FUNDAMENTALS OF TRIBOLOGY

FUNDAMENTALS OF TRIBOLOGY

Proceedings of the International Conference on the Fundamentals of Tribology

held at
The Massachusetts Institute of Technology
Cambridge, Massachusetts
June 1978

Edited by

N. P. Suh
and
N. Saka

The MIT Press
Cambridge, Massachusetts, and London, England

Printed and bound in the United States.

Library of Congress Cataloging in Publication Data

International Conference on the Fundamentals of Tribology, Massachusetts Institute of Technology, 1978.
Fundamentals of tribology.
Bibliography: p.
1. Tribology—Congresses. I. Suh, Nam P., 1936-
II. Saka, N. III. Title.
TJ1075.A2I64 1978 621.8′9 79-26144
ISBN 0-262-19183-0

DEDICATION

To
Professor Milton Clayton Shaw
a distinguished scholar,
a devoted teacher,
and a creative engineer,
who has enriched the field of tribology
with many scientific contributions.

TABLE OF CONTENTS

IX. WEAR IN PROCESSING

X. POLYMER WEAR

XI. WEAR MONITORING

PREFACE

Today, more than ever before, the intellect of mankind is being challenged by the aspirations of people everywhere for higher standards of living in the face of diminishing availability of natural resources. In order to meet these aspirations without violating the natural bounds of human habitat, we need to use materials and energy effectively and wisely. In this sense, the field of tribology takes on a particular significance, since the durability of a variety of goods and energy loss in machinery are largely determined by tribological factors.

Tribology—the science of friction, wear and lubrication—is a burgeoning field rich in opportunities and at the same time in need of critical assessment. Notwithstanding the enormous strides made during the past three decades, tribology is still one of the least developed scientific fields both because of the limited attention it has received and because of its complex and interdisciplinary nature. The field abounds with conflicting theories. There are only a few theoretical models that can be used as the basis for design of materials, and surfaces in relative motion.

The International Conference on Fundamentals of Tribology was organized to review the state of knowledge and to explore the exciting and important areas of research. To achieve these goals, leading scientists and engineers were invited to present definitive review papers and results of their recent research covering all aspects of tribology. They together with the other attendees of the conference made important contributions toward a better understanding of the subject through discussions.

In compiling these proceedings the papers were carefully read and edited for accuracy, continuity and readability. The discussions were transcribed from the recordings made at the time of the conference and then edited. An effort was made to convey the spirit as well as the content of the discussions. Our editorial task has been made bearable, and even enjoyable, by the thought-provoking ideas expressed by the discussors. It is strongly recommended that the discussions be read in addition to the reviews and contributed papers to gain a proper perspective on the subject matter.

Our current understanding of the tribological phenomena is incomplete. There is room for further inquiry even in areas where classical views are acknowledged. The recent recognition of the importance of tribology by scholars and policy-makers alike undoubtedly will have a significant impact on the future development of the field. We would consider our effort worthwhile if these proceedings could catalyze major breakthroughs in tribology.

Cambridge, Massachusetts
December 1979

N. P. Suh
N. Saka

ACKNOWLEDGEMENTS

The International Conference on Fundamentals of Tribology was made possible by many tribologists, as well as non-tribologists, who believe in the importance of the subject. They rendered unqualified support and provided the intellectual stimulus so essential to a conference of this magnitude.

The members of the Scientific Committee (F. T. Barwell, H. S. Cheng, N. H. Cook, I. Finnie, Y. Kimura, K. Kirschke, E. E. Klaus, I. V. Kragelskii, F. F. Ling, H. P. Martin, G. Mayer, R. S. Miller, M. B. Peterson, W. K. Petrovic, E. Rabinowicz, E. A. Saibel, L. E. Samuels, D. Scott, M. C. Shaw, D. Tabor, E. C. van Reuth, R. B. Waterhouse, V. C. Westcott, J. H. Zaat, and M. Zlotnick) were responsible for the selection of the speakers, papers and the format of the Conference. The Organizing Committee (M. P. Cleary, S. Jahanmir, N-H. Sung, M-K. Tse, A. P. L. Turner and F. Van Dyck) planned and organized the Conference.

The execution of the four-day conference was made possible by the enthusiastic assistance of Ms. Abigail Ryan, the Conference Secretary. Mrs. Young Suh introduced the rich history and delights of the Boston area to the spouses of the conferees.

The editorial assistance provided by Ms. Phyllis Brown in producing these proceedings was invaluable. We were fortunate to have had the support of the MIT Press and Science Press, Princeton, in the publication of the Proceedings. Mr. Frank Satlow and the late Mr. Carl Martinson were extremely helpful in this undertaking.

The Conference and the publication of these proceedings were made possible by the financial support of the Army Research Office, the Defense Advanced Research Projects Agency, the Department of Energy, the National Science Foundation, and the Office of Naval Research. The Massachusetts Institute of Technology kindly made its facilities available for the Conference.

CONTRIBUTORS

Ahearn, J. S., Martin Marietta Laboratories, Baltimore, MD 21227

Argon, A. S., Department of Mechanical Engineering, Massachusetts Institute of Technology, Cambridge, MA 02139

Bahadur, S., Engineering Research Institute, Iowa State University, Ames, IA 50011

Barwell, F. T., Department of Mechanical Engineering, University College of Swansea, Swansea SA2 8PP, U. K.

Beerbower, A., Energy Center, University of California, La Jolla, CA 92093

Bhateja, C. P., The Torrington Company, Torrington, CT 06790

Briscoe, B. J., Department of Chemical Engineering and Chemical Technology, Imperial College, London, SW7 2BY, U.K.

Buckley, D. H., NASA-Lewis Research Center, Cleveland, OH 44135

Burton, R. A., Department of Mechanical Engineering and Astronautical Sciences, Northwestern University, Evanston, IL 60201

Cadman, P., Edward Davies Chemical Laboratories, The University College of Wales, Aberystwyth SY23 1 NE, U.K.

Cheng, H. S., Department of Mechanical Engineering and Astronautical Sciences, Northwestern University, Evanston, IL 60201

Clerico, M., Instituto di Costruzione di Macchine, Politecnico di Torino, Torino, Italy

Czichos, H., BAM (Federal Institute for Materials Testing) D-1000 Berlin 45, West Germany

Devine, M. J., Naval Air Development Center, Warminster, PA 18974

Duquette, D. J., Materials Engineering Department, Rensselaer Polytechnic Institute, Troy, NY 12181

Evans, H. P., Department of Mechanical Engineering and Energy Studies, University College, Cardiff CF2 1TA, U. K.

Finkin, E. F., Allegheny Ludlum Industries, Inc., Pittsburgh, PA 15222

Gaul, D. J., Owens Corning Fiberglass Co., Granville, OH 43023

Glaeser, W. A., Battelle Columbus Laboratories, Columbus, OH 43201

Godfrey, D., Chevron Research Co., Richmond, CA 94802

Goldblatt, I. L., Exxon Research and Engineering Co., Linden, NJ 07036

Gossedge, G. M., Edward Davies Chemical Laboratories, University College of Wales, Aberystwyth SY23 1NE, U. K.

Gupta, P., Mechanical Technology, Inc., Latham, NY 12110

Hamrock, B. J., NASA-Lewis Research Center, Cleveland, OH 44135

Heyes, D. M., Vitreous State Lab, The Catholic University of America, Washington, DC 20064

Hirano, H. H., Sandia Laboratories, Livermore, CA 94550

Hirth, J. P., Department of Metallurgical Engineering, The Ohio State University, Columbus, OH 43210

Jahanmir, S., Sibley School of Mechanical and Aerospace Engineering, Cornell University, Ithaca, NY 14853

Johnson, J. H., Department of Mechanical Engineering-Engineering Mechanics, Michigan Technological University, Houghton, MI 49931

Kalivoda, D. A., Martin Marietta Laboratories, Baltimore, MD 21227

Kar, M. K., John Deere Product Engineering Center, Waterloo, IA 50704

Kimura, Y., Institute of Space and Aeronautical Science, University of Tokyo, Tokyo 153, Japan

King, T. G., Leicester Polytechnic, Leicester LE1 9BH, U.K.

Kirschke, K., BAM (Federal Institute for Materials Testing) D-1000 Berlin 45, West Germany

Knotek, O., Institut fur Werkstoffkunde, Techn. University Aachen, D-5100 Aachen, West Germany

Koury, A. J., Naval Air Systems Command, Washington, DC 20360

Kramer, B. M., Department of Mechanical Engineering, Massachusetts Institute of Technology, Cambridge, MA 02139

Ku, P. M., Southwest Research Institute, San Antonio, TX 78228

Lansdown, A. R., Swansea Tribology Centre, University College of Swansea, Seansea SA2 8PP, U. K.

Larsen-Basse, J., Department of Mechanical Engineering, University of Hawaii at Manoa, Honolulu, HI 96822

Latanision, R. M., Department of Materials Science and Engineering, Massachusetts Institute of Technology, Cambridge, MA 02139

Levy, A. V., Materials and Molecular Research Division, Lawrence Berkeley Laboratory, University of California, Berkeley, CA 94720

McCullagh, P., National Engineering Laboratory, East Kilbride, Scotland, U. K.

McNab, I. R., R&D Center, Westinghouse Electric Corporation, Pittsburgh, PA 15235

Mills, G., National Engineering Laboratory, East Kilbride, Scotland, U. K.

Mills, J. J., Martin Marietta Laboratories, Baltimore, MD 21227

Mizutani, Y., Toyota Central R&D Labs., Inc., Nagoya 468, Japan

Montrose, Vitreous State Laboratory, Catholic University of America, Washington, DC 20064

Murday, J. S., Surface Chemistry Branch, Naval Research Laboratory, Washington, DC 20375

Needelman, W. M., Pall Corporation, Glen Cove, NY 11542

Ohmae, N., Department of Precision Engineering, Osaka University, Osaka, 565, Japan

Peterson, M. B., Wear Sciences, Inc., Arnold, MD 21012

Pirvics, J., Technology Services Division, SKF Industries, Inc., King of Prussia, PA 19406

Quinn, T. F. J., Department of Physics, The University of Aston in Birmingham, Birmingham B4 7ET, U. K.

Rabinowicz, E., Department of Mechanical Engineering, Massachusetts Institute of Technology, Cambridge, MA 02139

Ramalingam, S., School of Mechanical Engineering, Georgia Institute of Technology, Atlanta, GA 30332

Rice, S. L., Department of Mechanical Engineering, University of Connecticut, Storrs, CT 06268

Rigney, D. A., Department of Metallurgical Engineering, The Ohio State University, Columbus, OH 43210

Ritchie, R. O., Department of Mechanical Engineering, Massachusetts Institute of Technology, Cambridge, MA 02139

Rohde, S. M., Engineering Mechanics Department, General Motors Research Laboratories, Warren, MI 48090

Ruff, A. W., Center for Materials Science, National Bureau of Standards, Washington, DC 20234

Saka, N., Department of Mechanical Engineering, Massachusetts Institute of Technology, Cambridge, MA 02139

Sakurai, T., Department of Chemical Engineering, Tokai University, Yokohama, Japan

Schnur, J. M., Optical Techniques Branch, Naval Research Laboratory, Washington, DC 20375

Schoen, P. E., Optical Techniques Branch, Naval Research Laboratory, Washington, DC 20375

Scott, D., Department of Mechanical and Production Engineering, Paisley College of Technology, Paisley PA1 2BE, U. K.

Shaw, M. C., Engineering Research Center, Arizona State University, Tempe, AZ 85281

Sin, H -C., Department of Mechanical Engineering, Massachusetts Institute of Technology, Cambridge, MA 02139

Singer, I. L., Surface Chemistry Branch, Naval Research Laboratory, Washington, DC 20375

Smith, R. A., University Engineering Department, University of Cambridge, Cambridge CB2 1PZ, U. K.

Snidle, R. W., Department of Mechanical Engineering and Energy Studies, University College, Cardiff, Wales, U. K.

Sproles, E. S., Bell Laboratories, Columbus, OH 43213

Stout, K. J., Leicester Polytechnic, Leicester LE1 9BH, U. K.

Suh, N. P., Department of Mechanical Engineering, Massachusetts Institute of Engineering, Cambridge, MA 02139

Tabor, D., Department of Physics, University of Cambridge, Cambridge, CB3 OHE, U. K.

Tamai, Y., The Chemical Research Institute of Non-Aqueous Solutions, Tohoku University, Sendai, Japan

Tessman, R. K., Fluid Power Research Center, Oklahoma State University, Stillwater, OK 74074

Tevaarwerk, J. L., Department of Mechanical Engineering, University of Waterloo, Waterloo, Ontario N2L 3GI, Canada

Tsukizoe, T., Department of Precision Engineering, Osaka University, Osaka, 565, Japan

Ura, A., Department of Mechanical Engineering, Nagasaki University, Nagasaki, 852, Japan

Waterhouse, R. B., Department of Metallurgy and Materials Science, The University of Nottingham, Nottingham NG7 2RD, U. K.

Watson, W., Leicester Polytechnic, Leicester LE1 9BH, U. K.

Werner, G., Department of Mechanical Engineering, Massachusetts Institute of Technology, Cambridge, MA 02139

Westcott, V. C., Foxboro/Trans-Sonics, Inc., Burlington, MA 01803

Westwood, A. R. C., Martin Marietta Laboratories, Baltimore, MD 21227

Whitehouse, D. J., Rank Taylor Hobson, Leicester LE5 3RG, U. K.

Wunder, S. L., Optical Techniques Branch, Naval Research Laboratory, Washington, DC 20375

Yust, C. S., Metals and Ceramics Division, Oak Ridge National Laboratory, Oak Ridge, Tenn 37830

I. INTRODUCTION

INTRODUCTION TO TRIBOLOGY

D. Scott

ABSTRACT

In the present and foreseeable future economic situation, material and energy conservation is becoming increasingly important. Since wear is a major cause of material wastage; friction, a serious cause of energy dissipation; and lubrication, the most effective means of controlling both - tribology, which is the science and technology of friction, lubrication and wear, plays a major role in conservation. The present state of the art is reviewed and the areas of inadequate knowledge identified.

Basic research studies can now be fully justified only if information is provided to industry to solve its immediate problems or produce speedy significant advances in technological progress. The trend may be to experiment less, and to measure and interpret more in order to utilize more fully the available information.

INTRODUCTION

In the present and foreseeable future economic situation, material and energy conservation is becoming increasingly important. As wear is a principal cause of material wastage, any reduction of wear can effect considerable savings. Friction is a serious cause of energy dissipation and considerable savings can be made by better control of friction. Lubrication is the most effective means of controlling wear and lowering friction. Thus, tribology (the science and technology of friction, lubrication and wear) plays a major role in material and energy conservation.

It is just over a decade since the word tribology entered the English language with the publication of the Lubrication Report(1) and, therefore, a convenient time to assess its current status.

IMPACT OF TRIBOLOGY

There is a continuously increasing awareness of the subject of tribology throughout industry. In the United Kingdom the National Centre of Tribology and Industrial Units of Tribology have been set up to provide expert advice to industry on the utilization of existing knowledge. These centers were provided with Government deficiency grants but are now viable establishments operating as contract research organizations selling their services at commercial rates.

In the United Kingdom, over 30 universities, polytechnics and technical collages have incorporated courses on various aspects of tribology into their syllabuses. A basic tribology module(2) for undergraduate mechanical engineering courses has been drawn up. Tribology is an elective subject for HNC in engineering, and a tribology content is included in some CNAA courses. The Iron and Steel, and Engineering Industry Training Boards include tribological topics in their recommendation for the training of craftsmen. Post-graduate research in tribology, leading to higher degrees, is carried out at several universities. Three universities have chairs in tribology. A comprehensive selection of courses and training programs is also available to industry.

Some 7000 papers on tribology are now published annually. Most of these report research directed toward a better understanding of the fundamental principles governing interacting surfaces; but, unfortunately, most of the information provided is not suitable for direct use by designers and engineers. This is partly because research workers generally find it more convenient to express results in terms of non-dimensional parameters rather than as the specific data required for design purposes. A tribology handbook(3) has been produced which presents tribological information to industry in a form that is readily accessible and easily understood by engineering designers, draughtsmen and works engineers. A synoptic journal(4) has also been introduced in an endeavor to save time spent in literature persual.

MECHANISMS OF WEAR

Progress in wear control and prevention can only be made when a better understanding of the mechanisms by which it occurs has been achieved and the dominant controlling factors have been established.

Wear is an interdisciplinary subject and its complex mechanisms are not easily elucidated. The progressive nature of wear destroys the evidence of its initial stages as cumulative action develops toward failure. The tendency of research workers has been to isolate and study specific mechanisms such as adhesion, abrasion, erosion and fatigue. Much research has been directed toward the study of surfaces in relative motion and changes brought about by their interaction and the effects of the lubricant and the environment. Until recently, comparatively little attention has been given to the wear particles generated by the interaction of relatively moving surfaces. Now attention to particle tribology(5) allows postulation of the mechanism of their formation which, together with refined techniques of surface investigation and the study of subsurface changes, aids the elucidation of the wear process.

Advances in understanding emerge only from a willingness to question accepted theories. Suh's(6) delamination theory of wear is a typical example of such progress. The delamination theory of wear introduced in 1973 to explain sliding wear departs completely from the classical adhesion theory. Extensive analytical and experimental work has confirmed the validity of the theory and many of the postulated mechanisms involved. The delamination theory is based on the behaviour of dislocations at the surface, subsurface void and crack formation and the subsequent joining of cracks by shear deformation of the surface. It predicts that the wear particle shape is likely to be thin, flake-like sheets (Figure 1), and that the surface layer can undergo considerable plastic deformation. It has been shown that bulk material hardness in itself is not the controlling factor in wear and that the delamination theory and not the adhesion theory satisfies the thermodynamic requirements of the frictional and wear behaviour of metals.

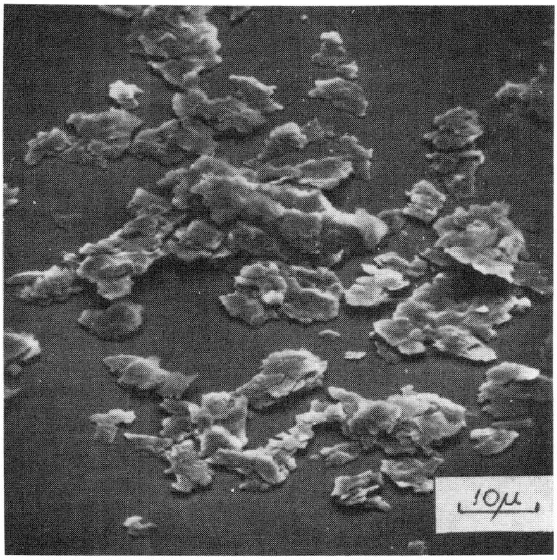

Fig. 1.—Plate like particles of rubbing wear.

The ultimate formation of the wear particle depends on two mechanisms, void formation and crack propagation. For materials such as medium tensile strength steels of good fracture toughness, where void nucleation can readily occur, crack propagation may be the controlling mechanism. However, for materials of high tensile strength and low fracture toughness, void nucleation can be difficult, but crack propagation can readily take place. Void nucleation may then be the controlling mechanism.

Surface examination in conjunction with wear particle analysis has led to the hypothesis[7] that the interaction of surfaces in relative motion polishes the surfaces (Figure 2) and creates a shear-mix layer of short crystalline order of almost superductile material which spreads over the surface. This was first proposed by Beilby[8]. Repeated rubbing causes the shear-mix layer to become fatigued and Beilby-type particles flake off.

Further work is required to provide a more complete description of the behaviour of material, surfaces, and wider application of the delamination theory of wear must await additional evidence. It may then be possible to predict the wear rate of materials based on first principles and fundamental properties. The application of systems analysis[9] to wear problems is extremely useful in the prediction of wear of materials.

Wear research has been almost non-existent in the United States except for a few institutions[10]. Perhaps the breadth of the subject of wear has prevented any technical society in the United States from being the recognized home of the subject. However, wear is at present the concern of about ten technical societies in the United States[11]. It appears that the era in which wear was considered a branch of studies in friction and lubrication is coming to an end. The success of the 1977 International Conference on Wear of Materials[11] encouraged the organizing committee to arrange a second conference, and thus establish wear as a subject of international importance in its own right.

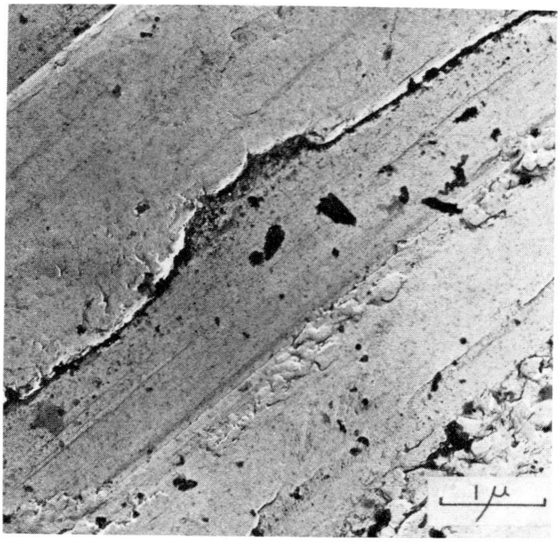

Fig. 2.—Flowed surface layer formed by running in.

SURFACE STUDIES

It has long been recognized that the friction and wear behaviour of materials is greatly dependent upon the topography, chemical composition, physical structure and mechanical properties of surfaces and surface material. Contact between surfaces by relative motion causes changes in these properties which may be of significance in subsequent contacts. However, detailed knowledge of what is happening at the interface when wear is occurring is difficult to acquire. It has been usual to study surfaces at various stages of wear to postulate the sequence of events. Besides this procedure, greater attention is now being given to the size, morphology, and structure of wear particles, as well as to the localized nature of damage to surface, interface and subsurface material. A number of comparatively new tools are available for the study of surfaces at atomic level, notably Auger electron spectroscopy, X-ray photoelectron spectroscopy, scanning ion spectroscopy and ion scattering spectroscopy; with complementary information from X-ray energy analysis in the scanning electron microscope and microprobe analysis, these tools aid the tribological elucidation of surface phenomena.

During the past decade, rapid strides have been made in the application of statistical techniques to the characterization of rough surfaces[12,13]. The entire statistical microgeometry of certain rough surfaces can now be completely described in terms of the number of peaks and mean line crossings counted on a single profile. These techniques are now being applied in tribology, and it appears that in instrumentation, three-dimensional mapping is now well established.

LUBRICATION

Since Reynolds produced his equation, the mathematical expression of the process of film formation between relatively moving surfaces has been fundamental to all lubrication theory. Equations have been derived and applied to the study of the various surface configurations used in industry. However, it was not until the intorduction of the high speed digital computer that a simultaneous solution of Reynolds equation, together with equations representing the variation of viscosity with pressure and the elastic deformation of the surfaces, became a practical proposition.

Optical studies of EHL films(14), infra-red temperature measurement(15) and the elucidation of the response of viscous liquids to high frequency shear(16) have greatly improved the understanding of elastohydrodynamic contacts. It is perhaps better to describe the lubricant in a highly loaded EHL contact as an elastic-plastic solid rather than as a viscous fluid. Based on the new understanding, a theory of EHL traction has been advanced(17) which may be applied to engineering components such as rolling bearings and variable speed drives.

Progress in hydrodynamic lubrication appears to be centered on detailed developments rather than improved fundamental understanding. Work on boundary lubrication seems to be oriented mainly to specific problems or problem areas such as elevated temperature and hostile environments.

Two centuries of study(18) have failed to unravel completely the mysteries of the lubrication problems most important to mankind, the mechanisms of human joints. From the engineer's point of view, the reciprocating engine and a walking human being have in common the need for similar bearings. As there are still questions unanswered regarding connecting rod bearings, it is readily appreciated that the more complicated human joints have as yet defied complete understanding. The tentative proposal of squeeze films made by Fein(19) a decade ago and the emphasis placed by Dowson(20) on the protective motion of the simple squeeze film is now supported as the mechanism which dominates the tribologist's desire for a full film of fluid between joints for loads of short duration. Higginson(18) considers that the prospect of EHL is good, but that the promising mode is squeeze film and not rolling/sliding.

LUBRICANTS

In the field of tribology, a disproportionate amount of research appears to be carried out on lubricants. This is understandable; when design is inadequate and failure in service occurs, it is commercially more acceptable to change the lubricant rather than the design. Thus, research and development work is continuously directed toward improved lubricants and additives to impart or reinforce desirable properties, and synthetic lubricants with unique properties. The more recent major developments in lubricant formulation appear to have been in cutting fluids, fire resistant hydraulic fluids and synthesized hydrocarbon fluids. The latter appear capable not only of allowing the formulation of specialized lubricants for extreme environments but also of replacing mineral oils in some traditional applications. Although these fluids may cost about four times as much as mineral oil based products, experience with them indicates that they may give an overall cost saving.

In some applications the complications caused by lubricants lead naturally to consideration of wear resistant materials which possess good frictional properties and which can operate without lubrication. Antipollution and conservation are placing emphasis on sealed, lubricated-for-life machinery using solid lubricants and surface treatments which lubricate. The general concept of metal plus lubricant is now giving way to refractory or similar materials plus coolant or parting compound. Under these conditions, techniques for studying and controlling surface structure become important, for as close control of structure and strength is achieved, the performance and wear resistance must depend increasingly on interfacial conditions. The absence of any recognized code of practice and well established tests for evaluating new materials for tribological design requirements means that suitability can be reliably established only by performance in practice, and accelerated service simulation testing is now replacing conventional test rigs.

Polymers and fluorinated polymers are finding increasing application, especially where chemical and thermal inertness are important. PTFE has become the standard solid lubricant in cryogenic applications. Its tendency to cold flow has been improved by suitable reinforcement. Newer polymers such as polyimides are being increasingly used where high thermal stability is required. Metal film lubricants are now finding use, and potential developments in solid lubricant technology may arise from the combination of solid and liquid lubricants to use the specific properties of each.

MATERIALS FOR TRIBOLOGICAL APPLICATIONS

We are living in the material age. A major incentive for the development of wear resistant materials and the acquisition of materials data is the emergence of new design concepts. For instance, the aero-space industry spawns new problems and solutions while each generation of nuclear reactors requires some new wear-resistant material. The thermal and stress problems associated with advanced tribo-engineering require materials of high strength, high elastic modulus, and light weight. Conventional materials have been improved by orthodox methods almost to the limit of their potential mechanical properties so that new types of material such as composites, synthetic diamond and sapphire, new graphites and carbides, borides and nitrides of certain metals which approach the hardness of diamond are being developed. Use of such materials requires new design concepts to utilize their specific properties, as substitution of such materials in existing designs can lead to problems and failures in service. Besides replacing metals, ceramics may be used as coatings to complement desirable metal characteristics by adding refractory properties, insulation and erosion, wear, oxidation and corrosion resistence.

In the field of plain bearings, it appears unlikely that any major development in soft metal bearings will take place in the immediate future, as the possible alloys of all commercially feasible softer metals have been explored thoroughly. Presently used materials come close to utilizing fully the potentiality of plain bearings of current designs and lubricating systems. In the field of plastics the development of bearing materials capable of being manufactured to achieve and maintain the close tolerances of metals may cause something of a revolution in the plain journal field.

Plastics and their composites dominate the dry bearing scene mainly due to the availability of design and performance data[21,22]. A significant advance in fundamental understanding of the wear of plastic composites has been the recognition of the dominant role of the counterface metal. The rate of wear depends on the counterface topography and composition generated by the sliding process and involving transfer of polymer or fillers, abrasion by fillers or corrosion by the environment or polymer degradation products. The development of vacuum deposition techniques such as sputtering, ion-implantation, ion plating and CVD processes appear potentially attractive for solid film lubricant solution to a wide range of dry bearing problems.

In the field of rolling bearings material, methods are being developed to improve contact fatigue resistance. Interstitial nitrogen content[23] can explain the inherent differences between different steel-making processes and the success of vacuum refining techniques which reduce gas content, as well as some lack of success by electro-slag refining which does not reduce nitrogen content. Figure 3 shows the effect of nitrogen content on rolling contact fatigue resistance of basic electric arc EN 31 ball bearing steels.

Developments in high speed tool steels for rolling bearings have centered around alloys produced by the powder route to provide a finer dispersion of carbides, but again gas content has proved to be deleterious[24]. The use of higher than normal alloying elements to provide marginally improved properties may not justify the increased cost. In the field of cutting tools there continues to be a steady but unspectacular development of ceramic materials such as silicon nitride[25,26].

Brakes and clutches are required to dissipate continually greater energies due to generally increasing loads and speeds, and improved materials are constantly being demanded to cope with the higher duties and temperatures. For over seventy years asbestos has been the most effective filler material for phenolic resins, both because of its fibrous nature and its heat resistance. Because of a possible health hazard, there is considerable pressure to replace asbestos with other fillers. Sintered metal matrices are used when the duty is severe, but attempts to introduce other organic and inorganic materials have not yet succeeded in displacing the more conventional materials except in highly specialized fields. More exotic materials have been successful in aircraft brakes. For example, Concorde uses carbon composites against themselves. Use of these materials with cheaper fibers and fillers may be one promising method of approach to the replacement of asbestos-filled phenolic resins and may help to contend with the other high energy dissipation situations such as advanced aircraft brakes.

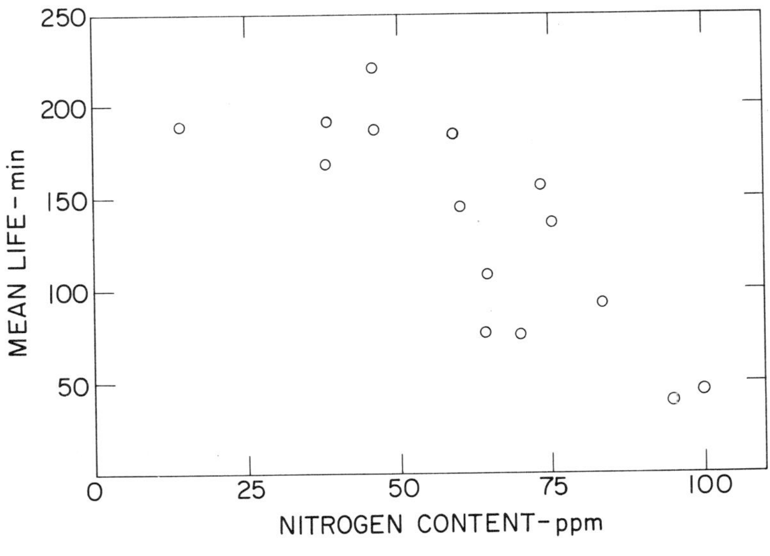

Fig. 3.—Correlation between mean rolling contact fatigue life and nitrogen content of basic electric arc EN 31 ball bearing steels.

SURFACE TREATMENTS

When confronted with a wear problem it is becoming increasingly difficult to make worthwhile innovations in the fields of design or lubrication, but considerable opportunities are now available in the realm of surface treatments[27,28].

Recently developed surface treatments involving thin surface films with specific properties are now finding increasing use and are proving to be advantageous as wear resistant coatings. The treatments include physical and chemical vapor deposition processes. The physical vapor deposition (PVD) processes are evaporation, ion plating and sputtering. Metal films deposited by ion plating are strongly bonded to the surface as the film is deposited on a surface cleaned by sputter etching. Due to the velocity of the evaporant, the surface is penetrated, and a graded interface is formed to give a strongly adherent film. Soft metal lubricant films may also be bonded to a metal surface by ion plating. Ion nitriding may be used to speed up the nitriding process.

The chemical vapor deposition (CVD) process deposits carbides of the transition metals on steel. The transition metal in the form of a volatile chloride is carried to the surface by a carrier gas together with a gaseous hydrocarbon. These react at the steel surface to form a carbide layer which is smooth, continuous and wear and corrosion resistant. TiC coatings on sintered carbide cutting tools considerably reduce tool wear[29]. Low temperature CVD processes and controlled nuclear thermo-chemical deposition have been developed. The latter produces equi-axed grains randomly oriented and with borides of hardness up to 8000 HV compared with about 2500 HV of CVD coatings[30]. Ion implantation is an interesting technique to improve wear resistance; its high cost may be a deterent to its wide usage in industrial applications, but its investigation may contribute to a better understanding of the behaviour of wearing surfaces.

COMPUTER AIDED DESIGN

Tribology stands between the pure sciences on the one hand, and the material and engineering requirements of technological progress on the other. It is only by the combination of improved scientific understanding and the most purposeful, speedy application of knowledge that rapid technological progress will be achieved. A major difficulty has been the delay or lack of feed back, the time taken from the inception of a good idea to its fruitful application. To eliminate such delays, increasing use is being made of the computer in design to enable almost instantaneous feed back. Designers in their offices in local industry or research laboratories can now communicate directly with the computer. For example, in the field of plain bearings with a suitably constructed program, the designer need only transmit information on bearing design, and within seconds he will be informed of the performance characteristics. In this way he can have instant feed back and can make use of the latest available research results without being an expert in the fields of tribology, computation or programming. In effect, he has at hand a universal testing machine in which he can plan his design, test its characteristics, modify the design and again measure its characteristics continuing this process until satisfied with the performance of the design before committing himself to production and service.

The use of computers for materials selection for optimum performance is approaching rapidly.

MACHINERY CONDITION MONITORING

Economic pressures are causing a switch from regular dismantling of machinery for maintenance to failure prevention maintenance. Increasing emphasis is thus being given to on-line monitoring techniques to detect deterioration of machinery so that remedial action can be taken before the breakdown point is reached.

Many techniques are available to monitor machinery condition. The technique chosen usually depends upon the specific information required and the cost of acquiring the information compared with the savings such information can effect.

Over the past few years there has been a gradual acceptance of vibration analysis, although this has proved to be neither the simplest nor the most effective method to use. There is the problem of interpreting the data caused in part by the fact that the vibration signal tells as much about the general dynamics of the machine as its damaged state. An inability to separate the two creates the need for trend analysis and a massive build up of data. Thus, the costs of complementary vibration analyses are high in relation to the information which it yields readily and it has found its most useful application in the monitoring of expensive rotating equipment.

Magnetic plugs and spectrographic oil analysis procedure (SOAP) are now extensively used to detect abnormal wear from the amount of debris which enters the lubricant from relatively moving surfaces. Although these techniques have proved effective in providing warning of changes in a system, they have some disadvantages. SOAP provides a knowledge only of the quantity of metal in the lubricant, but no information on the size or shape of the metallic particles. Some damage has usually occurred by the time the magnetic plug picks up debris large enough for observation.

Ferrography[7], a convenient method for the isolation and analysis of wear particles, has opened up a new dimension in wear detection and assessment in the form of particle tribology[5]. Each wear particle is unique, having distinctive characteristics bearing evidence of the conditions under which it was formed. Thus, the isolation and careful examination of the particles of wear can yield specific information concerning the condition of the surfaces from which the particles were produced. A Ferrographic oil analysis procedure by direct reading (DR) and analytical Ferrography has proved convenient for machinery condition monitoring[31,32]. Non-magnetic particles can also be isolated from lubricants so that lubricant degradation products can be identified to assess lubricant condition and performance. Recent developments have enabled the adaptation of Ferrography to bioengineering[32,33]. Fluids have been prepared which cause bone, cartilage and organic tissues to become magnetic for precipitation from synovial fluid. Ferrography is thus being applied to the study of wear of prosthesis joints and for condition monitoring of human joints to detect damage and the stages of degenerative arthritis.

Ferrographic trend analysis[34] from oil washed systems and the gas stream[35] appears to be a potentially attractive method of condition monitoring of gas turbines. The real time or automated Ferrograph has successfully undergone prototype testing. Ferrography looks at intermediate debris that is smaller than that collected by filters and magnetic plugs and larger than SOAP. Also as non-magnetic materials, man-made plastics, plant and animal tissues and living bacteria have been precipitated[36], it appears that the Ferrographic technique can be extended to the recovery of all materials for expanding applications in tribology.

CONCLUSIONS

In the short time since tribology was launched as a concept on its own, it has been described as the world's fastest growing applied science, though still in its infancy[34], and as a means of national wealth creation without commensurate capital investment[35]. So that tribology may reach maturity, still greater use must be made of existing knowledge. Most industrial tribological problems can be solved satisfactorily by the application of existing knowledge. Thus it appears that greater efforts are required to disseminate knowledge in a readily understood form to effect greater energy, materials and manpower savings at minimum cost. The trend may be to experiment less, and to measure and interpret more. Basic research studies may be justified only if they provide information to allow industry to solve its immediate problems and if they can produce significant advances in technological progress. The environmentalists may also play their role in demanding that tribologists strive for reduction of the noise level of mechanisms and the elimination of pollution from toxic lubricants.

REFERENCES

1. "Lubrication (Tribology)-Education and Research. A report on the present position and industry's needs," HM Stationery Office, London, 1966.
2. "A Basic Tribology Module," Dept. Trade and Industry, London, 1973.
3. Neale, M.J., ed., "Tribology Handbook," Butterworths, London, 1973.
4. *Synoptic Journal*.
5. Scott, D., *Institution of Mechanical Engineers. Proceedings,* Vol. 189, 1975, p. 623; NEL Report 627, National Engineering Laboratory, East Kilbridge, Scotland.
6. Suh, N.P., et al., *Wear,* Vol. 44, 1977.
7. Scott, D., Seifert, W.W. and Westcott, V.C., *Scientific American,* Vol. 230, 1974, p. 88.
8. Beilby, G., "Aggregation and Flow in Solids," Macmillan, London, 1921.
9. Czichos, H., "Tribology - A Systematic Approach to the Science and Technology of Friction, Lubrication and Wear," Elsevier, New York, 1977.
10. Peterson, M.B., in Proceedings of a Workshop on Wear Control to Achieve Product Durability, edited by M.J. Devine, Naval Air Development Center, Warminster, Pa., 1977, p. 111.
11. Ludema, K., "Wear of Materials, 1977," American Society of Mechanical Engineers, New York, 1977.
12. "Properties and Metrology of Surfaces," Proceedings of the Institution of Mechanical Engineers, Vol. 182, Part 3K, 1967-1968.
13. Thomas, T.R. and King, M., "Surface Topography in Engineering - A State of the Art Review and Bibliography," British Hydromechanics Research Association, Cranfield, 1977.
14. Cameron, A., et al., *Institution of Mechanical Engineers. Proceedings,* Vol. 184, Part 1, 1969-70.
15. Nagaraj, H.S., Sandborn, D.H. and Winer, W.O., *Wear,* Vol. 49, No. 1, 1978, p. 189.
16. Lamb, J., *Institution of Mechanical Engineers. Proceedings,* Vol. 182, Part 3A, 1967-68.
17. Johnson, K.L. and Tevarwerk, J.L., *Royal Society of London. Proceedings.* Series A, Vol. 356, 1977, p. 215.
18. Higginson, G., *Institution of Mechanical Engineers. Proceedings,* Vol. 191, 1977, Preprint 33/77.
19. Fein, R.S., in Proceedings Symposium on Lubrication and Wear in Living and Artificial Human Joints, Institution of Mechanical Engineers, London, Vol. 181, Part 3J, 1967, p. 125.

20. Dowson, D., in Proceedings Symposium on Lubrication and Wear in Living and Artificial Human Joints, Institution of Mechanical Engineers, London, Vol. 181, Part 3J, 1967, p. 45.
21. "A Guide to the Design and Selection of Dry Rubbing Bearings," Engineering Science Data Unit, Data Item 76029, 1976.
22. Lancaster, J.K., Tribology, Vol. 6, 1973, p. 219.
23. Scott, D. and McCullagh, P.J., *Wear*, Vol. 25, 1973, p. 339.
24. Scott, D. and Blackwell, J., *Wear*, Vol. 34, 1975, p. 149.
25. Scott, D. and Blackwell, J., *Wear*, Vol. 24, 1973, p. 61.
26. Dalal, H.M., et al., American Society of Mechanical Engineers/American Society of Lubrication Engineers Lubrication Conference, Montreal, 1977.
27. Scott, D., *Wear*, Vol. 48, 1978, p. 283.
28. Wilson, R.W., in Proc. 1st. Euro. Tribology Congress, Institution of Mechanical Engineers, London, 1975, p. 165.
29. Hintermann, H.E., *Wear*, Vol. 47, 1978, p. 407.
30. Holzl, R.A., *Wear*, Vol. 48, No. 2, 1978, p. 305.
31. Scott, D. and Westcott, V.C., *Wear*, Vol. 44, 1977, p. 173.
32. Scott, D. and Westcott, V.C., "Recent Developments in Ferrography," Preprint C42/78, Institution of Mechanical Engineers, Swansea, 1978.
33. Mears, D., et al., *Wear*, Vol. 50, No. 1, 1978, p. 115.
34. "Eurotrib '77," Bundesrepublik Deutschland Dusseldorf.
35. Jost, H.P., *Tribology International*, Vol. 2, 1978, p. 34.

DISCUSSION

P. K. DAS, John Deere: Mr. Scott should be complimented for his comprehensive presentation of tribological problems encountered in industry. I would like to comment on some difficulties that are faced in industry in relating the theories to practice. First, with the high speed computers that are available, now we have better methods of solving Reynold's Equation and calculate the parameters such as film thickness, film pressure, friction, etc. But it is not clear how to relate these parameters to failure criteria; how, for instance, the pressure relates to fatigue, S-N diagram, or any other material property, and how the film thickness relates to wear rate. Secondly, postulating criteria for bearing calculations through materials properties, even if primitive, is very much needed. It is very important to develop the relationships between fatigue, wear and other useful criteria. Finally, recommendation of test procedures that industry could follow and perhaps compare one bearing with another on the basis of these above criteria is very much needed. In practice one frequently finds that simple parameters such as the film thickness, pressure, friction, etc., are not the total story; i.e., a lower film thickness or high pressure does not necessarily relate to poor performance of a bearing. Work in this area will greatly aid the industry.

D. SCOTT: I think that you have emphasized my lecture theme. It is very difficult but essential to correlate basic research with practice and a greater effort is required to relate laboratory research results to service problems using language which the engineer and designer can readily understand. With test programmes there will be a trend towards service simulation testing using procedures which produce the same conditions and failure mechanisms as experienced in practice. This tends to reduce the time from advances in theory and fundamental research to its industrial utilization. Generally fundamental research at Universities is carried out using pure materials which are amenable to this type of work in order to establish principles but it is the speedy adaptation of their principles to practice which is the present grey area which requires attention. Dissemination of available knowledge and the correlation of test results with performance in practice will provide the greatest benefits at the lowest cost.

R. S. MONTGOMERY, U. S. Army: You mentioned that the interstitial nitrogen causes the surfaces to fail by rolling fatigue. What do you think about fatigue that occurs in gun barrels?

SCOTT: Work in rolling contact fatigue has shown that interstitial nitrogen contents of ball bearing steels is a dominant factor in their proneness to failure and can explain the considerable differences in fatigue life of otherwise apparently identical steels, the beneficial effect of vacuum refining techniques and the occasional lack of benefit of electro slag refining. If fatigue is a mechanism of gun barrel failure the interstitial nitrogen content of the steel used could be a dominant factor in fatigue resistance.

R. DASKIVICH, G. M. Research Laboratory: In addressing the topic of wear prevention, you made a comment that high hardness is not necessarily essential and you followed that statement addressing how implantation of soft materials builds in a state of equilibrium. Could you please elaborate on what would constitute this state of equilibrium?

SCOTT: The object of the research outlined is to produce failure and maintenance free machines for an economic life. Our object is to design the moving surfaces in a run-in state so that the interfacial layers are in a state of equilibrium for an economic service life.

To prevent wear it is often not necessary to use very hard materials. A systems analysis should be carried out to determine exactly the cause of wear and the necessary remedial action. In some instances sliding or vibratory movement may cause adhesion which may produce hard particles of debris which in turn initiate abrasive wear. Increasing the original surface hardness may produce harder particles and more severe abrasive wear. The correct solution would be to prevent the initial adhesive wear and generation of debris and a soft metal coating may do this.

N. P. SUH, MIT: I would just like to make a comment. Our work on the soft coatings has shown that about 0.1 μm thick, nickel and gold on steels can increase the life by more than a factor of ten.

SCOTT: This comment illustrates my point. There is evidence of a movement away from the concept of two hard surfaces separated by a film of lubricant to a designed interface between the surfaces.

S. CYTRON, U. S. Army: When you are applying very thin layers (0.1 μm) of soft metals for wear and erosion problem, do you have the difficulty of finding suitable mechanical testing techniques to evaluate the microhardness of these very thin films?

SCOTT: There can be problems but techniques are available and will be developed to overcome them. It has been quoted that industry gets things done and research finds out in the next ten years the principles of operation. However, advances in the understanding of the mechanisms of thin solid film lubrication should aid their application to wear control.

SUH: The reason may not be correct, but we hypothesized that a very thin layer of soft metal cannot retain dislocations and therefore cannot work harden. Based on that argument we have applied a very thin layer and it does work.

SCOTT: Hardness testing is used as it is a simple almost non-destructive test which can be carried out on the actual component or specimen to be used. Hardness testing is similar to the tensile testing of bridge material, based on past experience the test results can be correlated with suitability for service requirements.

DASKIVICH: But the very point is how can a hardness test be non-destructive to a 0.1 μm film. It cannot be done.

SCOTT: Microhardness may be determined at some non-critical position. Alternatively a thin film may be applied to a sample of a similar surface during the process for test purposes.

II. SURFACE TOPOGRAPHY

SURFACE TOPOGRAPHY AND QUALITY AND ITS RELEVANCE TO WEAR

D. J. Whitehouse

ABSTRACT

In the past two decades there has been a growing shift of emphasis from control of manufacture to prediction, or optimization, of performance. This has stimulated research into getting a better understanding of surfaces and how they interact in practical situations. From this research two important developments have taken place, one instrumental and the other theoretical. The instrumental advances have been digitizing the outputs of stylus instruments and using relocation methods. The theoretical advance has been the use of random process theory which enables the most suitable parameters for a given function to be identified. One principal objective of this paper is to show where and with what result these techniques have been employed in wear and related situations.

Particular attention has been given to looking at the fundamentals of the problem rather than listing what exists as empirical relationships in the hope that it will provide a more useful starting ground for researchers. For this reason a deliberately wide view of wear has been taken.

INTRODUCTION

The measurement of surfaces and the quantitative study of wear have the same timescale of about 50 years. This has meant that the researcher into topographic aspects of wear has rarely been in a position to investigate the relationship effectively because the necessary tools, the instrument and the theory have not been readily available; they have been about as advanced as his own understanding of the wear mechanism. However, this situation has changed dramatically in the past decade or so. With the advent of cheap digital computation and the increasing use of random process analysis, the tribologist, at last, has at his disposal all the power he needs to make new strides in understanding. It seems therefore that the purpose of this paper might best be served by setting out the recent advances in metrology and after this, by discussing how these have been applied and how they may be applied in the future to wear situations. The review is not exhaustive; it is more of a personal viewpoint of the subject.

Historical Background to the Metrology of Surfaces

The first serious investigator of surface texture both practically and theoretically was Abbott.(1) He devised the first instrument and he attempted to ascribe a meaningful parameter to the output. This parameter was in fact a function. It was called then the Abbott-Firestone curve and is now called the bearing ratio curve. It was devised as a curve describing how the ratio of metal to air changed as a lapping plate lapped the surface texture down from the highest peak to the lowest valley. This was the first functional parameter.

From this beginning the role of surface texture began to change. Attention began to focus on the use of surface texture to control manufacture. See for example Kayser(2) and Castro.(3) In this role the texture was used as an effective "go gauge" on the manufacture. It was the tight control of manufacture which ensured that a component worked as well as the previous one made to the same specifications.

For about two decades this system worked adequately, particularly during the war years, but due to economic pressures in the early 1950's it started to break down because production engineers demanded freedom to choose the manufacturing process. The result was that often the process was left off the drawing altogether - a state of affairs still with us today. This weakened the control and consequently caused a degradation of performance. To combat this trend more attention was given to controlling the function of a part directly and since the early 1960's much more effort has been put into discovering just how and why surface texture influences performance; the ultimate aim is to replace manufacturing process parameters with surface parameters. The next section will show how far surface metrology has progressed with just this aim in mind.

Modern Advances in Surface Metrology

At one time measurement of complicated surface parameters was out of the question. It was simply uneconomical for instrument makers to provide a wide range of multiparameter boxes. This situation changed dramatically when digital methods started to be used. It represented the biggest revolution in surface analysis; analogue devices being replaced by digital equivalents, for example filters,(7,8) with considerable gains in flexibility and accuracy. Examples of typical digital systems are given in references 9 and 10. Today, practically every researcher uses digital methods; a fact which might give rise to general rejoicing. Unfortunately, the increasing use of computer techniques has been paralleled by an equally big increase in the misuse of computers as will be made clear later.

The next big step in metrology was the use of relocation methods. In this, elaborate care is taken to ensure that during the progress of an experiment exactly the same part of the surface is measured. This enables a considerable saving in cost and an improvement in the ability to see small changes in surface texture. The latter point has proved to be exceptionally important in mild wear experiments.(11-13)

Digital methods and relocation have been combined to investigate parameter changes during run-in.(13) Another joint application has been in the mapping of surfaces. One of the biggest problems in wear and contact is that of "seeing" the surface and the way it contacts another. In a classic experiment Williamson digitized whole areas of a surface and from these he made contour maps (Figure 1). Subsequently he simulated "gap maps" of

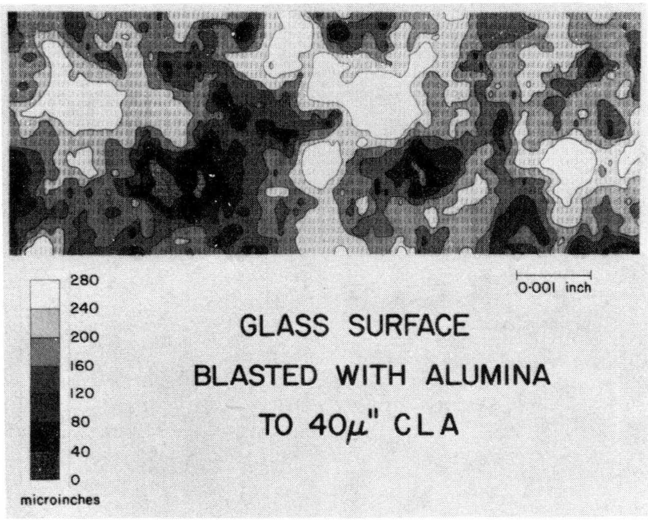

Fig. 1.—Schematic contour map of a machined surface obtained using a stylus instrument and scanning techniques. Scale relative to typical hills is about 10^7:1.

surfaces coming into contact. Using this technique, the shape and size of contacts could be assessed, all by computer. Others have used the same technique since.(12,15) In particular Sayles and Thomas have produced some spectacular pictures of surfaces(15) (Figure 2).

At about the same time the theoretical basis of surface metrology was undergoing its own upheaval; the old deterministic parameters were slowly being replaced by random process analysis. This had been used earlier with a limited application by a few workers.(16–18) It was principally because of the realization that surfaces could be regarded as Gaussian processes(17) which drew attention to statistical techniques. Peklenik began to classify surfaces by means of their autocorrelation function with the principal aim of better controlling manufacture.(19) He restricted himself to dealing with the profile graph - here referred to as a 1-Dimensional figure - rather than the surface as a whole.

Thus letting the surface height be $z=f(x)$ the autocorrelation function $c(\beta)$ can be defined as

$$c(\beta) = \lim_{L\to\infty} \frac{1}{L} \int_{-L/2}^{L/2} f(x)\cdot f(x+\beta)\, dx \qquad (1)$$

Its Fourier transform the power spectral density $P(w)$ is relatec to $c(\beta)$

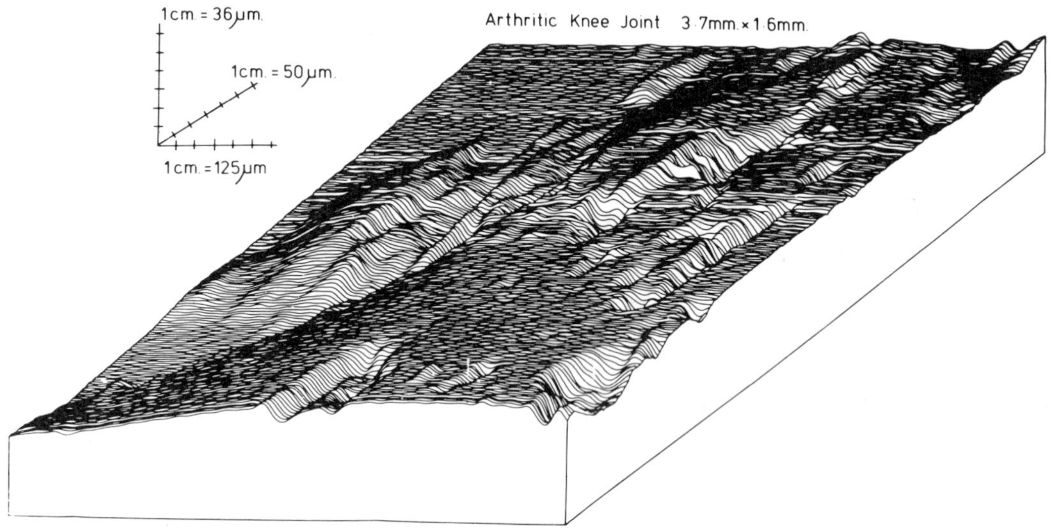

Fig. 2.—Topographic map of surface in isometric view again using a stylus instrument and computer scanning.

by the expression

$$P(w) = \frac{2}{\pi} \int_0^\infty C(\beta) \cos w\beta \, d\beta \tag{2}$$

Tallian et al.[20] used some aspects of random theory to determine the density of crossings of the profile with some arbitrary level. He did this in an attempt to predict asperity interactions through a thin oil film.

Greenwood[21] was the first to use a Gaussian distribution of peaks for his work on contact as we shall see shortly. Others[22] had also seen this possibility.

The breakthrough in the application of random process theory to tribology came with the realization that more than one feature of the texture could jointly be considered to be Gaussian. In its simplest form the slope g and the profile height z were both taken to be Gaussian by Kimura[23] but it was the application of the general case which produced the biggest impact.

Thus in general if $P_1, P_2, \cdots P_N$ are characteristics of the surface and $P_1, P_2 \cdots\cdots$ are obtained as a result of many random events,[24] then the joint probability density function (JPDF) of $P_1, P_2 \cdots$ is $f(P_1, P_2, \cdots)$ where

$$f(P_1, P_2 \ldots P_N) = \frac{1}{(2\pi)^{N/2} |M|^{1/2}} \exp\left[\frac{-\sum_{i,j=1}^{N} M_{ij} . P_i . P_j}{2|M|}\right] \tag{3}$$

Where $|M|$ is the determinant of M and M is given by the square matrix

$$M = \begin{pmatrix} d_{ii} & \cdots & d_{iN} \\ \vdots & & \vdots \\ d_{Ni} & \cdots & d_{NN} \end{pmatrix} \tag{4}$$

Where d_{ij} is the second moment of the variables P_i, P_j i.e., $E(P_i, P_j)$ and M_{ij} is the cofactor of d_{ij} in M.

It is the use of equation (3) which has dominated theoretical investigation of the tribological properties of surfaces.

The first application of it was due to Whitehouse and Archard[13,25,26] who linked the use of digital techniques with random process analysis and applied them, via equation 3, to derive expressions for many tribological properties of profile peaks in terms of three values of the autocorrelation function. The three points correspond to $c(\beta)$ for $\beta=o$, $\beta=\ell$, $\beta=2\ell$ where ℓ is the distance between digital measurements of z on the profile. The parameters $P_1, P_2 \cdots$ are taken as three consecutive distributions of ordinates of z, i.e., z_{+1}, z_0, z_{-1}

The variance-covariance matrix M then becomes

$$M = \sigma^2 \begin{pmatrix} 1 & \rho_1 & \rho_2 \\ \rho_1 & 1 & \rho_1 \\ \rho_2 & \rho_1 & \rho_2 \end{pmatrix} \tag{5}$$

and $f(P_1, P_2, \cdots)$ becomes $f(z_{+1}, z_0, z_{-1})$. ρ_1 is the correlation coefficient between successive ordinates and ρ_2 that between ordinates spaced by 2ℓ. Unity in equation (5) represents σ^2 (the variance of the surface). They made the crucial observation that all the relevant tribological parameters could be expressed in terms of just three numbers. This immediately meant a big saving in computational effort and made real-time measurements possible.

Because many surfaces exhibit exponential autocorrelation functions they constrained the general model to include only this case. Then $\rho_2 = \rho_1^2$.

Typical of the relationships are the mean peak height

$$\overline{\overset{*}{z}} = \frac{\sigma}{2N_p}\sqrt{\frac{1-\rho}{\pi}} \tag{6}$$

where N_p is the peak density given by

$$N_p = \frac{1}{\pi\ell}\tan^{-1}\sqrt{\frac{3-\rho}{1+\rho}} \tag{7}$$

Another important parameter is mean peak curvature $\overline{\overset{*}{c}}$ where

$$\overline{\overset{*}{c}} = \frac{(3-\rho)\quad\sqrt{1-\rho}}{2N_{p\ell}{}^{2}\quad\sqrt{\pi}} \tag{8}$$

Many other parameters are similarly derived.

Recent work by Whitehouse and Phillips[27] has reverted back to the general autocorrelation function where $\rho_2 \neq \rho_1^2$ yielding in place of equations 6, 7 and 8

$$\left.\begin{aligned} \overline{\overset{*}{z}} &= \frac{1}{2N_p}\frac{\sqrt{1-\rho_1}}{\pi} \\ N_p &= \frac{1}{\pi\ell}\tan^{-1}\left[\frac{3-4\rho_1+\rho_2}{1-P_2}\right]^{1/2} \\ \overline{\overset{*}{c}} &= \frac{(3-4\rho_1+\rho_2)}{2N_p\ell^2\sqrt{\pi(1-\rho_1)}} \end{aligned}\right\} \quad \sigma = 1 \tag{9}$$

One important relationship amongst the many others is the correlation between peak height and curvature

$$cor\ (c,\overset{*}{z}) = \frac{2(1-\rho_1)}{(6-8\rho_1 + 2\rho_2)^{1/2}} \tag{10}$$

which is always positive showing that high peaks are sharper than low ones.

Also they[20] proved that $\overline{N_p}\overset{*}{c}.\overset{*}{\sigma}$ is a constant ($\overset{*}{\sigma}$ is the standard deviation of the profile peak distribution). This demonstrated the dependence of some of the most useful parameters. Greenwood(21) regarded these as independent.

The other main value of this approach is that the tribological parameters associated with "each scale of size" can be investigated simply by changing ℓ. This changes ρ_1 and ρ_2 in a way dependent on the shape of the autocorrelation function and hence the surface. In section 4 some implications of this will be considered (Figure 3).

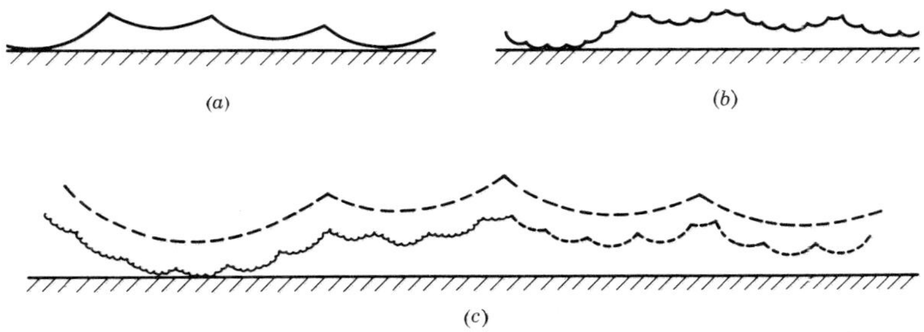

Fig. 3.—Schematic of a typical surface profile considered to be made up of different "scales of size" each with its own tribological function.

An alternative use of equation 3 in tribology was made by Nayak[28] whose results have greatly influenced later workers. His work is an extension to tribology of the pioneering work of Longuet-Higgins.[29,30] This work differs in three respects from that of Whitehouse et al.; it is based on the two dimensional surface, not the profile (2 D is used to represent the two independent variables x and y). It is based on continuous rather than discrete properties of random waveforms and it uses the power spectral density rather than the autocorrelation function.

Nayak uses differentials of the surface for $P_1, P_2 \cdots P_N$ in equation 3. Thus

$$P_1 = z,\ P_2 = \frac{dz}{dx} = g_x,\ P_3 = \frac{dz}{dy} = g_y,\ P_4 = \frac{d^2z}{dx^2} = c_x,\ P_5 = \frac{d^2z}{dxdy} = c_{xy}$$

$$\text{and}\quad P_6 = \frac{d^2z}{dy^2} = c_y \tag{11}$$

- a direct extension of the work of Rice[31] in communication theory.

By assuming that the surface can be represented by $z(x,y)$ where

$$z(x,y) = \sum_n F_n \ Cos\left(xk_{xn} + yk_{yn} + \varepsilon_n\right) \tag{12}$$

the second moments d_{ij} of the variance-covariance matrix are found to be m_{pq} where

$$m_{pq} = \frac{1}{2}\sum k_{xn}^{p} \cdot k_{yn}^{q} \; Fn \tag{13}$$

where F_n is the nth Fourier coefficient.

The variance-covariance matrix is expressible completely in terms of m_{pq}. Hence the tribological parameters can be evaluated simply by restricting the range of integration of $f(z,g_x,g_y,c_x,c_{xy},c_y)$.

Nayak took the specific case where the surface is isotropic in which case all parameters are in terms of m_o, m_2 and m_4 where m_o is the variance of the surface $(R_q{}^7 5\sigma^2)$, m_2 is the variance of the slopes of the surface and m_4 the variance of the second differential of the surface (assumed to be the curvature).

He defines a parameter $$\alpha = \frac{m_o m_4}{m_2^{\,2}} \tag{14}$$

which is meant to represent the bandwidth of the surface.

This is slightly different from others(32,33) but essentially all misinterpret the meaning of such a parameter. It is concerned with the shape of the spectrum and not the bandwidth as Sayles(34) rightly points out!

For comparison purposes the density of summits D_s not profile peaks D_p is given by the equation

$$D_s = \frac{1}{6\pi\sqrt{3}}\left[\frac{m_4}{m_2}\right]^{1/2} \tag{15}$$

This can easily be related to profile properties for isotropic surfaces because letting D_p be the density of profile peaks (N_p is discrete equivalent) then

$$D_s = 1.2\, D_p^{\,2} \tag{16}$$

This is surprisingly 20% different from the tribologists usual habit of simply squaring the profile density, $D_s = D_p{}^2$.

The mean summit curvature is $\frac{8\sigma''}{3\sqrt{\pi}}$ compared with $\sqrt{\frac{\pi}{2}}.\sigma''$ where σ'' is the rms curvature for the whole surface (m_4).

In fact Nayak showed that summits are on average about 0.75 higher than profile peaks. The shift actually depends on α but the dependence is small. Recent work(35) suggests that practical curvatures do not change as rapidly as theory predicts.

The relationships are not so simple for anisotropic surfaces. Nayak(36) has examined the requirement for assessing anisotropic surfaces and comes to the conclusion that five non-parallel profile traces would be needed. Sayles and Thomas reduce this to two at right angles for certain types of surface such as grinding. Further theory on anisotropic surfaces has been examined by Longuet-Higgins in detail but as yet no general theory for summit curvatures and such like parameters has been proposed; in principle it is similar to that already described but the calculations are more involved.

This represents the surface metrology situation as it is today when applied to tribology. It represents a considerable improvement on the situation fifteen years ago but there is a basic question which is as yet unanswered. The question is which of the two random process theories is best suited to the practical application in tribology.

Comparison of Theories

A number of people have set out to resolve this issue[35,37] because it is of vital importance. The main criticism levelled at Whitehouse and Archard[25] was that the exponential autocorrelation function was unsuitable for analysis because m_2 and m_4 were undefined. This has been overcome by making the autocorrelation function general.[27] It is still only valid for profiles but for isotropic surfaces the same relationships hold (equation 16 for example). Nayak provides a 2D analysis which only holds if there is a well defined limit to the detail on surfaces. Doubts have been expressed as to whether this is realistic.[35] The basic question underlying all this argument is whether or not peak curvatures, slopes, etc. are in fact intrinsic properties of the surface. Looking at surfaces through a scanning electron microscope would lead one to believe that they are not! It is becoming clear that the moments m_o, m_2, m_4, referred to earlier will have to be redefined in terms of upper and lower angular frequencies ω_h and ω_L. Thus

$$m_i = \int_{\omega_L}^{\omega_h} \omega^i P(\omega)\,d\omega \Big/ \int_{\omega_L}^{\omega_h} P(\omega)\,d\omega \qquad (17)$$

ω_h is usually a limit of the instrument, ω_L would be most probably restricted by the function. It could be the Hertzian zone width for instance. Whitehouse and Archard in their work restricted the bandwidth of interest to the main structure; that corresponding to spacings which made ordinates independent (Figure 4). This is not necessarily the best scale of size for

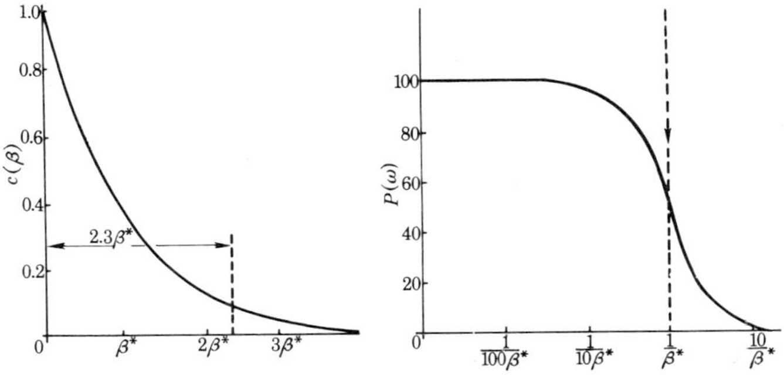

Fig. 4.—The relationship between the independence distance for sampling and an exponential autocorrelation function as well as the equivalent spatial frequency relative to the power spectral density.

all uses but it was a first step towards recognizing the problem. The distance over which the correlation fell to a low value was called the independence distance equal to $2.3\beta^*$ where β^* represented the decay time constant of the autocorrelation function. An equivalent length derived from the power spectrum and called the average or rms wavelength λ_a and λ_q was soon developed[36] for unifying surface metrology. In reference[36] the effect of filtering on parameters was discussed in detail. Nayak in later work[37] went into some detail in filtering and succeeded in showing that filtering the surface was not necessarily the same as filtering the profile.

The big advantages of discrete analysis[25,27] are that first it represents what is actually being measured on random surfaces and second that it can therefore be directly compared with the digital analysis of skewed distributions such as occur after run-in. This latter aspect is much more difficult using Nayak's approach.

The unpleasant truth is that researchers will have to use whichever theory fits better into practical measurement. Digital measurement is here to stay and already workers are meeting the digital analysis problems head-on. Using the discrete approach results can be verified, and the effect of filtering found directly.[39,40] As an example of what can be found peak curvature can change by 100:1 depending on *ℓ for the same profile* (Figure 5). This revelation casts serious doubts on many results displayed in the texts of recent papers.

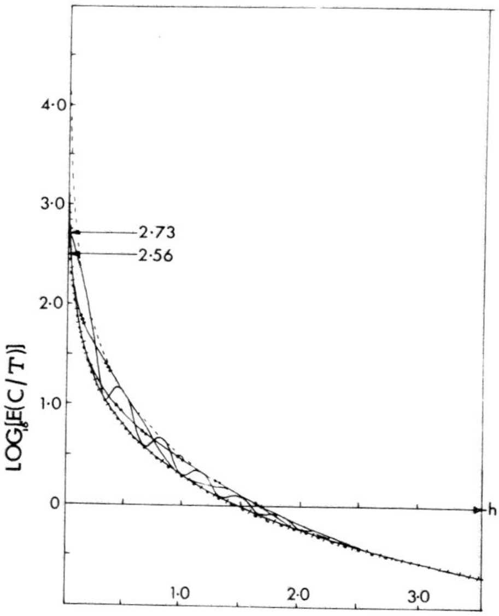

Fig. 5.—The effect of sampling on the measurement of mean peak curvature (E(C/T). The ordinate is in terms of the logarithm of the mean curvature and the abscissa has for a unit the value β for the exponential decay of the autocorrelation function. The spread of the curves shows results for different surfaces. The continuous line-is for example a turned surface the crossed line at the other extreme is for a ground surface.*

At present theoretical investigation tends to use Nayak's approach yet digital measurement implicitly implies Whitehouse's. In the limit as $\ell \to o$ they are the same so there is no theoretical conflict.[27,28] Time will tell which is of most practical benefit. Perhaps this will resolve itself more readily when the discrete approach is extended more completely to 2D. Theoretically[27] and practically[35] this is now being done.

Having established that the tools for theoretical and practical examination of surfaces are now adequate, we shall see what use has been made of them in tribology, particularly in wear. Friction, although very important, will be left out except incidentally. Contact mechanics, however, cannot be left out because contact is central to the whole issue. It will, therefore, be dealt with first.

CONTACT MECHANICS AND RELATED PROBLEMS

Mode of Deformation

Whether a surface in contact deforms elastically or plastically is of considerable importance. There is some doubt as to whether wear is confined to the removal of metal. It has been suggested by Summers-Smith[41] for example that metal movement is equally a wear mechanism; the surface has *changed*. This seems to provide a basis for a very much more comprehensive definition of wear. Plastic deformation is, under these conditions, very germane to the wear regime and so also is the boundary between elastic and plastic deformation. Functionally, this subject is important in determining the stiffness of joints, and in electrical and thermal conduction. Also in pure rolling, mechanical contact determines stresses and strains which produce pitting, spalling and scuffing.

As a historical fact early pioneering work by Bowden and Tabor[42] did not attach too much importance to texture. They argued that plastic flow would occur at the asperities when a surface was loaded until the load W was equal to the real area of contact A_r multiplied by the flow pressure of the softer material p_m. The shapes of the asperities were not regarded as especially important. Surface texture did not enter prominently into the contact mechanism except to ensure that the deformation was plastic. Problems soon arose, however, using this approach because it was not possible to explain running-in for example.

To answer this and similar problems attempts were made[43,44] to develop elastic models of contact. In particular criteria for the complete elastic compression of asperities were developed. One such criterion[43,44] related the profile absolute slope $\overline{g}$ to the hardness H and composite elastic modulus $\overset{*}{E}$. Thus if

$$\overline{g} < k\, H/\overset{*}{E} \qquad (18)$$

no plastic flow would occur. k is a constant taking account of asperity shape and having a range of values from 0.8 to 2.0.

Then came a conceptual breakthrough by Archard[45] who proved practically and theoretically that well-known contact laws would still be obeyed even for elastic deformation. He used for his analysis a "scale of size" argument (Figure 3). By postulating a profile as the superposition of different scales of asperity he was able to show for instance that it was possible to get a proportional relationship between real area of contact and load, a conclusion verified much later for random surfaces.(46) Once this elastic possibility was realized surface texture took on a much more important role, but the problem still remained of what determined the mode of deformation.

Since this work many researchers have derived indices with which to predict the mode of deformation of asperities on a surface under load. In what follows we shall attempt to extract the key topographic features. Perhaps the best known is due to Greenwood and Williamson(47) who were among the first to take into account the statistical nature of the asperity distribution on surfaces. Assuming that the peaks were distribution in a Gaussian manner in height on a surface profile they derived what they called a plasticity index ψ_1, where

$$\psi_1 = \frac{\overset{*}{E}}{H}\sqrt{\frac{\overset{*}{\sigma}}{R}} \tag{19}$$

In equation 19 $\overset{*}{\sigma}$ is the standard deviation of the peak distribution and R is the average radius of the peaks. In fact Greenwood and Williamson assumed that all asperities had constant radius. If $\psi_1>1$ a large part of the contact will be plastic whereas if $\psi_1<0.6$ the chances of plastic flow are remote.

A modification of ψ_1 based on random theory was due to Whitehouse and Archard(25) who extended the model to allow peaks to have a distribution of curvatures. In addition every parameter was related to profile parameters, the asperity model was dispensed with. Thus ψ_1 became ψ_2

$$\begin{aligned} \psi_2 &= 0.69\,\frac{\overset{*}{E}}{H}\,\frac{\sigma}{\overset{*}{\beta}} \qquad (25) \\ \psi_2^1 &= \frac{\overset{*}{E}}{H}\cdot\frac{\sigma}{\overset{*}{\beta}} \qquad (48) \end{aligned} \tag{20}$$

Where σ is the rms value of the surface and $\overset{*}{\beta}$ is related to what is known as the correlation or independence distance of the surface. The plasticity index ψ_2 was shown by Onions and Archard(48) to allow more plastic deformation than ψ_1 for the same numerical value. In other words ψ_1 under-estimates plasticity. This seems sensible because the Whitehouse and Archard theory(9) showed that the higher peaks, those which were more likely to make contact, had sharper radii than the lower ones.

Other indices include that by Mikic and Roca(49) ψ_3

$$\psi_3 = \frac{\overset{*}{E}}{H}\tan\theta \tag{21}$$

where $\tan\theta$ is the absolute mean slope of the surface. ψ_3 is such that at least 2% of asperities will deform plastically when $\psi_3 \geq 0.25$; for $\psi_3<0.25$ contact will be predominantly plastic. Notice that in this case the radius in ψ_3 has been replaced by a slope and notice also that $\sigma/\overset{*}{\beta}$ in ψ_2 could also be construed as a slope.(50)

Gupta and Cook(51) define a topographic index ζ essentially as a measure of the ratio of nominal to real contact area of two surfaces in contact.

Thus

$$\zeta = \frac{1}{\sqrt{2D_s}} \frac{1}{R_m} \left(\frac{\overset{*}{E}}{H} \right) \tag{22}$$

where D_s is the average number of peaks per unit area (or the smaller the two densities when the surfaces are different).

$$R_m = \frac{R_1 R_2}{R_1 + R_2}$$

where R_1 and R_2 are the average radii of asperities of the two surfaces. In their analysis the distribution of radii is assumed to be log-normal, and Gaussian.

Another index of plasticity due to Nayak was developed using full 2 dimensional random theory,[52] but was dependent on results reported by other workers which have since been seriously queried.[53] Thus

$$\psi_4 = \frac{E}{H} \sigma_1' \tag{23}$$

where σ_1' is the rms value of the differential of that component of a surface in contact which has a narrow spectrum. Nayak shows that if $\psi_4 > 5$ then most asperities will deform plastically.

Other workers have approached the problem somewhat differently.[54] Using the spectral moments of a profile m_o, m_2, m_4 described earlier they determine the probability density of peak curvature as a function of height. Then, assuming that a profile comprises of randomly distributed asperities having spherical tips, they estimate the proportion of elastic to plastic contact of the profile on a flat as a function of separation. As a basis for deformation they use the Hertz elastic deformation criterion.

Thus if R_p is the maximum peak height and h is the separation of the flat from the mean plane the probability of elastic deformation becomes P_{eh} where

$$P_{eh} = Prob\left[c < \left(\frac{H}{E}\right)^2 / (R_p - h)\right] \text{ given that } \overset{*}{z} \geq h \tag{24}$$

So that in this method the key surface parameter is in fact a function which describes how curvature c changes with asperity height $\overset{*}{z}$. In other words the joint probability density function between peak height and curvature, $f(c|\overset{*}{z} \geq h)$.

The basic problem with this technique is that of estimating R_p. It is not a limiting parameter; the longer the profile, the bigger R_p. Other aspects of this theory have been criticized[55] but it does not affect the validity of this approach. The next logical step would be to expand the treatment to 2D.

The most recent and comprehensive index yet devised is by Francis.[33] He defines what amounts to two indices ψ_5 and ψ_6.

where

$$\left.\begin{aligned}\psi_5 &= \frac{\sigma'}{\underline{k}^{1/2}}\ \frac{E_1}{P_m}\\ \psi_6 &= \frac{\sigma'}{\underline{k}^{1/2}}\ \frac{\overset{*}{\bar{E}}_1}{P_m}\end{aligned}\right\} \qquad (25)$$

where $\underline{k}$ is

$$\sqrt{1.5.\frac{\sigma'^2}{\sigma\,\sigma''}}.$$

and is a parameter similar to Nayak's α to describe the spectral nature of the surfaces. σ'' is the standard deviation of the second differential of the gap between two surfaces and σ' is the rms slope. σ is equivalent to R_q the rms value of the gap itself. $\overset{*}{E}$ is the harmonic mean of the moduli of the two surfaces P_m is the uniaxial yield stress. In general ψ_5 is the most important parameter; the larger the value of ψ_5 the greater is the deviation of the contact mechanics from totally elastic behaviour. For a given value of ψ_5 the departure from elasticity increases with decreasing strain hardening. For $\psi_6 < \sim 23$, ψ_5 is a better predictor than ψ_6. The elastic model is valid for any contact for which $\psi_6 < 4$ and the complex iterative plastic-elastic model must be used for $\psi_6 > 14$. Between 4 and 14 the behaviour depends on specific values of the parameters.

In equations 18 to 25 the physical parameters $\overset{*}{E}$ and H are common but there seems to be a variety of topographic parameters basically involving either first differentials ($\bar{g}$, $\tan\theta$, σ_1', σ', σ/β^*) second differentials (σ'', $\frac{1}{R}$) and sometimes a height parameter (σ_z^*) or peak density (D_p, $f(\sqrt{k})$). Sometimes, the index is related completely to peak behaviour (ψ_1, ζ), but more often to general properties of the profile, due largely to the properties of random signals which allows slopes to be taken as a measure of peak curvature. Of the indices mentioned most have been developed with restrictive assumptions. ψ_1 depends on constant curvature, ψ_2 has an exponential autocorrelation, ψ_3 assumes constant contact size, ζ makes the peak radii independent of height, ψ_4 depends on partitioning a profile into two components. Probably those due to Francis are least restrictive assumptions but are more difficult to interpret. From the point of view of measurement it is easier to measure slopes than curvatures and peak densities are difficult to measure realistically.(39) It seems unlikely that the best topographic parameter will be found for this important issue until it is possible to develop surfaces in which slopes and peak curvatures can be varied independently.(38) If the criterion for use is that of majority agreement, then ψ_2 and ψ_3 should be used because from the Whitehouse-Archard theory(25) $\tan\theta = 0.24\,\frac{\sigma}{\beta^*}$ which when inserted into equation (20) gives value of $\frac{E}{H}\tan\theta = 0.24$ for plasticity corresponding to $\psi'_2 = 1$ which agrees with the value 0.25 obtained by Mikic and Roca.

What emerges from these formulae is that characteristics of the surface texture determine the mode of deformation, the normal load has hardly any effect over a wide range of values.

Real Area of Contact - Asperity Persistence and Related Subjects

Other parameters deemed useful in contact studies include the bearing ratio curve because it has been shown by Bush et al.[46] that if under purely elastic conditions a random surface is loaded then the bearing ratio of real contact area to nominal of a flat pressed to the level z is given by

$$\frac{t_p(z)}{2} = \frac{A_r(z)}{A_n}$$

This means that if the load-compliance (stiffness) characteristics are known, the real area of contact could be found, or at least a good estimate of it made for a given load by referring to a bearing ratio curve derived from the surface. In the case of plastic deformation it is often acceptable to regard the bearing ratio as the ratio A_r/A_n at any height. The nominal area is taken to mean the physical area of the surface as defined by its dimensions of width and depth.

For plastic conditions the physical shape of the asperities is not important - only the homogeneity of the material;[53] but the shape does become more relevant for elastic contact, and spherical[47] or parabaloidal maxima[46] have so far been considered. In the latter case for large separations the normal force - area of contact relationship is linear; the constant of proportionality Ω being given by

$$\Omega = \frac{E^* \sigma'}{\sqrt{\pi}} \tag{26}$$

which highlights again the importance of the absolute slope in wear phenomena.

The question of whether two rough surfaces in contact can be considered to be equivalent to the contact of a composite rough surface on a flat has been discussed by many authors.[23,55-58] The general conclusion is that it can provided the surface slopes are low. Under these circumstances, there is only a small lateral component of either the force or the compliance; that is, contacts on shoulders rather than the peaks do not seriously affect the contact mechanics. Providing that those assumptions hold and that contacts and asperities are independent, the following relationships hold. See reference 55 for example.

$$\left.\begin{aligned} \sigma_T^2 &= \sigma_1^2 + \sigma_2^2 \\ C_T(\beta) &= C_1(\beta) + C_2(\beta) \\ \sigma_T^2 H_p^2 &= \sigma_1^2 H_1^2 + \sigma_2^2 H_2^2 \\ \sigma_T'^L &= \sigma_1'^2 + \sigma_2'^2 \\ \sigma_T''^2 &= \sigma_1''^2 + \sigma_2''^2 \end{aligned}\right\} \tag{27}$$

where H represents the density of crossings of the profile with the mean line. In the case where lateral effects are present and interaction can be taken into account O'Callahan and Cameron(56) have introduced, in an elastic regime, what they refer to as a "mutal insinuation" parameter v which is a measure of the interaction between contacting elastic asperities. In their analysis again we meet with the important slope parameter, because

$$v^2 = \lambda_2 u_2/(\lambda_2 + u_2) \tag{28}$$

where λ_2 and u_2 are the second moments (m_2) of the spectral density of the two surfaces in contact and therefore represent standard deviations of the slopes of the surfaces.

Interactive effects of asperities in plastic flow have been considered by Pullen and Williamson,(53) and others.(11,59,60) In these cases the interaction was studied when the bulk material was in a state of complete constriction. Under this condition the bulk could be considered to be in a state of hydrostatic stress. Using a slip line theory(60) Mitchell and Rowe(61) attempted to explain asperity persistence - the ability of small asperities to withstand tremendous pressures well above their flow pressure - by means of plasticity theory. They(61) also came to the conclusion that the most important surface parameter in all aspects of leakage was $\overset{*}{d}/\overset{*}{\sigma}$

where $\overset{*}{d}$ is the distance between mean peak and mean valley heights and $\overset{*}{\sigma}$ is the rms value of the peak distribution. They concluded that for a given apparent contact pressure $\frac{W}{A_n}$ the real area A_T is least for small $\overset{*}{d}/\overset{*}{\sigma}$ ratios, the most easily compressed surface would be one in which $\overset{*}{\sigma}$ is small and $\overset{*}{d}/\overset{*}{\sigma}$ is large. Because of the problem of "scale of size" they restrict their peaks to the most significant ones as determined from the autocorrelation function.

Earlier work on leakage had been done by Tsukizoe and Hisakado(62) who realized the importance of texture in determining the mean spacing between surfaces. They were also the first to bring back attention to the fact that the actual or nominal size of surfaces in contact could not be ignored.

Nuri and Halling, having carried out many significant experiments in contact,(63-66) emphasize two important contact features one physical and one geometrical. The physical parameter is strain hardening which will be referred to later; the geometrical parameter is the nominal area. Halling maintains that the population of peaks encountered by mating surfaces depends on the nominal size of the contact area. This changes the load-separation characteristic expected from continuous probablistic theory. The essential point is that not only do the number of peaks reduce as the area is reduced but the *nature* of the peaks also change; there is less chance of getting high peaks (which have the higher curvatures). Thus the average clearance and the minimum separation depend on the nominal area. The point about the nature of peaks is an interesting one because it illustrates one of the differences between surface phenomena and communication theory in which a waveform may well have the same shape as a profile. The effect cf sample length is to reduce the stability of the result, odd peaks are rarely important. In contact mechanics the high peaks can be a dominant factor. Halling and others(62,67,68) have helped to show this difference. Notice that this effect is not the same as the way the density of peaks contacting changes with load. Thomas(69) investigated the effect of correlation length[25] on clearance and mean gap width. He showed that as the correlation length increased, the clearance decreased. This conclusion makes sense because the correlation length divided into the nominal dimensions of a surface is a measure of the number of peaks. So for a constant size of surface one would expect, as Halling showed, that the actual number of peaks reduce.

Contact Size and Related Subjects

The important question of the distribution of contact size has continued to involve workers. Sometimes it is not enough to know the real area of contact, in thermal and electrical contact for instance. Holm[70] for example, showed that the conductance of a single spot is not proportional to its area but to its radius. It is therefore of importance to know how the area is broken down in terms of size of contact shape and density. This topic has been investigated by many people including Hisakado and Tsukizoe[71] who show that the number of contacts decrease as the range of slopes increase and that the mean radius increases with the range of slopes. The resultant effect, that the real contact area as a whole is not much influenced by slopes, is misleading because it hides the two very important opposing mechanisms which are occurring as the range of slopes is changed.

Microcontact sizes, however, do depend more on the surface slopes according to Jones et al.[72,73] They use Whitehouse and Archards' theory[25] to investigate microcontacts. They find, not surprisingly, that the scale of size is a vital consideration. What is more surprising is that the scale of most importance for this functional consideration is about one third of the spatial size of the main structure, which means sampling less than β^* apart.

Adhesion, too, depends considerably on the surface texture and is a vital factor in adhesive wear. It is so relevant that an "adhesion index" similar to the plasticity index could be developed. Early work involving the contact of a sphere on a flat[74] has been extended by Fuller and Tabor [75] to include surface finish using the results of Greenwood and Williamson (47).

The index showed that there would be negligible adhesion unless

$$\frac{E\sigma^{3/2}}{\sqrt{R}\,\Delta\gamma} < 10 \tag{29}$$

where $\Delta\gamma$ is the surface energy density, R is the peak radius as in (47). Later workers[76] have extended the treatment to include the complete 2D surface. They find that although the agreement between[75] and[76] is acceptable at small asperity extension, when it is large, the differences become serious.

This section has been looking at the influence of surface texture on some of the basic mechanisms upon which wear depends. In the next sections specific types of wear will be considered starting with those in which damage, at least in the initial condition is small. Later dry solid - solid sliding and similar subjects will be briefly examined.

LUBRICATED CONDITIONS

Pitting, Spalling Wear; Hydrodynamic, Elastohydrodynamic Lubrication

Early work in lubrication assumed smooth surfaces[77,78] but the importance of surface texture soon became apparent, especially in wear conditions. It soon became obvious that the mechanism of asperity contact through thin films lay provided the key to many of the failures.

Basically, when the oil film becomes of the same order of size as the roughness, the issue is: what proportion of load is carried by the film and what proportion is carried by asperities. This in turn determines the stresses transmitted to the bulk material and, by implication, the likelihood of failure. The problem is very complex because the film thickness of the oil depends on normal loading, spin-slide ratios and macrogeometry. It also depends on the roughness itself. Fatigue failure is further complicated by surface inclusions or flaws.

A well-known criterion relating fatigue wear such as pitting to the ratio of film thickness and roughness $\frac{d}{\sigma}$ has been found by Dawson.[79] This ratio has been called the D ratio; σ is the composite surface finish and d is the mean line separation between the surfaces. Onions and Archard[48] verified this ratio. Values of less than unity often lead to failure.

Tallian has been instrumental in developing both the theoretical and practical aspects of this problem particularly relating to ball bearing research. Theoretical advances in this field concerned the use of Rice type crossing theory[31] to establish the probability of meeting an asperity at a given height in an attempt to predict asperity contact for a given film thickness.[5] In terms of the surface theory mentioned earlier of prime interest is the *presence of a peak at a given height rather than its curvature*. This is the *essential difference* between the influence of surface texture in pure contact and contact through films. Interest changes from $f(c/\overset{*}{z},z)$ to $f(\overset{*}{z}/z)$. Obviously, it is only when pure rolling takes place that normal contact situations and hence $-(c/\overset{*}{z},z)$ should be considered first.

He and his colleagues later took up the 2D aspects of surfaces, but did not attempt to establish whether the simple addition of roughness to thin film e.h.l. for smooth surfaces is valid. However, he effectively demonstrated that in matters as complicated as e.h.l. the use of mathematical modelling to aid the physical understanding of the mechanism was invaluable. In fact he was able to use statistical mathematics to unravel the practical measurement of asperity interactions through the oil film. He used for his measurement the idea that when two asperities made contact, the electrical resistance between the rotating members was reduced to a low value. By monitoring the resistance, therefore, he was attempting to infer the degree and density of contacts. He seemed, however, sometimes to get low values. Greenwood[38] asserts that this is inevitable in a contact in which there is no slip; a good electrical contact always requires some degree of slide in order to breakdown any oxide film or high resistance deposit on the surface.

Perhaps the biggest advance in the relationship between texture and full film lubrication in the last few years was made by Christensen.[80] He developed a modified form of Reynolds equations which took account of the stochastic or random nature of surfaces. He split the geometry of the gap into two parts: one macroscopic - the deterministic component, and one microscopic - the random component. The texture he split up into two components which he investigated separately. In one case he made the lay of the texture parallel to the direction of motion and in the other case transverse. He therefore had two quite different hydrodynamic ecuations to solve. Normal Reynolds equations only contain the mean functions of film thickness and these are smooth. Christenson found that although the form of the Reynolds equation which are applicable to rough and smooth bearings are approximately the same, they cannot be brought into complete accord. It is therefore impossible to account for surface roughness by considering a geometrically smooth bearing with a suitably adjusted film thickness. Subsequently other workers have extended this to include rolling(81) and the general result.(82) Johnson et al.(81) showed that the separation of mean planes is greater than the smooth case for transverse texture and less for longitudinal. For isotropic surfaces they more or less balance.

One part of the treatment which still needs exploring is the effect of bandwidth of texture, i.e., the effect of nominal area of contact zone. Tallian developes many empirical models for pitting and smearing. He concluded that the surface roughness rms height and slope are the important parameters. Christensen on the other hand did not find the distribution of heights important.

Scuffing

Some work on e.h.l. supports the contention that the load on the inlet side of the contact is mainly supported by the film.(81) However, work by Dyson,(83) Cheng and Dyson,(84) suggests that surface roughness could be very much more important than hitherto thought possible. By direct computer simulation(81) of the Hertzian region he showed that an appreciable proportion of the load is carried by the asperities well before scuffing. He and others(84) have come to the conclusion that under heavy loading conditions the degree of asperity interaction is comparable with even the static loading conditions mentioned in the contact section. In fact it seems likely that the degree of interaction could provoke thermal instability in the inlet region which for scuffing would be a prelude to failure. The key mechanism seems to be that because of elastic asperity interaction the surfaces cannot get close enough to each other to generate the necessary hydrodynamic pressures to keep lubricant between asperities; the viscosity of the lubricant cannot increase enough because the pressure is too low. When this condition arises scuffing occurs. In his experimental procedure Dyson makes full use of digital methods. He takes digital records of many axial traces of two discs and runs them together in a computer to find the degree of interaction corresponding to a given separation. This technique was also applied by Onions and Archard.

The latest in the theory of scuffing uses, like Dyson, a mixture of real data and mathematical modelling.(85) Data obtained from surfaces or generated numerically are inserted into a modified Reynolds equation to obtain pressure and shear flow factors typical of that data set. These flow factors are expressed as empirical relationships in terms of $\frac{d}{\sigma}$ and the correlation length.(86) If $\frac{d}{\sigma} < 3$ the relationship between flow ∇ and finish becomes

$$\nabla = 1 - 0.9\exp\left(-0.56\,\frac{d}{\sigma}\right) \tag{30}$$

For $\frac{d}{\sigma} > 3$ the flow ∇ is the same as that for smooth surfaces $(\nabla = 1)$.

Anisotropy of the texture causes similar problems to flow as Christensen found.[80] The advantage of this technique is that, although it is somewhat laborious, it only needs to be done once for any particular geometrical configuration. As Dyson says it is capable of dealing in a general way with all the important influences of surface texture *at the same time* (in this instance hydrodynamic effects and interactive effects).

General Comment on Wavelength Dependence

Different degrees of importance have been attached variously between the small asperities and waviness, especially in pure rolling as in ball bearings. Here the function need not be fatigue wear but could equally be vibration and acoustic noise. Some workers dismiss track waviness as unimportant but it seems that its slope at least can be used as a process control parameter. Also, there seems to be evidence that a "scale of size" problem exists. It appears[87] that micropitting is caused by asperity interaction whereas spalling has dimensions comparable with the Hertzian contact zone.

RUNNING - IN AND MILD WEAR

General

Whereas the emphasis in the previous section was aimed at recognizing those surface features which contributed to damage, in this section the topographic changes which constitute running-in will be considered. This is to get some idea of the necessary topographic changes to achieve running-in and therefore opens up the possibility of manufacturing run-in surfaces.

Running-in is an effective way of *matching two components* in a functional situation. It is necessary largely because of our inability to make parts completely interchangeable; there is always some final fitting which has to take place in order to get sustained satisfactory performance. Basically, during running-in the surfaces conform geometrically and to some extent physically to each other. High spots are removed, voids filled and overall shape matched. Summers-Smith[41] distinguishes two mechanisms, one in which metal is moved by plastic deformation and one in which a small amount of metal is removed. The latter is called a "mild" or "normal" wear regime in which metal removal is confined to a depth which is not greater than the surface finish depth.

Plastic Flow

The first type of mechanism is similar to roller burnishing, the asperities literally get squashed down, the total metal remains the same. Surface features larger than the Hertzian zone width will not be correlated. One would expect some sort of smoothing to take place and this is so,[68] the width of the smoothing window is the same as the Hertzian contact zone see Figure 6(a). This has also been shown by Leaver et al. practically.[88]

Another effect on the geometry which tends to get produced in running-in, especially if sliding is also present, is directionality; slopes opposed to the direction of motion become steeper. As a result, a surface tends to

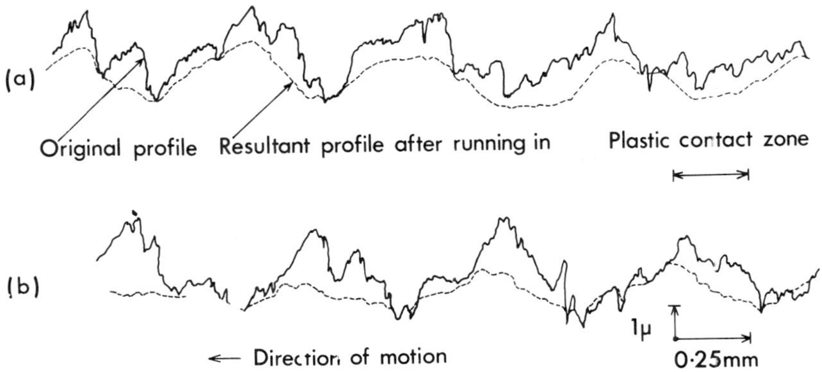

Fig. 6.—Shows (a) how the contact size acts as a smoothing window of the smaller wavelengths in a running-in experiment. (b) how when the running-in has a sliding component directionality is easily set up.

develop a sawtooth pointing in the direction of motion, Figure 6(b). This has been predicted by slip line analysis[89] and observed.[90] This mechanism produces a skew or assymmetry in the slope distribution anc as a result can cause profound frictional changes.[91] If, for instarce, the motion is reversed in direction, friction and wear increase and can degrade the bearing. A similar "directional" effect can be caused by machining.

The effect of this directionality is (because the total material is conserved) to reduce the R_a value of the surface. Ku and Li[92] have measured 40% changes in R_a but it is questionable whether this is solely due to plastic flow. Nevertheless, for a small change in R_a a large improvement in life can result, this is why so much importance is given to the influence of surface texture on running-in.

Mild Wear

Metal removal occurs as a result of adhesive and abrasive wear to an extent defined by Kaliszer[93] as "Normal." It fundamentally has the effect of rubbing off the higher asperities. This mechanism is called truncating or censoring the height distribution.[94] This wear produces a completely different profile change than does the former the bottom of the profile is retained while the upper gets removed progressively (Figure 7). It is a simplification to cut off the distribution sharply at any level. Archard has suggested that each plateau has a finite radius of curvature such as will support the normal load, and that the running-in process is merely the transition from plastic deformation condition to the elastic. There is evidence that the upper part gets much smoother, and yet retains the shape of a Gaussian distribution of smaller scale. Williamson[95] observed such characteristics of surfaces and described them as "transitional." They often occur on surfaces made by multiple processes and so justifies the use of multiprocessing of parts in order to short-circuit the lengthy running-in operation. The problem is, of course, that only certain aspects can be simulated because in the final analysis *the two mating surfaces are needed*. As Archard predicted the average peak curvature is the parameter which changes most rapidly during run-in (Figure 8). The common average

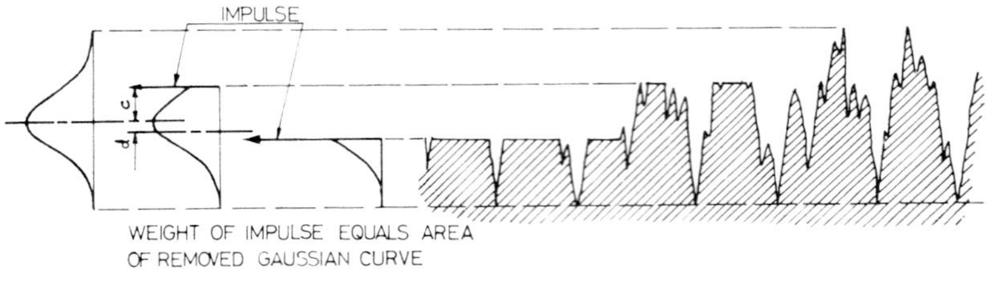

Fig. 7.—A simplified wear model showing how the surface can be imagined truncated at different levels.

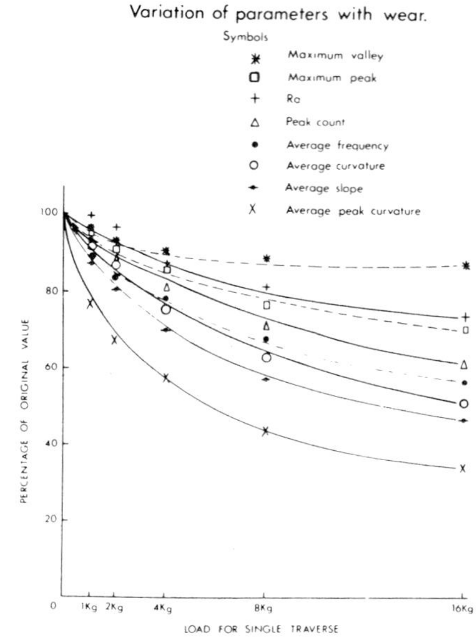

Fig. 8.—Shows how typical parameters of a particular profile change as a running-in experiment proceeds. The abscissa corresponds to accelerated loading; the origin represents the original profile. All values for the ordinate are measured relative to the original values.

type of height parameters (R_a,R_q) do not change quickly. In fact they are almost predictive of the run-in value. In some respects this is what is wanted - a parameter measured from the unworn surface which predicts the value of the run-in surface. Obviously, an opportunity is here to develop a suitable algorithm to simulate run-in.[68] Theoretical investigation on the surface texture parameters of a random wave can be used to get an idea of which mechanism is dominant in the run-in.[96] Take for example the autocorrelation function for a given lag β the original correlation coefficient ρ can be used to express the truncated one ρ_T Figure 9.

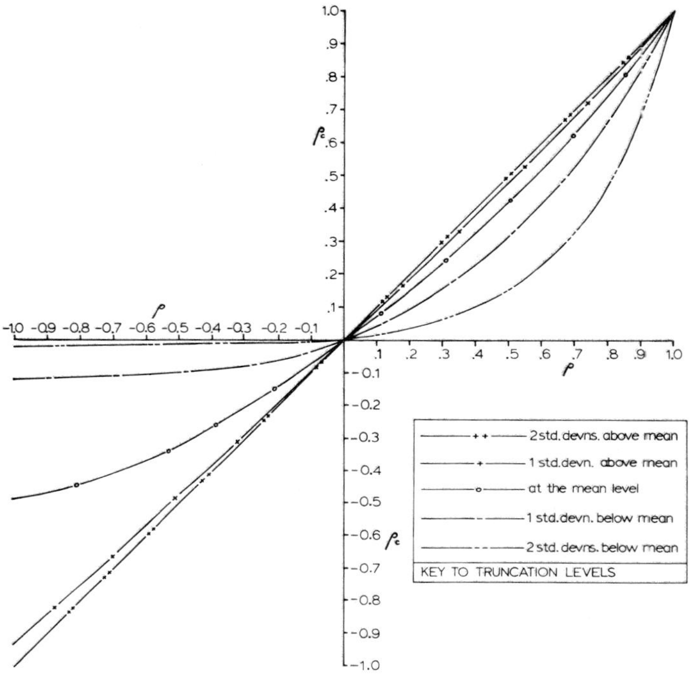

THE CORRELATION BETWEEN ORDINATES OF A TRUNCATED GAUSSIAN WAVE AS A FUNCTION OF THEIR UNTRUNCATED CORRELATION

Fig. 9.—Shows the relationship between the autocorrelation of a truncated waveform ρ_C *and that of the original* ρ. *It is obvious that there is not a big change until the truncated passes the mean line.*

$$\rho_T = \left(\rho \sin^{-1} + \rho\frac{\pi}{2} + \sqrt{1-\rho^2} - 1\right) \Big/ (\pi - 1) \tag{31}$$

This equation is interesting because it shows that the censored correlation coefficient is lower than the original. The spectral bandwidth *increases*. This is exactly opposite to the *decrease* in the spectrum caused by the other mechanism! This suggests that the autocorrelation function could be used to tell which is the dominant mechanism in a given situation. It could also be that the structure function $S(\beta)$ which is related to $C(\beta)$ could highlight the change.[68,97] (The structure function $S(\beta) = 2\sigma^2(1-\rho^2)$).

The use of common parameters to monitor running-in is not successful they all behave in the same sort of way. Plotting percentage change against amount of truncation often shows a Gaussian type of shape. This behaviour has been anticipated theoretically[96] and by simulation.[98] The use therefore, of more than one commonly used parameter such as R_a, R_t, R_q, etc. is a waste because it only gives redundant information.

In fact the shape of this truncation curve is not quite Gaussian

Figure 10(a). If c' is the level of truncation and d' is the truncated mean line level then for R_{qt} the relationship is

$$R_{qt} = R_q[(1-\Phi(c'))(1-2c'd' + c'^2) +\phi'(c'-2d') + d^2] \tag{32}$$

which for small truncations becomes $R_{qt} \sim R_q\ '(c')$

where Φ is the Gaussian cumulative Distribution function and ϕ' is the corresponding density function.

Essentially it is the shape of the amplitude distribution which changes most during run-in; therefore, one would expect a parameter or function which responds to shape to be most suitable. Indeed this is so; there are two possibilities, the bearing ratio curve and the skew. The former is good principally because it was designed by Abbot(1) to do just this. It is a measure of metal to air as the surface is truncated. The only problem is knowing at which points to take it. Skew on the other hand is a parameter and not a function and so the problem does not arise. It has the other advantage that it begins to change rapidly when the truncation is in the vicinity of the mean line Figure 10(b). But it can be somewhat unreliable.

Other mechanisms have been suggested to explain the typical run-in shape of surfaces. One(99) proposes that the valley eventually fill up with wear debris and another(100) that a thin layer of self generated glaze (in cast-iron) gets deposited over the surface. The evidence for these mechanisms especially the first(99) is not conclusive. Sreenath and Raman(99) show profile graphs, for example, taken throughout a running-in experiment on cylinder liners. Because they did not use relocation methods it is impossible from the graphs to tell whether the valleys fill up or not. Using relocation their secondary running-in mechanism would probably be easily seen.

Rowe et al.(101) and others(96,102) have used relocation to great effect in running-in experiments. The general idea is that accelerated wear tests can be dispensed with because relocation enables accurate data to be obtained from real bearing under normal loads. Indeed it is difficult to see how mild wear can ever be satisfactorily monitored without the use of relocation methods (Figure 11).

Wear and Boundary Lubrication and Other Topics

A significant experiment by Hirst and Hollander(103) into the boundary lubrication of steel balls in sliding showed some positive evidence of the usefulness of random theory in wear tests. The Whitehouse Archard plasticity Index, equation (20) was used as a damage criterion parameter. They showed (103) that σ and β^* could be used to predict failure of balls. In a computer simulation they showed that once the Hertzian contact width exceeds the main structure of the surface (determined by β^*) then the balls siezed. This is a good demonstration of the importance of "scale of size" in tribological situations.

Other surface texture effects have been studied in boundary lubrication. Anisotropic surfaces, those with a lay, have effects even in boundary lubrication. Bayer and Sirico(104) using sliding balls showed that movement across the lay increased wear for a given roughness. What they do not make clear is the direction of the machining marks when motion is along the lay. Wear rate differences could be expected depending on the direction of motion. Unfortunately, it is difficult to see directionality with a stylus instru-

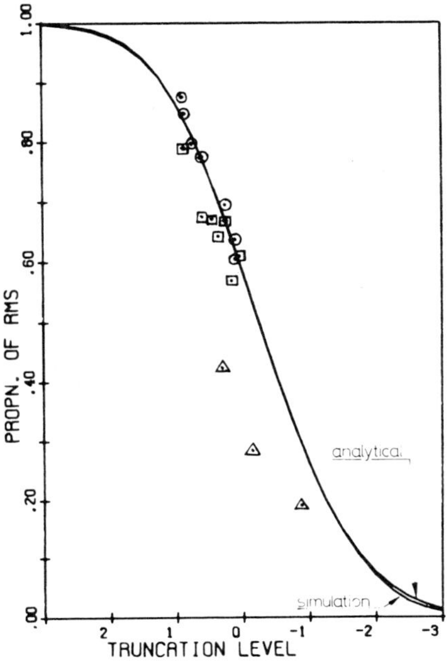

Fig. 10.—Shows (a) how a typical amplitude parameter Rq changes with truncation.

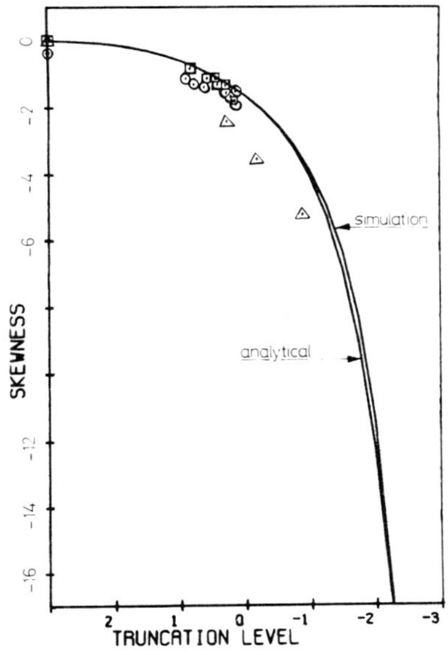

(b) how the skew value changes with truncation showing a much more severe rate of change after truncation beyond the mean line.

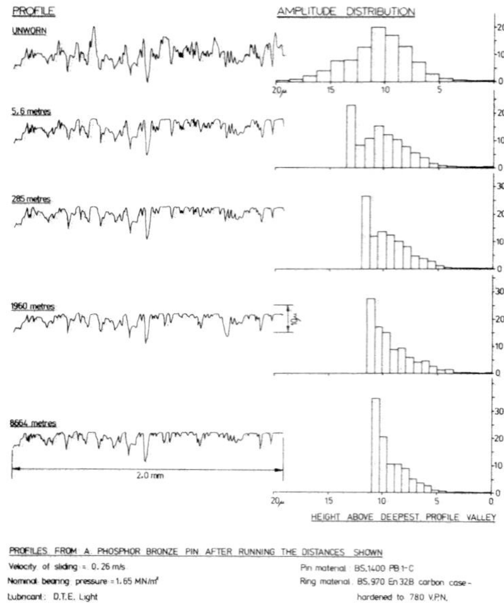

RELOCATED SURFACE PROFILES FROM A WEAR EXPERIMENT

Fig. 11.—Illustrates how the amplitude density function changes with running-in as seen by using relocation methods.

ment except in severe cases which probably explains why little work has been done on this aspect of wear.

SOLID CONTACT WEAR

Abrasive Wear, Adhesive Wear and Metrology

In many instances of the dependence of abrasive wear on the surface topography the problem has been that of finding whether or not there was any dependence rather than understanding the mechanism behind it. Earlier work such as Furey(105) and Davies(106) tended to belong to the former category as also is much of the wear work of Kragelski.(107)

In abrasive experiments the problem usually resolves itself into the movement of hard rough surfaces against soft smooth surfaces. One observation comes immediately to mind upon scanning the available literature. It is that the measurement of texture has been rather primitive in most cases, the specific topographic parameter influencing the wear has been the R_a value or the peak-valley. For example Beyer and Sirico(104) define a wear scar depth h' in a ball-plane configuration as

$$h' = a \times R_t \times n^i \qquad (33)$$

where a is a constant dependent on geometry, R_t is initial peak to valley height, n is the number or rubs and i is an index. In this situation the wear rate is linear with initial roughness up to a value of 0.4 μm R_t and then becomes constant.

Only recently has random process analysis been considered to identify the nature and extent of wear. Radhakrishnan[108] has tried to apply cross correlation measurement between the peak and the valley. In effect the symmetry of the peaks are compared with those of the valleys instead of using ordinary profiles. He maintains that by using this technique, re-location techniques do not have to be used to monitor wear because of the averaging property of correlation methods. Unfortunately, this very property means that small, specific detail such as occurs in mild wear cannot be observed. It may be, however, that the correlation methods would be useful in severe wear regimes.

An important point is worth making here concerning the use of either autocorrelation methods or spectral methods to investigate wear. A simple rule is this: if the mechanism is basically random then the autocorrelation should be used and if the mechanism is cyclic, then the power spectrum should be used. This leading statement is due to two reasons. The first is that if the process is nominally random the autocorrelation function reveals the unit mechanism.[68,108] Second, deviations from randomness are easily spotted; a well-known property of correlation methods.[24] This is the reason why correlation techniques have been used to good effect in the measurement of the statistical characteriwtics of friction.[68,109,110,111] The unit adhesive mechanism - corresponding to a single contact - can be seen.

Another example of the use of stochastic methods in wear experiments has been carried out by Burney and Wu[112] who, using the moments of the power spectrum defined in an earlier section, try to evaluate the wear on abrasive belts. Using digital methods they measure the autocorrelation function, the standard deviations of heights, slopes and curvature of profiles taken on the belts before and after wear. They find that the autocorrelation function decays more slowly after wear, indicating smoothing and not truncation. Also the standard deviations reduce with wear, again consistent with smoothing.

In many experiments in wear the importance of the initial surface is stressed.[79,68,104,118,101] It seems likely that there are two reasons for this, one primary and the other secondary. In the first case the initial surface finish itself might determine the mode of subsequent behaviour such as in scuffing where the finish on the inlet to the Hertzian zone sets off the temperature cycle,[80,83,84] or determines the rate of lubrication.[113] Secondary effects may well be due to strain hardening effects produced by asperity interaction (see for example references 114, 115, 116) or initial slope.[117] Note that whether or not these strain effects are absolutely valid is questionable; there is evidence that there is a considerable effect of "size" caused by the density of dislocations on the surface. The area of indentation of asperities during abrasive wear, for instance, is getting dangerously close to the average distance between dislocations; it may be that this is the mechanism behind effective strain hardening; it is a strain-area effect rather than a strain-penetration effect.[119]

Models of Wear Incorporating Texture

Although many workers have modelled contact and even friction[116,33] few have attempted wear because it is more complicated. Hisakado[120] assumes randomly-angled cones of a hard material abrading a soft surface.

He derives expressions for abrasive wear rate and the mean particle size in terms of asperity slope, which, as Archard has pointed out, is one of the basic unknowns in the wear mechanism. Probably the most important result that Hisakado gets is not quantitative wear parameters but a very interesting comment about the influence of random variables in wear studies. According to him using a model with single slopes overestimates wear rate and grit size compared with a *random* model having the same mean slope. This tends to support the contention made earlier that the main structure of the surface is often dominant.(13,103)

Halling in two papers using(63) the prow formation work of Cocks(120) and the concept of fatigue failure with Greenwood and Williamson statistics, relates material properties and surface topography(116) to wear. He comes to the conclusion in a simplified analysis that the surface finish is relatively unimportant in heavy wear situations.

Some other modelling has been carried out by Golding.(121,122) He gets some very important predictions of the wear mechanism using surprisingly simple assumptions! He simply ignores slopes and curvatures and concentrates on the degree of interpenetration of surfaces only. His mechanism assumes that wear is proportional, at each asperity, to this interaction and using the simple Gaussian distribution of peaks, he then predicts, quite convincingly, the presence of the two similar distributions of "transitional" surface finish often found at different heights on worn surfaces. Although not a quantitative analysis it does suggest that more insight might be gained using simplistic arguments.

Delamination and Texture

In cases of slow speed sliding Suh(123) and Suh et al.(124) have developed a theory of wear based on dislocation propogation. It is unusual in that the usual law linearly relating wear to sliding distance is refuted and evidence is given which suggests that wear debris particles are much longer than the asperity contact size(125) and that they are sheet-like in character. The details of the theory of this mechanism need not concern us here; what is of interest is the influence of surface texture. It seems that except in the initial contact, the texture is marginally important. It cannot be said that to have no influence because the stresses are communicated to the bulk through it, so the conclusion is that in situations involving considerable tangential shear at slow speeds the material properties become much more important than the geometric. This is a reversal of the usual order of importance between geometric and material properties in tribological situations as this paper seems to show.

DISCUSSION AND CONCLUSIONS

What do people want to know concerning the influence of surface texture on wear? They want to know two things. The first is whether surface texture has any influence on their particular wear situation. The second question follows if the answer to the first is yes. It is, what feature of the texture is important?

The evidence is that while there are many situations in wear and tribology where the answer is yes, the knowledge is very rudimentary and cannot be regarded as very useful. The reason for this has been the lack of the ability to measure surfaces adequately during experiments.

In the past two decades there has been a growing shift of emphasis from control of manufacture to prediction, or optimization of performance. This has stimulated research into getting a better understanding of surfaces and how they interact in practical situations. From this research two important developments have taken place, one instrumental and the other theoretical. The instrumental advances have been digitizing the outputs of stylus instruments and using relocation methods. The theoretical advance has been the use of random process theory which enables the most suitable parameters for a given function to be identified. One principal objective of this paper has been to show where and with what result these techniques have been employed in wear and related situations.

Particular attention has been given to looking at the fundamentals of the problem rather than listing what exists of empirical relationships in the hope that it will provide a more useful starting point for researchers. For this reason a deliberately wide view of wear has been taken.

Because the main problems seem to be that of measurement most emphasis has been placed on surface geometry.

Looking at some of the important parameters identified in the paper reveals a bewildering array of possibilities. The parameters to be used should depend on the function, and rarely is it possible to say that just one parameter is sufficient. This is not surprising considering the complexity of tribological behaviour. One could say that the average slope and mean curvature of peaks and bearing ratio would be useful additions to the arithmetic average R_a for normal contact and plastic flow situations, but this would have to be modified to include the way in which the curvature changed with height for light loads. Similarly when oil films are present, for instance as in e.h.l., it is the density of peaks at different heights rather than their curvature which determines the degree of interaction and thus susceptibility to fatigue wear or scuffing. Often the surface texture affects the performance in a secondary way for example by strain hardening the opposing surface. In some cases of slow speed working under high tangential stresses surface finish has little effect, but instances like this are rare. One major problem is that results obtained by different researchers are sometimes contradictory or specific to one small functional domain.

Another problem highlighted in this paper is that of the two different applications of random theory to tribology, that by Nayak and that by Whitehouse. This issue is of vital importance because it is mainly by means of theory that we can isolate the really important parameters and so cut down costly experiments and measurements. It therefore follows that we must know which theory is the best! The Nayak theory is two-dimensional and insists on finite values of curvature, slope, etc. for any surface. The Whitehouse theory on the other hand regards these parameters as effectively not intrinsic to the surface; the value you get will depend on the scale of size you need for your particular function; furthermore it is tied in closely with the digital analysis of surfaces (which is actually measured). Which of the two to use depends on the user deciding which is best for his practical application.

There is, however, one fundamental message that emerges from this paper: because of the multiplicity of parameters which can now be measured and because of the relative lack of reliable data it would seem sensible to look for *strategic parameters* from which the multiplicity can be derived.

This would stop the use of putting many parameters on drawings pending conclusive evidence of either one or the other being best. What could then happen is that as functional data based upon the many parameters accrue, it could be put into a *Functional Data Bank* similar to the now well established *Machinability Data Bank*.

The problem is first of finding such strategic parameters, and then making sure that they are *additive* from surface to surface. The last point is important because in tribology two surfaces are involved. It is the gap property which is very important (at least initially). One wants to be able to predict gap properties simply from parameters of each surface.

Fortunately there are parameters which satisfy both of these requirements. There are two possibilities, one derived from Whitehouse statistics and the other from Nayak Longuet-Higgins. In the former the strategic parameters are the three correlation points $c(o) = \sigma^2$, $c(\ell) = \sigma^2\rho_1$, $c(2\ell) = \sigma^2\rho_2$. If the two surfaces making contact are A and B, then the gap has three correlation points $c_j(o) = \sigma_A^2 + \sigma_B^2$, $c_g(\ell) = \sigma_A^2\rho_{1A} + \sigma_B^2\rho_{1B}$, $c_g(2\ell) = \sigma_A^2\rho_{2A} + \sigma_B^2\rho_{2B}$. Normalizing by $\sigma_A^2 + \sigma_B^2$ gives three strategic values $1, \rho_{1g}, \rho_{2g}$. To find the average peak gap, for example, these numbers are put in equation (9) and the value evaluated directly.

There is an alternative approach in Nayak's theory. Longuet-Higgins showed that the moments m_o, m_2, m_4 could be used to express tribological functions. Also he showed that $m_o\ m_2\ m_4$ could be expressed in terms of the zero crossing density H_p and the peak density D_p taken from a profile

$$H_p = \frac{1}{\pi}\sqrt{\frac{m_2}{m_o}}\ , \quad D_p = \frac{1}{2\pi}\sqrt{\frac{m_4}{m_2}} \tag{34}$$

Thus $\sigma_g^2 = \sigma_A^2 + \sigma_B^2$

$$H_g = \sqrt{\frac{\sigma_A^2 H_A^2 + \sigma_A^2 H_B^2}{\sigma_A^2 + \sigma_B^2}} \qquad D_g = \sqrt{\frac{\sigma_A^2 H_A^2 D_A^2 + \sigma_B^2 H_B^2 D_B^2}{\sigma_A^2 H_A^2 + \sigma_B^2 H_1^2}} \tag{35}$$

Hence making three strategic parameters σ^2, H_p and D_p.

There are good and bad points for both methods. The correlation points are more difficult to measure than H_p and D_p but they have the extra advantage that they can also be used to control the manufacturing process - the points could be used in both the Machinability and Function data banks. The peak and crossing densities are easier to measure from the output signal, i.e., the chart, but they are much more sensitive to instrument error;[33] they are not all additive and do not fit well into the control of finishing processes.

This concept could be taken further by incorporating other, physical features which are additive, for instance, elastic modulus E.

The criticism that these parameters relate to Gaussian surfaces is partly true, errors of skew of ±1 when using Whitehouse's statistics is allowable. Furthermore, in many cases, the problem is less severe when gap geometry is considered - the skew assymetry partially being compensated for by the other surface.

This approach is futuristic in concept; it would allow the designer to be able to look up functionability of components. The strategic parameter approach would not displace the research parameter; it would be used from the industrial point of view and would avoid the use of too many parameters. Obviously there would be exceptions, but in general it would simplify matters and avoid the state of surface metrology specification we find ourselves in today, where many similar parameters are specified on drawings, R_a, R_q, R_t, R_z, Rtm etc. - all height parameters, with absolutely no evidence of different functional significance. It is because people have been realizing the growing importance of surface texture without having sufficient knowledge that these problems have arisen. Now is the time to try to rationalize the position. It is suggested that we now have the necessary techniques and understanding to do it.

REFERENCES

1. Abbott, E.J. and Firestone, F.A., *Mechanical Engineering,* Vol. 55, 1933, p. 569.
2. Kayser, J.F., *Aircraft Engineering,* Vol. 16, 1944, p. 25.
3. Castro, H.L., *Tool Engineer*, Vol. 13, 1944, p. 84.
4. Reason, R.E., *Automatisme*, Vol. 9, No. 5, 1964, p. 177.
5. Tallian, T.E., et al., *American Society of Lubrication Engineers Transactions*, Vol. 2, 1964, p. 109.
6. Greenwood, J.A. and Williamson, J.B.P., Burndy Res. Rept. No., 1966.
7. Whitehouse, D.J. and Reason, R.E., "Equation of Mean Line of Surface Texture as Found by an Electric Wave Filter," Rank Organisation, 1965.
8. Whitehouse, D.J., *Institution of Mechanical Engineers. Proceedings*, Vol. 182, Part 3K, 1967-68, p. 306.
9. Kinsey, D. and Chetwynd, D.G., Proceedings, IMEKO VI, Dresden, 1973.
10. Thomas T.R., in 1st joint Polytechnic Symposium on Manufacturing Engineering, Leicester, June 1977, Leicester Polytechnic, England.
11. Williamson, J.B.P. and Hunt, R.T., *Journal of Institute of Physics*, Series E, Scientific Instruments, 2, Vol. 1, 1968, p. 749.
12. Grieve, D.J., Kaliszer, H. and Rowe, G.W., *CIRP Annals*, Vol. 17, 1967, p. 147.
13. Whitehouse, D.J. and Archard, J.F., in Symposium Surface Mechanics, Proceedings, Los Angeles, American Society of Mechanical Engineers, 1969, p. 36.
14. Williamson, J.B.P., *Institution of Mechanical Engineers. Proceedings*, Vol. 182, Part 3K, 1967-68, p. 1.
15. Sayles, R.S. and Thomas, T.R., *Journal of Institute of Physics*, Series E, Scientific Instruments, Vol. 9, 1976, p. 861.
16. Wormersley, J.R. and Hopkins, M.R., *Journal Etats Surface*, Vol. 135, 1945.
17. Linnik, Y. and Khusu, A.P., *Izhernernyi*, Vol. 20, 1954, p. 154.
18. Nakamura, T., *JSPMJ*, Vol. 25, 1959, p. 56; also Vol. 26, 1960, p. 226.
19. Peklenik, J., *CIRP Annals*, Vol. 12, 1963, p. 173.
20. Tallian, T., et al., "Lubricant Films in Rolling Contact of Rough Surfaces," American Society of Lubrication Engineers Lubrication Conference, Rochester, New York, 1963.

21. Greenwood, J.A., Paper 66, Lub., American Society of Mechanical Engineers, Vol. 10, 1965, p. 1.
22. Ling, F.F., *Journal of Applied Physics*, Vol. 29, 1958, p. 1168.
23. Kimura, Y., *Japan Society of Lubrication Engineers*, Vol. 11, 1966, p. 467.
24. Bendat, J.S., "Principles and Applications of Random Noise Theory," John Wiley and Sons, New York, 1958.
25. Whitehouse, D.J. and Archard, J.F., *Royal Society of London. Proceedings.* Series A, Vol. 316, 1970, p. 97.
26. Whitehouse, D.J., and Archard, J.F., *Institution of Mechanical Engineers. Proceedings,* Vol. 182, Part 3K, 1967-68.
27. Whitehouse, D.J. and Phillips, M.J., *Royal Society of London, Philosophical Transactions*, Series A, Vol. 290, 1978, p. 267.
28. Nayak, P.R., *Journal of Lubrication Technology*, July 1971, p. 398.
29. Longuet-Higgins, M.S., *Royal Society of London, Philosophical Transactions,* Series A, Vol. 249, 1957, p. 321.
30. Longuet-Higgins, M.S., *Royal Society of London, Philosophical Transactions,* Series A, Vol. 249, 1957, p. 966.
31. Rice, S.O., *Bell System Technical Journal,* Vol. 23, 1944, p. 282.
32. Cartwright, D.E. and Longuet-Higgins, M.S., *Royal Society of London. Proceedings.* Series A, Vol. 237, 1956, p. 212.
33. Francis, H.A., *Wear*, Vol. 45, 1977, p. 221.
34. Sayles, R.S., Ph.D. Thesis, Teeside Polytechnic, 1976.
35. Sayles, R.S. and Thomas, T.R., *American Society of Lubrication Engineers Transactions,* 1978, in press.
36. Spragg, R.C. and Whitehouse, D.J., *Institution of Mechanical Engineers. Proceedings*, Vol. 185, No. 47, 1971.
37. Nayak, P.R., *Wear*, Vol. 26, 1973, p. 165.
38. Greenwood, J.A., "Recent Advances in Surface Textures," Proceedings of the 4th International Conference on Tribology, Leeds-Lyon, Lyon, 1977.
39. Whitehouse, D.J., *Journal of Mechanical Engineering Science*, Vol. 20, No. 4, 1978, p. 221.
40. Chetwynd, D.G., *Journal of Mechanical Engineering Science*, Vol. 20, No. 2, 1978.
41. Summers-Smith, D., "Machinery and Production Engineering," February 1970, p. 169.
42. Bowden, F.P. and Tabor, D., "The Friction and Lubrication of Solids," Oxford University Press, London, 1950, 1954.
43. Block, H., *Royal Society of London. Proceedings,* Series A, Vol. 212, 1952, p. 480.
44. Halliday, J.S., *Institution of Mechanical Engineers. Proceedings,* Vol. 169, 1955, p. 177.
45. Archard, J.F., *Royal Society of London. Proceedings.* Series A, Vol. 243, 1957, p. 190.
46. Bush, A.W., Gibson, R.D. and Thomas, T.R., *Wear*, Vol. 35, 1975, p. 87.
47. Greenwood, J.A. and Williamson, J.B.P., *Royal Society of London. Proceedings.* Series A, Vol. 295, 1966, p. 300.
48. Onions, R.A. and Archard, J.F., *Journal of Physics D: Applied Physics,* Vol. 6, 1973, p. 289.
49. Mikic, B.B. and Roca, R.T., *International Journal of Heat and Mass Transfer,* Vol. 17, 1974, p. 205.
50. Tabor, D., *Wear*, Vol. 32, 1975, p. 269.
51. Gupta, P.K. and Cook, N.H., *Journal of Lubrication Technology*, Vol. 94, No. 1, 1972, p. 19.
52. Nayak, P.R., *Wear*, Vol. 26, 1973, p. 305.
53. Pullen, J. and Williamson, J.B.P., *Royal Society of London. Proceedings.* Series A, Vol. 327, 1972, p. 159.
54. Suratkar, P.T., Pandit, S.M. and Wu, S.M., *Wear*, Vol. 39, 1976, p. 239.

55. Tallian, T.E., *Wear*, Vol. 21, 1972, p. 49.
56. O'Callahan, M. and Cameron, M.A., *Wear*, Vol. 36, 1976, p. 79.
57. Greenwood, J.A. and Tripp, J.H., *Institution of Mechanical Engineers. Proceedings*, Vol. 185, 1971.
58. Yip, F.C. and Venart, J.E.S., *Journal of Physics D: Applied Physics*, Vol. 14, 1971, p. 1470.
59. Chivers, T.C., Mitchell, L.A. and Rowe, M.D., *Wear*, Vol. 28, 1974, p. 171.
60. Childs, T.H.C., *Wear*, Vol. 25, 1973, p. 3.
61. Mitchell, L.A. and Rowe, M.D., *Journal of Mechanical Engineering Science*, Vol. 11, 1969, p. 534.
62. Tsukizoe, T. and Hisakado, T., *American Society of Mechanical Engineers Transactions*, September 1965, p. 666.
63. Nuri, K.A. and Halling, J., *Wear*, Vol. 32, 1975, p. 95.
64. Nuri, K.A., *Wear*, Vol. 30, 1974, p. 321.
65. Halling, J., *Tribology International*, August 1971, p. 155.
66. El-Refaie, M. and Halling, J., *Institution of Mechanical Engineers. Proceedings*, Vol. 183, Part 3P, 1968-69, p. 116.
67. Whitehouse, D.J., *CIRP Annals*, Vol. 23, 1974, p. 1.
68. Whitehouse, D.J., Ph.D. Thesis, Leicester University, 1971.
69. Thomas, T.R., "Influence of Roughness on the Deformation of Metal Surfaces in Static Contact," 6th International Conference on Fluid Sealing, Munich, 1973.
70. Holm, R., "Electric Contacts Handbook," Springer-Verlag, 1958.
71. Hisakado, T. and Tsukizoe, T., *Wear*, Vol. 30, 1974, p. 213.
72. Jones, A.M., O'Callaghan, P.W. and Probert, S.D., *Wear*, Vol. 31, 1975, p. 89.
73. Jones, A.M., O'Callaghan, P.W. and Probert, S.D., *Wear*, Vol. 38, 1976.
74. Johnson, K.L., Kendal, K. and Roberts, A.D., *Royal Society of London. Proceedings*. Series A, Vol. 324, 1971, p. 301.
75. Fuller, K.N.G. and Tabor, D., *Royal Society of London. Proceedings*. Series A, Vol. 345, 1975, p. 327.
76. Bush, A.W., Gibson, R.D. and Keogh, G.P., *Wear*, Vol. 40, 1976, p. 399.
77. Grubin, A.N., Book 30, DSIR translation, Central Scientific Institute of Technical and Mechanical Engineering, Moscow, 1969, p. 377.
78. Michell, A.G.M., "Lubrication," Blackie and Son Ltd., London, 1950.
79. Dawson, P.H., *Journal of Mechanical Engineering Science*, Vol. 4, No. 1, 1962, p. 16.
80. Christensen, H., *Institution of Mechanical Engineers. Proceedings*, Vol. 184, 1969-70, p. 1013.
81. Johnson, K.L., Greenwood, J.A. and Poon, S.Y., *Wear*, Vol. 19, 1972, p. 91.
82. Berthe, D. and Godet, M., *Wear*, Vol. 27, 1973, p. 345.
83. Dyson, A., *Institution of Mechanical Engineers. Proceedings*, Vol. 190, 1976.
84. Cheng, H.S. and Dyson, A., *Lubrication Engineering*, 21, Vol. 1, 1976, p. 25.
85. Patir, N. and Cheng, H.S., ASLE/ASME Conf., Kansas, 1977.
86. Kubo, M. and Peklenik, J., *CIRP Annals*, Vol. 16, 1968, p. 235.
87. Michan, B., Berthe, D. and Godet, M., *Wear*, Vol. 28, No. 2, 1974, p. 187.
88. Leaver, R.H., Sayles, R.S. and Thomas, T.R., *Institution of Mechanical Engineers. Proceedings*, Vol. 188, 1974, p. 461.
89. Bey, N. and Wanheim, T., *Wear*, Vol. 38, 1976, p. 201.
90. Ostvik, R. and Christensen, H., *Institution of Mechanical Engineers. Proceedings*, Vol. 183, 1968-69, p. 59.
91. Myers, N.O., *Wear*, Vol. 5, 1962, p. 182.
92. Ku, P.M. and Li, K.Y., "Effect of Surface Topography on Sliding Rolling Disc Scuffing," Proceedings of 4th Leeds-Lyon Symposium, Lyon, 1977.

93. Kaliszer, H. and Rowe, G.W., Proceedings Society of Manufacturing Engineers, Pittsburgh, May 1973.
94. Thomas, T.R., "The Characterisation of Changes in Surface Topography During Running-In," Proceedings of 4th Leeds-Lyon Symposium, Lyon, Sept. 1977.
95. Williamson, J.B.P., Pullen, J. and Hunt, R.T., *American Society of Mechanical Engineers Transactions*, Vol. 24, 1969.
96. King, T., Whitehouse, D.J. and Stout, K., *CIRP Annals*, Vol. 1, No. 25, 1977, p. 351.
97. Sayles, R.S. and Thomas, T.R., *Wear*, Vol. 42, 1977, p. 263.
98. Thomas, T.R., *Wear*, Vol. 22, 1972, p. 83.
99. Sreenath, A.V. and Raman, N., *Tribology International*, 1976, p. 55.
100. Montgomery, R.S., *Wear*, Vol. 14, 1969, p. 99.
101. Kaliszer, H. and Rowe, G.W., Institution of Electrical Engineers publication, University of York, U.K., Vol. 103, 1973, p. 7.
102. Stout, K., King, T. and Whitehouse, D.J., 1st Joint Polytechnic Symposium on Manufacturing Engineering, Leicester, June 1977, Leicester Polytechnic, England.
103. Hirst, W. and Hollander, A.E., *Royal Society of London. Proceedings*. Series A, Vol. 337, 1974, p. 379.
104. Bayer, R.G. and Sirico, L., *Wear*, Vol. 35, 1975, p. 251.
105. Furey, M.J., *American Society of Lubrication Engineers Transactions*, Vol. 6, 1963, p. 49.
106. Davies, C.B., *New York Academy of Sciences. Annals*, Vol. 53, 1951, p. 935.
107. Kragelski, I.V., "Friction and Wear," Butterworths, London, 1965.
108. Hamed, S., Whitehouse, D.J. and Buttery, T., *CIRP Annals*, Vol. 27, 1978, p. 1.
109. Strang, C.D. and Lewis, C.R., *Journal of Applied Physics*, Vol. 20, 1948, p. 1164.
110. Rabinowicz, E., *Journal of Applied Physics*, Vol. 20, No. 11, 1951, p. 1373.
111. Nagash, H., *Bulletin of Japanese Society of Production Engineers*, Vol. 6, 1951, p. 124.
112. Burney, F.A. and Wu, S.M., *Wear*, Vol. 36, 1976, p. 225.
113. Wilson, W.R.D. and Delmaline, W.P., *Wear*, Vol. 29, 1974, p. 1.
114. Hisakado, T., *Wear*, Vol. 37, 1976, p. 41.
115. Hirano, F., Yamashita, N. and Kamitani, T., *Institution of Mechanical Engineers. Proceedings*, Paper No. C85, 1971, p. 151.
116. Halling, J., *Wear*, Vol. 34, 1975, p. 239.
117. Bey, N., Wanheim, T. and Peterson, A.S., *Wear*, Vol. 34, 1975, p. 77.
118. Asouros, G.M., Dimarogonas, A. and Lefas, K., *Wear*, Vol. 45, 1977, p. 375.
119. Hunt, R.T. and Whitehouse, D.J., Tech. Note T46, Rank Organisation, 1971.
120. Cocks, M., *Wear*, Vol. 9, 1966, p. 320.
121. Golden, J.M., *Wear*, Vol. 39, 1976, p. 25.
122. Golden, J.M., *Wear*, Vol. 42, 1977, p. 157.
123. Suh, N.P., *Wear*, Vol. 25, 1973, p. 11.
124. Suh, N.P., Jahanmir, S. and Abrahamson, E.P., II, Rept. 229-011, Massachusetts Institute of Technology, September 1974.
125. Abrahamson, E.P., II, Jahanmir, S., Colling, D.A. and Suh, N.P., *Scanning Electron Microscope Symposium. Proceedings*, Part 4, 1974, p. 889.

DISCUSSION

QUESTIONER: Are the tribologists now characterizing surfaces and then correlating those characterizations with wear properties?

D. J. WHITEHOUSE: Yes, that is absolutely what we are doing now.

SAME QUESTIONER: Without lubricant?

WHITEHOUSE: Without and with. But the problem of characterization when there is lubricant is easier than in the case of solid-to-solid contact. Solid-to-solid contact is purely to do with geometry and physical properties of the surface. However, one more element is introduced in the case of oil.

SAME QUESTIONER: I am intrigued about specifying the stresses.

WHITEHOUSE: Well, we can certainly specify the interaction.

JACK BOUCHARD, Northrop Corporation: I am a little concerned about how you read very narrow valleys. One would need a very sharp probe in order not to get reflection of a flat.

WHITEHOUSE: The problem with the stylus techniques of any sort is that you probably cannot measure better than 60 degrees included angle. Is that what you mean?

BOUCHARD: Yes, it is. Do you actually get down to four millionths point?

WHITEHOUSE: On the stylus, yes. You see there are two degrees of integration with the stylus. One is the physical size of the bottom. The second is the slope of the flank. With very rough surfaces you will find that the tip dimension is the main thing. But when you are talking about very smooth surfaces, for instance, those obtained by honing or polishing, very often measurement of the angle is still a problem even with sharp stylus. But we can definitely make styluses to four millionths. It is a very difficult task.

QUESTIONER: We had the same problem with stylus methods in our research. We used scanning electron microscope to follow this and we obtained a much better profile of the surface.

WHITEHOUSE: We have done extensive work with scanning electron microscopes, stylus methods, and inferometers. In brief, there is no correlation.

SAME QUESTIONER: We had difficulty relocating the area.

WHITEHOUSE: We indented a flat surface (say a flat surface for the sake of argument) with the same stylus that we used to make the profile. And then we made a track across to get the profile. Then we put the specimen in the scanning electron microscope (SEM) and we used and the Z modulation or scan modulation mode where the blacks and whites of intensity of the electron beam are transformed into peaks and valleys. This gives a scanning electron picture which normally looks like a profile and it was digitized. Similar procedure was carried on with an interferometer as well. And then we cross correlated the two sets of data and found that the correlation coefficient between a stylus instrument and a SEM is about 0.6 which is very low. The reason it is 0.6 is only because the spacings are correct, or very nearly. Instead of looking at the geometrical property of surface to the simple voltage of SEM, if we can differentiate the profile and do a cross correlation of the differential with the voltage from the SEM, we get a much better correlation; it goes up to about 0.8. Now, if you add on to that something like 10% of the second differential of the curvature, then we get up to 0.9, indicating that what you see on a SEM is a very complicated version of the geometry of the surface. It is not one to one at all.

THE EFFECTS OF SURFACE TOPOGRAPHY ON WEAR

T. Tsukizoe

ABSTRACT

The purpose of this paper is to discuss the mechanism of contact between metal surfaces and the surface topography effects on wear, both from theoretical and practical standpoints. Assuming that the distribution curve obtained from the profile curve of the surface has a Gaussian distribution, the mechanism of contact between metal surfaces has been deduced theoretically in the case of an ideal plastic flow at micro-contacts. The surface topography effects on adhesive wear, oxidative wear and abrasive wear are deduced theoretically using the above theory of contact mechanism. A comparison between the theoretical and the experimental results shows a good agreement.

INTRODUCTION

Wear is defined as an unwanted removal of solid materials from surfaces under relative movement. Four principal types of wear, namely, adhesive, abrasive, corrosive and surface fatigue, may be categorized.(1) The adhesive wear arises from a lump removal of materials due to a process of solid-phase welding, the abrasive wear may be the cutting caused by hard asperities or hard particles, the corrosive wear or the oxidative wear occurs in the form of a layer removal of corrosion products by mechanical action, and the surface fatigue is detachment of particles by fatigue which arises from cyclic stress variations. The first three occur under sliding motion and the last one under rolling motion.

The purpose of the present paper is to describe briefly what is known about the effect of surface topography on wear on the basis of the contact mechanism of metal surfaces.

In the review paper by Whitehouse,(2) recent advances in analytical procedure of surface topography were described. However, the review paper mentioned little about the quantitative analyses of the surface topographical effect on wear. The model of the surface employed by Whitehouse was, in fact, very precise, but it would seem difficult to predict or estimate wear volume with such a rigorous model. It seems likely that we may be able to evaluate and predict the effects of surface topography on wear with use of a rather simplified model.[3]

The assumptions made for the present analysis of the mechanism of contact are:

1) The distribution curve which is obtained from the profile curve of the surface has a Gaussian distribution.
2) The surface has the same profile curve in any direction.
3) The surface contains a large number of asperities in the form of cones of equal base angle.
4) The deformation of the metal occurring at the contact is plastic.

With these assumptions, the separation, the real area of contact, the number of the contact points, the summation of the radii of the contact points, and their average radius is calculated.

Simple laws for three types of wear, adhesive wear, oxidative wear and abrasive wear, are deduced. On the basis of these theoretical deductions the influnce of surface topography on wear are discussed. To check the validity of the theory, a comparison of theoretical and experimental results is carried out.

THEORY

Mechanics of Contact

Our analytical equations for the mechanism of contact between metal surfaces[4,5] are as follows (see Figure 1):

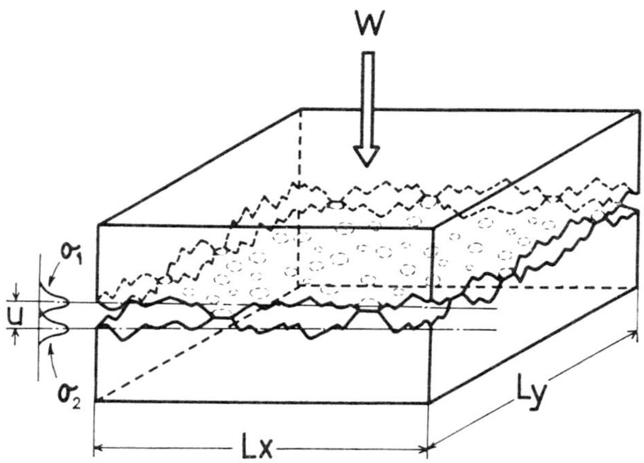

Fig. 1.—Schematic of two surfaces in contact.

(a) The separation $u=t\sigma$ is calculated from the following equation

$$\int_0^t \phi(t)dt = \left(1 - \frac{2W}{L_x L_y p_m}\right) \int_0^\ell \phi(t)dt \tag{1}$$

where
$$\phi(t) = \frac{1}{(2\pi)^{1/2}} \exp\left(-\frac{1}{2} t^2\right) \tag{2}$$

$t = u/\sigma$

$\delta = (\delta_1^2 + \delta_2^2)^{1/2}$ = standard deviation

W = load

$L_x L_y$ = apparent contact area

P_m = flow pressure of metal

$\ell\sigma$ = distance between two median planes of the surfaces at the beginning of contact (ℓ depends on $L_x L_y$).

(b) The real area of contact $A = \sum_{i=1}^{n} \pi a_i^2$ under the applied load W can be written as

$$A = \frac{1}{P_m} W \tag{3}$$

where n = number of contact points.

(c) The number of contact points between two metal surfaces n is given by

$$n = \frac{L_x L_y}{2\pi \int_{-\ell}^{\ell} \phi(t)dt} \frac{\tan^2\theta}{\delta^2} t\, \phi(t) \tag{4}$$

where θ = base angle of conical asperities.

The number of contact points n_x between two profile curves can also be expressed by

$$n_x = \frac{L_x}{\pi \int_{-\ell}^{\ell} \phi(t)dt} \frac{\tan\theta}{\sigma} \phi(t) \tag{5}$$

where L_x = length of profile curve.

(d) The summation of the radii of the contact points $\sum_{i=1}^{n} a_i$ is given by

$$\sum_{i=1}^{n} a_i = \frac{L_x L_y}{2\pi \int_{-\ell}^{\ell} \phi(t)dt} \frac{\tan\theta}{\sigma} \phi(t) \tag{6}$$

Dividing equation (6) by equation (4), we obtain

$$a_m = \frac{\sum_{i=1}^{n} a_i}{n} = \frac{\sigma}{t \tan\theta} \tag{7}$$

where a_m = average radius of contact points.

(e) The base angle of conical asperities θ can be expressed by

$$\tan\theta = \frac{\pi}{2} (\tan\theta)_{prof.} \tag{8}$$

where $(\tan\theta)_{prof.}$ = mean value of tangent of slope angle on profile curve of surface.

The empirical relation between maximum height of asperities $R_{max}(\mu m)$ and the value of $(\tan\theta)_{prof.}$ was derived from the experimental results shown in Figure 2 and written as

$$(\tan\theta)_{prof.} = \frac{R_{max}}{3.5R_{max} + 15} \tag{9}$$

where R_{max} = maximum height of asperities defined in JIS (Japanese Industrial Standard).

The empirical curve shown in the figure was deduced from equation (9).

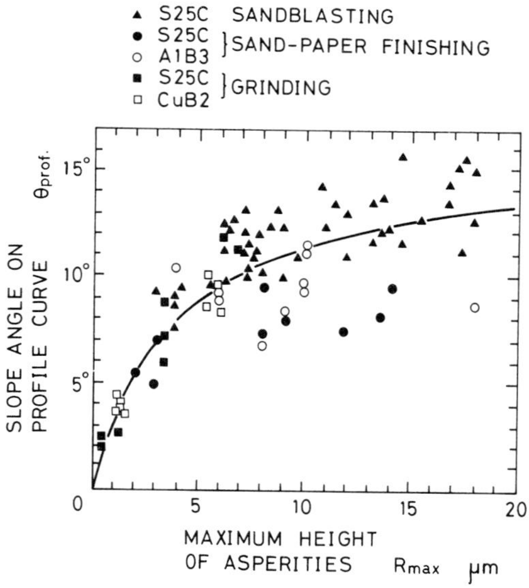

Fig. 2.—Relation between the slope angle on profile curve and the maximum height of asperities. S25C-0.25 percent carbon steel; AlB3-aluminum; CuB2-copper.

Adhesive Wear

When junctions are formed between metal surfaces, the shearing of them may occur in two different ways (Figure 3). If the junctions are formed in the absence of the oxide film, i.e., in the case of metal-to-metal contact, shearing will rarely occur at the interface but will take place within the bulk (Figure 3(a)). In this case a lump removal of metals will occur (adhesive wear). On the contrary, if the junctions are formed in the presence of the oxide film and if the oxide is not broken up by the deformation of the underlying metal, shearing will occur at the interface between the oxide film and the bulk of the metal (Figure 3 (b)). Under these conditions a layer removal of the oxide will occur (oxidative wear).

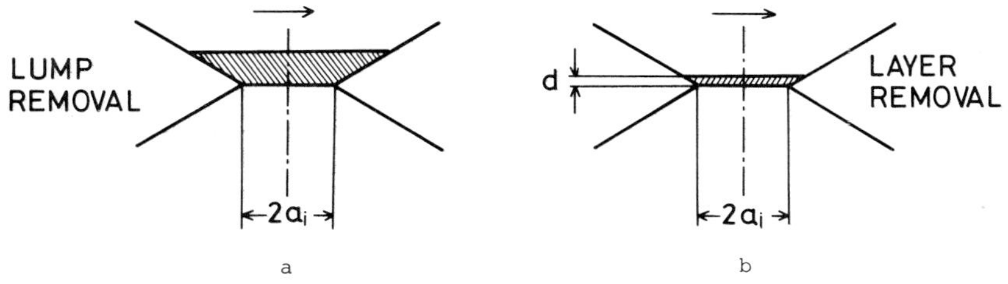

Fig. 3.—Idealized model of the two types of wear due to shearing of metallic junctions.

For the adhesive wear the most probable shape of the lump is the conical particle with a base angle θ. This idea can be supported by the fact that in the steady state of wear, the profile curves obtained from a wearing surface show a sequence of conical asperities of a base angle θ. In this case the volume ΔB of a given wear particle can be written as

$$\Delta B = \frac{1}{3} \pi a_i^3 \tan\theta \tag{10}$$

Assuming that the process of continuous formation and shearing of metallic junctions obeys the analysis by Rabinowicz,[6] the adhesive wear rate $Q_{adhe.}$, i.e., the total wear volume per unit sliding distance for the whole surface, is given by

$$Q_{adhe.} = \sum_{i=1}^{n} \frac{\Delta B}{2a_i} \tag{11}$$

Combining this equation with equations (10) and (3), we obtain

$$Q_{adhe.} = \frac{1}{6} \frac{\tan\theta}{P_m} W \tag{12}$$

Equation (12) is similar to the wear equation by Archard[7] and is obtained essentially by replacing Archard's concept of hemispherical wear particles by conical ones.

The main conclusions derived from equation (12) are:

1) The wear rate $Q_{adhe.}$ is proportional to the load W.

2) The wear rate $Q_{adhe.}$ is independent of the velocity of sliding.

3) The wear rate $Q_{adhe.}$ is inversely proportional to the flow pressure P_m (the metal with higher hardness shows a better wear-resistance).

4) The wear rate $Q_{adhe.}$ is proportional to the tangent of base angle of conical asperities $\tan\theta$ (the metal with smoother surface shows a better wear-resistance).

Oxidative Wear

From Figure 3 (b) the volume ΔB of a given wear particle, i.e., the amount of oxide removed from this asperity can be written as

$$\Delta B \simeq \pi a_i^2 d \tag{13}$$

where d is the thickness of the oxide film formed in time t during which this asperity moves a distance s. As for the distance s we may use the average value between two neighbouring asperities of contact, then we have

$$s = \frac{L_x}{n_x} - \frac{\pi}{2} a_m \tag{14}$$

and

$$t = \frac{s}{V} \tag{15}$$

where V = velocity of sliding.

The empirical relation for the initial stage of oxidation on a clean iron surface[8] is given by

$$\begin{aligned} d &= \eta \log_e\left(\frac{t}{\tau} + 1\right) \\ &\simeq \eta\frac{t}{\tau} \\ &= \frac{\eta}{\tau}\frac{s}{V} \end{aligned} \tag{16}$$

where η = constant concerned with material and temperature
τ = constant concerned with material
(τ = 2.6 ~ 10.4 s for iron[8]).

From equation (13) the oxidative wear rate $Q_{oxid.}$ can be written as

$$\begin{aligned} Q_{oxid.} &= \sum_{i=1}^{n} \frac{\Delta B}{2a_i} \\ &= \frac{\pi}{2}\sum_{i=1}^{n} a_i d \end{aligned} \tag{17}$$

Combining this equation with equations (16), (14), (5), (6) and (7), we obtain

$$Q_{oxid.} = \frac{\pi}{4}\frac{\eta}{\tau}\frac{1}{V}L_x L_y\{1-\frac{\phi(t)}{2t\int_{-\ell}^{\ell}\phi(t)dt}\} \tag{18}$$

The main conclusions to be derived from equation (18) are:

1) The wear rate $Q_{oxid.}$ is inversely proportional to the velocity of sliding V.

2) The wear rate $Q_{oxid.}$ is almost independent of such factors as the load W, the flow pressure P_m and the surface roughness (the second term in bracket of equation (18) is smaller than unity).

Abrasive Wear

Figure 4 diagrammatically illustrates the process of the abrasive wear produced on a soft flat surface by a hard conical asperity. The amount

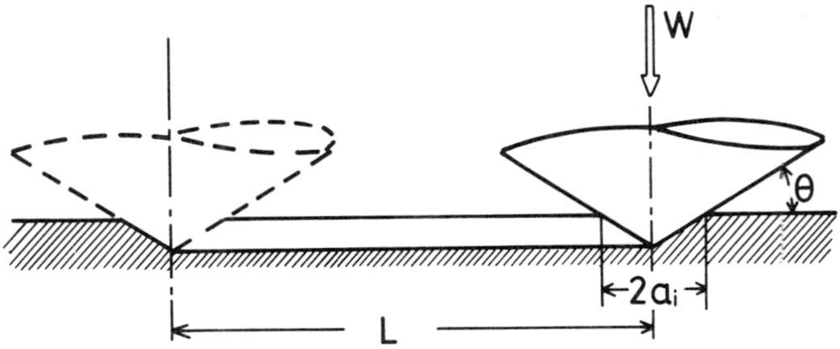

Fig. 4.—Idealized model of the process of abrasive wear.

of metal ΔB removed from the soft surface by this hard asperity is given by

$$\Delta B = \beta L a_i^2 \tan\theta \tag{19}$$

where β = constant ($\beta \lesssim 0.15$ for dry surfaces[9])
L = sliding distance.

Then the abrasive wear rate $Q_{abra.}$ can be written as

$$Q_{abra.} = \sum_{i=1}^{n}\frac{\Delta B}{L}$$

$$= \beta\tan\theta\sum_{i=1}^{n} a_i^2 \tag{20}$$

where $\sum_{i=1}^{n} a_i^2 = \frac{2}{\pi}\frac{W}{P_m}$ (21)

Hence $Q_{abra.} = \frac{2\beta}{\pi}\frac{\tan\theta}{P_m}W$ (22)

Equation (22) is similar to the equations by Goddard[10] and by Richardson[11] and leads to conclusions similar to those derived for $Q_{adhe.}$:

1) The wear rate $Q_{abra.}$ is proportional to the load W.

2) The wear rate $Q_{abra.}$ is independent of the velocity of sliding V.

3) The wear rate $Q_{abra.}$ is inversely proportional to the flow pressure P_m (the metal with higher hardness shows a better wear-resistance).

4) The wear rate $Q_{abra.}$ is proportional to the tangent of base angle of conical asperities $\tan\theta$ (the metal with smoother surface shows a better wear-resistance).

EXPERIMENTAL RESULTS AND DISCUSSIONS

Figure 5 shows the results of the wear curves obtained under various loads and velocities on the pin-on-cylinder testing machine. The wear represents the sum for the pin (12mm x 12mm square) and the cylinder (40mm dia.), made of 0.45 percent carbon steel (flow pressure P_m=500 kg/mm^2). Among these wear curves (A~I), B, C, D, E, F and I showed metallic wear particles (adhesive wear), and A, G and H oxide products (oxidative wear). The wear rates calculated from these curves are shown in the figure.

The wear rates for the velocity of 0.86 m/s (D, E, F and I) are plotted against the loads in Figure 6. The relation between wear rate and load appears to be a straight line, and it is concluded that the adhesive wear rate $Q_{adhe.}$ is proportional to the load W. Equation (12) thus holds valid. The theoretical curve shown in the figure was deduced as follows:

surface roughness of sliding surface $R_{max} \approx 25\mu m$

then $\tan\theta$ = 0.383 (from equations (8) and (9))
and P_m = 500 kg/mm^2

hence from equation (12)

$$Q_{adhe.} = 1.28 \times 10^{-4}\ W \text{ mm}^3/\text{mm} = 1.00\ W \text{ mg/m}$$

where W = load in kg.
Thus the agreement between the experimental and the theory is fairly good.

The oxidative wear rate for H in Figure 5 will be

$$Q_{oxid.} = 0.001 \text{ mg/m} = 1.27 \times 10^{-7} \text{ mm}^3/\text{mm}$$

Under this experimental condition, the theoretical value from equation (18) was

$$Q_{oxid.} = 1.13 \times 10^{-7} \text{ mm}^3/\text{mm}$$

where η/τ in equation (18) = $20\ \mathring{A}/s$.
The comparison between the experimental and the theoretical shows a good agreement.

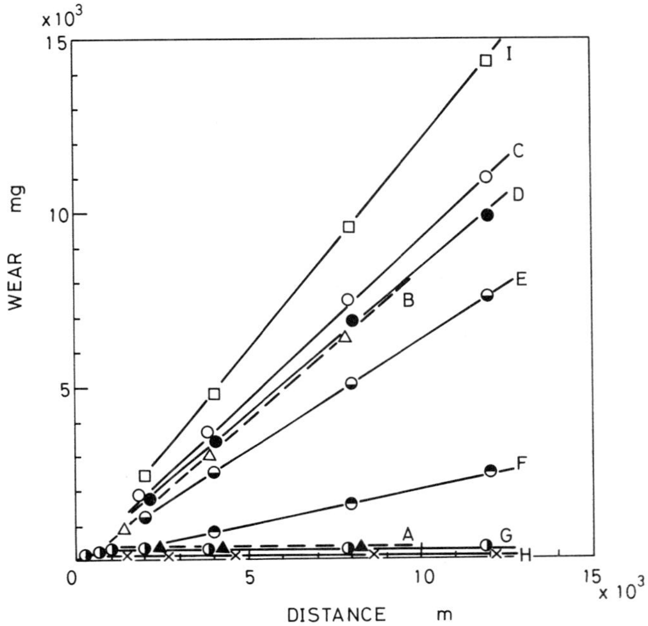

Fig. 5.—Dependence of wear on sliding distance.

	Load-Velocity	Wear Rate
A	1.0 kg-0.3 m/s	0.0010 mg/m
B	1.0 kg-0.6 m/s	0.85 mg/m
C	1.0 kg-1.0 m/s	0.92 mg/m
D	1.0 kg-0.86m/s	0.84 mg/m
E	0.6 kg-0.86m/s	0.63 mg/m
F	0.2 kg-0.86m/s	0.20 mg/m
G	1.0 kg-1.5 m/s	0.0012 mg/m
H	1.0 kg-2.0 m/s	0.0010 mg/m
I	1.4 kg-0.86m/s	1.20 mg/m

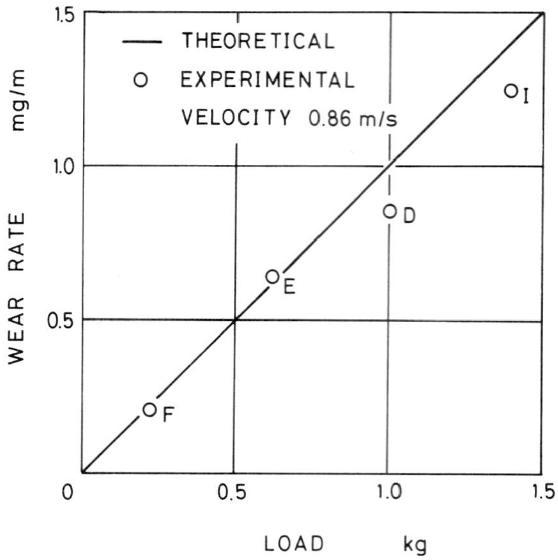

Fig. 6.—Dependence of wear rate on load.

CONCLUSIONS

Assuming that the distribution curve obtained from the profile curve of the surface has a Gaussian distribution, the mechanism of contact between metal surfaces has been deduced theoretically in the case of an ideal plastic flow at micro-contacts. Using the theory of contact mechanism, it has been possible to arrive at theoretical equations relating the amount of wear volume removed from the surfaces for three types of wear; adhesive wear, oxidative wear and abrasive wear. On the basis of these theoretical deductions, the influence of the operating conditions on wear, such as distance or time of travel, velocity, applied load, mechanical properties of metals, and surface topography, was analyzed, as shown in equations (12), (18) and (22).

REFERENCES

1. Burwell, J.T., *Wear*, Vol. 1, 1957-58, p. 119.
2. Whitehouse, D.J., These Proceedings.
3. Tsukizoe, T., et al., *Wear*, Vol. 1, 1957-58, p. 472.
4. Tsukizoe, T., et al., *Journal of Basic Engineering*, Vol. 87, 1965, p. 666.
5. Tsukizoe, T., et al., *Journal of Lubrication Technology*, Vol. 90, 1968, p. 81.
6. Rabinowicz, E., *Journal of Applied Physics*, Vol. 22, 1951, p. 1373.
7. Archard, J.F., *Journal of Applied Physics*, Vol. 24, 1953, p. 981.
8. Uhlig, H.H., *Journal of Applied Mechanics*, Vol. 21, 1954, p. 401.
9. Goddard, J., et al., *Nature*, Vol. 184, 1959, p. 333.
10. Goddard, J., et al., *Wear*, Vol. 5, 1962, p. 114.
11. Richardson, R.C.D., *Wear*, Vol. 14, 1969, p. 423.

DISCUSSION

QUESTIONER: I have a question for Dr. Whitehouse as well as the present speaker about the assumption that the characteristics of the surface are identical in all directions. In fact, only very few real surfaces are like that. I would like to know the amount of work that has been done to characterize the surface in two orthogonal directions.

D. J. WHITEHOUSE, Rank Taylor Hobson: Well, you are absolutely right of course. They are different. We have been doing work on the effect of the anisotropy to determine the validity of the approximations made. Only in high speed bearings we found some difference. The real feature of interest is not even whether performance is the same in one direction relative to that at 90 degrees to it. We find that there is one more unknown, rather than one less, to deal with: that is the directionality. The surface has different property depending on the direction (or sense) of sliding. This is simply because of the method of its manufacture. We are finding in fact that by taking a 2-dimensional picture of the surface and examining the different characteristics we get roughly something like 10 percent variation between peak curvatures and mean slopes, usually higher for the summits of 2-dimensional picture. But more important than that are the effects of the asymmetry of slopes from one direction to the other. And for instance, we did some work two years ago on bearings and we found that on ordinary hydrodynamic journal bearings we could get a difference in wear of 3 to 1, just depending on the direction of turning. I did not believe the data. So I asked some grinding machine manufacturers and they said that thet is well known in the trade. So I think before we start talking about the odd ten percent or fifteen percent that we get with the anisotropy problem, there is another problem that we might have to look forward to in the future. Certainly directionality is more important than I expected.

D. TABOR, Cambridge University: I would like to ask two questions which are related to what Dr. Whitehouse gave in his marvelous exposition earlier and in the paper by Prof. Tsukizoe. As a non-practical man, my heart bleeds for the practical man who wants to know what plasticity index to use. Dr. Whitehouse put up twelve equations and really one would like to know what he ought to use and whether in fact there is any real significant difference between them in practice except in exceptional circumstances. That is the first question. The second question really relates to this and it is on the paper by Prof. Tsukizoe. In the studies of contact between surfaces, in the estimation of areas of contact one is primarily concerned with the behavior of surfaces under the normal load and there you can use simple criteria of the transition from elastic to plastic deformation. However, as soon as tangential tractions are applied, we have the influence of combined stresses in producing plasticity at the asperities. I think in the work of Prof. Tsukizoe described here there was no reference to this, particularly for the case where the surfaces have little oxide and where adhesion is strong. What I would like to ask both the gentlemen is whether there is a possibility of coping with those problems analytically, digitally, or any other way so that we could get a better idea of what happens when relatively clean surfaces slide on one another, however much detail the initial surface topography may be.

N. OHMAE, Osaka University: Thank you for the comments, Prof. Tabor. Prof. Tsukizoe is presently analyzing the deformation using plasticity index and in the near future he will clarify this important problem.

WHITEHOUSE: Prof. Tabor has asked the most difficult question as usual. I do not think that people have done enough work yet to decide which plasticity index is the best one; they give two to one at the moment. Regarding the second question, under the conditions of very severe tangential forces the influence of the initial surface finish is relatively small; the initial surface finish is only a clue to the initial stresses set up in the surface. When we have essentially normal loading, the surface finish is vitally important. Obviously under extreme tangential loads it is a different question.

S. JAHANMIR, Cornell University: When we consider contact between two surfaces, are we also considering the cyclic forces at the contact? That will set limits on the plasticity conditions of the contacts and you have to consider the shakedown limits in the theory of plasticity. Is that considered in any of the criteria that have been used? I would like to know if that is possible if it has not already been used.

WHITEHOUSE: That is a very good question. I believe that in the pitting experiments of Dyson, he in fact attempts to do what you suggest. He is not doing it necessarily analytically, but he is physically getting a large number of tracks around a disc on a two-disc machine and putting these in a computer. Then literally turning them in a computer and seeing the interactions at given distances. He works out what he considers to be the cyclic variations. I believe he is the only one doing it. Prof. Cheng may be doing the same thing. I do not know if they have got any answers yet, but it is a new idea.

H. S. CHENG, Northwestern University: I am most delighted to hear the effects of directional properties on wear in Journal bearings. This work provides an incentive for us hydrodynamicists or elastohydrodynamicists to really work harder. So I would like to have a little confirmation. Is the longitudinal roughness more detrimental to wear?

WHITEHOUSE: Yes, longitudinal roughness.

CHENG: If so, can you provide the explanation for it?

WHITEHOUSE: I have no idea.

S. RAMALINGAM, Georgia Institute of Technology: In all this analysis, we seem to be purely concerned with the surface topography. Prof. Tabor raised the question of plasticity conditions. One might also examine in a little more detail. Any time a surface is produced, residual stress distribution, which is not symmetrical to start with is also produced. We are completely ignoring that. In addition, engineering materials are polycrystalline systems. When we polish the surface, we get surface relief with a variation in hardnesses. Those things are all being ignored. Therefore, when we talk about asperity contact we are not entitled to assume that all aspeities are identical. We have to take into account, in addition, the texture that is developed. Do the people in surface topography area pay attention to these details?

WHITEHOUSE: Excellent question, of course. The answer is, we are well aware of them. In fact what we have been finding is that in these plasticity indexes everybody tends to think that E and H are bulk properties. But they are not. It is true to say that the properties of the surface skin can be dominant although many people do not realize it.

If we measure hardness as a function of depth of indentation, we start to approach the surface skin which is a depth about equal to the surface texture we observe an important effect. As soon as you approach the surface to a depth of penetration equal to surface finish or thereabouts, you start to get great increases in the hardness. To our surprise we observed up to five to one increase in the hardness of the skin. This has nothing whatever to do with work hardening because when we do these experiments with indium we get exactly the same effects. I went to Cambridge to unravel this and found that it was due to the average distance between dislocations in the material. What happens in effect is that the stylus penetration is small compared with the average distance between dislocations which is what we have in all materials. Then we approach the theoretical strength of the material. And it is only when we have large indentations that we approach the bulk hardness of the material. And so you could argue that in a wear situation it is the skin hardness we should be looking at. I would very much agree with Prof. Ramalingam in that the values of the physical constants used presently for these plasticity indexes are not true. I understand that work is now going on to determine the effects of skin hardness. It is a very interesting subject, but I do not know the answer to it yet.

ON SURFACE TOPOGRAPHY AND QUALITY AND ITS RELEVANCE TO WEAR

P. Gupta

ABSTRACT

A discussion of several key points in the general area of surface interaction is presented. Some experimental data obtained by profilometric measurements is used to support the random process model for surface description and interaction. Asperity load sharing and partial elasto-hydrodynamic lubrication are discussed in the light of surface profile data on mating cylindrical surfaces. Some general conclusions regarding the estimate of real area of contact are cited and a distinction between the "mechanical" and "electrical" area of contact is made.

INTRODUCTION

An adequate description of surface topography and the interaction between mating surfaces has been a problem of great interest in a number of engineering areas. The complexities associated with this subject have imposed a number of limitations on the development of generalized theories in this area of extensive interest. Thus the question of universally supporting one model and rejecting another cannot be realistically asked at this point in time and it only seems justified to discuss further some of the key points. The purpose of this discussion is, therefore, to review a few points of general interest in the light of some experimental data.

RANDOM PROCESS MODELS FOR SURFACE DESCRIPTION

A stochastic description of rough surfaces has been a common feature of many random process models. The details and specific parameters used in the particular models generally vary. Whitehouse-Archard[1] model is dependent on digitized profilometric data and the sampling interval is an important parameter. In fact a small change in sampling distance could result in significant changes in surface description. Nayak's model[2] is free of such restrictions but it needs significant information about the power spectral density functions, which may be difficult to determine for actual surfaces. However, once these functions are determined the model seems fairly realistic.

The agreement between the above two models, in the limiting case (sampling distance $\ell \rightarrow 0$), is not clear to the discussor. It seems that as $\ell \rightarrow 0$, the author's model will result in infinite peak density, and peak radius of curvature will approach zero. Thus further elaboration of this point to clarify the apparent confusion is needed.

Since most surface profiles contain very low frequency components it is very difficult to determine the power spectral density function by using the conventional analogue techniques and, therefore, attempts have been made to use digital methods. Thus the sampling distance once again becomes significant. The only distinction now is that although the theory is sound and free of parameters such as sampling distance, the input to the theory may depend on such parameters. Primarily due to such difficulties, an acceptable experimental validation of the theoretical model due to Nayak[2] has not been possible.

One of the specific parameters in surface profile description is the correlation between peak height and curvatures. According to Nayak's model if the parameter α $(=m_0 m_4/m_2^2)$, where m_0, m_2 and m_4 are respectively the zeroth, second and fourth moment of the power spectral density function) approaches infinity, the peak height and curvatures become independent. The discussor has validated this to some extent.[3] In course of traction experiments with rolling discs, surface profiles on the cylinderical surfaces of the disc were obtained in the circumferential direction. These profiles were then digitized and the peak height and curvatures were computed. The statistical correlation coefficient for most curves was found to be less than 0.10. The autocorrelation and power density functions were also computed and typical curves are shown in Figures 1 and 2. From the spectral density curve the various moments were computed and α was found to be about 25, which would indicate very low correlation between peak height and curvature as observed. Beyond, this validation the discussor was unable to obtain any other parameters for the general validation of the Nayak's model. The primary difficulty was associated with the proper selection of the sampling interval used while digitizing the measured profilometric data.

The discussor agrees with the author in the fact that for most practical surfaces there has to be a lower limit on the sampling interval. In fact, this limit may be of the order of the stylus radius. Thus for all practical purposes, peaks with a radius of curvature below a certain limit may not have any significance and from this viewpoint the author's model may provide a realistic surface description. However, the question raised above concerning the apparent conflict between the two models must be resolved.

ASPERITY LOAD SHARING

Once the surfaces are defined and they are employed in practical applications where a lubricant is introduced between the surfaces, the next question raised concerns the load sharing between the mating asperities and the lubricant film. It is clear that if the film thickness is quite large compared to peak heights, then most of the load will be supported by the lubricant film. As the film thickness is reduced, the asperities will come into contact and they will eventually support all the load. In most cases, the transition or the shift of load from asperities to lubricant film has been found to be quite sudden.[3] The discussor confirmed this finding for the rolling disc surfaces[3] discussed above. An analysis developed earlier[4,5] was used for these surfaces. The lubricant film thickness or the nominal

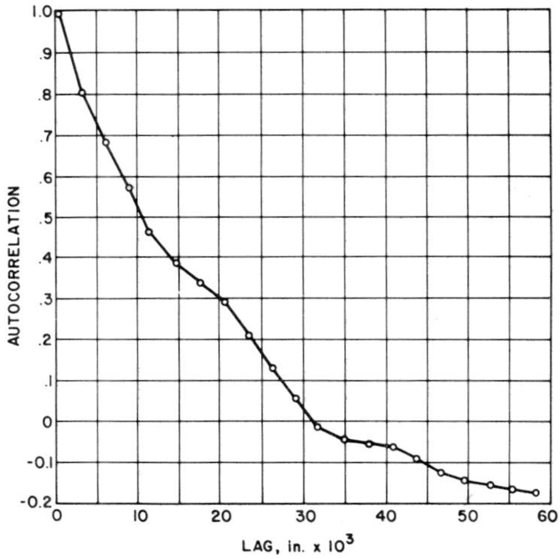

Fig. 1.—Autocorrelation Function for a surface profile obtained on the cylindrical surface of a disc specimen along the circumferential direction. RMS roughness ≈ 5 microinch.

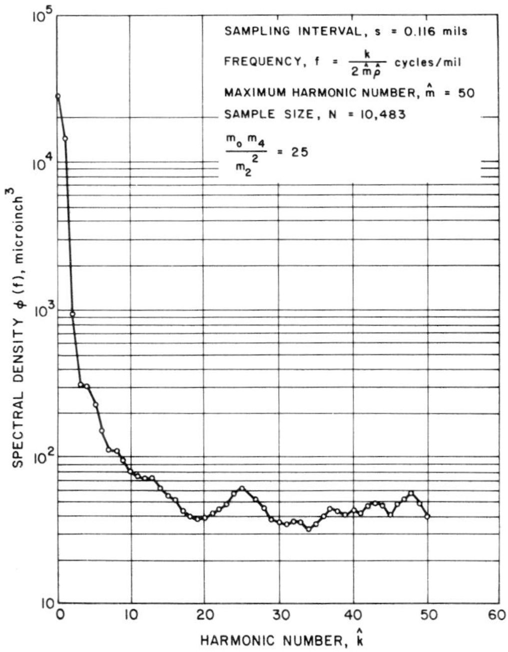

Fig. 2.—Power Spectral density function corresponding to the date shown in Figure 1.

separation between the mating surfaces was computed from the classical elastohydrodynamic solutions. The results of asperity loads as a function of the rolling velocity are shown in Figure 3. The steep gradient clearly demonstrates the sudden transition.

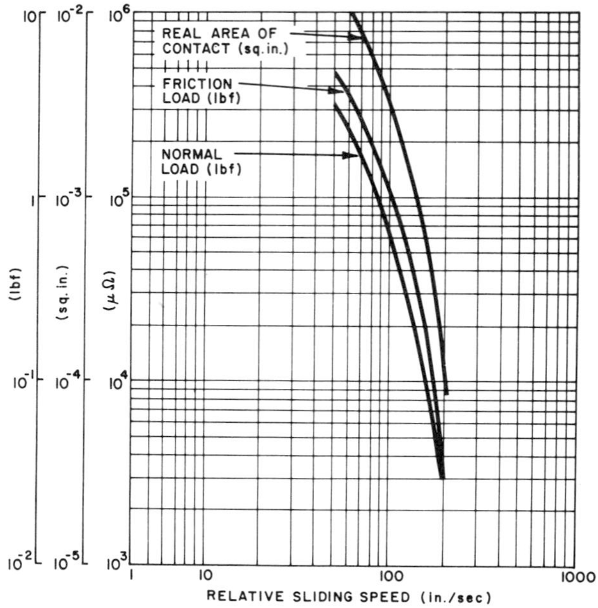

Fig. 3.—Asperity loads and real contact area as a function of rolling speed.

THE REAL AREA OF CONTACT

The real area of contact between mating surfaces is another parameter which has a substantial tribological significance. The lower bound estimate determined by the load to hardness ratio is often questionable. With a simple model of surface interaction the discussor computed the contact stress levels of a number of rough surfaces and the results are shown in Figure 4. The four experiments had varying roughness of the mating surfaces, the roughest and smoothest pairs being experiments #1 and #4 respectively. Although the surfaces in experiments #3 and #4 were prepared quite differently they demonstrated closely the same order of roughness. The general finding from these results is that the "real stress" could be up to half the hardness and hence the real area of contact can be twice the lower bound estimate.

Although the total real area of contact is of primary concern in friction and wear problems, the size distribution of contact spots becomes significant in thermal and electrical applications. In addition to this size distribution, Williamson[6], while dealing with electrical contacts, introduces the terms "mechanical" and "electrical" area of contact. Since all the mating asperities may not conduct due to the presence of some insulating contaminants the electrically conducting area may be quite small compared to the total mechanical area determined by the asperity deformations. Thus some caution should be observed while determining the real contact area with

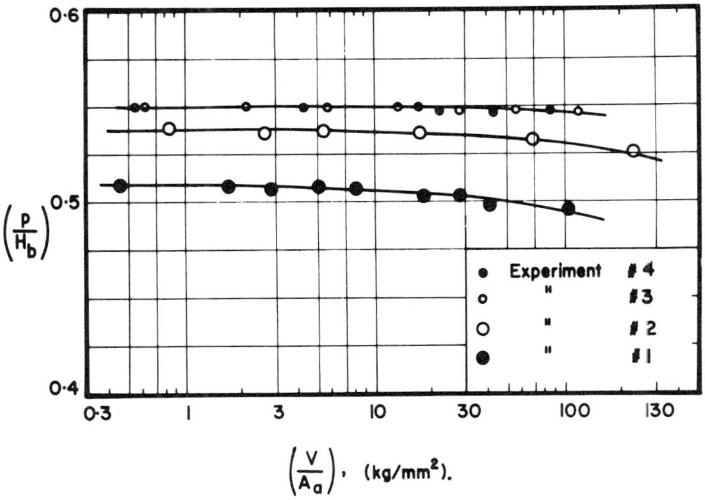

Fig. 4.—Variation of real contact stress as a function of normal load for several pairs of rough surfaces.

the use of electrical resistance measurements.

SUMMARY

The behavior of the Whitehouse-Archard model as the sampling interval goes to zero needs further elaboration. The discussor feels that the alternate model due to Nayak(2), though free of sampling interval problems, may lead to difficulties, somewhat indirectly, since the power spectral density functions are generally computed numerically. The peak radius and height independence as predicted by Nayak's model(2) seems to agree with some experimental data obtained by the discussor.

It is shown that the transition of load support from mating asperities and the lubricant film can be quite sudden and this is predicted with a simple surface interaction model.(4) The sharp transition has also been observed experimentally.(3)

For a number of rough surface pairs it is shown that the real area of contact can be twice as large as the lower bound value predicted by the load to hardness ratio. In electrical contact experiments the area determined by the mechanical interaction of asperities may not be truly relevant due to the presence of nonconducting contaminant films, and therefore an additional caution is necessary for such applications.

REFERENCES

1. Whitehouse, D.J. and Archard, J.F., *Royal Society of London. Proceedings*. Series A, Vol. 316, 1970, p. 97.
2. Nayak, P.R., *Journal of Lubrication Technology*, American Society of Mechanical Engineers Transactions, Vol. 93F, 1971, p. 305.

3. Smith, R.L., et al., "Research of Elastohydrodynamic Lubrication of High Speed Rolling-Sliding Contacts," U.S. Air Force Aero Propulsion Laboratory Technical Report AFAPL-TR-72-56, prepared by Mechanical Technology Inc. under Contract F33615-69-C-1305.
4. Gupta, P.K. and Cook, N.H., *Journal of Lubrication Technology,* American Society of Mechanical Engineers Transactions, 94F.
5. Gupta, P.K. and Cook, N.H., *Wear,* Vol. 20, 1972, p. 73.
6. Williamson, J.B.P., Private Communications.

DISCUSSION

D. J. WHITEHOUSE, Rank Taylor Hobson: Dr. Gupta, that was a very interesting discussion. I think that a number of points that you raise, and one in particular, need airing. It is possibly one of the most important factors in surface topography. That is, when do you use correlation methods and when do you use spectral analysis methods? Everybody says that there is a certain amount of information in correlation methods, as there is in spectral analysis techniques. That is absolutely true, but unfortunately it does nct quite work that way. In tribology, invariably we are dealing with surfaces that have been generated in a random way. Not necessarily, but usually. Under that circumstance, it is better to use a correlation method than spectral analysis method, for the simple reason that the correlation function will reveal the unit event of machining which in turn will give you a clue as to the unit event of function.

Three different things need to be looked for in the correlation function. One is the scale, that is the rate of decay of the exponent, the second is the shape and the third thing is the behavior at the origin. Now, if we imagine that a grit hits the surface, what happens is the surface characteristic is preserved in the correlation function. The decay in the correlation function is a direct result of the pile-up produced by that one unit event, the grit hitting the surface and leaving a pile-up. The extent of pile-up is exactly proportional to the depth of indentation of the correlation function and it is one to one. The width of indentation is proportional to the thickness of the indentation before we get to the plane so that by looking at the correlation function we can get an absolute replica of the machining process which is what gives the surface its properties anyway. And for that reason, in my experience, whenever we have a surface which has been manufactured by a random process (by that I mean a Poisson process in statistical terms), then we should use a correlation method.

Now there is another point and that is this: the reason you find spectral analysis in all the books is because they were written by communication people who are interested in the temporal behavior of surfaces. Very rarely, in any sort of communications, do you find anybody interested in the behavior that we are looking at. And for that reason alone you have to be a bit careful not to stray into spectroanalysis just because it is in the books. In tribology we have a rather different problem. We do not just have temporal behavior to take into account. But shall we say vertical behavior as well which we all know. Thus, there are at least two good reasons why we should think a little bit more carefully about using just spectral analysis. I hope this clears up the point about correlation which I think is absolutely fundamental.

On the other hand, if we are measuring turning, milling, or planing marks, obviously we would use spectral analysis, because these are basically cyclic events.

THE RELATIONSHIP BETWEEN SOME TRIBOFAILURES AND SURFACE ROUGHNESS

A. Ura

ABSTRACT

This paper discusses three separate topics related to tribology. The effect of surface quality on scuffing is discussed in terms of Λ*, the ratio of film thickness of a lubricant to the surface roughness followed by some interesting results on asperity interactions in rolling-sliding contact. Lastly, the effects of abrasive embedded layers on the wear of soft surfaces are presented.*

INTRODUCTION

As Whitehouse[1] has explained with some examples and analytical results in detail, the advance of the theoretical treatment of surface metrology is very noticeable. He has reviewed the merits and demerits of the applicability of the theory and studies done so far. He has also indicated the difficulties in choosing suitable parameters expressing surface topography because of their changeable character in practical running conditions.

In this paper, a few points about the surface roughness effects and the effects of an embedded layer in abrasive wear on surface quality will be discussed.

EFFECT OF SURFACE TOPOGRAPHY ON SURFACE FAILURES

In scuffing, it is difficult to state exactly the correlation between surface roughness parameters and failures. One of the reasons, as described by Whitehouse, is that the mating surfaces are likely to change at every moment with continued rubbing and it is difficult to keep the condition of the rubbing surface constant.

First of all, consider the well-known criterion to show the ratio of the film thickness to the roughness $\Lambda = h_o/\sigma$, where σ is the composite surface roughness and h_o is the mean film thickness. Λ is often used to indicate the extent of surface asperity interaction, since the breakdown of the film under elastohydrodynamic lubrication (EHD) may be due to the direct contact between asperities. Poon et al, for example, classified the condition of lubrication into three regimes, namely EHD, non-hydrodynamic lubrication, and the transient condition by means of Λ. Yamamoto et al. have reported[3] that Λ gives information about the contact condition during running by observing an indentation impressed on a surface previously.

Determining the relationship of the standard deviation of the profile peak distribution σ* to the correlation distance β*, an approach Whitehouse has tried, is very useful for indicating the relative difficulty of plastic deformation, similar to Plasticity Index by Hirst and Hollander[4] as shown in Figure 1. If equipment is developed to determine the parameter more easily, it will be very useful[5].

The thickness of the oil film is dependent upon the direction of the surface texture. As for the roughness parallel to the direction of motion, Chow and Cheng[6] have examined theoretically the oil film thickness and its load capacity as shown in Figure 2. Generally, it is reported that the transverse roughness has an excellent ability to resist scuffing and this could be explained by the fact that the oil film under EHD condition is formed easily owing to successive presence of asperities in the sliding direction.

RELATION BETWEEN SURFACE QUALITY AND FAILURES

Generally, Λ is considered as a criterion of a surface failure under lubrication condition[7]. The effect of a changeable surface quality on scuffing will be discussed here with the help of the experimental results by Hirano and Yamamoto[8]. They tried to establish the condition for scuffing at sliding velocities of 6 m/s and 4 m/s using a disk machine. They reported that scuffing did not occur even when Λ was less than unity. They could find oxide film over the entire surface and concluded that the oxide film of the surface is a necessary condition to prevent scuffing when the breakdown of oil film occurs.

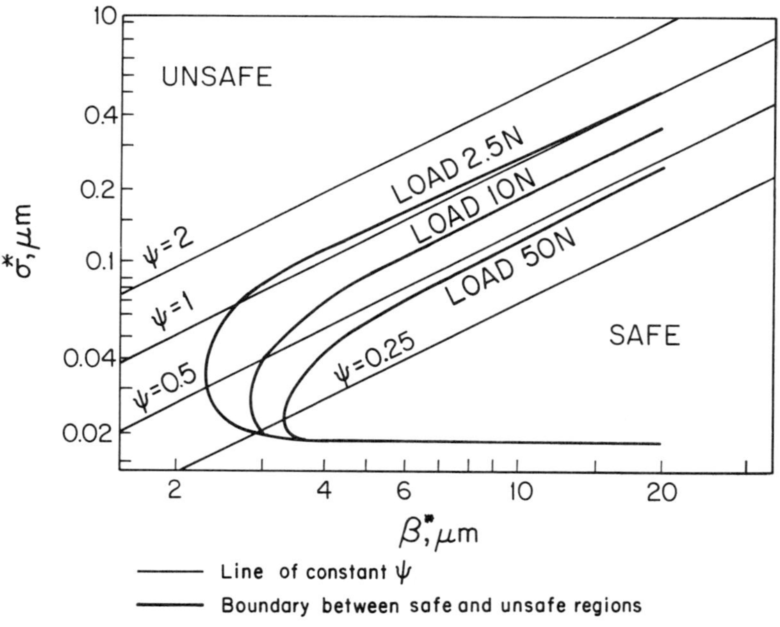

Fig. 1.—The effect of the surface asperity on scuffing[3].

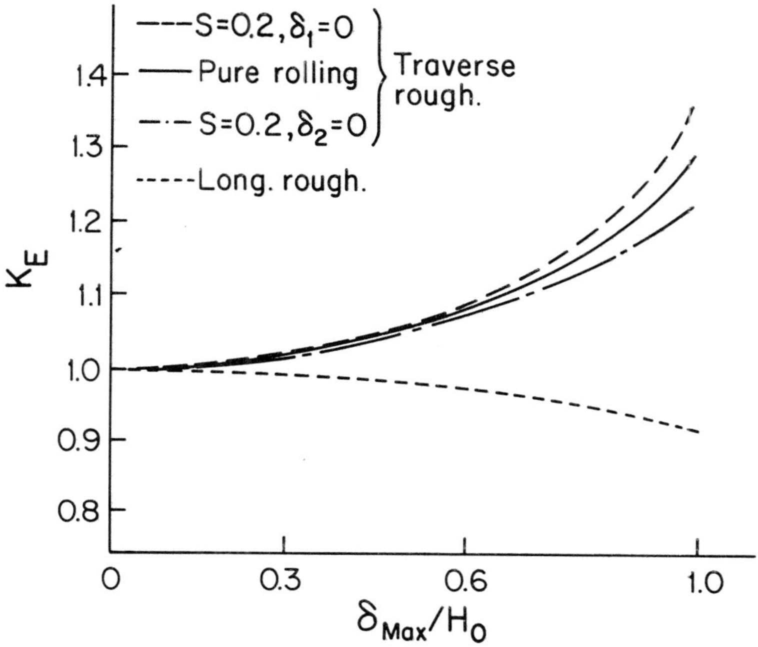

Fig. 2.—The effect of transverse and longitudinal roughness on K_E of an EHD contact[5].

Under a severe lubricated condition (e.g., high load, high sliding velocity and high temperature) the breakdown of the EHL film seems to be sufficient condition for scuffing. On the other hand, under mild conditions (e.g., low load, or slow sliding velocity) the oxide film is formed on the mating surface which alleviates the occurrence of scuffing when the breakdown of oil film occurs.

The effect of the amplitude and shape of the surface roughness is as follows. The amplitude of the surface roughness affects the deformation of asperities. When the amplitude is large, the plastic work done during the deformation increases the surface hardness. When this increase in hardness is accompanied by oxidation, scuffing resistance is further increased. When the asperity height is low, no scuffing occurs at low loads due to the lack of asperity contacts. However, occasionally, sudden scuffing occurs when severe surface contacts take place in the absence of any workhardening and a formation of an oxide film[6] as shown in Figures 3, 4.

The shapes of surface asperities can be compared using a machined surface and a ground surface. The surface machined by turning has a larger asperity than the ground one and, therefore, is likely to deform plastically. Accordingly, the running-in is not brought about effectively due to the slow change in roughness and consequently the surfaces are easily brought into scuffing even at low load as shown in Figure 5. By contrast, the surface finished by grinding has small asperities and the roughness is low. Based on these observations, we believe that tan θ, the absolute mean slope of the surface in an Index, ϕ (= $(E^*/H) \cdot \tan\theta$), should be a criterion for scuffing occurrence.

The hardness increases due to workhardening when the parameter Λ is less than two[8]. For surface with higher roughness, the hardness starts

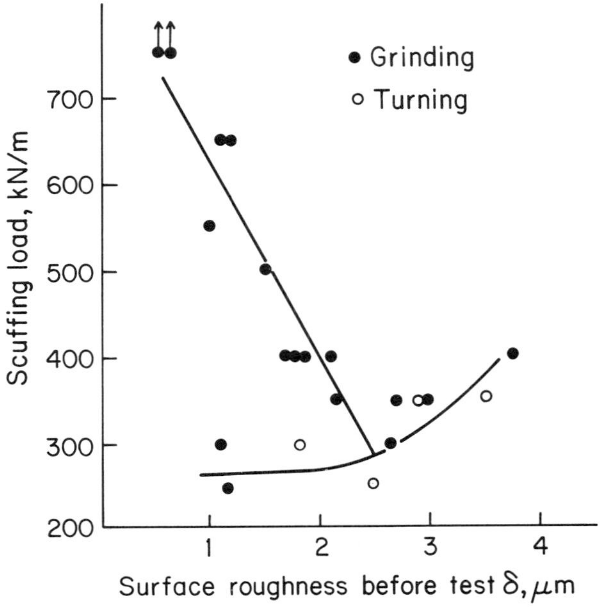

Fig. 3.—The effect of surface roughness on scuffing (S45C: Annealed)[7].

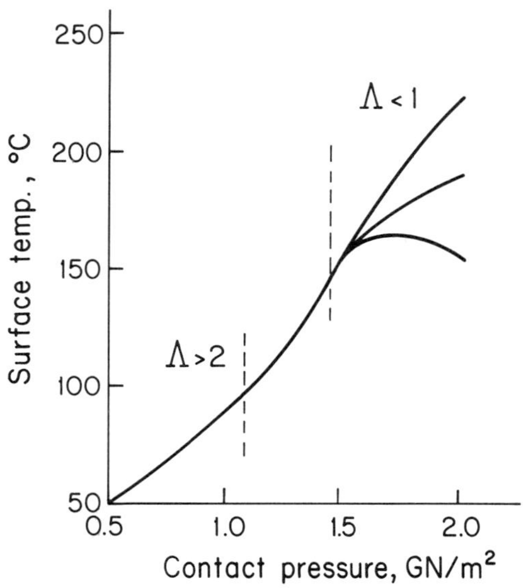

Fig. 4.—Contact temperature[6].

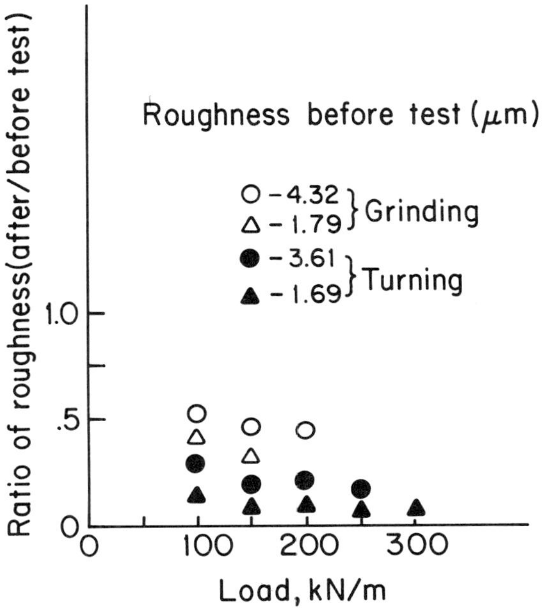

Fig. 5.—The effect of machining process on running-in.

to increase even when the applied load is low. However, when the surface has a comparatively smaller roughness, the hardness does not increase until the load reaches a high value. The load capacity against scuffing goes up linearly with the increase in the hardness. Based on these results, it may be concluded that the breakdown of the oil film comes about during running-in and causes a surface quality to change, resulting in activation of mechanisms which lead to scuffing.

Ichimaru et al.[8,9] reported interesting experimental results on the asperity interaction in a rolling-sliding contact. They carried out pitting tests using a soft and a hard roller finished by cylindrical grinding. Even when the initial roughness of the hard roller was much larger than the EHD film thickness, the pitting life was extremely long when there was no relative slip between the two rollers. The EHD film built up gradually, becoming fully established when the surface texture of the hard roller was replicated on the soft roller surface, usually after a long run. In the presence of sliding, however, they reported that pitting occurred at an earlier stage of running and the impression of asperities of the hard roller on the soft one was not evident.

The change of the surface quality during running affects wear. The alteration of the surface condition can yield a better abrasive wear resistance. Although it cannot be stated quantitatively, it can be discussed in terms of stress analysis[11,12].

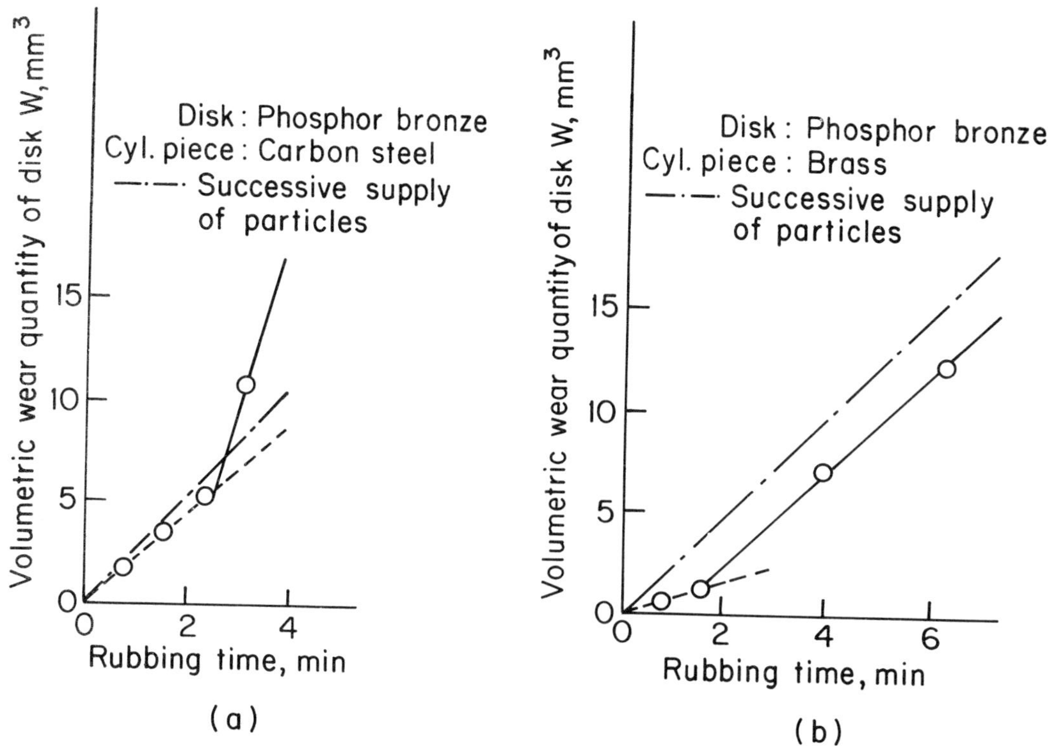

Fig. 6.—The effect of detaching process of particles from the embedded layer.

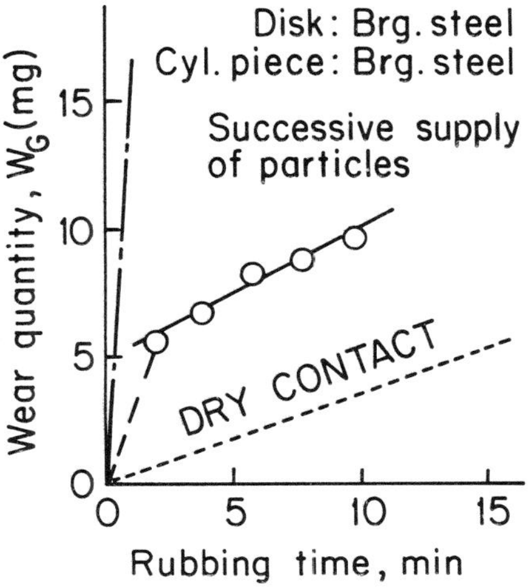

Fig. 7.—The abrading process of the embedded layer.

EFFECT OF EMBEDDED ABRASIVE PARTICLES ON WEAR

A soft surface impregnated with abrasive particles can improve the wear resistance. The protective action imparted by the embedded layers on a phospher bronze disk surface was examined. During the early stage of wear tests when the embedded layers are held on the surface, the wear rate was low, but it increased suddenly after a certain passage of time. The broken line of Figure 6a shows the wear rate in the case of a successive supply of abrasive particles. When the difference in hardness of the contacting surface is small, it is difficult to establish the abrasive embedded layer as shown in Figure 6b. In this case, the wear rate is as high as when the surfaces are slid against each other without any abrasive embedded layers.

Figure 7 shows the abrading abilities of the embedded particles. It proves that an embedded layer fails due to abrasive action. The embedded layer is not strong enough to withstand the rubbing force and therefore, the particles on a surface are detached after a few cycles of rubbing. These phenomena explain that when a successive supply of abrasives is maintained between two surfaces, the rate of the formation of an embedded layer and the rate of the detachment of particles from the surfaces are equal.

REFERENCES

1. Whitehouse, D.J., These Proceedings.
2. Poon, S.Y. and Haines, D.J., *Institution of Mechanical Engineers. Proceedings*, Vol. 181, Part 1, 1966-67, p. 363.
3. Yamamoto, Y., Hirano, F. and Hashimoto, M., *Japan Society of Mechanical Engineers*, Vol. 755, Part 2, 1977, p. 343.
4. Hirst, W. and Hollander, A.E., *Royal Society of London. Proceedings*. Series A, Vol. 337, 1974, p. 379.

5. Hirst, W., "Tribology Practical Reviews," Institution of Mechanical Engineers, London, 1975, p. 25.
6. Chow, L.S.H. and Cheng, H.S., *Journal of Lubrication Technology*, Vol. 98, 1976, p. 117.
7. Nagaraj, H.S., Sanborn, D.M. and Winer, W.O., *Journal of Lubrication Technology*, Vol. 99, 1977, p. 254.
8. Hirano, F. and Yamamoto, Y., in Institution of Mechanical Engineers Convention Tribology, 1978, Swansea, p. 43.
9. Ichimaru, K., Nakajima, A. and Hirano, F., *Japan Society of Lubrication Engineers. Journal*, Vol. 20, 1975, p. 799.
10. Nakajima, A., Ichimaru, K. and Hirano, F., *Japan Society of Lubrication Engineers. Journal*, Vol. 22, 1977, p. 255.
11. Hirano, F. and Ura, A., in Proceedings of the Conference on Tribology in Iron and Steel Works, Institution of Mechanical Engineers and Iron and Steel Institute, 1969, p. 163.
12. Ura, A. and Hirano, F., *Japan Society of Lubrication Engineers. Journal*, Vol. 22, 1977, p. 335.

DISCUSSION

H. DALAL, SKF Industries: The presentations of Dr. Whitehouse and Dr. Ura bring out an interesting conflict of interest. Dr. Ura presented a graph which shows that the load-carrying capacity of a surface depends on the transverse surface roughness as opposed to a longitudinal surface roughness. On the other hand Dr. Whitehouse presented that the longitudinal surface roughness is important to the wear of the surface. It seems to me that if this is the case then a Longuet-Higgins kind of charterization of the surface might be valuable in designing a surface for an application. Would Dr. Whitehouse like to comment on that please?

D. J. WHITEHOUSE, Rank Taylor Hobson: This is a valid point. I believe that there is no fundamental reason why directionality cannot be introduced into any random characteristic. However, the directionality of the surface is never taken into account in either my statistics or Nayak's statistics. The mathematical analysis assumes that things happen in time the same way in both directions. But we get completely different wear characteristics dependent on the way the surface is rotated. At any moment I know one way of characterizing directionality and that is by using the moments of the derivatives of the surface in two dimensions. Then you apply my statistics or Nayak's statistics. Without the differentiation you cannot account for directionality. I have tried this approach in one dimension, but unfortunately not yet in two dimensions. In principle it is possible to do by differentiating, and either taking correlation functions of the differentials or taking the moments of the differentials in 2-dimensions.

P. ENGLE, IBM: I would like to ask the speakers if they find a relationship between the macroscopic contact and the microscopic contact geometries. Obviously when contacting machine parts are designed, these relationships do come in, and I think there may be an interesting development if we correlate these two. The macroscopic contact conditions would also be helpful in finding the two dimensional contact conditions and the wear rates. I wonder if Dr. Whitehouse has a comment on that.

WHITEHOUSE: Again an interesting question. We have certainly tried to bring together the macroscopic and microscopic functions. Usually what we have found so far was that basically we cannot relate in terms of simple geometry. The factors that bring about the macroscopic errors have a different thermal characteristic from those that produce the microscopic ones. The result is that functionally the macroscopic effects are quite different, and necessarily so, from the microscopic ones. Presently it is difficult to relate the functional differences purely to the geometry although we can differentiate geometrically very easily as you know. Personally I do not believe it is possible.

THE MICRO-GEOMETRY OF LUBRICATED WEAR- CLASSIFICATION AND MODELLING

K. J. Stout, W. Watson and T. G. King

ABSTRACT

For the limited depths of wear which are allowable in many engineering applications the "worn" surface generally retains features of the original surface along with those generated by the wear process. When one of the rubbing surfaces is comparatively hard and lubrication conditions prevent metallic transfer, wear is confined principally to the softer surface and this exhibits a transitional topography.

There are occasions when all traces of the original surface profile disappear and this may be a result of the method of production of the original surfaces. To illustrate this point, a series of ground steel surfaces were prepared yielding differing amplitude and directional characteristics and their wear properties examined by running them against phosphor bronze pins. Surface profiles and wear scars produced on the pins were assessed. It is first shown that directional effects are of considerable significance for rougher surfaces, but become less important when surfaces of finer texture are considered.

To assist with the prediction of surface wear behaviour, attempts have been made to classify the distributions which best fit the amplitude characteristics. Finally, mathematical analyses are developed which model the topographical transitions which occur during "running-in" and these may be of assistance in evaluating the extent of wear under test conditions.

NOMENCLATURE

c = the truncation level

d = mean height of worn surface

$f(y)$ = the probability density function (amplitude distribution) of the worn surface

$g_u(y)$ = the probability density function (amplitude distribution) of the unworn surface

$g_t(y)$ = the probability density function (amplitude distribution) of the truncating function

p = the ratio of the R.M.S. roughness of the truncating function to that of the unworn surface

y = the general ordinate height of the worn surface

y_u = the general ordinate height of the unworn surface

y_t = the general ordinate height of the truncating function

$\Phi_u(y)$ = the cumulative probability density function of the unworn surface

$\Phi_t(y)$ = the cumulative probability density function of the truncating function

γ_K = covariance between two points y_o, y_K

μ = mean surface height

$\mu_n(y)$ = nth central moment of the amplitude distribution

INTRODUCTION

Whilst considerable work has been carried out on the compatibility of bearing pairs, less attention appears to have been given to the influence of surface finish and appropriate means for its specification. This is particularly true for the early stages of wear - "running-in", most wear rate evaluation being undertaken after initial running-in has been accomplished. To enable the designer to make effective estimates of bearing performance and life, there is a need for a more detailed understanding of the effects of surface roughness in its various forms.

Some recent work in which surface roughness effects have been considered has been carried out by Rowe et al.,[1] who investigated the performance of grease lubricated plain bearings using relocation profilometry techniques. They used steel shafts of a variety of surface finishes run against steel backed phosphor bronze bushes, and showed that surface roughness was of fundamental importance. They also noted that shaft end-float could have a dramatic effect on wear rate. Rowe et al. demonstrated convincingly the suitability of relocation profilometry for the study of wear. Another interesting paper has been presented by Bayer and Sirico[2] who studied the influence of the magnitude and directionality of surface texture on wear. They utilized 65Cu/35Zn brass balls run on lubricant flooded steel plates, both across and against the direction of lay of the surface. Their results showed that wear was most sensitive to surface roughness variations for finer surfaces, whilst coarser surfaces showed greater wear sensitivity to directionality.

Confining our considerations to the geometric aspects of wear, it is proposed that wear is affected not only by the scale (and shape) of the surface height distributions, but also by spatial considerations of the surfaces and their interaction. These include:

(i) The correlation functions of the surface profiles.

(ii) Directionality or anisotropy of the surface textures and its relation to the direction of relative motion.

(iii) The geometric constraints of the particular wear situation - allowing such effects as end float in journals or run out in thrust bearings.

In this initial study, it was considered important that the surfaces be produced by typical production methods. The great majority of bearing journals are produced by cylindrical grinding, either by plunge feeding or traverse grinding. In plunge grinding, form errors of the workpiece are principally controlled by the form of the wheel as imparted by dressing. Traverse grinding relates the form errors of the workpiece to the integrated errors of the grinding machine. This generally produces a more precise form but also imparts a helical lay to surface finish, the pitch of which depends on the traverse rate. It was, therefore, decided to produce specimens by both the above methods and contrast their wear performance.

EXPERIMENTAL AND TEST PROCEDURE

The wear experiments were carried out on a specialized pin and ring machine, employing an aerostatic support bearing yielding low inherent vibration levels and high rotational and axial stiffness. This rig allows relocated profiles of the pin and ring to be obtained without removal of either component from the machine, and has been described in detail elsewhere.[3,4] The rings used for these experiments were 90 mm in diameter and made from BS 970 En 32B steel case hardened to 750 VPN. The pins were produced from BS 1400 PB 1-C phosphor bronze and were 6 mm square. The rubbing surfaces of the pins were prepared by surface grinding them nominally flat, using an A46 wheel. The grinding direction was parallel to the direction of sliding, and the pin hardness was 140 VPN. The surface of the ring was lubricated by a contacting felt wiping pad and the lubricant used was Mobil D.T.E. light oil. The load for these tests was 10 kg and was applied by means of deadweights.

Initial alignment between pin and ring was made using Prussian blue to ensure nominally even bearing pressure across the width of the pin. The profile traces of the pin were taken at right angles to the wear direction, at a position central to the contact zone as indicated by this marking procedure.

In the first series of tests, the rings were ground using an A46 wheel. This was dressed using one traverse in each direction with 0.025 mm infeed at each pass and a traverse rate of 25 mm/sec. The grinding wheel was 300 mm in diameter and rotated at 1500 rev./min. Three specimens were prepared with this wheel. The first and second were plunge ground with a finishing infeed of 0.025 mm and allowed to spark out. The third was traverse ground at a feed rate of 2 mm/sec. with the same infeed. The results of the wear test with the first specimen are shown in Figure 1. As can be

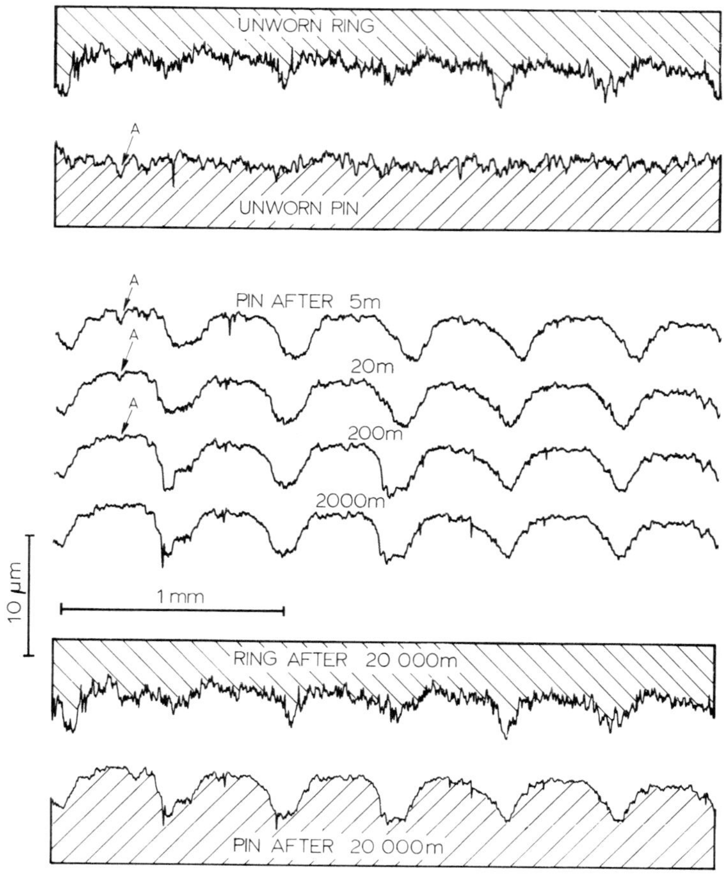

Fig. 1.—Surface Profiles during Wear Process - Plunge Ground Specimen with A46 Wheel, Coarse Dressed.

seen, the ring shows periodic features from the wheel dressing and these were rapidly reproduced on the pin after only 5 m of sliding. Further running results in relatively little modification to this established profile, although the general profile level drops. Some indication of the wear rate can be obtained from observation of the original profile feature marked "A" on the diagram. Note that the final pin profile conforms quite closely to the single ring profile shown illustrating the gross anisotropy of the ring. Examination of the initial and final ring profiles reveals that little change has occurred. The differences lie mostly in the peaks of the profile. A subsequent test using this ring with a new pin yielded the same pattern of events but resulted in a smaller wear scar, indicating that these minor changes may be significant. Three dimensional examination of the ring appears essential if the nature and extent of these profile changes is to be understood.

The second test utilized the traverse ground ring, and is shown in Figure 2. The short section of profile reproduced shows a periodic structure exists again, although a longer wavelength. In this case, this is

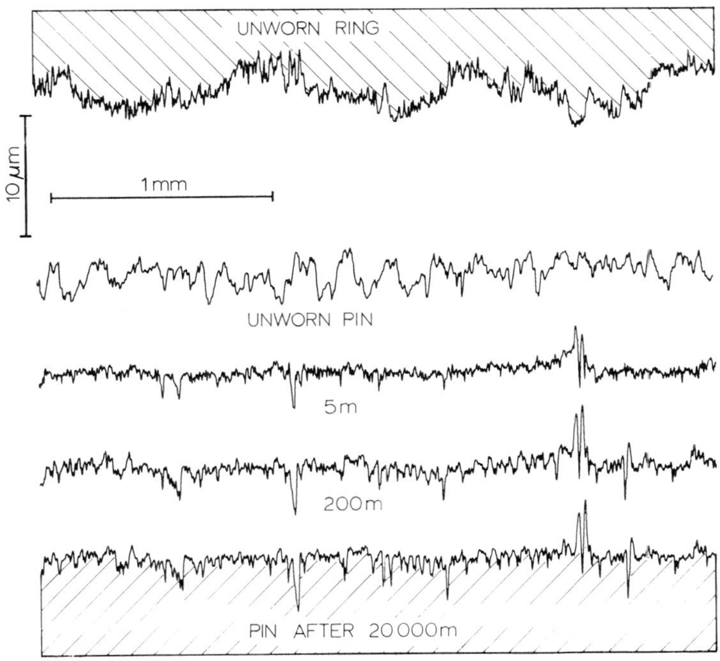

Fig. 2.—Traverse Ground Specimen with A46 Wheel, Coarse Dressed.

not transferred to the pin because of the helically structured surface texture of the ring. A further feature is that the worn surface profile of the pin is considerably flatter than the profile of the ring. An analogy may be drawn here to the process of traverse grinding itself in that an integrating process is occurring. The wear scar produced in this test was much larger than in the previous one. Despite the similarity of the later pin profiles, the scar width increased steadily throughout the test and large amounts of wear debris in the form of metallic particles were apparent.

Figure 3 shows the results of a test conducted using the second of the plunge ground specimens. The ring profile is closely similar to that of

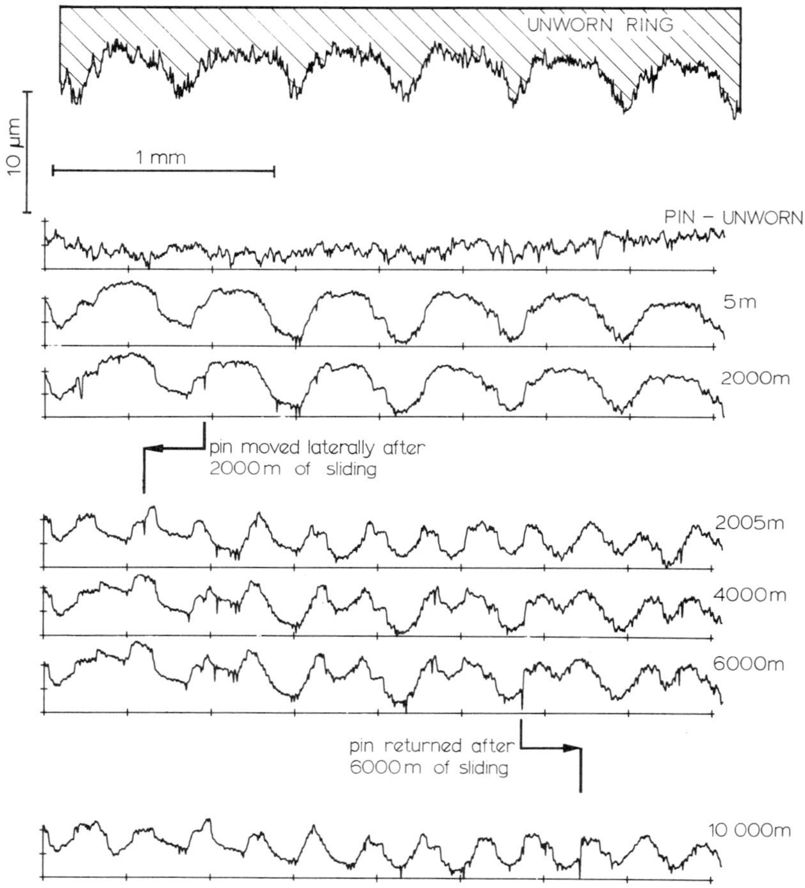

Fig. 3.—End Float Simulation - Plunge Ground Specimen with A46 Wheel, Coarse Dressed.

Figure 1, resulting from the same wheel dressing. The object of this test was to simulate relative profile movement as found with the helically structured surface or with shaft end float. The simulation involved lateral movement of the pin of one half wavelength of the profile periodicity.

This was performed whilst the surfaces were out of contact, without measurably altering the alignment of pin and ring. As can be seen from the figure, the early progress of the test conforms to the pattern seen in Figure 1, the ring rapidly imposing a new profile on the pin. The pin was moved laterally by 0.26 mm (estimated as the half wavelength) after 2000 m of sliding. A further 5 m of sliding showed the ring re-establishing its effective peak cutting profile and in doing so removing a significant volume of material. Inspection of the profiles after translation shows that adjacent pairs of profile valleys match closely, having been produced by interaction with the same features of the ring. After 4000 and 6000 metres of sliding, the new valleys have grown deeper relative to their displaced counterparts which experience no apparent modification. However, the rate of deepening decreases to a small fraction of that experienced in the first five metres after translation. Returning the pin to its original position after 6000 m resulted in further rapid wear on the original sites.

DISCUSSION OF EXPERIMENTAL RESULTS

The foregoing experimental observations serve to demonstrate the sensitivity of wear to spatial/directional nature surface geometrical features.

Figures 1 to 3 in conjunction with Table I give probably the clearest indication of the pattern of wear during early running. Whilst unavoidable

TABLE I.—WEAR SCAR DIMENSIONS AFTER 20 KM OF SLIDING

Test No.	Grinding Conditions	Mean Scar Width on Pin as assessed visually under X 10 Magnification	Mean Scar Depth as Measured by Talysurf
		mm	μm
1	Plunge/A46/Coarse Dressed	1.09	6
2	Traversed/A46/Coarse Dressed	3.32	35
3	As 1	1.72	10

differences exist between the amplitude characteristics of the three specimens as a result of the manufacturing process, the three tests taken together indicate that it is the directional nature of the surfaces which most significantly affect wear performance. Comparing the ring profile traces for Figures 1 and 2, it can be seen that on the basis of profile evidence alone, the traverse ground specimen (Figure 2) appears to offer better prospects for the achievement of conformity with its pin. In practice, by comparing the wear scar sizes given in Table I, the reverse is seen to be true, as shown by its wear scar being three times as large. That this is due to the directional nature of this surface is confirmed when consideration is given to the worn profiles of the pins. Figure 1 shows conformity between pin and ring leading to the proliferation of contact points and hence lower local bearing pressures. The worn pin profile of Figure 2 shows, by contrast, that conformity does not occur and thus high contact pressures are maintained leading to continued wear. The pin profile in this case becomes flatter. Further evidence is offered by the comparison of Figure 1 with Figure 3 where two very similar plunge ground surfaces are caused to have different wearing actions. The translation of the pin shown in Figure 3 results in a step increase in bearing pressure leading to an increased wear rate whilst conformity is re-established. It is interesting to note that the bulk of this wear takes place within 5 m of sliding after the translation. In situations where directional effects produced by the machining process (or by end float in journals and run-out in thrust bearings) effectively prevent profile conformity from being established, local bearing pressures can only be reduced to acceptable levels by increasing the conformity in the direction of sliding. This is likely to involve extensive wear leading to bearing play and inability to retain lubricants.

AMPLITUDE DISTRIBUTION MODELS

It has been noted by other workers[5,6] that some surface production techniques usually yield near gaussian amplitude distributions, whilst other production processes rarely provide gaussian surfaces. It is therefore of interest, in light of the above comments, to attempt to fit model distributions to both gaussian and non-gaussian surfaces; and to investigate analytically the effect of wear on the moments of the amplitude distribution. Attempts to fit model distributions including some non-gaussian surfaces as well as the usually considered gaussian case have been made. Figure 4 shows some typical surface profile height distributions for samples of 1500 heights. Using the Pearson system[7] for identification of distributions, appropriate models have been fitted to the data and can be seen plotted as the smooth curves on Figure 4. Further details of the types of density function used are given in Appendix I. The following analysis is related to the truncation of the gaussian amplitude distribution.

MODELLING THE MICROGEOMETRY OF LUBRICATED WEAR

This work extends some of the basic ideas on amplitude characteristics presented previously[8]; which allowed an estimation of the worn surface in which an initial surface profile was successively truncated by a straight line. This has been reported earlier in this Conference by Dr. Whitehouse.

Figure 5 illustrates the simplest truncation model, in which the wear generated surface features are considered to form a smooth plane. The normal distribution curve is progressively cut off as wear proceeds and an impulse representing all the observations of the plane is formed at the contacting surface level. The weight of this impulse is equal to the area of the removed portion of the normal curve since the total number of ordinate observations remains constant. This type of behaviour for a real surface is illustrated in Figure 6, which shows relocated profiles from a rough phosphor bronze pin, prepared with 120 grade SiCC paper to give an initially near gaussian surface, and then run against a hard steel ring. The ring had a ground surface finish of 0.2 μm Ra across the lay. Lubricant was Mobil DTE light applied by means of a felt wiping pad and the bearing pressure was 1.65 MN/m^2. Within the resolution of the histograms presented, it can be seen that the wear produces an impulse-like addition to the remaining portion of the original distribution.

It is more reasonable, in the wear context, to assume that the truncating function is not completely smooth, although this naturally renders the

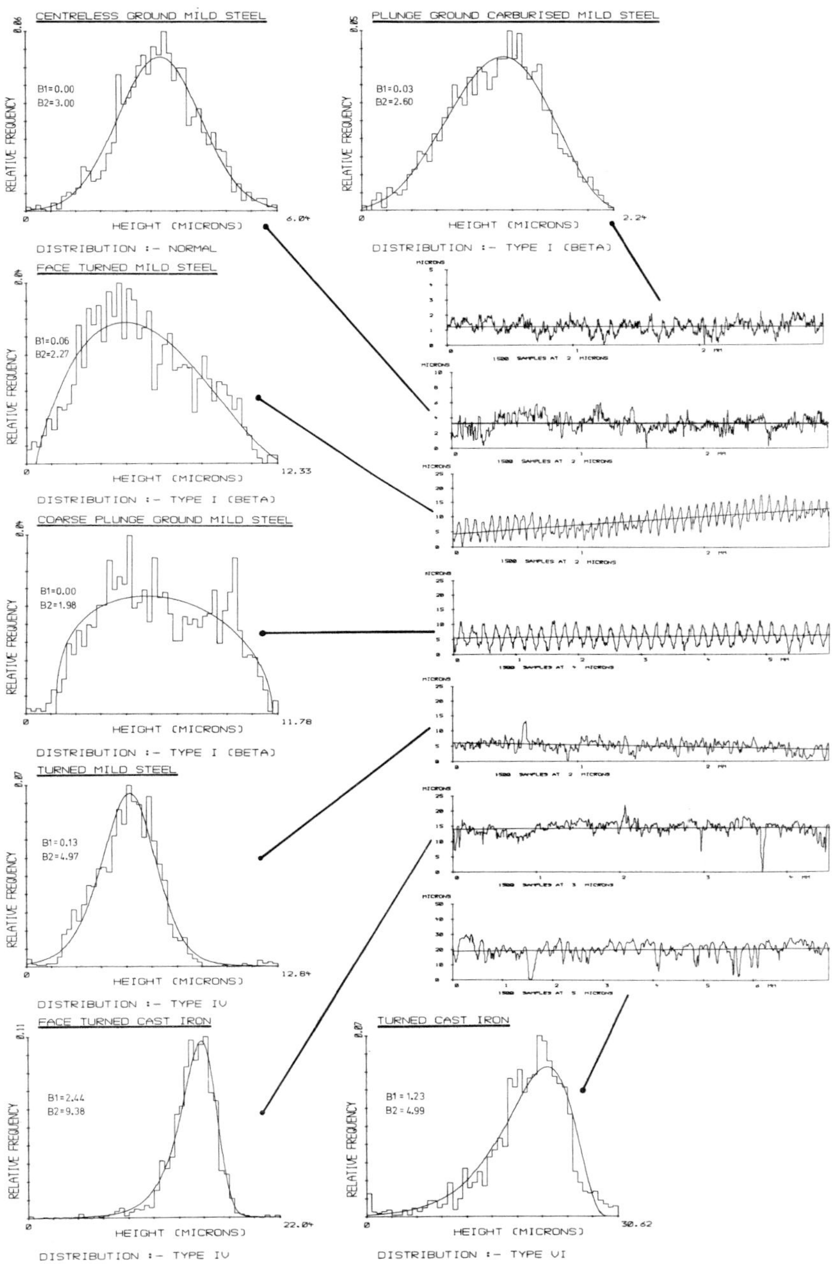

Fig. 4.—Some machined Surface Profiles with Amplitude distributions fitted by the Pearson System.

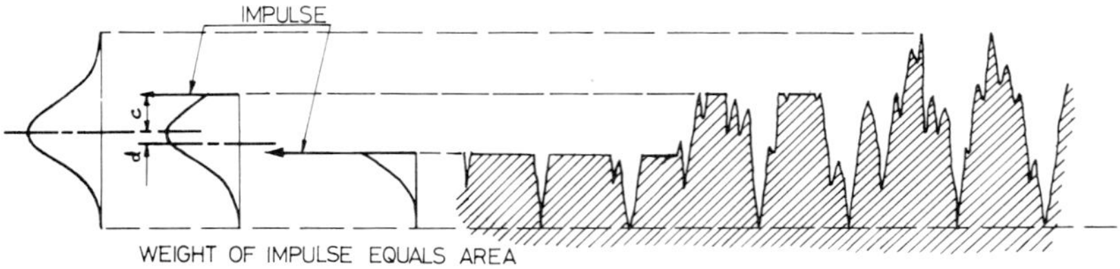

Fig. 5.—The Simple Wear Model-Surface Truncation by a Plane.

analysis more complex. The effects of truncation of a gaussian wave on its amplitude distribution is considered. Analytical results are presented for the amplitude distribution of the worn surface when the truncating function has a gaussian distribution, and this is treated by computer simulation.

The true shape of the worn surface amplitude distribution for the case where both the unworn surface and the truncating function have gaussian amplitude distributions has also been considered. Figure 7 shows a diagrammatic representation of this situation in which the surface being worn is taken to have mean, $y=0$ and standard deviation $\sigma=1$ and the truncating function to have mean $y=c$ and standard deviation $_t=p$. The general height of a point on the final worn surface is y, whilst the corresponding ordinates of the unworn surface and the truncating function are y_u and y_t respectively, then

$P[y<Y<y + \delta y] \simeq f(y)dy$ where $f(y)$

is the probability density function of the worn surface heights and thus

$$f(y)dy \simeq P\{[(y<y_u<y + \delta y) \cap (yt>y)] \cup [(y_u>y) \cap (y<y_t<y + \delta y)]\}$$

$$= g_u(y)dy \int_y^\infty g_t(y)dy + g_t(y)dy \int_y^\infty g_u(y)dy$$

$$\therefore f(y) = g_u(y)[1-\Phi_t(y)] + g_t(y)[1-\Phi_u(y)]$$

where $g_u(y)$, $g_t(y)$ are the probability density functions of the unworn surface and truncating function ordinate heights respectively, and $\Phi_u(y), \Phi_t(y)$ their cumulative density functions. The shape of the worn distribution thus obtained is illustrated by the solid line in Figure 8 for truncation at the mean level with $P=0.1$. Rather than deriving the worn surface distributions by this method and then obtaining their moments, it was decided that it would be simpler to more fully investigate this subtractive wear model, by computer simulation.

The surface truncating functions were each represented by 10,000 ordinates. Wear was simulated by offsetting the ordinates of the truncating function with respect to those of the unworn surface and then selecting the lower value of each pair or ordinates along the profile. The mean position and moment values for the resulting distribution at each level were then computed directly from these ordinate values using

$$u_n(y) = \int_{-\infty}^{\infty} (y-u)^n f(y)dy$$

and

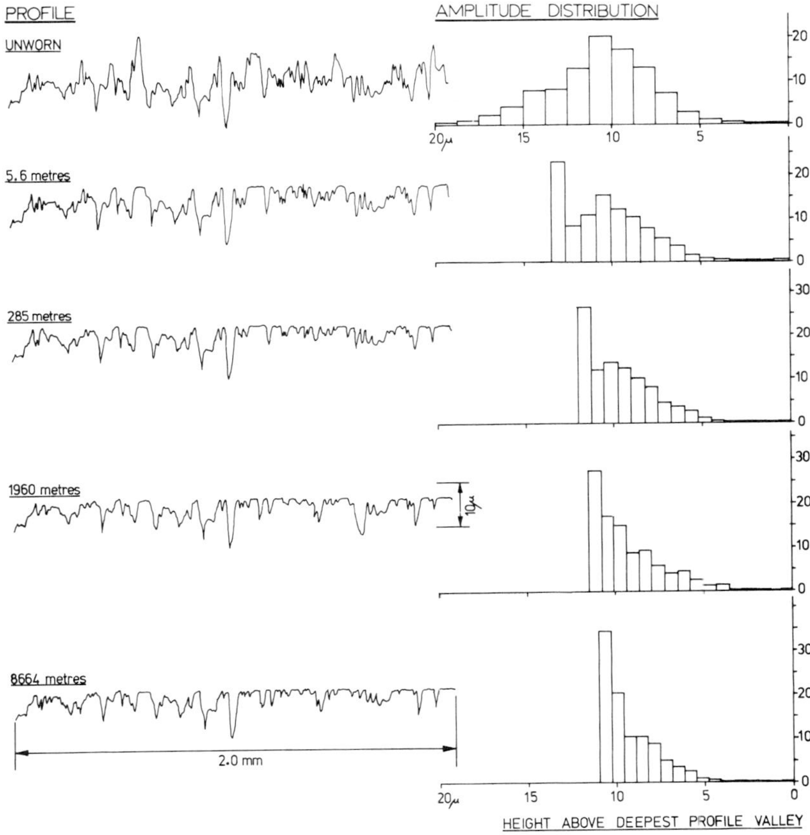

PROFILES FROM A PHOSPHOR BRONZE PIN AFTER RUNNING THE DISTANCES SHOWN

Velocity of sliding = 0.26 m/s

Nominal bearing pressure = 1.65 MN/m²

Lubricant: D.T.E. Light

Pin material: BS.1400 PB 1-C

Ring material: BS.970 En 32B carbon case-hardened to 780 V.P.N.

Fig. 6.—Re-located Surface Profiles from a Wear Experiment.

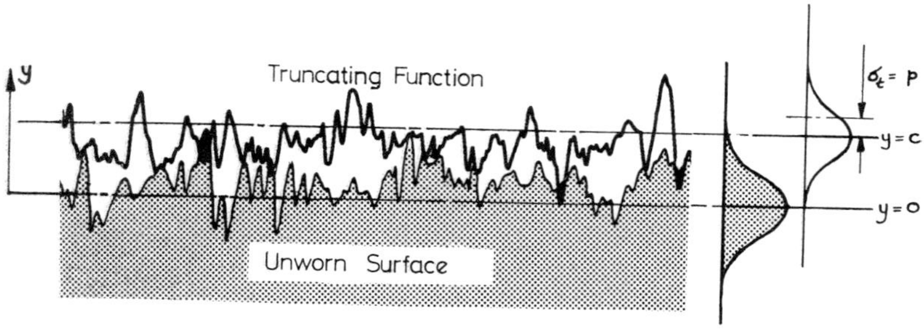

Fig. 7.—The Model for Truncation by a Wave with a Gaussian Distribution.

$$Skew = \frac{u_3(y)}{(u_2(y))^{3/2}} \qquad Kurtosis = \frac{u_4(y)}{(u_2(y))^2}$$

and the data sorted into 100 class intervals so that the shape of the resultant distributions could also be visually observed for different depths of wear. The typical distribution obtained for truncation at the mean with $p=0$ is illustrated as the dotted curve in Figure 8 and as can be seen, follows the previous result closely.

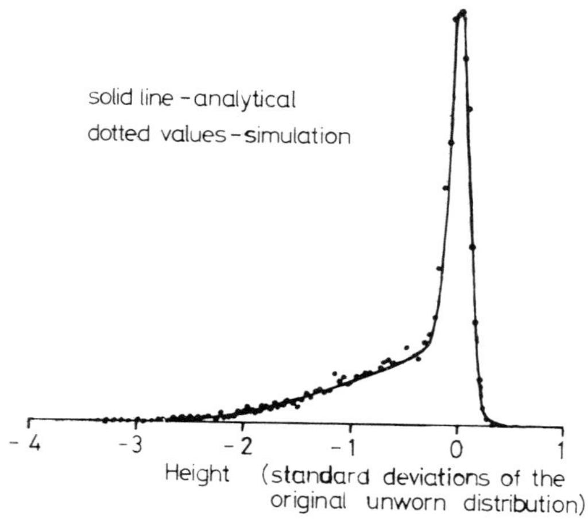

Fig. 8.—An Amplitude Distribution obtained for the Gaussian on Gaussian Truncation Model.

Figure 9 illustrates this model and also shows results obtained for various ratios, p, of the standard deviation of the truncating function to that of the unworn surface. It should be noted that since the amplitude

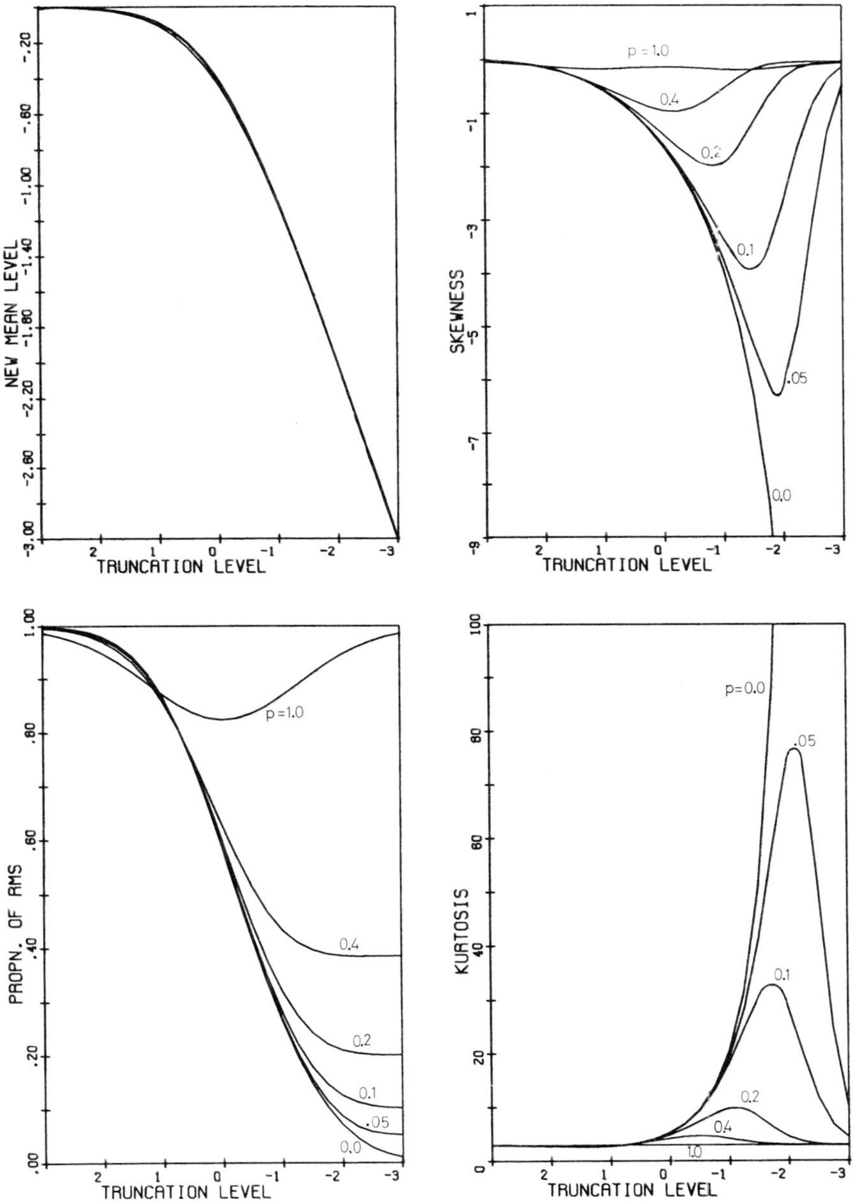

Fig. 9.—Amplitude Distribution Parameters for the Gaussian on Gaussian Truncation Model.

distribution of the truncating function is symmetrically disposed about the truncation level, the relationship between the level of truncation and the new mean position after truncation remains the same as for the impulse model. The RMS roughness for this model tends to the RMS value of the truncating function, as Figure 9 shows, whilst skewness and kurtosis reach maximum values dependent on the roughness of the truncating function alone, as this function effectively cuts its image into a solid medium when the depth of wear becomes large.

In the case of non-gaussian surfaces (representable by Pearson distributions) only the simple, impulse wear model has been applied so far to some Type I (Beta) distributions. The results are shown in Figure 10. This shows curves obtained (i) by simulated wear of a surface initially identified as being of Type I and (ii) be determination of the theoretical moments of such a Beta distribution, truncated by an impulse as in the gaussian case for $p=0$. The theoretical values were obtained using numerical integration methods to estimate appropriate moments of the truncated distribution with probability density function of the form given in Appendix I.

DISCUSSION AND CONCLUSIONS

Some impulse truncation models have been prepared for non-gaussian amplitude distributions, to establish the extent to which the worn surface moment values differ from the gaussian model, and therefore how sensitive these values are to the form of the unworn surface amplitude distribution.

The 'theoretical' and 'experimental' (simulated wear) results agree well. For the moderately skewed cases investigated, the results as shown in Figure 10 follow closely those previously obtained for gaussian surfaces. The wear of a broad category of engineering surfaces, whose amplitude distributions are statistically significantly non-gaussian, thus appears to follow a similar pattern of parameter changes as that of a gaussian surface.

From the foregoing analyses, it is possible to consider which statistical parameters might be most useful in the description of lubricated wear. For a parameter to be of use, it must be seen to vary significantly throughout the regions of wear acceptable in normal engineering applications. In many bearing applications, it may be considered that wear should not totally eliminate all traces of the original surface topography. The analyses presented in this paper have therefore considered changes giving rise to transitional topographies.[5] Within this wear region, it can be seen that measurement of the skewness of the amplitude distribution provides useful numeric information. Kurtosis values are somewhat less useful since the rate of change of kurtosis is almost zero in the early stages of wear, the rate of change being too great in the latter stages of wear. Kurtosis is only useful between truncation levels +1.0 to -0.5 standard deviations of the original unworn surface from its mean. An interesting feature of the kurtosis parameter is that it marginally reduces for all truncation models from truncation levels 3.0 to 1.0 .

ACKNOWLEDGEMENTS

The authors wish to thank Mr. T. A. Spedding and acknowledge his assistance in the development of this work as part of his final year project for the degree of B.Sc. (Maths.), Leicester Polytechnic (1978). The authors

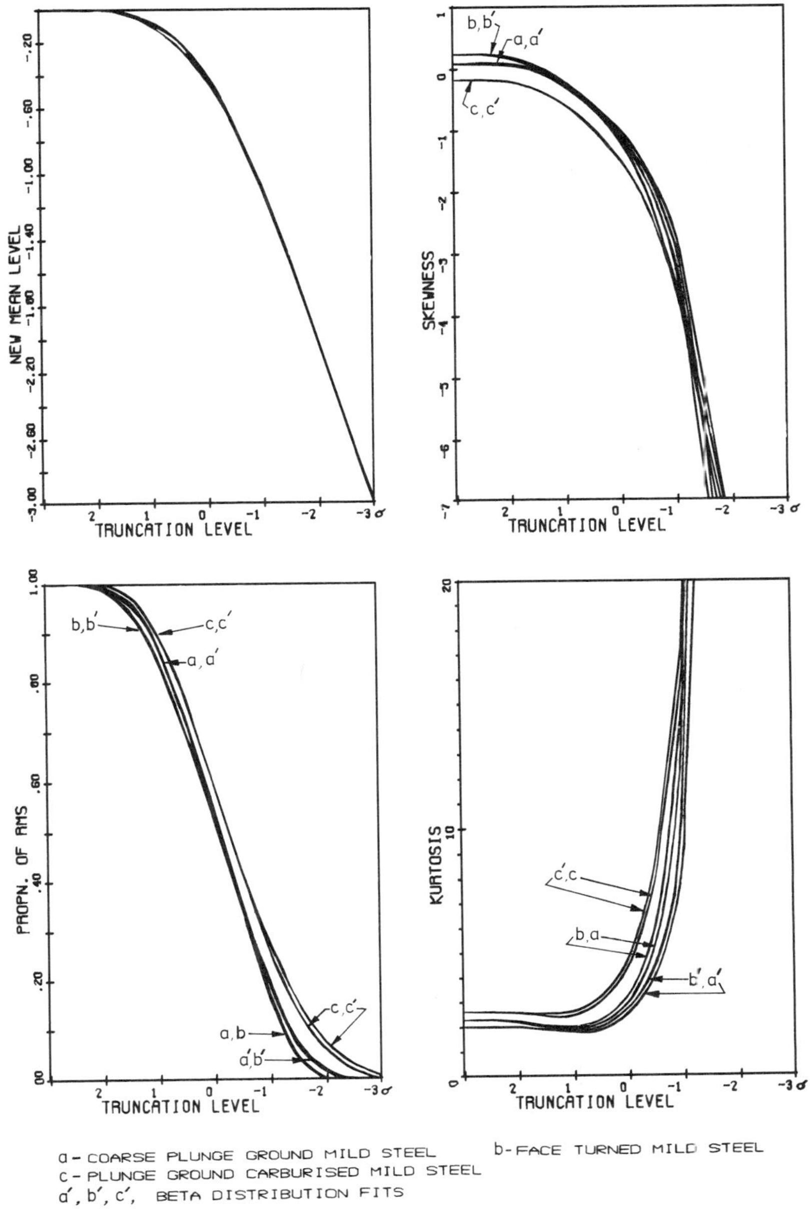

Fig. 10.—Amplitude Distribution Parameters for Three Non-Normal Distributions from Figure 4 under Impulse Truncation.

also are indebted to the Science Research Council for their sponsorship of the project, which has enabled a computer controlled digital surface metrology laboratory to be commissioned.

APPENDIX I

The Pearson system of distributions uses the basic differential equation

$$\frac{1}{f(y)}\frac{df(y)}{dy} = \frac{y + a}{c_o + c_1 y + c_2 y^2} \qquad (1)$$

to describe a family of probability density functions $f(y)$.

The differing types of Pearson distributions are defined by the various possible solutions to the defining equation (1) according to the roots of the quadratic equation $c_o + c_1 y + c_2 y^2 = 0$.

Pearson showed that the type for a particular set of observations may be identified by using a criterion k defined as

$$k = \frac{B_1 (B_2 + 3)^2}{4(2B_2 - 3B_1 - 6)(4B_2 - B_1)}$$

where

$$B_1 = \frac{(\mu_3(y))^2}{(\mu_2(y))^3} \quad \text{and} \quad B_2 = \frac{\mu_4(y)}{(\mu_2(y))^2}$$

Beta distributions (Type I) are indicated if $k<0$ and Type IV distributions if $o<k<1$. The other main type is Type VI ($k>1$). Pearson also distinguished various transition types of which the normal distribution ($k=0$, and $B_2=3$) is important.

The probability density functions are of the forms

$$y_o\left(1 + \frac{y}{a_1}\right)^{m_1}\left(1 - \frac{y}{a_2}\right)^{m_2} \quad \text{for} \quad -a_1<y<a_2 \quad \text{Type I (Beta)}$$

$$y_o\left(1 + \frac{y^2}{a^2}\right)^{-m} \exp\left(-V \tan^{-1}\left[\frac{y}{a}\right]\right) \quad -\infty<y<\infty \quad \text{Type IV}$$

$$y_o (y-a)^{m_1}\, y^{-m_2} \quad a<y<\infty \quad \text{Type VI}$$

$$y_o \exp\left(-\frac{y^2}{a}\right) \quad -\infty<y<\infty \quad \text{Normal}$$

(a, a_1, a_2, y_o, V Constant)

REFERENCES

1. Rowe, G.W., Kaliszer, R., Trmal, G. and Cotter, A., *Wear*, Vol. 34, 1975, p. 1.
2. Bayer, R.G. and Sirico, L., *Wear*, Vol. 35, 1975, p. 251.
3. Stout, K.J., Whitehouse, D.J. and King, T.G., *Wear*, Vol. 43, No. 1, 1977, p. 99.
4. King, T.G., Stout, K.J. and Whitehouse, D.J., Paper C1, in 1st Joint Polytechnic Symposium on Manufacturing Engineering, Leicester, June 14-15, 1977, Leicester Polytechnic, 1977.
5. Williamson, J.B.P., Pullen, J. and Hunt, R.T., "The Shape of Solid Surfaces," American Society of Mechanical Engineers Winter Annual Meeting, 1969.
6. Thomas, T.R. and Sayles, R.S., *Mecanique Materiaux Electricite*, No. 337, January 1978, p. 7.
7. Elderton, W.P. and Johnson, N.L., "Systems of Frequency Curves," Cambridge University Press, England, 1969, p. 35 et seq.
8. Stout, K.J., King, T.G. and Watson, W., *Mecanique Materiaux Electricite*, No. 337, Jan. 1978, p. 45.

DISCUSSION

D. J. WHITEHOUSE, Rank Taylor Hobson: The point I would like to make is that it is often difficult to characterize the amplitude distribution of a surface which is being truncated. It is difficult to introduce parameters to represent how the profile changes with running-in. Dr. Stout was talking about a Poisson distribution which is an attempt to do just that. Well, we have been carrying out a lot of work on this particular problem ourselves and we have found that the best function for this job is the β function, which is probably familiar to statisticians. It has some interesting properties. Basically, with this distribution, you can arrange to fit any sort of profile shape, for instance, a sinusoidal distribution. It is possible to get a β function to do just that - a case where the Pearson distribution would not work. And it has the advantage that it is very flat at the origin. It is now possible to specify completely all aspects of truncated amplitude distributions using the function which is an assistance to ordinary manufacturing processes.

III. PHYSICAL PROPERTIES

MECHANICAL PROPERTIES OF NEAR-SURFACE MATERIAL IN FRICTION AND WEAR

A. S. Argon

ABSTRACT

Friction and wear are inelastic deformation and fracture processes occurring on the surfaces of materials in sliding contact. Although only "skin-deep," such processes exhibit the same mechanisms of plastic flow or creep, and the same damage processes of crack formation and crack growth found in cyclically strained material in bulk. As in the case of such processes occurring in bulk, the key to the understanding of friction and particularly wear resides in obtaining the solution of the fundamental elasto-plastic-creep boundary value problems for cyclic deformation in a heterogeneous alloy, subject to the constitutive behavior of the separate phases, and terminated by the same critical processes of fracture - albeit all on a microscopic scale underneath asperity contacts. Account is taken of the recent progress in wear research that has followed along these lines.

INTRODUCTION

Problems of friction and wear have been traditionally treated as phenomena separate from plastic deformation and fracture familiar to engineers. Treatises and standard textbooks[1,2] on the subject often list a large number of separate mechanisms of wear each of which have their own advocates and possess their own lore. Although in many instances a differentiation between wear processes based on their special features may be justified to lay emphasis on an overriding key process that may call for a separate cure, more is to be gained by viewing the unifying features of the separate wear processes. In this manner it may be possible not only to perform better projections for control or containment of the problem but also to identify important omissions in research which must be rectified. It is reassuring to see that some wear research of recent years has already moved in this direction.[3]

In this paper we will take the view that friction and wear are microscopical inelastic deformation and fracture processes, often involving crack initiation and propagation under cyclic strains enforced in very small regions near the surface. Thus, our task is to elucidate the geometrical, physical, and mechanical properties and inelastic behavior of near-surface material that enter into the wear process. Detailed considerations of mechanisms of wear and effects of environments, including corrosion, will not be discussed here as they are topics handled by other authors.

FEATURES OF THE WEAR PROCESS

It is an inherent nature of the wear process that material is removed from surfaces sliding on each other by the action of sliding. It is generally known that once initial transients are over, the removal of matter in the form of wear particles is a linear function of the distance of sliding. Furthermore, it is also known as an elementary fact that in a material of a given hardness the rate of removal of material with distance of sliding is a linear function of the tangential resistance to sliding, i.e., the average shear traction produced at the sliding interface. This dependence - which is not *a priori* obvious - is held to be a result of the fact that rubbing during sliding actually occurs at minute asperities[4] of the contacting interface where the nature of the contacts at asperities remains largely unaltered, while their number per unit area changes in direct proportion to the change in normal traction to maintain overall force equilibrium. Hence

$$\frac{dV}{d\ell} = \alpha A_r \tag{1}$$

where V is the total volume of removed particles, ℓ the distance of sliding, and A_r the real area of contact at the rubbing interface with α being a constant of proportionality. Since the rate of wear at sliding interfaces can differ by many orders of magnitude, it is clear that the understanding of the entire phenomenon of wear is in the mechanistic understanding of what governs the level of the coefficient α.

It is clear that if rubbing at contacting asperities were a reversible elastic process, no wear would result. Hence, it is natural to expect that particle evolution from sliding interfaces must result from processes of inelastic deformation and fracture - albeit at a microscopic scale. Furthermore, since the more familiar phenomena of plastic flow and fracture in quasi-homogeneously strained parts can involve strain localization in heterogeneous plastic media, hole formation, hole growth, crack formation and propagation under monotonic or cyclic straining, and since these phenomena can be profoundly affected by the environment, it should be no surprise to find all these processes also present in the wear particle evolution. Hence, it is fruitful to take the view that wear is merely another problem of inelastic deformation and fracture occurring under more complicated microscopical boundary conditions and possibly involving more localized fields of stress and higher gradients of strain and damage but otherwise involving no new phenomena. This implies that as in the case of the mechanisms of fracture and fatigue, a microscopical view both in the literal sense and in the mechanistic sense is the only proper and profitable one as some recent wear research has so amply demonstrated.[3]

We view asperities as shown schematically in Figure 1, where medium 1, considered to be harder, is moving tangentially toward the right on a softer medium 2. The figure shows two asperities demonstrating two extremes of behavior in rubbing at a steady state after initial transients of wear have died away. The asperity at left caused by a large protrusion in medium 1 has plastically indented medium 2 and is plowing it, while the asperity on the right resulting from a smaller protrusion in medium 1 has

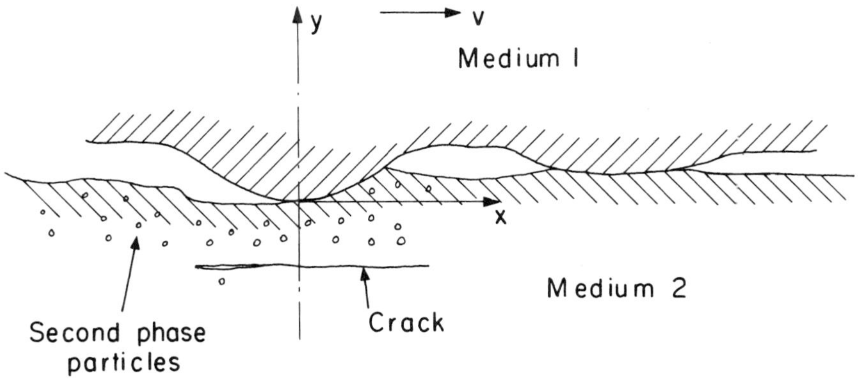

Fig. 1.—Schematic drawing showing two asperity contacts between a hard (1), and a soft (2) medium. Contact on left is plastically plowing while contact on right is largely elastic.

accommodated largely to medium 2 by *elastic flexing*. In contacts between a very hard and a very soft material, contacts of the plastic indentation type will dominate while between two materials where even the softer material has a high hardness on the scale of the modulus, a larger fraction of the contacts can be of the elastic flexing type. The latter case manifests itself by lower coefficients of friction, and lower wear rates where many contacts rub on each other in a more nearly reversible elastic manner. We will be primarily interested in the first type of contact involving the lateral translation of a plastic indentation. Although in some instances contacts at asperities can be of sub-micron dimensions, examination of wear tracks on initially machined surfaces indicates that contacts are often many microns in dimension and can therefore be treated by continuum concepts.* To deal with the problem of wear in a fully deterministic manner as a microscopic problem in deformation and fracture we require the distribution of stresses and strains in the heterogeneous material near its surface which is repeatedly being subject to plastic rubbing by moving contacts. A first order analysis of the problem of plastic deformation of asperity contacts subjected to combined compression and lateral shear has been given by Green[6] in the form of a set of slip line solutions between notches of arbitrary angle and tilt for varying degrees of adhesion at junctions. From such slip line solutions, a good understanding has evolved for the coefficient of friction for materials that approximate rigid plastic conditions, i.e., having yield stresses in shear k which are only a very small fraction of the shear modulus G such as $k/G \leq 10^{-3}$. For such cases in perfectly dry adhesive contact situations, the coefficient of friction should be between 0.4 and 1.0, depending upon whether the surfaces are very smooth or very

* The choice of when inelastic deformation can be treated by continuum concepts and when dislocation mechanics is necessary is never a simple one. A conservative rule of thumb is that whenever the largest characteristic dimension of a dislocation structure, resulting from the deformation, such as cells or sub-grains, is considerably smaller than the size of the region over which the imposed strain is uniform then a continuum approach is proper.[5] In many instances, however, where this rule cannot be satisfied a continuum approach may still be desirable as it indicates a smoothed out average behavior.

rough.[7] For high strength materials where $k/G=O(10^{-2})$ many contacts will have a substantially reversible elastic behavior and the coefficient of friction will be correspondingly lower. Real contacts at asperities as depicted by Figure 1 differ substantially from the idealized contacts analyzed by slip line solutions. In some instances, actual penetration of a hard contact may have occurred into the softer medium which may do considerable plowing when translated laterally. In many other instances of strong materials where $k/G=O(10^{-2})$, the elastic displacements at the contacts will be substantial, requiring an elastic plastic solution to determine the distribution of stresses and plastic strains in and around the contacts. In the latter instances, approximate elastic-plastic solutions for rolling contact incorporating multiple passes leading to a shake-down state[8] have proved to be useful for a better understanding of subsurface damage initiation in the wear problem.[9]

For purposes of study of subsurface crack propagation in high strength alloys even the classical elastic indentation solution of Hertz[10] may be appropriate and has led to useful results in the study of sub-surface crack propagation under conditions of rolling contact.[11]

FUNDAMENTAL NEAR-SURFACE PROPERTIES INFLUENCING WEAR

Surface Roughness

Since wear is a process of fatigue occurring in layers of material very near the surface, the initial roughness of the surface may be expected to be an important parameter governing the wear rate. Experiments by Jahanmir and Suh,[3] on surfaces having finishes ranging from 0.1 μm to 4.8 μm indicated that under very light contact stresses rough surfaces tend to wear less initially than smooth surfaces while the opposite appears to be true under high contact stresses. In any event, however, such differing behavior was observed to be limited to an initial transient which typically was confined to the first 100 meters of sliding during which the geometry of the initial asperities was altered radically by plastic flow and wear particle evolution into a common surface topography that is relatively smooth. The stages of how this is achieved have been studied in detail by Whitehouse.[12] After such a steady-state wear surface has evolved, no significant effect on the wear rate due to initial roughness remains.[3]

Cyclic Plastic Resistance of Surfaces

With very few exceptional cases, possibly involving coarse abrasive particles between very hard and brittle surfaces, wear particle evolution requires local plastic flow and fatigue. Hence, one of the most important properties of a surface is its cyclic plastic resistance. In the absence of detailed information, the cyclic plastic resistance can be approximated by the monotonic deformation resistance of a material. Clearly since wear may be of interest at any temperature and sliding velocity we need to consider the inelastic deformation resistance of a solid at all temperatures between 0°K and its melting point and at strain rates ranging from the common creep rates of c.a. 10^{-9} sec^{-1} to very high rates of the order of 10^3 sec^{-1}. It is common knowledge that different mechanisms of inelastic deformation can be active in different ranges of temperature, and strain rate, and that at any given temperature and strain rate the deformation resistance of different materials can be vastly different. Ashby and Frost,[13] however, have demonstrated that at similar homologous temperatures and at steady state strain rates that are in the same proportion to their atomic diffusivity, all crystalline materials of the same grain size exhibit deformation resistances that are roughly in the same proportion to their shear modulus. This is shown in Figure 2 for one characteristic normalized strain rate for a wide selection of crystalline materials with greatly different crystal structures and bonding while Figure 3 on the other hand, shows the temperature dependence of the normalized deformation resistance for a typical solid solution stainless steel alloy at a given grain size for a range of strain rates.[14] Figure 3 in particular shows that while the obstacle controlled plastic resistance at low temperatures is relatively strain rate insensitive, it becomes quite sensitive in the range where diffusion governs. In addition, in the latter range, grain boundaries begin to offer a lower resistance to deformation resulting in boundary sliding which makes deformation on the scale of a grain very inhomogeneous and accentuates critical fracture processes there. All such departures from the more familiar rate independent behavior will have important consequences on the wear process in the elevated temperature range.

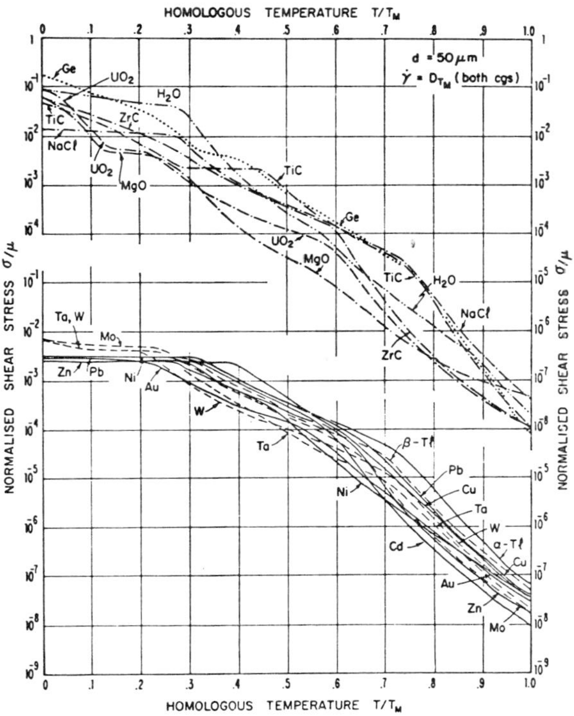

Fig. 2.—Dependence of normalized deformation resistance on homologous temperature, for a given normalized shear strain rate for a large group of crystalline materials (from Ashby and Frost,[13] courtesy of M.I.T. Press).

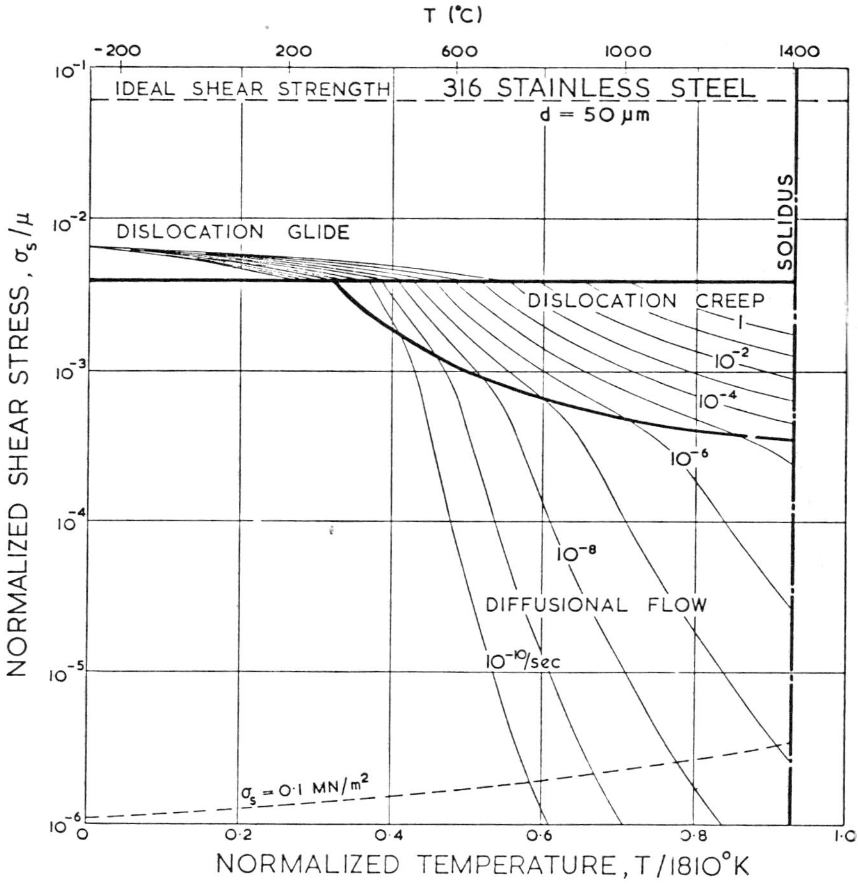

Fig. 3.—Deformation mechanism map for Type 316 stainless steel (from Frost and Ashby[14]).

Deformation mechanism maps of the type shown in Figures 2 and 3 are based on steady state deformation. At low temperatures where strain hardening prevents deformation at a steady rate, the plastic resistance considered in such deformation mechanism maps pertains to an asymptotic behavior where strain hardening has been saturated. Although saturation of strain hardening is rarely attainable in monotonic deformation, it is a common fact in cyclic deformation. Hence, idealized deformation mechanism maps as the one shown in Figure 3 for monotonic deformation are at least qualitatively more representative for cyclic deformation. The actual level of the saturated cyclic plastic resistance depends on the balance of cyclic strain hardening and recovery processes which are not yet clearly understood and can be described only qualitatively. Nevertheless, some effects are clear and are worth noting. While impenetrable particles of all sizes, at the same mean net spacing, perform equally well for the purpose of impeding dislocation motion and governing the plastic resistance, large particles tend to be more effective in influencing strain hardening. On the other hand, low temperature nondiffusive recovery in f.c.c. metals and other close packed structures is impeded by a low stacking fault energy that inhibits cross slip of screw dislocations and constrains slip to be planar. Similarly, at high temperatures where climb motion of dislocations adds another powerful mechanism of recovery, thermally stable particles can inhibit subgrain growth and maintain the creep resistance at high levels.

Initiation of Cracks

An intriguing feature of cyclic plastic deformation at low homologous temperatures in *single phase metals and alloys* is the formation of persistent slip bands throughout the volume but particularly near the free surfaces during roughly the first thousand cycles. Detailed studies of the dislocation content of persistent slip bands[15] has established that very well organized forward and backward dislocation motions occur inside these bands that produce extrusions and intrusions and eventually give rise to formation of cracks inside them by a mechanism that most probably involves surface grooving due to repeated in and out motion of slip on different crystallographic planes inside a persistent slip band.[16] Hence in this process the free surface plays a key role as the place from which grooving can start. In wear where there are intense strain gradients near the surface other mechanisms of crack initiation, less statistical and random in nature, such as surface grooving by fold-over events resulting in cold welds become more prominent. There are no known cases of fatigue crack formation in *single phase* alloys below the surface. At elevated homologous temperatures, back and forth sliding of grain boundaries and accompanying boundary migrations take the place of persistent slip bands in producing surface grooving and crack initiation as shown in Figure 4. Although the documentation in this case is far less complete than the low temperature process, again the existing evidence indicates that in *single phase* alloys cracks form at the surface.

In very *brittle materials* where plastic flow is very restricted or absent altogether, tensile stresses around indentations outside the zone of contact of the indenter can be high enough to propagate existing minute surface cracks that may have resulted from earlier chemical attack, local phase transformations, devitrification in glasses, or from damage due to contact with a particularly hard object with particularly small radius of curvature. Such cracking in brittle solids in the absence of environmental stress corrosion, however, is primarily of a monotonic nature and does not involve cumulative damage so characteristic of cyclic deformation.

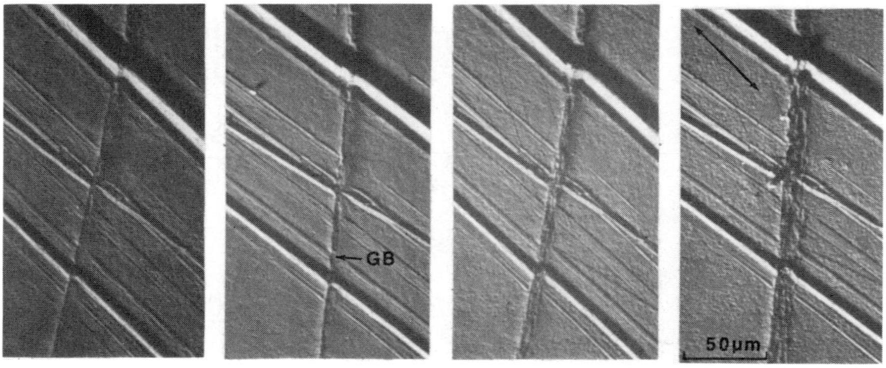

Fig. 4.—Cyclically sliding and migrating grain boundary (GB) initiating a surface groove and eventual crack in Fe 3% Si alloy at 900°C: a) after 10.5 cycles, b) after 20.5 cycles, c) after 30.5 cycles, d) after 40.5 cycles; arrow shows tension-compression direction.

The implications of the above statements are clear. In ductile *single phase alloys* or even brittle compounds and glasses, cracking that is necessary for wear particle evolution must start from the free surface as there are no other important weak sites or interfaces below the surface. The only exceptions to this rule might be the generation of very large tensile stresses bordering on the cohesive strength "focused" underneath the surface while the surface is protected by large compressive stresses *everywhere*. The asperity contacts do not fall into this category.

In *heterogeneous alloys*, fatigue cracks can be initiated from interfaces of second phase particles by decohesion or from cracked second phase particles. In monotonic straining, the process of particle decohesion or particle fracture occurs when interface tractions or stresses inside particles resulting from deformation incompatibilities between phases reach the interface strength or the cohesive strength of the particle.[17] The corresponding process in cyclic straining, where deformation incompatibilities are partially accommodated by rather effective cyclic recovery effects, many cycles are required to produce the same end result. The computation of the rate of production of such stresses involves elastic-plastic analysis incorporating repeated reversals, with residual stresses, which all require accurate knowledge of amplitude dependent cyclic stress strain behavior (to account for cyclic hardening, recovery and the Bauschinger effect) and highly efficient nonlinear numerical analysis programs. At the present time, neither the phenomenological information (let alone the mechanistic understanding) nor the required non-linear analysis techniques are available to handle such problems with any reasonable level of accuracy. Hence, the level of understanding of crack formation from second phase particles in cyclic straining is at best qualitative, but it has been established that cracks form by this mechanism after fewer strain reversals than would have been required in forming them from persistent slip bands. At elevated temperatures the places of action are still the sliding grain boundaries where cyclic sliding produces decohesion or fracture of grain boundary particles. Once again, the development of stresses on or in such particles due to incompatibilities resulting from grain boundary sliding and the attendant decohesion or cavity nucleation is reasonably well understood in monotonic deformation,[18-21] but not in cyclic deformation.

In crack formation during monotonic straining it has been reported that large particles are generally more effective than small ones.[22] It has now been established that there are two reasons for such behavior.[23] First, it turns out that for rigid particles of size less than about 100Å, embedded in a linearly hardening medium, there is not enough strain energy stored in the elastically strained lattice around the particles to provide for the surface energy of the cavity that is to form. Since stress concentrations are lower but have longer ranges in non-linearly strain hardening continua, the energy cut-off will occur at smaller particle sizes than that given above, the more non-linear the medium becomes.[20] Above the cut-off size, the necessary and sufficient condition for hole nucleation is the attainment of the cohesive strength at least over a small region of the interface or interior. What has been thought of as a second particle size effect has now been established to be an effect of particle interaction,[23] where larger stresses arise between two closely spaced particles due to the need to dissipate larger deformation incompatibilities in smaller regions and where the effect must be more correctly ascribed to random variations of local particle volume fraction from point to point.[24] Similar effects are likely to be present in cyclic straining.

The lesson for wear particle formation that can be derived from the above discussion is that holes and cracks in heterogeneous alloys can form interfaces between matrix and particle anywhere in the material where the conditions can be satisfied, and free surfaces play a far less important role.

Growth of Cracks

It was observed by Forsyth[25] that fatigue cracks have two growth regimes. In Stage I, cracks that have nucleated in persistent slip bands or by some initial coalescence of cavities around particles, grow inside the persistent slip bands or shear zones for some distance which in medium grain size polycrystalline metals is of the order of the grain size. This mode of crack growth is comparatively slow and is very difficult to study. It has been largely ignored or avoided by the vast majority of fatigue researchers on the weak argument that it may not make up much of the fatigue life or that it may be lumped together with the crack nucleation. In wear particle formation the state of stress at an asperity contact will in most instances mean that cracks need to grow in shear,[11] hence, by a Stage I mechanism. This should be an added stimulus for emphasizing research on Stage I Crack growth to supplement the meagre data available.[26,27] In Stage II, the growing crack switches into a direction nearly normal to the direction of maximum tension and after a period of rapid acceleration in a so-called threshold range settles down to a rate dc/dn that can be described by a simple power function of the amplitude of the stress intensity factor ΔK

$$\frac{dc}{dn} = \beta(\Delta K)^m \tag{2}$$

where β is a proportionality constant that can be temperature dependent. In this so-called Paris law region,[28] the exponent m depends on the plane strain fracture toughness K_{IC} and varies from an asymptotic value of 2 for materials with very high fracture toughness to rather large exponents as the fracture toughness decreases.[29] While the crack growth according to an exponent of 2 is readily explainable as resulting from accumulation of irreversible surface production due to the crack tip opening per cycle, the higher exponents cannot so readily be justified beyond stating that they often signify the presence of embrittlement effects that accelerate crack growth.[30] The growth of fatigue cracks in this region is well investigated and quite considerable data on it exists,[31] making it unnecessary for us to go into it further. It may suffice to add that at elevated temperature this mode of fatigue crack growth becomes also more strongly dependent on the frequency of reversals of stress,[32] as the plastic flow itself that is instrumental in the crack extension becomes strain rate dependent.

It has been well documented that the Stage II fatigue crack growth rate is accelerated by aggressive environments which tend to either inhibit undoing, during the compression phase of a cycle, an already generated free surface at a crack tip, or accelerate crack growth by embrittlement of the material.[33]

Some direct correlations have been reported between Stage II fatigue crack growth rates in tension cycling and wear rates in laboratory experiments.[34] An expanded study is likely to establish that the correlation will be between Stage I crack growth and the wear rate. It is interesting to observe in this connection that if certain environmental effects which accelerate fatigue crack growth rate are present and are observed to accelerate the wear rate, this would be a strong indication that the cracks producing wear particles are vented to the environment and may have initiated from the surface.

Hard Layers, Soft Layers

A controversy has raged for many years on whether or not surface layers of a plastically deforming body are harder or softer than the interior of the body (see e.g., (35,36)). Although this controversy is not readily settled by a short discussion here, it may be useful to take account of some facts.

Much has been made of Fisher's[37] suggestion that due to image stresses dislocation sources near the surface, in a Frank network of dislocations, require about half the stress to operate than similar sources in the interior. Although this is perfectly true, if the plasticity is not just to be restricted to the very surface layer but is to initiate deformation in the interior as well, the surface dislocation has to travel inward and cut through the remainder of the Frank network. Hence, there should be no reduction in the yield stress of a large part due to sources on the surface.[16]

Fourie[34] has shown in a set of very careful experiments that in single phase metals and alloys where strain hardening is a statistical process of dislocation storage by mutual interactions and blockage, surface layers are less efficient for such storage as they lack one half the dislocation cross-flux that interior points see. This evolutionary process of dislocation storage begins with the earliest phase of deformation in single crystals and persists into all later stages. Although many assertions have been made that under identical conditions of uniform strain, hard surface layers develop in the process of strain hardening,[35] no direct and unambigious evidence has ever been furnished to substantiate them.[38]

There are, however, numerous well documented cases of hard surfaces resulting from oxide layers, polycrystalline layers on single crystals, composition gradients, etc. which have far-reaching influences beyond what can be expected from their volume fraction. As in the case of the soft layers discussed above, the disproportionate influence of hard layers results from their "catalytic" role in accentuating strain hardening.(5) In the case of surface sliding and wear, where plasticity can be very intensive but only "skin deep," the cyclically deformed surface layers can retain very large densities of well accommodated dislocations in very tight multipolar clusters, which as a result should have little tendency to be extracted out by the free surface. Such dense multipolar clusters should present no long range stress but will pose strong obstacles to dislocations of the interior and act as moderately strong layers which block the passage of some interior dislocations before they give way and give rise to relatively sharp slip bands. If such heavily deformed thin surface layers are removed by electropolishing, it is found that the usual slip band or slip line structure that developes upon incremental deformation is far more diffuse.(39) This effect is well known in crystal plasticity by researchers who have learned never to perform fundamental experiments on slip distribution with mechanically polished single crystals. It is most likely that the Beilby layer, known to metallographers, is nothing more than this layer just discussed.

ACKNOWLEDGEMENT

The author's research on mechanisms of deformation and fracture is supported by DOE under Contract EG-77-S-02-4461.A000. I am grateful to Professor D. Rigney for catching an inaccuracy in an elementary equation appearing in an earlier version of the paper.

REFERENCES

1. Bowden, F.P. and Tabor, D., "The Friction and Lubrication of Solids," Clarendon Press, Oxford, 1954.
2. Rabinowicz, E., "Friction and Wear of Materials," John Wiley and Sons, New York, 1965.
3. Suh, N.P. and Coworkers, "The Delamination Theory of Wear," Elsevier Sequoia, S.A., Lausanne, 1977.
4. Bowden, F.P., Moore, A.J.W. and Tabor, D., *Journal of Applied Physics*, Vol. 14, 1943, p. 80.
5. Argon, A.S., in "Surface Effects in Crystal Plasticity," edited by R.M. Latanision and J. Fourie, Noørdhoff, Leyden, 1977, p. 383.
6. Green, A.P., *Journal of the Mechanics and Physics of Solids*, Vol. 2, 1954, p. 197.
7. McClintock, F.A. and Argon, A.S., "Mechanical Behavior of Materials," Addison-Wesley, Reading, Massachusetts, 1966, p. 665.
8. Merwin, J.E. and Johnson, K.L., *Institution of Mechanical Engineers. Proceedings*, Vol. 177, 1963, p. 676.
9. Jahanmir, S. and Suh, N.P., *Wear*, Vol. 44, 1977, p. 17.
10. Hertz, H., "Gesammelte Werke," Vol. 1, 1895, p. 155; Huber, H., "Annalen der Physik," Vol. 14, 1904, p. 153.
11. McClintock, F.A., in "Fracture 1977," edited by D.M.R. Taplin, Vol. 4, University of Waterloo Press, Waterloo, Ontario, Canada, 1977, p. 49.
12. Whitehouse, D.J., These Proceedings.

13. Ashby, M.F. and Frost, H.J., in "Constitutive Equations in Plasticity," edited by A.S. Argon, The M.I.T. Press, Cambridge, Massachusetts, 1975, p. 117.
14. Frost, H.J. and Ashby, M.F., "Deformation Mechanism Maps for Pure Iron, Two Austenitic Steels and a Low-Alloy Ferritic Steel," Cambridge University Engineering Department Report, July 1975.
15. Mughrabi, H., in "Constitutive Equations in Plasticity," edited by A.S. Argon, The M.I.T. Press, Cambridge, Massachusetts, 1975, p. 199.
16. Argon, A.S., in "Corrosion Fatigue: Chemistry, Mechanics, and Microstructure," edited by O.F. Devereux, et al., National Association of Corrosion Engineers, Houston, Texas, 1972, p. 176.
17. Argon, A.S., *Journal of Engineering Materials Technology,* Vol. 98, 1976, p. 60.
18. Crossman, F.W. and Ashby, M.F., *Acta Metallurgica,* Vol. 23, 1975, p. 425.
19. Raj, R. and Ashby, M.F., *Acta Metallurgica,* Vol. 23, 1975, p. 653.
20. Lau, W.C. and Argon, A.S., in "Fracture 1977," edited by D.M.R. Taplin, Vol. 2, University of Waterloo Press, Waterloo, Ontario, Canada, 1977, p. 595.
21. Argon, A.S. and Lau, W.C., to be published.
22. Palmer, I.G. and Smith, G.C., in Proceedings of the Second Bolton Landing Conference on Oxide Dispersion Strengthening, Gordon and Breach, New York, 1968, p. 253.
23. Argon, A.S., Im, J. and Safoglu, R., *Metallurgical Transactions,* 6A, 1975, p. 825.
24. Argon, A.S. and Im, J., *Metallurgical Transactions,* 6A, 1975, p. 839.
25. Forsyth, P.J.E., *Acta Metallurgica,* Vol. 11, 1963, p. 703.
26. Duquette, D.J. and Gell, M., *Metallurgical Transactions,* Vol. 2, 1971, p. 1325.
27. Neumann, P., Vehoff, H. and Fuhlrott, H., in "Fracture 1977," edited by D.M.R. Taplin, Vol. 2, University of Waterloo Press, Waterloo, Ontario, Canada, 1977, p. 1313.
28. Paris, P.C., Gomez, M.P. and Anderson, W.E., "The Trend in Engineering," University of Washington, Seattle, Washington, 1961, p. 9.
29. Ritchie, R.O. and Knott, J.F., *Acta Metallurgica,* Vol. 21, 1973, p. 639.
30. Wright, R.N. and Argon, A.S., *Metallurgical Transactions,* 1A, 1970, p. 3065.
31. Hertzberg, R.W., "Deformation and Fracture Mechanics of Engineering Materials," John Wiley and Sons, New York, 1976, p. 465.
32. Speidel, M.O., in "High Temperature Materials in Gas Turbines," edited by P.R. Sahm and M.O. Speidel, Elsevier, Amsterdam, 1974, p. 207.
33. Speidel, M.O., Blackburn, M.J., Beck, T.R. and Feeney, J.A., in "Corrosion Fatigue: Chemistry, Mechanics, and Microstructure," edited by O.F. Devereux, et al., National Association of Corrosion Engineers, Houston, Texas, 1972, p. 324.
34. Fleming, J.R. and Suh, N.P., *Wear,* Vol. 44, 1977, p. 57.
35. Kramer, I.R. and Demer, L.J., in "Progress in Materials Science," edited by B. Chalmers, Vol. 9, Pergamon Press, Oxford, 1961, p. 133.
36. Fourie, J.T., *Philosophical Magazine,* Vol. 17, 1968, p. 148; *Philosophical Magazine,* Vol. 21, 1970, p. 977.
37. Fisher, J.C., quoted in Cottrell, A.H., "Dislocation and Plastic Flow in Crystals," Clarendon Press, Oxford, 1952, 1953, p. 86.
38. Basinski, Z.S., in "Surface Effects in Crystal Plasticity," edited by R.M. Latanision and J.T. Fourie, Noørdhoff, Leyden, 1977, p. 433.
39. Brown, A.F., *Advances in Physics,* Vol. 1, 1952, p. 427.

DISCUSSION

S. JAHANMIR, Cornell University: I would like to offer some explanation for the strange behavior that was mentioned about the effect of load and surface roughness on wear. The effect of surface roughness is present only at the initial stages of wear. Once the original roughness of the surface is removed by the wear process, the steady-state wear does not depend on the initial surface roughness. As for the behaviors at different loads, we observed that at low loads the behavior was controlled by the gradual removal of the asperities. So the total wear of the surface which had the highest roughness was larger than the total wear of the surface with the smallest roughness. Of course, the steady-state wear rate of both surfaces was the same. When we increased the load, asperities of smooth surface could not support the load and thus were broken up very quickly in the sliding process starting the steady state delamination wear processes immediately. But for rough surfaces it took some time before the rough, hard asperities could be removed and the delamination process could start. In all cases, the steady state wear rate will not depend on the initial surface roughness.

A. S. ARGON: I believe that is really what I said. Once the surface has been worked into a steady state, the initial roughness is not very important. I am sorry if I failed to mention your research on surface roughness.

QUESTIONER: Most of the figures refer to the mechanical data obtained on bulk materials. Are you suggesting that we can use the bulk properties when the near surface properties might be quite different from the bulk material?

ARGON: In effect I am making partly that suggestion. If we know the hardness of the surface (this is not necessarily the bulk hardness of the material) after it has been rubbed, after it has been cyclically deformed, and after it has come to a cyclic steady-state of deformation resistance, presumably the temperature and strain rate dependence can be scaled by the deformation mechanism maps. That is really what I am saying. But we want to be cautious that we have some information about the state of deformation of that material at that particular strain. Whether or not we have that material in bulk or near the surface, I do not think it will matter all that much.

SAME QUESTIONER: It matters very much because there is a coupling reaction between the strain and the service mechanical properties.

ARGON: Well, I do not contest that, of course. I am just saying that as long as the microstructure of the material (the state of the dislocation arrangements, etc.) is the same in bulk and in the surface, the deformation resistance of that material will be the same both in the bulk and in the surface. We should be aware that the deformation behavior is not necessarily obtained from a tension test. But may have to be obtained from a torsion experiment, or under a great deal of compression, where we essentially get that state of strain by compressing the material. So far as the microstructure of the rubbed region is the same as the bulk, it should not make any difference. However, if you have an aggressive atmosphere, and if that has penetrated into the surface in some manner, then it may be different. Barring those things, it should make no difference.

E. F. FINKIN, Allegheny Ludlum Industries: In one of your figures you define the wear process as the fracture of the asperities. I think that is too confining because one can easily envision situations where the asperity rubs on a relatively smooth surface in a cyclic fashion, like in rotating specimens, and failure can take place due to the cyclic stresses in the smooth body, not by the failure in the asperity region of one of the rubbing asperities. To concentrate only on the fracture of the asperity is an unnecessarily restrictive approach.

ARGON: Don't you think though that once you have the steady-state roughness the asperities induce plastic deformation in the material by the rubbing action? Those things I call asperity. Maybe that definition is not common.

FINKIN: Yes, you will develop an equilibrium roughness,which depends on the dynamics of the system, but if you actually look at the sequence of events, the coming together and breaking of junctions like in Archard's model, is far too restrictive.

ARGON: No, I didn't mean that. I am just saying that when a surface of a given roughness rubs another surface and repeatedly indents it, at some stage during this process either on the surface or in the subsurface region a crack is formed. When that crack propagates wear particles come out. That is the picture I am sketching.

FINKIN: OK. I do not disagree with that; that is a broader view than just the tearing out of asperities. Failure can occur at either surface and when it occurs, it is not just an asperity failure.

ARGON: It looks as if I have used the word asperity in an overly general sense.

FINKIN: I think so.

WATERHOUSE, University of Nottingham: Could you please comment on the dissipation of energy in sliding, particularly where the amplitude is of the order of a few microns as in fretting? We seem to get some correlation between the wear damage and thermal conductivity.

ARGON: Well, I don't know really what to say there. The deformation work, particularly when the material attains a steady state hardness, goes into the specimen and it has to be lost in the form of heat. Depending on the rate of deformation, there may be a temperature rise. I think the small amplitude deformation may be particularly lossy and can raise the temperature quite high. Surface oxidation, oxide particle formation, etc. can take place. Am I right?

WATERHOUSE: Yes, we do see a similarity between say pure aluminum and pure copper which are of course widely separated if we think in terms of stacking fault energy.

ARGON: Well, I am sorry I don't have any bright answers.

WATERHOUSE: Thank you.

R. DASKIVICH, G.M. Research Laboratories: A major premise that you used in formulating the continuum mechanics approach is that the grain size be sufficiently small. Could you give some examples of materials where the continuum mechanics approach would work and where it would not. What I am looking for is a transition region.

ARGON: I think in heavily worked material the grain size is not important any more because the dislocation structure is on a finer scale than the grain. It is essentially the cell size which I emphasize. I think that will be the scale that we have to take above which a continuum mechanics solution is valid. Grain size will turn out to be an important parameter in initially annealed and then lightly deformed materials where plastic flow may go from one boundary to another boundary. Then the slip line length determines the scale. Once the material is sufficiently worked, the deformation resistance comes to a steady-state. Then the scale of the problem is very much finer which is fortunate because that enables us to use continuum mechanics. What that scale is I cannot say, although I have thought about it for a very long time. You can have upper and lower bounds for it which may be orders of magnitude apart. Surely that scale must be above any mean particle spacing, dislocation bow out spacing, and the cell size which is in the range of 0.5 - 0.1 μm. In annelled material it may be as large as the grain size. However, we can ignore those instances because they are not very important in the wear problem. Once the surfaces is properly worn in, then its deformation resistance is small and therefore the transition size should be about 0.2 μm.

ON THE MECHANICAL PROPERTIES OF NEAR-SURFACE MATERIAL IN FRICTION AND WEAR

D. A. Rigney

ABSTRACT

Professor Argon's paper is a welcome contribution to the literature on friction and wear. Investigators in these fields should be encouraged to emphasize inelastic deformation and fracture in their tribological investigations.

Comments in this discussion paper are designed to supplement those of Argon on such topics as the use of the term "plowing", the role of surface finish, the nature of break-in and steady-state regimes, stacking fault energy and microstructure, and controversies on crack origins and on hard vs. soft surface layers.

DISCUSSION

Professor Argon's paper[1] is a welcome contribution to the literature on friction and wear. The approach taken is one that is particularly appreciated by workers in materials-related fields. Though he is in Mechanical Engineering, he speaks our "language," and fluently.

At the outset, Argon suggests that wear processes which are often treated quite differently have certain features in common. In particular, he suggests that inelastic deformation and fracture are the important unifying features which should be considered if we are to advance in the understanding and control of friction and wear. I am in complete agreement on this point. In fact, our own work has been based on the same foundation[2,3].

I also agree with Argon that friction and wear do not really involve new phenomena compared with the deformation and fracture observed in systems subjected to simpler boundary conditions. The main differences arise from the highly localized applied stresses which in turn cause high local strains, large strain and temperature gradients, and changes in local microstructure. We can and should learn from the vast body of knowledge already published on deformation and fracture, and we should draw on this source more regularly for the specific conditions involved in friction and wear.

I agree with much of what Argon says in his paper. But there is one section near the beginning that I find rather puzzling. It begins with the following statement: "Furthermore, it is also known as an elementary fact that the rate of removal of material with distance of sliding is a linear function of the tangential resistance to sliding, i.e., the average shear traction produced at the sliding interface."(1) Thus, the wear rate is claimed to be simply proportional to the frictional force. I know of no experimental data which supports this claim, unless the author is simply referring to the well-known load dependence, which is related to the tangential force through the friction coefficient. The paper cited[4] does not clarify this question; therefore, further discussion by the author would be appropriate.

In discussing his Figure 1, Argon has used the term "plowing" to discuss the plastic deformation resulting when a hard asperity moves across a softer material. Traditionally, this term has been analyzed by assuming simple geometric models involving the projected area of the indenting asperity[5,6]. Thus, plowing has been treated by defining an effective yield pressure of homogeneous material. In this simple picture, local variations in structure and properties as well as deformation below the wear groove and to the sides have been ignored. A simple model of this approach would consist of a pre-machined groove filled with a soft homogeneous material like butter. The asperity simply pushes the soft material along the groove without deforming the material which constrains it. Such a model is obviously not adequate, but it closely approximates the usual treatment of plowing. Therefore, if the word plowing is to be used, I would prefer to see it carefully defined so as to allow for all the deformation work actually involved.

In addition to contacts which cause plastic deformation, Professor Argon has listed contacts which behave in nearly reversible elastic manner. I would simply like to add that if these were truly reversible, their contribution to friction would be zero, because there would then be no energy dissipated. Various inelastic effects would contribute to friction, but true elastic contacts would not[7].

Most of the early work on asperity contacts was restricted to situations involving normal forces. Green included tangential forces in his impressive analysis,[8] but Argon quite rightly calls it a first order analysis of the problem. Since Green used a model of adhesive junctions and since he neglected local differences in microstructure, texture, and properties, his analysis is unable to explain the observed topology of wear surfaces or the details which metallography reveals.

The role of surface finish in friction and wear depends very much on how long one runs a test. During an initial break-in period, the results are sensitive to surface finish, but after this transient period, the initial surface finish is irrelevant, as Argon indicates. Of course, in practical devices, especially those with close tolerances, surface finish can be critical. But for simple wear tests, if one waits for steady-state conditions, the initial surface finish and even the initial microstructure will not be important. Tests run during the break-in period (including any single-pass experiments) will give highly variable results which are difficult to interpret. Tests run under steady-state conditions should be more tractable.

We picture the break-in period as the time needed to deform surface asperities and to create a constant (in a statistical sense) surface topology and near-surface microstructure. Under certain conditions, this will include creation of a dislocation cell layer of constant average thickness[2].

When the microstructure near the surface remains constant, then the wear rate and the nature of the debris generated should remain constant. Conversely, if the microstructure changes character, the wear process changes accordingly. If input conditions such as temperature or strain rate produce changes in microstructure, such as cells going to sub-grains or recrystallized grains, then the wear rate can change as well.

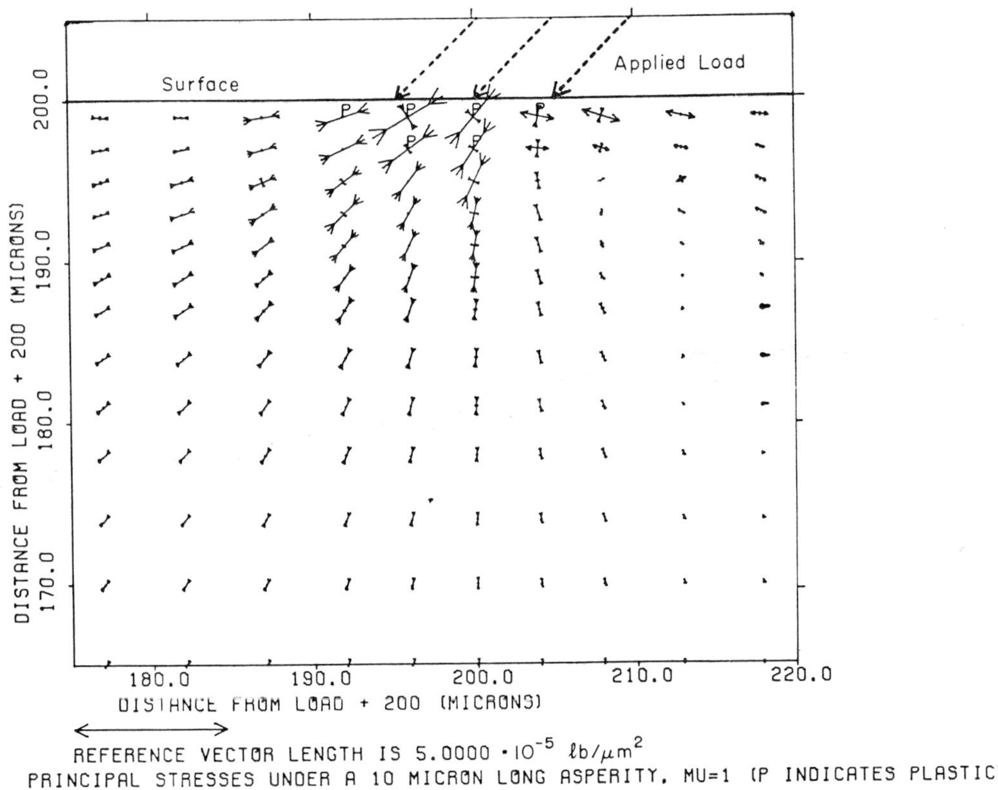

Fig. 1.—Plane strain stress distribution with tangential stress equal to yield and friction coefficient equal to one. (From N. P. Suh, Battelle Materials Science Colloquium on Fundamentals of Structural Alloy Design, ed. R. I. Jaffee and B. A. Wilcox, Plenum, N.Y. 1977.

As W. A. Glaeser and I pointed out at St. Louis last year,[3] the microstructure and the imposed conditions can be interrelated by the use of deformation mechanism maps, relatively new tools introduced by Ashby[9]. These devices may be considered as mechanical analogs of phase diagrams. By using deformation maps to correlate temperature, stress, strain rate, and microstructure, it should be possible to predict the wear mode for a given material subject to specific operating conditions. These maps should also be helpful for predicting wear rate transitions when these conditions are changed.

Argon has also called attention to the potential usefulness of deformation mechanism maps for friction and wear. He has, in addition, presented an interesting discussion on the applicability of these diagrams. He concludes that deformation mechanism maps are probably more representative for situations involving cyclic deformation than for monotonic deformation, because they are based on steady-state deformation, which in turn requires opportunities for recovery to balance work hardening.

Before going on to the subject of crack initiation, I should like to clarify the relationship between stacking fault energy and cross-slip. It is well known that cross-slip of screw dislocations can occur readily in materials with high stacking fault energy. Cross-slip in turn leads to cell formation at relatively low strains. On the other hand, low stacking fault energy metals such as silver or brass tend toward planar slip because of the difficulty of cross-slip. However, if the strains are sufficiently large, even these materials form well-developed cells[10-12]. This is important during wear, because the large strains needed for cell formation in these materials are almost certainly present. Thus, for wear, cell formation will generally occur near the surface. Other factors being equal, the stacking fault energy will affect how soon the cells form and how thick the cellular region will be.

Since wear generates loose debris, fracture processes are obviously involved. As Argon has noted, there is disagreement as to the origins of the wear cracks. Those who favor fatigue mechanisms believe that the cracks begin at the surface. In fact, as Argon points out, "There are no known cases of fatigue crack formation in single phase alloys below the surface"[1]. Of course, the usual fatigue tests involve more clearly defined and simpler loading conditions than those encountered with wear tests, so some exceptions may yet be found. The typical alternating stress patterns in most fatigue tests are replaced for wear by cycles of compressive stress followed by nearly complete unloading, without appreciable reverse loading.

A second group of workers includes Suh and his colleagues[13,14]. They have referred to Hertz stress calculations to rationalize cracks nucleating at some distance from the surface. In typical commercial alloys or other materials containing inclusions, interior cracks are to be expected because of plastic compatibility problems at the particle/matrix interfaces. These are the materials in which internal cracks are prominent. No one has clearly demonstrated for sliding wear that cracks nucleate at depths consistent with Hertz-type calculations in the absence of pre-existing voids or inclusions.

Suh has presented the results of a computer calculation of the stress field under a model asperity sliding on a homogeneous medium[15]. His results (Figure 1) do show some small tensile stresses behind the slider, but the stress orientations do not favor Suh's choice of the crack nucleation site. His plot shows that the tensile direction is roughly parallel to the sliding direction, and this of course favors fracture perpendicular to the wear direction.

Whether wear cracks begin *below* the surface and eventually propagate outwards or whether they nucleate *at* the surface and propagate inwards should not be considered as mutually exclusive alternatives. We may find that one mechanism operates to generate one wear particle and the other operates for a nearby wear event. A third possibility is that both occur simultaneously, with the crack tips eventually joining below the surface (see Figure 2). We should not be surprised to find such complications. After all, when any of these cracks are formed, the boundary conditions will be changed locally, and the stress patterns will differ from those predicted by simplified computer models.

One of the strongest arguments for preferring crack initiation at the wear surface is that involving environmental effects. Argon has correctly pointed out that the existence of such effects indicates strongly that cracks either initiate at the surface or vent there. In the first case, the aggressive environment could accelerate both the nucleation and propagation of fracture. In the latter case, it would mainly accelerate the crack propagation. If crack nucleation were known to be rate-controlling, the correct alternative would be clear.

The fact that various etchants clearly reveal the existence of a highly deformed near-surface region may offer a clue to the nature of environmental effects in many wear situations. Usually, the highly deformed region is revealed by preferential etching. This suggests that deformation has created highly defective regions, which, because of their relatively high energy, are readily attacked by the etch solution. Cell walls would seem to be reasonable candidates for such attack. Thus, it seems possible to include environmental effects in a natural way in the cell model described in reference[2].

After discussing the nucleation of cracks, Argon has focussed attention on the growth of cracks under fatigue conditions. The discussion of Stage I crack growth may be relevant for the deformation at some depth into the wearing material, but it seems unlikely that it will help us understand

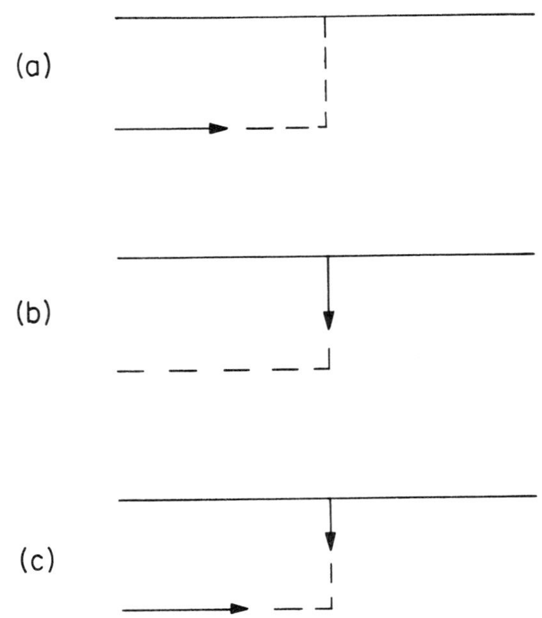

Fig. 2.—Three simple fracture paths for wear by delamination.

processes closer to the surface, where identifiable grains are absent. In that region, it is tempting to associate both nucleation and growth of cracks with the highly-textured and fine microstructure. Specifically, we have proposed a cell microstructure with fracture propagating parallel to (and probably within) cell walls[2].

It is interesting to note the discussion on whether the surface region is harder or softer than the bulk material. Hirth and I have recently discussed the same controversy[3], because it is indeed relevant to the subjects of friction and wear. I would like to suggest that part of the problem lies in what we mean by the words "hard" and "soft." One can question whether the near surface hardness, as measured by the usual penetration microhardness test, is a relevant parameter for wear, which seems to be associated with very high levels of highly directional strain. That is, the word "soft" is ambiguous for a case in which structural anisotropy is pronounced. Certain microstructures could exhibit very high microhardness values and yet they could be "soft" in the sense that stresses in a selected direction could easily give large strains.

Finally, I would like to refer to Argon's discussion of the dislocation structure of the heavily deformed surface region. He suggests that this region contains dense clusters of dislocation multipoles which are stable even near the surface and which have no long range stress field. I agree with this picture, but I would go further and suggest that at the high strains typically encountered, these multipoles are arranged in cell walls, and the traditional Beilby layer is probably a cellular region near the surface.

REFERENCES

1. Argon, A.S., These Proceedings.
2. Rigney, D.A. and Glaeser, W.A., *Wear,* Vol. 46, 1978, p. 241.
3. Rigney, D.A. and Hirth, J.P., *Wear,* Vol. 39, 1976, p. 133.
4. Bowden, F.P., Moore, A.J.W. and Tabor, D., *Journal of Applied Physics,* Vol. 14, 1943, p. 80.
5. Bowden, F.P. and Tabor, D., "The Friction and Lubrication of Solids," Clarendon Press, Oxford, England, 1954.
6. Walton, D., *Wear,* Vol. 6, 1963, p. 257.
7. Rigney, D.A. and Hirth, J.P., *Wear,* Vol. 53, 1979, p. 345.
8. Green, A.P., *Journal of the Mechanics and Physics of Solids,* Vol. 2, 1954, p. 197.
9. Ashby, M.F., *Acta Metallurgica,* Vol. 20, 1972, p. 887.
10. Bhargava, A.K., Moteff, J. and Swindemann, R.W., *Metallurgical Transactions,* Vol. 7A, 1976, p. 879.
11. Bailey, J.E., in "Electron Microscopy and Strength of Crystals," edited by G. Thomas and J. Washburn, Wiley-Interscience, 1963, p. 535.
12. Swann, P.R., in "Electron Microscopy and Strength of Crystals," edited by G. Thomas and J. Washburn, Wiley-Interscience, 1963, p. 131.
13. Suh, N.P., *Wear,* Vol. 25, 1973, p. 111.
14. Suh, N.P., et al., "The Delamination Theory of Wear," Elsevier Sequoia, S.A., Lausanne, 1977.
15. Suh, N.P., "Battelle Materials Science Colloquium on Fundamentals of Structural Alloy Design," edited by R.I. Jaffee and B.A. Wilcox, Plenum Press, New York, 1977.

DISCUSSION

S. GANESH, Bendix Research Laboratory: In actual sliding there will be a temperature gradient below the surface. Given that, it may be difficult to predict the dominant mechanism by deformation mechanism maps. Also because of the same reason the effect of stacking fault energy in controlling the wear may not be as important as you have emphasized. Could you please comment?

D. RIGNEY: I would like to comment on the temperature first. The temperature gradient has recently been shown to be not as steep as people normally thought it would be, the reason being that most estimates of the temperature and the temperature gradient have been based on the assumption that the temperature source was at the surface. That is, the surface was considered as a 2-dimensional heat source. As we all know, a localized heat source at one place will give a spike there so the temperature and the temperature gradient will be very high. Recent work at the Technion, published in Wear last year, analyzes the problem as an extended heat source over a volume, as Prof. Argon and I would expect, since plastic deformation is the source of heat. Then neither the temperature nor the temperature gradient will be high. I agree with you that high temperatures and temperature gradients will affect just about all the properties, but I want to point out that it is not as severe as many people think.

N. P. SUH, MIT: I would like to make a few comments. Certainly no one disputes the fact that there are dislocation cells. But some of the data you have shown should be analyzed carefully. If indeed mere dislocation cell formation creates internal cracks, some of the soft thin metallic coatings which were effective in preventing wear should never work. We have to realize the fact that a complex relationship between deformation mechanisms, state of stress (i.e., normal load and coefficient of friction) and the microstructure affects the wear process. Even in the case of stacking fault energy a number of processes such as crack propagation rate, crack nucleation rate, etc. depend on it. There is a danger in trying to collect a great deal of data from all sorts of materials and then trying to generalize in terms of one parameter. If you try to categorize there are eight different ways cracks can nucleate and propagate which I have shown elsewhere in these proceedings.

D. RIGNEY: I do not mean to say that stacking fault energy is the only important property -- certainly not. There are many other important properties. This was one correlation that was experimentally noted by many people in the past. John Hirth and I discussed why there might be such a correlation. The model I give here is just one reason why this stacking fault energy correlation might occur. I think that texture, which depends on stacking fault energy, also has some effect on wear. Certainly the great variety of cracking that occurs is also important.

You also cited the connection between friction and wear. There is no simple quantitative correlation between friction and wear. I think there is a very simple reason why that is so. Argon and I have both emphasized two factors: plastic deformation and fracture. I believe there would be a good correlation with the plastic deformation part if only we could separate it, but there is no reason why we should expect a correlation between friction and fracture. The fracture events that produce wear debris have very little to do with friction, except for the noise.

ON THE RELATIONSHIP BETWEEN DELAMINATION WEAR AND THE INITIATION AND GROWTH OF FATIGUE CRACKS IN ULTRAHIGH STRENGTH STEEL

R. O. Ritchie

ABSTRACT

According to the delamination theory of sliding wear, wear damage can result from cyclic shear deformation due to surface traction exerted by asperities leading to sub-surface initiation and propagation of fatigue cracks (under combined Mode I and II conditions). Estimates of the rate of such fatigue crack propagation under sliding wear conditions in steels have been found to be in the near-threshold regime (i.e., typically below 10^{-5}*mm/cycle). Accordingly, the object of this work was to investigate whether microstructures which are known to be highly resistant to Mode I near-threshold fatigue crack propagation in an ultrahigh strength steel (300-M) show similar resistance to sliding wear in environments of ambient temperature air. Effects of tempering temperature and hardness were specifically investigated. It was found that wear resistance in this steel bore no direct relationship to crack propagation resistance, whereas a direct correlation was found with resistance to crack initiation (or more correctly the initiation and growth of microcracks). The significance of these results are discussed in the light of the limiting mechanisms for delamination wear in ultrahigh strength steels and the size-scale of the events involved.*

INTRODUCTION

In recent years, the delamination theory of wear[1] has gained increasingly wide acceptance as one of the basic mechanisms of wear when sliding surfaces come into contact. The essence of the theory is that normal and tangential forces are transmitted through contact points at asperities between two surfaces by adhesion and plowing, such that the softer surface experiences cyclic loading (at asperity-to-asperity and asperity-to-plane contact points) when the asperities of the harder surface plow it. The resultant accumulated cyclic plastic shear deformation gives rise to sub-surface crack nucleation in the softer material, which on further deformation leads to crack propagation and coalescence parallel to the surface at a depth governed by material and friction characteristics. Removal of material through formation of wear particles is then achieved when cracks shear to the surface generating long, thin, "delaminated" wear sheets.

Whereas the rate controlling mechanism for delamination wear may be one of cyclic deformation, crack nucleation or crack propagation, recent evidence has suggested that in many materials the critical step is crack propagation.[1-4] Furthermore, estimates of the rate of growth of delamination cracks in steels have indicated that propagation rates are less than 10^{-5}mm/cycle,[5] and are thus in a regime of growth rates (the so-called near-threshold range) where the influence of microstructure becomes of paramount importance (regime A in Figure 1).[6] Accordingly, as noted by Argon,[7] attempts have been made to directly correlate wear rates with macroscopic fatigue crack propagation rates, measured under tensile opening Mode I loading, but to date results have been somewhat unconvincing.[4]

The object of the present work is to examine wear rates in a steel where fatigue crack propagation behavior, particularly at near-threshold rates (i.e., less than 10^{-5}mm/cycle) has been extensively characterized.[6,8] The rationale here is simply that, if near-threshold crack propagation is indeed a rate-controlling step in the development of delamination wear cracks, then microstructures which show superior resistance to fatigue at extremely slow growth rates should similarly show superior resistance to wear.

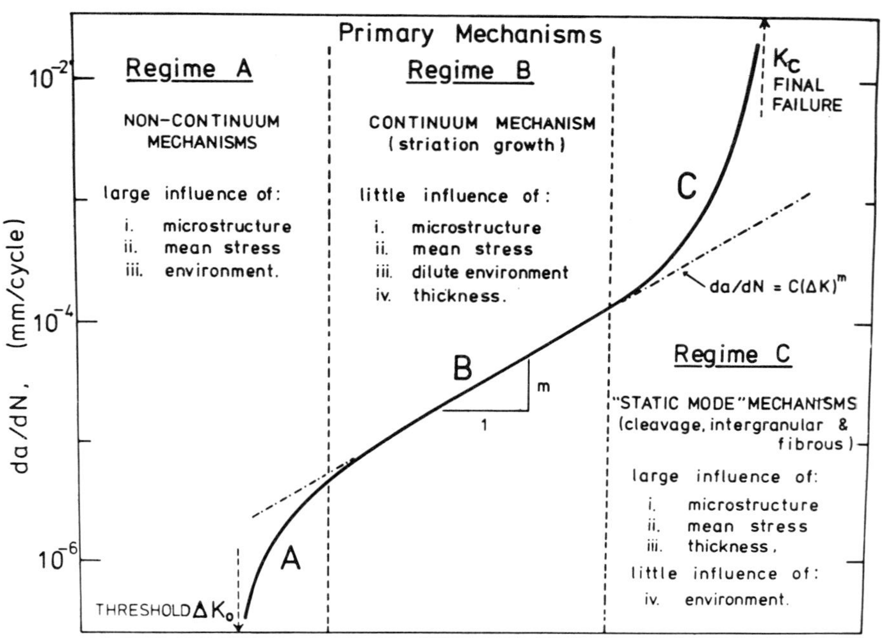

Fig. 1.—Schematic diagram showing the variation of fatigue crack propagation rate under tensile loading (da/dN) with alternating stress intensity (ΔK) indicating primary failure mechanisms. ΔK_O is the threshold stress intensity below which crack propagation cannot be detected, K_C the stress intensity at final failure.

EXPERIMENTAL PROCEDURES

The material chosen for study was an ultrahigh strength, aircraft-quality (vacuum-arc remelted) 300-M steel, of composition in weight percent shown below:

C	Mn	Cr	Ni	Mo	Si	S	P	V
0.42	0.76	0.76	1.76	0.41	1.59	0.002	0.007	0.10

The composition is essentially that of AISI 4340 modified with 1.3% silicon.

To investigate effects of microstructure on wear and fatigue characteristics in this material, specimens were austenitized for 1 hour at 870°C and quenched into agitated oil to form martensite, yielding a prior austenite grain size of 20 μm. Tempering was carried out for 1 hour at temperatures of 100°, 300°, 470° and 650°C to provide a wide range of properties. These treatments are hereinafter referred to as T100, T300, T470 and T650, respectively. The resultant microstructures were fully characterized with respect to ambient temperature mechanical properties from monotonic and cyclic uniaxial tensile tests, plane strain fracture toughness (K_{Ic}) tests, and fatigue crack propagation behavior over an extremely wide range of growth rates from 10^{-8} to 10^{-1}mm/cycle. Full details of the experimental procedures used are described elsewhere.(6,8) Unlubricated wear tests were performed on a lathe using a cylinder-on-cylinder geometry, where the cylindrical specimens of diameter 6.4 mm, were rotated against a similar sized stationary slider of AISI 52100 steel (R_C=69). A lathe tool dynamometer, connected to the slider, was attached to the carriage of the lathe and the normal force applied by the tranverse motion of the carriage. Normal and friction forces were monitored using a strain gauge assembly and displayed on a Sanborn recorder. Tests were carried out using a 1 kg load, for a duration of 30 minutes, at a surface sliding speed of 2m/min., corresponding to a sliding distance of 60 m. Wear rates were determined from weight loss measurements using a chemical balance.

All tests were performed in ambient temperature air (23°C) with relative humidity 45%, on specimens machined in the longitudinal L-T orientation.

RESULTS AND DISCUSSION

Ambient temperature mechanical properties of the structures tested are listed in Table I. All structures were typical of tempered lath martensite with some evidence of twinning. The T100 microstructure contained a high dislocation density, autotempered carbides and roughly 6% retained austenite distributed as thin, 200 Å thick films surrounding martensite laths. Evidence of ε-carbide was seen in T300 structures, the retained austenite content being reduced to ∿4%. Interlath cementite replaced ε-carbide in T470 structures, and the retained austenite content was below 1%. In the T650 structure, the cementite had spheroidized and all austenite had transformed.

Fatigue crack propagation data, in the form of growth rates (da/dN) versus the alternating stress intensity factor ($\Delta K = K_{max} - K_{min}$), are shown in Figure 2. The data presented represent behavior over the entire range of growth rates from the threshold stress intensity (ΔK_o), below which crack propagation cannot be detected, to final failure characterized by K_{Ic} (i.e., 10^{-8}-10^{-1}mm/cycle). Measurements were taken for a load ratio

TABLE I.—AMBIENT TEMPERATURE MECHANICAL PROPERTIES OF 300-M

Code	0.2% Offset Yield Stress		U.T.S.	Elongation	K_{Ic}	R_c
	Monotonic (MPa)	Cyclic (MPa)	(MPa)	(% of 1 in. gauge)	(MPa$\sqrt{m}$)	
T100	1497	2107	2338	12.4	36	55
T300	1737	1486	2006	11.9	65	50
T470	1497	1198	1683	12.1	69	45
T650	1074	861	1186	18.1	152	37

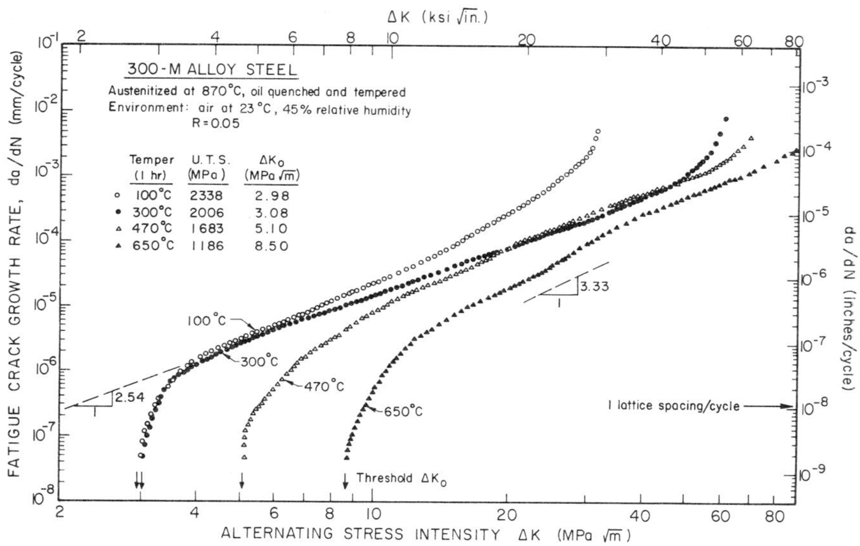

Fig. 2.—Variation of fatigue crack propagation rate (da/dN) with alternating stress intensity (ΔK) at R=0.05, for 300-M steel, quenched and tempered between 100 and 650°C. Data obtained on 10mm thick 1-T compact specimens at 50 Hz in ambient temperature air at 45% relative humidity.

($R=K_{min}/K_{max}$) of 0.05, under load control with the cyclic frequency maintained at 50 Hz.

It is apparent from Figure 2 that, whereas there is little difference in crack propagation resistance at intermediate growth rates (i.e., regime B in Figure 1 where $da/dN \approx 10^{-5}$-10^{-3}mm/cycle), tempering temperature has a marked influence on near-threshold rates less than 10^{-5}mm/cycle. At $\Delta K=9$ MPa$\sqrt{m}$, for example, the growth rate in the T470 condition is over two orders of magnitude less than in the T100 condition. As the tempering temperature is raised, the threshold ΔK_o increases from 3.0 to 8.5 $MPa\sqrt{m}$, concurrent with a two-fold reduction in strength. Clearly, as the tensile strength or cyclic yield strength of 300-M is decreased with increasing tempering temperature, there is a significant *increase* in resistance to near-threshold fatigue crack propagation.

Results from the wear tests are shown in Figure 3 as a function of tempering temperature and hardness. The decrease in hardness resulting from an increase in tempering temperature apparently does not affect the steady state friction coefficient (μ), which is observed to be approximately 0.5 for this steel. Wear rates, however, defined as the volume of material lost per unit sliding distance, are dramatically increased by nearly an order of magnitude as the tempering temperature is increased. Similarly, the wear coefficient (k), defined from Archard's wear equation[9] and representing the wear rate normalized with respect to hardness, is almost an order of magnitude larger in the softer T650 structure compared to the T100

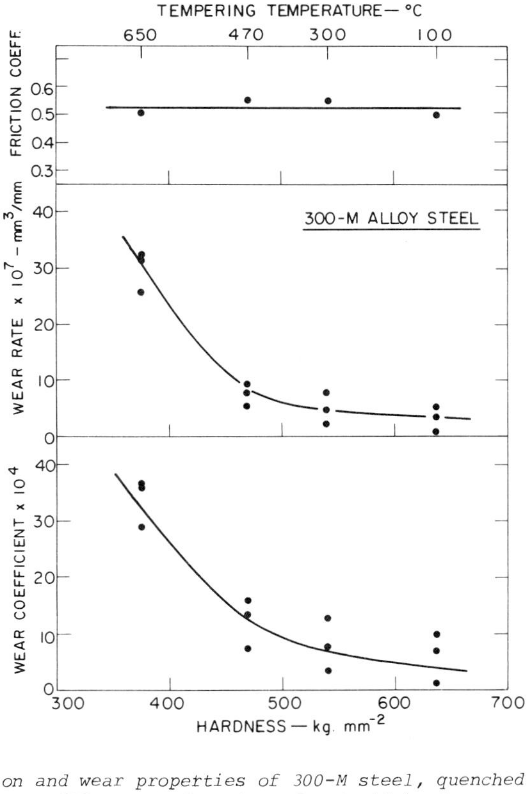

Fig. 3.—Friction and wear properties of 300-M steel, quenched and tempered between 100 and 650°C, as a function of Brinell hardness: a) friction coefficient, b) wear rate, and c) wear coefficient. The normal load was 1 kg, duration of tests was 30 minutes at a sliding speed of 2 m/min., in environment of ambient temperature air at 45% relative humidity.

structure. Thus as the tensile strength or hardness of 300-M is decreased with increasing tempering temperature, there is a significant *decrease* in resistance to unlubricated sliding wear.

A comparison of resistance to near-threshold crack propagation, characterized by the threshold ΔK_o, with wear resistance, characterized by the reciprocal of the wear coefficient (k^{-1}), in this steel (Figure 4) thus indicates a total lack of direct correlation between wear and macroscopic fatigue crack growth. In fact, microstructures which are most resistant to crack propagation are least resistant to wear. However, if the wear resistance is compared to the resistance to fatigue crack initiation, characterized by the smooth bar fatigue limit*, a direct correlation can be seen between wear and fatigue behavior as a function of tempering temperature (Figure 4).

Thus it appears that for ultrahigh strength steels, such as 300-M,

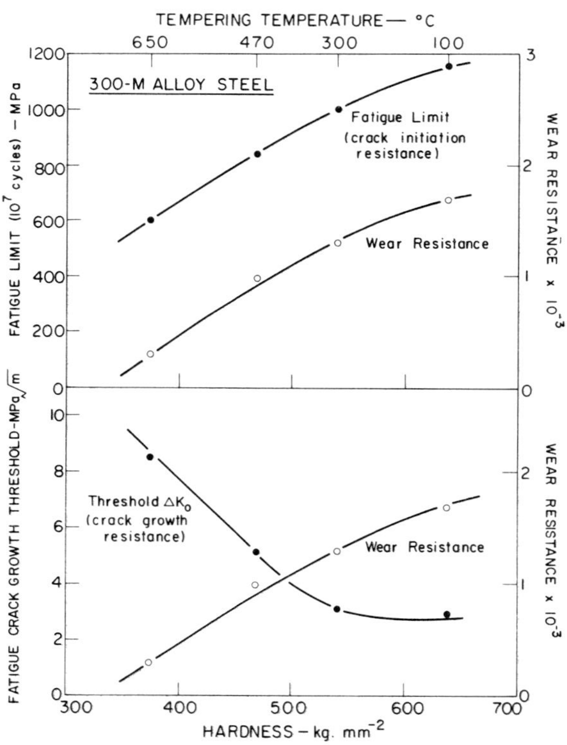

Fig. 4.—Correlation of fatigue and wear properties of 300-M steel, quenched and tempered between 100 and 650°C, as a function of Brinell hardness. Wear resistance characterized by reciprocal of wear coefficient, near-threshold crack growth resistance by threshold ΔK_o value, and crack initiation resistance by smooth bar fatigue limit.

wear rates cannot be directly correlated with representative Mode I fatigue crack propagation rates, but rather with crack initiation behavior characterizing the initiation and growth of microcracks less than a few grain diameters in size. A possible reason for this is apparent if one examines the size-scale of events involved. An estimate of the threshold defect size, below which linear elastic fracture mechanics is not applicable, can be obtained from

$$a_{th} = \frac{1}{\pi}\left(\frac{\Delta K_o}{\sigma_o}\right)^2 \tag{1}$$

where ΔK_o is the threshold stress intensity below which macrocracks do not propagate and σ_o is the smooth bar fatigue limit below which microcracks do not propagate and coalesce to initiate a macrocrack. For the T650 structure, for example, this threshold defect size (a_{th}) is of the order of 65 μm, which is significantly larger than the measured wear particle sizes in this steel of 20 μm. Thus, if wear damage and the resultant production of wear particles in this steel is assumed to be a consequence of sub-surface propagation of fatigue cracks, it is clear that such delamination cracks are too small to be treated as macrocracks and their growth rates will not correlate with macroscopic fatigue crack propagation behavior characterized by continuum mechanics parameters such as ΔK_o. Rather sub-surface wear damage in this steel must be related to the initiation and growth of microcracks, and resistance to such fatigue damage is best described by the fatigue limit. Since in ultrahigh strength steels microstructures which are *most* resistant to crack initiation (i.e., with maximum strength) are least resistant to early (macro) crack propagation,[8,10] it is therefore not surprising that wear rates do not show a direct relationship with fatigue (macro) crack propagation behavior.

We conclude then that the rate controlling step for delamination wear in this steel is not one of macrocrack propagation, but apparently one of initiation and growth of microcracks. Wear resistance in ultrahigh strength steels thus appears to be enhanced in microstructures which show good resistance to fatigue crack initiation rather than near-threshold macrocrack propagation and therefore correlation with macroscopic fatigue crack propagation behavior is not observed.

FINAL REMARKS

It is perhaps worth noting that any attempt to correlate wear rates to fatigue crack propagation rates in tension (Mode I) cycling (e.g., Ref. 3) is somewhat artificial since sub-surface delamination cracks appear to grow under additional shear (mixed mode) conditions.[1,11] However, whereas crack growth rates under mixed mode loading may differ in absolute magnitude to rates determined for pure Mode I loading,[12,13] it is unlikely that microstructural influences on crack propagation behavior would significantly differ. Hence, structures which show superior resistance to

---

*In strict terms, the fatigue limit represents the stress to initiate a macrocrack (i.e., a crack larger than a few grain diameters) in such high strength steels. In softer microstructures (i.e., T650) this occurs by the initiation, growth, and coalescence of several microcracks to form a "fatal" macrocrack, whereas in harder microstructures (i.e., T100) generally a single microcrack forms and propagates to form the macrocrack.[10]

fatigue crack propagation under tensile loading are likely to show similar superior resistance under mixed mode loading, and thus correlation procedures are not totally unrealistic.

CONCLUSIONS

Based on a comparative study of unlubricated sliding wear and fatigue behavior in an ultrahigh strength steel (300-M), quenched and tempered to a range of strength conditions, the following conclusions can be made:

1) As the hardness is decreased as a result of increasing tempering temperature, resistance to macroscopic near-threshold fatigue crack propagation is significantly *increased*, whereas resistance to unlubricated sliding wear is significantly *decreased*.

2) The decrease in wear resistance is consistent with a decrease in the smooth bar fatigue limit which infers easier microcrack initiation and growth.

3) Because of the size-scale of the events involved, wear rates cannot be directly correlated with bulk fatigue crack propagation rates (of macrocracks), but rather with the initiation and growth of microcracks.

ACKNOWLEDGEMENTS

Thanks are due D. Hanchar, S. Singer and G. Thomas, for experimental assistance and to Professor F. A. McClintock and Dr. N. Saka for helpful discussions. The work was supported by the Department of Mechanical Engineering, Massachusetts Institute of Technology.

REFERENCES

1. Suh, N.P. and Coworkers, in "The Delamination Theory of Wear," Elsevier Sequoia S.A., Lausanne, 1977.
2. Suh, N.P., *Wear,* Vol. 44, 1977, p. 1.
3. Fleming, J.R. and Suh, N.P., *Wear,* Vol. 44, 1977, p. 57.
4. Fleming, J.R. and Suh, N.P., *Wear,* Vol. 44, 1977, p. 39.
5. Jahanmir, S. and Suh, N.P., *Wear,* Vol. 44, 1977, p. 17.
6. Ritchie, R.O., *Metal Science,* Vol. 11, 1977, p. 368.
7. Argon, A.S., These Proceedings.
8. Ritchie, R.O., *Journal of Engineering Materials and Technology,* American Society of Mechanical Engineers Transactions, Series H, Vol. 99, 1977, p. 195.
9. Archard, J.F., *Journal of Applied Physics,* Vol. 24, 1953, p. 981.
10. Fine, M.F. and Ritchie, R.O., in "Fatigue and Microstructure," American Society for Metals, Ohio, 1979.
11. McClintock, F.A., in "Fracture 1977," edited by D.M.R. Taplin, Vol. 4, University of Waterloo Press, Waterloo, Canada, 1977, p. 49.
12. Iida, S. and Kobayashi, A.S., *Journal of Basic Engineering,* American Society of Mechanical Engineers Transactions, Series D, Vol. 91, 1969, p. 764.
13. Hourlier, F., McLean, D. and Pineau, A., *Metals Technology,* Vol. 5, 1978, p. 154.

EFFECT OF MICROSTRUCTURE ON FRICTION AND WEAR OF METALS

N. Saka

ABSTRACT

The effect of microstructure on dry friction and sliding wear of metals at room temperature is examined. In particular those structural parameters that control surface traction, subsurface deformation crack nucleation, and crack propagation are considered. The effects of crystal structure, grain size, solute atoms, precipitates, etc. on wear properties of metals and alloys are discussed. The need for further work on the cast structures and metastable phases is emphasized. The usefulness and limitations of deformation-mechanism and fracture-mechanism maps in the understanding of sliding wear of metals is critically examined.

I. INTRODUCTION

In his review paper[1] on the effect of near-surface mechanical properties on friction and wear, Professor Argon takes an instructive overview of the friction and wear behavior or metals in particular and of materials in general. The paper displays the characteristic elegance of Professor Argon's writings. It is a welcome contribution to the tribology literature.

Argon correctly points out that subsurface deformation crack nucleation, and crack propagation processes control the wear rates of materials. He also emphasizes that the key to the understanding of friction and wear processes lies in the solution of the elastoplastic boundary value problem. This approach originally proposed by Suh,[2] has been the basis for a detailed inquiry into the friction and wear behavior of materials by Suh and his coworkers over the years.[3] Although much progress has been made on the analyses of the subsurface deformation, crack nucleation and propagation processes, still much work needs to be done before the wear rates of materials can be described by analytical expressions.

Meanwhile, however, the task of designing wear-resistant materials cannot wait until such analytical expressions are obtained. Some practical guidelines, even if qualitative, are very much desired. The key to establishing these guidelines lies in the organization of the existing information on friction and wear as a function of the microstructure of materials. Such a microstructural frame work, once established, also serves as a basis for designing experiments and interpretation of results rationally.

Accordingly, this paper is aimed at a systematic description of the friction and wear behavior of metals as a function of their microstructure. In an earlier review of the subject by Suh,[4] certain aspects have already been covered. Only pertinent points of that review will be repeated here for the sake of continuity. Although many results reported here were obtained by the investigators at MIT, the results of other investigators will also be quoted and interpreted freely. Further, this paper is aimed at sliding friction and wear of metals only, although microstructure profoundly affects other modes of wear also. Moreover, the effects of environment and temperature are not discussed in detail here as they are covered adequately elsewhere in these proceedings.

II. CLASSIFICATION OF MICROSTRUCTURE

It is unnecessary to rigorously classify the microstructure of materials here. Standard works on quantitative metallography and stereology[5,6] have devoted considerable effort on such a classification. Fortunately, such a fine characterization is not warranted as friction and wear properties of metals are independent of the finer details of the microstructure.

A useful classification of metals and alloys can be carried out with the help of a binary phase diagram such as that shown schematically in Figure 1. Metals can be approximately classified as pure metals, solid solutions, two-phase alloys and composites (see Table 1). Pure metals in turn can be classified in terms of their crystal structure, substructure, grain structure, etc. Solid solutions can be further classified as interstitial and substitutional solid solutions. Two-phase metals can be classified in terms of the shape, size, connectivity, etc. of the second-phase. Composites are included as a separate group when the volume fractions of the various phases are comparable. Eutetics and eutectoids belong to this group.

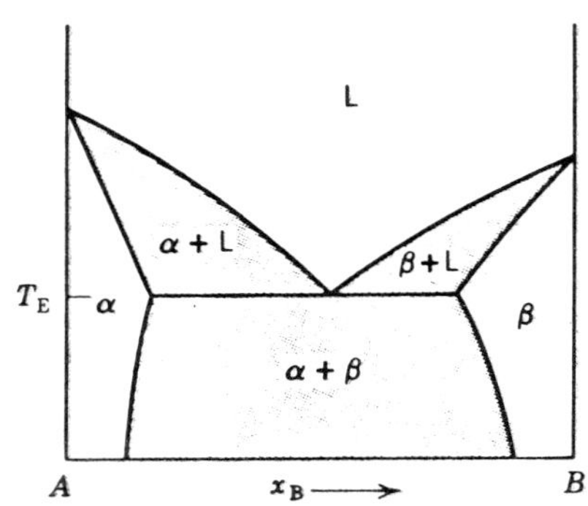

Fig. 1.—Schematic binary phase diagram.

The binary phase diagram includes many important structures as the amount of the solute is increased from 0 to 100 percent. Although many structures, which can be found in more complicated phase diagrams, cannot be represented by the binary phase diagram, this scheme still provides a reasonable microstructural framework for organizing the scattered information on the friction and wear properties of metals and alloys.

TABLE I.—APPROXIMATE CLASSIFICATION OF METALS AND ALLOYS

A. PURE METALS

- A.1 Crystal Structure
  BCC, FCC, HCP, etc.
- A.2 Substructure
  dislocation density, arrangement, etc.
- A.3 Grain Structure
  - A.3.1 grain size
  - A.3.2 grain shape
  - A.3.3 texture
  - A.3.4 grain boundary structure (high angle or low angle)

B. SOLID SOLUTIONS

- B.1 Ordered
  - B.1.1 Long Range Order
  - B.1.2 Short Range Order
- B.2 Disordered
  - B.2.1 Substitutional
  - B.2.2 Interstitial

C. TWO-PHASE (AND MULTIPHASE) ALLOYS

- C.1 Volume Fraction of Precipiates
- C.2 Size of Precipiates
- C.3 Shape of Precipiates
- C.4 Particle/Matrix Interface

D. COMPOSITES

- D.1 Size
- D.2 Shape
- D.3 Orientation
- D.4 Connectivity

} of various phases.

The objective of this discussion is to determine how the microstructure affects certain basic properties and how these basic properties in turn affect the friction and wear behavior of materials. Since friction and wear phenomena involve two bodies in contact, they are functions of the microstructures of both materials. While the number of combinations of microstructures to be considered is increased enormously, it also gives rise to a rich variety of phenomena that are not observed in areas such as elasticity, plasticity and creep. As a result it also offers many possibilities by which friction and wear can be minimized.

It should be recognized at the outset that many objections can be raised against this microstructural approach. An important one is that as friction and wear are not reversible processes, the initial microstructure is altered substantially during sliding to yield a completely different steady-state microstructure. The steady-state microstructure is a function of not only the initial microstructure but also the friction and wear behavior of materials in the transition period. Thus, one has to follow the microstructural changes from the time the materials are brought into contact until a steady-state is reached. Further, as friction and wear are essentially surface phenomena, large gradients in microstructure are possible. Such gradients may affect some basic properties during sliding. For example, migration of solutes can take place due to substructure gradients which may affect the surface energy which in turn may affect, as we shall see, the friction coefficient.

Despite these limitations, the task of relating friction and wear of materials to the initial structure is not hopeless. Results obtained by many investigators over the years indeed show a strong correlation between the initial microstructure and steady-state properties. Accordingly, an attempt is made here to summarize such correlations. Needless to say, this approach should be refined further to account for the microstructural variations during wear.

III. FRICTION

Friction is the resistance to motion (both initiation and continuation) when two contacting bodies are set into relative motion. Since it is the principal cause of energy dissipation and wear, the mechanism of friction has been studied extensively and many theories have been advanced. Surprisingly, the frictional behavior of metals as a function of microstructure is not fully understood yet.

The overwhelming effect of friction on wear can be seen from Figure 2 where the wear coefficient increases almost exponentially with the friction coefficient.(7) It is clear, therefore, that for a complete characterization of friction and wear, both phenomena have to be observed simultaneously. Indeed, the first step in minimizing wear, wherever possible, should be by the minimization of friction itself.

Argon points out that when a very hard material is in contact with a very soft material, plastic contacts are established. However, between two materials of high hardness (on the scale of the modulus) many contacts will be of the elastic type. In the latter case, he argues, the friction coefficients are expected to be lower than that in the former. When $s/G < 10^{-3}$ (s = the shear yield stress and G = the shear modulus), the friction coefficients will be high, whereas when s/G is of the order of 10^{-2}, the friction coefficients will be low. This argument oversimplifies the frictional behavior of metals. Accordingly a brief review of the theories of friction is in order.

The basic theories(8) of sliding friction proposed in the past are: roughness, plowing, and adhesion theories. According to the roughness theory, friction is caused by the interlocking of asperities of the contacting surfaces; i.e., the friction force is the tangential force necessary to lift the asperities of one surface over the asperities of the other. The roughness component of friction according to this model is given by

$$\mu_r = \tan\theta \tag{1}$$

where θ is the average slope of the asperities.

According to the plowing theory, friction is caused by the tangential motion of the hard and sharp asperities penetrated into the softer material.

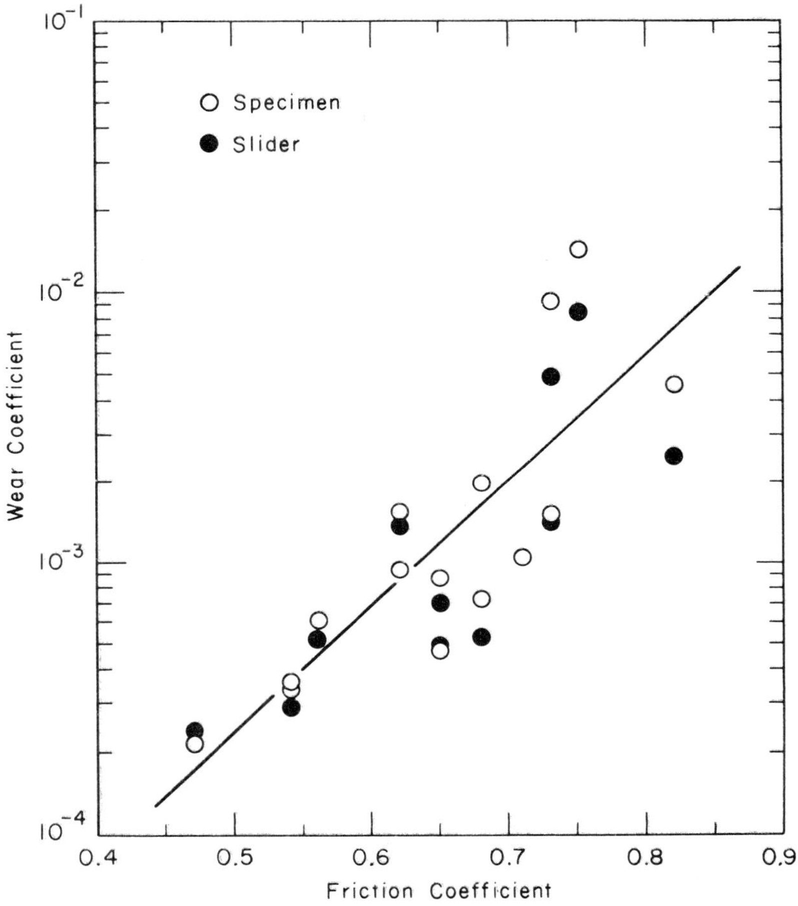

Fig. 2.—Wear coefficient of commercial steels as a function of friction coefficient.[7]

The frictional force is assumed to be the product of the cross-sectional area of the groove and the penetration hardness of the soft material. The plowing coefficient of friction then is given by

$$\mu_p = \frac{\tan\theta}{\pi} \tag{2}$$

The adhesion theory assumes that perfect adhesion occurs at the asperity contacts, and that the tangential force necessary to shear those adhered asperities when sliding occurs is the frictional force. If the average shear strength of the asperities is assumed to be the bulk shear strength, s, the softer of the contacting material, the adhesion coefficient of friction from a limit analysis is given by

$$\mu_a = \frac{s}{p} \tag{3}$$

where p is the penetration hardness.

Thus the total friction coefficient, μ, can be written as

$$\mu = \mu_r f_r + \mu_p f_p + \mu_a f_a \tag{4}$$

where f_r, f_p, and f_a are the fraction of the contacts in roughness, plowing and the adhesion modes respectively, and the f's are related by the equation

$$f_r + f_p + f_a + f_e = 1 \tag{5}$$

where f_e is the fraction of the contacts in the elastic mode. It is assumed in Equation (4) that the elastic contacts do not contribute to the friction coefficient.

For sliding contacts, the calculated values of μ_r, μ_p, and μ_a are very small (< 0.2) compared to the experimentally measured values ($0.3 - 1.2$). This situation has been improved by considering the details of plastic deformation by slip-line field analysis which leads to the following expression for the adhesion component of friction[9]

$$\mu_a = 1/(1 + \pi/2 - 2\theta) \tag{6}$$

where θ is now the maximum flank angle. Equation (6) predicts a friction coefficient in the range $0.4 - 1.0$ (depending on the value of θ) when all contacts are essentially in the adhesion mode.

It is interesting to note that the roughness, plowing and adhesion theories of friction described so far do not represent the microstructural effects of materials on friction coefficient. A more realistic theory from the microstructural viewpoint is due to Rabinowicz.[8] He refined the adhesion theory by proposing that the energy of adhesion, W_{ab}, influences the normal load supported by the plastic contacts at the asperities. According to this model, the friction coefficient is given as

$$\mu_a = \frac{s}{p} \cdot \frac{1}{1 - 2\,W_{ab} \cdot \cot\theta/pr} \tag{7}$$

where

$$W_{ab} = \gamma_a + \gamma_b - \gamma_{ab} \tag{8}$$

and γ_a, γ_b, and γ_{ab} are the surface energies and the interfacial energy of the contacting materials and r is the radius of the contact.

Now, Equation (7) represents the microstructural effects essentially through W_{ab}/p ratio, since both s/p ratio and the contact size are essentially constant for all materials. It is clear, therefore, that s/G ratio alone is insufficient to characterize the microstructural effects on friction.

IV. EFFECT OF MICROSTRUCTURE ON FRICTION

As mentioned earlier, the effect of microstructure on friction coefficient is essentially through its influence on W_{ab}/p ratio. The contact radius, r, and the slope of the asperities, θ, are assumed to be constant for many materials. In this section, the effect of various structural parameters, listed in Table I, on γ_a, γ_b, γ_{ab} and p and thus on the friction coefficient will be discussed.

A. Pure Metals

Rabinowicz and his coworkers(11-13) have conducted thousands of tests on commercially pure metals and reported friction coefficients for hundreds of pairs (see Table II). It appears from their publications that there is a systematic dependence of μ on the crystal structure. Both BCC and FCC metals essentially follow Equation (7) whereas HCP metals have markedly different behavior. In the latter case friction coefficient was found to be lower for c/a values larger than the ideal value. This is explained on the basis of the s/p ratio. It is argued that when c/a is larger than the ideal value, plastic deformation takes place at stresses below that given by the isotropic yield locus when the slip planes are favorably oriented.

Except for the deviation in hexagonal metals, Rabinowicz's results show a remarkably close agreement with his theory. This lead to the definition of the compatibility of metals. When the compatible parameter, defined as $W_{ab}/(\gamma_a + \gamma_b)$, is close to unity, the metals are compatible and the friction coefficients are high. (The compatible parameter of identical metals is unity.) The compatibility of metals is explained in terms of the solid solubility limits. Mutually soluble metals are compatible and, therefore, exhibit large friction coefficients.

The substructure effects do not seem to be important in pure metals. When metals are cold worked, the penetration hardness increases and hence W_{ab}/p should decrease. Accordingly the friction coefficient should also decrease (cold work should not affect W_{ab} term appreciably). Although no systematic investigations were carried out on the effect of cold work, it appears that the friction coefficient does not depend on dislocation density, arrangement, etc. In fact, the friction coefficient values obtained on well-annealed metals after a substantial amount of sliding should be strictly interpreted as that for cold worked metals because the material deforms substantially (perhaps reaching full hardness in a short time) during wear. In any case, since the W_{ab} values are usually not known to any reasonable certainty, it is difficult to separate surface energy and substructure effects.

Grain size and grain shape do not seem to have substantial effects on friction. This is understandable in view of the fact that both W_{ab} and p are essentially independent of these structural parameters.

Much work was done on the effects of crystallographic texture on friction.(14,15) It is known, for example, that the surface energy is orientation dependent. Further, it is also known that the yield stress, or hardness, of metals depends on the texture. It is not surprising, therefore, that friction coefficient depends on texture, especially in HCP metals.

B. Solid Solutions

When alloying elements are added to pure metals, two effects are en-

TABLE II.—FRICTION COEFFICIENTS FOR COMMERCIALLY PURE METALS.(11)

	W	Mo	Cr	Co	Ni	Fe	Nb	Pt	Zr	Ti	Cu	Au	Ag	Al	Zn	Mg	Cd	Sn	Pb	In
In	1.06	.73	.70	.68	.59	.64	.67	.79	.70	.60	.67	.67	.82	.90	1.17	1.52	.74	.81	.93	1.46
Pb	.41	.65	.53	.55	.60	.54	.51	.58	.76	.88	.64	.61	.73	.68	.70	.53	.66	.84	.90	
Sn	.43	.61	.52	.51	.55	.55	.55	.72	.55	.56	.53	.54	.62	.60	.63	.52	.67	.74		
Cd	.44	.58	.58	.52	.47	.52	.56	.59	.50	.55	.49	.49	.59	.48	.58	.55	.79			
Mg	.58	.51	.52	.54	.52	.51	.49	.51	.57	.55	.55	.53	.55	.55	.49	.69				
Zn	.51	.53	.55	.47	.56	.55	.50	.64	.44	.56	.56	.47	.58	.58	.75					
Al	.56	.50	.55	.43	.52	.54	.50	.62	.52	.54	.53	.54	.57	.57						
Ag	.47	.46	.45	.40	.46	.49	.52	.58	.45	.54	.48	.53	.50							
Au	.46	.42	.50	.42	.54	.47	.50	.50	.46	.52	.54	.49								
Cu	.41	.48	.46	.44	.49	.50	.49	.59	.51	.47	.55									
Ti	.56	.44	.54	.41	.51	.49	.51	.66	.57	.55										
Zr	.47	.44	.43	.40	.44	.52	.56	.52	.63											
Pt	.57	.59	.53	.54	.64	.51	.57	.55												
Nb	.46	.47	.54	.42	.47	.46	.46													
Fe	.47	.46	.48	.41	.47	.51														
Ni	.45	.50	.59	.43	.50															
Co	.48	.40	.41	.56																
Cr	.49	.44	.46																	
Mo	.51	.44																		
W	.51																			

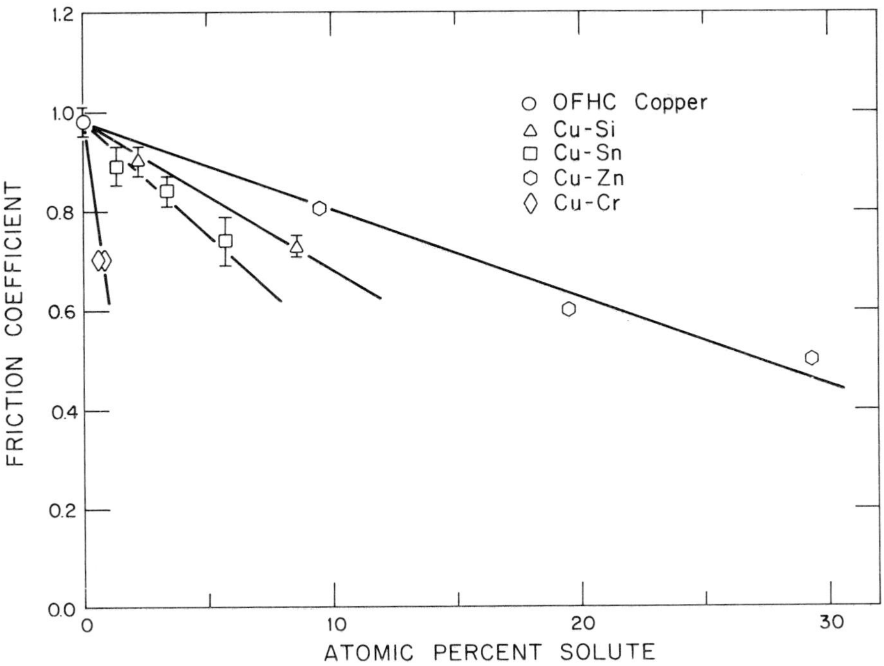

Fig. 3.—Friction coefficient as a function of atomic percent of solute of copper-based alloys.[18]

countered. First, the hardness of the metal increases by the substitutional or interstitial hardening mechanisms.[16] Secondly, the surface energy is decreased as the alloying content is increased.[17] Since it is known that many impurities segregate to the grain boundaries and interfaces, the interfacial energy is also expected to decrease with alloying content. Although it is not clear whether W_{ab} remains the same or decreases, W_{ab}/p is expected to decrease and hence the friction coefficient should also decrease. Indeed, experimental results on substitutional solid solutions obtained by Teixera, Saka and Suh[18] show that the friction coefficient decreases when the concentration of the alloying elements is increased (Figure 3). Recent data on a variety of steels obtained by Tohkai[7] and some results on oxide dispersion strengthened alloys[19] also indicate that the friction coefficient decreases when the solute content is increased. The solute effect should be much more predominant in the interstitial alloys because of the high hardening rates obtained in interstitial solid solutions.

Data on ordered alloys is scanty. Some unpublished work of Saka indicates that short range order (SRO) does not have a large effect on friction coefficient. Again this can be explained on the basis of W_{ab}/p. No systematic work was done yet on the long range order (LRO). It may be speculated, based on the above arguments, that LRO alloys exhibit lower friction coefficients than both solid solutions and SRO alloys, at least at the beginning of the test.

Slider			Specimen: Armco Iron	AISI 1020 Steel	AISI 1045 Steel	AISI 1095 Steel	AISI 52100 Steel
Slider	Armco Iron	μ_i	0.82 ± 0.04				
		μ_s	0.71 ± 0.13				
		k_a	$(4.60 \pm 3.72) \times 10^{-3}$				
		k_b	$(2.51 \pm 1.75) \times 10^{-3}$				
	AISI 1020 Steel	μ_i	0.56 ± 0.04	0.75 ± 0.03			
		μ_s	0.55 ± 0.04	0.68 ± 0.06			
		k_a	$(6.10 \pm 2.75) \times 10^{-4}$	$(1.43 \pm 0.88) \times 10^{-2}$			
		k_b	$(5.23 \pm 2.60) \times 10^{-4}$	$(8.54 \pm 4.84) \times 10^{-3}$			
	AISI 1045 Steel	μ_i	0.54 ± 0.01	0.62 ± 0.09	0.73 ± 0.01		
		μ_s	0.52 ± 0.02	0.53 ± 0.03	0.71 ± 0.03		
		k_a	$(3.59 \pm 1.69) \times 10^{-4}$	$(1.56 \pm 0.59) \times 10^{-3}$	$(9.42 \pm 1.78) \times 10^{-3}$		
		k_b	$(2.89 \pm 2.37) \times 10^{-4}$	$(1.39 \pm 0.72) \times 10^{-3}$	$(4.94 \pm 0.75) \times 10^{-3}$		
	AISI 1095 Steel	μ_i	0.47 ± 0.04	0.65 ± 0.03	0.68 ± 0.04	0.73 ± 0.08	
		μ_s	0.47 ± 0.04	0.54 ± 0.02	0.58 ± 0.03	0.67 ± 0.05	
		k_a	$(2.11 \pm 0.89) \times 10^{-4}$	$(8.83 \pm 4.04) \times 10^{-4}$	$(7.32 \pm 5.10) \times 10^{-4}$	$(1.51 \pm 0.76) \times 10^{-3}$	
		k_b	$(2.35 \pm 0.95) \times 10^{-4}$	$(4.97 \pm 2.72) \times 10^{-4}$	$(5.24 \pm 3.78) \times 10^{-4}$	$(1.47 \pm 1.03) \times 10^{-3}$	
	AISI 52100 Steel	μ_i	0.54 ± 0.04	0.62 ± 0.05	0.68 ± 0.03	0.71 ± 0.03	0.65 ± 0.04
		μ_s	0.54 ± 0.03	0.58 ± 0.03	0.66 ± 0.01	0.70 ± 0.02	0.61 ± 0.07
		k_a	$(3.41 \pm 0.87) \times 10^{-4}$	$(9.43 \pm 3.46) \times 10^{-4}$	$(1.97 \pm 1.11) \times 10^{-3}$	$(1.05 \pm 0.69) \times 10^{-3}$	$(4.78 \pm 2.05) \times 10^{-4}$
		k_b	—	—	—	—	$(6.99\ 3.65) \times 10^{-4}$

μ_i : friction coefficient at the initial stage

μ_s : friction coefficient at 36 m of sliding distance

k_a : wear coefficient of specimen

k_b : wear coefficient of slider

Fig. 4.—Friction and wear coefficients of heat-treated commercial steels.[7]

C. Two-Phase Alloys

While the hardness increases enormously with the amount of second-phase particles, the surface energy should remain constant when the solubility limit is exceeded. Thus, even though the surface and interfacial energies of two-phase alloys are not substantially different from the corresponding values of saturated solid solutions, according to Equation (7) the friction coefficient should decrease as the volume fraction of the second-phase particle is increased. Further, for a given volume fraction, the hardness can be increased by decreasing the particle size and interparticle spacing. This process again should not change either s/p, or W_{ab}. Thus, the friction coefficient should decrease substantially in two-phase alloys when the particle size and spacing are decreased for a given volume fraction of the second phase.

Unfortunately, the experimental results on a variety of commercial steels,[7] precipitation hardened alloys,[20] and oxide dispersion strengthened alloys[19] do not confirm this hypothesis. As can be seen from Figures 4-6, the friction coefficient is essentially independent of the volume fraction of the second phase and hardness of the alloys. Thus Rabinowicz's theory, which is so successful in explaining the friction of pure metals, breaks down in the case of two-phase alloys. At present, there is no theory that can explain this anomaly.

It may be speculated that as metals are subjected to precipitation hardening, dispersion strengthening, etc., hard particles are produced, and these particles may plow the soft metals and contribute to plowing coefficient of friction. However, Figures 4-6 indicate that the friction coefficient does not abnormally increase as the volume fraction of the second-phase particles is increased. Although plowing tracks are observed in some instances, the plowing component of friction calculated for these submicron size particles is extremely low. This is due to the fact that these submicron particles act as inefficient abrasives.[21] Further, as the volume fraction of the second-phase particles is generally less than 10 percent, the plowing contribution can be essentially ignored. This fact is further demonstrated by the observation that even highly oriented structures like pearlite in steel do not unreasonably increase the friction coefficient.

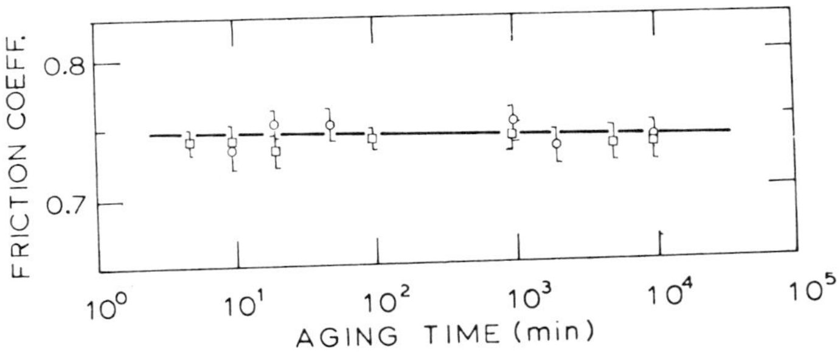

Fig. 5.—Friction coefficient of precipitation hardened Cu-Cr alloys. Note that the friction coefficient remains constant as a function of the aging time. The hardness increases initially, reaches a peak value and then decreases.[20] *See also Figure 18.*

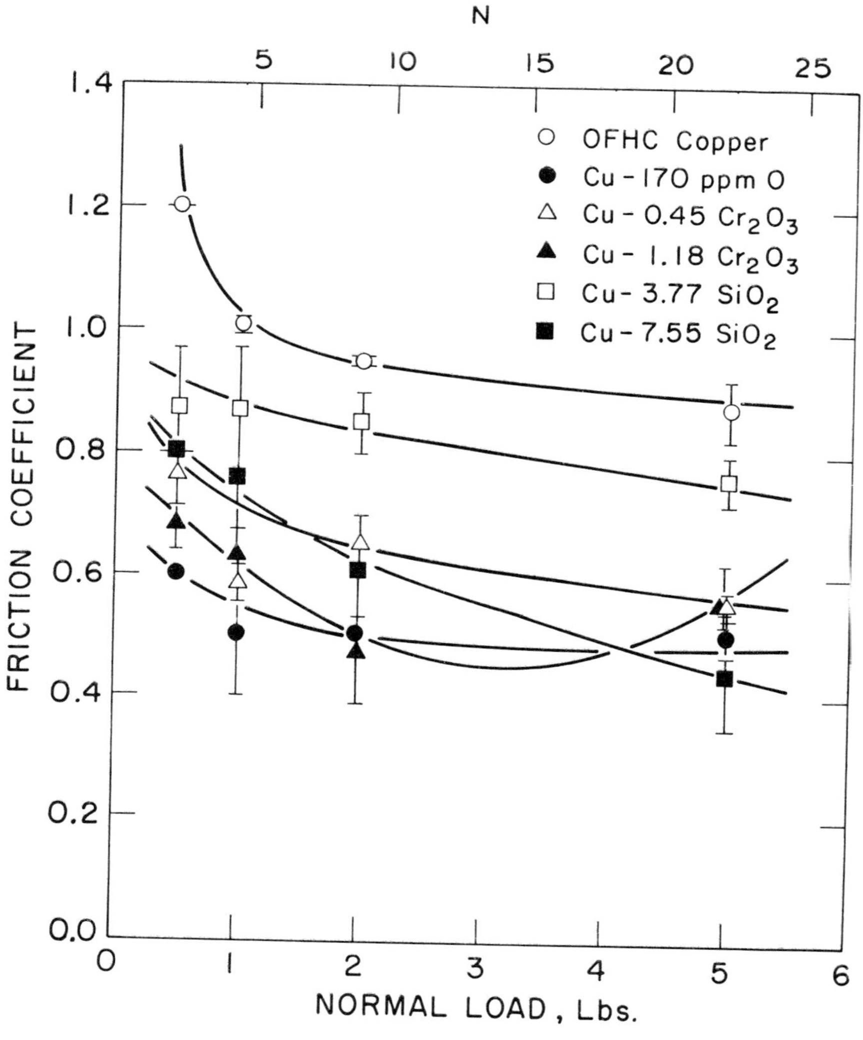

Fig. 6.—Friction coefficient versus normal load.[19]

D. Composites

Eutetics and eutectoids have not been studied extensively because of to their brittleness. However, a substantial amount of information exists on fiber reinforced polymer-based composites slid against metals. It has been shown that the friction coefficient is minimum when the sliding direction coincides with the fiber direction and maximum when the fibers are normal to the sliding surface.[22,23] Further, the friction coefficient

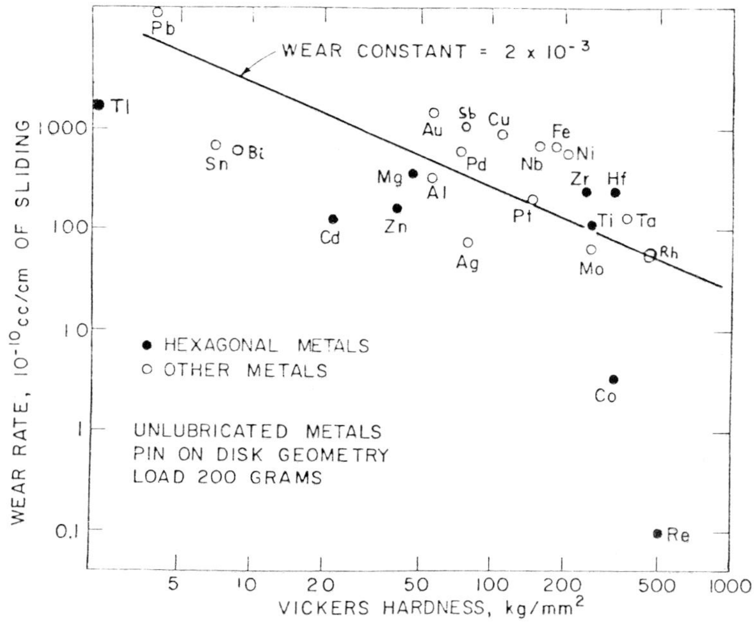

Fig. 7.—Wear rate of metals as a function of hardness. Note that the two metals showing very low wear rates have HCP structure.[10]

depends largely on the nature of the reinforcing phase. Surface energy considerations may be used for the fiber itself as the fiber generally gets exposed during sliding. However, as the hardness of the fiber is much larger than the matrix phase, Equation (7) cannot be used for interpreting the results. In fact, plowing and elastic contact problems may have to be considered depending on whether the mating surface is soft or hard.

Summarizing the microstructural effects in friction, Rabinowicz's theory of friction, which takes into account the surface and interfacial energies and the hardness of metals, appears to explain the friction behavior of pure metals and solid solutions reasonably well. However, the theory needs further improvements in the case of two-phase metals and composites.

V. WEAR

The wear volume, V, is generally expressed by

$$V = k \frac{P \cdot \ell}{3p} \tag{9}$$

where P is the normal load, ℓ the sliding distance, and k is the wear coefficient. Experimental evidence indicates that the wear rate of metals, $dV/d\ell$, is approximately proportional to the normal load P. However, the inverse dependence of $dV/d\ell$ on the hardness, p, is questionable. In the case of pure metals, for example, the inverse dependence is approximately obeyed as shown in Figure 7. However in the case of solid solutions, the wear rate

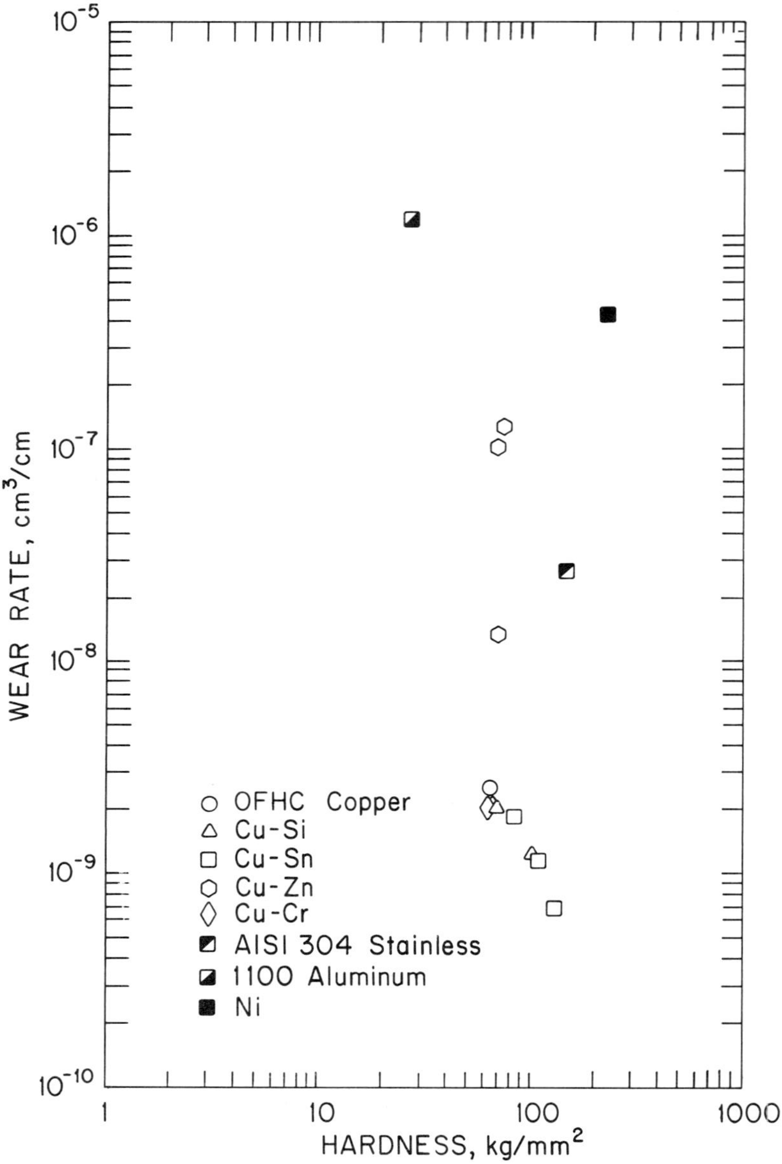

Fig. 8.—Wear rate versus hardness for some commercially pure metals and copper-based solid solutions.[3]

was not found to correlate well with hardness as can be seen from Figure 8. As we will see later, the disagreement in the case of precipitation-hardened and dispersion-strengthened alloys is even more pronounced.

It should be recognized that only when k is a constant (i.e. approximately independent of microstructure), the wear rate $dV/d\ell$ would be inversely proportional to the hardness, p. Then the microstructural dependence of the wear rate is determined completely by the microstructural dependence of hardness, and hard metals will be more wear resistant than soft metals. On the other hand, if the wear coefficient, k, is a strong function of microstructure, the wear rate, $dV/d\ell$, would be a strong function of microstructure, too. Then the wear rate of materials will be determined not only by p but also by k. As suggested rightly by Argon, "the understanding of the entire wear phenomenon is in the mechanistic understanding of what governs the levels of the coefficient k."

This writer agrees with Argon in that wear takes place (under normal conditions of load, speed, etc.) by three sequential processes: subsurface deformation, crack nucleation, and crack propagation. Therefore, we can write symbolically

$$k = k\ (\textit{subsurface deformation, crack nucleation, crack propagation})$$

or

$$k\ (S) = k\ [\dot{\varepsilon}\ (S),\ \dot{\phi}\ (S),\ \frac{dc}{dn}\ (S)] \qquad (10)$$

where $\dot{\varepsilon}$, $\dot{\phi}$, $\frac{dc}{dn}$, represent the subsurface deformation, crack nucleation, crack propagation rates respectively and s the microstructure.

It is well known that a variety of microstructural parameters affect all the three processes and therefore the expression for the wear coefficient, k, is expected to be a complex function of microstructure. Fortunately, however, the above three processes take place at any given location sequentially and the slowest of the three controls the overall wear rate and hence the level of k. Further, as the temperature affects these processes profoundly, Equation (11) can be rewritten as

$$\begin{aligned} k\ (S,T) &= k\ [\dot{\varepsilon}\ (S,T);\ \dot{\phi}\ (S,T);\ \frac{dc}{dn}\ (S,T)] \\ &= k_1\ [\dot{\varepsilon}\ (S,T)] && \textit{subsurface deformation controlled} \\ &= k_2\ [\dot{\phi}\ (S,T)] && \textit{crack nucleation controlled} \qquad (11) \\ &= k_3\ [\frac{dc}{dn}\ (S,T)] && \textit{crack growth controlled} \end{aligned}$$

The key to an understanding of the microstructural effects on k, then, lies in the understanding of the deformation, crack nucleation, and propagation aspects as a function of the microstructure, at least separately if not together.

A. Subsurface Deformation

The mechanical aspects of subsurface deformation during wear have been adequately discussed earlier by Suh and Sridharan(24) and Jahanmir and Suh.(25) These analyses emphasize that it is the cyclic plastic deformation and not the uniaxial stress-strain behavior that should be considered in wear. However, detailed information on cyclic deformation of materials is not available as a function of the microstructure of metals at present. In the absence of such information, we have to be content with the uniaxial stress-strain data. Accordingly, the objective here is to investigate the

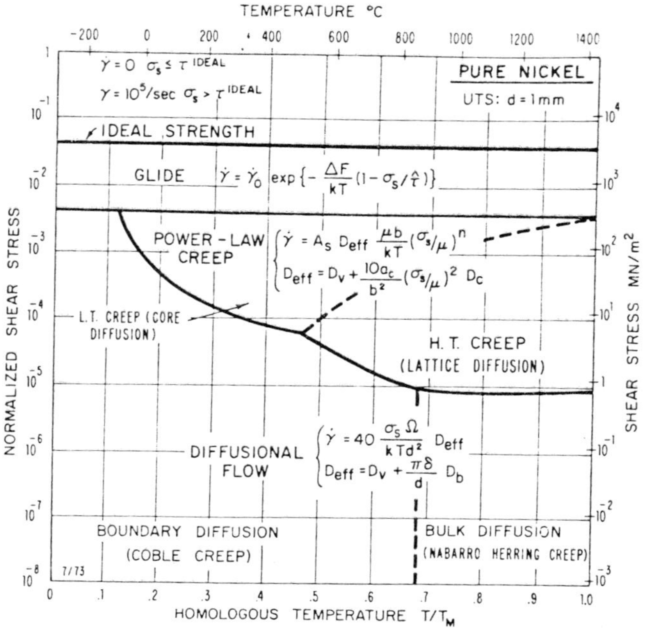

Fig. 9.—Deformation-mechanism map for nickel.[26]

influence of microstructure on the uniaxial deformation on the behavior of metals.

Professor Argon points to an excellent source of such information: the deformation mechanism maps published by Ashby and his coworkers.[26] According to Ashby, different deformation modes operate at different temperatures and strain rates, and these mechanisms in turn are microstructure dependent. The common mechanisms are dislocation glide, dislocation creep, diffusional flow (lattice and grain boundary), and grain boundary sliding (Figure 9). Complete maps are available for a variety of BCC, FCC and HCP metals, solid solutions, and even many engineering alloys. From these maps, normalized shear stress, s/G (in Ashby's notation σ_s/μ), can be obtained if we know the strain rate, $\dot{\varepsilon}$, at a given temperature or vice versa. As the normalized shear stress can be converted to the hardness of the materials, subsurface deformation rates can be calculated as a function of the hardness.

Now, it is necessary to emphasize a few aspects of these maps which were not considered in Argon's paper. First, as subsurface deformation rates (and even crack nucleation mechanisms as we shall see shortly) are dependent on the exact mechanism of deformation, it is absolutely necessary to identify the mode of deformation under wear conditions. The mechanism of deformation can be identified as follows. The size of a typical asperity is about $10\mu m$, and the sliding speed is about $1\ cm/s$ (which is typical of speeds employed in testing). Therefore, the contact time is about $1x10^{-3}s$. If the asperity deforms by a few percent (which is a conservative estimate), the strain rate is of the order of $10/s$. From the deformation mechanism maps

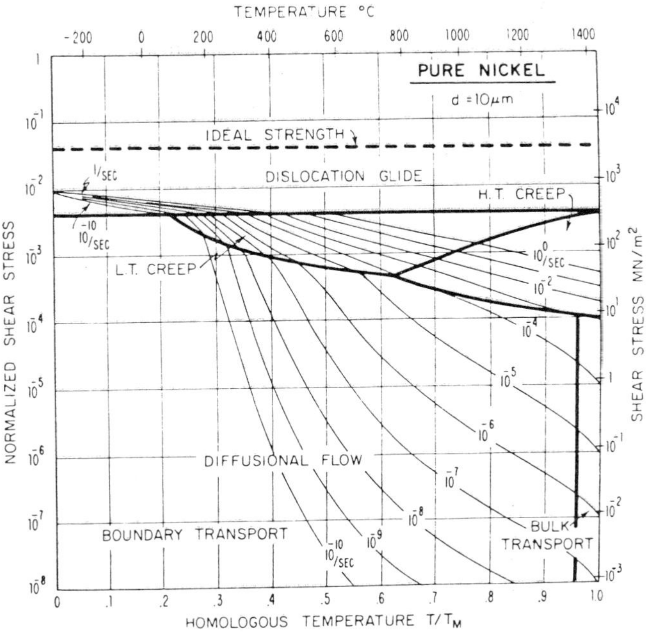

Fig. 10.—Deformation-mechanism map for nickel with strain rate contours. [26] *Note that the field boundaries are shifted when grain size is decreased.*

for nickel (Figure 10), for example, it is clear that for a strain rate of *10/s* only two mechanisms are possible. At temperatures far below the melting point, the dislocation glide is the dominant mechanism; in the temperature range 0.5TM - TM dislocation creep takes place. In fact, considering that the estimate is extremely conservative, even the operation of dislocation creep mechanism is doubtful.

The implication of the above conclusions are clear. In wear, nonlinear dislocation glide controls the subsurface deformation rates making the analysis extremely complex. And, to minimize subsurface deformation, the material should be strengthened to impede dislocation glide by methods that are used for room temperature strengthening such as solid solution strengthening, precipitation hardening, etc. Methods employed for high temperature strengthening such as grain coarsening will not be effective in reducing wear.

Another important effect to be considered is the grain size effect. It has been well-documented that small grain size metals exhibit high flow stress by the Hall-Petch effect. Accordingly, the deformation mechanism maps show large flow stress (at low temperatures) for small grain size materials. Although this effect is observed in large macroscopic tensile specimens, since the grain size of many engineering materials is larger than the asperity size, such an effect can never be realized in wear.

Finally, the texture effects are not included in these maps. (These maps are given for randomly oriented polycrystalline samples.) As remarked earlier, substantial texture develops during sliding and, therefore, texture effects are very important in HCP metals.

Nevertheless, the many advantages cited by Argon make the deformation-mechanism maps extremely useful in unifying the subsurface deformation characteristics of a variety of microstructures from $0°K$ to the melting point for a wide range of strain rates. The usefulness of these maps will become even more apparent when we consider the fracture mechanism maps shown in the next section.

B. Crack Nucleation

Argon observes that crack nucleation in ductile single-phase materials and brittle solids takes place at the surface unless the stresses below the surface reach theoretical cohesive strength of the materials. He also maintains that in heterogeneous materials, cracks can be initiated at the interfaces of the second-phase particles. At elevated temperatures, decohesion and fracture of grain boundary particles can take place.

Crack nucleation at the hard second-phase particles and even fracture of hard cementite plates have been metallographically demonstrated by Jahanmir, Abrahamson and Suh.[27] There is no question about the location of the crack nucleation sites in heterogeneous alloys: cracks nucleate at the particle matrix interface. Although subsurface cracks have been observed in commercially pure metals and solid solutions, the mechanism of crack nucleation in these materials is not clear at present. It appears that, at least in ductile metals, crack nucleation still takes place below the surface whereas in brittle materials the surface may act as a nucleation site.

In any case, recently published fracture-mechanism maps can be used to clarify some of the above issues. Fracture-mechanism maps,[28] similar to the deformation mechanism-maps, provide information about the mechanism of fracture as a function of the homologous temperature, T/T_M, and the normalized tensile stress σ_n/E. As shown in Figures 11 and 12, several mechanisms of fracture take place in metals depending on the temperature and the applied normal stress and the microstructure. These mechanisms include cohesive failure, cleavage, ductile fracture, transgranular creep fracture, intergranular creep fracture, etc. Although all these mechanisms have been observed in a uniaxially loaded tensile specimen, they probably will not operate in a wear specimen. For example, rupture is a characteristic of flow instability under tensile stresses. Therefore it should not operate in wear where the stresses are predominantly compressive. Nevertheless, the fracture mechanism maps help us to identify the various modes of fracture and hence the location of crack nucleation sites.

However, some modes of fracture mechanisms may be eliminated using the ideas presented in the section on subsurface deformation. For example, diffusional creep fracture, intergranular creep fracture due to grain boundary sliding, are not possible because these modes of deformation do not take place as discussed earlier. However, crack nucleation at grain boundary can still take place due to the deformation incompatibility even when dislocation glide and transgranular creep are the dominant mechanisms of deformation. It appears, therefore, that three fracture mechanisms are possible during wear: cleavage, ductile fracture, and intergranular creep fracture.

The energetics of crack nucleation has been discussed in Argon's paper and will not be repeated here. However, a point worth noting is that crack nucleation takes place readily in two-phase alloys which contain second-phase particles.[30] Even in the so-called pure metals, ductile fracture takes place due to the presence of small amounts of inclusions. It is not possible at present to speculate the crack nucleation mechanism in high purity metals. However, such metals may not be used in tribological systems, although in the form of coatings they have been found to be extremely wear resistant.[29] In brittle metals, cleavage is another mechanism by which fracture can take place.

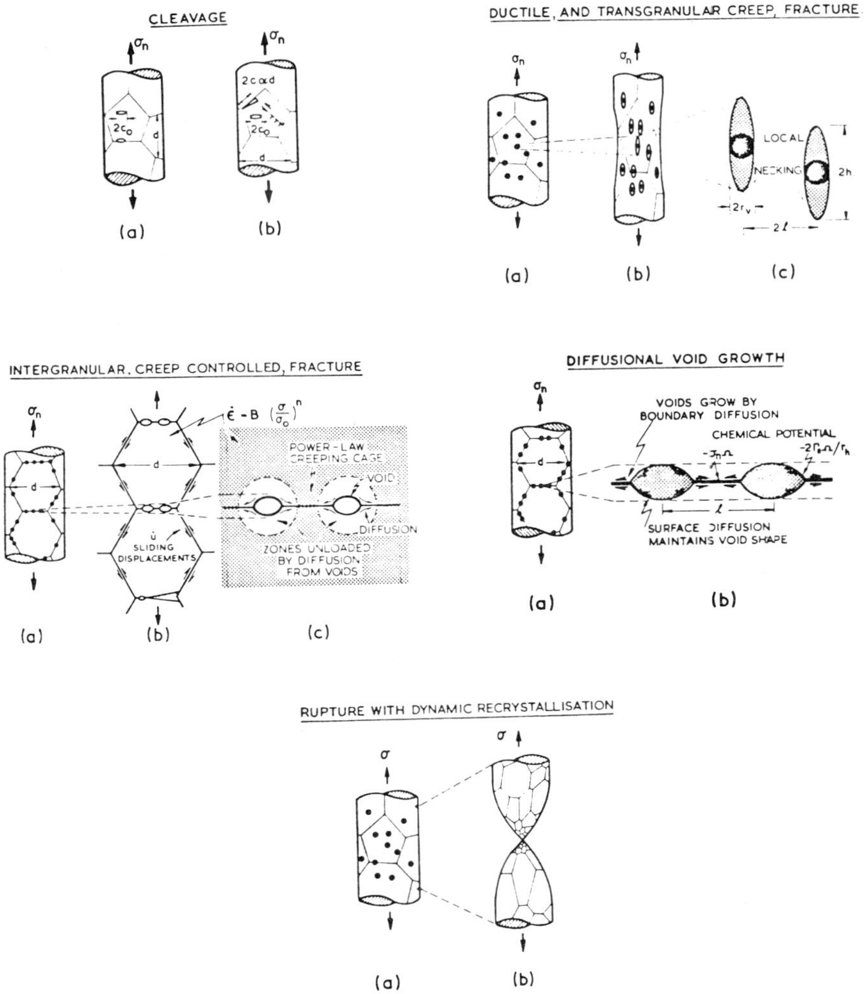

Fig. 11.—Schematic of fracture mechanisms.[28]

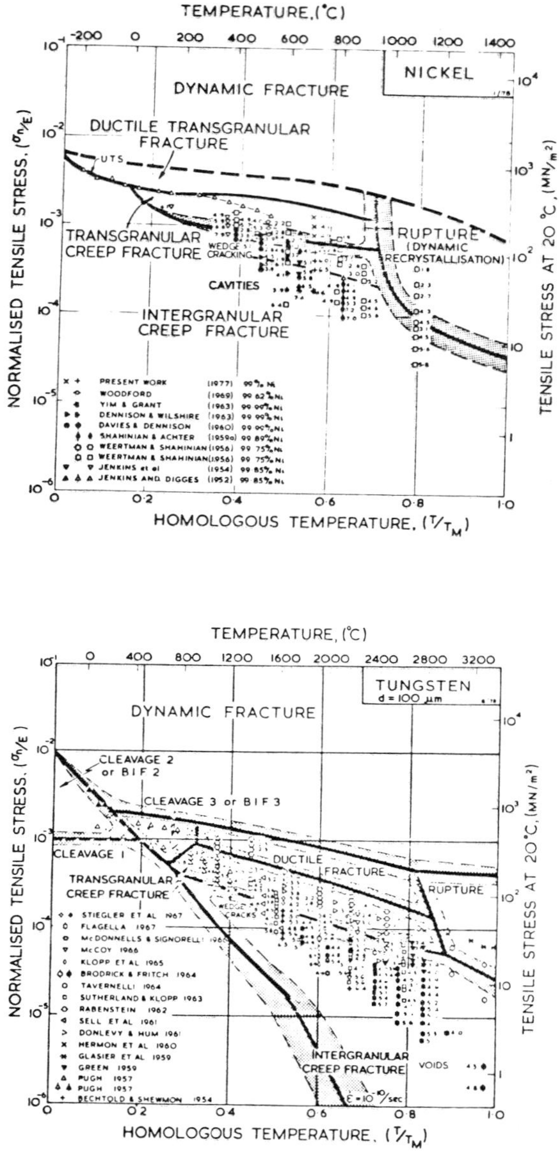

Fig. 12.—Fracture-mechanism maps for nickel and tungsten.[28]

C. Crack Propagation

Although the mechanism of crack advancement on the microscopic scale is not clearly known, a wealth of accumulated literature suggests that a power-law relation is obeyed in crack growth.

$$\frac{dc}{dn} = \beta\ (\Delta K)^m \tag{12}$$

where $\frac{dc}{dn}$ is the crack growth rate, β and m are constants, and ΔK is the change in the stress intensity factor. The crack growth aspects have been adequately treated by Argon and Ritchie in these proceedings and are not considered in detail here.

It is surprising to note that the crack growth rates are structure insensitive for ΔK values that lie in the range where Equation (12) is valid. If that is true, the crack growth rates in wear should also be structure insensitive, leaving the subsurface deformation and crack nucleation mechanisms reflect the microstructural dependence. However, as Ritchie points out, the crack growth rates are structure sensitive when the ΔK values are either too low (in the threshold range) or too high (near K_{IC}). Although some progress has been made in the calculation of ΔK values for wear,(30) much needs to be done to compute exact ΔK values. Unless exact ΔK values are known, it is difficult to point out the effect of microstructure on crack growth rates.

At high ΔK values, the crack advance mechanisms should be similar to the static fracture mode. Cleavage, void growth in ductile materials, and intergranular crack growth should be possible. Favorably oriented cleavage planes, impurity and particle loaded grain boundaries should enhance crack growth rates and the wear rates of materials. Again the fracture mechanism maps can be used to identify the crack advance mechanisms as shown schematically in Figure 13.

Even for low ΔK it has been reported that microstructure has some influence on the crack growth rates. For example, Ishii and Weertman(31) have shown that the crack growth rates depend on the stacking fault energy (SFE) of *Cu-Al* alloys. Similar effects have been observed in titanium alloys.(32)

VI. EFFECT OF MICROSTRUCTURE ON WEAR

It is clear from the previous discussion that the wear rate (or wear coefficient) is a function of three sequential processes (subsurface deformation, crack nucleation and propagation) which in turn are very complicated functions of the microstructure of materials. More often than not, many microstructural factors affect all of the three processes in the same way, thus making it impossible to isolate the exact rate controlling mechanism. Further, the friction coefficient (which is strongly dependent on the microstructure) affects all the three processes very significantly. Nevertheless, certain gross effects of structural parameters on wear can still be identified.

A. Pure Metals

Extensive work was done in the past on the friction and wear of commercially pure metals by Rabinowicz and his associates (Figure 7). It

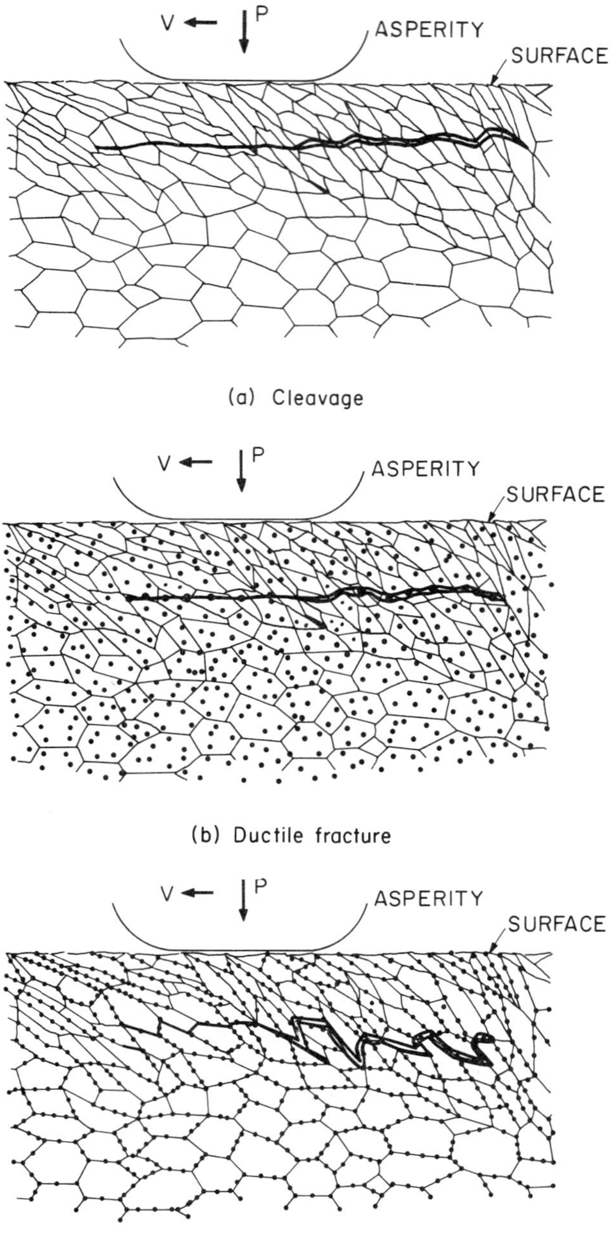

Fig. 13.—Schematic of crack advance mechanisms in wear.

appears that the hardness and friction coefficients are the basic properties that affect the wear rate of pure metals.[11-13]

As pure metals are softer than solid solutions and two-phase metals, subsurface deformation is easier. Since BCC metals are in general harder than FCC and HCP metals, the wear rate exhibited by BCC metals is generally lower than the other two crystal classes. However as the wear coefficient is a function of the friction coefficient, some hexagonal metals, due to very low friction coefficients, exhibit low wear rate as shown in Figure 14.

An important effect in pure metals, at least in FCC metals, is the effect of substructure on wear. The development of substructure depends on the stacking fault energy of metals. Low stacking fault energy metals exhibit planar slip whereas high stacking fault energy metals exhibit cross-slip and cellular dislocation structure. It has been speculated by Hirth and Rigney[33] that the cell boundaries would act as potential crack nucleation sites and propagation paths. Although these speculations seem reasonable, there is no direct proof to confirm them. Nevertheless, it has been shown experimentally by Suh and Saka[34] that high stacking faulty energy materials indeed exhibit high wear rates. (More about this will be discussed in the section on Solid Solutions.)

Grain size and grain shape do not have a large influence on the wear rate. However, texture may have some influence on the wear rates of BCC metals when the cleavage planes are favorably oriented to promote crack propagation.

In summary the overall wear rate, is determined by the subsurface deformation rate which in turn is controlled essentially by the hardness and friction coefficients. The traditional view that hard metals are wear resistant is in fact justified by pure metals only.

B. Solid Solutions

Addition of the solute atoms to pure metals generally increases the hardness without enhancing the tendency of crack nucleation. It has been remarked earlier that the friction coefficient decreases with the increase in solute content. Thus, the overall effect is that the wear rate decreases as the solute content is increased as shown in Figure 15. Both interstitial and substitutional solid solutions are found to exhibit superior wear properties.

The effects of ordering, clustering, etc. have not been studied in the past in detail, and these studies are in progress in our laboratory. The preliminary results indicate that SRO has no influence on the wear rate of materials.

As pointed out earlier, the stacking fault energy (SFE) decreases the wear rate of materials. In addition to raising the hardness of metals, substitutional solutes reduce the SFE of FCC metals substantially (Figure 16). As shown in Figure 17, the wear coefficient of metals decreases substantially by reducing the SFE. This effect is essentially due to the reduction of crack propagation rates as discussed earlier.

Embrittlement and high wear rates may be observed when the impurities segregate to grain boundaries, cell boundaries, etc. However, no definitive work has been done to date on these effects.

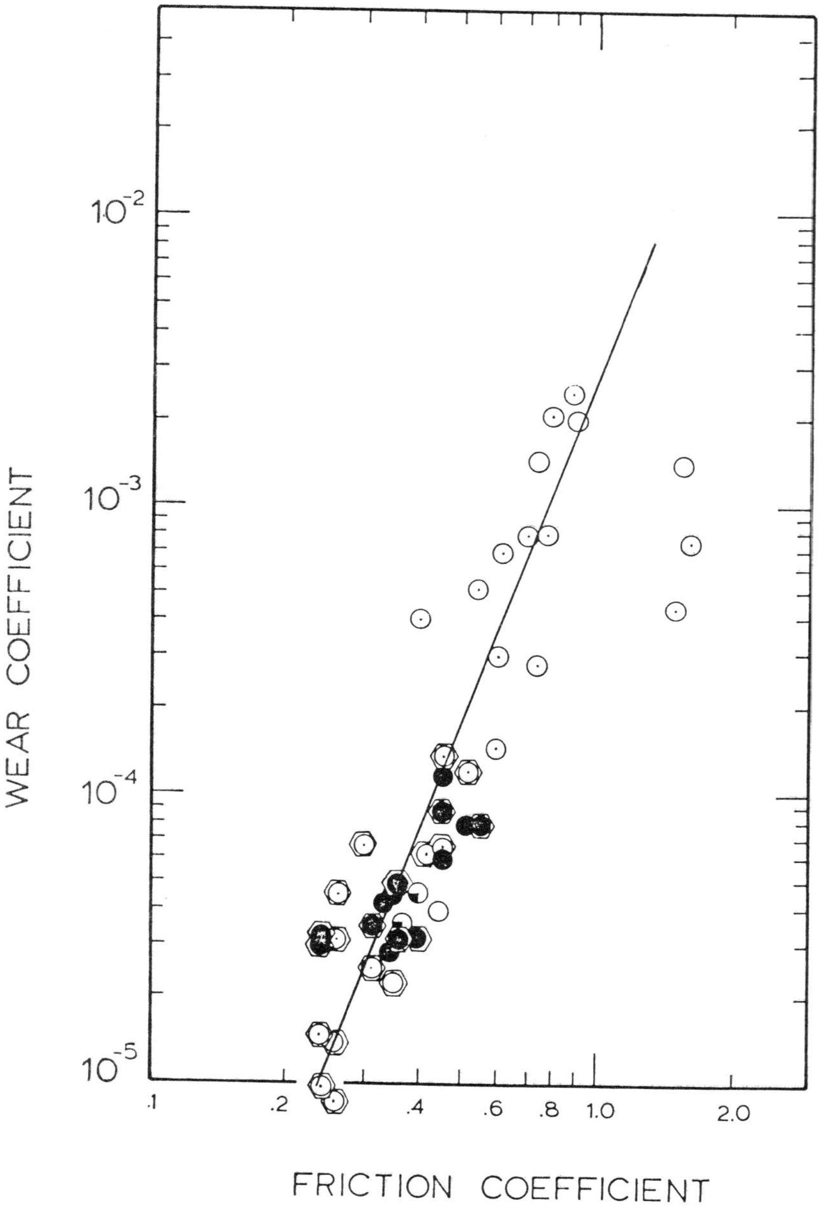

Fig. 14.—Plot of the mean wear coefficient as a function of the friction coefficient.(13)

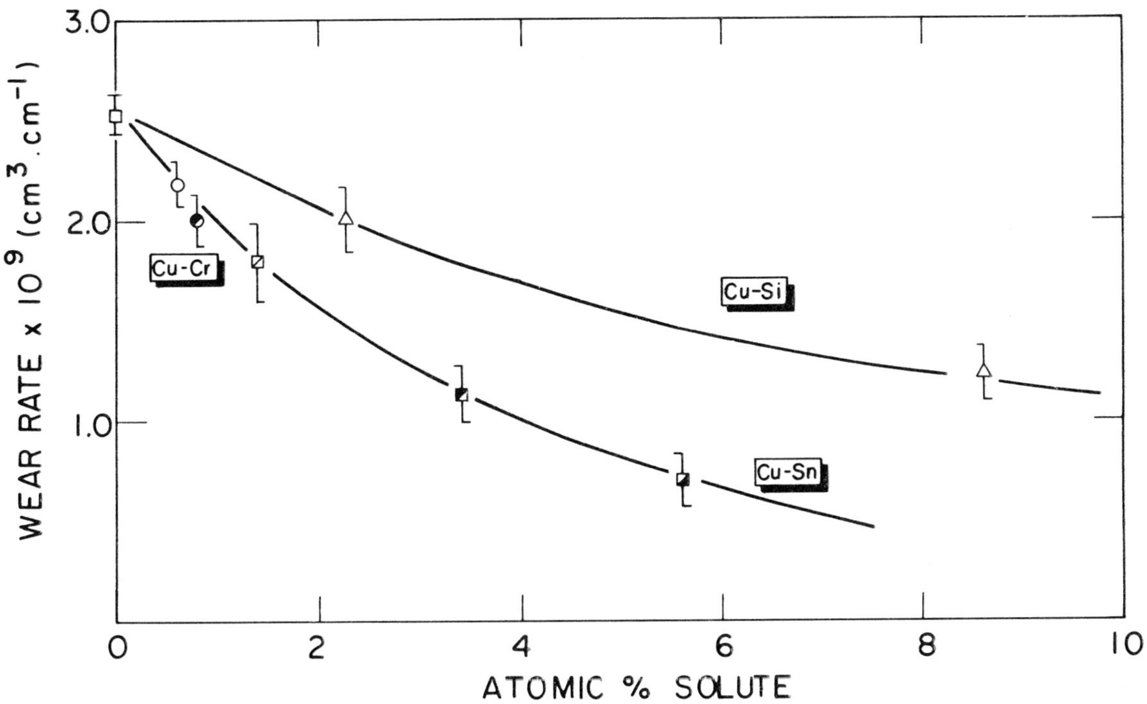

Fig. 15.—Wear rate as a function of the atomic percent solute in copper-based solid solutions.[18]

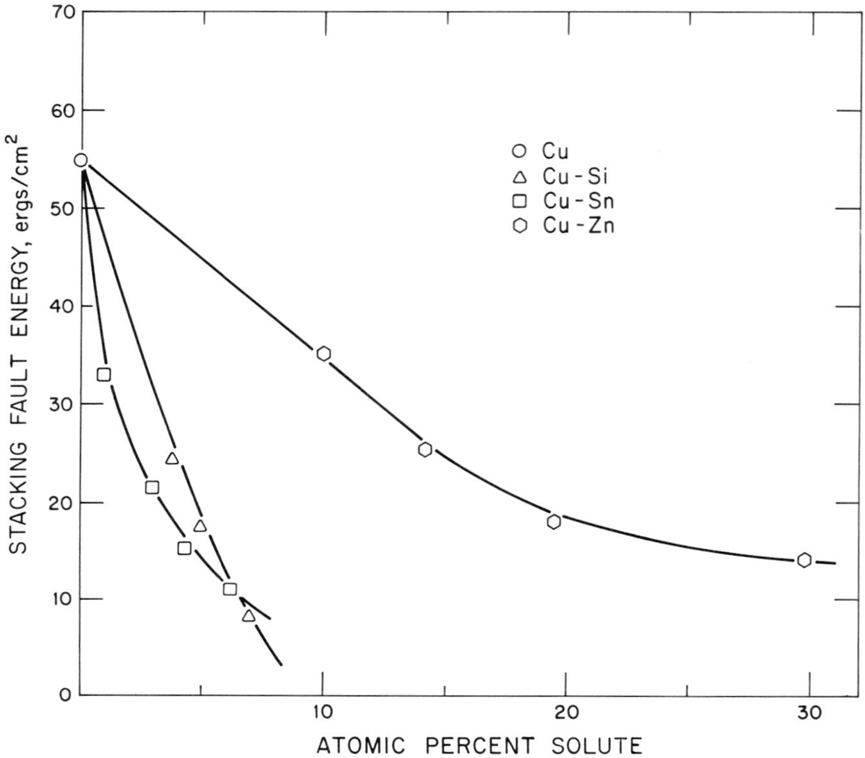

Fig. 16.—Stacking fault energy (SFE) as a function of solute conent in copper-based solid solutions.[3]

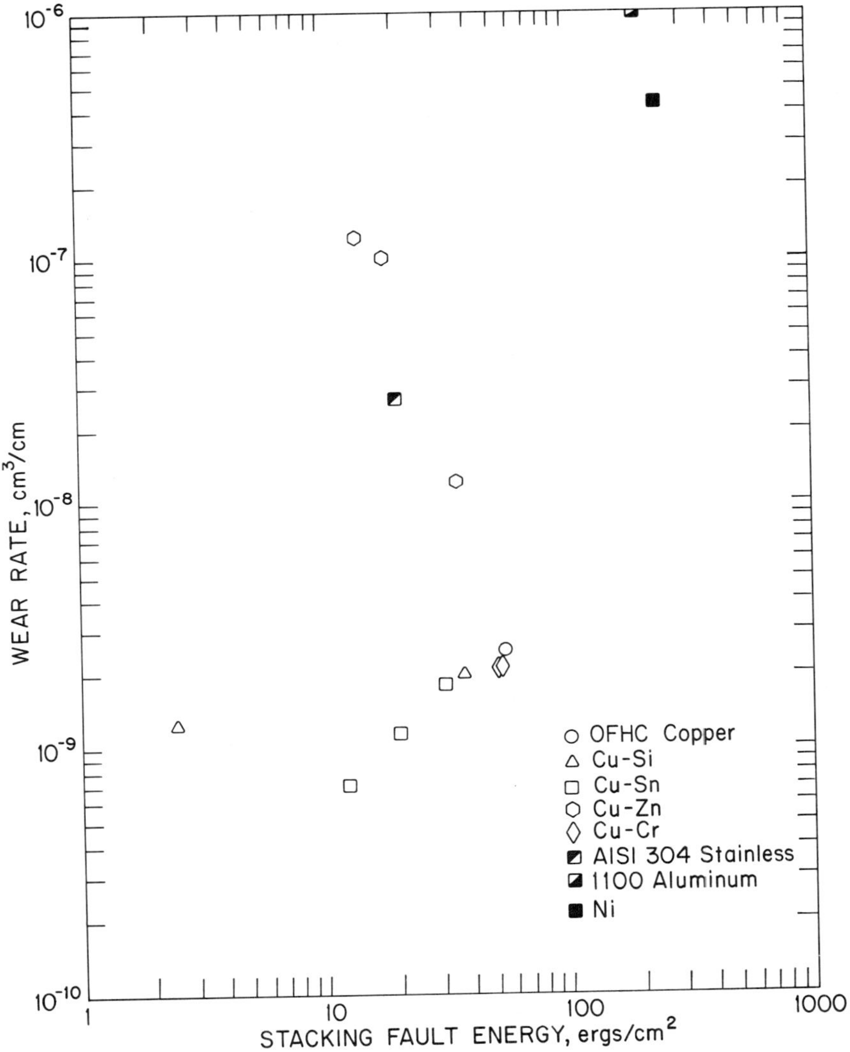

Fig. 17.—Wear rate as a function of SFE.[3]

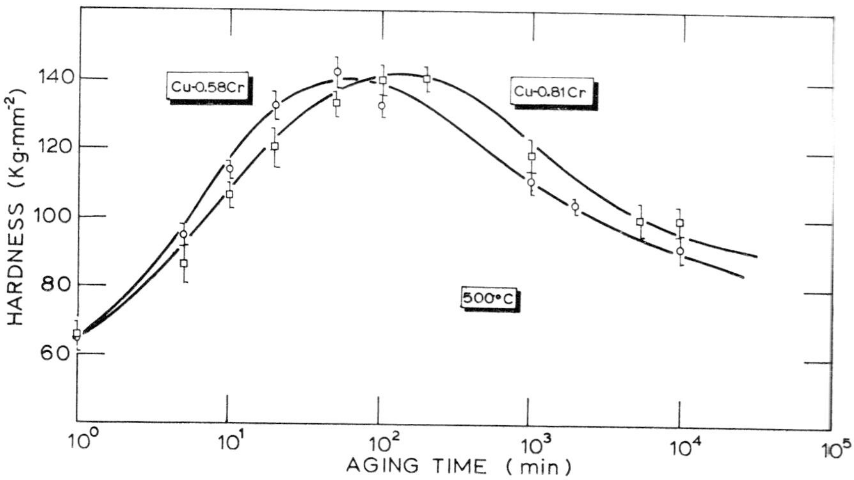

Fig. 18.—The Vickers Microhardness as a function of the aging time.[20]

C. Two-Phase Alloys

As pointed out earlier, the friction coefficient of two-phase alloys is independent of the volume fraction, size, and shape of the second-phase particles. This is an important effect, for it enables us to clearly identify the rate controlling mechanisms in wear. The mechanics of crack nucleation has been worked out by Jahanmir and Suh.[25] Basically, from a microstructural point of view, we need to describe the effects of particle morphology only.

When the mean free path of the particles is decreased, the hardness increases and the subsurface deformation rate decreases. But the wear rate may increase or decrease depending on the particle size and coherency because these parameters affect crack nucleation rate. The coherency and size effects have been investigated by Saka, Teixeria and Suh[20] in *Cu-Cr* alloys and by Jahanmir, Suh and Abrahamson in Steels.[27] As shown in Figure 18, as the aging time increases, the hardness increases due to a decrease in mean free path, and the wear rate and wear coefficient decrease (Figure 19). This was interpreted as due to the difficulty of crack nucleation at the particle-matrix interface when the particles are coherent. However, when the aging time is increased, the wear rate increases even though the hardness increases; and with further aging, the wear coefficient remains constant. These effects are interpreted as due to loss of coherency and increase in the size of the particles.

When the particles are large (*>100Å*), the coherency is lost and the wear rate of two-phase metals increases even if the hardness increases. Age-hardenable *Cu-Cr* alloys (Figure 20), spherodized steels (Figure 21), and oxide dispersion strengthened copper alloys (Figure 22) indicate that as the volume fraction of the second-phase particles is increased, the wear

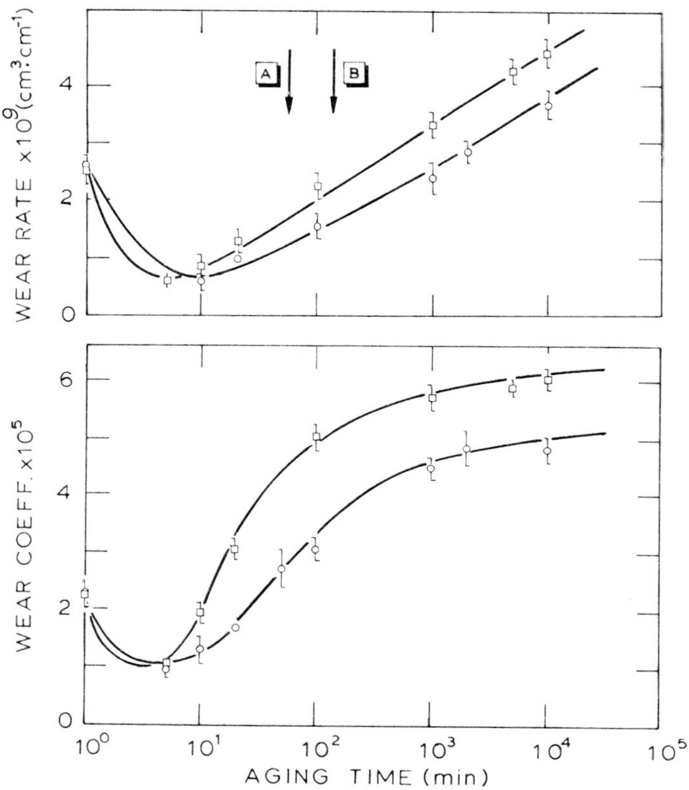

Fig. 19.—Wear properties of precipitation-hardened Cu-Cr alloys as a function of aging time.[20] *The arrows indicate the aging time at peak hardness.*

resistance decreases exponentially. In fact, the variation of the wear resistance (Figures 20-22) is exactly similar to the variation of ductility with respect to volume fraction of the particles[35] (Figure 23). However, this does not imply that ductility always correlates well with wear rates of metals.

The effects of particle shape on the wear rate is not clearly known at present although experimental evidence suggests that cracks can be nucleated by particle fracture when the aspect ratio is very large (such as cementite plates in steels).

D. Composites

Although much work has been done on polymeric composites,[22,23] experimental data on metal matrix composites is scarce. However, certain generalizations can be made based on the evidence available for polymeric composites. In the case of particulate composites, crack nucleation should

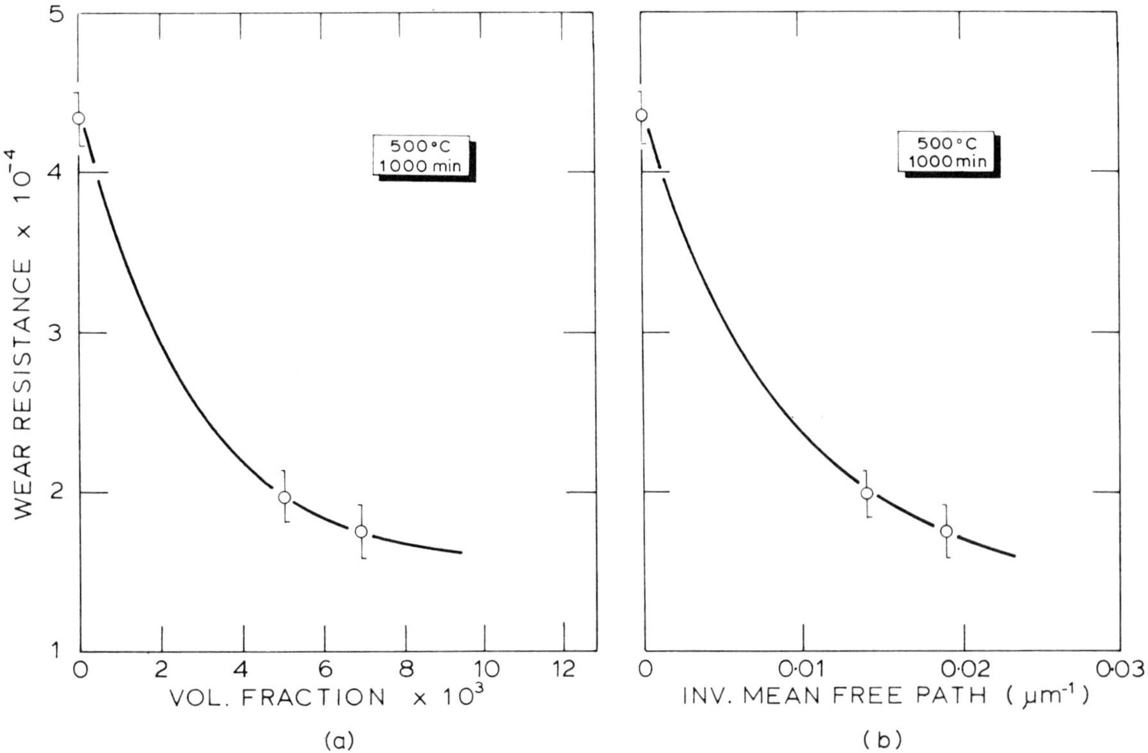

Fig. 20.—Wear resistance (reciprocal of wear coefficient) vs. the volume fraction and the inverse mean free path of Cu-Cr alloys.[20]

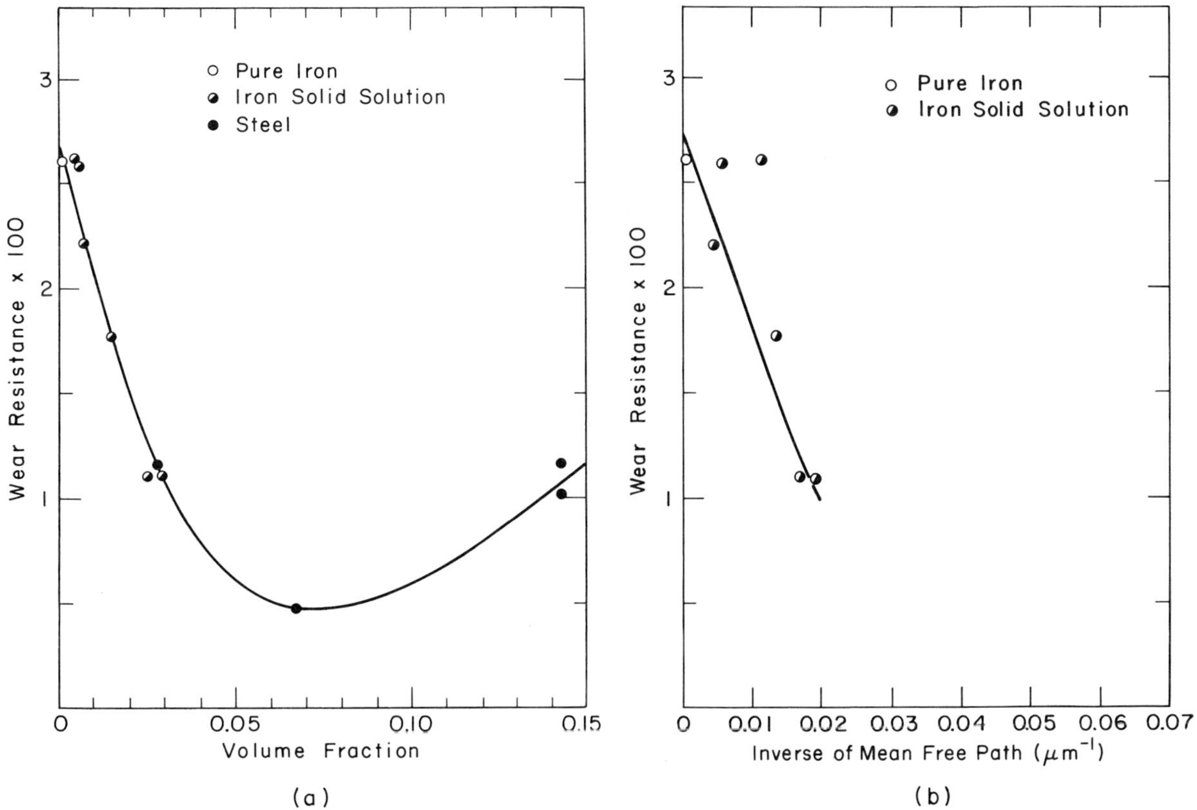

Fig. 21. —The effect of volume fraction and mean free path on the wear resistance of iron and spherodized steels.[27]

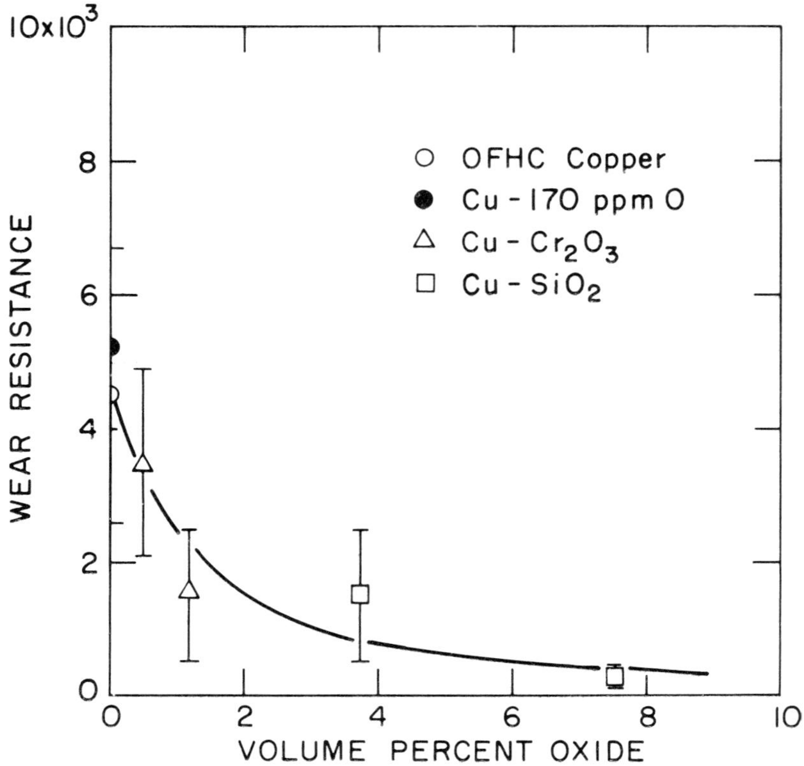

Fig. 22.—Wear resistance as a function of the volume percent oxide in internally oxidized copper-based alloys.[19]

take place readily, but the wear coefficient can either be high or low depending on the hardness.

Experimental results of boron reinforced aluminum composites[36] indicate that the wear rate is dependent on the orientation of the fibers with respect to the sliding direction. When the fibers are oriented normal to the contact interface, the wear rates are found to be very low. When sliding takes place in the plane of fibers and normal to the fiber axis, the fibers are pulled out by debonding. As debonding, fiber pull-out, etc. are crack nucleation and growth processes, the fiber-matrix interface and the orientation of the fiber with respect to sliding direction are important aspects. Abrasion may also take place during the sliding wear of composites because the fibers are generally hard and have high aspect ratio.

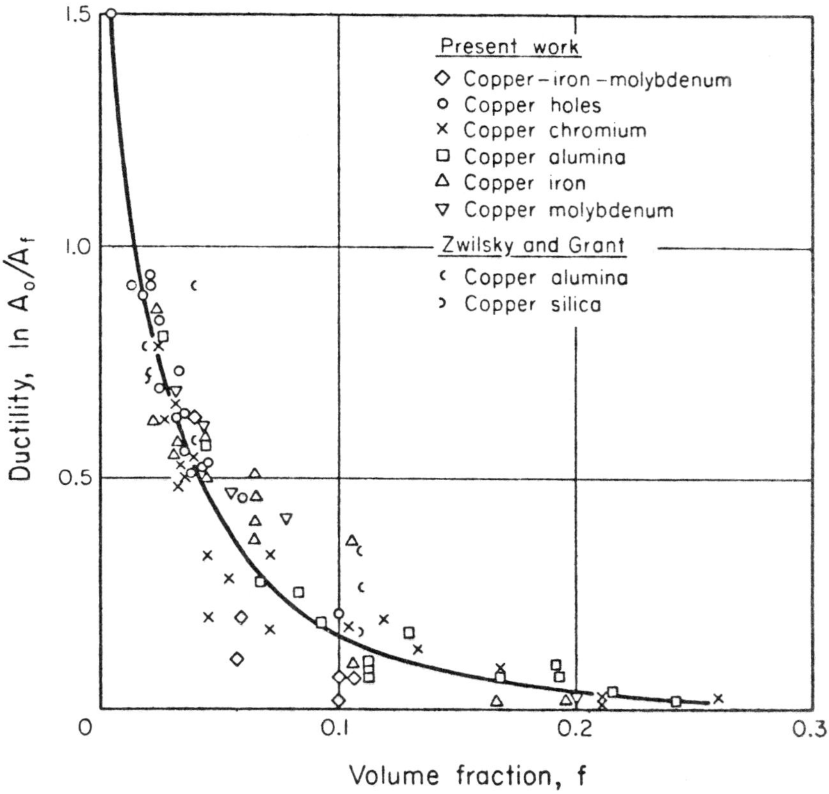

Fig. 23.—Effect of volume fraction of second-phase particles on ductility for various copper alloys.[35]

VII. OTHER MICROSTRUCTURES

As mentioned earlier, the classification of microstructures based on the binary phase diagram excludes many important engineering materials from consideration. Metallic glasses, metastable phases, cast alloys, etc. belong to this category. One complication with these structures is that they may transform to stable microstructures during testing because of frictional heating. Nevertheless, a study of these structures is important as some of them may possess the optimum microstructure for wear minimization. Structures produced by laser and electron beam melting should also be studied as these methods have many advantages over the conventional induction hardening and flame hardening processes.

VIII. SUMMARY

The objective of this paper is to emphasize the importance of the microstructure of metals on friction and wear. So far, the microstructural effects have been largely ignored by many tribologists due to the strict adherence to the adhesion theory of wear. The delamination theory now points the way to the study of tribological behavior of metals as a function of the microstructure.

The effects of crystal structure, grain structure, solute atoms, second-phase particles, etc. have been discussed in terms of friction, subsurface deformation, crack nucleation, and crack propagation processes that lead to wear particle generation. Although some important effects have been uncovered in recent years, much is desired to be understood -- especially in the case of composites and metastable phases. Only then it would be possible to design wear-resistant materials for sliding applications on a rational basis.

IX. ACKNOWLEDGEMENTS

Many results quoted in this paper were obtained through the research sponsored by the Defense Advanced Research Projects Agency (DARPA) and the Office of Naval Research (ONR). The author is grateful to Drs. Arden L. Bement and Edward C. Van Reuth of DARPA, and Dr. Richard S. Miller and Commander Harold P. Martin of ONR for their support and encouragement. The author is also grateful to Mr. Ming Tse for many helpful comments on the manuscript.

REFERENCES

1. Argon, A.S., These Proceedings.
2. Suh, N.P., *Wear,* Vol. 25, 1973, p. 111.
3. Suh, N.P. and Coworkers, "The Delamination Theory of Wear," Elsevier Sequoia S.A., Lausanne, 1977.
4. Suh, N.P., in "Fundamental Aspects of Structural Alloy Design," edited by R.I. Jaffee and B.A. Wilcox, Plenum Press, New York, 1977, p. 565.
5. Dehoff, R.T. and Rhines, F.N., eds., "Quantitative Microscopy," McGraw-Hill, New York, 1968.
6. Underwood, E.E., "Quantitative Stereology," Addison-Wesley, Reading, Massachusetts, 1968.
7. Tohkai, M., S.M. Thesis, Department of Mechanical Engineering, MIT, 1978.
8. Rabinowicz, E., "Friction and Wear of Materials," John Wiley, New York, 1965, pp. 66, 68.
9. McClintock, F.A. and Argon, A.S., "Mechanical Behavior of Materials," Addison-Wesley, Reading, Massachusetts, 1966, p. 665.
10. Rabinowicz, E., "Friction, Wear and Lubrication, Study Guide," MIT, Cambridge, Massachusetts, 1977.
11. Rabinowicz, E., *American Society of Lubrication Engineers Transactions,* Vol. 14, 1971, p. 198.
12. Ohmae, N. and Rabinowicz, E., Preprint No. 78-LC-2C-1, the ASLE/ASME Lubrication Conference, Minneapolis, Minnesota, October 1978.
13. Rabinowicz, E., in the Proceedings of the International Conference on Wear of Materials, St. Louis, Missouri, 1977, edited by W.A. Glaeser, K.C. Ludema and S.K. Rhee, American Society of Mechanical Engineers, New York, 1977, p. 36.
14. Buckley, D.H., *American Society of Lubrication Engineers Transactions,* Vol. 11, 1968, p. 11.
15. Buckley, D.H. and Johnson, R.L., *Wear,* Vol. 11, 1968, p. 405.

16. Fleischer, R.L., in "The Strengthening of Metals," edited by D. Peckner, Reinhold, New York, 1964.
17. Murr, L.E., "Interfacial Phenomena in Metals and Alloys," Addison-Wesley, Reading, Massachusetts, 1975.
18. Pamies-Teixeira, J.J., Saka, N. and Suh, N.P., *Wear,* Vol. 44, 1977, p. 65.
19. Saka, N. and Suh, N.P., *Journal of Engineering for Industry,* Trans. ASME, Vol. 99, Series B., No. 2, 1977, p. 289.
20. Saka, N., Pamies-Teixeira, J.J. and Suh, N.P., *Wear,* Vol. 44, 1977, p. 77.
21. Suh, N.P., Sin, H-C. and Saka, N., These Proceedings.
22. Lancaster, J.K., *Tribology International,* Vol. 5, No. 6, 1972, p. 249.
23. Sung, N-H. and Suh, N.P., *Wear,* Vol. 53, 1979, 129.
24. Suh, N.P. and Sridharan, P., *Wear,* Vol. 34, 1975, p. 291.
25. Jahanmir, S. and Suh, N.P., *Wear,* Vol. 44, 1977, p. 17.
26. Ashby, M.F. and Frost, H.J., in "Constitutive Equations in Plasticity," edited by A.S. Argon, MIT Press, Cambridge, Massachusetts, 1975, p. 117.
27. Jahanmir, S., Abrahamson, E.P., II and Suh, N.P., in Proc. 3rd North American Metal Working Research Conf., Carnegie Press, 1975, p. 854.
28. Ashby, M.F., Gandhi, C. and Taplin, D.M.R., *Acta Metallurgica,* Vol. 27, 1979, p. 699.
29. Jahanmir, S., Abrahamson, E.P., II and Suh, N.P., *Wear,* Vol. 40, 1976, p. 75.
30. Fleming, J.R. and Suh, N.P., *Wear,* Vol. 44, 1977, p. 39.
31. Ishii, H. and Weertman, J., *Metallurgical Transactions,* Vol. 2, 1971, p. 3441.
32. Stoloff, N.S. and Duquette, D.J., *CRC Critical Reviews in Solid State Sciences,* Vol. 4, 1974, p. 615.
33. Hirth, J.P. and Rigney, D.A., *Wear,* Vol. 39, 1976, p. 133.
34. Suh, N.P. and Saka, N., *Wear,* Vol. 44, 1977, p. 135.
35. Edelson, B.I. and Baldwin, W.M., Jr., *American Society for Metals Transactions,* Vol. 55, 1962, p. 230.
36. Saka, N., Unpublished Work.

DISCUSSION

B. WILCOX, National Science Foundation: I am interested in your plot which shows how the volume fraction of the second-phase particles influences the wear behavior. Is it a ductile fracture mechanism?

N. SAKA: Yes, it is ductile fracture essentially. The cracks nucleate at the second-phase particles. The more the number of particles, the more will be the crack nucleation sites. The cracks do not have to travel too far before they link up with the neighboring cracks to form large wear sheets.

WILCOX: In an another figure you showed the old Edelson-Baldwin plot of ductility as a function of volume fraction. Is it basically the same mechanism?

SAKA: Yes, it is the same mechanism. That is why the wear rate of dispersion strengthened alloys correlates very well with the Edelson-Baldwin plot.

D. RIGNEY: I just have one brief comment. You showed some correlations including one with stacking fault energy. I would just like to say that whenever one shows a simple correlation he has to be extremely careful in that he holds all other properties constant. On that stacking fault energy plot, for example, there was tremendous scatter. All kinds of materials are there, some of which, for example, have very different hardnesses. There have been very few works in the wear field in which a series of materials were compared making some effort to hold several things approximately constant. I can think, for example, of work by Moore on abrasion and also some recent work by Boas and Rosen in which they tried to hold structure and hardness approximately constant. More of that type of work is needed. The single correlations are fine, but you have to be careful to see that all the other factors are held constant.

SAKA: That is absolutely true and nobody would dispute it. In fact I have mentioned in the paper that when solute atoms are added to change stacking fault energy, changes in surface energy and substantial changes in friction coefficients may be brought about. Further, some times these solute atoms can segregate, for example, to the grain boundaries. That can be a source of large scatter in the wear coefficient as a function of stacking fault energy. Nevertheless, there appears to be a direct correlation between the stacking fault energy and the wear coefficients of materials. However, it still remains to be investigated whether crack nucleation or propagation controls wear in these materials.

Investigation of wear behavior of materials that have nearly constant microstructure and properties while changing one parameter at a time is almost an impossible task. There is no way that one can alter the microstructure for one property without altering other properties. After all, the properties we are dealing with are all structure dependent. Understanding wear should be pursued not just by careful design of experiments only but by clever interpretation of the results also.

IV. CHEMICAL PROPERTIES

DEFINITION AND EFFECT OF CHEMICAL PROPERTIES OF SURFACES IN FRICTION, WEAR AND LUBRICATION

D. H. Buckley

ABSTRACT

Much of the data on the properties of surfaces that have been used in the past in analyzing, interpreting and predicting adhesion, friction and wear behavior for solid surfaces is now suspect. With the advent of analytical surface tools, careful and complete characterization of surfaces indicate that very frequently the outermost layers of solid surfaces are markedly different in chemistry than had previously thought. These layers, as will be shown, are extremely important in adhesion, friction and wear behavior. Some of the properties to be discussed in the paper relative to their role in adhesion, friction, wear and lubrication will include: (1) adsorption, both physical and chemical; (2) orientation of the solid as well as the lubricant; (3) surface energy; (4) surface segregation; (5) surface versus bulk metallurgical effects; (6) electronic nature of the surface; and (7) bonding mechanisms.

INTRODUCTION

The properties of surfaces are extremely important in the adhesion, friction, wear and lubrication of materials. This fact was recognized by Sir William Hardy(1) over a half a century ago! It has, however, only recently been possible to define and fully characterize these surfaces. The advent of special surface tools has assisted in the characterization. These include: field ion microscopy (FIM), the atom probe, low energy electron diffraction (LEED), Auger emission spectroscopy analysis (AES), electron spectroscopy for chemical analysis (ESCA), ellipsometry and scanning electron microscopy (SEM).

Through the years from Hardy to the availability of these tools, considerable research has been conducted relative to the physics, chemistry, and metallurgy of solid surfaces, in general, and more specifically as related to tribological systems. In light of the current identification of the real nature of solid surfaces many of the concepts, mechanisms and theories previously held may have to be modified or discarded. This may be particularly true where bulk properties have been used to predict surface behavior.

The objective of this paper is to review the importance of surface effects and the need for a careful definition of solid surfaces in friction, wear and lubrication. While in the past the physicist, chemist, and metallurgist had well defined areas of activity with regard to the behavior of materials it will become evident that the chemistry of surfaces to which this paper is devoted will involve, of necessity, physics and metallurgy. Some of the properties to be discussed in the paper relative to their role in adhesion, friction, wear and lubrication will include, adsorption, both physical and chemical, orientation of the solid as well as the lubricant, surface energy, surface segregation, surface versus bulk metallurgical effects, electronic nature of the surface and bonding mechanisms.

REAL SURFACES

Until approximately ten years ago it was very common to find in classical texts on surface chemistry a nearly complete absence of the characterization of solid surfaces with which gases and liquids would interact.(2) A wealth of literature has been developed through the years concerning adsorption to solid surfaces, particularly chemisorption. Again, very little attention has been paid to the nature of the surface of the solid involved. Frequently M was used to designate the metal involved where adsorption studies were conducted with metals.(3,4)

In the field of tribology, because of the importance of surfaces, the presence of oxides and adsorbates has been recognized.(5) Despite this recognition it has been a common practice to consider solid surfaces, which are usually metals, as reacting directly with the lubricants, indicated by way of example in References 6 to 10. Frequently metal powders, particularly iron have been used as the adsorbing surface for the lubricant.(6,7) Metal powders have, however, been identified as being poor relative to gas adsorption; they are poor with respect to both surface cleanliness and characterization.(11) The same may be said with regard to liquids.

If the surface of high purity iron (vacuum zone refined) is examined with Auger emission spectroscopy it is found to contain more than simply iron as indicated in Figure 1(a). In Figure 1(a) in addition to iron, oxygen, carbon, and sulfur are detected on the surfaces. These elements are stable on the surface and resist techniques such as heating to 500°C to achieve their removal. The only method found effective in accomplishing the removal of all the extraneous elements was argon ion bombardment.

Figure 1(b) is an Auger emission spectrum for the iron surface after argon ion sputter cleaning. The only element detected is iron. The surface atoms of the iron however have been strained by the ion bombardment. If the clean iron surface is then heated to 500°C to anneal the surface, carbon from the bulk diffuses to the surface and contaminates the surface. A LEED pattern is presented in Figure 2(a) for this surface.

In Figure 2(a) the four brightest spots in a rectangular array are due to the iron (011) crystal surface. The diffuse and less intense spots appearing in a circular pattern are due to the carbon contamination. The Auger spectrum of Figure 3 indicates that the surface contaminant is carbon.

The removal of the carbon by argon ion bombardment results in the LEED pattern of Figure 2(b). The diffraction spots are diffuse and elongated. Heating at 200°C for a very short period of time results in the LEED pattern of Figure 2(c) for the clean annealed iron (011) surface.

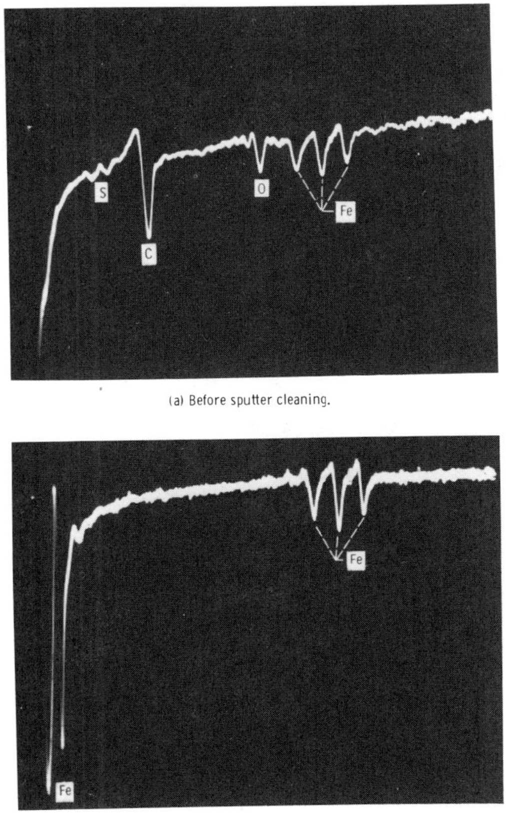

(a) Before sputter cleaning.

(b) After sputter cleaning.

Fig. 1.—Auger spectra for an iron surface before and after sputter cleaning. High purity iron vacuum zone refined.

The interaction of even the simplest of elements with the clean iron surface becomes very involved and must be carefully followed to understand the mechanisms of interaction. This is accomplished with the aid of surface analytical tools. It has been very effectively done in the study of the chemisorption and chemical reaction of oxygen with iron.[12-14]

The surfaces of most metals are not too different from that observed for iron and usually there are a number of elements present on the surface in addition to those of the metal. The role of these elements on the interaction of lubricants with the solid surface is not fully understood but some of these reactions will be discussed later in reference to gas-solid surface interactions.

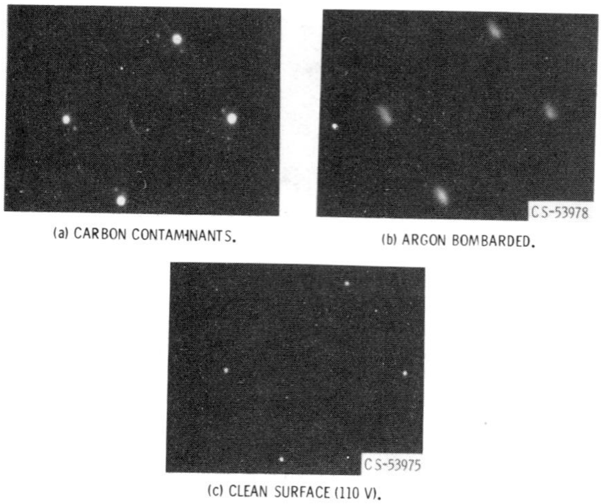

Fig. 2.—LEED patterns of iron (011) surface with carbon present and after argon iron bombardment.

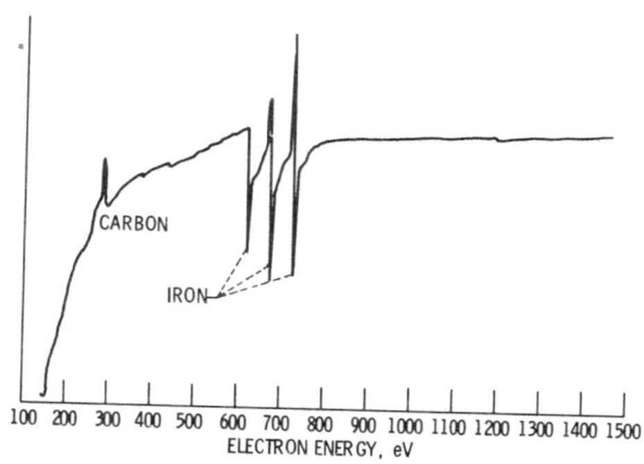

Fig. 3.—Auger spectrometer trace of iron (011) surface with carbon present on surface.

SURFACE ENERGY

If one cleaves a crystalline solid along its cleavage plane, two highly chemically active surfaces are generated. The cleavage process causes the fracture of cohesive bonds across the cleavage interface and these fractured bonds leave the surface in a highly energetic state. The energy of the surface will be dependent on both the elemental nature of the bonds broken and the coordination number of the atoms in the resultant surface layers. As a result, surface energy will be a function of the material[15] as well as the surface orientation.

There is no question but that surface energy is important in the tribological behavior of materials. It will influence adhesive bonds for solids in contact and hence friction and adhesive wear. In addition it will determine the nature of the interaction of lubricants with solids. The lubricant may either: (1) physically adsorb; (2) chemisorb; or (3) undergo decomposition, as has been observed for some hydrocarbons with a clean metal surface.[22] Surface energy has been used in the formulation of an adhesive wear mechanism.[23]

While surface energy can be very helpful in understanding the adhesion, friction, wear and lubrication behavior of materials, its present usefulness is very limited. The principle restriction has been the inability to obtain accurate experimental surface energy values.

An examination of the surface energy literature reveals wide disparities in reported values for any one material. Table I indicates the minimum and maximum surface energy values which can be found in the literature for some of the elemental metals. These data were taken from a summary by Wawra.[15]

*TABLE I.—VARIATION IN VALUES OF REPORTED SURFACE ENERGIES**

Element	Surface energy, ergs/cm^2 max	min	Temp., °c
Cu	4258	950	-273
Ag	2493	600	-273
Au	2540	590	-273
Fe	5267	1980	-273
Ti	2730	1330	25, -273
Cr	4061	1515	-273
W	9410	1497	3370

* H. H. Wawra, Radex R. H., 4(1973)602.

While the broad range of values obtained are of concern, the fact that, for example, the minimum to maximum for some metals such as iron and chromium fall within the range found for tungsten are of even greater concern. It would be difficult, based upon reported experimental data, to identify differences in the surface energy for iron, chromium and tungsten.

As has already been indicated, the surface energy of solids such as metals is sensitive to crystallographic orientation. Most researchers conversant in the subject of surface energy readily agree that this is the case. Differences arise, however, when actual results are compared. The research results of three different investigators who have measured the surface energies for various planes of face centered cubic metals are presented in Table II. The results are presented as the ratio of the surface energies for the various planes over that for the (111) surface. The results of Table II indicate that not only does the value vary with the investigator, but more importantly, the relative order of the metals as well.[24]

TABLE II.—STRUCTURAL DEPENDENCE OF SURFACE ENERGY ON FACE-CENTERED CUBIC METALS

Plane	Au, Ag, Cu, Ni	Winterbottom & Gjostein[21] Au	Mykura[18] Ni
(111)	1.00	1.00	1.00
(100)	1.047	1.072	0.95
(311)	1.119	1.065	1.00
(110)	1.15	1.047	1.01
(210)	1.16	1.055	1.00

One of the most significant reasons for the wide disparity in the surface energy values reported by various investigators has been inadequate control over the impurities in the materials. Small concentration of impurities in the bulk of a metal can markedly alter the measured surface energy of a material. This is indicated in the data of Figure 4 for sulfur in iron. With an increase in concentration of sulfur, there is an accompanying decrease in surface energy, Figure 4.

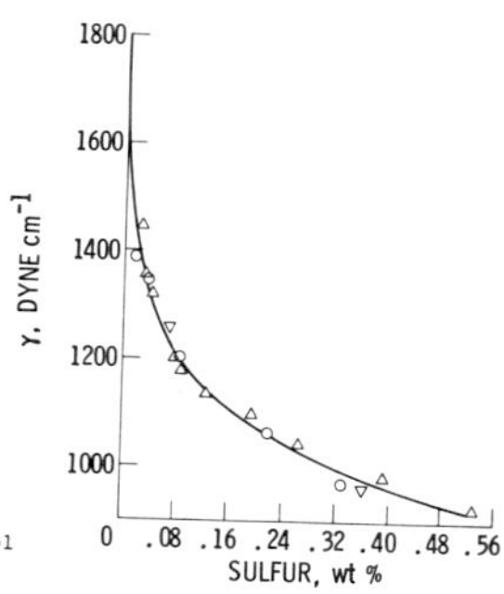

CS-77-151

Fig. 4.—Dependence of surface energy of liquid iron on sulfur content (ref. 17).

Extremely small concentrations of bulk contaminant in a metal such as iron can have a pronounced effect in contaminating a surface. For example, as little as 8 ppm of carbon in iron will diffuse to the surface, segregate there and contaminate it.[25] This segregation will undoubtedly affect measured surface energies.

The use of high purity materials and the careful characterization of solid surfaces should result in the acquisition of meaningful surface energy values in the future. Surface analytical tools are currently being used for the needed surface characterization. One which has proven especially useful in this regard is the field ion microscope.[20,26,27] When used in conjunction with the atom probe, its contribution will be enhanced. It gives the atom by atom structural arrangement on a solid surface, and with the atom probe, an atom by atom chemical analysis.

ALLOY CHEMISTRY

In practical lubrication systems the mechanical components in solid state contact are most frequently alloys rather than elemental metals. The composition of these alloy surfaces are important in considering the chemical interactions of such solids with other solids, with gases and with lubricants. Even where elemental metals are used, the surfaces of these metals may have compositions entirely different from the bulk which results from impurity segregation.

The field of tribology contains a number of excellent texts. Many of these books do not, however, discuss the real nature of the surface to be lubricated (see, for example, Refs. 23, 28, 29 and 30). Much attention is given to the chemistry of the lubricant, but little to the chemistry of the alloy surface to be lubricated. In fairness to the authors of these texts the identification of these surfaces with surface tools just began to emerge at the time these texts were written. Future texts on the subject should, however, not neglect the importance of the metal or alloy surface chemistry and the interaction of the lubricant with that surface chemistry.

With elemental metals, the effect of small concentrations of impurities such as parts per million carbon in iron have been shown to affect surface chemistry. This was discussed earlier in this paper. Similar effects have been observed with other impurity elements in a number of different metals. Surface segregation has been noted for oxygen in platinum,[31] phosphorus in iron,[32] sulfur and carbon in nickel,[33] sulfur in molybdenum,[34] carbon in nickel,[35] and even sodium in lithium.[36] The foregoing are only a few examples. Other impurity effects have been studied and still other systems, it is certain, will be studied. With some of these impurities the metal surface is completely covered by the contaminant which can diffuse from the bulk and even form compounds with the metal itself at the test surface. These compounds do not exist in the bulk.[34]

In lubrication systems, the rubbing off of surface oxides and adsorbates because the solid surfaces are in contact under relative motion, is a common occurrence. Frictional heating of the surface layers can promote the diffusion of impurities to the surface under such conditions. The presence of impurities in metals is therefore important to surface chemical behavior of the materials in solid state contact. A foreign atom which may simply be an impurity in the bulk can be an alloy constituent of the surface or compound with it.[34]

When elements are alloyed, the segregation of one element to the surface can occur and its concentration at the surface can exceed that in the bulk alloy. Thus, for a given alloy the surface metallurgy can differ appreciably from the bulk metallurgy. Such surface enrichment has been observed with a host of systems. A few examples of such systems include nickel in iron,[37] silver in palladium,[38] gold in copper,[39] copper in nickel,[40] silver in gold,[41] aluminum in copper,[42] tin in copper,[43] aluminum in iron,[43] and platinum in osmium.[44]

The amount of material which can be present on the surface of alloys relative to the bulk concentration of the alloying element can be appreciable. This is indicated in the data of Table III for aluminum and tin segregating to the surface of copper and aluminum to the surface of iron. The surface concentration ranges from approximately 3 to 15 times the bulk concentration of the alloying element.

TABLE III.—MAXIMUM COVERAGE OF MINOR CONSTITUENT ON ALLOY SURFACES

Alloy	Ratio of surface concentration to bulk concentration	Atomic size from lattice nearest neighbor distance	
Cu-1 a/o Al	6.5	Cu-2.556 Angstrom	-f.c.c.
Cu-5 a/o Al	4.5	Al-2.862	-f.c.c.
Cu-10 a/o Al	3.1	Sn-3.022	-Tetragonal
Cu-1 a/o Sn	15.0±2	Fe-2.481	-b.c.c.
Fe-10 a/o Al	8.0		

Note: Atomic size gives a rough measure of the amount that the alloy atom strains the parent lattice.

One theory relevant to the reason for the segregation of the solute to the surface of the solvent metal is that involving lattice strain. It has been postulated that if the solute atom is larger in size than the solvent atom, it will strain the solvent lattice and therefore there exists the tendency to squeeze the solute out of the solvent lattice. From the atomic sizes presented in Table 4 it appears that the experimental results agree with the theory. Still another theory involves the concept of surface energy reduction. The difficulties in using surface energies has, however, already been discussed.

Surface segregation of alloy constituents has a very definite effect upon the adhesion, friction, wear and lubrication of alloy surfaces. The effect on the adhesion of copper base alloys can be seen in the data of Figure 5. A fivefold increase in the adhesion of copper occurs with the addition of as little as one atomic percent aluminum. Further increases in the aluminum concentration beyond one percent do not produce any further change in adhesion behavior. In fact, the adhesion data for pure aluminum in Figure 5 is approximately the same as that obtained for the copper containing one atomic percent aluminum. The data of Figure 5 were all obtained with metal and alloy single crystals having a (111) surface orientation in order to eliminate orientation as a possible variable.

Both LEED and Auger emission spectroscopy were used to identify the segregation of the aluminum to the copper surface. The surface structure resulting from this identification is presented in Figure 6. From an examination of Figure 6 it can be seen that the outermost layer of the solid consists of a layer of aluminum atoms. The density of aluminum atoms in

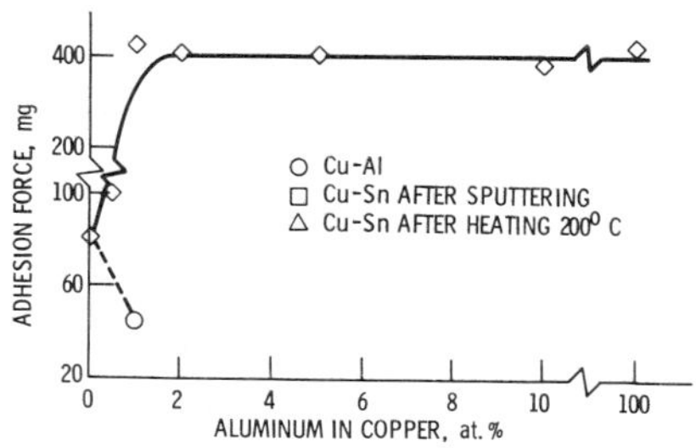

Fig. 5.—Adhesive force of (111) gold to (111) surface of copper and copper alloys as a function of bulk concentration.

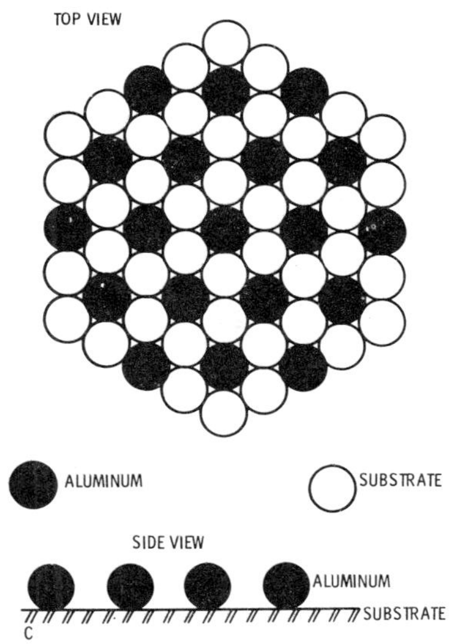

Fig. 6.—Surface segregation of aluminum in copper-aluminum alloys.

this layer will vary with bulk concentration but the layer is always aluminum. It is for this reason that no difference in adhesion behavior was detected from one atomic percent to pure aluminum.

The presence of some alloying elements can, upon segregation, promote surface chemical activity and thereby adhesion. Such behavior was observed for aluminum in copper while with other elements chemical surface activity and adhesion are reduced. Such a reduction effect is observed for tin in copper.[43] If an alloy of tin (one atomic percent) in copper is sputter cleaned, the adhesion behavior of the alloy is comparable to that for pure copper as indicated in the data of Figure 5. The reason for this is that sputtering removed the surface segregated tin. However, heating to 200°C causes the tin to segregate at the surface and brings about a reduction in adhesion (Figure 5).

Among ferrous base alloys, both aluminum[43] and silicon[45] have been found to segregate to the surface of iron. Aluminum in iron segregates to the surface and increases adhesion. It also causes, in the clean state, an increase in friction and wear over that observed for iron without aluminum.[46] Alloy surface chemistry is therefore important to friction and wear as well as adhesion. It should be indicated that, while increased surface chemical activity of the aluminum-iron alloy produces an increase in adhesion, friction and wear for dry metal contact, it also results in increased activity with lubricants which can be beneficial.[46]

Silicon alloyed with iron behaves in a rather unusual manner. If a silicon-iron alloy is heated, silicon will segregate to the surface.[45] When, however, the alloy is cooled to room temperature the silicon returns to the bulk. It is a reversible segregation. This is unlike the behavior of other alloy systems where the segregation is irreversible. If, however, the silicon is allowed to react with oxygen while on the surface, the formation of silicon oxide prevents the return of the silicon to the parent lattice from which it came.

In Figure 7 friction coefficient is plotted as a function of oxygen exposure for iron and for an iron 3.5 percent silicon alloy. Prior to the

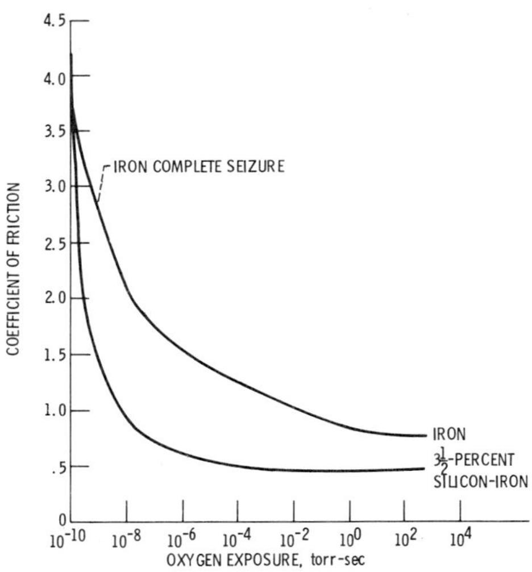

Fig. 7.—Coefficient of friction for iron and 3.5 percent silicon-iron as function of oxygen exposure. Sliding velocity, 0.001 centimeter per second; ambient temperature, 20°C; ambient pressure, 10^{-10} torr.

admission of oxygen, the friction coefficient is extremely high for the alloy and the pure iron seizes completely. As the surfaces are exposed to oxygen, the friction for both the alloy and the elemental iron decrease. This decrease occurs, however, much more rapidly for the alloy than for the iron. The difference is due to the segregated silicon and its interaction with oxygen at the surface. The sliding process is capable of generating sufficient frictional heating to cause the silicon to segregate at the alloy surface. Auger analysis confirmed its presence.[45]

GAS-SOLID INTERACTIONS

Specie and Concentration

Almost all surfaces involved in tribological systems, with the exception of those operated in a good vacuum (e.g., 10^{-10} torr), are exposed for interactions with gaseous constituents of the environment. This is true for lubricated as well as unlubricated systems.

Oxygen is probably the best "lubricant" available. It will reduce friction coefficients for clean metals from complete seizure to values of less than 1.0 (see data of Figure 7). Liquid lubricants will recuce the dry sliding value from less than 1.0 to a value of approximately 0.1.

Extremely small concentrations of surface films can influence adhesion and friction. For example, fractions of a monolayer of adsorbed oxygen or chlorine on a clean iron, copper or steel surface will reduce static friction as seen from the data of Figure 8. In Figure 8 static friction

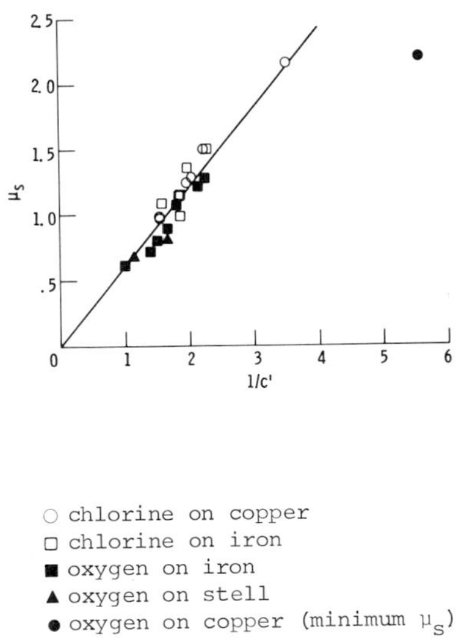

Fig. 8.—Static coefficient of friction μ_s as a function of inverse of adsorbate concentration.

is plotted as the inverse of surface coverage to the point where the surface is covered by a monolayer.[47] From these data the importance of gas-solid interactions at the surface of the solid is readily apparent.

Interaction Mechanisms

There are three basic types of interactions between a gas and solid surface: (1) physical adsorption; (2) chemisorption; and (3) chemical reaction. Physical adsorption involves weak bonding forces (van der Waals) and are not specific. Because of the highly energetic state of clean metal surfaces it is seriously doubted that such adsorption occurs on these surfaces other than with inert gases. The forces involved are comparable to those involved in liquification.[48]

Chemical adsorption or chemisorption involves very strong chemical bonds, comparable to those for chemical reaction, and further are highly specific. They play a very important role in adhesion and friction as indicated in the data of Figure 8.

Great care must be taken in using chemisorption data which appears in the literature. The current use of surface tools in the study of adsorption has indicated that earlier findings are incorrect in many instances. For example, earlier studies have indicated that gold does not chemisorb oxygen.[49] More recent studies have, however, with the aid of LEED and AES analysis, indicated that gold in fact does chemisorb oxygen.[50,51] Reference 49 indicates that nitrogen does not chemisorb to platinum while Ref. 52 reports that it does. (The reason for the earlier study not observing adsorption of nitrogen to platinum can be found in Ref. 52). Nitrogen adsorption becomes significant only after any carbon contamination is removed from the platinum surface by heating in oxygen and removing the CO that forms from the system. Such results indicate the importance of using analytical surface tools to characterize the adsorbing surface.

Orientation of Solid

Adhesion and friction are extremely sensitive to surface character as already indicated. With various adsorbed gases, friction is not only a function of the adsorbed gas but the surface orientation as well. In Tables 4 and 5 friction coefficients are presented for various gases chemisorbed to three different atomic planes of tungsten (see data for oxygen and hydrogen sulfide, Table 4).

An examination of Table 4 indicates that even hydrogen will reduce the friction of tungsten. This occurs on all three planes of tungsten with only the magnitude of the reduction varying with the plane. Oxygen is more effective than sulfur in reducing the friction of tungsten.

Molecular Structure

The data of Table 5 indicates that even the degree of bond unsaturation with hydrocarbons has an effect on friction. The greater the degree of bond unsaturation, the lower the friction coefficient for any given plane. It is difficult to interpret these results in light of adsorption mechanisms proposed for the adsorption of these gases to a tungsten surface. For example, with the adsorption of ethylene whether the adsorption is a single or two step decomposition process is in dispute.[52] Disagreement exists even as to the mechanism for the adsorption of the simple hydrocarbon methane to a tungsten surface.[53] Hopefully future surface studies with well character-

*TABLE IV.—INFLUENCE OF VARIOUS CHEMISORBED GASES ON FRICTION COEFFICIENT OF TUNGSTEN IN VACUUM**

Chemisorbed gas	Coefficient of friction (110) plane	(210) plane	(100) plane
None	1.33	1.90	3.00
H_2	1.25	1.33	1.66
O_2	0.95	1.00	1.30
CO_2	1.15	1.15	1.40
H_2S	1.00	----	1.35

* Rider specimen, (100) atomic plane of tungsten; load, 50 g; sliding velocity, 0.001 cm/sec; temperature, 20°C; pressure, 10^{-10} Torr.

*TABLE V.—INFLUENCE OF BOND SATURATION OF CHEMISORBED GASES ON FRICTION COEFFICIENT OF TUNGSTEN IN VACUUM**

Chemisorbed gas	Coefficient of friction (110) plane	(210) plane	(100) plane
Ethane (H_3C-CH_3)	1.10	1.10	1.25
Ethylene ($H_2C=CH_2$)	0.88	0.85	1.20
Acetylene (HC-HC)	0.70	0.66	1.00

* Rider specimen, (100) plane of tungsten; load, 50 g; sliding velocity, 0.001 cm/sec; temperature, 20°C; ambient pressure, 10^{-10} Torr.

ized surfaces will resolve this conflict.

From the data of Table 5 there is no question that what is present on the tungsten surface varies either in composition or structural arrangement, otherwise friction differences would not be observed. This is true not only with the adsorption of the various gases on a single plane but with different planes of tungsten as well.

Much of the difficulty in the study of adsorption of gases, and liquids as well, to solid surfaces is the pronounced effect small concentrations of bulk impurities can have on the surface. This is true with metals such as tungsten and even more so with metals like iron. Concentrations of 10-100 ppm of carbon, nitrogen or sulfur in bulk iron will segregate to the surface and will have an effect.[54] When gases are adsorbed to well characterized iron, pronounced tribological effects are seen.

Frictional Energy Effects

An area to which very little attention has been paid is the effect of the interfacial frictional energies of two surfaces in rubbing contact on gaseous adosrption. Some studies by the present author indicate that there is an effect but it is a function of the chemistry of the adsorbate. For example if methyl mercaptan is adsorbed onto a clean iron surface in the presence and absence of sliding, differences in the quantity of sulfur adsorbed are observed. These differences are indicated in the Auger spectroscopy data of Figure 9.

More sulfur is observed on the iron surface in the absence of sliding (Figure 9). Frictional heating at the interface can promote desorption of the sulfur; this and wear could account for the lesser amount of sulfur observed with sliding. These data indicate that care should be taken in applying static adsorption results to tribological systems.

If sulfur dioxide is adsorbed to a clean iron surface either in the presence of or in the absence of sliding, no difference in adsorption behavior is observed. The Auger spectroscopy data of Figure 10 indicate the same surface concentration of sulfur and oxygen under both conditions.

The data of Figure 9 indicate a sensitivity of adsorption to rubbing with methyl mercapton while the data of Figure 10 indicate an absence of such sensitivity. If sliding in Figure 9 causes desorption of sulfur, then oxygen bonded to the sulfur must assist in resisting desorption, Figure 10. This implies stronger bonding of oxygen to iron than exists for sulfur to iron.

The relative stabilities of sulfur and oxygen on an iron surface can be demonstrated with the aid of the data from Figure 11. The data of Figure 11 were obtained in experiments in which a well characterized clean iron surface was first exposed to 10,000 langmuirs of hydrogen sulfide by the present author. The hydrogen sulfide dissociatively adsorbs on the iron surface leaving a surface saturated with a sulfide film. If that surface is then exposed to oxygen, the oxygen will nearly completely displace the sulfur. This phenomena of displacement is evidenced by the data of Figure 11.

An oxidized iron surface was exposed to hydrogen sulfide. Sulfur did not displace the oxygen on the iron surface. Thus, from these results one could infer that iron oxide is more stable than is iron sulfide. It should be indicated that the oxide is thermodynamically unstable relative to the sulfide in hydrogen sulfide when no oxygen is present but the activation energy hump must be overcome.

Effect of Mechanical Parameters

Mechanical effects other than simply sliding of the surfaces can have an effect on adsorption and correspondingly on such tribological properties as friction behavior. For example, increasing loads for surfaces in solid state contact as well as increasing sliding velocity between such surfaces will increase the generated interfacial energy. It is reasonable to assume that such changes in energy will alter adsorption behavior.

Sliding friction experiments were conducted by the present author in which an iron surface containing a normal surface oxide was operated in a vinyl chloride atmosphere at a pressure of 10^{-6} torr. In a series of experiments, the load was increased and both friction force and Auger peak intensities for chlorine monitored. The results obtained are presented in Figure 12.

An examination of Figure 12 indicates that the concentration of chlorine

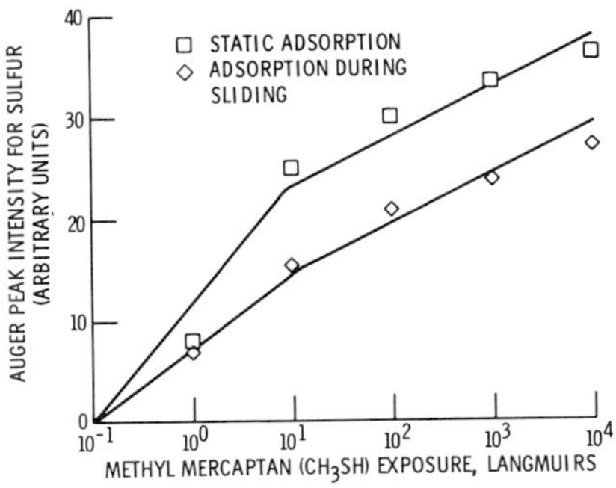

Fig. 9.—Auger spectroscopy detection of sulfur adsorbed on clean iron surface exposed to methyl mercaptan under static conditions and during sliding friction. Sliding velocity, 30 centimeters per minute; load, 100 grams; ambient temperature, 23°C.

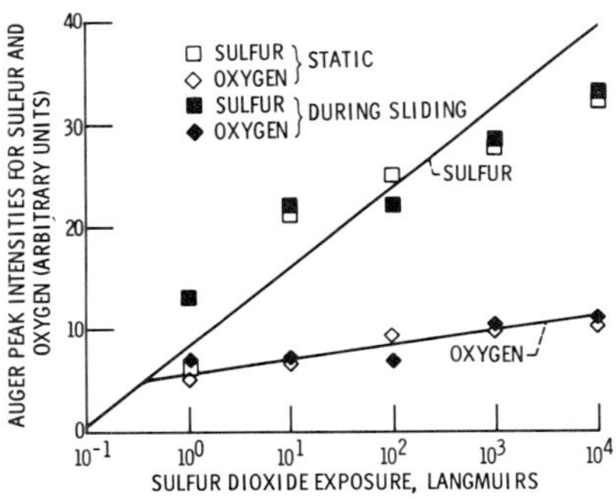

Fig. 10.—Auger spectroscopy detection of sulfur and oxygen adsorbed on a clean iron surface exposed to sulfur dioxide under static conditions and during sliding friction. Sliding velocity, 30 centimeters per minute; load, 100 grams; ambient temperature, 23°C.

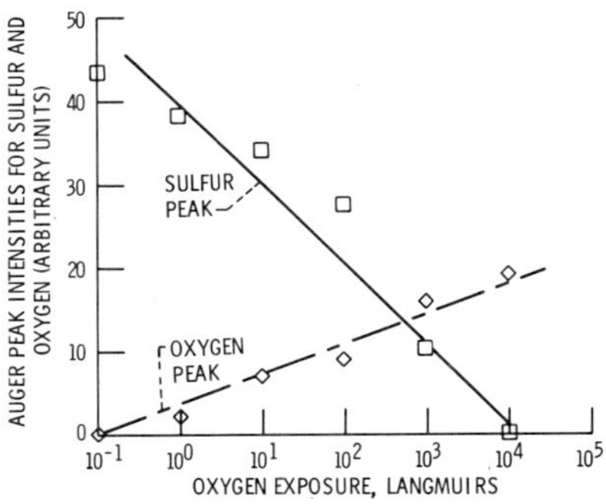

Fig. 11.—Auger spectroscopy evidence for the displacement of sulfur from an iron surface by oxygen. Initial sulfide film formed by exposure of iron surface to 10,000 langmuirs of hydrogen sulfide at 23°C.

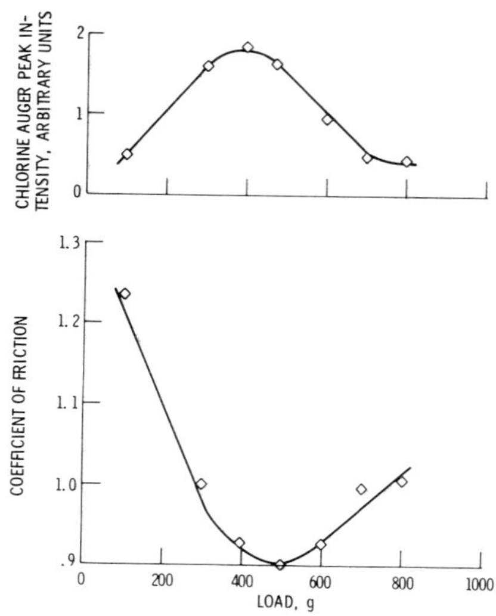

Fig. 12.—Coefficient of friction and Auger chloride peak intensity as function of load for vinyl chloride on iron surface. Ambient pressure, 10^{-6} torr of vinyl chloride; rider specimen, aluminum oxide; sliding velocity, 30 centimeters per minute; temperature, 23°C; normal oxides present on iron surface.

in the wear contact zone is a function of load. Chlorine concentration first increases with load as the friction coefficient decreases. An optimum surface coverage is achieved at which point the friction coefficient is at a minimum. Beyond this point surface coverage by chlorine decreases and this decrease is accompanied by a corresponding increase in friction coefficient. Very little wear occurs to the surfaces presented in Figure 12. The concentration of chlorine at the surface is 10^{-6} torr and consequentially corrosion or corrosive wear is not involved.

A considerable amount of data in the literature dealing with adsorption of lubricating species to solid surfaces are the result of static exposures. More data on the dynamic interfacial effects on adsorption are needed.

LIQUID-SOLID INTERACTIONS

In lubrication systems the liquid to solid interface and the chemical interactions that take place at the liquid to solid interface are extremely important. As a consequence considerable research has been expended in understanding these interactions. The adsorption of liquids and the effects of hydrocarbon chain length, functional groups, etc., were explored back at the time of Hardy.

An area which has been neglected by the tribologist is the effect of surface liquids on the mechanical behavior of solids. The materials scientist is very familiar with these effects. For example, some surface films produce mechanical strengthening of surface layers of the solid while other produce a weakening or softening.[55] These effects must play a role in the behavior of mechanical components of tribological systems.

Figure 13 illustrates surface effects in a typical stress-strain diagram. With certain surface films (e.g., oxides) a surface mechanical strengthening occurs called the Roscoe effect because it was first observed by him in the strengthening of cadium crystals by its oxide.[56]

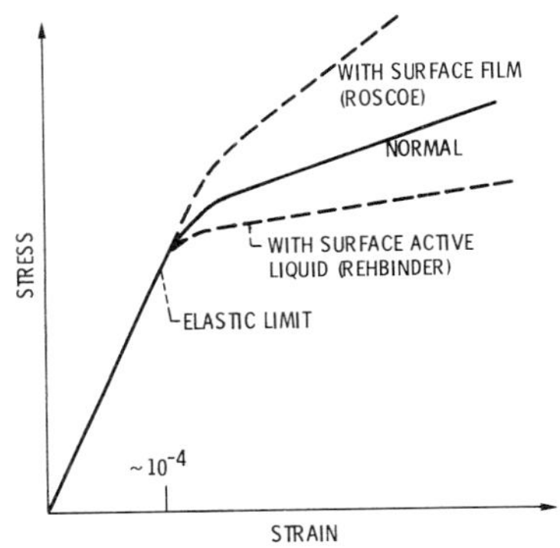

Fig. 13.—Schematic illustration of the principal extrinsic surface effects (ref. 55).

In contrast the presence of certain liquids on the surface of solids produce a softening effect. Many of these liquids are lubricants (e.g., oleic acid.)[57] Rehbinder observed this surface softening on many solids in a number of different liquids and it is therefore called the Rehbinder effect.

From the stress-strain curves of Figure 13 it is apparent that surface chemistry can influence the mechanical behavior of solids. Films such as oxides can strengthen the material while certain lubricating types of films can increase plasticity. Such effects are important and should not be overlooked in attempting to understand lubricated systems. These effects are demonstrated in Figure 14 with data from friction and wear experiments.

The data of Figure 14 are from friction and wear experiments in which films were examined on the surface of the basal (0001) plane of a single crystal of zinc. Three surface states of the zinc were studied, a clean surface generated by cleavage in liquid nitrogen, an oxidized surface and a cleaved surface containing a layer of 5 percent hydrochloric acid in water. The cleaved clean surface would be analogous to the normal surface of Figure 13; the oxidized surface, strengthening by the Roscoe effect and the acid solution, would be a manifestation of the Rehbinder effect of Figure 13.

The friction coefficient as a function of load in Figure 14 is greater for the oxidized surface than it is for the surface lubricated with the acid

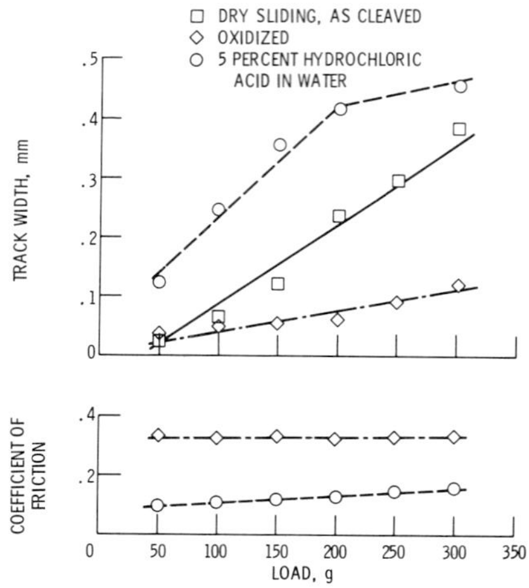

Fig. 14.—Width of wear track and coefficient of friction produced with ruby ball sliding on zinc single crystal (0001) surface in [10$\bar{1}$0] direction. Sliding velocity, 1.4 millimeters per minute; temperature, 23°C; dry argon atmosphere.

film. These results are as might be anticipated because, in a sense, the acid layer is a lubricant. Chemical reactions with the surface leads to the formation of zinc chloride.

The wear track widths are presented in Figure 14 for all three surface states. These and the friction results were obtained some years ago by the present author.[58] The greatest amount of surface deformation occurred with the acid solution, the substance which produces surface softening (Rehbinder effect).

The clean, "as cleaved" crystal surface produced intermediate amounts of surface deformation with the least surface deformation taking place in the presence of the surface oxide. Thus, surface chemistry is important in the mechanical behavior of tribological surfaces.

The mode of deformation as well as the amount is affected by the presence or absence of surface films. In Figure 15 the zinc single crystal surfaces are examined after sliding experiments with hexadecane present on the surface. Figure 15(a) is for the surface without an oxide layer preformed and Figure 15(b) is the wear track generated in the presence of an oxide film. Without the oxide, deformation is completely plastic via slip. When the oxide is present deformation twinning is observed as shown in the photomicrography of Figure 15(b). A "ladder" of deformation twins is detected in the wear track of Figure 15(b). These twins are completely absent in Figure 15(a).

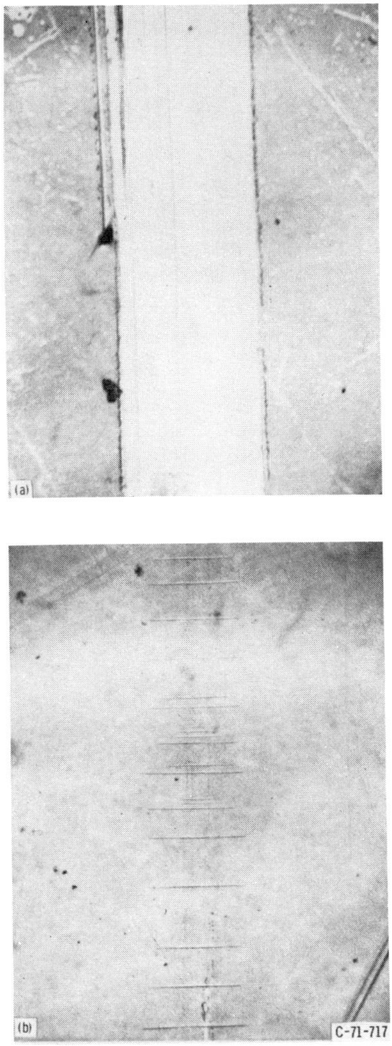

Fig. 15.—Deformation tracks developed on a zinc (0001) surface in sliding contact with a ruby ball under a 200 gram load and in hexadecane.

SOLID-SOLID INTERFACE

The surface chemistry and bonding across an interface for two solids in contact is extremely important in understanding the adhesion, friction and wear behavior of materials. With metals and alloys, the materials which have received the greatest attention with respect to chemistry, various properties have been found to affect their surface chemical activity. Such properties include: (1) crystal structure; (2) crystal orientation; (3) solid solubility; (4) surface segregation; (5) surface energy; and (6) d bond character.

Crystal Structure and Orientation

The manner in which metallic atoms bond to one another in the bulk will determine the crystal structure that results. Interfacial bonding of hexagonal metals generally results in strong bonding at the interface but easy shear along basal planes which allows for limited growth in the real contact area, low adhesive bonding forces (because of easy separation along basal planes) and low friction.[59]

Crystallographic orientations at metal surfaces are extremely important in surface chemistry and in solid to solid interactions. Generally in any crystal system (e.g., C.P.H., F.C.C. or B.C.C.) the high atomic density planes are the low surface energy planes and correspondingly the least chemically active. Thus, when {111} planes are brought into contact for face-centered cubic metals the adhesive bonding is less than when two {110} planes are brought into contact. Likewise, with hexagonal metals, bonding is weaker between {0001} planes than it is between {1010} planes. This orientation influences bonding effects, adhesion, friction, and adhesive wear.[59]

The foregoing discussion applies to metals in clean solid state contact. When a metal is adsorbed on a solid surface of the same metal the same interaction characteristics exist. Thus, for example, the bonding energy for tungsten atoms is less on the (110) plane than it is for the (111) plane.[60-61] In Ref. 60 it is 121 kcal g atom^{-1} for the (110) plane and 139 kcal g atom^{-1} for the (111) plane. Similar observations have been made in adhesion and friction studies with tungsten;[62] that is, adhesive bonding (and correspondingly friction) are less on the (110) than on the less dense atomic planes.

Solid Solubility

Through the years a number of attempts have been made to correlate the solid solubility of metals (bulk) with adhesion, friction and wear of metal surfaces.[63-67] Adhesion, friction and adhesive wear are largely the result of surface properties and consequently surface chemistry, rather than bulk chemistry should be considered. Ample evidence exists to establish differences between surface and bulk bonding of, for example, dissimilar metals.

Tin, copper, and gold are completely insoluble in tungsten.[68] Despite this bulk insolubility, these elements bond very strongly to the surface of tungsten. With tin, there are two binding energies, a weaker one comparable to the binding energy in bulk tin and a second energy which is much stronger and occurs in the first monolayer of tin where each tin atom contacts four tungsten atoms.[60] This bond is unusually strong. With gold to tungsten, the bond is intermetallic in nature at the interface.(70) Copper alloys with the tungsten surface.(71)

Adhesion experiments in the field ion microscope have revealed strong adhesive bonding and transfer of gold to tungsten with surface compound formation.(72) Adhesion and transfer of copper to tungsten has been observed in sliding friction experiments.(59)

The foregoing caveat against using bulk properties to predict surface behavior not only applied to metal-metal contacts but to metals contacting nonmetals also. For example, gold has extremely limited solid solubility in both silicon and germanium (less than 10^{-5} atomic percent).(68) Despite this very limited solubility, gold bonds very strongly to silicon and germanium in adhesion experiments.(73)

Figure 16 is a photomicrograph of a silicon (111) surface after adhesive contact with gold. Heavy transfer of gold to the silicon is observed.

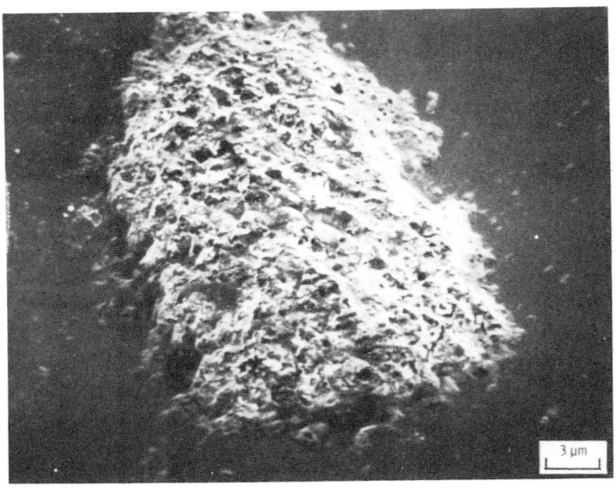

Fig. 16.—Gold transferred to a clean silicon (111) surface after adhesive contact. Load, 30 grams; temperature 23°C; pressure, 10^{-8}N/m^2.

served. The amount of transfer is greater than is frequently seen for metals in contact with other metals where complete solid solubility exists.(59) The gold cohesive bonds were weaker than the gold to silicon interfacial adhesive bonds, thus resulting in fracture in the gold.

Strong interfacial bonding was also observed (Ref. 73) for gold to germanium. Separation of the adhesive junction, however, fracture occurred in the germanium rather than in the gold. The cohesive binding energies of gold and germanium are nearly equal. Thus, where the interfacial binding is stronger than the cohesive binding in the elements fracture can occur in either. In the studies, of Ref. 73, it takes place in the germanium. The cohesive binding energy of silicon, however, is stronger than gold and consequently, fracture occurs in the gold for this particular couple.

The foregoing discussion indicates the importance of avoiding the use of bulk properties to predict surface behavior. On the basis of the solid solubility theory, very little adhesion and transfer should occur for the particular couples discussed, yet the opposite result is observed. Conversely, beryllium is soluble in cobalt yet very little adhesive transfer is observed.

Valence Bonding

The importance of surface segregation and surface energy in solid state contacts has already been discussed. The role of d valence bond character of metals on bimetallic adhesion and friction can be useful in predicting friction coefficients for various metals in contact.(74) This concept gives an indication of the relative amount of electron energy available at a free surface for bonding based on that which is committed to cohesive bonding within the metal itself. Recently, the concept has been applied to metal to nonmetallic bonding, adhesion, and friction of metal to nonmetal couples.(75)

SUMMARY

A critical review of the chemistry of surfaces indicates that much of the interpretation of the data generated in the past may now be suspect because of inadequate characterization of the surfaces. Analytical tools are now available to assist in this characterization. They have already revealed that many metal surfaces do not consist of metallic atoms but rather atoms of bulk impurities which diffuse to the surface and contaminate it. It is now apparent that the wide disparity in the values of surface energy may be a result of these surface contaminates. These surface films are important because even fractions of a monolayer can affect adhesion and friction.

Bulk alloy chemistry cannot be extended to the surface because equilibrium segregation will result in a different surface alloy chemistry. The adsorption of lubricating species on the surface of solids is highly specific and not only dependent upon chemistry but orientation as well. Mechanical surface activity influences the adsoprtion process in some instances and mechanical properties, such as plastic deformation are altered by the presence of adsorbates.

Strong interfacial bonding between dissimilar materials which are insoluble in the bulk occurs and in some instances the bonding and material transfer observed is greater than seen with those materials which are completely soluble, one in the other. Such observations argue against the use of bulk properties to predict surface behavior.

Adhesion, friction, wear and lubrication are highly dependent upon surface chemistry and, therefore, surface rather than bulk behavior should be used in studying mechanisms which are clearly surface related. Assuming that bulk properties are also surface properties can be very misleading.

REFERENCES

1. Hardy, W.B., "Collected Papers," Cambridge University Press, England, 1936; see Papers no. 37 (1919), no. 39 (1920), no. 40 (1920), no. 41 (1922), no. 42 (1922), no. 43 (1923), no. 44 (1925), no. 46 (1925), no. 50 (1926), and no. 57 (1928).
2. Adamson, A.W., "Physical Chemistry of Surfaces," 2nd edition, Interscience Publishers, New York, 1967, p. 263.
3. Hayward, D.O. and Trapnell, B.M.W., "Chemisorption," 2nd edition, Butterworths, Washington, 1964, p. 135.
4. Little, L.H., in "Chemisorption and Reaction on Metallic Film," edited by J.R. Anderson, Vol. 1, Academic Press, London and New York, 1971, p. 489.
5. Bowden, F.P. and Tabor, D., "The Friction and Lubrication of Solids," Oxford Press, London, 1950, p. 145.
6. Forbes, E.S. and Reid, A.J.D., *American Society of Lubrication Engineers Transactions,* Vol. 16, No. 1, 1973, p. 50.
7. Barber, R.I., *American Society of Lubrication Engineers Transactions,* Vol. 19, No. 4, 1976, p. 319.
8. Cottington, R.L. and Ravner, H., *American Society of Lubrication Engineers Transactions,* Vol. 12, No. 4, 1969, p. 280.
9. Sakurai, T. and Sato, K., *American Society of Lubrication Engineers Transactions,* Vol. 9, No. 1, 1966, p. 77.
10. Prutton, C.F., Turnbull, D. and Dloughy, G., *Institute of Petroleum. Journal,* Vol. 32, No. 266, 1946, p. 90.
11. Hayward, D.O., in "Chemisorption and Reactions on Metallic Films," edited by J.R. Anderson, Vol. 1, Academic Press, London and New York, 1971, p. 229.
12. Simmons, G.W. and Dwyer, D.J., *Surface Science,* Vol. 48, 1975, p. 373.
13. Sewell, P.B., Mitchell, D.F. and Cohen, M., *Surface Science,* Vol. 33, 1972, p. 535.
14. Brucker, C.F. and Rhodin, T.N., *Surface Science,* Vol. 57, 1976, p. 523.
15. Wawra, H.H., *Radex Rundschaw,* Vol. 5, 1972, p. 351.
16. Tyson, W.R., *Journal of Applied Physics,* Vol. 47, No. 2, 1976, p. 459.
17. Averbach, B.L., in "Fracture, An Advanced Treatise, Vol. 1: Microscopic and Macroscopic Fundamentals," edited by H. Liebowitz, Academic Press, New York, 1968, p. 441.
18. Mykura, H., *Acta Metallurgica,* Vol. 9, No. 6, 1961, p. 570.
19. Sundquist, B.E., *Acta Metallurgica,* Vol. 12, No. 1, 1964, p. 67.
20. Kumar, R. and Grenga, H.E., *Surface Science,* Vol. 50, 1975, p. 399.
21. Winterbottom, W.L. and Gjostein, N.A., *Acta Metallurgica,* Vol. 14, No. 9, 1966, p. 1041.
22. Roberts, R.W., *New York Academy of Sciences. Annals,* Vol. 101, 1963, p. 766.
23. Rabinowicz, E., "Friction and Wear of Materials," John Wiley and Sons, Inc., New York, 1965, p. 161.
24. Ehrlich, G., "Surface Phenomena of Metals," Society for Chemical Industry Monograph No. 28, 1968, p. 13.
25. Buckley, D.H., "Absorption of Ethylene Oxide and Vinyl Chloride on Iron (011): Surface and Effect of These Films on Adhesion," NASA TN D-5999, 1970, p. 5.
26. Müller, A. and Drechsler, M., *Surface Science,* Vol. 13, 1969, p. 471.
27. Brenner, S.S., *Surface Science,* Vol. 2, 1964, p. 496.
28. Halling, J., "Introduction To Tribology," Wykeham Publications, London, 1976, p. 21.
29. Cameron, A., "Principles of Lubrications," Longmans Green and Company, Ltd., London, 1966, p. 412.

30. Akhmatov, A.S., "Molecular Physics of Boundary Friction," translated from Russian by N. Kaner, Israel Program for Scientific Translation, Jerusalem, 1966, p. 25.
31. Ralph, B., in "Field Ion Microscopy," edited by J.J. Hren and S. Ranganathan, Plenum Press, New York, 1968, p. 157.
32. Shell, C.A. and Rivière, J.C., *Surface Science,* Vol. 40, 1973, p. 149.
33. Sickafus, E.N., *Surface Science,* Vol. 19, 1970, p. 181.
34. Kunimori, K., Kawai, T., Kondow, T., Onishi, T. and Tamaru, K., *Surface Science,* Vol. 46, 1974, p. 567.
35. Isett, L.C. and Blakely, J.M., *Surface Science,* Vol. 47, 1975, p. 645.
36. Powell, G.L., Clausing, R.E. and McGuire, G.E., *Surface Science,* Vol. 49, 1975, p. 310.
37. Wandelt, K. and Ertl, G., *Surface Science,* Vol. 55, 1976, p. 403.
38. Wood, B.J. and Wise, H., *Surface Science,* Vol. 52, 1975, p. 151.
39. McDavid, J.M. and Fain, S.C., Jr., *Surface Science,* Vol. 52, 1975, p. 161.
40. Yamashina, T., Watanabe, K., Fukuda, Y. and Hashiba, M., *Surface Science,* Vol. 50, 1975, p. 591.
41. Overbury, S.H. and Somorjai, G.A., *Surface Science,* Vol. 55, 1976, p. 209.
42. Ferrante, J. and Buckley, D.H., "Auger Electron Spectroscopy Study of Surface Segregation in Copper-Aluminum Alloys," NASA TN D-6095, 1970, p. 4.
43. Ferrante, J. and Buckley, D.H., *American Society of Lubrication Engineers Transactions,* Vol. 15, No. 1, 1972, p. 18.
44. Rivière, J.C., *Journal of the Less Common Metals,* Vol. 38, 1974, p. 193.
45. Ferrante, J., "Auger Electron Spectroscopy Study of Surface Segregation in the Binary Alloys Copper-1 Atomic Percent Indium, Copper-2 Atomic Percent Tin and Iron-6.55 Atomic Percent Silicon," NASA TN D-6982, 1973, p. 9.
46. Buckley, D.H., "Influence of Aluminum on Friction and Wear of Iron-Aluminum Alloys Dry and Lubricated," NASA TN D-6359, 1971, p. 6.
47. Wheeler, D.R., *Journal of Applied Physics,* Vol. 47, No. 3, 1976, p. 1123.
48. Young, D.M. and Crowell, A.D., "Physical Adsorption of Gases," Butterworths, Washington, 1962, p. 4.
49. Hayward, D.O. and Trapnell, B.M.W., "Chemisorption," 2nd edition, Butterworths, Washington, 1964, p. 75.
50. Chesters, M.A. and Somorjai, G.A., *Surface Science,* Vol. 52, 1975, p. 21.
51. Schrader, M.E., *Journal of Colloid and Interface Science,* Vol. 59, No. 3, 1977, p. 456.
52. Barford, B.D. and Rye, R.R., *Journal of Vacuum Science and Technology,* Vol. 9, No. 2, 1972, p. 673.
53. Yates, J.T. and Madey, T.E., *Journal of Vacuum Science and Technology,* Vol. 9, No. 2, 1972, p. 672.
54. Grabke, H.J., et al., *Surface Science,* Vol. 63, 1977, p. 377.
55. Gilman, J.J., *Philosophical Magazine,* Vol. 6, No. 61, January 1961, p. 159.
56. Roscoe, R., *Philosophical Magazine,* Vol. 21, 1936, p. 399.
57. Likhtman, V.I., Rebinder, P.A. and Karpenko, G.V., "Effect of Surface-Active Media on the Deformation of Metals," Chemical Publishing Company, New York, 1960, p. 38.
58. Buckley, D.H., *American Society of Lubrication Engineers Transactions,* Vol. 15, No. 2, 1972, p. 96.
59. Buckley, D.H., in "Adhesion or Cold Welding of Materials in Space Environments," American Society for Testing and Materials Special Technical Publication, edited by D.V. Keller, No. 431, 1967, p. 248.

60. Ehrlich, G. and Kirk, C.F., *Journal of Chemical Physics,* Vol. 48, No. 4, 1968, p. 1465.
61. Plummer, E.W. and Rhodin, T.N., *Journal of Chemical Physics,* Vol. 49, No. 8, 1968, p. 3479.
62. Buckley, D.H., *Journal of Applied Physics,* Vol. 39, No. 9, 1968, p. 4224.
63. Keller, D.V., *Wear,* Vol. 6, 1963, p. 353.
64. Sikorski, M.E., *Wear,* Vol. 7, 1964, p. 144.
65. Roach, A.E., Goodzeit, C.L. and Hunnicutt, R.P., *American Society of Mechanical Engineers Transactions,* Vol. 78, No. 11, 1956, p. 1659.
66. Ernst, H. and Merchant, M.E., "Chip Formation, Friction, and Finish," The Cincinnati Milling Machine Co., 1940, p. 3.
67. Rabinowicz, E., in "Wear of Materials, 1977," edited by S.K. Rhee, K.C. Ludema and W.A. Glaeser, American Society of Chemical Engineers, New York, 1977, p. 36.
68. Elliott, R.P., "Constitution of Binary Alloys, First Supplement," McGray-Hill Book Company, New York, 1965, p. 107.
69. Nishikawa, O. and Saadat, A.R., *Surface Science,* Vol. 60, 1976, p. 301.
70. Jones, J.P., "Surface Phenomenon of Metals," Society of Chemical Industry Monograph No. 28, London, 1968, p. 263.
71. Taylor, N.J., *Surface Science,* Vol. 4, 1966, p. 161.
72. Brainard, W.A. and Buckley, D.H., "Preliminary Studies by Field Ion Microscopy of Adhesion of Platinum and Gold to Tungsten and Iridium," NASA TN D-6492, 1971, p. 39.
73. Buckley, D.H. and Brainard, W.A., "Adhesion and Friction of Iron and Gold in Contact with Elemental Semiconductors," NASA TN D-8394, 1977, p. 4.
74. Buckley, D.H., *Journal of Colloid and Interface Chemistry,* Vol. 58, No. 1, 1977, p. 36.
75. Miyoshi, K. and Buckley, D.H., "Friction and Wear of Single-Crystal and Polycrystalline Manganese-Zinc Ferrite in Contact with Various Metals," NASA TN E-9168, 1977, p. 5.

DISCUSSION

QUESTIONER: Do we use the right combinations of metals today? Are there better combinations we could use such as soft metals against hard metals?

D. H. BUCKLEY: The answer is yes. There are a lot of properties that can guide us in selecting materials. Many years ago it used to be a common expression among people in lubrication, and I have heard it from designers many times myself, that dissimilar materials in contact should be used, never like-on-like. That is not necessarily so. For example, I can take beryllium-on-beryllium and cobalt-on-cobalt in the clean state and get markedly lower friction coefficients with complete absence of adhesion and seizure of those materials. Any other couple that we might select would give adhesion and seizure in the clean state. That does not necessarily mean that unlike combination cannot be used. There are certain basic properties that influence adhesion, friction and wear behavior. For example, through the years people have established that crystal structure, order - disorder reactions and textures have a profound influence on wear. Certain surface textures of a metal can give markedly lower friction and wear than other textures. There are many other properties of solid surfaces such as cohesive binding which can be used as criteria. Let us face that adhesion is adhesive wear. Every time we pull surfaces apart, even electrical contact, we lose material from one surface. There are some basic properties that we can use in selecting materials. Of course, we are hoping that as more people get involved, more of these properties can be identified and someday we can catalogue various properties. Some day we can select materials on a rational basis instead of simply saying, "Don't use like on like."

B. BRISCOE, Imperial College: Do you find any correlation between the magnitude of the adhesive force and friction force and whether the factors which influence one in general will influence the other in a similar way?

BUCKLEY: That is a good question. In the case of contact between clean surfaces there is a correlation; the friction behavior and the adhesion behavior correlate fairly well. The materials that adhere strongly at the interface will give very high friction and seizure. In the presence of surface films, however, it is very difficult. Once the surface is contaminated, adhesion does not take place very effectively. The adhesive force drops off rapidly with the monolayer from the contaminant film. We still measure friction, no matter what is present on the solid surface; but we may not be able to measure adhesion any longer. It is difficult to draw a correlation between adhesion and friction because the adhesion, so to speak, drops out of the picture. Of course, it has a small value but it is very difficult to measure. I guess it is a qualified answer.

H. CZICHOS, Federal Institute of Materials Testing, Berlin: You made some experiments sliding iron against germanium first in dry state and then in the lubricated state and you got different patterns. First you observed fracture and then more plastic deformation. My question is, is it due to the different tangential forces or is it due to the change in the chemical nature?

BUCKLEY: It actually is a combination of both. When you bring clean surfaces into contact, they are going to bind as soon as you put them in close contact. Therefore, the tangential motion is resisted because of interfacial bonding. In the lubricated case there will be very little resistance to tangential motion, and therefore friction forces will be relatively low, and we should not see much fracturing. In one case we get fracture with large amounts of energy and in the other case is the complete absence of fracture, but we still have frictional energy involved because we observe plastic deformation of the surface. But whatever frictional energy is involved, it is remarkably less than the case of the clean system.

D. GODFREY, Chevron Research: Regarding the diffusion of alloying elements, what is the driving force that would cause aluminum for example, to diffuse to the surface of copper during sliding?

BUCKLEY: As we observed this effect in tribology, several others have observed the same effect in other areas. Surface physicists call this effect surface segregation. There is a controversy going on now about the driving force for diffusion. We took a naive approach. If you look at the appropriate table you see the lattice dimensions of the various species. It is an indicator of lattice strain that is produced in the crystal lattice when the impurity atom is present. Tin produces the greatest amount of lattice strain, and its concentration is the highest on the solid surface. Therefore, we proposed that it is a simple matter of lattice strain. The greater the amount of lattice strain the solute porduces in the solvent, the higher will be the driving force to diffuse to the surface. Since then a number of other theories have been proposed and it seems that, in addition to our own, there are about three other theories. No one was really proved.

A. W. RUFF, NBS: I think it is also true that the energy of a surface is a function of composition in many systems and therefore if you have a system that minimizes the surface energy with a composition change, diffusion to the surface may take place.

BUCKLEY: Dr. Ruff has just expounded on one of the other hypotheses. And there are two others.

QUESTIONER: The wear rates we measure are very sensitive to the presence of oxide film. I have great difficulty in maintaining a stable oxide film. Were you going over the same wear track in your experiments?

BUCKLEY: In some experiments it was a single pass, and in others they were multiple passes over the same wear track. It depends on the object of the experiment. If the object of the experiment is to look at something which one pass will establish, then single pass tests are used. If, for example, absorption of lubricant species during sliding are studied as a function of the number of passes, we use as many as 200-250 passes and then compare the absorption of different species on the solid surface.

TRANSMISSION ELECTRON MICROSCOPE STUDY OF THE INTER-RELATIONSHIP BETWEEN FRICTION AND DEFORMATION OF COPPER SINGLE CRYSTALS

N. Ohmae

ABSTRACT

A high voltage transmission electron microscope study on friction-deformed copper single crystals was conducted to clarify the relationship between friction and deformation, since friction is dependent not only on surface chemical or physical properties but also on crystallographic orientation. It was found that the friction process resulted in a complex deformation, i.e., a tensile deformation under compression. The work hardened layer due to friction consisted of (i) {110} <211> texture, (ii) a distorted region dependent on sliding direction, (iii) compressed zone, and (iv) undisturbed matrix. The type of slip systems which acted during texturing determines the friction property.

INTRODUCTION

The paper 'Definition and Effect of Chemical Properties of Surfaces in Friction, Wear, and Lubrication' by Buckley emphasizes the importance of surface chemistry in tribology.[1] The use of surface analytical tools pioneered by Buckley has provided a new insight into tribology by characterizing the surface.[2-4] It is widely accepted now that the tribological properties are very sensitive to surface characteristics, both chemical and physical. The present author's adhesion and friction experiments on f.c.c. metals directly inside the Auger emission spectrometer have also showed the importance of surface cleanliness, temperature, load, velocity, and crystallographic orientations.[5-10] In addition to these factors, this paper deals with the influence of crystallographic orientation on friction. In addition, dislocation behavior in the deforming subsurface material cannot be ignored. A number of research publications reported anisotropic friction due to crystallographic orientations both in air[5-8] and in vacuum.[9,10] The discussion on the effect of dislocation may repeat the other review paper, but this paper aims to clarify the mechanism of friction from the standpoint of crystallography.

One of the most powerful methods for observing microstructure is transmission electron microscopy. However, the difficulties in preparing a thin foil as well as the problem of electron transmission capability do not make such an observation easy. In addition, the region observable with a conventional transmission electron microscope is so limited that the obtained results cannot be extended to an overall phenomenon. In the present study, the use of the high voltage transmission electron microscope operating at 2000 kV enabled an observation of dislocation structure both in the wear groove and on the ridges. Observations were carried out at relatively low magnifications to cover a wide range of wear track. Disk speciments employed were 99.99% pure copper single crystals grown by Bridgman method[11] and 99.999% pure aluminum single crystals obtained by strain-annealing.[12] Pin specimens were 99.98% copper polycrystals and 99.99% aluminum polycrystals with a shape of 120° cone angle. Unlubricated single pass sliding tests between similar metals (relative humidity 60±2%, temperature 20±1°C) were conducted in air at a sliding velocity of 10 μm per second. The results obtained at 1.5×10^{-1} N normal load for copper and at 7.4×10^{-2} N for aluminum are presented in this paper.

ANISOTROPIC FRICTION AND CRYSTALLOGRAPHIC ORIENTATION

It is well known that f.c.c. metals exhibit anisotropic friction behavior.[13] Table I summarizes the friction coefficients between similar f.c.c. metals in air. Evidently, an anisotropy exists in the friction of copper and aluminum. When friction is considered on the basis of plastic deformation, the influential factors can be: (1) the type of slip system, (2) the ease of crystal rotation, and (3) the degree of work hardening. The interaction of the above mentioned factors is so complicated that the mechanism of friction cannot be clarified simply. Friction, in an essential sense, is a resistance to sliding. To rupture adhesive junctions, fracture should occur either at interface or within bulk depending on the strength of materials.

Friction is sensitive not only to surface conditions but also to bulk deformation. When materials are brought into contact, adhesion takes place. Then under relative movement, material at the surface undergoes large plastic deformation. This plastic deformation further increases adhesion. Thus strain in the contact region continues to increase, and when the movement of dislocations is not able to ease the strain, fracture will occur at the weakest point, such as pre-existing cracks, cell boundary, voids due to dislocation pile-ups, precipitates and grain boundaries. Although surface cleanliness, load, temperature and sliding velocity also influence the friction behavior, friction primarily is a function of material deformation. This will be discussed later.

One of the most important factors in the friction-deformation relationship is the type of slip system at the surface of friction track. In order to examine the orientational effect, friction experiments were conducted with various aluminum single crystals slid in varied directions against aluminum cones. The relationship between friction coefficient and the direction of stress axis is shown in the stereographic standard triangle (see Figure 1). In this case, the mean stress axis was assumed to be along the resultant of the normal load and the friction force. Figure 1 clearly sets the problem which should be studied in this paper. Friction coefficients higher than 1.2 were obtained in the neighborhood of the sides of the standard triangle, while friction coefficients less than 1.0 were mainly observed in the inner region of the triangle. This crystallographic orientational effect indicates that the slip system which acted on the surface of friction track is the basic factor in the mechanism of friction. Multiple slip easily occurs at the sides of the triangle, while a single

TABLE I.—ANISOTROPIC FRICTION FOR Cu AND Al (REFS. FROM 5 TO 9)

	Disk	Pin	Atmosphere	Load(N)			Friction Coeff.
Bailey &		Sapphire				[001]	0.23
Gwathmey	Cu	Hemisphere	Air	0.25	(001)		
		(22 μm R)				[110]	0.67
		Diamond				[100]	0.27
Steijn	Cu	Hemisphere	Air	0.10	(001)		
		(17.8 μm R)				[110]	0.38
		Cu				[100]	1.0
Roshon	Cu	Hemisphere	Air	3.9	(001)		
		(1.59mm R)				[110]	0.8
					(001)		40
			1.3×10^{-9}	0.49	(110)	<110>	40
		Cu	Pa		(111)		21
	Cu	Hemisphere			(001)		0.60
		(0.8mm R)	Air	0.49	(110)	<110>	0.40
Buckley					(111)		0.21
				0.08-	(001)		1.3
		Al		0.59	(110)	<110>	1.0
		Hemisphere	Air		(111)		0.7
		(0.8mm R)		0.29-	{110}	<100>	0.55
				0.59		<110>	1.0
						[100]	0.8
		Cu			(001)	[210]	1.0
	Cu	Cone	Air	0.74-		[110]	1.25
Present		(120°)		8.8	(001)		1.25
Author					(110)	<110>	0.9
					(111)		0.4
		Al					
	Al	Cone	Air	0.074	Random		0.8-
		(120°)					1.4

slip occurs inside the triangle because of a high resolved shear stress of the primary slip system. The surface texture caused by friction of f.c.c. metals, is the {110} <211> texture, which is to be shown later for copper single crystals. However, the type of slip systems which acted during the formation of texture determines the relationship between friction and deformation.

Slip Systems

Slip bands formed on the copper single crystals were observed by optical microscopy and optical interferometry. As slip system for f.c.c. metals is {111} <110>, the active slip systems for varied sliding conditions were identified as shown in Table II. A schematic illustration of the slip model is shown in Figure 2 for the slidings on the (001) surfaces in the

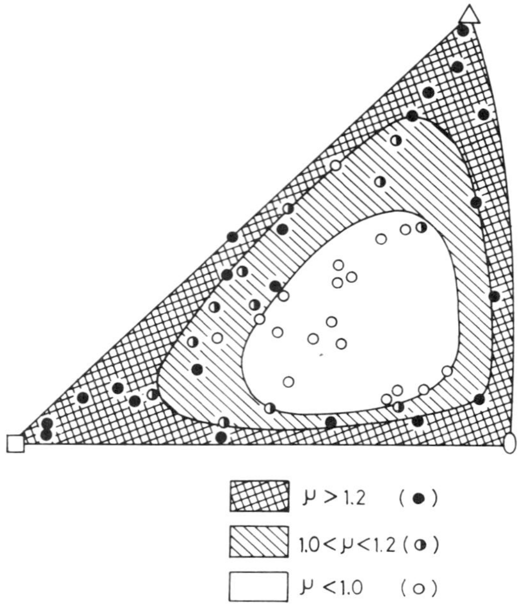

Fig. 1.—Relation between friction coefficient and direction of stress axis shown on the standard triangle, the results obtained from friction experiments with randomly oriented Al single crystals.

[100] direction, the [110] direction, and the [210] direction. It should be noted that the slip systems tabulated in Table II mainly contribute to the deformation of ridges. In some cases, a different slip system governs the deformation inside the groove.

Optical microscope observations also showed that the slip bands flowed in the direction of sliding particularly in the vicinity of friction track. This suggests that a crystal rotation occurred around the axis normal to the friction surface along sliding direction. In other words, dislocations with screw component were introduced. Under such a complicated stress condition, one of the techniques for the analysis of active slip systems is the Schmid factor.(14-16) By calculating the relation between Schmid factors and measured friction coefficients,(17) it is possible to estimate the most active slip system during the formation of friction-deformed structure.

MICROSTRUCTURE ON FRICTION TRACK

The difference in the microstructure due to varied conditions of friction was studied by a high voltage transmission electron microscope. The dislocation structure of the sliding on the (001) surface in the [100] direction is shown in Figure 3. In the figure, the direction of stress axis, which might act during friction process, is indicated. A high density of dislocations was observed on the friction track, and an elongation of cell structure to the direction of sliding was predominant. The appearance

TABLE II.—SLIP SYSTEMS WHICH CONTRIBUTED TO THE DEFORMATION ON RIDGES

	[100]		[110] ; [210]			
(001)	Downward	Upward	Downward	Upward		
Lefthand	$(111)\ [01\bar{1}]$	$(\bar{1}\bar{1}1)\ [011]$	$(\bar{1}11)\ [01\bar{1}]$	$(1\bar{1}1)\ [011]$		
Side	$(\bar{1}11)\ [01\bar{1}]$	$(1\bar{1}1)\ [011]$	$(\bar{1}11)\ [\bar{1}0\bar{1}]$	$(1\bar{1}1)\ [\bar{1}01]$		
Righthand	$(1\bar{1}1)\ [0\bar{1}\bar{1}]$	$(\bar{1}11)\ [0\bar{1}1]$	$(1\bar{1}1)\ [10\bar{1}]$	$(\bar{1}11)\ [101]$		
Side	$(\bar{1}\bar{1}1)\ [0\bar{1}\bar{1}]$	$(111)\ [0\bar{1}1]$	$(1\bar{1}1)\ [0\bar{1}\bar{1}]$	$(\bar{1}11)\ [0\bar{1}1]$		
(110)	[001]		[$\bar{1}$10]		[$\bar{1}$11]	
					Downward	Upward
Lefthand			$(111)\ [0\bar{1}1]$	$(\bar{1}\bar{1}1)\ [0\bar{1}1]$	$(\bar{1}\bar{1}1)\ [1\bar{1}0]$	$(111)\ [1\bar{1}0]$
Side	$(1\bar{1}1)\ [\bar{1}\bar{1}0]$	$(\bar{1}11)\ [\bar{1}\bar{1}0]$	$(111)\ [\bar{1}01]$	$(\bar{1}\bar{1}1)\ [\bar{1}01]$	$(111)\ [\bar{1}10]$	$(\bar{1}\bar{1}1)\ [110]$
Righthand			$(\bar{1}\bar{1}1)\ [\bar{1}0\bar{1}]$	$(111)\ [01\bar{1}]$	$(\bar{1}\bar{1}1)\ [1\bar{1}0]$	$(111)\ [1\bar{1}0]$
Side	$(\bar{1}11)\ [\bar{1}\bar{1}0]$	$(1\bar{1}1)\ [\bar{1}\bar{1}0]$	$(\bar{1}\bar{1}1)\ [0\bar{1}\bar{1}]$	$(111)\ [10\bar{1}]$	$(111)\ [\bar{1}10]$	$(\bar{1}\bar{1}1)\ [110]$
(111)	[$\bar{1}$10]					
Lefthand	$(\bar{1}\bar{1}1)\ [0\bar{1}\bar{1}]$	$(\bar{1}\bar{1}1)\ [011]$				
Side	$(\bar{1}\bar{1}1)\ [\bar{1}0\bar{1}]$	$(\bar{1}\bar{1}1)\ [101]$				
Righthand	$(\bar{1}\bar{1}1)\ [0\bar{1}\bar{1}]$	$(\bar{1}\bar{1}1)\ [101]$				
Side	$(\bar{1}\bar{1}1)\ [\bar{1}0\bar{1}]$	$(\bar{1}\bar{1}1)\ [101]$				
(112)	[$\bar{1}$10]					
Lefthand	$(\bar{1}\bar{1}1)\ [0\bar{1}\bar{1}]$	$(\bar{1}\bar{1}1)\ [011]$				
Side	$(\bar{1}\bar{1}1)\ [\bar{1}0\bar{1}]$	$(\bar{1}\bar{1}1)\ [101]$				
Righthand	$(\bar{1}\bar{1}1)\ [0\bar{1}\bar{1}]$	$(111)\ [011]$				
Side	$(\bar{1}\bar{1}1)\ [\bar{1}0\bar{1}]$	$(\bar{1}\bar{1}1)\ [101]$				

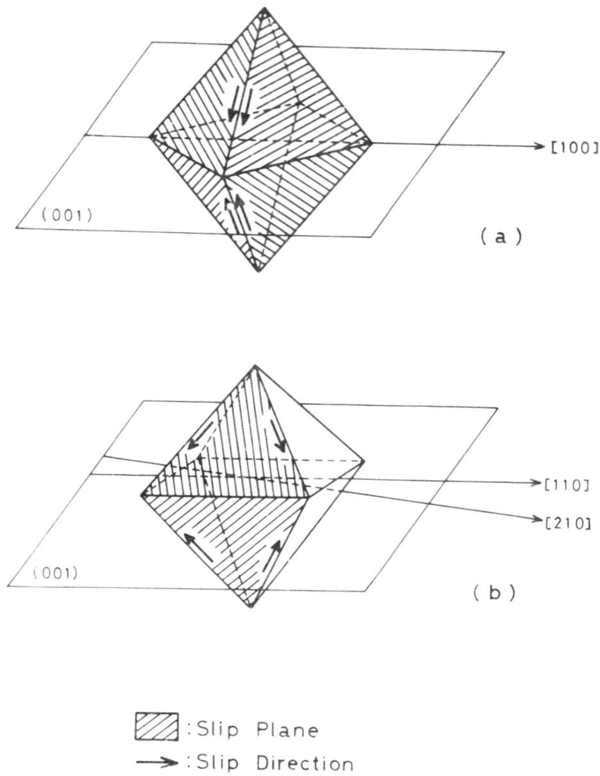

Fig. 2.—Schematic view of active slip systems for the slidings on Cu (001) in the [100], [110] and [210] directions.

of the dislocation structure is not similar to that of copper single crystals subjected to uni-axial tensile deformation or compressive deformation[18] but is similar to that of rolled copper at a heavy processing of about 60%.[19] The slip system which first acted on the friction track was estimated by the Schmid factor to be a double slip of (111) $[0\bar{1}1]$ and $(1\bar{1}1)$ [011]. However, the fact that the elongated cell structure was obtained signifies that the crystal rotation by a single slip on the primary slip plane must take place. This can be explained by the effect of latent hardening.[20] Thus the primary slip system preferentially acted. This is also clear from the selected area diffraction pattern which shows that the plane on the friction track is very near to {110}. A cell boundary was found to be parallel to <211>. Therefore the {110}<211> texture was formed on the friction track. This result is consistent with the rolling texture reported by Wilman et al.[21] The cell size perpendicular to sliding is small (on the order of 0.5 μm). This value is small compared to that of copper deformed in uni-axial tension. An isometric cell structure was found on the ridges. It appears that this cell structure is similar to the tensile deformed structure of copper at Stage II or Stage III.[22] The

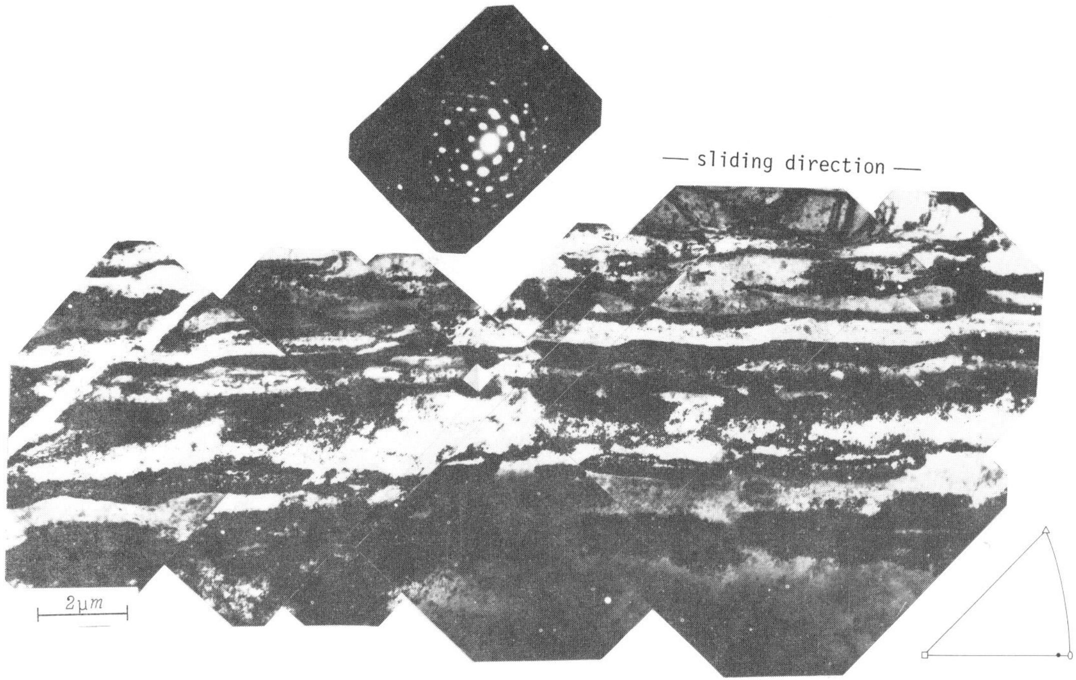

Fig. 3.—Transmission electron micrograph showing a deformation structure for the sliding on the Cu (001) surface in the [100] direction. Note that the pattern shows {110} diffraction and that the cell boundary is nearly along <211>.

isometric cell structure may be caused either by $(111)\ [01\bar{1}]$, $(\bar{1}11)\ [01\bar{1}]$, $(\bar{1}\bar{1}1)\ [011]$ and $(1\bar{1}1)\ [011]$ slip systems or by $(1\bar{1}1)\ [0\bar{1}\bar{1}]$, $(\bar{1}\bar{1}1)\ [0\bar{1}\bar{1}]$, $(\bar{1}11)\ [0\bar{1}1]$ and $(111)\ [0\bar{1}1]$ slip systems shown in Table II.

Figure 4 shows a dislocation structure on the (001) surface with a sliding in the [110] direction. Cell structure developed on the two slip planes forming a grid of parallelograms. The angle between the two cell boundaries on the two slip planes was approximately 80°, and was nearly consistent with the geometrical angle between the two slip planes on {110} surface, 70.5°. Since the direction of stress axis was on [001]-[111] boundary, it is reasonable to mention that the primary slip system of $(1\bar{1}\bar{1})$ [011] and the conjugate slip of $(\bar{1}1\bar{1})\ [10\bar{1}]$ occurred simultaneously. Thus such a layered cell structure was accomplished. The elongated cell boundary was nearly along the <211> direction. Densely tangled dislocations were visible. The dislocation cell structure on ridges was isometric, which might be caused by $(\bar{1}11)\ [01\bar{1}]$, $(\bar{1}11)\ [\bar{1}0\bar{1}]$, $(1\bar{1}1)\ [011]$ and $(1\bar{1}1)\ [\bar{1}01]$ slip systems or by $(1\bar{1}1)\ [10\bar{1}]$, $(1\bar{1}1)\ [0\bar{1}\bar{1}]$, $(\bar{1}11)\ [101]$ and $(\bar{1}11)\ [0\bar{1}1]$ slip systems.

Figure 5 shows the dislocation structure on the (001) surface in which a sliding was conducted along the [210] direction, the intermediate angle between the [100] and the [110] directions. As is similar to the case of the (001) [100] sliding, the elongated cell structure to the sliding direction was predominant. It is therefore considered that the primary slip system of $(111)\ [01\bar{1}]$ preferentially acted. The cell size perpendicular to the sliding direction was as small as 0.5 μm, which is similar in size to that of rolled copper.(19)

The dislocation structures on (110) surface are shown in Figures 6, 7 and 8 for slidings in the [001] direction, the $[\bar{1}10]$ direction, and the $[\bar{1}11]$ direction respectively. Figure 6 shows a predominant cell structure obtained for the sliding in the [001] direction. It is considered from this micrograph that a rotational slip on (110) plane occurred. The cell structure showed a slightly elongated appearance which was parallel to <211>. The cell structure on the ridges might be formed by $(1\bar{1}1)\ [\bar{1}\bar{1}0]$ and $(\bar{1}11)$ $[\bar{1}\bar{1}0]$ slip systems.

The elongated cell structure obtained with a sliding in the $[\bar{1}10]$ direction is shown in Figure 7. A fine cell structure developed. Since the stress axis is close to [100]-[111] boundary, a double slip consisting of the $(111)\ [01\bar{1}]$ primary slip and the $(\bar{1}1\bar{1})$ [011] conjugate slip would occur. Either $(111)\ [0\bar{1}1]$, $(111)\ [\bar{1}01]$, $(\bar{1}\bar{1}1)\ [0\bar{1}1]$ and $(\bar{1}\bar{1}1)\ [\bar{1}01]$ or $(\bar{1}\bar{1}1)\ [\bar{1}0\bar{1}]$, $(\bar{1}\bar{1}1)\ [0\bar{1}\bar{1}]$, $(111)\ [01\bar{1}]$ and $(111)\ [10\bar{1}]$ slip systems may form an isometric cell structure on the ridges.

When sliding was conducted in the $[\bar{1}11]$ direction on the (110) surface, a high density of dislocations and non-uniform cell structure were observed as shown in Figure 8. Friction track appears to consist of elongated cells and uniformly distributed dislocations. This structure may be obtained by active slip systems of $(\bar{1}11)$ [110] and $(111)\ [1\bar{1}0]$. An isometric cell structure, resulting from slip systems of $(\bar{1}\bar{1}1)\ [1\bar{1}0]$, $(111)\ [\bar{1}10]$, $(111)\ [1\bar{1}0]$ and $(\bar{1}\bar{1}1)$ [110], is visible on the ridges.

Figure 9 shows the dislocation structure in the case of friction on the (111) surface in the $[\bar{1}10]$ direction. A high density of dislocations and elongated cell structure were found on the friction track, while an isometric cell structure predominated on the ridges. The former might be

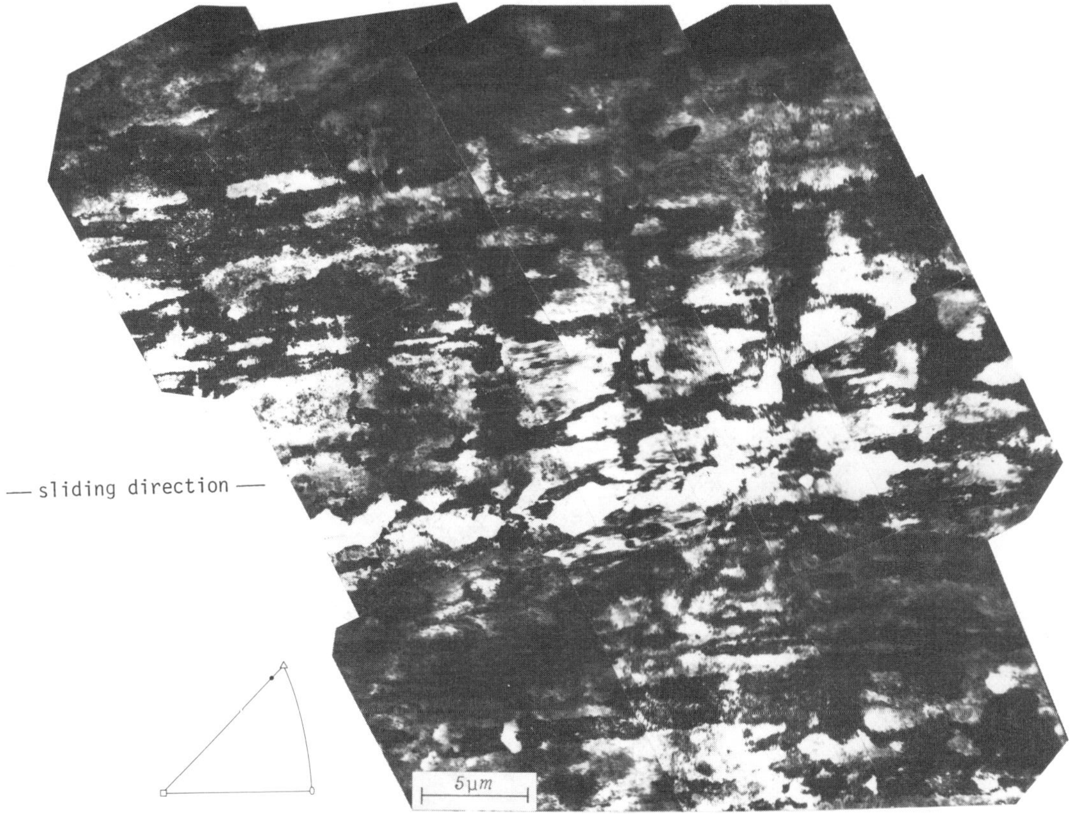

Fig. 4.—Layered cell structure resulting from a sliding on the Cu (001) surface in the [110] direction.

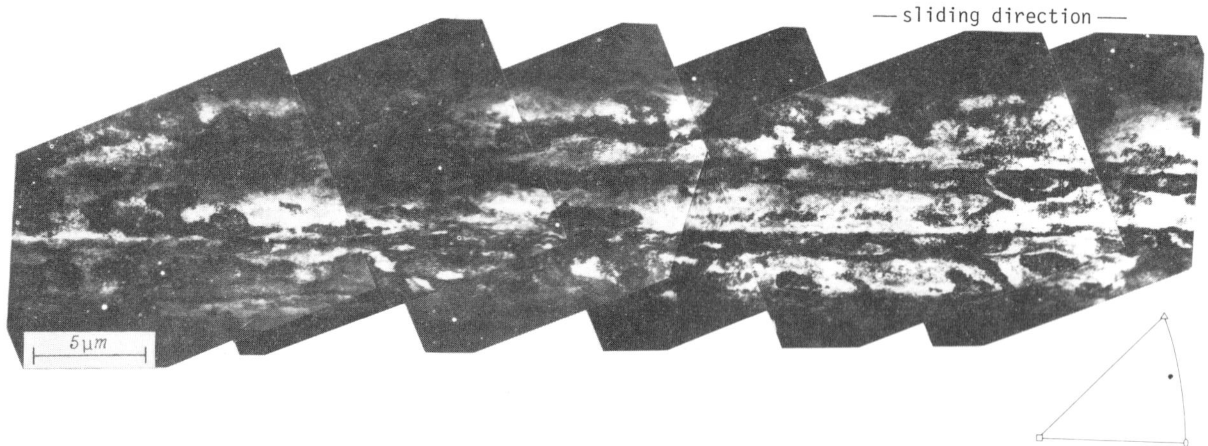

Fig. 5.—Elongated cell structure observed for a sliding on the Cu (001) surface in the [210] direction.

Fig. 6.—Transmission electron micrograph showing cell structure; sliding on the Cu (110) surface in the [001] direction.

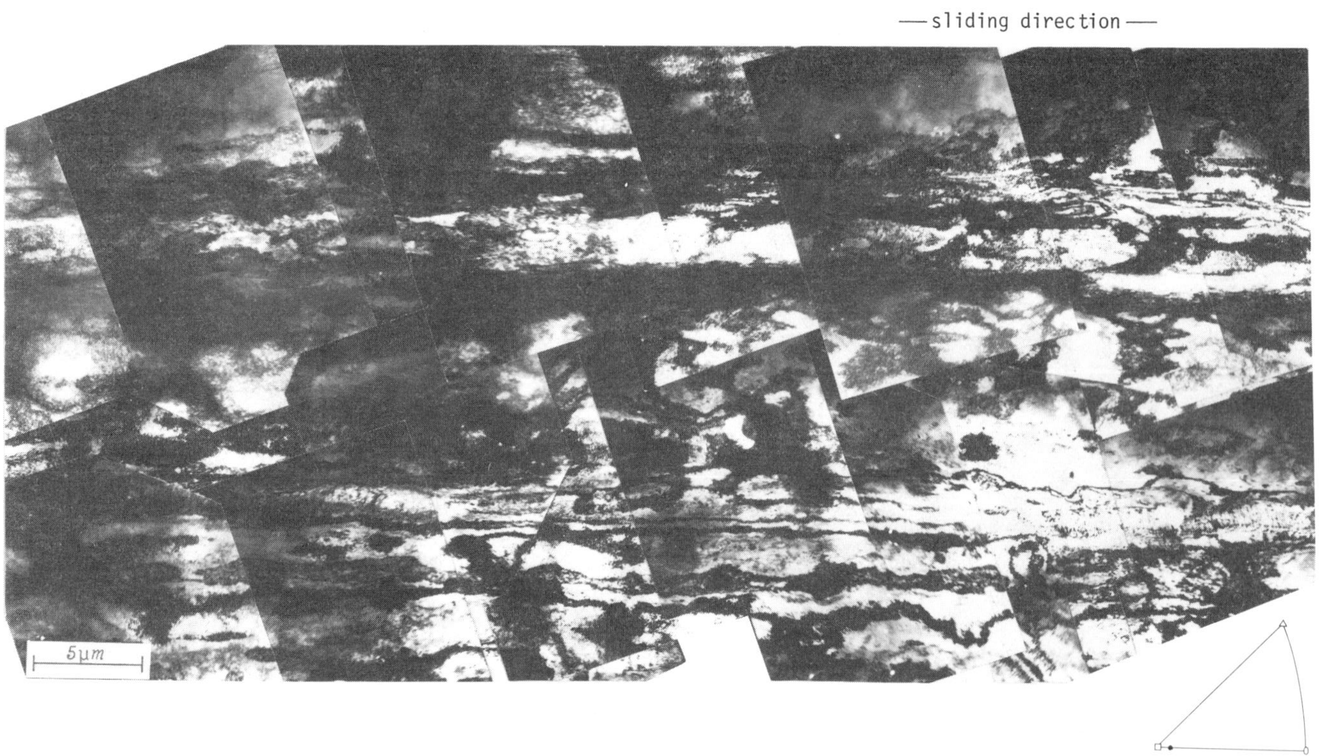

Fig. 7.—Elongated cell structure resulting from a sliding on the Cu (110) surface in the [$\bar{1}$10] direction.

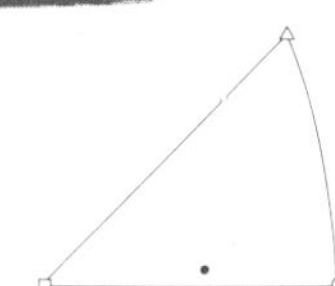

Fig. 8.—A high density of dislocations and well developed cell structure observed for a sliding on the Cu (110) surface in the $[\bar{1}11]$ direction.

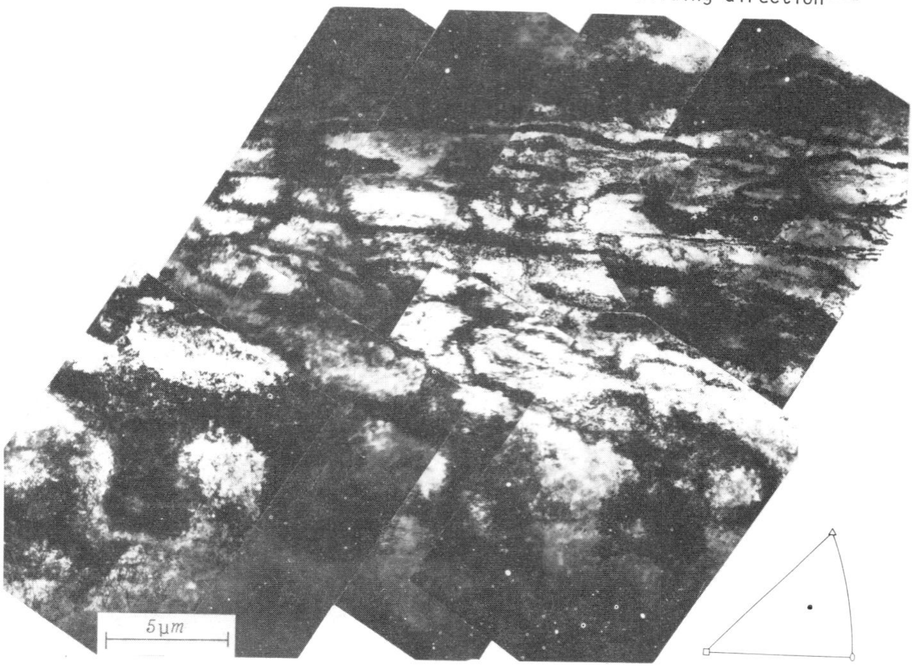

Fig. 9.—Elongated cell structure resulting from $(\bar{1}11)$ [110] single slip; sliding on the Cu (111) surface in the $[\bar{1}10]$ direction.

caused by the single slip of $(\bar{1}11)$ [110]. A wide cell boundary and a uniform distribution of dislocations were observed on ridges.

Figure 10 shows an elongated cell structure when sliding was made on the (112) surface in the $[\bar{1}10]$. The single slip of $(\bar{1}11)$ [101] may occur to form this elongated cell structure, and the cell boundary was nearly parallel to the $\langle 21\bar{1}\rangle$ direction. The structure on the ridges suggests the actions of $(\bar{1}\bar{1}1)$ $[0\bar{1}\bar{1}]$, $(\bar{1}\bar{1}1)$ $[\bar{1}0\bar{1}]$, $(\bar{1}\bar{1}1)$ [011] and $(\bar{1}\bar{1}1)$ [101] slip systems.

Figure 11 is a plot of cell size obtained for each sliding condition. In this case, an assumption was made that the shape of the cell was an ellipse. For the slidings on the (001) surface, the largest cell size was obtained in the [100] sliding, an intermediate one in the [210] sliding and the smallest in the [110] sliding. It is accepted that the cell size is inversely proportional to the degree of plastic deformation.[23] Therefore the degree of plastic deformation will be described as a simplified expression of [110]>[210]>[100]. Similarly on the (110) surface, it is possible to describe the degree of plastic deformation in the order $[\bar{1}11]$>$[\bar{1}10]$>[001]. It is clear therefore that the greater the number of active slip systems which contributed to a rotational slip, the greater the degree of plastic deformation. Attention should be given to the fact that the relationship in the order of plastic deformation is exactly consistent with that of fric-

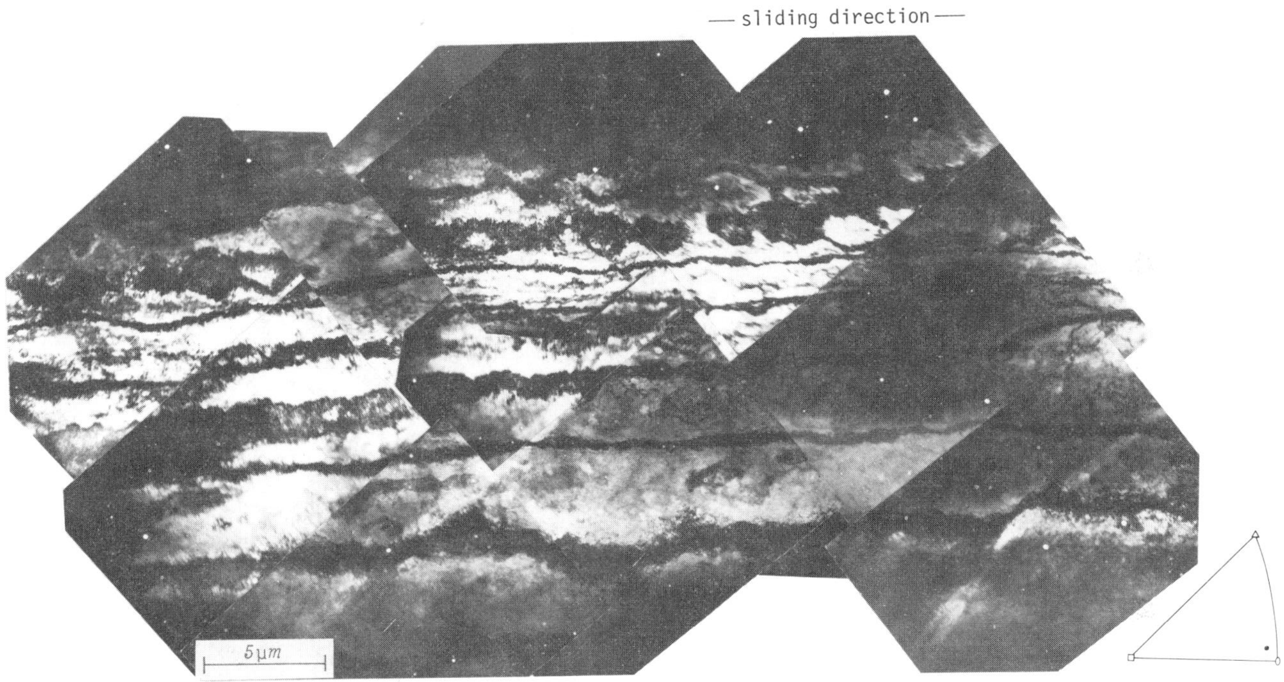

Fig. 10.—Elongated cell structure resulting from $(\bar{1}11)$ [101] single slip; sliding on the Cu (111) surface in the $[\bar{1}10]$ direction.

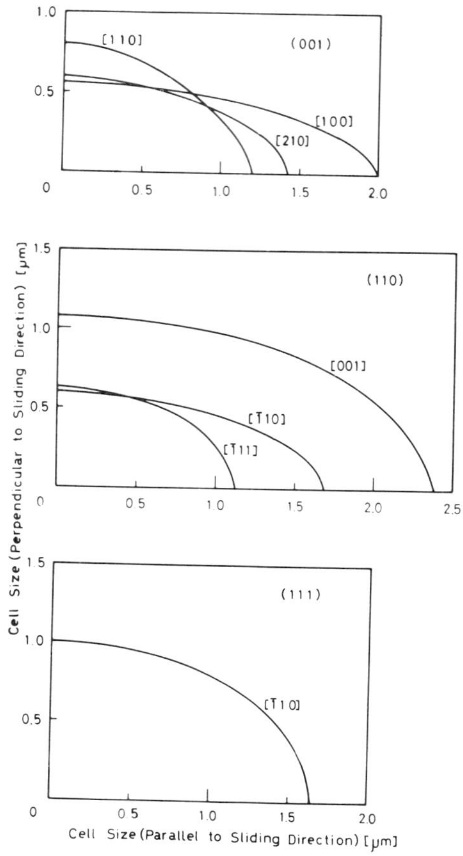

Fig. 11.—Plot of cell size for varied sliding conditions. Note that the cell size on each plane is inversely proportional to friction coefficient shown in Table III.

tion coefficient shown in Table III. It is therefore reasonable to mention that the process of plastic deformation determines the magnitude of friction.

MICROSTRUCTURES ON THE CROSS-SECTIONS OF FRICTION TRACK

The mechanism of deformation due to friction could accurately be studied by observing the cross-sections. The results shown in Figures 12 and 13 were obtained with a sliding on the Cu (001) surface in the [110] direction. Figure 12 shows the dislocation structure on the cross-section normal to the sliding direction. A very fine cell structure was observed at the region approximately 1~2 μm below surface. The appearance of cell structure in this region is rather isometric with a mean cell diameter of 0.4 μm. Therefore the cell structure at the surface may be cylindrical in

Fig. 12.—Dislocation cell structure on the cross-section normal to the sliding direction; sliding on the Cu (001) surface in the [110] direction.

TABLE III.—A SUMMARY OF FRICTION COEFFICIENTS OBTAINED FOR THE CONDITIONS SHOWN IN FIGURES FROM 3 TO 10.

Plane	Direction	Friction Coefficient
(001)	[100]	0.9
(001)	[110]	1.25
(001)	[210]	1.0
(110)	[001]	0.8
(110)	$[\bar{1}10]$	0.95
(110)	$[\bar{1}11]$	1.4
(111)	$[\bar{1}10]$	0.4
(112)	$[\bar{1}10]$	0.5

shape, when considering the cell structure shown in Figure 4. Figure 12 also suggests that the texture is limited to the surface region and that the compressive deformation is predominant in the inner region. As the depth from surface increases, the cell size becomes large and the misorientation between each cell becomes large too. The cell structures were also observed around the ridges. However, they are large in diameter compared to that around the friction track. Thus the degree of deformation is relatively low on the ridges. At far below the surface, a dense clustering of stacking faults was visible.

Figure 13 shows the dislocation structure on the cross-section parallel to the sliding direction. The dislocation density below surface was extremely high and the formation of cell structure was predominant. The cell size becomes large, as the depth from surface increases. A misorientation between cells was large at the region of several micrometers below surface. The appearance of cell structure exhibits a slight elongation to the sliding direction in the region of approximately 5 μm below surface. In contrast, an isometric cell structure was observed in the region away from surface. The region of low dislocation density due to image force, postulated by Suh,[(24)] was not observed in the present study.

DEFORMATION DURING FRICTION

Figure 14 is a scheme of dislocation cell structure based on the transmission electron microscope study in the foregoing sections. It has become evident that a difference in the dislocation structures exists between the friction track (groove) and the ridges, and that friction is primarily dependent on the former. At the very surface of the groove, the {110} <211> texture resulted independently of the crystallographic planes and directions. For the sliding on the (001) surface in the [100] direction, the primary slip system predominated due to the latent hardening, although a double slip of $(1\bar{1}1)$ [011] and (111) $[0\bar{1}1]$ occurred. The texture was caused by the rotational slip. For the slidings on the (001) surface in the [110] direction, a double slip of $(1\bar{1}\bar{1})$ [011] primary slip and $(\bar{1}1\bar{1})$ $[10\bar{1}]$ conjugate slip occurred. In the case of the [210] sliding on the (001) surface, a single slip of $(1\bar{1}\bar{1})$ $[01\bar{1}]$ acted first, then a double slip combined with a conjugate slip occurred. As for the slidings on the (110) surface, the double slips of the combinations; primary slip and conjugate slip, primary slip and conjugate slip, and primary slip and conjugate/critical slips occurred in the [001], $[\bar{1}10]$, and $[\bar{1}11]$ directions, respectively.

Fig. 13.–Dislocation cell structure on the cross-section parallel to the sliding direction; sliding on the Cu (001) surface in the [110] direction.

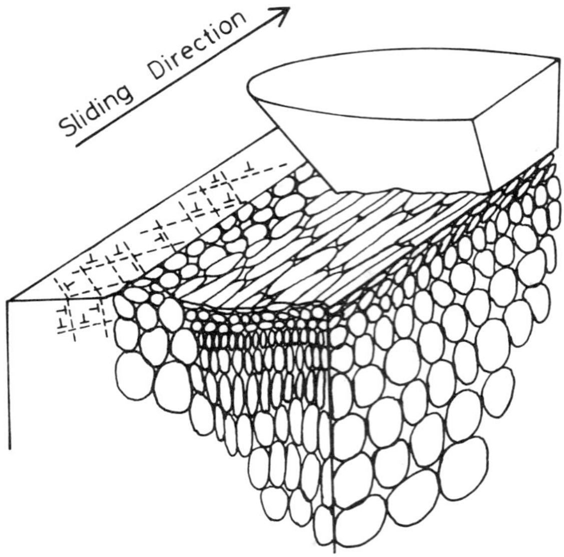

Fig. 14.—A model showing deformed structure due to friction, as it is based on the transmission electron microscope studies.

The slidings on the (111) surface in the [$\bar{1}$10] direction and on the (112) surface in the [$\bar{1}$10] direction caused a crystal rotation by a single slip. Thus higher friction coefficients were obtained when a double slip occurred, while lower friction coefficients were frequently obtained when a single slip took place. In the case where the latent hardening led to a predominant primary slip in spite of the action of double slip, intermediate values of friction coefficients resulted. As shown earlier for aluminum, the types of slip system do determine the friction.

The second layer is caused by a crystal rotation, the axis of which lies on the surface and normal to the sliding direction. In other words, the bending axis of the second layer is dependent on the sliding direction. This layer is consistent with the deformation structure which has been clarified by the existence of Laue asterism.[25]

The third layer consists of the compressed zone. Although the deformation mode is a relatively simple compression, both the deformed region and the degree of work hardening depend on the dislocation interactions on each active slip system.

The fourth layer is undistorted matrix.

CONCLUSIONS

It is concluded from the present transmission electron microscope study that the friction property primarily is dependent upon the plastic deformation at the very surface of friction track. The anisotropic friction behavior was explained by the relationship between the active slip systems and the direction of stress axis. The deformation mode due to friction was defined as a complicated tensile deformation under compression. Friction caused a work hardened layer consisting of the {110} <211> texture, a distorted region depending on the sliding direction, a compressed zone, and an undisturbed matrix. In view of the experimental results, the point to be emphasized is that friction depends both on the crystallographic variables and on the chemistry of solid surfaces.

ACKNOWLEDGEMENTS

I would like to thank Fusao Akiyama for carrying out much of the transmission electron microscopy. The use of the high voltage transmission electron microscope was by the courtesy of Professor H. Fujita. His valuable suggestions on this work are greatly acknowledged. Thanks are also due Professor T. Tsukizoe, S. Tanaka and T. Okuyama for helpful comments. This work was supported, in part, by a Grant-in-Aid for Scientific Research (A) from the Ministry of Education, Japan.

REFERENCES

1. Buckley, D.H., These Proceedings.
2. Buckley, D.H., *Wear,* Vol. 20, 1972, p. 89.
3. Buckley, D.H., *Journal of Vacuum Science and Technology,* Vol. 13, 1976, p. 88.
4. Buckley, D.H., in Proceedings, Wear of Materials '77, edited by W.A. Glaeser, K.C. Ludema and S.K. Rhee, American Society of Mechanical Engineers, New York, 1977, p. 12.
5. Bailey, J.M. and Gwathmey, A.T., *American Society of Lubrication Engineers Transactions,* Vol. 5, 1962, p. 45.
6. Steijn, R.P., *Wear,* Vol. 7, 1964, p. 48.
7. Roshon, D.D., *Journal of Applied Physics,* Vol. 35, 1964, p. 1262.
8. Buckley, D.H., "The Effect of Orientation and the Presence of Surface Active Materials on the Friction, Deformation and Wear of Aluminum," NASA TN D-4994, 1964.
9. Buckley, D.H., *American Society for Testing and Materials. Special Technical Publications,* No. 431, 1968, p. 248.
10. Buckley, D.H., *American Society of Lubrication Engineers Transactions,* Vol. 11, 1968, p. 89.
11. Young, F.W., *Journal of Applied Physics,* Vol. 32, 1962, p. 963.
12. Buckley, H.E., "Crystal Growth," John Wiley and Sons, New York, 1951, p. 93.
13. Buckley, D.H., "Friction, Wear, and Lubrication in Vacuum," NASA SP-277, 1971, p. 86.
14. Bowden, F.P. and Brookes, C.A., *Royal Society of London. Proceedings.* Series A, Vol. 295, 1966, p. 244.
15. Tsuya, Y., *Wear,* Vol. 14, 1969, p. 309.
16. Sato, J., *Japan Society of Lubrication Engineers. Journal,* Vol. 15, 1970, p. 751 (in Japanese).
17. Akiyama, F., M.Sc. Thesis, Osaka University, 1978 (in Japanese).
18. Steeds, J.W., *Royal Society of London. Proceedings.* Series A, Vol. 292, 1966, p. 343.
19. Kawasaki, Y., *Physical Society of Japan. Journal,* Vol. 36. 1974, p. 142.
20. Jackson, P.J. and Basinski, Z.S., *Canadian Journal of Physics,* Vol. 45, 1967, p. 707.
21. Goddard, J., Harker, H.J. and Wilman, H., *Journal of Physics B: Atomic and Molecular Physics,* Vol. 80, 1962, p. 771.

22. Essman, U., *Acta Metallurgica,* Vol. 12, 1964, p. 1467.
23. Keh, A.S. and Weissman, S., "Electron Microscopy and the Strength of Crystals," edited by G. Thomas and J. Washburn, Wiley-Interscience, New York, 1963, p. 231.
24. Suh, N.P., *Wear,* Vol. 25, 1973, p. 111; *Wear,* Vol. 44, 1977, p. 1.
25. Nakajima, K. and Kawamoto, J., *Wear,* Vol. 11, 1968, p. 21.

DISCUSSION

D. H. BUCKLEY, NASA: Did you look at the effect of impurities in the copper? The impurities alter dislocation cell structure in copper.

OHMAE: That is true. We obtained the copper single crystals by the Bridgman method. The Auger analysis showed impurities like carbon. In these observations by TEM at low magnification, it is rather difficult to show the impurity pinning of dislocations there. Surely the impurity has an influence on dislocation cell structure. That is a good point.

S. RAMALINGAM, Georgia Institute of Technology: In one of your figures of dislocation cell structure you show a crack. Quite far away from the crack, you show clustering of stacking faults. It is surprising that you could observe that in copper at these low strain rates, would please clarify?

OHMAE: Well, it is rather surprising to me, too. The misorientation between the cell is very large in that area. Probably Frank Partials are responsible for the clustering effect.

A. W. RUFF, NBS: Was the speed so fast that there might be heating at the interface?

OHMAE: The experiments were conducted at relatively low sliding velocity (10 μm per sec) to avoid heating.

RUFF: All right. Was there any evidence of structural changes near the surface? I know you saw fairly consistent patterns of cells from the interior material right up to the deformed surface. Although the size and the shape of the cells are changing, neither deformation twinning nor voids can be seen near the surface.

OHMAE: We saw voids. The dislocation of density is very high, around $10^{11}/cm^2$ or so. In these micrographs it is very difficult to distinguish void because it is not clear. But by direct observation in the microscope it is possible to detect voids especially in the cross-section.

RUFF: Finally, was there an oxide film on the surface and was that oxide worked into the copper specimen during the sliding?

OHMAE: That is important. Using the electron diffraction technique in the TEM we could not detect the copper oxide, but in some cases we did. We are not sure whether this oxide is formed during the wear test or during the preparation of thin foils.

QUESTIONER: What are the size and the misorientation of the cells of the top layer?

OHMAE: The cell size perpendicular to the sliding is about 0.5 μm. The misorientation is three degrees as obtained by the TEM diffraction technique.

QUESTIONER: Were the sliding experiments done with single pass?

OHMAE: Yes.

STRUCTURAL CHANGES IN THE SURFACE LAYER DUE TO SLIDING AND THEIR EFFECTS ON FRICTION AND WEAR OF MATERIALS

Y. Mizutani

ABSTRACT

Segregation, phase transformation, plastic deformation and oxidation in solid surface layers during sliding are discussed on the basis of the results obtained by a combined use of AES, EPMA, X-ray and EM methods. It is emphasized in this paper that such changes play an important role in the friction and wear behaviors of metals and alloys. This view is consistent with that proposed by Buckley.

INTRODUCTION

Very careful investigation of the sliding surface of solids is needed to understand the fundamental aspects of tribology, since the nature of the surface may strongly affect the friction, wear and lubrication of materials. Most of the subjects reviewed by Buckley(1) are concerned with the basic surface science, i.e., studies on the outermost layer of solid surfaces in which highly purified single crystals were examined in a well-controlled atmosphere under very low sliding speed and load. The use of surface analytical tools has gradually revealed the basic phenomena on the sliding surfaces of solids.

Detailed behavior of the solid surfaces at sliding contacts, even of simpler types of sliding in conventional sliding tests and practical systems is very complicated. A full understanding of what happens at the contact interfaces has not yet been attained. It is therefore desirable to correlate the basic knowledge with test results and investigate the changes in structure and composition of solid surface layers due to friction, from both macroscopic and microscopic points of view. For the microscopic observation, it is very important to use a proper combination of analytical tools such as scanning electron microscopy (SEM), Auger electron spectroscopy (AES), electron probe microanalysis (EPMA) and X-ray diffraction techniques.

In this paper, some topics on surface segregation, phase transformation, plastic deformation and oxidation in solid surface layers due to friction will be discussed in relation to the subjects reviewed by Buckley.(1)

STRUCTURAL CHANGES DUE TO FRICTION

As a result of relative motion of two surfaces in contact, some of the external work is dissipated, resulting in generation of heat, plastic deformation and wear in their surface layers. The amount of energy dissipation depends on the surface and bulk materials in contact and the environmental conditions.

Figure 1 illustrates the structural changes in the surface of metals and alloys due to friction.(2) This indicates a mechanism by which frictional energy dissipation results in structural changes; plastic deformation, surface segregation, phase transformation, precipitation and corrosion (or oxidation). Thus, a careful investigation of the sliding surface layer is very important for the understanding of friction, wear and lubrication.

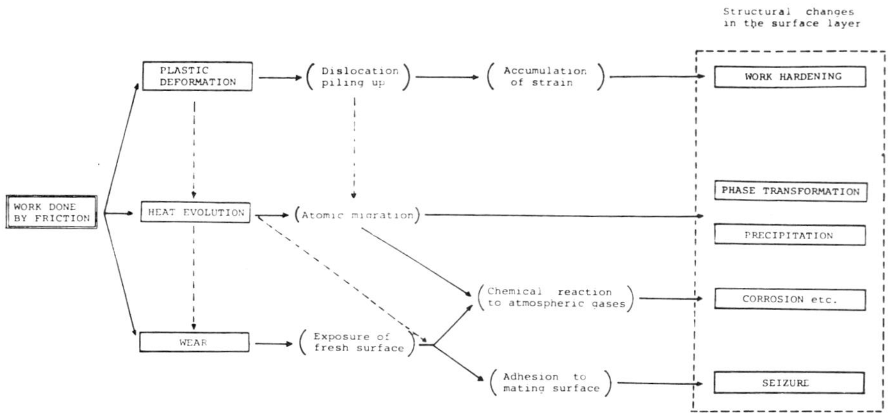

Fig. 1.—Illustration of structural changes in the surface layer of metals and alloys due to friction. (Mizutani, Ref. 2)

CRYSTAL ROTATION AND PLASTIC DEFORMATION

The surface orientation of crystalline solids may change due to crystal rotation or plastic deformation during the rubbing process. Such change may be associated with the change in surface energy, because the surface energy is sensitive to the crystallographic orientation.(1)

The axis of crystal rotation due to rubbing is dependent on the rubbing direction.(3) Ito et al.(4) utilized X-ray diffraction and AES techniques to study the effect of gentle rubbing of MoS_2 sputtered film. They showed that in the surface layer the texture parallel to the surface changed from $(10\bar{1}0)$ to (0002) during the rubbing process (see Figure 2). They also reported that the intensity of Auger spectra of the sputtered MoS_2 film changed with the crystal rotation and the Mo/S intensity ratio coincided with that obtained from a cleaved surface of MoS_2 single crystal.(4) The formation of the (0002) preferred orientation in the surface layer might lead to a decrease in the coefficient of friction. Such a rotation in crystalline solids during friction should arise from the rearrangement of

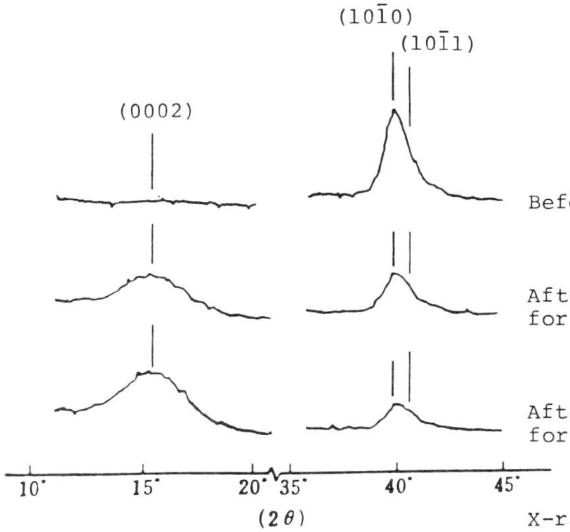

Fig. 2.—Change in X-ray diffraction profiles of sputtered MoS_2 film with rubbing. Thickness of the film; 0.7 μm, rubbing velocity; 5cm/sec., apparent pressure; 13 g/cm^2. (Ito, Ref. 4)

the surface layer so as to minimize the energy of deformation.[(5)]

It has been usually observed in metals and alloys that large plastic deformation occurs in the sliding surface layer. Figure 3 shows a sliding surface layer of SUS 304 Stainless steel (18Cr-8Ni) in which a plastic flow is observed along the sliding direction.[(6)] In this case, the thick-

Fig. 3.—Cross section through the surface of SUS 304 stainless steel (18% Cr-8%Ni) after sliding. Sliding velocity; 311 mm/sec., load; 3.15 kg (6.3 kg/cm^2). (Shimura, Ref. 6)

ness of the deformed layer is closely related to the particle size of wear debris (see Figure 4).

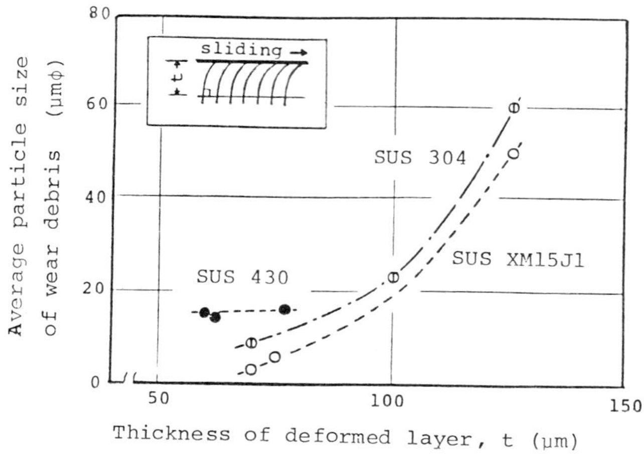

Fig. 4.—Relation between thickness of deformed layer and particle size of wear debris in stainless steels: SUS 430 (16%Cr), SUS 304 (18%Cr-8%Ni) and SUS XM15J1 (16%Cr-12%Ni-4%Si). Sliding velocity; 311 mm/sec., load; 3.15 kg (6.3 kg/cm^2). (Shimura, Ref. 6)

Pile-up of dislocations has been observed in the surface layer of metals and alloys after sliding.(7,8) In 0.3% carbon steel, the dislocation density varied with the depth from the sliding surface as shown in Figure 5.(3) According to Cottrell,(9) impurity and solute atoms may interact with dislocations, and Cohen(10) indicated that atomic migration may be promoted by plastic deformation, owing to mutual interaction between moving dislocations and diffusing atoms. Such atomic migration toward the surface may cause a change of chemical composition in the surface layer, that is, "surface segregation."

SURFACE SEGREGATION

Recently, surface segregation of minor components of metals and alloys has been investigated by using surface analytical tools such as AES, LEED (Low energy electron diffraction) and ESCA (Electron spectroscopy for chemical analysis). However, the data on tribological properties is not extensive despite the importance in understanding fundamental phenomena of friction, wear and lubrication of materials.

Of earlier works, the following results are very instructive. Nakajima and Isogai(11) showed by using EPMA that the solute content in the surface layer of *Cu-Al, CuZn* and *Fe-Si* single crystal tends to increase due to friction. They(12) also observed with the aid of electron diffraction technique that during the break-in period, fine particles of graphite were produced on the surface of a spinning frame ring (a plain carbon steel), rotating around a ring traveler at about 30 m/sec., resulting in a good lubrication at the sliding interface. On the other hand, Ferrante and

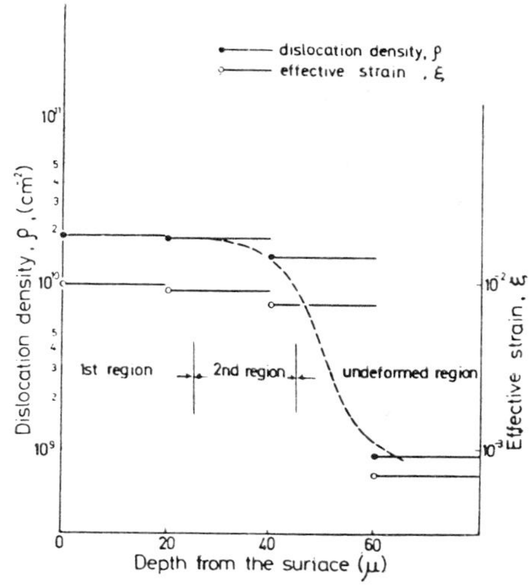

Fig. 5.—Dislocation density and effective strain vs. depth from surface of specimen Fe-0.3%C abraded at 7 cm/sec. under 700 g/mm^2. (Nakajima, Ref. 3)

Buckley[13] indicated that the adhesion force between *Cu-Al* and *Au* single crystals was dependent on the composition of the surface layer, rather than on the bulk composition of *Cu-Al* crystal. The surface layer was enriched with *Al*, due to surface segregation.

Compositional changes in the sliding surface layers of metals and alloys may be caused by friction-induced diffusion due to the generation of heat and plastic flow during friction.[2,14] The diffusing atoms may behave so as to form an equilibrium state, i.e., a state of the lowest free energy in the surface.[2,13,15]

As is shown in the review paper,[1] the effect of heating on the surface segregation is significant. For example, the intensity change of Auger spectra of *Sn* on the surface of *Cu-5%Sn* alloy with temperature is given in Figure 6. It is seen that the intensity of the Auger peak increased with increasing temperature and reached a maximum after heating at 300°C for 30 minutes, which corresponds to about 8 times larger than the original intensity. The present author[15,16] also observed a remarkable surface segregation of *Mo* and *S* in *Fe-Mo-S* alloys, forming a layer of $MoS_{1.8}$, a defective modification of MoS_2 which was identified by X-ray diffraction and EPMA. It is seen in Figure 7 that the diffraction intensity of $MoS_{1.8}$ in the surface layer varies with temperature and time in the vacuum heating. The maximum layer thickness of 0.3 μm was obtained for 800°C, at which an optimum balance was maintained between the rates of layer formation and evaporation. According to Buckley et al.,[17] the rate of evaporation in MoS_2 increases with increasing temperature and decreasing atmospheric pressure. It has been concluded that the formation of the

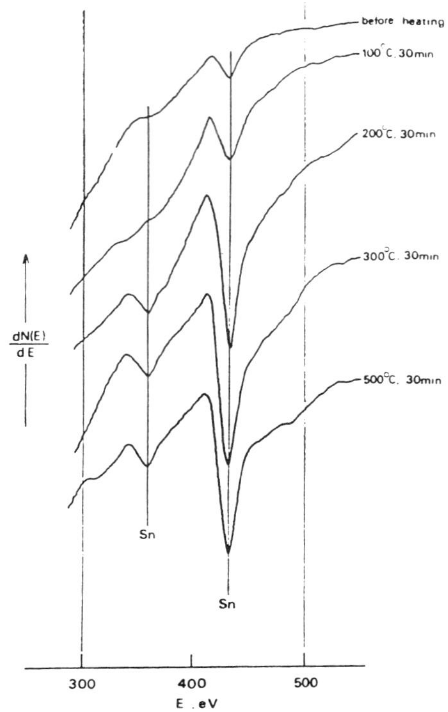

Fig. 6.—Change of Auger spectra of Sn in the surface of Cu-5%Sn as a function of heating temperature. Ambient pressure; 7 x 10^{-10} Torr. (Nakajima, Ref. 14)

molybdenum sulfide layer depends on heating temperature, heating time, ambient pressure and degree of metastability; i.e., the more unstable the initial structure is, the more the formation occurs. Figure 8 is an electron micrograph of the surface of $MoS_{1.8}$ layer prepared by the heating of *Fe-Mo-S* alloy in vacuum, showing a mosaic-like structure.

Rubbing caused surface segregation of *Sn* in *Cu-Sn* alloy as observed by AES(14) and the intensity of Auger spectra was found to increase with rubbing and reach a value, about 3 times larger than the original one.

Figure 9 shows the effect of surface segregation in a *Fe-Mo-S* alloy on the friction coefficient. In specimen A, the coefficient of friction decreases with sliding time, which corresponds to the formation of $MoS_{1.8}$ layer during sliding, as shown in Figure 10. In specimen B, the thin layer of $MoS_{1.8}$ formed before sliding shows a good lubrication at the beginning of the test. These results suggest that heat generation during sliding may play an important role in forming the molybdenum sulfide layer on the surface.

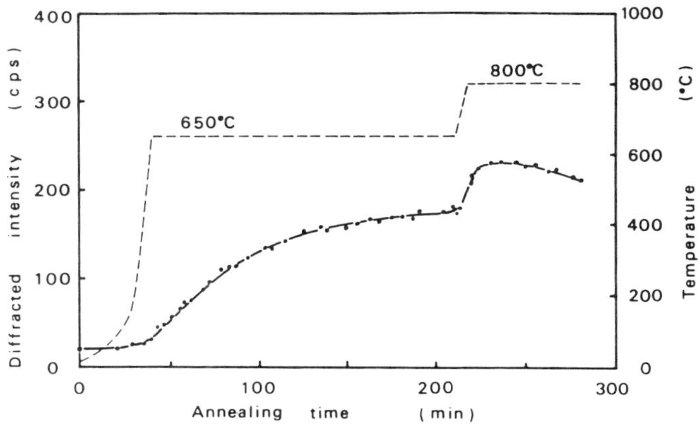

Fig. 7.—Intensity change of molybdenum sulfide ($MoS_{1.8}$) with isothermal annealing in the surface of Fe-16.6%Mo-2.34%S (as cast). Ambient pressure; 3∿5 x 10^{-5} Torr. (Mizutani, Ref. 15)

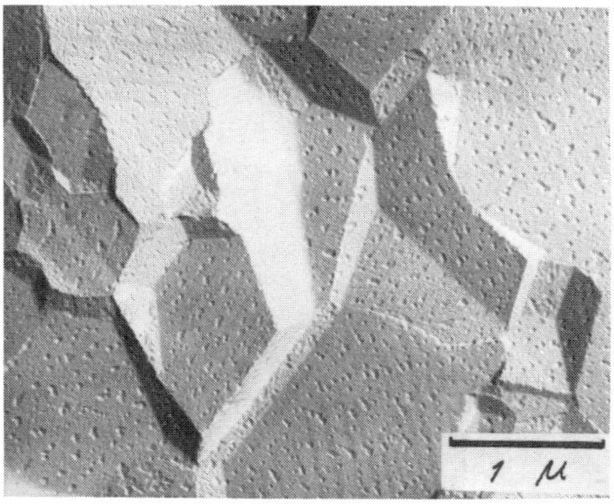

Fig. 8.—Electron micrograph obtained from the surface of Fe-22.1%Mo-2.61%S alloy annealed in vacuum at 900°C for 2 hours. (Mizutani, Ref. 16)

PHASE TRANSFORMATION

Surface segregation during friction is sometimes accompanied by phase transformation or precipitation.

Oda et al.[18] investigated the sliding surface of a multi-layered alloy bearing, consisting of over-lay (*Pb-8%Sn*), second layer (*Cu-28%Pb-7%Sn*) and back metal (a low carbon steel), which was against a cast iron. It was found that the content of *Sn* in the second layer was remarkably enriched (*25%Sn*), resulting in the formation of intermetallic compound, ε-Cu_3Sn, when the bearing was scuffed seriously after wearing off the over-lay. The formation of ε-phase might be caused by friction-induced diffusion of *Sn* from the over-lay into the second layer. A similar phenomenon was

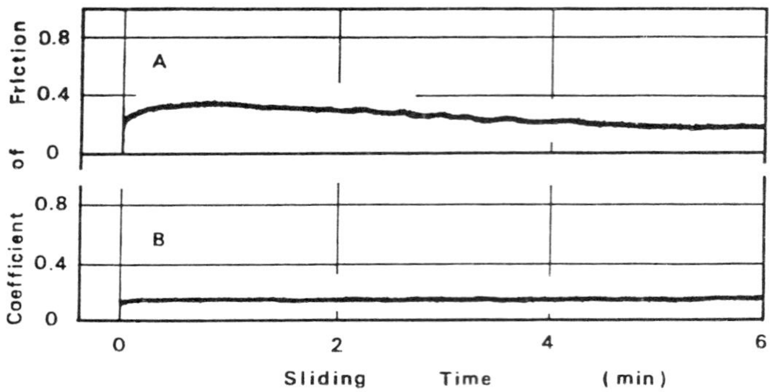

Fig. 9.—Variation of coefficient of friction with sliding time for Fe-22.1% Mo-2.61%S on hardened 0.5% carbon steel. Sliding velocity; 7.9 m/sec., load; 18.75 kg (3 kg/cm²). Specimen A; as cast, Specimen B; annealed at 900°C for 2 hours in vacuum to form a thin layer of molybdenum sulfide on the surface. (Mizutani, Ref. 15)

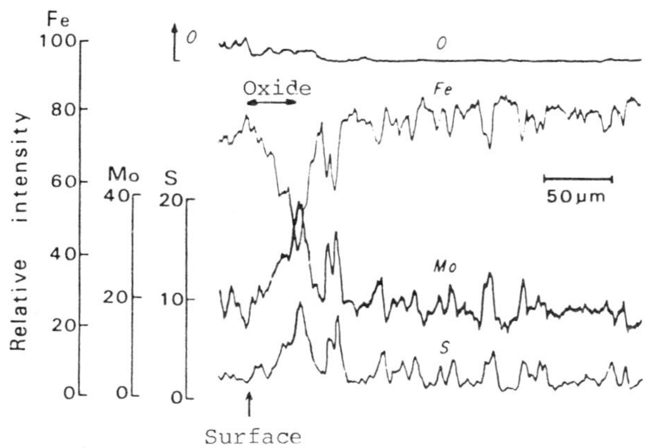

Fig. 10—EPMA line analysis of Mo, S, Fe and O along the depth from the surface in the taper section of Fe-22.1%Mo-2.61%S alloy (as cast) after sliding. Sliding velocity; 7.9 m/sec., load; 18.75 kg (3 kg/cm²). (Mizutani, Ref. 15)

observed in the second layer after heating of the bearing at 170°C for 120 hours.

Figure 11 shows the result of line analysis by EPMA obtained from the surface of η-Cu_6Sn_5 after sliding. This represents occurrence of partial decomposition of η-Cu_6Sn_5 into ε-Cu_3Sn during sliding process. A similar experiment was carried out with ε-Cu_3Sn, but no change could be observed in the surface composition. The friction coefficient of ε-Cu_3Sn is much higher than that of *Cu*, *Sn* or η-Cu_6Sn_5.[14]

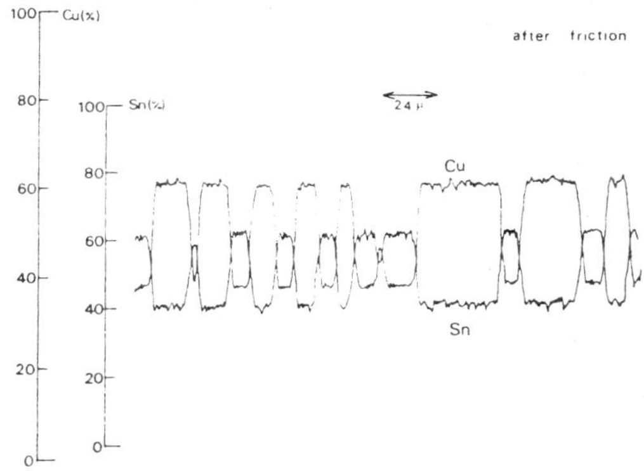

Fig. 11.—EPMA line analysis of η-Cu_6Sn_5 surface after sliding, showing partial decomposition of η-Cu_6Sn_5 into ε-Cu_3Sn. Sliding velocity; 78 cm/sec., load; 1026 g. (Nakajima, Ref. 14)

In carbon steels, martensitic transformation has been often observed in the surface layer after sliding. This transformation is an allotropic one without any change in the chemical composition, which is different from that described above. The formation of the martensite suggests that the temperature at the surface has risen enough to transform ferrite to austenite, since the martensitic transformation results from quenching the austenite.[19] Although the hardness of the martensite structure is very high, the effect of its hardness on the friction and wear property may disappear with the temperature rise during friction, because of its retransformation to austenite.

An attempt has been carried out to investigate a strain-induced martensitic transformation in austenitic stainless steels due to sliding,[6] but no meaningful correlation between the transformation and the frictional property was found.

It is thus concluded that the generation of heat due to friction plays an important role in phase change in the surface and that, in some cases, its change strongly affects the friction and wear properties of metals and alloys. Much more data on this, however, is desired.

OXIDATION

It is a very interesting fact that oxygen is probably the best "lubricant" available,[1] because oxidation does occur, more or less, on the surface of metals and alloys during sliding in ambient air.

Figure 12 shows a well-known wear curve with sliding velocity in cast iron and steel.[20] This wear curve consists of four regions; 1) Oxidative wear region in which oxide wear debris such as Fe_2O_3 is formed, 2) Abrasive wear region in which metallic wear debris predominates, 3) Oxidative wear region in which Fe_3O_4 or FeO is formed and 4) Adhesive wear

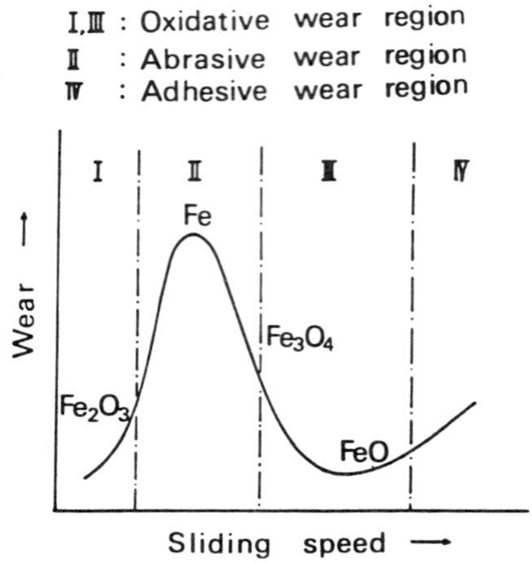

Fig. 12.—Variation of wear loss with sliding velocity in cast iron and steel. Main constituent of wear debris in each wear stage is also described. (e.g., Okabayashi, Ref. 20)

region. It is seen from this figure that formation of oxide on the sliding surface can reduce severe wear of cast iron and steel.

Figure 13 indicates an interesting result to show the effect of oxide on wear.[6] The result was obtained from the sliding tests of some steels at the speeds less than 311 mm/sec. This graph reveals an obvious dependence of wear on the amount of oxide. This oxide was identified by X-ray diffraction as α-Fe_2O_3.

The friction coefficients of the oxides as indicated in Figure 12 are different from each other. Both Fe_3O_4 and FeO have a low coefficient of friction, while Fe_2O_3 has a high value, ~0.8, which is 2 or 3 times larger than that of Fe_3O_4.[21]

According to Figure 12, the formation of oxide on the sliding surface depends on the sliding conditions such as sliding velocity. This finding suggests that the oxide once formed under a given condition may change into another type of oxide under another condition. As a result of this change, the friction and wear property may vary to some extent. Figure 14 is an example of this change which was obtained in a sliding test of Fe_3O_4 film against a hardened tool steel.[22] The coefficient of friction of Fe_3O_4 is initially low, but with sliding time it increases rapidly and then reaches a constant. It is found by careful X-ray diffraction that this increase corresponds to the change of Fe_3O_4 to α-Fe_2O_3. It is also noted in this diagram that such change can be accelerated by increment of applied load. Thus, it is pointed out that the oxidation phenomenon is closely related to the frictional heat which is a function of applied load and sliding velocity.[23]

As long as Fe_3O_4 or FeO is formed during the sliding process, a good lubricating interface can be produced without the need for any solid lubricant. In such a case, oxide can be the best lubricant available for use, as proposed by Buckley.[1]

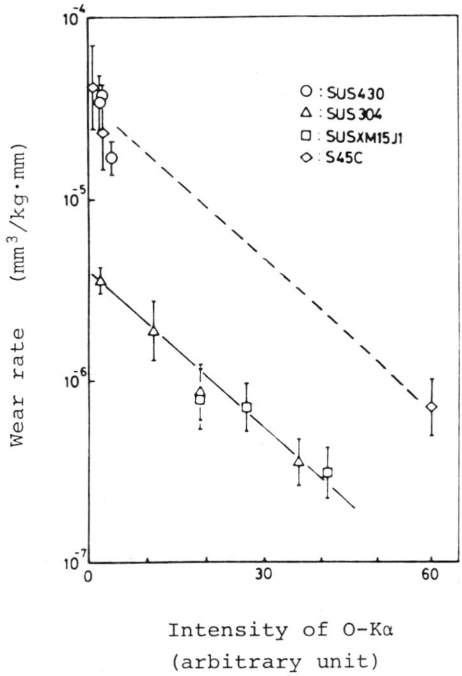

Fig. 13.—Average intensity of O-Kα radiation from the sliding surface vs. wear rate in SUS 430, 304 and XM15J1 stainless steels and 0.45% carbon steel. Sliding velocity; 69, 138 and 311 mm/sec., load; 1.15 kg (2.3 kg/cm²) and 3.15 kg (6.3 kg/cm²). (Shimura, Ref. 6)

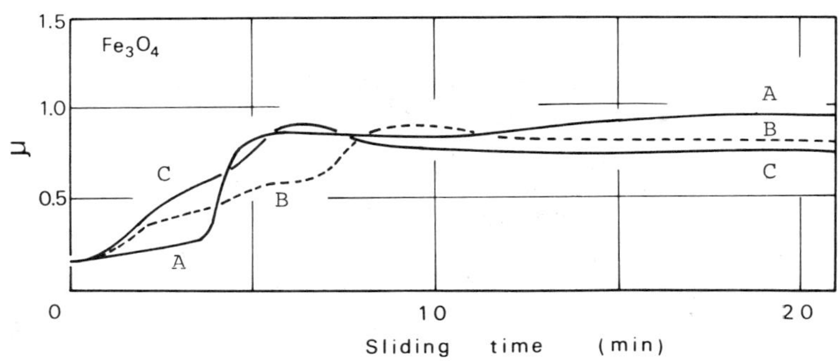

Fig. 14.—Variation of coefficient of friction with sliding time in Fe_3O_4 film against hardened tool steel. Sliding velocity; 1.13 m/sec., load; (A) 227 g, (B) 527 g and (C) 1027 g. (Mizutani, Ref. 22)

REFERENCES

1. Buckley, D.H., These Proceedings.
2. Mizutani, Y., Ph.D. Thesis, University of Nagoya, 1977, p. 8.
3. Nakajima, K. and Kawamoto, J., *Wear,* Vol. 11, 1968, p. 21.
4. Ito, T., Taga, Y. and Nakajima, K., *Journal of Japan Society of Lubrication Engineers,* Vol. 20, 1974, p. 355.
5. Shimura, Y., Ito, T., Taga, Y. and Nakajima, K., *Wear,* Vol. 49, 1978, p. 179.
6. Shimura, Y., Mizutani, Y. and Nakajima, K., *Japan Society of Lubrication Engineers,* Preprint of the 22nd Meeting (Tokyo, Japan), May 17, 1978, p. 41.
7. Nakajima, K. and Mizutani, Y., *Wear,* Vol. 13, 1969, p. 283.
8. Bill, R.C. and Wisander, D., *Wear,* Vol. 41, 1977, p. 351.
9. Cottrell, A.H., "Dislocations and Plastic Flow in Crystals," Oxford Press, London, 1961, p. 133.
10. Cohen, M., *Bulletin of the Japan Institute of Metals,* Vol. 9, 1970, p. 271.
11. Nakajima, K. and Isogai, A., *Wear,* Vol. 10, 1967, p. 151.
12. Nakajima, K., Isogai, A. and Kitagawa, F., *Wear,* Vol. 11, 1968, p. 233.
13. Ferrante, J. and Buckley, D.H., *American Society of Lubrication Engineers Transactions,* Vol. 15, 1972, p. 18.
14. Nakajima, K., Isogai, A. and Taga, Y., *Japanese Journal of Applied Physics,* Supplement 2, Part 1, 1974, p. 309.
15. Mizutani, Y., Imada, Y. and Nakajima, K., in Proceedings of the JSLE-ASLE International Lubrication Conference, June 9-11, 1975; edited by T. Sakurai, Elsevier Press, Amsterdam, 1976, p. 161.
16. Mizutani, Y., Imada, Y. and Nakajima, K., *Journal of the Japan Institute of Metals,* Vol. 38, 1974, p. 205.
17. Buckley, D.H., Swikert, M. and Johnson, R.L., *American Society of Lubrication Engineers Transactions,* Vol. 5, 1962, p. 8.
18. Oda, Y., Kimura, M. and Nakajima, K., *Wear,* Vol. 20, 1972, p. 159.
19. Nishiyama, Z., "Marutensaito Hentai, Kihon-Hen," Maruzen Press, Tokyo, 1971, p. 1.
20. For example, Okabayashi, K., Kawamoto, M. and Notani, H., *Journal of the Japan Foundrymen's Society,* Vol. 38, 1966, p. 501.
21. Bisson, E.E. and Anderson, W.J., "Advanced Bearing Technology," NASA SP-38, 1964, p. 54.
22. Mizutani, Y. and Nakajima, K., *Journal of Japan Society of Lubrication Engineers,* Vol. 22, 1977, p. 177.
23. Bowden, F.P. and Tabor, D., "The Friction and Lubrication of Solids," Oxford Press, London, 1950; translated edition by Soda, N., Maruzen Press, Tokyo, 1961, p. 31.

DISCUSSION

K. TRIPATHI, IIT Research Institute: I am very impressed by your work on surface oxide formation. Dr. Buckley used ESCA which detects monolayers or fractions of monolayers, but you are using X-ray diffraction which penetrates thousands of monolayers. How do you correlate these two?

Y. MIZUTANI: I studied the outermost layer by using Auger and ESCA also. Our data agrees with Dr. Buckley's data.

D. H. BUCKLEY, NASA: Are you talking about the rotation of molybdenum disulfide single crystal that you were sliding on, or a film of molybdenum disulfide?

MIZUTANI: This molybdenum disulfide is not single crystal.

BUCKLEY: I have a comment to make which is related to one of the comments that were made on my paper. This is about the experimental conditions such as load and speed, and other mechanical parameters. Many years ago, early in the days of tribological research, practical systems were simulated in the laboratory using high loads and high speeds. But one began to observe that it was very difficult to identify what was happening at the surface because many changes were produced by mechanical parameters such as load and speed. As you point out, for example, just by changing load we can get recrystallization at the surface. I have been able to recrystallize tungsten surface by changing the load and sliding speed. If you can recrystallize tungsten, almost any other metal can be made to recrystallize by raising load and speed. We also can bring about phase transformations; cobalt can be transformed from hexagonal to cubic form by changing velocity or load, and as a consequence, the friction coefficient can b e changed markedly. It is not just due to the change in the sliding speeds or the load, but it is due to the metallurgical transformation. Therefore, if we conduct the tests at low speed and light loads and understand what properties affect friction at those conditions, then we can increase the speed or load and see whether any properties changed. When we have done that and seen, for example, that we can follow the what change in crystal structure, order/disorder reactions will bring about changes in friction processes. If you run at very high speeds and high loads you will not be able to understand what is going on.

S. RAMALINGAM, Georgia Institute of Technology: I would add that in addition to the mechanical conditions one should worry about the environmental conditions also. We have, from what we have seen of the previous papers, very fine cell structure being formed, cell structure as small as 1000 $\mathring{A}$. If we therefore calculate the mobility of the ambient species through the grain boundaries, as opposed through the lattice, we find extremely high mobility through the grain or cell boundaries. Consequently the partial pressure of the ambient gases is very important when the pressure is very high, oxidation may take place; at low pressure we may observe only absorption.

ON THE DEFINITION AND EFFECT OF CHEMICAL PROPERTIES OF SURFACES IN FRICTION, WEAR AND LUBRICATION

A. R. Lansdown

Dr. Buckley has emphasized in his paper the new picture of surface composition which has emerged in recent years from the use of new techniques such as SEM, ESCA and Auger spectroscopy. There can be no dispute about the data which he presents, and which has in fact been confirmed by a great deal of work in Swansea.

I would suggest, however, that it is possible to exaggerate the influence of this new data on our understanding of what takes place in sliding contacts. It is only our theories about mechanisms which need to be reassessed; our empirical data on the nature and magnitude of friction and wear still remain valid. Since our theories about physical and chemical mechanisms have generally never been considered convincing or satisfactory, it follows that this re-assessment is itself only part of a continuing process of assessment and re-assessment.

During the past forty years there has developed a general awareness of the chemical and physical complexity of surfaces[1,2] from the first evidence of the importance of the surface oxide film[3] to the effects of adsorbed films.[4,5] In recent years when specialists have talked about, for example, the relative sliding of two steels, it has been recognized that they meant two steels with their more or less undefined oxide films and adsorbed contaminants. Now we find that we must also include the presence of carbon which we didn't realize was present, and surface enrichment or impoverishment in one or more elements which we can much more accurately describe.

But to discount any of our previous understanding because of the new information is to run the risk of throwing away the baby with the dirty bath-water. The baby is alive and well, even if the skill of the dermatologist has improved.

Perhaps the greatest hope from this new information is that we may be able to improve our design of lubricants and additives in a much more scientific way than in the past. One difficulty will be in deciding how far to simplify the surfaces which we are studying in order to achieve an understandable picture. The simplified surface (e.g. argon bombarded) may be easy to describe and understand, but its behavior is likely to bear no resemblance to that of the dirty, complicated real bearing. We may find a sort of tribologist's Heisenberg Uncertainty Principle operating.

In our experience one major benefit from these new surface examination techniques is in understanding the changes which have taken place in dry contacts, such as in shears for polymers or slitters for paper or fabric, where the surface is a little less complicated than with lubricated systems. Simply by monitoring these changes, geometrically and chemically, we are sometimes able to make useful progress in improving design and material selection.

REFERENCES

1. Lansdown, A.R., in "Interdisciplinary Approach to Liquid Lubricant Technology," NASA SP-318, 1973.
2. Lansdown, A.R. and Hurricks, P.L., *Institute of Marine Engineers Transactions,* Vol. 85, 1973, p. 157.
3. Bowden, F.P. and Tabor, D., "The Friction and Lubrication of Solids," Parts I and II, Clarendon Press, Oxford, 1964.
4. Godfrey, D., in "Interdisciplinary Approach to Friction and Wear," NASA SP-181, 1967.
5. Fein, R.S. and Kreutz, K.L., in "Interdisciplinary Approach to Friction and Wear," NASA SP-181, 1967.

INVESTIGATION OF LUBRICATED BEARING SURFACES BY X-RAY PHOTOELECTRON AND AUGER ELECTRON SPECTROSCOPIES

I. L. Singer and J. S. Murday

ABSTRACT

Surface analytical tools such as X-ray photoelectron spectroscopy (XPS or ESCA) and Auger electron spectroscopy (AES) reveal the species present at the outermost layers of solids. These tools can therefore be used to examine the effects of various chemical and mechanical treatments on the composition of solid surfaces. These same tools can also help in analyzing more complex films formed on lubricated bearing surfaces in an air environment. In this presentation, we review recent surface analytical studies of bearing steels which have been wear tested in lubricants containing several widely used antiwear formulations. The consensus is that a sulfide is the predominant sulfur specie formed on surfaces worn in most organosulfur additives. Other surface species suggested are B-O and P-O complexes from boron- and phosphorous-containing additives, respectively. In several experiments the concentration of certain surface species could be correlated with the amount of wear, although the role a given species plays in reducing wear is not yet understood. The influence of the oxide layer and other surface species on the wear rate has also been examined. In view of these studies and advances in surface science, the capabilities and limitations of surface analysis on tribological surfaces are assessed.

INTRODUCTION

Surface science has been making rapid progress in its development of new analytical tools and concepts which can be applied to tribological studies. As with any analytical tools, the most complete characterizations can be made on relatively simple surfaces, i.e., those with the least amount of chemical or spatial heterogeneity. For the past ten years, Buckley has studied the tribological properties of surfaces cleaned, i.e., stripped of the native oxide layer, in an ultrahigh vacuum system and subsequently modified in a carefully controlled manner. Since the chemical and physical modifications are kept at a relatively simple and fairly well defined level, surface sensitive analytical tools can contribute a good deal of interpretable information. His comprehensive review of these studies, which appears elsewhere in this volume,(1) shows one way in which surface analytical tools can contribute to tribology: by elucidating a group of fundamental interactions between sliding surfaces. An understanding of such interactions will ultimately lead to improved friction and wear performance in practical situations.

If these fundamental interactions are to be related to practical situations, it will be necessary to determine what chemical and physical modifications of bearing materials take place under wear conditions in the real world. Bearings have metal components that are coated with a predominantly oxide film which can incorporate other species as well. Lubricants, additives and ambient gases react with this surface, resulting in a very complex film which buffers the interaction between two opposing surfaces. The chemical properties of this film can also be probed by the surface-sensitive analytical tools. While the information so gathered may be difficult to fully interpret, it can provide valuable insight into the role of these films in friction and wear.

Given the fact that the tribological roles of surface films, as well as their composition, are complex, the initial application of surface analysis to their study is more likely to be profitable when the film composition and its change play a major role. This leads one naturally to boundary lubrication as an initial tribological subject for surface analytical study. Several other articles in this volume recognize this same point.[(2)]

Boundary lubrication occurs most often when bearings are run at high load or low speed. Most, if not all, of the load is carried by the contacting surfaces, not by a fluid film. Certain reactive compounds containing sulfur, phosphorous or chlorine are known to reduce wear under boundary lubrication conditions. These compounds are often called either antiwear (AW) additives or extreme pressure (EP) additives (which technically depends on their effectiveness at moderate or high loads, respectively) to distinguish them from those organic compounds which operate effectively only at lower temperatures. Although the wear effectiveness of these additives has been thoroughly investigated over the past 50 years[(3,4)] little could be learned about how these additives modify the bearing surfaces until the advent of analytical tools capable of examining the structure and composition of surfaces on a fine scale (microns or less).

In 1974, Coy and Quinn[(5)] presented results of a multi-analytical study which sought to identify surface films formed on steel by organosulfur compounds such as organic disulfides and dialkyldithiophosphates used as EP additives. Using the tools available--glancing angle X-ray diffraction, the electron microprobe and the scanning electron microscopy--they were able to identify the presence of crystalline FeS and small amounts of α-Fe_2O_3 and Fe_3O_4 in the EP region of wear. They were unable to identify surface film composition in the antiwear region. It was suggested at that time that surface analytical tools such as X-ray photoelectron spectroscopy (XPS, ESCA) and Auger electron spectroscopy (AES) could play a major role in examining the films in the antiwear region. We will examine the XPS and AES work on lubricated steel surfaces that has been reported in the intervening years.[(6-14)]

One major point made by Coy and Quinn was the need for more than one analytical approach to the study of boundary lubrication. As we shall see, the use of XPS and AES, coupled with depth profiling, do add a significant new approach but one which has not been fully exploited both because the spectroscopies themselves are having growing pains (they are effectively only ten years old) and because they really need to be used together rather than separately.

CHEMICAL CHARACTERIZATION OF WEAR RELATED SURFACES

The chemical characterization of a surface involves identifying the elements present, establishing their stoichiometry, mapping out valence electron distributions, and locating the atomic geometries. In addition to establishing the chemical identity of the surface species, it is also necessary to locate these species relative to the surface topography in a three-dimensional picture. According to the above requirements, the complete description of a surface composition involves a tremendous amount of information, and no one analytical tool can provide all of it.

In their multi-analytical study, Coy and Quinn(5) utilized electron microprobe analysis (EPMA) for elemental identification and stoichiometry and scanning electron microscopy (SEM) to locate these species relative to topography. Glancing angle X-ray diffraction probed the atomic geometries, at least for crystalline forms. Both the diffraction and microprobe analysis sampled depths on the order of a micron. Identification of chemical species located within this sampling depth had to be inferred from stoichiometry and crystal structure since the valence electron distribution could not be examined directly.

In contrast, AES(15-17) and XPS(15,18,19) are more surface sensitive techniques and do give directly chemical information about surface species. AES and XPS typically probe 1-4 nm. Auger spectroscopy is the surface sensitive analog to the electron microprobe. It detects all elements except H, He, and Li, and can be made to give elemental images with a lateral spatial resolution of 0.2 μm, which is slightly better than the electron microprobe resolution. An added advantage of AES as compared to EPMA is that, whereas the two are roughly equally sensitive to heavier elements, the electron microprobe loses sensitivity for such key lighter elements as carbon, nitrogen, and oxygen. AES also provides a limited amount of chemical information, none of which is accessible with commercial electron microprobes.

Chemical information as well as elemental identification can be extracted from XPS measurements. Further, XPS is a bit easier to quantify than is AES and thereby lends itself better to establishing stoichiometry. On the other hand, XPS presently requires a fairly large area for analysis and cannot be related to finer scale topographic features. There are also electron diffraction techniques, LEED and RHEED, which are the surface sensitive analogs of X-ray diffraction but unfortunately these techniques are not as readily used as are AES and XPS. A comparison of AES, XPS, and EPMA features relevant to tribological studies is made in Table 1.

The new analytical dimension which is added by the use of XPS, and to a lesser but still significant extent by AES, is the ability to probe the valence electron distributions. Different valence states of an element result in small but frequently distinguishable shifts in the kinetic energies of emitted Auger or photoelectrons. In XPS, this "chemical shift" is a direct measure of the change in binding energy of the photoelectron; in AES, it is a convolution of binding energy shifts of the three electrons involved in the Auger de-excitation process. The spectral densities of the valence electrons can also be observed directly with XPS and indirectly by means of lineshape changes with AES. The two spectroscopies thereby provide a method for chemical analysis which does not rely on the crystallinity of the surface.

TABLE I.—COMPARISON OF AES, XPS AND EPMA CAPABILITIES

Features	AES	XPS	EPMA+
	Techniques		
elemental identification (Atomic No. Z)	Z > 3	Z > 2	Z > 4[a]; Z > 10[b]
sensitivity (Atomic %)	0.1	1	0.1
chemical information	possible	yes	very difficult
lateral resolution	0.2 μm	1 mm	0.5 μm
depth resolution	0.5-3 nm	0.5-3 nm	0.5-3 μm

+ X-ray emission spectroscopy (XES); often available with a scanning electron microscope (SEM).
a Wavelength dispersive analyzer.
b Energy dispersive analyzer.

If XPS is to provide useful chemical information, then it must be capable of resolving the various chemical states of the elements of interest. For bearing steels, the surfaces will be basically iron oxides, and the expected active elements in the lubricants will be P, S, Cl, and O. Surface films which form under wear conditions might therefore include complexes of, for example, Fe-O-S, each element being in one or many valence states. Since XPS chemical shifts are sensitive to the valence state, different ligand species or different stoichiometries will lead to slight differences in the recorded electron binding energies. The measurement of binding energies with sufficient accuracy for chemical identification is, however, not easy, and often inaccurate binding energy values for chemical states have been reported in the literature. Their use in one boundary lubrication study[6] led to an identification of an iron oxide specie as a ferrous oxide when it was most likely a ferric oxide.

Fortunately, several very careful XPS studies of the iron-oxygen system have been reported recently.[(20-22)] These results are presented in Table 2 to show that there can now be agreement to withon 0.4 eV between different laboratories as to the iron binding energies in metallic iron, FeO, Fe_2O_3 and FeOOH. When sufficient care is not taken these binding energies can vary considerably more than 0.4 eV as illustrated by other measurements found in the literature[(23-25)] and quoted in boundary lubrication studies.[(6, 10)] Since these groups variously choose the $FE_{2p3/2}$ or the Fe_{3p} line to characterize the chemical shift, the relationship between the binding energies of the two lines is presented in Figure 1 to facilitate comparisons.

Whereas X-ray photoelectron spectra for the iron-oxygen system are well characterized, the literature results[(6,9,23,26,27)] for iron-sulfur systems, presented in Table 3 leave those binding energies open to considerable debate. When it is considered that conclusions are often drawn from binding energy shifts of 1 eV, the differences reported in Table 3 are quite substantial. It is not safe to assume, as did Wheeler,[10] that the reported binding energies are good to ± 0.3 eV. Nor is it possible to determine the valence

TABLE II.—REPORTED Fe ($2p_{3/2}$) BINDING ENERGIES (IN eV) FOR IRON-OXYGEN COMPOUNDS

Compound	McIntyre and Zetaruk[20]	Asami and Hashimoto[21]	Brundle et al.[22]
Fe	706.9	706.8	707.0
FeO	709.5	-----	709.7
Fe_2O_3	710.8	711.0	711.2
FeOOH	711.6	711.5	711.2
Spectrometer Calibration Lines			
Au ($4f_{7/2}$)	84.0	84.1	84.0
C (1s)	285.0	-----	-----
Cu ($2p_{3/2}$)	932.5	932.5	-----

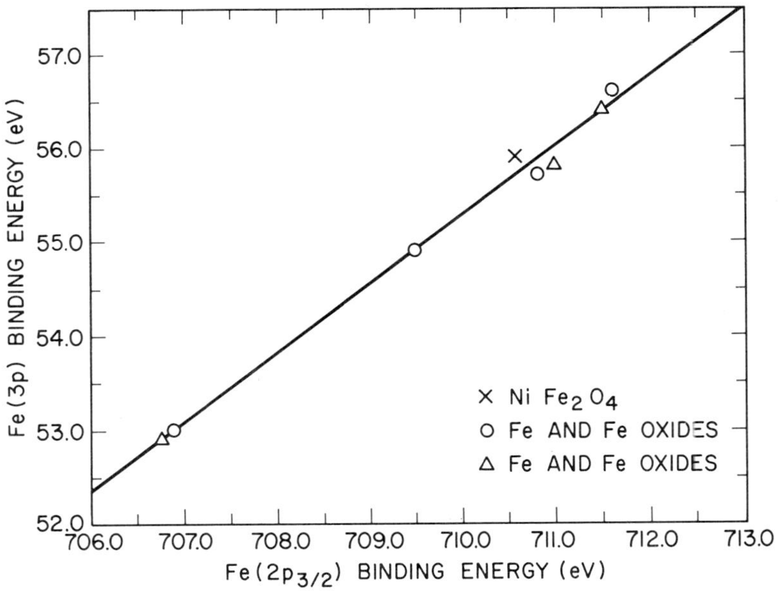

Fig. 1.—Fe(3p) vs. Fe($2p_{3/2}$) binding energies. Data taken from Ref. 20 (O, X) and Ref. 21 (Δ).

TABLE III.—REPORTED S(2p) BINDING ENERGIES (IN eV) FOR IRON-SULFUR COMPOUNDS

Compound	*Lindberg et al.*[26]	*Kramer and Klein*[23]	*Bird and Galvin*[9]	*Baldwin**[6]	*Furuyama et al.*[27]
FeS	-----	160.7	162.8	161.8	161.3
FeS_2	-----	161.5	-----	162.6	-----
DBDS	163.8	-----	-----	163.8	-----
S_8	164.2	162.8	164.3	163.4	-----
$FeSO_4$	168.9	-----	169.6	169.3	-----
$Fe_2(SO_4)_3$	169.3	-----	169.7	-----	-----
Spectrometer Calibration Lines					
C (1s)	285.0	-----	285.0	285.0	-----
Au ($4f_{7/2}$)	-----	-----	-----	-----	84.0

* $C_{(1s)}$ taken as 284.0 in Ref. 6; data adjusted here by 1 eV to reflect the accepted value of $C_{(1s)}$ = 285.0 for the carbon overlayer (see Ref. 26).

state of a specie from its binding energy alone. For example, for iron oxides and iron sulfides, the binding energy of Fe in the ferrous (2+) state is several eV below that in the ferric (3+) state; yet, ferrous sulfate and ferric oxide have binding energies within a few tenths of an eV of each other.[23] Ionic compounds like the iron sulfates have higher binding energies than do covalent compounds like iron oxides for the same nominal oxidation state.

Although XPS provides more easily differentiated chemical information in most cases, AES can often be very useful in discerning chemical states. For instance, Phillips et al.[11] observed that Auger spectra from wear surfaces with a ZDTP-containing oil showed new peaks in the vicinity of the phosphorus LVV Auger line. They correctly assigned these peaks to a non-phosphide form of phosphorus; we now know that the lineshape they observed is characteristic of P-O ligands.[14] Furthermore, many elements, including the antiwear species Cl, S, P, C, and B, show characteristic Auger lineshapes related to oxygen ligands.[28] A comparison of XPS and AES lines from several sulfur-oxygen compounds is shown in Figure 2. Note that while the

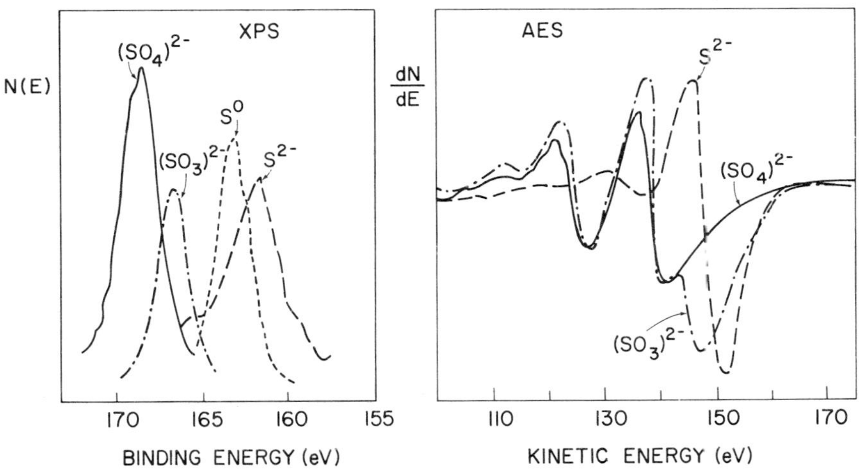

Fig. 2.—Sulfur lineshapes of several sulfur-oxygen compounds obtained by XPS and AES. XPS spectra are taken from Ref. 7; the binding energy scale is adjusted to the C(1s) value of 285.0 eV. Note the expanded energy scale of the XPS spectra vis-a-vis that of the AES spectra.

XPS measurements would be a more convenient basis from which to ascertain mixture compositions, the Auger lineshapes are sufficiently different to provide an alternative. This capability would be important, for instance, if small surface areas were to be examined. In other cases the AES kinetic energy shifts are more sensitive to the chemical environment than are XPS shifts. Bird and Galvin[9] took advantage of this fact in their analysis of Zn chemistry in wear surfaces exposed to an additive containing Zn.

The surface sensitivity of XPS and AES, coupled with the ability to etch away surface layers by ion beam sputtering,[29] allows one to form a three dimensional picture of surface composition. There are many concerns associated with sputter etching, especially if one hopes to extract chemical information. The ion beam can be destructive to surface composition as has been shown by Kim and Winograd[30] and by Brundle, et al.[22] for ion oxides and suggested by Bird and Galvin[9] for SO_4^{2-} species. It is not really known whether the ion beam disruption of oxide ligands will accumulate to a significant level if only thin films of these materials are to be etched away. On the positive side, McIntyre and Zetaruk[31] recently reported that ion bombardment of an oxidized Inconel alloy (Ni-Cr-Fe) does not result in chemical decomposition sufficient to distort a composition profile of a thin oxide film grown on the alloy.

The use of XPS with sputter profiling also leads to problems if the surface is heterogeneous on the scale of the area analyzed. For example, Figure 3 shows scanning Auger micrographs of a severe wear area which has been lightly sputtered. Some areas have been etched through to metallic iron while other areas are still covered by an oxide. An XPS study of this surface would show a composition averaged over both areas and would mask the heterogeneity. The heterogeneity depicted by the SAM images is believed to be characteristic of the surface composition in the example shown in Figure 3, but shadowing of the incident ion beam by rough surfaces, or angular dependent etch rates, might also have led to compositional variations.[29]

It should be noted, however, that in order to use any of these electron spectroscopies, wear samples have to be cleaned and inserted into the spectrometer's vacuum chamber. At each stage, many of the materials in contact with the wear surfaces may be removed; the remaining species are only those sufficiently bound to the surface so as not to be removed by solvent washing or volatilization. If these species play a significant role in the boundary lubrication process, they should correlate in some way with the wear rate.

SURFACE COMPOSITION AND WEAR

There have been only a handful of studies of boundary lubrication which have utilized the surface analytical tools. A summary of this work is presented in Table 4; note that a comprehensive multianalytical study in the spirit of Coy and Quinn[5] has not yet been tried. The role of sulfur compounds has been the subject of most of the work with B, P and Cl being examined to a lesser degree.

The main conclusion to be drawn from the studies to date is that the lubricant-additive-surface interaction results in a limited number of detectable surface species. Metallic sulfide (S^{2-}), probably FeS, was found on steel surfaces worn in oils containing organosulfur compounds.[6-10] The detection of a sulfide specie is consistent with the identification of FeS by X-ray diffraction,[5] albeit the surface sensitive tools detect the presence of sulfide before X-ray diffraction patterns do so. A zero-valent sulfur specie (S°) was detected on surfaces worn in several of the least effective sulfur-containing antiwear additives.[6-8] Sulfate species (SO_4^{2-}) were also detected, but only in immersion tests with selected sulfur compounds and over a limited temperature range.[9,10] Wear tests with a variety of boron compounds resulted in the formation of a single surface boron specie, likely a B-O complex.[8] The rather limited number of species formed during wear is all the more interesting when one considers the variety of bonding configurations assumed by the load-carrying elements. Examples of sulfur containing additives which form sulfide surface species during wear

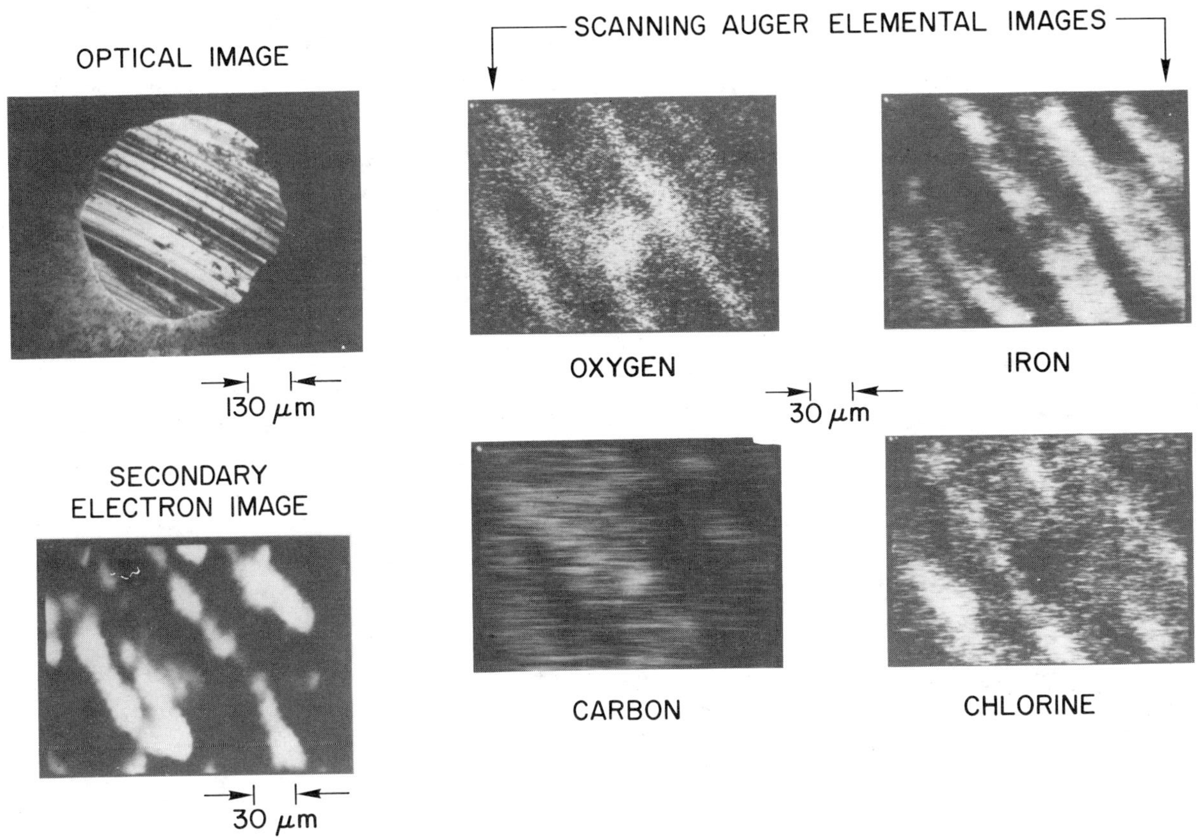

Fig. 3.—Optical, secondary electron and scanning Auger micrographs of a sputter-etched surface in a severe wear area.

*TABLE IV.—REPORTED XPS/AES STUDIES INVOLVING ANTIWEAR ADDITIVES**

Source	Additive†	XPS	AES	SAM	Sputter Etching
Baldwin[6,7]	S (11)	W			
Baldwin[8]	S + B (7)	W			
Bird and Galvin[9]	"S_8, DBDS, ZDTP"	W/I			√
Wheeler[10]	"DBDS"	W/I			√
Phillips, et al.[11]	"ZDTP"		W		√
McCarroll[12]	S (2), Cl (2)		W		√
Shafrin and Murday[13]	"ZDTP and TCP"		W/I	√	√
Shafrin and Murday[14]	"TCP"		I		√

* Wear (W) or Immersion (I) experiments on steel surfaces.
† Load-carrying elements and number of compounds examined (in parenthesis) or compounds examined (in quotes).

are: Zinc Dialkydithoiphosphate (ZDTP), dibenzyl disulfide (DBDS), T-Butylsulfenyl-N, N-Dimethyldithiocarbamate, and Benzyl mercaptan.

One class of sulfur containing compounds studied by several groups(6,9,11,13) are the ZDTP's. The ZDTP's are known for their antiwear properties. XPS and AES are particularly useful for surface modification studies in this wear regime where the techniques of Coy and Quinn are rather insensitive. Analyses of ZDTP-promoted surface films showed Zn, S and P present, but not with the stoichiometry of the original compound. XPS studies indicated that the sulfur is principally a sulfide,(6,9) but the phosphorous is observed by both Auger(11,13) and XPS(6,9) to have oxygen ligands. Phillips, et al.(11) found that the phosphous became a phosphide during sputter etching into the depth of a wear area; on the other hand, Shafrin and Murday(13) found phosphorous retained its oxygen ligand lineshape with depth. These two observations are not considered contradictory since the sample preparations and wear conditions were very different.

The actual compounds contained in the ZDTP promoted surface film could not be positively identified. X-ray photoelectron spectra of the film, however, could be likened to spectra of the original ZDTP compound depleted of some sulfur. Bird and Galvin(9) suggested that some sulfur reacts separately with the steel surface, a possibility consistent with depth profiles from immersion samples. Further comparison of the surface film spectra with the reference compounds provides strong evidence that several compounds whose presence had been suspected are in fact absent.(9)

A second phosphorous containing additive known for its antiwear properties is tricresyl phosphate (TCP). Shafrin and Murday(14) examined with AES films formed on bearing steels soaked in heated TCP (neat) and found evidence for both phosphide and phosphate-like (P-O) species. This work, however, did not examine any wear surfaces.

Several studies in which wear tests were performed showed clearly that the concentration of detected surface species does correlate with the amount of wear. The type of correlation, however, seems to depend on loading conditions and, perhaps, on wear regime. For wear tests conducted at a fixed load, Baldwin[7] observed that the amount of surface sulfide, S^{2-}, correlated inversely with the amount of wear. Similar concentration vs. wear curves were obtained for an oil containing 11 different sulfur compounds and for an oil diluted to several concentrations with a single sulfur compound. (Catastrophic wear resulted when no compounds were added.) McCarroll et al.[12] also found that surface sulfur (and chlorine) concentrations were inversely related to wear scar diameter at a fixed load. For wear tests in which the additive concentration remained fixed but the load was increased, Wheeler[10] found a significant and increasing S^{2-} concentration only when the loading conditions passed from the AW (mild scar) regime to the EP (severe scar) regime.

More intricate correlations were found by Baldwin for wear surfaces on which more than one additive-promoted surface specie is present.[6-8] When amines were present in oils containing sulfur compounds, he observed a significant concentration of surface nitrogen, a decrease in the amount of S^{2-} and an increase in the amount of wear.[7] He[8] also determined the relative antiwear efficiencies of three surface species (S^{2-}, S°, and B-O) found after wear testing of boron-sulfur compounds. For equal concentrations, S^{2-} was approximately 50 times more effective than the B-O species. The zero valent sulfur, S°, on the other hand, appeared to have prowear behavior, associated perhaps with corrosive wear. It would appear that antiwear properties can be associated with the concentration of a single surface species even in the presence of other additive-promoted surface species.

These correlations, and the inability of the native oxide layer to prevent wear, led Baldwin[6,7] to suggest that the concentration of the antiwear species at *the outermost atomic layer* should be the most important in preventing metal-to-metal contact. In addition, on the basis of some tenuous interpretations of XPS data, Baldwin[6] concluded that a sulfide film forms on top of the original oxide layer. This double layer picture was not, however, supported by the more direct observations of composition vs. depth obtained by sputter etching.[9-13] Results of XPS depth profiles on both worn [10] and immersed[9] samples suggested that additive-promoted films probably consist of mixed sulfides and oxides and that some of the sulfur may migrate into the bulk of the sample. Scanning Auger micrographs, coupled with depth profiles, should be useful to see if sulfide is uniformly spread across the surface or if islands form which would give profiles with the appearance of a mixed sulfide/oxide layer.

An accurate three-dimensional picture of additive-promoted films would be very useful in determining their role in resisting wear. Baldwin[6-8] suggested that a sulfide film reduces metal-to-metal contact either directly or by providing adsorption sites for other molecules. Alternatively, the sulfur may promote the formation of a thicker oxide layer. On the basis of his depth profile studies, Wheeler[10] concluded that, under mild wear conditions, the native oxide film is thickened whereas under severe wear conditions it becomes thinner. The results of Shafrin and Murday[13] and of Phillips, et al.[6] support the concept of thickening under mild wear. Here, again, this conclusion must be tested by careful SAM work to better define how surface heterogeneity affects this observation.

SUMMARY AND CONCLUSIONS

XPS and AES in conjunction with sputter etching can provide new and useful chemical information on modifications of surfaces resulting from wear. The techniques probe the electronic structure of the outermost atoms on a solid surface and are therefore more sensitive to composition changes associated with early or mild wear than the analytical techniques (EPMA, SEM, glancing angle X-ray diffraction) used previously. Recent studies on steel surfaces lubricated with oils containing antiwear additives have provided several examples of the capabilities of these techniques.

The composition of surface films, promoted by additives containing a variety of sulfur, boron, and phosphorous compounds, have been examined. It appears that only a limited number of surface species are formed, despite the variety of bonding configurations assumed by the load-carrying elements in the additive compound. The composition of the film, however, has not yet been identified due, in part, to difficulties in interpretating spectra from heterogeneous surfaces.

Correlations have been made between the presence and concentration of additive-promoted surface species and the amount of wear. A metallic sulfide, probably FeS, appears to be associated with efficient antiwear performance. Other species containing boron or nitrogen, which may compete with sulfide for surface sites, appear to be less efficient. One specie, zero valent sulfur, has been associated with prowear behavior. The role of these surface species, or more generally, the additive compounds has not yet been established, although there is some indication that the species themselves or some precursor specie may assist in the formation of oxide layers.

In future studies of wear surfaces, we expect to see more use of the multi-analytical approach; in particular, SAM is called for on wear-scarred surfaces. More careful analysis is also needed to establish the relationship of XPS and, especially, AES lineshapes to the chemical compounds of interest. Although XPS and AES are powerful tools they cannot be expected to adequately examine the roles of the organic compounds in surface films. For these molecules, other surface sensitive techniques such as secondary ion mass spectroscopy (SIMS),[32] and attenuated total reflection, Fourier-transform infra-red spectroscopy (ATR/FTIR)[33] may be useful.

REFERENCES

1. Buckley, D.H., These Proceedings.
2. Godfrey, D., Tamai, Y. and Tripathi, K., These Proceedings.
3. Bowden, F.P. and Tabor, D., "The Friction and Lubrication of Solids," Part 1, Oxford Press, Cambridge, 1950, Chapter 11.
4. Riesz, C.H., in "Metal Deformation Processes," edited by J.A. Schey, M. Dekker, Inc., New York, 1970, Chapters 3 and 4.
5. Coy, R.C. and Quinn, T.F.J., *American Society of Lubrication Engineers Transactions,* Vol. 18, 1975, p. 163.
6. Baldwin, B.A., *Lubrication Engineers,* Vol. 32, 1976, p. 125.
7. Baldwin, B.A., *American Society of Lubrication Engineers Transactions,* Vol. 19, 1976, p. 335.
8. Baldwin, B.A., *Wear,* Vol. 45, 1976, p. 345.
9. Bird, R.J. and Galvin, G.D., *Wear,* Vol. 37, 1976, p. 143.
10. Wheeler, D.R., *Wear,* Vol. 47, 1978, p. 243.

11. Phillips, M.R., Dewey, M., Hall, D.D., Quinn, T.F.J. and Southworth, H.N., *Vacuum,* Vol. 26, 1976, p. 451.
12. McCarroll, J.J., Mould, R.W., Silver, H.B. and Sims, M.L., *Nature,* Vol. 266, 1977, p. 518.
13. Shafrin, E.G. and Murday, J.S., *Journal of Vacuum Science and Technology,* Vol. 14, 1977, p. 246.
14. Shafrin, E.G. and Murday, J.S., *American Society of Lubrication Engineers Transactions,* Vol. 21, 1978, p. 329.
15. Carlson, T.A., "Photoelectron and Auger Spectroscopy," Plenum Press, New York, 1975.
16. Joshi, A., Davis, L.E. and Palmberg, P.W., in "Methods of Surface Analysis," edited by A.W. Czanderna, Vol. 1, Elsevier Scientific Publishing Company, New York, 1975, Chapter 5.
17. Chang, C.C., in "Characterization of Solid Surfaces," edited by P.F. Kane and G.B. Larrabee, Plenum Press, New York, 1974, Chapter 20.
18. Riggs, W.M. and Parker, M.J., in "Methods of Surface Analysis," edited by A.W. Czanderna, Vol. 1, Elsevier Scientific Publishing Company, New York, 1975, Chapter 4.
19. Hercules, S.H. and Hercules, D.M., in "Characterization of Solid Surfaces," edited by P.F. Kane and G.B. Larrabee, Plenum Press, New York, 1974, Chapter 13.
20. McIntyre, N.S. and Zetaruk, D.G., *Analytical Chemistry,* Vol. 49, 1977, p. 1521.
21. Asami, K. and Hashimoto, K., *Corrosion Science,* Vol. 17, 1977, p. 559.
22. Brundle, C.R., Chuang, T.J. and Wandelt, K., *Surface Science,* Vol. 68, 1977, p. 459.
23. Kramer, L.N. and Klein, M.P., *Journal of Chemical Physics,* Vol. 51, 1969, p. 3618.
24. Castle, J.E. and Durbin, M.J., *Carbon,* Vol. 13, 1975, p. 23.
25. Kishi, K. and Ikada, S., *Chemical Society of Japan. Bulletin,* Vol. 46, 1973, p. 341.
26. Lindberg, B.J., et al., *Physica Scripta,* Vol. 1, 1970, p. 286.
27. Furuyama, M., Kishi, K. and Ikeda, J., *Journal of Electron Spectroscopy and Related Phenomena,* Vol. 13, 1978, p. 59.
28. Bernett, M.K., Murday, J.S. and Turner, N.H., *Journal of Electron Spectroscopy and Related Phenomena,* Vol. 12, 1977, p. 375.
29. Wehner, G.K., in "Methods of Surface Analysis," edited by A.W. Czanderna, Vol. 1, Elsevier Scientific Publishing Company, New York, 1975, Chapter 1.
30. Kim, K.S. and Winograd, N., *Surface Science,* Vol. 43, 1974, p. 625.
31. McIntyre, N.S. and Zetaruk, D.G., *Journal of Vacuum Science and Technology,* Vol. 14, 1977, p. 181.
32. Benninghoven, A., *Surface Science,* Vol. 53, 1975, p. 596.
33. Shida, H.I. and Koenig, J.L., *American Laboratory,* Vol. 10, 1978, p. 33; see also Tompkins, H.G., in "Methods of Surface Analysis," edited by A.W. Czanderna, Vol. 1, Elsevier Scientific Publishing Company, New York, 1975, Chapter 10.

DISCUSSION

D. H. BUCKLEY, NASA: I think we have to recognize that with Auger and the X-ray spectroscopy there is a problem of degradation of the surface film which you are actually analyzing. For organic films, there are other techniques such as photoacoustic spectroscopy that we are working with now to identify organic films. The acoustic techniques do not damage the organic films. The structure of the organic molecules on the surface can be analyzed without destroying them.

The other thing I would like to comment on is the work of Wheeler who works at NASA with me. His work shows that the kind of compounds formed in dynamic studies are not the same as those that form by immersion studies; they are markedly different.

I. SINGER: It is valuable information.

J. F. HUTTON, Shell Research, UK: In your talk you mentioned the work of G. D. Galvin. I should just like to say that this group, of which I am a member, is continuing the work with ESCA, EPMA and other analysis. I would like to point out that indeed the whole situation is very complicated. We are not generally in favor of trying to solve the chemistry of the whole apparatus. We try to vary the contact condition by a roller on a flat plane, but even then the distribution of material on the surface is quite complicated. And averaging techniques, that is techniques like XPS which give you an average composition of the surface may be a bit suspect. But on the whole the analyses we find are very similar to the analyses we find at concentrated area with EPMA. Yet the SEM shows quite clearly that some of the EP additives are not all that uniform. The distribution of the film material is extremely patchy, and whether or not this is linked with the surface topography is yet to be decided. Whether it is due to chemistry or perhaps due to melting and spreading of the material and so on will have to be decided also.

SINGER: I would like to point out that one of the researchers that I admire is Galvin, who used several techniques, including the use of TEM to examine the fine structure of the deposits on the surface.

D. GODFREY, Chevron Research: I would like to elaborate on what you said regarding the use of XPS. I think the disadvantage is that it covers a large area, and this consists of troughs which are the real load bearing areas and valleys which are covered by junk. Observations of degradation of the oil and the additives are not applicable to load carrying capacity. And so, focusing on the real load bearing areas with small beams has much more importance in tribology.

SINGER: Obviously, it would be nice to have an X-ray laser. I think also that we have to appreciate that these are new terrains that we are trying to cover here. We have to be clever in the kinds of techniques we apply. For example, it may be useful to have a sample rotating and examine it with a sputter beam so as to errode away the junk constantly -- junk that might not be erroded in the case of a stationary sample.

QUESTIONER: I would like to ask the speaker, in view of the many ways in which lubricant can decompose on a surface, whether it occurred to the tribologists to take a simple system like the single crystal and strain it in some way or another and put the lubricant on it and develop a whole new chemistry that is peculiar to tribology with its own rules and regulations so that we do not have to refer to the principles of general chemistry which we have been doing for so many years?

SINGER: I think some very good experiments can be planned along these lines.

SAME QUESTIONER: Do you plan any?

SINGER: I would like to. I think that is a very good approach -- to take a well-characterized surface, strain it and examine the chemistry for that amount of strain. I think very good experiments can be done that way, especially using these sensitive surface techniques.

V. SURFACE AND INTERFACIAL PHENOMENA

SURFACE EFFECTS IN CRYSTAL PLASTICITY

R. M. Latanision

ABSTRACT

This paper includes discussion of not only the phenomenology and current mechanistic understanding of manifestations of surface effects in crystal plasticity, but also the application of these phenomena to technology. At the outset, certain aspects of the surface chemistry and physics of various crystalline inorganic solids are examined in order to illustrate that the surfaces of such solids are likely to be structurally, chemically and electronically distinct from the bulk. This is followed by brief discussion of surface effects in uniaxial plastic deformation, the influence of surface films on mechanical properties, chemisorption-induced variations in the plasticity and fracture of metals and nonmetals, and surface effects in embrittlement phenomena. Throughout these discussions emphasis will be placed on areas of controversy and an attempt will be made to invoke some of the principles of surface chemistry and physics in describing each topic. The remainder of the paper is intended to demonstrate that our understanding of surface effects in crystal plasticity, imperfect though it may be, has been applied by various investigators with remarkable success to such technological problems as metal cutting, the machining of ceramics, the rapid excavation of hard rock, improvement in the fatigue life of various alloys, and, potentially, even to the control of earthquakes.

1. INTRODUCTION

The fact that the mechanical properties of solids are sensitive to the environment to which these solids are exposed is, of course, not newly discovered. I am reminded by Westwood(1) that more than 100 years ago, Reynolds(2) first associated certain detrimental effects on the ductility of iron with the presence of hydrogen. Not only is hydrogen embrittlement still a major industrial problem, it now seems that in the case of nearly every major engineering alloy system there are some researchers who believe that hydrogen is responsible, at least in part, for serious mechanical degradation. It seems safe to say that in a mechanistic sense we still do not know what hydrogen does on an atomic scale to induce this degradation. It was almost 50 years ago that the late academician Rebinder(3) in the Soviet Union first reported that the plasticity and fracture of solids were remarkably affected by surface-active media. Here, too, years of disagreement and uncertainty in interpretation have followed that first report. More recently, an interesting and often spirited debate has developed over the role of the free surface, in the absence of specific environments, in the plasticity of crystalline solids. Despite some very clear indications that the free surface does play a role in determining bulk mechanical properties, it is nevertheless true that surface effects have by and large been ignored in modern theories of work hardening and in general discussions of crystal plasticity.

In what follows, an attempt will be made to establish objectively a general sense of perspective, with due regard given to the interdisciplinary character of these proceedings. Therefore, some of the phenomenology of surface effects in crystal plasticity will be described pointing out areas of controversy. Subsequently attention will be directed to a brief summary of applications of this subject to areas of technology. Where possible emphasis will be placed on areas in which an interdisciplinary surface science/materials science approach might be useful in providing new insight into the mechanism of some manifestations of surface effects in crystal plasticity. So broad a treatment cannot hope to be comprehensive. My detailed views on many of the issues treated here - as well as others that are not - are given in more lengthy accounts.[4-7]

The logical starting place for this discussion is the surface itself. There follows a brief account of some of the properties of surfaces, particularly those which may well affect mechanical behavior.

2. SOME ASPECTS OF THE PHYSICS AND CHEMISTRY OF SURFACES IN RELATION TO MECHANICAL PROPERTIES*

The simple view of the surface or interface between a solid and its environment as a sharp discontinuity in an otherwise infinite continuum is adequate for many purposes. In many cases, this simplification is invoked in order to arrive at solutions to otherwise intractable analytical problems. Perhaps the classic example here is the concept of an image dislocation which has an electrostatic analogue. In fact, however, it must be recognized that there is a region of finite thickness between any two media that may differ structurally, chemically, and electronically from either. (For recent general treatments, see References 8 and 9.)

Now, it is also recognized that the mechanical behavior of a crystalline solid is determined principally by the generation, motion, and interaction of dislocations. The fact that the plasticity and fracture behavior of such solids can be affected by the environment suggests that the surface and its condition must somehow affect dislocation behavior. Indeed, it has been proposed over the years that the surface or near-surface layers may serve as regions in which dislocations are easily generated or as barriers to the escape of dislocations from within the solid and that these functions may be modified by the presence of adsorbed surface-active species, charge double layers, non-equilibrium concentration of solute atoms or vacancies, solid surface films, corrosive (solvent) environments, and the like. In a simplified form, the formidable path that a dislocation must travel as it approaches the surface of a crystalline solid is suggested schematically in Figure 1. Structural, chemical, and electronic perturbations at the surface are indicated, for the purpose of illustration, as lattice parameters variations, solute concentration gradients, and an electrical double layer, respectively. By and large, the impact of these perturbations, particularly surface structure, on plasticity is poorly understood. In the next section, this situation

* An extended version of this discussion is given in Reference 5.

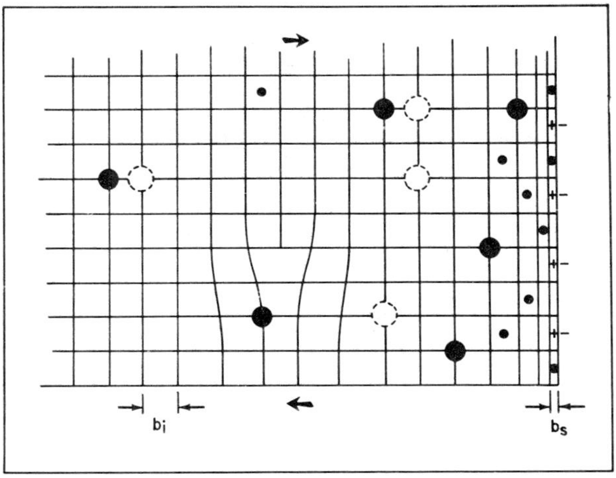

Fig. 1.—Schematic illustration of some of the structural, chemical and electronic perturbations seen by a dislocation as it approaches a crystal surface (after Latanision).[5]

is discussed briefly.

2.1 Surface Structure

The model that is often used for the structure of crystalline surfaces derives from the ideas of Kossel,[10] Stranski,[11] and Burton, Cabrera, and Frank.[12] This model which has been referred to as the terrace-ledge-kink model (TLK) has been used extensively in discussions of crystal growth, evaporation, crystal dissolution, and surface diffusion. The TLK model, Figure 2, includes low index terraces and monatomic height ledges which are occasionally displaced by atomic distances at kink sites. In considering such models, however, it should be remembered that they represent an ideal surface. No allowances have been made, for example, for the relaxation of surface atoms which seems analytically likely, particularly in the direction perpendicular to the surface and, perhaps, as well, parallel to the surface. On this point, there is a vast literature of relatively recent origin related to the structure of crystalline surfaces. There is also a considerable uncertainty in the interpretation of some information collected over the years by surface analytical tools such as LEED.[13] Some of this uncertainty appears to have been resolved by refined methodology and diffraction theory. At any rate the general consensus based upon observations seems to be that atoms in the topmost layer of many clean metal surfaces have the same arrangement in the plane of the surface as atoms on parallel planes in the body of the crystal, with perhaps only slight adjustment of the spacing of atoms normal to the free surface. The magnitude of this variation in lattice parameters seems uncertain and the many claims - some in excess of 10% perturbations - are too numerous to mention here. Duke's opinion[13] is that this variation (expansion or contraction) is not likely to be more than ±5% of the bulk lattice parameters for most clean single crystal surfaces - not an insignificant variation. There are exceptions, of course, and indeed crystal periodicity and/or symmetry are sometimes reported to be perturbed at the surface of presumably clean metals.[8,9] For example, phase transformations have been observed on platinum surfaces as a function of temperature though

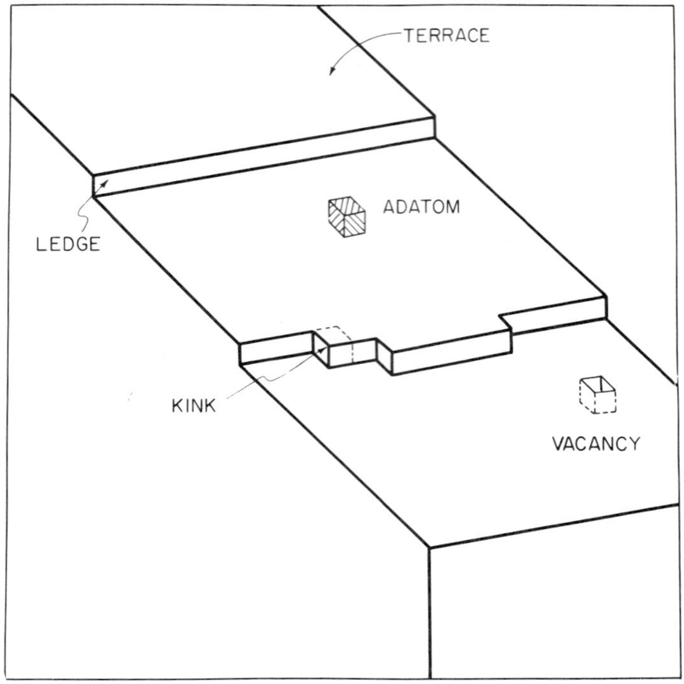

Fig. 2.—Schematic of the terrace-ledge-kink (TLK) model for a crystal surface (after Latanision).[5]

similar transformations are not observed in the bulk. Likewise a hexagonal surface structure has been reported on the cube face of fcc gold. Although there is some debate as to whether some of the surface strucutres reported for metals are the property of clean metal surface or associated with surface impurities, recent evidence tends toward the former view. In contrast, it is apparently well agreed that constructive rearrangements (reconstruction) of varying complexity do, in fact, occur on covalently-bonded semiconductor surfaces (Si, Ge, etc.); and surface superlattices with unit cells ranging from 2 to 8 times the unit cell dimension of the substrate have been reported.

One of the most surprising and exciting observations to come from studies of surface structure is the large effect small amounts of adsorbed gas atoms appear to have on the structure of otherwise clean surfaces. Indeed, it seems that in many cases, adsorbed gasses dislodge substrate atoms from their normal positions leading to reconstructed surfaces composed of both substrate atoms and chemisorbed gas atoms in periodic arrays. Apparent reconstructive interactions following chemisorption have been observed in many circumstances and are summarized by Somorjai.[9]

Despite all that is known about the structure of crystalline surfaces, very little consideration appears to have been given to the influence of surface structure on mechanical properties. Fleischer[14] has suggested that the gradual change in lattice spacing over 4-5 atom distances from the surface may lead to the development of an obstacle to the emergence of dislocations from the crystal. It seems not unlikely that ledges or kinks on otherwise flat surfaces may act as stress concentrators[15] and increase the probability that dislocations sources may operate preferentially in the near-surface layers. The potential catalyzing effect of vicinal surfaces on dislocation nucleation[16] may also play a role in determining the mechanical behavior of crystalline solids. While there are some other possibilities as well, the fact remains that, qualitatively and quantitatively, the influence of surface strucutre on mechanical properties is virtually unexplored. It is difficult to say now how significant these aspects of surface structure are relative to mechanical properties. It does seem, however, that there exist several good analytical problems worthy of attention.

2.2 Chemical Segregation at Free Surfaces

In addition to the contamination of the surface by adsorption of impurities from an external source, which was discussed briefly in the preceding section, impurities which originate in the bulk of the solid may also accumulate at the free surface. In short, the segregation of bulk impurities may occur at the free surface as at internal interfaces (grain boundaries, twin boundaries, etc.) or at dislocations. The driving force for solute segregation at a free surface may be thought of in terms of a relaxation of near-surface elastic distortion by equilibrium or Gibbsian adsorption of solutes of appropriate size. Alternately, in a thermodynamic sense, any solute which lowers the surface energy of the solvent might be expected to segregate to the free surface. In either case, this would lead to an enrichment or depletion of solute over a few atom layers from the surface, since lattice distortion is thought to drop off rapidly with distance into the bulk and essentially extinguishes by the fourth or fifth atom layer from the surface as indicated schematically in Figure 1.

Aust, Westbrook, and co-workers[17,18] have also pointed out that non-equilibrium solute segregation may occur because of the development of a gradient in vacancy concentration near free surfaces, since the surface is an effective sink for vacancies. In those cases where solute-vacancy interactions are strong, it is suggested that the vacancies will diffuse toward the surface dragging impurities with them, Figure 1. The vacancy-solute couples dissociate at the surface, vacancies are annihilated, and a net excess of solute remains at or near the free surface. Extrapolating measurements of segregation at internal interfaces to the free surface, one may predict that non-equilibrium segregation near the latter may occur to depths on the order of several tens of microns depending upon the system involved.

Solute segregation near the free surface may well exert an influence on every aspect of surface properties and behavior, not the least of which is mechanical behavior. It is perhaps more obvious, however, that thermionic emission, surface diffusion, surface energy, surface structural trarsformations, catalytic efficiency, etc., are more likely to be affected by perturbations in surface composition. On the other hand, evidence[17,18] for non-equilibrium segregation is considered to account for significant increases (or decreases) in microhardness near grain boundaries and, in some cases, near the free surface over distances covering about 100 μm. Hardening is presumably due to the formation of solute clusters[17] or secondary vacancy defects,[18] respectively. Considerable importance has been attached to the equilibrium segregation of impurities at internal interfaces (grain boundaries) in understanding various embrittlement phenomena.[19] In most cases, embrittlement is thought to be due to solute-induced decrease in cohesion. More recent work by Latanision and Opperhauser,[20] however, has indicated that segregation alone is not always sufficient to lead to embrittlement but that embrittlement may be a function of the subsequent interaction between the segregated impurities and the environment to which the solid is exposed - i.e., to impurities and microchemistry.

It should also be appreciated that near-surface concentration gradients may also affect the flow stress and tensile behavior of metals. Indeed, in view of recent indications that yielding of metal monocrystals begins in the near-surface layers (see Section 3.1.1) recognition of the potential effects of surface contamination becomes even more important. An indication of sensitivity of tensile behavior to solutes may be seen by studying the effects of cathodically-produced hydrogen on the behavior of (impure) nickel monocrystal electrodes.[21] In Figure 3, the behavior of a crystal deformed in air is indicated by Curve C, while B shows the effect of precharging with hydrogen followed by simultaneous charging and deformation. The important point to note here is that serrated yielding occurs when the specimens are deformed and cathodically charged simultaneously, even if precharging is eliminated. Serrated yielding implies interaction between mobile dislocations and solute atoms, and it is significant that this occurs even though calculations indicate that the depth of penetration of hydrogen should not have been greater than 15 microns and must have been considerably less in the early stages of deformation. The implications of this work are, therefore, that slip begins in the near-surface layers of the crystal and that the resultant stress-strain behavior is dependent on the composition of the near-surface regions of the solid. Other observations consistent with this view have been summarized recently.[4,7]

2.3 Electronic Properties of Surfaces

In addition to structural and chemical singularities, it should be recognized that at the surface there is in general an electrostatic potential difference associated with some kind of charge double layer.[22] In the case of a metal surface in a vacuum, for example, we know that quantum mechanically there is a finite probability of finding an electron outside the metal surface. Hence, there is an electron excess just beyond the surface and a deficiency in the near-surface region of the solid. The result is an electrical double layer or dipole with its negative side outward - i.e., the double layer creates an electrostatic potential that is more positive inside than outside as indicated schematically in Figure 1.

While the above simplified illustration of a complex problem serves to indicate that electrical double layers are characteristic of the interface between almost any two phases, our major interest for the purpose of this discussion is the double layer present at the solid-electrolyte interface. I will return to this point later. It should be sufficient at this stage to point out that because of the high density and mobility of conduction electrons in a metal, interfacial electric fields cannot penetrate more than an angstrom or so into the bulk of a metal crystal. For nonmetals, however, because of the lower density and mobilities of the charge carriers involved, the influence of such fields may be sizable and much greater depth (for example, in excess of 1 μm for insulator materials). As we shall see later (Section 3.3.2), changes in the space charge distribution appear to exert a significant influence on mechanical behavior.

3. THE PHENOMENOLOGY OF SURFACE EFFECTS IN CRYSTAL PLASTICITY

In this section I will briefly describe some of the issues of importance with regard to the phenomenology of surface effects in crystal plasticity, throughout attempting to invoke some of the principles of surface chemistry

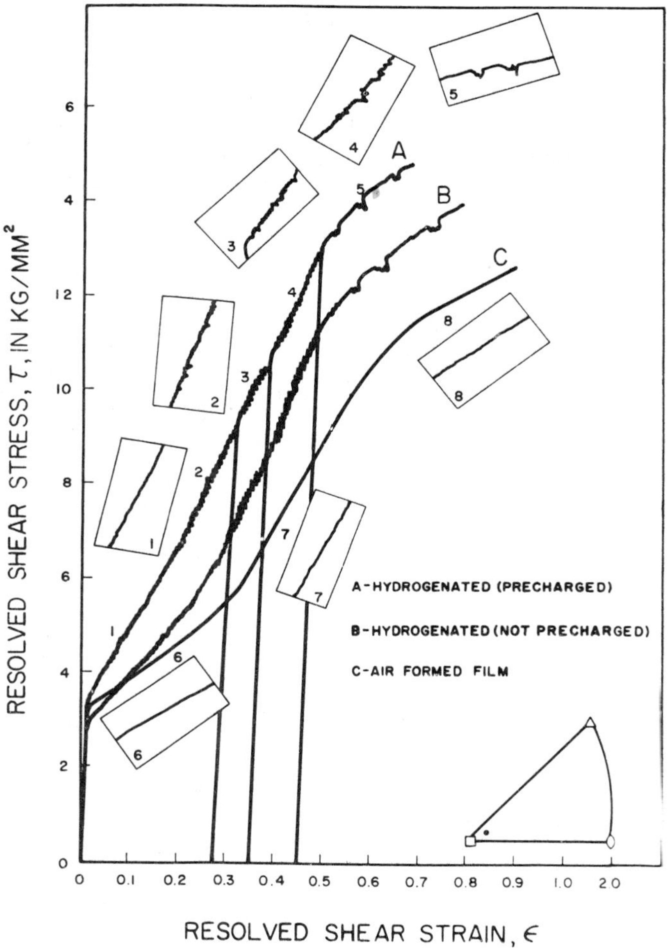

Fig. 3.—Effect of hydrogen charging on the deformation on nickel monocrystals. Insets show detailed recorder tracings (load-extension)(after Latanision and Staehle).[21]

and physics in understanding these phenomena.

3.1 Surface Effects in Uniaxial Tension

The role of the free surface in the plastic deformation of homogenous and presumably clean* crystalline solids has been the subject of considerable

---

* It is now very obvious indeed that the mechanical properties of crystals with clean surfaces in a surface analytical sense have never really been examined.

interest, particularly recently. The details of these many studies are complex and often confusing and need not be treated here. These details have been discussed elsewhere[4,7] and were presented throughout the course of the NATO Advanced Study Institute (ASI).[4] In essence the centers of controversy focus on two main issues: (1) the influence of the surface on dislocations multiplication, and (2) the distribution of dislocations and consequent flow stress gradients which might be expected after large scale yielding has occurred - i.e., the hard-versus-soft surface layer controversy.

3.1.1 The surface and dislocation sources

A carefully prepared and undeformed metal monocrystal typically contains abous 10^6 cm of dislocation line per cm^3, although wide variations are possible. This dislocation density rises with increasing deformation, perhaps to a value as high as 10^{12} cm/cm^3 in heavily cold worked specimens. The means by which dislocation multiplication occurs is not surprisingly an important aspect of the theory of dislocations.[23,24] From the point of view of tribology, the effect of the surface on dislocation multiplication is also a central issue. In short there are at present three models which consider the role of the surface in this respect, each of which tries to answer the question of whether the surface, in fact, performs in the role of an easy source of dislocations, an obstacle to the emergency of dislocation, or perhaps both.

The differences between each model are shown in Figure 4. In each case the mechanism of dislocation generation is essentially that proposed by Frank and Read in 1950[25] (and which now remains the most likely multiplication mechanism) although the location of the source relative to the surface differs in each model. Frank-Read sources have in fact been observed in operation, the classic observation being that of Dash in 1956.[26]

In the debris layer model, Figure 4(a), first suggested by Kramer[27] in 1961, some dislocations leaving the crystal in the early stages of deformation are presumed to be trapped in the near-surface region, forming a zone of high dislocation density - the debris layer - which introduces a back stress opposing the motion of the other dislocations in the surface region. The near-surface layers thus act as an obstacle to dislocation egress. Typically this debris layer is considered to be on the order of 50-100 μm in thickness. It seems implicit in this model that dislocations are generated by sources far removed from the surface - i.e., internal Frank-Read sources. Presumably these sources lie outside the debris layer and, hence, are elastically well removed from the range of influence of image forces which are considered important in Figure 4(b) and (c). Dislocations are thus presumably generated somewhere in what might be considered the core of the crystal and move toward the surface.

Others have invoked the suggestion by Fisher[28] that the free surface provides an easy source of dislocations. The basis for this view is that a dislocation line terminating unpinned at the surface, but being pinned somewhere in the interior (at a distance L from the surface), behaves elastically as a dislocation line in a uniform medium which is anchored by the internal pinning point and its image above the surface. Thus, the effective length of this surface Frank-Read source may be twice its actual length as shown in Figure 4(b). Since the stress required to operate such sources is inversely proportional to the length of the dislocation segment between pinning points, a surface $(2L)$ should operate at some stress lower than that required to operate an internal source of the same length (L) and, in fact, at half the stress if the dislocation line happens to lie normal to the

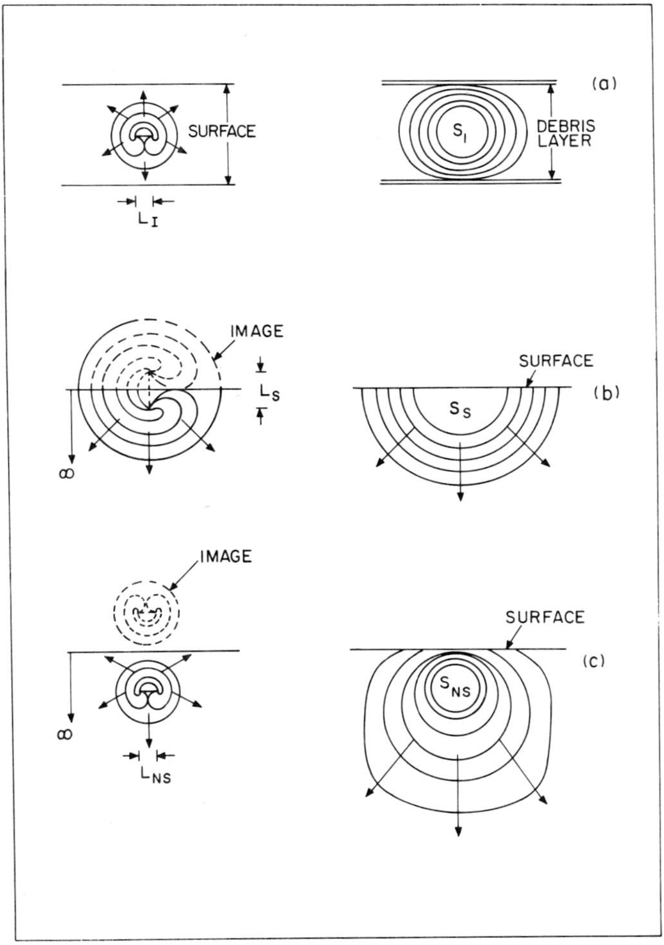

Fig. 4.—Schematic illustration of (a) the "debris-layer" model, (b) a surface source of dislocations, and (c) dislocation generation by a near-surface source (after Latanision).[5]

surface. Half loops produced by such a source would then propagate from the surface into the interior.

The third alternative, namely that the surface acts simultaneously as both a source and obstacle was first suggested by Kuhlmann-Wilsdorf.[29] Latanision and Staehle[30] extended this argument and proposed that the most likely dislocation generator in the early stages of deformation is a near-surface source. This is pictured, Figure 4(c), as a source of the Frank-Read type, which happens to lie near enough to the surface (on the order of a micron away) so that it operates in the presence of its own image; but, unlike the Fisher course, does not presume to have an arm swinging freely at the surface - i.e., both pinning points are within the crystal. The image force diminishes as the reciprocal of the distance from the surface, and, hence, the segments near the surface are strongly attracted toward the surface while the motion of segments running into the interior is more weakly resisted by a similar but proportionately smaller attractive image force. Sumino[31] had considered similar sources analytically in 1962 and showed that these sources produced dislocations more easily than otherwise equivalent internal sources. Such sources produce loops just as do internal sources, but because of the proximity of the surface, the half loops nearest the surface may either slip readily out of the crystal or be held up (i.e., the surface acts as an obstacle) depending upon environmental conditions, while the remaining half loops move into the interior. There are sensible reasons for believing that under normal atmospheric conditions the surface of a crystal should act as an obstacle. Among others, surface films are almost certainly present. Likewise, composition gradients (segregation) are likely to occur near the surface of annealed single crystals, hence, affecting the motion of dislocations into such regions (see References 4 and 7 for summary). One might therefore suspect that dislocations may accumulate at the surface, much as Kramer[27] suggests, producing a back stress which limits the operation of initially active near-surface sources. On the other hand, if the surface is slowly but actively dissolved (i.e., in the absence of surface films), potential barriers may be removed to an extent; and the development of a back stress which would otherwise limit the operation of a given near-surface source would be delayed. One might expect that in the latter case a given source might generate a relatively unlimited number of loops which upon emerging from the crystal would produce slip steps with much greater than ordinary step height. Precisely this observation has been reported by Latanision and Staehle.[3] They found that in Stage II, for example, step heights on 99.8% nickel monocrystals actively dissolved during deformation were about 2000 $\underline{b}$ - more than an order of magnitude larger than the crystals deformed in the absence of surface dissolution.

There is increasing evidence that in monocrystalline metals[30,32-40] as well as nonmetals[41-43] yielding does in fact begin somewhere near the free surface. There is reason to suspect that slip processes in the surface grains of a poly-crystal are relatively less restricted compared to grains in the interior[44] and etch pitting studies by Vellaikal and Washburn[40] indicate that preferential multiplication of dislocations occurs near the external surface of copper polycrystals. It appears, therefore, that one must decide whether the Fisher or near-surface models best account for these observations. My own view is that the Fisher surface source, while it is attractive in principle, is not likely to operate in practice.[4,7] In the first place, the mechanical properties of atomically clean surfaces have yet to be examined. Specimens examined in the laboratory ambient are unavoidably contaminated with adsorbed gas or, in most cases, oxide films. In such situations one would expect that the free end of any potential Fisher source would be pinned. Likewise, one must wonder about the possible influence of segregates near the free surface. Moreover, even if it were possible to produce and maintain a clean free surface on tensile specimens, surface atomic relaxation and reconstruction may conceivably interfere with the operation of such sources. In short, the surface is not the simple discontinuity that it is presumed to be in the Fisher source model; and it seems to me, therefore, that the operation of classical surface sources is unlikely under ordinary circumstances.

Fourie and Dent[33] have invoked the near-surface model in explaining flow stress gradients in *Cu-Al* monocrystals, and others[38,40] have proposed explanations for preferential yielding in the surface layers by models similar to that in Figure 4(c), but taking into account the possibility that the active segment may be inclined to this surface rather than parallel to it. In either case, the source operates in the presence of its image with both pinning points just inside the crystal surface. At any rate, it seems safe to conclude that the details of dislocation multiplication and the influence of the surface have yet to be clarified. Much of the debate on this point has been summarized elsewhere(4) and need not be treated further here. The view does emerge, however, that in the early stages of plastic deformation the net flow of dislocations is expected to be into the crystal from the surface.

3.1.2 The hard-versus-soft surface layer controversy

The quesiton of what role the surface plays after large-scale yielding has occurred is much debated. Again, this has been summarized in detail elsewhere.(4) In essence, Kramer(27[b]) has made the striking observation, Figure 5, that the initial yield stress and work-hardening behavior of an aluminum monocrystal could be recovered by chemically removing a 1 mm envelope from the surface, the implication being that work-hardening in Stage I is concentrated in the near-surface region. Moreover, continuous surface removal during deformation was shown to affect the work-hardening characteristics of fcc monocrystals, and similar observations have been reported by Latanision and Staehle.(20) Kramer has interpreted this in terms of the formation of a dislocation-rich (and, hence, hardened) debris layer in the surface regions of plastically deformed crystals, as explained earlier. The debris layer is considered to impede the motion of subsequent dislocations and serves as a barrier against which they pile up. The consequent debris layer (back) stress is reportedly dissipated by dissolution or by allowing

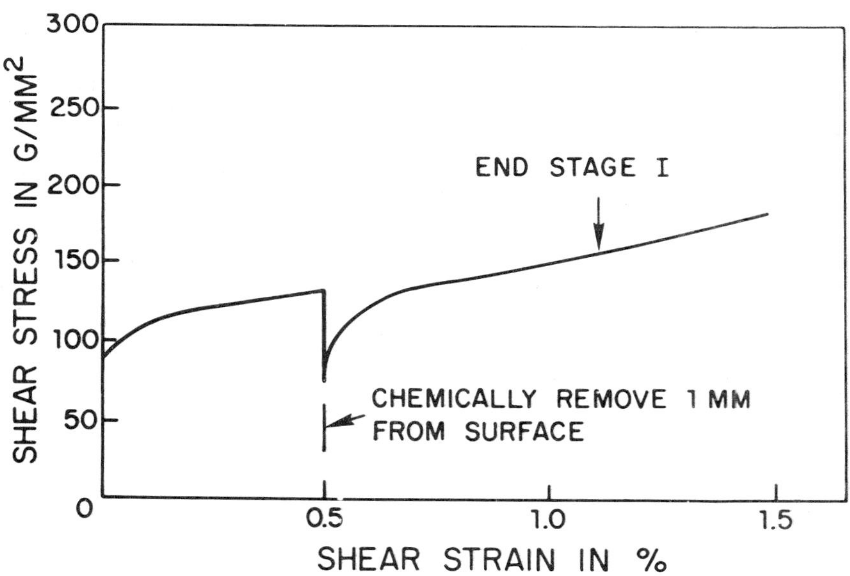

Fig. 5.—Effect of electrochemically removing 1 mm from the surface of an aluminum monocrystal following 0.5% strain. Note recovery of the original yield stress upon reloading (after Kramer and Demer).(27[b])

specimens to relax without dissolution for various periods of time. This layer is presumed to extend to depths of about 60 μm in aluminum and 100 μm in gold and iron, for example. Kramer has presented a recent review of his work in Reference 45.

Kitajima et al.[34] also suggested that a hardened subsurface layer develops just after yielding, consistent with his etch pit observation that the dislocation density within 70 μm of the surface in lightly deformed copper crystals is higher than at a point further into the interior. By virtue of the geometry of their crystals, however, Kitajima et al. conclude that this result is not likely to derive from the accumulation at the surface of dislocations emitted from internal sources, i.e., the debris-layer model. Rather, they prefer the view that dislocations are generated by surface sources and that the advance of primary glide dislocations into the crystal is hindered by the creation of barriers developed through interacting near-surface primary and secondary dislocations, evidence for preferential activation of sources on conjugate and critical slip planes near the surface also having been observed.

It is interesting that, based on the observation that work-hardening in Stage I can be completely eliminated by removing the surface layers, Kramer[46] suggests that in Stage I additional internal dislocation obstacles are not formed in sufficient numbers to affect mechanical behavior - i.e., the Stage I work-hardening is confined to the surface region. Moreover, it seems reasonable on this basis to suppose[46] that secondary slip systems may be operative in Stage I deformation but only in the surface region. This implies that Stage II deformation should begin first at the surface and propagate inward since the transition to Stage II, on the above basis, seems more likely in the surface than in the interior. What is interesting is that this is just opposite to the point of view developed by Fourie,[47] namely, that the transition to Stage II occurs first in the core of deformed crystals. Indeed, while Kramer proposes a hard surface-soft core model, Fourie imagines just the opposite. As is also true of Kramer's view, considerable evidence is available to support this view. In essence, Fourie has observed that flow stress gradients develop during plastic deformation and that, indeed, the flow stress near the surface is less than in the core. These dynamical flow stress measurements are made by slicing larger (14 mm wide in the direction of the primary slip vector) predeformed crystals into thin component crystals (ranging from 0.065 - 0.6 mm in thickness) and then restraining these components. One result is shown in Figure 6. Notice that in initially large crystals, the soft surface layer appears to extend to distances into the bulk on the order of 2 mm - comparable to the mean free path of edge dislocations at the beginning of Stage I. Mughrabi,[49] on the other hand, has observed evidence for similar gradients in crystals with diameters not much larger than the mean free path. Fourie[47] initially accounted for this flow stress gradient on the basis of a model that assumed a uniform distribution of dislocation sources throughout the crystal, but later with regard to *Cu-Al* monocrystals[33] preferred a model based on near-surface sources.

In essence, it is expected that plastic deformation gives rise to a disparity of dislocation signs between the surface and interior, Figure 7.[48] Assuming that work-hardening occurs by a mechanism in which the transition from Stage I to Stage II is initiated by the creation of dipole bundles, Fourie suggests[47] that a deficiency of dislocations of either sign near the surface would delay the onset of Stage II hardening there. In contrast, in the core or central region, the density of dislocations of both signs might be expected to be about equal. Hence, on this basis, the interior of such crystals should work-harden more rapidly than the surface, and one might

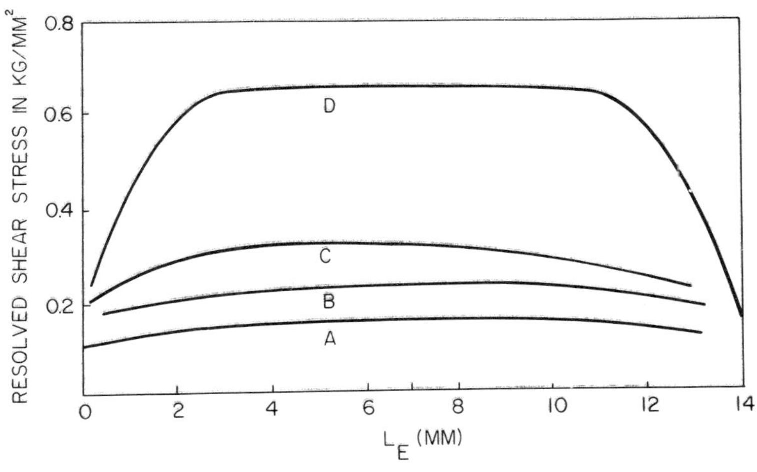

Fig. 6.—The flow stress distribution in copper monocrystals plotted as a function of the length of the glide path of edge dislocations, L_E, which has the value 0 and 14 mm at the original surfaces. Curve A is for an as-grown crystal: Curves B, C, and D are for prestrains of 0.02, 0.029, and 0.058 respectively (after Fourie).[47]

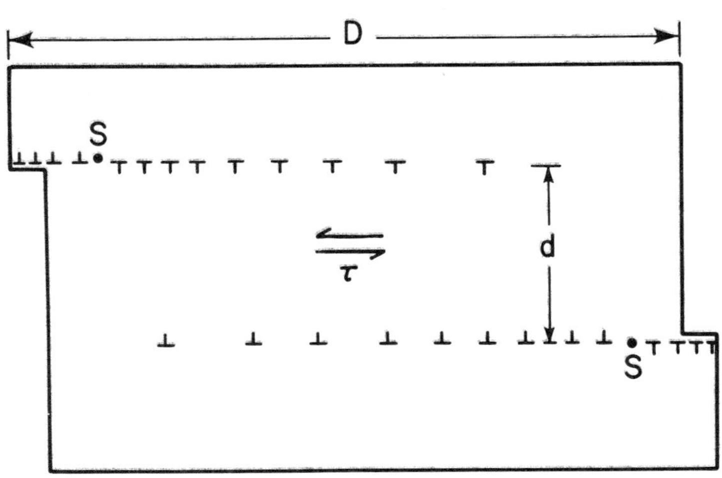

Fig. 7.—Model of near-surface source operation illustrating the action of the free surface as a barrier to dislocation emergence as well as the disparity of dislocation signs developed by the glide of dislocations into the bulk from the near-surface layers (after Latanision).[48]

envision that even in a crystal deformed macroscopically into Stage II, the surface layer might still be deforming in a Stage I mode. Indeed, by means of transmission electron microscopy Fourie[47] and later Mughrabi[49] have observed precisely that. For a crystal deformed into Stage II, the dislocation distribution near the surface is still typical of Stage I processes; Figure 8(a) - more single primary dislocations and isolated edge dipoles - whereas the characteristics of Stage II hardening - cell formation, higher dislocation density, etc. - are observed in the interior, Figure 8(b).

There is much more to be said regarding the hard-versus-soft surface layer controversy, but I believe the above represents the essence of the problem. This controversy has been much discussed recently (see discussion to References 45 and 50, as well as References 7, 51, and 52). While there are clearly great differences in opinion, what is necessary now is to determine if apparently divergent experimental observations might be understood from some rational perspective. More often than not, the experimental bases for the various views which have been developed have been so dissimilar and vague in the literature that direct comparisons have been quite impossible. Various questions have been raised about the possible influence of (a) specimen bending in Fourie's dynamical flow stress measurements (see discussion to Reference 50), (b) dislocation loss and rearrangement in thin foil searches for the debris layer,[51] (c) solid surface films present on the surface of electrodes during electropolishing in regard to Karmer's surface removal experiment,[7] (d) the injection of vacancies by dissolution and their effect on mechanical behavior during surface removal experiments,[52] (e) size effects,[32] and many others that will surely be aired during these deliberations. In the light of our improving understanding of the nature of crystal surfaces, one must also now consider, for example, the role of near-surface solute and vacancy concentration gradients in crystals deformed under ambient conditions and, similarly, the removal of such gradients by dissolution prior to or during deformation. While it is true that surface removal experiments such as those performed by Kramer over the years and by others as well[30,53] are complex in the sense that they do represent a moving frame of reference,[47,32] those experiments which are done under carefully controlled anodic conditions[30,53,54] probably most closely approach the deformation of a solid with a "clean" surface. It is interesting to note in passing that the increased plasticity observed in such experiments is similar in some respects to the Joffe effect.[55] In the classic demonstration of this phenomenon, a salt crystal is shown to be weak and brittle if deformed in air but remarkably ductile and stronger if deformed in water - i.e., while being slowly dissolved. Though still not yet well understood, the Joffe effect has been attributed to a combination of factors - the dissolution of pre-existing notches or cleavage cracks (Joffe's original explanation) or brittle surface films, removal or at least decrease in the active length of surface sources,[42] etc. In any case, it seems reasonable to expect that the near-surface regions of a crystal may act as both a preferential source of dislocations as well as a barrier to the exit of dislocations at the surface. What remains in view of the conflicting observations, I think, is to decide the circumstances under which one of these competitive processes suppresses and dominates the other.

3.2 The Influence of Surface Films on Mechanical Properties

In 1934, Roscoe[56] discovered that thin oxide films on the surface of cadmium crystals caused a significant increase in their yield stress. Similar observations have been reported many times in various solid-film combinations. Often such films have been found to increase not only the yield points but also the rate of work hardening and in many cases the three-stage

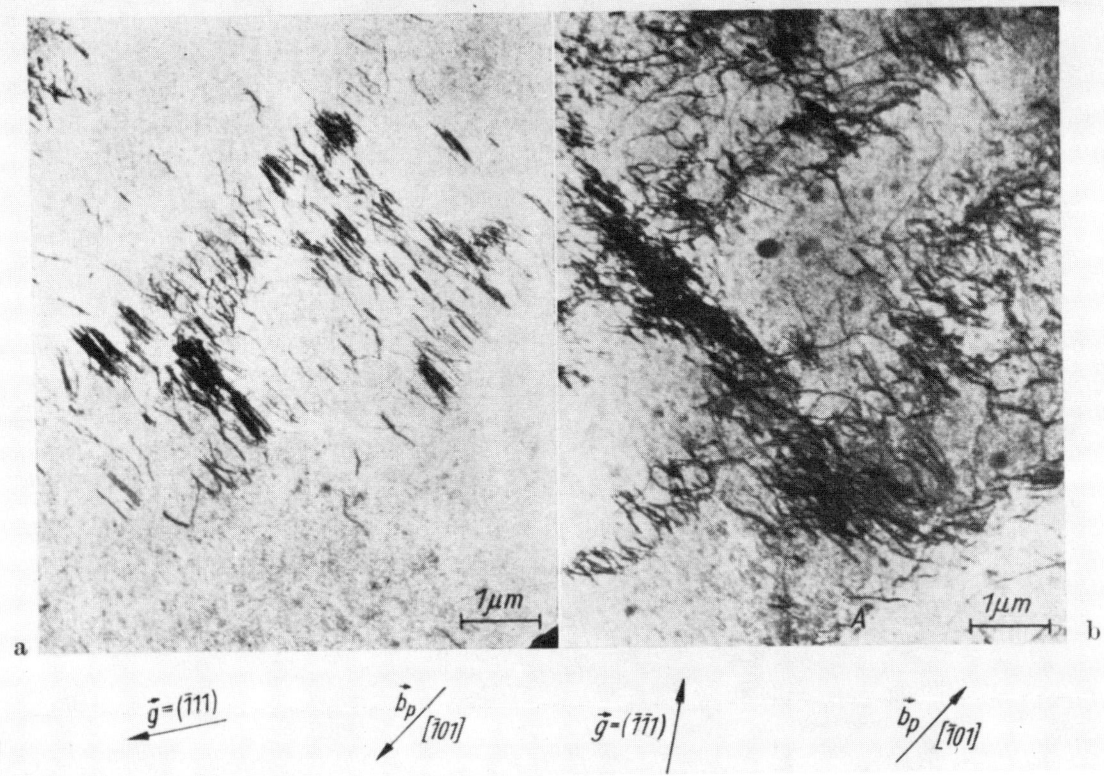

Fig. 8.–Dislocation arrangements in a copper crystal deformed into Stage II (a) 0.5 mm from the surface, (b) center (after Mughrabi).[49]

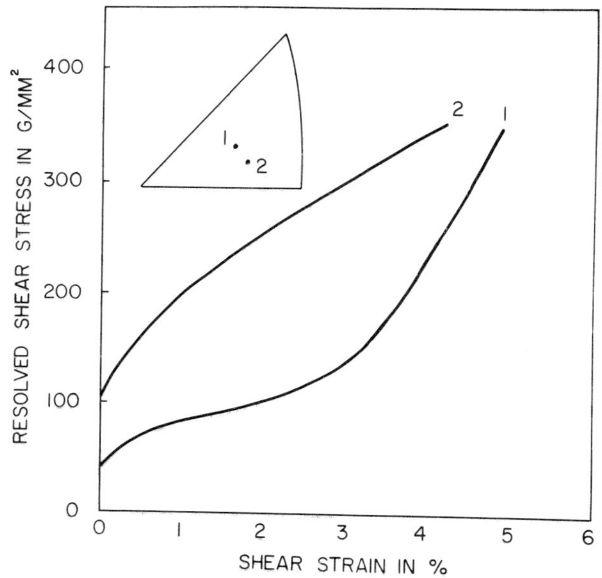

Fig. 9.—Effect of nickel-chromium film on the stress-strain curve of similarly oriented copper monocrystals: (1) unplated, (2) plated (after Garstone et al.).[57]

work hardening behavior of fcc metals has been totally suppressed by films. For reference, Figure 9 shows the influence of a 2 μm nickel-chromium coating on a copper monocrystal oriented for easy glide.[57] There is now no doubt that the presence of oxide layers, electrodeposited or evaporated metal films, or alloyed layers can profoundly affect the mechanical behavior of a variety of crystalline metals and nonmetals. There have been many explanations for such phenomena - the pinning of surface or near-surface sources, etc. - and many controversies remain. Over the years it has been considered sufficient in many instances to imagine that such films act as barriers to dislocation emergence and that this accounts for some measure of strengthening. The means by which a surface film might serve as a barrier in a mechanistic sense continues to be debated. Moreover, there are more recent indications that surface films can, under certain circumstances, induce softening rather than hardening of some solids (see below).

The mechanisms by which films affect near-surface dislocation behavior in solids have been related to the elastic interaction of dislocations with a film of shear modulus different from that of the substrate, the formation of accommodation dislocation networks at the film-substrate interface, and the existence of surface damage caused by film cracking due to residual plating stresses. In what follows, I will briefly consider some indications of the relevance of these parameters.

3.2.1 Elastic interactions

Dislocations theory (see Reference 58) predicts that if a dislocation approaches a surface that is covered with a film whose modulus, μ_F, is less than that of the crystal substrate, μ_S, then the dislocation will experience

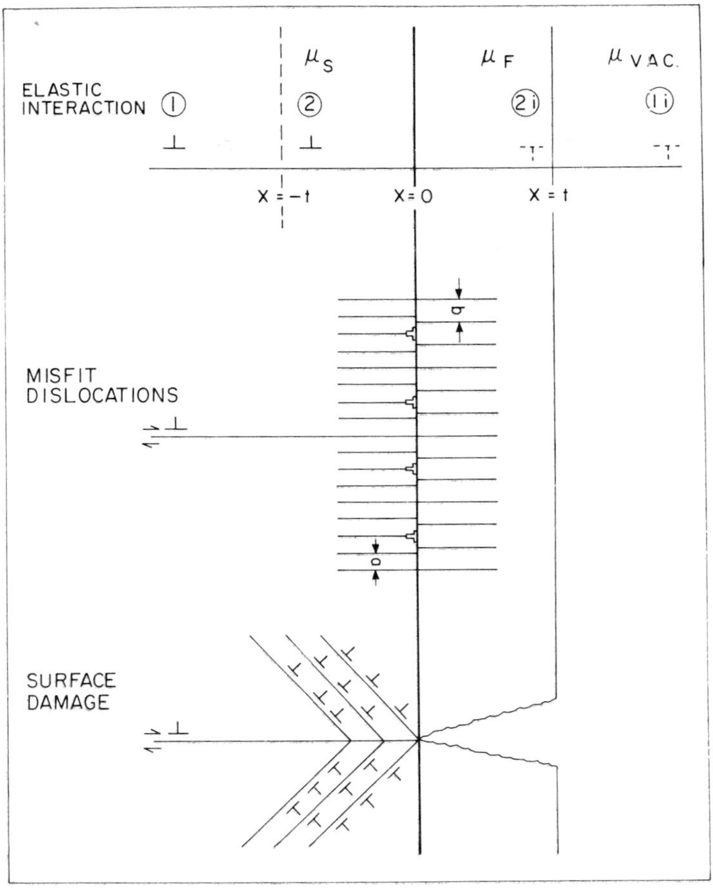

Fig. 10.—Schematic illustrating the possible means by which a surface film may act as an obstacle to dislocation emergence.

an image force attracting it to the surface. Conversely, if $\mu_F > \mu_S$ the dislocation will feel a long-range attraction and a short-range repulsion (over a distance comparable to the film thickenss, t), the consequence is that a position of stable equilibrium is expected at a distance of about t from the film-substrate interface, Figure 10. The force between the dislocation and image in the film is given by

$$F = \frac{\mu_S b^2}{4\pi r} \left(\frac{\mu_F - \mu_S}{\mu_F + \mu_S}\right)$$

where r is the distance of the dislocation from the interface and b the Burgers vector. When no film is present, the ratio of the moduli becomes unity, and the equation represents the attractive image force experienced by a dislocation approaching a "clean" free surface. Notice that when $\mu_F < \mu_S$ the image force is attractive but of a lesser magnitude than when no film is present. For a dislocation to emerge from the crystal in the case where $\mu_F > \mu_S$, sufficient stress must be applied in order to overcome the repulsive force discussed above (neglecting all other factors). Hence, the film acts as a barrier or resistance to dislocation emergence, and one might expect to find a dislocation accumulation just under the interface.

An elastically stiffer surface film should, thus, induce a certain strengthening in the softer substrate as observed, for example, by Roscoe.[56] On this basis, the strengthening observed should be a systematic function of the modulus of the film. However, in a relatively few cases where this prediction has been examined[59,60] no such systematic correlation has been observed. Moreover, it is difficult to imagine how softening effects observed in some cases[61,62] could be accounted for on this basis. In short, some other factors must also play a role in determining the influence of surface films on mechanical properties.

3.2.2 Interfacial mismatch or misfit

Evans and Schwarzenburger[63] have pointed out that the transfer of dislocations from a substrate to a surface film of the same crystal structure but different orientation is difficult because of the misorientation of slip planes across the interface. This mismatch may thus serve as an obstacle to slip. If the film and substrate are of different crystal structures, or if one is a metal and the other a ceramic, the transfer of dislocations becomes even more difficult since changes in Burgers vector, stacking fault energy, etc., must be accommodated. Even in a less complex case where the film and substrate lattices are of the same structure and occur in parallel orientations, the atomic misfit in lattice spacing between the film and substrate may be accommodated by a two-dimensional grid of dislocations in the interface, the mesh size of the network decreasing as the degree of mismatch increases. The accommodation of such misfits, sometimes in part by dislocations and in part as well by homogenous strains, is described in detail by Van der Merwe.[64] An example[65] of such a misfit dislocation network is shown in Figure 11. This interfacial dislocation network may serve as an effective barrier to glide dislocations approaching the surface from the substrate, Figure 10, and one might expect that the strength of this barrier would decrease as the lattice mismatch decreases. There is some evidence to support this view.[66,67] On the other hand, there are cases[61,62] where coatings exert a distinct softening, not hardening, influence on mechanical properties. For example, Ruddle and Wilsdorf[61] report that oriented copper monocrystals coated with 600 Å electrodeposits of nickel yield in tension at approximately half the stress required for unplated crystals. In this case, it is presumed that the misfit network acts not as an obstacle to dislocations approaching the surface but rather as a source of dislocations and, importantly, a source of stresses less than the macroscopic yield point of uncoated copper crystals.

3.2.3 Surface damage due to residual stresses in films

Johnson and Block[59] report that chromium and rhodium films markedly strengthen copper monocrystals. They also conclude that strengthening on the basis of either a lattice misfit or elastic effects is incompatible with their observations. Instead, they propose that strengthening is associated with cracking of these coatings, either prior to or during deformation of the composite, and the consequent local generation of dislocation-dense regions

Fig. 11.—Transmission electron micrograph showing long, straight misfit dislocations in a deposit of platinum on gold. The plane of the figure is (100) and its vertical and horizontal boundaries are parallel to <110> directions. Misfit, 3.9% magnification about 250,000X (after Mathews and Jesser).[65]

through the stress pulses accompanying film cracking. Strengthening then occurs as a result of interacton between dislocations approaching the surface and the dislocation-dense regions near the interface associated with film cracking as suggested schematically in Figure 10.

It is well known that electrodeposits and evaporated metal coatings as well as oxide films often contain significant residual strains after growth. While there has been disagreement on this point,[68] it seems not improbable that such stresses and their release could, in fact, contribute to surface film strengthening as described above. On the other hand, it is not clear why the injection of dislocations via the release of residual stresses in the coating could not just as well lead to a softening effect - i.e., to large-scale dislocation motion at stresses lower than the yield point of the uncoated substrate. Indeed, Sethi and Gibala[62] find substantial decreases in the flow stress and stress-strain behavior of niobium and tantalum single crystals coated with anodic oxides and account for this observation in just such a manner.

In summary, the influence of solid surface films on the mechanical behavior of crystal substrates is likely to depend, in a complex and interrelated way, on such factors as the elastic properties of both components; the degree of atomic misfit at the interface; residual stresses in the film; etc. But while all of these factors (and probably others as well) contribute to the effects of surface films on mechanical behavior, the important interrelations between them remain to be established. It should be important, for example, to determine under what conditions during the deformation of a film-substrate composite slip is transferred from the substrate to the film, in which case the film may act as a barrier, or vice versa, in which case the film may provide an easy source of dislocations.

3.3 Adsorption-Induced Changes in the Plasticity and Fracture of Metals and Nonmetals

Adsorption-induced reductions in the hardness of nonmetallic solids were first reported in 1928 by Rebinder(3) and are known as Rebinder effects. In the period of time since Rebinder's first announcement, many others have reported that surface-active media (long-chain organic compounds, liquid metals, etc.) affect the plasticity and/or fracutre of a variety of solids, including metals, covalent and ionic crystals, molecular crystals, polymers, amorphous glasses, etc. These are far too numerous to summarize here, but there are several reviews which might serve as useful guides(4,69-72) including those by Shchukin(71) and Macmillan.(72)

While there have been a variety of explanations for Rebinder effects, I will limit discussion in this section to the two that are best supported by observation. The point of view developed by Rebinder and his colleagues is that adsorption-induced softening and strength reduction occur as a result of the lowering of the specific free surface energy of the solid, i.e., the work of formation of new surfaces during deformation and fracture. On the other hand, the view of Westwood and his co-workers(73) is that such phenomena - adsorption-induced hardening as well as softening - may be understood in a conceptual way from consideration of the type, concentration, mobility, and adsorption-induced redistribution of charge carriers in the solid. In this context such phenomena might well be called Westwood effects. While these views are at first glance quite different, they have in common a basis for understanding which is derived from colloid- and/or electrochemistry. In order to better understand this basis, the following section begins with a discussion that itself began in Section 2.3, namely, a description of the charge double layer present at a solid/electrolyte interface.

3.3.1 Solid/electrolyte interface

Let us first distinguish the metal/electrolyte interface from the semiconductor/electrolyte interface. The structure of the double layer in the former case is likely to take the form illustrated in Figure 12(a) for the case of a dilute aqueous electrolyte. The compact double layer consists of an excess or deficiency of electrons at the metal surface and a layer of ions of charge oppostie in sign to that at the surface of the metal adsorbed at the interface. This system of charges is known as the Helmholtz double layer (HDL). The locus of electrical centers of hydrated ions in contact with the electrode surface is known as the outer Helmholtz plane (OHP) while the locus of centers of unhydrated specifically adsorbed ions in contact with the electrode surface is known as the inner Helmholtz plane (IHP). The Gouy-Chapman layer (GCL) represents a distribution of charges in a space charge layer with an excess of ions similar in sign to that in the outer Helmholtz plane smeared out over a distance of up to about 1 μm. Beyond this point, the ions are present in concentrations typical of the bulk electrolyte. A schematic illustration of the potential drop expected across the double layer is shown in Figure 12(b). There is a linear potential drop across the HDL and then a more gradual exponential decay in the diffuse double layer. The difference in potential between the OHP and the bulk electrolyte, i.e., the contribution of the diffuse double layer, is called the electrokinetic or

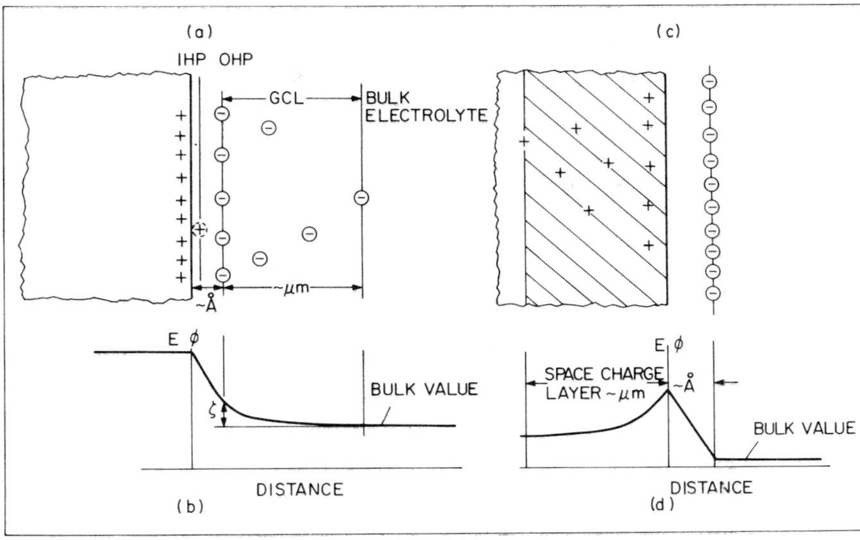

Fig. 12.—Distribution of excess charge and corresponding variation of electron energy in the solid and potential of ions in solution at interfaces between a metal and a dilute electrolyte (a and b) and between a semiconductor and a concentrated electrolyte (c and d) (after Latanision).[6]

zeta (ζ) potential.

The relation between the charge distribution in the space charge layer near the surface and the electrostatic potential change at the surface is given by a Poisson-Boltzmann equation, the consequence of which is that the thickness of the space charge layer is essentially a function of the density of mobile charge carriers in each phase, i.e., solid or electrolyte. In essence, as the carrier concentration increases the thickness of the space charge layer decreases. In the case of a metal, the density of free electrons is very high, so that all of the charge behaves like a surface charge. Hence, the charge redistribution extends to only a few angstroms at the most beneath the metal surface. The diffuse double layer becomes essentially compressed into the OHP when the electrolyte concentration is sufficiently high. In contrast, in semiconductors the density of charge carriers (electrons and holes) is orders of magnitude lower than in a metal. In pure germanium, for example, the concentration of electrons and holes at room temperature is on the same order of magnitude as the concentration of H^+ and OH^- in pure water. Consequently, the space charge layer in a typical semiconductor is on the order of a micron - 10^4 times thicker than in a metal. The double layer present at the interface between a semiconductor and a concentrated electrolyte is shown in Figure 12(c). A hybrid model showing the electron energy levels in the solid and the potential of ions in solution[74] is given in Figure 12(d).

3.3.2 The space charge layer and the mechanical properties of nonmetals

It is possible to change the distribution of charge at the electrode surface by several means - by the application of an external field, by illumination with light, or of interest in the present context by the chemisorption of charged ionic or polar molecules, for example. Now, is it possible that adsorption-induced changes in the space charge distribution can affect mechanical behavior? There are many good indications of this, perhaps the best available illustration being the ζ-potential correlation studied extensively by Westwood and his colleagues at Martin Marietta Laboratories and summarized by Macmillan.[72] The ζ-potential correlation, which has been observed in a variety of inorganic solids including Al_2O_3, MgO, quartz, as well as a variety of other minerals, is shown schematically in Figure 13, which shows the variations of ζ-potential with solution concentration (a), along with the corresponding variation in hardness (b) and dislocation mobility (c). The ζ-potential is considered to be related in sign and magnitude to the surface charge on the solids, i.e., as ζ changes sign it must pass through a zero point, and it is presumed that the surface is uncharged when $\zeta = 0$, although this is perhaps only rigorously true when specific adsorption is absent. The interesting correlation in Figure 13 is that the hardness is a maximum when $\zeta = 0$ and decreases with increasing magnitude of the ζ-potential regardless of its sign.

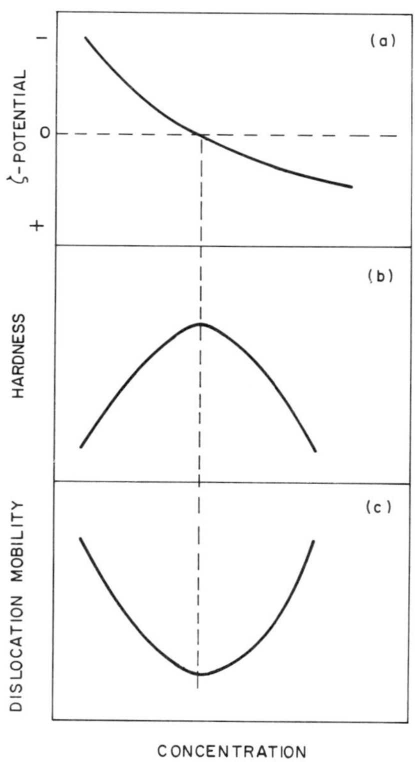

Fig. 13.—Schematic of the apparently generic correlation between ζ-potential, hardness, and dislocation mobility observed in solids such as MgO, Al_2O_3, SiO_2, as well as a variety of rocks and minerals.

Just how the ζ-potential correlation may be understood is not now clear. In a general sense, the view developed by Westwood et al.[75] in order to account for adsorption-induced changes in mechanical behavior is that these phenomena are a consequence of the influence of adsorbed species on the mobility of near-surface dislocations. In short, adsorption-induced changes in a near-surface charge distribution are thought to alter the state of ionization of point and line defects in the near-surface region. Such changes induce variations in the mutual interactions between dislocations, between dislocations and point defects, between dislocations and the lattice, etc., and these variations are in turn reflected in changes in near-surface dislocation mobility or microhardness. Specifically, however, what one must account for is the symmetry that is observed in the hardness about the isoelectric point (i.e., when $\zeta = 0$). One possibility has been suggested by Westwood and Mills in the ASI proceedings.[76] Another which also has its origins at Martin Marietta Laboratories has been proposed by Swain and Latanision.[77] This is shown schematically in Figure 14. In essence, one expects that the charge distribution in the space charge layer of the solid must respond to the charge (in sign and magnitude) populating the solution side of the double layer, Figure 12. For simplicity, it is assumed that this charge is localized in the outer Helmholtz plane. Electron energy levels in the near-surface region must bend up or down depending on the nature of the adsorbed charge. An electronegative adsorbate would induce upwards band bending while the converse would be true in the presence of an electropositive adsorbate, Figure 14(a) and (c), respectively. Now, as mentioned earlier, the ζ-potential tells us something about the surface charge. If one associates the flat band condition of Figure 14(b) with $\zeta = 0$, one might then consider the possibility that the hardness maximum (minimum dislocation mobility) observed in relation to ζ-potential measurements may be typical of the flat band condition in the space charge layer. Recognizing that dislocations may be assigned energy levels in the band gap,[78] as shown in Figure 14, it is tempting to suggest that dislocations in the space charge layer may also acquire an excess charge (i.e., bands bend up or down) in response to the presence of adsorbates. If this were true, then in addition to the force imposed upon near-surface dislocations mechanically by an indenter, electrostatic repulsion between adjacent and like-charged dislocations - the charge acquired by adsorption - would add a further driving force to their motion. Admittedly, this neglects electrostatic image forces, and other factors which should be considered. Nevertheless, except at a flat band condition, the mechanical and electrostatic forces would combine to give greater apparent dislocation mobility - decreased hardness - as is suggested by the ζ-potential correlation.

While this explanation may not be the correct one - it is true that the relation between the ζ-potential, surface charge, and electronic band structure is not well understood - the above model is susceptible to concerted interdisciplinary examination by materials scientists and surface scientists. The critical experiment in terms of the above explanation would be to examine dislocation mobility while at the same time monitoring the distribution of charge in the space charge layer. To an extent, this has been done via the ζ-potential measurements. But, although the ζ-potential may be determined by a relatively simple electrokinetic measurement, its interpretation is not always very direct or unambiguous. On the other hand, it is possible to change the distribution of charge in the space charge layer by several means as described earlier, one of which is by illumination with light. In principle it would seem that if minimum near-surface dislocation mobility in semicondictors (and presumably insulators as well) is associated with the flatband condition, the means by which the flat band is reached should be of no consequence. In this context I hope to soon begin a study of the mobility of dislocations in cadmium sulfide monocrystals examined in such a manner

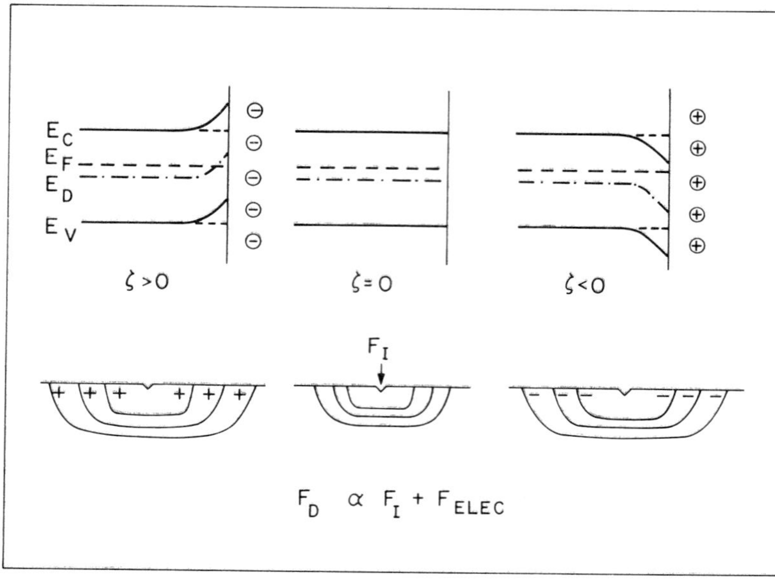

Fig. 14.—Schematic illustrating a possible rationale for the symmetry observed in the ζ-potential correlation shown in Figure 13 based upon the contribution of electrostatic interactions between near-surface dislocations.

that the space charge distribution may be controlled by sub-band gap illuminatnon and monitored by means of surface photo-voltage spectroscopy (SPS) which has been extensively studied by Gatos and co-workers at MIT.[79] CdS is a seemingly appropriate choice; the electronic characteristics of the basal and prismatic surfaces have been recently studied and the band theory for dislocation energy states in CdS has been evaluated by Elbaum and Holmes.[80] If chemisorption-induced changes in dislocation mobility may be related to coincident changes in band structure, the above surface analytical approach is attractive, since through it both elements of this problem become accessible. A crucial experiment will be to determine the dislocation mobility under flat band conditions, as identified by SPS, and to compare this with the dislocation mobility observed when band bending is known to occur.

3.3.3 The double layer and the mechanical properties of metal electrodes

At this point, we might return to consideration of the double layer in order to understand a possible rationale for the ζ-potential correlation based on the Rebinder school's view of adsorption-induced changes in mechanical behavior. This will then be extended to the case of metal electrodes which are also known to be sensitive to changes in the surface charge density.

One hundred years ago Lippmann[81] observed the following relation between the surface energy, γ, the surface charge density, q, and the applied potential, ϕ, for an ideal polarized electrode:

$$\left(\frac{\partial \lambda}{\partial \phi}\right)_{T,P,\mu_i} = -q$$

The Lippmann equation tells us that the surface energy of such an electrode - for example, *Hg* in aqueous alkali halide solutions - passes through a maximum when the surface is uncharged, i.e., when $q = 0$, and, indeed, this has been demonstrated experimentally on many occasions. The potential at which the surface becomes uncharged is known as the potential of zero charge, pzc. The above relationship may be understood physically by recognizing that because of electrostatic repulsion between the like charges in the surface charge layer, the work required to expand the interface is smaller in the presence of a net surface charge density of either sign than in the absence of electrostatic interaction. Because many of the early experiments in this field were done with a Lippmann capillary electrometer, the above is often described as the electrocapillary effect.

The Lippmann equation in the form given above is valid only for liquid electrodes (e.g., *Hg*). In the case of a solid electrode, it is the surface stress (the stress required to deform the surface) which passes through a maximum when the surface is uncharged.[82] In essence, because the much reduced mobility of near-surface atoms in a solid, as compared to liquid, prevents the process of creating new surface from occurring under anything like thermodynamically reversible conditions in a finite period of time, some extra work must be done, and it is the sum of this extra work and the surface energy which is maximized at the pzc. Hence, even in the case of a solid electrode, one may cautiously conclude that, except when the surface is uncharged, electrostatic repulsion between like charges in the surface charge layer will reduce the total effort required to produce unit area of new surface.

It is possible to argue, therefore, that the symmetry of the dependence of hardness on ζ-potential in Figure 13 could conceivably be attributed to the electrocapillary effect as has been acknowledged by Westwood et al.[83] and Shchukin.[71] Indeed, Shchukin et al.[84] suggest that dislocation mobility (hardness) becomes sensitive to adsorbed environments only when two conditions are met: (1) surface steps are formed during dislocation motion and (2) the environment causes large reductions in the energy of surface slip steps formed during deformation. In short, hardness would be greatest and dislocation mobility least when the surface step energy is a maximum.

While there are clearly many more facets to this difference of opinion regarding the basis for understanding adsorption-induced changes in mechanical behavior - i.e., electronic effects versus surface energy reduction - the above represents the essence of this difference. Recently, Macmillan, Huntington, and Westwood[83] have shown that in *MgO* and *LiF* the mobilities of both edge dislocations (which did not produce surface slip steps in the particular hardness test performed) and screws (which did) were environment sensitive. This would seem to argue strongly against the surface energy concept presented above, at least in the case of nonmetallic solids. On the other hand, it seems possible that the contribution of the surface step energy may be more significant in some other instances, particularly in the case of metal electrodes. In contrast to semiconductors and insulators in which the distribution of charge in a space charge layer that extends to depths on the order of microns may well influence the mobility of near-surface dislocations, the same argument cannot be invoked in the case of metal electrodes. As explained earlier, because of the high density of mobile charge carriers (electrons) in a metal, the charge in the solid consists of an excess or deficiency of electrons within at most a few angstroms of the surface. Correspondingly, any effect of charge density on mechanical behavior must result from events occurring right at the metal surface.

It has in fact been demonstrated the mechanical behavior of metal electrodes is a function of charge density on the metal surface.[85-88] The first reported studies of electrocapillary effects on mechanical behavior were reported by Rebinder and co-workers.[85] They reported that the pendulum hardness of metal electrodes (determined from the amplitude damping of an oscillating pendulum) passed through a maximum at the pzc. However, Bockris and Parry-Jones[89] later observed precisely the opposite effect in pendulum experiments, using a smooth fulcrum rather than the ground glass fulcrum used by Rebinder, and suggested that friction between the fulcrum of the pendulum and the metal, rather than hardness, may be the underlying physical property that is most affected by the charge double layer in such experiments. While there is evidence to show that friction in such cases is potential sensitive,[90] others have shown[86,87] that the creep rate of metals passes through a minimum at the pzc, and it is clear that this could not be related to frictional effects. The latter investigators have considered a contribution to the activation energy for creep arising from the energy required to create surface steps. Recent work by Latanision et al.[88] on zinc monocrystal surfaces has confirmed the notion that electrocapillary changes in the charge density can indeed affect the motion of dislocations producing slip steps. In these experiments the diamond pyramid hardness and (etch pit) dislocation distribution about the indentation were monitored as a function of applied electrode potential on the basal $\{0001\}$ and prism $\{10\bar{1}0\}$ planes of zinc monocrystal electrodes. As shown in Figure 15, observable slip about indentations on the basal plane occurs on second order pyramidal planes $\{11\bar{2}2\}$ and in $\langle 11\bar{2}3 \rangle$ directions. Hence, screw dislocations on the pyramidal planes that intersect the basal plane have components of their Burgers vector normal to the surface and slip around the hardness impression exposes $\{11\bar{2}2\}$ steps. The motion of such dislocations is resisted in part by the work required to produce the slip steps left in their trails, and one expects that at the pzc, where more effort is required to create surfaces, the extent of glide of dislocations about a hardness impression would be a minimum. Conversely, the hardness would be a maximum as observed, Figure 16. It is therefore seemingly possible to harden or soften zinc surfaces by changing charge density. Note that the hardness maximum at -1200 mV on Figure 16 corresponds well with the anticipated value of the pzc for the $\{11\bar{2}2\}$ slip plane as should be expected, since the new area being formed by slip is that of the $\{11\bar{2}2\}$ plane. On the other hand, if one indents the prism surface, in which observable pyramidal dislocations have their Burgers vector in the plane of the surface, then such dislocations do not produce steps and are expected to show no sensitivity to potential variation. This, too, has been observed, Figure 16.

We now have the situation that in the case of ceramics like *MgO*, the mobilities of both edge and screw dislocations are environment sensitive, whereas in metals like zinc,[88] gold,[86] and lead[87] only the mobilities of those dislocations which trail slip steps appear to be sensitive. One explanation for this difference between such metals and nonmetals may arise because dislocation mobility in the latter is principally a function of electrostatic interaction between dislocations and point defects - which vary similarly with environment for both edge and screw dislocations - and the additional resistance, if present, due to slip step formation is so small as to be negligible. Alternately, one might compare the energy consumed in creating slip steps with the energy consumed in sweeping out the area traveled by the dislocations loops which produce those steps as a consequence of plastic deformation beneath an indenter. In this case one would find[91] that only in materials with relatively high surface energies and low yield stresses (some pure metals) would the surface energy contributions be significant and, hence, reflect changes in surface energy, however

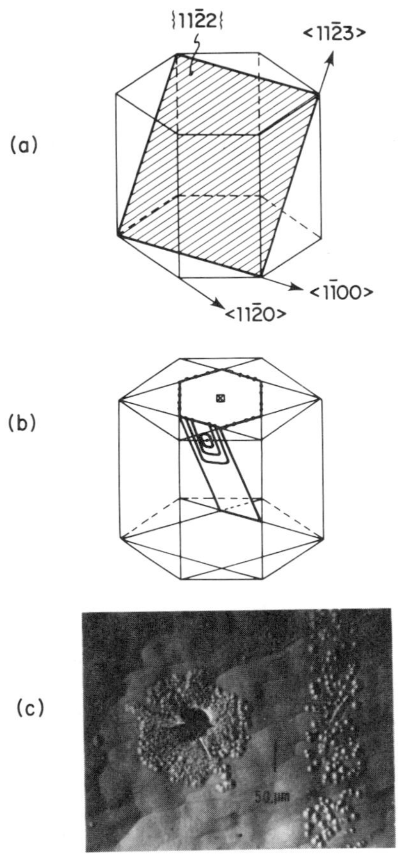

Fig. 15.—Nonbasal slip about a microhardness indentation on (0001) zinc surfaces showing second-order pyramidal slip systems and a schematic of the dislocation distribution at an indentation (after Latanision et al.).[88]

induced, in mechanical testing. Conversely, in materials of high flow stress (ceramics, metallic alloys, etc.) or relatively low surface energy (*LiF, NaCl, MgO,* etc.) or relatively low surface energy might not be reflected in dislocation mobility. In short, only if the surface energy and plastic work terms were of the same magnitude could a change in one significantly affect the other. This seemingly can only happen in the case of relatively pure metals. At any rate, while it is true that the correlation between hardness (dislocation mobility) and surface energy is poor for most solids, it may be that this is more a matter of degree than a failure in principle.

3.4 Surface Effects in the Environment-Sensitive Fracture of Solids

The Joffe effect, mentioned earlier (see Section 3.1.2) is one example

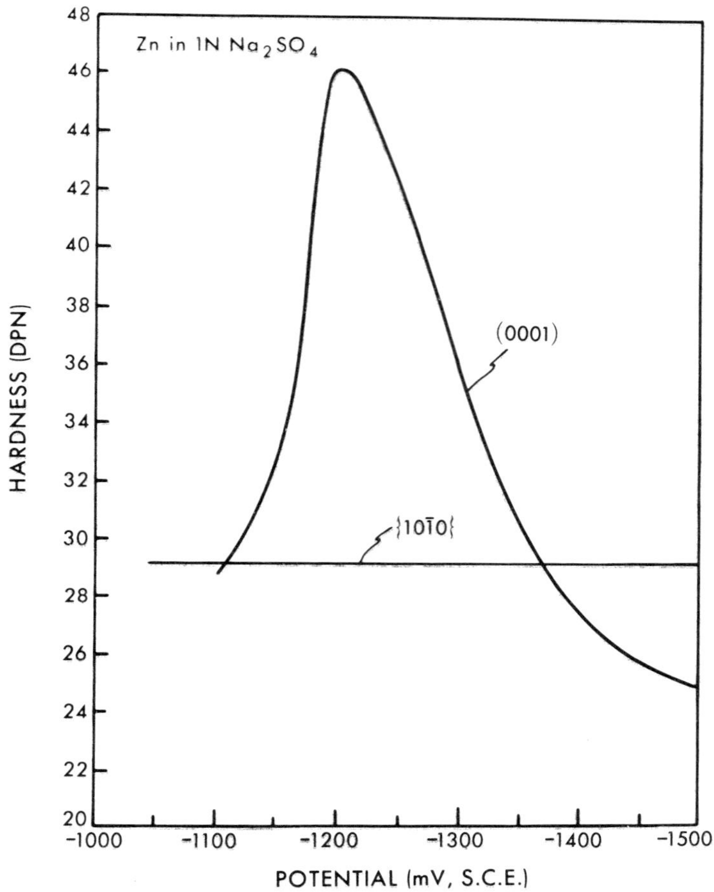

Fig. 16.—Potential dependence of hardness on (0001) and {10$\bar{1}$0} surfaces of zinc (after Latanision et al.).[88]

of the remarkable influence that environments - sometimes very innocuous ones - may have on the fracture behavior of solids. In the case of the Joffee effect, the presence of ordinary water has the staggering effect of turning a typically weak and brittle material (e.g., *KCl*), into one with increased strength and almost unbelievable ductility. This is a relatively pleasant and useful surprise. Often the surprises are not so happy. Small amounts of *Hg* on the surface of otherwise (relatively) ductile zinc polycrystals lead to a catastrophic loss of strength and ductility; hydrogen has a similar effect on alloys of iron, nickel, etc.; certain mildly corrosive electrolytes can lead to the premature cracking of certain (often normally ductile) alloys (304 stainless steel in Cl^- environment, etc.) at stresses far below their normal fracture stresses, and so on. All of these embrittlement phenomena - liquid metal embrittlement,[92] hydrogen embrittlement[93] stress corrosion cracking[94] - and others[95] - have been the subject of recent reviews and conferences and will not be discussed in detail here. Some of the issues of concern, particularly in relation to dissolution-related embrittlement, were treated by Engell[96] at the ASI.

In all of the above, the action begins at the interface between the solid and the environment which surrounds it, and it should not be too surprising to find that many of the mechanistic models for these phenomena involve film formation, adsorption of critical species from the environment, interfacial solute and vacancy concentration gradients, etc. However, we know little about the possible influence of yielding initiation in the surface layer on the embrittlement of polycrystals.[20,97] Nor, for example, do we seem to know very much about the influence of surface charge on crack propagation in metal electrodes, or about the detailed atomic-order interactions between embrittling species and strained crystal surfaces, or about the atomistics of decohesion or about the electron distribution at a crack tip (which has, however, received some attention).[98] Some of these will be discussed briefly in this section.

In the previous section, evidence was presented to show that changes in the surface charge density on metal electrodes appear to affect their plasticity - the electrocapillary effect. It would also seem possible that changes in the charge density on a metal electrode may affect its fracture behavior. For example, an equilibrium crack in a solid subject to an increasing force, Figure 17, will either propagate by cleavage or grow slowly by shear, depending on whether the tensile fracture stress, σ, for the atom-atom bond or the shear stress, τ, to cause dislocation motion on a favorable slip system is achieved first. Hence, as the ratio of σ/τ decreases, cleavage becomes more likely and, conversely, shear failure becomes more probable as the ratio increases.[99] As discussed earlier, the micrchardness of metals (and, hence, crystal plasticity, τ) may be affected by changes in the charge density, provided that glide dislocations produce surface steps. On this basis, one anticipates that the motion of such dislocations, Figure 17, away from sources near the crack tip may be inhibited (i.e., increasing τ) at the pzc encouraging cleavage. Likewise, because of the repulsion between like charges, the bond strength or cohesion between atoms in the surface layer of the crack tip, Figure 17, may also be affected by the charge density in the double layer. In this case one expects that at the pzc, where the surface charge is extinguished, cohesion of surface atoms would be maximized - i.e., cleavage would be discouraged.

Although it is difficult to predict whether the expected influence of charge density on τ or on σ should be most significant, recent straining electrode experiemnts[100] suggest that in the case of zinc monocrystal electrodes, the former may be quite important. In these experiments, cylindrical electrodes with their tensile axis normal to the basal plane were deformed at (a) a constant strain rate ($10^{-4}sec^{-1}$) or (b) a fixed load of 90% of the yield stress and their fracture strain and time-to-fracture, respectively, were measured at a function of applied potential over the same range of potentials as shown in Figure 16. In this orientation, the limited slip which does occur takes place on second-order pyramidal planes allowing correlation of fracture behavior with the known[88] plastic response of pyramidal slip to electrode potential. Both the fracture strain and time-to-failure were observed to pass through sharp minima at about -1200 mV, SCE, i.e., at the pzc for the $\{11\bar{2}2\}$ plane. The latter is shown in Figure 18. This coincides with the hardness maximum - i.e., decreased mobility of dislocations (plasticity) - which was observed in Figure 16. This suggests that once a crack is initiated, it may be expected to propagate most rapidly at the pzc, since crack blunting, via dislocation emission from the crack tip,

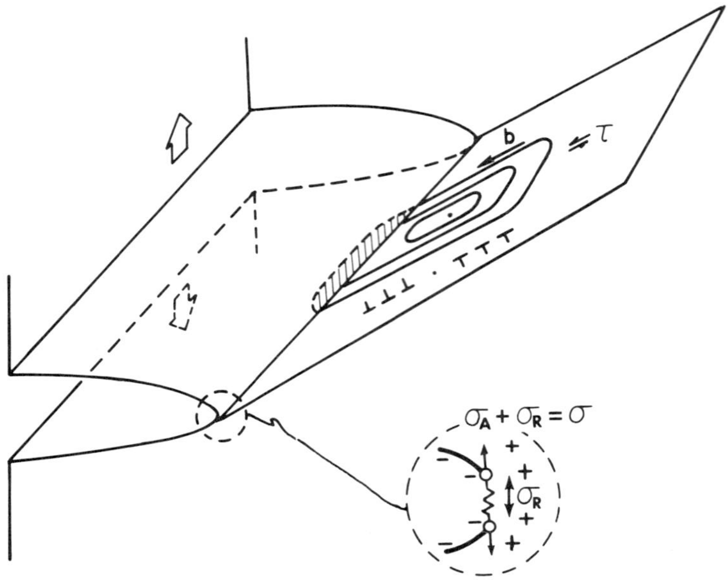

Fig. 17.—Crack tip showing the possible influence of the double layer on σ and τ (after Latanision).[6]

would be most restricted (i.e., the ratio of σ/τ is decreased), provided dislocations produce surface steps as shown in Figure 17. This increases the tendency for cleavage and should be reflected in reduced time-to-fracture and fracture strain.

It should be noted in passing that we really cannot say whether the variation in σ had any effect on the observed results. While one would argue that a maximum in τ would occur at the pzc for the $\{11\bar{2}2\}$ slip plane, the maximum in σ would occur at the pzc for the plane of the free surface which in this case cannot be specified. Since the pzc is anisotropic, there is no reason to expect singularities in τ and σ to occur at the same potential. Hence, by inference, we presume to have seen a dependence of the initiation and propagation of cleavage cracks in zinc by a shear and not a bond strength mechanism.

It may be worth noting as well that the model proposed in Figure 17 would allow for the possibility that adsorbates may significantly affect the fracture of metals, since the propagation of a surface-initiated crack

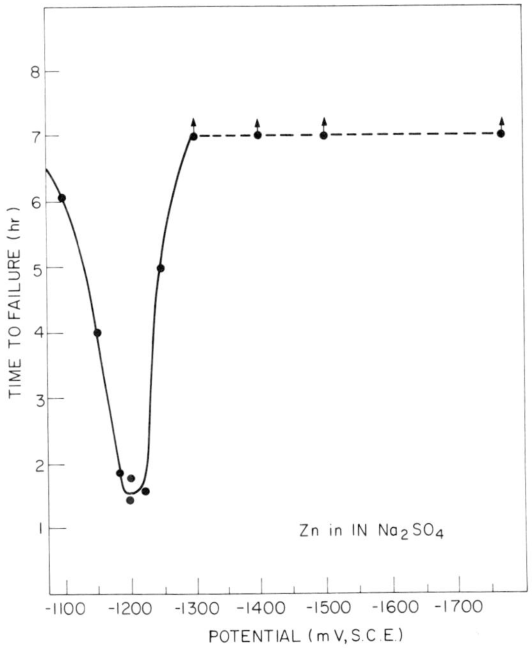

Fig. 18.—The potential dependence of the time-to-failure of zinc monocrystal electrodes.

involves the consecutive rupture of surface bonds, and chemisorption, which is likely to involve some change in the distribution of charges in the double layer, may thus affect the bond strength or cohesion (perhaps as described above) between the atoms constituting the crack tip. Liquid metal embrittlement is a classic example of this.(92) Note that in liquid metal embrittlement, the embrittling species is considered to be adsorbed directly on the solid surface and, hence, strongly affects the strength of bonds between surface atoms. Thus, the consecutive rupture of surface atomic bonds, induced by reduction in the cohesive strength of surface atoms, leads to catastrophic fracture. In contrast, in aqueous electrolytes and in the absence of specific adsorption, ions present in the outer Helmholtz plane, Figure 11, are effectively shielded from the surface by water molecules, and are less likely to significantly influence cohesive strength (i.e., fracture). If it occurs, specific adsorption, on the other hand (for example, of Cl^-), may lead to a situation in an aqueous electrolyte approximating that of liquid metal em-embrittlement. Adsorption models based on this premise have been proposed for stress corrosion cracking.(94) At any rate, it should be appreciated that σ and in some cases τ may both be affected by variations in the charge density in the electrical double layer, regardless of whether the variations are due, for example, to the application of external potential or to the chemisorption of surface-active species. It should be mentioned that Gilman(101) has also recently considered some aspects of surface effects in embrittlement phenomena.

While it is possible to describe liquid metal embrittlement, hydrogen embrittlement, etc., on the basis of environmentally-induced decohesion, the atomic scale interactions between the adsorbate (*Hg*, *H*, etc.) and strained metal surfaces are not well understood. In this context, one interesting electronic model of hydrogen embrittlement proposed in 1960 by Troiano but never really examined experimentally, may now be tractable through the use of available surface analytical tools. In this model, Troiano presumes that hydrogen behaves in an electropositive sense and donates its electron to the unfilled d-bands of the metallic cores as shown schematically in Figure 19. The increase in electron density leads to an increase in the repulsive force between adjacent metal cores, or, in other words, a decrease in the cohesive strength of the lattice, i.e., in a manner not unlike that described in Figure 17. This may be studied by means of chemical shifts in the Auger spectrum for iron, for example, due to the presence of adsorbed hydrogen. If hydrogen does in fact behave in an electropositive sense, iron's Auger peaks should undergo positive chemical shifts. The converse is true if hydrogen-behaves in an electronegative sense, as is suggested in fact by work function measurements.[103] Of course, there may be other ways of understanding the atomistics of hydrogen embrittlement based on knowledge of the charge transfer process which occurs when hydrogen is adsorbed on a suitable metal surface. For example, recognizing that the great cohesive strength of the transition metals may be associated with their band structure, particularly the filling of d-orbitals, it is tempting to suggest[104] that an electropositive adsorbate on iron, for example, donates its electrons to iron surface atoms, allowing them to behave, at least insofar as cohesion is concerned, rather like cobalt atoms. Conversely, in the presence of an electronegative adsorbate, iron surface atoms might behave as their neighbor in the periodic chart with one less d-electron, namely, manganese. It is interesting that cobalt has a reportedly[105] lower cohesive strength than iron, suggesting in this context that the former rather than the latter is more likely to occur, i.e., hydrogen behaves as an electropositive adsorbate. At any rate, the nature of the charge transfer process which occurs should be accessible and would provide valuable insight into the mechanism of hydrogen embrittlement as well as other presumably adsorption-induced embrittlement phenomena such as liquid metal embrittlement. In short, we are now in a position to

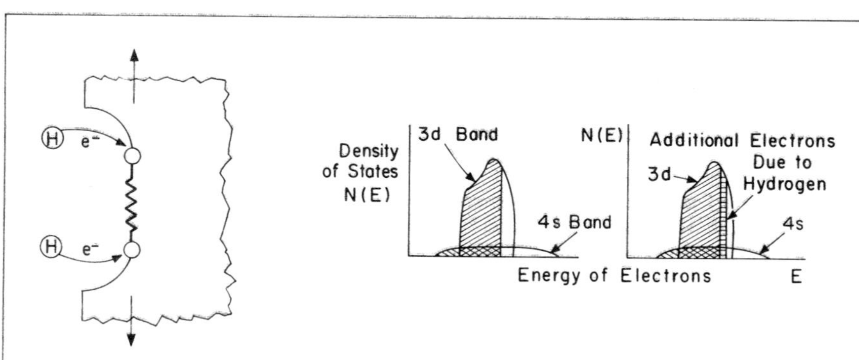

Fig. 19.—An electronic model for hydrogen embrittlement (after Troiano).[102]

use surface analytical tools to study the atom-scale interactions between strained surfaces and embrittling species (adsorbates).

In this brief commentary, I have not attempted to discuss the specific and often exciting areas of controversy that have arisen in the studies of environmentally induced embrittlement - for example, the role of hydrogen (embrittlement), if any, in stress corrosion cracking.[94] What I have attempted to do, however, is to suggest some relatively unexplored means of viewing environment-sensitive fracture, and an example of a surface analytical approach to studying the atomic-order processes which occur when embrittling species interact with crystal surfaces. Perhaps by means of such experiments we may better understand the age-old question of why normally ductile metals become embrittled when exposed to certain environments.

4. TECHNOLOGICAL APPLICATIONS OF SURFACE EFFECTS IN CRYSTAL PLASTICITY

Generally, the effects of environments on mechanical behavior are considered to be adverse, and to be sure, often that is the case. Stress corrosion cracking, corrosion fatigue, hydrogen embrittlement, liquid metal embrittlement, etc., are all examples of failure phenomena which take on catastrophic consequences. Sometimes, however, the effects of environments on the mechanical behavior of solids are more beneficial and can be used to advantage. For example, we have already seen that through variations in the surface or near-surface charge density on various metals and nonmetals, the hardness of such solids may be controlled - if, however, for different reasons. Moreover, we also know that the coefficient of friction may be affected by changes in the charge density. Hence, it should not be surprising to find that much attention has recently been directed toward making application of research on the surface- and environment-sensitive mechanical behavior of solids in metal cutting, the machining of ceramics, rapid excavation of hard rock, comminution, and even in the control of earthquakes, all of which are related to such parameters as friction, flow, and the fracture behavior of solids. A detailed account of these effects in both the US and in the USSR was presented by Westwood,[76] Pertsov,[106] and others at the NATO Advanced Study Institute. A superb review in this context has recently been prepared by Westwood.[1] In any case, a brief commentary on some of these follows. What is important to recognize now, however, is that not only are there many exciting academic problems to tackle but, despite an admitted lack of detailed understanding, we are able to apply this understanding with remarkable success to technological problems.

Let's first consider metal cutting. In conventional metal cutting, the removal of material is considered to be governed principally by two parameters: friction at the tool-chip and tool-workpiece interfaces and the strength or workhardening capacity of the workpiece. Significantly, both the frictional and mechanical behavior of metals may be made controllable in the presence of a charge double layer affect mechanical behavior, but the presence of anodic surface films, surface dissolution, and the generation of the various gases on the electrode surfaces are also known to affect mechanical and frictional behavior. Hence, one expects that machining of metals should be significantly affected by making the workpiece an electrode in a suitable electrolyte and then applying an electrode potential which imparts appropriate physical and mechanical properties to the electrode. Thus, in this approach metal removal still occurs by contact of a tool with the workpiece; however, frictional and mechanical behavior of the tool-workpiece system are controlled electrochemically. This technique, developed particularly for use in machining hard-to-cut alloys and patented at Martin Marietta Laboratories[107] under the name of electromechanical machining (EMM), is already being examined in a pilot-scale production operations.

The use of surface active environments in the machining of ceramic materials has its origins in the book "Hardness Reducers in Rock Drilling," published in the USSR.[108] Today surfactants are being examined for various uses in the cutting, grinding, and drilling of ceramics, minerals, and rocks - both on the very large scale, as when driving a railroad tunnel into the side of a mountain, and on the very small scale, as when manufacturing the Al_2O_3 substrates for modern microelectronic components. By maximizing the surface hardness through ζ-potential control (i.e., by changing the composition of the surface active environments surrounding the solid), researchers at Martin Marietta Laboratories have been able to increase the rate of drilling by factors of 6-10 in the case of alumina, up to 14 in the case of quartz, and 3-4 for various types of granites. From a practical viewpoint, the importance of such improvements in drilling rate - which potentially represent an enormous reduction in drilling costs - is that they may be achieved merely by adding a very small amount of a relatively cheap and innocuous chemical to the drilling water normally employed for cooling, lubrication, and flushing away of drilling debris.

A very different application for environmental control of the coefficient of friction may be a tool for reducing earthquake hazards.[109] In simple terms, an earthquake occurs when the shear forces resulting from movements in the earth's crust exceed the frictional resistance along a plane of crustal weakness (a fault), and this frictional resistance varies directly with the coefficient of friction between the fault surfaces and inversely with the pressure of any liquid present in the fault. One can therefore imagine influencing an earthquake either by pumping natural pore water (usually brine) out of the fault, or by pumping in some man-made fluids of suitable chemistry, thereby changing the pressure tending to force the fault surfaces apart and/or varying the coefficient of friction between them. In this context it might be possible to "defuse" a large and potentially destructive earthquake by environmentally releasing its stored strength and energy as a succession of small and harmless tremors rather than as a less frequent but large-scale and catastrophic release of strain energy, i.e., a major earthquake.

The above are just a few of several exciting and potentially useful applications of the phenomenology of surface effects in crystal plasticity to real technological problems.

5. CONCLUDING REMARKS

My objective in this paper was to provide a general overview of the science add technological applications of surface effects in crystal plasticity. I have not discussed adhesion, oxidation, catalysis, etc., all of which are dependent as well on the surface and its condition. This effort has been confined to inorganic crystalline solids, thereby excluding vitreous and polymeric solids whose mechanical properties are also environment-sensitive.

I believe that materials scientists have made considerable progress in understanding surface effects in crystal plasticity, but I also believe that the detailed, atom-scale understanding that will lead to definitive treatments of these phenomena and to application on a broad scale demands the concerted action of materials scientists and surface scientists alike.

6. ACKNOWLEDGEMENTS

This review was begun during a visit to Max-Planck-Institut für Eisenforschung in Düsseldorf as a recipient of a Senior US Scientist Award administered by the Alexander von Humboldt Foundation. I am grateful to the Foundation and to Professor H.-J. Engell, Director of MPI, for providing me with the facilities and resources to pursue this activity. I am also particularly pleased to acknowledge many helpful and informative discussions over a number of years with my friends and colleagues, A. R. C. Westwood, N. H. Macmillan, and R. W. Staehle.

REFERENCES

1. Westwood, A.R.C., *Journal of Materials Science,* Vol. 9, 1974, p. 1871.
2. Reynolds, O., *Manchester Literary and Philosophical Society. Memoirs and Proceedings,* Vol. 13, 1874, p. 93.
3. Rebinder, P.A., Reports to the VI Congress of Physicists, Moscow, 1928, p. 29.
4. Latanision, R.M. and Fourie, J.T., eds., "Surface Effects in Crystal Plasticity," Noørdhoff International Publishing, Leyden, 1977.
5. Latanision, R.M., "Corrosion Fatigue," National Association of Corrosion Engineers, Houston, 1972, p. 185.
6. Latanision, R.M., in Proceedings of the International Conference on Surface Technology, Society of Manufacturing Engineers, Dearborn, 1973, p. 1.
7. Latanision, R.M., Sedriks, A.J. and Westwood, A.R.C., "Structure and Properties of Metal Surfaces," Honda Memorial Series on Materials Science, Maruzen Co., Tokyo, 1973, p. 500.
8. Blakeley, J.M., "Introduction to the Properties of Crystal Surfaces," Pergamon Press, New York, 1973.
9. Somorjai, G.A., "Principles of Surface Chemistry," Prentice-Hall, Englewood Cliffs, 1972.
10. Kossel, W., *Nach. Ges. Wiss.,* 1927, p. 135.
11. Stranski, I.N., *Zeitschrift fuer Physikalische Chemie,* Vol. 136, 1928, p. 259.
12. Burton, W.K., Cabrera, N. and Frank, F.C., *Royal Society of London. Philosophical Transactions.* Series A, Vol. 243, 1950, p. 299.
13. Duke, C.B., *Advances in Chemical Physics,* Vol. 27, 1974, p. 1.
14. Fleischer, R.L., *Acta Metallurgica,* Vol. 8, 1960, p. 598.
15. Marsh, D.M., "Fracture of Solids," Wiley-Interscience, New York, 1963, p. 119.
16. Hirth, J.P., "Relation Between Structure and Strength of Metals and Alloys," Her Majesty's Stationery Office, London, 1963, p. 218.
17. Aust, K.T., Niessen, P., Hanneman, R.E. and Westbrook, J.H., *Acta Metallurgica,* Vol. 16, 1968, p. 291.
18. Aust, K.T. and Westbrook, J.H., *Acta Metallurgica,* Vol. 19, 1971, p. 521.
19. McMahon, C.J., Proceedings 4th Bolton Landing Conference: Grain Boundaries in Engineering Materials, 1974.
20. Latanision, R.M. and Opperhauser, H., Jr., *Metallurgical Trans.,* Vol. 5, 1974, p. 483; also Vol. 6A, 1975, p. 233.
21. Latanision, R.M. and Staehle, R.W., *Scripta Metallurgica,* Vol. 2, 1968, p. 667.
22. Herring, C., "Metal Interfaces," American Society for Metals, Metals Park, Ohio, 1952, p. 1.
23. Nabarro, F.R.N., "Theory of Crystal Dislocations," Clarendon Press, Oxford, 1967.

24. Hirth, J.P. and Lothe, J., "Theory of Dislocations," McGraw-Hill, New York, 1968.
25. Frank, F.C. and Read, W.T., in Symposium on Plastic Deformation of Crystalline Solids, Carnegie Institute of Technology, Pittsburgh, 1950, p. 44.
26. Dash, W.C., *Journal of Applied Physics,* Vol. 27, 1956, p. 1193.
27. Kramer, I.R. and Demer, L.J., *Progress in Materials Science,* Vol. 9, 1961, p. 133; also *Trans. AIME,* Vol. 221, 1961, p. 780.
28. Fisher, J.C., *Trans. AIME,* Vol. 194, 1952, p. 531.
29. Kuhlmann-Wilsdorf, D., "Environment-Sensitive Mechanical Behavior," Gordon and Breach, New York, 1966, p. 681.
30. Latanision, R.M. and Staehle, R.W., *Acta Metallurgica,* Vol. 17, 1969, p. 307.
31. Sumino, K., *Physical Society of Japan. Journal,* Vol. 17, 1962, p. 454.
32. Mughrabi, H., *Physica Status Solidi,* Vol. 44, 1971, p. 391.
33. Fourie, J.T. and Dent, N.C.G., *Acta Metallurgica,* Vol. 20, 1972, p. 1291.
34. Kitajima, S., Tanaka, H. and Kaieda, H., *Japan Institute of Metals. Transactions,* Vol. 10, 1968, p. 12.
35. Vol'shakov, V.I. and Orlov, I.G., *Soviet Physics - Solid State,* Vol. 12, No. 3, 1970, p. 576.
36. Vesely, D., *Physica Status Solidi,* Vol. 29, 1968, p. 685.
37. Young, F.W. and Sherrill, F.A., *Canadian Journal of Physics,* Vol. 45, 1967, p. 747.
38. Lohne, O. and Rustad, O., *Philosophical Magazine,* Vol. 25, 1972, p. 529.
39. Lohne, O., *Physica Status Solidi (A) - Applied Research,* Vol. 18, 1973, p. 473.
40. Vellaikal, G. and Washburn, J., *Journal of Applied Physics,* Vol. 40, 1969, p. 2280.
41. Tsunekawa, Y. and Weissman, S., *Materials Science Engineering,* Vol. 17, 1975, p. 51.
42. Mendelson, S., *Journal of Applied Physics,* Vol. 33, 1962, pp. 2175, 2182.
43. For review, see: Westwood, A.R.C., "Environment-Sensitive Mechanical Behavior," Gordon and Breach, New York, 1966, p. 1.
44. Kolb, K. and Macherauch, E., *Philosophical Magazine,* Vol. 7, 1962, p. 415.
45. Kramer, I.R. and Kumar, A., "Corrosion Fatigue," National Association of Corrosion Engineers, Houston, 1972, p. 146.
46. Kramer, I.R., *Trans. AIME,* Vol. 233, 1965, p. 1462.
47. Fourie, J.T., *Philosophical Magazine,* Vol. 17, 1968, p. 735.
48. Latanision, R.M., in Proceedings International Conference on Strength of Metals and Alloys, American Society for Metals, Metals Park, Ohio, Vol. 2, 1970, p. 446.
49. Mughrabi, H., *Physica Status Solidi,* Vol. 39, 1970, p. 317.
50. Fourie, J.T., "Corrosion Fatigue," National Association of Corrosion Engineers, Houston, 1972, p. 164.
51. Kramer, I.R., *Scripta Metallurgica,* Vol. 8, 1974, p. 1231.
52. Revie, R.W. and Uhlig, H.H., *Scripta Metallurgica,* Vol. 8, 1974, p. 1235.
53. Duquette, D.J. and Uhlig, H.H., *Transactions of the American Society for Metals,* Vol. 61, 1968, p. 445; also Vol. 62, 1969, p. 839.
54. Duquette, D.J., Hahn, H. and Andresen, P., "Surface Effects in Crystal Plasticity," Noødhoff, Leyden, 1977, p. 469.
55. Joffe, A., Kirpitschewa, M.W. and Lewitsky, M.A., *Zeitschrift fuer Physik,* Vol. 22, 1924, p. 286.
56. Roscoe, R., *Nature,* Vol. 133, 1934, p. 912.

57. Garstone, J., Honeycombe, R.W.K. and Greetham, G., *Acta Metallurgica*, Vol. 4, 1956, p. 485.
58. Bullough, R., "Surface Effects in Crystal Plasticity," Noørdhoff, Leyden, 1977, p. 321.
59. Johnson, R.M. and Block, R.J., *Acta Metallurgica*, Vol. 16, 1968, p. 831.
60. Jemian, W.A. and Law, C.C., *Acta Metallurgica*, Vol. 15, 1967, p. 143.
61. Ruddle, G.E. and Wilsdorf, H.G.F., *Applied Physics Letters*, Vol. 12, 1968, p. 271.
62. Sethi, V.K. and Gibala, R., *Scripta Metallurgica*, Vol. 9, 1975, p. 527.
63. Evans, T. and Schwarzenburger, D.R., *Philosophical Magazine*, Vol. 4, 1959, p. 889.
64. van der Merwe, J.H., "Surface Effects in Crystal Plasticity," Noørdhoff, Leyden, 1977, p. 301.
65. Matthews, J.W. and Jesser, W.A., *Acta Metallurgica*, Vol. 15, 1967, p. 595.
66. Brame, D.R. and Evans, T., *Philosophical Magazine*, Vol. 3, 1958, p. 971.
67. DeJonghe, L.C. and Greenfield, I.G., *Acta Metallurgica*, Vol. 17, 1969, p. 1411.
68. Pridans, J., Berkowitz, B. and Billelo, J.C., *Scripta Metallurgica*, Vol. 5, 1971, p. 701.
69. Rebinder, P.A. and Shchukin, E.D., *Progress in Surface Science*, Vol. 3, No. 2, 1972, p. 97.
70. Westwood, A.R.C. and Macmillan, N.H., "Science of Hardness Testing and Its Research Applications," American Society for Metals, Metals Park, Ohio, 1973, p. 372.
71. Shchukin, E.D., "Surface Effects in Crystal Plasticity," Noørdhoff, Leyden, 1977, p. 701.
72. Macmillan, N.H., "Surface Effects in Crystal Plasticity," Noørdhoff, Leyden, 1977, p. 629.
73. Westwood, A.R.C., Preece, C.M. and Goldheim, D.L., "Molecular Processes on Solid Surfaces," McGraw-Hill, New York, 1969, p. 591.
74. Boddy, P.J., *Journal of Electroanalytical Chemistry and Interfacial Electrochemistry*, Vol. 10, 1965, p. 199.
75. Westwood, A.R.C., Goldheim, D.L. and Lye, R.G., *Philosophical Magazine*, Vol. 16, 1967, p. 505; Vol. 17, 1968, p. 951.
76. Westwood, A.R.C. and Mills, J.J., "Surface Effects in Crystal Plasticity," Noørdhoff, Leyden, 1977, p. 835.
77. Swain, M.V. and Latanision, R.M., Unpublished Work, Martin Marietta Laboratories, 1974.
78. Haasen, P. and Schröter, W., in "Fundamental Aspects of Dislocation Theory," Special Publication 317, National Bureau of Standards, 1970, p. 1231.
79. Lagowski, J. and Gatos, H.C., *Surface Science*, Vol. 30, 1972, p. 491.
80. Elbaum, C. and Holmes, R.P., in "Fundamental Aspects of Dislocation Theory," Special Publication 317, National Bureau of Standards, 1970, p. 1293.
81. Lippmann, G., *Ann. Chim. Phys.*, Vol. 5, 1875, p. 494.
82. Latanision, R.M., Macmillan, N.H. and Lye, R.G., *Corrosion Science*, Vol. 13, 1973, p. 387.
83. Macmillan, N.H., Huntington, R.D. and Westwood, A.R.C., *Philosophical Magazine*, Vol. 28, 1973, p. 923.
84. Shchukin, E.D., Savenko, V.I., Kochanova, L.A. and Rebinder, P.A., *Dokl. Akad Nauk SSSR*, Vol. 200, No. 2, 1971, p. 406.
85. Rebinder, P.A. and Venstrem, E.K., *Acta Physica et Chemica*, Vol. 19, 1944, p. 36.

86. Pfutzenreuter, A. and Mazing, G., *Zeitschrift fuer Metallkunde,* Vol. 42, 1951, p. 361.
87. Likthtmann, V.I., Kochanova, L.A., Leykis, E.J. and Shchukin, E.D., *Elektrokhimiya,* Vol. 5, 1969, p. 729.
88. Latanision, R.M., Opperhauser, H., Jr. and Westwood, A.R.C., "Science of Hardness Testing and Its Research Applications," American Society for Metals, Metals Park, Ohio, 1973, p. 432; also in Proceedings 5th International Congress on Metallic Corrosion, National Association of Corrosion Engineers, Houston, 1974, p. 111.
89. Bockris, J. O'M. and Parry-Jones, R., *Nature,* Vol. 171, 1953, p. 930.
90. Bockris, J. O'M. and Sen, R.K., *Surface Science,* Vol. 30, 1972, p. 237.
91. Latanision, R.M., Unpublished Work.
92. Westwood, A.R.C., Preece, C.M. and Kamdar, M.H., "Fracture," Plenum Press, New York, 1970, Vol. 3, p. 589.
93. Bernstein, I.M. and Thompson, A.W., eds., "Hydrogen in Metals," American Society for Metals, Metals Park, Ohio, 1970.
94. Staehle, R.W., "Theory of Stress Corrosion Cracking," NATO Scientific Affairs Division, Brussels, 1971, p. 271.
95. Staehle, R.W., McEvily, A.J. and Devereux, O.F., eds., "Corrosion Fatigue," National Association of Corrosion Engineers, Houston, 1973.
96. Engell, H.-J., "Surface Effects in Crystal Plasticity," Noørdhoff, Leyden, 1977, p. 749.
97. Liu, H.W., in Proceedings of the First International Conference on Fracture, 1965, p. 191.
98. Tiller, W.A., *Scripta Metallurgica,* Vol. 8, 1974, p. 487.
99. Kelly, A., Tyson, W.R. and Cottrell, A.H., *Philosophical Magazine,* Vol. 15, 1967, p. 567.
100. Latanision, R.M., Opperhauser, H., Jr. and Westwood, A.R.C., Unpublished Work, Martin Marietta Laboratories, 1974.
101. Gilman, J.J., *Philosophical Magazine,* Vol. 26, 1972, p. 801.
102. Troiano, A.R., *Transactions of the American Society for Metals,* Vol. 52, 1960, p. 54.
103. Tompkins, F.C., "Solid-Gas Interface," Marcel Dekker, New York, 1967, p. 765.
104. Nabarro, F.R.N., Private Communication, April 1975.
105. Watson, R.E. and Ehrenreich, H., *Comments on Solid State Physics,* Vol. 4, 1970, p. 109.
106. Pertsov, N.V., "Surface Effects in Crystal Plasticity," Noørdhoff, Leyden, 1977, p. 863.
107. Latanision, R.M., "Electromechanical Machining Method," U.S. Patent No. 3,873,512, March 25, 1975.
108. Rebinder, P.A., Schreiner, L.A. and Zhigach, K.F., "Hardness Reducers in Rock Drilling," English Translation, CSIRO, Melbourne, 1948.
109. Westwood, A.R.C., Paper TP-386, Martin Marietta Laboratories, 1971.

DISCUSSION

J. MECHOLSKY, NRL: The layers are very thin and yet they seem to affect the bulk hardness.

LATANISION: The surface layer in the case of a semiconductor or insulator is about a micron thick. If the environment is continually available as new surface is being exposed, I would argue then that it can affect bulk properties.

E. SAIBEL, Army Research Office: As I recall, Joffe gave the explanation that he was removing defects, stress concentrations and so on from the surface of the crystal. Is this still valid?

LATANISION: I think it is a valid explanation. But others, for example Mendelson, looked at this question from the point of view of surface sources. The argument that developed from the dislocation etchpit studies by Mendelson is the following: first, plastic deformation takes place by the action of the Fisher surface sources, one arm pinned and the other free to move in the solid; and secondly, as the surface is removed by dissolution, the arm length of that source became shorter. We would expect two things to happen. The first is that as we remove the flaws the material is no longer brittle. Secondly, as the length of the Fisher source decreases, the stress required to operate the source increases. Therefore we observe tremendous plasticity and an increase in fracture strength. Joffe's explanation is still a part of the phenomenology, but more sophistication is involved in the understanding of the whole picture.

SAIBEL: It seems to me that there is a confusion. For example, if we give a light wash with hydrofluoric acid to a piece of glass we strengthen it because a lot of the imperfections, surface scratches and so on are removed. But at the same time people who cut glass would scratch it and then immediately wet the glass to make it fracture easily. The effect of wetting appears to facilitate stress concentration.

LATANISION: Not necessarily. In fact there is a phenomenon known as static fatigue in glasses which is comparable to the familiar stress corrosion cracking of metals. Water is a stress corrosion agent for glass; it appears to increase the sharpness of notches (similar to the chloride ions in the stress corrosion cracking of austenitic stainless steel). Uniform surface dissolution which occurs when we put stainless steel in sulphuric acid, or glass in hydrofluoric acid is different from localized dissolution which occurs when we put a stainless steel in a chloride environment or glass in water. The localized phenomenon leads to stress corrosion cracking in the case of stainless steels but static fatigue in the case of glasses.

MECHOLSKY: Uniform dissolution can take place in glasses that are hygroscopic. It is actually rate dependent. You can either dissolve the cracks or sharpen them depending on the rates of reaction. In fact, there have been experiments in which glass that is susceptible to stress corrosion cracking was put under water for a year and it came out stronger presumably due to the rounding of the cracks.

LATANISION: You are absolutely right. Actually, it is composition dependent. My comments would apply to soda lime glass which have a substantial soda content and perhaps not the others. We find from the literature that the activiation energy for crack propagation corresponds to the activation energy for the diffusion of sodium ions in glass. It seems that sodium is somehow involved in the process.

D. A. RIGNEY, Ohio State University: The experiment of Fourie and Kramer were for fairly simple loading conditions on very simple systems. The conditions we have for typical wear experiment are more complicated. But a hard versus soft controversy can arise in wear as well.

I think part of the problem is that the definitions "hard" and "soft" are somewhat ambiguous. The argument for hard layer uses microhardness measurements. These measurements indicate that the near surface material is harder than the bulk. On the other hand the soft argument uses response to the shear stress. We see evidence of large shear strain which indicates that we have a fairly soft material in the sense that it is ductile. We are talking about two very different things here when we apply the "hard" and "soft" ideas to the case of wear.

LATANISION: There have been people who measured the hardness as a function of depth and observed softening. On the other hand, Kramer has removed surface material and has shown that the core is softer. There really is a major dichotomy.

RIGNEY: That could well depend on the amount of strain which would change the structure. In one case it could be soft and in the other it could be hard.

LATANISION: Yes.

D. TABOR, Cambridge University: May I just make a point. We carried out some very sensitive experiments using an indentor that has a very small radius of curvature and the scanning electron microscope. We found that when we did the experiments very quickly, we did not get a hardness much higher than the bulk hardness. We left the specimen in the electron microscope which is a relatively dirty instrument and gradually formed a thick film of carbonaceous material, and we found that the elastic stresses were perhaps five or ten times the bulk hardness. We were not happy with that discrepancy. So we did this again with one of those ultra high vacuum microscopes (10^{-10} Torr) with devices for cleaning the surface. We observed that when the surfaces were clean we did not get any marked effect. The point that surprises me is that the surface film is so thin and it still shows a large increase in hardness.

LATANISION: An important point is that none of the experiments that I described has ever been done under the conditions which would satisfy a surface physicist as being really clean. When you think about the Kramer-Fourie arguments you must remember that Kramer did many of his experiments with metals like iron and aluminum which are easily oxidized. No one has taken the time yet to look at really clean surfaces. In any case, I am not surprised by the fact that very thin films may have large effects. I would say that people had not done experiments quite as clean as yours before. I would guess that the surfaces were much dirtier and had much thicker films.

INFLUENCE OF SURFACE CHARGE ON THE HARDNESS OF ZnO

J. S. Ahearn, J. J. Mills, A. R. C. Westwood and D. A. Kalivoda

ABSTRACT

Past work has shown that surface-active environments can influence the hardness of such non-metallic solids as silver chloride, magnesium oxide, alumina, quartz, and soda-lime glass. Specifically, a maximum in hardness occurs when the ζ-potential of these solids is zero. This "ζ-correlation" has been taken to imply that surface charge can markedly influence hardness. To examine this possibility directly, therefore, the microhardness and size of dislocation rosettes on the (0001) and {10$\bar{1}$0} surfaces of ZnO were measured as a function of applied potential, electrolyte pH, and time in an electrolytic cell -- all of which alter surface charge. The results indicate that, for both ZnO surfaces, a maximum in hardness is produced not when the surface charge is zero -- as expected -- but rather when the surface is slightly positively charged (downward band bending). Since earlier interpretations of the cause of the "ζ-correlation" now appear to be inappropriate for ZnO, an alternative mechanism, involving charge exchange between donor levels and the conduction band near moving dislocations, is suggested.

INTRODUCTION

In the previous review paper[1] Dr. Latanision has discussed the ζ-potential correlation* observed in insulating solids, such as AgCl[2], Al_2O_3[3], MgO[4], quartz[5-6], soda-lime glass[7-11], and Si[12]. He has correctly pointed out the importance of this correlation as a unifying principle in the field of chemomechanical effects. Although several suggestions have been offered[1] to explain the "ζ-correlation," no detailed understanding has yet been achieved. However, one feature common to recent theories is the concept that surface active species affect the charge on the solid surface, this being revealed by changes in the ζ-potential. This alteration of surface charge, σ, is then considered to change either the charge on point defects or dislocations or the Peierls stress. Since these factors influence dislocation behavior, near-surface dislocation mobility is altered[4].

*The occurrence of an extremum in some mechanical property at zero ζ-potential.

To examine the validity of this basic concept it would be best if we could control, vary, and measure σ directly. In practice, this is not readily accomplished. Moreover, the ζ-potential, which is readily determined, does not directly measure σ, but only the potential at the slipping plane in the liquid double layer surrounding the solid(13). Furthermore, the relationship between ζ-potential and σ is not always straightforward. For instance, when the ionic strength of an aqueous electrolyte at a fixed pH is increased, the ζ-potential will decrease towards zero -- even though the surface charge increases (see, for example, Parks and de Bruyn(14)). Thus, while the ζ-correlation *implies* that surface charge is intimately involved in altering near-surface dislocation behavior, the direct link between ζ-potential, surface charge, and a mechanically different surface region is lacking.

With this problem in mind, therefore, recently we have examined the variation of hardness, H, with σ for the model material, ZnO, for which the relationship between σ and near-surface electronic band structure is reasonably well understood(15,16). In this contribution, we summarize the results of this work.

APPROACH AND EXPERIMENTAL PROCEDURES

In choosing a model system for study, we were guided by two considerations. First, we wanted to study the effect of σ on both polar and non-polar dislocation glide systems, and on point defect-dislocation interactions, dislocation-dislocation interactions, and the lattice friction (Peierls) stress. Second, we wanted to control σ directly and quantitatively and, thereby, the extent and direction of band bending at the surface.

Evidence in the literature(17) suggests that basal slip (polar) in ZnO is controlled by the interaction of dislocations with point defects. For example, crystals oriented for basal slip exhibit a relatively slow decrease in yield stress (τ) with increasing temperature up to ~100°C, a temperature independent behavior between 100 and 300°C, and a further decrease at higher temperatures (Figure 1(a)). Crystals oriented for prismatic slip behave quite differently, however (Figure 1(b)). When this data for prismatic slip is replotted as log τ vs 1/T, a linear relationship is revealed, (Figure 1(c)), characteristic of kink-controlled (i.e., Peierls stress controlled) dislocation motion(18). Thus, two types of deformation controlling mechanisms can be studied in ZnO by simply choosing crystal orientations favoring basal or prismatic slip.

To control σ, the test crystal was made a working electrode in an electrolytic cell, and the applied potential varied appropriately. For n-type, wide-band-gap semiconductors such as ZnO in concentrated electrolytes, the electrical picture is particularly simple (Dewald(19)). Under sufficient anodic bias (semiconductor positive), the charge is negative at the surface, the bands bend up and the space charge region is depleted (Figure 2(a)). The surface charge may be increased to zero (Figure 2(b)) and then made positive (Figure 2(c)) by applying less anodic bias. Thus, band bending is controlled by the applied potential and this, in turn, allows controlled alteration of the carrier density in the conduction and valence bands, and occupancy of dislocation and defect levels in the space charge layer. The influence of these changes on mechanical properties can then be studied via microhardness measurements.*

---

*The high ionic strength of the 1M KCl electrolyte compresses the double layer to the point where the slipping plane coincides with the outer Helmholtz plane. Hence, $\zeta = 0$ at all values of pH.

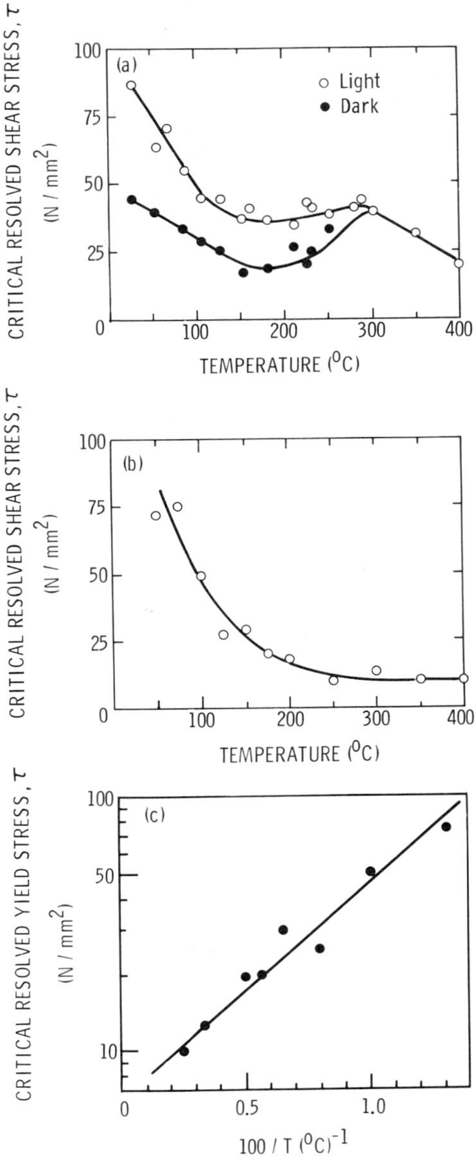

Fig. 1.—(a) Critical resolved shear stress of ZnO for basal slip under dark and illuminated conditions as a function of temperature. (b) Critical resolved shear stress of ZnO for prismatic slip as a function of temperature (from Ref. 17). (c) Critical resolved shear stress of ZnO for prismatic slip as a function of 1/T (data from Ref. 17).

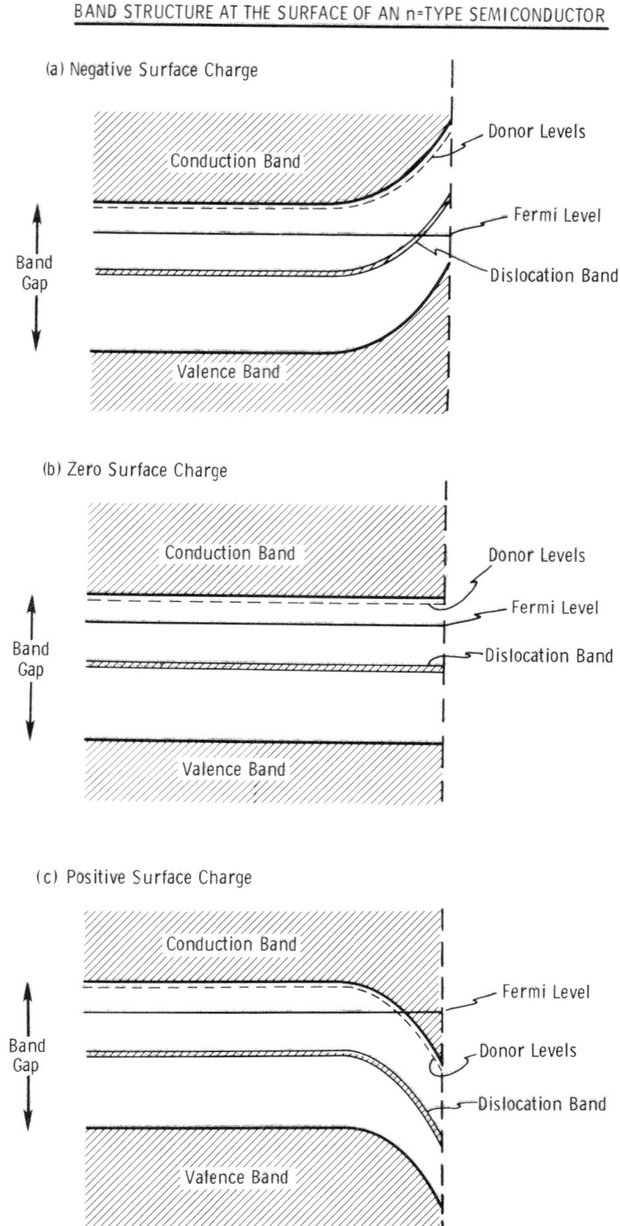

Fig. 2.—Schematic diagram of band bending with negative, zero, and positive surface charge.

To determine the applied voltage necessary to produce $\sigma = 0$, i.e., the flat band potential, FBP, the capacitance-voltage (C-V) technique used originally on ZnO by Dewald[19] was employed. The FBP was determined by measuring C, plotting $1/C^2$ vs V, and extrapolating the straight line to $1/C^2 = 0$. (For further details, see references 19 and 20.)

The detailed experimental arrangement has been reported previously[15,16] and so will be discussed only briefly here. The single crystals of ZnO were either flat disks with (0001) and $(000\bar{1})$ surfaces or hexagonal rods with the rod axis in the [0001] direction and $\{10\bar{1}0\}$ side surfaces. To prepare these specimens for hardness measurements, an ohmic contact was created by evaporating indium onto one surface of the sample and mounting that surface on a brass plate with conducting silver paint. Indentations of 10-s duration were then made on the opposite parallel face of the crystal using a Vickers diamond indenter with a 10-g load. The capacitance of the crystal-electrolyte interface was measured on samples similarly mounted. The position of dislocations around the hardness indentations was revealed by etching in a saturated chromic acid solution for 30s at 50°C followed by immersion in phosphoric acid for 60 s.

In the experiments to be discussed, we have concentrated on measurements at and near $\sigma = 0$ (the FBP) since we particularly wanted to clarify the mechanistic factors responsible for the ζ-correlation.

OBSERVATIONS AND DISCUSSION

Hardness of (0001) and $(000\bar{1})$ Surfaces

The microhardness of both the (0001) and $(000\bar{1})$ surfaces was measured as a function of voltage and pH. Although variations in H with voltage were observed on both surfaces, consistent data could be obtained only on (0001) probably because of the increased chemical activity of the $(000\bar{1})$ surface compared to (0001) as noted in the earlier etching studies of Heiland et al.[21]. Data from (0001) surfaces are presented in Figure 3 for $pH = 8.6$ and 12.2. Another set of hardness measurements taken at $pH = 8.5$ and 11.5 over the limited voltage range (producing the hardness maximum at A in Figure 3(a)) is given (Figure 4). Each time a maximum in hardness is observed at ~-100 mV. Note that the maximum for $pH = 11.5$ is more pronounced than that for $pH = 12.2$. This may be related to dissolution and/or adsorption kinetic effects, as will be discussed below.

A more sensitive measure of environmental effects on mechanical behavior than hardness is the extent of dislocation motion, L*, around a hardness indentation (Figure 5(a)). And, data obtained from a plot of L vs V (Figure 5(b) indicate an overall change in L of 50% with changes in applied potential. It is significant that the L-V curve exhibits a minimum in L at about -100 mV, approximately the same potential that produces a hardness maximum in the H-V curves since this suggests that the maximum in H and the minimum in L are caused by the same effect. The minimum in L does appear broader than the hardness maximum, H_{max}, found on the H-V curves. This may be caused by the higher sensitivity of L to changes in dislocation mobility, as compared with H.

*L is determined by measuring the extent of glide of the leading dislocations on the glide plane away from the hardness indents.

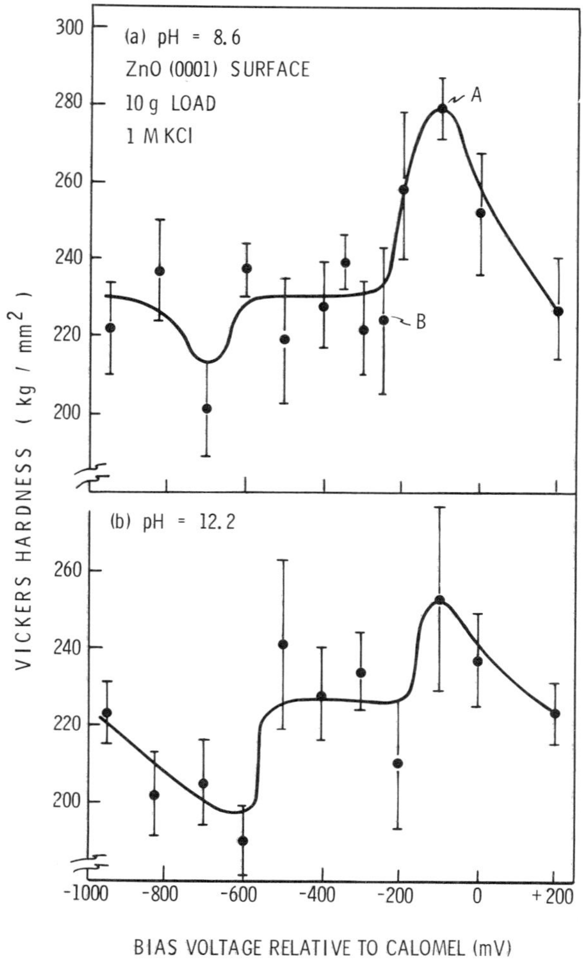

Fig. 3.—Hardness of the (0001) ZnO surface vs bias voltage in an electrolytic cell. Indentation time = 10 s; load, 10 g. (From Ref. 15).

Hardness of $(10\bar{1}0)$ Surfaces

The variation of hardness of the $(10\bar{1}0)$ surface with bias voltage in a 1M KCl electrolyte buffered to *pH = 8.5* is shown in Figure 6(a). The peak in hardness at ~-300 mV is illustrated in more detail in Figure 6(b), which presents data obtained using smaller voltage intervals than those used for Figure 6(a) (i.e., 50 mV instead of 100 mV).

Measurements were also made using solutions at *pH = 6.5, 9.5, and 11.5* (Figure 7). These data may also imply the existence of hardness maxima as indicated by the arrows in Figure 7, but reproducible measurements were difficult to obtain, probably because of either (1) the increased solubility

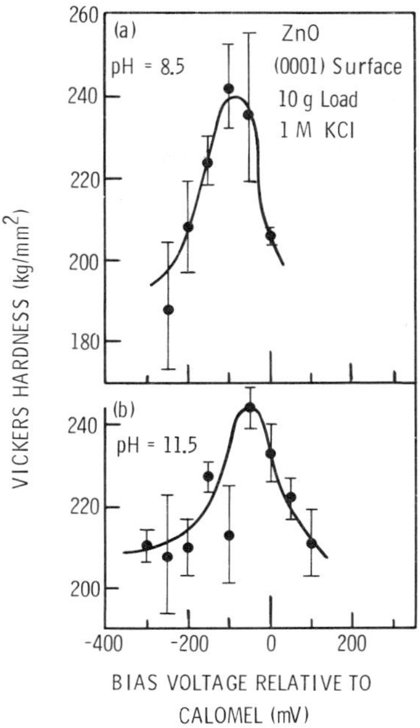

Fig. 4.—Hardness of the (0001) ZnO surface vs bias voltage in an electrolytic cell. Indentation time = 10 s; load 10 g.

of $(10\bar{1}0)$ surfaces in solutions of *pH* < *8* and > *10*[22] or, (2) adsorption kinetic effects (discussed below). As an example of the first, evident dissolution of the strained region around the indentation was observed after a ZnO crystal had been immersed in solutions of *pH = 6.5* for 30 min, the time typically involved in performing a series of measurements. Such dissolution would alter the size of the indentation and, therefore, the measured hardness value. Because of the higher stability of the $(10\bar{1}0)$ surface, however, no significant dissolution effects were observed on the results in Figure 6.

To verify the presence of hardness maxima in solutions with pH different than 8.5, data similar to that shown in Figure 6(b) were obtained at *pH = 9.5* and *11.5* (Figure 8). These data do appear to indicate maxima in hardness over particular voltage ranges. In fact, it is possible to fit the data at *pH = 8.5, 9.5,* and *11.5* (Figure 8) on one master curve by suitable shifting of the voltage and hardness axes (Figure 9).

A shift in the FBP and, presumably, the applied potential producing a hardness maximum, V_{max}, of -59 mV for each pH unit is expected from electrochemistry theory[7]. Although the voltage shifts necessary to produce the master curve of Figure 9 are in the direction predicted by the theory, they are also about twice as large as expected (~-100 mV/pH unit). This apparent discrepancy is not understood.

The shift of the hardness scale is also of importance. To produce Figure 9, it was necessary to shift the curves obtained at *pH = 8.5* and *11.5* down relative to that obtained at *pH = 9.5*. Thus, there appears to be a maximum in overall hardness at *pH = 9.5*.

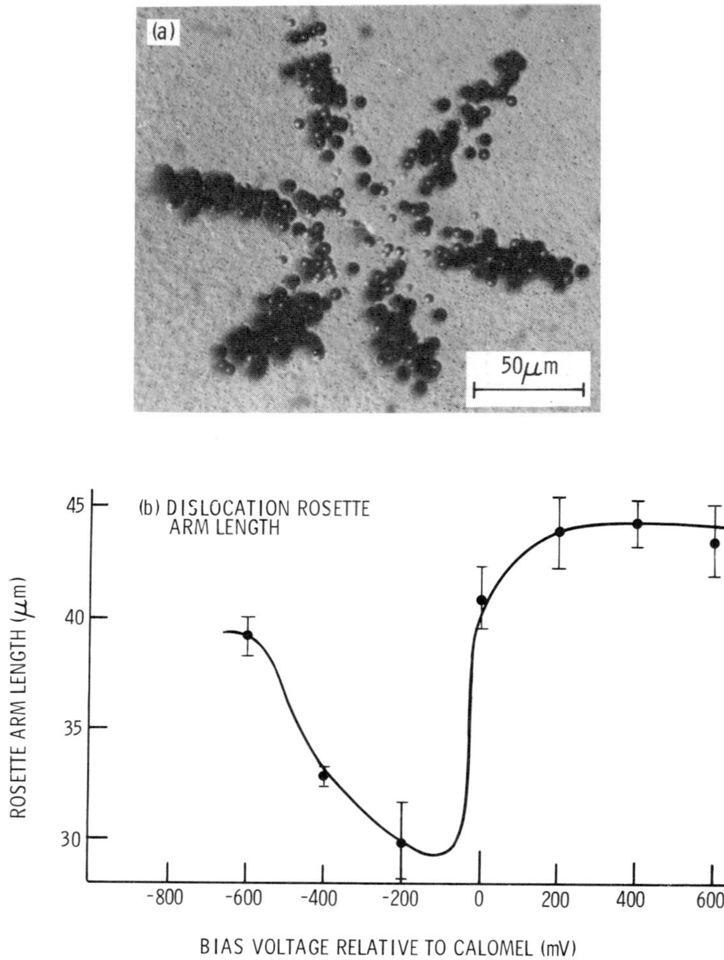

Fig. 5.—(a) Position of dislocations around a hardness indentation in ZnO revealed by etching in a saturated chromic acid solution for 30 s at 50°C followed by immersion in phosphoric acid for 60 s. (b) Rosette arm length, L, on the surface of (0001) ZnO vs bias voltage in an electrolytic cell. Vickers indentation produced with a load of 10 g applied for 10 s.

This result may be rationalized from considerations of the dependence of ZnO solubility on pH. Dissolution may be enhanced near a hardness impression, due to local mechanical strain, enlarging the impression and lowering the apparent hardness. The pH controls the dissolution rate, and minimum solubility is expected at $pH = 9.5$, the zero point of charge of ZnO[22]. Thus, dissolution effects should be least in solutions of this pH, and greater in solutions with $pH = 8.5$ and 11.5. In short, the "averaged hardness" will appear to be less for ZnO surfaces exposed to environments of $pH = 8.5$ and 11.5 than 9.5, as observed.

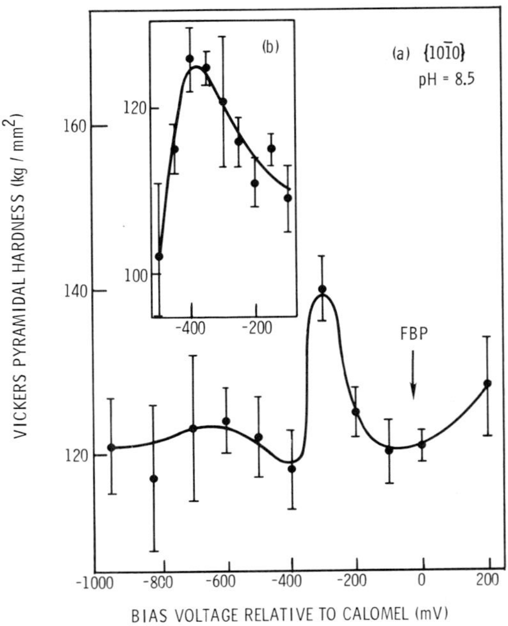

Fig. 6.—Hardness of a (10$\bar{1}$0) ZnO surface vs bias voltage in 1M KCl. Indentation time = 10 s, load 10 g. (From Ref. 16).

It should be emphasized that the dissolution effect cannot be responsible for the observed maxima in hardness on the H-V curves since this would require an unrealistically abrupt discontinuity in dissolution at a particular applied voltage. Dissolution may be important under cathodic potentials, but should result only in a lower overall hardness level, and/or a gradual increase in the measured hardness as V is changed from cathodic to anodic potentials.

In characterizing the solid/electrolyte interface with current capacitance versus voltage measurements, we found that the equilibrium between the solid and liquid was not achieved instantaneously. Accordingly, adsorption kinetics also may be a factor in hardness measurements. Thus, in addition to the dissolution problem mentioned above, poor reproducibility of hardness data in electrolytes at pH's other than 8.5 might be caused by a time-dependent surface charge at fixed applied voltage. This could result in time-dependent hardness values.

To examine this possibility, we assumed that there is some surface charge density, σ_{max}, established by applying a particular potential, V_{max}, which does produce a hardness maximum. We also assumed that to maintain σ_{max}, it may be necessary to alter the applied potential (i.e., V_{max} changes with time). For example, for (10$\bar{1}$0) ZnO in electrolytes at *pH = 9.5*, we assumed that V_{max} ~-500 mV immediately after the cell is switched on, but that it shifts slowly toward -600 mV over a 5-10 min period. Accordingly, we measured H vs time for three time and voltage histories (Figure 10). For Figure 10(a), the cell was switched on at -500 mV, and hardness indentations were immediately produced. After 5- and 10-min intervals, further hardness

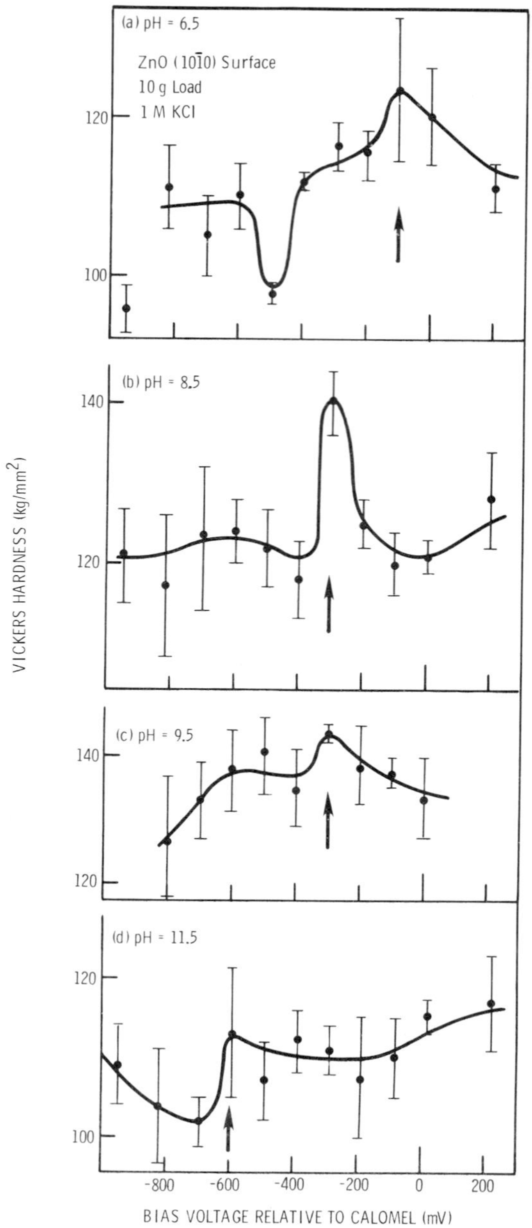

Fig. 7.—Hardness of the (10$\bar{1}$0) surfaces vs bias voltage in 1 M KCl. Indentation time = 10 s, load 10 g. a) pH = 6.5, b) pH = 8.5 (Ref. 16), c) pH = 9.5, d) pH = 11.5.

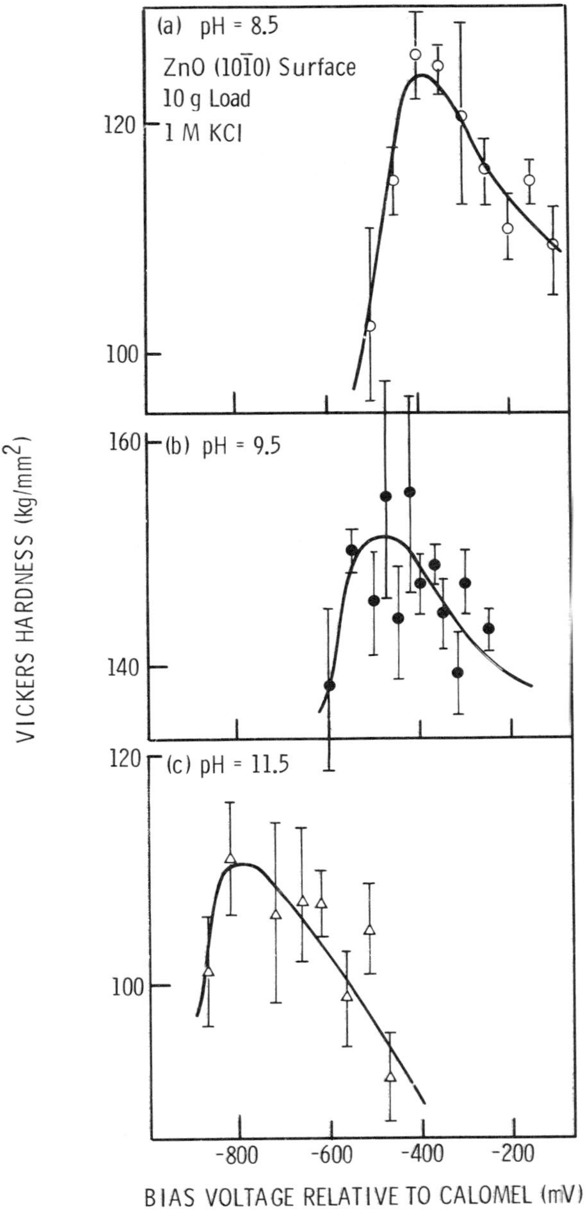

Fig. 8.—Hardness of the (10$\bar{1}$0) ZnO surface vs bias voltage in 1 M KCl. Indentation time = 10 s, load 10 g. a) pH = 8.5, (Ref. 16), b) pH = 9.5, c) pH = 11.5.

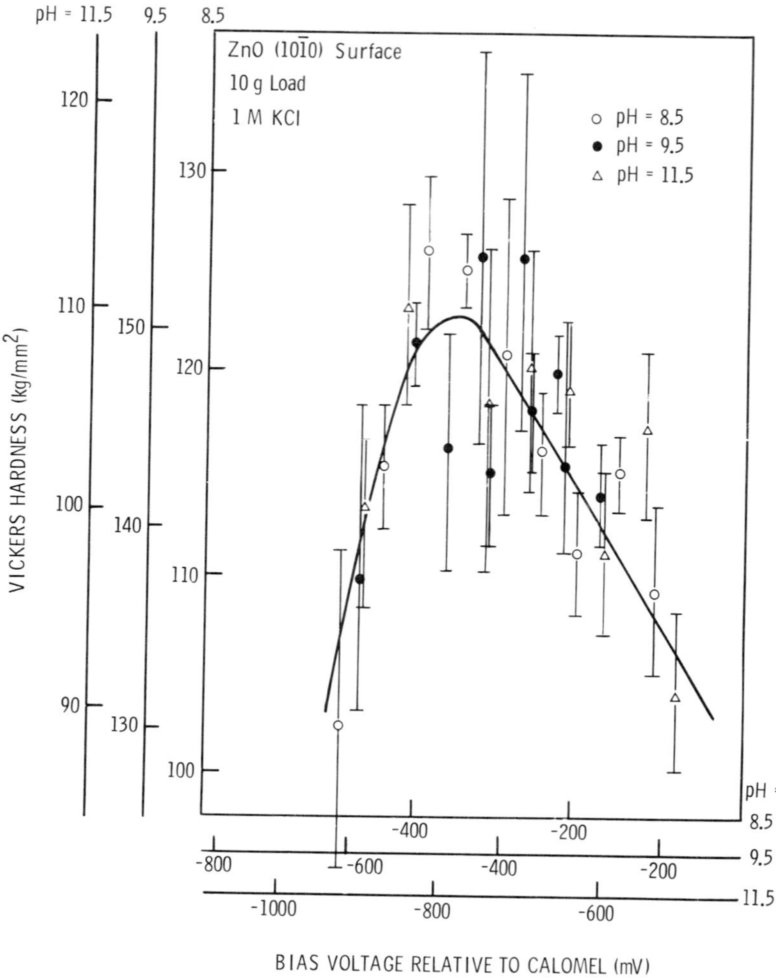

Fig. 9.—Master curve of hardness data from Figure 10.

indentations were made, the cell was switched to -400 mV, and further indentations were produced at 5-min intervals. These data indicated a progressive decrease in hardness when the cell was switched on at -500 mV, but an almost constant hardness after it was switched to -400 mV. These results agree with the model, since we would expect hardness obtained with the cell set at -500 mV applied potential to decrease with time if V_{max} shifts initially from -500 mV toward -600 mV. The hardness at -400 mV is constant since we would not expect dislocation mobility to be sensitive to changes in surface charge in this potential region.

In the second experiment (Figure 10(b)), the voltage was set to -600 mV, and hardness indentations produced immediately and at 5-min intervals. The voltage was then switched to -500 mV, and a final set of indentations were

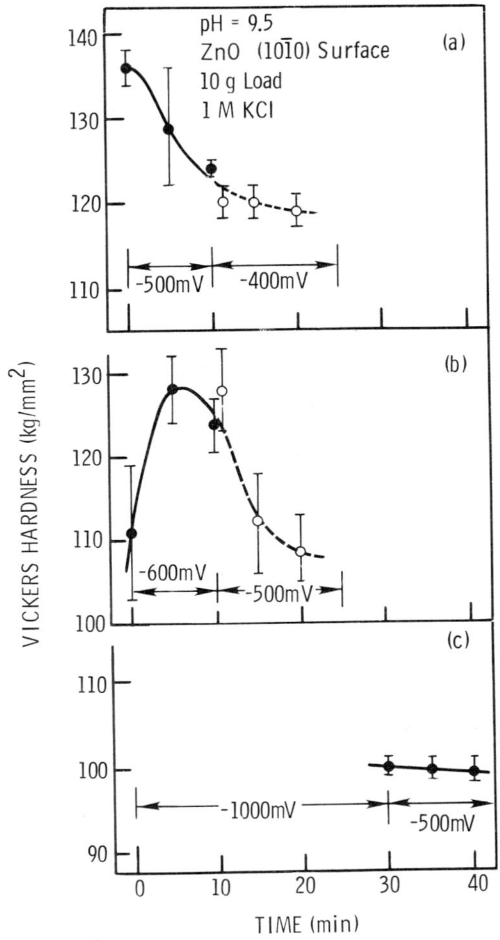

Fig. 10.—Hardness vs time of (10$\bar{1}$0) ZnO surface in 1 M KCl under three voltage histories. Indentation time = 10 s, load 10 g.

produced. These data indicated an initial increase in H with time, followed by a small decrease at -600 mV, and a further decrease in H with time after the voltage was switched to -500 mV. Again, these results agree with the model since we would expect hardness obtained with applied potential set at -600 mV to initially increase with time as V_{max} shifts from -500 mV toward -600 mV. Hardness decreases with time after the applied voltage is switched to -500 mV because the maximum had shifted past -500 mV and apparently continued to shift in the cathodic direction.

In the final experiment (Figure 10(c)), the cell was held at -1000 mV for 30 min and then switched to -500 mV. Hardness indentations were produced immediately and after 5 and 10 min. The hardness was nearly constant.

This result confirms that, after a suitable time interval, V_{max} had shifted well away from -500 mV.

In summary, the results of these experiments on $\{10\bar{1}0\}$ ZnO conform with the model that a hardness maximum exists in the voltage range $-600 < V_{max} < -500$, and that V_{max} shifts slowly with time to more cathodic potentials when the applied potential is held constant.

Current and Capacitance Measurements

The carrier density of the ZnO crystal, calculated from the slope of the $1/C^2$ -V graphs was, as expected, independent of the voltage sweep-rate and pH over a limited pH range. But the flat band potential was found to depend on a number of parameters, including sweep rate, direction of sweep (from positive to negative or negative to positive bias), and pH. Also, for pH's other than those giving minimum solubility (pH 8-10), C-V curves were difficult to obtain under certain circumstances. For example, linear $1/C^2$ vs V were not always obtained at $pH = 12$ for $(10\bar{1}0)$ surfaces. Moreover, it was difficult to record a C-V curve at pH 6.5 for $(10\bar{1}0)$ surfaces because of non-equilibrium changes in the capacitance and current.

Another important observation is that the FBP on the (0001) and $(000\bar{1})$ surfaces did not shift with pH under fast sweep rates, but did when sweeping more slowly. For instance, for a (0001) surface at pH 8.5 and at a sweep rate of 1 mV/s, the FBP occurred at ~0 mV. At 0.2 mV/s, however, the FBP occurred at ~-110 mV. Similarly, for the same pair of sweeping conditions and pH 11.5 the FBP occurred at 0 and ~-170 mV, respectively.

In the case of the $(10\bar{1}0)$ surfaces, a shift in the FBP of -50 mV was found when the sweep rate was changed from 1 to 0.5 mV/s. These variations were examined by measuring the cell capacitance and current as a function of potential and time. In the potential range from -1000 mV to -200 mV, both C and current, I, varied slowly over a period of minutes, the direction (increasing and decreasing) and the rate of change depending on the history of the sample in the solution. These shifts in FBP, and the changes in C and I with time, suggest that equilibrium between the liquid and the space charge region of the solid is not reached instantaneously, and that kinetic effects are important when determining the FBP, and hence, presumably, the hardness.

The unexpected variation in FBP with sweep rate and sweep direction presented a problem when trying to relate the C-V measurements to the hardness results. This difficulty was overcome via the following procedure. As noted above, the H-V curves were obtained by altering the voltage from (1) -1000 mV to +200 mV in 100-mV increments at an average rate of 1 mV/s, or (2) over a limited voltage range in 50-mV increments at an average rate of 0.5 mV/s. Therefore, to match the conditions for the capacitance measurements to those of the hardness, the voltage was swept from -1000 mV to +1000 mV at a rate of 1 or 0.5 mV/s, and the part of the C-V curve from 0 to +1000 mV was used for the $1/C^2$ -V plots (Figure 11). Values of the FBP obtained in this fashion, along with the carrier density and Fermi level (derived from the carrier density), are presented in Table I for (0001) and $(10\bar{1}0)$ surfaces.

IMPLICATIONS OF THE RESULTS TO THE MECHANISM OF THE ζ-CORRELATION

The results described demonstrate that the microhardness of a ZnO crystal can be significantly changed by altering its surface charge via an applied potential. Therefore, the existence of a relationship between

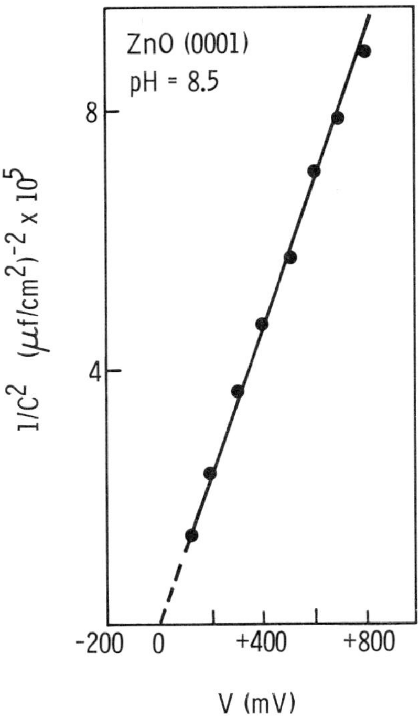

Fig. 11.—Typical C^{-2} vs V plot for the ZnO-aqueous electrolyte system employed here.

surface charge and hardness, presumed in earlier work on chemomechanical effects has now been verified for the ZnO-aqueous electrolyte system. In particular, the data of Table I reveal that a maximum in hardness, H_{max}, occurs at potentials producing a slightly positive surface charge on both (0001) and ($10\bar{1}0$) surfaces. Thus, although the results confirm the importance of surface charge in chemomechanical effects, they do not exactly confirm the ζ-correlation as earlier conceived (namely, that hardness maxima occur when $\zeta \equiv \sigma = 0$), at least for ZnO.

The positions of the maxima have implications for the mechanism of chemomechanical effects in ZnO. According to previous understanding[5], active environments influence dislocation mobility, and hence hardness, by changing -- through surface charge -- the nature of the interaction of dislocations with (1) the lattice (i.e., changing the Peierls stress), (2) point defects, or (3) themselves.

As a basis for discussing these possibilities, consider the energy level diagram in Figure 12. This figure was constructed using the value of the FBP and the Fermi level, derived in the present work, the Zn^+ interstitial donor levels[23], the conduction and valence band energies, and an assumed location of the dislocation energy level[15]. From this diagram, it can be seen that the valence band and the dislocation level cannot change occupancies under the conditions of our experiments (applied voltages near the FBP). Thus, mechanisms involving changes in interionic forces, i.e.,

TABLE I.—POSITION ON THE VOLTAGE AXIS OF THE MAXIMUM IN HARDNESS, H_{max}, AND THE FBP OF ZnO AS A FUNCTION OF SURFACE ORIENTATION AND pH OF 1M KCl ELECTROLYTES. ALSO LISTED ARE THE CARRIER DENSITIES AND THE POSITION OF THE FERMI LEVEL BELOW THE CONDUCTION BAND FOR THE ZnO SAMPLES, AND THE PARAMETER ΔE.

Surface	Electrolyte pH	Position of H_{max} (mV)	Predicted Range of V_{max} from Charge Exchange Model (mV)	FBP (mV)	Carrier Density (cm^{-3})	Fermi Level (meV)	ΔE[a] (meV)
$(10\bar{1}0)$[b]	8.5	-350	-150-350	-50	5×10^{13}	275	+200
(0001)[c]	8.6	-100	-75-125	0	1×10^{14}	250	-50
(0001)[c]	12.2	-100		0	1×10^{14}	250	-50

[a] ΔE is the assumed distortion of the energy bands due to the presence of dislocations (see Figure 13).

[b] Voltage sweep rate 0.5 mV/s from Ref. 16.

[c] Voltage sweep rate 1.0 mV/s from Ref. 15.

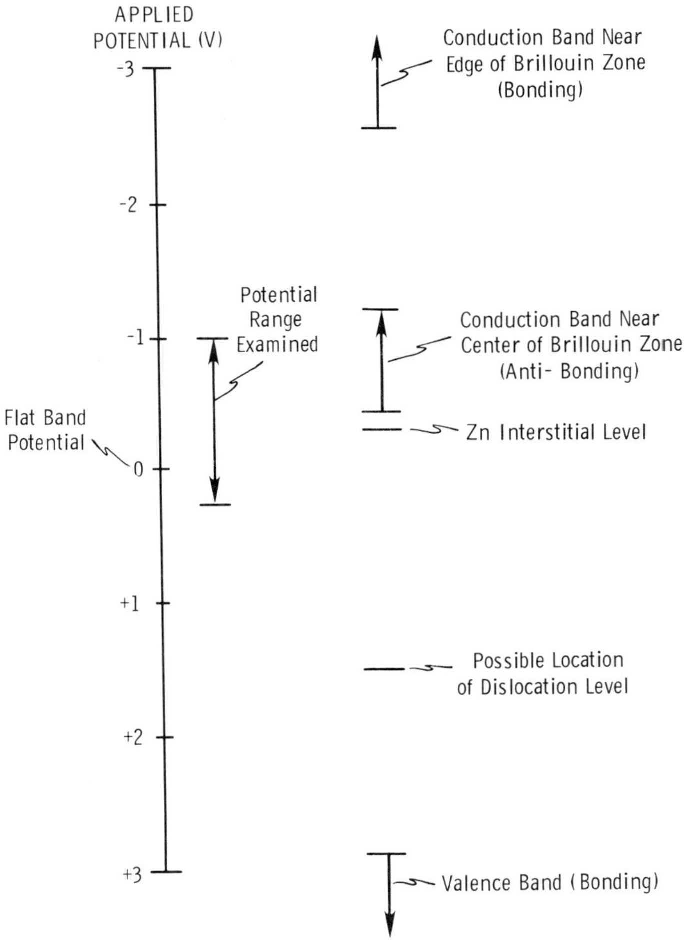

Fig. 12.—Schematic locating electronic levels of interest relative to the applied potential (in volts) expected to produce alteration in electron occupancy of these levels.

changes in the Peierls stress, cannot cause the observed hardness maxima. However, Figure 12 does reveal that both the donor levels and the conduction band near the center of the Brillouin zone are expected to change occupancy under specific applied voltages near the FBP. Accordingly, the following "charge exchange" hypothesis, involving donor levels and the conduction band, has been developed[15,16] to explain the results.

First, assume that both conduction and valence bands are distorted near a dislocation (Figure 13(a)). This distortion might result from: (1) a positively or negatively charged dislocation with a surrounding screening cloud of conduction electrons or interstitial zinc ions; (2) coupling of the

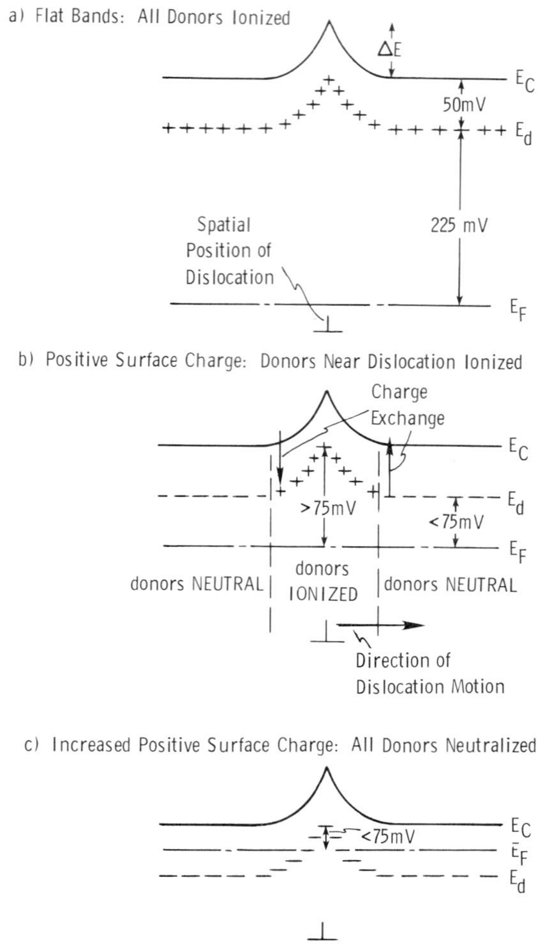

Fig. 13.—Schematic of the distortion in electronic levels in the neighborhood of a basal dislocation near the edge of the conduction band.

strain field around the dislocation and the energy bands through the deformation potential, or (3) the piezoelectric effect. To interpret the results of Table I on the "charge exchange" hypothesis when dislocations induce, for example, an upward band distortion, it should first be appreciated that the motion of a dislocation also requires movement of the band distortion. Under flat band conditions, only rearrangement of the conduction electrons is involved when the dislocation moves; thus, motion is relatively easy, and the crystal is (relatively) soft. As the bands bend downward, however, the Zn^+ donors away from the dislocation become neutralized. Therefore, dislocation motion now necessitates excitation of electrons from the neutral donor levels to unoccupied levels in the conduction band (Figure 13(b)). Such excitation requires about 50 meV/donor (the energy between the conduction band edge and the Zn^+ donor level). The consequence should be a decrease in dislocation mobility, and concomitant increase in hardness because this amount of additional energy must be supplied before dislocation motion can occur. With further downward band bending, however, essentially all of the donors in the space charge layer become neutralized, so that the dislocation may again move without exciting donors. Hence, dislocation mobility will be less for the situation illustrated in Figure 13(b) than that shown in Figure 13(a) or 13(c). Similar arguments can also be used when the band distortion near the dislocation is downward.

The degree of band bending necessary to produce a maximum depends on the local distortion of the conduction bands, ΔE (Figure 13(a)). For a positive ΔE (upward band distortion), a hardness maximum is produced by bending the bands downward so that the donors far from the dislocation are neutralized, but the donors near the dislocation are still ionized -- that is, when the band bending, Ψ, falls in the range (as measured from the edge of the conduction band):

$$E_F - (E_d + \Delta E + 3\ kT) > \Psi > E_F - (E_d + 3\ kT).$$

When ΔE is negative (downward band distortion near the dislocation), a hardness maximum is produced by bending the bands downward at the surface so that the donors near the dislocation are neutralized, but the donors far from the dislocation are still ionized -- that is, when the band bending, Ψ, falls in the range (as measured from the edge of the conduction band):

$$E_F - (E_d + 3\ kT) > \Psi > E_F - (E_d + \Delta E + 3\ kT).$$

To predict the position of H_{max} on the potential axis, it is necessary to assume a value for ΔE. Agreement between the predicted potential range and the range on the potential axis in which H_{max} is observed to occur can be achieved by selecting $\Delta E = 200$ meV for the dislocations controlling the hardness of the $(10\bar{1}0)$ surface or -50 meV for those controlling the (0001) surfaces (see Table I for a summary of this comparison).

These values of ΔE for the different types of dislocations can be rationalized from the expected properties of these types of dislocations. For example, prismatic dislocations (the type of dislocation thought to control the hardness of the (0001) surface) are probably not charged, because of the non-polar nature of the dislocation core.* Hence, there is probably no band distortion caused by charge on prismatic dislocations. Indeed, Carlson[17] explains the absence of a photoplastic effect on the prismatic slip system in ZnO as due to the uncharged nature of prismatic dislocations. Band distortions near prismatic dislocation may, however, be caused by the coupling of the strain field and the bands through the deformation potential. Calculations for Ge[25,26] predict a downward band distortion of 50-100 meV near screw or 60° dislocations. Although this effect has not been studied for ZnO, if a similar distortion is present around prismatic dislocations,

*Although kinks and jogs along the dislocation line may carry positive or negative charge, Hirth and Lothe[24] have pointed out that in divalent solids such as ZnO, such charges may easily be neutralized by free electrons or holes.

then the assumed ΔE for the (0001) surface and that caused by coupling of the dislocation strain field and the electronic structure through the deformation potential are comparable.

For basal dislocations (the type of dislocation thought to control the hardness of $\{10\bar{1}0\}$ surfaces), the charge on the dislocation due to the polar dislocation core should be the major contributor to band distortion. One problem in interpreting the results on the $(10\bar{1}0)$ surface is that indentation on this surface is expected to generate both α- and β-type basal dislocations, which are expected to be oppositely charged. However, the hardness results we obtained on $(10\bar{1}0)$ surfaces cannot be explained by a charge exchange mechanism involving downward band bending near basal dislocations -- as expected about α-dislocations -- but only by upward bending -- expected about β-dislocations. As a result, this conclusion tends to imply that β-dislocations, because they are so much less mobile than α-dislocations[27], control the hardness of the $(10\bar{1}0)$ surfaces -- as might be expected.

ACKNOWLEDGEMENTS

It is a pleasure to acknowledge helpful discussions with Professor P. Mark of Princeton University, Professors P. Haasen and W. Schröter, of the University of Göttingen, and colleagues Dr. J. D. Venables, Dr. N. E. Byer, Dr. Jar-mo Chen, Dr. L. G. Boxall and Dr. J. A. S. Green. This work was supported in part by the U.S. National Science Foundation under Grant DMR 75-05443.

REFERENCES

1. Latanision, R.M., These Proceedings.
2. Heins, R.W. and Street, N., *Society of Petroleum Engineers Journal,* Vol. 5, 1965, p. 177.
3. Westwood, A.R.C., Macmillan, N.H. and Kalyoncu, R.S., *American Ceramic Society. Journal,* Vol. 56, 1973, p. 258.
4. Macmillan, N.H., Huntington, R.D. and Westwood, A.R.C., *Philosophical Magazine,* Vol. 28, 1973, p. 923.
5. Westwood, A.R.C. and Macmillan, N.H., in "Science of Hardness Testing and Its Research Applications," edited by J.H. Westbrook and H. Conrad, American Society for Metals, Metals Park, Ohio, 1973, p. 377.
6. Ryncarz, A. and Laskowski, J., *Powder Technology,* Vol. 18, 1977, p. 179.
7. Westwood, A.R.C. and Huntington, R.D., in Proceedings of the 1971 International Conference on Mechanical Behavior of Materials, Society of Materials Science, Kyoto, Japan, Vol. IV, 1972, p. 383.
8. Fox, P.G. and Smith, A.J., presented at Symposium on Strength of Glass, Society of Glass Technology, University of Sussex, U.K., March 1974.
9. Malin, M. and Vedam, K., *Journal of Applied Physics,* Vol. 48, 1977, p. 1155.
10. Adams, R., Private Communication, University of Warwick, U.K., April 1978.
11. Cuthrell, R.E., *Journal of Applied Physics,* Vol. 49, 1978, p. 432.
12. Yost, G.B. and Williams, W.S., *American Ceramic Society. Journal,* Vol. 61, 1978, p. 139.
13. Ahmed, S.M., in "Oxides and Oxide Films," edited by J.W. Diggle, Vol. 1, Marcel Dekker, New York and Basel, 1972, p. 486.
14. Parks, G.A. and de Bruyn, P.L., *Journal of Physical Chemistry,* Vol. 66, 1962, p. 967.
15. Ahearn, J.S., Mills, J.J. and Westwood, A.R.C., *Journal of Applied Physics,* Vol. 49, 1978, p. 96.

16. Ahearn, J.S., Mills, J.J. and Westwood, A.R.C., *Journal of Applied Physics,* Vol. 49, 1978, p. 614.
17. Carlsson, L., *Journal of Applied Physics,* Vol. 42, 1971, p. 676. See also Carlsson, L. and Svensson, C., *Journal of Applied Physics,* Vol. 41, 1970, p. 1642 and *Solid State Communications,* Vol. 7, 1969, p. 177.
18. Alexander, H. and Haasen, P., in "Solid State Physics," Vol. 22, edited by F. Seitz, D. Turnbull, and H. Ehrenreich, Academic Press, New York, 1968, p. 28f.
19. Dewald, J.F., *Bell System Technical Journal,* Vol. 39, 1960, p. 615.
20. Pettinger, B., Schoppel, H.R., Yokayama, T. and Gerischer, H., *Berichte der Bunsengesellschaft fur Physikalische Chemie,* Vol. 78, 1974, p. 1024.
21. Heiland, G., Kunstman, P. and Pfister, H., *Zeitschrift fur Physik,* Vol. 176, 1963, p. 485.
22. Blok, L. and de Bruyn, P.L., *Journal of Colloid and Interface Science,* Vol. 32, 1970, p. 518.
23. Kroger, F.A., "The Chemistry of Imperfect Crystals," North-Holland, Amsterdam, 1974, p. 691.
24. Hirth, J.P. and Lothe, J., "Theory of Dislocations," McGraw-Hill, New York, 1968, p. 376.
25. Celli, V., Gold, A. and Thomson, R., *Physical Review Letters,* Vol. 8, 1962, p. 96.
26. Winter, S., *Physica Status Solidi (b),* Vol. 79, 1977, p. 637.
27. Michara, M. and Ninomiya, T., *Physica Status Solidi,* Vol. 32, 1975, p. 43.

DISCUSSION

H. E. HINTERMAN, LSRH, Switzerland: Do adsorbed species influence the measurement of capacitance?

J. S. AHEARN: They do if they contribute to the surface charge. The measurement of the capacitance in strong electrolytes is only over the space charge in the solid. The layer in the electrolyte is so thin that it contributes essentially nothing. We assume that we have H^+ or OH^- as the controlling charge species on the surface. Now if you have specifically adsorbed surface charge impurities they will just affect the surface charge there but they do not affect the capacitance measurement. The capacitance measurement is blind whether the charge is H^+, OH^-, or impurities. It is only measuring the semiconductor properties and not the electrolyte properties. We are fairly confident about this technique. Originally it was used on zinc oxide to study electronic exchange processes between redox couples and the solid.

J. SHEASBY, University of Western Ontario: Could you tell me what would be the magnitude of current densities required to change the potentials?

AHEARN: In the upper band bending conditions the currents are minimal. You get a barrier. In other words you bend the bands up and what you have is a negative surface charge and a depletion layer. You repel the electrons and you get upward band bending. The Zn interstitial donor levels supply the space charges essentially and there is very little current going through the cell. In the lower band bending configurations you can get charge exchange across the interface. The important thing is that you have to have something in the electrolyte to pick up the electrons. The current does increase but generally not until you get down to about -1.2 V when you start to get gas evolution. The benefit of zinc oxide is that it is already oxidized and you will have nothing to oxidize at the surface.

R. M. LATANISION, MIT: At the meeting in Germany it was suggested that there may be another explanation for displacement of the hardness maximum relative to the zero point charge. That had to do with the notion that dislocations in ionic crystals are charged.

AHEARN: That is particularly relevant in univalent solids like LiF and NaCl. In divalent solids like zinc oxide the kinks or jogs should carry a plus or minus charge and they are easily compensated. We do not have any information on whether the kinks and jogs in ZnO are charged. From the fact that we do have electrons and holes available to compensate those charges on a dislocation I would assume that they are not charged by that mechanism.

W. NEEDELMAN, Pall Corporation: Has there been any work on the hardness of particles in non-aqueous solutions?

AHEARN: A lot of work was done in the alcohols which I would qualify as non-aqueous solutions. Usually hardness was measured on macroscopic bodies like single crystals, but not on particles. This effect will be much more significant for small particles than for bulk material.

APPLICATION OF CHEMOMECHANICAL EFFECTS TO HARD-ROCK DRILLING

J. J. Mills and A. R. C. Westwood

ABSTRACT

Adsorbed, surface-active species have long been known to influence the microhardness, frictional properties, and laboratory-scale drilling behavior of rocks and other nonmetallic solids. But the reliable application of such chemomechanical phenomena to the field-scale, diamond-bit drilling of hard rocks requires that due consideration also be given to other less-obvious factors, e.g., the mechanism likely to be involved (namely, chemomechanically induced reductions in bit wear), the polarity and concentration of the adsorbed species, adsorption dynamics, bit rotation speed, diamond failure mode, etc. Given such perspective, drilling fluids can be formulated which can increase bit life by a factor of four and substantially reduce drilling costs.

INTRODUCTION

The use of additives with cutting fluids to enhance bit penetration rates has been studied and attempted in practice for over 40 years[1,2]. Routine field application has been limited, however, because of the unpredictable or unreliable results obtained. This situation has arisen, we believe, because the number and complex interactions of the various factors involved have not been generally appreciated. Table 1, for example, presents some of the functions the circulating fluid ("coolant") must perform during the diamond-bit drilling of a hard rock. The Table also lists some of the *physical* properties of the fluids on which these useful functions depend. Clearly, if an additive to the fluid causes any change in these properties, then the efficiency of one or more of the functions of the liquid also will change, and drilling performance may be affected. Conceivably, opposite influences on different functions can also occur, e.g., on bit cooling and debris removal, leading to bit performance which is difficult to predict or explain.

The *chemical* properties of a coolant can also markedly affect drilling performance, and, while a high concentration of additive usually is required to change the physical properties of a drilling fluid (say 10-20% by volume), the relevant chemical properties can be markedly altered by the presence of surfactant concentrations as low as 10^{-5} mol/liter. For example, fatty acid additives can increase drilling efficiency by enhancing bit

TABLE I.—INFLUENCE OF PHYSICAL PROPERTIES OF CIRCULATING FLUID ON FACTORS INVOLVED IN DIAMOND-BIT DRILLING OF HARD ROCKS

Function†	Relevant Physical Property*		
	ρ	η	κ
1. Drill string lubrication	yes	yes	no
2. Bit lubrication	yes	yes	no
3. Bit cooling	yes	yes	yes
4. Debris removal up the annulus	yes	yes	no
5. Debris removal from under bit face	yes	yes	no
6. Corrosion inhibition	no	no	no
Influence On:			
1. Rock properties	none	none	none
2. Bit life	yes	yes	yes

* ρ = density of fluid

η = viscosity of fluid

κ = thermal conductivity of fluid

†In drilling for oil in soft rocks, other functions, such as hole wall stabilization and control of subsurface pressures, are also involved.

life[3]. Presumably, they adsorb to provide improved boundary lubrication, and hence improve functions 1 and 2 in Table 1. Chemically active additives can also alter the surface charge on the particles created by drilling, changing their agglomeration characteristics and the ease with which they are flushed out of the hole. Another possibility is that of altering the flow, fracture, and frictional characteristics of rocks by the use of chemomechanical effects[2,4-7].

In this contribution, we consider how chemically induced changes in rock mechanical properties can beneficially influence diamond-bit drilling in granitic rocks. We also discuss some of the non-obvious factors involved in the practical application of chemomechanical effects to rock drilling.

CHEMOMECHANICAL EFFECTS IN DIAMOND-BIT DRILLING

In earlier reports from our Laboratories[5-7], it was suggested that chemisorbed species can facilitate fracture during drilling by directly influencing the near-surface flow processes that control hardness and brittleness. While this possibility is still viable for softer, nonmetallic solids, recent studies with hard rocks indicate that often the more important environmental influence is a reduction in the wear rate of the diamonds in the bit[8]. To illustrate this point, consider the data in Figure 1(a). In this experiment, the penetration obtained with a freshly dressed bit, drilling into granite under water, is compared to that from the same bit after redressing but drilling under an active fluid, namely, 10^{-3} mol/liter aq. DTAB (dodecyl trimethyl ammonium bromide). Note that the total penetration obtained using the active fluid is greater than that for water. Next, using one bit throughout, the drilling environment was alternated (without redressing) between water and the aq. 10^{-3} mol/liter DTAB solution (approximately the concentration that produces a zero ζ-potential on the rock phases). The data, Figure 1(b), reveal no abrupt change in drilling rate with change of environment such as might be expected if the cause of the differences in Figure 1(a) were environmentally induced changes in rock fracture behavior. On the other hand, the results presented in both Figures 1(a) and (b) can be understood if the primary role of the active fluid is to reduce the wear rate of the diamonds in the bit. The argument is as follows: when drilling is conducted in solutions that produce $\zeta \simeq 0$, the hardness of the rock surface is a maximum, and so the coefficient of ploughing friction between the diamonds and the rock is a minimum. Consequently, the heat generated by their impingement also is a minimum, and the diamonds run cooler. Since temperature is a prime factor in determining wear of the diamonds, they now wear less and stay sharp longer. Therefore, both the rate of penetration (averaged over some common time period) and the life of the bit are enhanced[8].

FACTORS INVOLVED IN SCALING-UP FROM LABORATORY EXPERIMENTS

Polarity and Concentration of Adsorbed Species

The correlation between the ζ-potential of a rock or mineral, its hardness, and its ease of drilling with small diameter (<0.25 in. o.d.) diamond bits, shown in Figure 2, has been termed the "ζ-correlation"[2]. Since most hard rocks are negatively charged in water, this correlation implies that cationic surface-active agents should be most effective in chemomechanically increasing drilling performance.

On the other hand, many surfactants are long-chain organic molecules known to enhance boundary layer lubrication and so intrinsically likely to improve bit performance. And, it might be thought that lubrication would be the principal origin of their influence. However, it now appears that a maximum in bit performance will occur whenever the ζ-potential of the rock is zero, regardless of the type of chemical used to achieve that condition. For example, a maximum in bit performance similar to that shown in Figure 2(c) has been observed by Appl[9] when using $AlCl_3$ solutions to influence the single point diamong cutting of granite. Appl found that concentrations of Al^{3+} - an inorganic ion - close to those that produce $\zeta = 0$, maximized normal and tangential forces on the diamond and minimized diamond wear, incidentally reproducing the results obtained by Rebinder et al. in their earlier work[1]. Of course, $AlCl_3$ is not a practical additive, being corrosive, but the results obtained with this inorganic ion at low concentrations (10^{-5} mol/liter) emphasize the generic nature of the "ζ-correlation."

Such *chemical* effects should be distinguished from other effects caused by changing the *physical* properties of the fluid, for example, those reported by Selim et al.[10] from experiments using anionic surfactants, glycerine and ethylene glycol, and by Unger et al.[11] from studies using anionic, non-ionic, and cationic surfactants. Since the latter effects

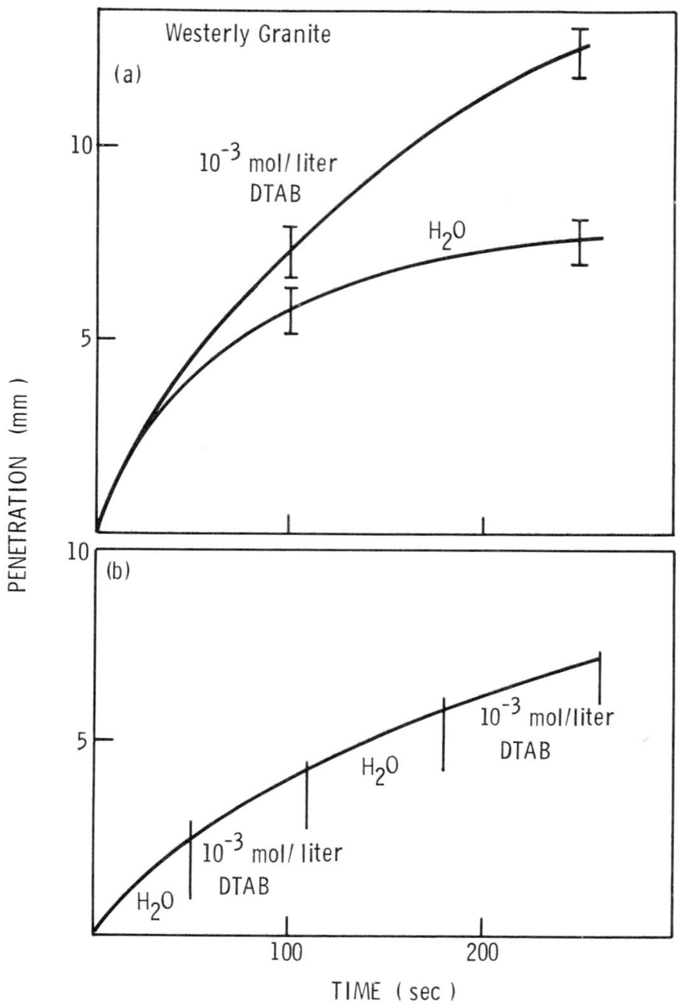

Figure 1.—(a) Penetration vs time curves for freshly redressed bits drilling into Westerly granite under water or 10^{-3} mol/liter DTAB. (b) Penetration vs time for a 0.25 in. o.d. diamond-impregnated bit rotating at 200 rpm under a thrust of 9 lb into Westerly granite. Note that on changing from water to a 10^{-3} mol/liter aq. DTAB environment, no significant change in slope occurs.

occur at much higher additive concentrations, they are less likely to be cost-effective. To exhibit the potential for producing chemomechanical effects during drilling, therefore, an additive should be of the correct polarity and should be added in concentrations sufficient to reduce the ζ-potential to zero under the *dynamic* conditions present during drilling (see below).

Type of Test

The test strategy adopted to reveal any influence of chemomechanical effects in rock drilling must take into account the fact that such effects

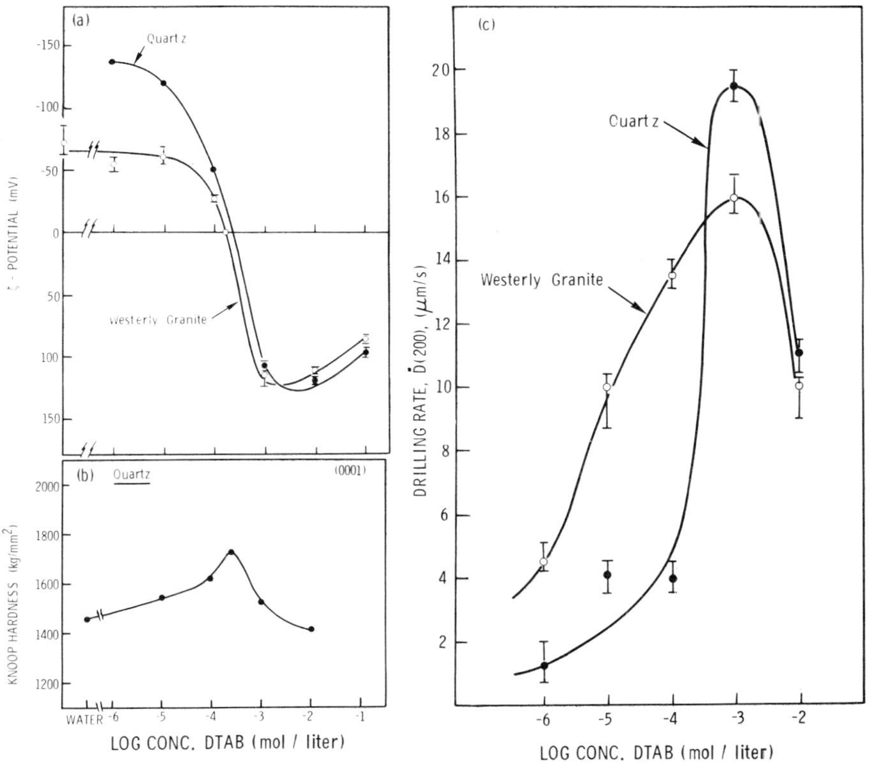

Fig. 2.—Plots of (a) the ζ-potential of Westerly granite and its major constituent, quartz; (b) Knoop hardness of the (0001) surface of quartz; and (c) drilling rate after 200 sec, D(200), in Westerly granite and quartz as a function of DTAB concentration. Note that both hardness and drilling rate are greatest when ζ ≃ 0, the slight shift to higher concentrations for drilling rate probably being associated with the dynamics of adsorption in a non-equilibrium system.

probably occur primarily via reductions in the rate of bit wear. Thus, as was often the case in earlier work, experiments which measure drilling rates over only a small fraction of the total bit life (as an extreme example, over the first 10 sec in Figure 1(a)), will *not* reliably reveal differences in the effectiveness of various environments. Tests must be conducted over a significant fraction of the life of the diamonds. Of course, such extended experiments could be very expensive if full-size bits were used in field-scale tests. Fortunately, plots of bit penetration rate vs. penetration distance can provide a useful early indication of the influence of drilling environment[12,13]. For example, consider the data presented in Figure 3. Note that, after a settling-in distance of about 0.3 in., the rate of decay of penetration rate with distance is markedly less in the Aerosol C-61 solution than in water; but the initial drilling rates are not significantly different.

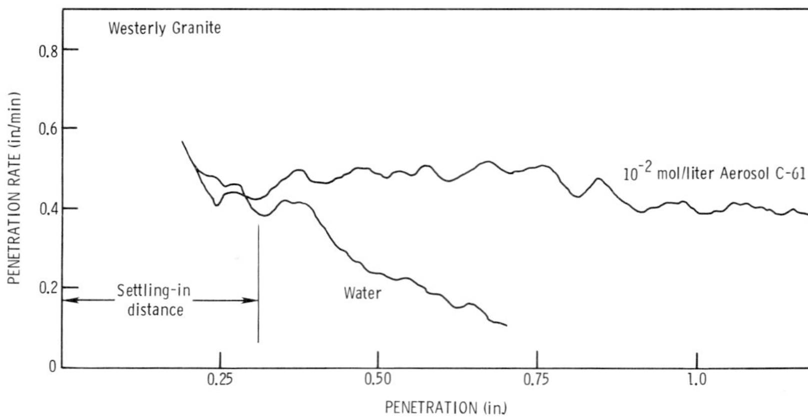

Fig. 3.—Penetration rate of a 0.25 in. o.d. diamond-impregnated bit into Westerly granite as a function of penetration, illustrating the reduced wear rate induced by a 10^{-2} mol/liter Aerosol C-61 solution from that in water; 13 lb thrust; 2000 rpm rotational speed.

In general the rate of penetration ($\dot{D}$) of a bit is related to the distance drilled (D) by the expression[12,13]:

$$\dot{D} = V_o - WD \tag{1}$$

where V_o is the initial drilling rate and W a wear rate parameter. As Figure 1 and 3 illustrate, V_o is not significantly environment sensitive, but W is[8]. Thus, it is possible to define a parameter R, where

$$R = W(water) \;/\; W(surfactant\ solution). \tag{2}$$

Because V_o is essentially environment independent, R can also be defined as

$$R = D_o(surfactant\ solution) \;/\; D_o(water) \tag{3}$$

where $D_o = V_o/W$, and is termed the "life" of the bit. Thus, R can be taken as either the factor by which the wear rate of a bit is decreased, or as that by which life of a bit is increased. This quantization of the environmental influence is especially useful because, empirically, R appears to be largely independent of both the flow rate of the circulating fluid and the size of the bit[14].

The importance of bit wear has not been recognized in previous work, as can be seen by comparing the results of Selim et al.[10] with those of Unger et al.[11], or earlier work from our Laboratories[5-7] with later studies by Tweeton et al.[15], Cooper and Berlie[16], and ourselves[8]. In Selim's work, plots similar to those in Figure 3 were employed, and the chemicals used produced significant increases in bit life, largely due to changes in the physical properties of the drilling fluid rather than to any chemomechanical effect. Unger et al., on the other hand, measured the average penetration rate over a significant, but not large, fraction of the life of the diamonds. That is, as illustrated in Figure 4, they compared the slopes of the lines 0-P(2) and 0-P(1). Consequently, the effects they observed were smaller than those reported by Selim and coworkers.

In their early lab-scale studies, Westwood and coworkers[5,6] investigated the effects of various environments on the rate of bit penetration after 200 sec, $\dot{D}(200)$, see Figure 4. Because their bits wore out shortly thereafter, this parameter proved to be a useful indicator of environmental influences... under their *particular* drilling conditions. In contrast, although Tweeton et al.[15] and Cooper and Berlie[16] used fluids capable of reducing the ζ-potential of their selected materials to zero, they observed little influence on drilling rates into marble, microcline, or serpentine under drilling conditions apparently similar to those employed by Westwood et al. However, since these minerals are relatively soft and non-abrasive, the life of diamond bits used to drill them was relatively long. Since the bits effectively suffered no wear during the tests, no

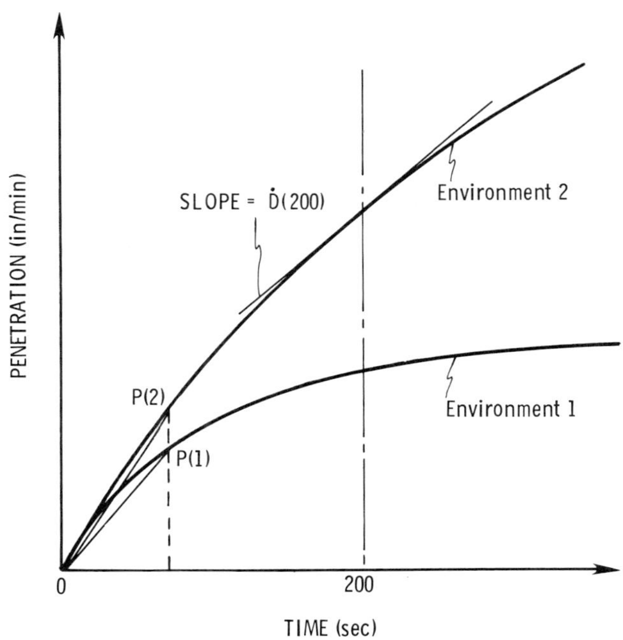

Fig. 4.—Characteristic form of relationship between penetration distance and time for diamond bits drilling in different environments.

significant chemomechanical influences were revealed. On the other hand, when Tweeton et al. drilled into the much harder and more abrasive mineral quartz using a solution producing $\zeta \cong 0$, a small but beneficial effect was observed[15]. The improvement observed was smaller than those reported by Westwood et al.[5,6] because drilling rates were compared relatively early in the life of the bits.

Cooper and Berlie[16] demonstrated yet another effect when drilling Bohus granite in various n-alcohols. In their experiments, changing from water to an alcohol produced an immediate and irregular increase in drilling rate. This behavior suggests the possibility that the n-alcohols caused some redressing of their bits, i.e., some physical property of the alcohols caused the metal matrix holding the diamonds to wear away more rapidly than when water was used. Accordingly, fresh, sharp diamonds were continually exposed. Under such conditions, the n-alcohols would certainly increase drilling rate, but perhaps only temporarily because bit life (defined in this case as the number of dressings possible before all diamonds have been released from the matrix) is likely to have been shortened. Bit redressing also probably masked any chemomechanical effects that might have been observed under other circumstances. For example, when Westwood et al.[5,6] drilled Westerly granite under n-alcohol environments, it is presumed that the light loads used and the shallow holes produced did not cause bit redressing; thus, the chemomechanical influences of the n-alcohols were more clearly revealed.

In recent studies of the effects of dilute solutions of cationic surfactants on the drilling of granite - i.e., concentrations sufficiently low ($\lesssim 10^{-2}$ mol/liter) that changes in density, viscosity, or conductivity of the cutting fluid were unlikely - fourfold reductions in diamond-bit wear rate have been observed[8].

Type of Diamond Drilling Bit

The two types of diamond bit in common use are termed impregnated and surface-set. Impregnated bits use many small diamonds embedded throughout a metal matrix. When these bits cease to cut at an acceptable rate, they are redressed, i.e., resharpened either by drilling them a short distance into an abrasive medium or by sand blasting to remove a layer of the matrix metal and expose fresh, sharp diamonds. The data in Figure 5 indicate that the interval between redressings for impregnated bits can be substantially increased by chemomechanically active fluids. The curves in this figure were produced using the same 2.375 in. o.d. bit, freshly redressed before each test. If we assume that redressing would be called for whenever the penetration rate fell below, say, 0.5 in./min., then it can be seen that the interval permitted when drilling under 10^{-2} mol/liter Aerosol C-61 is more than twice that for water.

However, impregnated bits are now frequently fabricated with a matrix that wears at about the same rate as the diamonds so that new, sharp diamonds are continually exposed. These are called self-sharpening bits, and for these, even if diamond wear rate is decreased chemically, bit life is not substantially improved because the wear rate of the metal matrix is not affected. To use the chemomechanical effect with these bits, they must be fabricated with a more abrasion-resistant matrix to match the chemomechically enhanced diamond life.

Surface-set bits, on the other hand use relatively few, large diamonds embedded in the surface of a metal matrix. When these diamonds are worn

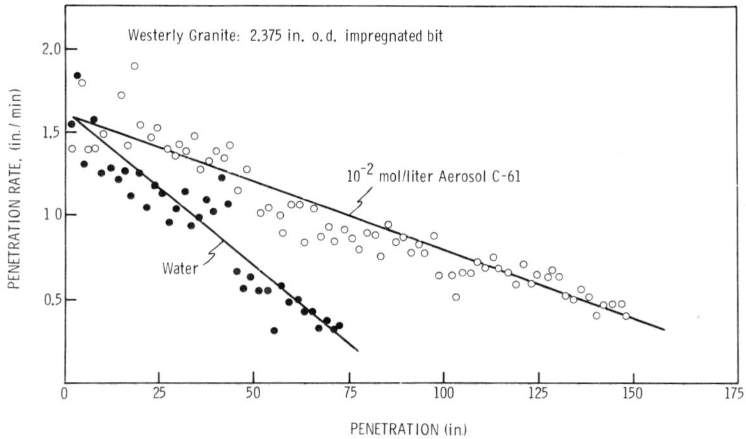

Fig. 5.—Influence of a 10^{-2} mol/liter Aerosol C-61 solution on wear rate of a BX impregnated bit rotating at its design speed of 300 rpm; 3000 lb thrust.

out, the bit is returned to the manufacturer to have the diamonds reset in a new matrix. Chemomechanically active fluids can have a profound influence on the life of such bits. For example, the results presented in Figure 6 were obtained with two comparable surface-set bits of 2.375 in. o.d., one drilling under water, the other under a 0.25 w/o solution of Marvansoft FBH, a cationic surfactant. Note that in these full-scale tests, the initial drilling rates of the two bits were similar (~1.5 in./min) and typical of the rates employed in the field for this rock. However, the FBH solution enhanced the life of the bit 2-3 fold over that when using water (i.e., from ~4 ft to ~10 ft). Further, after drilling 3-4 feet, the rate of penetration was some three times greater under the FBH solution than under water.

To eliminate the possibility that the differences illustrated in Figure 6 are due to using different bits, the FBH solution was replaced by water at point A without altering any of the other drilling parameters. The penetration rate promptly decreased at a more rapid rate than under FBH solutions, and the bit drilled only a few inches more instead of the several feet that may have been achieved by continued drilling in the FBH solution.

Adsorption Dynamics

It is possible to select a surfactant concentration producing $\zeta = 0$, to carry out the appropriate test with the right kind of bit, and yet still not observe any significant influence on drilling behavior either because the surface-active molecules do not have time to arrive at a freshly fractured surface before the next cutting edge arrives, or because, having arrived, they do not chemisorb quickly enough.

The importance of this factor was first appreciated when it was observed that, although certain very long-chain organic surfactants can, under equilibrium conditions, produce $\zeta = 0$ and hardness maxima at concentrations of 10^{-5} - 10^{-6} mol/liter, the concentrations necessary to maximize drilling efficiency might be as large at 10^{-3} mol/liter[17]. This shift was

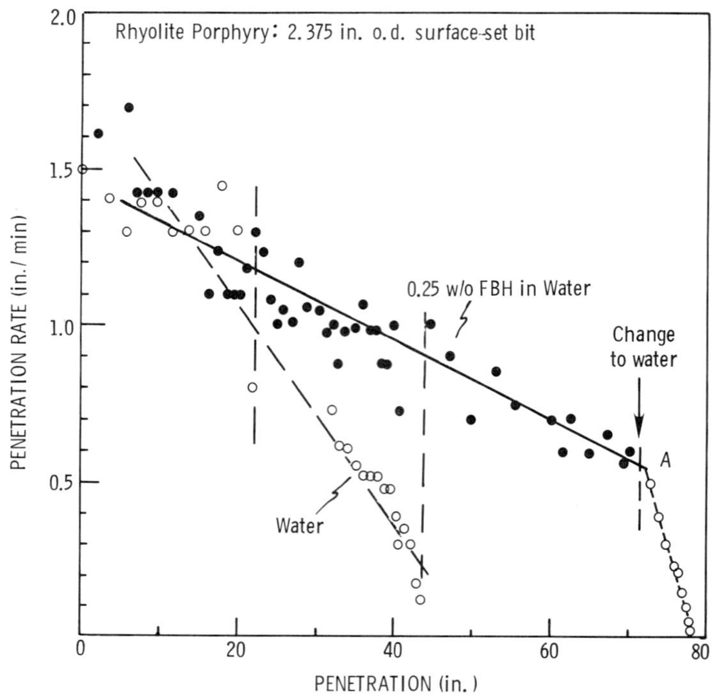

Fig. 6.—Penetration rate vs penetration distance for BX surface-set bits drilling into a rhyolite porphyry rock. Test conditions were: 500 rpm; 15 gpm of either water or 0.25 w/o Marvansoft FBH solution; 3000 lb thrust.

attributed to the inability of the 10^{-5} - 10^{-6} mol/liter solution to produce $\zeta = 0$ under the dynamic and distinctly non-equilibrium conditions of drilling.

The influence of adsorption dynamics was later studied(18) by drilling Westerly granite with bits of three sizes at different rotational speeds and under various concentrations of Aerosol C-61 bracketing the concentration producing $\zeta \simeq 0$ under equilibrium conditions, namely, 3.3 x 10^{-4} mol/liter. It was found that when the bit life parameter R (see eq. 2) was plotted against the peripheral speed of the diamonds, a "master" curve could be obtained for each surfactant concentration by shifting the curves for each bit along the abscissa, Figure 7. The shapes of the curves can be interpreted as follows(18):

For granite in 10^{-4} mol/liter C-61, ζ is <0, but is not as negative as it is in water. Hence, for slow drilling speeds (<10 in./min for a 0.25 in. o.d. bit), $R > 1$ but not as large as it would be if $\zeta = 0$. Now, as the bit peripheral speed, v, increases (plotted towards the left in Figure 7), the time available for adsorption before the next cutting edge arrives, t_e, decreases (in fact, $t_e \propto v^{-1}$). Thus, fewer C-61 molecules adsorb in the time available, and R decreases to its value in water, i.e., to 1.

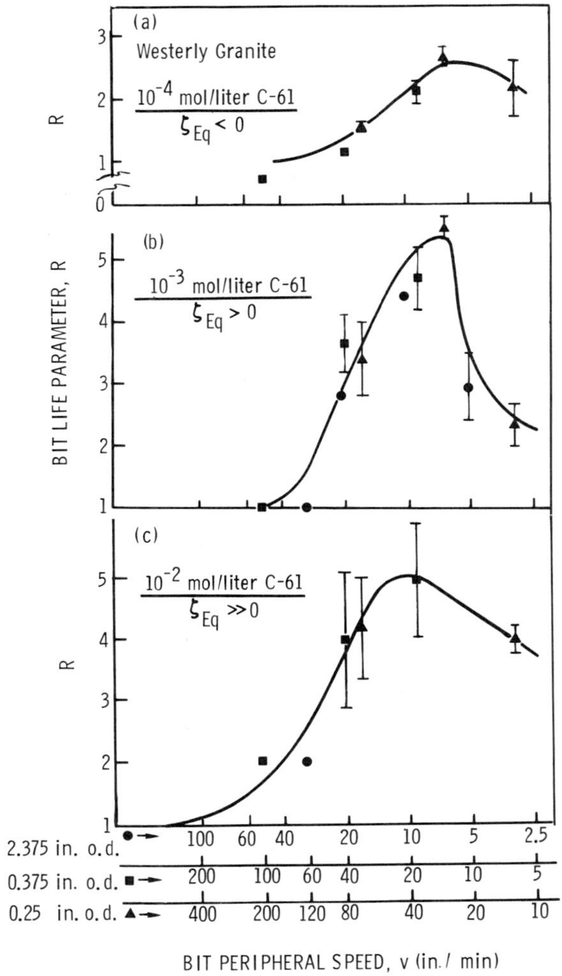

Fig. 7.—Variation in bit life parameter, R, for diamond bits drilling into Westerly granite induced by three different concentrations of Aerosol C-61 as a function of bit peripheral speed.

For granite in 10^{-3} mol/liter C-61, ζ is slightly >0 under equilibrium conditions, i.e., when v is low. Thus, $R > 1$, but again is less than it would be if $\zeta = 0$. As v increases and t_e decreases, however, the number of C-61 molecules arriving in time to affect cutting behavior decreases. In effect, the concentration of C-61 available on the rock surface decreases, initially back to that which would produce $\zeta = 0$ under equilibrium conditions ($\sim 10^{-3.5}$ mol/liter). For the 0.25 in. o.d. bit, this occurs when $v \simeq 30$ *in./min*, and R is then at its maximum value of 4-5. When $v > 30$ *in./min*, however, ζ again becomes <0, and R decreases towards 1. Similar behavior occurs for the 0.375 in. o.d. and the 2.375 in. o.d. bits, except that in this case the optimum values of v are $\approx$ 15 in./min and $\approx$ 7 in./min.

For the 10^{-2} mol/liter C-61 environment, $\zeta >> 0$ when v is low, but, as with the 10^{-3} mol/liter C-61 environment, as v increases, the effective value of $\zeta \to 0$, and R increases to its maximum value. The maximum appears to be located at a somewhat higher value of v than for the 10^{-3} mol/liter solution, however, as might be expected in view of the greater available concentration of surfactant molecules. Finally, as v increases beyond ~60 in./min for a 0.25 in. o.d. bit, ζ becomes increasingly negative, and R again decreases towards 1.

These experiments reveal, therefore, (i) the importance of bit rotational speed in optimizing the use of chemomechanical effects in diamond-bit drilling, and (ii) the existence of characteristic or "master" curves for given rock-environment combinations. The significance of the latter finding is its implication that lab-scale screening tests with small diameter bits may be used to simulate in a reasonably reliable fashion full-scale drilling tests. Thus, surfactant solutions that can produce $\zeta = 0$ and also meet acceptable criteria for toxicity, corrosivity, and biodegradability could be rapidly and inexpensively screened for effectiveness in the field.

Such an evaluation procedure has been used to select a drilling fluid for a rhyolite porphyry, which is inexpensive (less than 3¢ per drilled foot, assuming total loss of fluid), biodegradable, corrosion inhibiting, and which increases the life of surface-set bits up to a factor of four. Figure 6 presented some of the results obtained with this fluid, 0.25 w/o Marvansoft FBH in water. The data of Figure 8 also reveal that the thrust required to maintain a constant penetration rate of 1 in./min is only about one half of that necessary when using water. The trends of the data indicate that the bit operating in the FBH solution would have drilled at least twice as far as the one in water before the thrust required to maintain penetration rate reached 3,000 lb, the figure used to indicate that the bit has worn out.

Diamond Failure Mode

Another factor which, if not given due consideration, may confuse the proper interpretation of drilling data is the mode of diamond failure. During drilling, diamonds can fail either by wear, a process involving abrasion, oxidation, and graphitization, or by fracture and disintegration. If the diamonds fail by fracture, chemomechanically active fluids will have only limited influence on bit life. For example, compare the data presented in Figure 6 with those in Figure 9. In the former experiments, the bits were used to drill about 50 in. into the porphyry under conditions of gradually increasing thrust before the test thrust of 3000 lb was applied. This procedure rounded off any sharp diamond points, allowing them to bear better the 3000-lb load without fracture. In contrast, the bits used to obtain the data presented in Figure 9 were first used to drill only about 6-10 in. under a thrust of 1000 lb before the bit load was increased to 4000 lb. After only a short settling-in, this high load caused shattering of the diamonds, and a rapid, environment-insensitive, nonlinear decrease in drilling rate.

SUMMARY

It follows from the foregoing that drilling strategy -- which includes all the factors discussed above -- should be carefully planned if the

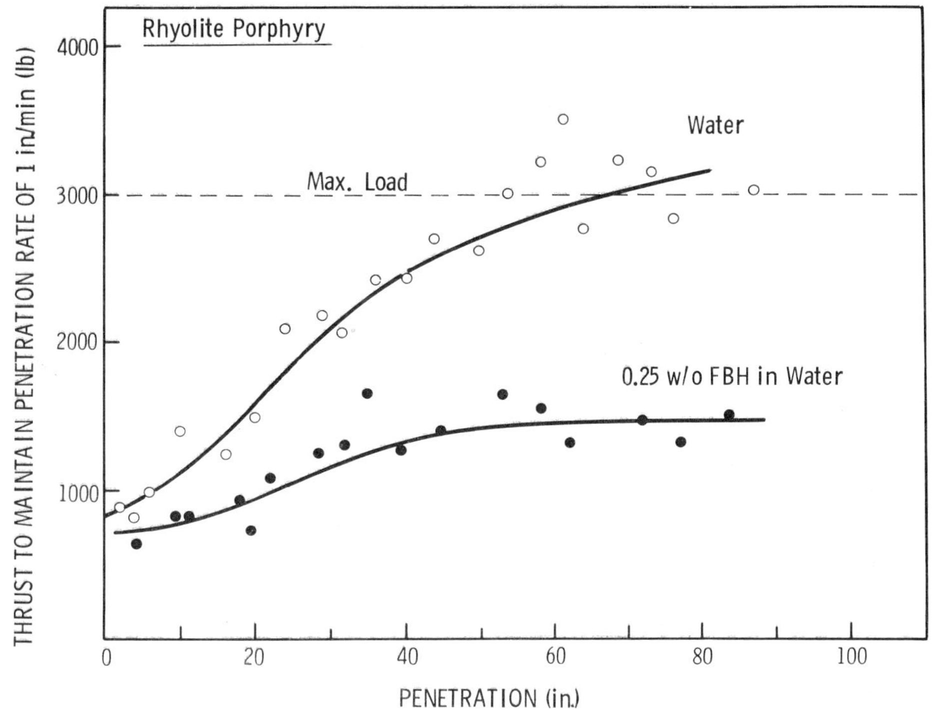

Fig. 8. Thrust necessary to maintain BX surface-set bits penetrating at a constant rate of 1 in./min as a function of penetration distance into a rhyolite porphyry rock. Test conditions were: 500 rpm; 15 gpm of either water or a 0.25 w/o solution of Marvansoft FBH.

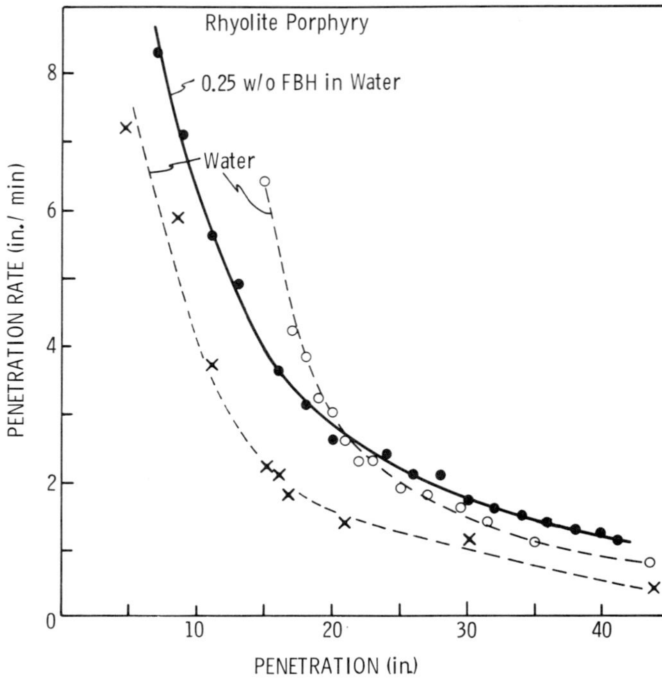

Fig. 9.—Penetration rate vs. distance for BX surface-set diamond bits drilling into a porphyry rock. Test conditions were: 500 rpm; 4000 lb thrust; water or a 0.25 w/o solution of Marvansoft FBH; 15 gpm flow rate.

maximum beneficial influence of chemomechanically active fluids on diamond-bit life is to be reliably achieved. Hill's(19) analysis of the relative costs of drilling 3000 ft-deep exploratory holes into hard rock indicates that the savings resulting from such carefully planned use of chemomechanical effects should be worthwhile. Given the assumptions that (i) fluid chemistry influences bit life only (not particle removal, for example), (ii) bit life is increased by a factor of four (as is the case in Figure 7, for example), and (iii) the unrecirculated fluid flow rate is 10 gpm, then the potential cost savings are on the order of 50%.

We believe that such savings could be achieved provided due consideration is given to the somewhat less than obvious factors discussed in this paper, together with others already well known to the experienced drilling engineer.

ACKNOWLEDGEMENTS

This work was supported in part by the U.S. National Science Foundation under Grant APR 73-07787.

REFERENCES

1. Rebinder, P.A., Schreiner, L.A. and Zhigach, S.F., "Hardness Reducers in Rock Drilling," Academy of Sciences, Moscow, 1944, CSIRO Translation, Melbourne, 1948.
2. Westwood, A.R.C., *Journal of Materials Science,* Vol. 9, 1974, p. 1871.
3. Wicklund, P.C., AMAX Corp., Private Communication, December 1976.
4. Ahearn, J.S., Mills, J.J. and Westwood, A.R.C., These Proceedings.
5. Westwood, A.R.C., MacMillan, N.H. and Kalyoncu, R.D., *Transactions AIME,* Vol. 256, 1974, p. 106.
6. MacMillan, N.H., Jackson, R.E., Mularie, W.M. and Westwood, A.R.C., *Transactions AIME,* Vol. 258, 1975, p. 278.
7. Westwood, A.R.C. and Mills, J.J., in "Surface Effects in Crystal Plasticity," edited by R.M. Latanision and J.T. Fourie, Noørdhoff International Publishing, Leyden, 1977, p. 835.
8. Mills, J.J. and Westwood, A.R.C., *Journal of Materials Science,* Vol. 13, 1978, p. 2712.
9. Appl, F.C., Private Communication, Kansas State University, May 1978.
10. Selim, A.A., Schultz, C.W. and Strebig, K.C., *Society of Petroleum Engineers Journal,* Vol. 9, 1969, p. 425.
11. Unger, H.F., Snowden, B.S. and Englemann, W.H., *Transactions AIME,* Vol. 258, 1975, p. 185.
12. Tsoutrelis, C.E., *Transactions AIME,* Vol. 244, 1969, p. 365.
13. Mills, J.J. and Westwood, A.R.C., *Industrial Diamond Review,* August 1977, p. 264.
14. Mills, J.J., Unpublished Work, Martin Marietta Laboratories, 1976.
15. Tweeton, D.R., et al., "Influence of Surface-Active Chemicals on Drilling and Fracturing Rock," Report No. 8186, U.S. Bureau of Mines, 1976.
16. Cooper, G.A. and Berlie, J., *Journal of Materials Science,* Vol. 11, 1976, p. 1771.
17. Westwood, A.R.C. and Huntington, R.D., Third Six-Monthly Report to the National Science Foundation on Grant AI-38114, November 1974.
18. Mills, J.J., Seventh Six-Monthly Report to the National Science Foundation on Grant 73-07787, MML TR 76-90c, November 1976.
19. Hill, B.S., *Industrial Diamond Review,* August 1975, p. 282.

DISCUSSION

G. F. HARDY, Celanese Research Company: Did you attempt to measure the ζ-potential under actual operating condition?

J. J. MILLS: No. We did a little calculation and found out that we would have to measure within a millisecond. I could not see any way of measuring ζ-potential or any form of adsorption in such a short duration.

B. WILCOX, NSF: Did you measure the thermal conductivity?

MILLS: We did not measure the thermal conductivity, but we have measured density and viscosity. We are talking about concentrations that are so small that neither of those two change. Typical concentrations are lower than one percent, and can get as low as 0.01 percent. Since viscosity is related through kinetic theory to thermal conductivity, we assumed that the thermal conductivity did not change either.

J. LARSEN-BASSE, University of Hawaii: Have you looked into the possibility that the properties of the tool are changing?

MILLS: We did not do that because of other evidence. For example, the correlation between hardness and drilling rate always appears at the zero $- \zeta$ potential of the material being drilled. We found this to be true for percussive bit, carbide bit and diamond bit. We did not do that experiment because the evidence strongly suggested to us that the properties of the rock are being changed and not the drill bit.

QUESTIONER, Dow Corning Corporation: We did a study a number of years ago on the effects of viscosity and surface tension. We took a diamond saw and did a number of cutting tests by changing the surface tension of the cutting fluids. As we lowered the surface tension of the cutting medium, we got increased cutting rates and also better blade lives. We also changed the viscosity of the silicone fluids. When we decreased the viscosity, the cutting rates were lower.

MILLS: It sounds as if you are changing the physical properties. A lot of work is being done on that sort of thing. What I am trying to show here is that we can also get the same kind of things at much lower concentrations by using these chemomechanical effects.

SAME QUESTIONER: The fact is, you do alter the surface tension.

MILLS: But aluminum chloride does not. And you do see the chemomechanical phenomenon with simple aqueous solutions. I am trying to show that there is a distinction between the two kinds of phenomena.

ELECTRIC CURRENT TRANSFER ACROSS SLIDING SURFACES

I. R. McNab

ABSTRACT

During recent years, considerable interest has been shown in developing an improved understanding of the phenomena involved in the low loss transfer of electric current across sliding surfaces. The requirement that a substantial electric current be transferred across a sliding interface significantly complicates the more conventional processes of friction and wear. Thus, in addition to frictional heating, electrical heating takes place at the transiently contacting asperities, and within the body of the sliding contact materials. The necessity for low electrical losses precludes the use of an electrically conducting film of significant thickness at the interface. However, the requirement for low frictional losses and low wear rates generally ensures that metal-to-metal contact is unacceptable. Recent work shows that an effective compromise can be achieved when a thin (2 nm or less) film is present on the sliding surfaces. This prevents the occurrence of high adhesive forces but permits low loss electrical current transfer by tunneling. Modern diagnostic techniques (SEM, Auger, etc.) are presently being used to evaluate the topography and composition of suface films formed by sliding metal-graphite composites against metal surfaces in controlled atmosphere environments.

INTRODUCTION

Recent developments in advanced electrical machinery[1-3] have led to a demand for material combinations which will permit the efficient transfer of high currents across sliding surfaces while simultaneously providing low frictional losses and low wear. In the terminology of the subject, the static component is generally called the "brush" (for historical reasons) and the counterface against which it operates is the "slip ring" or "commutator" for a rotating machine, or the "rail" for a linear system. In traditional practice, the brush is made of, or based on, carbon and the slip ring or commutator is made of copper, and operation is in the air. However, many material combinations have been investigated and a number of interesting new approaches are presently under development.

Two major features distinguish modern electric current power transfer systems from conventional friction and wear situations.* The first of these is the requirement to transfer significant electric currents (tens of kilo-amps or more) from the static brush to the moving slip ring. This condition generally prevents the use of lubricant or oxide films which, although beneficial for friction and wear, may provide a significant barrier to the transfer of current. That is, the *total* (electrical plus frictional) losses at the interface have to be considered. To accommodate the capability to transfer tens of thousands of amps in a reasonable space also requires that a very high operating current density should be used (1.5 to 15 MA/m^2). These electrical aspects of the sliding contact bring with them a whole range of interesting, but complex, phenomena that are not present in conventional friction and wear situations. The range of literature available on this subject is considerable, and only an outline of some of these effects is attempted in this paper. The books by Holm(4) and Shobert(5) provide in-depth data.

The second feature that is present in modern electrotechnological applications of sliding contacts is not exclusively related to electrical apparatus, but is relatively uncommon in purely mechanical situations. It is that very high sliding speeds may be required(2). Thus, in rotating machinery, sliding speeds up to 280 m/s have been achieved on a continuous basis(6), with short-term operation up to 360 m/s(7). These values, and any apparatus operating at speeds beyond this level, require very high strength-to-weight ratio materials, and careful attention to mechanical design criteria. In linear electromagnetic accelerator systems ("rail guns"), sliding speeds over an order of magnitude higher than the above (up to 5800 m/s) have been achieved on a transient basis, with simultaneous current transfer(8).

The requirements for the simultaneous transfer of high currents with low electrical and frictional losses and low wear rates at these ultra-high speeds presents an exciting challenge which is leading to a re-evaluation of the fundamental aspects of sliding contacts. A related aspect is the availability, in recent years, of greatly improved techniques for characterizing surfaces, such as scanning electron microscopes, Auger electron spectroscopy, etc.

INTERFACE PHENOMENA

The fundamental phenomena that take place at the sliding interface are complex, and by no means fully understood at present. Some of the contributory features are outlined below. Note that the great majority of traditional experiments relating to current transfer have been carried out in air, although recent interest for high power applications has focused on controlled environments where quite different processes may dominate the sliding contact.

Brush and Slip Ring Materials

For many traditional brush applications, it is adequate to use a brush made entirely of carbon or carbon-graphite. However, for high power applications, the relatively high electrical resistivity of this material (typically 1000 μΩ cm or higher) would cause unacceptably high electrical

*This paper does not discuss instrument type slip rings where only small (< 1 amp) currents are transferred.

losses. It is, therefore, more common to employ brushes which are composed of a metal-graphite mixture. Such materials are generally made by powder-metallurgy techniques, and metal mass fractions up to 0.96 are available. Molten metal infiltration techniques have also been used at lower metal mass fractions, the limiting factor in this case being the strength of the base material to be infiltrated. Figure 1 shows surface micrographs (X200 to X250) of several typical brush materials with metal mass fractions from zero to 0.80. The significant variation in surface morphology is an

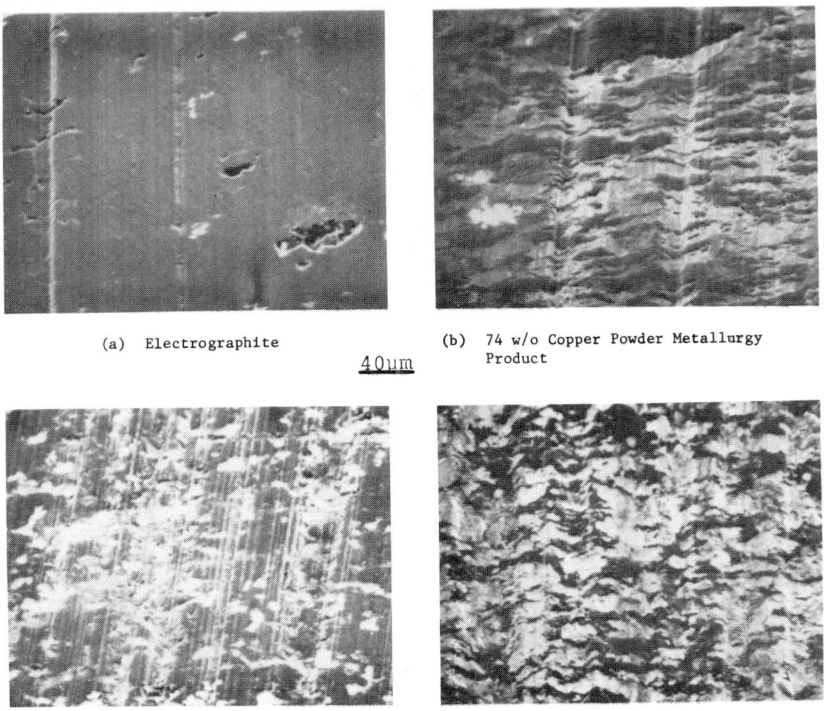

(a) Electrographite

(b) 74 w/o Copper Powder Metallurgy Product

40μm

(c) 41 w/o Copper Infiltrated

(d) 80 w/o Silver Powder Metallurgy Product

Fig. 1.—Graphite-metal mixture brush materials.

indication of the wide variation in brush materials. This is undoubtedly a major factor leading to the observed large differences, for example, in brush wear.

The metal phase of conventional metal-graphite brushes is most commonly copper, although silver offers better electrical performance and may be used if cost considerations are not prohibitive. The natural or electro-graphite component may be partially or entirely replaced by a dichalcogenide for operation in dry atmospheres or in space. The dichalcogenide most commonly used is molybdenum disulphide, although niobium diselenide has been used and offers a lower electrical resistivity. A very wide range of brush materials is produced, and many different material combinations, methods of processing, and additives are available for use in particular cases.

Copper or a copper alloy is generally chosen as the slip ring material, although steel is also fairly widely used. A limited but useful amount of research has been undertaken on noble and refractory metals (such as silver, palladium, rhodium, ruthenium), mainly for low or zero current applications[9], and further research is required in this area. In general terms, the harder ring materials have an associated higher electrical resistivity but lower frictional losses, so that a choice of materials may be optimized around the specific applications and loss requirements.

It is apparent from these studies that mutual solubility and the lattice structure of the two metals plays an important role for purely metallic systems. The influence of graphite in such a situation has yet to be fully explored.

Area of Contact

Application of a load force (F) to the brush, either by means of a spring or some other form of actuator, brings it into mechanical contact with the slip ring. The resulting area of mechanical contact is determined by a complex balance of elastic and plastic deformation of the metal component of the brush, and elastic and crushing deformation of the brittle graphite brush constituent at the microscopic asperities on the brush and slip ring surfaces. For typical brush operating conditions (5-20 newtons), the true area of contact appears to be about one hundredth that of the apparent area (the brush face area). Whether the elastic, plastic, or crushing deformation regime will be dominant depends on the brush composition and factors such as the brush history, period of operation, and surface films. The initial contact may be controlled by plastic yielding of the softer (generally the brush) material, and for this case, Holm[4] and Shobert[5] recommend the use of the following equation to evaluate the true area of mechanical contact:

$$A_t = 3F/H$$

where H is the Meyer (or contact) hardness of the brush[10]. Over a longer period of operation, when the brush and slip ring contours have become matched, elastic effects will control the behavior. In that case, it is much more difficult to evaluate the true contact area, since a knowledge of the surface geometry is required that is seldom available[11]. With carbon and graphite materials, there is significant evidence to show that film build-up occurs on the surface of the base material[12]. The properties of this thin and relatively soft film contribute to the determination of the area of contact, although if the film is sufficiently thin, the elastic modulus of the substrate will be the controlling factor. The extent to which this effect is present with metal graphite materials is undetermined at this time.

The area of contact is usually considered to be subdivided into a relatively small number of randomly distributed contact spots which, as the brush wears, move around the brush surface, covering all of the available area over the brush life time.

Under circumstances of high speed and/or high current transfer, the contact spot distribution may become unstable, with one spot growing thermally from the substrate to such an extent that it lifts the remainder of the brush surface from the slip ring. Once initiated, this situation will persist, with one spot taking all the current and generating all the frictional heating, until it becomes worn away. Such phenomena have been observed experimentally in high current brushes by Marshall[13], and analyzed in detail recently by Burton and Chen[14].

Constriction Resistance

Depending on the material properties and the operating environment, significant films may be present on the surface of the brush and the slip ring. Depending on the electrical properties of these films, all or part of the area of mechanical contact may be available for the transfer of current. Common operating experience is with carbon brushes on copper slip rings in air, and, in this case, most of the mechanical area of contact is covered with a film which is largely composed of cuprous oxide and carbon wear debris. Such a film is electrically insulating and conduction takes place only at localized areas where the film is disrupted either by electrical breakdown or by mechanical abrasion. In such situations, the area available for current transfer may be as small as a tenth or a hundredth of the mechanical contact spot area. In contrast, the elimination of a substantial surface film by operation in a controlled atmosphere (e.g., non-oxidizing) environment will permit the electrical contact area to approach more nearly the mechanical contact area.

In most sliding contacts, two fundamental contributions to the electrical resistance are present: (1) the constriction resistance, which is caused by the channeling of all the brush current through the few small contact areas and (2) the interface film resistance.

Solution of Poisson's equation in oblate spherical coordinates (first derived by Maxwell) yields the constriction resistance for a single symmetric contact:

$$Rc_1 = \frac{\rho}{2a}$$

where a is the radius of the contact spot (assumed circular in this simple case and ρ is the material resistivity). For n multiple interacting spots, Greenwood[15] gives

$$Rc_n = \frac{\rho}{2a}\left[\frac{1}{n} + \frac{1.081\ a}{\alpha}\right]$$

where α is the so-called Holm radius. Constrictions occur both in the brush and slip ring, the resistivities of which are assumed equal in the above derivation. If the resistivity of one material is markedly greater than the other, it will dominate the total resistance, and the other may be ignored. This is the case, for example, for a carbon brush on a copper slip ring, although when a metal-graphite composite brush is used its resistance is sufficiently small that the slip ring resistance has to be included.

In the case of sliding contacts, the contact areas are more likely to be elliptical than circular; in extreme cases a line contact is a good approximation[5].

Film Resistance

The second major contribution to the electrical resistance of the sliding contact is provided by the film that is always present at the sliding interface. Compared with some other aspects of sliding contacts, relatively little is known about this film, although it is clear that it is of a complex nature.

Where chemical reactions can occur with the brush or slip ring materials, reaction by-products such as oxide or sulphide films will be present. In the case of copper, the oxides are harder than the base material. In addition, there is known to be transferred material from the brush, both graphitic and metallic in nature, and the slip ring wear debris. With electrographitic brushes, it has been shown[12] that a carbon film is formed first on the brush and, subsequently, transferred to the copper slip ring. Initially, the graphite crystallites in the brush are not broken down, but eventually a structure with a preferred orientation and a characteristic detail of less than 6 nm was observed. The extent to which the graphite can bond to the slip ring depends on the nature of the metal and the oxide, either through physical or chemical attachment. There is some evidence to suggest that graphite crystallites embedded in a soft copper slip ring surface have a shingle-like structure aligned at a small angle to the slip ring surface[16]. Over all and/or dispersed within this will be physisorbed and chemisorbed molecules from the surrounding gas atmosphere, both gases and vapors.

The general expression for the film resistance is $R_f = \sigma/A_e$ where σ is the film resistivity and A_e is the area available for electric current transfer.

When insulating films of significant thickness (of the order of 10-100 nm) are present at the interface, conduction will take place only when the applied voltage becomes high enough to raise the local electric field strength in the film to the breakdown value. When this occurs, metallic "bridges" are formed through the film and, once formed, these conducting spots may persist for many hundreds or thousands of slip ring revolutions before oxidation or wear causes the current to be transferred to another location[5]. Disruption of the surface film by mechanical abrasion can also occur, especially if the oxide film is harder than the underlying substrate, as in the case of copper.

The persistance of such spots is, in part, due to a process (called B-fritting by Holm[4]) in which the local electric field distribution at the edges of the contact spot acts to enlarge the spot radius, opposing the oxidative processes that tend to cover the spot with an insulating film.

The general characteristics of thick films of this type, mostly encountered in air operation of copper slip rings, are a non-linear current voltage characteristic, generally with a significant difference between positive and negative polarities as a result of ion migration effects. The non-linearity is a consequence of the breakdown characteristics.

In the absence of a significant film (or on stationary contacts) a more ohmic characteristic is found. If the film thickness at the contacting spots on the interface can be maintained at 2 to 3 nm, or less, electronic conduction by quantum tunneling can be effective in transferring large current densities with low losses. Recent experiments with Auger techniques appear to offer direct evidence for such thin films at some parts of the slip ring track (see Figure 2). At the same time, based on an estimated sputtering time - depth calibration, film thicknesses up to about 1 μm are present in other areas, even on silver slip rings operated in non-oxidizing (CO_2) atmospheres.

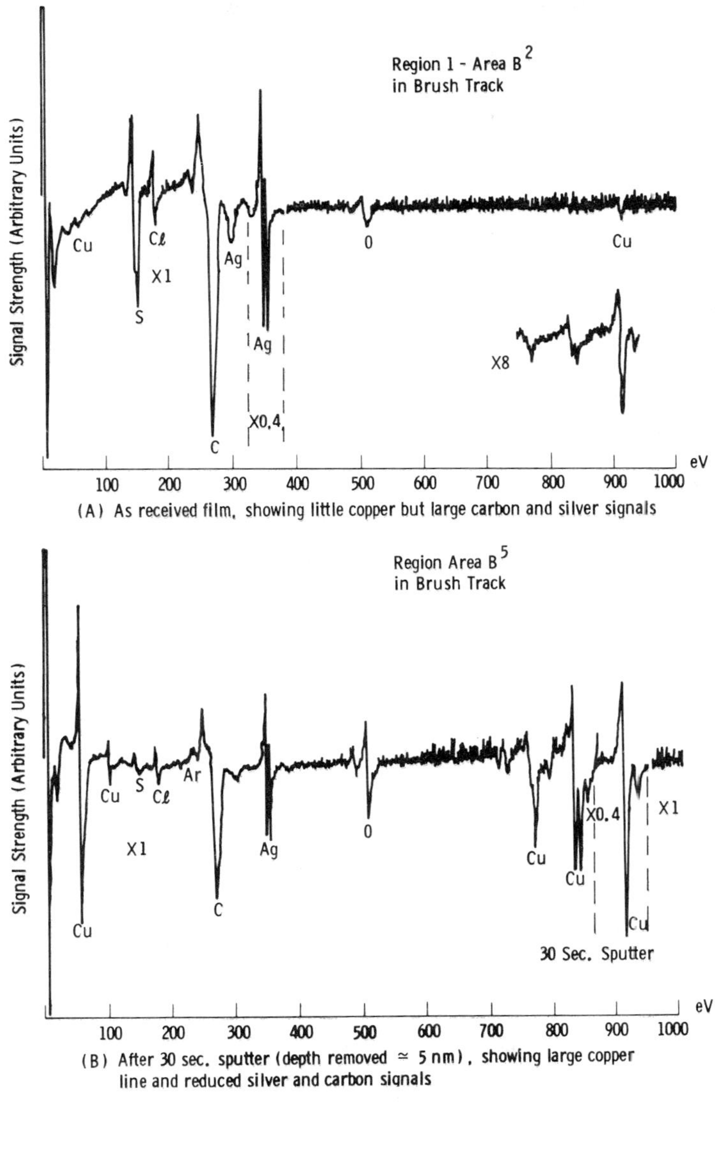

Fig. 2.—Auger spectra of film on OFHC copper slip ring after operation under an 85 w/o silver-graphite brush.

Friction and Wear

The friction and wear behavior of brush and ring materials is dependent not only on the materials themselves but also, and probably to a larger extent, on the solid and adsorbed surface films that are present at the interface. Since these are complex, depend on external circumstances, and can vary widely from one brush material to the other, the friction and wear may also be subject to similar variations.

Factors such as the atmospheric composition are of major importance in the behavior of brushes containing graphite or dichalcogenides, at least in certain operating regimes.

In general, the frictional behavior of brushes is not subject to such a large variation as the wear rate. Thus, despite the many phenomena that may be involved (shearing of graphite platelets, shearing of metallic conduction bridges, adhesive forces, viscous effects in adsorbed gas or vapor films) it is relatively uncommon to find friction coefficients below 0.08 or above 0.35. In general, the pure graphites, or low metal content brushes fall toward the lower end of this range, and the high metal content materials toward the upper end. However, depending on the specific material, there may be significant variations.

There is evidence[12] to show that the frictional behavior of non-graphitic carbons is virtually the same as that of graphite, at least after a wearing-in period. This is a strong indication that the weak intercrystallite bonding in the surface film, rather than shearing of the graphite base material crystallites, controls the frictional behavior. As described below, the ambient atmosphere in which the brushes operate plays an important role in the friction and wear process.

In contrast to friction, the brush wear rate is a sensitive parameter, which may vary by two or more orders of magnitude with relatively minor changes in conditions. Figure 3 shows that a variation by a factor of two in the silver content of a metal-graphite brush can cause a 3 to 1 change in the friction coefficient, but up to a 100 to 1 increase in the wear rate[17]. In part, it is possible that the surprisingly small variation in the friction coefficient may arise as a result of compensating factors, such as the action of the relatively large quantities of brush debris as ball or roller bearings at high wear rates. Some authors[18] consider that the conduction process may be dominated by this debris, although this is not a common view.

Microhardness measurements indicate that despite the relatively easy shearing, the individual graphite crystallite is hard and brittle. Thus, on rough surfaces (e.g., in initial wearing-in) high wear may occur as local stresses cause brittle fracture of the crystallites. However, if allowed to continue, this process produces a smoother surface, both by destroying asperities and by producing a surface film from the retained wear debris. Under these conditions of low surface irregularities, elastic deformation of the surface occurs and some authors[12] have observed evidence for a surface fatigue mechanism in which the sequence: polishing → formation of a film → film blistering → wear flake removal → polishing, etc., is repeated indefinitely.

A significant contribution to the friction and wear of carbon- or graphite-containing brushes operating in oxidizing atmospheres is the

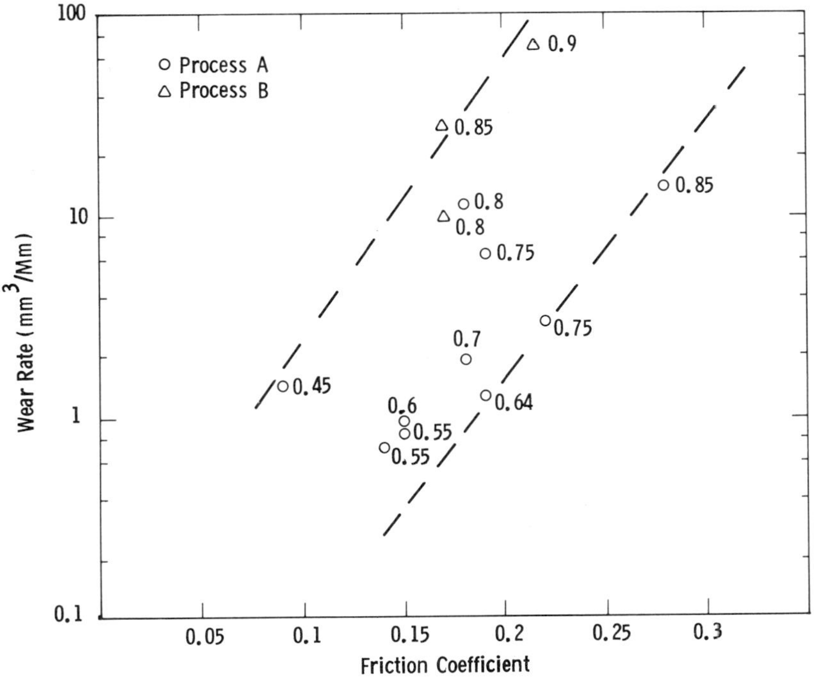

Fig. 3.—Variation of wear rate and friction coefficient for silver-graphite brushes operating on a copper slip ring in CO_2 atmospheres at a current density of 0.78 MA/m^2 (silver mass fraction adjacent to data pcint).

catalytic oxidation of the carbon films or base material by metal oxides. Certain metals, present in impurity quantities ($\gtrsim$ 0.25 w/o) in the brush may, therefore, exert a strong influence on wear processes. Leac oxide is particularly bad in this context, causing the graphite ignition temperature to be reduced by 358°C[19].

Temperature and Atmosphere

The influence of temperature and atmospheric environment are closely interrelated through the adsorption of gaseous layers on the slip ring and brush surfaces. Up to about 180°C, increasing the bulk brush temperature of a brush operating in a humidified atmosphere generally causes a modest decrease in the friction coefficient, whether because of reduced material shear strength, reduced "viscosity" of adsorbed gas or vapor films, or both. However, at or near 180°C, a marked change in the wear rate of graphite-based brushes occurs, the change being by a factor of 1000 or more. This relates to the desorption of gases and/or vapors from the edge sites of the graphite crystallites, thereby allowing strong pi-electron bonds to be developed at the (now) unsaturated edge sites. Under these conditions, wear rates can increase up to levels of cm per minute, a phenomena known as dusting because of the production of large amounts of very fine (6 nm) brush debris.

In the case of graphite, a minimum amount of a condensable vapor or gas is necessary to prevent the formation of these strong bonds and ensure easy shearing. About 60 vppm of water vapor appears to be necessary in air(20), for silver-graphite brushes, although values up to 6000 vppm have been found necessary under other conditions (speed, material, temperature, etc.)(21), but lower concentrations of normal hydrocarbons having long $-CH_2-$ chains are also effective in preventing wear. Figure 4 summarizes the results of some recent experiments undertaken at Westinghouse on the effects of hydrocarbon vapors on brush friction and wear(22). The greater

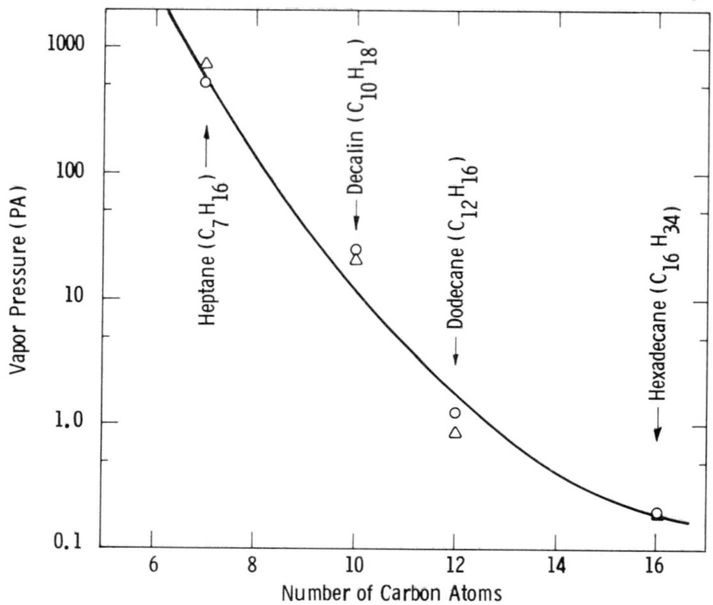

Fig. 4.—Vapor pressure of hydrocarbon vapor required to ensure a contact drop of 0.08 V (o) and wear rate of 11 mm^3/Mm (Δ) for silver-graphite brushes sliding on a copper slip ring in CO_2.

efficacy of the longer chain length vapors apparently relates to the greater bonding strength of the chain onto the graphite lattice, both of which have a hexagonal structure, with similar lattice constants.

The extent to which water vapor is effective in preventing dusting-type wear is dependent on the atmosphere in which the brushes are operated, probably through a mechanism of competition or adsorptive activation of the surface sites on the graphite crystallites, as discussed by Savage(23) and found experimentally by Pardee(20).

With dichalcogenides, which are added to some brush materials in moderate quantities (up to 10%), the reverse situation applies, namely, that the materials shear most readily in the absence of an atmosphere. Aparin et al.(24) have suggested that the role of water molecules and hydroxyl groups on MoS_2 surfaces is to increase the adhesion of MoS_2 and by accumulation at the grain boundaries during sliding prevent the orientation of crystallites required for low friction. The choice of dichalcogenide for good lubricity is related to the crystal structure/lattice structure of the lamellar solid[25]. Of the available materials, MoS_2 is most widely used, although the stoichiometric $NbSe_2$ (63 w/o Se) has been used on a number of occasions and is much preferable for brushes from an electrical conductivity point of view. (Note that a change to 59 w/o Se increases the resistivity by a factor of 50.)

Supertemperatures

The temperatures mentioned above relate to the bulk brush and slip ring conditions. At the actual contact spots, much higher temperatures undoubtedly exist on a transient basis. Methods for evaluating these "hot spot" or "super"-temperatures were first developed by Blok[26] for frictional heating. Holm[4] and Shobert[5] have extended that theory to allow an estimate to be made for the additional electrical heating that occurs under brushes. The usual, though approximate, method used to solve the heat flow equations for steady-state, and apply a correction factor to determine the fraction of the steady-state temperature that may be achieved in a transient contact.* The frictional and electrical supertemperatures are added to yield the total supertemperature.

This technique permits an estimate to be made of how the asperity supertemperature compares with the softening and/or melting temperatures for the contact. With very high speeds or currents, the rate of rise of temperature may be extremely high (10^8 °C/sec) but, since the period of contact is limited to microseconds, the actual temperature rise may be quite modest.

Since the contacting and current-transferring asperity contacts experience these supertemperatures, their influence may be of major importance in friction, wear, and electrical current transfer processes, although direct evidence on this topic is very limited.

Speed Dependent Effects

At low or moderate sliding speeds, the heat produced at the interface by frictional or electrical heating is equipartitioned between the two bodies on the basis of their thermal diffusivities. However, as Jaeger showed[28], at higher speeds an increasing fraction of the interfacially produced heat is transferred into an increasingly thin layer on the (relatively cold) incoming slip ring. Assuming uniform bulk contact, Jaeger also showed that the temperature profile under a sliding contact was modified with increasing speed, with the maximum temperature shifting towards the rear of the brush.

Aerodynamic forces and surface irregularities become increasingly important at high sliding speeds. Although only limited experimental evident is available[29], aerodynamic gas "wedges" under the brush face may be the cause of observations in which low friction and high contact voltages are simultaneously observed. Such effects may be reduced, if not eliminated by the use of a spiral groove cut in the slip ring surface. A second important purpose of such a groove is to reduce selectively at the brush face by forcing all parts of the brush to carry current in turn.

---

*Note that the heat flow equations have the same form as the electrical current flow equations *only* for steady-state conditions[27].

Brushes operating on high speed slip rings are also sensitive to the frequency and amplitude of slip ring imperfections and rotor vibrations. Such effects are likely to be responsible for relatively rapid changes in brush characteristics which have been observed over relatively restricted speed ranges[30]. An estimate of the conditions at which such effects become important may be obtained by equating inertial terms with slip ring accelerating forces[5,31]. For very small disturbances, the brush elasticity and speed of response (i.e., speed of sound) become limiting factors.

Electrical Effects

The transfer of electric current across the sliding interface brings into play a range of phenomena which are not ordinarily present in conventional friction and wear situations. These effects include:

(a) A mutual repulsive force at each point contact caused by the magnetic field distribution associated with the passage of current through the constriction.

(b) Capacitative and inductive effects in any circuits in which switching of current occurs. On a macroscopic scale, this clearly occurs in commutative circuits; on a microscopic scale, it may occur each time asperities make and break contact.

(c) Thermoelectric effects at the junction of dissimilar metals, including Peltier and Seebeck effects.

(d) In some applications, for example involving superconducting machines, substantial ambient magnetic fields may be present in the vicinity of the brushes. Not only can this give rise to substantial $\underline{J} \times \underline{B}$ forces on the body of the brush, but second order effects, such as the Hall effect, could also become important.

(e) Polarity effects may be present due to a variety of electrical effects. The most common is ionic migration due to the local electric field in, or near, the brush-ring interface film. Such effects are of major importance in causing the breakdown of insulating films on the slip ring and establishing metallic bridges. Measurements on thin silver sulphide films indicate that a field strength of 2 MV/m is necessary for breakdown and silver bridge formation[32], although values up to 100 MV/m have been quoted[4]. For certain species, the electron wind can cause ionic species to be moved in the opposite direction to that predicted from electrostatic effects[33]. Jansen et al.[34] have remarked that electromigration effects may be beneficial in removing impurities or lattice defects from the contact region.

(f) The quantum tunneling process that appears to permit electron current transfer through thin films has been referred to earlier.

CONCLUDING REMARKS

Stringent demands are made on current-transferring sliding contacts by the requirements of modern steady-state and pulsed homopolar machines and electromagnetic rail guns. To ensure efficient machine operation and ease the machine design, the total (i.e., mechanical plus electrical) losses, and wear, have to be minimized. This necessitates consideration of all of the complex and interrelated aspects of sliding contact behavior to ensure that optimum solution is achieved for each particular situation. As a consequence of these considerations, the operating constraints are appreciably more severe than in normal friction and wear situations. For example, the use of substantial lubricant films to reduce friction and wear is unacceptable.

Although a fairly significant body of knowledge exists in some areas of sliding current transfer, much still remains to be done to improve the understanding of friction, wear, and electrical processes with metal-graphite mixture brushes. The imaginative use of new surface analysis techniques (SEM, Auger, etc.) offers the prospect of improving our understanding, and work along these lines has been initiated at the Westinghouse R&D Center under Defence Advanced Research Projects Agency/Office of Naval Research sponsorship. Material combinations and techniques other than the conventional metal-graphite monolithic brushes offer the prospect of significant improvements, although further research and development is required.

ACKNOWLEDGEMENTS

Much of the work on which this paper was based was performed under Advanced Research Projects Agency Contract N00014-76-C-0683, monitored by the Office of Naval Research. Thanks are due J. J. Schreurs for the data shown in Figure 2 and to J. L. Johnson for the data shown in Figures 3 and 4.

REFERENCES

1. Satkowski, J.A. and Seng, W.R., in Australian-U.S. Seminar on Energy Storage, Compression, and Switching, Canberra, Australia, November 1977, edited by E.K. Inall, ANU Press, Canberra, 1978.
2. Mole, C.J. and Mullan, E., "Design of a 10 MJ Fast Discharging Homopolar Machine," in Australian-U.S. Seminar on Energy Storage, Compression, and Switching, Canberra, Australia, November 1977, edited by E.K. Inall, ANU Press, Canberra, 1978.
3. Mole, C.J. and Mullan, E., "Design Trends in Homopolar Machines Since the mid-1960's," in Australian-U.S. Seminar on Energy Storage, Compression, and Switching, Canberra, Australia, November 1977, edited by E.K. Inall, ANU Press, Canberra, 1978.
4. Holm, R., "Electrical Contacts," 4th edition, Springer Verlag, New York, 1967.
5. Shobert, E.I., "Carbon Brushes - The Physics and Chemistry of Sliding Contacts," Chemical Publishing Company, New York, 1965.
6. Marshall, R.A. and Slepian, R.M., in Ninth International Conference on Electric Contact Phenomena, Chicago, September 1978, Illinois Institute of Technology, p. 513.
7. Robson, A.E., et al., "An Inductive Energy Storage System Based on a Self-Excited Homopolar Generator," 6th Symposium on the Engineering Problems of Fusion Research, San Diego, November 1975.
8. Rashleigh, S.C. and Marshall, R.A., *Journal of Applied Physics*, Vol. 49, 1978, p. 2540.
9. Angus, H.C., "Instrument Practice," March 1966, p. 241.
10. Holm, E., Holm, R. and Shobert, E.I., II, *Journal of Applied Physics*, Vol. 20, 1949, p. 319.
11. Clark, W.T., Connelly, A. and Hirst, W., *Journal of Physics D: Applied Physics*, Vol. 14, 1963, p. 20.

12. Clark, W.T. and Lancaster, J.K., *Wear*, Vol. 6, 1963, p. 467.
13. Marshall, R.A., *Wear*, Vol. 37, 1976, p. 233.
14. Chen, C.P. and Burton, R.A., in Ninth International Conference on Electric Contact Phenomena, Chicago, September 1978, Illinois Institute of Technology, p. 571.
15. Greenwood, J.A., *Journal of Physics D: Applied Physics*, Vol. 17, 1966, p. 1621.
16. Porgess, P.V.K. and Wilman, W., *Journal of Physics A: General Physics*, Vol. 76, 1960, p. 513.
17. McNab, I.R. and Johnson, J.L., in Ninth International Conference on Electrical Contact Phenomena, Chicago, September 1978, Illinois Institute of Technology, p. 493.
18. Mayeur, R., *Revue Generale de L'Electricite*, Vol. 66, 1957, p. 207.
19. McKee, D.W., *Carbon*, Vol. 8, 1970, p. 623.
20. Pardee, R.P., *Institution of Electrical and Electronic Engineers Transactions: Power Apparatus and Systems*, Vol. 86, 1967, p. 616.
21. Dobson, J.V., *Electrical Journal*, Vol. 32, 1935, p. 527.
22. Lee, P.K. and Johnson, J.L., *Institution of Electrical and Electronic Engineers Transactions: Components, Hybrids and Manufacturing Technology*, Vol. 1, 1978, p. 40.
23. Savage, R.H., *New York Academy of Sciences. Annals*, Vol. 53, 1951, p. 862.
24. Aparin, V.I., et al., *Soviet Physics - Doklady*, Vol. 232, 1977, p. 65.
25. Jamison, W.E. and Cosgrove, S.L., *American Society of Lubrication Engineers Transactions*, Vol. 14, 1971, p. 62.
26. Blok, H., *Institution of Mechanical Engineers. Proceedings*, Vol. 2, 1937, p. 222.
27. Greenwood, J.A. and Williamson, J.B.P., *Journal of Physics D: Applied Physics*, Vol. 11, 1960, p. 389.
28. Jaeger, J.C., *Royal Society of New South Wales. Journal and Proceedings*, Vol. 56, 1942, p. 203.
29. Schwab, A., Dissertation, Technische Hockschule, July 13, 1957.
30. McNab, I.R., *Institution of Electrical and Electronic Engineers Transactions: Components, Hybrids and Manufacturing Technology*, Vol. 1, 1978, p. 30.
31. McNab, I.R., in Seminar on Energy Storage, Compression, and Switching, Canberra, Australia, November 1977, edited by E.K. Inall, ANU Press, Canberra, 1978.
32. Holm, E., *Institution of Electrical and Electronic Engineers Transactions: Power Apparatus and Systems*, Vol. 84, 1965, p. 404.
33. Huntington, H.B. and Grone, R.R., *Journal of Physics and Chemistry of Solids*, Vol. 20, 1961, p. 76.
34. Jansen, A.G.M., Mueller, F.M. and Wyder, P., *Science*, Vol. 199, 1978, p. 1037.

DISCUSSION

QUESTIONER: What estimates of temperatures have you made and by what methods have you estimated?

I. R. McNAB: We only made theoretical estimates of the temperature based on what we felt was the likely number of asperities or contact points present under the brush. Attempts have been made to measure temperatures not only by us but by other investigators also. I believe some have been successful and some have been not too successful. In general, it is felt that the local temperatures of the contact points should stay below what is called the softening temperature for those materials. If they go above that it may not be catastrophic, but one certainly would not like to go beyond the melting point of the materials. And, of course, that depends on the materials involved.

I. GOLDBLATT, Exxon Research: Did you observe any of these organic films breaking down and forming a laquer and thus affecting conductivity?

McNAB: I would say that we have them. On the other hand, we have not made a detailed study of those films. There is always some film that is present except for the conditions I mentioned in the paper for which very light films are formed. The influence of the hydrocarbon vapor is not noticeable up to this point in time. It is not a marked effect even if it is present.

GOLDBLATT: Have you been able to distinguish between the wear that occurs due to sliding and that which might be due to the making or breaking of the electrice contact?

McNAB: Well, we have done some work on that although not as much as we would have liked. There is no doubt that the passage of current does increase the wear rate (see references 30 & 31 of paper). I think that has to be the case because the local temperature of contact spots increases substantially with high current densities. The extent to which the wear is increased depends upon the particular circumstances. I do not have the numbers. I think for most operating conditions one sees increases in the wear rate between one and ten for increased current transfer.

J. J. MILLS, Martin Marietta Laboratories: What is the effect of water vapor on wear rate?

McNAB: I think with the water vapor we are only seeing the bottom part of the curve. If it is continued we would see that the wear rate would start to rise dramatically into an area known in brush terminology as dusting, when one gets into wear rates of inches per minute. I think with the hydrocarbon vapors we still have not gotten to that level. If we continue to put lower and lower amounts of the vapor, sooner or later we would reach the point at which the insufficient hydrocarbon vapor cannot maintain the lubricity of graphite.

The water vapor and the hydrocarbon vapor operate by saturating the bonds on the surface of the graphite crystallites, as we understand it at present. If that does not happen, suddenly the edge sites of the graphite crystals become very active and particles can stick to one another and this gives rise to a kind of abrasive situation. The polar molecules such as water adsorb on the edge sites. The hydrocarbon vapors seem to adsorb on the basal planes. But in the end they seem to have the same influence.

S. GANESH, Bendix Research: Have you observed the slip rings after tests for the nature of damage?

McNAB: We are trying to do that now. By looking at the slip ring after a typical period of operation, which is many thousands or many tens of thousands of revolutions, we see the damage, but we do not really understand all the mechanisms that are involved in causing that damage. We are trying to set up a system which could be used both in an Auger spectrometer and an electron microscope so that we can actually operate brushes on a slip ring and watch the film build up stage by stage. This is undoubtedly going to be a bit tedious because one has to avoid operation in vacuum conditions. That means alternately filling and purging, but we hope that in this way we can actually see how his film is laid down on these different surfaces. This is the most interesting and the least understood area. Although some of the work that Dr. Buckley has done on the films may be very relevant here, I do not think we have a good enough understanding of that yet.

GANESH: Did anybody study the graphite films on metal substrates?

McNAB: Lancaster, for example, has done quite a lot of work on films and with some interesting results. It turns out that wear rates under certain conditions seem to be the same with quite different graphites or carbons. And the implication is that it is not the shearing of the base material that is important as far as the friction is concerned. Probably it is the shearing that takes place within the interfacial layer that is formed which has characteristics that are significantly different from those of the base materials. At the same time one needs to have some graphitic material present to keep good operation.

VI. FRICTION

FRICTION—ESPECIALLY LOW FRICTION

E. Rabinowicz

ABSTRACT

Although there have been a number of recent reviews of the friction phenomenon, its causes, and the equations governing the magnitude of the friction coefficient, these have tended to emphasize situations in which high friction coefficients are obtained. However, practical interest is mainly confined to cases where the friction is moderate or low. It is shown that the lower limit to the friction of any sliding system is the quotient of the shear strength of the interfacial layer, at a hydrostatic pressure equal to the hardness of the softer contacting material, to the softer material hardness. An examination of measured values of this quotient suggest that friction coefficient values under 0.02 for boundary lubricated or solid film lubricated sliding systems should be quite rare, whereas in fact such values are regularly reported in the literature. In some cases, this discrepancy may arise from errors in measuring the friction coefficient; however, in other cases, lubricants giving very low friction deserve close investigation as their more general application may greatly reduce energy consumption during sliding.

INTRODUCTION

The phenomenon of friction is one which ancient man encountered very commonly, and it is no surprise that some of his earliest activities involved the reduction of friction when it was inconvenient or the use of friction when this might be beneficial. In the first category we might place such developments as the use of vegetable oils and animal fats as lubricants, as well as the use of rolling motion to take advantage of the low coefficient of rolling friction. In the second category belongs the use of twigs rubbed together to start a fire, and, of course, this use survives today in the striking of matches.

Much later, there came various explanations of friction and the long controversies between the roughness and adhesion theories, concluded in the decade 1940-1950 by the almost universal acceptance of the adhesion theory, although the roughness theory is still encountered, and often in unexpected places.[1,2] The distilled essence of the adhesion theory is the expression for the friction coefficient in the form

$$f = \tau/p \qquad (1)$$

where τ is the shear strength of the interface between the two sliding solids and p is the penetration hardness of the softer of them.

In practice, this expression has proved to be a mixed blessing. First, it predicts the wrong value for the friction coefficient (approximately 0.17 for unlubricated surfaces) whereas such surfaces generally give friction coefficients in the range 0.3 to 1.0. To make things worse, a more elaborate consideration of the combined stresses prevailing at a junction (involving Mohr's circle) indicates that the actual friction coefficient could be expected to be considerably less than 0.17,[3] whereas of course actual friction values are found to be greater than 0.17.

This then has led to what might be regarded as a 'second round' of friction theories. In these theories, it is hypothesized that when sliding is initiated, the first slip occurs when the friction coefficient is indeed low, but friction then increases as slip develops. One prominent theory is the surface film theory of Tabor,[4] which gives the friction coefficient f in the form

$$f = .33\ (k^{-2}-1)^{-1/2} \qquad (2)$$

k is the ratio of the shear strength of the interface to the shear strength of the softer material in contact.

This can give very high friction coefficients if k is close to 1.

Another theory along these lines is the Rabinowicz surface energy theory,[5] which gives the friction coefficient in the form

$$f = \frac{\tau}{p} \cdot (1 - 2\ W_{ab}\cot\theta/rp)^{-1} \qquad (3)$$

where W_{ab} is the surface energy of adhesion

θ is the roughness angle of the surface

r is the radius of the junctions.

Again, this can give very high friction values if 2 W_{ab} $\cot\theta$ is close to rp.

The various theories of the friction coefficient are discussed in two fine reviews, to which the reader is referred for further details.[6,7]

It should be noted that this theoretical activity has been mainly concerned with explaining why clean surfaces give such high friction and indeed it has been shown that high friction values can arise rather easily in such circumstances. Moreover, the models predict, and practical experience confirms, the fact that these friction values can change very drastically as the surface conditions are changed slightly. And yet, from a practical point of view this concern with the high friction of clean surfaces has been largely of theoretical interest, quite far removed from practical engineering significance. As I think of my involvement with friction problems arising in industry in the past twenty years, they break down somewhat as follows.

1. Very high friction (about 1.0) ~ 2%. (A manufacturer of toy racing

cars who needed a friction coefficient above 1.5 between tire and track, and achieved it by using tires made of a very soft rubber).

2. Very low friction (below 0.10) ~ 18%. These were generally deformation processing situations, in which a low friction allows optimum deformation to occur. Also, a manufacturer of battery powered equipment has expressed a concern for achieving very low friction. Third, occasionally I get a telephone inquiry from someone involved with winter sports.

3. Medium-high friction, around 0.25 (20%). These have mainly been traction problems associated with slippage involving people or automobiles.

4. Interest in constant friction values, irrespective of magnitude (20%). These were often situations involving brakes, clutches or the like, in which case variations in friction would give poorly controlled motion.

5. Interest in avoiding stick-slip oscillations (~40%). This is a persistent problem, and, of course, is related to the slope of the friction-velocity curve rather than any specific value of the friction coefficient itself.[8]

Note that there has been little interest in achieving low friction in application involving steady sustained sliding. It is precisely in these situations, of course, that friction dissipates such large amounts of energy (overall perhaps 5% of all the energy generated by mankind) and one might have anticipated enormous concern in this area. The reason there has been so little interest in low friction is doubtless associated with the fact that energy has until recently been cheap, and furthermore that devices sliding continuously are generally full fluid lubricated, in which case there is much greater concern with wear occurring during start-up, which from an economical point of view may cost an order of magnitude more than friction.

Recently, however, thanks to the energy crisis, interest in low friction has increased. Accordingly, it seems sensible to consider in this review

a) How we can achieve low friction in systems in which the sliding materials contact, i.e. those that operate in the regime of boundary lubrication.

b) What is the lowest friction that can be produced in such systems.

SLIDING OF BOUNDARY LUBRICATED SURFACES

There are a few substances like graphite, ice and teflon which give inherently lower friction, for reasons that are more or less adequately understood. But in most practical situations in which neither a full fluid film nor rolling contact can be used, low friction is achieved by the use of a boundary lubricant or of a solid lubricant.

The equation for the friction force arising from boundary lubricated solids is in the form

$$F = \alpha A s_m + (1 - \alpha) A s_\ell \qquad (4)$$

where α is the proportion of the surface where material-material contact exists, A is the real area of contact, and, of course, $(1 - \alpha)$ is the proportion of the surfaces for which the lubricant separates the surface.[9]

S_ℓ is the shear strength of the lubricant film, while s_m is the shear strength of the material-material interface.

Dividing this equation by the relationship for the normal load L in terms of the hardness and the area of contact, namely

$$L = Ap \tag{5}$$

we have the familiar two-term relationship for the friction coefficient

$$f = \alpha s_m/p + (1 - \alpha)s_\ell/p \tag{6}$$

We may use this relationship to discuss the experimental fact that a solid lubricant like a metal stearate gave a friction coefficient of about 0.05 for metals ranging from soft ones like lead ($p = 4\ kg/mm^2$) to hard steel ($p = 1000\ kg/mm^2$). Since s_ℓ is the shear strength of the metal stearate (a constant), the second term in eqn. 6 will vary greatly, and so will f, unless the first term is larger. Thus we are led to the conclusion that α has a constant value of about 0.1 for all metals, which, combined with a uniform s_m/p value of about 0.5, will produce a friction coefficient value of 0.05 for hard metals, and a somewhat greater value (about .07) for soft metals like lead.

Later work with radiotracers showed that, with metal stearates as lubricants, α values were likely to be less than .01, and thus the first term in eqn. 6 was likely to be negligible in comparison with the second. Thus, for well lubricated metals we have, putting $\alpha = 0$.

$$f = \frac{s_\ell}{p} \tag{7}$$

According to this relationship, a constant value of f with soft and hard materials could only be produced if s_ℓ was a function of p, specifically if s_ℓ was proportional to p.

Consideration of the results obtained during high pressure studies make this indeed a likely possibility. The argument, which dates back to Rabinowicz and Tabor[10] is that, if we make a junction by pressing together two asperities separated by a thick film of metal stearate, the film will tend to become squeezed out, and initially each junction will look rather like figure 1, in which the lubricant near the center of the junction is under a hydrostatic pressure equal to the hardness of the softer of the two contacting materials, and the peripheral regions are at lower pressure, down to zero at the outside of the junction. If for simplicity we assume that the central region is appreciably larger than the peripheral region, we have

$$f = (s_\ell)_p/p \tag{8}$$

where the notation $(s_\ell)_p$ indicates that the shear strength of the lubricant is to be measured under the hydrostatic pressure p.

Two points should be made about eqn. 8. First, the value of friction coefficient described there is a minimum. If there are sizeable numbers of material-material contacts f will increase, and if a sizeable amount of the lubricant is under a pressure less than p, f will also increase. Thus, eqn.

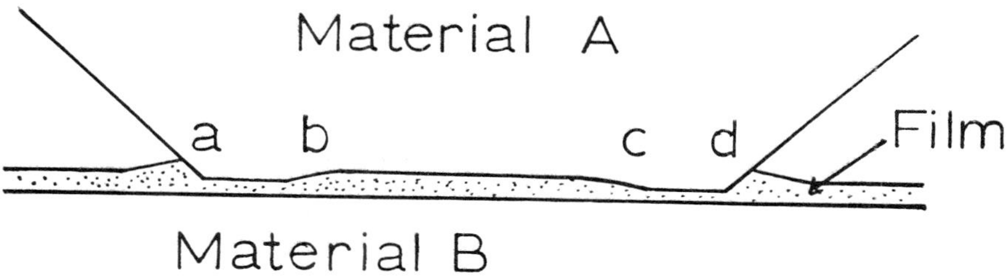

Fig. 1.—Schematic illustration of a junction involving a hard material B, a somewhat softer material A of hardness p, and a very soft solid film at the interface. In the region bc the film is under a hydrostatic pressure equal to p while in the region ab and cd the pressure is between p and zero.

8 represents a rock-bottom value for friction coefficient.

Second, it should be mentioned that the idea of postulating a shear strength of lubricant that varied with hardness of substrate never appealed to many eminent tribologists. Thus Holm[11] produced a special section of his monograph to argue against the idea, while Rowe[12] has produced a calculation in which only a numerical error hid the fact that the shear strength of the lubricant must vary as p increased.

If we assume that eqn. 8 is correct, then in the search for low friction values we are led to a search for low $(s_\ell)_p/p$ values.

MAGNITUDE OF THE QUANTITY $(s_\ell)_p/p$

Comprehensive measurements of the shear strength of various substances as a function of the hydrostatic pressure have been made by Bridgman,[13,14] by Boyd and Robertson[15] and values obtained by other investigators are tabulated by Godfrey.[16] My interpretation of this data is Figure 2, which shows what appears to be the typical relationship between s_ℓ and p. At low p values s_ℓ is independent of p (this indeed is the assumption contained in the two normally used yield criteria, the Mises and the Tresca in applied mechanics),[17] but at high pressures s_ℓ becomes proportional to p. However, this relationship is not obtained experimentally when s_ℓ is measured in the normal way, by confining a small amount of the lubricant between two hard anvils, applying a high normal pressure, and then measuring the shear force. In that case, at low pressures the shear occurs between the lubricant and one of the surfaces (the dotted line in Figure 2) and only at high pressures does shear within the lubricant layer occur. Thus, the experimental data follow the dotted line, at low p values, and the solid line at high p values.

For the solid line shown in Figure 2, Figure 3 gives the s_ℓ/p ratio as a function of p, starting with a value of p equal to 6 s_ℓ, which is the value that would be obtained when one of the two sliding surfaces is made of a material with the same hardness as the lubricant layer. In this case, there are high friction coefficient values when p is small, and a steady decrease as p is increased until eventually a constant s_ℓ/p ratio is reached.

This finding is indeed paralleled by friction coefficient measurements that have been made on effective boundary lubricants applied on a variety of

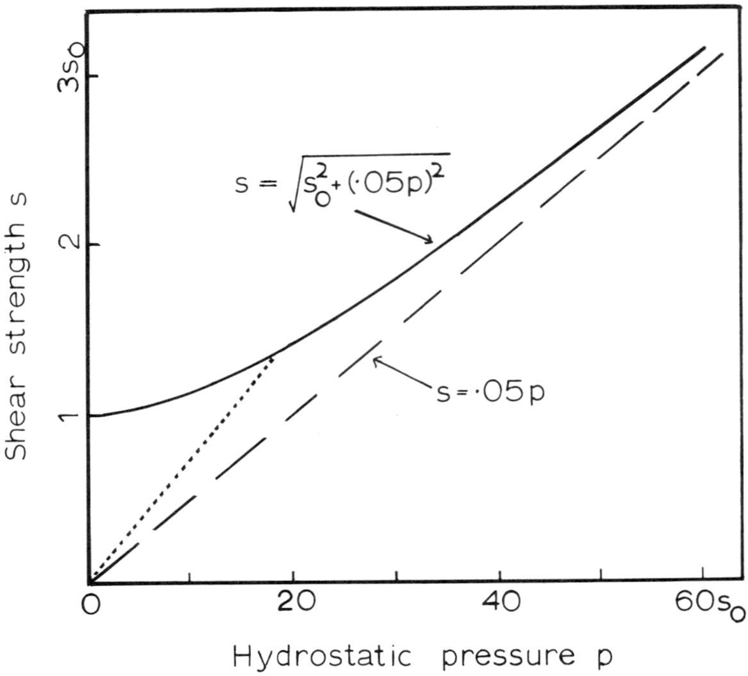

Fig. 2.—Solid line-schematic representation of the shear strength of organic substances, whose shear strength at zero pressure is s_o, as a function of the hydrostatic pressure p. Dotted line - s/p ratio obtained in an anvil shear test in which there is slippage when p is low. Dashed line - the relationship s - .05p, which the experimental curve approaches asymtotically.

metal substrates (Figure 4). Typically, when the substrate is soft, the friction coefficient tends to be high, but when the substrate is hard, the friction coefficient levels off to a nearly constant value.(18) This agreement between Figure 3 and Figure 4 suggests that we can use the high pressure work of Bridgman and Boyd and Robertson in looking for potential low-friction boundary lubricants.

Table I gives a listing of a few of the many s_ℓ/p values, derived from the work of Bridgman, and obtained at his highest pressure of .500 kg/mm^2. Table II is a similar compilation from the work of Boyd and Robertson, who worked at pressures to 280 kg/mm^2, supplemented by three data points for polymers, obtained by Towle.(19) It will be seen that these data provide a fairly consistent picture, namely that the s_ℓ/p values for organic substances are all in the range 0.05 to 0.10, while among inorganic materials, only a few, mainly the soft metals, can go lower.

Perhaps some additional comments on the data shown in Tables I and II are in order. The few substances that were tested in both series of investigations, namely boron, graphite, lead iodide, and silver sulphate, did not

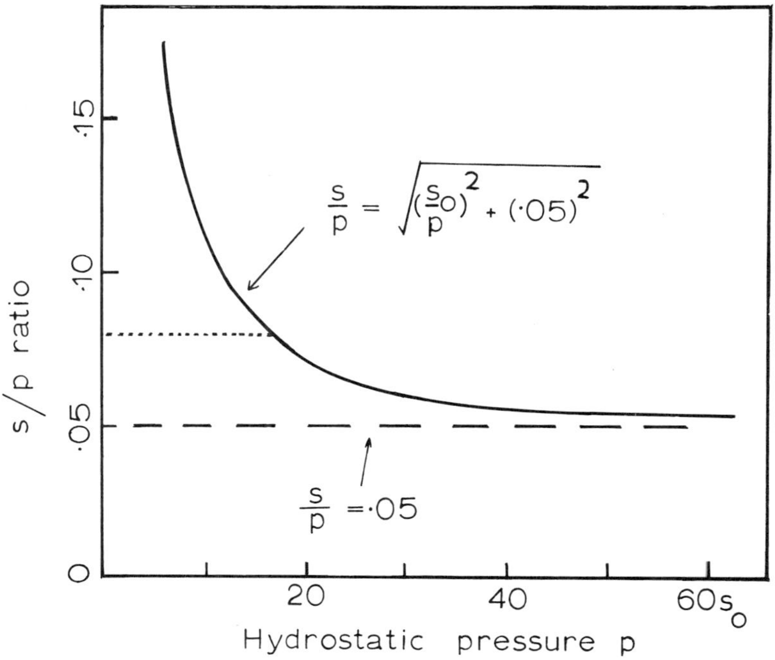

Fig. 3.—Plot of the s/p ratio derived from Figure 2. As p is increased s/p drops to a value of .05.

give very similar results, perhaps because the pressures used were not the same, and some phase transformations may have occurred in the Bridgman tests which were conducted at higher pressures. Then, the measured shear strengths were to some extent functions of the rate of shear and the thickness of the sheared layer. However, it is significant that the two investigations rated the four substances mentioned above in almost the same order, and they agree in finding that organic substances generally give (s_ℓ/p ratios between .05 and .10, with extreme values of .029 and .119 except for one isolated data point of .020 obtained by Towle.

The above evaluation of the limiting values of the s_ℓ/p ratio achievable with organic lubricants is very similar to that of Briscoe, Scruton and Willis,[20] although there is some disagreement in the assumed slope of the s_ℓ/p function at low values of p.

APPLICATION TO ACTUAL FRICTION VALUES

It should be realized that, while the s_ℓ/p values listed in Tables I and II are convertible to friction coefficient values by the use of eqn. 8, they represent minimum attainable friction coefficients. Only if the substrate is sufficiently hard, and if there is no material-material interaction, will these values be achieved. Otherwise, the friction coefficients will be

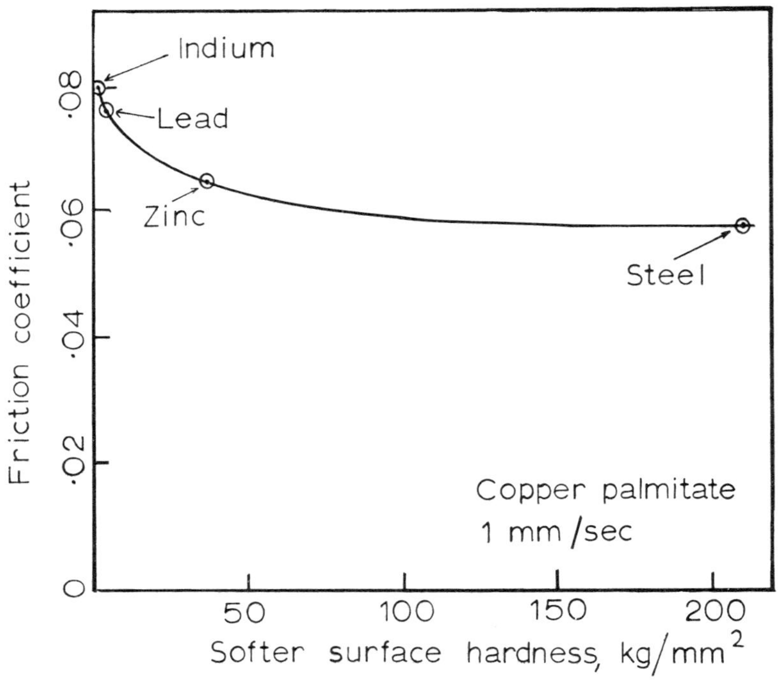

Fig. 4.—Experimental friction values obtained with a copper palmitate solid lubricant on metals of various hardness. The shape of the curve resembles Figure 3.

greater.

In fact, there is reasonably good agreement between these friction coefficient values and those reported by most workers in the field. For example, Bowden and Tabor[21] give comprehensive tables of friction coefficients covering a wide range of sliding conditions. They list 229 friction values in all, and Figure 5 is a histogram of these values, which, as it happens, are a reasonably close approximation to the friction values which I have obtained over the years for a wide variety of sliding systems. As will be seen in Figure 5, there are five friction coefficient values of below 0.04, and all of these refer to systems involving ice. This anomalous behavior of ice can readily be explained. Since water expands when it freezes, ice has an s_ℓ value which diminishes as p increases, and consequently the s_ℓ/p ratio does not approach a limit, but continues to decrease.

The fact that Bowden and Tabor cite no friction coefficient value under 0.04 for clean and boundary lubricated surfaces is paralleled by my own experience. I don't believe I have ever measured a friction coefficient value under 0.03. And yet, it must be stated that lower friction values are regularly encountered in the literature. For example, at a recent international conference on Wear of Materials,[22] three papers reported friction coefficient values of less than 0.02. At an international conference on Solid

TABLE I.—DATA OBTAINED BY BRIDGMAN[13,14]

A. Organic substances

Name	s_ℓ/p at highest p
Acetanilide	.056
Alizarin	.051
Amidol	.078
Aniline hydrochloride	.062
Anthraquinone	.066
Dinitrotoluene	.058
Guanidine	.050
Picric acid	.050
Tartaric acid	.100

B. Inorganic substances

Name	s_ℓ/p at highest p
Aluminum	.068
Boric acid	.112
Boron	.360
Cadmium	.038
Cadmium chloride	.062
Cadmium iodide	.014
Graphite	.198
Indium	.015
Lead	.014
Lead iodide	.059
Lead monoxide	.048
Lithium fluoride	.216
Magnesium	.174
Silver sulphate	.011
Sodium	.104
Tin	.015
Zinc	.037

Lubrication[23] some years ago, four papers were presented which cited friction coefficients of below 0.02. Beerbower,[24] in a search for anomalous effects in boundary lubrication, cites two other cases of friction coefficients of 0.02 or below.

I can think of several reasons why such low values might be obtained

a) The experiment might involve a thick lubricant layer, which would allow hydrodynamic lubrication, elastohydrodynamic lubrication, 'weeping lubrication' or the like.

b) The experimenter might be reporting minimum friction values achieved during an experiment, rather than average friction values, and have confused "noise" of his friction measuring apparatus with fluctuation in the friction coefficient.

c) The apparatus might involve a sliding contact besides that at the interface being tested, so that the actual reported friction value represents a difference or a sum involving a measured value and an applied

TABLE II.—DATA OBTAINED BY BOYD AND ROBERTSON[15] AND FOWLE[19]

A. Organic substances	s_ℓ/p at highest p
Capric acid	.109
Castor oil	.081
Grease (calcium base)	.082
Oleic acid	.119
Palm oil	.075
Sperm oil	.085
Stearic acid	.029
Turbine oil	.108
B. Inorganic substances	
Boron	.710
Graphite	.058
Lead iodide	.071
Mica	.305
Molybdenum disulfide	.033
Silver sulphate	.054
Soapstone	.306
Tungsten disulfide	.037
C. Polymers	
Polyethylene	.020
Polyvinylidene fluoride	.113
Polytetrafluoroethylene	.051

correction. If the correction value is wrong, the reported friction value may be too low (Figure 6).

d) The load in the experiment might be applied through some linkage having friction, so that at light loads uncertainty in the actual load present led to uncertainty in the friction (Figure 7).

e) There might possibly be unusual effects involving non-isotropic substances, such as layer-lattice materials, although I have seen few signs of them.

DISCUSSION

As I have mentioned above, at least one investigator of unusual phenomena associated with friction tests has mentioned low friction coefficients, but low friction values as such appear never to have been regarded as a separate category involving anomalous friction behavior. However, I think that low friction coefficients are unusual and anomalous. Equation 8, combined with Tables I and II, suggests that only in systems involving very soft metal lubricants and very hard substrates are low friction values likely to be encountered, and in fact there have been few reports of low friction observed with such sliding systems.

If it turns out that friction coefficient values below these derived

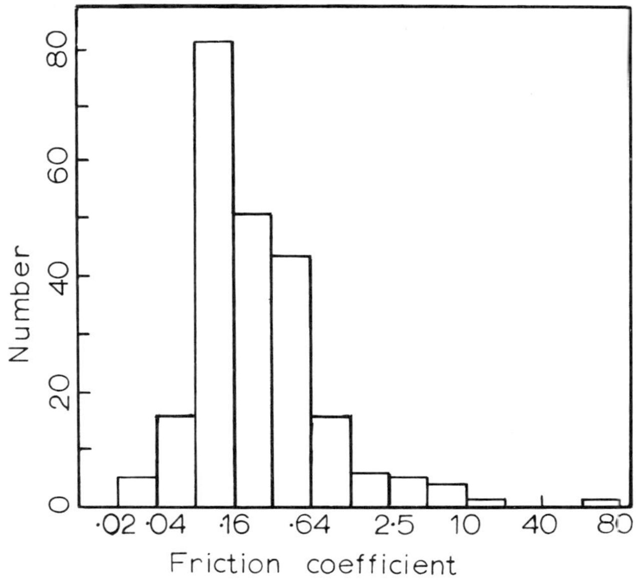

Fig. 5.—Histogram of the 230 friction coefficient values tabulated by Bowden and Tabor.[21] *Five values were under 0.04, and all had ice as one of the contacting materials.*

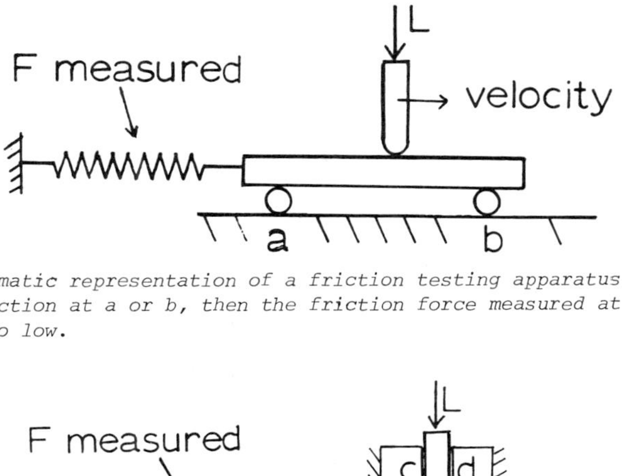

Fig. 6.—Schematic representation of a friction testing apparatus. If there is any friction at a or b, then the friction force measured at the spring will be too low.

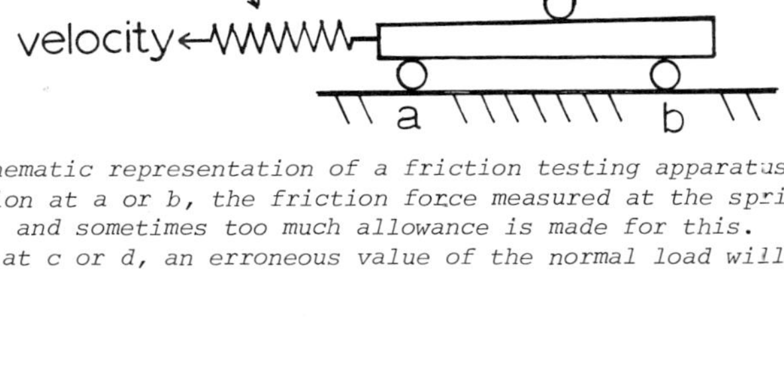

Fig. 7.—Schematic representation of a friction testing apparatus. If there is friction at a or b, the friction force measured at the spring will be too high, and sometimes too much allowance is made for this. If there is friction at c or d, an erroneous value of the normal load will be obtained.

from eqn. 8 can be obtained in boundary lubricated or solid film lubricated sliding systems, then this important information should be widely documented, not only for its great theoretical interest but also for its possible practical importance. However, I should reiterate the fact that I personally have never observed friction coefficient values of below 0.03 (except perhaps when slipping on a patch of ice) and am skeptical about their prevalence.

Perhaps a comment is in order about the curious fact that limiting s_ℓ/p ratios, certainly for organic materials, seem to cover such a narrow range, mostly between 0.05 and 0.10. Cameron[25] has tried to explain this by an atomic model involving attractive and repulsive forces. Perhaps this should be followed up.

ACKNOWLEDGEMENTS

I wish to thank the U. S. Army Research Office for sponsorship of this work under contract DAAG 29-76-C-0058, and Dr. Nobuo Ohmae for helpful discussions.

REFERENCES

1. French, A.P., "Newtonian Mechanics," W.W. Norton and Co., New York, 1971, p. 188.
2. Harris, W.H. and Levey, J.S., eds., "New Columbia Encyclopedia," Columbia University, New York, 1975, p. 1016.
3. McFarlane, J.S. and Tabor, D., *Royal Society of London. Proceedings.* Series A, Vol. 202, 1950, p. 244.
4. Tabor, D., *Surface and Colloid Science,* Vol. 5, 1972, p. 245.
5. Rabinowicz, E., "The Friction and Wear of Materials," John Wiley and Sons, New York, 1965, Section 4.4.
6. Merchant, M.E., in "Interdisciplinary Approach to Friction and Wear," NASA SP-181, 1968, p. 181.
7. Tabor, D., *Surface Physics of Materials,* Vol. 2, 1975, p. 475.
8. Rabinowicz, E., "The Friction and Wear of Materials," John Wiley and Sons, New York, 1965, Section 4.16.
9. Bowden, F.P., Gregory, J.N. and Tabor, D., *Nature,* Vol. 156, 1945, p. 97.
10. Rabinowicz, E. and Tabor, D., *Royal Society of London. Proceedings.* Series A, Vol. 208, 1951, p. 455.
11. Holm, R., "Electric Contacts Handbook," 3rd Edition, Springer, Berlin, 1958, p. 216.
12. Rowe, G.W., in "Principles of Lubrication," edited by A. Cameron, John Wiley and Sons, New York, 1966, p. 451.
13. Bridgman, P.W., *Proceedings of the American Academy of Arts and Sciences,* Vol. 71, 1937, p. 387.
14. Bridgman, P.W., *Proceedings of the American Academy of Arts and Sciences,* Vol. 72, 1938, p. 227.
15. Boyd, J. and Robertson, B.P., *American Society of Mechanical Engineers Transactions,* Vol. 67, 1945, p. 51.
16. Godfrey, D., in "Interdisciplinary Approach to Friction and Wear," NASA SP-181, 1968, p. 335.

17. Crandall, S.H., Dahl, N.C. and Lardner, T.J., "An Introduction to the Mechanics of Solids," 2nd Edition, McGraw-Hill, New York, 1972, p. 317.
18. Rabinowicz, E., in "Friction and Wear," edited by R. Davies, Elsevier, Amsterdam, 1959, p. 149.
19. Towle, L.C., in ASLE Proceedings - International Conference on Solid Lubrication, Denver, American Society of Lubrication Engineers, Park Ridge, Ill., 1971, p. 202.
20. Briscoe, B.C., Scruton, B. and Willis, F.R., *Royal Society of London. Proceedings*. Series A, Vol. 333, 1972, p. 99.
21. Bowden, F.P. and Tabor, D., "Friction and Lubrication of Solids," Clarendon Press, Oxford, 1950, p. 322.
22. "Wear of Materials - 1977," American Society of Mechanical Engineers, New York, 1977.
23. "ASLE Proceedings - International Conference on Solid Lubrication," American Society of Lubrication Engineers, Park Ridge, Illinois, 1971.
24. Beerbower, A., "Boundary Lubrication," Report AD 747336 to the Army Research Office, Washington, D.C., 1972.
25. Cameron, A., "A Theory of Boundary Lubrication," *American Society of Lubrication Engineers Transactions,* Vol. 2, 1960, p. 195.

DISCUSSION

QUESTIONER: Would you expect lower friction coefficient with pin-on-disc apparatus or with flat-on-flat? Would you also comment on the interfacial slip.

E. RABINOWICZ: The difference between having two large flat surfaces with a thin film in between and a pin on the disc is as follows. In the case of two flat surfaces the end region where the pressure drops from the hydrostatic pressure to atmospheric pressure is a small part of the total area. In the case of a hemisphere on the flat, the transition region tends to be a larger proportion of the whole region. When we work out the consequence of that in terms of expected friction coefficient, it would be higher for the pin-on-disc than for the two flat surfaces. At the transition region you get higher effective S/P ratio and therefore, in practice they are either the same or we would expect higher friction for the pin-on-disc.

Now as to the question of interfacial slip, it is a fact that in the early stages of these high pressure tests when slip takes place at the interface, the friction coefficient is generally higher; it is about 0.1 as opposed to 0.05. That is why we get the characteristic knee in the curve which all the investigators comment on having observed. I suppose that it turns out to be a lower net force in the initial stage, and I assume that this is no longer true when the shear shifts to the bulk film. In other words, the apparatus is always trying to do as little work as possible, which means there is interfacial slip initially and then slip within the film later on.

SAME QUESTIONER: How does anisotropy affect the friction coefficient?

RABINOWICZ: Structures of materials like graphite and cadmium iodide which Bridgman studied are, of course, different from normal materials. But the friction properties were either the same or similar, especially for graphite. In fact, if Bridgman had an anomalous effect, it was observed in some of the soft metals like indium. The possibility that the anisotropic material would behave differently is just a possibility and is by no means established. There is no experimental data to suggest that they will give friction coefficients below 0.02.

SAME QUESTIONER: Would you please comment on the women's heel problem? Is the friction coefficient of nylon higher or lower than rubber?

RABINOWICZ: Lower. The friction of nylon against typical ground materials is about 0.3, for typical elastomer it is about 0.6, and that provides far more margin of safety during walking when you hit, say, a grease layer because the minimum amount you need for walking actually is about 0.20. You do not have much safety margin with nylon.

SAME QUESTIONER: What happens with water film?

RABINOWICZ: Well, if you have a water film on the surface and you are moving fast, then you can very easily get an effective friction coefficient under 0.20.

R. DASKIVICH, G. M. Research Laboratories: With respect to the s/p ratio, do you know the values of the shear strengths of films such as sulphides which might be formed on metal substrates?

RABINOWICZ: I don't offhand. I think Bridgman tested a few sulfides. He did a whole series of tests on minerals and those sulfides that happen to exist as minerals. But I am not so sure that they would behave the same way as a sulfide layer formed on a metal surface which tends to include a lot of oxides as well and have quite different composition. I do not know of any measurements that have been made of that kind of film.

QUESTIONER: You have been talking about repetitive sliding on the same wear track for measuring friction. Is that indeed the best way of measuring it? Once you deform the track, any subsequent measurement of friction coefficient would be in error by the amount of deformation the surface has seen. It is not indeed responsible for the scatter that you get? By using identically similar surfaces would it not be possible to minimize the scatter in the data?

RABINOWICZ: I suppose it is the faith of the scientist that identical measurements made on identical surface will give identical results. Since friction values are not identical, obviously something is not identical. There is, of course, an indication that repeated sliding has some effect on friction. I doubt that that accounts for the major discrepancy in the curve. We had occasion to conduct an elaborate study of surfaces lubricated by water for high density nuclear storage pools, and I have seen very little effect between the initial measurements and after some sliding. Clearly it is an effect, but I don't think it is a major effect.

FRICTION WITHOUT WEAR

A. Beerbower

ABSTRACT

The status of non-adhesion theories of friction is briefly examined, leading to the conclusion that the only promising one is the direct stimulation of surface molecules into thermal vibration. A means for implementing this model, using the surface free energy and compressibility of the solid film lubricant, is developed and illustrated for stearic acid. The results compare favorably with experimental data from several sources, and also with the quadratic relation of friction to lubricant shear strength and bearing metal hardness postulated by Rabinowicz.

NOMENCLATURE

β_S - Isentropic secant compressibility (Pa^{-1})

β_T - Isothermal secant compressibility (Pa^{-1})

δ - Solubility parameter $(\text{Joules/meter}^3)^{0.5}$

E_1, E_2 - Energy of attraction of two molecules (Joules)

F - Coefficient of friction

γ - Surface free energy (Joules/meter^2)

γ_S - Surface free energy of solid (Joules/meter^2)

ΔH_V - Heat of vaporization (Joules/mole)

N_a - Avogadro's number, 6.024×10^{23} (molecules/mole)

P - Pressure; hardness of softer metal in pair (Pa)

R - Gas constant (Joules/mol°K)

S - Shearing stress (Pa)

S_o - Shearing stress at ambient pressure (Pa)

V - Molar volume (cubic meters/mole)

X - Distance between molecules (meters)

INTRODUCTION

The title of this paper is lifted from the subtitle of Eudier's[1] study on brake and clutch materials. The point is to emphasize that while all wear causes friction, the converse "all friction is caused by wear" is definitely not true. The present purpose is at the opposite end of the friction spectrum from Eudier's, but if his conclusions are sound, they should apply to that minimum part of friction which remains when all kinds of wear are eliminated--and help us see how large it is.

It is important to bear in mind that friction is a thermodynamically irreversible process, converting work (free energy) into heat at essentially ambient temperature. This is the principle used by Count Rumford to demonstrate the mechanical equivalence of heat, which led to the first law of thermodynamics. In more mechanistic terms, this process involves conversion of unidirectional motion (kinetic energy) into random molecular motion. This can arise from:

- Viscous drag in a fluid
- Vibration due to low frequency stick-slip
- Vibration due to asperity interactions
- Molecular vibration due to shear-plane sliding

(Two other tribological processes which result in stored free energy are, in that sense, trivial. These are the increased surface free energy due to the new surface of wear particles, and energy stored in elastic surface strains.)

THE CONVENTIONAL PROCESSES

In the present context, we may exclude fluid friction, though we must be cautious not to accept "mixed film" lubrication as "boundary." In the former, the troughs full of liquid between the asperities contribute enough bydrodynamic lift to partially de-load the true boundary contacts.

Rabinowicz[2] has studied stick-slip as a part of the friction process, and concluded that it can be eliminated by stiffening the system, without doing much to reduce the energy conversion. Kilburn[3] made elaborate experiments on the frequency spectrum of dry sliding, and found that the vast majority of friction was in the steady, unidirectional mode. Of course, his equipment would not detect the *molecular* vibration at the 50 GHz level. Rabinowicz[4] disparages the "roughness" theory, and rightly so. A case was made for elastic asperity interaction as the basic cause for fatigue wear and delamination by Beerbower,[5,6] but such interactions are reversible unless suddenly released, to twang like tuning forks. It is hard to get much twang out of asperities with only an 8° slope.

It appears that gross vibration is not responsible for much friction.

Indeed, a well-known but little used effect is that vibration can reduce friction. We all have seen how a vibrating motor causes tools to walk off a workbench. The only use made of this effect was in early jet aircraft, where the cockpit meters tended to stick until a vibrator was installed to simulate a reciprocating engine.

Eudier[1] used a composite material to deliberately induce vibration as an energy sink, which is not very relevant to ordinary tribology.

VIBRATION DUE TO SHEAR-PLANE SLIDING

Since neither the low nor intermediate frequency range serves to explain the energetics of friction, we reach the fundamental molecular frequency. This is the "relaxation oscillation," common to all molecules regardless of phase. From this viewpoint, the gas, liquid and solid phases differ only in the number of degrees of freedom with which they can exhibit thermal perturbation. Gases have six or more degrees; they can translate and rotate on three axes, plus internal oscillations between their various atoms. Liquids are less free, and solids have lost all except the oscillatory modes.

The direct stimulation of molecules into thermal motion by unidirectional motion is taken so much for granted in fluid mechanics that it seems surprising so little attention has been given to its counterparts in solid friction. Even a valid macroscopic friction theory would eventually have to result in this conversion, because matter has no other way to contain thermal energy. Its only other recourse is to emit infrared, and in most tribological situations the difference from ambient temperature is so small that radiation output and input are equal.

Rabinowicz[4] suggested further study of the model by Cameron,[7] but it did not stand up very well to criticism by van Battum.[8] In brief, Cameron's approach was to quantify the variation of London-van der Waals energy of attraction between molecules on the two sides of the shear plane. E_1 was the energy at closest approach of opposing molecules and E_2 that when the moving molecule was half-way between two fixed ones. This difference, divided by the average distance (X) between these two points, gave the tangential force. Since E_2 was considered to be very small with respect to E_1, the shear stress was given by

$$S = E_1/X \tag{1}$$

remembering that E_1 and hence S are functions of the load pressure P.

There were three criticisms by van Battum.[8] The first took exception to Cameron's method of numerical integration to obtain E_1. This was admittedly cumbersome and inaccurate, and could be done much better today with a modern computer. In fact, Macmillan[9] has recently computed the materials constants and shear plane orientation for solid argon and NaCl, with admirable accuracy as evaluated by Tyson.[10]

The second criticism was more serious, but more difficult to assess as it concerned the question of whether E_2 was really negligible. I do not completely understand the argument, nor is it fully resolved by Macmillan, who has not yet been able to assign a meaningful yield stress in shear in either of his cases.

The third objection was a perfectly logical one, which Cameron failed to answer. It concerned his practice of using a constant coefficient of compressibility. This is in direct conflict with the Lennard-Jones model, in which the intermolecular repulsion increases inversely with the 12th power of the intermolecular spacing. It also conflicts with numerous experiments which show the coefficient decreases rapidly with increasing pressure.

Though he made no direct us of it, Cameron offered an alternative to the integration problem. This was to replace the controversial $E_1 - E_2$ by the "cohesion energy," more properly the work of cohesion, which is equal to twice the surface free energy. It is not an easy matter to explain, plausibly and concisely, why it is appropriate to use a *free* energy to compute a heat. It hinges on the generally accepted fact that tensile failure is a *reversible* process, indeed, the thermodynamic definition of "surface tension" is based on tensile separation to infinity of a column of unit cross section. In vacuum, the column would reheal on re-approach with recovery of all the work. In shear, however, the surfaces move in a tangential direction, and in rotary shear are not separated at all. Though the input work is the same, there is no way to recover any work. Remembering that each molecule is suspended in a double network of repulsion and attraction "springs," it is necessary to visualize each molecule being agitated by the alternating passage of zones of attraction and repulsion on the other side of the slippage plane. This results in vibrational energy quanta called "phonons"(10) which propagate into the substrate and environment as heat.

METHODS OF COMPUTATION

The first step is to obtain the surface free energy (γ_S) of stearic acid. Extrapolation of the values in Quayle's compilation(11) gives 0.027J/m^2 (27 ergs/cm^2) at 20°C for the liquid (γ). Solidification involves a reduction in volume and a consequent increase in van der Waals forces; this averages 14% increase over γ for the liquid according to work reported by Beerbower.(6) (It has since been learned that this factor does not apply to ionic solids, which have a larger contraction on freezing.) The resulting value is 0.030 J/m^2 for γ_S.

The next step is assignment of an average sliding distance (X) by which this energy must be divided to obtain the shearing stress. Cameron used the square root of the area of the chain, but then illogically divided it by 2 to obtain "the distance between the positions of maximum and minimum attraction energy." He should have used the distance for a complete cycle. The X-ray data he used for the area was from 1945, and more recent sources give other values. It is more general to compute this from the molecular weight and density. Beerbower(5,6) used a spherical model, so that

$$X = (6V/\pi N_a)^{1/3} \qquad (2)$$

where V is molar volume (m^3/mole), and N_a is Avagadro's number (6.024 x 10^{23}). Molar volume is molecular weight divided by density. For stearic acid, this is 0.2845/9.37 x 10^2, or 3.03 x 10^{-4}m^3/mol, from which X = 9.9 x 10^{-10}m.

From these, the shear strength at ambient pressure is 2 x 0.030/9.9 x 10^{-10} or 6.1 x 10^7 Pa. This falls within the scatter of results by various methods. Briscoe(12) showed about 4 x 10^6 Pa as the result of combining his data with that from other sources. However, some unpublished experiments using the ASTM D217 penetrometer as a test instrument indicated a yield stress of around 7.5×10^7 Pa in bulk.

The third problem is the compression under high normal loads. No data could be found on stearic acid, but Wright[13] published extensive charts for the bulk modulus of all hydrocarbons with densities between 0.7 and 1.0×10^3 kg/m^3 at working temperature (-18 to 260°C), for pressures of 0 to 7×10^8 Pa gage. Since stearic acid is 84% hydrocarbon by weight (even more by volume) and it is that portion which is involved in shear, it does not seem too far-fetched to apply Wright's correlation to it.

Cameron used the *linear* compressibility, apparently feeling that a rigidly confined column was involved. This seems quite unlikely, so the *volume* compressibility (reciprocal of Wright's bulk modulus) was used. Since tribology is a rapid process and stearic acid a poor heat conductor, it seemed necessary to convert the isothermal modulus ($1/\beta_T$) to the isentropic ($1/\beta_S$). This involves multiplying by the ratio of specific heat at constant pressure to that at constant volume. Based on many similar organic materials, the ratio 1.30 was used. The Wright charts produce the "secant" modulus; that is, the average from ambient to maximum pressure, exactly what is needed.

Use of these compressibilities to adjust surface free energies involves an equation discussed in detail by Beerbower[6]

$$\gamma = 0.0715\ V^{1/3}\delta^2 \tag{3}$$

where δ is the Hildebrand[14] solubility parameter

$$\delta = \sqrt{(\Delta H_V - RT)/V} \tag{4}$$

where ΔH is the heat of vaporization at temperature T, and R is the gas constant. The effect of volume change is

$$d(\ln\delta) = -1.25\ dV/V \tag{5}$$

We are interested in dS/dP. Combining Equations (1), (2) and (3), cancelling $V^{1/3}$, taking the natural log and differentiating:

$$d(\ln S) \propto 2d(\ln\delta) = -2.50\ dV/V \tag{6}$$

Integrating this over the intervals S to S + ΔS and P to P + ΔP,

$$\Delta S/S = -2.50\ \beta_S P \tag{7}$$

where β_S is the isentropic secant compressibility ($\Delta V/V\Delta P$). With these tools, it is easy to put together an expression for the shear stress

$$S = 2.28\gamma(1 + 2.5\ \beta_S P)/X \tag{8}$$

Dividing through by pressure puts this into the "two part friction" form

$$F = 2.28\gamma/XP + 5.70\beta_S\gamma/X \tag{9}$$

In this and subsequent expressions, P should be understood to represent both the load pressure and the penetration hardness of the softer bearing metal,[4] since the latter limits the former in tribological contacts. Up to 3.5×10^8 Pa, β_T is read directly from Wright's chart, and 7×10^8 Pa can be safely extrapolated as the chart is linear in that respect. Pressures above that require non-linear extrapolation.

It should be kept in mind that not all the above parameters enter the calculation with the same effect on the precision and accuracy. Thus, uncertainty in the assumed ratio $\beta_T/\beta_S = 1.30$, or in β_T, is worse than in V, but less serious than in γ or the ratio $\gamma_S/\gamma = 1.14$. The combined uncertainties seem to add up to no more than $\pm$ 20%. Any problems beyond that are presumably at the conceptual level.

COMPARISON WITH EXPERIMENTAL DATA

Briscoe's[12] experimental study and data compilation is a standard by which theories may be tested. His data for stearic acid are shown in Figure 1. His Figure 5, upper line, representing "monolayer retracted from the melt," was plotted. The lower line, "monolayer deposited from solution" is considered in a later section.

The Rabinowicz[4] Figure 4 data on copper palmitate are plotted for comparison. This curve will be discussed later.

MacGregor[15] edited into a handbook form most of the data obtained at the IBM Corporation by R. G. Bayer and T. C. Ku. These included coefficient of friction under boundary conditions for 418 combinations of platten alloy, ball alloy and lubricant. Of these, 60 were run on a highly refined oil containing 0.2% stearic acid. All experience indicates that this blend will lay down a film containing at least 40% stearic acid.[6] The data are displayed in Figure 1, plotting the penetration hardness of the softer alloy as P. In all the 26 points with the hard 52100 steel, the platten was softer. The brass ball was softer for all 10 of those points, but the 320SS ball was harder for 15 points and softer for 9. The beryllium copper platten acted in a quite anomalous way, with $F = 0.35$ against 52100 and 0.32 against 302SS, so those two points were not plotted. The results are somewhat scattered, but confirm in a general way Rabinowicz'[4] prediction that the hardness of the softer material does indeed affect the coefficient of friction.

The predictions from Equation (9) are plotted in Figure 1. The shape of the curve appears quite promising, but all values are well above the experiments. It is evident that both numerical coefficients need to be decreased. Probably those derived above relate to static rather than kinetic friction.

FITTING AN EMPIRICAL QUADRATIC EQUATION

Rabinowicz[4] proposes an empirical equation, which can be expressed as

$$F^2 = (S_o/P)^2 + F_\infty^2 \qquad (10)$$

where S_o is the yield stress of the lubricant at ambient pressure, and F_∞ is the ratio of shear stress to pressure at very high pressures (approaching infinity). He generalized $F_\infty = 0.05$, but for specific cases it can be determined by regression analysis.

Simple two-point fitting of the points on Figure 1 leads to some interesting conclusions:

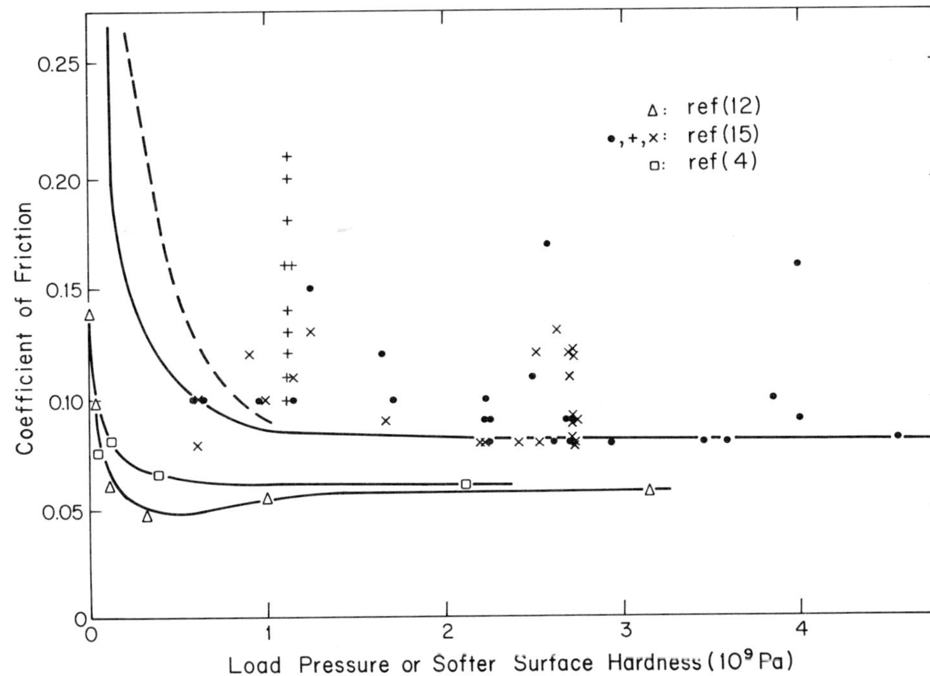

Fig. 1.—Comparison of calculated coefficients of friction with experimental results. Points show the data of Briscoe[12] and MacGregor[15] on stearic acid, and of Rabinowicz[4] on copper palmitate. These data are fitted by Equation (10) (solid lines) but not by Equation (9) (broken line).

- The Briscoe[12] data are fitted rather well by

$$F^2 = (1.62 \times 10^6/P)^2 + (0.050)^2 \qquad (11)$$

- The Rabinowicz[4] Figure 4 data cannot be fitted by this type of equation. There is reason to suspect that the P value of indium is wrong. If the γ_S value indium of 0.63 J/m^2 from Tyson's[16] compilation is plotted in Rabinowicz'[2] Figure 2.14, the hardness is predicted to be about 5 x 10^7 Pa (5 kg/m^2). When this value is used instead of the $\sim$ 1.7 x 10^7 Pa he plotted,

$$F^2 = (2.2 \times 10^6/P)^2 + (0.06)^2 \qquad (12)$$

gives an acceptable fit.

- The MacGregor[15] data are too scattered to provide much confidence, so no regression analysis was made. A visual fit to the lowest points in Figure 1 gives

$$F^2 = (3.5 \times 10^7/P)^2 + (0.08)^2 \qquad (13)$$

- The predictions of Equation (9) were surprisingly well fitted by

$$F^2 = (6.1 \times 10^7/P)^2 + (0.065)^2 \qquad (14)$$

The $S_o = 6.1 \times 10^7$ value was calculated from γ/X. It fits the predictions so closely that plotting in Figure 1 would show only one line.

EFFECT OF SURFACE FREE ENERGY

Briscoe[12] also shows 5 other solid lubricants, and also calcium and copper stearates. The latter resemble stearic acid too closely to be of interest. The data are all plotted in his Figure 7. Since compressibilities were not available on most of these lubricants, the frictions at the lowest load used on all of them (3.16 x 10^7 Pa) were scaled from his graph. The following sources were used for γ:

Anthracene: Quayle[11] does not list γ values for this, so his data for phenanthrene were extrapolated to 25°C.

Polythene: Values for γ_S and V were taken from Wu.[17] No distinction could be made between molecular mass 30,000 and 10,000 in parameters, but the higher would tend to be less liquid-like. The γ_S value was divided by 1.14 to put γ on a common basis.

Polytetrafluoroethene (PTFE): Data on γ_S and V from Wu, handled as for polythene.

Sebacic Acid: Estimated γ by methods outlined by Beerbower[6] must be considered quite rough as there was no previous experience with dibasic acids.

The results are shown in Table I and they are not at all reassuring. If the concept borrowed from Cameron were true, FX/γ would be a constant. However, it appears to be a function of (γ/X). As shown by Equation (3), this group is proportional to δ^2, the "cohesive energy density."

*TABLE I.—COMPARISON OF SURFACE FREE ENERGY WITH FRICTION**

Lubricant	Figure No.*	$10^{10}X$(m)	γ (J/m^2)	F	$10^{10}FX/\gamma$
Stearic Acid	5 Upper	9.9	0.027	0.10	36.7
Stearic Acid	5 Lower	9.9	0.027	0.07	25.7
Anthracene	7	7.6	0.041	0.44	81.5
Polythene 30000	7	4.6	0.031	0.35	51.9
Polythene 10000	7	4.6	0.031	0.18	26.7
Sebacic Acid	7	8.0	0.044	0.23	42.0
PTFE	7	5.3	0.017	0.05	15.6

* Briscoe[12] at 3.16 x 10^7 Pa and 20°C.

The information in Table I was log-log plotted two ways--F versus (γ/X), and XF versus γ. Both showed a similar trend but there were not enough points to be decisive. They were supplemented by the 22 metals for which Rabinowicz[2] gave S_0 values and Tyson[10] listed experimental γ_S values. These were at or near the melting point, so they were adjusted to 25°C, adding 4 more from recent literature. These were plotted as above, S_0 versus (γ_S/X) and S_0X versus γ_S. Both showed about the same moderate scatter and identical slopes of 2.5. A plot of S_0 versus Tyson's H(0)/V was very definitely inferior, so a useful relation to δ^2 was not established. (Tyson's H(0) is equivalent to ΔH_V - RT in Equation (3).)

The plots for organic solids were both well fitted with lines of 2.5 slope, but the better was

$$XF = 9.63 \times 10^{-7}\gamma^{2.5} \qquad (15)$$

The fact that this is dimensionally unbalanced is an unpleasant reminder that the model and all that follows from it are definitely oversimplified. It also destroys any hope of salvaging the theoretical derivation of the numerical coefficients in Equation (9), but does provide a less scientific approach.

AN EMPIRICAL ENGINEERING MODEL

To give due weight to all the lubricants in Table I, it would be necessary to have their individual compressibilities. Only the polythene can be read off Wright's[13] charts; by coincidence, it has the same density (~0.95) as stearic acid. The charts do permit a guess at anthracene compressibility, but nothing on sebacic acid or PTFE. It seemed best to use the stearic acid data to evaluate the coefficients. The development was essentially as for Equation (9), with the further assumption that the energy barrier (E_1 - E_2) depends on ($\gamma_S^{2.5}$/X). In its simplest form, this results in

$$F = (\gamma^{2.5}/X)(4.80/P + 1680\ \beta_S) \qquad (16)$$

The constants were evaluated from the stearic acid film retracted from the melt (upper line in Briscoe's Figure 5), at pressures of 10^7 and 3.16×10^8 Pa. It was assumed that these *two* data points could be extended to cover the entire range of lubricants in Table I. The results are shown in Table II, in comparison with the experimental data from Briscoe's Figures 5 and 7. It is surprising to see how closely this rather crude model predicts actual behavior. The compressibility of an oil of the same density as stearic acid and polyethylene was used for all lubricants, though anthracene is surely less compressible.

One feature of the stearic acid data not simulated by Equation (16) is the tendency for the friction to rise again at very high pressures. This is shared by calcium stearate (Briscoe's Figure 2, 0.1 and 40mm sliders only), but none of the other lubricants. It seemed worthwhile to elaborate Equation (17) to simulate this behavior, by assuming that the 2.5 power "law" extends to the compression term:

$$F = \frac{\gamma^{2.5}}{X}\ (1200/P + 88200\ \beta_S^{2.5} P^{1.5}) \qquad (17)$$

This showed excessive dip between the two reference points of 10^7 and 3.16×10^8 Pa, so was abandoned for

$$F = \frac{\gamma^{2.5}}{XP}\ (2.41 + 63.0\ \beta_S P)^{2.5} \qquad (18)$$

which seemed to have about the required properties. The results are shown in Table II, under those for Equation (16). One additional "experiment" was tried--the compressibility of anthracene was lowered to 58% that of stearic acid, based on the free volume theory. That led in the wrong direction, but is shown in Table II as a warning.

Equations (16) and (18) seem to fit the data about equally well. However, (18) seems more logical, since at very high pressure lubricants tend to become glassy and the friction should be high.

TEMPERATURE DEPENDENCE OF FRICTION

A severe test of any model is its ability to predict correctly the temperature behavior of the real system. Briscoe[12] provided the data in his Figure 8. All lubricants were tested from 20° to 80°C, so those were taken as the key points. The γ values were obtained as before: stearic acid and phenanthrene from Quayle,[11] polythene and PTFE from Wu.[17] The Briscoe data on sebacic acid are quite anomalous, so no attempt was made to compare them with an estimate of γ at 80°C. No adjustment was made on X, since it changes by less than 1%, but β_S was raised by 21%.

The results of Equation (16) and (18) are compared with the actual data in Table III. The stearic acid from solvent data showed an anomaly above 50°C and were omitted. Again, it is surprising to see how well these models respond to a situation not even contemplated when the 2.5 exponent was derived from Table I. It is disappointing that the predicted decreases are all less than the actual data, and ironic that the fit would have been much better if the value of β_S had been left uncorrected.

TABLE II.—PREDICTED VS. ACTUAL FRICTION AT 20°C

Pressure (Pa)	10^7	3.16×10^7	10^8	3.16×10^8	10^9	3.16×10^9
Compressibility ($1/10^{10}$ Pa)	3.98	3.72	3.19	2.23	1.35	-
Stearic Acid						
(12) Figure 5, from melt	0.140	0.097	0.064	0.047	0.057	0.058
(12) Figure 5, from solvent	0.100	0.073	0.048	0.033	0.036	0.058
Predicted (16)	0.140	0.094	0.071	0.047	0.028	-
Predicted (18)	0.141	0.076	0.056	0.054	0.055	-
Anthracene						
(12) Figure 7	0.73	0.44	0.31	0.22	-	-
Predicted (16)	0.51	0.35	0.26	0.18	-	-
Predicted (18)	0.58	0.28	0.21	0.20	-	-
Predicted (β_s= 58%)	0.53	0.22	0.12	0.09	-	-
Polythene						
(12) Figure 7, 30000 MW	0.54	0.35	0.29	0.24	0.17	-
(12) Figure 7, 10000 MW	0.32	0.18	0.14	0.12	0.07	-
Predicted (16)	0.42	0.29	0.21	0.14	0.09	-
Predicted (18)	0.47	0.23	0.17	0.16	0.17	-
Sebacic Acid						
(12) Figure 7	0.30	0.23	0.21	0.20	0.13	0.11
Predicted (16)	0.58	0.39	0.30	0.20	0.12	-
Predicted (18)	0.65	0.32	0.24	0.23	0.23	-
PTFE						
(12) Figure 7	-	0.05	0.03	0.02	0.01	-
Predicted (16)	-	0.06	0.04	0.03	0.02	-
Predicted (18)	-	0.04	0.03	0.03	0.03	-

*TABLE III.—PREDICTED VS. ACTUAL EFFECT OF TEMPERATURE ON FRICTION**

Lubricant	Coefficient of Friction		%
	20°C	80°C	Decrease
β_s ($1/10^{10}$ Pa) =	3.50	4.20	-
Stearic Acid - γ =	0.027	0.022	-
Retracted from melt	0.088	0.035	68
Predicted (16)	0.083	0.058	30
Predicted (18)	0.056	0.039	30
Anthracene - γ =	0.041	0.038	-
Actual	0.39	0.25	36
Predicted (16)	0.31	0.30	3
Predicted (18)	0.21	0.20	5
Polythene - γ =	0.031	0.028	-
30000	0.301	0.120	60
10000	0.163	0.065	60
Predicted (16)	0.252	0.229	9
Predicted (18)	0.170	0.153	10
PTFE - γ =	0.017	0.014	-
Actual	0.025	0.008	68
Predicted (16)	0.049	0.035	29
Predicted (18)	0.033	0.024	27

* Briscoe,(12) Figure 8 at 5×10^7 Pa.

CONCLUSIONS

The following conclusions may be drawn from this study:

- A very good case can be made for friction arising from a non-wear (non-adhesive) mode of sliding.

- Rabinowicz' proposed two part friction equation can be fitted to much of the data, but its quadratic structure does not relate well to molecular theory.

- Cameron's suggestion on surface free energy does not work out as stated, but has interesting implications for further work.
- The empirical $\gamma^{2.5}$ model, in either of two forms, comes close to fitting all the data at 20°C, but is less satisfactory on the temperature dependence.
- This limited success provides an answer to Rabinowicz' question--any coefficients of dry friction below 0.02 at 20°C should be considered suspect.
- There is still incentive to perform Macmillan types of integrations on these lubricants to produce a dimensionally balanced equation.

REFERENCES

1. Eudier, M. and Yousseff, H., *Powder Metallurgy,* Vol. 12, 1969, p. 462.
2. Rabinowicz, E., "Friction and Wear of Materials," John Wiley and Sons, New York, 1965, p. 94ff.
3. Kilburn, R.F., *Journal of Lubrication Technology,* Vol. 96, 1974, p. 291.
4. Rabinowicz, E., These Proceedings.
5. Beerbower, A., *American Society of Lubrication Engineers Transactions,* Vol. 14, 1971, p. 90.
6. Beerbower, A., "Boundary Lubrication," Report AD 747336 to the Army Research Office, Washington, D.C., 1972, pp. 253ff.; 267.
7. Cameron, A., *American Society of Lubrication Engineers Transactions,* Vol. 2, 1960, p. 195.
8. van Battum, C.M. and Broeder, J.J., *American Society of Lubrication Engineers Transactions,* Vol. 5, 1963, p. 285.
9. Macmillan, N.H. and Kelly, A., *Royal Society of London. Proceedings.* Series A, Vol. 330, 1972, p. 291.
10. Tyson, W.R., *Journal of Applied Physics,* Vol. 47, 1976, p. 459.
11. Quayle, O.R., *Chemical Reviews,* Vol. 53, 1953, p. 439.
12. Briscoe, B.J., Scruton, B. and Willis, F.R., *Royal Society of London. Proceedings.* Series A, Vol. 333, 1973, p. 99.
13. Wright, W.A., *American Society of Lubrication Engineers Transactions,* Vol. 10, 1967, p. 349.
14. Hildebrand, J.H. and Scott, R.L., "The Solubility of Nonelectrolytes," 3rd edition, Dover, New York, 1950, p. 129.
15. MacGregor, C.W., "Handbook of Analytical Design for Wear," Plenum Press, New York, 1964, p. 60ff.
16. Tyson, W.H., *Canadian Metallurgical Quarterly,* Vol. 14, 1975, p. 307.
17. Wu, S., *Journal of Macromolecular Science, Part C. Reviews in Macromolecular Chemistry,* Vol. C 10, 1974, p. 1.

DISCUSSION

D. TABOR, Cambridge University: I would like to congratulate Mr. Beerbower for his attempt to formulate frictional properties of polymeric surface films in terms of thermodynamic models. And I think that one of the problems here is really not so much the calculation as the model. It is the model that worries me.

You consider that the energy required to pull the surfaces one atomic spacing across is 2 γ. It seems to me that that model cannot be right because in fact we are not removing the molecules to infinity. We are moving the molecules only a small distance away from the surface, sliding them over, and then we put them back. Therefore one would think that the energy expended in the sliding process is not 2 γ but only some fraction of making it 1/4 or 1/10 of γ. And where this becomes relevant in a different sphere is, for example, in surface diffusion. If we look at the energy involved in the movement of molecules over the surface, it is only 20 percent of the energy required to remove the molecules from the surface. Clearly, the energy required to remove the surfaces from one another is much greater than the energy to move in a potential field of another surface. From that point of view, I think, using 2 γ is a complete over-statement of the energy involved. Now the amount you actually require may depend on the geometry of molecules.

The second aspect that worries me is the application of thermodynamic concepts to strength properties. This is by far the greatest weakness. Certainly in crystalline solids, we cannot describe the strength properties of materials in terms of its interatomic forces. Indeed we do need to know whether dislocations are playing a part in the sliding process. If layered materials slide one over another, the forces involved will depend not only upon intermolecular forces but upon imperfections in the layer also.

We have been doing a study at Cambridge on the shear strength of monolayers and the conditions where we know, first of all, the true area of molecular contact, and secondly, the precise thickness of the area that is being sheared. This is done by depositing a monolayer of stearic acid and then measuring the thickness by multiple beam interferometer. Since the optical resolution of the technique used is better than 3Å, we can measure the thickness of the sandwich and it comes out to be 48Å; and it is not a bad guess for two of the hydrocarbon chains put in contact. We can then load the surfaces together, pull them and measure the shear force as a function of contact pressure. We can monitor the thickness of the film while the shearing is taking place. And on those conditions we are able to do the following: We determine the real area of material that has been sheared so there is no question of trying to do calculations based on topographical studies and assumptions. It is a direct measurement. The second thing is, we can measure the thickness of the film so we know whether the film is being destroyed or distorted and, on the whole, the film remains at its original thickness to within about 10% which means, at the most, the chain may be tilting over by a few degrees.

The results suggest that the difference between stearic acid and fluorinated stearic acid is the amount of disorder in the monolayer and the role of, if I may say, dislocations that affect the shear properties of the material. First of all, one needs direct experimental evidence on which to base a theory — the more direct, the better. The second point is that if one is trying to apply thermodynamics to strength properties, he should think whether he really needs to know the role of disorder in shear properties of what is essentially a slightly disordered crystalline material.

A. BEERBOWER: Your point is certainly a good one about the applicability of thermodynamics to shear properties of materials. However, I would take some comfort in the fact that Tyson correlates the strength of materials to γ . If you take the Macmillan integration you can get a very good Young's modulus for solid argon. You can even get the direction of the shear plane. You can do everything except get the shear strength. This was a warning but I did not heed it. I also took some comfort in the fact that we should be able to predict the tensile strength from γ and we can generally guess the shear strength from the tensile strength. But I know these are not by any means rigorous. As I said at the end I hope somebody will take Macmillan's integrations for the stearic acid. Frankly, I do not even trust the value of Young's modulus for stearic acid. The ones I got for teflon and polythene looked plausible compared with what I calculated. With stearic acid it is anybody's guess.

QUESTIONER, C. S. Draper Laboratories: I wonder if I can ask Professor Tabor to elaborate a little more on the fluorinated stearic acid. Is that a mixture of fluorine-containing stearic acids?

TABOR: The bottom half of the molecule was stearic acid and the top half is fluorinated material. It did not have a fully fluorinated molecule. It is a question of what one can get.

D. H. BUCKLEY, NASA: I have a comment to make on the presentations of both Professor Rabinowicz and Mr. Beerbower. Both advance surface energy concepts; Professor Rabinowicz uses hardness terms. When I look for the hardness value of a particular material, I get six different hardness values from six different references.

Now about the surface energy, I would like to indicate a word of caution. Much of surface energy data that has been obtained in the past is useless so to speak. This is not only my opinion; it is the opinion of the people who are conducting current research on the surface energy values. For example, the results of Wawra published a few years ago clearly indicate that we should exercise caution in using surface energy values in our calculations. It may be that we are using values that are not reliable trying to explain friction and wear data. Part of the reason for this may be impurities. Many years ago when people did experiments like these, lot of attention was paid to the purity of the material. In all frankness, the high purity materials just were not available in those days and it was very common to assume 99% purity as high purity. Since then people have found that very small concentrations of impurities in the metal can markedly alter surface energy.

Alan Beerbower said that he was correct in using the liquid surface energies and obtaining the solid surface energy by extrapolation from the liquid surface tension. Data for iron shows that the surface energy falls off drastically with small concentrations of sulfur in iron as discussed in my paper in these proceedings. I am sure some of you are aware that small concentrations of impurities can settle on the surface and vastly alter surface energies. We have to be very careful in using these values in making calculations.

BEERBOWER: I would like to explain something about Wawra's data. His table represents calculated values from the 1920's and 30's for maximum values, and other data from obsolete handbooks for the minimum values. In addition his correlation based on the elastic constants of the crystals, which he assumes to be a single constant value and applies to everything, is not valid. Compressibility depends on three other variables so that anything that Wawra says is not suspect; it is wrong. What Dr. Buckley neglected to say is that by the time we have so much sulfur we no longer have iron, but iron sulfide on the surface. And I would suppose a value of 900-1000 $ergs/cm^2$ would be the surface free energy of iron sulfide.

In Tyson's paper that I mentioned, we find screened values representing the consensus of perhaps a hundred recent investigations on the surface free energy of solids. I have checked these against the interfacial energy of the melt against the solid, and the surface free energy of the liquids. As far as I am concerned, they are quite consistent. However, the data on organic solids should be further refined. The surface energy values of the alkaline halides are also in excellent condition. If we look at the recent data instead of the data from the 20's and 30's, we find that the surface energy values are really in good shape.

BUCKLEY: I disagree with that statement. Surface physicists believe presently that there is no reliable surface energy data available in the literature and they are now trying to find ways of generating it. We will see it generated in the coming years, but presently we do not have good surface energy measurements for metals.

FRICTION OF BOUNDARY FILMS

W. A. Glaeser

ABSTRACT

Boundary lubrication conditions exist in many practical engineering situations, as Rabinowicz points out. Therefore, low friction values are often encountered in solid surface contact. The friction levels encountered appear to be associated with the plastic flow properties of very thin surface films. Recent work at Battelle has shown that solid film friction can be detected when clean metal surfaces are wet with a dilute solution of a fatty acid. The implications of these results are that boundary lubrication by fatty acids is not based on monolayers but on disordered multilayered thin films. These thin films show an increase in shear strength with contact pressure, a decrease in shear strength with increasing surface temperature and sensitivity to sliding velocity - or a quasi-viscous response to shearing. Behavior of these films is linked to the combined effects of pressure, shear rate, surface temperature, and film growth/dissolution characteristics.

INTRODUCTION

Low friction is often encountered under lubricated conditions and is accompanied by continuous mild wear. These "boundary lubrication" conditions presume that a solid film of soft material supports a large portion of the load and that friction forces arise from plastic deformation of the film during sliding. Even if some load support occurs by contacting asperities, the shearing of these junctions contributes an almost negligible amount to the total friction force at the interface. This fact becomes evident if one calculates the force to shear a very small total asperity contact and compares it to the shear force to yield the remaining solid film area to be sheared.

If one accepts the predominance of the solid film in boundary lubricated friction phenomena, then the mechanical properties of the boundary films involved are important. Research involving the measurement of mechanical properties of solid organic films and soft metallic films on hard substrates has been under way for many years. Briscoe, Scruton and Willis,[1] for instance, have measured the effect of pressure on the yield strength in shear of soft films deposited on glass substrates and found that the yield point increases with pressure. The rate of increase with pressure varies with material species--like pressure-viscosity phenomena.

BEHAVIOR OF BOUNDARY LUBRICANTS AS SOLID FILMS

Research at Battelle[(2)] has shown that solutions of fatty acids in long chain hydrocarbon solvents will act much like solid films deposited on metal surfaces. Using an iron-coated glass surface, the shear strength of surface films generated from the solution have been measured. The shear strength-pressure characteristics of fatty acid solutions have been compared with similar data for solid films and similar trends were found. These findings are summarized in Figure 1. The two lines labeled "Briscoe Stearic

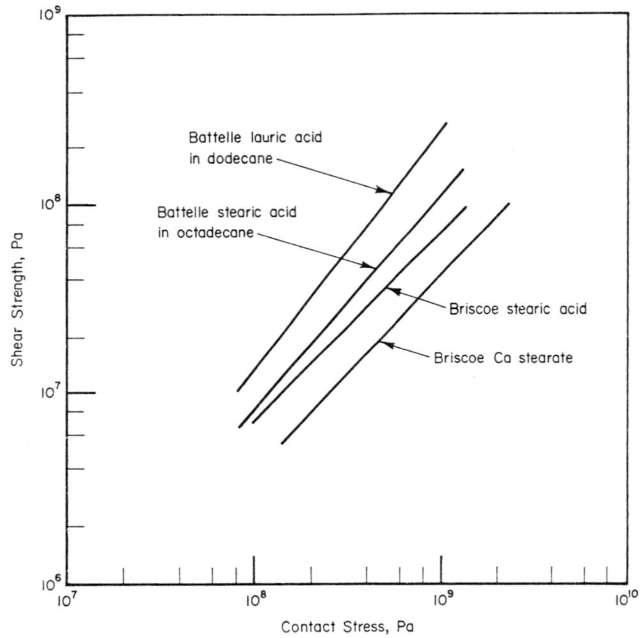

Fig. 1.—Comparison of Room-Temperature Shear Strength of Fatty Acid Films as a Function of Pressure for Battelle's Data (Liquid Solution) and Briscoe's Data (Retracted Solid Films).

Acid" and "Briscoe Ca Stearate" are from data on shear strength of solid films deposited on glass by the Langmuir-Blodgett technique from work by Briscoe and co-workers.[(1)] The two remaining lines are from Battelle data for liquid solutions of fatty acids. The data in Figure 1 show that there is a similarity in shear strength-pressure characteristics for both the solid films and the liquid solutions. Note the close similarity for stearic acid solid film and stearic acid in Octadecane. In addition, Figure 1 shows that the presence of a solvent and advent species appear to influence shear strength. The Battelle research has indicated that boundary films developed from the fatty acid solutions are more than monolayers of adsorbed polar molecules. The boundary films appear to be molecularly-oriented, multi-layered films several layers thicker than monolayers. Evidence for multi-layered ordered films on metal surfaces covered with liquid solutions of fatty acids has been found in the Battelle experiments using multiple-internal-reflection infrared spectroscopy.[3] Very pure iron and chromium substrates were used to form in situ boundary films from liquid solutions containing fatty acids.

The "low" friction characteristic of boundary lubrication essentially can be related to the low shear strength of the organic film formed from the liquid solution. The only contribution made by the substrate material in boundary lubrication is asperity contact. The total area of asperity contact is very small compared to the total area of contact involving a soft organic film; however, asperity contact is presumed to be the source of the wear that usually accompanies boundary lubrication. Although the extent of surface film action at the interfaces between asperities is at question, it may not be significant since the asperity deformation at concentrated contacts (plowing or asperity-asperity clashes) is the source of wear particle generation.

Friction coefficient should be related to the shear properties of the solid boundary film. What happens when the contact pressure is increased and the shear of strength of the film increases? The approach of Briscoe and Tabor[4] for thin polymer films can be used to answer that question. Assuming that the film separates two contacting surfaces, that the elastic deformation of the substrate determines the true area of contact, and that friction derives from shearing of the separating film, the coefficient of friction may be related to material properties as

$$\mu = F/w$$

$$= \frac{A\tau}{w} = \frac{\tau_o}{H} + \alpha$$

where μ = *coefficient of friction*

w = *normal load*

τ_o = *shear strength of the film at atmospheric pressure*

α = *pressure coefficient*

A = *contact area*

H = *plastic yield pressure of substrate*

F = *Frictional force*

Briscoe and Tabor assume for most metals that H is much greater than τ_0 and therefore, the friction coefficient is approximately equal to the pressure coefficient, α. Therefore, μ is independent of load. This may be true so long as the yield point of the substrate is not exceeded (this limits the maximum pressure on the boundary film). In addition, temperature will reduce the film yield strength and Battelle data indicates that shear properties of fatty acid boundary films have some quasi-viscous properties and therefore, may be influenced by sliding velocity. All of these effects can influence α and therefore change friction coefficient.

The dynamic response of very thin organic boundary films during sliding is not well understood. The Battelle model assumes an ordered multilayered structure for a quiescent film in the presence of a fatty acid solution. Once this film is stressed, it may become highly disordered and may, to some extent, disintegrate. In recent experiments at Battelle, the wear debris of copper lubricated with a solution of 1 per cent stearic acid in dodecane was found to contain about 20 per cent copper particles and 80 per cent platelets of an organic solid, presumably the metal soap formed during contact of the lubricant with the copper surface.

REFERENCES

1. Briscoe, B.J., Scruton, B. and Willis, F.R., *Royal Society of London. Proceedings*. Series A, Vol. 333, 1973, p. 105.
2. Allen, C.M., et al., "Aircraft Propulsion Lubricating Film Additives: Boundary Lubricant Surface Films," AFAPL-TR-73-121, Vol. 3, May 1976.
3. *Ibid.*, Volume II, 1972.
4. Briscoe, B.J. and Tabor, D., *Wear*, Vol. 34, 1975, p. 29.

SOME PROBLEMS IN THE ADHESION THEORY OF FRICTION

Y. Kimura

ABSTRACT

Although a majority of practical friction problems are concerned with the case of moderate or low friction, as Rabinowicz pointed out, high friction also poses some problems in practice (which appear as catastrophic failure of sliding surfaces) and a few remarks on our understanding of general friction phenomena are in order. As is well known, the adhesion theory proposed by Holm succeeded in explaining the so-called dry friction. It was further developed to elucidate a wide range of values the coefficient of friction can take, where the junction-growth concept was introduced. It is interesting to note that the early experimental evidence supported the classical theory. However, as the techniques to obtain well-defined clean surfaces were developed, it has been found that extremely strong adhesion takes place between clean surfaces which cannot be predicted from simple mechanical considerations. A further investigation is clearly needed to provide a quantitative representation of the frictional behavior.

INTRODUCTION

As usual, Professor Rabinowicz has written very clearly. It is so clear that we are tempted to believe that the two-term relationship (i.e., Eq. 6 in his text) will suffice in explaining the frictional behavior ordinarily encountered in practice.

Indeed it may be the case. Ordinary friction values of practical interest are relatively low, and are not so widely diffused. In his paper(1) Professor Rabinowicz showed that some 80 percent of 210 dry friction values, obtained from various combinations of 20 elemental metals, fell between 0.4 and 0.6. For these cases it can be said that the two-term relationship may provide an approximation of a sufficient accuracy, and that the adhesion theory could successfully explain friction phenomena. This might have resulted in a situation in which the wild enthusiasm for study into friction mechanisms almost cooled down. So the discusser would like to ask whether there are no important problems left awaiting solution. One should find the answer if this conference is to accomplish its major purpose.

The discusser does not lack appreciation of the general validity of the adhesion theory. Nevertheless, he thinks some problems should be clarified further for a systematic understanding of friction phenomena.

ADHESION THEORY OF FRICTION

The adhesion theory of friction is well known and it might not be necessary to state it here. However, a brief description of the discusser's understanding would serve to confirm our common knowledge.

Research on the mechanisms of friction originated in an attempt to elucidate the values of the coefficient of friction which we commonly obtain from measurements in air, the so-called dry friction. The concept of dry friction has considerable practical importance. For example, we refer to its value when some mechanism is designed, and conveniently we can find them in tables and in handbooks. It is also widely known that the values usually differ unlike other physical quantities such as density or elastic constants. However, it would be important to recognize that they are rarely higher than 1.0 or lower than 0.04 as suggested in the review. A reasonable explanation by the adhesion theory has been provided for these values in the past, which may be regarded as a rough approximation to the problem.

The history of tribology would have been plain but for the development of techniques to obtain cleaner and well-characterized surfaces. Seemingly, however, history does not like such plainness. When Holm first introduced an equation for the coefficient of friction, such as Eq. 1 in the review paper(1), along adhesion theory, it was already known that the friction coefficient increased infinitely if the surfaces were properly cleaned(2). A series of experiments conducted with "clean surfaces" have shown that the coefficient of friction can actually exceed 100 in some cases, and that a slight change in surface states can cause a marked change of its values even with a specified pair of materials(3-8). Table 1 presents a partial listing of these results obtained for Cu/Cu. This behavior was explained when the junction growth theory was introduced by Tabor(9) in the "second round," which reduced to the classical adhesion theory, if the junction growth was prevented by contamination (Figure 1). Probably from such a point of view, dry friction is defined as "an imprecise term frequently used to indicate friction under nominally unlubricated conditions" by the OECD group(10).

TABLE I.—COEFFICIENT OF FRICTION BETWEEN CLEAN COPPER SURFACES.

Authors	Shape of Specimens	Preparation	Atmosphere	Coeff. of Friction
Bowden and Hughes(3)	cylinder/cylinder	heated in vacuo	vacuum (10^{-6} torr)	4.8
Simon et al.(4)	cylinder/V-shaped groove	heated in vacuo	N_2 and He	1.3
Gwathmey et al.(5)	sphere/plane	reduced in H_2	H_2	>100
Takagi and Tsuya(6)	sphere/plane	reduced in H_2	H_2	31~147
Kimura(7)	cone/plane	reduced in H_2	H_2	20~140
Brown and Burton(8)	sphere/plane	heated in vacuo	vacuum (10^{-9} torr)	3.5~4.8

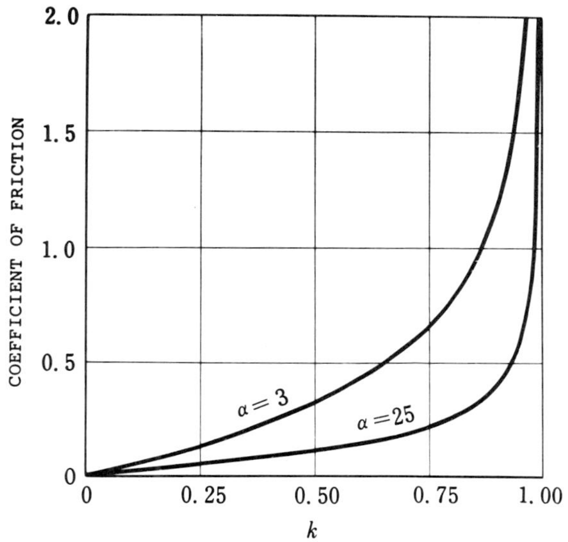

Fig. 1.—Coefficient of friction as a function of k, after Tabor[9]. α is a constant appearing in the criterion of plastic flow.

THEORETICAL PROBLEMS

An important consequence of the junction growth theory is that the coefficient of friction can be given as a function of k, the ratio of the shear strength of the interface to that of the materials in contact. It is no more a quantity peculiar to the material combinations. Thus, a table of dry friction should be regarded as one showing the values of k.

Then our interest should naturally be focused on the parameter k for various sliding systems; i.e., we must know what surface films are formed on sliding surfaces, and their mechanical properties. This is not so much a problem of friction as boundary lubrication. Unfortunately, however, researchers in the field of boundary lubrication are mostly chemists who have little interest in mechanics so that surface films are hardly discussed in connection with friction. As a consequence, little is known about mechical behavior of surface films. A mechanism proposed by Rabinowicz for the low friction associated with a change in the shear strength of lubricating film has an important implication in this respect. When it is interpreted in terms of k, the discusser questions whether the pressure also has an effect on the shear strength of the substrate. Further cooperative work is required on chemical and mechanical aspects of this problem.

EXPERIMENTAL PROBLEMS

There is abundant evidence which shows that adhesion indeed takes place at contacts. Further, it plays a dominant role in friction. Nevertheless, there is still a problem in its quantitative aspect.

Once McFarlane and Tabor reported that a linear relationship was observed between normal load and force of adhesion in air[11], Figure 2. They described that the surfaces were free from greasy matter but would be

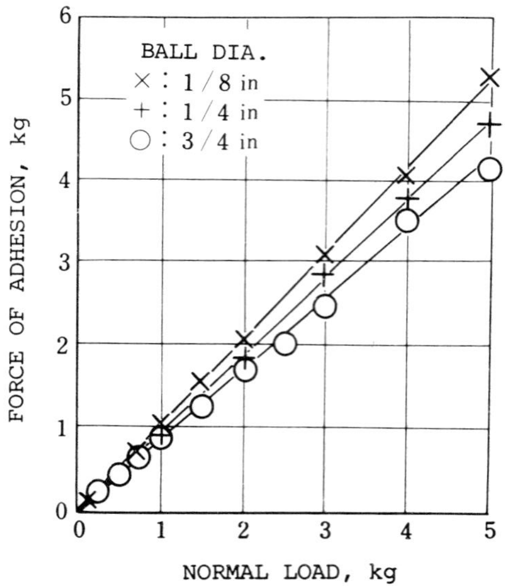

Fig. 2.—Adhesion of steel spheres on indium, after McFarlane and Tabor(11).

covered with thin films of oxide or absorbed gas from the atmosphere. From our present knowledge they cannot be said to be "clean." It is very interesting that such a linearity was found with the surface of, strictly speaking, uncertain conditions. This gave a strong support for an explanation of dry friction with the adhesion theory.

Recently various techniques were developed by which we can obtain atomically clean surfaces and can identify clearly the states of the surfaces. Buckley conducted a series of experiments in high vacuum. He prepared clean surfaces, examined their state with LEED and AES, and obtained results as shown in Figure 3(12). He reported that simple touch contact of the clean iron was sufficient to give extremely strong adhesion which exceeded the measuring capabilities of the apparatus. If we define the coefficient of adhesion as the ratio of force of adhesion to normal load, the foregoing will imply that it tends to infinity. Even in the case of slightly contaminated surfaces, the linearity can no longer be found. When no foreign surface films exist, interaction between the atoms lying on both surfaces would need no normal load. It is geometry, not mechanics, that determines the force of adhesion. Nevertheless, we lost a quantitative foundation on which the frictional behavior should be interpreted.

CONCLUDING REMARKS

As Professor Rabinowicz stated, a majority of practical friction problems are concerned with the case of moderate or low friction. Friction is also important in other tribological phenomena such as wear, scuffing, seizure; occasionally they cause catastrophic failure of sliding systems. Nevertheless, the possible roles of friction in such phenomena have not been properly analyzed, in terms of fundamental understanding of friction. Further investigation is clearly needed to explain and to control friction.

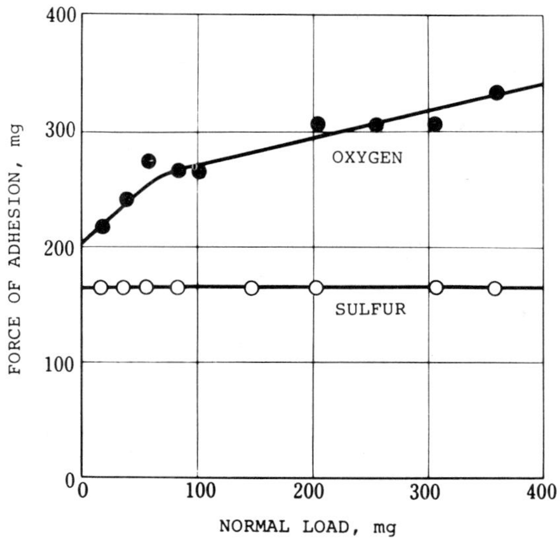

Fig. 3.—Adhesion of iron single crylstals, after Buckley[12]. *Data for atomically clean surfaces are not shown (far above). Oxygen and sulfur make partial monolayers of 6.25 atomic percent in surface concentration.*

REFERENCES

1. Rabinowicz, E., These Proceedings; also *American Society of Lubrication Engineers Transactions,* Vol. 14, 1971, p. 198.
2. Holm, R. and Kirschstein, B., *Wissenschaftliche Veröffentlichungen aus den Siemens-Werken,* Vol. 15, 1936, p. 122.
3. Bowden, F.P. and Hughes, T.P., *Royal Society of London. Proceedings,* Series A, Vol. 172, 1939, p. 263.
4. Simon, I., et al., *Journal of Applied Physics,* Vol. 22, 1951, p. 177.
5. Gwathmey, A.T., et al., *Royal Society of London. Proceedings,* Series A, Vol. 212, 1952, p. 464.
6. Takagi, R. and Tsuya, Y., *Japan Society of Lubrication Engineers, Journal,* Vol. 6, 1961, p. 81.
7. Kimura, Y., *Japan Society of Lubrication Engineers, Journal,* Vol. 9, 1964, p. 475.
8. Brown, R.D. and Burton, R.A., *Journal of Lubrication Technology,* Vol. 89, 1967, p. 425.
9. Tabor, D., *Royal Society of London. Proceedings.* Series A, Vol. 251, 1959, p. 378.
10. Research Group on Wear of Engineering Materials, OECD, "Glossary of Terms and Definitions in the Field of Friction, Wear and Lubrication," OECD, Paris, 1969, p. 29.
11. McFarlane, J.S. and Tabor, D., *Royal Society of London. Proceedings.* Series A, Vol. 202, 1950, p. 224.
12. Buckley, D.H., NASA TN D-5689, 1970, p. 1.

DISCUSSION

D. TABOR, Cambridge University: I would like to have heard a little more about the things that are taking place at the interface. Bascially you were describing results and not telling us very much about what is going on at the contact. For example, the junction growth equation is irrelevant in a way. It is the physical picture that is important. When junctions are formed and are subjected to tangential stress, the combined stress produces increased plasticity and increase in area of contact, whether we can express this process in that simple equation or not is of secondary importance. Certainly the clean surface is pulled together by the tangential and normal force. That is why we get large friction coefficients which will depend on the geometry of the surface and for very clean surfaces you will get coefficients of friction which may range from 1.0 to 100. The result is not confusing in terms of the physics of what is taking place at the interface. If we have surfaces which are contaminated, the interface has different physical properties. Probably it lacks ductility, and therefore we get much clearer failure criterion than in the case of clean surfaces and extended plasticity. But there again the physics of what is going on at the interface is important.

I enjoyed your talk very much, I would have liked to have heard a little more about what you think is really taking place at the interface.

Y. KIMURA: Thank you very much for your generous comments. You may say that the numerical value is of secondary importance, but we have observed this consistently.

TABOR: I agree. And this is really the difference between the practical man who has to make things work and people like me who just try to find out why.

May I just make another point. Some time ago a paper was published in the Soviet Union on the friction of molybdenum sulfide surfaces in vacuum where the surfaces were irradiated and then exposed to air, and then evacuated again. Coefficients of friction of the order of 0.002 were quoted and this seems to me is very hard to understand. Tsuya has published some work in which she has got not only coefficients of friction of 100 for clean surfaces but also in her PhD thesis she has some coefficients of friction of sulfide surfaces of 0.005 or so. There are systems where very low friction occur, and I would like to know why.

KIMURA: I am aware of Tsuya's results. I do not know how to interpret such low values.

D. H. BUCKLEY, NASA: I think that the concept of clean surfaces is what matters. To use an equation or to try to put numbers on, I think, is meaningless. Between any two well-characterized clean surfaces in contact adhesion always occurs instantaneously. I think, as Professor Tabor rightly indicated, the magnitude of the contact area is unimportant. What is going on at the interface is important. At the interface you can have adhesive bond of like metal to itself or dissimilar metals. For example, with two perfectly matched single crystals of copper with the planes and directions identical across the interface, we no longer have two separate single crystals but one continuous crystal. There is no such thing as a boundary separating them. It will be one continuous piece of copper. When we talk of very clean surfaces, the numbers become meaningless.

KIMURA: I thought that phenomenologically we can understand the result.

D. A. RIGNEY, Ohio State University: When we take ductile materials and put them together instantaneously, we get some plastic deformation also.

BUCKLEY: Of course, we cannot match two crystals; it is impossible. However, if we could do that we would lose the interface, and cohesive binding takes place across the interface. In the real situation, the equivalent at best may be a grain boundary. There will be lattice distortions when we try to match the surfaces.

E. F. FINKIN, Allegheny Ludlum Industries: A paper published by K. L. Johnson a year ago offers an alternative explanation why when junction growth takes place, the calculated coefficient of friction by the adhesion mechanism is very low. A possible fundamental reason is that the surface energy is responsible for substantial increase of the contact force beyond the applied force. There will be a magnification of real area of contact because of the surface energy. It is not that junction growth does not count. We can get four or five times an increase in the real area of contact by this effect. Furthermore this effect is critically dependent on the surface energy which explains why the surface films are so effective in decreasing friction. If we put all these together we have a much better explanation. The reason for the confusion in the literature is that not enough of the major factors have been identified. This is a new major factor that should be considered.

B. J. BRISCOE, Imperial College: The paper which you state is about elastomers of low surface energy and low modulus. It is true that the surface energy can be correlated with an increase in the area of contact. For soft elastomers, one can predict and measure that there will be an area of contact even at zero load. And you can extend the argument further by taking the magnitude of the adhesive force to pull the material away from the surface. If you look at the literature, there are precedents dating back to 1930.

PLASTIC DEFORMATION AND SLIDING FRICTION OF METALS

D. A. Rigney and J. P. Hirth

ABSTRACT

A model is described for the source of friction during the steady state sliding of metals. It focusses on the plastic work done in the near surface region, described in terms of work hardening, recovery, and the microstructure existing during steady state sliding. The model is closely related to a model for sliding wear which has been presented earlier.

PLASTIC DEFORMATION AND SLIDING FRICTION OF METALS

Rigney and Glaeser have described a wear model[(1)] in which attention is focussed on steady state wear. They emphasized plastic deformation near the surface, particularly in the highly deformed region which has a very fine microstructure and a high degree of preferred orientation. In metals and in some ceramic materials, this near-surface microstructure consists of dislocation cells developed during an initial break-in period. Under steady-state conditions the average cell structure at a given distance from the surface remains constant, and the average thickness, t, of the cell region is a constant that depends on material properties and on the details of the sliding wear test.

In this preliminary report,[(2)] the ideas applied earlier to wear are extended to apply to friction as well. In each case we are suggesting an alternative to traditional theories, such as those which emphasize the local shear of adhesive junctions.

With the assumptions that the frictional force arises from plastic deformation and that most of the deformation work is confined to the cell region, one can develop an expression for the friction coefficient using a model that is consistent with the wear model of reference,[(1)] and with the steady state microstructures described in references.[(3-7)]

Since steady state friction is assumed to depend on a steady state microstructure, all of the plastic work is dissipated as heat and none is added to the energy already stored in the microstructure.

The simplest case is that in which one component of a sliding pair is much harder than the other, so that plastic deformation is essentially confined to the near-surface region of the softer material. Then the plastic work per unit volume is given by the product of the shear stress, τ, in the cellular region and in the sliding direction, and the average strain per cycle, ε.

The total work is then given by (Volume, V) ($\tau\varepsilon$). For a virtual displacement, δx, the frictional work is just $F\delta x$, where F is the frictional force, and these two expressions for frictional work must be equal:

$$V\tau\varepsilon = F\delta x \qquad (1)$$

The volume is equal to $wt\delta x$, where w is the width of the highly deformed region (approximately equal to the wear track for a pin on disk geometry). Since the friction coefficient is defined by $\mu=F/L$, the following expression for the friction coefficient is obtained:

$$\mu = \frac{wt\tau\varepsilon}{L} \qquad (2)$$

All four terms in the numerator are material properties or are closely related to material properties. Three of the terms (w, t, L) in equation (2) are directly measureable and readily available for a given sliding situation. The shear stress τ is not the bulk value that might be available from simple stress-strain curves. However, it could be estimated from appropriate tests on severely cold rolled material with similarly textured cell microstructure. The average strain per cycle, ε, is somewhat more difficult to measure. Perhaps a chemical marker technique could be used together with Auger analysis to measure the variation of strain with depth in the near surface region.

At first sight, equation (2) seems to give the result that μ depends on the load. However, both w and t should increase with the load. If the product wt is roughly proportional to load, then Amontons' familiar first law is obeyed: i.e., the friction coefficient is independent of the load. If the relationship between wt and L is not quite linear, then there would be some dependence of the friction coefficient on the load.

Since stress falls off rapidly near the surface, one would expect that the thickness of the highly deformed (cell) layer would vary with load by some power less than one. That is, it would increase with load, but in such a way that strain is concentrated near the surface and t does not reach large values. Limited data for copper[8] indicate that $t \sim L^n$, where $n \approx \frac{1}{2}$.

It is reasonable to expect that the width w would also be $\sim L^m$, with $m<1$. In fact, data on the Vickers Hardness Number, VHN,[9] which is roughly proportional to load/area, would indicate that $m \approx \frac{1}{2}$, with perhaps a somewhat higher value at very light loads. However, one can do better than to use this plausibility argument, since Tsuya[8] has also measured track width vs. load for copper, and she finds $m \approx \frac{1}{2}$.

Since both $n \approx \frac{1}{2}$ and $m \approx \frac{1}{2}$, equation (2) describes a friction coefficient that is approximately independent of load. When $(n+m) \neq 1$, there is some dependence of μ on load, as is sometimes observed experimentally.[10,11]

The preceding derivation and discussion were based on the simple case of a hard material sliding on a softer material, in which plastic work is largely confined to the cell layers in the softer material. However, since contributions to friction from more than one region are additive, one can extend equation (2) to other cases of practical significance. In fact, whenever plastic deformation is largely confined to a well-defined region of a material in a sliding system, an expression like that of equation (2) applies for that region.

EXAMPLES

(1) If metal A slides against a piece of the same metal A, each develops a cell region of thickness t, and the friction coefficient is given by $\mu = w(2t)\tau\varepsilon/L$. This is the case treated by Tsuya.[8]

(2) If a thick metal film is applied, and the substrate does not deform at all, then $\mu = wt\tau\varepsilon/L$ applies, but t, τ, and ε characterize the cell region of the metal film.

(3) If the metal film is fully cellular, and the substrate also develops a cell layer, then a 2-term expression is needed for friction:

$$\mu = \frac{wt_s\tau_s\varepsilon_s}{L} + \frac{wt_f\tau_f\varepsilon_f}{L},$$

where s refers to the substrate and f to the film. Solid lubricants could be treated by a similar 2-term expression, one term representing the lubricant layer and the other term representing the substrate.

(4) A 2-term expression also results if two dissimilar materials slide against each other:

$$\mu \doteq \frac{wt_1\tau_1\varepsilon_1}{L} + \frac{wt_2\tau_2\varepsilon_2}{L}.$$

(5) For a composite with a lamellar structure, if the lamellae become aligned parallel to the sliding surface, then a separate term can be written for each lamella in which plastic shear is appreciable. A multi-term expression is then appropriate for the friction coefficient:

$$\mu = \frac{wt_1\tau_1\varepsilon_1}{L} + \frac{wt_2\tau_2\varepsilon_2}{L} + \cdots\cdots + \frac{wt_n\tau_n\varepsilon_n}{L}.$$

In all these cases, the contributions to friction from deformation below the "easy" shear region are neglected. The work of Tsuya[8] shows that this approximation is reasonable, since most of the deformation is localized near the surface. Therefore the error involved in using this simplified model is small. Of course, during the break-in period, these contributions would be more important, but the model described here and in references[1,2] is not meant to apply before steady-state conditions are established.

ESTIMATES OF FRICTION COEFFICIENTS

The quantities w, t, and L can be measured directly for a given test system. The quantities τ and ε are not directly available. However, it is possible to estimate these quantities from data published in the literature. Therefore, it is possible to estimate friction coefficients.

As an example, we have selected the case of an OFHC copper sample tested in a LFW machine with a 440 C ring in argon. The load was 20.4 kg (45 lb.) and the sliding speed was 0.05 m/sec. Since the 440 C alloy is much harder than the copper, most of the deformation should be in the copper, and the simple one term expression can be used for the friction coefficient, i.e.,

$\mu \approx (wt\tau\varepsilon)/L$. The resulting track width was the width of the sample block, or 6.3(5) mm. The value of t was determined from a transverse section of the copper sample, so that a proper average could be obtained. The approximate value for $t_{av} \approx 12 \mu m$.

Since the textured microstructures generated by large reductions during cold rolling[6] and those generated during sliding[7] are similar, one can use tabulated strength data on cold rolled material to estimate τ.[12] Assuming that the preferred orientation of the near surface material is that which allows easiest shear, then τ should be approximately half the ultimate tensile strength (UTS). For copper, the appropriate UTS is about 416 MPa (60,000 psi); therefore $\tau \approx 208$ *MPa (30,000 psi)*. Since dislocation cell size is known to vary with distance from the sliding surface,[7,13] τ probably varies also. Therefore, the value used here is an estimate for an average value of τ.

Suh and co-workers[14] have used an intercept method described by Dautzenberg and Zaat[15] to obtain approximate strain profiles for copper samples. While it is not clear that this method can be used directly when the reference region has a qualitatively different microstructure from the deformed region, Suh's value of $\varepsilon \approx 10$ will be used here. The resulting calculated value of the friction coefficient is $\mu \approx 0.78$. For comparison, the measured value was 0.7(5) $\pm$ 0.1. The agreement is much better than might be expected from the use of so many approximate values.

It is interesting to estimate the range of friction coefficients that might be expected for a wide variety of metal systems. If one excludes, for now, the low melting point metals which recrystallize readily near room temperature, then a reasonable range for τ is 70 to 1000 MPa (10,000 to 150,000 psi). Then, using 5 to 20 for ε and 5 to 50 μm for t, one predicts a range of friction coefficients from roughly 0.1 to 30 for unlubricated metal-to-metal sliding. Smaller values of μ could probably be traced to surface films of reaction products or lubricant. The higher values observed in a high vacuum environment have been explained by citing the absence of such films.[16] Surface films in our model affect the way in which stress is transferred from one surface to another in a sliding system.

FRICTIONAL HEATING

For the steady state model of friction described in this paper, most of the deformation energy is dissipated in a region of thickness t below the wear surface. This means that the frictional heat source is *not* a two-dimensional heat source localized at the surface. Instead, the heat source is dispersed over a volume. This has important implications for estimating surface temperatures, which are often difficult to measure directly.

It is well known that solutions of the heat flow equation with localized heat sources indicate higher maximum temperatures than solutions for more dispersed heat sources. Therefore, it is not surprising that evidence for high surface temperatures cannot always be found in sliding systems.[17] Of course, under appropriate conditions, high temperatures can result from either type of heat source, but the dispersed heat source must always give a lower surface temperature than a localized heat source if other conditions are comparable.[18,19]

CONCLUDING REMARKS

The existence of a steady state regime in friction and wear tests indicates that the material near the surface has a low net work hardening rate. It seems reasonable that an ideal material for low friction would be one which has a low work hardening rate near the surface and a high work hardening rate in the interior. This would force deformation toward the surface, reduce friction, delay fracture, and minimize the size of wear particles. The highly deformed layer which develops near the surface appears to be the result of nature's attempt to reach that set of conditions.

The model presented in this paper involves friction as a continuous process. Wear, on the other hand, is more complicated than friction, because it involves plastic deformation *plus* localized fracture events. These discrete events contribute in a random manner to friction, but they are not required for friction. That is, one can have friction between wear events or even without wear. Thus, friction and wear are related, but not in a simple way. It is no surprise that dependable correlations of friction and wear have eluded those who have been tempted to draw them too closely together.

ACKNOWLEDGEMENTS

An early version of this paper was presented on February 27, 1978, at the Annual Meeting of AIME, Denver.

It is a pleasure to acknowledge project support under contracts administered by R. J. Reynik (National Science Foundation), G. Mayer (Army Research Office), and R. S. Miller (Office of Naval Research).

REFERENCES

1. Rigney, D.A. and Glaeser, W.A., *Wear,* Vol. 46, 1978, p. 241.
2. A more comprehensive version of this paper by D.A. Rigney and J.P. Hirth is published: *Wear,* Vol. 53, 1979, p. 345.
3. Wheeler, D.R. and Buckley, D.H., *Wear,* Vol. 33, 1975, p. 65.
4. Bill, R.C. and Wisander, D., *Wear,* Vol. 41, 1977, p. 351.
5. Stobbs, W.M., Kallend, J.S. and Williams, J.A., *Acta Metallurgica,* Vol. 24, 1976, p. 1083.
6. "Metals Handbook," Vol. 8, American Society for Metals, Metals Park, Ohio, 1973, p. 220.
7. Ohmae, N., These Proceedings.
8. Tsuya, Y., "Microstructure of Wear, Friction, and Solid Lubrication," Technical Report No. 81, Mechanical Engineering Laboratory, Igusa Suginami-ku, Tokyo, 1976.
9. Shimadzu Seisakusho, Ltd., Kyoto, Japan, microhardness tester literature.
10. Bowden, F.P. and Tabor, D., "The Friction and Lubrication of Solids," Oxford University Press: Clarendon Press, Oxford, 1964.
11. Buckley, D.H., NASA Report TM X-71781, 1975.
12. "Metals Handbook," Vol. 1, American Society for Metals, Metals Park, Ohio, 1961.
13. Ahn, T.M., Unpublished Research, Ohio State University.
14. Suh, N.P., Jahanmir, S. and Abrahamson, E.P., II, "The Delamination Theory of Wear," Report to U.S. Advanced Research Projects Agency, 1974, p. 125.

15. Dautzenberg, J.H. and Zaat, J.H., *Wear,* Vol. 23, 1973, p. 9.
16. Buckley, D.H., "Friction, Wear and Lubrication in Vacuum," NASA SP-277, 1971.
17. Sproles, E.S., Jr. and Duquette, D.J., *Wear,* Vol. 47, 1978, p. 387.
18. Carslaw, H.S. and Jaeger, J.C., "Conduction of Heat in Solids," Oxford University Press, Oxford, 1959.
19. Malkin, S. and Marmur, A., *Wear,* Vol. 42, 1977, p. 333.

DISCUSSION

Y. KIMURA, University of Tokyo: I think your model is a reasonable one. How do you explain the variation of the friction coefficient with time?

D. A. RIGNEY: I am talking about average friction here, leaving out the noise. There are all kinds of things that are responsible for the variations including the wear debris when it comes loose.

N. OHMAE, Osaka University: Do you have any idea how to distinguish the cell layer from the deformed layer in an experiment because the deformed layer has some cells in it too.

RIGNEY: Yes, I agree. It is certain that if we go deep enough we will find the bulk structure. It may or may not have cells in it depending on the prior preparation. Now as we move toward the surface, we will get to higher and higher strain levels and in some regions cells will come in earlier than others depending on the local microstructural details. As we approach the surface the whole region will be cellular, and if we get close to the surface it will be finer. I agree that it will be cellular below that layer in some places. The cells will not be as well-formed, and they will not be as highly oriented (the preferred direction will not be as striking).

VII. WEAR MECHANISMS

WEAR OF MACHINE ELEMENTS

F. T. Barwell

ABSTRACT

The common machine elements--pistons, crankshafts, gears, cams, valves, electrical commutators, ball and roller bearings are classified from the tribological viewpoint; and the forms of wear encountered are described and quantified where possible.

Various manifestations of wear, for example scuffing, pitting, fretting and abrasion are discussed in relation to their causes and preventive measures. The effect of environment, the relevance of the mode of lubrication and the selection of materials is reviewed. The state of the art is assessed and an attempt is made to identify those areas where fundamental knowledge is insufficient to enable the achievement of practical results.

The modes of wear known to occur in actual machines are then related to fundamental concepts. Thus the adhesive wear concept, whilst relevant in extreme cases, has only limited application to machines which are required to have a normal life under the influence of naturally occurring environments.

The surface is regarded as being stressed mechanically and thermally, as well as being prone to chemical and electrical attack. The nature and intensity of these aggressive agencies determines the nature and extent of wear. The dimensional scale of the various actions is contrasted. Thus degradation due to small-scale stresses on the same scale as asperities may lead to delamination, whereas more intense stress, related to the Hertzian contact area, can lead to pitting failure which may be greatly accelerated by the effects of corrosion. However, some forms of oxidation may be beneficial.

Most tribological situations involve the concentration of a great deal of energy into a small quantity of matter so that thermal considerations are particularly important. Such phenomena as scuffing and thermal stressing can be critical under high speed transient conditions.

Electrical wear can take place simultaneously with the chemical wear and recent advances, notably the effect of micro-arcing, are described.

The implications of progress in tribology for design are discussed with particular reference to elasto-hydrodynamic lubrication and to the availability of wear-resistant materials.

DEFINITION AND CLASSIFICATION OF WEAR

Wear has been defined as "the progressive loss of substance from the operating surface of a body occurring as a result of relative motion at the surface."[1] Examination of machine elements which have suffered wear in industrial use indicates, however, that the action is rather more complex than the phrase 'loss of material' would indicate. In some cases the material may be lost in such a way that, although the size of the machine element may change, its shape and its fundamental effectiveness may be very little affected by the removal of a considerable quantity of material. In other circumstances the change in shape of a component, accompanied by the removal of a very small amount of material, may seriously affect the functioning of an appliance.

Whilst wear can be regarded as the result of the surfaces being stressed mechanically, thermally, chemically or electrically, the actual mode of removal of the material therefrom may take one of a number of forms dependent on the composition of the surface and the nature of the stress system. Thus it may be expected that different types of machine element may give rise to different manifestations of wear.

Wear may be characterized in two ways. It can be dealt with phenomenologically; that is, the terminology of the workshop may be used to describe the appearance of worn parts. Or alternatively, the basic mechanism whereby material is removed from the surface can be identified and the wear system designated accordingly.

The most important classification of interacting surfaces arises from the definitions derived from the science of mechanics of machines.[2] Interacting surfaces, known as 'pairs,' fall into two categories--'lower pairs' and 'higher pairs.' The lower pairs are characterized by sliding such as may occur when two planes are in contact or when two closely-fitted cylinders move one relative to another. Apart from an intermediate form identified as the screw pair, all other pairs are 'higher pairs' insofar as they permit relative rotation of the two components irrespective of whether or not there is relative sliding. This implies nominal 'point' or 'line' contact. Of course, any attempt to transmit a force through a point or line having zero area will result in infinite stress. Accordingly, all bodies made of real materials will deform either elastically or plastically so as to give rise to a finite area of contact. Naturally, if a component is intended to retain its shape it must be made sufficiently hard for the load to be distributed without permanent deformation, that is elastically. The first calculations quantifying the contact of semi-infinite bodies were published by Hertz and these are, therefore, known as Hertzian contacts.

The essential difference between the lower and the higher pairs is, of course, that it is possible in a lower pair for the two surfaces to conform together whereas in the higher pair the surfaces are non-conforming or counter-formal. The nature of practical surfaces is that they are never atomically smooth and, in the case of the lower pairs, contact takes place between many asperities. Because the two surfaces are macroscopically conforming, the points of contact are distributed over a wide area, and, therefore, can be referred to as 'dispersed contact.' Because of the relatively large area, contact stresses may be low and the soft materials can be used, preferably for only one of the two elements of the pair. In these circumstances asperities will deform plastically thereby enhancing the degree of conformity.

In most common arrangements of bearing systems one of the elements of the pair is stationary and the other moves. It is, therefore, generally preferable for the moving component of a journal bearing to be made of hard material and the stationary part of a soft material so that the latter becomes bedded-in to suit the shape of the cylindrical portion.

In addition to the shape and the intensity of loading characteristic of the interacting surfaces, it is also important to consider their relative motion. When surfaces are closely conforming but not quite perfect, the combined effect of the motion, geometry and viscosity of the fluid can cause the generation of hydrodynamic pressures sufficiently high to separate the surfaces. This is known as 'hydrodynamic lubrication' and the force which can be generated to separate the surfaces is proportional to the viscosity times the effective velocity, together with geometrical factors characteristic of the bearings. When the two surfaces move relative to some reference plane (often for convenience taken as the direction of the forces applied to the plane) the effective velocity is the sum of the two velocities measured in the same direction. This is also a measure of the amount of fluid induced to flow between the surfaces.

A critical factor in almost all tribological applications is the derivation and dissipation of frictional heat. Frictional heat is quantified as the product of the load, the coefficient of friction and the relative velocity between the two surfaces (where both surfaces are moving, this represents their difference). In the absence of fluid lubrication, the heat generated at any interface has to be conducted through the material. When the zone of contact in an element of a kinematic pair moves over the face of that element, the thermal conditions are not as severe as when it is stationary because it moves over a relatively cool material. Thus the system can be characterized from a thermal point of view by a ratio known as the 'slide sweep ratio' defined in Equation 1 as follows:

$$\frac{U1 - U2}{U1} \text{ for body 1 and } \frac{U1 - U2}{U2} \text{ for body 2} \qquad (1)$$

The numerator represents the heat input to the surface represented by the difference in velocities and the denominator represents the speed at which new material is passed to the surface. Thus the simple journal bearing wherein the housing is stationary gives a figure of unity for the shaft and infinity for the stationary bush. This signifies the high thermal loading associated with the stationary part from which all heat must be removed by conduction. Figure 1 shows a bearing which has failed by overheating.

The configuration of a bearing system, i.e., whether counter-formal or conformal, will not only determine the selection of materials used in its construction but will also influence the mode of lubrication, if any, as well as ultimately determining the form to be taken by wear. In the case of counter-formal contacts, the intensity of pressure in the region of contact is such that classical concepts of hydrodynamic lubrication cannot apply. Nevertheless components such as spur gears have operated for many years and are quite clearly effectively lubricated. It appeared from consideration of the viscosity of the lubricant that some other property was responsible for the effectiveness of the lubrication and this led to the introduction of the term 'boundary lubrication.' Recent work[3] has shown

Fig. 1.—Bearing seized due to overheating.

that the effective lubrication under these conditions may indeed be hydrodynamic but is made possible by two factors not taken into account by the early workers. The first is that, due to the deformation locally under the action of the Hertzian stress, the surfaces conform to a greater degree than assumed in the simple hydrodynamic theory. The second is the fact that the viscosity of the petroleum lubricant is increased many times as a result of the applied pressure. Indeed it has been shown recently(4) that in many circumstances the lubricant may assume a glassy state.

MODES OF WEAR COMMONLY ENCOUNTERED IN MACHINE ELEMENTS

Figure 2 attempts to display the design characteristics and the form of wear encountered by a selection of common machine elements.

Plain Bearings

These occupy pride of place in most texts on lubrication. When allowed to operate eccentrically the journal bearing develops a hydrodynamic film which separates the surfaces and promotes a copious flow of lubricant which prevents overheating. Similarly the plane thrust bearing embodies a wedge action. In these circumstances wear does not take place. This ideal situation is not always achieved. Sometimes misalignment (either inherent in the way the machine is assembled or of a transient nature arising from thermal or elastic distortion) may cause metal-to-metal contact to occur, rubbing may occur at the instant of starting before the hydrodynamic film has had an opportunity to develop, the bearing may be overloaded from time to time and most probably particles of foreign matter will be introduced into the film space. In some appliances, diesel-engines for example, acids and other corrosive substances may be formed during combustion and transmitted by the lubricating oil so as to induce a chemical attack on a surface. The continual application and removal of hydrodynamic stress on the shaft may dislodge loosely-held particles. It is, however, the particles of foreign matter which are responsible for most wear. Sometimes these can exhibit attrition

CONDITIONS OF APPLICATION

NATURE OF CONTACT	STRESS SYSTEM	TYPICAL APPLICATION	SLIDE/SWEEP RATIO
Lower pair (Conformal)	Dispersed	Plain bearings	1 for journal ∞ for lining
		Piston and piston rings	∞
		Cylinder bores	1
		Worm gears	High
		Brakes and clutches	1 for moving ∞ for fixed part
		Commutators and slip rings	1 for rotor ∞ for brush
		Fixtures subject to vibration	High
Higher pair (Counterformal)	Hertzian	Rolling contact bearings	0
		Involute-gears	Real, but passes through zero at pitch line
		Hypoid gears	High
		Cam and followers	0
		Friction drives	0
		Wheel on rail	0 nominally
		Electrical switchgear	Indeterminate

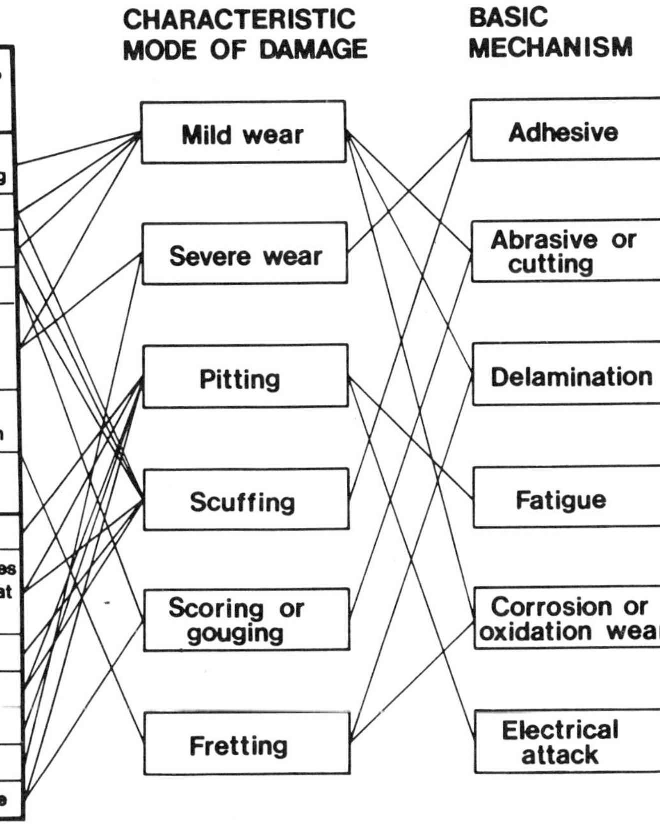

Fig. 2.—Classification of Tribological Nature of some machine elements.

in the vicinity of the duct through which oil is supplied to the bearing. Most commonly, however, the particulate matter is trapped between the journal and the bearing. Indeed it is a characteristic of white metal that it can allow particles to become embedded, thereby relieving the situation. However, it has been shown[(5)] that the hard shaft of an engine often wears more rapidly than the nominally soft bearing metal. This is attributable to the embedment of hard particles within the soft material to constitute a lapping system. Generally, however, the wear of hydrodynamic bearings can be regarded as mild and attributable to occasional abrasive action. Chromium plating of crankshaft bearings is successful in combating abrasive and corrosive attack.[(6)]

Piston, Piston Rings and Cylinder Bores

One of the most common machine elements is the piston within the cylinder which normally forms part of an engine although similar arrangements are encountered in pumps, hydraulic motors, gas compressors and vacuum exhausters. The primary function of a piston assembly is to act as a seal and to counter-balance the action of fluid forces acting on the head of the piston. In the majority of cases the sealing action is achieved by the use of piston rings although these are sometimes omitted in fast running hydraulic machinery finished to a high degree of precision.

Pistons are normally lubricated although in some cases, notably in the chemical industry, specially formulated piston rings are provided to function without fluid lubrication. It has been shown[(7)] that lubrication is hydrodynamic, the viscosity of the lubricant is therefore critical from the point of view of maintaining and lubricating film and of carrying out its main function which is to act as a sealing element.

Lloyd[(8)] has shown that piston-rings tend to wear-in so as to assume a parabolic profile and Hamilton and Moore[(9,10)] have measured film thickness and pressure in running engines. They confirmed the existence of hydrodynamic action but measured film thickness of 1μm when theory indicated 10μm.

Failure of a piston system to function properly is evidenced by the occurrence of blow-by and eventual loss of compression. Design must be a matter of compromise insofar as a very effective piston assembly (operating with a thick oil film, low friction and no 'blow-by') could produce undesirably high oil-consumption. On the other hand it has shown that most wear occurred in the vicinity of the top dead-centre where the combination of pressure, velocity and temperature are least favorable to the operation of a hydrodynamic film. Conditions in the cylinders of an internal combustion engine, for example, can be seriously corrosive due to the presence of sulphur and other harmful elements present in the fuel oil. Corrosion can be particularly harmful before an engine has warmed up when the cylinder walls are below the dew-point of the acid solution. Wakuri and Ono[(11)] have reported the occurrence of abnormal wear which was attributed to conditions established during the initial running-in process. Serious edge-loading of piston rings caused scuffing which was accompanied by accelerated wear occasioned by large-size worn particles scraped from the cylinder liner.

The normal running-in process can be completed during the period of the works trial after which the wear rate tends to fall as time goes on. High alkaline oil is more apt to cause abnormal wear and this is attributed to a lack of spreadability at high temperature. Machined finishes are regarded as having more resistance to scuffing than ground finishes because of the oil retaining characteristics of the roughened surfaces. The use of taper face-rings is effective in preventing scuffing by relieving the edge load in the earliest stages of the process. A high-phosphorous lining is somewhat better than a vanadium lining in preventing scuffing and Wakuri and Ono[11] recommend the use of a rotating piston mechanism to enhance the resistance to scuffing.

Some difference in viewpoint exists regarding the prevention of scuffing of engines, some designers maintaining that this is no problem as far as they are concerned whereas others admit to having difficulties. It is general experience that, provided an engine type can be run-in satisfactorily, scuffing is unlikely to occur during normal running.

Being a term used to describe the appearance of failed components there is some ambiguity regarding the meaning of the term 'scuffing.' The OECD glossary[1] describes it as follows:

"Scuffing -

Localised damage by the occurrence of solid phase welding between sliding surfaces without local surface melting.

1. in the UK scuffing implies local solid-phase welding only;
2. in the USA scuffing may include abrasive effects;
3. in the USA the term 'scoring' is sometimes used as a synonym for scuffing."

If we accept the definition based on the concept of solid-phase welding then we must classify scuffing as being a manifestation of adhesive wear. It is probably the most frequent example of this mechanism encountered in ordinary engineering practice.

The incidence of scuffing in engines must be associated with the breakdown of hydrodynamic conditions. It is the writer's view that insufficient attention has been given to the importance of temperature of the ring belt and it is understood that some manufacturers have removed this to a position lower down the piston where it is cooler and have thereby avoided all difficulties. In a comprehensive review of piston-ring scuffing Neale[12] proposes two main methods by which the problems may be attacked as follows:

1. to improve the design of the rings and their operating environment so that they operate with increased film thickness at the critical periods of the cycle, and
2. to leave the ring design and operating conditions basically unchanged so that the film thickness remains low but to use materials and finishes for the surface of the rings and cylinders which have great resistance to scuffing.

As previously indicated, the enhancement of the film thickness as envisaged in Method 1 may very well result in a prohibitive increase in oil consumption. Nevertheless, a reduction in temperature at the ring-belt might produce an effective compromise between film thickness and oil consumption.

As regards materials there is considerable evidence[13] that scuffing is associated with drastic metallurgical changes. Micro-hardness determinations of scuffed material show it to be very hard, and taken in conjunction with the very fine structure revealed by the electron microscope, indicate that any process of fusion and solidification has taken place with great rapidity.

Ting and Mayer[14] conclude that the cylinder bore wear pattern for reciprocating engine can be predicted analytically by considering the hydrodynamic lubrication of the piston rings, gas blow-by through the ring-pack, piston ring elastic characteristics and the lower limit of the film lubrication.

Whilst the ring pack dominates the situation, the piston side thrust force contributes to the uneven wear along the major and minor thrust sides of the cylinder wall. The largest bore wear occurs at the upper end of the cylinder and is evidenced in two peaks. The first peak arises from the high gas pressure behind the top ring at the onset of the power stroke and the second peak results from the accumulation of wear due to each individual ring rubbing over that point. The wear tapers off at the middle of the stroke and then increases at the bottom of the stroke.

Wiln and Brett[15] conclude from experiments in a mechanical rig that scuffing failure is sensitive to speed, load and temperature and that for a particular temperature condition failure levels are related by $PU^{7/16}$. A test on actual engines indicated that the metallurgical characteristics of cast-iron lining materials such as graphite structure, phosphorous content and free-ferrite content had a relatively small influence on performance. Special ring and liner materials and surface treatment could, however, give up to 75% improvement in performance. Bore finish was particularly critical and a diamond-honed finish gave a result which was only 35% of that of the normal finish treatment.

Phosphate treatment of the bores of cylinders is beneficial particularly during the running-in period.

Chromium plating of cylinder bores leads to reduced wear rate particularly under corrosive conditions. Surface texture is important and two variants are available. The bores may be provided with a dense chromium layer which is then honed and 'honey-combed' using eddy-current etching or the initial plating may be carried out so as to provide a porous surface.

When Jones et al[16] heated the ferrograms from a series of tests taken with a diesel engine during running-in they obtained the surprising result that initially only 40% of the larger particles were cast iron and 60% were carbon steel. This percentage gradually changed until after five hours test running 20% of the particles were cast-iron and 80% carbon-steel. During the period 74 and 113 hours only 10% of the particles were cast iron. This result appears to suggest that a high proportion of the wear damage during running-in arose from components made of steel such as the crankshaft, gear train and oil pump rather than within the cylinder bore itself. It was characteristic of the cast iron that it generated small fragmented particles in the sub-micron size range indicating that the normal mode of wear can best be described as 'mild wear.' Further investigation is necessary to elucidate the mechanism of this mild wear but it is probably based on fatigue action analagous to that proposed in the delamination theory of wear.

Worm Gears

The worm gear is somewhat anomalous because of the degree of conformity which is greater than any other type of gear, and it is best classed as representing a screw pair within the family of lower pairs. However, it represents a fairly critical situation in view of the very high degree of relative sliding. Practically the only suitable combination is phosphor-bronze with hardened steel. Good finish and accurate, rigid positioning is essential. Lubrication is critical and Wilford[17] has shown that either a castor oil or a castor-oil synthetic ester blend was necessary for effective operation. Wear is mild and probably corrosive as a result of the action of the boundary lubricants.

Auger electron micrographs of the bronze pins which have been rubbed against steel in order to simulate the action of worm gears have been published by Jones in 1976.[18] An interesting feature of this method is that progressive removal of the surface layers by bombardment with Argon ions whilst in the Auger spectrometer reveals how the composition of the surface film may vary through the thickness. The value unity in Figure 3 represents the composition of the bulk material which in this case was commercial quality 5% tin phosphor bronze. When plain mineral oil was used as the lubri-

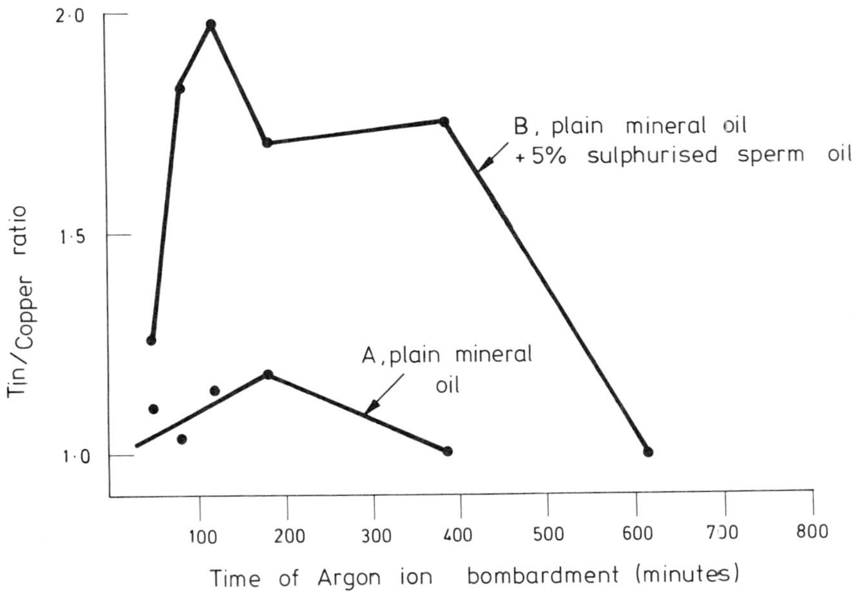

Fig. 3.—Ratio of Tin/Copper concentration peaks, bulk material taken as unity.

cant there was little variation in film content. When, however, 5% by volume of commercial quality sulphurised sperm oil was added to this oil the results were as shown in curve B. The reasons for this behavior are not yet fully understood, but it is clearly associated with a differential rate of reaction of the sulphurised oil with the tin and copper respectively. This is clearly an area which warrants further investigation, both from the point of view of practical application and for the elucidation of the fundamental mechanism of certain classes of extreme pressure lubricants.

Brakes and Clutches

Brakes and clutches are usually constructed so that a continuous metal member is in sliding contact with another member which is sometimes discontinuous. The continuous member is usually made of steel and the secondary member of a non-ferrous material; or, if it is made of a ferrous metal, the secondary member is of a very different composition from the continuous member. Thus in railway braking, blocks of cast iron are forced against a continuous tyre made of carbon steel.

Because the function of a brake is to convert mechanical work into frictional heat, a first requirement of brake lining material is heat resistance. Consequently, the most common form of brake linings for automobile use are based on asbestos which is held together in various resin formulations. In order that the rate of heat generation must be distributed evenly over the surface, brake lining materials require a high degree of compliance. Even so, the tendency exists for the distribution of temperature to become unstable because of the thermal expansion of the heated asperity regions.(19) Although aircraft braking materials are subject to a less complex duty cycle than those used for automobiles, conditions during relatively infrequent brake applications are very much more severe and the materials used for liners are usually copper-based composites which contain silicon oxide and aluminum oxide to resist wear, and graphite to maintain constant frictional properties. Ting Long Ho(20) has studied the wear of frictional pads of conventional copper-based brake-material contained in a steel cup. This composite material contained by volume about 31% copper, 22% mullite, 32% graphite and 15% friction modifiers such as iron, lead or tin. Wear debris from this material was metallic and heavily oxidized and consisted of flat plates of average size about $4x3x0.3x10^{-4}$m.

The formation of these particles resulted from three distinct modes of crack generation. The generation of the three cracks was attributable as follows:

1. cracks across the sliding direction were due to tensile failure in the surface layers as a result of the frictional force.

2. Longitudinal cracks in the operating surface arose through the thermoelastic and thermo-plastic instabilities in sliding as a result of what is referred to as the 'Barber' effect.

3. A third kind of crack which arose in the sub-surface region initiated by some stress raiser such as a large particle embedded in the surface layer and accelerated by oxidation.

The effect of termperature on the wear rate is therefore two-fold. In the first place it can set up fatigue action due to volume instability and secondly, it can increase the rate of oxidation. Further studies are necessary in order to assess the effects of various possible inclusions in the wear resistance of the brake material.

Commutators and Slip Rings

Essential features of most electrical machines, these components impose the conflicting requirement of provision of a low-friction low-wear surface with intimate contact. The most common material for the continuous element is copper although stainless steel has been successfully used for slip rings. The stationary discontinuous elements known by the traditional name of 'brushes' now consist of carbon blocks containing various additive materials. Successful operation depends on the development of a suitable patina on the metallic commutator or ring surface which consists of an admixture of metal oxide and carbon flake. Electrical contact and conduction only occurs over a small proportion of the area and the mechanism involved is different according to whether the current passes from the ring to the brush or vice versa. It is believed that metallic spots are necessary in both cases but in the case of flow of electrons from the ring to the brush, a contact spot is formed on the moving part which is traversed across the face of the brush. In the converse case it is believed that detached metallic particles become embedded in the stationary brush.

The effect of the ambient atmosphere has been shown to be of critical importance, attempts to operate in a vacuum leading to a disintegration of the protective surface film accompanied by high rates of wear. Special additives are incorporated in brushes which are required to operate in the partial vacuum conditions, for example, in aircraft required to operate at high altitudes. In addition to ohmic contact through minute metallic asperities as indicated, Llewellyn Jones[21] points out that an effective resistance of less than 10^{-3} ohms is required when kilo amps are required to be passed. This is made possible by a quantum mechanical tunnel effect in which electronic conduction takes place over distances of one nano-metre under the influence of electric fields of the order of giga-volts per metre through films of matter which in bulk would be insulating. The tarnish oxide layers, which have been found necessary for the satisfactory tribological operation of commutators, fall within this category. A naturally occurring film which can also protect the elements from excessive wear and yet conduct by the tunnel effect is a layer of absorbed gas molecules. Under vacuum conditions supply may be maintained by occlusion from the bulk graphite which can absorb a high content of gas.

It has also been noted that the temperature of commutator segments on machines which are run without electric power, i.e., railway traction equipment during a period of 'coasting' is higher than when current is flowing. This may suggest the possibility that metallic contact areas may melt and provide an element of hydrodynamic lubrication. Indeed the sliding electric contact represents a tribological system about which a great deal has still to be learned.

Fixtures Subject to Vibration

A form of wear frequently encountered between components which are not intended to move one relative to the other is 'fretting corrosion.' In addition to the degradation of the surfaces resulting from this phenomena, its occurrence may cause the initiation of fatigue cracks.[22] Fretting corrosion is very much affected by environmental condition. There appears to be no way of eliminating it apart from preventing the relative motion of the interacting surfaces by better fitting or the introduction of interference fits. Palliatives are the avoidance of metals which form hard abrasive oxides, phosphating the surface and providing sufficient lubricant to prevent the ingress of oxygen.

Rolling Contact Bearings

These bearings typify the widest class of tribological system, that which embodies Hertzian contact. From the practical point of view they are divided into two broad classes--ball and roller bearings, although the nature of the contact and the laws governing friction and wear behavior are common to both classes. Although the contact is basically a rolling action, most bearings involve an element of sliding and this is particularly the case with certain types of roller bearing, notably the taper rolling bearing where sliding occurs at point A, Figure 4. Any rolling contact bearing is characteristic numbers, 'the static load rating' and the L_{10}.

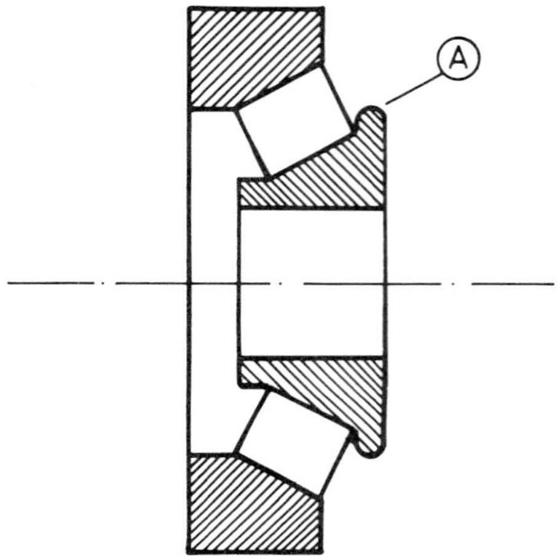

Fig. 4.—Critical tribological region in tapered roller bearing.

Static load rating

The concept of 'the static load carrying capacity' is used to denote the load that can be applied to a bearing which is either stationary or subject to slight swivelling motion without impairing its running qualities for the subsequent rotation. In practice this is taken as the load which results in the combined deformation of the rolling element and race-ways at any point not exceeding 1000th of the diameter of the rolling element. Most manufacturers' rate permissable static load on a maximum Hertzian stress of 2.6 GPa (380,000 psi). Hartnett[23] maintains that these ratings are extremely conservative. It would appear to be a matter of some importance to determine what the rating should be in the light of modern materials and manufacturing methods.

Running load capacity

It is accepted that the lives of rolling contact bearings are subject to a wide dispersal due to the random incidence of fatigue failure. The L_{10} life is defined as that which can be stood by 90 out of 100 bearings without failure. Thus a designer has to determine the expected life of a bearing and also to specify the reliability, i.e., the probability that the machine will continue to operate for that life without its performance being impaired by bearing failure. It is customary to obtain the appropriate information from manufacturers' catalogues. In recent years better understanding of rolling element bearing design, materials, processes and manufacturing techniques has permitted general improvement in bearing performance as measured by a higher bearing reliability or longer life in a given set of circumstances. The implication of these improvements has been studied by the Rolling Elements Committee of the Lubrication Division of the ASME and recommendations have been published in an engineering design guide.(24)

The expected bearing life may be related to the calculated rating life by L_{10} or to the concept C, the basic load rating, as shown in Equation 2.

$$L_a = (D)(E)(F)(G)(H)L_{10}$$
$$= (D)(E)(F)(G)(H)\left[{}^{C}/_{P}\right]^n \qquad (2)$$

where (D) to (H) are life adjustment or bearing design factors

C	basic load rating
P	static equivalent load
n	load-life exponent; either 3 for ball bearing or $^{10}/3$ for roller bearing.

Materials factor (D)

A number of investigations into the dispersal of life characteristics of rolling contact bearings have been carried out and the results are such as to lead the ASME Committee to assign the value 2 as a materials factor for the AISI 52100 steel.

Processing factor (E)

Five methods of control of the melting process were considered as follows: Air, Vacuum Induction, Consumable Electrode Vacuum Remelt, Vacuum De-Gassing and Electro-Slag Remelt. Consumable Electrical Vacuum Remelt in particular enabled bearing life to be extended beyond the present-day catalogue value. Some investigators have shown an improvement of up to 13 times, but the Committee recommends a factor of 3 as a conservative figure.

Lubrication factor (F)

Under the intense Hertzian conditions characteristic of rolling contact bearings, elasto-hydrodynamic lubrication may be expected to occur. In this system the Dowson-Higginson formula(26) indicates the effect of speed is significant, but that variation in loads leads to very little variation in film thickness (additional load leading to a commensurate increase in area of contact) as follows:

$$H_m = {}^{h}m/_R = 2.65\,\frac{U^{0.7}\,G^{0.54}}{W_1^{0.13}}$$

h_m = minimum film thickness

U = speed parameter $\eta_o \quad (U1 + U2)/E^1R$

W_1 = *load parameter* P/E^1R

G = *materials parameter* αE^1
where α *is the index in the expression* $\eta = \eta_o \varepsilon^{\alpha P}$

$$\frac{1}{E^1} = {}^1/_2\left(\frac{1-\gamma_1^2}{E_1} + \frac{1-\gamma_2^2}{E_2}\right) = \frac{1-\gamma^2}{E} \quad \textit{for similar materials}$$

where E_1 and E_2 and γ_1 and γ_2 are the Young's moduli and Poissons ratios of the materials of the two interacting bodies respectively. Curves are provided in the design guide to enable the factor (E) to be determined for any given set of circumstances.

Speed effect factor (G)

This may be an adverse effect due to the importance of the centrifugal force in modern bearings which may be required to operate at much higher speeds than heretofore. Tabulated data is provided in the design guide.

Misalignment factor (H)

Misalignment is regarded as being more detrimental to the operation of roller bearings than of ball bearings. The design guide suggests that misalignment in radians should be limited to the following values:

spherical bearings	0.0087
ball bearings	0.0035-0047
cylindrical and taper roller bearings	0.001

Provided misalignment does not exceed these limits, it can be ignored in calculations.

In view of the importance of roller contact bearings in engineering practice it would seem to be essential that the possible improvement of ratings should be kept under continuous review.

Cleanliness in lubricant

In a recent investigation(26) special precautions were taken to ensure cleanliness of the lubricant. This resulted in the absence of failures within a period which in normal circumstances would have been appropriate to produce a number of failed bearings. It is therefore clear that debris of various types in the oil can result in the initiation of failure and quantification of this effect would appear to be an important subject for future study.

Involute Gears

At the instant where the line of contact crosses the common tangent to the pitch circles, involute gear teeth roll one over the other without sliding, but during the remainder of the interaction, i.e., when the contact zone lies in the addendum and dedendum, a certain amount of relative sliding occurs. Therefore, pitting is mostly to be found on the pitch line whereas scuffing is found in the addendum and dedendum region. A certain amount of clarification is still required to elucidate the precise conditions for the commencement of scuffing. Some work on pitting of metallic and non-metallic gears is reported by Krishna Murthy.[27]

Rozeanu and Grosberg[28] show that with good quality hardened gears, scuffing occurs at the point where deceleration and overload combine to produce the greatest disturbance. However, before reaching the scoring stage, another type of damage is observed which is located in the vicinity of the tip of both pinion and gear teeth. This type of damage is believed to be due to abrasion by hard debris detached from the tip wedge.

Brigger, Daus and Schultz[29] provide data on the fatigue life of gears and provide clear evidence of sub-surface fatigue due to Hertzian stress. They draw attention to the growth of fatigue cracks in relation to the effect of lubricant trapped in an incipient crack during successive cycles.

Hypoid Gears

Hypoid gears are mainly encountered in right-angle drives associated with the axles of automobiles. Tooth actions combine the rolling action characteristic of spiral-bevel gears with a degree of sliding which makes this type of gear critical from a point of view of surface loading. Towle and Graham[30] list five modes of failure by surface damage by automobile gears, notably 'rippling,' 'ridging,' 'pitting,' 'scratching' and 'scuffing.' Successful operation of hypoid gear is dependent on the provision of the so-called extreme pressure oils. U. S. military specifications Mil.L2105 and Mil.L.2105b define the properties required for acceptable axle oils. Although these are all based on military requirements they are also used for civilian purposes. There are several types of additives for compounding hypoid lubricants. Lead-soap, active-sulphur additives may prevent scuffing in final drives which have not yet been run-in, particularly when the gears have not been phosphated. They are not satisfactory under high torque but are effective for high speeds. Lead sulphur chlorine types are generally satisfactory under high-torque low speed conditions but are sometimes less so at high speeds. Figure 5 illustrates examples of 'scuffing,' 'scoring,' and 'rippling.'

Friction Drives

Friction drives which are being used increasingly in infinitely variable gears, are the converse of hypoid gears insofar as it is the intention that two smooth machine elements should roll together without sliding whilst being capable of transmitting a peripheral force from one to the other. Problems of friction drive have led to a number of investigations of the drag forces in elastohydrodynamic lubrication[31,32] and a good deal of work was reported by Crook[33] who, in common with the majority of workers in the field of elastohydrodynamic lubrication, used two cylinders to represent the interacting elements. With this arrangement, slide-roll ratio can be varied and more elaborate arrangements embodying three or more cylinders may be adopted.

If frictional traction is plotted against relative sliding three principal modes may be identified as in Figure 6. First there is the linear mode in which traction is proportional to the relative velocity of sliding; then there is the transition zone during which a maximum is reached, and

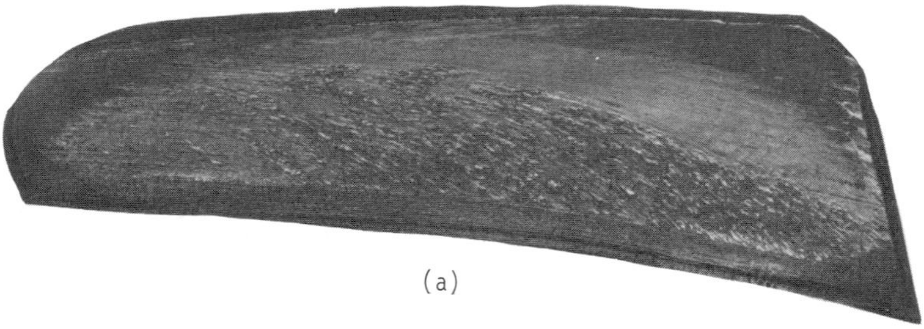

(a)

(b)

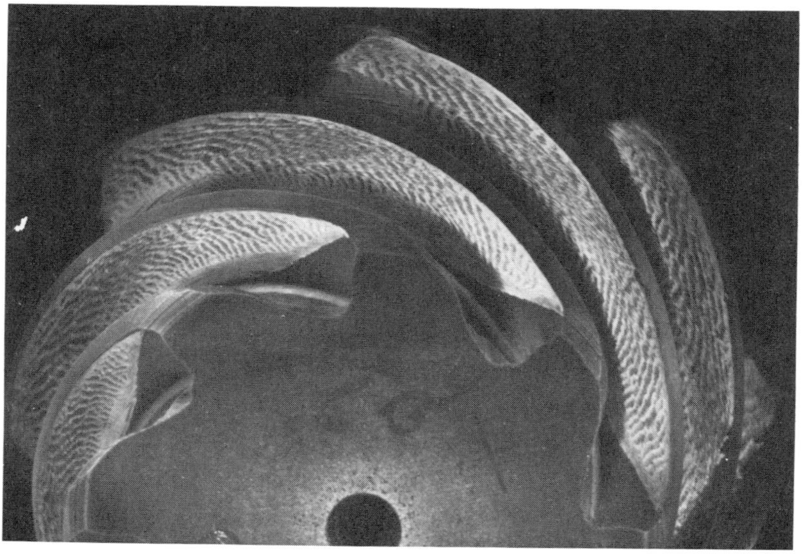

(c)

Fig. 5.—Hypoid Gears:
(a) Scuffed gear flank (coast side)
(b) Scored gear flank (drive side)
(c) Rippled gear

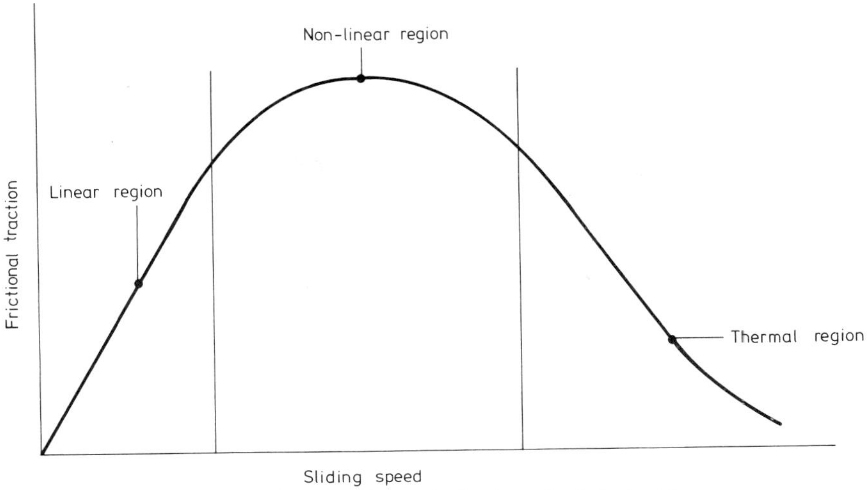

Fig. 6.—Frictional traction in elasto-hydrodynamic lubrication.

finally a third zone with a falling characteristic. The initial region can be shown to relate to the rheological properties of the oil, viscosity being of predominant importance.[34] However, a certain mystery surrounds the fact that a maximum value is observed in the second zone. It is now believed that under appropriate circumstances a lubricant within a film under the intense pressures of the Hertzian contact becomes a glass-like solid, which in common with other solids, has a limiting strength corresponding to the maximum value of traction[4] Regarding the third zone, the falling off in traction is attributed to the fall in its viscosity associated with increase in temperature of the lubricant. Odi-Owei and Roylance[35] investigated contacts in which one of the surfaces was stationary. They applied Crook's theory to the situation and showed that if appropriate allowance was made for variation in the thickness of the film occasioned by diminution of viscosity with temperature, excellent agreement was to be found between theory and experiment (Figure 7). A difference in traction performance of different lubricants was noted and was shown to be directly related to lubricant properties notably the base viscosity and the temperature coefficient of viscosity.

Cam and Cam Followers

An often quoted instance of the benefit resulting from use of elastohydrodynamic theory was published by Muller.[36] He indicated that the calculation of Hertzian stress alone could be misleading and that it was necessary to take the mean rolling speed into account. He showed that a reduction in nose radius of a cam which increased Hertzian stress, also increased the relative velocity and thus the oil film thickness. This was shown to be an advantage because the cam with the thicker film operated satisfactorily in service whereas the one with the thinner film failed prematurely.

Temperature limitations are likely to be important in the case of cams required to operate under intense conditions, and it has been shown by Watson and Milkins[37] that scuffing is most likely to occur at speeds removed

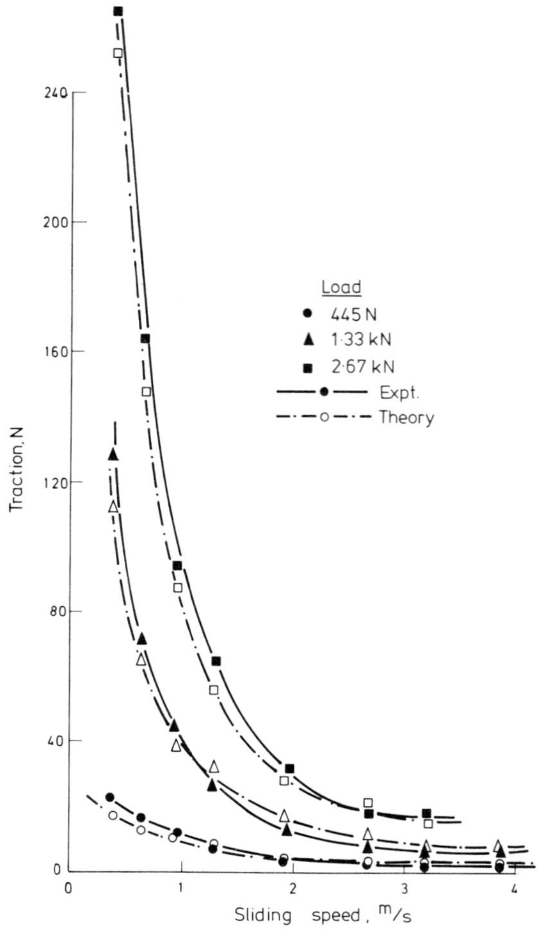

Fig. 7.—Correlation of theoretical traction with experimental data (HVI 650).

from the design speed. The loading of cams is never steady and an important development of the elastohydrodynamic theory must be towards the evaluation of transient load conditions. Vichard and Godet(38) and Vichard(39) have demonstrated that a 'sqeeze' component may be beneficial in this respect. Beeze and Clark(40) have reviewed the selection of materials for use in cams. Relatively thick coatings are envisaged in order to embrace the region of maximum Hertzian shear stress.

Wheel on Rail

In railway practice Hertzian stresses are high (in the order of 10^9 N/m^2) and a certain amount of work hardening occurs in the early life of the component. When a typical wheel load of $10x10^4N$ is applied, deflection of the contact zone will be .1 of a millimetre so that the contact region can itself be regarded as a spring. Association with the mass of the axle can give rise to vibration and, as demonstrated in a disc machine,[41] to a form of wear known as 'corrugation.' Experience of corrugation on the track does not necessarily correlate with laboratory experience. In Johnson's laboratory tests the pitch of the corrugations was approximately 50mm, whereas on the track, the wave length varies because of the resilience of the rail and of its support on sleepers and ballast. Johnson and Gray[41] consider that friction and wear on the rail head are more significant in the generation of corrugation than plastic flow of metal.

Other forms of rail wear are associated with oxidation and for many years it was considered that this was the predominant mode.[42] Recent work by Clayton[43] has shown the importance of severe plastic working particularly where route alignments are sharply curved. The evaluation of the relative importance of corrosion and plastic deformation waits further investigation.

Electrical Switchgear

Electrical switchgear involving the making and breaking of electrical circuits by bringing conducting components into contact with or without a degree of relative sliding (wipe) is of importance, not only because of the limitation in the life of equipment due to the wear of the contacts, but also in the interests of safety. If, for instance, contacts became welded together so that when it was desired to switch off current, the circuits did not separate, disaster would result. The physics of electrical contacts has been studied recently, notably by Llewellyn Jones[44] who defines the functions of a contact as being to carry out the following sequence:

a. to close the circuit

b. to allow the current to pass for a specified time

c. to open the circuit and terminate the current.

This sequence must be repeatable on demand.

Continued operation leads to deterioration of the surface or loss of material so that eventually the system must be taken out of service. The underlying cause of this deterioration is the electrical process rather than ordinary mechanical wear.

When two charged conductors approach each other, high electrostatic forces are set up in the vicinity of microscopic protruberances. Thus conduction may commence even before physical contact is made. Once the circuit has been completed, contaminating films or rough surfaces may concentrate the areas of true contact so that melting occurs. If, for any reason, current ceases to flow before an attempt is made to open the switch it may become welded. Normally, however, as contact pressure is reduced, so is the number of contact areas, until the high current densities operate from a very small area. This leads to the formation of molten bridges as described in Section--Electrical Damage.

SOME WEAR MECHANISMS

Mild and Severe Wear

The terms 'mild' and 'severe wear' were first introduced by Hirst in order to describe the sudden transitions which were observed in wear experiments.(45) As such they denoted qualitative differences and the use of quantitative terminology to describe a qualitative difference is in the view of the writer somewhat unfortunate.

Kragelskii(46) refers to 'internal' and 'superficial' friction. Superficial friction occurred when the interface between the rubbing surface was weak so that sliding could take place therein with relatively little damage and comparatively no friction. When the bond between the surfaces (whether produced by penetration of asperities or by adhesion) was stronger than the underlying pairs, failure would occur in the bulk of the material. This internal friction will affect larger volumes of material causing considerable roughening and damage to the surface. Lancaster(47) explains the transition from mild to severe wear as arising from competition of two opposing dynamic processes. One of these is the formation of fresh metal surface and the other is the formation of surface oxide film by reaction with the surrounding atmosphere. The transition from one mode to another, ostensibly due to changes in load of speed can generally be attributed to a change in temperature. Severe wear conditions are usually intolerable in machinery and when they occur the general deterioration in the surfaces may lead to a catastrophic worsening of the situation and eventually to seizure. Mechanisms which leave the surface relatively unaltered such as delamination wear (see Section--Delamination Wear) can lead to an adequate life in practical machines.

Effects of lubrication

If the heights of asperities are randomly distributed, some interactions may be expected to be so severe as to cause intense but localized surface damage whereas others will be accommodated by elastic deformation.

Whether or not deformation will be primarily plastic will be determined by the value of the plasticity index λ as defined in equations 4(a) and 4(b).

$$\lambda = \frac{E^1}{H} \left(\frac{\sigma}{r}\right)^{\frac{1}{2}} \qquad 4(a)$$

or

$$\lambda = 0.6 \frac{E^1}{H} \frac{\sigma}{L} \qquad 4(b)$$

where

$$\frac{1}{E^1} = \frac{1 - \gamma^2}{E_1} + \frac{1 - \gamma^2}{E_2}$$

and E_1 and E_2 and γ_1 and γ_2 are the Youngs moduli of elasticity and Poissons ratios of the materials of bodies 1 and 2 respectively.

σ = standard deviation of asperity height distributions

r = asperity radius

H = hardness

L = correlation distance.

The effect of λ on the occurrence of mild or severe sliding conditions has been investigated by Hirst and Hollander(48) using a stainless steel ball of hardness 500 vpn and 12.7mm diameter which was arranged to slide over flat plates made of the same material but of hardness value 180 vpn. Different surface finish characteristics of the flat plates were obtained by using different grades of abrasive paper in the finishing process. Values of σ ranged from 0.01μm to 2μm and of L from 2μm to 30μm. Lubricant was a 1% solution of stearic acid in white oil.

The tests revealed a sharp transition from smooth sliding to irregular motion accompanied by high friction values. For any given load it was possible to determine 'safe' and 'unsafe' regions as indicated in Figure 8.

From Equation 3(b) for a given combination of materials and a given value of λ, a plot of σ against L will represent a straight line. A series of such lines, each relating to a particular value of λ, is shown in Figure 8. For a large part of the diagram these lines are parallel to lines representing the transition from safe to unsafe conditions for loads of 2.5, 10 and 50 N respectively. The variation for a twentyfold increase in load is not great. Thus it can be concluded that the transition from smooth to rough sliding is dependent on the value of λ, that is, whether asperity deformation is primarily plastic or elastic.

Another important conclusion to be drawn from Figure 8 relates to the limiting value of σ (just below 0.02μm) below which all surfaces failed readily under test indicating that surfaces may be too smooth to be lubricated effectively. It is also apparent that L must be above a certain value for smooth sliding to occur.

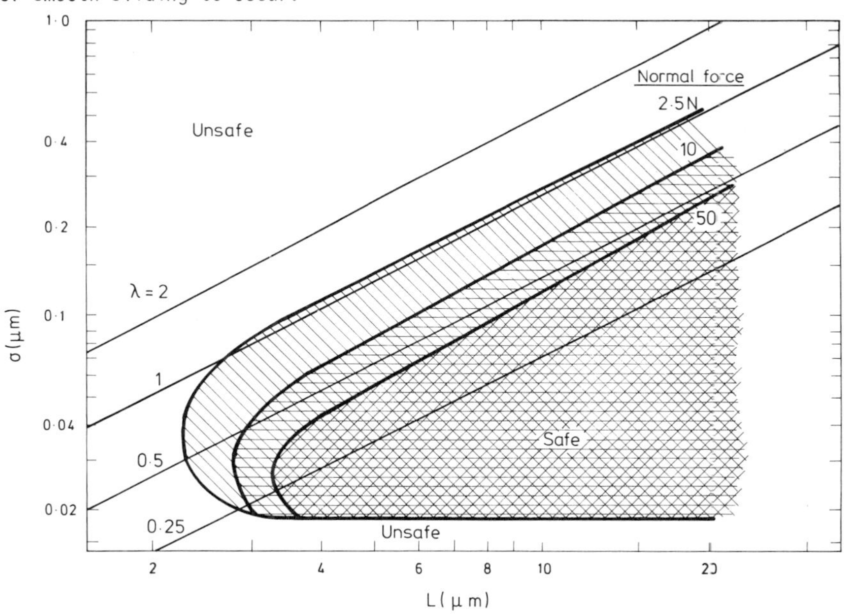

Fig. 8.—Effect of surface texture on study of lubricated stainless steel (after Hirst and Hollander)

Pitting

This is the characteristic mode of failure of rolling contact bearings as illustrated in Figure 9 and also occurs in other Hertzian contacts such as the faces of involute gears particularly in the vicinity of the pitch line. Compressive Hertzian stress falls off rapidly with distance from the

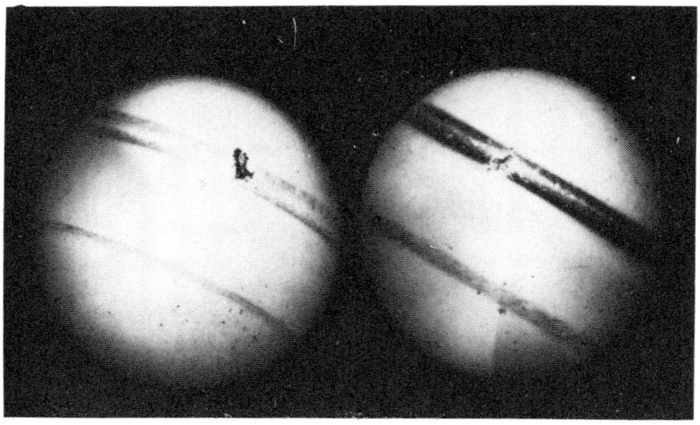

Fig. 9.—Pitting failure of 0.5 in. diameter steel balls.

surface in contrast with the shearing stress which increases to a maximum within the bulk of the compacting material. Maximum shear stress depends on the maximum compressive stress. The depth below the surface at which maximum shear stress occurs is related to the major semi-axis of the elliptical contact by a factor which varies between 0.78 for a line of contact to 0.475 for point contact. The maximum shear stress changes sign within the contact for every passage of a loaded element so that the loading cycle may be compared with the direct stress test in which the cycles consist of compressive and tensile stress symmetrically above and below zero stress. The onset of plastic yielding will be determined by the Von Mises - Henky criterion but for practical purposes the 'maximum shear stress' criterion is more convenient and gives virtually the same result. Thus elastic limit is reached when maximum Hertzian stress equals 1.7 times the yield stress in compression. There is thus a region within the rolling element which is particularly subjected to fatigue and cracks may be expected to start from any inclusion acting as a stress raiser. These cracks may spread parallel to the surface until other points of weakness are encountered so enabling a flake of material to become detached so as to leave a pit in the surface. There is some controversy as to whether or not the initiation of the cracks is at the surface or in the region of maximum Hertzian shear stress. Evidence is conflicting.

A number of investigators have observed spherical particles about 1μ in diameter which appear to be formed within the growing fatigue crack. Further work is necessary to elucidate the mechanism of formation of these spherical particles (Figure 10).

Evidence exists that the presence of lubricant affects the initiation of pitting failure and the slide/roll ratio may have an important effect. Further clarification of this aspect is also important.

Scuffing

Scuffing (sometimes known as 'scoring' in the U.S.A.) is a character-

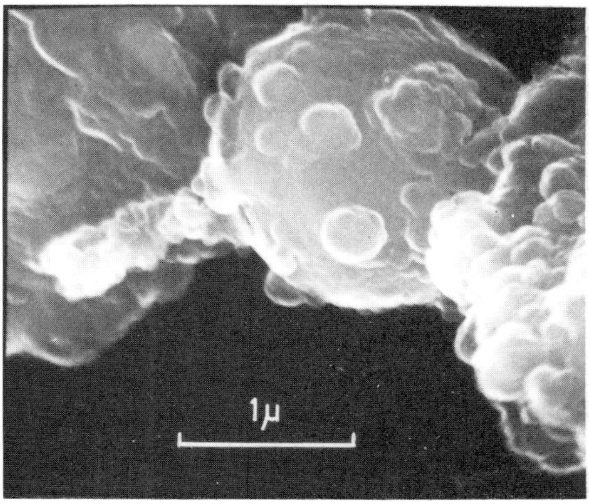

Fig. 10.—Spherical wear particle.

istic form of damage usually associated with a combination of stress concentration and relative sliding. The following definition adopted by the Institution of Mechanical Engineers in 1957 namely "gross damage characterized by the formation of local welds between sliding surfaces" is not entirely adequate because it presupposes a mechanism. Similarly the O.E.C.D. defines scuffing as "localized damage caused by the occurrence of solid-phase welding between sliding surfaces without local surface melting."

Scuffing is usually characterized by the transfer of material between the interacting surfaces, in general material being transferred from the faster moving surfaces to the slowing moving surface. Whether or not melting and welding occur, there is ample evidence of drastic metallurgical change. Phase transformations occurring in the vicinity of the surface consistent with materials having been heated and rapidly quenched.[(49)] An important aspect of scuffing failure is the relative ease with which this can be prevented by the use of extreme pressure lubricants.

These usually contain sulphur, halogens or phosphorous and can be shown to react chemically with the surface to leave a non-metallic layer. Prior phosphating of the surface of gears has been shown to have similar results to extreme pressure lubrication. The temperature of operation appears to be of crucial importance in determining the incidence of the scuffing. It also appears to be the triggering agent initiating the action of extreme pressure lubricants.

A tentative theory of scuffing failure is put forward for discussion as follows:

"There will be occasional points of intimate contact between the sliding surfaces particularly if particles of a size greater than the minimum film thickness exist in the lubricant. At these points of interaction the pressures and the coefficient friction will be high and considerable heat will be generated. At low loads and speed (depending on the coefficient of friction and other factors) the relative heat generation would increase so as to bring about an unstable thermal system wherein the temperature rose until the melting point of one of the materials was attained."

Figures 11 to 14 show initiation and development of scuffing damage.

Scoring or Gouging

Gouging wear is associated with the intense ploughing action of individual particles or asperities as shown in Figure 15. This is obviously the result of intense contact stresses giving rise to plastic deformation or cutting of material. It is more often encountered on the working surfaces of material as opposed to the elements of the mechanism of the machine, particular examples being the effect of flint being embedded in the ground interacting with a plough share or blade of earth-moving equipment. It may be expected to be resisted by materials such as manganese steel which possesses good work hardening capability.

Fretting

This form of wear results from very small oscillatory displacement between surfaces. It represents the interaction of several forms of wear, being initiated by adhesion, amplified by corrosion and having its main

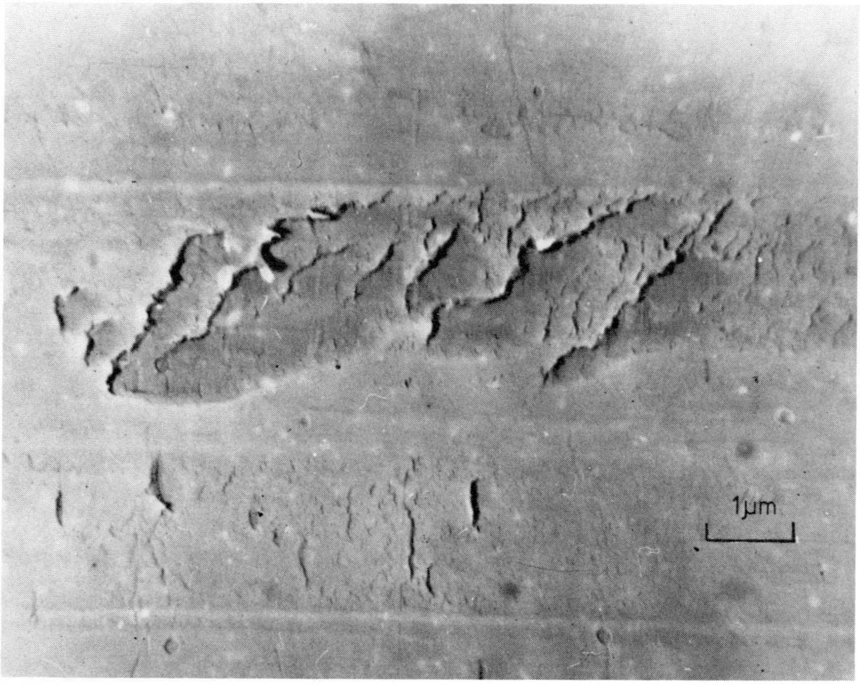

Fig. 11.—Electron-micrograph of initiation of scuffing.

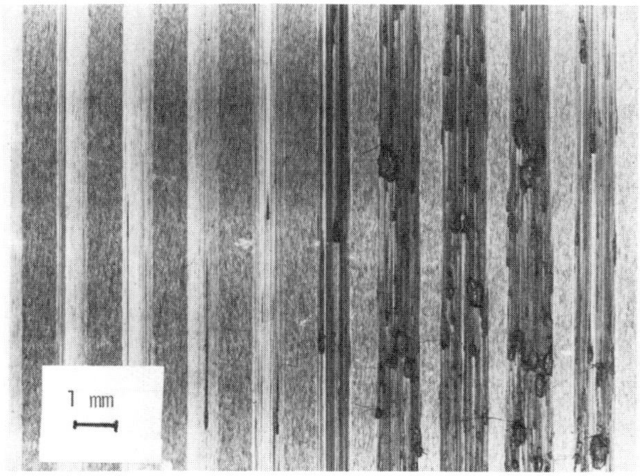

Fig. 12.—Growth of scuffing in crossed cylinder machines.

Fig. 13.—Initial scuffing.

effect by abrasion fatigue. The corrosion effect was illustrated by Wright[50] who demonstrated the effects of relative humidity. Hurricks[51,52] also stressed the effect of environment and considered that the main agency of surface degradation was micro-fatigue.

At temperatures above 200°C the fretting of mild steel diminished with temperature until a second transition temperature was reached between 500-600°C above which wear rate increased. Above 380°C the proportion of Fe_3O_4 to Fe_2O_3 increased with a corresponding reduction in the rate of wear. FeO appears to be the most harmful form of oxide. There appears to be no way of preventing fretting apart from eliminating the relative motion by improved fitting, use of interference fits, etc. Palliatives are the avoidance of metals which form hard abrasive oxides, phosphating the surface or providing a sufficient lubricant to add as a barrier to the supply of oxygen

Fig. 14.—Developed scuff.

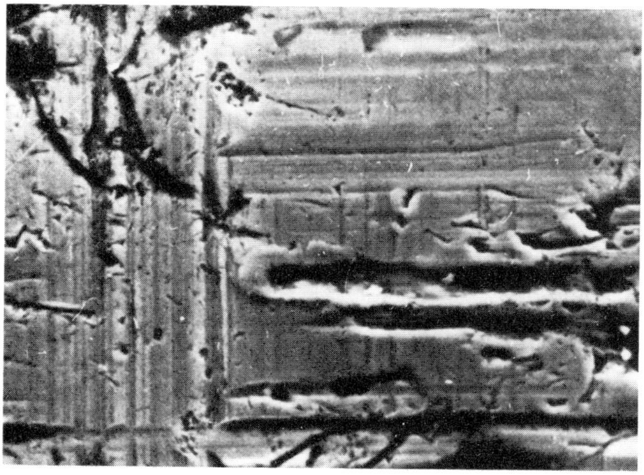

Fig. 15.—Gouging type wear.

to the fretting surfaces.

RELATIONSHIP BETWEEN THE DEGRADATION OF MACHINE PARTS AND THE FUNDAMENTAL CONCEPTS OF WEAR

Adhesive Wear

When clean metal surfaces are brought into contact a strong chemical bond is formed between them. Ferrante and Smith[53] have reported calculations of adhesive energy and conclude that the range of strong chemical bonding is 0.2nm.

At greater separations than this bodies in close proximity are attracted to each other by Van der waal's forces. Depending on the atomic structure of the interacting bodies two groups of forces exist known respectively as the 'retarded' and the 'non-retarded' forces. When the distance between the interacting bodies is below 10^{-8}m the retarded forces operate and the attraction forces are determined by the square of the distance between the surfaces. When the spacing is in the range 10^{-8}m to 10^{-7}m a transition takes place. The 'non-retarded' forces act at separations above 10^{-7}m when the van der Waal's forces fall off according to a cube law (See Figure 16).

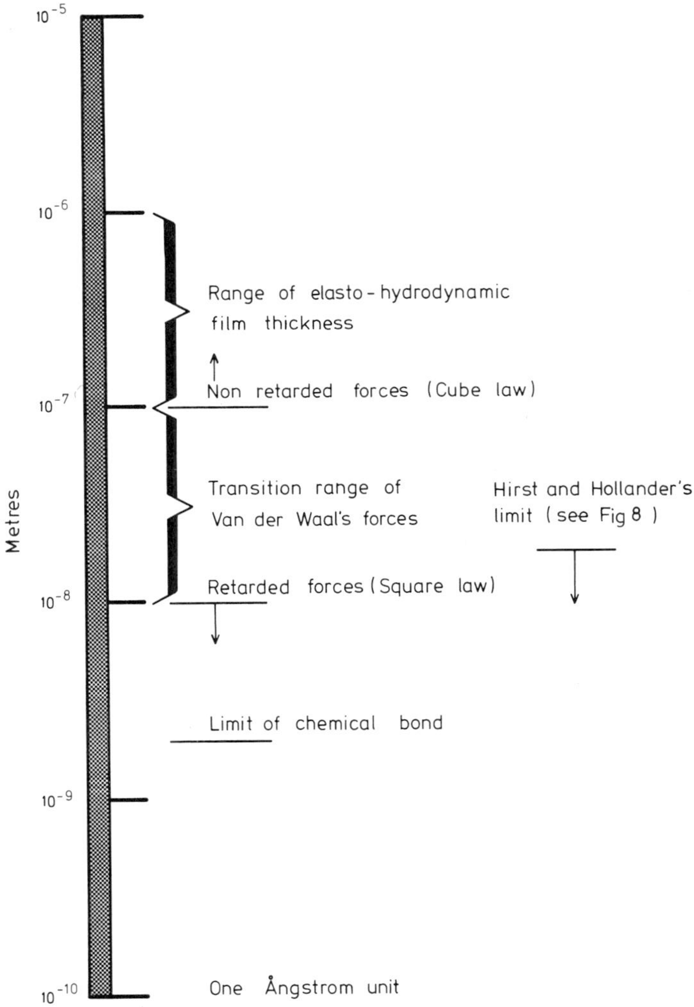

Fig. 16.—Range of operation of surface forces.

Strong experimental evidence has been obtained of the importance of van der Waal's forces, notably by Israelachvilli and Tabor[54] who studied the interaction of cylindrical surfaces of clean mica, Johnson et al[55] who experimented with rubber and gelatine and Kohno and Hyodo[56] who brought stylii with tips of small radius made of tungsten or fused quartz into contact with optically flat steel surfaces. All these investigators showed evidence of powerful adhesion between the contacting surfaces. Sometimes the forces required to separate the surfaces were orders of magnitude higher than those initially applied in forcing the surfaces together. There is thus a powerful argument in favor of an adhesive mechanism of wear in which material can be torn from one interacting body and made to adhere to another. This implies a rather special form of wear where material is transferred from one surface to the other rather than being removed completely from the interacting system. Whilst examples of this form of wear can be observed in, for example, systems operating at high vacuum, it is not usual in lubricated machine systems for two reasons. Firstly, the surfaces are covered by oxide layers to say nothing of absorbed gases or lubricant molecules of the same order of magnitude as the range of the van der Waal's forces and secondly, because the irregularities of the surface texture are usually at least of an order of magnitude higher. Thus the intimate contact necessary for powerful adhesion does not exist.

Pollock et al[59] record high values of adhesion for clean surfaces and noted that sometimes separation took place at the interface and sometimes within the bulk of one of the samples. In the latter case, evidence was provided of ductile extension of the material.

It is possible for intimate contact to occur as a result of the interaction of asperities representing a tiny fraction of the nominal area and it is here that the junctions envisaged in the classical adhesion theory of wear would occur. The area of junction was assumed to be proportional to the applied load divided by the flow pressure of the solid and frictional force to the area of the junction multiplied by the shear strength of the material. In isontropic materials having normal properties this would have resulted in a coefficient of friction of about 1/6th for unlubricated metals, a value which is much lower than anything experimentally observed. The theory was refined by assuming that, as tangential forces were applied, the area of the junction grew until friction corresponded with the observed value.

The writer is not convinced that this mechanism is important in relation to the mechanical engineering systems required to operate for considerable periods of time. The mechanism has a great deal in common with scuffing and, if the frictional heat is taken into account, it could well lead to an unstable system. Thus as the interacting metals became heated, their flow strength would be reduced and the junction area increased. It is a matter for discussion whether in fact adhesive wear and scuffing are not the same phenomena. Steady state adhesion wear appears to be the exception rather than the rule because of the inherently unstable thermal system involved.

Abrasive Wear

Writers in the English language restrict use of the term 'abrasion' to those situations wherein material is removed from one body by the interaction with another body which is harder or by the intervention between two bodies of harder particulate material. However, in the Preliminary Standard DIN 50320(November 1953) the German term "Verschleiss" is translated as "Abrasion" (Mechanical Effects). The term "Wear" is used therein as the translation of "Abnutzung" which is employed as a generic term covering chemical and/or electrochemical effects (corrosion), thermal and other effects in addition to mechanical effects. In this article the word "Abrasion" will be used in the restricted sense defined above and the German word "Verschleiss" will be translated as "Wear." This is consistent with the usage of the provisional revision of this standard dated November 1976.

Abrasive wear may be defined as damage to a surface by a harder material. This harder material may have been introduced between two surfaces from outside (three body wear), it may be formed between them by some chemical process such as oxidation or it may be part of the material forming the second surface. Tests of abrasive wear under realistic practical conditions are extremely difficult and many investigators have used test arrangements embodying abrasive paper or cloth. Careful work by Kruschov[58] reveals some anomalies in the effect of hardness in resisting abrasive wear. When the hardness of the test materials had been obtained by a work-hardening process prior to test, a simple relationship between hardness and wear rate did not exist. A possible explanation of this result may be that the relative wear resistance is a measure of the hardness of the material in a fully work-hardened condition, it being immaterial whether the work-hardening action took place during the abrasive process or prior to the commencement of the test. Similar difficulties arose in the case of heat-treated steel. In the annealed condition a range of steels gave results which were entirely in accordance with the linear relationship between the wear rate and hardness but when these steels were quenched and tempered a different coefficient of proportionality resulted. In no case was increase in wear resistance resulting from quenching and tempering commensurate with the enhanced hardness which resulted from these processes.

The development of materials with enhanced resistance to abrasive wear is of obvious economical importance. It is very difficult, however, to devise test methods which are representative of the conditions in industry and yet at the same time are amenable to rigorous analysis. Whilst it can be argued that abrasion is simply a cutting process which should be amenable to stress analysis it is nevertheless a dynamic process during which the properties of the materials are changing as a result of strain and temperature. No doubt progress will be made in the immediate future by various forms of empirical testing; nevertheless, the evolution of a sound theory of abrasive wear is well overdue.

Delamination

Friction must be considered as a means of converting mechanical energy into thermal energy. However, the adhesion theory which presupposes that all action takes place on a plane of cleavage does not suggest a very powerful means for absorbing the work of friction. Indeed the only energy required in this theory is that involved in the creation of a new surface. However, if we accept that shear forces are applied to the surface of an order of magnitude comparable with the shear strengths of the junctions, there is no reason to assume that the plastic deformation resulting from these forces should be confined to the surface. We may expect strain to take place within the body of the material, albeit in the immediate vicinity of the material in contact. Microscopic evidence of surfaces which have been sheared provide evidence of plastic strain to a considerable depth and Figure 17 shows evidence of extensive slip-lines in a friction test. It is, therefore, the writer's view that plastic deformation plays a more important part in the mechanism of friction than adhesion.

Fig. 17.—Slip-lines indicating plastic working of substrate.

Rigney and Glaeser[59] state that the repeated ploughing of asperity contacts over a mating surface can produce high dislocation densities. This brings about an eventual change in the microstructure which assumes a cellular pattern characteristic of heavily deformed metals. This structure presents many favorable sites for sub-surfac- crack generations and the eventual release of thin wear flakes.

If it is assumed that much frictional energy is dissipated by plastic strain within the material then it is necessary to presuppose some other mode of wear to take the place of adhesive wear in the general theory of our subject. Such a theory has been put forward by Suh and his co-workers known as 'delamination theory of wear.' A quotation from a recent paper by Suh[60] is as follows:

"The delamination theory of wear described the following sequential (or independent, if there are pre-existing sub-surface cracks) events which lead to wear particle formation:

1) When two sliding surfaces come into contact normal and tangential loads are transmitted through the contact points by adhesive and ploughing actions. Asperities of the softer surface are easily deformed and some are fractured by the repeated loading action. A relatively smooth surface is generated, either when these asperities are deformed or when they are removed. Once the surface becomes smooth, the contact is not just as asperity-to-asperity contact, but rather an asperity-plane contact; each point along the softer surface experiences cyclic loading

as the asperities of the harder surface plough it.

2) The surface traction exerted by the harder asperities on the softer surface induces plastic shear deformation which accumulates with repeated loading.

(The increment of permanent deformation per asperity passage is small compared to the total plastic deformation per passage which is almost completely reversed. This accounts for the dissipation of energy).

3) As the subsurface deformation continued, cracks are nucleated below the surface. Crack nucleation very near the surface is not favored because of a triaxial state of highly compressive stress which exists just below the contact regions.

4) Once cracks are present (either due to the crack nucleation process or pre-existing voids), further loading and deformation causes cracks to extend and to propagate, joining neighboring ones. The cracks tend to propagate parallel to the surface at a depth governed by material properties and the coefficient of friction.

5) When these cracks finally shear to the surface (at certain weak positions), long and thin wear sheets 'delaminate.' The thickness of a wear sheet is controlled by the location of subsurface crack growth, which is controlled by the normal and the tangential loads at the surface."

Evidence in support of this theory is the production of thin plate-like particles of wear debris. The wear rates associated with delamination wear are low enough to be tolerable in most machinery and it can be assumed that crack propagation in the surface region is the predominant mechanism of mild wear.

When conditions become too severe for this mild form of wear due to high stress leading to plastic flow or to high temperature leading to oxidation, other forms of wear emerge. It is possible to arrange forms of wear in sequence of intensity starting with delamination wear, passing on to oxidation wear and finishing with severe wear or scuffing, the detailed sequence depending on the application and, in particular, on the environment.

Fatigue

In most sliding systems it is possible to identify elements in the surface which are subject to periodic variation in stress. These may be on the macro scale, for example rotating shaft of a hydrodynamic bearing will be subject to the hydrodynamic pressures generated in the oil film. In the case of steady load these will be fixed in space so that each element on the shaft surface passes through a cycle of stress on each load. Alternatively they might be on a micro-scale wherein a succession of asperities might pass across an element in the stationary surface. In a real system there may well be a number of stress systems operating simultaneously which can lead to more than one fatigue system. The mode of failure which actually then occurs will be determined by competition between the respective systems. Thus Berthe[61] distinguishes between the circumstances giving rise to micropitting and those leading to the conventional pitting failure which occurs in conformity with the Hertzian stress applied to the system. Table I displays his results.

TABLE I.—RELATIONSHIP BETWEEN MICRO AND MACRO PITTING (AFTER BERTHE)[61]

Lubrication Factor	Stress Factor	
	Po/K <1	Po/K >1
$\frac{E1 + E2}{h_m} < 1.4$	No failure	Subsurface crack
$1.4 < \frac{E1 + E2}{h_m} < 2.5$	Surface cracks	Surface and sub-surface cracks
$\frac{E1 + E2}{h_m} > 2.5$	Micro-pitting	Micro and macro-pitting

E_1 and E_2 are half peak to valley heights of the two surfaces respectively.
h_m = minimum thickness of lubricant film.
Po = maximum Hertzian compression stress.
K = 'shake-down' value of yield stress of material as defined by Johnson and Jefferies.[62]

The micro-pitting may well be related to the stresses arising from asperities at the surface whereas the macro-pitting clearly relates to the Hertzian stress. However lubricated context the quality of the surface finish becomes of importance and can be related to the nominal film thickness by Equation 5:

$$\Delta = \frac{h}{\sqrt{\sigma_1{}^2 + \sigma_1{}^2}} \qquad (5)$$

where h is the film thickness and σ_1 and σ_2 the relative respective r.m.s. roughnesses of the interacting surfaces.

There is evidence[63] that the onset of pitting is related to this factor. If, as propounded in Section-Worm Gears, pitting originates in the zone of maximum Hertzian stress which is well within the surface it is very difficult to see how surface roughness can affect pitting. This problem can be resolved if it is assumed that different modes of failure are in competition. In some circumstances of speed, load, and lubrication, the failure may originate in the subsurface region; whereas in other circumstances, a low value of Δ, for example, a superficial mechanism may intervene to cause earlier failure. Some support for this view is provided by Vaessen and Gee[64] who used a disc machine to investigate the effect of rolling contact fatigue on maraging steels.

Because of the difficulty of experimenting with actual gears most published work on fatigue relates to tests on disc machines. It is unfortunate, as evidenced by Onions and Archard[65] that the life of actual gears may be considerably less than that assumed from tests on discs. This difference may arise from the difference in character of load as between the tooth flank and the disc systems. In the first case the load will be dynamic in character, impulsively applied and of short duration, as contrasted with the steady state. The difference may also be due to the transient nature of the film forming process and the conditions of lubricant supply may also be of importance. In this connection there is some evidence that slide-roll ratio may also have some effect on the incidence of pitting-type failure because such a sharp transition occurs as this ratio passes from negative to positive values. As long ago as 1935 Way[66] produced evidence of interaction between the lubricant and the growing crack which raises problems which have not yet been fully resolved.

Kragelskii has put forward a general equation of fatigue wear based on the depth of penetration of asperities and their radii.[67] The equation is as follows:

$$I = \frac{\Delta \ell}{S} = \frac{\ell}{R}\,\frac{Pa}{P_2}\,\frac{K_1}{m} \qquad (6)$$

S - distance of sliding. $\Delta\ell$ - height of worn layer.

ℓ - depth of penetration. R - radius of asperity.

$P_a P_2$ - nominal and real pressure.

K_1 - parameter of distribution of asperity (usually 1.12-0.15).

m - number of cycles $m = \frac{\sigma_\ell \; t}{k_2 f p_2}$

σ_ℓ - strength of material.

K_2 - coefficient of tensile stress acting on surface layer (usually for elastic material - 3, for brittle - 5).

m - $\frac{\sigma_\ell}{\sigma_x}\; t$

t - 5, 6 or 7

Corrosion

Some form of chemical attack must be expected on all interacting surfaces, oxidation being the most probable effect. The oxidative wear process can have a dynamic character as evidenced by the work of Kerridge[68] who showed that an equilibrium situation could be established between transfer of material and its removal. The mechanism suggested by these results is as follows:

Metal is transferred from the soft pin to the harder ring. This metal was oxidized rapidly until the oxidized film limited the further transfer of material. As this oxide film was rubbed off a further steady amount of transfer wear became possible.

In the case of automobile cylinders the wear can be accelerated by the presence of acid-bearing moisture condensed onto the cylinder walls when the engine is operating at low temperature, particularly during start-up periods when the cylinders walls may be starved of oil. A number of acids, formed as a result of the combustion of hydro-carbons, may act as corrosive agents, (also products of combusiton of sulphur).

As indicated in Section-Fretting the action of oscillatory motion is greatly accelerated by corrosive wear, not only by removal of oxide debris but in providing a grinding mechanism where this oxide is harder than the basic material.

Oxidation under mild wear conditions has been quantified by Quinn and Sulivan.[69] They describe the system where oxidation first occurs when an area of contact cracks and breaks off to leave a virgin metal. During the early stages of its life this area will oxidize at surface temperature until it becomes one of a few new areas of contact. Such areas are determined by plastic deformation of the materials constituting the interacting surfaces. The oxidation of the area of contact will proceed at an increasing rate until a critical oxide thickness is reached. The film then cracks up due either to dissimilarities in the crystal of the metal or its oxide or dissimilarities in their expansion coefficients.

Oxidative wear may occur by spalling of oxide flakes from a substrate which shows little evidence of deformation. This is to be contrasted to metallic wear which may occur by plastic deformation and fracture on a massive scale. Shivaneth and Eyre[70] have investigated the wear of binary aluminum alloys containing up to 20% silicon. They confirm the existance of the two mechanisms described above and, in a series of pin-and-disc machine experiments, demonstrate that a transition from oxidative to metallic wear occurs when the load is increased beyond a critical value. This value is increased in accordance with an increase in the silicon content of the aluminum alloy.

Electrical Damage

As an electrical contact begins to open, the current streamlines become concentrated at fewer and fewer points of contact until they are finally concentrated at a single microscopic area. Finally the resistance across the contact becomes almost entirely due to the constriction resistance. The temperature can therefore rapidly reach the boiling point of the metal. This provides a molten globule which extends as the two surfaces separate until it evaporates. It may even explode. Llewellyn Jones[71] set out two important problem areas embodied in this general process. These are (1) the mode of formation development and final rupture of this molten metal bridge and (2) the consequence of the conditions thus produced as far as the initiation and subsequent development of micro-plasma. Both these sets of phenomena have important practical consequences as far as electrode-wear is concerned. Detailed analysis of the fracture of the micro-bridge yields the following succession:

a. A small gap (10^{-6}m) is set up between the two electrodes.
b. Each contact spot is at a high temperature and is therefore a high thermionic emitter.
c. There is a gaseous atmosphere containing metal vapor.
d. The metal vapor may exceed atmospheric gas density and its ionization potential must be even lower than that of the atmosphere.

A seld-inductance of the local circuit can set up a pulse of voltage sufficient to produce ionization of the gas or metal vapor. The occurrence of a micro-arc is therefore practically certain in most cases of practical importance and may be a primary source of electrical wear.

QUANTITATIVE MEASUREMENTS OF WEAR

A number of possible ways of quantifying wear is enumerated in DIN 50321 dated March 1961. These comprise Linear absolute amount of wear, Linear rate of wear, Specific reference linear amount of wear, Line ratio amount of wear, Volumetric absolute amount of wear, Volumetric rate of wear, Specific reference rate of wear, Volumetric ratio amount of wear, gravimetric absolute rate of wear and gravimetric rate of wear.

Whilst the selection of the most suitable measure must depend on the method of measurement and the application, it appears to the writer that the most generally applicable is the ratio of the loss in dimension of the test piece measured in the direction perpendicular to the face of interaction to the total amount of relative motion.

The value of this result would depend on the stress acting at the interface, the shape of the interacting bodies as well as the temperature and other environmental factors. The wear resistance of a material might then be computed by an expression such as Equation 7.

$$\text{Wear resistance} = \frac{S}{\Delta \ell} + \frac{Po}{K} \times A \times B$$

where A = shape factor

" B = environmental factor

Kharach and Kragelskii[72] define a quantity which they designate as 'Wear Intensity' as follows:

$$I = \frac{Dh}{DS} = \frac{DV}{A_R DS} = \frac{\lambda}{\rho} \frac{Dg}{A_A DS} = \lambda r \frac{DV}{DW} \qquad (7)$$

where (in our notation)

I = Wear intensity

h = linear wear

V = Volume wear

g = loss of mass (gravitational wear)

D = operator $\frac{d}{dt}$

S = distance of sliding

ρ = density of material of worn component

A_R = real (frictional) area of contact

A_A = apparent area of contact

λ = $\frac{A_A}{A_R}$

F = Force of friction

r = specific nominal force of friction = $\frac{F}{A_A}$

W = Work of friction = SF

They define Wear Class as given in Table II.

TABLE II.—SCALE OF WEAR RESISTANCES

Class	Linear Wear Intensities = $\frac{\text{height of worn layer}}{\text{sliding distance}}$
0	10^{-12} - 10^{-13}
1	10^{-11} - 10^{-12}
2	10^{-10} - 10^{-11}
3	10^{-9} - 10^{-10}
4	10^{-8} - 10^{-9}
5	10^{-7} - 10^{-8}
6	10^{-6} - 10^{-7}
7	10^{-5} - 10^{-6}
8	10^{-4} - 10^{-5}
9	10^{-3} - 10^{-4}

Some practical cases are given in Table III.

TABLE III.

Machine	Machine Part	Class
Excavator	Chain gear shovel teeth	1 9
Automobile Engine	cylinder liner piston ring crank pin	1-0 1 1
Lathe	guides	3
Shaper	tool slides	1-0
Disc brakes	friction parts	5 to 3
Automobile tire	tread	4 to 2

TECHNIQUES OF INVESTIGATION

The advancement of a knowledge of wear can take place on three inter-related fronts. (1) the development of theory (such as that of elastohydrodynamic lubrication), (2) the improvement of instrumentation enabling us to make observations which were not previously possible and (3) developments in practice.

It is the view of the writer that theory should not run too far ahead of experiment. Conversely attempts to explain experimental results can be a very fruitful source of inspiration to theoretical workers.

Rapid progress may be made in idealized situations, both mathematical and experimental, leading to results which whilst true are not relevant to any particular industrial situation.

A greater need is for the study of wear of actual machines operating under representative industrial conditions. It is fortunate that this demand coincides with the economic demand for 'condition-monitoring' by various means of which perhaps spectograph oil analysis and ferrography are likely to be the most revealing from the point of view of increasing our knowledge of the wear processes. Thus in addition to engineers using the data for the purpose of making maintenance decisions from day to day, provision should be made for collecting data for the benefit of research and development scientists, whose objective is to understand the wear process and to devise means by designs or materials to reduce the incidence of wear in the interests of the economy as a whole. As was pointed out by Professor Blok many years ago, this is particularly important for developing countries whose limited rate of capital formation must be devoted to expansion rather than replacement of worn-out machinery.

CONCLUSION

In order to keep this review within a manageable size, attention has been confined to machine elements which operate under moderate environmental conditions. The importance of different modes of wear is, of course, determined by the application. Thus adhesive wear may be expected to occur more frequently in mechanisms operating in high vacuum or in inert gases which preclude the regeneration of oxide films.

Another important range of tribological problems arises with machine components, usually characterized as tools, where contact is made with material which is constantly renewed or whose shape is changed during the period of contact. Operations such as cutting, rolling, forging and wire-drawing impose severe requirements for wear resisting materials and machines for handling minerals such as mining or earth-moving machinery require the application of a wide variety of wear resistant techniques. For treatments of these wider implications of wear, reference must be made to other reviews in this conference and to a collection of reviews edited by Scott.[73] Nevertheless, even within the restricted field selected for this review numerous topics for further investigation have been identified, and it is contended that progress will be more satisfactory if this research can be pursued as part of a consistent body of theory of wear rather than as isolated studies of particular machines and situations.

REFERENCES

1. O.E.C.D. Research Group on Wear of Engineering Materials, "Glossary of terms and definitions in the field of Friction, Wear and Lubrication.-Tribology," Paris, 1969.
2. Morrison, J.L.M. and Crossland, B., "An Introduction to Mechanics of Machines," Longman, London, 1970.
3. Dowson, D. and Higginson, G.R., "Elasto-Hydrodynamic Lubrication," Pergamon, Oxford, 1966.
4. Winer, W., "Lubricant Rheology Applied to Elastohydrodynamic Lubrication," Contractors Report, NASA-Lewis Laboratory, R-2837, Cleveland, Ohio, 1977.
5. Baker, D.W.C. and Brailey, E.D., in Proceedings of the Institution of Mechanical Engineers Conference on Friction and Wear, London, 1957, p. 720.
6. Byer, K., Dallimore, B.J. and Lowe, C.B., *Institute of Marine Engineers Transactions,* Vol. 87, 1975, p. 49.
7. Eilon, S. and Saunders, O.A., *Institution of Mechanical Engineers. Proceedings,* Vol. 171, 1957, p. 427.
8. Lloyd, T., *Institution of Mechanical Engineers. Proceedings,* Vol. 183, Part 3P, 1969, p. 28.
9. Hamilton, G.M. and Moore, S.L., *Institution of Mechanical Engineers. Proceedings,* Vol. 180, 1974, p. 253.
10. Hamilton, G.M. and Moore, S.L., Proceedings of the Institution of Mechanical Engineers Conference on Piston Ring Scuffing, London, 1975, p. 61.
11. Wakuri, Y. and Ono, S., "Experimental Studies of the Abnormal Wear of the Cylinder Liners and Piston Rings in a Marine Diesel Engine," University of Kyshu, Preprint.
12. Neale, M.J., "Tribology Practical Reviews," Institution of Mechanical Engineers, 1975.
13. Milne, A.A., Scott, D. and Macdonald, D., Proceedings of the Institution of Mechanical Engineers Conference on Lubrication and Wear, London, 1957, p. 735.
14. Tring, L.L. and Mayer, J.E.H., *Journal of Lubrication Technology,* Vol. 96, 1974, pp. 258; 305.
15. Willn, J.E. and Brett, P.S., *Institution of Mechanical Engineers. Proceedings,* Vol. 191, 1977, p. 241.
16. Jones, M.H., Sastry, V.R.K. and Youdan, G.H., in Proceedings 4th Leeds-Lyon Symposium, 1977, Mechanical Engineering Publications Ltd., Bury St. Edmunds, U.K.
17. Wilford, A.T., *Institution of Mechanical Engineers. Proceedings,* 1957, p. 524.
18. Jones, M.H., "Element Concentration Analysis of Films Generated on a Phosphor Bronze Film Worn Against Steel Under Conditions of Boundary Lubrication," American Society of Lubrication Engineers Paper No. 76-LC-2B-3, 1976.
19. Barber, J.R., *Royal Society of London. Proceedings.* Series A, Vol. 312, 1969, p. 381.
20. Ting Long Ho, in Conference on Wear of Materials, St. Louis, Missouri, American Society of Mechanical Engineers, 1977, p. 70.
21. Llewellyn Jones, F., World Electromechanical Congress, Moscow, 1977.
22. Fenner, A.J. and Field, J.E., *North East Coast Institution of Engineers and Shipbuilders Transactions,* Vol. 76, 1960, p. 76.
23. Hartnett, M.J., to be published, 1975.
24. "Life Adjustment Factors for Ball and Roller Bearings, An Engineering Design Guide," American Society of Mechanical Engineers, New York, 1971.

25. Dowson, D. and Higginson, G.R., in Proceedings of Lubrication and Wear Conference, Boummouth, Institution of Mechanical Engineers, 1963, p. 216.
26. Dalal, H., et al., Final Report, SKF Report No. AL74 T002, SKF Research, King of Prussia, Pennsylvania, 1974.
27. Krishnamurthy, R., in Conference on Wear of Materials, St. Louis, Missouri, American Society of Mechanical Engineers, 1977, p. 317.
28. Rozeanu, L. and Grosberg, J., in Congrès Mondial des Engrenages, Paris, France, 1977, p. 657.
29. Brugger, H., Draus, G. and Schultz, in Congrès Mondial des Engrenages, Paris, France, 1977, p. 467.
30. Towle, A. and Graham, R., "Tribology Practical Reviews," Institution of Mechanical Engineers, London, 1975.
31. Trachmann, E.G. and Cheng, H.S., in Symposium on Elastohydrodynamic Lubrication, Institution of Mechanical Engineers, 1972, p. 142.
32. Hirst, W. and Moore, A.J., *Royal Society of London. Proceedings.* Series A, Vol. 337, 1974, p. 101.
33. Crook, A.W., *Royal Society of London, Philosophical Transactions.* Series A, Vol. 255, 1963, p. 281.
34. Dyson, A., *Royal Society of London, Philosophical Transactions.* Series A, Vol. 266, 1970, p. 1.
35. Odi-Owei, S. and Roylance, B.J., *Journal of Lubrication Technology,* American Society of Mechanical Engineers Transactions, Vol. 100, 1977, p. 115.
36. Muller, R., *Motortechnische Zeitschrift,* Vol. 27, 1966, p. 58.
37. Watson, H.C. and Milkins, E.E., in Proceedings Conference on Mechanisms, Institution of Mechanical Engineers, London, 1974, p. 39.
38. Vichard, J.P. and Godet, M.R., *Institution of Mechanical Engineers. Proceedings,* Vol. 182, Part 2, 1968, p. 109.
39. Vichard, J.P., *J. Mech. Eng. Sci.,* Vol. 13, 1971, p. 173.
40. Beese, J.G. and Clark, M., in Proceedings Conference on Mechanisms, Liverpool, Institution of Mechanical Engineers, London, 1974, p. 87.
41. Johnson, K.L. and Gray, C., *Institution of Mechanical Engineers. Proceedings,* Vol. 189, 1975, p. 45.
42. Dearden, J., *Wear,* Vol. 3, 1960, p. 43.
43. Clayton, P., in Proceedings Convention on Tribology, Swansea, Institution of Mechanical Engineers, 1978, p. 95.
44. Llewellyn Jones, F., "Institution of Electrical Engineers Colloquium on Electrical Contacts," 1976.
45. Hirst, W., in Proceedings Conference on Lubrication and Wear, Institution of Mechanical Engineers, London, 1957, p. 674.
46. Kragelskii, I.V., "Friction and Wear," Butterworths, London, 1965.
47. Lancaster, J.K., *Journal of Physics D: Applied Physics,* Vol. 13, 1962, p. 468.
48. Hirst, W. and Hollander, A.E., *Royal Society of London. Proceedings.* Series A, Vol. 337, 1974, p. 379.
49. Milne, A.A., Scott, D. and Macdonald, D., in Proceedings Conference on Lubrication and Wear, Institution of Mechanical Engineers, London, 1957, p. 735.
50. Wright, K., *Institution of Mechanical Engineers. Proceedings,* Vol. IB, 1952, p. 556.
51. Hurricks, P.L., *Wear,* Vol. 19, 1972, p. 207.
52. Hurricks, P.L., *Wear,* Vol. 30, 1974, p. 189.
53. Ferrante, J. and Smith, J.R., *Solid State Communications,* Vol. 20, 1976, p. 393.
54. Ishraelachvilli, J.N. and Tabor, D., *Royal Society of London. Proceedings.* Series A, Vol. 331, 1972, p. 19.

55. Johnson, K.L., Kendal, K. and Roberts, A.D., *Royal Society of London. Proceedings*. Series A, Vol. 324, 1977, p. 301.
56. Kohno, K.C. and Hyodo, S., *Journal of Physics D: Applied Physics*, Vol. 7, 1974, p. 1243.
57. Pollock, H.M., Shufflebottom, P. and Skinner, J., *Journal of Physics D: Applied Physics*, Vol. 10, 1977, p. 127.
58. Kruschov, M.H., in Proceedings Conference on Lubrication and Wear, Institution of Mechanical Engineers, London, 1967, p. 655.
59. Rigney, D.A. and Glaeser, W.A., in Conference on Wear of Materials, St. Louis, Missouri, American Society of Mechanical Engineers, 1977, p. 41.
60. Suh, N.P., *Wear*, Vol. 44, 1977, p. 1 (see the original paper: Suh, N.P., *Wear*, Vol. 25, 1973, p. 111 - Eds. Note).
61. Berthe, D., Thesis No. 216, L'Universite Claud Bernard, Lyon, France, 1974.
62. Johnson, K.L. and Jefferies, S.A., in Symposium on Fatigue in Rolling Contact, Institution of Mechanical Engineers, London, 1963, p. 54.
63. Dawson, P.H., *Journal of Mechanical Engineering Science*, Vol. 4, No. 1, 1962, p. 16.
64. Vaessen, G.H.C. and de Gee, A.W.S., in Symposium on Elastohydrodynamic Lubrication, Institution of Mechanical Engineers, 1972, p. 40.
65. Onions, R.A. and Archard, J.F., *Institution of Mechanical Engineers. Proceedings*, Vol. 188, 1974, p. 673.
66. Way, S., *Journal of Applied Mechanics*, Vol. 57, 1935, p. 49.
67. Kragelskii, I.V., *Journal of Basic Engineering*, Vol. 87, 1965, p. 785.
68. Kerridge, M., *Physics Society Proceedings*, Vol. 68B, 1955, p. 400.
69. Quinn, T.F.J. and Sullivan, J.L., in Conference on Wear of Materials, St. Louis, Missouri, American Society of Mechanical Engineers, 1977, p. 110.
70. Shivaneth, R., Sengupta, P.K. and Eyre, T.S., in Conference on Wear of Materials, St. Louis, Missouri, American Society of Mechanical Engineers, 1977, p. 120.
71. Llewellyn Jones, F., in 4th International Conference on Gas Discharges, Swansea, IEE, 1976, p. 429.
72. Kharach, G.M. and Kragelskii, I.V., in Japan Society of Lubrication Engineers-American Society of Lubrication Engineers International Conference, Tokyo, 1975, p. 129.
73. Scott, D., ed., "Wear," Academic Press, New York, 1979.

DISCUSSION

H. CZICHOS, BAM, Berlin: You gave a very good compilation of various modes of wear. One of your tables includes the term fretting. Could you say a few words about the machine elements that are prone to fretting?

F. T. BARWELL: My first experience with fretting was on high-speed rolling contact bearings wherein the inner race was secured to the shaft by a press fit. Because these bearings operated faster than had been the previous normal practice centrifugal force caused a loss of grip between race and shaft with consequential fretting. This was overcome by applying hard chromium plating and increasing the interference fit. Another case of fretting commonly occurs between the backs of steel-backed sintered bearings and their housings due to vibration if they are not fitted properly. A third and very serious case may occur on railway axles where the wheel is pressed on. If this is not done correctly fretting takes place which can lead to fatigue, derailment and disaster.

D. GODFREY, Chevron Research: One mechanism of wear that was not shown in your table is cavitation damage. In my experience, especially with hydraulic systems, this is a very significant kind of wear. Would that fall into some other group in your table?

BARWELL: Cavitation erosion is not included in Fig. 2 and this is a serious omission. Thank you for drawing my attention to it. I have not come across it very often which may explain its omission from the table. It sometimes occurs in the wet liners of internal combustion engines when vibration forces predominate. Hydrodynamic bearings subject to loads which fluctuate rapidly in magnitude or direction depend on the existence of a squeeze film which may sometimes cavitate with consequential damage of the bearing lining. Of course in the design of hydraulic machinery cavitation is on of the primary design parameters which limits the output of equipment.

GODFREY: At the wear conference in 1977 at St. Louis I was impressed with the example on cavitation that occurred to the sluiceways in dams. Tremendous quantities of concrete are removed due to the cavitation of the water coming down the sluiceway at very high velocity. Another case which is quite serious is the cavitation damage to the diesel cylinder liners on the coolant side. I know of a case in which the cavitation of the cylinder lead to the squirting of water into the piston area.

T. F. J. QUINN, University of Aston: I would like to ask Prof. Barwell what forms of wear he thinks are involved in the wear of rails.

BARWELL: In the Standard Handbook on Lubrication I had cited the wear of rails as an example of the interaction of abrasion and corrosion. This was based on the work of Loach who drew attention to the great effect of environment on wear as evidenced by the greatly different rates of wear in tunnels depending on whether steam or electric traction was employed.

However, in recent years there has been a considerable increase in the intensity of operation of railways which has introduced problems of side wear in curved track in particular. In these circumstances rails have to be changed after several months in service as compared with a normal life of twenty years in straight track. Severe wear conditions occur when there is relative sliding between rail and tyre in association with high Hertzian stress. It appears that the best criterion for judging the performance of steel under these circumstances is that based on the cumulative plastic strain derived from a monotonic tensile stress strain curve.(1) The form of wear which occurs resembles plastic extrusion which can be included in the Chart under the heading 'Severe Wear'.

---

(1)Clayton, P. (1978) 'Lateral Wear of Rails on Curves', Proc. J. Med. E. Convention on Tribology, Swansea, April, 1978.

WEAR MECHANISMS: AN ASSESSMENT OF THE STATE OF KNOWLEDGE

N. P. Suh

ABSTRACT

The literature on friction and wear is often less than clear in elucidating the fundamental mechanisms. The causality for friction and wear is often misunderstood, the rate determining process in wear is not explicitly identified, the transition from one wear mechanism to another is not clearly comprehended, and even the scientific terminology varies from publication to publication. An attempt is made in this discussion paper to establish the state of understanding of various wear mechanisms and outline the areas that require further research.

INTRODUCTION

Tribology is still a burgeoning field of science. As such it offers challenges and at the same time suffers from the over-abundance of unfiltered and undigested information. Often the uninitiated reader forms an impression that all the wear mechanisms cited in the literature are equally important in a given situation. He also tends to believe that the wear behavior of materials is independent of the frictional behavior. This confusion is a reflection of the current state of knowledge that is unable to discriminate the validity of various claims and counterclaims.

In many situations there is often only one rate controlling mechanism, although under some conditions several wear mechanisms may operate simultaneously. The identification of the rate controlling mechanism from all the plausible wear mechanisms is of prime importance in tribology. This enables us to isolate the most critical wear mechanisms and eliminate them from being operative in a given application. It has been possible in the past to minimize the wear by simply dealing with these rate controlling mechanisms. Therefore, it is imperative that the future advances in tribology enable us to identify and quantify the rate controlling mechanisms from a complex set of possibilities.

Professor Barwell has made a comprehensive and critical review of a very difficult subject.(1) His comments on the validity and limitations of various theories provide a new insight into the problems of tribology. The purpose of this paper is to supplement the review paper by Professor Barwell.

In particular, an attempt will be made to identify specific rate determining wear mechanisms, the relationship between delamination and abrasive wear, and the wear caused by chemical dissociation of materials at high temperatures.

THREE ASPECTS OF TRIBOLOGICAL PROBLEMS

The phenomena occurring between two contacting bodies in relative tangential motion may be divided into the following three aspects:

1) Chemical and physical interactions of the surface with lubricants and other constituents of the environment.

2) Transmission of forces at the interface through asperities and loose wear particles.

3) The response of a given pair of solid materials to the forces at the surface.

These three aspects are interrelated. The understanding of this interrelationship and the ability to describe each one of the three aspects quantitatively are the prime requisites of tribology. Changes in any one of these aspects can drastically alter the wear process.

The chemical and physical interactions of the surface with the environment are important because they affect the surface tractions which, in turn, control the response of the materials. It is well known that the same pair of sliding materials wear differently in vacuum than in air. The major effect of the interactions of the surface with the environment is, in many cases, the reduction of surface traction by eliminating metal-to-metal contact and by redistributing the applied normal load. Even oxidation, which is the best known chemical reaction occurring at the surface of most metals, affects the wear process by changing the surface traction. The wear of metals is rarely a direct consequence of surface oxidation under normal sliding conditions.(2)

It is important to recognize that, depending on the magnitude of the surface traction, the rate controlling wear mechanism may, for example, change from delamination wear to other types of wear. The magnitude of the forces transmitted and the contact area across which the force is transmitted determine the state of stress in the material being worn, which in turn activate different wear mechanisms. Since the surface interaction affects the traction, the sliding velocity can affect the wear process by changing the surface temperature and the species present at the surface.(3)

Surface topography is important in characterizing the nature of the forces transmitted and thus the resulting wear mechanism. For example, when the force is transmitted through sharp abrasive particles, the induced stress field may be such that the cutting and plowing components may predominate the wear process. However, the surface topography under the actual sliding conditions has not usually been characterized and, therefore, the asperity contact size and the load distribution are not exactly known. The elaborate studies of asperity contacts in the past were normally made under static or quasi-static conditions. These measurements do not provide information on the actual load transmission, because the soft asperities undergo plastic deformation under the load exerted by the sliding action and even the hard asperities undergo a continuous shape change. Only through a realistic characterization and modeling of the surface interaction can we hope to understand the genesis of frictional force.

Although the surface topography is important, because it determines the surface traction, the appearance of the worn surface should not be considered as the indicator of the basic rate controlling wear mechanism. For example, both the so-called mild wear and severe wear may be consequences of the delamination wear mechanisms; the differences in the number of surface craters being due to the difference in loading conditions for a given pair of metals.

Due to the forces transmitted at the surface contacts, the material experiences cyclic loading as the asperity contact moves along the surface.[4] Depending on the mechanical behavior of materials, the material will deform plastically and/or fracture under the influence of the cyclic load. The cumulative damage experienced by the material eventually leads to loose wear particle formation.[5,6] In this sense, the wear phenomenon is a consequence of materials response to the applied cyclic load at the surface and theoretical prediction of the wear process can be made only when the stress and the velocity boundary conditions are known at the surface contacts. It was stated that when the sliding velocity and the applied load are low, the temperature rise is small[3] and its effect on wear is mainly through surface traction.[2] However, as the sliding velocity increases, chemical stability of materials and diffusion of elements across the interface may become the rate controlling process due to the high temperature at the interface. For example, when cemented carbide tools are used to cut steel, the wear rate of carbide tools is controlled by the chemical stability of the tool materials.[7] Under exactly identical conditions chemically stable aluminum oxide tools wear by mechanisms controlled by its mechanical behavior.

The important points to be reiterated are:

1) It is useful to separate tribology problems into three interrelated aspects in identifying the rate controlling mechanism.

2) There are only a few (often one) rate controlling wear mechanisms under a given condition and, therefore, it is important to identify them in a specific situation.

THE EFFECT OF SURFACE TRACTION AND SLIDING VELOCITY ON FRICTION AND WEAR

The genesis of the frictional force is not clearly known at present. Although it is commonly assumed that adhesion and plowing are major contributors to the surface traction, the frictional force cannot be predicted based on existing theoretical models. As Barwell[1] pointed out in his review paper, the adhesion theory predicts values which are much smaller than those measured experimentally. Even a more serious defect of the adhesion model is that the prediction becomes increasingly worse as the surface is made cleaner and cleaner to satisfy the conditions assumed by the theory.

The plowing component of surface traction is also poorly understood. An understanding of plowing will also enhance our understanding of the adhesion component of the surface traction by clarifying the nature of the metal-to-metal contact. For example, if plowing enables a cleaner metal-to-metal contact, the frictional force will be affected. In fact, it is likely that the forces generated at asperity contacts may be different from contact to contact; the freshly-generated surfaces due to wear and plowing give rise to much larger surface traction. Certainly it is a great challenge to be able to predict the frictional force from the first principles.

The effect of surface traction on wear is different depending on the behavior of materials, temperature, and sliding velocity. They may be classified as follows:

Case 1) When the material is perfectly elastic with little resistance to crack propagation, the cracks formed will be simple ring cracks perpendicular to the surface when the coefficient of friction, μ, is zero. When μ is greater than zero, the cracks will form perpendicular to the surface behind the slider (Figures 1 and 2).

Case 2) When the elastic solid of Case 1 has internal cracks and voids and when μ is greater than zero, sub-surface cracks can propagate parallel to the surface (Figure 3). These brittle materials can also wear by generating small wear particles due to the large number of cracks running in all directions, especially when there are many defects in the material.

Case 3) When an elasto-plastic solid is slid against another surface with a coefficient of friction larger than zero, plastic deformation of the surface layer will occur which will in turn cause sub-surface crack nucleation and propagation leading to delamination wear.[8] Cracks can also form at the surface perpendicular to the surface. These cracks will be blunted when plastic deformation occurs. Also these cracks cannot penetrate too deeply into the material since they will soon be in a compressive state of loading (Figure 4). Both the plastic deformation and the state of stress may change the crack propagation direction, when the metal cannot heal the crack. It should be noted that the crack propagation can only occur when the stress intensity factor at the crack tip is larger than the threshold value of the material.

Case 4) The special case of the elasto-plastic solid with $\mu=0$ leads to the case of the so-called shake-down load when the normal stress at the contact is equal to four times the shear flow stress (Figure 5). In this case the plastic deformation may occur at the maximum shear stress (Hertzian stress) zone below the surface. However, it quickly reaches an elastic state as a consequence of the residual stress remaining in the material after initial plastic loading and unloading.[9] When the normal stress is larger than four times the shear flow stress, there is always a finite plastic deformation.

Case 5) When an elasto-plastic solid is subjected to a small finite oscillatory displacement at the surface, cracks normal to the surface can form, especially when the material has low fracture toughness (Figure 6). When the displacement is large, sub-surface cracks can nucleate and propagate parallel to the surface. This is often referred to as fretting.

Case 6) When hard carbide tools are used to cut steel at high speed, the wear can occur by chemical dissociation of carbides.[10] Wear occurs when products of dissociation are transported away by the chip through the formation of a solid solution (Figure 7).

Case 7) Although it does not occur with Group IV-B and Group V-B carbides while cutting steel, wear at high temperature can be controlled by diffusional processes when the temperature is sufficiently high and/or if the change in the free energy of dissociation is very small (Figure 7).

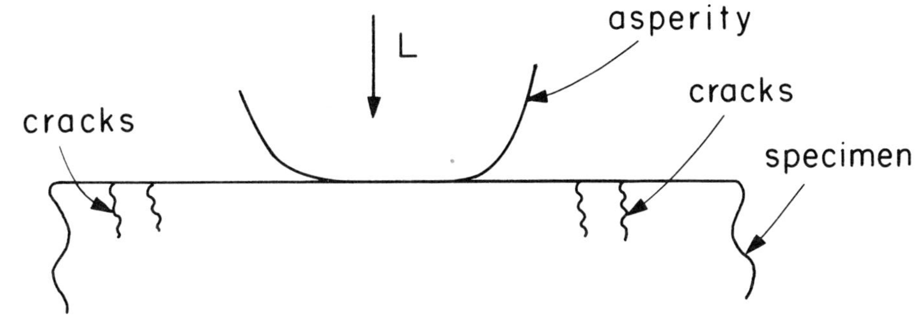

Fig. 1.-Perfectly Elastic Solid with $\mu=0$.

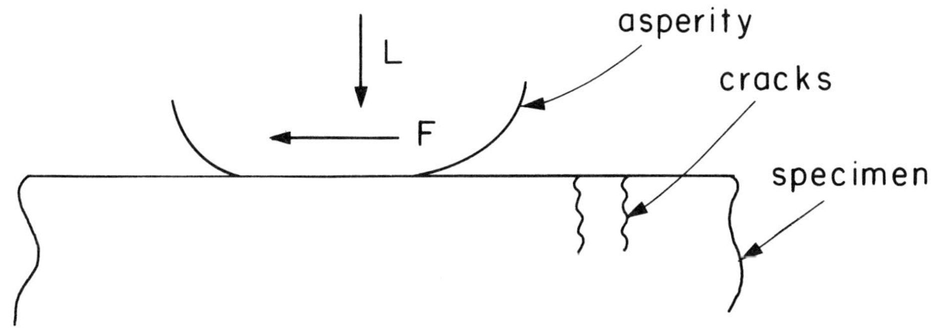

Fig. 2.-Perfectly Elastic Solid with $\mu>0$.

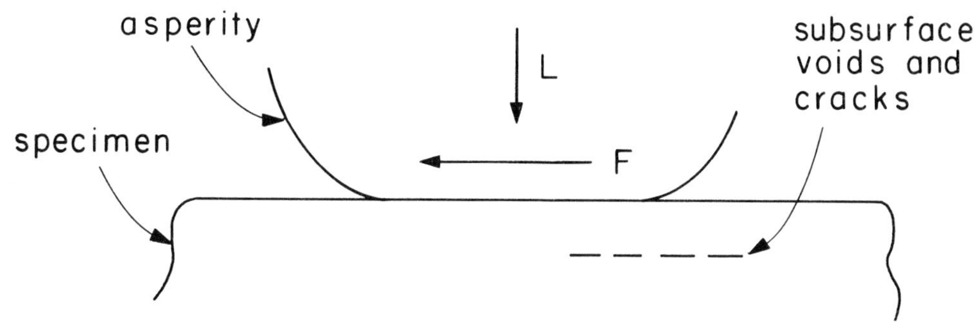

Fig. 3.—Elastic Solid with Internal Voids and Cracks, μ>0.

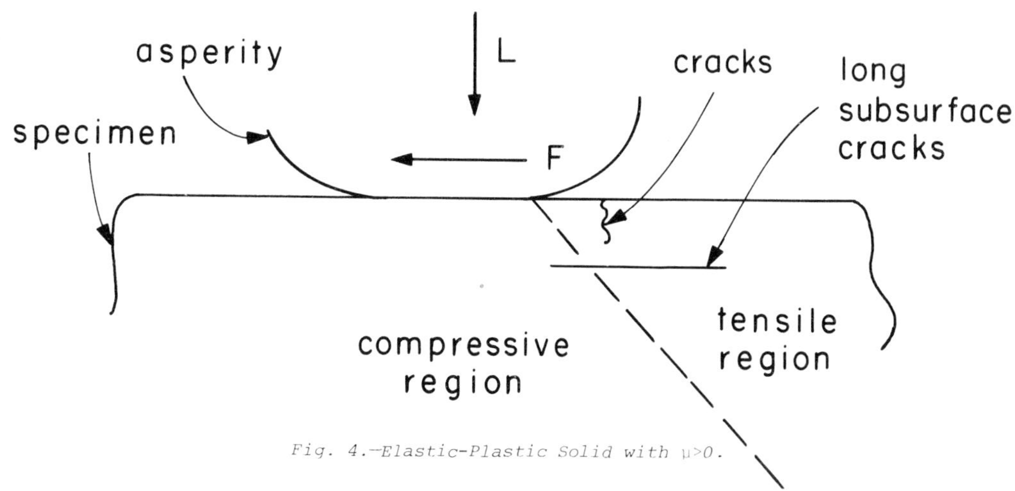

Fig. 4.—Elastic-Plastic Solid with μ>0.

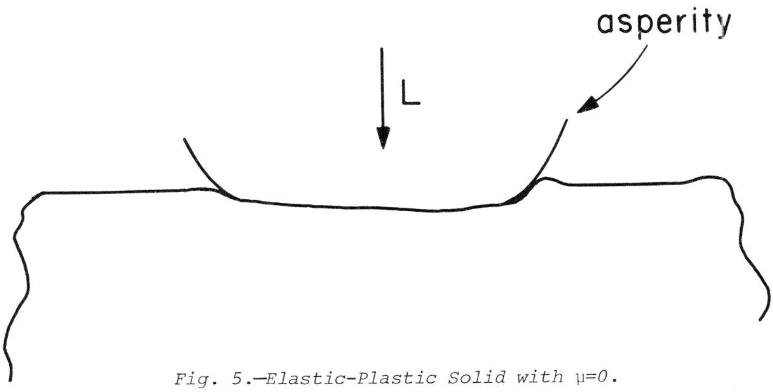

Fig. 5.—Elastic-Plastic Solid with μ=0.

Case 8) When abrasive particles are present on the surface of elasto-plastic solids, wear can occur by cutting and plowing actions when the abrasive particles are much larger than about 1 μm, but it occurs by delamination mechanism when the particle is less than about 1 μm (Figure 8).

ON WEAR DICTATED BY MECHANICAL BEHAVIOR OF MATERIALS

The discussion of the previous section indicates that the wear mechanism is likely to be controlled by the mechanical behavior of solids at moderate sliding speeds. The so-called adhesive wear mechanism does not seem to control the sliding wear in most practical cases. The presence and absence of adhesion has significant effects on the surface traction which in turn controls the delamination wear mechanisms, when the coefficient of friction is relatively high.

Barwell stated in his paper that the mild wear is caused by the mechanisms postulated by the delamination theory of wear, whereas other wear phenomena such as severe wear are not. That may not be true. The difference between the severe wear and mild wear is simply the difference in the rate of removal of wear sheets for a given pair of materials. In the case of mild wear the number of wear sheets coming off at a given time is small due to smaller numbers of sub-surface cracks propagating at a slow rate. When the applied load and the frictional force are increased, the number of wear sheets coming off at a given instant will significantly increase. Since the number of cracks and the sub-surface crack propagation rate depend strongly on the micro-structure, the transition from "mild" to "severe" wear depends on material properties as well as the loading conditions.[7]

The delamination wear mechanisms were operative even in carbon steels when the sliding speed and load were very high.[2] Because of the high interface temperatures, the wear rate was high with a great deal of metal transfers. However, the cause for wear was again by the nucleation and propagation of sub-surface cracks. The only possible exception to the wear caused by sub-surface crack nucleation and propagation was when phase transformation and corrosion were present. In the latter case, pitting and the change in crack propagation rate in the presence of corrosives also affect the wear rate.

The delamination wear occurs in many materials. There is sufficient evidence that it occurs in a variety of metals,[7] polymer/glass fiber com-

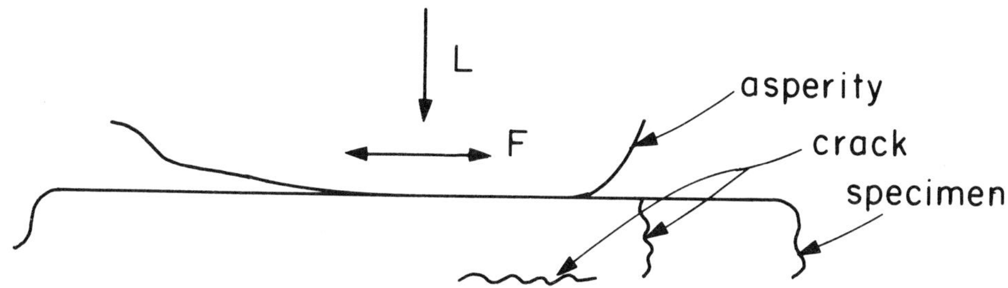

Fig. 6.—Elastic-Plastic Solid with $\mu>0$ and Oscillatory Motion.

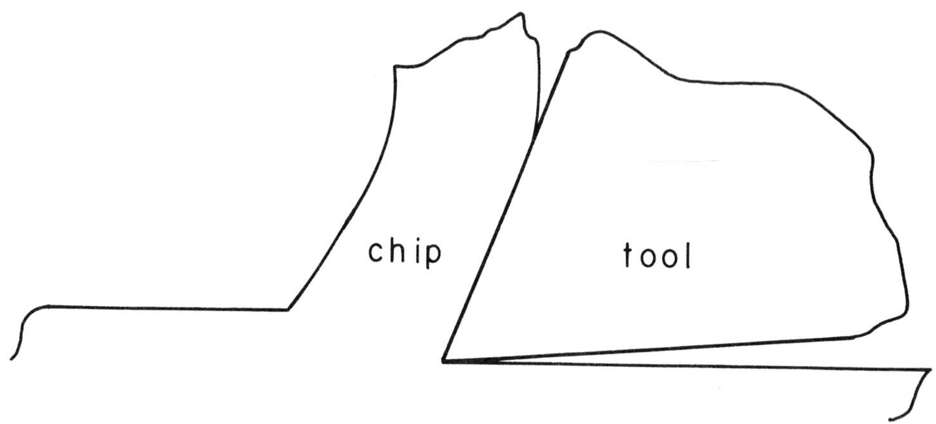

Fig. 7.—Hard Material with Low Chemical Stability at High Temperature and $\mu>0$.

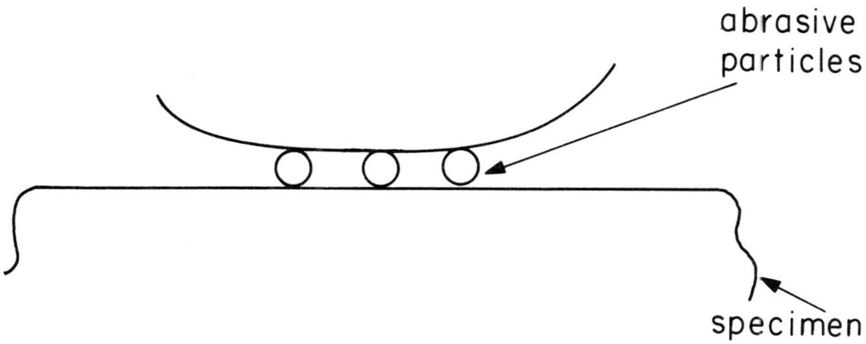

Fig. 8.—Elastic-Plastic Solid Loaded Through Abrasive Particles.

posites,[11] and even in aluminum oxide.[12]

A few words of caution on delamination wear are in order. Delamination wear does not always involve crack propagation parallel to the surface. When the surface traction is so small that the stress intensity factor is less than the threshold value, crack propagation cannot occur. In this case, the wear process will be extremely slow since the wear rate is controlled by the crack nucleation mechanism; a large number of cracks must be nucleated for loose wear particle formation when the surface traction exceeds the strength of the material connecting these cracks. This may be the case when the coefficient of friction is very small.

Another commonly recognized wear that is controlled by the mechanical behavior of solids is abrasive wear. Abrasive wear has been a popular subject of wear research as indicated by an extensive literature. However, the work done in the past was incomplete and could not explain some of the critical experimental observations. The two most critical deficiencies are:

(a) The theoretically predicted wear coefficient for abrasive wear is approximately 1. (The wear coefficient may be interpreted as a ratio of the work done to remove metal in the form of chips to the total external work.[13]) This value of wear coefficient is at least an order of magnitude larger than the experimentally observed values.

(b) The dependence of the wear rate on grit size cannot be predicted by the existing theories.

Recent MIT work[14] seems to explain the basic mysteries of abrasive wear. The reason that the experimentally observed values are smaller than the theoretical value is because of the plastic work done to deform the substrate and the repeated plastic deformation the cut chips are subjected to before they separate as a loose particle. That is, the plowed or cut material does not come off loose from the surface until the cumulative strain (imposed by the asperities) exceeds a critical fracture strain. This argument leads to the conclusion that the abrasive wear coefficient should be a function of both ductility and hardness. Hardness alone is an insufficient characterization of the material.

The grit size effect in abrasive wear is due to the changes in the "apparent" sharpness of abrasives as the size of abrasive particles decreases.[14] When the grit size is large, the grit resembles a conical cutter, whereas small grits resemble a spherical cutter. In the latter case, a greater portion of the external work done is consumed to deform the surface. When the particle is very small, the relative ratio of the depth to the width of penetration is so small that all the external work is consumed in deforming the surface.

There are still many unknowns related to delamination and abrasive wear which need further clarification. They are:

1. Fracture mechanics involved in wear, especially under a combined state of loading

2. The wear behavior at low coefficients of friction

3. Crack nucleation and propagation process in single phase metals

4. The mechanical behavior of materials very near the surface

5. The prediction of the wear coefficient as a function of microstructure, loading conditions, and the environmental interaction

6. The wear mechanism of very lightly loaded cases.

WEAR DOMINATED BY CHEMICAL BEHAVIOR OF MATERIALS

At high velocity and high loads, the interface temperature eventually exceeds a critical temperature at which different wear mechanisms control the wear rate. One such wear mechanism is that controlled by chemical stability of materials. In the case of cemented tungsten carbide tools the carbide dissociates and the dissociated elements form solid solutions with the chip which is then transported away by the chip from the interface by convection.[10] Based on this model the wear rates of various Group IV-B and V-B carbides were theoretically predicted based on the kinetics of chemical dissolution at the interface. Oxide tools do not wear due to chemical instability; their wear is controlled by mechanical processes due to their extremely low free energy of formation.

In highly oxidative environments the oxidation may be the rate controlling wear mechanism at extremely high temperatures. The oxide layer may flake off if the oxide formed is not mechanically stable at the surface or may be scraped off by the sliding action. Despite numerous papers on this subject, there is no evidence that at moderate temperatures oxidation is an important wear mechanism. Its effects are mainly manifested through surface traction when the temperature rise is moderate. Oxides normally lower the surface traction.

Other forms of corrosion can occur and affect the wear process.[15] However, presently very little is known about the synergistic effect of corrosion and wear. Further research must be done to understand this very important subject.

CONCLUSIONS

1) In understanding and solving wear problems, it is more instructive to think in terms of rate controlling wear mechanisms under a given set of

surface traction and displacement boundary conditions, rather than in terms of gross wear phenomena (e.g., abrasive wear, mild wear, etc.).

2) The basic mechanisms postulated by the delamination theory of wear are operative in a number of seemingly different situations such as mild wear, severe wear, and some cases of abrasive wear.

3) One of the major tasks of tribology is the quantitative prediction of friction and wear behavior of solids based on the relationship between surface traction, temperature, materials properties and wear mechanisms.

4) At high sliding speeds wear can occur either due to the chemical instability of materials involved or due to the mechanically controlled wear mechanisms. The dominating mechanism is indicated either by the solubility or by diffusivity of elements at the sliding temperature.

ACKNOWLEDGEMENT

The research on delamination wear has been sponsored by the Defense Advanced Research Projects Agency. The stimulating discussions and the support of Drs. Arden L. Bement, Richard S. Miller, and Edward Van Reuth and Commanders Kirk Petrovic and Harold P. Martin are gratefully acknowledged.

REFERENCES

1. Barwell, F.T., These Proceedings.
2. Saka, N., Eleiche, A.M. and Suh, N.P., *Wear,* Vol. 44, 1977, p. 109.
3. Cook, N.H. and Bhusan, B., "Sliding Surface Interface Temperatures," Paper 7-Lub-34, American Society of Mechanical Engineers, 1974.
4. Jahanmir, S. and Suh, N.P., *Wear,* Vol. 44, 1977, p. 39.
5. Suh, N.P., *Wear,* Vol. 44, 1977, p. 1.
6. Fleming, J.R. and Suh, N.P., *Wear,* Vol. 44, 1977, p. 39.
7. Suh, N.P., "Surface Treatment and Coating Techniques for Cemented Carbide Tools," Proceedings of the North American Metal Working Research Conference, May 1973.
8. Suh, N.P., *Wear,* Vol. 25, 1973, p. 111.
9. Merwin, J.E. and Johnson, K.L., *Institution of Mechanical Engineers. Proceedings,* Vol. 197, 1963, p. 676.
10. Kramer, B.M. and Suh, N.P., These Proceedings; also Kramer, B.M., Ph.D. Thesis, Massachusetts Institute of Technology, 1979.
11. Clerico, M., These Proceedings.
12. Suh, N.P. and Fillion, P.D., to be published.
13. Suh, N.P., in Proceedings of Conference in Scientific Problems of Coal Utilization, University of West Virginia, 1977, p. 168.
14. Suh, N.P., Sin, H-C. and Saka, N., These Proceedings.
15. Tse, M.K. and Suh, N.P., *Wear,* Vol. 44, 1977, p. 145.

ON THE WEAR MECHANISMS AND THE WEAR EQUATIONS

S. Jahanmir

ABSTRACT

Various types of wear such as delamination, adhesion, abrasion, fretting, corrosion, erosion, impact wear and surface fatigue wear are discussed. It is shown that the fundamental mechanisms of all types of wear are similar in nature. In general, either one or a combination of delamination, adhesion or abrasion mechanisms controls the wear behavior. The basic nature of these wear mechanisms are discussed and their similarities are indicated. Based on these similarities a general wear equation is proposed in which the worn volume is inversely proportional to the specific wear energy. (The specific wear energy is a material property and it is the amount of energy expended for wear of one unit volume of material.) The specific wear energy is composed of a number of components; mainly, the plastic strain energy, surface energy, cutting energy and plowing energy. The contribution of each component to the specific wear energy depends on the wear mechanism, but in most cases the plastic strain energy for subsurface deformation predominates.

INTRODUCTION

One of the least understood and yet important branches of tribology is that dealing with the wear of materials. Wear may be defined as the removal of material from solid surfaces as a result of relative sliding motion at the surface. Wear is controlled by the type of interacting materials, the environment, the loading condition and the type of sliding interaction. Various terminologies have been used to characterize different types of wear. Sliding, adhesive, delamination, scoring, gouging, scuffing, pitting, mild, severe, abrasive, erosive, fretting, surface fatigue, corrosive and diffusive wear are only some of the terms which are used. This type of terminology is rather unfortunate since their use is often arbitrary.

In this paper a number of wear situations are considered and their underlying mechanisms are described. Basically, most types of wear are similar in nature since they occur as a result of an interaction between two surfaces. In each wear process, one of the surfaces applies a set of normal and tangential loads to the wearing surface. This surface traction deforms a surface layer elastically or plastically and causes wear particle generation. A number of mechanisms may occur independently or simultaneously to cause material loss in each wear process. The purpose of this paper is to describe and identify the mechanisms which occur during each type of wear.

It is proposed that an energy analysis may be a useful tool in identifying the wear mechanisms.

WEAR TYPES AND WEAR MECHANISMS

Wear may be characterized based on the appearance of the worn parts or based on the mechanisms and conditions which prevail during material removal. Such classifications as mild wear, severe wear, scoring and scuffing are based on the observation of worn parts. However, other terminologies such as adhesive wear, delamination wear, erosive wear, etc., are based on the mechanisms of material removal or the conditions which exist during the wear process. The phenomenological characterization of wear may be useful for workshop but it does not provide a fundamental understanding of the wear mechanism or the previous history of the worn part.

In this paper wear is classified according to the wear mechanisms and conditions. Based on this classification the wear types are,

i) adhesion
ii) delamination
iii) fretting
iv) abrasion
v) erosion
vi) impact wear
vii) surface fatigue
viii) corrosive wear
ix) diffusive wear
x) electrical contact wear

The first seven types occur by mechanical interactions, whereas the last three types are largely influenced by special environments or conditions. The wear types which occur as a result of mechanical interactions are considered here. The other wear types are described elsewhere in these proceedings.

Adhesion

Adhesive wear occurs as a result of relative sliding between two surfaces under a normal contact load. It is assumed that when asperities come into contact, they adhere strongly to each other and form asperity junctions. Subsequent separation of the surfaces occurs in the bulk of the softer asperities. This process generates particles of the softer material which adhere to the harder surface.

There are a number of important criticisms of the assumptions of the adhesion theory of wear:

1. For generation of transferred particles, complete adhesion should occur at the junctions, and the interface must be stronger than either material. This interface cannot be stronger than the contacting materials since it may contain vacancies, impurities, or oxides. Therefore, complete adhesion and material transfer is a very rare event, except for some selected materials under very special conditions (e.g., seizure in vacuum or writing with a chalk).
2. The adhesion theory implies that the harder surface should not wear at all. This is contrary to the experimental evidence. The theory is modified to explain the wear of harder surfaces by stating that the hard material has some "weak spots" inside some of the asperities, and occasionally fracture takes place in these weak regions rather than in the softer metal(1). This explanation, however, is overly simplified.

3. The adhesion model suggests that transferred particles are generated only through asperity-asperity interactions. This suggests that the steady state wear rate should depend on the surface roughness, which is contrary to experimental observations[2].
4. The adhesion theory only considers the formation of adherent particles. It is a matter of general observation that wear occurs by loss of materials from the system. Rabinowicz[1] introduced a model to account for generation of loose wear particles. According to this model, loose particles are generated from the adherent particles when the stored elastic energy in the particle becomes greater than the adhesion energy of the particle to the surface. This model, however, implies that only particles larger than about 1 μm become loose.
5. The maximum work required to generate a given volume of wear particles by the adhesion mechanism as given by the upper bound solutions of plasticity (assuming a Green's[3] type of solution) is two to three orders of magnitude smaller than the external work done. This point is explained by the proponents of adhesion theories by considering a very low probability factor (i.e., wear coefficient) or by stating that the particles transfer back and forth between the surface a number of times before becoming loose[1].
6. The adhesion theories ignore the controlling effect of friction force on the wear rate; yet wear depends strongly on the friction force[4].
7. The adhesion theories do not consider the structure and the mechanical properties of materials. It is known that the behavior of materials under load depend sensitively on their microstructure and the metallurgical parameters. Therefore, these factors cannot be ignored in a physically realistic model for sliding wear of metals.

These remarks indicate that the adhesion models have neglected to include many important factors. Nevertheless, the experimental evidence[1,5] is sufficient to conclude that adhesion may occur in some sliding wear situations. The adhesive wear is perhaps the controlling wear mechanism under vacuum, inert atmospheres or when the asperity flash temperature is large to promote asperity adhesion. Under many practical situations the presence of contaminates on the surface or in the environment retards adhesion. Under these conditions adhesion may occur but it cannot be the controlling wear mechanism.

Delamination

Delamination wear takes place as a result of relative sliding between two surfaces under a normal contact load. The delamination process occurs by the following set of mechanisms[6]:

a) When two sliding surfaces come in contact, normal and tangential loads are transmitted through the contact regions by adhesive and plowing actions. The asperities of the softer surface are easily deformed and some are fractured by the repeated loading. A relatively smooth surface is generated, either when these asperities are deformed or when they are removed.
b) Once the surface becomes smooth, the contact is not just an asperity-to-asperity contact, but rather an asperity-plane contact; each point along the softer surface experiences cyclic loading. The surface traction exerted by the harder asperities on the softer surface induces plastic deformation which accumulates with repeated loading.

c) As the subsurface deformation continues, voids are nucleated below the surface. Void nucleation below the contact is controlled by two factors. Deformation by the shear component of stress promotes void nucleation, while the triaxial state of compressive stress opposes nucleation. Since the compressive stress is a maximum at the surface, voids cannot nucleate at the surface. However, below a certain depth, where the deformation induced stress exceeds the compressive stress, void nucleation becomes possible. At depths far away from the surface void nucleation cannot occur since the stresses die off very quickly. In materials which contain hard particles (e.g., second phase or inclusions) void nucleation is preferentially initiated at the particle-matrix interface.

d) Upon further deformation and repeated loadings, cracks extend and propagate, joining the neighboring ones. Presence of voids or small cracks, however, does not guarantee crack propagation. Cracks can only propagate in the regions where criteria for propagation can be satisfied. For example, the tensile state of stress existing below the surface right behind each moving asperity could be responsible for subsurface crack propagation. Cracks tend to propagate parallel to the surface since the state of loading is repeated along the surface.

e) When these subsurface cracks reach a critical length they become unstable and propagate to the surface, generating thin wear sheets. The thickness of the wear sheets depends on the material properties and the magnitude of the normal and tangential loads at the surface, since they control the magnitude and the depth of the maximum tensile stress behind the moving asperities.

Since the delamination process is controlled by deformation, void nucleation and crack propagation, the microstructure of the material and the state of stress are two important parameters of the wear behavior. Such microstructural parameters as hardness, number of second phase particles or inclusions and the morphology of hard second phase particles affect the wear process since they control the mechanical behavior of materials. Jahanmir et al.(7), Pamies-Teixiera et al.(8) and Saka et al.(9) have shown that as the hardness of metals without hard particles or with coherent second phase particles is increased, the wear rate decreases because of the reduced plastic deformation and void nucleation. Jahanmir et al.(7) and Saka and Suh(10) have shown that the wear rate increases as the number of incoherent second phase particles or inclusions is increased. The increase in the number of hard particles which act as the void nucleation sites, increases wear because of the accelerated crack nucleation and ease of crack propagation. Therefore, a solid-solution metal with a high hardness or a metal containing small coherent second phase particles is the most suitable material for sliding applications.

The effect of second phase particle morphology on wear has been investigated by Jahanmir et al.(7). Among the various structures studied for AISI 1095 steel, a coarse pearlite and bainite had the lowest and the highest wear rates, respectively. It was also found that the spheroidized steels had larger wear rates than peralitic steels. The wear rate depends on the morphology of second phase particles because the mechanism of void formation depends on the shape of the particles. Voids nucleate from equiaxed particles by particle-matrix separation. However, for the case of plate-like particles, such as pearlite, subsurface voids generate by particle fracture(7).

The processes of void nucleation and crack propagation have been analyzed by Jahanmir and Suh(11) and Fleming and Suh(12), respectively. In both studies a continuum mechanics approach was used to investigate the mechanics of void nucleation and crack propagation. In the void nucleation model, it was assumed that voids nucleate when the tensile stress normal to the particle-matrix interface reaches the bond strength. This void nucleation criterion was found to be satisfied only at characteristic depths below the surface. This depth increases with friction coefficient and normal load. It was also shown that void nucleation in metals with partially coherent or incoherent particles can occur at early stages of the delamination process.

In the study of mechanics of crack propagation during delamination, Fleming and Suh(12) showed that the tensile stress which exists below the surface and behind each moving asperity can cause subsurface crack propagation. The calculations showed that there is a characteristic crack propagation depth below the surface which increases with increasing coefficient of friction. In addition, the change in stress intensity factor, and hence crack propagation rate, increases with increasing coefficient of friction.

Fretting

Fretting wear arises when contacting surfaces undergo oscillatory tangential displacement of small amplitude. This type of wear occurs by interaction of several wear mechanisms. Fretting wear is initiated by adhesion and when wear particles are oxidized, abrasive wear is combined with the adhesive process. Corrosion has a strong influence on fretting wear, especially on oxidation of wear particles which produce abrasive action. According to Suh(6) and Waterhouse(13) delamination is responsible for the final generation of fretting particles. In some material systems, adhesion may be absent and the delamination process controls the fretting wear particle generation(14). It is not surprising to find that subsequent fretting particle generation is by delaminations, since in fretting similar to sliding wear a set of tractions is exerted on the surface. This surface traction can result in subsurface deformation, void nucleation and crack propagation(6).

Abrasion

Abrasive wear occurs when a surface is in contact with harder particles. These particles may be loosely held between two sliding surfaces, or they may be a part of the second surface. The mechanism of abrasive wear has been shown to be microcutting of small chips by the harder particles acting as cutting tools(15,16). However, consideration of the input energy and the energy needed for metal cutting(17) indicates that microcutting is not the only mechanism. Portions of the input energy may be used up in surface or subsurface deformation of the wearing surface. Some of the abrasive particles do not have the required cutting angle for producing microchips(18). These particles slide and deform the surface and the subsurface by plowing. The subsurface deformation could result in the process of subsurface void and crack formation similar to delamination. The delamination process may not lead to material removal in the form of wear sheets, but it could damage the surface layer such that the microcutting action can proceed faster. Some experimental evidence has been published on the existence of subsurface damage in abrasive wear(19,20). However, the effect of subsurface damage on the abrasive process has not been tested experimentally.

Erosion

Erosive wear by solid particle impinigement arises from impact of hard abrasive particles on a target surface. Erosion is believed to occur by a combination of two mechanisms, mainly cutting or abrasive wear and deformation wear[21,22]. The cutting action by impacting abrasive particles dominates at smaller impact angles; whereas the deformation mechanism dominates at larger angles of impact. Erosion by the cutting action takes place similar to the abrasive wear process. The exact nature of the deformation mechanism has not been explicitly explained in the literature. Some proposals have been made such as cold working and fracture after repeated impact, fatigue crack generation and final fatigue failure of wear particles, etc.[21], but none have been shown to be correct.

Recently Suh[19] suggested that the process of delamination which occurs in sliding wear of metals may take place in erosion. This suggestion was supported by subsurface examination of eroded samples which showed evidence of subsurface cracking. Some recent experimental evidence obtained elsewhere[23] has also shown the possibility of delamination in erosive wear. The requirement for initiation of the delamination process is the formation of subsurface voids and cracks after repeated deformation of the surface layer. This is possible in view of the following concepts. First, only some of the impacting particles are capable of removing material by the cutting action, because they may not have the correct cutting angle or a sufficient energy. These particles will slide on the surface, after impact, and plow the surface material. Secondly, the particles which remove material by cutting, slide on the cut surface before leaving the surface. Therefore in both cases the impacting particles slide on the surface and plastically deform the surface layer. These concepts are, however, valid at shallow impact angles, lower than 50°, since at large angles most impacting particles rebound after impact.

The possibility of subsurface void nucleation in erosion has been checked analytically[24]. The results indicate that void nucleation from inclusions or second phase particles occurs in specific regions below the target surface, and that the voids nucleate more readily at lower impact angles. It should be pointed out that the effect of subsurface void nucleation in erosive wear is not on the generation of wear particles similar to delamination wear, but rather subsurface void nucleation weakens the surface material and makes its removal easier by the cutting action.

At larger impact angles where the abrasive action of the impacting particles is reduced the wear process occurs by the deformation mechanism[22]. The material removal mechanism at these impact angles is probably by generation of subsurface cracks at locations of maximum shear stress (in ductile metals) and propagation of these cracks by a fatigue mechanism. Presence of inclusions or hard second phase particles could accelerate the erosion rate, since crack nucleation becomes possible by matrix/particle separation. This process is similar to the delamination mechanism, the difference being the motion of the interacting surfaces. In delamination wear, the two surfaces are in a sliding interaction, whereas in large impact angle erosion the two surfaces are in an impact interaction.

Impact Wear

Impact or percussive wear arises from repetitive impact of two surfaces. Impact wear differs from erosive wear in the sense that erosive wear occurs by impact of small solid particles on the surface, whereas impact wear arises from repetitive impact of two solid surfaces. The most common mechanism of impact wear is by formation and propagation of subsurface cracks in ductile materials and by formation and propagation of surface cracks in brittle materials[25]. This mechanism is similar to erosion mechanism at large impact angles.

Surface Fatigue Wear

Surface fatigue wear is generally associated with surfaces in a rolling contact where the friction coefficient is negligible. The wear mechanism is by formation of surface or subsurface cracks and fatigue crack propagation[26]. The subsurface crack nucleation is associated with inclusions located at regions of maximum shear stress. Especially, in overloaded contacts the crack nucleation occurs in the plastic deformation zone which is located at a characteristic depth below the surface. The crack nucleation process in surface fatigue wear is similar to the delamination process. The main difference is that in sliding wear the plastic deformation zone extends to the surface (at friction coefficients larger than 0.35) but in rolling contacts the plastic zone is surrounded by elastic material.

The foregoing description of the various types of wear indicates that these types of wear are basically similar in nature. In all cases the wearing body undergoes a cyclic type of deformation which is the result of the cyclically applied surface tractions. The deformation is either elastic as in the case of surface fatigue or elastic-plastic as in the case of delamination. In all the wear types the wear particles are separated from the surface by a fracture process which is controlled by the material properties, loading condition and the environment. The various wear types differ from each other in the shape and type of contact (i.e., elastic or plastic), surface tractions (i.e., the magnitude of normal and tangential loads) and the motion of the surfaces (i.e., rolling, sliding, or impact). These conditions are summarized in Table 1. The differences in conditions of the types of wear give rise to the different wear mechanisms which control each type of wear. A close examination of each type of wear indicates that only three mechanisms are responsible for wear: adhesion, delamination and abrasion. In some types of wear, one of these mechanisms is the controlling process, in some others wear occurs as a result of interaction between these mechanisms. For example, in delamination wear, the controlling mechanism is delamination; whereas in erosion a combination of abrasion, delamination and perhaps some adhesion interact to cause wear particle generation. The wear mechanism for the types of wear considered here are summarized in Table 1.

TABLE I.—THE MECHANISMS OF VARIOUS TYPES OF WEAR

Type of Wear	Type of Contact	Motion	Wear Mechanism
Adhesive wear	Elastic/Plastic	Sliding	Adhesion
Delamination wear	"	"	Delamination
Fretting wear	"	Oscillating	Adhesion, delamination abrasion
Abrasive wear	"	Sliding	Abrasion, delamination
Erosive wear	"	Impact/Sliding	Abrasion, delamination
Impact wear	"	Impact	Delamination (fatigue)
Surface fatigue wear	Elastic	Rolling	Fatigue

Sliding wear may be defined as the wear occuring between two sliding surfaces in the absence of abrasion. The adhesion and delamination are the basic mechanisms for sliding wear. The adhesion mechanisms may be the rate controlling process under conditions where strong adhesion between two sliding surfaces exists, such as inert atmosphere, heavy contact loads or high sliding speeds. Under normal sliding conditions the adhesion tendency is low and the adhesion mechanism may not control the sliding wear process. Under these conditions, delamination may be the controlling mechanism. The adhesion mechanism can occur under normal conditions only on surface regions where the subsurface has been weakened by the delamination process and the weak adhesive forces are sufficient to complete the final fracture process and wear particle generation. Therefore, a few transferred wear particles are expected to be observed on wear surfaces, but the adhesion mechanism is not the controlling process under normal sliding conditions.

The removal of material from metallic surfaces by any wear mechanism occurs by deformation and fracture of wear particles from the wearing material. Therefore, the wear rate must depend on such mechanical properties as the flow stress, hardness, strain hardening capability, ductility and toughness of the wearing material. The effect of hardness has been the subject of many experimental investigations. As a general rule, the adhesion, delamination and abrasion rates are inversely proportional to the hardness. This relationship, however, only holds if the increased hardness does not grossly affect other mechanical properties. It has been observed that increasing hardness by introducing many inclusions or hard second phase particles may increase delamination rate, because the hard particles act as void nucleation sites[7]. The effect of ductility and toughness has not been studied extensively. It may be expected that an increase in ductility should decrease the wear rate. However, this may not be observed since by increasing the ductility, the hardness of the alloy decreases, which may increase the wear rate.

The lack of a simple relationship between the wear rate and some of the mechanical properties is partly due to the interdependence of mechanical properties and partly due to the difference in the state of stress during wear and during mechanical property testing. Most mechanical properties are obtained in simple tension, compression or shear; whereas the state of stress in a wearing body is fairly complex and consists of tension and shear or compression and shear plus hydrostatic stresses[11]. Moreover, the stress distribution in a wearing material is not uniform and may contain sharp stress or strain gradients[11]. Therefore, a simple and independent relationship may be absent between wear rate and some mechanical properties. The other factor which complicates the issue is that in most wear processes the surface traction is applied cyclically. A closer relationship may exist between cyclically determined properties and wear rates. This has been substantiated experimentally for delamination wear[27] by the dependences of the wear rate on the fatigue crack growth rate.

WEAR EQUATIONS

Wear arising from the three basic mechanisms of adhesion, delamination and abrasion has been found to be proportional to the applied normal load and the sliding distance. In the case of abrasion and for some materials under adhesion and delamination, an inverse relation has been found between wear and hardness of the wearing body, i.e.:

$$W = K\frac{NS}{CH} \tag{1}$$

where: W = *worn volume*
N = *Normal load*
S = *Sliding distance*
H = *Hardness*
K = *Wear constant*
C = *geometrical constant, equals 1 for abrasion and equals 3 for adhesion.*

Since both hardness and wear constant depend on the microstrcuture of materials, they should not appear independently in the equation. Therefore, the wear equation can be written as:

$$W = k\ N.S \tag{2}$$

where k = *wear factor.* Although both equations 1 and 2 can be derived by considering the basic mechanisms of adhesion, delamination or abrasion(1,6), the constants must be determined experimentally. A wear equation must be capable of predicting wear based on the basis mechanisms in each wear situation. Such an equation is not available and it appears to be in a distant future.

In any wear situation a specific amount of energy is expended through the interaction of surfaces. It may be assumed that all of this energy is used in wear particle generation, and that for any unit volume to be worn a specific amount of energy is required. Therefore, the worn volume can be found by:

$$W = \frac{\text{Energy input}}{\text{Wear energy/unit volume}}\ . \tag{3}$$

The energy input is the total energy expended in the surface interaction. For adhesion, delamination or abrasion the energy input is equal to the external work done; i.e., (Tangential force) x (sliding distance). The wear energy per unit volume or the specific wear energy, is the energy expended in removal of one unit volume of material. The magnitude of the specific wear energy depends on the wear mechanism involved in a particular wear situation, the material properties and the environment. The specific wear energy is composed of energy components required for each energy dissipation mechanism involved in a given removal process. These components are: specific energy for adhesive wear, specific energy for cutting (abrasion), specific energy for plowing, specific energy for subsurface deformation (delamination) and specific energy for generation of new surfaces.

The specific energy for adhesive wear particle generation can be calculated based on the adhesion mechanism. Since wear particle generation by the adhesion mechanism occurs as a result of shear in the softer asperities, the specific energy for the process must be equivalent to the toughness of the material under shear fracture or,

$$u_a \simeq \tau\gamma_f \tag{4}$$

where u_a = *specific adhesion energy*, τ = *shear fracture stress*, and γ_f = *shear fracture strain.* In normal torsion testing γ_f is generally less than one for most metals. However, under large hydrostatic pressures which exist under contacts γ_f can be as large as 10. The shear fracture stress is on the order of the hardness of the material. Therefore,

$$H \leq u_a \leq 10\ H \tag{5}$$

where H = *hardness of the wearing material.*

The specific energy for cutting by abrasive particles can be understood by analogy to the machining of metals. In metal machining chip formation occurs by shear in the shear zone, and the specific cutting energy is the plastic strain energy expended in the shear zone[28]. During cutting with sharp cutting tools it has been found that the specific cutting energy is on the order of the hardness of the material being cut[28]. Therefore,

$$u_c \simeq H \tag{6}$$

where u_c = *specific cutting energy*. In abrasive wear process the cutting action is similar to metal machining, but it takes place on a much smaller scale.

In the plowing process the abrasive particles form grooves on the surface without any material removal. It has been found that the width of the groove is proportional to the hardness of the material and the specific plowing energy is approximately equal to twice the hardness of the material. Therefore,

$$u_p \simeq 2H \tag{7}$$

where u_p is the specific plowing energy.

The specific energy for subsurface deformation or delamination is equal to the total plastic strain energy expended in removal of a unit volume of material by the delamination process. This energy component may be approximated by,

$$u_d \simeq m\bar{\sigma}_0\bar{\varepsilon}^p_{ave} \tag{8}$$

where u_d = *specific subsurface deformation energy*, m = *ratio of total equivalent strain per cycle to the net plastic strain per cycle*, σ_0 = *the equivalent yield stress*, and $\bar{\varepsilon}^p_{ave}$ = *the average subsurface equivalent plastic strain*. The values of m for cyclic tension-compression testing is generally between 100 to 500[29]. The yield stress is approximately equal to one third of the hardness. The average subsurface equivalent plastic strain depends on the surface tractions and material properties. The value of the $\bar{\varepsilon}^p_{ave}$ is much larger than the tension fracture strain value due to the large hydrostatic pressure under the contact of two surfaces. The average subsurface equivalent plastic strain may range from 1 to 50[30]. Therefore, by assuming that m equals 300,

$$100\ H < u_d < 5{,}000\ H \tag{9}$$

The specific energy for the generation of new surfaces is in general much smaller than the other energy components considered for wear particle generation. Therefore, the effect of surface energy on wear particle generation may be neglected. Table 2 summarizes the value for the specific energy components for various metals. The hardness of most engineering metals ranges from 50 to 5,000 MNm^{-2}. In determining the specific energies, the largest possible strain has been considered for the softest material and the smallest strain has been considered for the hardest materials, to be consistent with the mechanical behavior of materials. The table shows that the specific energy for subsurface deformation has the largest value; this is to be expected since the subsurface deformation is a cyclic process and a large amount of energy can be dissipated in a cyclic deformation process[29].

TABLE II.—CALCULATED VALUES OF SPECIFIC ENERGIES

Mechanism	Specific Energy MN m/m^3
adhesion	$5 \times 10^2 - 5 \times 10^3$
cutting	$5 \times 10 \;\; - 5 \times 10^3$
plowing	$1 \times 10^2 - 1 \times 10^4$
deformation	$2 \times 10^5 - 5 \times 10^5$

The experimentally determined values of specific energeis for delamination, abrasion and erosion for various metals are listed in Table 3. The mechanism of material removal in each of these wear situations may be understood by comparing Tables 2 and 3. The specific energies in Table 3 have been calculated for various metals tested by different investigators. It is observed that the specific energy for delamination is in the range of expected values for subsurface deformation. This indicates that the primary mode of energy dissipation in delamination is subsurface deformation. In fact, Suh and Sridharan(30) have shown that the delamination wear rate can be related to the energy dissipated by subsurface deformation.

Comparison of the specific energies for abrasion and for cutting indicates a one to three orders of magnitude difference. This large difference suggests that cutting is not the only mechanism which occurs in abrasion of metals. It was discussed in the previous section that some abrasive particles are not involved in the cutting action and cause surface plowing. The plowing specific energy is one to two orders of magnitude smaller than the abrasion specific energy. The other mechanism for the dissipation of energy in abrasion is subsurface deformation by the abrasive particles. This process is similar to deformation during delamination with the exception that during abrasion some of the deformed layer is removed by the cutting action. Therefore, the total accumulated subsurface strain is much less in abrasion than in delamination. In fact, subsurface examinations of abraded samples have shown a very small value of strain(19,20). Therefore, equation (8) can be used for the case of abrasion, since the deformation in this case is also a cyclic process. If the average subsurface strain of 0.5 to 1 is used for abrasion the specific energy for deformation in abrasion ranges from 5×10^3 to 2×10^4 MN m/m^3. Addition of cutting and plowing specific energies to the deformation energy, increases the specific energy of abrasion to the range of experimental values.

TABLE III.—EXPERIMENTALLY DETERMINED VALUES OF SPECIFIC ENERGIES

Type of Wear	Specific Energy MN m/m^3	Reference
Delamination	$10^5 - 10^6$	7-9, 31
Abrasion	$10^4 - 10^5$	1, 16, 32
Erosion	$10^5 - 10^6$	20, 23

The specific energy for erosion, in Table 3, is on the same order of magnitude as the specific energy of subsurface deformation. This indicates that cutting or abrasive action alone cannot be responsible for energy dissipation under erosive conditions, and that the subsurface deformation is the main energy dissipation process. This conclusion agrees with subsurface observations of eroded samples(23). The subsurface deformation and the number of cracks observed in eroded samples is, in fact, much larger than in the abraded samples.

CONCLUSIONS

The energy point of view appears to be a useful tool in understanding the wear mechanism involved in various wear situations. Even though the analysis was very approximate, the preceeding discussion indicated that the main wear mechanisms are adhesion, delamination and abrasion. In each type of wear one or several of these mechanisms are responsible for wear particle formation. The specific energy for wear is composed of energy components for each mechanism involved in the particular wear situation. The contribution of each component to the specific wear energy depends on the wear situation, but in most cases the plastic strain energy for subsurface deformation predominates.

REFERENCES

1. Rabinowicz, E., "Friction and Wear of Materials," John Wiley and Sons, New York, 1965.
2. Jahanmir, S. and Suh, N.P., *Wear,* Vol. 44, 1977, p. 87.
3. Green, A.P., *Journal of Mechanics and Physics of Solids,* Vol. 2, 1954, p. 197.
4. Rabinowicz, E., *Product Engineering,* Vol. 29, 1958, p. 71.
5. Bowden, F.P. and Tabor, D., "The Friction and Lubrication of Solids," Clarendon Press, Oxford, 1954.
6. Suh, N.P., *Wear,* Vol. 25, 1973, p. 111.
7. Jahanmir, S., Abrahamson, E.P., II and Suh, N.P., in Proceedings of the Third North American Metalworking Research Conference, Carnegie-Mellon University, edited by M.C. Shaw, Carnegie Press, Pittsburgh, 1975, p. 854.
8. Pàmies-Teixeira, J.J., Saka, N. and Suh, N.P., *Wear,* Vol. 44, 1977, p. 65.
9. Saka, N., Pàmies-Teixeira, J.J., and Suh, N.P., *Wear,* Vol. 44, 1977, p. 77.
10. Saka, N. and Suh, N.P., American Society of Mechanical Engineers Paper No. 26-WA/Prod. -29.
11. Jahanmir, S. and Suh, N.P., *Wear,* Vol. 44, 1977, p. 17.
12. Fleming, J.R. and Suh, N.P., *Wear,* Vol. 44, 1977, p. 39.
13. Waterhouse, R.B., and Taylor, D.E., *Wear,* Vol. 29, 1974, p. 337.
14. Waterhouse, R.B., *Wear,* Vol. 45, 1977, p. 355.
15. Aghan, R.L. and Samuels, L.E., *Wear,* Vol. 16, 1970, p. 293.
16. Hirano, H.H. and Levy, A.V., These Proceedings.
17. Suh, N.P., "Scientific and Technical Problems in Erosive Wear," Institute on Scientific Problems Relevant to Coal Utilization, West Virginia University, 1977.
18. Doyle, E.D. and Samuels, L.E., in Proceedings of the International Conference on Production, Tokyo, Japan, 1974, p. 45.
19. Suh, N.P., in "Fundamental Aspects of Structural Alloy Design," edited by R.J. Jaffe, Plenum Press, New York, 1976.

20. Doyle, E.D., *Wear,* Vol. 24, 1973, p. 249.
21. Finnie, I., in Proceedings of the 3rd United States National Congress on Applied Mechanics, 1958, p. 527.
22. Bitter, J.G.A., *Wear,* Vol. 6, 1963, pp. 5; 169.
23. Brass, L.L., Lawrence Berkeley Laboratory Report No. LBL-6277, 1977.
24. Jahanmir, S., to be published in *Wear,* 1980.
25. Engel, P.A., "Impact Wear of Materials," Elsevier, New York, 1976.
26. Syniuta, W. and Corrow, C., *Wear,* Vol. 15, 1970, p. 137.
27. Fleming, J.R. and Suh, N.P., *Wear,* Vol. 44, 1977, p. 57.
28. Cook, N.H., "Manufacturing Analysis," Addison-Wesley, Reading, Massachusetts, 1966.
29. Morrow, J., Department of Theoretical and Applied Mechanics, Report No. 277, University of Illinois, 1965.
30. Suh, N.P. and Sridharan, P., *Wear,* Vol. 34, 1975, p. 291.
31. Jahanmir, S., Ph.D. Thesis, Massachusetts Institute of Technology, 1976.
32. Larsen-Basse, J. and Mathew, K.G., *Wear,* Vol. 14, 1969, p. 199.

A REVIEW OF WEAR MECHANISMS AND RELATED TOPICS

S. L. Rice

ABSTRACT

This paper discusses Barwell's review of wear of machine elements and related topics. Barwell's paper is a comprehensive survey of wear mechanisms arising in machine elements, and well reflects the complexity of this subject.

The discussion paper focuses on the nature of contact and resultant stress systems, and on the importance of material temperatures occasioned in machine elements. It further elaborates upon Barwell's taxonomy, including discussions of characteristic modes of damage and on basic wear mechanisms.

The significance of material substrates in wear is highlighted, and illustrative experimental results from impact wear studies are reported. Suggestions are offered for a more unified framework from which future wear research might be conducted.

INTRODUCTION

Professor Barwell's review[1] of the wear of machine elements is comprehensive and well reflects the complexity of this subject. While details of his taxonomy will be debated, there are no substantive ideas with which this reviewer takes issue. Accordingly, this paper will present reactions to those of Barwell's points which prompted the same, and will then propose some lines of investigation which may lead to further progress under a more unified framework.

NATURE OF CONTACT

The classification of machine elements as kinematic pairs is useful in establishing a context for discussion. Barwell notes that, theoretically, conformal pairs give rise to dispersed contact while counterformal surfaces lead to Hertzian contact. He further points out that in lubricated systems, a higher pair can be effectively transformed to a lower one. Thus, while this classification is potentially useful, it likewise is potentially misleading. This is so because the essential determinants of wear are not whether surfaces are theoretically conforming but rather include the specifics of the stresses applied to the material and the environmental factors surrounding their application. In most contacting systems leading to wear, both "dispersed" and Hertzian stress systems operate. That is, in either dry or lubricated systems, individual asperities are subject to different stress levels than is the material which constitutes the relatively gross Hertzian contact zone.

SLIDE/SWEEP RATIO

The concept of slide/sweep ratio is helpful in categorizing the thermal "loading" to which machine elements are subjected. Unfortunately, no correlations between this quantity and wear are suggested. Presumably this is because the essential determinant of wear is not relative surface velocity or fixity but rather the surface and subsurface temperatures which the stress-affected material achieves. Often, these temperatures are not easily calculated or measured, but if correlations with wear behavior are sought, such analysis and experimentation should be performed.

CHARACTERISTIC MODES OF DAMAGE

Barwell cites six "characteristic modes of damage": mild wear, severe wear, pitting, scuffing, scoring or gouging, and fretting. He pays special attention to the definitions of most of these terms. In particular, he notes that two definitions which have been advanced for scuffing presuppose a mechanism. He then discusses scuffing in some detail, and sets forth an explanation with which this reviewer agrees.

There is a weak point, however, in Professor Barwell's utilization of the terms mild and severe wear. He correctly points out that these terms denote qualitative differences observable in some wear studies. He then suggests, however, that these words constitute "quantitative terminology". In this writer's view, the terms are qualitative descriptors. But more importantly, if these terms are to be utilized to identify wear damage arising in a variety of circumstances, adequate definitions should be put forth. For example, mild wear might be defined as that which occurs to produce no observable surface degradation. This definition would allow for subsurface damage (not observable as surface damage) and also for a polishing or smoothing of surfaces (not degradation). Accordingly, this term would then adequately describe the condition of most machine elements following run-in and following service under normal operating conditions. As such, it assumes no greater (or lesser) significance than terms such as pitting or gouging. That is, it allows for a qualitative characterization of wear, without reference to any mechanism. In concert with this definition, severe wear might be described as that which occurs to produce observable whole-field surface degradation. The modifier "whole-field" is included to differentiate from localized (yet severe) surface damage as in cases of pitting and scuffing.

BASIC MECHANISMS

In discussing adhesive wear, Barwell makes the statement that this form of wear implies material transfer from one surface to another without material being removed from the interacting system. Adherents to the adhesion point of view would argue that enactment of this mechanism does not preclude material removal. This simply *follows* the adhesive action, and can be due to subsequent abrasion of adhesively transferred particles, etc.

Barwell's comments concerning the role of frictional heat in adhesive action are well taken. Many discussions of adhesive phenomena seem to overlook the importance of thermal effects. However, the suggestion that adhesive wear and scuffing are the same phenomenon is somewhat misleading. Barwell's explanation of scuffing, which is quite plausible, goes only so far as to suggest that melting of at least one of the materials occurs.

Following this, adhesive action could serve to transfer material. But explaining scuffing in terms of a sequence of events which includes adhesion does not serve to make adhesive wear and scuffing into the same phenomenon. Indeed, scuffing is best reserved as a term which characterizes a mode of wear damage; adhesion is a basic physical mechanism by which material transfer may occur.

In discussing delamination, Barwell makes further reference to adhesion, stating that the adhesion theory presupposes that all action takes place on a plane of cleavage. Again, this is not a necessary supposition of a theory of adhesion, although some workers have made the cleavage plane assumption in attempting to account for frictional work.

As to delamination itself, the now well known work of Suh is briefly referenced[2]. While the importance of this theory cannot be too strongly emphasized, it is the writer's view that delamination is better classified as a characteristic mode of damage than as a basic mechanism of wear. Thus, delamination is identified by flat, sheet-like particles and by subsurface crack propagation. The basic mechanism is fatigue.

With these observations, Figure 1 is offered as an alternative classification to that presented in Barwell's Figure 2.

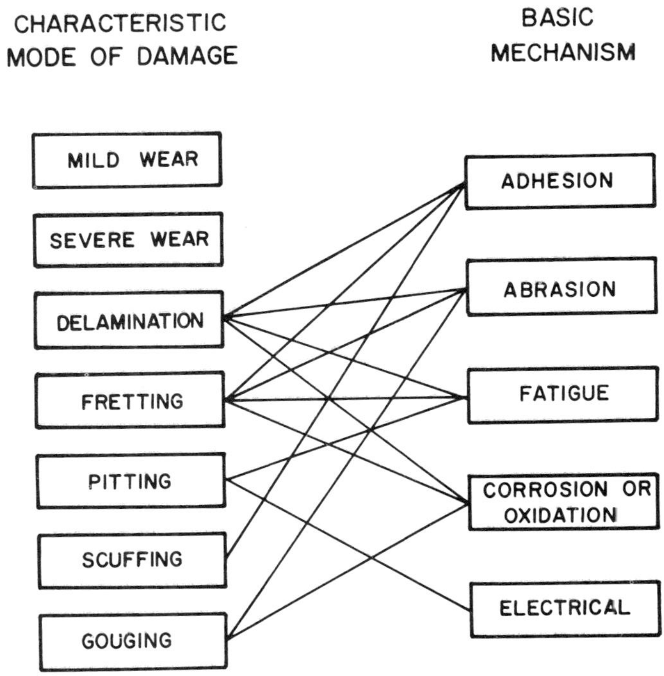

Fig. 1.—Modes of damage and basic mechanisms in wear.

THE SUBSURFACE

One of the major contributions of the delamination theory of wear is the attention it focuses on the subsurface of wearing elements. By now, considerable evidence has been accumulated to substantiate the occurrence of subsurface plastic deformation, void and crack nucleation and propagation.

In our own work in the area of impact wear, we observe strikingly different subsurface features between different materials and under various experimental conditions. Figures 2 - 6 show subsurface features from five different metals. In each case, the material specimen has been repeatedly impacted against a transversely moving counterface (compound impact). The counterface material is 17-4 PH stainless steel of initial surface roughness of approximately 0.2 μm CLA. The impacting specimen is a flat-ended cylinder so that planar strikes are achieved. With the impulse measured by means of a piezoelectric force transducer, peak normal impact stress levels are monitored and maintained constant over a given experimental run[3]. Thus, on each figure, the transverse (relative) sliding velocity (V) is given, as is the nominal peak normal impulsive stress (σ) and the number of impact load cycles (N).

While differences between materials are apparent in these figures, there appear to be (at most) three characteristic zones beneath the surface. The zone farthest from the wear surface ranges from unaffected microstructure toward observably deformed "grains". The intermediate zone consists of material which has been subject to exceedingly high plastic deformation, but the basic "grain" structure is still observable, although deforming and refining into flow lines. The near-surface zone is still more plastically deformed, and probably consists of material both from the specimen and that deposited from the counterface. It appears to be highly refined and homogenized, and preliminary TEM results indicate that it is crystalline.

Investigations by other workers have recently been conducted on such near-surface layers arising in sliding wear. Bill and Wisander[4] and Van Dijck[5] find this zone to be crystalline, as distinct from the amorphous Beilby layer which results from polishing and is often noted in the literature. In particular, Van Dijck finds the smallest grains ever measured, ranging in thickness from 200 to 600 Å. Moreover, the plastic deformations in this layer are likewise the largest ever measured, ranging up to strains of 230.

Thus, considerable work remains to be undertaken to characterize and understand the formation of subsurface zones. When the subsurface contains zones of markedly different character, is the delamination model adequate to describe the wear process? Subsurface voids and cracks are found in some materials, but not in others. Yet all materials contain regions of high plastic deformation, and most contain a near-surface zone with apparently markedly different properties from the base material. Further research is needed to elucidate the similarities and differences between subsurface zones in materials subject to various types of surface loading. Where possible, stress levels and surface and subsurface temperatures should be quantified in concert with such research.

CONCLUSIONS

As Professor Barwell notes, progress in future studies will best be made if pursued from a consistent framework rather than as isolated studies on particular machines and situations. It is this writer's view that future work should include the following:

1) experimental studies should be conducted with systems in which stress levels are determinant; these stress levels should be reported in concert with any results on wear resistance or with any discussion on wear mechanisms.

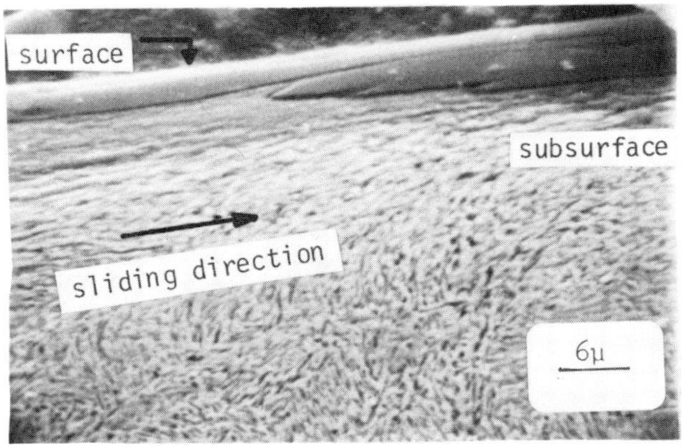

Modified HY180 High Strength Steel
V = 3.7 m/s
σ = 42.0 MPa
N = 150,000 cycles

Fig. 2.—Subsurface zones in impact wear of modified HY 180 high strength steel.

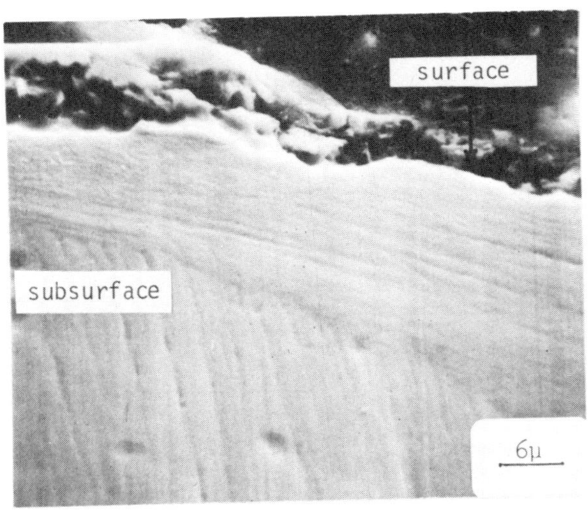

Titanium Alloy, IMI 685
V = 5.3 m/s
σ = 10.6 MPa
N = 500,000 cycles

Fig. 3.—Subsurface zones in impact wear of IMI 685 titanium alloy.

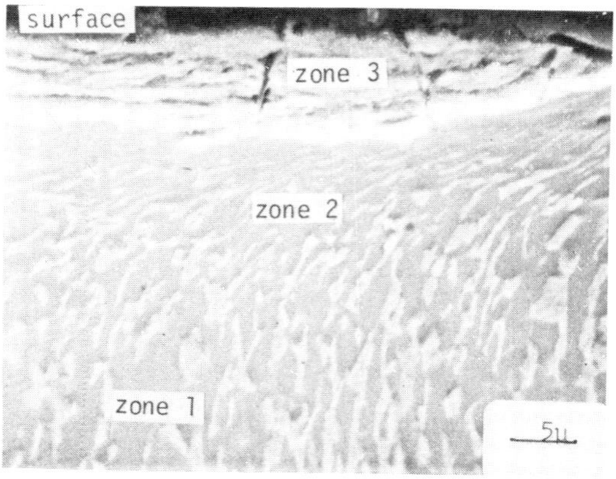

Titanium Alloy, RMI 5522S
V = 0.43 m/s
σ = 6.8 MPa
N = 200,000 cycles

Fig. 4.—Subsurface zones in impact wear of RMI 5522S titanium alloy.

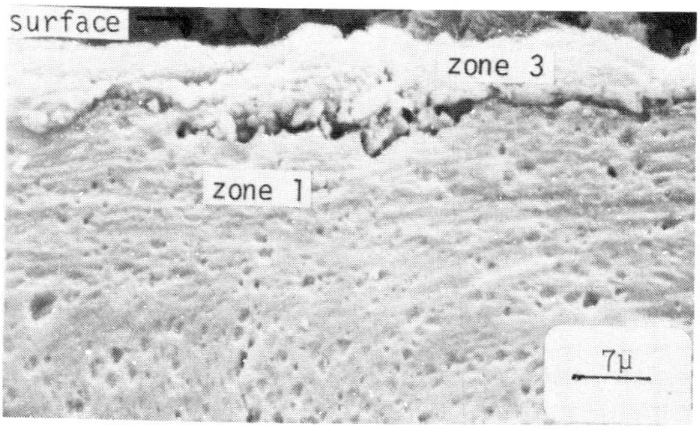

Aluminum Alloy, 2124
V = 5.3 m/s
σ = 4.2 MPa
N = 150,000 cycles

Fig. 5.—Subsurface zones in impact wear of 2124 aluminum alloy.

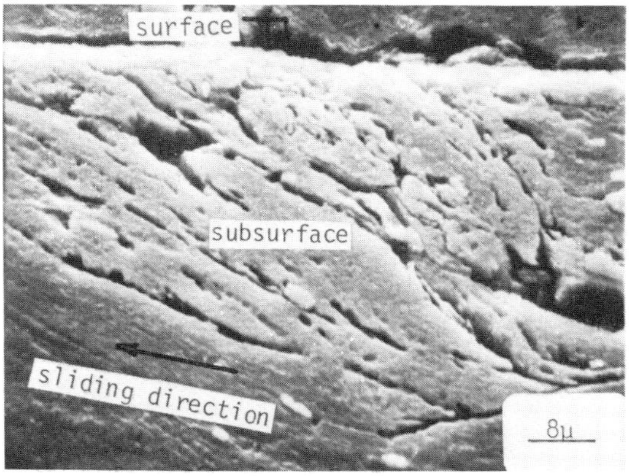

Aluminum Alloy, 2011-T3
V = 2.7 m/s
σ = 16.9 MPa
N = 250,000 cycles

Fig. 6.—Subsurface zones in impact wear of 2011-T3 aluminum alloy.

2) experimental and analytical studies should be conducted to determine surface and subsurface temperatures occasioned in various wear systems.

3) experimental and theoretical work should focus on the formation and characterization of subsurface zones.

ACKNOWLEDGEMENTS

The results reported herein concerning impact wear are a part of research sponsored by the Air Force Office of Scientific Research, Air Force Systems Command, USAF, under Grant No. AFOSR-76-3087. The United States Government is authorized to reproduce and distribute reprints for governmental purposes notwithstanding any copyright notation hereon.

The author is indebted to Mr. William J. Walker of AFOSR and to Mr. Steven F. Wayne of The University of Connecticut.

REFERENCES

1. Barwell, F.T., These Proceedings.
2. Suh, N.P., *Wear,* Vol. 44, 1977, p. 1.
3. Rice, S.L., *Wear,* Vol. 45, 1977, p. 85.
4. Bill, R.C. and Wisander, D.W., NASA Tech. Note D-7840 (1974); also Bill, R.C. and Wisander, D.W., *Wear,* Vol. 41, 1977, p. 351.
5. Van Dijck, J.A.B., *Wear,* Vol. 42, 1977, p. 109.

DISCUSSION

F. T. BARWELL, University College of Swansea: I have always had some difficulty with the terms "mild" and "severe" wear. The example I quoted from Kragelskii was, I think, a good example of the description of "mild" wear -- a situation in which we have an oxide film on the surface as tested by electrical resistance methods. In the other case we get metal-to-metal contact with low resistance. Recently a lot of our thinking has been conditioned by the study of wear particles, but again one recognizes the mild wear condition where thin flake-like particles come off as opposed to more intensive conditions in which we get many more much thicker particles.

Now when it comes to quantitative k factors and so forth, I have never been able to use a k factor. The best I have ever been able to do (and I think Prof. Czichos mentions this in his systems approach) is to set up in a laboratory a test rig which represents a real machine which simulates the important process. I think Prof. Czichos makes his check when he gets the same appearance of surface under the electron microscope in the laboratory as he gets in the field for a meaningful test. All I have been able to do (having got many variables, different steels, etc.) is to have a standard. Then I report the relative wear resistance on the different materials and I do not think the k factor means very much.

I would like to add one other point about your comment on the delamination wear. I may have put it in the wrong place out of respect for the chairman. When Prof. Suh published this, it did seem something of importance in the physical world which one should put alongside of adhesion and abrasion. I knew it was a fatigue mechanism. I think we could argue for a long time, but the reason I think I put it where I did is because I give considerable importance to it. I think that a lot of the mild wear takes place by the delamination process.

THE CLASSIFICATIONS, LAWS, MECHANISMS AND THEORIES OF WEAR

T. F. J. Quinn

ABSTRACT

This paper critically examines the current classifications, laws, mechanisms and theories of wear, in an attempt to provide a cohesive appraisal of the whole field of the fundamentals of wear. It is shown that the "mild" and "severe" wear classifications proposed by Archard and Hirst in 1956 are sufficient for most purposes, being based only on observable phenomena such as contact resistance, surface topography and analysis of the size and composition of the wear debris. The Laws of Wear are compared with the Laws of Friction, the aim of this comparison being to show how still incomplete are the current "explanations" of the Laws of Friction. Under the heading "Mechanisms of Wear," the mild (oxidational) wear mechanism is discussed in some detail, since it is proposed that "severe wear" is what happens when "mild wear" processes cannot occur. It is suggested that the most prevalent form of severe wear mechanism is the delamination mechanism recently proposed by Suh and his co-workers.

Finally, the oxidational theory of mild wear and the delamination theory of severe wear are discussed in terms of their ability to provide an expression for the K-factor of the so-called "Archard Wear Law" (w = KA). It is pointed out that both theories contain some ill-defined parameters and further work is needed to elucidate these parameters. Some details are given of the recent experimental verification of the mild (oxidational) wear theory by work being carried out at the author's laboratory.

INTRODUCTION

The aim of this proceedings is to establish the state of fundamental knowledge in Tribology and to assess the major program areas for future work in this subject. It would seem that the fundamentals of lubrication and friction are fairly well covered by Professors Cheng and Rabinowicz and by Mr. Godfrey. The review paper by Professor Barwell is concerned with how the practicing engineer can classify and deal with his particular wear problem rather than fundamentals of wear. Whilst this is a very important aspect of wear, there is surely a need for a review in which most of the current classifications, laws, mechanisms and theories of wear are critically examined from the point of view of how much is known about the fundamentals of wear. The aim of this paper is to attempt to satisfy this need.

DEFINITIONS

Before starting on the main theme of this paper, it will be worthwhile to consider what is meant by the words "Classifications," "Laws," "Mechanisms" and "Theories."

A "classification" is a method for grouping together items which have at least one feature in common. In crystallography, for example, it is possible to classify all the many possible shapes and forms of crystals into just seven *crystal systems*, using their basic minimum symmetry properties as a means of classification. Similarly, a classification of wear should be made on the basis of observable quantities, such as contact resistance or surface topography. Quite often, classifications are made in situations where there is insufficient experimental or theoretical knowledge to propose laws or theories. Classification is the first step towards the eventual scientific description of a phenomenon.

The word "law," when applied to scientific research, refers to the concise description of the general behavior of a large group of subjects under certain well-defined conditions. Typically, the word relates to a statement connecting dependent and independent variables, all of which are measurable quantities. For instance, the friction and wear laws are statements about the dependence of the frictional force and the wear rate upon the normal load and the independence of these quantities of the *apparent* area of contact. The so-called "Archard Wear Law"[1] is not truly a law, since it involves the dependence of the wear rate upon the *real* area of contact, a parameter not susceptible to experimental measurement in all but the most artificial circumstances.

A "mechanism" is literally the way a machine works, but when applied to scientific research, it is clearly intended to provide a framework upon which to base an explanation of the "Law" discovered (usually) by empirical means. A friction or wear mechanism is one which explains the friction or wear laws. It should lead directly to a friction or wear "theory" (see next paragraph). It should not be included in friction or wear "classifications," since these should be based on observable characteristics.

In scientific research, a "theory" is typically a mathematical expression from which the experimental "laws" can be seen to be derived. It normally involves some mechanism (which is normally a gross over-simiplification of what probably happens) without which the theory could not have evolved. Sometimes, the law and the theory become inextricably entwined, especially when the theory is shown to be successful over a wide range of conditions. Sometimes, theories emerge without any basic laws having first been discovered, but this is a rare occurrence in the complex world of surface interactions.

Although these definitions may not be of *universal* validity, the author intends to use them for the remainder of the paper, since he is convinced of the need for some clarification of these terms when applied to Tribology.

CLASSIFICATIONS OF WEAR

Although there have recently appeared some new classifications of wear according to the type of wear particles (Bowen and Westcott[2]), these are strongly related to the more general classifications proposed by Burwell and Strang[(3)] in 1952 and Archard and Hirst[(4)] in 1956. These classifications have existed independently for the past 20 years or so, with most researchers favoring the Burwell and Strang classifications. More recently, however, it has become evident to the author that some attempt should be made to reconcile these two classifications of wear, as a necessary preliminary step before formulating a definitive wear theory. This section describes the author's own ideas on how to make this reconciliation.

Essentially, it is maintained that the Archard and Hirst classifications[4] are the most basic and easy to apply to any given wear situation, since they are entirely phenomenological. On the other hand, it is maintained that most of Burwell and Strang's[(3)] classifications are special cases of the simpler classifications proposed by Archard and Hirst,[(4)] namely those of *mild wear* or *severe wear*. It is not always appreciated that these classifications are based on (i) measurements of contact resistance (severe wear is characterized by low contact resistance whereas mild wear provides surfaces which give mainly high contact resistance), (ii) analysis of the wear debris both as regards size and composition (severe wear debris normally consists of large, i.e., ~ 10^{-2} mm diameter, metallic particles, whereas mild wear debris consists of small, i.e., ~ 100 nm diameter particles which have been produced by reaction with the ambient atmosphere or fluid) and (iii) microscopic examination of the surfaces (severe wear produces rough, deeply torn surfaces whilst mild wear produces extremely smooth surfaces, often much smoother than the original surface finish). These classifications do *not* specify a range of wear rates for each class of wear--in fact, it is often possible for mild wear processes to occur at a rate equal to severe wear processes. In general, however, the severe wear processes are much more "efficient" at removing material from the sliding surfaces.

Let us now consider the classifications proposed by Burwell and Strang in 1952. Their first class was assigned the name "Adhesive Wear." This title clearly implies a *mechanism*, namely adhesion between the surfaces, and was probably proposed with the Bowden and Tabor "cold welding" mechanism of friction (see The Laws of Wear and the Laws of Friction) in mind. Burwell and Strang[3] envisage parts of one surface being pulled off to adhere to another. These fragments then either break away to become wear particles or may be transferred back to the original surface. It is pertinent to ask, how does one know that a given pair of surfaces have been worn in this way? Optical microscopy will reveal rough surfaces exist, but not whether transfer or back transfer has occurred. Radioactive tracers have been used to test whether back transfer has occurred, but this is only discernible after much experimentation. As far as this author can find out, the "Adhesive Wear" classification of Burwell and Strang is no different from "Severe Wear" as defined by Archard and Hirst, except for the difficulty with which one can assign the classification "Adhesive Wear" to any given situation.

"Corrosive Wear" is another classification proposed by Burwell and Strang in 1952. Again, a *mechanism* seems to be envisaged by the authors, since they suggest that in this type of wear, in which sliding takes place in a "corrosive" environment, the products of corrosion form a film on the surface which slows (or prevents) further corrosion. However, the action of sliding removes the film, thereby allowing corrosive attack to continue.

The implication here is that the sliding has a deleterious effect on the formation of protective surface films. It is not clear whether or not the action of sliding, and the temperatures involved due to frictional heating, have any effect on the type of film formed. In fact, if the "corrosive environment" were the air (or a lubricant, in which air is always found in large quantities) it is possible that the interaction of the surface with the oxygen in that air will have beneficial effects (see The Mechanisms of Wear).

It seems that Burwell and Strang[3] had a severely corrosive condition in mind when they proposed the classification of "corrosive wear" i.e. one in which "severe wear" is maintained by removal of the surface film formed by reaction with the corrosive environment. If, however, they had in mind the more beneficial conditions of "mild wear," in which protective oxide films are formed due to interaction at the sliding temperatures of the interfaces (or protective films are formed by interaction with the "extreme-pressure" additives often found in lubricants) it is not clear from their publication. This author suggests that the "corrosive wear" classification is so difficult to assign, that it would be better to use the "mild wear" classification for the beneficial form of wear whilst sliding under a corrosive environment and the term "severe wear" for the deleterious form in which the environment is truly "corrosive" in the normal meaning of that word.

Burwell and Strang[3] also propose "surface fatigue wear" as another classification. This is the form of wear which occurs after many millions of repeated loading and unloading cycles at a surface asperity, when sub-surface cracks eventually reach the surface and form large pits in the surface. Although the words "surface fatigue wear" imply a mechanism, it is clear that this form of wear is characterized by observable features, namely the pits which can be seen in the surfaces of roller and ball bearings, in gear teeth and in cams and tappets. Clearly, the authors had the "pitting wear" of such practical geometries in mind when they proposed this type of wear. It is suggested, however, that surface fatigue is a *mechanism* by which "mild wear" occurs (e.g. see the crack formation in Figure 2.13 of reference 5) and could be involved in "severe wear" also (if one assumes that Suh's "Delamination Wear Mechanism"[6] relates mainly to this form of wear). It is a mechanism, which happens (in the case of geometries in which *both* sliding and rolling occurs) to manifest itself in the form of characteristic surface cracking and eventual pitting, features which are, of course, readily discerned in the surface. It does not replace "mild" or "severe" wear classifications.

"Fretting," sometimes also known as "Fretting Corrosion," was thought by Burwell and Strang[3] to be yet another type of wear. In fact, it is clearly a form of "oxidational wear," occurring under conditions where small oscillatory tangential motion occurs between two surfaces which are held together in such a way that the oxide wear particles cannot escape, thereby causing severe stresses to be set up in the constraining mechanism (e.g. in the bolt holding together two plates subjected to vibrations with extremely small amplitude (of the order of microns). As will be seen in a later section, "oxidational wear" is a form of "mild wear" occurring when sliding takes place in a oxidizing environment. Hence, "Fretting" can be classified as a special form of "mild wear."

Burwell and Strang[3] also maintain that "Abrasive Wear," "Erosion" and "Cavitation" are separate wear classifications. It is the view of the author that these are all forms of wear which occur when a rough hard surface (or hard particle for "Erosion," or bubbles of liquid for "Cavitation") slides (or impinges at high velocities) upon a softer surface and causes a series of grooves (or large holes) to be formed in that surface. They are all forms of "abrasion" which in itself is a mechanism rather than a classification! Abrasive mechanisms can occur in different stages of both "mild" or "severe" wear (see later section on "wear mechanisms").

Having shown that the classifications of "mild wear" and "severe wear" as proposed by Archard and Hirst[4] can be readily applied to actual wear situations by observations of the topography, contact resistance and the structure of the wear debris, *without recourse to any wear mechanisms*, let us now deal with the experimental laws which both "mild" and "severe" wear have been found to obey. It will be instructive, however, to deal with the Laws of Wear in conjunction with the Laws of Friction, since some very interesting conclusions can be drawn regarding their development since discovery and their general acceptance by the scientific and technical world.

THE LAWS OF WEAR AND THE LAWS OF FRICTION

The laws of wear are a comparatively recent (1950) description of the general wear behavior of surfaces which interact during sliding. They should be compared with the well-established laws of friction, which were first discovered by Leonardo da Vinci in the 15th Century. In fact they are so similar that both the wear rate and the force of friction during sliding are proportional to the load applied normal to the direction of sliding, and both are independent of the apparent area of contact. The proportionality of wear rate with load is only true provided the nature of the surfaces does not change. Apart from this, the similarity is most striking, leading one to believe that, since the friction laws have been "explained" in terms of elastic and plastic deformation of the surfaces, it will only be a matter of time before the wear laws are also "explained" in these terms. However, it should be clear from Professor Barwell's review (in so far as it relates to the fundamentals of wear) that the wear laws are a long way from being explained to everybody's satisfaction, especially the practicing engineer. From the abundance of investigations of the wear of various materials under various conditions of load, speed, distance of sliding, surface hardness, surface roughness, ambient temperature, lubricating oil (and its additives) and specimen geometries, a very confused picture of the wear behavior of sliding systems has emerged. The only work which has revealed any systematic wear behavior[3,4] has been carried out under conditions often far removed from conditions existing in any practical wearing system. For this reason, the laws of wear which describe this systematic behavior have not been so widely accepted as the laws of friction.

Why are the laws and mechanisms of friction so much *less* controversial than those of wear? To answer this question, we must review the sequence of events which led up to the current theories of friction. It is interesting to note that this was an instance in which the "laws" came first. (In fact, as far as the author is aware, there are no "classifications" of friction in the same way as there are classifications of wear, which is strange, when one recalls the tenacity of the belief that friction and wear *must* be related). These laws have been shown to be valid (to a first approximation) over an extremely long period (about 500 years). During that time, several hypotheses have been proposed regarding the mechanisms of friction. Amontons (1663-1705), for example, produced 3 possible mechanisms. One involved the work done in lifting one surface over the roughness of the other, another related friction to the bending down of the roughnesses, whilst the third hypothesis proposed that friction was due to the breaking off of the roughness. Other mechanisms include the Adhesion mechanism of Desaguliers (1683-1744) and the mechanism proposed by Leslie (1766-1832) in which the frictional force was related to the work done in deforming the surface of one body by the roughness of another. It turns out that all these hypotheses had elements of truth in them, in much the same way as all the various wear hypotheses currently being proposed probably have some relevance to some situations. However, it was not until the 1940's that a mechanism was proposed which led to a "satisfactory" theory of friction. This was the "cold welding" mechanism of friction proposed by Bowden and Tabor,[7] who also maintained that the coefficient of friction (μ) between metals was given by the ratio of the shear strength of the softer material of a sliding pair to the yield pressure of that material. This theory was supported by results with many combinations of materials. It should be noted, however, that the values of the coefficients of friction did not change widely with changes in material or speed. Bowden and Tabor explained this by pointing out that the shear strength and the yield pressure will vary in the same way as the temperature varies (due to sliding). It should also be noted that many of the experiments were carried out on a Bowden-Leben (linear sliding) machine at fairly low speeds over short friction paths.

In the 1950's, the work of Archard and Hirst[4] indicated that, for many combinations of materials and speeds, the surfaces of the specimens showed no signs of welding as would be expected from the Bowden and Tabor Theory. Archard and Hirst[4] used pins loaded against rotating disks, so their results are probably more relevant to the repeated contact conditions found in rotating machinery. One might suggest that Bowden and Tabor's experiments related to the *initial stages* of sliding whereas Archard and Hirst's experiments related to *equilibrium conditions*. Archard[8] has attempted to explain the Friction Laws in terms of purely elastic deformation and a surface asperity model which he claimed approached the complexity of real surfaces. This explanation was not widely received due mainly to its inability to give a clear picture of what causes friction between surfaces suffering elastic deformation only. The surface model has sometimes been criticized as being unrealistic. In fact, however, Archard must be credited with being one of the first people to propose a surface asperity distribution (in fact, a uniform distribution of surface asperity heights with depth) and use this distribution to obtain important information about the real areas of contact due entirely to elastic deformation.

Apart from Archard's attempt to explain the frictional force in terms of completely elastic deformation of the surface asperities, there have also been modifications proposed to Bowden and Tabor's[7] original theory due to the increased interest, in recent years (i) in the friction and wear properties of polymers sliding against metal counterfaces and (ii) in the effects of surface films formed during sliding. The interest in polymers has led to a revival of interest in the "ploughing component" of friction, which is actually related to the force required to move an asperity of the harder material tangentially through the asperities of the softer material. The simple theory of Bowden and Tabor[7] assumed that this component could be ignored for most typical metal-metal combinations. Any attempts to include this *mechanism* leads to a Theory of Friction which involves other (less well-defined) parameters than the two strength properties of the softer material. The increased interest in "extreme-pressure" and oxide films, formed on surface asperities during sliding by interaction with the lubricant or the atmosphere, led Bowden and Tabor[9] to modify their theory such that the coefficient of friction was said to be the ratio of the critical shear stress at the interface (s_i) and the plastic yield pressure of the underlying metal

(p_m). The inclusion of the *mechanism of film formation during sliding* into the Bowden and Tabor[7] simple theory of friction also leads to an apparently unresolvable situation since at any given time, it is almost impossible to know whether to use (s_i) or the strength of the underlying material (s_m) for any given asperity at which the interface may be a thin film of boundary lubricant, an oxide film, or possibly a film formed by interaction between the "extreme-pressure" additives in the lubricant and the metal exposed by the sliding processes.

Thus, we see that the *simple mechanism of friction* in which "cold welds" are made and broken during sliding has limited application when considering situations outside those simulated in the early Bowden and Tabor experiments. Nevertheless, the simple theory of Bowden and Tabor[7] has been retained in the literature of friction because it indicates how to affect friction (by altering the yield pressure of the underlying material or the critical shear strength of the interface) rather than how to predict the absolute value of the friction coefficient. Another reason for its retention is that it is very difficult to obtain much variation in friction coefficients beyond the range $0.1 < \mu < 1.0$, so that any deficiencies in the predictions made on the basis of the theory are never very far from the actual measured values. Finally, it has been retained because it gives a physical interpretation, which the elastic deformation mechanism of Archard has failed to give, although private communication with that author reveals that he envisages the frictional energy to be related to the work done in taking asperities through a hysterisis loop in the stress-strain curve associated with the elastic deformation of real materials.

From all the foregoing, we see that the fundamentals of friction are not so "satisfactorily" explained in terms of current mechanisms and theories as the text books would have us believe. In fact, the position in Friction is almost as confused as that in Wear. This should not prevent us from perservering in our attempts to produce *definitive* theories of both friction and wear, based on mechanisms with some factual foundations. In the next section, some of the more important wear mechanisms will be discussed, mechanisms which, in the opinion of the author, have some chance of leading to definitive theories.

THE MECHANISMS OF WEAR

In this section, we will describe only those mechanisms relevent to "mild wear" and "severe wear," since we have already seen that these are the basic classifications in respect of which all other classifications may be described. "Mild Wear" mechanisms will be discussed first, since it appears that "severe wear" is what happens when "mild wear" processes cannot occur.

"Mild Wear" clearly involves reaction with the environment, in particular with the oxygen in the environment. It is not always appreciated that "mild wear" can occur even under lubricated conditions due to the large amounts of air entrained in all typical lubricants. If the conditions of sliding involve lubricants containing "extreme-pressure" additives, it is probable that the dominant reaction will be between the newly-exposed metal and the additive (or, more likely, the reactive parts of the additive molecules). Most typically, however, "mild wear" will involve reactions with oxygen, i.e. "mild wear" occurs mainly through an "Oxidational Wear Mechanism." This mechanism, which has gradually evolved over the past 16 years or so (see references 10, 11, 12, 13, 14, 15), can best be summarized as follows:

After the initial (severe wear) stages, the surfaces will have achieved a measure of conformity, so that large areas of both surfaces come into contact during sliding. At any given instant one of these areas will bear most of the load. This area then thermally expands in a direction perpendicular to the plane of contact between the specimen (in a manner similar to that proposed by Barber)[16] so that one has a plateau of contact which will tend to remain in contact until it is removed by wear. If the sliding speed is too low, however, the frictional heating will be insufficient to produce the required expansion, so the mechanism cannot proceed. Given a sufficient amount of frictional heating, however, there will occur a certain amount of oxidation of the surface of the particular plateau in contact, this oxidation mainly occurring at the temperature of the interface between the areas of contact on the plateau and similar areas on the opposing surface. These plateaux have been observed by many investigators (for example, see references 6, 12, 17). They are extremely smooth with fine wear tracks parallel to the directions of sliding. Typically they have an area approximately equal to about 10^{-2} mm^2 and heights of about 2 or 3 x 10^{-3} mm. The plateaux often show surface cracks perpendicular to the direction of sliding. Occasionally, it is possible to find an area within a plateau where clearly a flake of material has been removed between adjacent cracks, showing the virgin surface beneath. Sometimes, at the trailing edge, flakes can be seen which are about to become detached (e.g. see Figure 4 of reference 6 and Figure 2.9 of reference 5). The surface surrounding the plateaux is rough and packed with wear debris and contains no wear tracks. It can only be the interface between plateaux which became detached at an earlier stage and the bulk of the metal. There is very little wear debris on the surface of the plateaux.

The area of these plateaux (A) has been shown to be about what would be expected if all the load (W) were borne at one of these plateaux and the area was formed by a stationary plastic contact envisaged by Bowden and Tabor,[7] the hardness (p_m) of the bulk metal being used to provide the value for the area from the equation $A = W/p_m$. It seems unlikely that every plateau/plateau interaction between opposing surfaces will take place along the whole of the surface area of a plateau. It is much more probable that there are several sub-areas of contact on each plateau and that these are the regions of actual contact at which oxidation occurs. Under the oxidational wear mechanism, the areas of contact are the sites for oxidation at the temperature generated by frictional heating (T_c). The oxide film builds up until it reaches a critical thickness (ξ_c), at which time the film becomes unstable and is removed. When all the sub-areas of contact have been removed from the plateau, then another plateau elsewhere on the surface becomes the operative one. The virgin surface beneath the original plateau is now clear of the contacting surfaces and hence can only oxidize at the general temperature of the surface (T_s). Without external heating, the amount of oxidation at a typical value of T_s of (say) 80°C is orders of magnitude less than the amount at the typical T_c values of (say) 400°C. Hence, the original plateau sub-surface region will not oxidize significantly until it becomes the dominant area of contact once more.

This mechanism is based on actual microscopic evidence. Before one can use it to predict wear rates, one must know the oxidation characteristics of the metal, the value of T_c (which is assumed to be not very different from T_0, the temperature at which the contacts oxidize), ξ_c, and the number of contacts (N). The oxidational wear theory, which is based on this mechanism, will be discussed in the next section. Let us now briefly deal with "severe wear" mechanisms.

Before "mild wear" starts, there is always a period of "severe wear" known as "running-in." Clearly, this is the time when the surfaces are being made to conform to each other. This must be through some adhesive, abrasive, or ploughing process. If the abrasion or adhesion is so severe that conformity (on the microscopic scale revealed by electron microscopy of "mild wear" surfaces) cannot occur, there is no opportunity for an area to become the dominant one. Similarly, if the ploughing is severe (such as in metal-cutting processes), removal of metal will dominate over any oxidation process. Obviously, for an unreactive material, such as a polymer, there is no opportunity for mild wear to occur unless, of course, one of the sliding pair is a metal which readily oxidizes. In effect, this paragraph says that if one has microscopically non-conforming surfaces, or if one has unreactive surfaces, then severe wear must ensue by plastic deformation and subsequent removal, perhaps in the manner described in the "Delamination Mechanism" proposed by Suh and his co-workers.[(6)] It would seem that most of their work relates to "severe wear," although this is not explicitly stated in their publications. This is just one more example of how the basic classifications of Archard and Hirst[(4)] seem to have been overlooked.

THEORIES OF WEAR

In much the same way as the friction theories depend on relating the frictional force to the real areas of contact, theories of wear must relate the volume removed per unit sliding distance (i. e. w, the wear rate) to the real area of contact (A). The so-called "Archard Wear Law"[(1)] is just an attempt to express the wear rate in terms of the real area of contact, namely,

$$w = KA \tag{1}$$

Archard[18] interprets K as the probability of producing a wear particle at any given asperity encounter. If K is independent of load, then one could say that Eq. (1) leads to the Wear Laws, since $A = W/p_m$ (where p_m is normally taken to be the hardness of the bulk of the material beneath the areas of contact and is, of course, independent of the load). Using this value for A, Archard[(19)] showed that K can vary from 10^{-2} down to 10^{-5} dependent on whether one used mild steel upon mild steel (for $K = 10^{-2}$) or Stellite upon tool steel (for $K = 10^{-5}$), all other operating parameters being maintained constant. It is this wide variation of K-factors (which sometimes even occurs merely by changing the load by a few per cent above or below a transition load, see reference 20) which makes wear results more difficult to analyze than friction results.

It is clear to this author that the understanding of *the factors influencing the K-factor should be given the highest priority in our research into the fundamentals of wear*. For mild wear, the oxidational wear theory[15] leads to the following expression for the wear rate (w):

$$w = \left[\frac{d.A_p.\ \exp -(Qp/RT_o)}{\xi_c^2\ \rho_o^2\ f^2\ U} \right] .\ A \tag{2}$$

where the portion inside the brackets is the expression for the K-factor. ρ_o is the density of the oxide, f is the fraction of the oxide which is oxygen, U is the linear sliding speed, d is the distance of a sliding contact during which oxidation occurs at a temperature T_o, Ap is the Arrhenius Constant for parabolic oxidation during wear, Qp is the Activation Energy for oxidation and R is the Gas Constant.

For "severe wear" it would seem that the delamination theory is the most relevant one. Engel[21] shows how Suh's mechanism leads to the expression:

$$w = \left[\frac{B_1 h_1}{d_{c_1}} + \frac{B_2 h_2}{d_{c_2}} \right] A \qquad (3)$$

where again, the expression within the brackets is the K-factor; h_1 is the removed layer thickness from one surface and h_2 is the thickness removed from the other surface; d_{c1} and d_{c2} are the critical plastic displacements for each surface; B_1 and B_2 depend mainly on topography and are not very well defined.

It must be emphasized that *neither* of these 2 equations can be used to predict wear rates without making further assumptions about a surface model. It is suggested, however, that Eq. (2) and (3) could form the basis of a concentrated attack on the fundamentals of wear. There have been other wear theories proposed over the past few years, but none has received the amount of substantiative research as these two theories. It is time to concentrate our efforts on trying to relate our new results to these 2 expressions for "mild" or "severe" wear, or at least to make some attempt to relate to previous work in this field. This subject seems to have suffered from a lack of really definitive papers over the past few years, so possibly the time is ripe for such endeavors. In the final section of this paper, some details are given of the efforts of the Tribology Group in the Physics Department of the University of Aston in Birmingham in searching for a definitive treatment of the oxidational theory of mild wear.

EXPERIMENTAL VERTIFICATION OF THE OXIDATIONAL THEORY OF MILD WEAR

All theories stand or fall by their ability to predict experimentally obtained results. The theories relating to mild (oxidational) wear and severe (delamination) wear are not yet sufficiently developed for this to be done. However, it is clearly worthwhile to use these theories, together with the experimental wear results, to predict the surface model which is most compatible with these results. This has been done for mild (oxidational) wear and a full report will be given at the Wear Conference in Dearborn in 1979 (see reference 21). The essence of that report will be given in the remainder of this paper.

Essentially, the heat flow along a steel pin has been measured whilst the pin was sliding against a rotating disk of the same material. The division of heat at the interface (δ_{expt}), the surface temperature (T_s) and the wear rates of the pin (w_{expt}) have all been determined by suitable measurements. If one now supposes that on the operative plateau there will be N areas of contact, each of area πa^2 (where 'a' is the radius of the area of contact which, for convenience, is assumed to be circular), then we can express d (the distance of each wearing contact) as being equal to 2a, where $N\pi a^2 = A = W/p_m$, i.e. $d = 2\sqrt{W/N\pi p_m}$. Furthermore, it is assumed that on each of these contacts, a critical oxide film thickness (ξ_c) builds up during the time for (1/K) contacts to occur. This step is based on Archard's[1] interpretation of the K-factor, namely that (1/K) encounters are needed (on the average) for a wear particle to be produced. The presence of this oxide film will, of course, affect the theoretical division of heat (δ_{theory}) at the sliding interface, as also will the assumption that there are N areas of contact. Since the excess temperature (θ_m) over the surface temperature (T_s) also depends on δ_{theory}, this means that T_o (which is assumed to be equal to ($T_s + \theta_m$)) also depends on N and ξ_c.

From the last paragraph, we see that Eq. (2) does *not* indicate a monontonically increasing wear rate with increasing speed (as stated by Saka, Eleiche and Suh)[22] since the speed (U) appears explicitly in the denominator and implicitly in T_o (see Equation (15), (14), (13) and (12) of reference 14). The effect of speed upon the wear rate is not readily deduced, without evaluating the constants upon which T_o depends. Furthermore, it is likely that increases in speed will take the oxidation temperature T_o beyond the range for which Ap, Qp, ρ_o and f will apply for a given type of oxide. Clearly, there is no point in discussing how Eq. (2) will vary with operating parameters U and W, without getting some ideas regarding the appropriate values of N, ξ_c, Ap, Qp, ρ_o and f to use.

We have just carried out some experiments on a low carbon steel (EN8) at 2.0 $m.s^{-1}$ in which we have measured the wear rate, the division of heat, the average friction coefficient and the surface temperature for a range of loads from 4 newtons up to 40 newtons.[21] We have also analyzed the debris by X-ray diffraction. The graph of wear rate versus load (not shown in this paper) reveals definite transitions at about 15 and 30 newtons, indicating changes in K-factor at these loads. The 15 newton transition appears to be connected with the first appearance of Fe_3O_4 (magnetite) in the debris, whereas the 30 newtons transition seems to correlate with the appearance of FeO (Wustite).

It was decided to use the computer search technique mentioned in reference 14 to deduce the appropriate values of N, ξ_c and T_o which will satisfy Eq. (2) for the wear rate and Eq. (14), of reference 14, for the division of heat, remembering, of course, that T_o appears in Eq. (2) of this paper and is related to the division of heat through the expression

$$T_o = (\delta_{theory}\ Tp) + T_s \tag{4}$$

where Tp is given by Eq. (10) of reference (14) and is the "fictitious" flash temperature, assuming all the heat generated at the sliding interface is supplied to the pin.

There was a real problem involved in finding the *appropriate* values of Ap and Qp to insert into Eq. (2). Although the various publications relating to bulk (static) oxidation agree about the value to be assigned for Qp (the Activation Energy) for iron oxidation, there are large differences in the values for Ap (the Arrhenius constant). These differences are, of course, due to many factors, some of which relate to the experimental procedure used by the various authors. We decided to use the Qp values of Caplan and Cohen[23] and insert these values into the computer search programme used in reference 14, except that now we were looking for consistent Ap values instead of consistent Qp values. This reference relates to medium-carbon steel specimens which had been worn on the same apparatus, so it was considered these Ap values would also be appropriate to our low-carbon steel experiments. The "appropriate" values are given below:

$$\text{For } T_o < 450°C,\quad Ap = 10^{16}\ kg^2.m^{-4}.s^{-1},\quad Qp = 208\ kJ,\ mole^{-1}$$

For 450°C < T_o < 600°C, Ap = 10^3 $kg^2m^{-4}s^{-1}$ Qp = 96 kJ. $mole^{-1}$

For T_o > 600°C, Ap = $10^8 kg^2.m^{-4}$. s^{-1} Qp = s10 kJ. $mole^{-1}$

The changes in the values of Ap and Qp are clearly related to changes in the structure of the oxide. For T < 450°C, the oxides would be expected to be mainly α - Fe_2O_3 (the rhombohedral oxide). For 450°C < T < 600°C, we would expect the oxides to be both α- Fe_2O_3 and Fe_3O_4. For T > 600°C, one could expect the oxides would be mainly FeO. Using the appropriate values of f and ρ_0 for each of these ranges, and rejecting any T_0 values thrown up by the search technique lying outside the range of temperatures for which Ap Qp f and ρ_0 are valid, it is possible to get 3 sets of values of N, ξ_c and T_0 for each experiment.

It is necessary now to choose which set of N, ξ_c and T_0 are relevant to each experiment. The criterion to be used is the analysis of the wear debris. Accepting only those sets which correlate with the expected oxides, it is possible to produce a table of results (Table 1) in which only those values of N, ξ_c and T_0 consistent with the wear debris analysis appear. Although these, and other analyses are discussed in great detail in reference 21, it is interesting to note that before the transition in the wear rate versus load graph occurs at about 15 newtons, the wear is mainly through α-Fe_2O_3, with N (the number of wearing contacts) remaining approximately constant at about 120, ξ_c approximately constant at about 1.5μm, and T_0 also approximately constant at about 280°C. The value of ξ_c = 1.5μm is consistent with electron microscope evidence regarding typical plateaux heights.[12,17] Between the transitions at 15 and 30 newtons, wear seems to be through the formation of *both* α-Fe_2O_3 and Fe_3O_4. The value of ξ_c is about the same as before, but N has reduced to about 60 and T_0 has increased to about 420°C. Above the 30 newtons transition, wear is through mainly FeO formation, with N now reduced to 30, ξ_c about 1μm and T_0 about 600°C.

The above values for N, ξ_c and T_0 are somewhat different from the values quoted in reference 14. Although that reference relates to the wear of a medium carbon steel, the differences are probably due to the fact that *inappropriate values of Ap and Qp* were used, that is, values inappropriate to the temperature at which the oxidation of the wearing contacts was occurring rather than inappropriate to the type of steel. Unfortunately, without any definite analysis of the wear debris formed in the experiments in reference 14, there is no way in which the computer analysis can be repeated.

The above results obviously need to be discussed in relation to other experiments with different steels at different speeds. This is, in fact, done in reference 21. For our purposes, however, it is sufficient to point out how complex the analysis of mild (oxidational) wear can be. Unless one has all the measurements of surface temperatures, division of heat at the sliding interface, wear rates, friction coefficients and the analysis of the wear debris, one cannot hope to understand mild-wear processes. It is suggested that a similar all-inclusive set of measurements and analyses should be obtained before attempting to use the delamination mechanism to explain "severe wear" in an equally self-consistent theory. Too much tribological research, including the early work of the present author, has been concerned with just one particular aspect of the surface interactions, for example, the wear rate. This paper will have served a useful purpose if the only "message" received by the reader is that he (or she) should always aim to provide facilities for measuring *all* the parameters involved in sur-

TABLE I.—RESULTS OF COMPUTER SEARCH FOR N, ξ_c AND T_o

Expt No	Load (Newtons)	Wear Rate (m^3/m)	T_s (^oC)	N	ξ_c (μm)	T_o (^oC	Oxides in Debris
1	3.94	2.063×10^{-13}	40.1	113	0.89	277	
2	6.89	1.943×10^{-13}	39.2	65	2.66	290	Mainly
3	9.85	4.030×10^{-13}	52.5	184	1.17	278	α-Fe_2O_3
4	9.85	3.039×10^{-13}	39.3	140	1.69	282	
5	14.77	6.676×10^{-14}	65.7	60	1.43	368	
6	14.77	1.487×10^{-13}	66.5	82	1.47	406	α-Fe_2O_3
7	19.70	3.003×10^{-13}	77.2	76	2.04	440	and
8	24.62	4.405×10^{-13}	87.2	87	1.98	444	Fe_3O_4
9	29.55	4.418×10^{-13}	84.9	47	2.63	442	
10	34.47	1.045×10^{-12}	164.8	34	0.94	583	Mainly
11	39.40	1.789×10^{-12}	176.9	28	1.38	612	Fe_3O_4

face interactions *before* embarking on a particular tribological research project, rather than have to repeat the experiments to obtain the data which only became of obvious relevance at a later date.

ACKNOWLEDGEMENTS

The author is indebted to his colleagues, Mr. J. L. Sullivan and Dr. D. M. Rowson, for their criticisms of the script of this paper and for their help and advice throughout, especially in respect of the last two sections relating to Wear Theories.

REFERENCES

1. Archard, J.F., *Journal of Applied Physics,* Vol. 32, 1961, p. 1420.
2. Bowen, E.R. and Westcott, V.C., "Wear Particle Atlas," Final Report on Contract No. N00156-74-C-1682 for Naval Air Engineering Center, Lakehurst, New Jersey, 1976, p. 12.
3. Burwell, J.T. and Strang, C.D., *Journal of Applied Physics,* Vol. 28, 1952, p. 18.
4. Archard, J.F. and Hirst, W., *Royal Society of London. Proceedings.* Series A, Vol. 236, 1956, p. 397.
5. Quinn, T.F.J., "The Application of Modern Physical Techniques to Tribology," Newnes-Butterworths, London and Van Nostrand Reinhold, Cincinnati, Ohio, 1971, p. 55.
6. Suh, N.P., *Wear,* Vol. 44, 1977, p. 1.
7. Bowden, F.P. and Tabor, D., "The Friction and Lubrication of Solids," Part 1, Clarendon Press, Oxford, 1954.
8. Archard, J.F., *Royal Society of London. Proceedings.* Series A, Vol. 243, 1957, p. 190.
9. Bowden, F.P. and Tabor, D., "The Friction and Lubrication of Solids," Part 2, Oxford University Press, Oxford, 1964.
10. Quinn, T.F.J., *Journal of Physics D: Applied Physics,* Vol. 13, 1962, p. 33.
11. Quinn, T.F.J., *American Society of Lubrication Engineers Transactions,* Vol. 10, 1967, p. 158.
12. Quinn, T.F.J., *Institution of Mechanical Engineers. Proceedings,* Vol. 182, 1968, p. 201.
13. Quinn, T.F.J., *Institution of Mechanical Engineers. Proceedings,* Vol. 183, 1969, p. 129.
14. Quinn, T.F.J., *American Society of Lubrication Engineers Transactions,* Vol. 21, 1978, p. 78.
15. Quinn, T.F.J. and Sullivan, J.L., "A Review of Oxidational Wear," First International Conference on Wear of Materials, St. Louis, Missouri, April 1977.
16. Barber, J.R., *Royal Society of London. Proceedings.* Series A, Vol. 312, 1969, p. 381.
17. Quinn, T.F.J. and Wooley, J.L., *Lubrication Engineering,* Vol. 26, 1971, p. 312.
18. Archard, J.F., *Journal of Applied Physics,* Vol. 24, 1953, p. 981.
19. Archard, J.F., *New Scientist,* Vol. 5, 1959, p. 1299.
20. Welsh, N.C., *Royal Society of London, Philosophical Transactions.* Series A, Vol. 257, 1965, p. 31.
21. Quinn, T.F.J., Rowson, D.M. and Sullivan, J.L., "New Developments in the Oxidational Theory of the Mild Wear of Metals," Second International Conference on Wear of Materials, Dearborn, Michigan, April 1979.
22. Saka, N., Eleiche, A.M. and Suh, N.P., *Wear,* Vol. 44, 1977, p. 109.
23. Caplan, D. and Cohen, M., *Corrosion Science,* Vol. 6, 1966, p. 321.

DISCUSSION

H. CZICHOS, BAM, Berlin: I would like to make a comment. I think the various aspects discussed by Professors Quinn and Barwell should be combined to give a classification of wear. Let me explain this by considering analogy between strength characteristics and wear characteristics of materials. In strength characteristics we have to distinguish first the external type of stress -- bending tension and so on. Then we have a fracture mechanism, ductile fracture or brittle fracture. Finally, a measure, say, yield stress. If we extend the analogy to wear, we should distinguish first the type of external attack; this can be the type of motion -- rolling, sliding, impact, oscillation, etc. Secondly, we should distinguish between the modes or mechanisms such as adhesion, delamination and abrasion. Third, we need a measure. This is where the classification "mild" and "severe" comes in. To be unambiguous, these three things should be combined as it is done for strength characteristics of materials. In my opinion, such an approach should be used to characterize wear.

T. F. J. QUINN: It is very difficult with wear because we do not have just one material; it is an interaction of two materials. This is a very great difference and this is what is causing all the problems. The constant of proportionality in wear depends also on the interaction of surfaces.

CZICHOS: All right. We can take specification of the material combination as a prerequisite and then go on further.

QUINN: I am familiar with your approach to the subject and I agree that it is important to take these factors into account. But I am interested in trying to be quantitative about this from a fundamental point of view. I feel that we should reduce the number of classifications and mechanisms and this is moving in the right direction.

A. W. RUFF, NBS: There is a considerable body of information on the oxidation of metals, for example steels under fairly static conditions. The problem is, of course, extremely complex because of the environmental effect. However, one can determine the type of oxides that are produced, the rates of oxidation, etc. and then get activation energies and pre-exponential coefficients.

Have you been able to obtain a quantitative connection between the activation energies, pre-exponential factors, etc. in the oxidative wear data that you presented and the static oxidation data? Or do you think it is possible to make that kind of a simple connection?

QUINN: That is a good question. We found it very difficult to establish such a simple connection between the static oxidation data and wear data. We are convinced that the oxidation that occurs in a wear situation is nowhere near the static case. So we have the two parameters Ap and Qp which we use to fit theory to our experiments.

In short, we cannot use all of the static bulk oxidation data, but we think we can use part of it.

I. L. GOLDBLATT, Exxon Research: I would like to pursue what Prof. Czichos was saying a while ago by pointing out one aspect with regard to wear studies. In simulated wear test, we usually have a pin sliding on a flat disc. The slider is continuously loaded whereas the disc is loaded cyclically. The kind of wear phenomenon which we would expect under cyclical conditions is different from the monotonically loaded case. An analysis of the wear debris (whether it is oxide, what type of oxide, etc.) without any distinction of where it is coming from is not going to give an indication of the mode of wear. Therefore, anykkind of wear classification has to be done with the recognition that the test members themselves should be looked at very carefully. The type of wear each member is subjected to has to be evaluated. Only by doing this, will one be able to get a wear mechanism which is descriptive of the actual phenomenon occurring at the surface.

QUINN: Yes, that is a good point. I do agree that we should take into account the wear rates of both members. In our work we measured the weight loss of both members and found that quite often the wear rates are approximately the same.

FUNDAMENTAL ASPECTS OF ABRASIVE WEAR

N. P. Suh, H-C. Sin and N. Saka

ABSTRACT

The classical theory of abrasive wear differs from the experimental observations in two significant aspects: (a) the predicted wear coefficient is one to two orders of magnitude larger than the experimentally observed values and (b) it cannot predict the effect of the abrasive grit size on the wear coefficient. A semi-quantitative model is presented which takes into account the geometric shapes of the abrasive grains, the cutting and deformation aspects, and the ductility of the abraded materials. It is found that the wear mechanism undergoes a transition from cutting to sliding wear as the grit size is decreased; when the grit size is less than about 1μm, wear is controlled by sliding wear mechanisms. The implications of this research on abrasive wear are that in lubricated systems abrasion caused by hard particles can be eliminated by continuously filtering particles larger than 1μm and that in designing two-phase materials for abrasive wear resistance the size of the hard particles should be kept below 1μm for minimum wear.

INTRODUCTION

As mentioned in Barwell's review paper[1], abrasion is generally idealized as a simple cutting process, but in reality it is far more complex. The classical theories of abrasive wear are oversimplified and cannot fully explain many details of the abrasion process. The purpose of the present paper is to review briefly various theories of abrasive wear and discuss the reasons for the discrepancies between theories and experimental results. Also, a semi-quantitative model is presented to explain the dependence of abrasive wear rate on the abrasive grit size.

ABRASION THEORIES

Traditionally, abrasion has been idealized as a simple cutting process with the assumption that abrasive particles have such simple shapes as sphere, cone, or pyramid. Using the simple model of a rigid conical asperity (Figure 1), Rabinowicz and coworkers[2,3] derived quantitative expressions for abrasive wear. For an asperity carrying a load L, the force

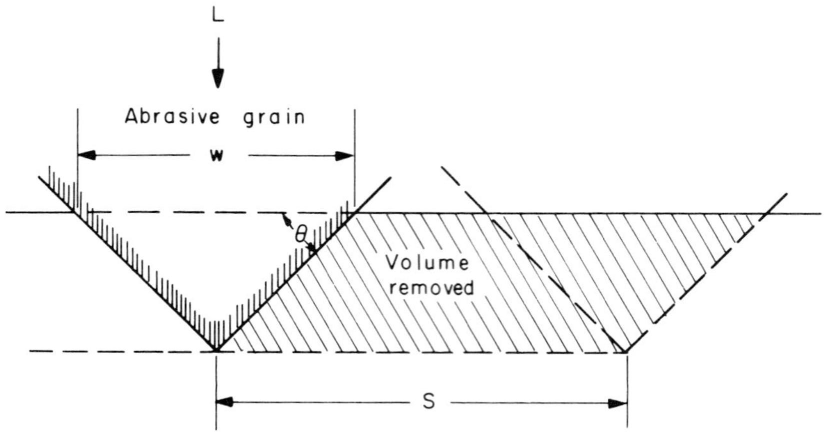

Fig. 1.—Idealized abrasion model in which a cone removes material from a surface.

equilibrium requires that

$$L = p \cdot \frac{\pi}{4} w^2 \tag{1}$$

where p is the hardness of the abraded material based on the projected area of indentation and w the diameter of contact of the cone. The groove area A_g, which is the projected area of the penetrating cone in the vertical plane, is given by

$$A_g = \frac{1}{4} w^2 \tan\theta = \frac{L \tan\theta}{\pi p} \tag{2}$$

where θ is the slope angle of the cone measured from the plane of the surface. Thus, when the cone moves through a distance S, it will sweep out a volume V given by

$$V = \frac{L \cdot S \cdot \tan\theta}{\pi p} \tag{3}$$

Comparing Equation (3) with Archard's wear equation, an expression for the wear coefficient K is derived as

$$K = \frac{3Vp}{LS} = \frac{3 \tan\theta}{\pi} \tag{4}$$

The total friction coefficient μ for this conical model is expressed in terms of the adhesion component μ_a and the plowing component μ_p as

$$\mu = \mu_a + \mu_p = \frac{s}{p} + \frac{\tan\theta}{\pi} \tag{5}$$

where s is the shear strength of the abraded material. For the simple cutting model the friction coefficient can be approximated as

$$\mu \cong \mu_p = \frac{\tan\theta}{\pi} \tag{6}$$

From Equations (4) and (6) we get

$$K \cong 3\mu \qquad (7)$$

The specific energy, u, defined as the energy expended in removing a unit volume of material, according to this model, is given by

$$u = \frac{\mu \cdot L \cdot S}{V} = \frac{\mu \cdot \pi p}{\tan\theta} \cong p \qquad (8)$$

Although the conical model is oversimplified, the above expressions provide a reasonable theoretical basis for abrasion in the absence of any sub-surface plastic deformation.

In abrasion it is reasonable to assume that the real area of contact is minimum when the normal stress at each asperity is equal to the hardness of the abraded material. This has been well confirmed experimentally by Larsen-Basse[(4)] (see Figure 2), and recently by the authors.[(5)] Figure 3 shows that the real area of contact estimated from the grooves formed on *PMMA* specimens agrees with Equation (1) within 50 percent for a variety of grit sizes.

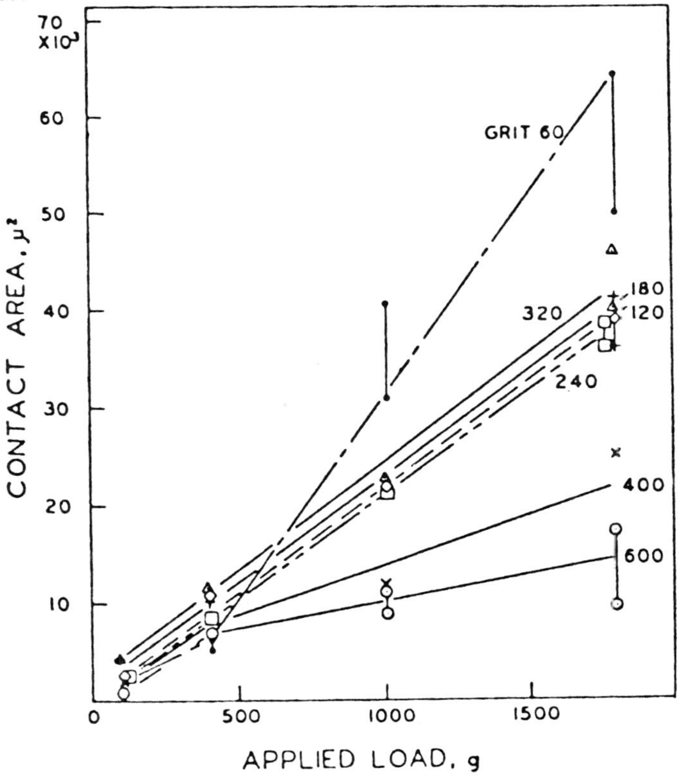

Fig. 2.—The contact area, calculated as $\Sigma\,(w^2)$ where w is the groove width, as a function of the applied load for various grit sizes.[(4)]

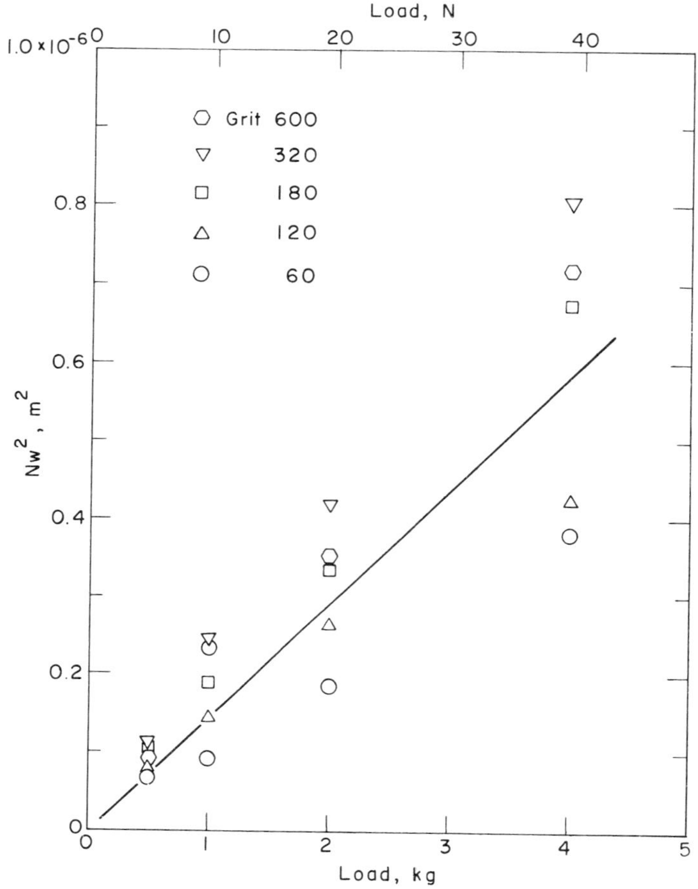

Fig. 3.—Number of contacting particles times the square of the average groove width as a function of the applied normal load (Nw^2 is proportional to the real contact area).

When abrasive particles slide over a specimen, it may be argued that the indentation must become deeper to support the same load because the contact interface is only on the front half of the particle.[(6)] Then the groove area A_g should be twice as large as that given by Equation (2). However, for most materials a raised ridge is formed around the abrasive particle during sliding as shown by Tsukizoe and Sakamoto.[(7)] The geometry of this ridge depends on material and the shape of the slider. Nevertheless, it may be assumed that the groove area given by Equation (2) still holds good within a factor of two. For particles with positive rake angle, however, Sedriks and Mulhearn[(8,9)] have shown that experimental results agree well with the expressions for the groove area derived from the mechanics of cutting (see Figure 4).

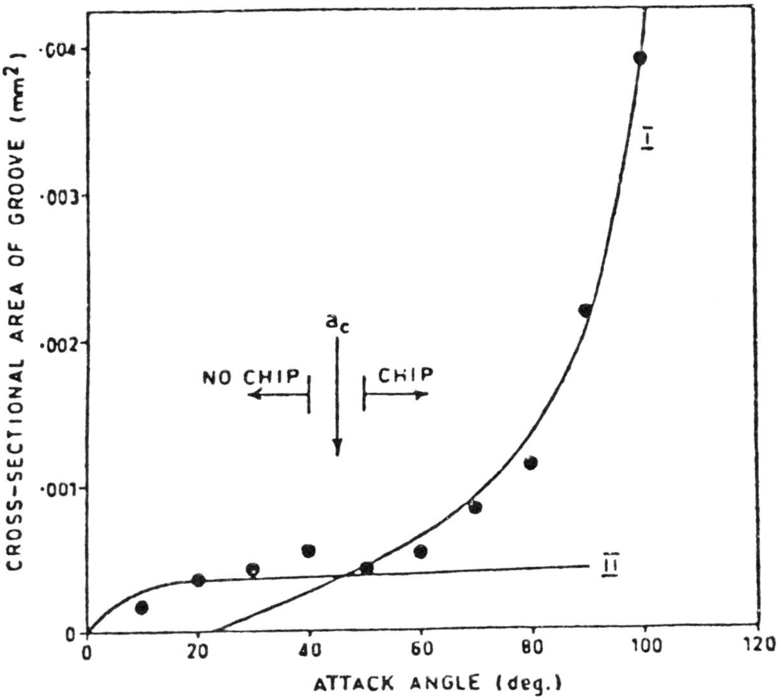

Fig. 4.—Transition from rubbing to cutting. Curves I and II represent the theoretically calculated cross-sectional area of groove with attack angle for cutting and rubbing respectively. The black circles represent experimental values for copper.[9]

As long as Equation (2) is valid, the groove volume can be obtained by multiplying the groove area by the sliding distance s. Therefore, Equation (3) represents only the groove volume and not the wear volume, unless all material from the groove is removed and becomes loose. Contrary to the assumption in this theory, only a fraction of the groove volume becomes loose, and the remaining is plowed to ridges at the sides of the groove. Stroud and Wilman[10] measured the real groove area after a short distance of sliding and have shown that only 10-40 percent of the groove volume is removed. On the other hand, it may be argued that only particles whose cutting face is aligned at an angle larger than a critical attack angle do cut. Mulhearn and Samuels[11] obtained an expression for the removal volume through this critical rake angle approach, in terms of the fraction f of the particles that cut as

$$V = \frac{f \cdot L \cdot S}{k_p \, \phi \, p} \tag{9}$$

where k_p is a constant which relates the contact area to the square of the half width of the groove, and ϕ the ratio of the half width to the depth of the groove. They have shown that the fraction of cutting particles is relatively small (~15%).

Considering the randomness of shape and orientation, there are difficulties in determining the fraction *f*. If the cutting face is not perpendicular to the sliding direction, only a portion of the groove volume will become loose even when the angle is greater than the critical attack angle. As shown in Figure 5, the results of cutting tests with diamond conical tools indicate that indeed only a fraction from the groove is removed and, therefore, the cutting efficiency of the individual particle should also be considered.

Equation (3) predicts that the wear volume is directly proportional to normal load and sliding distance, and inversely proportional to the hardness of the abraded material. These predictions have been well confirmed experimentally, and the results summarized in recent review papers.[12,13] Khrushchov[14] has shown for pure metals that the abrasive wear rate is inversely proportional to the hardness, although work-hardened and heat-treated metals exhibit some anomalous effects. (This anomalous behavior may be due to changes in plastic properties brought about by heat treatment.[15,16])

The wear coefficient κ given by Equation (4) is only dependent on the slope angle, θ, of the abrasive. The actual wear coefficient is much less than that predicted by Equation (4) since the wear volume is not equal to the groove volume as pointed out earlier. Results of cutting tests with conical tools shown in Figure 6 clearly illustrate the point. Because the particle shape does not depend on grit size (as will be seen later), the wear coefficient given by Equation (4) also predicts that κ is independent of the grit size. Further κ should be independent of the properties of the abraded material.

As for the friction coefficient, there are several models. Three typical theories are compared with the results of cutting experiments in Figure

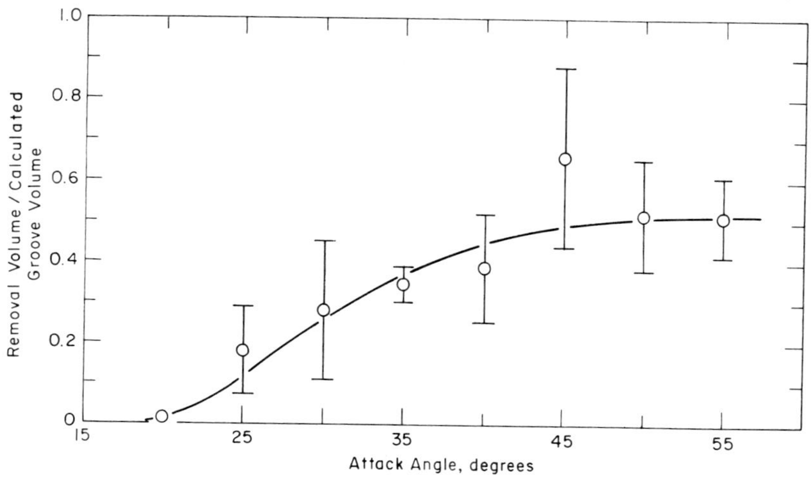

Fig. 5.—Ratio of the volume removed to the calculated groove volume as a function of attack angle. Material: AISI 1095 steel, applied normal load: 4.4 N, cutting speed: 1.6 cm/s.

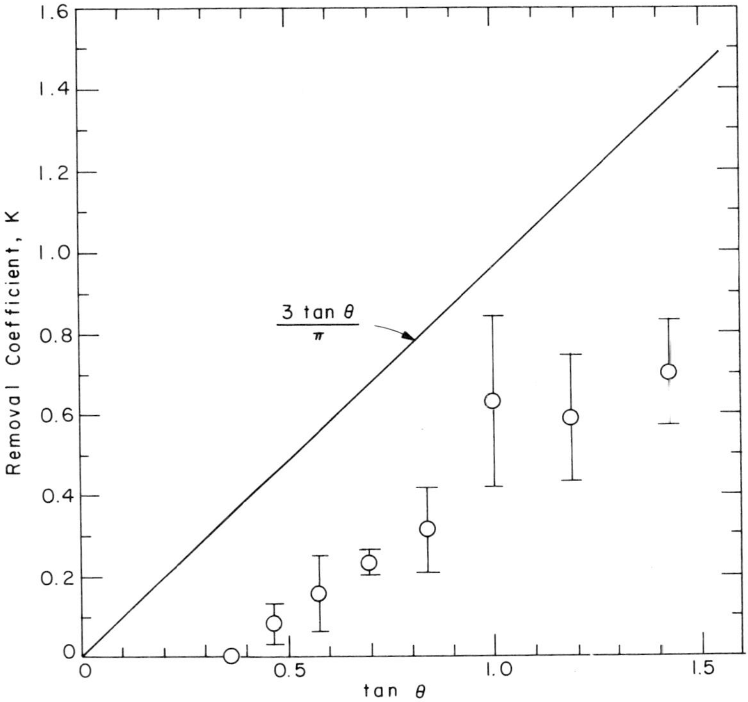

Fig. 6.—Removal coefficient as a function of the slope of the cone in cutting tests of AISI 1095 steel.

7. From this figure it can be seen that the friction coefficient predicted either by Equation (5) or by Equation (6) is too low, and that experimental results agree with theories of Goddard and Wilman[6] and Hisakado[17] within a limited range. Equations (5) and (6) predict too small values of friction coefficient and also ignore the grit size effect.

It is well established that when a metal is cut by very sharp cutting tools which leave little plastic deformation on the workpiece, the specific energy u is equal to the hardness of the workpiece.[18] This is also true in abrasion when the material from the groove is removed without leaving any plastic deformation. Using Equations (4), (7) and (8), it can be easily shown that, for the model considered, the external work is completely expended in removing material. However, in an actual case, abrasive particles leave large plastic deformation at and below the surface during abrasion.[19] Figures 8 and 9 show the plastic deformation of plowing and subsurface deformation, respectively. For this reason, the specific energy is much higher than the hardness of the abraded material. From Figure 10 it can be seen that the specific energy is at least ten times higher than the hardness of the material tested.

In summary, although the classical theory provides a simple understanding of abrasion process, several important experimental results still

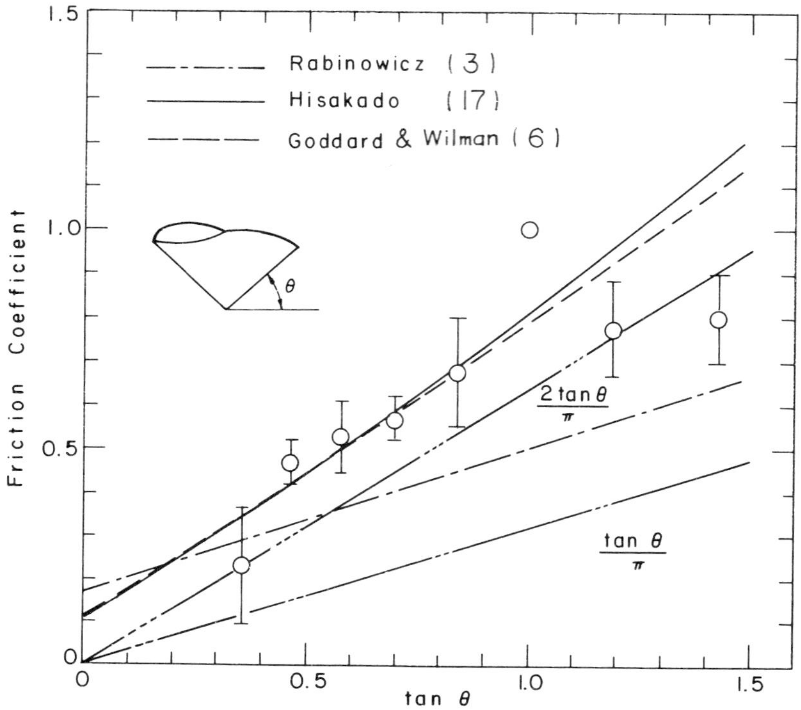

Fig. 7.—Friction coefficient as a function of the slope of the cone surface in cutting tests of AISI 1095 steel. The value of $\mu_a = 0.1$ was used in Hisakado's expression.

cannot be explained. First, experimentally observed values of wear coefficient are at least an order of magnitude smaller than the predicted values. Second, the dependence of friction and wear coefficients on the abrasive grit size (see Figures 11 and 12) cannot be predicted. Third, since the wear equation contains only hardness as a material property, the effect of hardness has been overemphasized. And finally, attention has not been paid to other properties such as ductility.

A model is developed here to take into consideration some of the above factors to predict the abrasive wear rates of materials.

WEAR COEFFICIENT - COARSE GRIT

As discussed earlier, if all material ahead of the abrasive particle becomes loose without leaving any surface and subsurface plastic deformation, the external work is completely expended in removing material. However, Mulhearn and Samuels(10) and Stroud and Wilman(11) have shown that only 15 percent of the total groove volume is removed for cold-drawn steel, 10-40 percent for silver. In addition, abrasive particles leave large plastic deformation at and below the worn surface. These effects seem to be

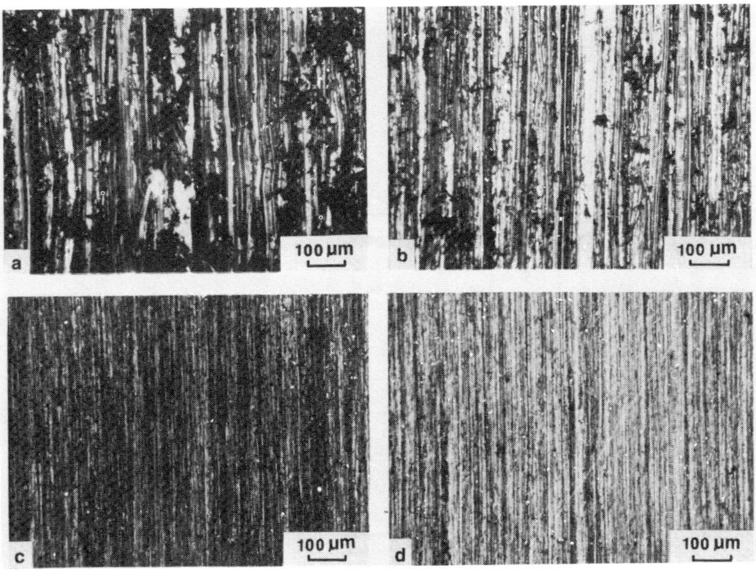

Fig. 8.—Surfaces of worn OFHC copper specimen for different grits: (a) 60, (b) 180, (c) 600, (d) 4/0. The normal load was 39.2 N.

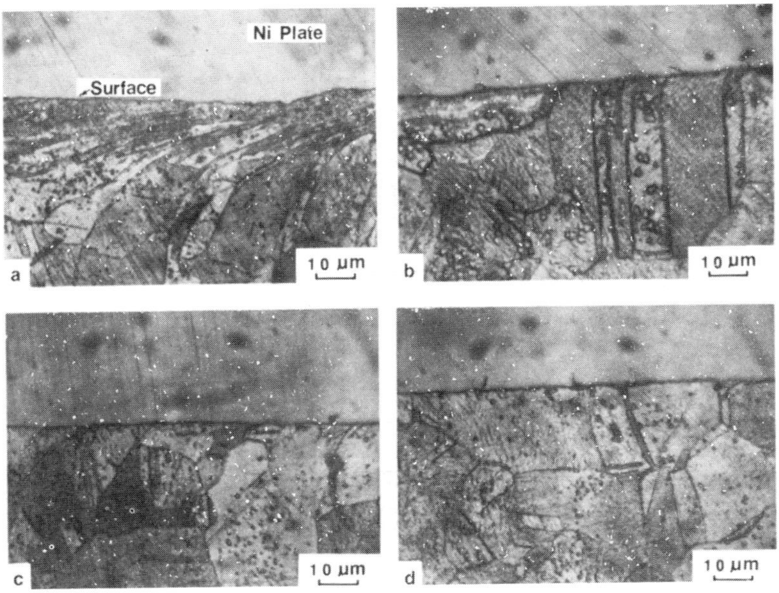

Fig. 9.—Optical micrographs of the OFHC copper subsurface for different grits: (a) 60, (b) 180, (c) 320, (d) 600 grit. The normal load was 39.2 N.

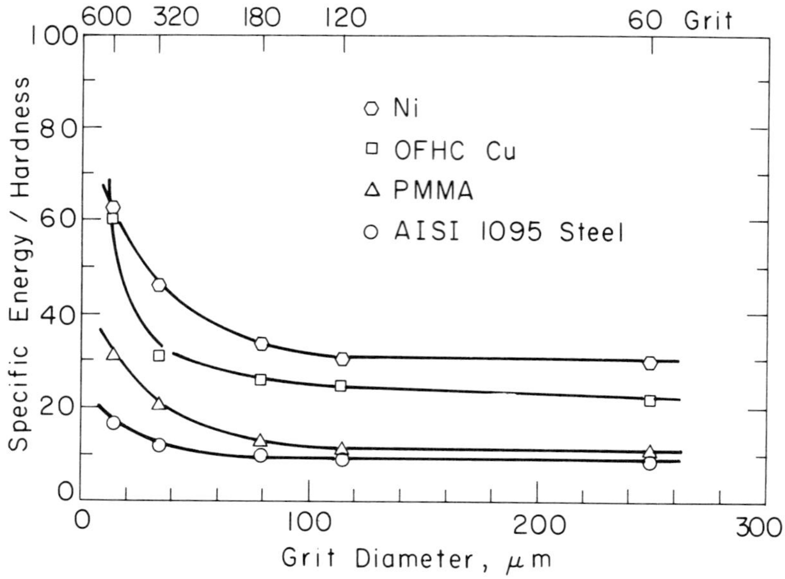

Fig. 10.—Ratio of the specific energy to the hardness as a function of the abrasive grit size.

dependent on material and need further systematic study.

At present it is impossible to treat abrasion in a purely analytical manner even if the abrasive is assumed to have such simple shapes as sphere, cone, etc. The randomness of the shape and orientation of abrasive particles makes the problem even more difficult. However, a dimensional analysis approach makes it possible to quantify the wear problem approximately. In abrasion the wear coefficient K may be written as

$$K = \frac{3Vp}{LS} = \frac{3\mu Vp}{\mu LS} \cong \frac{Vp}{\mu LS} \cong \frac{Vu}{\mu LS} \qquad (10)$$

Since $3\mu \simeq 1$, it is then seen that the wear coefficient K is the ratio of the work done to generate wear particles in the form of chips to the total external work done.[20] Therefore, when the external work is completely consumed to cut the surface, the wear coefficient should be nearly equal to 1. However, the experimentally observed wear coefficients are much less than 1 (0.2 for PMMA and AISI 1095 steel, 0.06 for nickel, and 0.07 for OFHC copper), indicating that abrasive wear particles are not generated by a simple cutting mechanism and that energy is dissipated through other mechanisms such as plastic deformation of the surface material.

In fact, the pattern of plastic flow of material around the abrasive particle will determine the chip volume removed. If the material is ductile, it flows easily around the abrasive particle without appreciable wear and forms ridges which are subsequently removed after extensive repeated deformation. For this reason, the wear coefficient depends on the ductility

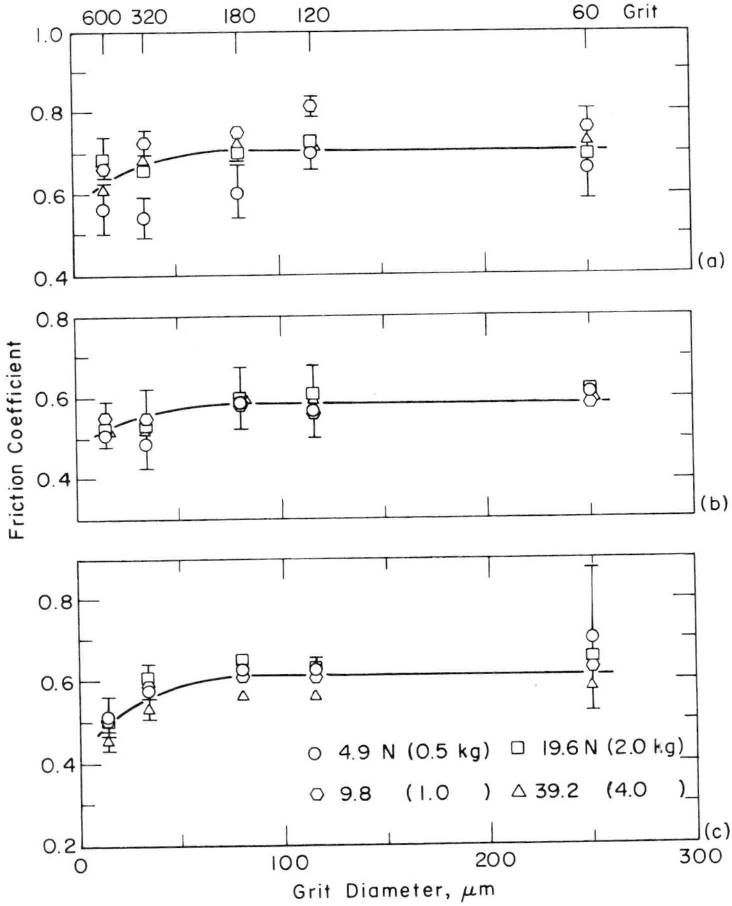

Fig. 11.—Friction coefficient as a function of the abrasive grit diameter for different normal loads: (a) PMMA, (b) nickel, (c) AISI 1095 steel.

of the material as shown in Figure 13. It is seen that less ductile materials such as PMMA and AISI 1095 steel have generally higher wear coefficients than more ductile materials such as commercially pure nickel and copper.

It can be concluded, therefore, that the experimentally observed values of the wear coefficient are at least an order of magnitude smaller than the theoretical values, since only a small fraction of material from the groove becomes loose wear particles. The work done by abrasive particles is expended as large plastic deformation, and the ductility of materials also has an important influence in determining the wear coefficient.

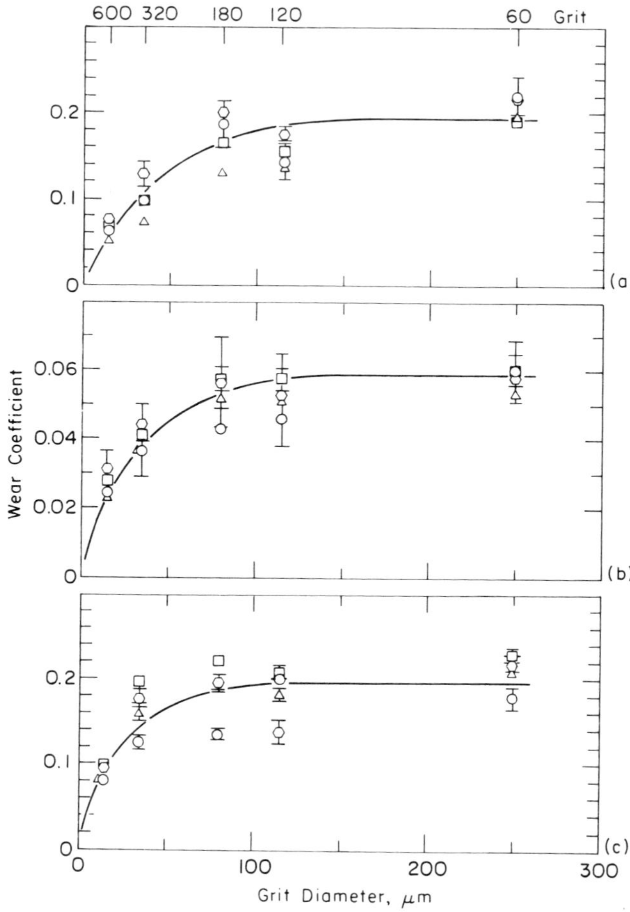

Fig. 12.—Wear coefficient versus abrasive grit diameter for different normal loads (symbols are same as in the previous figure): (a) PMMA, (b) nickel, (c) AISI 1095 steel.

GRIT SIZE EFFECT ON FRICTION

Experimental results of friction coefficient (Figure 11) show that initially it increases slightly as the grit size is increased, up to about 80 μm, and then levels off becoming essentially constant. Investigations have been made in the past to determine the grit size effect on friction coefficient. The decrease in the coefficient with a decrease in grit size has been attributed to clogging of the interstices(21) and an increase of plowing component of friction with decreased grit size.(22)

Theories of friction for such simple shapes as spheres, cones and pyramids(6,17) show that the plowing component of friction is strongly

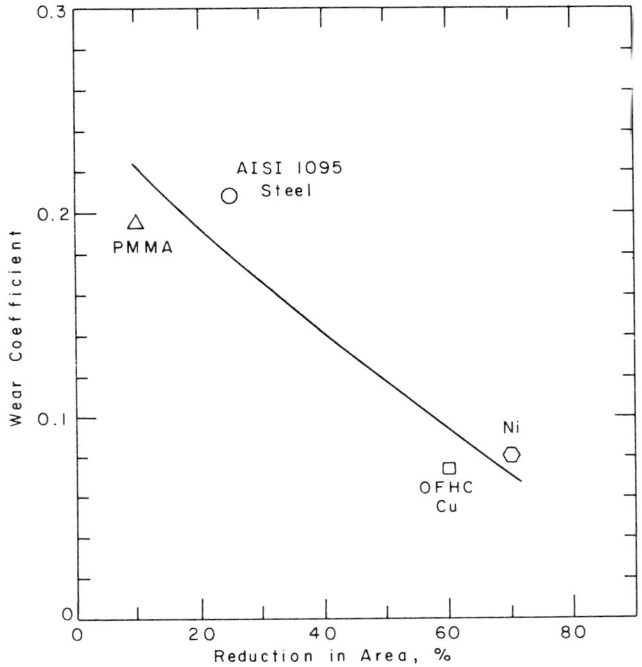

Fig. 13.—Wear coefficient as a function of ductility for 60 grit. Applied load was 39.2 N.

dependent on the shape of the abrasive, and experimental results agree fairly well with these theories.[17,23-25]

For a spherical asperity, the adhesion component μ_a and the plowing component μ_p of the friction coefficient can be expressed as[6]

$$\mu_a = \frac{4s}{\pi p} \left(\frac{D_g}{w}\right)^2 \left\{1 - \left[1 - \left(\frac{w}{D_g}\right)^2\right]^{1/2}\right\}$$

$$\mu_p = \frac{2}{\pi} \left\{\left(\frac{D_g}{w}\right)^2 \sin^{-1}\left(\frac{w}{D_g}\right) - \left[\left(\frac{D_g}{w}\right)^2 - 1\right]^{1/2}\right\} \qquad (11)$$

where D_g is the diameter of the sphere and w is the groove width. Figure 14 shows the variation of the friction coefficient as a function of the ratio w/D_g with the value of 1/6 for s/p, and the experimental results of Hisakado[17] and Lal and Shaw.[25] In abrasion the ratio w/D_g is about 0.3 and the friction coefficients are 0.5 - 0.7 (Figure 11), while this model predicts the friction coefficient of about 0.24 for this value of w/D_g. Therefore, the abrasive particles cannot be modeled as spheres; they are angular, as clearly seen from Figure 15.

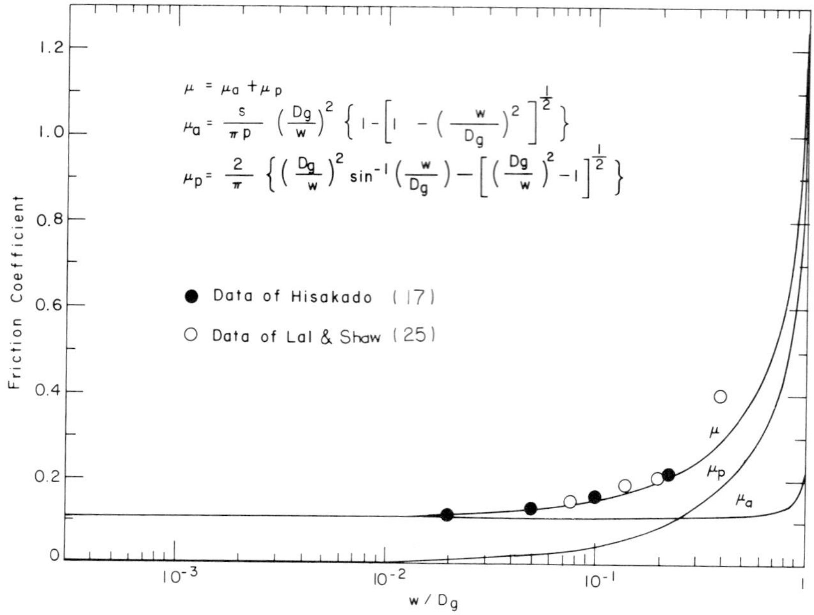

Fig. 14.—Friction coefficient as a function of the ratio of groove width to diameter of sphere.

For conical particles the friction coefficient can be calculated using theories of Rabinowicz(3), Goddard and Wilman(6) and Hisakado.(17) The Hisakado equation is expressed as

$$\mu = \frac{\tan\theta + \frac{\pi}{4}\mu_a(\sec\theta + 1)}{\frac{\pi}{2} - \mu_a \tan\theta} \tag{12}$$

and the Goddard-Wilman equation is given as

$$\mu = \frac{2}{\pi}\left[\tan\theta + \frac{s}{p}\sec\theta\right] \tag{13}$$

In Figure 7 results of simulated cutting tests with conical diamond tools are compared with these theories, and they show agreement only for a limited range of θ values. For the range of $\theta = 25 - 40$ degrees, the friction coefficient varies from 0.5 to 0.7 which is similar to the friction coefficient in abrasion tests (Figure 11). Therefore, it may be assumed that abrasive particles are conical in shape with this range of slope angles.

These theoretical models predict that the friction coefficient is independent of grit size when the grits are geometrically similar. But the experimental results show that initially the friction coefficient slightly increases with the increase of the grit size (up to about 80 μm) and then levels off to essentially a constant value (Figure 11). This clearly indicates that the abrasive grains can neither be treated as spheres nor as

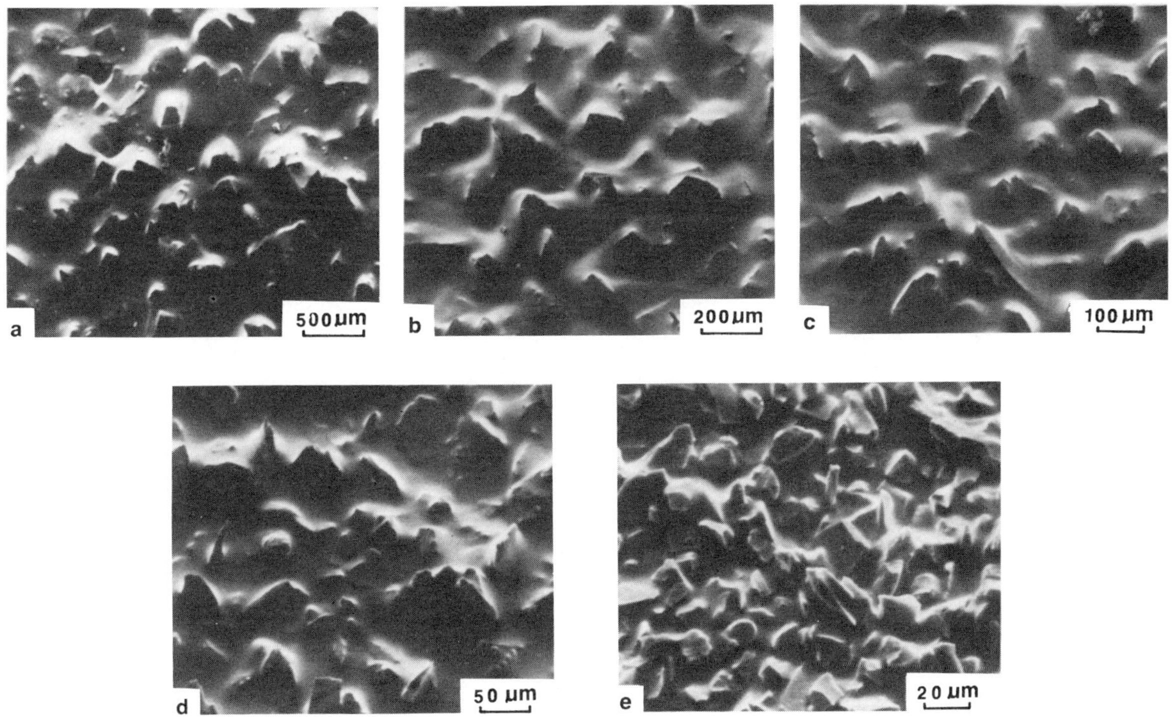

Fig. 15.—Scanning electron micrographs of the abrasive surface: (a) 60, (b) 120, (c) 180, (d) 320, (e) 600 grit.

cones. Assuming that the particles are conical with hemispherical tip of radius r, the friction coefficient can be calculated using the method of Goddard and Wilman.[6] The adhesion and plowing components of the friction coefficient are expressed respectively as

$$\mu_a = \frac{4s}{\pi p}\left(\frac{w}{2r}\right)^{-2}\left\{1 - \left[1 - \left(\frac{w}{2r}\right)^{-2}\right]^{1/2}\right\}$$

$$\mu_p = \frac{2}{\pi}\left\{\left(\frac{w}{2r}\right)^{-2} \sin^{-1}\frac{w}{2r} - \left[\left(\frac{w}{2r}\right)^{-2} - 1\right]^{1/2}\right\} \tag{14}$$

when $w \leq 2r\sin\theta$, and

$$\mu_a = \frac{2s}{\pi p}\left(\frac{w}{r}\right)^{-2}\left[8\,(1-\cos\theta) + \frac{\left(\frac{w}{r}\right)^2 - 4\sin^2\theta}{\cos\theta}\right]$$

$$\mu_p = \frac{2\tan\theta}{\pi}\left[1 - \frac{4(\tan\theta-\theta)}{\tan\theta}\left(\frac{w}{r}\right)^{-2}\right] \tag{15}$$

when $w \geq 2r\sin\theta$.

Figure 16 shows the friction coefficient according to this model as a function of w/r, the ratio of the groove width to the tip radius. Assuming that r is about 1-5 μm (i.e., the radius of a very sharp machining tool), the ratio w/r is 1-5 for 600 grit, 3-15 for 320 grit and 14-70 for 60 grit (Figure 17). It can be seen from Figure 16 that for 320 and 600 grits it falls on the transition region (sphere + cone); therefore, the friction

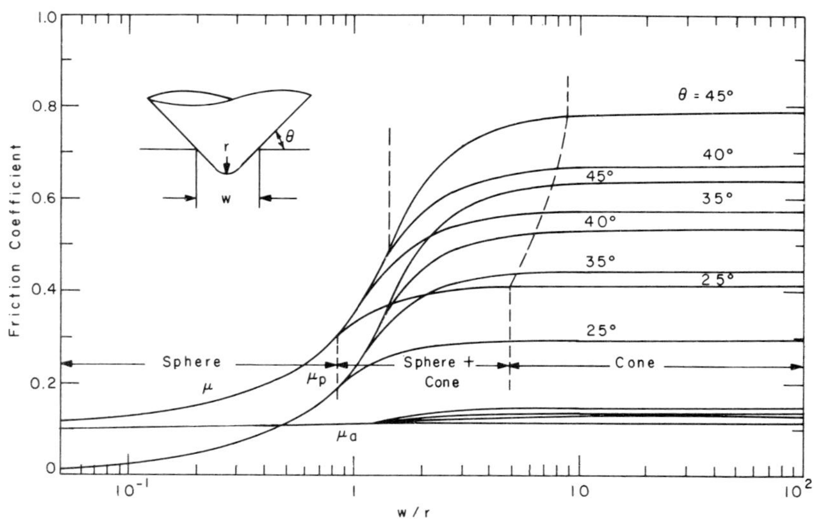

Fig. 16.—Friction coefficient as a function of the ratio of the groove width to the tip radius of conical particle for different cone angles.

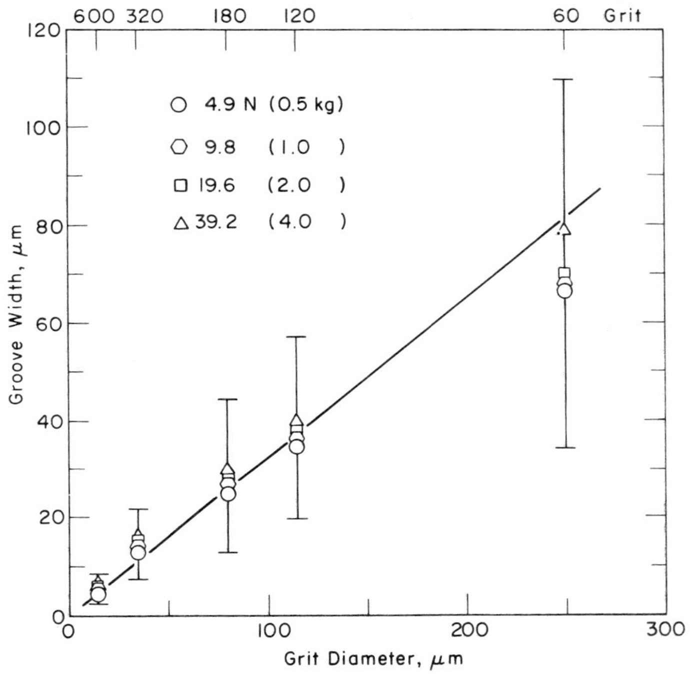

Fig. 17.—Groove width formed on PMMA specimen versus abrasive grit diameter.

coefficient decreases with decreased grit size. Since the ratio w/r is a measure of relative dullness at the contact region, it can be concluded that the increasing dullness of the particle due to decrease of the depth of penetration is the main reason for the decrease of friction coefficient with grit size in abrasion.

GRIT SIZE EFFECT ON WEAR

Several arguments have been proposed in the past for the grit size dependence of the abrasive wear rate of materials. They include clogging of the interstices between fine abrasive grits(3,23), deterioration of the abrasive(26), pickup of abrasive particles by wearing surfaces(27) and elastic contact(4) theories. Although those arguments provide plausible explanations for the size effect, no definitive evidence has been given to support the conjecture.

During abrasion chip formation, plowing and subsurface deformation occur simultaneously; therefore, the relative significance of these mechanisms will determine the wear rate. However, the estimation of the wear coefficients is difficult based on the available knowledge. The magnitude of the plastic work done in the subsurface layer can be estimated only approximately. The extent of the plastic work depends on the load exerted by an abrasive particle on the surface and the abrasive particle size.

Although the actual plastic deformation of the subsurface under a given set of surface tractions exerted by an abrasive particle can be determined approximately, the plastic deformation associated with chip formation and plowing cannot be predicted analytically. Nevertheless, the variation of wear coefficient as a function of grit size can be explained semiquantitatively by using the analytical result of Kragelskii[28] and geometric reasoning similar to those used in discussing the friction coefficient.

Kragelskii[28] has proposed that, when a spherical asperity of radius R slides on a metal surface and indents it to a depth h, various types of asperity-surface interactions occur depending on the value of h/R. In the case of ferrous metals, for $h/R<0.01$ only elastic deformation takes place; for $0.01<h/R$ 0.1 plastic displacement, and when $h/R>0.1$ cutting occurs. This criterion can be generalized for an arbitrary geometry using Shaw's interpretation of the wear coefficient.[29] From Figure 18 the plastic area A_2 is proportional to the area A_1 and, therefore, the wear coefficient K is expressed as

$$K \sim \frac{\text{Volume worn away}}{\text{Plastically deformed volume}} \sim \frac{A_g}{A_2} \sim \frac{A_g}{A_1} \tag{16}$$

Thus, the wear coefficient is proportional to the ratio of the groove area A_g to the contact area A_1. Therefore, this ratio characterizes the interaction for arbitrary shapes as h/R does for a sphere.

Assuming that the abrasive particle is conical with hemispherical tip of radius r, as in the case of friction, the groove area A_g can be calculated as a function of the groove width w.

For $w \geq 2r\ \sin\theta$, the groove area A_g is

$$A_g = \frac{w^2 \tan\theta}{4} \left[1 - \frac{4(\tan\theta - \theta)}{\tan\theta} \left(\frac{w}{r} \right)^{-2} \right] \tag{17}$$

and for $w \leq 2r\ \sin\theta$,

$$A_g = w^2 \left\{ \left(\frac{w}{r} \right)^{-2} \sin^{-1} \frac{w}{2r} - \frac{1}{2} \left(\frac{w}{r} \right)^{-1} \left[1 - \left(\frac{w}{2r} \right)^2 \right]^{1/2} \right\} \tag{18}$$

In Figure 19 contact geometry, mechanical interaction, and wear mode are shown as a function of w/r for different slope angles. With the assumed value of r of about 5 μm, this model clearly explains the grit size dependence of abrasive wear as in the case of friction. This combined model also shows that for extremely small grit sizes wear is controlled by the sliding wear mechanisms. Therefore, the grit size dependence can be regarded as a transition from abrasion to sliding wear. Rough estimation by Kragelskii's criterion shows that the sliding wear regime is below the grit size of about 1μm.

This model also explains semi-quantitatively the variation of the energy components -- cutting, plowing, subsurface deformation -- due to the change of groove geometry with grit size.

From Equation (11) the wear coefficient K can be expressed as

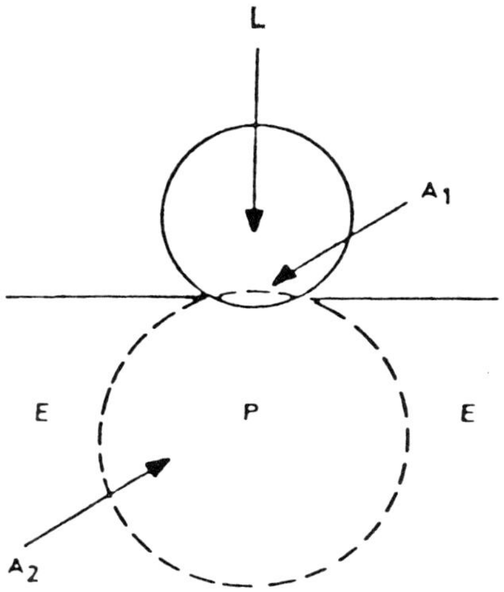

Fig. 18.—Indentation of plastic material: E = elastic region, P = plastic region.[29]

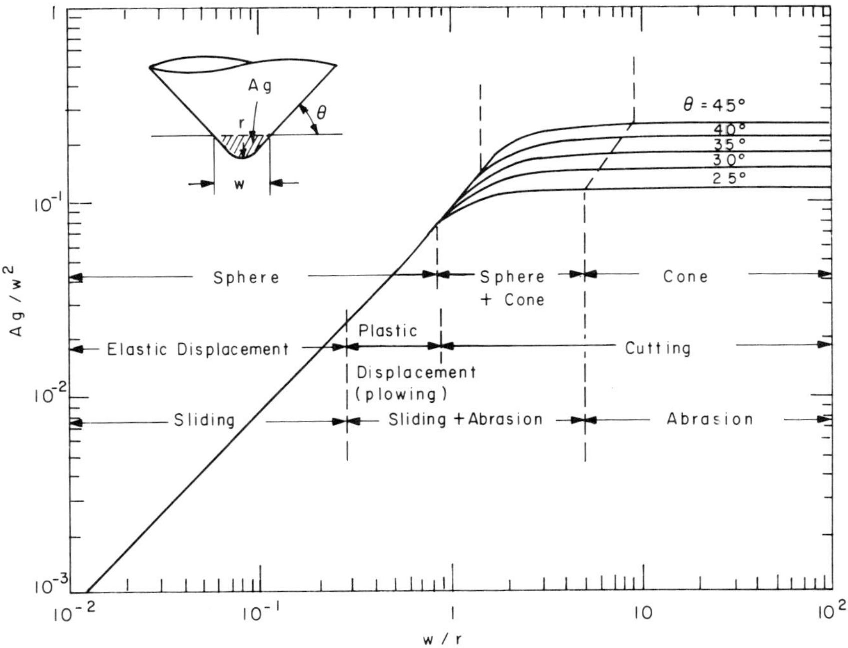

Fig. 19.—Ratio of the groove area to the square of the groove width as a function of the ratio of the groove width to the grit tip radius of conical particle.

$$K \approx \frac{W_c}{W_c + W_p + W_s} \tag{19}$$

where W_c is the cutting energy, W_p the plowing energy and W_s the subsurface deformation energy. This implies that the relative values of energy components in abrasion determine the wear coefficient. Since the friction coefficient changes only 20 percent with the grit size, the total external work done can be regarded as independent of the grit size, i.e., $W_c + W_p + W_s \approx constant$.

By applying the method of Suh and Sridharan[30] to abrasion, the subsurface deformation energy can be calculated approximately. As shown in the Appendix, this energy per unit sliding distance is given as

$$\frac{W_s}{S} \simeq \frac{1}{12} m L \int_0^{\infty} \Delta \bar{\gamma}^p \, d\left(\frac{y}{a}\right) \tag{20}$$

where m is the ratio of the total equivalent strain to the net plastic strain per cycle, L the applied load, a the half width of contact, and $\Delta \bar{\gamma}^p$ the incremental plastic strain per cycle at y. The integral can be approximately calculated by using the result of Jahanmir and Suh.[31] This calculation shows that the subsurface deformation energy is independent of abrasive grit size and therefore, $W_c + W_p \approx constant$.

As discussed above, not all groove volume is removed in the form of chips. The fraction of this removal volume is dependent on the contact geometry and determines the relative amounts of energies; the larger the fraction, the more the cutting energy. At present, no rigorous solution for this removal fraction exists, but results of cutting experiments with spherical tools[32] show that it increases with the increase of the ratio h/R, and so does the cutting energy. The ratio w/r for the model of a conical tool with hemispherical tip is more or less equivalent to h/R so the cutting energy will decrease with the decreased w/r and, therefore, with the grit size as shown schematically in Figure 20. Accordingly, the wear coefficient will decrease with the decrease of grit size.

OTHER ASPECTS

The grit size effect in 3-body abrasion can be explained in terms of the relative bluntness of abrasive grits and the particle motion on the effective "rake" angle. In addition to being "relatively blunt," the abrasive particles of a given size have a much larger negative rake angle due to their rolling motion, i.e.. small θ, since the effective θ is the angle between the plane of the surface and the abrasive-workpiece interface.

The relative hardness between the abrasive and the specimen also determines the critical grit size. As shown by Richardson[33], the wear resistance of the material increases when its hardness exceeds about 80 percent of the abrasive hardness. If the hardness of the wearing material is comparable to the abrasive hardness, the wear rate is no longer controlled by abrasive mechanism. Under a given load, the particles cannot penetrate deep and, therefore, the contacting geometry becomes duller even for coarse grits. Therefore, increasing hardness of material or decreasing

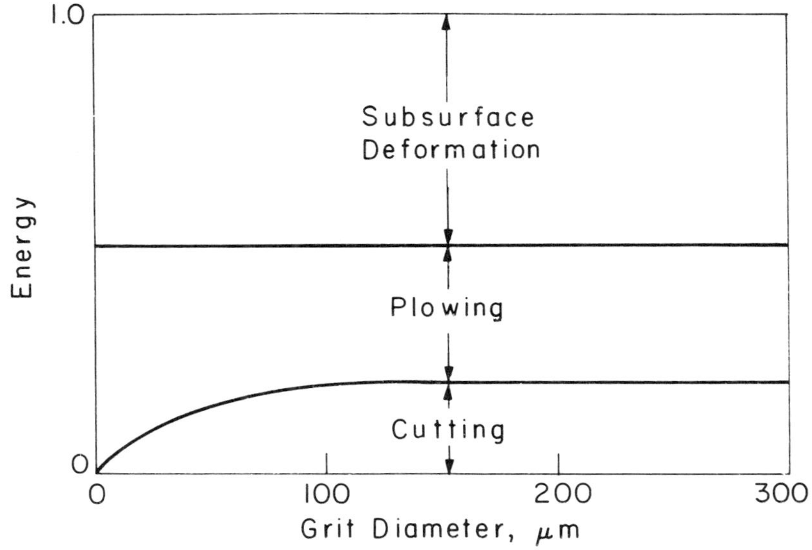

Fig. 20.—Schematic of energy components in abrasion.

hardness of the abrasive will accelerate the transition from abrasion to sliding wear.

The dependence of the abrasion behavior of materials on the grit size has practical significance for wear prevention. Since the small abrasive particles are less harmful than larger ones, the entrapment of large particles is more desirable in oil filtration. Also it is obvious that very small wear particles (smaller than 0.1 μm) may not act as abrasive particles, eliminating the need to filter these particles. From the metallurgical point of view, it is imperative that the hard second-phase particles be less than 80 μm in the event that they are dislodged from the matrix if the material is to be used in wear applications.

CONCLUSIONS

1) The wear coefficient predicted by classical theories of abrasive wear is of the order of unity, whereas the experimentally measured values are at least an order of magnitude smaller, because a large fraction of the external work is expended for plastic deformation and little in actually removing the material by a simple cutting process.

2) The abrasive wear coefficient of brittle materials is larger than that of ductile materials, indicating that ductility controls the rate of loose chip formation.

3) The relative bluntness of the abrasive particle is the major cause for the dependence of friction and wear coefficients on grit size. An idealized model of the abrasive particle as a cone with hemispherical tip can describe semi-quantitatively the variation of both friction and wear

coefficients with grit size, including a transition from abrasion to sliding wear with the decrease in abrasive grit size.

4) The grit size effect should be utilized in designing filters for lubrication systems and materials for wear resistant applications.

ACKNOWLEDGEMENTS

This study was supported by the Defense Advanced Research Projects Agency through the Office of Naval Research under Contract No. N00014-76-C-0068. The authors are grateful to Dr. E.C. van Reuth, Dr. Arden L. Bement, Commander H.P. Martin, and Lt. Commander W.K. Petrovic for their support and interest.

APPENDIX

Energy Dissipation by Subsurface Deformation in Abrasion

As in the case of sliding wear, deformation is accumulated in abrasion by cyclic loading, each cycle corresponding to the passage of an abrasive grit. Compared to sliding wear, the rate of removal of the surface layer is much faster. This makes the computation of energy dissipation by subsurface deformation extremely difficult. However, a very approximate analysis can be given as in the case of sliding[29] by neglecting the material removal aspect.

The stress-strain history of an infinitesimal volume of material below the surface depends on the depth. For the elastic semi-infinite solid under a moving concentrated load, the Boussinesq solution states that the material very near the surface experiences a reversed compression-tension stress while the material below it may only experience compressive loading and unloading. Even for the elastic-plastic case, the material undergoes a similar loading history. A plausible stress-strain history is shown schematically in Figure A.

The energy dissipated by subsurface deformation, W_S, is obtained by summing the work done over many cycles for the entire loading history as

$$W_s = nb\ell \sum_{i=1}^{N_f} \int_0^{\infty} \int_0^{\bar{\varepsilon}_i} \bar{\sigma}\, d(\bar{\varepsilon}_i)\, dy \qquad \text{(A-1)}$$

where n is the number of grooves, b the average width of the groove, ℓ the length of the groove (or approximately the diameter of the specimen), $\bar{\sigma}$ the equivalent stress and $\bar{\varepsilon}_i$ the equivalent strain in the ith cycle.

As a first approximation to the problem, one may assume that

$$\bar{\sigma} = s$$

and (A-2)

$$\bar{\varepsilon}_i = \bar{\gamma}_i$$

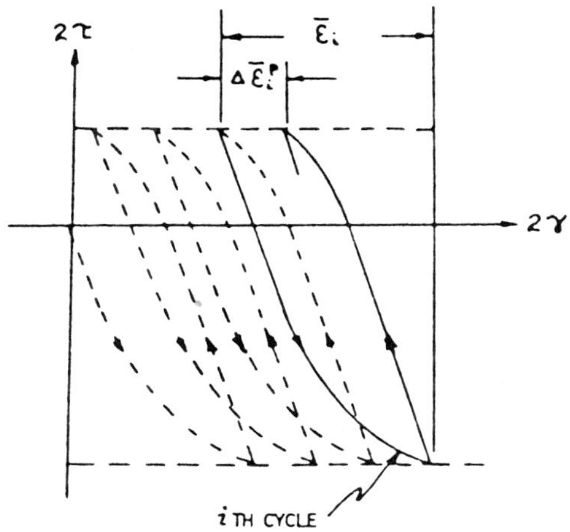

Fig. A.—Stress-strain history of a cyclically loaded metal; the ith cycle is shown by the solid line.(30)

where s is the shear yield stress and $\overline{\gamma}_i$ is the shear strain. Substituting these into Equation (A-1), it may be written as

$$W_s = nb\ell \sum_{i=1}^{N_f} \int_0^\infty \int_0^{\overline{\gamma}_i} s \, d(\overline{\gamma}_i) \, dy$$
$$= nb\ell \sum_{i=1}^{N_f} \int_0^\infty s \, \overline{\gamma}_i \, dy \qquad \text{(A-3)}$$

Let the ratio m_i of the total equivalent strain $\overline{\gamma}$ to the incremental plastic strain per cycle $\Delta\overline{\gamma}_i{}^p$ be defined as

$$m_i = \frac{\overline{\gamma}_i}{\Delta\overline{\gamma}_i{}^p} \qquad \text{(A-4)}$$

As an approximation let $m_i = m$, where m is a constant independent of both y and i. Substituting Equation (A-4) into Equation (A-3), we obtain

$$W_s = nb\ell m \, N_f \int_0^\infty s \, \Delta\overline{\gamma}^p \, dy$$
$$= nb\ell \; a \; m \, N_f \int_0^\infty s \, \Delta\overline{\gamma}^p \, d(\tfrac{y}{a}) \qquad \text{(A-5)}$$

where a is the half width of contact.

From Figure B, it can be seen that $\int_0^\infty \Delta\overline{\gamma}^p \, d(\frac{y}{a})$ is constant for a given friction coefficient μ. Since μ does not change much with the grit size, it may be assumed that $\int_{t}^{\infty} \Delta\overline{\gamma}^{p}\, d(\frac{y}{a})$ is approximately constant. If this integral is denoted by I_o, then Equation (A-5) becomes

$$W_s = nb\, \ell\, a\, m\, N_f\, s\, I_o \tag{A-6}$$

From the relationship between the real contact area A and the number of contacting particles N_c, the number of cycles N_f is given by

$$N_f \simeq \frac{S}{\ell}\,(N_c)^{1/2}$$

$$\simeq \frac{S}{2a\ell}\,(A)^{1/2} = \frac{S}{2a\ell}\,(\frac{L}{p})^{1/2} \tag{A-7}$$

where S is the sliding distance, L the applied load and p the hardness of the abraded material. Substituting Equation (A-7) into Equation (A-6) gives

$$\frac{W_s}{S} \simeq \frac{nbsmI_o}{2}\,(\frac{L}{p})^{1/2} \simeq \frac{1}{12}\, m \cdot I_o \cdot L \tag{A-8}$$

since $n \simeq (N_c)^{1/2}$, $b \simeq 2a$ and $s/p = 1/6$. Therefore, the ratio of this energy to the external work W_e is obtained as

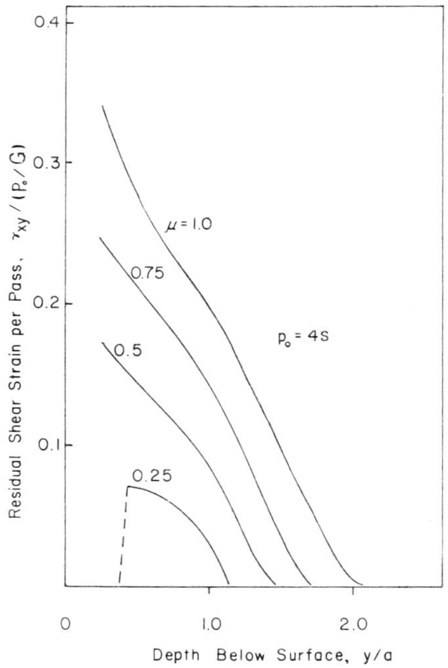

Fig. B.—The steady state residual shear strain per pass, for different friction coefficients, normalized with respect to the maximum applied normal stress p_o = 4s, divided by the shear modulus G (from Reference 31).

$$\frac{W_s}{W_e} \simeq \frac{1}{12} \cdot \frac{m\, I_o\, LS}{\mu LS} = \frac{m\, I_o}{12\mu} \qquad \text{(A-9)}$$

which states that the fraction of the external work dissipated by subsurface cyclic plastic deformation is independent of the grit size.

A numerical example may be given using the abrasion data and Figure B with the relationship $p_o = 4p/\pi$. For OFHC copper under the applied load of 39.2 N, the friction coefficient is approximately 0.5. Using $G = 4.5 \times 10^{10}$ N/m^2 and $p = 430\ MP_a$, $I_o \simeq 2 \times 10^{-3}$ from Figure B. For $m = 500$, the ratio is approximately 0.2, indicating 20 percent of external work is expended in the subsurface deformation. Although the value of m is for AISI 1045 steel, the analysis seems to predict the right order of magnitude value of the subsurface deformation energy.

Of course, this order of magnitude type analysis does not consider many important details of abrasion process. For instance, the effects of the material removal from the surface layer and the variation of contact geometry with the grit size, etc., have not been considered. Further, it was assumed that the history of cyclic deformation is similar for all y. As for the friction coefficient, the separation of the component responsible for the subsurface deformation from the total friction coefficient is required. A rigorous analysis which takes these details of abrasion process into consideration is very much desired.

REFERENCES

1. Barwell, F.T., These Proceedings.
2. Rabinowicz, E., Dunn, L.A. and Russell, P.G., *Wear*, Vol. 4, 1961, p. 345.
3. Rabinowicz, E., "Friction and Wear of Materials," John Wiley and Sons, New York, 1965, p. 168.
4. Larsen-Basse, J., *Wear*, Vol. 11, 1968, p. 213.
5. Sin, H.-C., Saka, N. and Suh, N.P., *Wear*, Vol. 55, 1979, p. 163.
6. Goddard, J. and Wilman, H., *Wear*, Vol. 5, 1965, p. 114.
7. Tsukizoe, T. and Sakamoto, T., *JSME Bulletin*, Vol. 17, 1974, p. 1637.
8. Sedriks, A.J. and Mulhearn, T.O., *Wear*, Vol. 6, 1963, p. 457.
9. Sedriks, A.J. and Mulhearn, T.O., *Wear*, Vol. 7, 1964, p. 451.
10. Stroud, M.F. and Wilman, H., *Journal of Physics D: Applied Physics*, 1962, p. 173.
11. Mulhearn, T.O. and Samuels, L.E., *Wear*, Vol. 5, 1962, p. 478.
12. Finkin, E.F., in "Evaluation of Wear Testing," ASTM STP 446, American Society for Testing and Materials, 1969, p. 55.
13. Moore, M.A., *Wear*, Vol. 27, 1974, p. 1.
14. Krushov, M.M., in Proceedings of the Conference on Lubrication and Wear, The Institution of Mechanical Engineers, London, 1957, p. 655.
15. Larsen-Basse, J., *Transactions of the Metallurgical Society of the American Institute of Mining, Metallurgical, and Petroleum Engineers*, Vol. 236, 1966, p. 1461.
16. Larsen-Basse, J. and Mathew, K.G., *Wear*, Vol. 14, 1977, p. 199.
17. Hisakado, T., *JSME Bulletin*, Vol. 13, 1970, p. 129.
18. Cook, N.H., "Manufacturing Analysis," Addison-Wesley, Reading, Massachusetts, 1966, p. 41.
19. Moore, M.A. and Douthwaite, R.M., *Metallurgical Transactions A: Physical Metallurgy and Materials Science*, Vol. 7A, 1976, p. 1833.

20. Suh, N.P., "Scientific and Technical Problems in Erosive Wear," Presented at Institute of Scientific Problems Relevant to Coal Utilization, West Virginia Institute, Morgantown, West Virginia, 1977.
21. Goddard, J., Harker, H.J. and Wilman, H., *Nature*, Vol. 184, 1959, p. 333.
22. Spurr, R.T. and Newcomb, T.P., in Proceedings of the Conference on Lubrication and Wear, The Institution of Mechanical Engineers, London, 1957, p. 269.
23. Avient, B.W.E., Goddard, J. and Wilman, H., *Royal Society of London. Proceedings*. Series A, Vol. 258, 1960, p. 159.
24. Maan, N. and Broese van Groenou, A., *Wear*, Vol. 42, 1977, p. 365.
25. Lal, G.K. and Shaw, M.C., *Wear*, Vol. 29, 1974, p. 153.
26. Date, S.W. and Malkin, S., *Wear*, Vol. 40, 1976, p. 223.
27. Johnson, R.W., *Wear*, Vol. 16, 1970, p. 351.
28. Kragelskii, I.V., "Friction and Wear," Butterworths, London, 1965, p. 25.
29. Shaw, M.C., *Wear*, Vol. 43, 1977, p. 263.
30. Suh, N.P. and Sridharan, P., *Wear*, Vol. 34, 1975, p. 291.
31. Jahanmir, S. and Suh, N.P., *Wear*, Vol. 44, 1977, p. 17.
32. Malkin, S., Wiggins, K.L., Osman, M. and Smalling, R.W., in Proceedings of the 13th International Machine Tool Design and Research Conference, Birmingham, 1972, edited by S.A. Tobias and Koenigsberger, MacMillan, London, 1973, p. 291.
33. Richardson, R.C.D., *Wear*, Vol. 11, 1968, p. 245.

DISCUSSION

J. LARSEN-BASSE, University of Hawaii: I notice that you say some of the previous theories are not fully valid. I agree with you. However, I wonder if you can extrapolate wear coefficient to 1 μm. In my work using the abrasive paper, I always found that there was a good deal of scatter. The work done in Australia shows that there is chip formation even below 1 μm grit size.

N. P. SUH: Obviously there is scatter because we are not using grits of uniform sharpness. The important point here is: what size particle should be removed from filtration systems? The ideal thing is to remove everything. Once you remove particles below 1 μm, or thereabouts, abrasive wear will not be a serious problem. Then wear takes place by subsurface deformation, crack nucleation and crack propagation processes.

LARSEN-BASSE: In our experiments we found that the abrasive grit size effect was really a test effect. It depends on the size of the specimen.

AN INVESTIGATION OF TWO-BODY ABRASIVE WEAR

H. H. Hirano and A. V. Levy

ABSTRACT

An investigation of two-body abrasive wear utilizing a pin specimen on a rotating SiC abrasive paper test device is described. The test materials were 7075 aluminum and 4340 steel which were thermally treated to attain a range of hardness, fracture toughness and yield strength values. Wear resistance is seen to correlate directly with hardness and yield strength for both materials. For the 4340 steel the wear resistance is inversely proportional to the fracture toughness; for the 7075 aluminum, the inverse proportionality is dependent upon the microstructure. The effect of an increase in hardness by heat treatment of an alloy is much less pronounced than when the wear resistance of pure metals of differing hardness are compared. Scanning electron microscopy of abraded surfaces was used to examine the removal mechanisms involved in abrasive wear. The meaning of the observed mechanisms in terms of a model are assessed.

INTRODUCTION

Many methods for producing liquid or gaseous hydrocarbons from coal operate continuously above atmospheric pressure. The development of dry coal screw feeders has shown that abrasive wear problems may be a limiting factor in design. In addition, abrasive wear is a problem in coal transportation equipment as well as in mining machinery. These aspects were the motivation for the present study of abrasive wear.

This paper is devoted primarily to a study of two-body abrasion as a first step in developing a more complete understanding of abrasive wear in coal conversion equipment. However, some comparisons are made with three-body wear results.

Considerable amount of work has been done in the area of two-body abrasion, much of it involving testing a wide range of materials and correlating the wear rate with other mechanical properties. The principal contribution of this type is the work of Kruschov.[1] For pure metals and annealed steels, he found that the wear resistance was directly proportional to the Vicker's hardness of the material. As a steel was heat-treated to increase its hardness, the wear resistance also increased. However, the increase in wear resistance was not as dramatic for a heat-treated steel as was the increase in wear resistance of pure metals or annealed steels for the same increase in hardness.

In the case of cold-worked materials, little or no change was discerned in the wear resistance although the hardness increases due to cold-working. Kruschov concluded that the degree of strain hardening which occurs during the abrasion process is much greater than any initial work hardening and that this value of work hardening corresponds to the maximum level of hardness preceding destruction of the material.

Rabinowicz[2] developed a simple expression for the volume of material displaced by idealizing abrasive grains as being conical in shape. He simply calculated the amount of material which would be displaced if the abrasive grain is dragged along the metal at some depth of penetration. Then, by noting the similarity between his expression and that for adhesive wear, he derived the following relation:

$$V = \frac{k\ L\ x}{3H}$$

where V = *Volume displaced*
L = *Load*
x = *Distance travelled*
H = *Hardness of base metal*
k = *Dimensionless abrasive wear coefficient*

While this equation is based on a simplified model, it does express some of the basic observations of both two-body and three-body abrasive wear. The volume of material removed is directly proportional to the load and distance travelled by the harder abrasive particle and is inversely proportional to the hardness of the softer abraded material.

Rabinowicz used this expression to compare the results obtained by a number of experimenters for two-body and three-body abrasion. The values of the abrasive wear coefficients were approximately an order of magnitude less for three-body than for two-body abrasion.

In addition to the simplification of the shape of the abrasive grain, the expression describes the total amount of material which is displaced from its original position in the softer material as being removed from the surface. This displaced material is not necessarily detached from the base material, as under some conditions it may be plastically deformed.

Mulhearn and Samuels[3] investigated the manner in which material is displaced. They identified ploughing and cutting as the primary modes of material displacement. The term "ploughing" is applied when material is plastically deformed by the abrasive grain. "Cutting" describes the situation where a chip is formed, as would be if the abrasive particle were replaced by a machine cutting tool. As a method of material removal, ploughing has a low efficiency, whereas almost all the displaced material in cutting is removed in the form of a chip.

Mulhearn and Samuels modeled the abrasive particle with a cutting tool and studied the effect of the orientation angle of the tool to the base metal. The critical value of this angle, which they termed the "critical attack angle," determined whether cutting or ploughing occurred. Angles greater than the critical value resulted in ploughing; whereas cutting occurred for smaller angles. From their tests, the critical attack angle measured from the plane of the abraded surface to the tool face was found to be 90°.

Mulhearn and Samuels also spent a good deal of time characterizing abrasive grains and the abrasive papers used in two-body abrasion tests. Among their results is the number of abrasive grains which contact the surface of the abraded material and actually cut. They determined this number to be 12% for the loading and particle size distribution they studied.

Numerous other investigations of abrasive wear include studies of the hardness ratio of the abrasive to the abraded material;[1,4] the effect of specimen size;[5] and the effect of abrasive grain size[6,7] and shape.[8]

EXPERIMENTAL PROCEDURES

Wear Tester

The abrasion tester used in this work is the simple pin on disc machine. Abrasive, normally in the form of abrasive paper, is fixed onto the top surface of the rotating disc. The head of the pin specimen is held against the abrasive by dead weight loading. As the disc is rotated at a fixed speed, the head of the specimen is abraded. In most cases the pin is allowed to traverse radially inward during the course of the test. This allows the pin to be continually exposed to fresh abrasive. A strain gauge ring was also included to allow measurement of the friction coefficient during sliding abrasion. For most of the testing, Carbimet Silicon Carbide Abrasive Papers with Pressure-Sensitive Adhesive backing were used.

Unless otherwise noted, the specimens were dead weight loaded with 500 gm against 120 grit silicon carbide abrasive paper. All of the tests comprising this investigation were performed at the same, constant speed of rotation of the disc (20 rpm) for 45 seconds.

The mechanism which controls the radial traversing of the specimen was adjusted so that all of the specimens travelled from 1.8 inches radius to .54 inch radius in the 45 seconds. Thus, all of the specimens were abraded for the same distance which was determined to be 2800 mm. The average velocity of the abrasive relative to the specimen was 6.2 cm/sec. The abrasive paper was changed before each test and examination of the wear track on the paper indicated that the radial traversing speed was adequate to insure that the specimen was continually exposed to fresh abrasive.

Wear rate was reported as volume of metal lost per unit distance travelled, in mm^2. The inverse of this number was used in the curves to designate erosion resistance.

Wear Test Specimen

The specimens were 0.25 in. diameter by 0.70 in. long. The end of the specimen which was to be abraded was machined hemispherical to discourage any initial effects due to misorientation. After machining, the specimens were lightly buffed with 600 grit silicon carbide abrasive paper.

The primary materials of interest for this investigation were 7075 aluminum and 4340 steel. Both of these materials were heat-treated to various levels of hardness and fracture toughness to study the effect of these mechanical properties on abrasive wear resistance.

The 7075 Al was obtained from two previous investigations of the effect of heat treatment on the fracture toughness of the alloy.[9,10] The fracture toughness specimens used in these investigations were 20 in. x 10 in. x 1 in.

plates which were differentially heat-treated by means of strip heaters. In this manner, heat-treated conditions in both underage and overaged from the optimal T-6 condition were obtained.

The wear specimens made from the eight overaged conditions tested in this program were machined from the grip ends of broken tensile specimens. The values of fracture toughness and yield strength used for this investigation were as reported by Pujari.[9]

In the case of the underaged material, four conditions were selected. Tensile and wear specimens were machined from the original fracture specimen. The specimens were machined such that the cylindrical axes were transverse to the rolling direction of the plate, in order to be consistent with the overaged material. The fracture toughness values used were as reported by Shenai,[10] while the yield strength values were obtained from the tensile specimens. To perform hardness tests on both overaged and underaged material, a 0.25 inch thick slice of material was removed from the wear specimens. The mechanical properties of interest for the 7075 Al wear specimens are listed in Table I.

The 4340 steel was obtained in the form of Charpy specimens from a previous investigation. These notched bar specimens were first heat-treated at 675°C so that they could be machined. The wear specimens were then machined from this material to 0.040 inches oversize on the diameter. The oversize specimens were austenitized at 870°C and then oil quenched. Four tempering treatments were selected: 200°C, 350°C, 500°C, and 650°C. The tempering treatment was performed in a neutral salt bath followed by a water quench. Following the tempering treatment, the wear specimens were ground to the final dimensions. The mechanical properties of interest for the 4340 steel are listed in Table II.

TABLE I.—MECHANICAL PROPERTIES OF 7075 ALUMINUM WEAR SPECIMENS

	Aging Temperature (°F)	Vicker's Hardness (Kg/mm^2)	Yield Strength (Ksi)	Fracture Toughness K_C ($Ksi\text{-}in^{\frac{1}{2}}$)
Underaged:	202	158.0	66.8	61.0
	209	163.0	68.6	58.0
	223	167.0	69.1	51.0
Overaged:	250	180.5	78.5	23.0
	254	181.0	77.1	23.0
	264	179.5	76.8	23.5
	305	167.0	73.5	25.0
	318	134.0	70.4	25.0
	332	122.5	67.5	26.0
	359	106.0	50.2	29.0
	371	102.0	46.2	31.0

Yield strength and fracture toughness data for overaged material obtained from Reference 9.

Fracture toughness data for underaged material from Reference 10.

TABLE II.—MECHANICAL PROPERTIES OF 4340 STEEL WEAR SPECIMENS

Tempering Temperature (°C)	Vicker's Hardness (Kg/mm^2)	Yield Strength (Ksi)	Fracture Toughness K_{Ic} ($Ksi-in^{\frac{1}{2}}$)
200	602	233	39
350	498	208	40
500	407	164	89
650	318	111	160

Yield strength data from Reference 11.

Fracture toughness data from Reference 12.

RESULTS AND DISCUSSION

Load and Abrasive Particle Size Effect

Several tests were conducted to compare the wear data obtained in this investigation with values observed by other experimenters. The wear rate was, as expected, directly proportional to the load applied on the specimen (Figure 1). The graph connecting the data points does not pass through the origin. This is attributed to the specimen support arm of the wear tester not being in perfect balance. While attempts were made to reduce this effect, all of the later testing on the heat-treated 7075 Al and 4340 steel were conducted at a large enough load level so that the error due to the un-

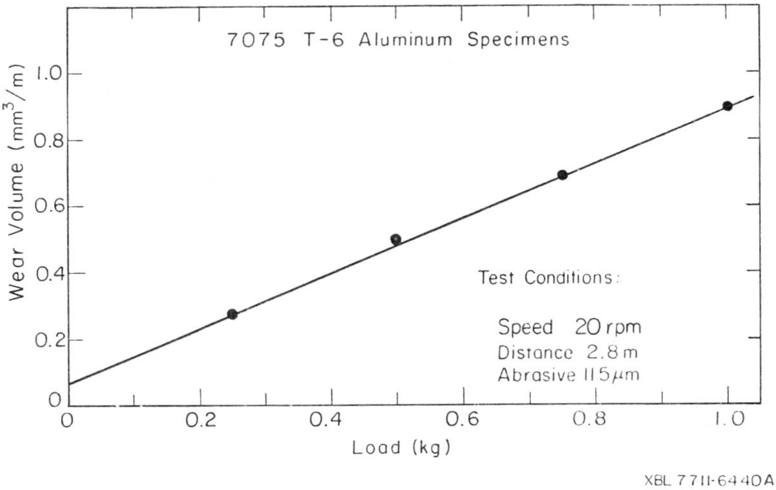

Fig. 1. —Effect of Load on the Wear Rate of 7075-T6 Aluminum.

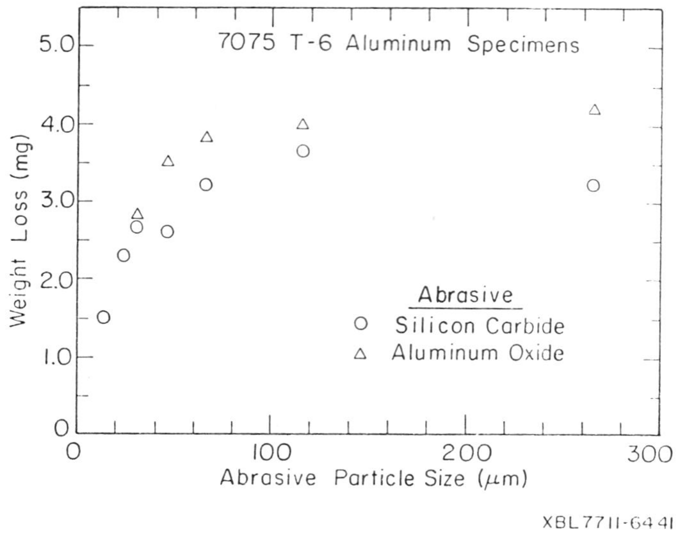

Fig. 2.—Effect of abrasive particle size on wear rate of 7075-T6 Al.

certainty in load was less than ten percent.

The effect of the abrasive particle size is an interesting one which has received some attention. The particle size above which no significant effect is seen was ~ 100 μm (Figure 2). This is the same size which has been noted by others for two-body abrasion.[7] The same result has also been reported for erosion.[13] The weight loss measured when using ~ 265 μm *SiC* abrasive particles may have been affected by the bonding technique used in the abrasive paper. Therefore, that data point was neglected in establishing the general trend discussed above.

This is an important effect to consider and warrants an investigation of its own. For the purpose of investigating the effect of the mechanical properties of the base metal, it was decided to simply use a fixed abrasive size for all the tests. This size (~ 115 μm) was chosen to lie in the region for which no grain size effect is expected.

Effect of the Mechanical Properties on Abrasive Wear

The wear resistance of 7075 Al is shown as a function of hardness, yield strength and fracture toughness in Figures 3, 4 and 5. The same information is shown for the 4340 steel in Figures 6, 7 and 8.

For the purpose of establishing any kind of correlation, hardness and yield strength seem to be good measures of the wear resistance. For 7075 Al, there was less scatter in the general trend of the data with yield stress than with hardness. However, one would expect the general trends associated with

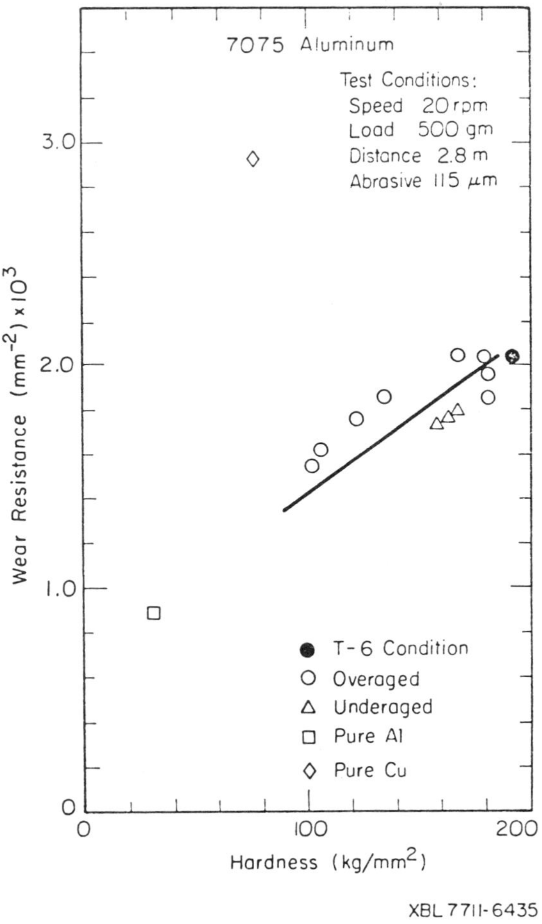

Fig. 3.—Wear resistance of 7075 Al vs. hardness.

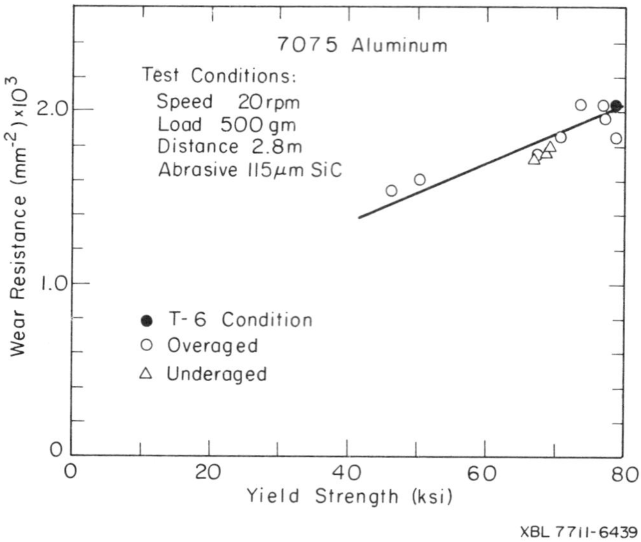

Fig. 4.—Wear resistance of 7075 Al vs. yield strength.

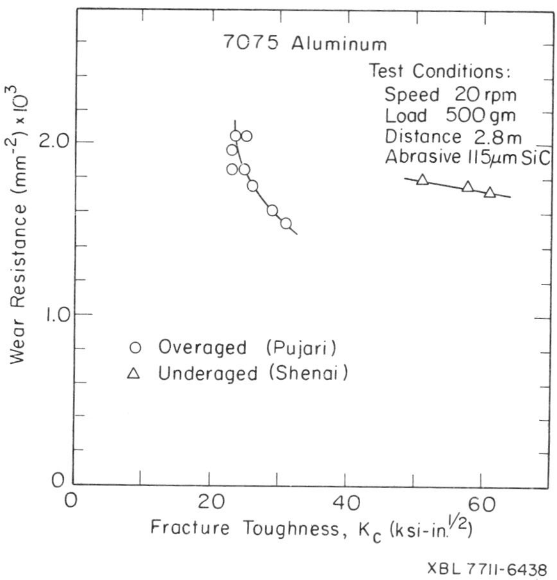

Fig. 5.—Wear resistance of 7075 Al vs. fracture toughness.

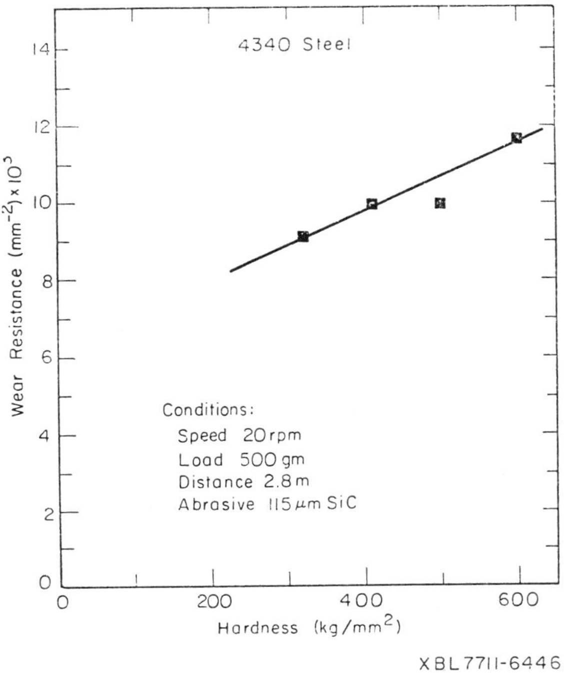

Fig. 6.—Wear resistance of 4340 steel vs. hardness.

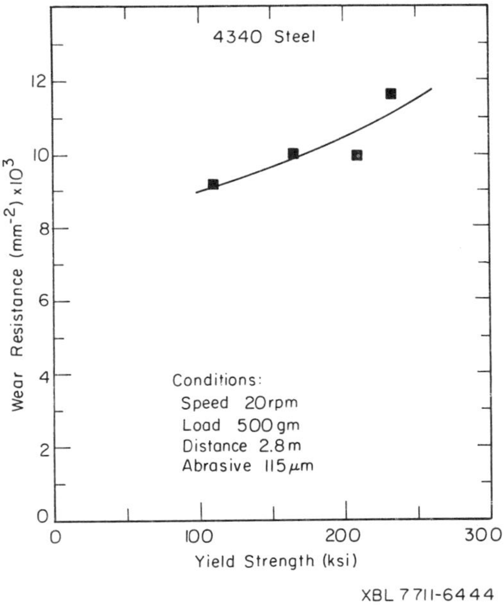

Fig. 7.—Wear resistance of 4340 steel vs. yield strength.

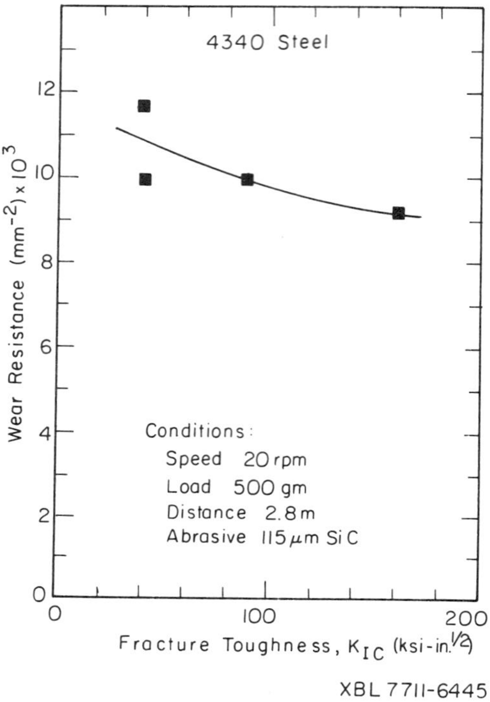

Fig. 8.—Wear resistance of 4340 steel vs. fracture toughness.

hardness and yield strength to be similar since both quantities are measures of the resistance to plastic flow in a metal. No single correlation could be seen as a function of the fracture toughness for the 7075 Al. It appears from Figure 5 that the overaged and underaged structures each has a different relation between fracture toughness and wear. The wear resistance of the 4340 steel also has a correlation with the fracture toughness. It is interesting to note that the wear resistance decreases with increasing fracture toughness, the opposite effect of what one would suppose. The implications of this require further study. The difference in hardening mechanism and microstructure between the two types of alloys is significant and could account for the fracture toughness correlation difference.

Figure 9 shows the wear resistance of the 7075 Al and 4340 steel as a function of hardness. From this figure it can be seen that the general effect on the wear resistance due to increases in hardness as a result of heat treatment for the 7075 Al is remarkably similar to that for the 4340 steel. This effect is consistent with that reported by Kruschov for heat treated steels. While the wear resistance does increase with hardness, the amount of increase is much less than for the same increase in hardness in pure metals. In the case of the 7075 Al, doubling hardness resulted in an increase

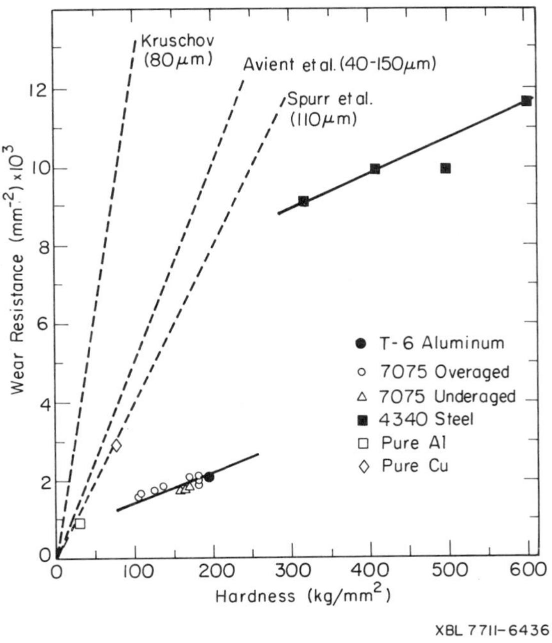

Fig. 9.—Comparison of wear resistance of 7075 Al and 4340 steel.

in the wear resistance of ~ 35%. For the 4340 steel, doubling the hardness increased the wear resistance ~ 30%.

Using Rabinowicz's model and the values of the abrasive wear coefficient which he calculated using the data which other workers reported for many materials (Table III), lines representing this data are also included in Figure 9. For comparing to the data of this investigation, the line representing Spurr should be better since the abrasive particle size is above the critical size for no size effect, and the size is approximately the same as the 115 μm particles used in this investigation. From these lines, it is seen that for the many materials studied by Spurr a two-fold increase in hardness would result in an increase in the wear resistance of approximately the same magnitude.

TABLE III.—ABRASIVE WEAR CONSTANTS REPORTED BY RABINOWICZ[2]

Investigator/Material Studied	Abrasive Size (μm)	k(x10^{-3})
Kruschov/Several	80	24
Avient, et al./Several	40-150	120
Spurr, et al./Several	110	150

SEM Examination of Wear Specimens

Aghan and Samuels[14] compared abrasive polishing (fine abrasive smaller than 8 μm) to abrasion (abrasives larger than 10 μm). Among their conclusions was that the mechanisms of material removal were similar, only the scale differed. From Figures 10 and 11, it is easy to justify expanding this conclusion to two-body abrasion in general. The specimens in both figures were 7075-T6 Al tested under identical conditions with the exception of the abrasive particle size. The specimen in Figure 10 was abraded against ~ 14 μm particles while that in Figure 11 was abraded against ~ 115 μm abrasive. In Figures 10 and 11 the horizontal lines represent wear tracks which start at the juncture of those lines and the vertical polishing marks.

Figure 12 shows a 4340 steel specimen whose hardness is approximately three times that of the 7075 Al. The test conditions for the 4340 sample were the same as that of the 7075 Al specimen in Figure 11. From close examination of these three figures it can be seen that the basic material removal process is the same. Returning for the moment to Rabinowicz's simple model, decreasing the abrasive particle size would indicate the same material removal mechanism but on a smaller scale. This is consistent with the observations of Figures 10 and 11. Increasing the hardness of the base metal would mean that the abrasive particle should penetrate less into the base metal for the same loading of the pin. This also indicates that the same type of material removal mechanism occurs, but on a smaller scale.

Rabinowicz's model assumes that when material is displaced, it is cleanly cut away, with no chips remaining and no plastic deformation. However, from Figure 13, at least three very obvious mechanisms of material displacement can be identified. These are: plastic flow, cutting and gouging. The plastic flow mechanism is indicated by the raised ridges found at the edges of the grooves. The presence of cutting chips indicates that a cutting process occurs and the large pit in the lower portion of the photo testifies that a gouging mechanism is responsible for at least some of the material removal.

Figure 14 is of a specimen abraded under the same conditions as that of Figure 13 with the exception that the load was doubled to 100 gm. Both of these photographs were taken at the trailing edge of the specimen. Because of the location it is probable that the gouging which is seen occurs when abrasive grains are fractured or pulled out of the paper, embedded in the softer material and pulled out by succeeding particles which are still attached to the paper. Figures 10, 11 and 12 were taken at the leading edge of the specimen and do not show signs of gouging. Generally, gouging does not appear until somewhat past the leading edge.

Plastic flow at first appears to be the only mechanism present if one examines the leading edge of a specimen. Figure 15 shows the leading edge of the same specimen as shown in Figure 14. Any cutting chips which may be formed at the leading edge are either removed or continued to the trailing edge.

It is also interesting to compare Figure 15 with Figure 11. Both specimens were abraded under the same conditions with the exception that the load of that in Figure 15 is twice that of Figure 11. Figure 15 shows much overlapping of the grooves, possibly due to the increased number of abrasive grains contacting the surface and actually deforming it. Since there are more contacting grains over the same area, the grooves are much closer and overlap to a much greater extent than for lighter loads.

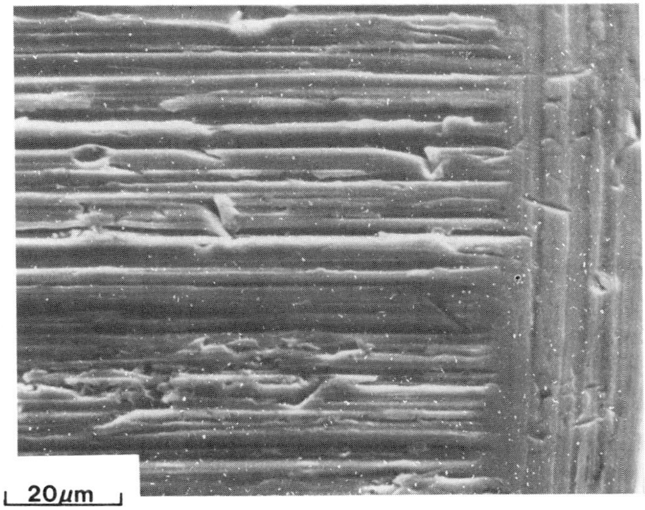

Fig. 10.—7075-T6 Al abraded on ~ 14 μm abrasive.

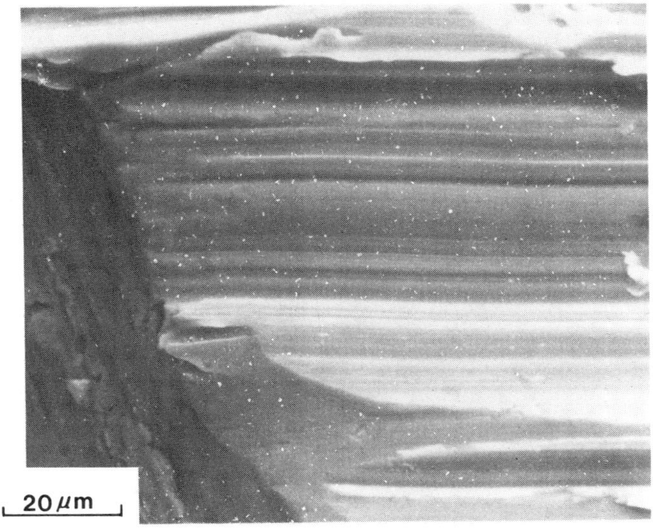

Fig. 11.—7075-T6 Al abraded on ~ 115 μm abrasive.

Fig. 12.—4340 steel abraded on ~ 115 μm abrasive.

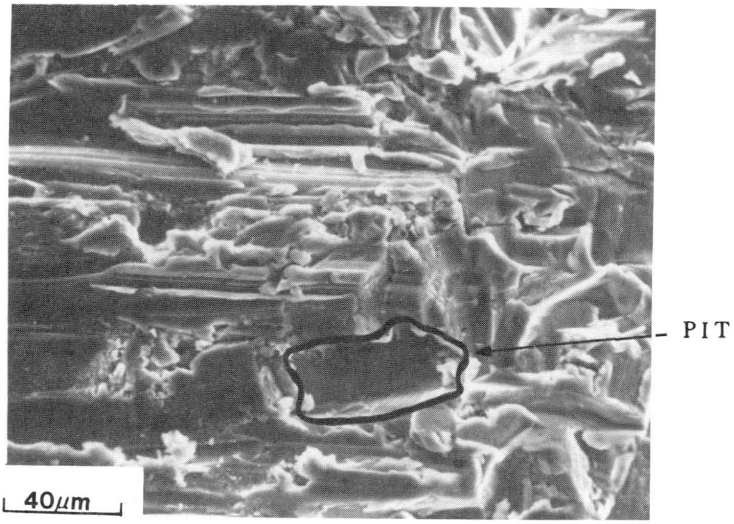

Fig. 13.—Evidence of cutting, plastic flow and gouging on 7075-T6 Al specimen loaded by 500 gm.

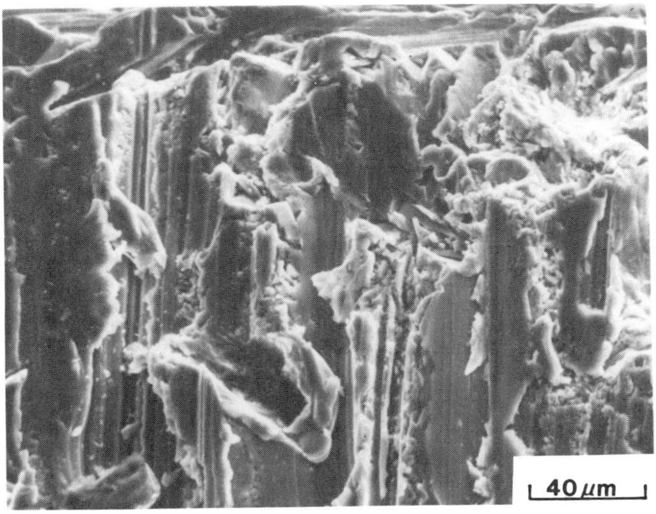

Fig. 14.—7075-T6 Al (loaded by 1000 gm) at the trailing edge.

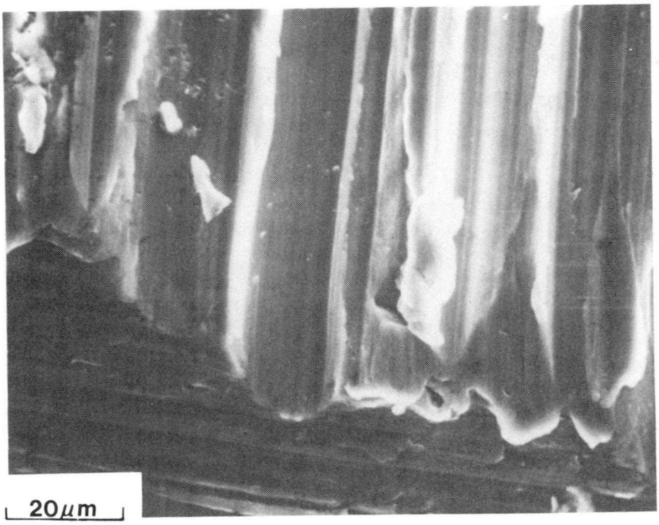

Fig. 15.—7075-T6 Al (loaded by 1000 gm) at the leading edge.

Figures 16 and 17 are of a 7075 Al specimen which has been cross-sectioned transverse to the abrading direction. The figures show cutting, as seen in the V-shaped grooves, and ploughing where the material from one groove has been extruded over the top of a neighboring groove.

Figure 18 shows a *SiC* abrasive particle which is embedded in the specimen. Johnson[15] has studied this and found that some embedded particles may lie beneath the surface. If the embedded particle protrudes above the surface, it will be exposed to other abrasive grains still attached to the paper. As it is pushed along, large pits may form (Figure 19). Although this may appear to be strictly a three-body wear phenomena, this type of interaction is bound to occur in two-body abrasion as the abrasive particles are fractured or pulled loose.

Figures 20 and 21 show another form of degradation caused by loose particles. In this case, what probably occurred is that the loose abrasive particle did not become embedded in the soft material as in the case where deep gouges were formed. Rather, the loose abrasive particle rolled along causing a large plastically distorted region. The abrasive grain size for this specimen was ~ 50 μm, which is approximately the width of the pit. By comparing the size of this pit to the surrounding grooves, one can see that only the tip of an abrasive particle contacts the base metal in a two-body situation.

Figures 22 and 23 are examples of the chip cutting mechanism which Mulhearn and Samuels concluded was the most effective removal process.

Modelling of Two-Body Wear

The development of an analytical model for two-body abrasive wear requires that a mechanism (or mechanisms) be defined that is conducive to analytical representation such as the concept of micro-cutting developed by Finnie[13] to account for low impingement angle erosion behavior. The mechanisms of material movement that have been observed in two-body wear are:

1. ploughing or extrusion of material ahead of and alongside the track of abrasive particles through the material being abraded (Figure 15);

2. micro-machining of chips of material from the surface of the abraded material (Figure 23);

3. embedment of abrasive particles that have separated from the abrasive disc in the surface of the abraded material protecting the material under it (Figure 18);

4. gouging out of vulnerable partially embedded abrasive particles by collision with subsequent particles still bonded to the abrasive disc, removing abraded material with them to form pits (Figures 13 and 19);

5. abraded material removed upstream adhering to downstream surfaces, adding to its surface and protecting it (Figure 14);

6. adhered upstream material taking welded-on downstream materials with it when it is abraded off the surface (Figures 13 and 14).

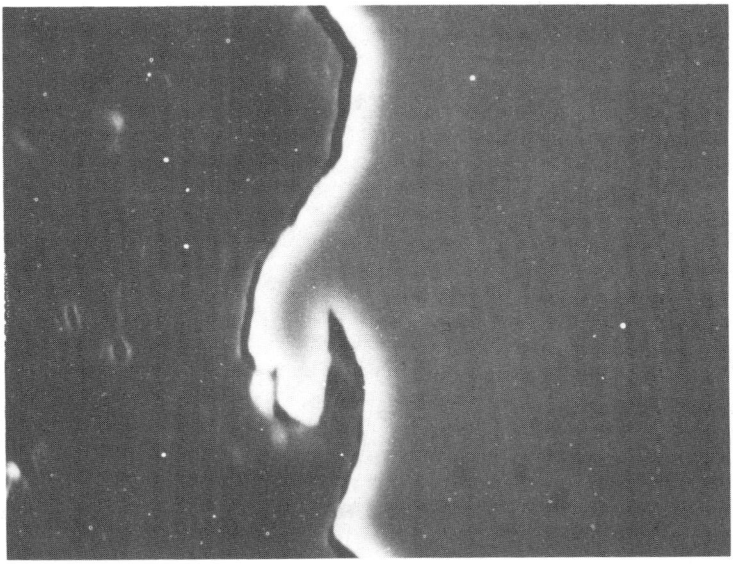

Fig. 16.—7075-T6 Al cross-sectioned transverse to abrading direction.

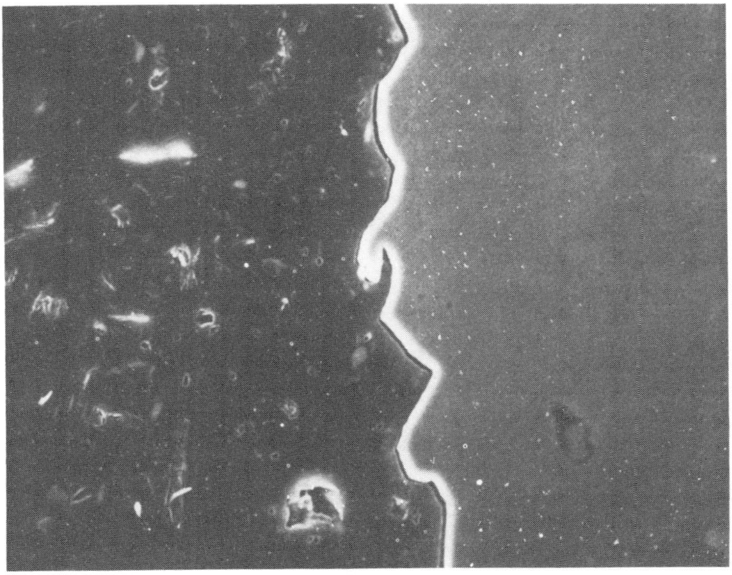

Fig. 17.—7075-T6 Al cross-sectioned transverse to abrading direction.

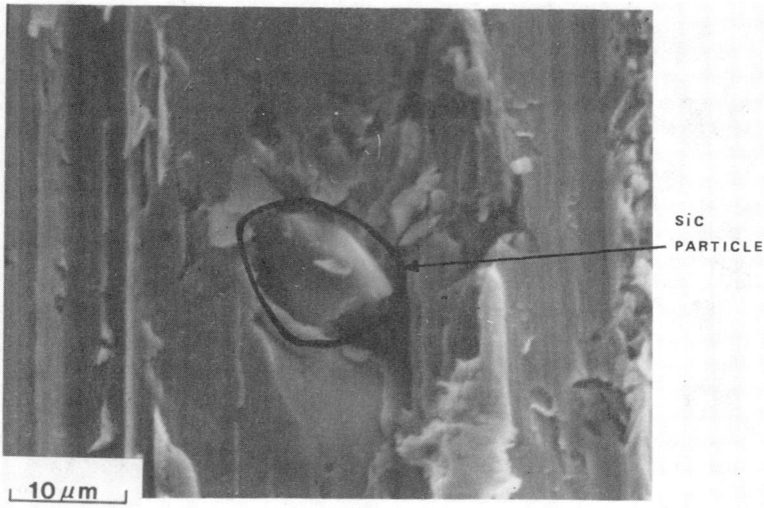

Fig. 18.—Embedded SiC abrasive particle.

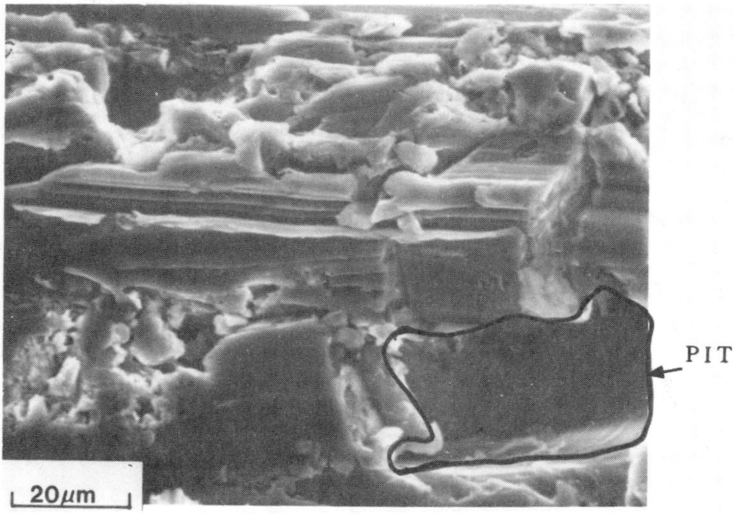

Fig. 19.—Large pit formed in 7075-T6 Al.

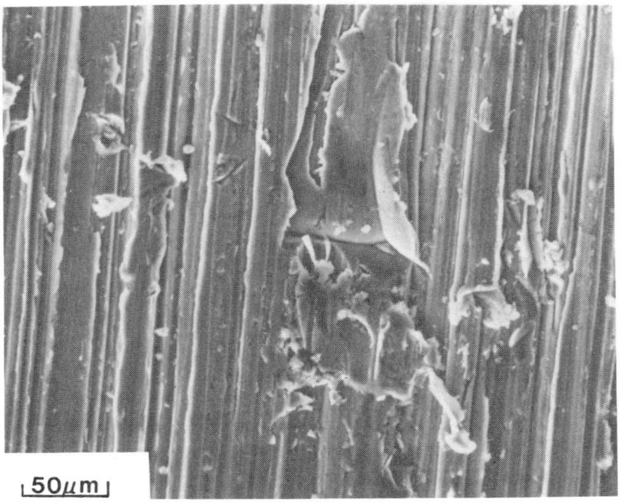

Fig. 20.—Evidence of plastic flow in gouging.

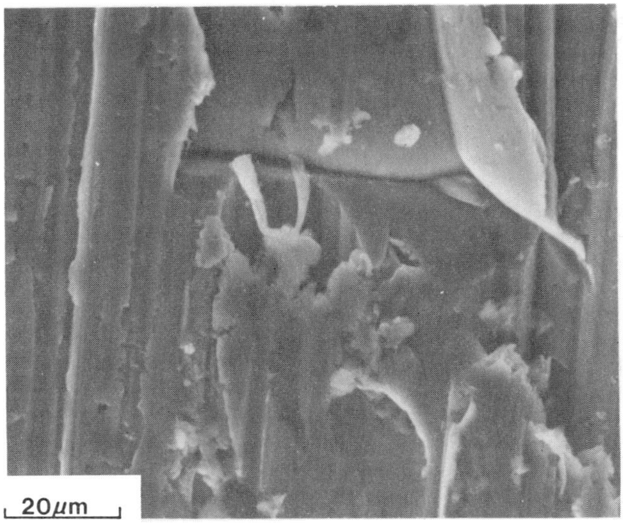

Fig. 21.—Larger magnification of Figure 20.

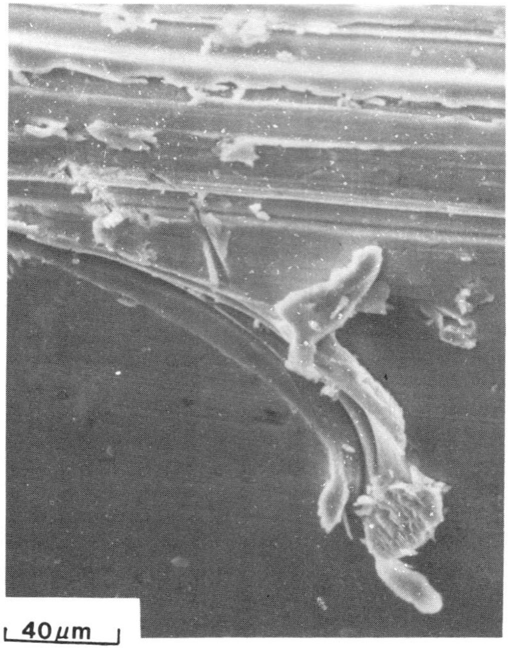

Fig. 22.—Cutting chip formation on 7075-T6 Al.

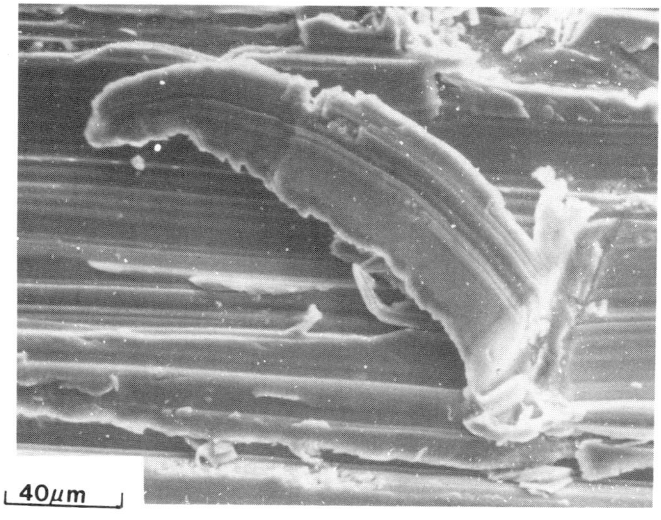

Fig. 23.—Example of cutting chip in 7075-T6 aluminum.

It is difficult to establish a single mechanism or even a number of mechanisms that can be addressed analytically that would adequately represent all of the above mechanisms.

As an alternative, one can look for material property variations that directly relate to wear rate as has been done in this investigation, i.e., hardness, yield strength, and fracture toughness; and establish a simple dependence ratio embodying the material property that shows a regular relationship to the wear rate. The use of a coefficient of proportionality completes the analytical representation of the relationship. Kruschov did this for hardness and the modulus of elasticity E in reference,[1] developing simple and effective equations for each property for relating pure or annealed metal behavior to wear resistance.

$$\varepsilon = bH \quad \text{and} \quad \varepsilon = C_1 E^{1.3}$$

where ε = *relative wear resistance*
b = *coefficient of proportionality for* H
H = *indentation hardness*
C_1 = *coefficient for proportionality for* E
E = *modulus of elasticity*

Similar expressions can be used with the data obtained in this investigation.

It has been observed in solid particle erosion of metals[16] that the wear rate could be related to the relative proportion and distribution of hard second phase material in the softer matrix phase. The greater the concentration of the hard phase, resulting in more continuous ductile matrix phase, the more resistant to metal loss the alloy was. In that investigation *Al-4.75%Cu* alloy was used; its microstructure is somewhat similar to that of 7075 Al used in this study.

From the curve of fracture toughness vs. wear resistance of 7075 Al (Figure 5) it can be seen that there are two distinct behaviors, one for the overaged material where the harder phase precipitate is consolidating with increasing aging temperature, and the other for the underaged material, which is progressing towards maximum distribution of the hard phase with a minimum of soft, matrix phase between hard precipitate material. The overaged material has the largest amount of ductile matrix phase between the hard precipitate. It can be seen in Figure 5 that differences in the fracture toughness of the overaged material cause significantly greater changes in the wear resistance than do changes in the underaged material.

CONCLUSIONS

As a means of estimating the abrasive wear resistance of a material by its mechanical properties, hardness and yield strength seem to offer the best correlations. However, the Vicker's hardness test is a measure of flow properties of a material at low strains whereas abrasion involves high strains. For annealed, pure, FCC metals the stress-strain curves should be similar; therefore, one would expect a good correlation between wear resistance and the Vicker's hardness. For cold-worked or heat-treated materials such as 4340 steel, the stress-strain curves are much flatter than for annealed metals. This implies that the flow pressure at high strains would be overestimated by the Vicker's hardness. Thus, the increase in wear resistance for a given increase in the Vicker's hardness is less for cold-worked or heat-treated materials than for pure annealed metals. This is seen as the flatter slope in Figure 9.

The 4340 steel demonstrates a level of correlation of the wear resistance to the fracture toughness. The 7075 Al has two distinct relationships between wear resistance and fracture toughness, each related to the type of microstructure.

The observed mechanisms of abrasive wear are plastic flow, cutting and gouging. All three mechanisms are likely to exist during two-body abrasion; however, one may dominate. The extent of gouging depends on the fracture properties of the abrasive and the manner in which it is fixed to the abrasive surface in the two-body condition. Whether cutting or plastic flow occurs depends largely on the angle of attack of the abrasive particles. Therefore, for situations where the abrasive is very much harder than the material being tested, the dominant wear mechanism probably depends primarily on the characteristics of the abrasive (i.e. angle of attack, fracture behavior, bonding technique).

The development of an analytical model for two-body abrasive wear will require the integration of several mechanisms of material movement-removal with pertinent material properties and morphologies.

ACKNOWLEDGEMENTS

The guidance and support received from Professor Iain Finnie during the course of this investigation is gratefully acknowledged.

Appreciation is also expressed for assistance received from V.K. Pujari, M.S. Bhat, R.M. Horn, Don Krieger, Lee Johnson and Richard Lindberg in carrying out the investigation.

This work was supported by the Division of Materials Sciences, Office of Basic Energy Sciences, U.S. Department of Energy.

REFERENCES

1. Kruschov, M.M., *Wear,* 1974, p. 69.
2. Rabinowicz, E., "Friction and Wear of Materials," John Wiley and Sons, New York, 1965.
3. Mulhearn, T.O. and Samuels, L.E., *Wear,* Vol. 5, 1962, p. 478.
4. Bowden, F.P. and Tabor, D., "The Friction and Lubrication of Solids," Oxford University Press, London, 1964.
5. Larsen-Basse, J., *Wear,* Vol. 19, 1972, p. 27.
6. Larsen-Basse, J., *Wear,* Vol. 11, 1968, p. 213.
7. Date, S.W. and Malkin, S., *Wear,* Vol. 40, 1976, p. 223.
8. Goddard, J. and Wilman, H., *Wear,* Vol. 5, 1962, p. 114.
9. Pujari, V.K., Ph.D. Thesis, University of California, Berkeley.
10. Shenai, K.P., M.S. Thesis, University of California, Berkeley.
11. Horn, R.M. and Ritchie, R.O., paper submitted to *Metallurgical Transactions.*
12. Throop, J.F. and Miller, G.A., in "Achievement of High Fatigue Resistance in Metals and Alloys," ASTM STP 467, American Society for Testing and Materials, 1970, p. 157.
13. Finnie, I., *Wear,* Vol. 19, 1972, p. 81.
14. Aghan, R.L. and Samuels, L.E., *Wear,* Vol. 16, 1970, p. 293.
15. Johnson, R.W., *Wear,* Vol. 16, 1970, p. 351.
16. Brass, L.M., M.S. Thesis, University of California, Berkeley, 1977.

DISCUSSION

S. A. KARPE, DTNSRDC, Annapolis: When you increase fracture toughness, the wear resistance is going down. Do you have a plot of the friction coefficient versus fracture toughness?

A. V. LEVY: This is just the first data. I really wanted to show you that we were trying to get the mechanical property relationships, but the thing that intrigues me most was the microstructure. Presently I do not have any data other than what I showed.

KARPE: What do you plan to do with the model? Are you going to specify the material properties?

LEVY: I hope to relate the model to alloy design or use it as predictive means; but it appears to me that it is much more complicated than erosion which is complicated enough.

QUESTIONER: I am wondering what you can do with this model. I do not know anyone who would deliberately operate without lubrication. You are ignoring the effect of delamination, chemical reaction, and so on. All these numerous chemicals we have include phosphates, esters, silicones, and each of them operates differently. They are not all the same and I am just wondering what relationship this is going to have in actual practice.

LEVY: Now that I have gotten this far and seen how complex it is I have the same question. That is why I made the comment that the challenge is to develop a theory. It is a great challenge to find the way to be able to predict from a set of operating conditions what kind of wear rates we would have, and to know a combination of properties of a metal that would produce very low wear rates and to be able to design a metal to reach that goal.

J. LARSEN-BASSE, University of Hawaii: On the point of lubrication, I do feel that there are a lot of unlubricated situations such as mining, cement handling, and so forth, where abrasive wear is a serious problem.

QUESTIONER: We mold dry silicon carbides at pressures of several tons per in^2. We observe abrasion on steel and tungsten carbide under these conditions. I would very much like to have a model or a basis for selecting materials to improve abrasive wear conditions under these circumstances.

LEVY: I am attempting to do this, but I am also looking for help.

R. A. DISKIVICH, General Motors Research Laboratories: On one of those figures that you showed (and I also seen in my own work), there is an apparent imbedment of a silicon carbide particle in the workpiece. Would you make a conjecture as to whether or not that particle was imbedded in a void, or was it already there, or did it create a void of its own?

LEVY: The one that is appearing in that picture looks like it was imbedded in a void that was already there because the void dimensions are different in size and configuration from the particle that is caught in it.

C. S. YUST, Oak Ridge, National Laboratory: In some of your early ploughing and cutting photomicrographs of the abraded surface, you made a distinction that is not clear to me. Would you define it more precisely?

LEVY: In the interpretation that I give, "cutting" is the smooth portion of the wear track; the "ploughing" is the material that was lifted on either side of an area that looks smooth down the center of it.

R. A. BURTON, Northwestern University: Professor Shaw published a paper some time ago on grinding. According to him when the tool makes an angle larger than a critical angle (I can't remember the exact value, but something like 60 degrees) with the surface, we would have a little chip going up. Below that angle we would get just ploughing and the material is pushed aside.

P. A. ENGEL, IBM: I would like to know about the abrasive particle size effect. People have found threshhold mechanisms in the past.

LEVY: I have shown that in the paper. We get a lower abrasion rate (or erosion rate) depending on the size of the particle.

UNIFYING FACTORS IN EROSION AND WEAR OF MACHINE ELEMENTS

C. S. Yust

ABSTRACT

Erosion is a complex phenomenon which may involve the simultaneous action of other basic wear mechanisms, i.e., abrasion, adhesion, fracture, and corrosion. Although these wear mechanisms appear to be diverse, they have many common features. A study of these common features in various wear systems may lead to the necessary background for the development of the unified wear theories as suggested by Barwell.

This paper presents evidence for the operation of various wear mechanisms during the erosion of ceramics. Abrasion is recognized by the ploughed paths and tool-like cutting action of impinging particles. Adhesion is demonstrated by the evidence of bonding of the impinging particle and the surface, and the removal of surface material with the particle. Fracture is observed over a wide range of impingement angles and evidence is presented which demonstrates that the fracture mechanism varies with erosion angle. Plastic deformation of the surface and the bulk material surrounding the impact site is also observed in these results.

The features of erosion discussed in this work confirm that identical mechanisms are operative in erosion and in the wear of many machine elements. This may contribute to the development of a unified theory of wear.

INTRODUCTION

It is indeed important, as suggested by Barwell, that research on wear "be pursued as part of a consistent body of theory of wear rather than as isolated studies of particular machines and situations."[1] It is in this spirit of seeking common factors in wear processes that this paper presents some observations of erosion in ceramics and discusses these observations in terms of basic wear mechanisms. In particular, the damage introduced at single particle impacts on several ceramics will be presented and discussed in terms of the operative wear mechanism(s).

Typically, erosive wear rate is measured in terms of mass loss per unit time or mass loss per unit mass of eroding particles directed at the target. Wear rates measured in this way not only yield data on the specific wear resistance of given materials under particular circumstances, but also have been used to infer the mechanism(s) by which material is removed from the damaged surface. By means of observations of typical impact pits rough estimates can be made of wear depth and sliding distance in a single erosion incident. Assuming a worn layer 0.25 μm deep (2500 Å) corresponding to a sliding distance of ∿25 μm, a typical wear intensity (i.e., the ratio of the height of the worn layer to the sliding distance) derived is 10^{-2}. This corresponds to a wear intensity value of class 9, the maximum shown in Table II, by Barwell.(1) Although the very short sliding distance (as compared to sliding distances in other wear processes) over which an impinging eroding particle acts is responsible for such large wear intensity numbers, this estimate does demonstrate the magnitude of the wear problem in erosion. This wear intensity evaluation can also be applied to the total process of erosion which is the summation of many individual class 9 impact events. The only application cited by Barwell as having a comparable wear intensity is that of excavator shovel teeth, which are subject to the same type of impact damage as that experienced in erosion circumstances. By this kind of intensity measure, the significance of erosion as a wear process is readily recognized.

THE COMMON MECHANISMS

All the wear mechanisms cited by Barwell as contributors to the degradation of machine elements may also be active in the erosion process. This has been noted by others; for example, by Rabinowicz(2) in his description of erosion as a complex wear phenomenon which may involve simultaneous abrasion, adhesion, fracture, and corrosion contributions. The delamination and fatigue mechanisms mentioned by Professor Barwell can be categorized as fracture mechanisms.

The operation of the various wear mechanisms during erosion can be observed by close investigation of specific impact sites. The examples which will be presented and discussed are taken from the author's recent studies of erosion of commercial refractory oxides, selected polycrystalline oxides, and single crystal alumina. The refractories and the polycrystalline specimens were eroded by a steady stream of 240 μm diameter silicon carbide particles carried by argon gas at a velocity of 24 m/s. The single crystal alumina sample was exposed to a single batch of 240 μm diameter silicon carbide particles accelerated by a compressed air gun to a velocity of 55 m/s. All the specimens were eroded at room temperature, and corrosion was not a factor in any of these experiments. The impact sites were examined by scanning electron microscopy.

A significant proportion of the material removal in erosion occurs by abrasion, although particles which cause erosion damage are not always harder than the target material, as in the case of liquid droplet erosion. As examples of hard particle abrasive damage, consider the damage produced by *SiC* particles on polycrystalline *MgO* (Figure 1) and on glass (Figure 2). The photomicrographs show a distinct ploughing path due to the passage over the surface of a sharp corner of an impinging particle which acted as a cutting tool. The specific mechanism responsible for material removal must be comparable to that involved in machining processes; note the apparent "chip" of glass in Figure 2. The abrasion process is active at moderate to low impingement angles.

Particle impact sites are often observed to be regions of extensive surface shear. Hutchings et al.(3) have demonstrated that spherical particles impacting a metal surface obliquely will shear the surface layers, and that adhesion of the surfaces is significant in this process. Figure 3 is

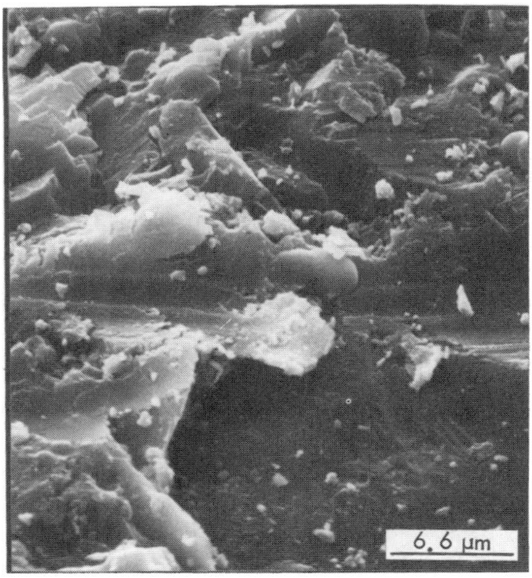

Fig. 1.—Polycrystalline MgO eroded at 470° C and 45° impingement angle by 240 µm diameter silicon carbide particles. The particle velocity was 24 m/sec. Note gouging of the central part of the area shown.

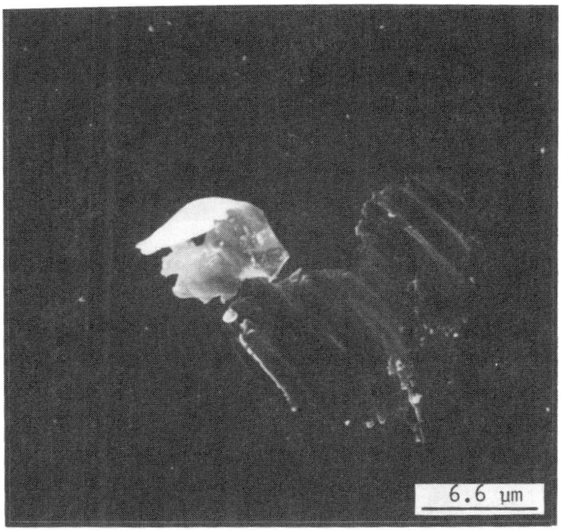

Fig. 2.—Glass eroded at room temperature by 100 µm diameter silicon carbide particles. The particle velocity was 55 m/sec; the impingement angle was 15°. Note the chip machined from the surface.

evidence for the operation of such a mechanism in crystalline nonmetals, in this instance in polycrystalline mullite ($3Al_2O_3 \cdot 2SiO_2$). The surface shear may also be accompanied by deformation of the volume of material surrounding the impact site, as shown by transmission electron microscopy studies of erosion impact pits in alumina (Al_2O_3) by Hockey et al.[4]

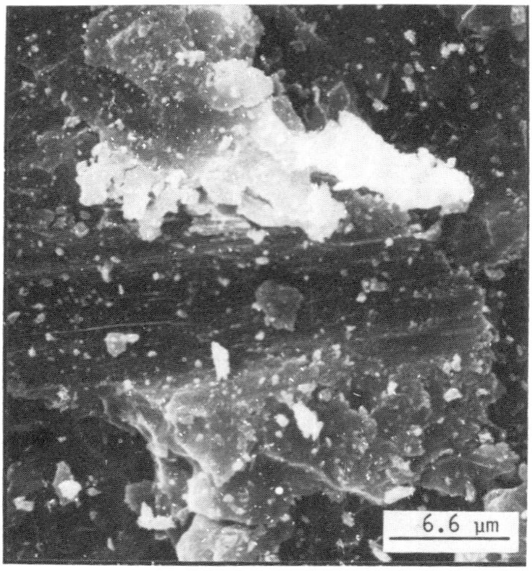

Fig. 3.—Surface shear deformation at an impact site in mullite ($3Al_2O_3 \cdot 2SiO_2$). Eroded at room temperature by 240 µm diameter silicon carbide particles at 25° impingement angle. The particle velocity was 24 m/sec.

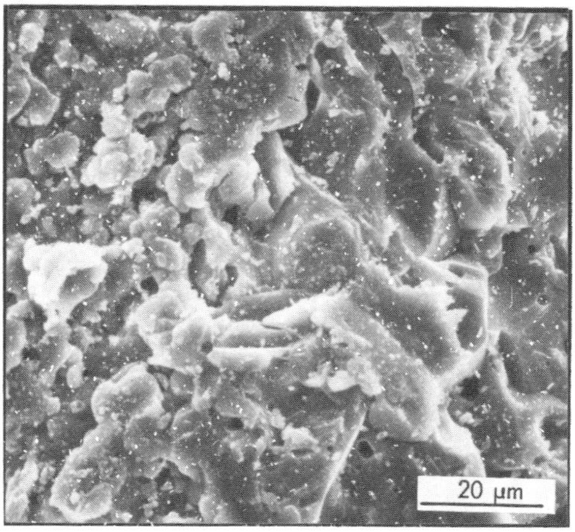

Fig. 4.—The eroded surface of a high alumina (Al_2O_3) brick consists principally of fracture surface. The sample was eroded at 90° impingement angle at room temperature by 240 µm diameter silicon carbide particles. The particle velocity was 24 m/sec.

Fracture is a commonly observed and anticipated feature at particle impact sites in brittle, ceramic materials, Figure 4. At high impingement angles, i.e. 75-90°, particle edges act as surface indenters. Tensile components of the stress field generated beneath the indentation lead to cracking around and beneath the impact surface.[5] Fracture is not limited to high impingement angles, however, and is not always due to the indentation stress field alone. Cracking often accompanies low-angle impacts, and, in addition to the indentation stress field, may be due to secondary deformation processes. (Evidence for deformation associated fracture processes will be presented in the next section.)

Finally, corrosion is always a possible contributor to the erosion process, just as it is to the wear of machine elements, although it was not a factor in the experiments described here. Erosion-corrosion is a very complex wear process, which has not been closely studied, and about which much remains to be learned.

It is clear, therefore, that erosive wear and the wear of machine elements occur by the operation of common mechanisms and our understanding of both depend on a more detailed comprehension of the specific mechanisms.

IMPACT SITE OBSERVATIONS

It is well known that the rate at which material is removed from a surface by erosion varies with the angle at which the eroding particles strike the surface. In ideally ductile materials the rate is maximum at 20-30° impact; in this angular range the magnitudes of the shear and normal forces maximize the removal of material by shear processes. In ideally brittle materials, the wear rate is greatest at 90° impact, when the shear force is zero, the normal force component is maximized, and the energy available for crack formation and propagation is at a maximum. Most materials exhibit behavior intermediate to these extremes. Oxide refractories subject to erosion, for example, do not behave in a completely brittle manner, despite the limited ductility of the bulk form. The eroded surface of alumina and mullite refractories show evidence of both brittle and plastic responses to erosion impacts, Figure 5. In some areas of this photomicrograph fracture surface is evident, while the central portion shows the characteristic surface shear associated with plastic deformation processes.

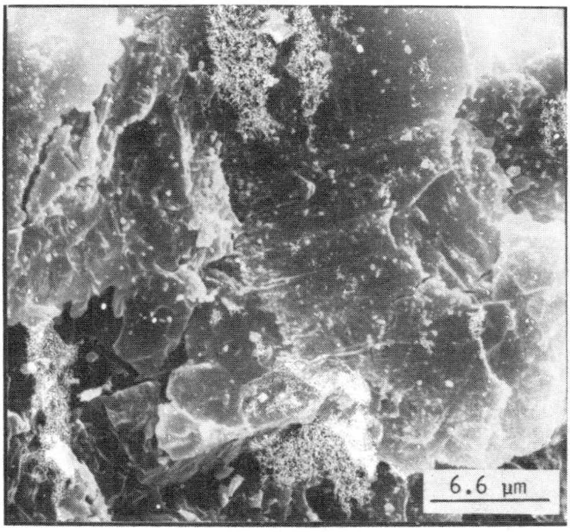

Fig. 5.—Both shear deformation (central area) and brittle fracture are evident on the eroded surface of a mullite refractory brick. The sample was eroded at room temperature by 240 μm diameter silicon carbide particles at 25° impingement angle. The particle velocity was 24 m/sec.

Fig. 6.—The eroded surface of an inhomogeneous refractory body becomes very irregular, and the particle impact angle depends on the specific impact site.

The simultaneous operation of both brittle fracture and plastic deformation in oxide refractories is due, in part, to the irregularity of the surface shape developed during erosion of these inhomogeneous bodies. Typically, the refractory body is composed of dense aggregate particles of varying size bonded together with a second phase. As the body is worn by the impinging particles, the bonding material may be selectively removed and/or void spaces between the aggregate particles opened to the surface, Figure 6. Subsequently, the eroding particles will strike the surface at a wide range of impact angles. Since the operative damage mechanism varies with impact angle, each of the anticipated mechanisms may be expected to operate concurrently over the face of the refractory.

The generation of heat is a critical factor in most tribological systems, and such systems can be thermally characterized by means of a "slide sweep ratio."[1] Application of this concept to the single particle impact shows that heat generation at an impact site occurs almost entirely in the eroding surface, very little temperature rise occurs in the colliding particle. If enough energy is exchanged at a sufficiently high rate in a body of relatively low-thermal conductivity, local temperatures at the impact site may become quite elevated. Such an "adiabatic shear" process has been described and analyzed by Recht.[6] Our recent observations[7] have shown that in alumina single crystals eroded by *SiC* particles at room temperature, melting occurs at particle impact sites, Figure 7. Further evidence of local melting has been detected in single crystal alumina, Figure 8, where ejected molten material is apparent at the edge of the impact pit. Electron microbeam analysis confirms that the content of the ejected material is aluminum (and possibly species of atomic number lower than 12). In these cases, the particle and the eroding surface were in close contact for the duration of the impact event, and material was removed by the colliding particle. Although this dynamic situation produces only a transient contact, the nature of the contact and the removal of material by the particle satisfies the criteria expressed by Barwell for classification of this material removal process as adhesion.

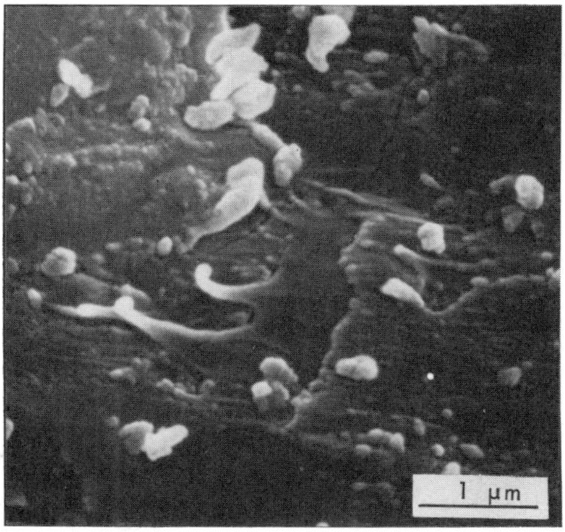

Fig. 7.—Melting has occurred at this particle impact site on mullite. Eroded at room temperature by 240 μm diameter silicon carbide particles. The particle velocity was 24 m/sec and the impingement angle was 25°.

Fracture may also be observed at low impingement angle (<30°) impact sites in single crystal alumina. Under the low-contact angle conditions fracture appears to develop from two mechanisms, one of which is related to the high impact angle mechanism. (At high impact angles, the stress field generated in the eroded body at the point of impact is very similar to that resulting from a single point indentation.) A particle striking the surface at a low angle develops a surface stress state which is better simulated by a moving indenter. The stress field of the moving indenter contains amplified tensile stresses in the path ahead of the moving object which are oriented in such a way that cracks will be initiated perpendicular to the course of travel, Figure 9(a), and these cracks may propagate into the surrounding material, Figure 9(b).

A second mechanism for low-angle fracture is associated with the deformation of the lattice volume adjacent to the impact. The deformation must occur by the motion of dislocations on lattice planes and if the crystal structure does not provide a sufficient number of independent slip systems to permit a general shape change, the insufficiently accommodated slip may lead to severe stress concentrations which result in crack generation. An example of this behavior is shown in Figure 10. In this instance, slip has occurred on the basal planes of the hexagonal alumina lattice. The stress and temperature conditions have permitted slip on only this single plane. As a consequence, cracks are created to accommodate the inhomogeneous plastic displacement.

TOPICS FOR ADDITIONAL RESEARCH

The observation of local melting at impact sites in alumina and other ceramics suggests further topics for study. The shear process which precedes melting proceeds by means of a deformation mechanism which is not well defined. This is especially true for the strain rates imposed by typical particle impacts, which are several orders of magnitude higher than those usually considered in deformation studies. The influence of compositional,

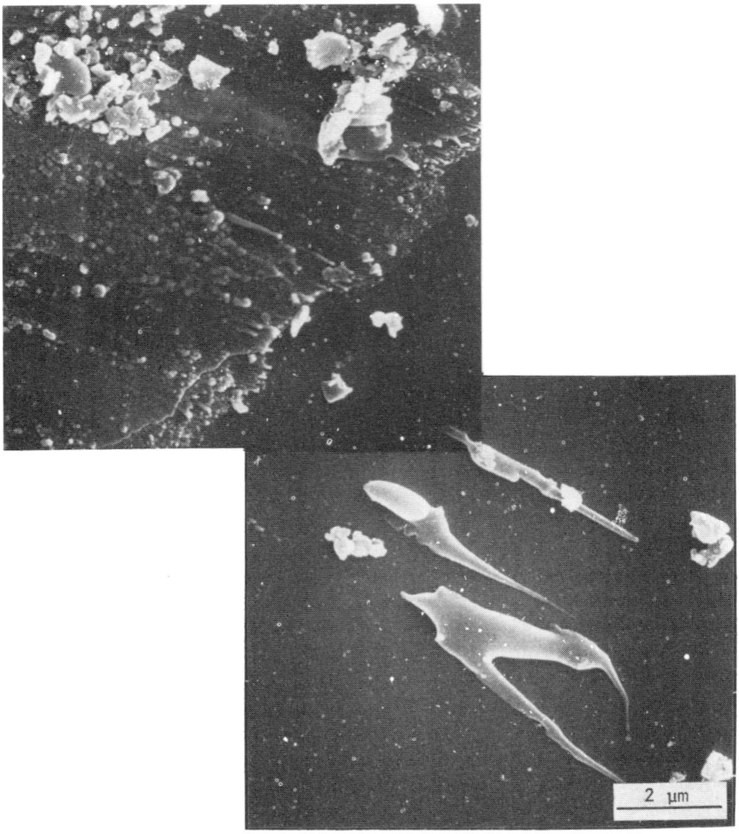

Fig. 8.—A particle impact site in single crystal alumina. Melting was induced at the impact site and caused the ejection of molten alumina at the exit side of the pit. Eroded at room temperature by 240 µm diameter particles; particle velocity was 55 m/sec; impingement angle was 15°.

and consequently microstructural, variations on the erosion process is also a largely unanswered question. It is evident that careful and extensive study will be required to resolve all the unknowns concerning specific erosion mechanisms under various erosion conditions.

CONCLUSIONS

Erosion is a complex process, which has not as yet been fully analyzed. It can, however, be treated as a wear phenomenon having many features in common with other wear processes. The results discussed in this paper support Barwell's suggestion that progress will come only by the development of broad and consistent theories of wear.

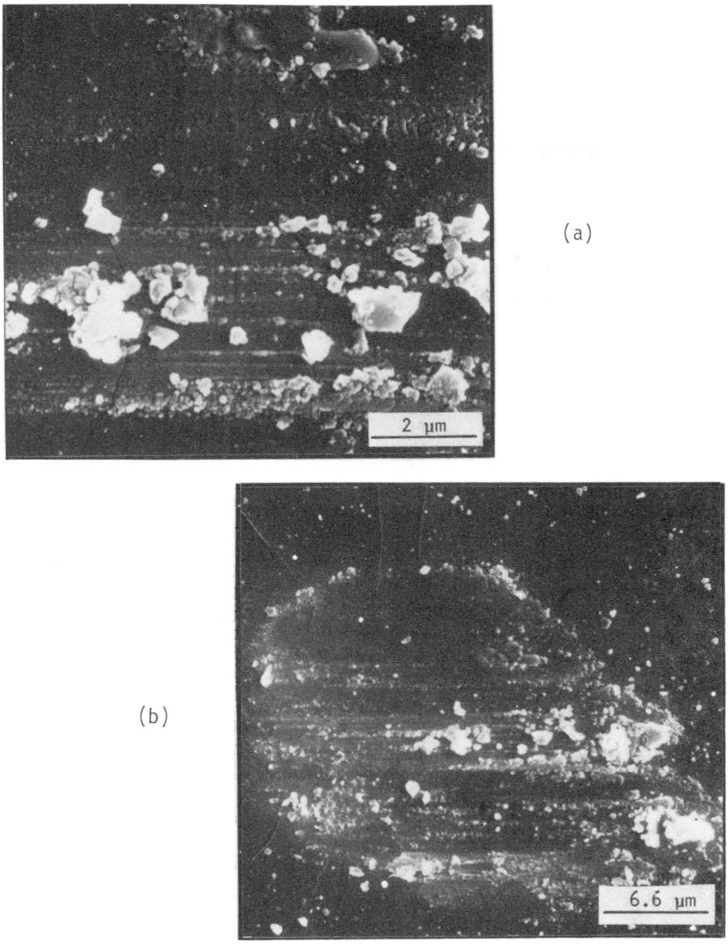

Fig. 9.—A particle impact site in single crystal alumina. In (a) the origin of the cracks in the sheared material is evident; in (b) the cracks are seen extending into the surrounding material. Sample eroded at room temperature by 240 μm diameter silicon carbide particles. The particle velocity was 55 m/sec; the impingement angle was 15°.

ACKNOWLEDGEMENT

This research was sponsored by the Division of Materials Sciences, U.S. Department of Energy, under contract W-7405-eng-26 with the Union Carbide Corporation.

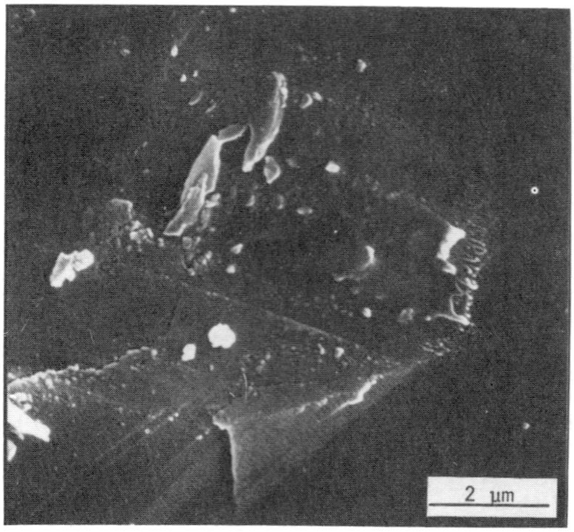

Fig. 10.—Slip traces adjacent to a particle impact pit in single crystal alumina. Slip on only one set of planes leads to crack generation. Note melting at exit edge of pit. Eroded at room temperature by 240 µm diameter particles. The particle velocity was 55 m/sec and the impingement angle was 15°.

REFERENCES

1. Barwell, F.T., These Proceedings.
2. Rabinowicz, E., *Materials Science and Engineering,* Vol. 25, 1976, p. 23.
3. Hutchings, I.M., Winter, R.E. and Field, J.E., *Royal Society of London. Proceedings.* Series A, Vol. 348, 1976, p. 379.
4. Hockey, B.J., Wiederhorn, S.M. and Johnson, H., "Fracture Mechanics of Ceramics," Plenum Press, New York, 1978, p. 379.
5. Lawn, B. and Wilshaw, R., *Journal of Materials Science,* Vol. 10, 1975, p. 1049.
6. Recht, R.F., *Journal of Applied Mechanics,* Vol. 86, 1964, p. 189.
7. Yust, C.S. and Crouse, R.S., *Wear,* Vol. 51, 1978, p. 193.
8. Hamilton, G.M. and Goodman, L.E., *Journal of Applied Mechanics,* Vol. 88, 1966, p. 371.

DISCUSSION

N. P. SUH, MIT: We also observed what appears to be melting type phenomenon when we did erosion experiments on aluminum. We saw in some cases an indication that something flew up in the liquid form and we tried to understand what happened. Have you taken any pictures of what is happening underneath the surface?

C. S. YUST: We are planning to do that. That is not a trivial problem. Professor Waterhouse referred to that in his paper about the TEM samples from the fretted samples. We are dealing with rather friable specimens which do not lend themselves to the mechanical processes of grinding, thinning, etc. in preparation for the ion milling which we ultimately want to do.

I am interested in what happens immediately beneath the surface. I think we can deduce some of it in terms of the slip traces, and we have indications that there is clear plastic deformation associated with the impact. We need to be concerned with the strain rate effects also. I do not know whether the deformation processes that we observe here are comparable to what we are accustomed to on the basis of normal strain rate experiments in bulk ceramic samples. I would like to look at that. So I endorse your suggestion that we should investigate the subsurface damage in some detail.

J. J. MECHOLSKY, NRL: I think your work supports previous models on impact erosion of materials in that you can explain chipping, cracks at various angles, etc. Basic indentation models, although modified, are also applicable to machining, scribing, etc. Does the same model work for glasses as well as polycrystalline materials? It is hard to explain slip or dislocations in glasses, but you still get the same process just from the stress field.

YUST: That is right, but we are also talking about the temperature rise at these impact sites. There is nothing that says that a glass that is warmed up will not flow into form as a viscous liquid.

MECHOLSKY: I am not arguing about the flowing.

YUST: As a matter of fact, in the written version of the paper I have included a photo-micrograph of an impact pit on a piece of glass as an example of an abrasion type of mechanism. I think that the particle edge has come along and cut out a piece of glass and left the chip on the surface right at the edge of the impact. I do not know precisely what the mechanism is.

MECHOLSKY: I was referring to the shear track.

YUST: I hope to distinguish between abrasion and adhesion. In many instances, these things are not quite clear; one blends into the other.

QUESTIONER: I would like to know what the relationship is between the single impact event and multi-particle erosion mechanism where you dig out the cement from around the aggregate particles and the particles fall out.

YUST: I think the quickest thing to say about it is that we are dealing with two different phases there and each behaves differently. Two different materials are eroded there. In my view one erodes much faster than the other, and that leaves the more erosion-resistant phase behind. In some circumstances some of those phases will be removed as bulk pieces.

SAME QUESTIONER: Are single events primarily in the binder?

YUST: Yes. After all that is coming out first. When the hard phase protrudes, it will be eroded away.

P. A. ENGLE, IBM: Do you have wear rates for the ceramics?

YUST: Yes, I have some wear rate numbers for the commercial ceramics.

ENGEL: As a function of velocity, and so forth?

YUST: Yes, it is limited data. I have some data for mullite, alumina and commercial refractories.

FLUID PARTICULATE CONTAMINATION, COMPONENT WEAR AND PERFORMANCE

W. M. Needelman

ABSTRACT

The relationship between wear of component surfaces and the service life performance of components is discussed. Since fluid particulate contamination is present in all lubrication, pneumatic and hydraulic systems, it can lead to component wear in these systems by a variety of mechanisms. The manner by which suspended particle contamination related wear influences component performance is then examined, with several illustrative examples cited from the literature.

INTRODUCTION

The major wear mechanisms responsible for deterioration of component surfaces in mechanical systems have been discussed previously[1,2]. These surveys are in basic agreement with the review by Barwell[3]. A Wear Chart found to summarize this information conveniently is included in Reference 2, a simplified version of the chart is shown in Figure 1. It is the purpose of this presentation to study the various modes of wear in fluid wetted components from the viewpoint of component operation, as well as to examine an operational parameter having a major influence on component wear: fluid particulate contamination.

FLUID PARTICULATE CONTAMINATION AND WEAR

The effects of particulate contamination on the wear of fluid-wetted component surfaces were highlighted in previous reviews[1,2]. Some of the major features of these surveys are condensed here.

1. There are three classes of particle interaction producing mechanical wear. In three-body wear a loose particle makes simultaneous contact with two moving surfaces. Two-body wear involves an abrasive particle embedded in or adsorbed on one surface making contact with a second component surface. In particle erosion a particle carried in the fluid impacts at high velocity against a component surface.

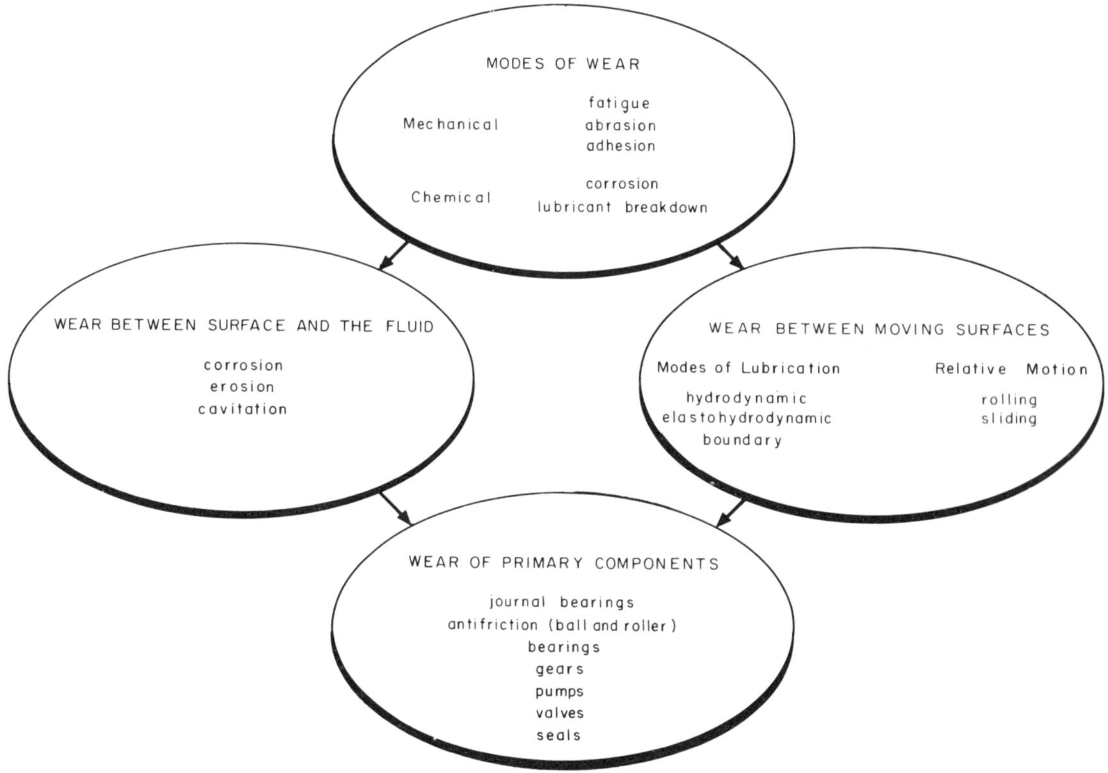

Fig. 1.—Wear of component surfaces.

2. Two mechanisms of particle produced wear have been established. The first is severe abrasive wear. It is the surface deterioration associated with abrasion, and passes under various labels including polishing, cutting, lapping, and abrasive rubbing. It is characterized by a rigid particle penetrating a softer component surface, cutting away material on a single pass (Figure 2). The wear rate of this mechanism is proportional to the number of particles

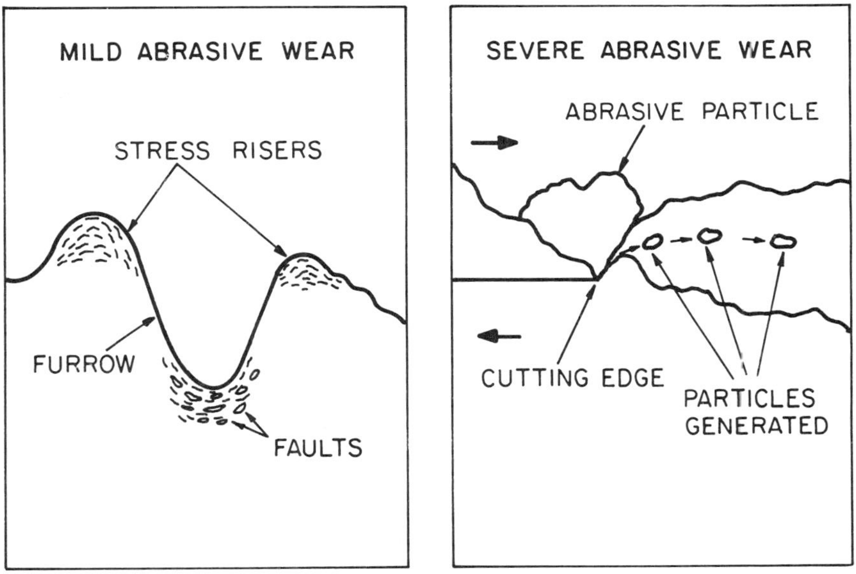

Fig. 2.—Two mechanisms of abrasive wear.

coming into contact with the surface, particle hardness, and particle size (the width of the wear scar). The second wear mechanism occurs when particles plough through or dent a surface producing substantial amounts of plastic deformation, without the removal of surface material. The large number of dislocations generated makes a significant contribution toward reducing the binding of the surface material, actually a form of accelerated surface material fatigue. Furthermore, the raised material along the edges of the furrow or pit are subjected to increased stresses on subsequent contact with another surface, causing surface fatigue of the adjacent component material.

3. Both mechanisms can occur in the lubricated contacts between surfaces. Larger particles entering the contact zone between lubricated surfaces will cause more surface damage than smaller particles of similar composition. The wear rate will also increase with particle hardness.

In order for particles to cause surface deterioration in these contacts they must be at least the size of the minimum lubricant film thickness. The surface damage produced by both the severe and the milder wear mechanisms will increase with increasing concentration of suspended particles greater than this minimum lubricant film thickness. As the minimum film thickness decreases, the percentage of the suspended particle population meeting this crition increases.

4. Particle erosion involves the impact of a surface by particles possessing kinetic energies larger than the forces binding the surface material. The rate of erosion increases with the particle kinetic energy. Thus, the rate of wear increases with particle weight, and with the square of the flow velocity of the carrier fluid. The erosion rate also increases with the ratio of particle hardness to component surface hardness.

Soft ductile particles release a pressure pulse into the solid upon impact. This sudden release of energy causes material deformation and the generation of dislocations. This is a case of wear by the surface fatigue mechanism.

In addition to producing dislocations, hard particles can also cut away surface material by the severe wear mechanism. Thus the rate of erosion is maximum when hard particles impact at glancing angles, during which long furrows are cut through the surface material.

COMPONENT PERFORMANCE TIME-LINES

Two types of component failures account for most component losses. A catastrophic failure occurs when a component is operated under intolerable conditions, rapidly causing failure. Examples are complete lubricant starvation of a bearing, or a rock passing through a gear pump. The second type, deterioration failure, is characterized by a period of component operation (service life) during which wear gradually leads to component failure. Only deterioration failures will be discussed here.

Component life may then be considered as a succession of distinct phases of operation. The initial phase usually involves a break-in period. This is typically followed by a phase of normal (satisfactory) performance. The terminating point is component failure. Component failure is defined here as a loss of performance to a level below that considered satisfactory. The criterion of failure may be the inability of a component to complete its required function, the necessity of component removal during scheduled maintenance (overhaul), or perhaps a monitoring device reaching a predesignated output.

The performance behavior of operating components suggests that component performance can be conveniently grouped into two categories. Each category of performance may be represented by the Component Performance-Time lines, as shown in Figure 3. Furthermore, each category of component performance is associated with characteristic types of component wear.

CATEGORY I - PROGRESSIVE PERFORMANCE DETERIORATION

The characteristic feature is component performance deteriorating progressively with operation, Figure 3(a). Wear mechanisms leading to this type of performance decay would be expected to produce a continuous attrition of surface material. Performance declines monotonically with the wear parameters responsible for the wear mechanism. For example, the increase in leakage rate (performance decay) by the removal of material from the metering edge of a servo-valve by hard particle erosion (wear mechanism) would depend on the concentration of suspended hard particles (wear parameter).

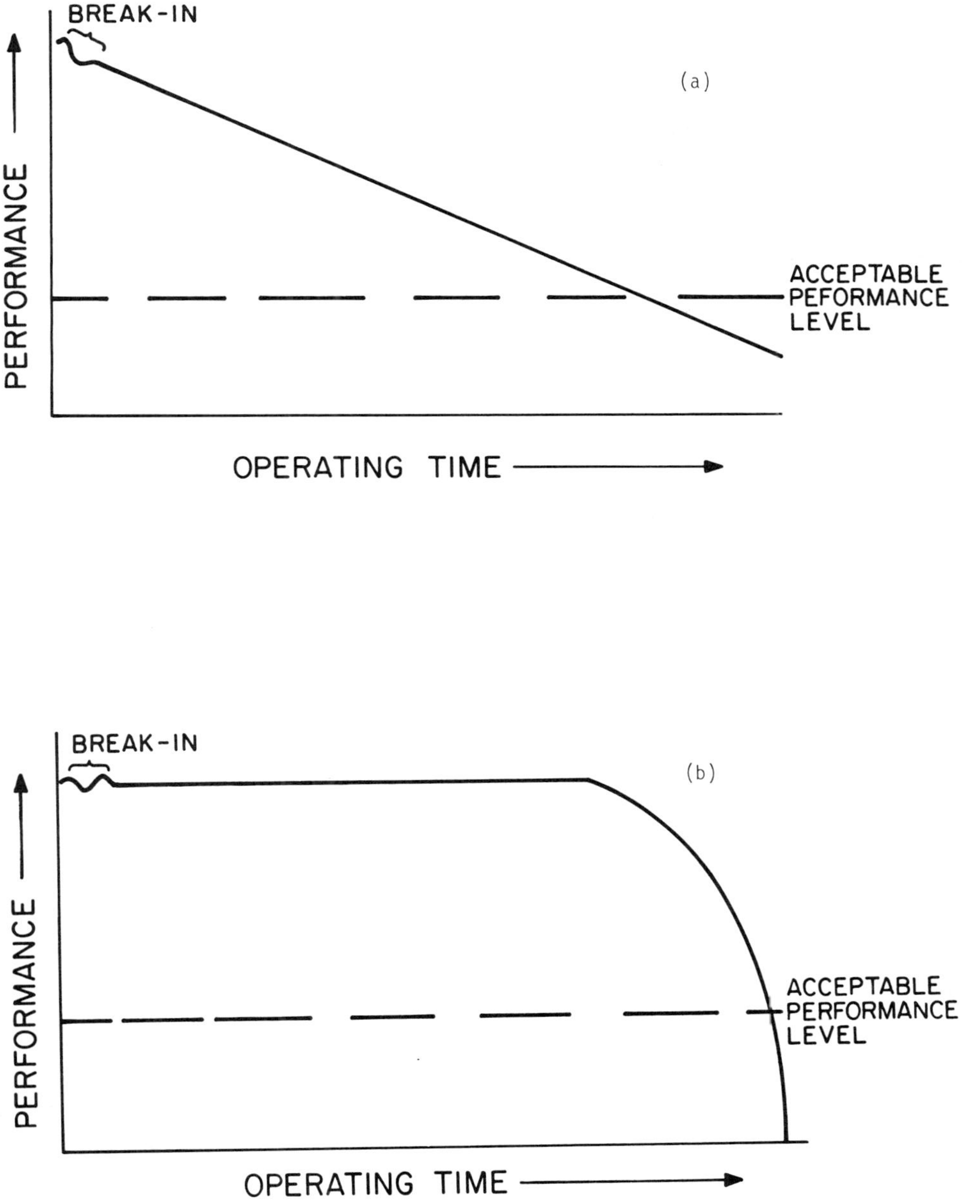

Fig. 3.—Wear of Component Surfaces. (a) Component performance-time lines. (b) Cumulative performance deterioration.

Although performance decay is related to the dominant wear parameters, the functional dependence may be a complex relationship.

Using the following definitions

P = Component performance

x = Wear parameter

d P = Component performance decay

We can formally state that component performance and performance decay are functions of the wear parameter.

$$P = P(x)$$

$$dP = dP(x)$$

A Taylor expansion of component performance in terms of a wear parameter shows to a first approximation that small changes in performance will be linearly related to the wear parameter.

$$P(x) = P(xa) + \left[\frac{d\ P(x)}{dx}\right]_{x=a} dx + \frac{1}{2}\left[\frac{d^2\ P(x)}{dx^2}\right]_{x=a} dx^2 + \ldots$$

Therefore for small changes,

$$dP(x) \approx \left[\frac{d\ P(x)}{dx}\right]_{x=a} dx = constant \cdot \Delta x$$

for any arbitrary point a on the Component Performance-Time line.

CATEGORY II - CUMULATIVE PERFORMANCE DETERIORATION

The essential feature is component performance showing negligible decay over an extended period of operation, followed by a phase of rapid decline in performance until failure. This is shown by the Component Performance-Time line of Figure 3(b).

Surface wear under these conditions will likely occur by the creation of surface and subsurface defects during operation. When these defects have accumulated to a critical level, gross and rapid surface deterioration occurs leading to component failure. The wear parameters responsible are those processes leading to surface and subsurface defects.

Since only negligible changes in performance are observed during Phase 2 of the time-line, performance does not have a meaningful functional relationship to wear parameters. However, component life (operating time until failure) does depend on these wear parameters. Therefore component life can be expressed as a cumulative survival probability having the general form of a Weibull distribution.

$$S = exp\ (-A_i\ m)$$

S = *percent population surviving*

where

m = Weibull slope

A_i = parameters showing life depending on operating variables, material, and wear factors

The component life will be related to the wear parameters through the Weibull parameters m and A_i.

COMPONENT WEAR, PERFORMANCE, AND LIFE

Several examples demonstrating the relationship between both types of Performance-Time lines and various wear mechanisms will now be examined. Because fluid particulate contamination is present in all operating fluid-wetted systems, and may contribute to both categories of component performance behavior, examples will be cited which show the effects of particulate contamination.

EXAMPLES OF CATEGORY I - PROGRESSIVE PERFORMANCE DETERIORATION

A recent study by Fitzsimmons and Clevenger[4] shows several aspects of Category I component performance. The removal of material from the edge of a tapered rolling bearing was measured. As material is removed, the amount of play in the bearing increases, ultimately leading to the need for replacement. The mechanism of surface wear was abrasive removal of material from the roller edge during sliding contact, Figure 4(a). The wear parameter studied was lubricant particulate contamination. When the particulate contamination was maintained at a given level, as is typical of operating systems, material was progressively removed from the bearing edge - characteristic of Category I behavior. Furthermore, the study showed a linear increase in the amount of material removed with an increase in particulate contamination over a wide range of contamination levels, Figure 4(b). As also shown in Figure 4(b), the rate of material removal increased with particle hardness.

Different size ranges of particles did not lead to significant differences in wear rate[5]. This was probably because most particles in any of the size ranges examined would have to be ground down before they could enter the edge lubricant film of about 0.2 μm (8 μ in.). A parameter that was found to accelerate edge wear was the concentration of anti-wear additives. It is believed this was a synergistic effect of surface chemical attack by increasing amounts of additives and abrasive attack by particles.

Several other examples of progressive performance decay can be found in a study by Black[6] on hydraulic servo-valves. One case involves an increase in spool leakage. The wear mechanism was abrasive removal of material from the sliding spool surfaces analogous to the bearing edge wear previously discussed. The wear parameter was particulate contamination in the 0-10 micrometer size range.

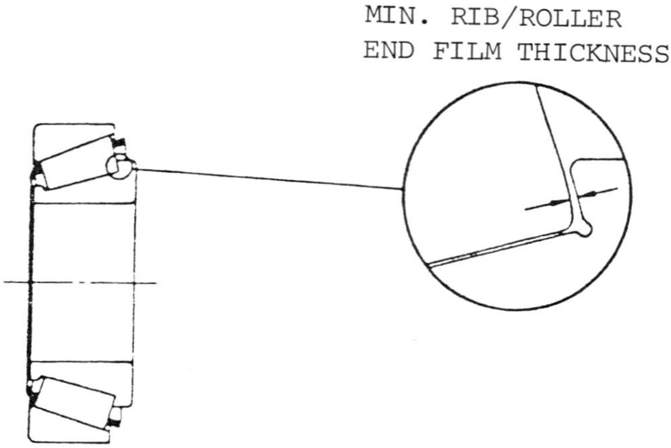

4a.

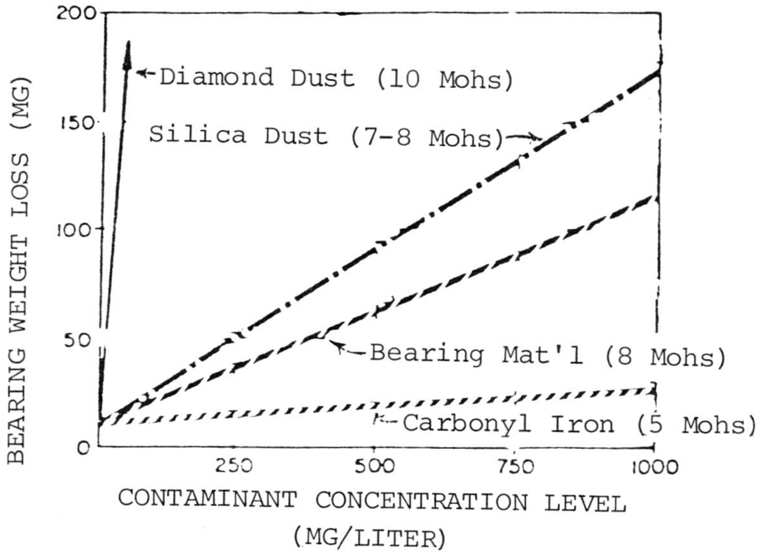

4b.

Fig. 4.—Particle contamination related roller bearing edge wear.
(a) Tapered roller bearing showing rib/roller lubricated contact.
(b) Relationship of bearing wear to contaminant particle concentration and hardness.

Another case examined in this study was the increase in null leakage flow due to contaminant particle erosion of metering orifices. The wear parameter was the size of contaminant particles. Figure 5 shows that the performance declined more rapidly when larger particles were present in the contaminated hydraulic fluid. This is expected since the kinetic energy

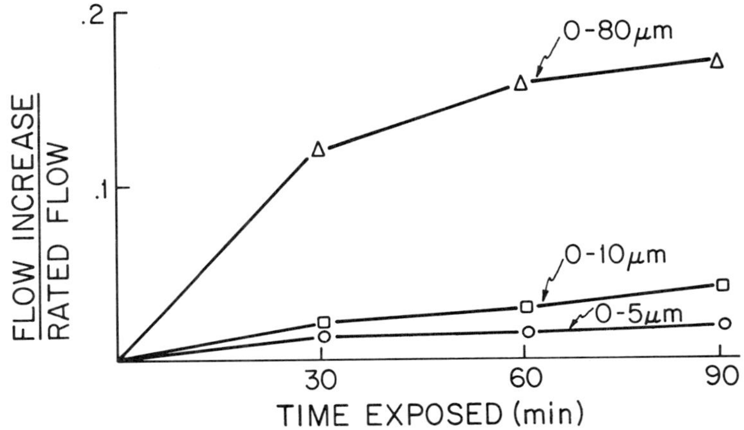

Fig. 5.—Hydraulic servovalve static spool erosion test.

released on impact by different size particles of the same composition carried at the same velocity in a fluid stream is linearly proportional to particle mass.

As a final example of Category I behavior, some results from a program in progress at the Basic Fluid Power Research Program at Oklahoma State University will be briefly described. Hydraulic pump performance is measured as the output flow at a prescribed pressure. When operated with high concentrations (300 mg/liter) of suspended hard particles composed of various size cuts of AC Fine Test Dust pump performance is found to decrease rapidly when operated with increasing larger size particles(7). The wear mechanism is probably abrasive removal of material increasing clearances between moving pump surfaces. Tests at lower contamination levels show a progressive loss of pump performance with operating time(8), as shown in Figure 6.

EXAMPLES OF CATEGORY II - CUMULATIVE PERFORMANCE DETERIORATION

Ball bearing fatigue life has the Performance Time-line of Figure 3(b). Negligible performance change occurs for most of the bearing operating life, as may be measured by load bearing ability or vibration monitoring. Near the termination of service life spalling commences to lead rapidly to bearing failure. A recent series of papers(9,10) indicates that surface damage originating from suspended particles entering the elastohydrodynamic film at the contact zone has a major effect on fatigue life under typical operating conditions. In Reference 9 several models are presented which give possible associations between lubricant particulate contamination and rolling contact fatigue life:

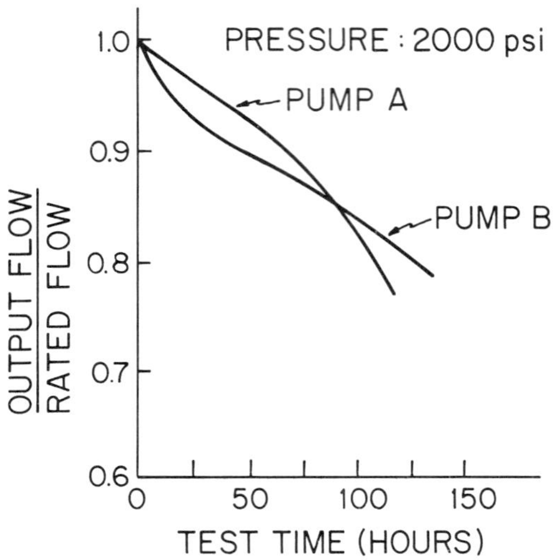

Fig. 6.—Hydraulic gear pump performance loss.

Case I: Pre-existing number of defects on roller bearing contact surface which does not change throughout operating life.

Case II: No initial defects; a pre-existing and constant population of solid contaminants in the lubricant, which produce microscopic surface "dents" throughout operating life.

Case III: No initial defects; no pre-existing contamination; a steady increase in population of contaminant particles which produce microscopic "dents" throughout operating life.

Case I would occur if contaminant particles did not affect fatigue life. Case II and Case III could occur if contaminant particles reduce fatigue life by causing surface damage leading to spalling. The contamination level described in Case II is representative of the equilibrium number of contaminant particles greater than size = x maintained in a system by a filter controlling particles of size x. Case III is typical of particle contaminant build-up when the filter cannot control the particle size range of interest, or when no filter is present. Case I is, by hypothesis, insensitive to the presence of contaminant particles.

Reference 10 compared a number of bearing tests with these models. No good correlations were found with Case I. Good correlations were obtained between Case II and ball bearing L_{10} fatigue life in recirculating filtered lube systems. Grease lubricated (no churning) ball bearing L_{10} life correlated well with Case III.

This strong relationship found to exist between ball bearing fatigue life and lubricant particulate contamination was convincingly demonstrated by a third paper of this study[11]. When deep-groove ball bearings were operated under "ultraclean" lubricant conditions, they lasted for more than 40 times the theoretical L_{10} life without any failures. The tests then had to be terminated because of time limitations.

Another example of Category II behavior, a rapid performance decline after a period of service, is the breakage of gear teeth. One commonly accepted mechanism is the direct (severe) abrasive removal of tooth material by large particles or high asperities in the contact region. Over a period of operation enough material is removed from the tooth to appreciably reduce its load-carrying ability, leading to breakage(13,14).

It is proposed here that several other gear tooth failures can be produced by a related wear mechanism. Smaller and/or softer particles entering the contact region may not be able to remove abrasively the base material of the gear. However there is evidence that particles bridging the lubricant film between opposing gear teeth can break up the protective boundary layer material formed by the action of dissolved oxygen or antiwear additives commonly found in gear oils. For example, two helicopter transmissions operating with different contamination levels were examined after 100 hours testing at high power spectrums. Bevel gears operated at the higher contamination level had complete removal of the gear teeth parco lubrite coating, while the transmission bevel gears operated at the lower contamination level (maintained with three micrometer absolute filtration) had "less than normal wear as evidenced by the incomplete removal of the parco lubrite coating"(15).

Deterioration of the protective boundary layer exposes fresh metal(12), which can lead to microscopic welding, adhesion, and scoring. Furthermore, over an extended period of operation enough tooth material could be lost by this mechanism to cause gear tooth breakage.

CONCLUSIONS

1. Fluid particulate contamination can produce both severe abrasive wear and milder accelerated material fatigue wear of component surfaces.

2. The operating performance of a component may be described in terms of one of two types of Component Performance-Time lines:

   (a) Category I: Progressive Performance Deterioration - progressive loss of performance until failure.

   (b) Category II: Cumulative Performance Deterioration - negligible performance loss followed by a phase of rapidly decreasing performance until failure.

3. Fluid particle contamination can be responsible for loss of component performance by continuously abrading away substantial amounts of component surface material (Category I).

4. Fluid particle contamination can lead to a variety of surface defects responsible for Category II performance behavior, significantly reducing the service life of components in operating systems.

REFERENCES

1. Needelman, W.M., *National Conference on Fluid Power. Proceedings*, Vol. 31, 1977, p. 106.
2. Bishop, F.E. and Needelman, W.M., "The Effects of Fluid Contamination on Component Wear," Field Service Report No. 56, Aircraft Porous Me-

dia, Inc., 1977.
3. Barwell, F.T., These Proceedings.
4. Fitzsimmons, B. and Clevenger, H.D., *American Society of Lubrication Engineers Transactions,* Vol. 20, No. 2, 1977, p. 97.
5. Fitzsimmons, B., Private Communication.
6. Black, R.L., *National Conference on Fluid Power. Proceedings,* Vol. 27, 1973, p. 291.
7. Fitch, E.C., Paper P74-42, 8th Annual Fluid Power Research Conference, 1974.
8. Bensch, L.E. and Tessman, R.K., Paper P76-5, 10th Annual Fluid Power Research Conference, 1976.
9. Tallian, T.E., Paper No. 75-Lub-37, American Society of Mechanical Engineers, 1975.
10. Tallian, T.E., Paper No. 75-Lub-38, American Society of Mechanical Engineers, 1975.
11. Dalal, H., et al., "Final Report on Progression of Surface Damage in Rolling Contact Fatigue," U.S. Navy Contract No. N00014-73-C-0461, 1974.
12. Fowle, T.I., *Lubrication Engineering,* Vol. 32, No. 1, 1976, p. 17.
13. Ku, P.M., *American Society of Lubrication Engineers Transactions,* Vol. 19, No. 3, 1976, p. 296.
14. Will, R.J., *Machine Design,* Vol. 49, No. 20, 1977, p. 116.
15. "Product Improvement Program Item 0012 Increased Service Life Sub-Item 0012A0 Transmission 3-Micron Filter," Final Report, Hughes Tool Company-Aircraft Division, 1972.

DISCUSSION

J. W. WILSON, Mobil Research and Development Corporation: What methods did you use for establishing your particle size distribution?

W. NEEDELMAN: They were done with both optical microscope counts where we looked just at the longest dimension and with an automatic particle size analyzer that used the light scattering property, and therefore we were looking at the cross sectional area. I think that is close enough.

WILSON: We have some serious doubts about the way these are done. I think a lot of people have lost some very early work done by St. George back in 1938-1940 when he gave some very definite particle size ranges for abrasive and non-abrasive silicas, and other abrasives. I think we tend to generalize too much about silica as an abrasive. There are certain sizes and types of silica that are not abrasive. I think we have to be careful in defining the abrasiveness of silica.

NEEDELMAN: Several studies were done that used silicon dioxide. Freshly crushed contaminant was used that had sharp edges and certainly it has been found to produce a lot of wear. There is another point that has to do with the size ranges. We would expect two or three-body interactions to occur when the particle size is at least the thickness of the lubricant film and when the particle is able to get into the lubricant film. There are some situations in which the big particles do not seem to be able to work their way in. In the case of a ship journal bearing which has 25 μm thick film, 10 μm particles may not affect it. High performance components tend to have a tenth of a micron film thickness.

WILSON: From the practical viewpoint, I think we have to be somewhat discerning about what is considered abrasive and what breaks the lubricant film. Particles smaller than the film thickness can be abrasive. There are other types that are very small, but they will never be abrasive.

THE EFFECT OF ENVIRONMENT IN WEAR PROCESSES AND THE MECHANISMS OF FRETTING WEAR

R. B. Waterhouse

ABSTRACT

Barwell's paper deals comprehensively with problems of wear in machine elements. One of the factors which is mentioned several times in his review is the influence of the environment on wear processes. This aspect of wear is considered in more detail in the present paper, or more correctly, its antithesis, i.e., the effect of the mechanical action of wear on the chemical processes occurring on the bearing surfaces. This is related to the specific topics of wear of mechanical seals operating in sea water and high temperature fretting wear of gas turbine components. The wear mechanisms dealt with in Professor Barwell's paper are discussed in relation to fretting wear together with recent evidence of dislocation structures and formation of subsurface voids in the fretted region.

INTRODUCTION

Professor Barwell's paper is a comprehensive review of the occurrence of wear in practical situations, in particular that of machine elements, and forms a valuable compilation of existing knowledge for the practicing engineer. He also reviews various mechanisms that have been proposed to explain the wear process and details the factors which are of importance in contributing to their application. It is apparent that a particular mechanism may be relevant to some situations and not to others. For example, the process which produces metallic debris in the more catastrophic, and thus intolerable, forms of wear such as "wire-wool" failures, is obviously very different from the process occurring in the milder forms of wear in which the debris is a metal oxide. The latter is much more common in practice and represents a slow process of attrition. One area where this can be seen is in the wear of railway lines; the supporting sleepers and ballast are covered with a brown dust. In underground railways the walls of the tunnels and the cables running along them are similarly covered. The production of debris in the form of a compound indicates that the environment has a role to play in the wear process, and this factor is mentioned several times in Barwell's paper particularly relating to corrosive products in lubricants and the effects of lubricants in limiting access of the atmosphere. The purpose of this paper is to look at this aspect of wear particularly in relation to fretting wear, and to offer some evidence of the possible mechanisms of wear based on my own experimental observations.

FRETTING WEAR

Fretting wear is a form of mild wear in which oxide debris is produced. The main difference between it and other wear situations is that one is concerned with microslip, i.e., the relative movement is occurring over only part of the areas in contact. The movement is oscillatory and the amplitudes are relatively small, generally below 25 μm. At higher amplitudes where slip occurs over the whole of the contact area, conditions appear to change, the wear rate increases and can be represented by one of the well known equations relating wear rate to distance travelled and normal load.[1] In the early stages intimate metallic contact undoubtedly occurs resulting in very severe damage which is, however, on a minute local scale. Figure 1 shows a typical example of a ruptured local weld with evidence of metal fracture. Obviously such a defect would be a likely source of initiation

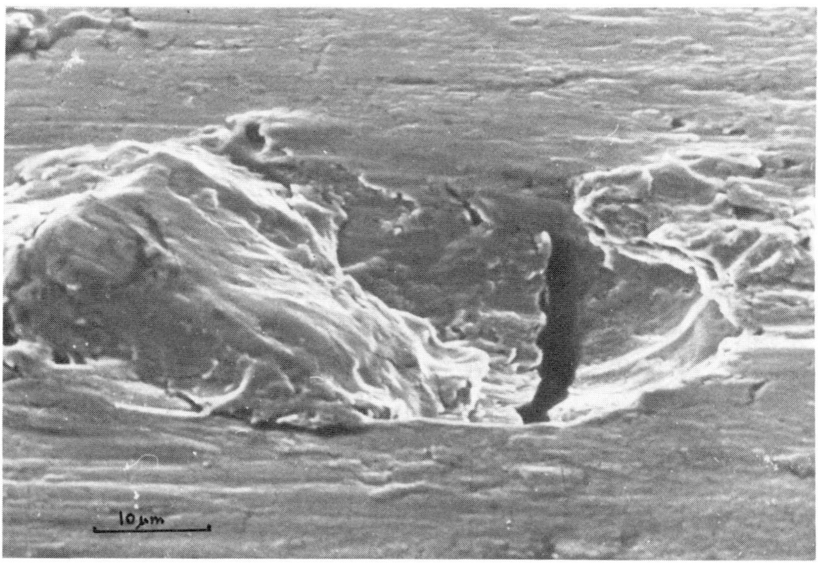

Fig. 1.—Adhesive damage in the early stages of fretting on a 0.20C steel surface--amplitude 12 μm.

for a fatigue crack if, as in this case, the movement were caused by fatigue. There is ample evidence that fatigue cracks are initiated early in the process.[2] However, in circumstances where fretting wear predominates, fatigue failure is usually not a serious problem. Measurements of the progress of fretting wear with number of cycles show that after an initial accelerated wear rate, corresponding to the initial adhesive wear, the process settles down to a steady state.[3] At this stage measurements of electrical resistance across the contact show a very high value indicating that metal-to-metal contact is no longer occurring.[4] It was therefore thought that the continued formation of debris was by abrasion by the existing debris, which being an oxide is invariably harder than the metal from which it originated.

This would argue an accelerating rate of wear unless debris were able to escape from the contact and cause no further damage. In this case a steady wear rate could be achieved. However, examination of debris and damaged surfaces under the scanning electron microscope (SEM) shows that the shape of the debris is plate-like[5] and this is thought to indicate that the removal of material is by delamination. The delamination theory of wear is based on the formation of subsurface dislocations by a shear stress transmitted across the surface. The presence of a compact layer of debris on the surface would still allow this to occur. Further evidence of the role of delamination in fretting wear is presented later in the paper.

ENVIRONMENTAL EFFECTS

Unidirectional Sliding

Since corrosion and tribology are surface phenomena, the occurrence of both of them simultaneously leads to interesting possible interactions; the presence of non-metallic corrosion products will have considerable influence on frictional behavior, while the sliding action is likely to disrupt otherwise protective oxide films and stimulate further chemical attack. The latter effect can be so marked in atmospheric air that it has merited the term tribo-oxidation.[6-8] In experiments where slip occurs between a driven and a braking roll in contact, considerable plastic deformation of the surfaces occurs, resulting in the stretching and rolling of the surface grains of the material which in this case was a variety of steels and an aluminum alloy. Enhanced oxidation occurs by penetration of oxygen down the grain boundaries, which are much increased in area, and the greater reactivity of the deformed material. This is a case of severe wear. Another example is the sliding wear of titanium in various gaseous environments.[9] The wear rate in dry hydrogen was greater than in dry nitrogen or dry air. SEM observation revealed that in hydrogen there was considerable surface cracking, which was attributed to a reduction in the fatigue properties of the surface material due to hydrogen penetration. In this case the environment under the influence of the sliding action has a deleterious effect on the mechanical properties of the material rather than increasing chemical activity. The presence of water vapor can result in the combination of both effects since the more reactive metals such as aluminum and titanium react with water molecules forming an oxide film and liberating hydrogen. Under normal static conditions the oxide film acts as a barrier to hydrogen penetration of the metal but this is not so under sliding conditions. The wear rate of titanium in moist gases is the same irrespective of the nature of the gas and is greater than in dry hydrogen.

Environmental effects are also manifest in aqueous solutions. There is the added interest in the case of an electrolyte of the possibility of varying the electrochemical conditions. Ijzermans[10] found a low wear rate of stainless steel guides in sliding contact with yarn in an aqueous electrolyte under conditions of cathodic protection, but at higher potentials in the passive region the wear rate increased due to continual scraping off of the normally-protective passive oxide film. Similar behavior has been found in the wear of stainless steels sliding in contact with an alumina block in dilute sulphuric acid.[11,12] The corrosion current in the passive region increases directly with the applied load as does the wear, assessed gravimetrically. This is to be expected since the real area of contact is directly proportional to the applied load and hence the area over which the oxide is being scraped off and reformed as the surface repassivates (giving rise to the corrosion current) is also proportional to the applied load.

Secondary effects such as the stirring action which reduces concentration polarization have much less influence. However, in a case where a hydrodynamic fluid film is developed between sliding surfaces the stirring effect assumes greater importance. This has been found to be the case in our work on the corrosive wear of bonded carbide surfaces sliding against graphite in artificial sea water. An annulus of graphite containing a small proportion of white metal was loaded in contact with a disc which was coated with a highly polished nickel bonded tungsten carbide layer in aerated 3% sodium chloride solution. Figure 2 shows polarization curves obtained with the graphite annulus static and rotating at 1500 rpm. The stirring action

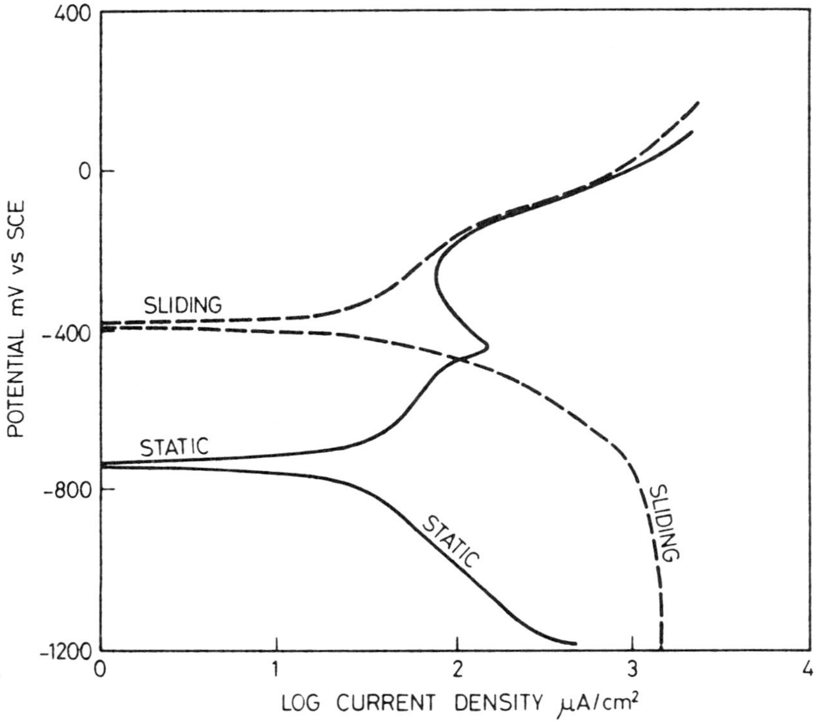

Fig. 2.—Polarization curves of a nickel bonded tungsten carbide surface in contact with a graphite slider in aerated 3% sodium chloride solution-- static and sliding.

caused by the sliding movement results in a large increase in the cathodic current density (the lower parts of the curves) and a consequent rise in the corrosion potential from -730 to -300 mV vs SCE. The upper parts of the anodic curves are identical. This material shows no evidence of passivation. The peak on the static curve at -460 mV arises from the graphite material. A polarization curve constructed with the graphite annulus rotating out of contact with the bonded carbide surface gives essentially the same curve as the sliding curve which indicates that the two surfaces are separated by a hydrodynamic liquid film. The stimulation of the cathode reaction by sliding results in increased pitting corrosion of the binder phase in the bonded carbide.

Fretting

Fretting in most practical engineering situations occurs in air at room temperature, where the major corrosive agents are oxygen and water vapor. Their respective influences on the fretting corrosion of mild steel have been well investigated in the past twenty years.(13) In recent years cases of fretting have been reported in more exotic materials and in more corrosive environments, for example, high temperature fretting in gas turbine engines, fretting between components of orthopaedic implants in the human body, and fretting at high temperature in carbon dioxide in the advanced gas cooled reactor.

There is much evidence that fretting wear and fretting fatigue damage is less in a protective atmosphere or vacuum than in an oxidizing atmosphere.(13) A recent example which illustrates how reactive a fretted surface is given in Figure 3, which shows the damage suffered by a vibrating contact between iron-nickel alloy components after one million cycles at 50 Hz in a nominally pure hydrogen/nitrogen gas mixture. Despite the presence of mere traces of oxygen the formation of oxide in the contact zone is considerable.

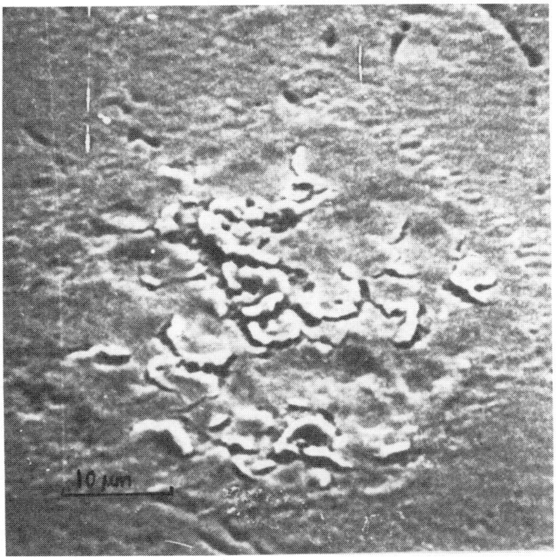

Fig. 3.a.—Fretting damage to an iron-nickel alloy surface after 1 x 10^6 cycles in nominally pure hydrogen/nitrogen gas.

In some cases, however, an oxide film may be formed which has adequate thickness and mechanical properties to withstand the fretting action and act as a protective coating. This was found in the case of fretting mild steel at elevated temperatures in air.(14) The wear rate fell to approximately one tenth of its room temperature value at temperatures above 140°C. The author has found similar behavior in a nickel base alloy, Inconel 718, although in this case the effect of fretting was assessed by its effect on the fatigue strength. The results are shown in Table 1. The conclusion is that the

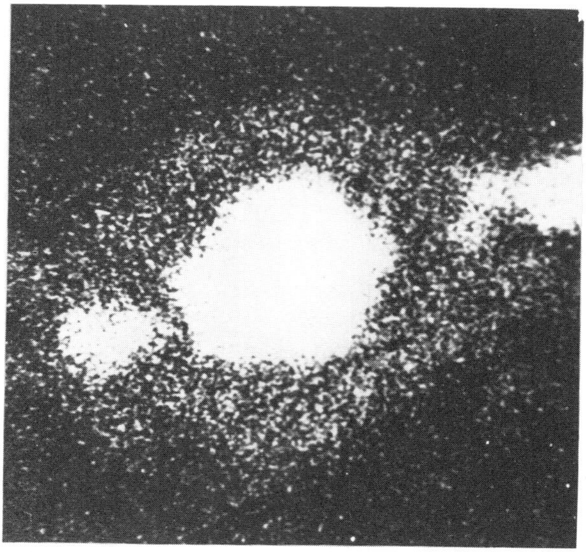

Fig. 3.b.—Electron microprobe picture of the area in Figure 3a showing oxygen distribution.

TABLE I.—FATIGUE STRENGTHS IN MN/m^2 AT 10^7 CYCLES AND STRENGTH REDUCTION FACTORS (SRF) OF INCONEL 718 TESTED WITH A MEAN STRESS OF 550 MN/m^2

Temperature	No fretting	Fretting	SRF
R.T.	275	120	2.29
280°C	325	120	2.71
540°C	325	275	1.29

much improved fatigue strength under fretting conditions at 540°C is due to the protective action of the oxide film. Figure 4 shows that over much of the surface a smooth glaze oxide is formed.

Another possible hazard associated with fretting of normally corrosion-resistant materials at high temperatures is that the disruption of the protective oxide layer leaves the underlying surface depleted in the particular element which confers oxidation resistance and the material is likely to oxidize more rapidly. This has been found to be the case in fretting of austenitic stainless steels in carbon dioxide atmospheres at 600°C.[15] Figure 5a shows the damage to the surface oxide film. Figures 5b and 5c are electron microprobe pictures of the same area showing the chromium and iron distributions respectively. The protective oxide is chromium rich (Cr_2O_3)

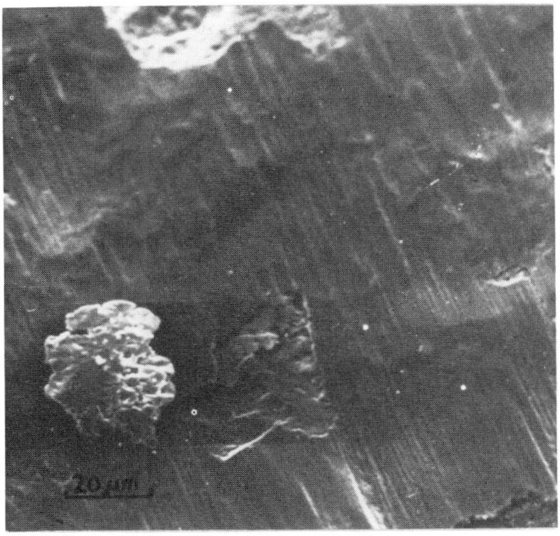

Fig. 4.—Surface of nickel base alloy, Inconel 718, after 1 x 10^6 cycles of fretting at 540°C, showing glaze formation--amplitude 10 μm.

whereas the underlying material is deficient in chromium and has inferior corrosion properties to the material as a whole.

Fretting of metal surfaces in aqueous electrolytes has been shown to result in fall of surface potential which in the case of the reactive metals such as aluminum and titanium can be several hundred millivolts with consequent large increases in corrosion current.[16] This has consequences in the construction of certain orthopaedic devices within the human body. Metal-to-metal contact occurs between screws and bone-plates and in some all-metal hip replacements. The change in potential which occurs when the austenitic stainless steel 316L, much used in orthopaedic applications, is fretted against itself in simulated body fluid, is recorded in Figure 6. This increase in activity accelerates the removal of material frcm the surface and may initiate a fatigue crack.

In many designs of an artificial hip, the femoral component is cemented into the femur with polymethylmethacrylate bone cement. The bond between the cement and the metal stem is not particularly strong and if debonding occurs local movement causes rubbing marks on the stem. Such areas in the scanning electron microscope show an increased thickness of oxide which is cracked and delaminating, Figure 7. Samples of 316L stainless steel were fretted against bone cement in simulated body fluid and the potential drop measured. The results are shown in Figure 8. It was found that the potential dropped on switching on, but soon recovered during continued fretting, in contrast to the metal-on-metal case. These specimens had a similar appearance to Figure 7 in the scanning electron microscope. The rubbing action of the bone cement appears to stimulate the growth of oxide leading to the rise in potential without producing severe surface damage. It had little effect on the fatigue strength of the material. However, if the growth of oxide becomes sufficiently thick that it starts to crack and flake

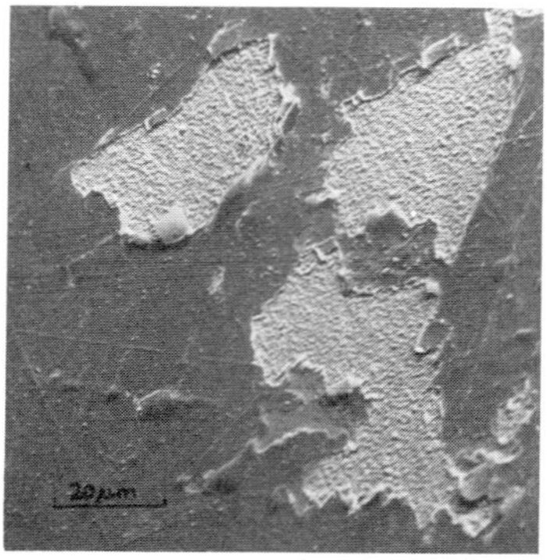

Fig. 5.a.—Surface damage on 18Cr-10Ni stainless steel after 1.8 x 10^6 cycles of fretting in carbon dioxide at 600°C - amplitude 25 μm.

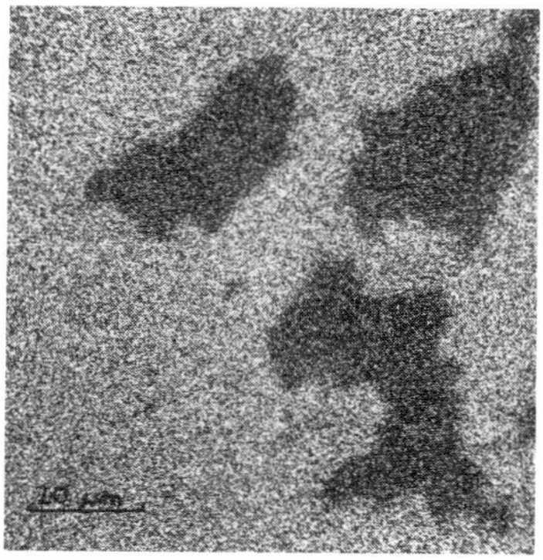

Fig. 5.b.—Electron microprobe picture of area in Figure 5a showing chromium distribution.

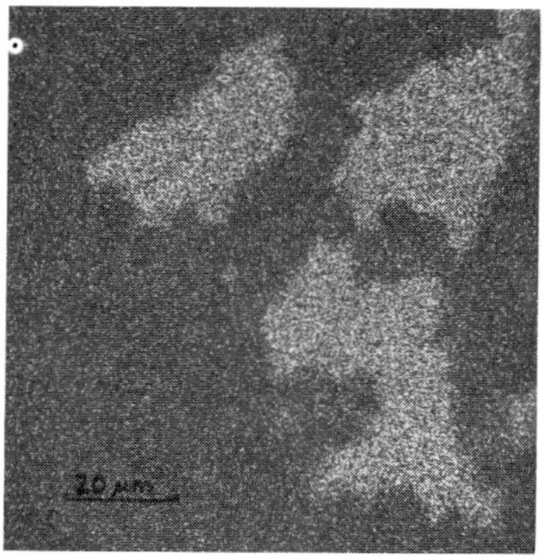

Fig. 5.c.—Electron microprobe picture of area in Figure 5a showing iron distribution.

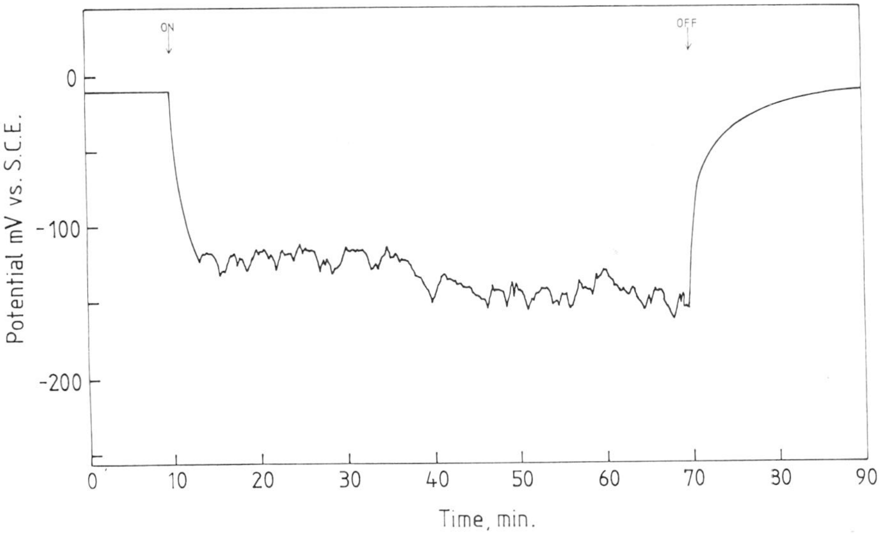

Fig. 6.–Potential change on fretting 316L stainless steel against itself in Hanks' solution - amplitude 25 µm.

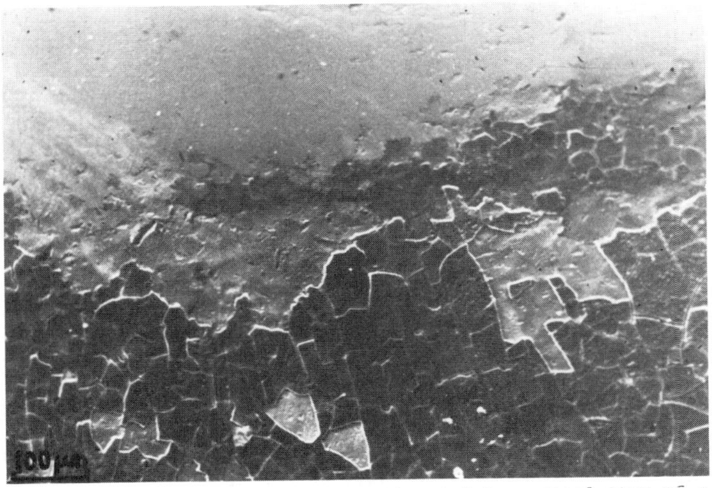

Fig. 7.–SEM picture of a rubbed spot on the stainless steel stem of a femoral component of a total hip replacement which was removed from the patient after four years.

off, it may result in an increased wear rate. It should be said that the device shown in Figure 7 had been in the patient for four years and surface breakdown appeared to be just beginning.

The conclusion to be drawn from this section is that unidirectional

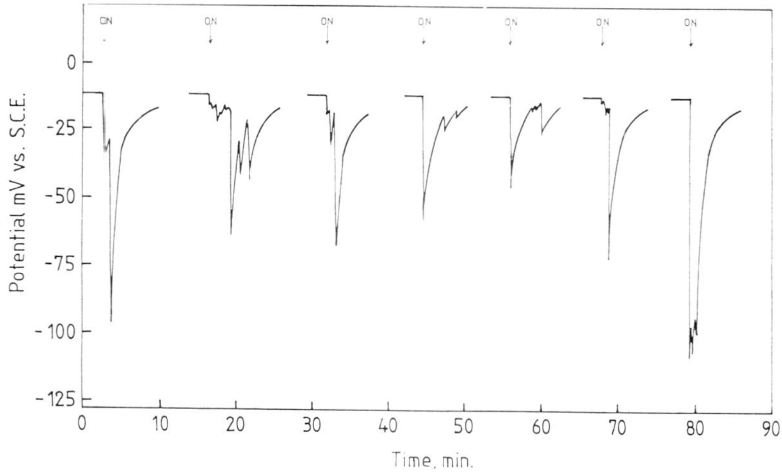

Fig. 8.—Potential changes on fretting 316L stainless steel against Simplex RO AKZ bone cement in Hanks solution - amplitude 25 μm.

sliding and fretting stimulate chemical reaction at surfaces by disruption of oxide films leading to an increased rate of oxidation with possible penetration of the surface by other species such as hydrogen. In some circumstances the oxide film can confer some protection on the underlying surface, e.g., in the high temperature wear of some alloys and rubbing of a non-metal against a metal.

THE MECHANISM OF WEAR

Professor Barwell details six specific types of wear in his paper. It is proposed now to see how certain of these apply to the case of fretting.

(a) Adhesive wear

Intimate intermetallic contact resulting in surface roughening as local welds are formed and broken, undoubtedly occurs in the early stages of fretting producing the type of damage shown in Figure 1. In some cases, particularly in a protective atmosphere, measurable adhesion develops between the surfaces.(17) With the more noble metals, such as copper, measurable adhesion can be developed in quite corrosive environments such as sodium chloride solution. With most alloy systems fretting in air macroscopic adhesion soon falls off as debris begins to form and accumulate. This initial adhesive wear period is not thought to contribute seriously to the overall wear rate although it may have a part to play in initiation of fatigue cracks propagating into the interior of the material.

(b) Abrasive wear

There is some controversy as to whether the continuing and most important phase of fretting wear, after adhesive wear has ceased, is the result of abrasion by the developing oxide debris. In one of the earlier studies of fretting of mild steel[18] the formation of debris resulted in the coefficient of friction falling to the low value of 0.05. The debris was described as having a "ball-bearing" action between the surfaces. In the author's experience abrasive wear is only found in situations where the amplitude of movement is rather high, i.e., between 75 and 150 μm. It can be recognized in the scanning microscope by a repeated "scallop" pattern. In low amplitude fretting abrasive wear is not thought to be a significant factor.

(c) Delamination

When surfaces which have undergone many hundred thousands of cycles of fretting are examined in the scanning electron microscope a very common picture is seen. It resembles crazy paving with the occasional stone missing, as in Figure 7. The wear products appear to be initially formed by the release of plate-like particles some 50 μm in diameter and 1 to 2 μ in thickness. Subsequent rolling to and fro between the surfaces may significantly alter their shape before they escape from the contact area, and so examination of debris may not always reveal a plate-like shape. The scanning electron microscope observations have strongly suggested a delaminated type of mechanism and additional evidence is continually being sought. Figure 9a shows a fretting fatigue failure produced by a Ti-6Al-4V alloy bridge clamped on to a Ti-6Al-4V specimen. The lower part of the picture is the

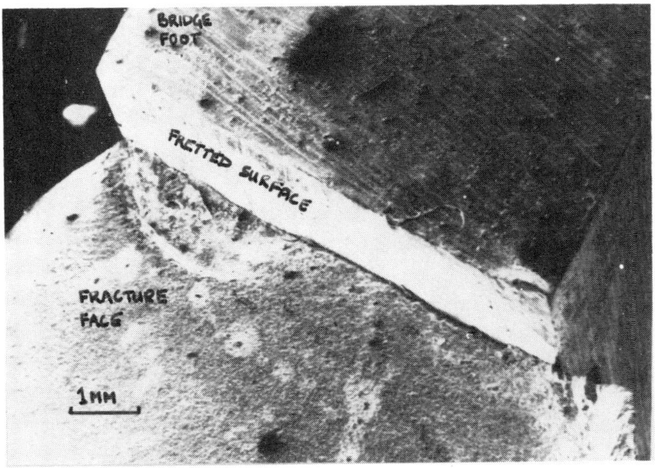

Fig. 9.a.—Ti-6Al-4V bridge in contact with fractured Ti-6Al-4V after 1.1 x 10^6 cycles of fretting.

fracture surface photographed with the bridge still clamped on to it. Examination of the interface, Figure 9b, shows a plate-like particle in process of formation. Some fretted surfaces have been examined by the taper technique developed by Dr. T. Eyre of Brunel University. It is possible by this means to examine in the scanning electron microscope both the damaged surface and the sectioned material below the surface which is polished and etched in the normal metallurgical manner. Such an area is seen in Figure 10a, which is of a 0.2C steel after 100,000 cycles of fretting. The upper part is the fretted surface and the lower the underlying structure. Grain

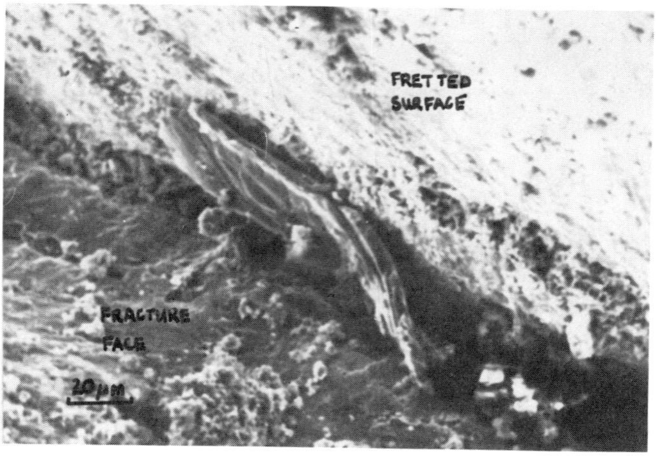

Fig. 9.b.—Part of the interface between bridge foot and specimen shown in Figure 9a at higher magnification showing the formation of a plate-like particle.

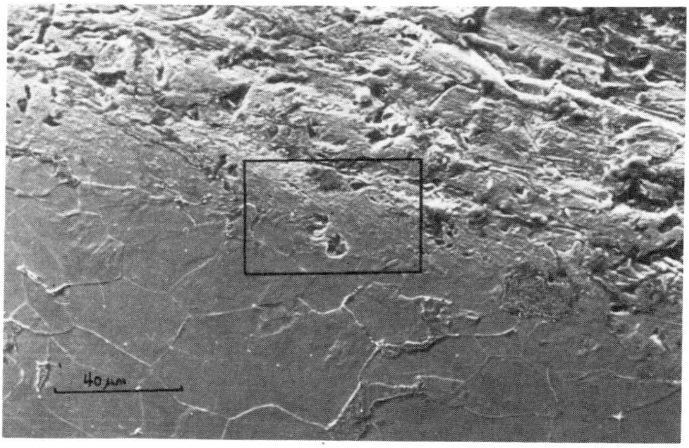

Fig. 10.a.—Taper section showing fretting damage to 0.2C steel surface after 10^5 cycles of fretting.

boundaries and areas of pearlite are visible. The area outlined is shown at higher magnification in Figure 10b. At point A a sub-surface crack is visible. There is a possibility this may be part of a crack initiated in the surface and propagating below the surface, although there do not appear to be any such cracks visible in the surface in Figure 10a. It is thought that this could be a delamination type of crack. The two large pits below the surface are probably inclusions torn out during the polishing process.

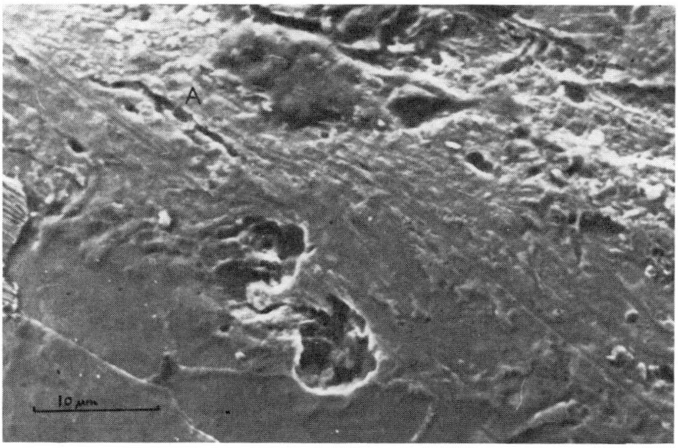

Fig. 10.b.—Enlarged area of Figure 10a showing sub-surface crack A.

The other approach which has been used is to thin down specimens by electropolishing and ion bombardment which have been taken from fretted areas. One technique is to cut a thin slice of material at right angles to the fretted surface by spark machining and to thin it by ion bombardment. Although no specimens have yet been produced for examination by transmission electron microscopy they have been examined in the scanning electron microscope. Figure 11 shows a section through the fretted surface of a Ti-6Al-4V alloy showing a subsurface crack running parallel to the surface. Other

Fig. 11.—Section through the fretted surface of Ti-6Al-4V alloy fretted for 10^4 cycles.

specimens have been prepared parallel to the surface by cutting off a thin slice of the surface and thinning by electropolishing from the back face. This means that any pits in the fretted surface go into holes first. It was possible with such specimens to examine areas in the fretted region by transmission electron microscopy. Such an area is shown in Figure 12, which is of a Ti-6Al-5Zr-1/2Mo-1/4Si (IMI 685) fretted for 2 x 10^6 cycles in air.

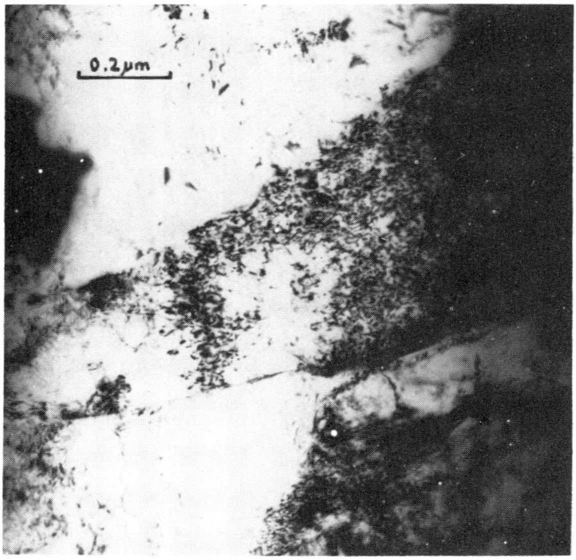

Fig. 12.—TEM picture of thinned area in the fretted surface of Ti-6Al-5Zr-1/2Mo-1/4Si alloy after 2 x 10^6 cycles of fretting showing twinning and areas of high density of dislocation.

The wedge-shaped area across the middle of the picture is a prominent twin. Areas of high dislocation density are visible but since this is a view through a section parallel to the surface, subsurface cracks would not be visible. An interesting feature was observed at the edges of the holes. The polishing technique left surface oxide protruding into the holes. These were observed in many cases to have a lath-like structure as in Figure 13. Electron diffraction of such a lath indicated strong preferred orientation, Figure 14. It should be said that the specimen when examined in the scanning microscope showed the normal fretted surface apart from the few holes produced in the thinning process. The polishing technique had not modified the surface in any way. Surface oxide from areas away from the fretted region show the usual polycrystalline ring pattern. This indicates that fretting either results in developing preferred orientation in the outer surface layer of the metal which is preserved on subsequent oxidation, or that it encourages the epitaxial growth of oxide.

The results shown in this section clearly support a delamination process in fretting wear, but there is as yet no indication as to how the subsurface cracks initiate and propagate. It is hoped in the near future that satisfactory specimens of sections through the surface for examination in transmission will be prepared, in order to determine the dislocation arrange-

Fig. 13.—TEM picture of lathlike oxide penetrating into a hole polished in the thin film of the specimen shown in Figure 12.

Fig. 14.—Electron diffraction pattern of the lathlike oxide shown in Figure 13.

ments and their relation to the surface and features such as inclusions and grain boundaries, and hence to verify the ideas in Suh's original theory.[19]

The question of corrosion and wear has been dealt with earlier in the paper. Local surface fatigue obviously plays a part in fretting, particularly in the development of propagating fatigue cracks which result in ultimate fatigue failure. In the wear process delamination may be a form of fatigue failure. As indicated above this has yet to be established.

ACKNOWLEDGEMENTS

The author wishes to express his thanks to many of his research students upon whose work this paper has been written, in particular Messrs. Baskerville, Doeser, Hamdy, Lamb and Mudge, and Drs. Taylor and Wharton.

Figure 3 is reproduced by kind permission of the Wolfson Institute of Interfacial Technology, and the author is indebted to Dr. Wanhill of the National Aerospace Laboratory in Amsterdam for thinning the specimens which form the subject of Figures 12 and 13.

REFERENCES

1. Stowers, I.F. and Rabinowicz, E., *Journal of Lubrication Technology,* Vol. 95, 1973, p. 65.
2. Waterhouse, R.B., *Wear,* Vol. 45, 1977, p. 355.
3. Feng, I.M. and Uhlig, H.H., *Journal of Applied Mechanics,* Vol. 21, 1954, p. 395.
4. Fenner, A.J., Wright, K.H.R. and Mann, J.Y., in Proceedings International Conference on Fatigue of Metals, Institution of Mechanical Engineers, 1956, p. 386.
5. Waterhouse, R.B. and Taylor, D.E., *Wear,* Vol. 29, 1974, p. 337.
6. Krause, H., *Schmiertechnik - Tribologie,* Vol. 17, 1970, p. 76.
7. Krause, H., *Praktische Metallographie,* Vol. 6, 1969, p. 167.
8. Krause, H., *Metall,* Vol. 28, 1974, p. 895.
9. Jones, J.W. and Wert, J.J., *Wear,* Vol. 32, 1975, p. 363.
10. Ijzermans, A.B., *Wear,* Vol. 14, 1969, p. 397.
11. Wiegand, H., et al., *Werkstoffe und Korrosion,* Vol. 23, 1972, p. 87.
12. Wiegand, H., et al., *Zeitschrift für Werkstofftechnik,* Vol. 3, 1972, p. 75.
13. Waterhouse, R.B., "Fretting Corrosion," Pergamon, Oxford and New York, 1972.
14. Hurricks, P.L. and Ashford, K.S., *Institution of Mechanical Engineers. Proceedings,* Vol. 184, Part L, 1969-70, p. 165.
15. Waterhouse, R.B., in Treatise on Materials Science and Technology, Vol. 13, "Wear," edited by D. Scott, Academic Press, New York, 1979, p. 259.
16. Taylor, D.E. and Waterhouse, R.B., *Corrosion Science,* Vol. 14, 1974, p. 111.
17. Bethune, B. and Waterhouse, R.B., *Wear,* Vol. 12, 1968, p. 289.
18. Halliday, J.S. and Hirst, W., *Royal Society of London. Proceedings.* Series A, Vol. 236, 1956, p. 411.
19. Suh, N.P., *Wear,* Vol. 25, 1973, p. 111.

DISCUSSION

S. GANESH, Bendix Research: To support the delamination theory you showed some micrographs that contain fretting cracks running parallel to the surface. I have seen numerous cracks running perpendicular to the surface in Ti-6Al-4V. How would you explain that?

R. B. WATERHOUSE: We found that the initiation of fatigue cracks that propagate across the specimen occurs in the very early stages of the fretting process and in particular they seemed to form at the slip-nonslip boundary. In actual contact slip does not occur over the entire contact area and where you get the boundary between the slip and the nonslip areas you get the formation of a crack and that never runs parallel to the surface; it is always inclined to the surface. Even when it gets out of the influence of the fretting action, it goes across the specimen. But the parallel cracks are always associated with the wear process. They become visible in the later stages of the process when it is mostly wear. We have never seen one of those turn into a fatigue crack.

GANESH: What concerns me is that too much emphasis is given to the delamination theory of wear. It is a good theory and does explain a lot of phenomena. I have seen instances where cracks run parallel to the surface and emerge at the surface. However, it cannot explain that the crack started below the surface and then propagated to the surface; it could be the other way around.

WATERHOUSE: Yes, I think in the case of fretting the cracks definitely start in the surface. We see them opening up; they are usually wider at the surface and then narrow down to a crack. Certainly the surrounding atmosphere can have quite an influence on the fretting behavior.

A. W. RUFF, NBS: I want to ask a general question about the passive film forming metals in aqueous saline solutions. We have been looking at stainless steel, titanium and several titanium alloys. The nature of the film, i.e., the thickness and properties, depends on the pH of the solution and the oxygen level for those systems. In fretting there seems to be a competition between the mechanical processes and the chemistry. In the case of restricted access, for instance, we find that low pH and low oxygen levels that develop in crevices severely affect the corrosion characteristics of titanium. I would think that the fretting geometry would lead to restricted access to the solution so that when you do a fretting corrosion experiment you are really looking at both the mechanical and the local chemical problems. Perhaps there is a whole range of chemical conditions that you have to study before you can really assess the fretting behavior. When you vary the pH and perhaps the oxygen level, do you get a different fretting corrosion response?

WATERHOUSE: Certainly with stainless steel we have looked at the effect of lowering the pH and there was a large increase in the corrosion current when we got down to pH of 3 and lower. The fretting situation is usually also the crevice situation where we would expect crevice corrosion and I should say that the crevice corrosion depends on oxygen depletion. There is a certain amount of movement in fretting which could reduce that polarization and enable oxygen to migrate perhaps more easily to the site.

S. A. KARPE, DTNSRDC, Annapolis: What parameters do you use in selecting materials to prevent fretting?

WATERHOUSE: Presently this is very much a matter of looking at the situation and trying it out. Some one asked, "What would be a good material for having contact with titanium and sea water?" - a question of condenser tubes passing through a baffle where there may be vibration. Well, I can't say offhand what will be the best combination. I think it has to be tried out. Usually you have very little choice; if it is corrosive environment both surfaces should be corrosion resistant and that is the main limitation I would say.

KARPE: The problem I am particularly concerned with is about bronzes used in a controllable pitch propeller hub system running against steel in an oil environment which may not be corrosive in the sense that you are talking about.

WATERHOUSE: Well, in a situation like that you might be able to get over it by altering the surface with some suitable coating. There is no doubt that coatings are one of the answers. We found, for instance, that sprayed molybdenum coatings can be very beneficial. But there is no one answer to this. I think one has to look at all the possibilities; probably Prof. Czichos' method should be applied to this to come up with an answer.

A NEW INTERPRETATION OF THE MECHANISM OF FRETTING AND FRETTING CORROSION DAMAGE

E. S. Sproles, Jr., D. J. Gaul and D. J. Duquette

ABSTRACT

Fretting experiments were conducted in air at room temperature on 4130 steel in the annealed (spheroidized) and quenched and tempered (martensitic) conditions. The mechanism of material removal is similar in both conditions. Examination of the fretted surfaces revealed that metallic material is removed from the fretted surfaces in the form of flakes. Cross sections through the fretted regions revealed severe plastic deformation of the surfaces. The observed mode of metallic material removal suggests wear by a delamination mechanism rather than by an abrasive wear mechanism or by a welding and material transfer mechanism. Oxide debris formation resulting from oxidation of metallic debris or from the formation and subsequent scraping away of a thin oxide film from the metallic surface is consistent with experimental results. No evidence was found to support the in situ formation of thick surface oxide films such as those which form by oxidation at elevated temperatures.

INTRODUCTION

Debris generated by fretting of metals or alloys may alter the physical properties of the surfaces in contact. For example, simple discoloration of the surfaces may occur, contact resistance may be increased, seizing of machinery parts may result, or, in the extreme, premature fatigue crack initiation may develop if cyclic stresses are imposed. A graphic example of the latter case is sometimes observed under rivet heads in aircraft wings where the cyclic stresses are imposed by elastic deformation. In order to devise methods to reduce fretting damage, the mechanism of material removal from the fretting surfaces must be understood.

In oxidizing atmospheres, fretting of most metals results in both metallic and oxide debris, and thus distinctions between the two must be made since different mechanisms of debris formation are most likely involved. A number of possible mechanisms of debris generation in the fretting of metals have been proposed. Adhesion, welding, or mechanical interlocking of surface asperities have been suggested as processes which cause fracture of surface layers of the metal, and thus result in the formation of loose metallic wear particles.[1-6] Oxidation of these loose metallic particles has been suggested as an origin of oxide debris.[1-7] Abrasive wear of metal surfaces by this oxide debris has also been suggested as a possible mechanism of metallic material removal.[3,4,6,7] The scraping of oxide from one surface by asperities on the opposite surface followed by the rapid reoxidation of the freshly exposed metal has been proposed as still another possible mechanism of material removal and oxide debris formation.[8]

The basic concepts in the mechanisms described above were proposed in the 1950's and little further refinement has occurred. There is, however, a need to reassess the validity of such mechanisms, both to increase understanding of the fretting process and to develop useful engineering data on the performance of specific materials. The purpose of the present study is to evaluate the applicability of the various proposed mechanisms to the fretting of steel in air, and to extend or supplement such models as required by experimental observations.

EXPERIMENTAL

Experiments on the mechanism of fretting are part of a general study of the role of fretting on premature fatigue failure. Fretting was induced in the interface of a flat fatigue specimen and a fretting "pad" pressed against the side of the specimen with a measured normal force. The fretting fatigue test fixture is illustrated schematically in Figure 1. Fretting occurred at the flat specimen-pad interface due to the movement of the specimen relative to the pad, the specimen movement being caused by elastic deformation of the specimen during the cyclic loading. By varying the position of the pads along the specimen it was possible to vary the interfacial slip independent of the normal load on the pads and the cyclic load on the specimen. Slip was measured optically, employing a strobe lamp, microscope and scale with 10 micron divisions. Normal load was measured by measuring the length of a co-linear array of disc springs through which the normal load was applied.

The material used in the experiments was commercial 4130 sheet steel of 2.7 mm thickness. The material was tested in two conditions to investigate the influence of microstructure on the fretting mechanism. The first microstructure consisted of an annealed spheroidized structure, i.e., spheroidal carbide particles in a ductile ferrite matrix. The second microstructure was produced by annealing at 950°C for 20 minutes, quenching in water and tempering at 450°C for 60 min, i.e., tempered martensite.

The experiments were conducted in laboratory atmosphere at room temperature at 15 cycles per second until the fatigue specimen failed, or until an arbitrary number of cycles was reached, generally 5 x 10^6. The fretted areas of the fretting fatigue specimen and the fretting pads were examined by optical and scanning electron microscopy. Cross sections through fretted areas were prepared by plating the sample with copper or nickel, then sectioning the sample and preparing a metallographic specimen by standard techniques.

RESULTS

The free surfaces of (a) annealed and of (b) quenched and tempered steel specimens fretted in air at room temperature are shown in Figures 2 and 3, respectively. No significant differences were observed between the specimen and pad surfaces. Relatively large metallic flakes are distributed

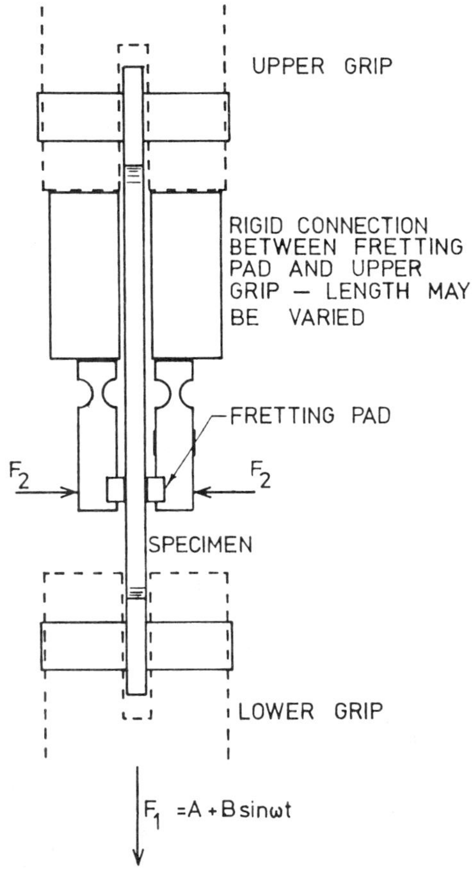

Fig. 1.—Schematic diagram of fretting fatigue test fixture.

at random orientations on each of the surfaces of both specimens. Further investigation has shown that these flakes can be easily dislodged by probing with tweezers or by stripping plastic replicas from the surface.

Typical cross sections through the fretted regions are shown in Figures 4 and 5. The distortion of grains near the surface is indicative of the large amount of plastic deformation sustained by the material immediately below the fretted interface. Other evidence of plastic deformation of the surface material was found in microhardness measurements with the region within 15 microns of the surface being measurably harder than the material deeper below the surface. Examination of metallic debris particles in the SEM disclosed extensive secondary cracking in particles removed from the annealed surfaces.

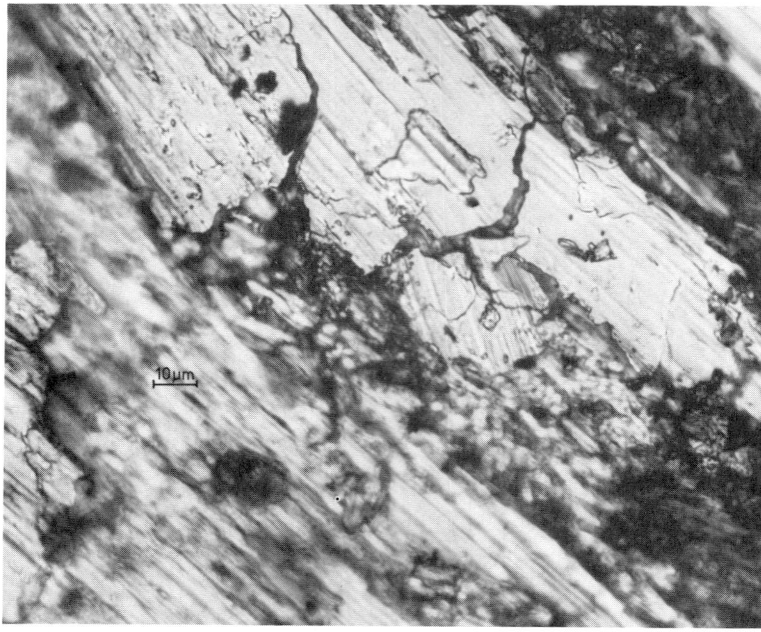

Fig. 2.—Free surface of annealed 4130 steel subjected to fretting fatigue for 5 x 10^6 cycles showing dislodged platelets, secondary cracks and non-parallel surface grooving.

Fig. 3.—Free surface of quenched and tempered 4130 steel subjected to fretting fatigue for 5 x 10^6 cycles showing platelets and surface cracks. Note similarity with Figure 2, independent of microstructure.

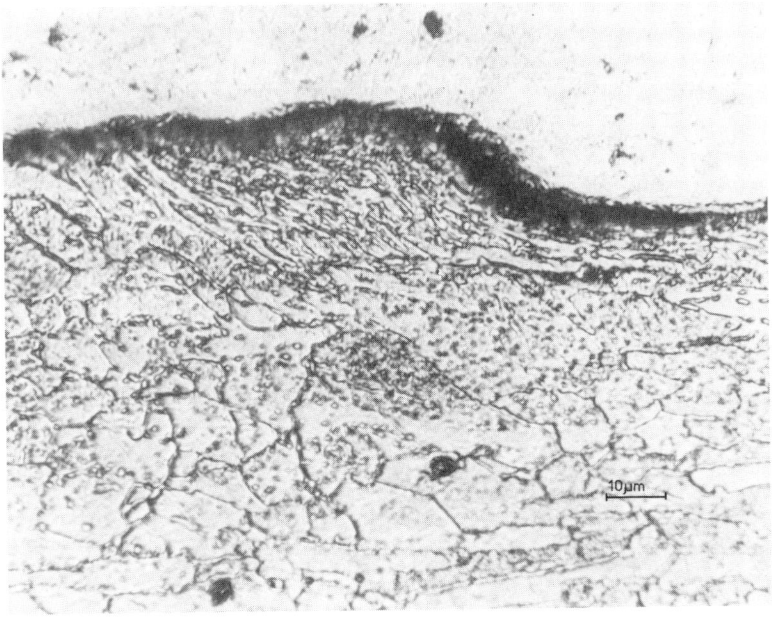

Fig. 4.—Cross section of annealed 4130 steel subjected to 5 x 10^6 cycles of fretting fatigue showing the boundary between fretted and non-fretted areas. Note extensive plastic deformation, removal of metal and oxide on surface.

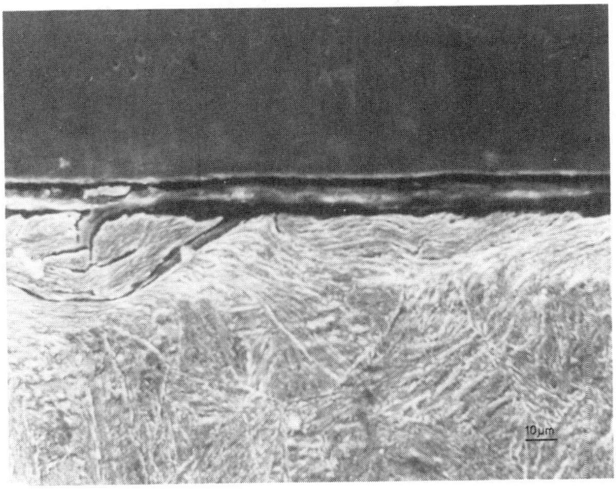

Fig. 5.—Cross section of quenched and tempered 4130 steel subjected to 5 x 10^6 cycles of fretting fatigue. Note extensive plastic defcrmation of martensite platelets, surface connected cracking and oxide wedging of cracks.

In addition to metallic flakes, copious amounts of oxide are generated at the fretted surfaces. The majority of this oxide is a loose, non-adherent powder, which is exuded from the rubbing surfaces. A small proportion of the oxide adheres to the surfaces in a random fashion and appears as dark areas in Figures 2 and 3.

The thicker oxide layer appears to have resulted from agglomeration of finely divided oxide and metal debris, and appears to be made up of many small grains, predominantly dark in color. A few bright particles are present and are assumed to be metal.

An area of a fretted surface is also shown in Figure 6 where the surface condition varies from practically bare metal to oxide of sufficient thickness to be opaque. Of particular interest is an area of oxide where

Fig. 6.—Surface of annealed 4130 steel subjected to 5 x 10^6 cycles of fretting fatigue showing areas of rubbed bare metal and areas of adherent oxide.

a crack in the metallic surface is visible through the oxide layer. Since the oxide film appears to be continuous even across the crack and contains metallic particles, the film must have formed by agglomeration.

DISCUSSION

The results of the fretting experiments include several observed features which relate directly to the mechanism of the fretting process. These features include:

1. A layer of heavy plastic deformation below the fretting surface.
2. Metallic flakes which cover extensive areas of the fretting surface.
3. Opaque oxide layers which appear to have been formed by agglomeration of debris rather than by *in situ* oxidation.
4. Closely interspersed areas of exposed metallic surface and surface covered with an opaque oxide layer.

Metallic Debris

There are two possible ways that the fretted surface can become covered with metallic flakes: a) the flakes could have formed *in situ* by the breakup of the metallic surface or; b) they could have been transferred from the opposing surface as a result of one of the adhesion or welding wear mechanisms. Experimental results indicate strong support for the former case. The evidence is at least twofold. First, the flakes show grooves parallel to the slip direction which continue from one flake to another, indicating that several flakes were once a continuous surface. If the flake-like material is transferred between surfaces, it must be transferred as a large flake and then break up into many smaller flakes after transfer, otherwise the cracks and grooves would not match between flakes. Although such a mechanism is possible, it requires the assumption of a more complicated mechanism than the breakup of an existing surface into flakes. The second argument for the *in situ* formation of flakes is still more compelling. If material is transferred from one surface to the other, some of the observed surface must be the reverse surface of transferred material. Some of the flakes were dislodged by probing with tweezers or by stripping replicas from the surface, and the reverse side of the flakes examined. The reverse side lacks regular parallel grooves and has more relief. This surface morphology is not observed on the fretted surface except in the bottom of pits, thus the flakes are not the result of material transfer. It is concluded, on the basis of experimental observations, that the primary mode of metallic material removal in the fretting process is the breakup of the surface into flakes and the subsequent removal of those flakes from the fretting surfaces.

Additional support for flake formation, *in situ*, is provided by the following observations. First, metallographic cross sections through the fretted region indicate that the surface material is severely plastically deformed. Second, microhardness measurements on such cross sections indicate some hardening of the surface region, as would be expected by the introduction of cold work. Third, plastic replicas stripped from the fretted surface remove some of the flakes, indicating that they are very weakly attached to the surface. In some areas, repeated stripping of replicas from the same area resulted in the removal of several layers of weakly attached metallic material. In other areas of the fretted surface, no material was removed by the replicas, indicating that only certain areas had reached the point where metallic flakes could be easily removed.

The experimental observations fit more closely with wear of the metallic material by the delamination theory of wear than with any other proposed mechanism.[9-11] The delamination mechanism may be applied to the fretting situation in the following manner. During any given period in the fretting cycle, there are some small areas of contact between the surfaces. As the surfaces move back and forth, the material in those areas of contact undergoes plastic deformation and eventually breaks into flakes. The flakes are gradually worked out of the contact areas by the motion of the surfaces, allowing other areas of the surfaces to come into contact. Thus, the delamination mechanism accounts for the flake-like morphology of the metallic wear particles. The delamination theory, however, does not address the formation of several layers of loose flakes in a single area. This observed departure from the single layer delamination theory is undoubtedly a result of the cyclic nature of fretting. Repeated wear of the surfaces leads to a kind of contact fatigue where multiple cracks are initiated in the near surface regions. These multiple cracks can accordingly link up to form a three-dimensional flake configuration.

Oxide Debris

There are two possible modes of formation of the oxide debris: (a) oxide may form by the *in situ* oxidation of the fretting surfaces and/or; (b) loose metallic wear debris may oxidize. Although experimental results indicate (Figure 6) that the thicker oxide layers are formed by agglomeration of smaller particles, the origin of these particles is not certain.

For example, oxide may form as a thin film which is subsequently scraped away, in a manner similar to that proposed by Uhlig.[8] The Uhlig model proposes that an asperity contact scrapes away a thin oxide film, and the bare metal so exposed quickly reforms a thin oxide film which is again scraped away the next time that an asperity contact rubs this region of the surface. The oxide scraped from the surface could agglomerate into the thicker oxide films or could migrate out of the interface and appear in the loose debris. Oxide formed by the oxidation of loose metallic particles could also agglomerate into thicker oxide films or could migrate out of the interface and appear in the loose debris. Similarly, oxide formed by the oxidation of loose metallic particles could agglomerate into the observed oxide layers.

The evidence in support of the formation and subsequent scraping away of a thin oxide film follows. First, all areas where the metallic surface is visible in the fretted area are slightly discolored in comparison to the free surface of the fretting fatigue specimen. This discoloration is slightly yellow or red, and probably is due to an oxide layer. Second, the wear rate can readily be explained by a simple model of the asperity scraping process. If it is assumed that during each half cycle of fretting, one monolayer of iron oxide is scraped from the entire fretting surface, and another monolayer forms and is scraped away during the next half-cycle, the calculated wear rate is much greater than the observed wear rate. If it is assumed that the monolayer is removed only from an area equal to the calculated area of contact, the calculated wear rate is slightly less than the observed wear rate. Since much of the wear occurs as removal of metallic flakes, the simple model just described is surprisingly accurate in predicting the quantity of oxide produced in fretting.

There is little evidence available to either prove or disprove the concept of oxide debris originating from the oxidation of metallic debris. The morphology of the debris which falls from the interface suggests certain limitations on the way in which such a mechanism could operate. This debris is a mixture of metal and oxide particles. The metallic portion consists of bright flakes and smaller particles of sizes ranging down to the resolution limits of optical microscopy. The morphology of the oxide portion is best described as a network of very small particles joined to each other in such a way that spongy structure is produced. The individual particles in the debris are generally much smaller than the metallic flakes seen on the fretting surface. As a result of the presence of metal and oxide in the debris, a mechanism involving oxidation of metallic debris must account for the fact that some metal particles oxidize while others do not. As a result of the relative particle sizes, such a mechanism must include the breaking of the metallic material into smaller pieces before or during the oxidation process. The experimental observations suggest that if oxidation of metallic debris is important, most probably the metallic flakes are broken into finer metallic particles and these smaller particles undergo oxidation. Such a mechanism could account for the presence of larger metallic particles mixed with the oxide and for the extremely small dimensions of the oxide particles. The mixture of sizes of metallic particles observed and the fact that cracks are observed in the metallic particles would support a mechanism requiring the mechanical breakup of the larger metallic particles.

The results of the fretting experiments do not conclusively prove or disprove either oxide debris formation mechanism. They do, however, place some limitations on the details of the mechanisms, especially on the mechanism of oxide debris formation by the oxidation of metallic debris. Such limitations must be carefully considered in the construction of a model of the entire fretting process.

Based on the observations of the present experiments, a model of the mechanism of material removal in the fretting of steel in air may be constructed. Figure 7 schematically shows a cross section of an asperity contact in a fretting situation and the breakup of the contacting surfaces.

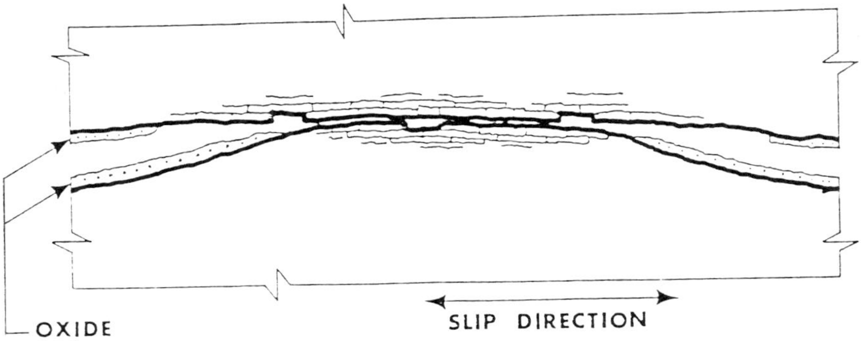

Fig. 7.—Schematic model of fretting phenomenon showing production of metallic platelets, bare surfaces and surface oxidation.

In the region of sliding contact, the surfaces break up into flakes, and some of the flakes become detached from the surfaces and escape from the sliding interface, leaving pits in the surfaces. Beyond the region of contact, oxide and metal debris collect into a relatively thick layer. In the region of sliding, little oxide can collect, thus accounting for the appearance on the fretted surface of closely interspersed areas of metallic surface and thicker opaque oxide.

CONCLUSIONS

Detailed examination of the surfaces of 4130 steel specimens subjected to fretting in air and cross sections through these surfaces has yielded significant new information on the mechanism of fretting. The metallic surface is severely deformed and eventually fails by breaking into metallic particles of a platelike morphology. It is concluded that wear by a delamination mechanism is the predominate form of metallic material removal in fretting. Abrasive wear or wear by adhesion, welding and material transfer are at best of minor importance in the fretting case studied in these experiments.

No evidence has been found to support the *in situ* oxidation of the surface to form a thick oxide film. It is found, however, that oxide debris formation either by the formation and subsequent scraping away of thin, translucent oxide films or by the oxidation of metallic wear debris is consistent with the experimental results. It is also found that oxide debris agglomerates into relatively thick layers in some areas of the surface.

The microstructure of the steel has been shown to have little effect on the mechanism of material removal in fretting. Both the ferrite with spheroidized carbides and tempered martensite structures fail in fretting by a delamination mechanism.

REFERENCES

1. Godfrey, D. and Bisson, E.E., *Lubrication Engineering,* Vol. 8, 1952, p. 241.
2. Godfrey, D. and Bailey, J.M., *Lubrication Engineering,* Vol. 10, 1952, p. 155.
3. Feng, I-Ming and Rightmire, B.G., *Lubrication Engineering,* Vol. 9, 1953, p. 134.
4. Feng, I-Ming and Rightmire, B.G., *Institution of Mechanical Engineers. Proceedings,* Vol. 170, 1956, p. 1055.
5. Halliday, J.S. and Hirst, W., *Royal Society of London. Proceedings.* Series A, Vol. 236, 1956, p. 411.
6. Waterhouse, R.B., *Institution of Mechanical Engineers. Proceedings,* Vol. 169, 1955, p. 1159.
7. Wright, K.H.R., *Institution of Mechanical Engineers. Proceedings,* Vol. 1B, 1952-53, p. 556.
8. Uhlig, H.H., *Journal of Applied Mechanics,* Vol. 21, 1954, p. 401.
9. Suh, N.P., *Wear,* Vol. 44, 1977, p. 1.
10. Suh, N.P., et al., *Journal of Lubrication Technology,* Vol. 96, 1974, p. 631.
11. Suh, N.P., *Wear,* Vol. 25, 1973, p. 111.

DISCUSSION

S. JAHANMIR, Cornell University: I would like to offer an explanation for the multiple layers that you observed on the surface. In general the formation of subsurface cracks, or the propagation of these cracks does not occur at only one depth below the surface; it occurs within a region below the surface. In many materials we observed cracks at different layers which were very close to each other. When a wear particle is formed there is a possibility of stripping off a number of layers at the same time rather than by the removal of just a single layer.

E. S. SPROLES: While reading the papers on delamination theory, it was not clear to us how that could occur. I think in fretting we may have a very special situation in that the two regions of contact remain in close contact many cycles and therefore, it may be a case in which we usually remove a pre-existing layer. We can easily form a number of layers there.

QUESTIONER: In our study of fretting corrosion from the standpoint of formulating lubricants to prevent fretting corrosion or reduce it, we found that the mechanism was different for different machines. The end result might have been the same, but in our study we found that the lubricants that reduce fretting and corrosion in one instance would not do so in another case. For example, rolling contact fretting corrosion was different from sliding contact. We concluded that although the end result was probably the same, the mechanism by which it reached that end result could definitely be different.

SPROLES: Did you come to any conclusions as to whether adhesion or delamination appeared to be occurring?

SAME QUESTIONER: I think adhesion was probably the primary mode of failure in the rolling contact bearing. Presently we are conducting some research and it will be published soon.

SPROLES: I think the point is very well taken. Each case has to be studied individually and I think that is where a careful examination of the surfaces and the microstructure is really very important in understanding fretting in a given situation.

R. B. WATERHOUSE, University of Nottingham: What is the amplitude of slip in your experiment set-up and also what was the effect of fatigue strength on fretting?

SPROLES: The amplitude of slip in one case was about 60 μm and the normal load was 2000 psi. In another case the slip was somewhat like 50 *μm*. I realize that you believe that the slip amplitude should be less than 5μm for it to be considered as fretting.

WATERHOUSE: By increasing the amplitude, which of course gives you more wear debris, it is convenient to measure wear by weight loss. Some of the experimental investigations have gone to rather high amplitudes to take advantage of that. However, when you have these higher amplitudes and increase the amount of wear, actually you can improve the fatigue situation and you find that it does not have any effect on the fatigue strength. According to one much publicized idea you initiate a fatigue crack and then wipe it out in the next moment never giving an opportunity for the fatigue crack to get going.

SPROLES: In our experience with fatigue we found that the most damaging slip region is in the neighborhood of 25 μm for both structures of AISI 4130 steel. We did not observe great differences in the appearance of the surfaces although I showed the microstructure for 70 μm slip. We believe it is very similar even at 25 μm, although it appears to be on a somewhat finer scale and thus it is harder to take a micrograph.

S. A. KARPE, DTNSRDC, Annapolis: I hear many people talking about wear as weight loss or volume loss. Some of the systems that I work with really do not have weight loss; but we observe large globs of material being transferred, large cavities in the bearing surface and transfer of a lot of material to the rigid counterface. I do not understand why everyone talks about weight loss. In other words, wear does not have to occur by weight loss; destruction of the original geometry without weight loss can be considered as wear.

Did you use like or unlike combinations in your experiments, i.e., tempered vs. tempered, or tempered vs. annealed?

SPROLES: Most of our experiments were conducted with annealed against annealed, or tempered vs. tempered. We did some specific exploratory experiments with other combinations and did not find significant differences in fatigue life. There were differences in the appearance of the fretted surface in that we get more wear from the softer surface.

D. GODFREY, Chevron Research: This is just a comment and I thought that some people might be interested in it. We made a study of the fretting corrosion in the presence of mineral oil and there we observed something completely different. It results in a highly polished but undulating surface free of cracks and pits. The properties of the oil that affect this fretting are its oxidation characteristics. An artificial oxidant in the oil will induce more wear with a greater area of pitting and conversely an oxidation inhibitor reduces wear. I interpreted this to be that the fretting debris in this case is actually finely divided Fe_3O_4 of the order of 100Å and it is sufficiently abrasive to polish off the tarnish caused by the acids that are formed in the decomposing oil.

SPROLES: We did some exploratory experiments in which we used an argon atmosphere and noticed the characteristics of the surface which indicated that perhaps adhesion was a signficiant problem. The explanation obviously would be that the oxide particles are holding the pieces apart and, as you say, we got very polished surfaces.

R. DASKIVICH, G. M. Research Laboratories: Is the normal pressure on the pad 2000 psi?

SPROLES: For the particular microstructure I showed that was the case, but the pressure varies from test to test.

DASKIVICH: Did you look at the surface topography before the test, or by what process were the pads prepared?

SPROLES: Both the specimens and the pads were surface ground by the machine shop and then they were wet polished on 600 grit paper with the scratches parallel to the slip direction. The original surface was quickly lost, long before the structures shown in the figures were obtained.

FRACTURE THEORY OF WEAR

Y. Kimura

ABSTRACT

Wear can be considered as surface fracture and, therefore, the fracture theory of wear involves the analysis of actual forces acting on the surface, evaluation of accumulated damage in the surface and subsurface, together with material properties which represent the resistance to fracture. A preliminary approach to this procedure is outlined.

INTRODUCTION

Professor Barwell presented an impressive and comprehensive review of a complicated subject(1). However, many sections of the review of the wear phenomena close with expressions such as "further investigation is necessary." It shows clearly the limits of our present understanding and the need for further investigation of the wear phenomena. Paradoxically, wear assumes increasing complexity rather than simplicity as invesitgation proceeds.

Barwell also states that theory should not run too far ahead of experiment. On the contrary, it must be said that the theory at present lags too far behind experience, although there are a number (indeed a surplus number) of theories which allegedly explain the mechanisms of wear. Therefore, there is a need to examine the validity of the current theories of wear.

Many authors claim the validity of their theories by explaining some specific experimental results, qualitatively of quantitatively. Although it is a necessary step to show the usefulness of a theory, it is not the sufficient condition for its general applicability to wear phenomena. Usually the design of an experiment is simplified to avoid effects of factors which are not considered essential. However, there are a number of factors which may affect wear in practical situations, and the factors which have the critical influence may vary from system to system. The simplification may sometimes exclude these critical factors and results in placing emphasis on less important factors.

Then we may ask, what is a reliable criterion of validity? We should also ask whether practical wear problems can be solved by a logical extension of a theory. From such a point of view, the current theories are quite insufficient. Practical lubrication, for example, has almost nothing to do with the fundamental theories of wear.

ADHESIVE WEAR CONCEPT

As one of the predominant modes of wear in moderately clean environmental conditions the so-called adhesive wear has long been under discussion. It seems, however, that there is some confusion about the implications of the term. In a wider sense of the term, a wear process is generally called "adhesive" when it is neither abrasive nor corrosive. Metallic wear, in not-intentionally lubricated systems (or even in systems lubricated with non-added oils) falls into this class. In such cases, the Holm's law, or the Archard's law, is approximately obeyed. One feels relieved to classify a phenomenon as adhesive wear when he finds it obeys those laws, although nothing is learned by doing so.

Barwell made an important statement on this point. A question is raised about the adhesive mechanism of wear in which material can be torn from one interacting body and made to adhere to another. He states that this implies a rather special form of wear, and that this mechanism is not important in relation to the mechanical systems required to operate for considerable periods of time. Clearly he defines the term "adhesive wear" in a narrow sense.

The author is in agreement with this understanding of adhesive wear. One should avoid the unlimited use of the term. This does not mean that adhesion plays no role in the wear process; it can assume a limited role under moderate environmental conditions. Such a view seems to gain increasing support. Typically, it is interesting to note that no "adhesive wear particles" were found in Ferrographic analyses. Consequently we must look for an alternative explanation for wear mechanisms.

FRACTURE THEORIES OF WEAR

Wear is defined as the gradual removal of material from solid surfaces. Removal must be the result of some breakage, or fracture. In other words, an explanation of wear must include a fracture process. This is a promising starting-point for an alternative theory of wear.

Basically, two approaches are available to understanding the wear process which was once considered as adhesive. One possible way is to modify the adhesive theory of wear where some physically meaningful interpretations are introduced. The other is to construct entirely new theories incorporating all the steps from contact to removal of the wear particles. Before discussing the fracture theory of wear in detail, a brief review of the former approach will be made.

Feng(2) criticized that the adhesion theory failed to explain the presence of the loose wear particles. According to the theory, shearing of adhesive junctions results in either breakage at the original interfaces or transfer of material. He postulated that the interface of contacting asperities was roughened due to plastic deformation, and its interlocking effect led to fracture along the weakest section giving rise to loose particles, Figure 1.

Kraghelsky(3) classified the wear process into three categories: microcutting, elastic and plastic fatigue. According to him, wear results from stresses and deformation, its value depending on the geometrical factors and adhesion forces. He also related the probability factor in the linear wear law with the expressions for fatigue curves in the latter two classes.

This fatigue concept was adopted by several investigators, and recently a considerably sophisticated model was proposed by Halling(4), who tried to predict wear behavior coupling the Manson-Coffin low cycle fatigue relation, the Archard's contact model, and Greenwood and Williamson's probabilistic

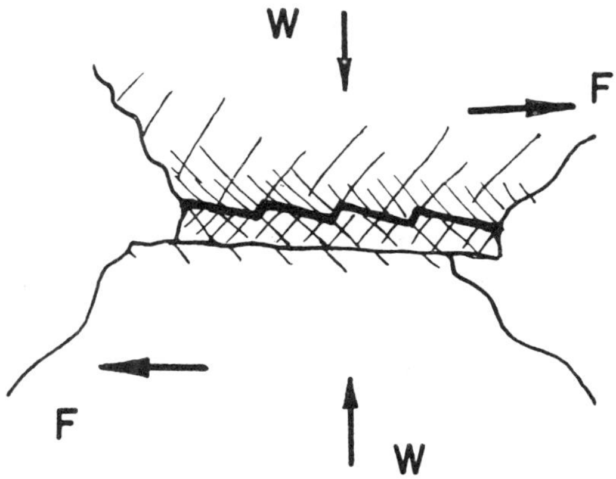

Fig. 1.—An asperity interaction model of Feng[1]. *Fracture occurs along the weakest section rather than the original interface resulting in formation of a loose wear particle.*

model of asperity interaction. This treatment was further developed by Finkin[5], who replaced the rather restrictive assumption of Archard's model with a generalized one, and took into account the individual contributions of elastic and plastic contacts.

CONTACT PROBLEMS

There is a common way of studying fracture and wear problems. Let us compare a wear experiment with material testing, Figure 2. In material testing, the shape and the size of specimens are defined as well as the nature and magnitude of the load to be applied. Fracture can be discussed only when these variables are properly defined. This is true in the case of wear also, but the parameters that control wear are not generally known.

One might argue that even in wear experiment the shape and the size of specimens and the load to be applied are well-defined. It is important to note here that the significance of shape and size in wear are completely different. In the case of material testing the conditions for expected fracture are almost completely described by macroscopic variables. By contrast, the macroscopic variables in wear experiment can never define such conditions. It is the magnitude of the forces and the contact mechanics that govern where and how fracture occurs. The "true specimen' in wear experiment is not a wear specimen itself. It can be envisaged that a wearing surface carries on it a number of "true specimens" as shown in Figure 2(c). What we perceive as wear particles are "broken specimens." Thus, it is clear that contact problems are of prime importance in wear processes.

We may find a basic agreement as to the importance of asperity contact. Though it loses a direct connection with the formation of wear particles as in the Archard's model, the forces which cause fracture are no doubt concentrated on it. Then we must know the magnitude of forces acting on individual contacts, and the scale of fracture to take place which determines the size of wear particles. These quantities are closely related with

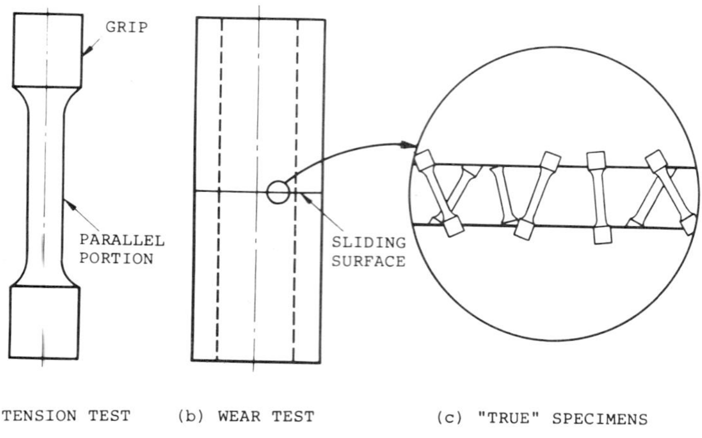

Fig. 2.—Specimens in material and wear testing showing the difference in what they signify.

surface topography, but the topography is often generated through the wear process itself. This makes the process more complicated. However, in some cases it is observed that the wear behavior is predominantly governed by the contact mechanics. Figure 3 shows an example in which force on contacts determines the thickness of deformed layer which has a definite correlation with the mean size of wear particles, and the change of wear amount is due to the change in the particle size[6,7].

The role of lubrication also must be discussed in this context. Its primary effect is on the coefficient of friction which alters the magnitude of actual forces. Another important effect of lubrication on contact mechanics, in the author's opinion, is that it results in a smoother topography on sliding surfaces which minimizes the size of individual contacts, thereby minimizing the magnitude of forces acting on them. It is well known that only minute wear particles are usually observed in wear process under proper lubrication.

RESPONSE OF MATERIALS

When the contact mechanics is learned we can inquire how the materials behave in response. This phase of the wear process has received more attention than the contact problem. The delamination theory[8] is a typical example of the analytical work in this phase.

A characteristic feature of wear process is that forces act on asperities repeatedly. As sliding continues damage accumulates in the material which leads to eventual fracture. This is nothing but fatigue failure. Here we should invoke the pioneering work of Kraghelsky[3]. However, his theory must be extended to incorporate an important fact that the substrate is also fatigued, which requires more sophisticated formulation[9] than seen in Archard's theory.

It seems clear, therefore, that the property representing the resistance of materials to wear is the resistance to fatigue failure, no matter

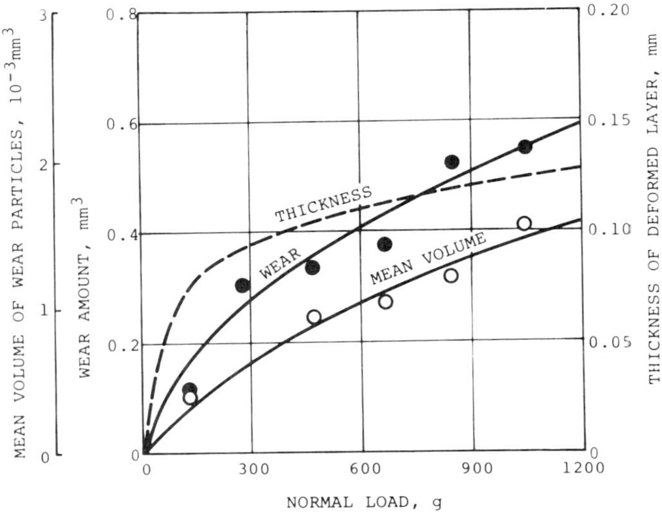

Fig. 3.—A mechanical aspect of wear[5,6]. *Normal load changes thickness of deformed layer, mean volume of wear particles, and thereby wear amount. Sliding of Ni/Ni in dry air. Sliding speed 168 mm/s, sliding distance 20 m.*

what model is employed. It is generally known that fatigue resistance of materials is more sensitive to environment when compared with other properties such as tensile or shear strength. Probably this contributes partly to marked effects of lubrication on wear. Figure 4 provides an example in which a clear correlation was observed between fatigue resistance and the rate of formation of wear particles having substantially constant mean size[10].

It is interesting to note the similarity of the process of wear particle formation to rolling contact fatigue (pitting, flaking or spalling). The difference between rolling contact fatigue and wear seems rather arbitrary. The particle size and the marks left on the surfaces are far larger in the rolling contact fatigue necessitated by its contact mechanics. In the case of wear, minute particles usually allow the system to operate further on. This difference has, however, nothing to do with their basic mechanisms. Much can be learned from analyses of rolling contact fatigue which have been conducted under more clearly defined conditions.

CONCLUDING REMARKS

The necessity for further investigation of wear phenomena, as suggested by Professor Barwell, implies that the approaches which consider the whole wear mechanisms as a single black box are insufficient. We must try new approaches in which plural black boxes are introduced to take the details of every stage of wear process into account, and thereby allow for the inherent complexity of wear process.

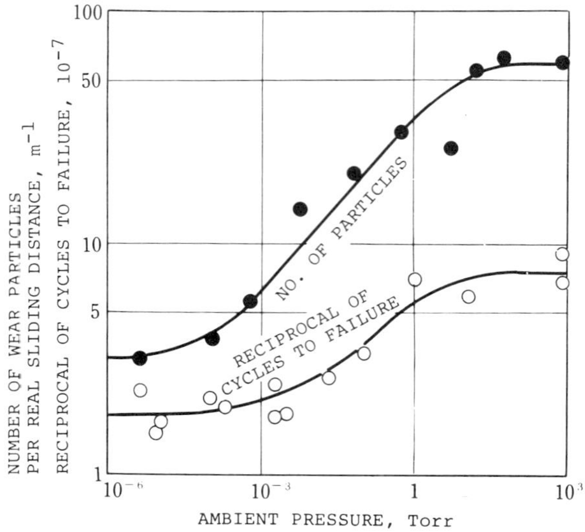

Fig. 4.—Response of materials(9)*. Increasing atmospheric pressure reduces fatigue resistance of Ni, and thereby increases the formation rate of wear particles in Ni/Ni sliding. Wear experiment: normal load 260 g, sliding speed 168 mm/s, sliding distance 50 m. Fatigue experiment: reverse bending at 16.7 Hz, initial maximum strain 0.077%.*

The fracture theory of wear as outlined above will present considerable complexity. However, as long as wear process includes fracture, it cannot be expected to be less complex than fracture itself which has its own complexity.

REFERENCES

1. Barwell, F.T., These Proceedings.
2. Feng, I.-Ming., *Journal of Applied Physics*, Vol. 23, 1952, p. 1011.
3. Kraghelsky, I.V., *American Society of Mechanical Engineers Transactions*, Vol. D87, 1965, p. 785.
4. Halling, J., *Wear*, Vol. 34, 1975, p. 239.
5. Finkin, E.F., *Wear*, Vol. 47, 1978, p. 107.
6. Soda, N., Kimura, Y. and Tanaka, A., *Wear*, Vol. 35, 1975, p. 331.
7. Soda, N., Kimura, Y. and Tanaka, A., *Wear*, Vol. 40, 1976, p. 23.
8. Suh, N.P., *Wear*, Vol. 44, 1977, p. 1.
9. Kimura, Y., in JSLE-ASLE International Lubrication Conference, Proceedings, edited by T. Sakurai, Elsevier, 1976, p. 67.
10. Soda, N., Kimura, Y. and Tanaka, A., *Wear*, Vol. 43, 1977, p. 165.

DISCUSSION

QUESTIONER: I was impressed by your correlation between regular or severe wear, wear and atmospheric pressure. I wonder what the correlation would be for mild wear. Did you try such a correlation?

Y. KIMURA: I have not tried it, but I think it is completely different. In the case of mild wear, there will be predominant effect or surface films as it may change the actual forces on the asperity. In the results shown the actual force is almost similar to that at atmospheric pressure and therefore such a correlation is observed. But in the case of mild wear I am not sure.

N. P. SUH, MIT: Professor Rabinowicz and Dr. Ohmae were doing experiments with pure metals and they got good correlation between wear resistance and hardness also. And the point I would like to make here is that if you start with multiphase metals with different microstructure, they are not going to give you good correlation with hardness. This is very important to recognize. Just because there is some kind of correlation with single phase metals which are unique (many of the engineering materials are not single phase), one can easily be mislead. If you plot wear rate as a function of hardness or fatigue life you can get a correlation. Some times these correlation schemes are very simplistic because the process is very complex.

KIMURA: Why doesn't a correlation exist in the case of two-phase metals? Do you think it is a completely different mechanism from fatigue?

SUH: Well, you are asking a question which would take some time to answer. Let me simply say the following. The extrusion and intrusion ideas Professor Argon metnioned in his paper in these proceedings do not occur in the case of wear. The kind of things that happen in uniaxial fatigue are very different from those that happen in the wear. Wear, as we all know, is very complex.

In the case of two-phase materials, much internal stress develops around inclusion which is responsible for crack formation. To believe that single phase materials, which do not have inclusions, behave in the same way as the two-phase materials, is not correct.

The other point I would like to make is that the problem is not simply correlating wear with ductility. In steels treated to yield pearlitic and spheroidal structure, we find that uniaxial tension tests give different ductility; the wear resistance is exactly opposite to the uniaxial ductility. The point I am trying to make is that one cannot take uniaxial tension test results and try to predict wear rate. That is a very difficult thing to do.

KIMURA: There are two problems. The first is the contact mechanics. If the contact mechanics is varied, the correlation would not be the same. But when the contact mechanics is the same, then I do not agree with your statement. It is a problem of relative size of cracks in the substrate to that of the inclusion. If the second phase is very small compared to the expected damage such as a crack, I think basically the same reasoning will hold.

S. GANESH, Bendix Research Laboratory: I think we are talking about fatigue crack initiation and propagation. We accept fatigue as one of the mechanisms for the wear damage. In pure material you have to initiate cracks and propagate it. The fatigue life is perhaps longer. In multiphase materials with inclusions and other phases, we have built-in cracks, and we do not have to initiate a fatigue crack but we have to make that crack grow. In that sense the fatigue process is still there in multiphase material, but we are possibly eliminating the initiation process. Otherwise they are both the same.

SUH: We have done a substantial amount of work on crack nucleation and propagation in wear. One of the things we looked at first was the number of cycles needed to nucleate a crack, both in single-phase and two-phase materials. We then asked how a crack propagates in wear and how does that differ from the fatigue case. One difference that comes out very obviously is that in the case of normal fatigue, the stress intensity factor increases as the fatigue test proceeds. In wear test that does not happen because of the geometric configuration and loading conditions. There are some basic differences. There are differences in crack nucleation mechanisms as well because of the very high hydrostatic pressure. One cannot simply look at the uniaxial tensile data and predict wear behavior from that. If we could do that it would be great, but we cannot.

KIMURA: I do not think that the direct comparison is always possible. Of course, fatigue should be compared on the basis of mechanical conditions, that is, the same stress intensity factor. But in my opinion this difference does not prevent correlation of wear data with fatigue data.

INTERFACES OF WEAR AND FATIGUE

R. A. Smith

ABSTRACT

Wear, leading to surface deterioration, in components subjected to heavy concentrated loads, is examined as a fatigue process.

The background of fracture mechanics and its application to fatigue crack propagation is reviewed. Some of the theories proposed to predict the direction of growth of cracks subjected to mixed mode loading are discussed. New experimental results on fatigue crack growth under compressive loading are presented. The cracks are shown to extend in a shear mode, whilst spherical wear debris is produced by the repeated sliding action.

Previous reported attempts to examine crack propagation in contact type stress fields are discussed and the areas of uncertainty of this type of analysis are outlined.

INTRODUCTION

Wear in some types of components may be regarded as a fatigue process if the load is applied repeatedly and the mechanism of wear involves some degree of crack propagation. The review by Barwell[1] cites examples of failure in roller contact bearings, gears and rails subjected to wheel loads.

For some time the generally held view has been that the mechanism of failure involves the initiation of cracks (at a sub-surface stress concentrating feature) by the maximum local shear stresses generated by the applied Hertzian type loading.[2,3] Subsequent propagation of these cracks towards the surface leads to flakes of material becoming detached, and the rapid onset of failure. In other cases cracks might be initiated on the surface and their progress accelerated by entrapped oil. Traditional approaches to fatigue assessment of ball bearings have been based on semi-empirical endurance predictions, typically by combining a stress/life material relationship with the Hertz shear stress, to yield a strong power law relationship between bearing load, P and life of the type, N_f:

$$N_f P^3 = const \tag{1}$$

However, more recently the emphasis of fatigue research has tended towards the study of crack propagation. By using the tools provided by this more quantitative approach, we may hope to predict growth rates and directions of crack advance in cracked components subjected to repeated Hertzian (perhaps modified by rolling, slipping and plasticity effects) loading conditions. This paper aims to discuss and review the basis of fracture mechanics and to discuss recent theoretical and experimental approaches to the Hertzian cracking problem.

PRINCIPLES CF FRACTURE MECHANICS

The so called 'Fracture Mechanics' approach depends on the simplified description of the stresses near a crack tip. The linear elastic solution for the stress field around a crack shows that the stress components σ_{ij} are always of the same form; in terms of polar co-ordinates r, θ from the crack tip, they are

$$\sigma_{ij}(r,\theta) = \frac{K}{\sqrt{2\pi r}} f_{ij}(\theta) + \text{'other terms'} \qquad (2)$$

If the point (r,θ) is sufficiently close to the crack tip then the singular nature of the stress field dominates and the *'other terms'* can be neglected. The stress intensity factor, K, is a function of the loading applied to the cracked body and its geometry. For a loading system tending to open the crack surfaces (Mode I), we define the stress intensity factor as:

$$K_I = \alpha \sigma \sqrt{\pi a} \qquad (3)$$

where α is a geometry correction factor which modifies the solution of a central crack in an infinite plane solution to account for more realistic boundaries ard loading configurations, σ is the remote applied stress and α the work length. An analogous factor K_{II} is used to describe loadings tending to cause the crack surfaces to slide in the plane of the crack (Mode II). Physically, the stress intensity factor K_I may be identified with the 'force' being applied which tends to cause cracking.

It is assumed, and confirmed experimentally, that a given material is only able to sustain a limited stress intensity factor before the crack advances catastrophically. This maximum value, the critical stress intensity factor, associated with plane strain deformation, is called the fracture toughness of the material, K_{IC}. A suitable review of these concepts can be found in a report by Pook[4], whilst the techniques used to measure fracture toughness are described in several other publications.[5-7] It will be realised that real materials are unable to sustain the infinite crack tip stresses predicted by Eqn. 2 and some plastic deformation occurs. We are still able to use the simplified description given above if this plastic deformation is limited compared with the crack length and other specimen dimensions. Full discussion of these limitations is given in the references.

APPLICATION TO FATIGUE CRACK GROWTH

At values of applied stress intensity factor less than the fracture toughness of a material, sub-critical extension can occur by fatigue and/or corrosion mechanisms. It has been well demonstrated experimentally that the *range* of stress intensity factor, $\Delta K = \alpha(\sigma_{max}-\sigma_{min}) \sqrt{\pi a}$, can be used to correlate crack growth rates. A typical relationship is shown in Figure 1 - the majority of data can be represented in the form:

$$\frac{da}{dN} = C(\Delta K)^m \qquad (4)$$

where C and m are material constants, m for steels being typically 4, although somewhat higher values have been reported for harder materials.[8] A minimum value of stress intensity factor, below which fatigue crack growth does not occur, and associated with the minimum amount of crack tip elasticity needed for crack growth, is termed the threshold stress intensity factor, ΔK_{TH}.

We might note in passing that the observations outlined above indicate that in ball bearings the traditional, $N_f P^3 = constant$, relationship, Eqn. 1, can only arise if the life is initiation rather than propagation controlled, otherwise from the integrated form of Eqn. 4, a relationship much less dependent on load, $N_f P^{4/3} = const$, would result. On the other hand, propagation might well dominate the situation of growth from existing flaws in rail heads.

PREDICTION OF THE DIRECTION OF CRACK GROWTH

Most of the experimental data referred to so far has been collected for pure Mode I loading, i.e. tensile stress perpendicular to the crack surfaces.

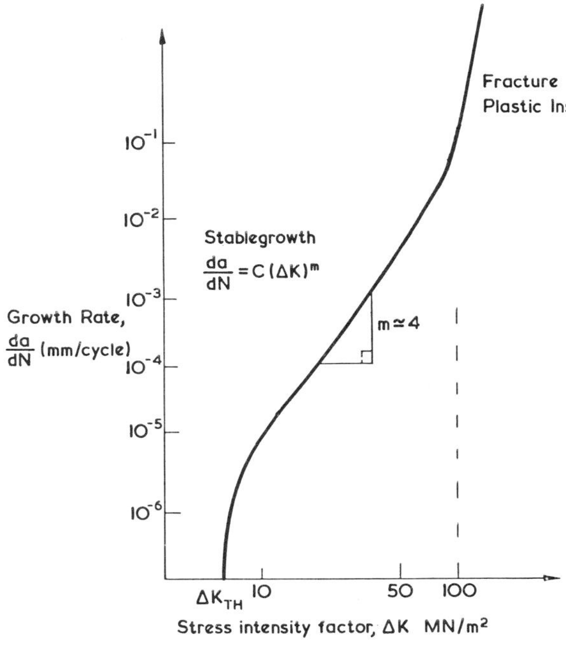

Fig. 1.—Generalised relationship between fatigue crack growth rate and stress intensity factor range for metals.

The question of the *direction* of crack advance does not arise, as the crack is assumed to extend in a self similar manner. However, when the loading is pure Mode II or a combination of Modes I and II, which is of particular interest in contact fatigue problems, the prediction of the direction of growth is important. Much recent work has centered on this problem without yet reaching a satisfactory conclusion.

Briefly, the main approaches, reviewed in detail by Swedlow(9), have involved the calculation of either local stress or energy quantities on contours surrounding the crack tip. The simplest criterion is to calculate the direction of maximum hoop stress, e.g. Erdogen and Sih(10), the crack then extending under local Mode I control. Most efforts to obtain pure Mode II loadings on cracks, Jones and Chisholm(11), Pook(12), have resulted in cracks tending to orientate themselves in a Mode I direction. An alternative approach, in some respects an extension of the Griffith energy method, is to calculate the direction in which strain energy density is a minimum, Sih.(13,14) In the case of applied *tensile* principal stress, both methods lead to similar results; all the local field quantities have very shallow maxima or minima with respect to angular position from the crack tip, a fact which is reflected in the considerable scatter of experimental results. No clear distinction between the rival criteria can therefore be made on the basis of experimental comparison.

Very little data exists for the compression principal stress case. The most serious objection to Sih's hypothesis here is that no distinction is made between tension and compression strain energy density. Other difficulties arise in the modelling of the contact between the crack surfaces; a slender ellipse has been used to avoid this problem, otherwise some coefficient of friction must be added to a line crack model and the Mode I stress singularities dropped. Finally, most of the debate has centered around *monotonic* loading extension of a crack in brittle materials. Perspex or *PMMA* have been the materials used in experimental investigations. Little attention has been paid to the problem of mixed-mode sub-critical fatigue extension under *cyclic* loadings.

EXPERIMENTS ON ANGLED FATIGUE CRACK GROWTH

Tanaka(15) has tested angled cracks in commercially pure, near isotropic, aluminium plate of 3.2 mm thickness. Fatigue cracks were grown from initial saw cut slots, by loads applied normal to the slot direction. Several mm of crack extension at either end of the slot resulted in an overall slot + crack length of about 15 mm. Low fatigue loads were used to give growth rates in the order of 7×10^{-6} *mm/cycle*. New smaller specimens were then cut at an angle to the initial crack direction (for details see figure 2), and the specimens were then annealed to remove crack tip damage for 2 hours at 270°C. The resulting angled crack specimens were then subjected to tensile fatigue loads. The angles of crack extension, θ_0, measured as a function of the inclination of the crack, ϕ, are shown in Figure 3.

A further set of experiments on the same material, using the same preparation procedure and testing machine were performed by the present author. For these experiments a wholly compressive fatigue load was applied. A balance had to be made to apply sufficient load range to cause crack extension, yet not enough to buckle the specimens. The main difference between the tension and compression cases was that the compression cracks all began to extend in their own direction, then stopped. This is a case of pure

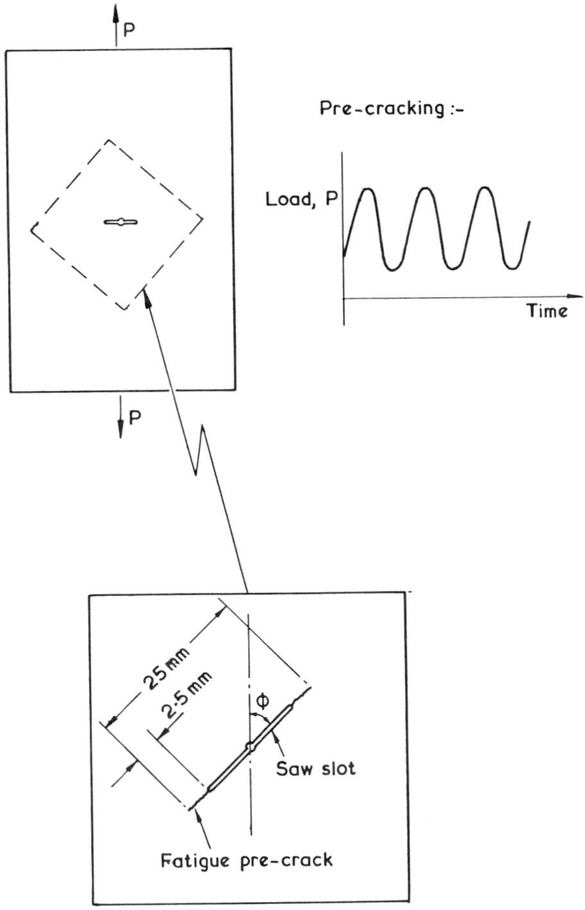

Fig. 2.—Details of pre-cracked inclined angle fatigue specimens. (Material - 3.2 mm thick, Aluminium plate).

Mode II extension before arrest, presumably caused by an increasing friction retardation as a greater length of crack surface came into contact. On continued cycling, angled cracks grew from the ends of the saw slots. The angles are shown in Figure 3.

Too few results have been obtained to arrive at firm quantitative conclusions. That there is a considerable amount of scatter is obvious, as is the need for some free surface movement for the start of the angled crack growth, which in the compression case was from the ends of the saw slot, rather than the longer fatigue cracks. It must be realised that fatigue crack growth is a consequence of the magnitude and direction of the irreversible plastic deformation at a crack tip. Although the elastic stress intensity factor approach gives us a good correlating parameter for growth

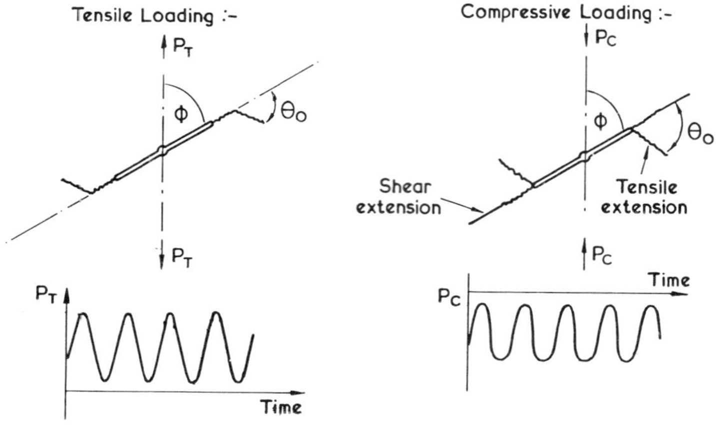

Initial crack angle ϕ (degrees)	Extension angle, θ_o (degrees)	
	Tensile load	Compressive load
90	0	—
72	28	—
60	—	56
45	49	76·5
30	{23, 52}	{30, 63}

Fig. 3.—Details of the extension directions of angled cracks subjected to tensile and compressive fatigue loadings.

rates, its major achievement is that it de-focusses attention from the crack tip by using bulk parameters. Work is now in hand to investigate the nature of crack tip *plastic fields* subjected to various loading combinations, and hence to predict the direction of crack growth.

Examination of the fracture surfaces under a S.E.M. showed that the angled cracks demonstrated features typical of fatigue cracks grown by tensile stresses across the crack surfaces, see Figure 4. The Mode II cracks however, showed considerable deformation typical of rubbing surfaces and a surprisingly large amount of wear debris in the form of spherical particles, Figure 5.

SPHERICAL WEAR PARTICLES

The presence of these particles was gratifying evidence of the relevance of these experiments to the rolling contact problem. As mentioned by Barwell[1], a number of investigators have observed these particles[16-19],

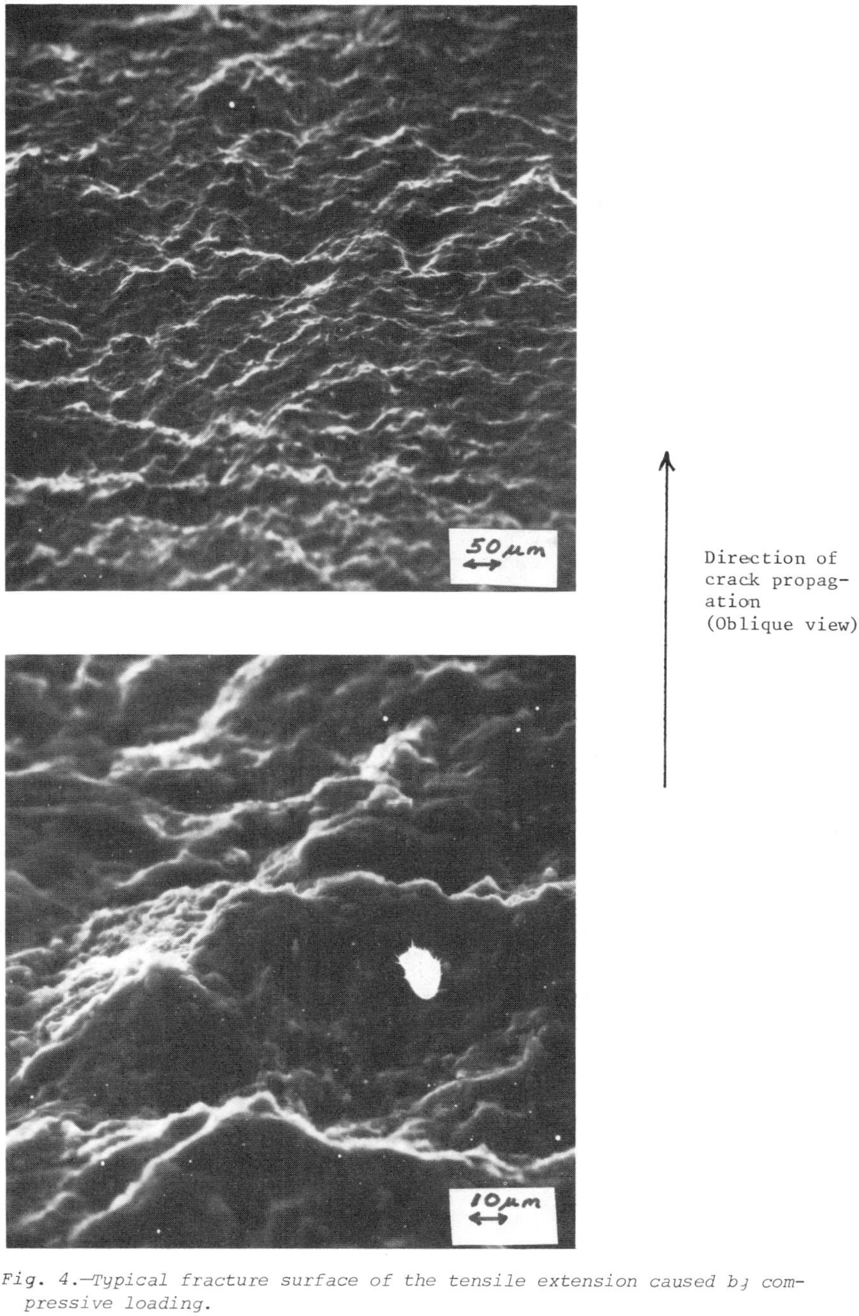

Fig. 4.—Typical fracture surface of the tensile extension caused by compressive loading.

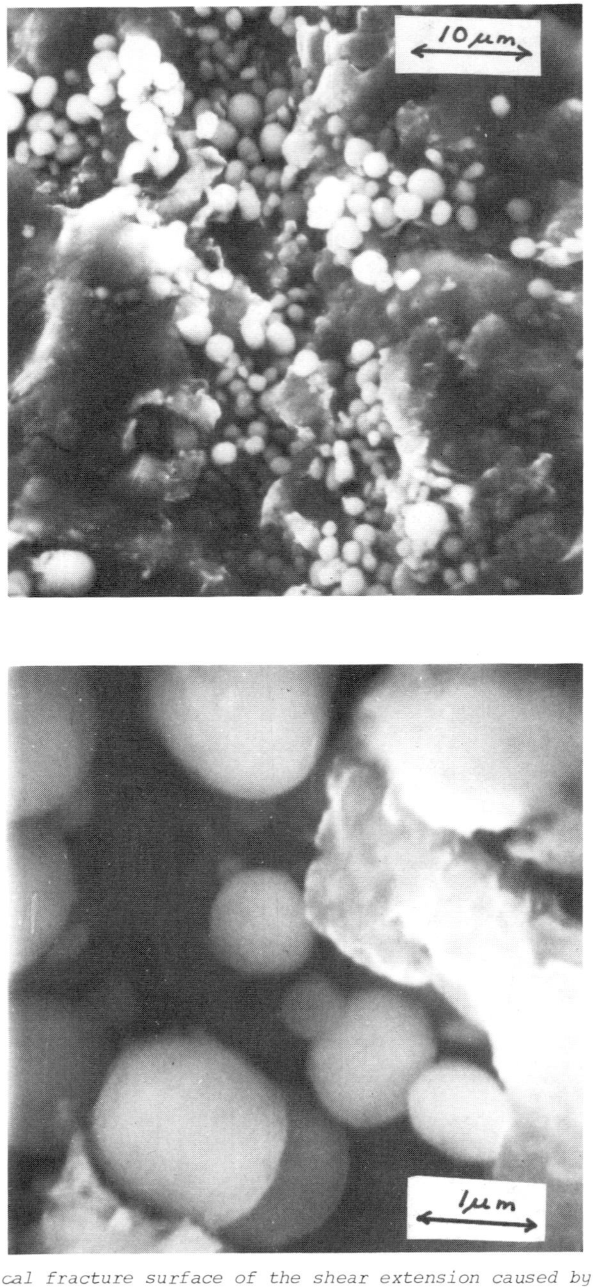

Fig. 5.—Typical fracture surface of the shear extension caused by compressive loading.

which are so characteristic of rolling contact fatigue, that their presence in lubricating oil can form the basis of a powerful diagnostic tool for the detection of ailing bearings.[20]

Various mechanisms have been suggested for the formation of these particles - sub-surface deformation between two parallel cracks[21], deformation of slivers of metal caused by build-up of lubricant pressure in the fatigue crack[20] and local high temperatures sufficient to cause melting and spheroidization.[22,23] Rabinowicz[24] has suggested that adhesive wear particles are trapped in cavities in the sliding surfaces and become smooth by burnishing. He derived a criterion:

$$d \leq x.k_b \tag{5}$$

where d is the maximum diameter of the spherical wear particle, x is the total sliding distance and k_b is the wear coefficient for burnishing, typically 10^{-6} for unlubricated metals. By using the near crack tip stress field/displacement equations (Appendix) for the geometry of our aluminium speciman subjected to a compressive load range of 1.5 tons, then for a 45° angle crack, the displacement per cycle is in the order of $2\ \mu m$. The surfaces were examined after 10^6 loading cycles, giving a total movement of $2\ m$. Thus $x.k_b$ is of the order $2 \times 10^{-6}\ m$ and the particles were of $1\ \mu m$ diameter, in agreement with the inequality of Eqn. 5.

It would seem likely that the melting mechanism is not applicable in this case, since the sliding velocities are very low; certainly there is no lubricant interaction, so a mechanical rolling action is the most probable candidate for the production of the particles. The lack of cylindrical or cigar-shaped particles amongst the debris is remarkable - given the one dimensional nature of the relative movement of the surfaces. One must assume that the highly irregular crack surfaces force the debris to follow an almost random path during the rounding process. From the S.E.M. evidence, very few particles seem to be restrained by 'sockets' as required by the Rabinowicz[24] model.

Summary of Reported Work

Finally, a brief summary will be made of some of the recent work reported in the literature which is of direct relevance to this problem.

Gervais and McQueen[25] simulated the fatigue of rails by repeatedly pressing a hardened steel cylinder against a steel plate. The sub-surface dislocation rearrangement was studied as a function of number of loading cycles, but even after 80×10^6 cycles, no sub-surface micro cracks were observed. The same authors later published[26] a numerical analysis of the same problem which predicted the location of maximum shear stress amplitude to be the same as the regions of maximum structural change of the experimental specimens.

Shieh[27] has specifically designed a specimen to produce compressive maximum shear failure in Mode II. Both the fracture initiation stress and the crack propagation rate, with respect to stress increment (*not* fatigue loading cycles) were determined for a 52100 bearing steel. Tail and parallel multiple cracks, features often noted in rolling contact problems, were observed.

Suh's delamination theory of wear[28] depends upon the multiple initiation and propagation of sub-surface cracks on paths parallel to the surface. Sheets of material delaminate when the cracks finally shear to the surface (at certain weak positions). Jahanmir and Suh[29], Fleming and Suh[30], analyse respectively the mechanics of initiation and propagation of such cracks. An approximate form of the stress intensity factors generated by an elliptical distribution of loads at the surface was used. A model was constructed for the correlation of crack propagation and wear rates which was shown to be in reasonable agreement with experimental values in aluminium alloys.[31] However, although the growth rates were shown to fit an equation of the form of Eqn. 3, $\frac{da}{dN} = C(\Delta K)^m$, it was not clear in what mode these cracks were propagating.

On the rail cracking problem McClintock[32] has reported a boundary integral relaxation method used to calculate the plasticity at the tip of a small horizontal crack in the rail head. A wheel passage was shown to give initial sliding, followed by locking, squeezing of the plastic zone, reversed sliding, locking and finally unloading. The flow was shown to be primarily shear along the plane of the crack indicating the likely direction of growth to be the same as the crack direction. It was concluded that the problem could be approximated with reasonable computational economy.

CONCLUSIONS

Many features of the problem of contact fatigue still await investigation by fracture mechanics methods. The main areas in which further knowledge is required are:

(i) The details of the initiation of cracks from second phase particles or inclusion either on or below the surface.

(ii) Mode II (sliding) type of crack growth, particularly when the crack surfaces are simultaneously subjected to compressive loads.

(iii) Nature of the criteria governing the direction of extension of cracks subjected to mixed mode fatigue loading.

(iv) Solutions for the stress intensity factors of arbitrarily inclined sub-surface cracks subjected to the passage of Hertzian type loading.

(v) The limiting cases as these cracks approach the free surface, since this event will mark the rapid approach of failure.

REFERENCES

1. Barwell, F.T., These Proceedings.
2. "Fatigue in Rolling Contact," Institution of Mechanical Engineers, London, 1963.
3. Littmann, W.E., in NASA SP-237, edited by P.M. Ku, 1970, p. 309.
4. Pook, L.P., "Linear Elastic Fracture Mechanics-What It Is, What It Does," NEL Report No. 465, Glasgow, 1970.
5. ASTM, STP 381, American Society for Testing and Materials, 1965.
6. ASTM, STP 410, American Society for Testing and Materials, 1966.
7. ASTM, STP 463, American Society for Testing and Materials, 1970.
8. Pook, L.P., *Journal of Strain Analysis,* Vol. 10, 1975, p. 242.

9. Swedlow, J.L., in ASTM STP 601, American Society for Testing and Materials, 1976, p. 506.
10. Erdogen, F., and Sih, G.C., *Journal of Basic Engineering*, Vol. 85, 1963, p. 519.
11. Jones, D.L. and Chisholm, D.B., *Engineering Fracture Mechanics*, Vol. 7, 1975, p. 261.
12. Pook, L.P., *International Journal of Fracture*, Vol. 13, 1977, p. 867.
13. Sih, G.C., *Engineering Fracture Mechanics*, Vol. 5, 1973, p. 365.
14. Sih, G.C. and MacDonald, B., *Engineering Fracture Mechanics*, Vol. 6, 1974, p. 361.
15. Tanaka, K., *Engineering Fracture Mechanics*, Vol. 6, 1974, p. 493.
16. Scott, D. and Mills, G.H., *Wear*, Vol. 16, 1970, p. 234.
17. Stowers, I.F. and Rabinowicz, E., *Journal of Applied Physics*, Vol. 43, 1972, p. 2485.
18. Scott, D. and Mills, G.H., *Wear*, Vol. 24, 1973, p. 235.
19. Hurricks, P.L., *Wear*, Vol. 27, 1974, p. 319.
20. Scott, D. and Mills, G.H., *Nature*, Vol. 241, 1973, p. 115.
21. Loy, B. and McCallum, R., *Wear*, Vol. 24, 1973, p. 219.
22. Brosseit, E. and Hess, F.J., *Wear*, Vol. 17, 1971, p. 314.
23. Doyle, E.D., *Metallurgical Forum*, Vol. 19, 1974, p. 276.
24. Rabinowicz, E., *Wear*, Vol. 42, 1977, p. 149.
25. Gervais, E. and McQueen, H.J., *Iron and Steel Institute. Journal*, March 1972, p. 189.
26. Gervais, E. and McQueen, H.J., *Canadian Metallurgical Quarterly*, Vol. 15, 1976, p. 97.
27. Shieh, W.T., *Engineering Fracture Mechanics*, Vol. 9, 1977, p. 37.
28. Suh, N.P., et al., "The Delamination Theory of Wear," Elsevier Sequoia, S.A. Lausanne, 1977.
29. Jahanmir, S. and Suh, N.P., *Wear*, Vol. 44, 1977, p. 17.
30. Fleming, J.R. and Suh, N.P., *Wear*, Vol. 44, 1977, p. 39.
31. Fleming, J.R. and Suh, N.P., *Wear*, Vol. 44, 1977, p. 57.
32. McClintock, F.A., in "Fracture 1977," edited by D.M.R. Taplin, Vol. 4, University of Waterloo Press, Waterloo, Canada, 1977, p. 49.

DISCUSSION

N. P. SUH, MIT: We all agree that crack propagation is a major factor in wear. I would like to suggest one thing. It may also steer your research in a way that we could made use of the crack propagation data. When we calculate the crack propagation rate per cycle,it is almost in the threshhold regime of the stress intensity factor. My colleagues tell me that microstructure, environment, etc. are very important in that regime. If you could run an experiment at small ΔK, it would be very interesting.

The reason spherical particles are observed in ball bearings and not in sliding situations is because of the following reason. In normal sliding situations, the way the compressive loading is applied does not slip very much along the crack. I think mode II is important and I agree with you. The question is what happens when we have mode I and mode II combined.

R. A. SMITH: That was the object of using angle cracks because that is an easy way of combining mode I and mode II loading. What we really want to do is to try to go for the simplest geometry we can and really understand the simple situation with uniform loading conditions. We cannot get much simpler geometry than an inclined crack; but even that is quite difficult.

S. GANESH, Bendix Research Laboratory: You showed spherical particles and they looked a little different in terms of the color compared to the fractured steel surface. Did you do any X-ray spot analysis on those particles? Were they oxide particles?

SMITH: We did think that they might be oxide particles. If they were oxide particles we might expect the same particles to be in other cracks as well. But they were completely absent from the other cracks. The difficulty about the microprobe analysis is that the oxygen is a very light element and it is difficult to distinguish the trace oxygen from the aluminum.

D. SCOTT, Paisley College of Technology: You mentioned that you consider crack initiation in rolling contact fatigue is more important than crack propagation. Yet from service data and experimental results we find that lubricants in the form of environment can reduce bearing life by considerable amounts (about 1/5). This clearly shows that propagation is really the important process in rolling contact fatigue. Secondly, you talk about crack initiation and crack propagation, but how do you differentiate between when the crack initiates and when it propagates? In other words, when is a crack not a crack? The third point is that if you get the spherical particles forming at the cracks below the surface and then when it opens up, you get them out where do these particles come from below the surface and what is the original shape of the particles before they are rolled into balls?

SMITH: The first point was about initiation. That can be affected by the environment very severely. The cracks are coming from the surface and the lubricant will affect them. The second point or question -- when is a crack not a crack? It is just impossible to give you any sensible definition. In some simple cases we can say initiation is finished when slip bands have developed and the cracks start propagating; they always seem to be propagating to begin with in this mode II shear growth before the tensile stresses take over. Obviously that is not a useful definition for this sort of experiment. The spherical particles come from the rubbing surfaces. I can not see where else they can come from.

GANESH: How do they form into spherical shape?

SMITH: By mechanical working. That is the only reason I can think of in these experiments. If you take two flat plates with a piece of plasticine in between and roll them around it becomes a sphere.

J. L. TEVAARWERK, University of Waterloo: Do you have any idea of the relative motion between the two surfaces right at the crack tip? Is it about two or three diameters of the particles or less than that? I mean it must roll at least once.

SMITH: Right at the crack tip, the relative displacement is zero.

TEVAARWERK: Therefore, they form behind the crack tip.

SMITH: Yes, they form at some distance from the crack tip. About 25 to 50 μm depending on the load.

TEVAARWERK: When you broke your specimen apart, did you notice any particles very near the crack tip or was it too small a dimension to observe?

SMITH: We have some information on that. They spread fairly well over the whole fracture surface with a tendency to get high density away from the crack tip, but I wouldn't be too specific, we have not yet done enough experiments.

VIII. THERMOMECHANICAL EFFECTS

THERMOMECHANICAL EFFECTS IN SLIDING WEAR

R. A. Burton

ABSTRACT

When sliding occurs with significant frictional heating, thermoelastic deformation may lead to a transition from smoothly distributed asperity contact to a condition where the surfaces are supported by a few thermal asperities. This circumstance may be associated with a transition to a condition of severe wear because of the elevated contact pressure and temperature, and also because of production of tensile stresses. This second stress component may lead to heat checking whereupon the rough checked surface acts to abrade the mating material.

The factors influencing transition are discussed, including wear, cooling, and hydrodynamic lubrication. The transition state is also discussed as to stress distribution, rate of movement of the contact patches and temperatures.

NOMENCLATURE

- b parameter in heat equation
- c speed of movement of contact patch along surface
- E Young's modulus
- E^* modified modulus for two body contact: $1/(1/E_1 + 1/E_2)$
- h axial length of sealing ring
- $\bar{h}$ film thickness in seal
- k thermal diffusivity
- K thermal conductivity
- ℓ half-length of contact patch
- p pressure
- q heat flow through unit area of surface
- Q total heat flow through contact patch
- R radius of curvature of surface
- t time

T	temperature
v	surface normal displacement
V	sliding speed
V^*	critical sliding speed
w	wear coefficient
w^*	critical wear coefficient
x	coordinate along direction of sliding
y	coordinate normal to surface
Pe	Peclet number
$Pe\ell$	Peclet number of contact patch
α	coefficient of thermal expansion
ζ	heat transfer parameter
η	viscosity
κ	wave number
λ	wave length
μ	friction coefficient
ξ	dummy variable
$(\hat{\ })$	implies amplitude of wave
$(\overline{\ \ })$	implies mean value

INTRODUCTION

The process of transition from nominally flat to highly deformed surfaces may be spoken of as *thermoelastic transition.*(1-9) It may sometimes occur in a sequence of stable, continuously related states as operating conditions are changed. At other times, however, the stably evolving behavior of the sliding system crosses a threshold, whereupon a sudden change of contact conditions occurs as the result of an instability. This involves a feedback loop which comprises: localized elevation of frictional heating, resultant localized thermal bulging, localized pressure increase as the result of the bulging and further elevation of frictional heating as the result of the pressure increase. This process, when it leads to an accelerated change of contact stress distribution, is spoken of as *thermoelastic instability* (TEI).

The ultimate result of growth of the thermal disturbance is the parting of the surfaces in some portions of the nominal contact area, with gap height being several times the roughness-asperity height. As a consequence of reduction of the nominal contact area the remaining contact patches acquire elevated stress.

Contact patch formation can occur in lubricated as well as dry contact and is influenced by wear, cooling, materials properties, and macroscopic constraints on the contacting bodies. The physics of thermoelastic transition, thermoelastic instability and contact patch behavior have begun to receive serious attention only in the past few years, and the explanation of many important effects is not generally well known. For this reason space must be devoted here to a brief introduction of the phenomena involved, before discussion may proceed to problem areas needing further research.

HISTORICAL BACKGROUND

Long before the introduction of the concepts of thermoelastic transition Ling and Mow[10] developed an influence function for surface displacement for high-Peclet* number sliding of bodies in plane strain, and outlined the procedure for treating a moving contact patch on the surface of a slab, also for high Peclet number. Mow and Cheng[11] have examined the companion problem of thermal stress in plane elastohydrodynamic contact. Early investigations of thermoelastic effects on lubricated sliding were reported by Korovchinsky[12] for the change of contact stress of a ball with frictionally heated contact, and by Nica[13] for radius change of journal bearings with fixed external radius. The bearing work has been broadened[14,15] to include cooling effects and more realistic boundary conditions for fluid-film bearings, and also to show that similar phenomena exist for rolling contact bearings.[16] Most interesting here is the role of thermal expansion in the catastrophic chain of events of seizure of the bearing.

Sibley and Allen[17] carried out a series of experiments on seal materials, developing a criterion for thermal checking and showing photographic evidence of systematically moving hot patches in the contact zone.

Interest in contact instabilities and patch formation was accelerated by the work of Barber[18-20] who has demonstrated the phenomenon experimentally and has provided analyses which partially explain his observations, as well as fundamental contributions to the field of thermoelasticity.

His explanations draw upon the modification of asperity contact by frictional heating and wear. His initial interest was in explaining hot-spot effects in railroad brakes.

Dow[21,22] addressed the problem of a scraper sliding perpendicular to its edge on a conductive slab and showed that (1) instability would be predicted in the absence of wear and (2) would be modified by wear which would: (a) raise the instability threshold and (b) sometimes give rise to oscillating pressure. He carried out a numerical simulation[23] which predicts the formation of contact patches, and their translation along the edge of the scraper as the result of wear. More recently he has carried out experiments which vividly display patch formation.[24] The onset of such patches is close to his theoretical critical sliding speed for instability for dry contact; but, although qualitatively similar, it is only poorly predicted for wick lubricated contact.

Kennedy and Ling attacked the problem of severely loaded aircraft brakes drawing upon numerical analysis,[25] later followed by experiments.[26] They postulated a wear model which suggests the mild-wear/severe-wear transi-

* Peclet number is a dimensionless measure of speed of movement of a heat source, being of the form $c\ell/k$, where c is speed of movement, ℓ is a characteristic length (patch length) and k is thermal diffusivity.

tion, assuming that no wear occurs until a critical shear stress is reached in the material. For sandwiches of moving disks, alternated with stationary ones, they find contact reducing to a band at the outer radius, then moving inward, only to repeat the sweep again several times in a typical "stop." The confinement of the disk brake causes material properties and operating parameters to interact differently than for the "floating" scraper.. Furthermore the appearance of patch contact in the brake may be a transient phenomenon rather than an instability in that there is some evidence that the sweep would ultimately die out as the brake wore-in. This cannot be tested since the overall temperature rises severely even in the brief runs reported, and thus limits the time of operation.

Nerlikar[27] has addressed the problem of ring contact in an idealized face seal sliding tangentially, and has shown that the relative conductivities of the two bodies strongly influence instability. The least stable is the thermal conductor on insulator; and the most stable is a material sliding on its own kind. This has been extended[28] to show that extremely thin contaminant, solid-lubricant or oxide films can play a major role in determining instability. Nerlikar has also explored the influence of roughness on instability,[29] and has calculated the gap width and contact temperatures for seals in the contact-patch configuration.

Lebeck[30] has improved the model for sealing-ring contact, allowing for thin-beam bending, and showing how to account for cooling from the sides of the rings.

Kilaparti has addressed the problem of patch contact in the absence of wear[31] and with wear[32,33] developing an improved influence function similar to that of Ling and Mow,[34] and discovering that there is a critical wear coefficient above which instability will not occur.

Banerjee has treated the idealized seal with liquid lubricant in the hydrodynamic short-bearing regime.[35] He has predicted a range of instabilities, modified at the thin film extreme by elastic deformation under contact pressure, and at the thick extreme by convection of heat in the film. He has carried out experiments (see Figures 1 and 2) which show the instability to occur where expected.[36] More recent work[37] has modified his model but does not alter the basic conclusions.

Closely related to this work is that of Hahn and Kettleborough,[38] Ettles[39] and Tanaguchi[40] on thrust washers, where only thermoelastic deformation can explain the load support of notched rings in lubricated, wide-bearing contact. Why these rings (and for that matter sliding systems such as engine pistons) do not show thermoelastic instability is a question yet to be answered. We can only note here that Banerjee predicted a stabilizing effect from increased face width.

Heckmann[41] has reassessed the ring and scraper instability problems from the point of view of controlling-dimensionless-groups, there being w/w^* a wear measure, and H and ζ two groups for cooling. He has also extended the seal and scraper problem to line contact on a slab[42] thus bringing three-dimensionality in to replace the earlier two-dimensional models. An axially symmetric contact patch on a slab has also been treated elsewhere.[43] Heckmann's studies show that, except for quantitative differences, a cylinder sliding in line contact with a slab is substantially the same as Dow's scraper. Indeed, Dow's experimental pieces,[24] simulating Wankel seals, lie somewhere between scrapers and cylinders.

Wangkrajong[44] has carried out high speed sliding experiments for carbon-graphite on mild steel, where at 400 ft/sec and higher the evidence suggests extremely small contact area. Work in progress on high speed turbine blade and labyrinth seal rubs not only suggests patch contact, but calls for analyses to treat bouncing and vibratory compliance of the contacting members.[45] Although this high speed (1000 to 1500 ft/sec) contact represents a spectacular application where thermoelastic instability is inevitable, one should remember that the phenomenon is also found at speeds as low as 10 ft/sec in some geometries.[46]

THERMOELASTIC TRANSITION OF NOMINALLY FLAT SLIDING CONTACT

Face-seals represent one of the best definable examples of thermoelastic transition. Such seals are ordinarily in the form of rings meeting at a plane perpendicular to their axis, with contacting surfaces lapped to quarter-light-wave smoothness, and supported by bellows or O-ring arrangements to make them self-aligning. At times they may operate with no lubricant, or they may incorporate solid lubricants, boundary lubricants or even liquid lubricants which provide hydrodynamic support.

An experiment[36] which simulates the frictional contact of a face seal is illustrated in Figure 1. A cylindrical cup of metal is inverted and supported by an axial stem in the chuck of a drill press. The circular edge of

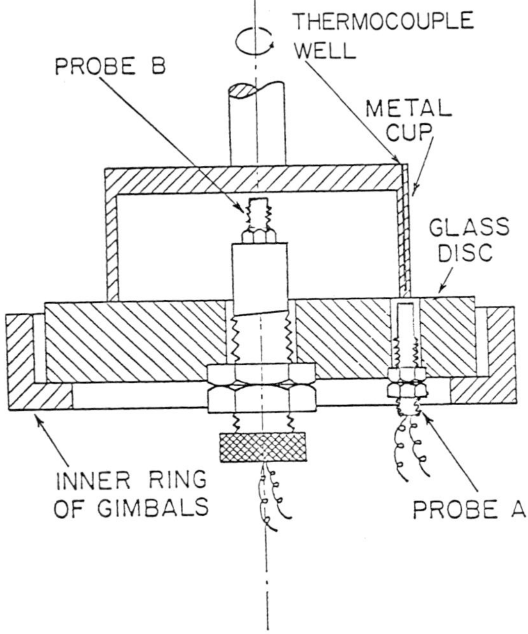

Fig. 1.—Cross-section of apparatus where ring contact occurs between an inverted cup and a glass plate. Probe A reads surface profile of the cup, while B indicates overall rising or bouncing.

the cup is pressed against a flat glass plate which is supported in gimbals to permit a self-aligning, almost-uniform contact. Whether the surfaces are lubricated or dry, as the cup turns about its axis, one will find a temperature rise which is determined by speed and load but is typically only about 100°C. As the speed is increased, one may observe the interface from below, looking through the glass plate; and at a characteristic, somewhat reproducible speed he will observe red light flashing from the contact.

If the glass and cup are inverted, so that the glass turns and the cup is stationary, the event of light evolution will be found to correspond to the appearance of two or three discrete spots on the cup, which glow orange-red and twinkle slightly. When operation is continued the spots move slowly from their original position at a rate of about 10° per minute. Thus even though the glass may be turning at 2000 rpm the hot spots will take roughly half an hour to make a circuit around the circumference of the cup. Typically they will remain as two spots 180° from one another, although for short times more spots may appear and later vanish.

Examination of the metal cup immediately after the formation of a hot spot shows evidence of some plastic flow and a darkened or "burned" patch on the metal. If examination follows a longer period of operation, it will show tracks on the surface over which the spots have traversed.

Although the spots are red-hot the general temperature of the cup remains moderate. Damage is not catastrophic because only a shallow layer of metal is affected. Nevertheless there is evidence that prolonged operation in this regime may significantly affect the wear rate and can produce surface cracking.

The first explanation that comes to mind is that the contact is at the peaks of roughness asperities. However, when one measures the contour of the metal surface in the deformed state he finds the hot spots may be associated with pimples on the surface, as high as 10^{-3} cm while the roughness is about 2.5×10^{-5} cm and initial waviness is 10^{-4} cm or less (see Figure 2). More interesting is the fact that if speed of sliding is reduced, the pimples shrink back into the surface and in some cases are replaced by slight depressions.

To explain this we may postulate a thermal expansion which produces the pimples. Then, because there is concentrated contact on their peaks, frictional heating is concentrated there so as to maintain the local thermal expansion. Since frictional heating is speed dependent, it would be reduced upon reduction of speed, and would ultimately be insufficient to maintain the height of the pimples, hence the collapse back to flat surface contact.

CONDITIONS FOR SELF-SUSTAINING PATCH-CONTACT

To show the factors which permit the existence of patch contact, let us assume that the edge of a plate rubs against a rigid thermal insulator, and that the contact patch is 2ℓ wide on the edge of the plate. Not, at this point, knowing the contact pressure distribution we assume it to be the uniform $\bar{p}$; thus if the sliding speed is v and friction coefficient is μ, the heat input per unit of area is

$$q = \mu \bar{p} v \tag{1}$$

It has been shown that the curvature of the edge of a plate is related to heat input such that[48]

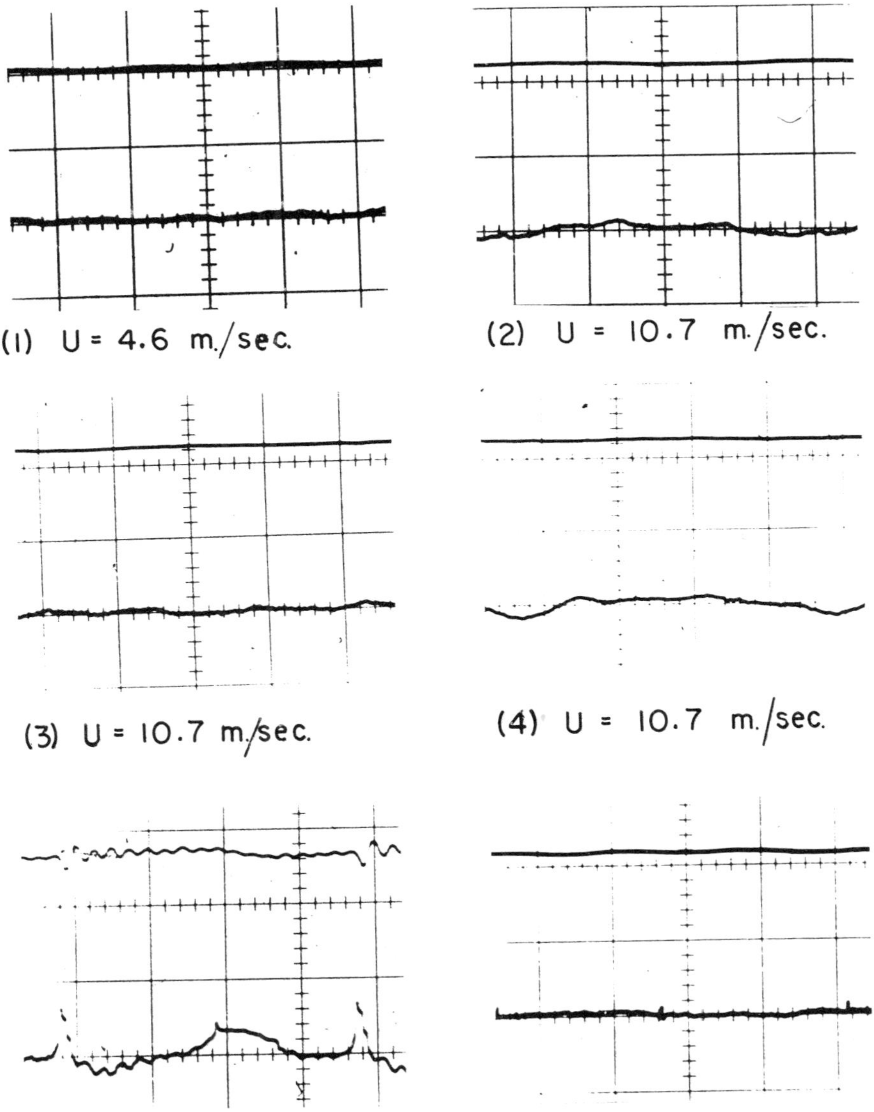

(1) U = 4.6 m./sec.

(2) U = 10.7 m./sec.

(3) U = 10.7 m./sec.

(4) U = 10.7 m./sec.

(5) U = 13 m./sec.

(6) U = 4.6 m./sec.

Fig. 2.—Oscilloscope traces of probe outputs. Upper trace is probe B and lower is probe A. As speed is increased moving waves grow in cup face and at 13 m/sec there is a sudden formation of a thermal asperity. Slight bouncing is indicated as this goes over the probe indentation. Upon lowering speed (plate 6) the thermal asperity disappears except for a tiny wear scar.

$$\frac{d^2v}{dx^2} = \frac{1}{R} = \frac{\alpha q}{K} \tag{2}$$

where R is radius of curvature, v is normal displacement and K is thermal conductivity. At this point we may ask what contact pressure would be required to flatten this curvature across the contact zone, and turn to the well known relationship for Hertzian contact

$$\ell/R = 2.3\ \bar{p}/E^* \tag{3}$$

Eliminating R between Equation 2 and Equation 3 and substituting for q from Equation 1

$$\ell = \frac{2.3K}{E^* V\mu\alpha} \tag{4}$$

More exact studies bear out the essential correctness of this finding, which shows two interesting features:

(1) the sliding speed required to sustain a patch of a particular ℓ is independent of the loading.
(2) there is nothing to prevent there being one, three or any number of contact patches on a given sliding edge, thus posing a problem of uniqueness.

Commenting on one first, we note that a load influence could appear in the physical problem if μ, E, α, K depend on temperature or stress level. Commenting on the second, there is a problem of uniqueness which disappears when a second consideration is introduced, this being the stability of the solution under small disturbances. A test for this can easily be introduced at the present level of argument.

Nerlikar's work(29) showed that if the contact load on a patch is increased the surfaces will move further apart in the region between patches. This paradoxical effect causes leakage to increase with increased face load on a seal. It also gives rise to an instability which causes the number of contact patches to reduce to the minimum number consistent with mechanical equilibrium. For a self-aligning seal the minimum is *two* patches; for one where tilting is constrained *one* patch is the minimum.

To visualize this we note that the seal can be developed into an infinite plate, where contact patches are cyclically repeated. If alternate patches are reduced in load the others must experience load increase. These load shifts give rise to displacements which further increase the load changes until ultimately the patches with decreasing load break contact.

THERMOELASTIC INSTABILITY

The question arises now as to how smooth contact evolves to the patch configuration. This may be posed as a stability problem, where the conditions may be found for run away growth of a small departure from uniform pressure. If the edge of the cup is allowed to have a circumferential waviness, and if the waviness is pressed flat by axial loading, the contact pressure may be represented by

$$p = \bar{p} + \hat{p}\cos\theta = \bar{p} + \hat{p}\cos\kappa x \tag{5}$$

where θ is a measure of angular position and x is a curvilinear coordinate along the edge of the cup, the pressure $\hat{p}$, will be spoken of as the "pressure perturbation." Frictional heating will occur according to Equation 1, but the uniform portion produced by $\mu\bar{p}V$ will not affect waviness, only causing the cup to flare out a small amount. The perturbation will give rise to a distribution: $\hat{q} \cos \kappa x$. This may be inserted into Equation 2 and integrated to produce a waviness:

$$v = \frac{\alpha}{K} \frac{q}{\kappa^2} \cos \kappa x \tag{6}$$

From elasticity theory it is known that the pressure required to press flat a sine wave of displacement v is

$$p = \frac{E\kappa v}{2} \tag{7}$$

Hence, if the displacement increment from the heating is held flat it will supplement the initial perturbation by an amount

$$p_{th} = \frac{E\alpha}{2K\kappa} \hat{q} \cos \kappa x' \tag{8}$$

But the heating perturbation will now be determined by

$$\hat{q} = \mu V(\hat{p} + p_{th}) \tag{9}$$

Inserting this into Equation 8 yields

$$p_{th} = \frac{E\alpha\mu V}{2K\kappa} (\hat{p} + p_{th}) \tag{10}$$

Let us draw upon the convenient grouping of variables V^* to simplify Equation 10; where $V^* = (2K\kappa)/(E\alpha\mu)$

$$p_{th} = \frac{V\hat{p}}{(V - V^*)} \tag{11}$$

It is now apparent that if $V = V^*$, $p_{th} \rightarrow \infty$ irrespective of the size of $\hat{p}$. This suggests that V^* is the critical sliding speed for transition; since ultimately the negative portion of the p_{th} wave would exceed $\bar{p}$ and the surfaces would part, thus initiating patch contact.

Table 1 is a list of thermal and elastic properties of some typical materials. When the material aluminum, for example, is selected for the cup, one finds that for $\mu = 0.1$, $\kappa = 2$ rad/cm.

$$V^* = 0.64 \text{ m/sec} \tag{12}$$

This corresponds to about 150 rpm for a 75 mm diameter cup; yet no major transition has been observed in actuality up to about 2500 rpm. Clearly the analysis is incomplete. Indeed, wear and other factors will be shown to have dominant effects upon the growth of waviness.

Although this analysis predicts a critical speed at which waviness is

TABLE I.—PROPERTIES OF REPRESENTATIVE MATERIALS

	Aluminum	Cast Iron	Silicon Carbide	Graphite	Glass
E(MPa)	6.8×10^5	1.2×10^5	8×10^5	0.68×10^8	8×10^5
α(1/°C)	17×10^{-6}	11×10^{-6}	4.7×10^{-6}	4.7×10^{-6}	5.4×10^{-6}
K(N/sec °C)	227	50.4	18	13.5	0.9
k(cm^2/sec)	83×10^{-2}	11×10^{-2}	6×10^{-2}	7.9×10^{-2}	0.31×10^{-2}

greatly multiplied in amplitude it also predicts that at higher speeds the amplitude will drop; and extremely high speeds would be associated with improved smoothness. This would at first appear to be something like the resonance phenomenon of synchronous whirl, where machines regularly operate above the critical speed; but more careful analysis will show an important difference.

CONDITIONS FOR EXPONENTIAL GROWTH OF SURFACE WAVINESS

The existence of steady state solutions as in the previous section does not assure that the system will be bound to these if they are themselves unstable to small disturbances. A test for such an instability of a solution may be made in several well established ways, one of which is to hypothesize an exponentially growing waviness and to test for the circumstances under which this car exist. The argument behind this would be that if (1) the system were disturbed and if (2) the disturbance contained a time dependent component corresponding to exponential growth, then such growth would continue no matter how small the initial disturbance. Thus a grain of dirt or a temperature fluctuation could trigger the growth process.

Before proceeding with such a test, let us review the initial analysis, noting that sinusoidal waves of contact pressure were accompanied by sinusoidal waves of frictional heating and thermal expansion. Further examination of the equation for heat transfer in the tube would suggest also that there is a corresponding sinusoidal temperature perturbation, proportional to $\hat{p}$ and having its peak where $\hat{p}$ is maximal.[22,35] Let us therefore postulate a temperature perturbation of the type

$$T = \hat{T}e^{-by} e^{i\beta t} \cos \kappa x = T_s e^{-by} \tag{13}$$

Then contact pressure will be given by

$$p = E\alpha T_s \left(\frac{\kappa}{\kappa + b} \right) \tag{14}$$

Here it is again assumed that the axial load is sufficient to assure full surface contact.

Returning to Equation 13 we find the heat flux through the surface will be

$$q = -K \left(\frac{\partial T}{\partial y}\right)_{y = 0} = K\, bT_s \tag{15}$$

Let us require now, that for a self-sustained wave to exist the frictional heating μVp must correspond to the heat passing through the surface as given in Equation 11. It follows that

$$KbT_s = \mu VE\alpha T_s \; (\kappa/b + \kappa) \tag{16}$$

Drawing upon the definition of V^*, this reduces to

$$(1 + b/\kappa)\; b/\kappa = 2V/V^* \tag{17}$$

If the temperature distribution in Equation 13 is to satisfy the Fourier heat flow equation,

$$\nabla^2 T = \frac{1}{\kappa}\frac{\partial T}{\partial t} \tag{18}$$

then, the following relationship must prevail, where b must be positive to satisfy distant boundary conditions.

$$b = \sqrt{\kappa^2 + i\beta/k} \tag{19}$$

Substituting this into Equation 17 and simplifying, one finds

$$\sqrt{1 + i\beta/\kappa^2 k} = -\tfrac{1}{2} \pm \sqrt{\tfrac{1}{4} + 2\frac{V}{V^*}} \tag{20}$$

Examination of this equation will show that $i\beta$ may be replaced by a *real, positive* number when

$$V/V^* \geq 1$$

Hence the condition for a self sustained exponentially growing perturbation is that the speed of sliding exceed V^*.

Again we note that the rationale behind this test is that if such a condition *can* exist, it may be triggered by fluctuations in operating conditions during extended periods of nominally steady operation. Obviously if such a pressure perturbation continues to grow it will reach a condition where the negative lobes exceed the initial contact pressure provided by the axial loading. This will lead to regions of parting between the surfaces. As has been shown this condition ultimately stabilizes the contact in a new configuration, where contact is in patches spaced around the edges of the cup.

Returning to the definition of V^* it is seen that the transition speed is dependent upon wave number κ, which is π/λ, and λ is half the wave length of the perturbation. Since small κ (or large wave length) has the lowest critical speed, it follows that upon advancing speed up from zero, instability will first be reached for the longest permissible wave length. If the glass and cup are stiffly mounted to prevent tilting, the critical wave length will correspond to one cycle around the cup. If, however, the glass is held in gimbals it will tilt so as to make the existence of $p = \hat{p} \cos \theta$ impossible. The next longest wave length would be that for $p = \hat{p} \cos 2\theta$ and would have two pressure maxima 180° apart. In experiments such a disturbance tends to dominate, in that hot spots (or contact patches) 180° apart are frequently observed.

As a word of caution, these derivations apply to the cup described, or to relatively stiff rings. Lebeck[30] has pointed out that slender rings, which obey beam theory, are more compliant and provide a stabilization of long wave length disturbances. They may also give rise to more contact patches than the minimum mentioned above.

Returning to Equation 11 the equation for critical speed may be rewritten as

$$V^* = 4\pi\kappa/E\alpha\mu\lambda$$

where λ is the wave length. For very thin bearings the longest wavelength permissible is 3.7 h, where h is the axial length of the ring. As speed is raised, longer wave lengths or spacings between contact patches will be tolerated, the length rising as $V^{1/3}$, until the number of patches reaches the above stated minimum permissible number.

EFFECTS OF HEAT CONDUCTION AND WEAR ON INSTABILITY

Because uniform heat flow down the axis of a ring or tube does not lead to wavy displacement in the axial direction, flat contact instability is independent of this factor. Patch contact is, however, influenced by the overall heat flow. In both cases the relative conductivities of the contacting bodies are of major importance. For the tube-on-tube geometry, Nerlikar[27] has carried out a theoretical study allowing both bodies to have the properties of aluminum, with one having hypothetically altered conductivity K_h (the effect of this on diffusivity is also accounted for). Results for critical sliding speed are as shown in Figure 3, as a function of the ratio of conductivities, and the friction coefficient. Except for high friction, the curves turn upward at $K_h/K = 1$, and the critical sliding speed goes to infinity for material sliding on its own kind. Another effect of differing conductivities is that the perturbation wave is not stationary in either body, but moves more slowly relative to the more conductive body. Even when both bodies are of the same material, the asymmetry necessary for instability can arise from thin insulating films on one or either of the two bodies.

Such a film will offer much less resistance to heat flow for low-speed disturbance movement than for high-speed movement. Consequently a condition can arise where a disturbance is nearly stationary in one body and has high relative speed on the second body, and the apparent resistances of the bodies can be such as to lead to instability at low sliding speeds. Calculations for aluminum-on-aluminum with a glassy film of thickness λ show that for infinite film thickness on one body the critical speed approaches that of conductor on insulator whereas films of reduced thickness lead to increased critical speed (Figure 4). There is however no critical speed for zero film thickness or zero friction coefficient.

Figure 4 is based on calculations where frictional heating is assumed to take place at the interface. It is possible, however, that some of the heating may be due to plastic deformation below the surface, and thus the film would not impede the flow of this heat component into the material. This question would bear further investigation. It does not, however, obscure the fact that even if only half the generated heat is affected by the film, instability would be called for; yet it would be absent without the film.

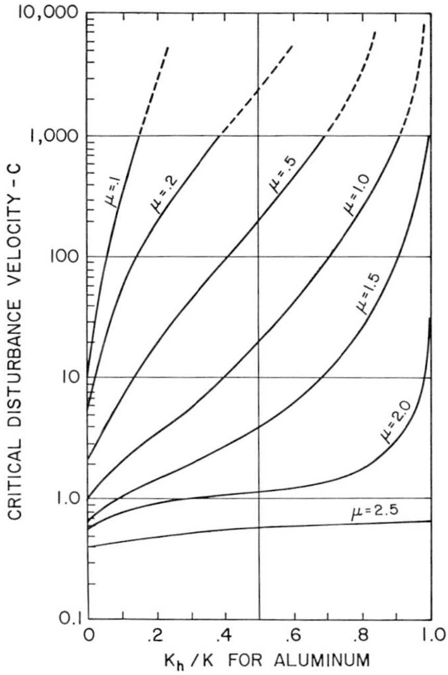

Fig. 3.—Critical disturbance velocity in the less conductive body for a range of friction coefficients and for aluminum with hypothetical reduced conductivity sliding on aluminum.

When wear is present[42,22] it may be treated in terms of Archard's wear law[47] in modified form, $dv/dt = wpV$, where p is contact pressure, V is sliding speed. The quantity w is an empirical wear coefficient which is influenced by material properties, temperature and other environmental factors. The wear coefficient may be nondimensionalized upon dividing by w^*, a group of variables which may be called critical wear coefficient.

$$w^* \equiv 2\alpha\mu k/K \qquad (21)$$

For conductor sliding on insulator, when $w/w^* \geq 1$ in the conductor, instability will not occur for finite sliding speed. The critical sliding speed in the presence of wear, V_w^*, is approximately

$$V_w^* = V^* \,[1/(1 - 1.2\, w/w^*)]\, \exp\,(w/w^*) \qquad (22)$$

Dow[22] has treated the case of a scraper running against a conductive drum, with wear. Heckmann[41] has generalized these results in terms of dimensionless quantities as shown in Figure 5. The critical sliding speed is set in ratio to V^* for the scraper for several values of the conduction parameter ζ. The physical quantities incorporated into ζ are those of the scraper except for k, the diffusivity of the drum. Cooling from the sides of the scraper (and also circular seals) was also investigated; and for reasonable values of convection heat transfer coefficient the effects were found to be small.

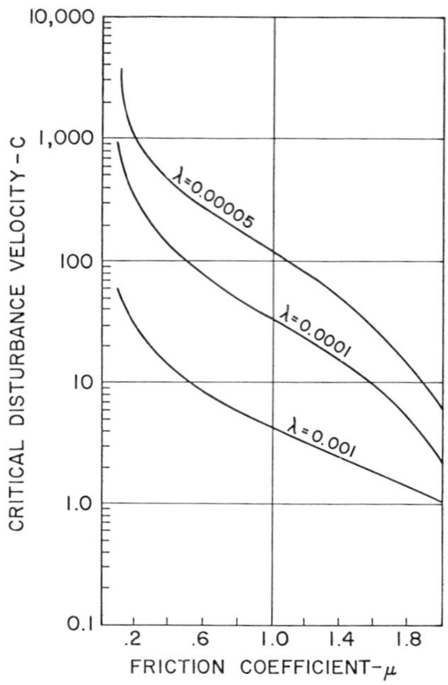

Fig. 4.—Effect of thin glass films on Al as to critical disturbance velocity.

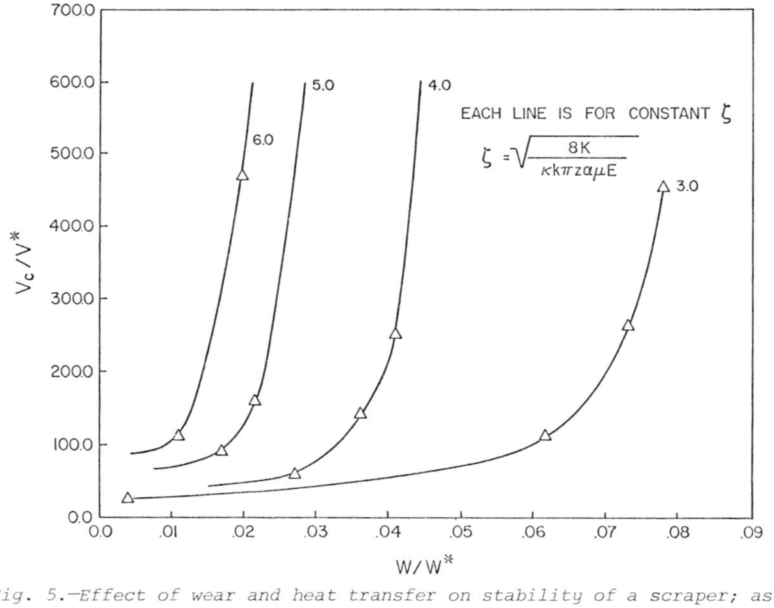

Fig. 5.—Effect of wear and heat transfer on stability of a scraper; as ζ increases, heat transfer increases to the surface the scraper slides against.

INSTABILITY IN A HYDRODYNAMICALLY LUBRICATED SEAL

The instability illustrated in Figure 2 took place in a seal with sufficient oil present to assure hydrodynamic, short-bearing support for the faces, which were self aligning and supported a fixed axial load. Banerjee[35] has investigated the stability of seals with constant mean film thickness and found the critical sliding speed to be

$$U_{crit} = (K_m/\alpha_m \eta)^{\frac{1}{2}} \bar{h}\kappa \qquad (23)$$

where metal runs against rigid insulator as in the previous illustrations. Strikingly good experimental confirmation of this result was obtained. An improved model, where the constant $\bar{h}$ condition is relaxed, has been investigated and has led to the paradoxical conclusion that quasi-equilibrium states of the system approach, but do not cross the stability threshold. Because of the approach to the conditions of Equation 23 it is not surprising that the equation is satisfied. Recent experiments confirm that steady operation, with great care, can be carried out without transition.

Turning now to Dow's results for a scraper running on a drum, Table 2 compares the measured wear coefficient w and that required to predict the observed transition speed, w_c. Note that for combinations run dry, agreement is good except for the steel on steel which may have been influenced by overall heat transfer through the contact *after* the transition. For the lubricated runs prediction is poor but the transition was observed to be qualitatively the same as for the dry contact. Actually, it was the deformed patch-contact state that was observed as evidence of transition. In view of this it is possible that the friction coefficient and wear coefficient on the hot patches would be nearer to dry contact than for the smooth sliding of lubricated material. If that is so, then the similarity to dry contact behavior can easily be explained.

PATCH CONTACT WITH HEAT TRANSFER AND WEAR

Both the experiments of Dow and Banerjee, cited above, suggest that if the deformed state *can* exist, it *may* exist. In the earlier studies of conductor on insulator the condition for patch contact to exist was shown to be

TABLE II.—MEASURED AND CALCULATED WEAR COEFFICIENTS

Drum	Blade	Friction Coef., μ	V*(m/sec)	Measured w	Calculated w_c
Al_2O_3	Al	0.38	7.11	14×10^{-10}	13×10^{-10}
Al_2O_3	Brass	0.16	8.13	9×10^{-10}	1.7×10^{-10}
Al_2O_3	Steel	0.26	5.08	2.5×10^{-10}	1.1×10^{-10}
Steel	Steel	0.25	12.20	13×10^{-10}	0.5×10^{-10}
		Lubricated			
Steel	Alum. Graphite	0.026	32	0.26×10^{-10}	<0
Steel	Alloy Steel	0.051	23.90	0.54×10^{-10}	1×10^{-13}

the same as for thermoelastic instability. When more realistic models are considered, however, the formation of patches is strongly influenced by overall heat transfer from one body to another, whereas idealized flat surface instability is not, being dependent only on the growth of zero-average waves - at least for the case of the axisymmetric ring configuration. This can be demonstrated by referring back to Equation 2 and adapting it to the case of heat flow from one body to another. If both are of the same material the one losing heat will curve inward and the one receiving heat will bulge outward the same amount. If they are different there will be a relative curvature change (see also Reference 49)

$$\left(\frac{d^2v}{dx^2}\right)_{rel} = \left(\frac{\alpha_1}{K_1} - \frac{\alpha_2}{K_2}\right) q \tag{24}$$

Depending upon the sign of the heat flow and the materials, this influence may greatly alter the patch size. It may also give rise to patch contact for normally flat surfaces even in the absence of sliding. For these reasons, patch contact observations require knowledge of heat partitioning for their full interpretation.

Although study of the stationary patch with or without wear has provided considerable physical insight, it can easily be shown that the patches must be moving relative to both contacting surfaces when wear is present, and each body is a thermal conductor. Even at low Peclet number the nature of the stress distribution is strongly affected by this movement. Kilaparti[32] has examined the case of the uniformly heated patch ($q = constant$), and has shown that, as the patch moves, wear causes a reduction of outward displacement v from leading edge to trailing edge. At the same time receipt of heat causes a rise in v over the same distance. Hence, if the patch is in contact with a rigid body these two effects must cancel one another. It follows that speed of traversal goes up with wear rate, and must exceed infinity for wear coefficient exceeding w^*. Even cases of so-called severe wear of several abradable materials lie below $w/w^* = 1$. A more detailed study[50] has shown that the pressure distribution on the contact becomes nearly triangular as wear rate is increased; but the study was limited to modestly high Peclet number because the numerical scheme required an excessively fine grid for higher values. A third study directed toward higher Peclet number[33] has reviewed the work of Ling and Mow[10] and suggested a simplified relationship for surface displacement:

$$\int_x^{\infty} \frac{2\alpha k}{cK} q(\xi)d\xi + \int_{-\infty}^{x} \frac{2}{\sqrt{\pi}} \left(\frac{k}{c}\right)^{3/2} \frac{\alpha}{K} \frac{q(\xi)d\xi}{\sqrt{x - \xi}} = \delta(x) \tag{25}$$

Derivations drawing upon this function have supported the earlier prediction of nearly triangular contact pressure distribution.

CONTACT STRESS

Recent calculations show that a tensile stress may appear at the edge of a moving contact patch with uniformly distributed heat input, this being given by

$$\sigma_{xth} = \frac{E\alpha Q}{K(1-\upsilon)\varepsilon} (0.8\ Pe\ell) \tag{26}$$

where $Pe\ell = c\ell/k\pi$. In one example where the contact patch was virtually stationary in the carbon of a carbon/cast-iron contact, the rate of movement c relative to the iron would be approximately the sliding speed. For both bodies close to ambient temperature in the bulk, almost all of the frictional heat would be expected to pass into the iron, hence $Q \approx \mu PV$. Under these circumstances Equation 26 becomes *(for* $\nu = 0.3$*, and* $\ell/\varepsilon = 1$*)*

$$\sigma_{xth} = \frac{1.14\ \mu E\alpha P}{\pi K k}\left(\frac{\ell}{\varepsilon}\right)V^2 \qquad (27)$$

and for V in m/sec and σ_{xth} in bars one finds upon substituting the properties of cast iron

$$\sigma_{xth} = 26{,}394V^2$$

obviously this can lead to large stresses.

On the one hand this gives an easy explanation of surface heat checking where patch contact passes over the surface. At the same time it raises the question as to why heat checking is not observed more often. One possible explanation may be that the depth affected is shallow. It is possible also that the numerical magnitude may be reduced by substituting a different distribution of heat input on the patch. Whatever the modifying factors may be they offer an attractive problem for further investigation.

An illustration of such heat checking is shown in Figure 6, which is an enlarged photograph of the band of contact of a large seal. The light colored band is over 2 mm wide and lies on the face of a hard brittle alloy which has been run against a carbon ring with a raised nose. The tiny hairline cracks run across the contact band and are believed to be the result of tangential tension. Evidence suggests that the thermal asperity forms and moves slowly on the carbon, and that the above described tension arises in the metal as it passes through the contact patch. There is no evidence of gross heating, and such heating is not necessary to produce the stresses necessary for cracking.

DIRECTIONS FOR FURTHER INVESTIGATION

One of the troublesome problems left unsolved in the above reported work involves delineating the path from stable operation to patch contact. When this is understood we will be in a better position to avoid the formation of patches with their severe effects. Undoubtedly, along the way we shall learn a great deal more about the mechanisms of lubrication and wear in general.

In particular this is of interest in the case of *lubricated contact* where predictability is the poorest. We should find out when flat sliding surfaces (such as the sides of pistons) develop the thermal asperity, or else discover the mechanisms which prevent this.

Although journal bearings have been investigated as to radial clearance loss due to thermal deformation, this writer knows of no investigation of ellipticity or multilobar deformation patterns which might be thermally excited and could lead to catastrophic failure.

In other bearing configurations such as tilted pads one can visualize a thermal asperity forming at high speed. Again, we should determine whether

Fig. 6.—Radial hairline cracks across the contact band (light area) on the metal face of a seal, after running against a carbon ring at high peripheral speed.

or not this actually happens in practice.

It is this writer's opinion that even the minimum film thickness criterion of bearing failure should be re-investigated with a thermoelastic analysis of the thin film region to determine the possibility of a load-concentrating temperature rise and thermal breakdown of the film.

It is clear that thermoelastic effects distort gas bearings, but as far as this writer can determine no stability analysis has been made of such systems. Furthermore, although ball bearings have been shown to be possibly subject to thermoelastic seizure processes, the study of thermoelastic effects in catastrophic failures is in the most primitive stage of development and should be extended.

Thermoelastic deformation of supposedly non-contacting jet engine shaft seals appears on initial investigation to have a strong likelihood of occurring, and could lead to thermal asperities on the seal face or concentration of load on two or more pads with attendant increased leakage, and possibly failure. The mating rings may also deform significantly.

The high-speed rubs of turbomachines in the form of blade tip contact and labyrinth seal contact are already under investigation. Further investigation of thermoelastic deformations brought about by such contact may help to provide guides to improve designs.

Granting that thermal asperities do appear in important applications such as seals, it is essential to understand the thermoplastic behavior of the patches and to estimate the stresses, metal working, and fatigue mechanisms in such contacts.

In conventional face seals, where the effects are most easily demonstrated, the complex interactions of tilt, nutation and other motions associated with thermal deformation can only be conjectured; yet experiments which have been made, suggest that these interactions are complex and closely inter-coupled. Stress analysis for the thermoelastic regime for hard materials should be undertaken and the conditions for stress cracking (heat checking) should be defined.

Turning now to physically different problems, Marshall[51] has reported contact patch formation in high-current electrical brushes. It appears that current as well as friction can produce the deformed state, and the combined effects of these should be studied to determine their influence upon wear as well as the limits of operation of brushes.

Although no publications have appeared on the subject, this writer has been told that thermoelastic instability, involving roller deformation, sets the limit for speed of rolling of thin sheet. Rollers may be undercut to compensate for "barreling out" as a result of heating. However, at some limiting speed this is no longer effective in assuring uniform sheet.

Recently Quinn[52] has given evidence for the thermoelastic oscillation of asperities in a concentrated, rider-on-drum contact, and has provided estimates of the number of asperities in contact at one time. On the basis of earlier theories thermal augmentation of asperities would not have been expected in the small dimensions involved; but the evidence is compelling.

Summarizing, it is this writer's opinion that thermal deformation and the formation of the thermal asperity are not isolated phenomena to be encountered in special circumstances. They appear and sometimes become predominant in quite ordinary applications. Indeed, it is believed that neglect of accounting for them has needlessly complicated the interpretation of many friction and wear observations. Increased understanding has led to improved tools for use in studying these effects. To apply these to some of the obvious problem areas listed above should provide, in some cases, improved physical understanding; in others, design guides; and in a few, a clearing of the air by showing that such effects are under control.

REFERENCES

1. Holm, R., Wiss, Veroeff. a. d. Siemens Konzern, Berlin, Vol. 7, No. 2, 1929, p. 217.
2. Bowden, F.P. and Tabor, D., "The Friction and Lubrication of Solids," Oxford University Press, 1950.
3. Merchant, M.E., *Journal of Applied Physics*, Vol. 11, 1940, p. 230.
4. Belidor, M., "Architecture Hydraulique," C.A. Jambert, Paris, 1737.
5. Coulomb, C., "Memoires de Mathématique et de Physique," Académie des Sciences, Vol. 10, 1785.
6. Blok, H., *Institution of Mechanical Engineers. Proceedings*, Vol. 2, 1937, p. 26.
7. Holm, R., *Journal of Applied Physics*, Vol. 19, 1948, p. 361.
8. Archard, J.F., *Wear*, Vol. 2, No. 6, 1959, p. 438.
9. Ling, F.F. and Simkins, T.E., *American Society of Mechanical Engineers Transactions*, Series D, Vol. 35, 1963, p. 481.

10. Ling, F.F., "Surface Mechanics," John Wiley and Sons, New York, 1973, p. 125.
11. Mow, V.C. and Cheng, H.S., ZAMM, Vol. 18, 1967, p. 500.
12. Korovschinsky, M.V., *American Society of Mechanical Engineers Transactions*, Series D, Vol. 87, 1965, p. 811.
13. Nica, A., ibid., p. 781.
14. Burton, R.A., *Wear*, Vol. 8, 1965, p. 157.
15. Hsu, Y.C., *American Society of Mechanical Engineers Transactions*, Series F, Vol. 89, 1967.
16. Staph, H.E., *American Society of Lubrication Engineers Transactions*, Vol. 10, 1967, p. 408.
17. Sibley, L.B. and Allen, C.M., American Society of Mechanical Engineers Paper 61-Lubs-15.
18. Barber, J.R., Dissertation, St. John's College, Cambridge, England, 1968.
19. Barber, J.R., *Royal Society of London. Proceedings.* Series A, Vol. 312, 1969, p. 381.
20. Barber, J.R., *Wear*, Vol. 10, 1967, p. 155.
21. Dow, T.A. and Burton, R.A., *Wear*, Vol. 19, 1972, p. 315.
22. Dow, T.A. and Burton, R.A., *American Society of Mechanical Engineers Transactions*, Series F, Vol. 95, 1973, p. 71.
23. Dow, T.A., Dissertation, Northwestern University, 1972.
24. Dow, T.A. and Stockwell, R.D., *American Society of Mechanical Engineers Transactions*, Series F, Vol. 261, 1977, p. 359.
25. Kennedy, F.E. and Ling, F.F., *American Society of Mechanical Engineers Transactions*, Series F, Vol. 96, 1974, p. 496.
26. Santini, J.J. and Kennedy, F.A., *Lubrication Engineering*, Vol. 31, 1975, pp. 402, 413.
27. Burton, R.A., Nerlikar, V. and Kilaparti, S.R., *Wear*, Vol. 24, 1973, p. 169.
28. Burton, R.A., *Wear*, Vol. 24, 1973, p. 189.
29. Burton, R.A. and Nerlikar, V., *American Society of Mechanical Engineers Transactions*, Series F, Vol. 97, 1975, p. 539.
30. Lebeck, A.O., *American Society of Mechanical Engineers Transactions*, Series F, Vol. 98, 1976, p. 277.
31. Burton, R.A., Kilaparti, S.R. and Nerlikar, V., *Wear*, Vol. 24, 1973, p. 199.
32. Kilaparti, R., *American Society of Lubrication Engineers Transactions*, Vol. 20, 1977, p. 64.
33. Kilaparti, R., *American Society of Mechanical Engineers Transactions*, Series F, Vol. 100, 1978, p. 65.
34. Ling, F.F. and Mow, V.C., *American Society of Mechanical Engineers Transactions*, Vol. 87, 1966, p. 814.
35. Banerjee, B.N. and Burton, R.A., *American Society of Mechanical Engineers Transactions*, Series F, Vol. 98, 1976, p. 157.
36. Banerjee, B.N. and Burton, R.A., *Nature*, Vol. 261, 1976, p. 399.
37. Banerjee, B.N., Dissertation, Northwestern University, 1977.
38. Hahn, E.J. and Ketteleborough, C.F., *American Society of Mechanical Engineers Transactions*, Series F, Vol. 20, 1968, p. 233.
39. Rockwell, K., Ettles, C. and Stokes, M., Paper No. 12, Institution of Mechanical Engineers, Brighton, Vol. 184, Part 3L, 1970, p. 82.
40. Tanaguchi, S. and Ettles, C., *American Society of Lubrication Engineers Transactions*, Vol. 18, 1975, p. 299.
41. Heckmann, S.R. and Burton, R.A., *American Society of Mechanical Engineers Transactions*, Series F, Vol. 99, 1977, p. 247.
42. Heckmann, S.R. and Burton, R.A., *American Society of Mechanical Engineers Transactions*, Series F, Vol. 100, 1978, p. 136.
43. Burton, R.A., "The Mechanics of Contact between Deformable Bodies," Delft University Press, Delft, 1975, p. 191.

44. Wankrajong, C., M.S. Thesis, Northwestern University.
45. Burton, R.A., Kilaparti, S.R. and Heckmann, S.R., *American Society of Mechanical Engineers Transactions,* Series A, Vol. 98, 1976, p. 435.
46. Wetenkamp, H.R. and Kipp, R.M., American Society of Mechanical Engineers Paper 75-RT-2.
47. Archard, J.F., *Journal of Applied Physics,* Vol. 24, 1953, p. 981.
48. Dundurs, J., *Mechanics Research Communications,* Vol. 1, 1974, p. 121.
49. Dundurns, J. and Panek, C., *International Journal of Heat and Mass Transfer,* Vol. 19, 1976, p. 731.
50. Kilaparti, R. and Burton, R.A., *American Society of Mechanical Engineers Transactions,* Series F, Vol. 98, 1976, p. 596.
51. Marshall, R.A., *Wear,* Vol. 37, 1976, p. 233.
52. Quinn, T.F.J., *American Society of Lubrication Engineers Transactions,* Vol. 21, 1978, p. 78.

DISCUSSION

S. GANESH, Bendix Research Laboratories: What kind of probe did you use to detect thermal asperity?

R. A. BURTON: We reviewed almost everything possible and ended up with a Bentley probe. We were able to get a very small Bentley probe made for us. It was about 1/32 thick. It worked well for the dynamic passage of the surface waves; but, in the measurement of static film thickness it was subject to drift because it did not expand at the same rate as the glass. However, we were able to run the apparatus, get it stabilized at some operating condition and then if we wanted to know what the mean film thickness was, we would turn off the rotor very quickly and see it descend rapidly. We would see a step in the output of the probe before it changed temperature and we could get quite reproducible results. We added half of the waviness to the descent to get the mean film thickness, and it seemed to be correct.

QUESTIONER, Dow Corning Corporation: Did you see any difference between concave, convex and flat surfaces?

BURTON: No. Experiments done a long time ago at Battelle by Sibley and Allen are interesting in this respect. They pressed a disc brake puck against a disc to study seal materials and they came up with a thermal cracking criterion. Unfortunately, that paper was never published. It appeared as a pre-print for some reason. We could see the hot disc material passing out from under the puck and we could see streaks of light which everyone thought were due to thermal asperity formation. These would sweep back and forth. Sometimes there would be more than one, and sometimes they would be stationary. The original purpose, therefore, was to explain the flat surface contact. When Dow started his work, he couldn't handle three-dimensions and he decided to do it as a two-dimensional scraper, with a line contact. Later Heckman looked at these scrapers and asked what if the scraper was part of a cylinder just touching in Hertzian line contact? He found that it would go unstable.

In the brush contact problem Dr. Marshall of Westinghouse Research Laboratory has observed little blue patches on the surfaces of brushes that indicate that under certain conditions of pulse current, they go to this kind of contact. It is easy to see how it could happen.

A. BEERBOWER, UCSD: Professor Blok was convinced that this was the mode by which gears went into sudden failure -- a thermal spike followed by vaporization, and running in a bubble which leads to scuffing.

P. A. ENGEL, IBM: Is it possible to contaminate the surface to cause deliberately such thermoelastic instability?

BURTON: Well, there are cases. For example, as I showed you, metals might be stable; but when we put a thin surface film on them that insulates, it could make them go unstable. A paradoxical case is one in which a lubricant film on the surface might actually lower friction, but by affecting the partitioning of the heat it causes instability. Within the instability phenomenon, the interactions are complicated and we cannot really determine anything at the outset. We have to look at each case and ask what is going on. Some of these are almost completely beyond my ability to visualize.

GANESH: You mentioned thermal spikes moving around. What is causing that? I find it hard to visualize.

BURTON: When we turned the metal cup and we looked up through the glass, all we could see were red flashes. Then we changed things around. We turned the glass and held the metal cup still. When the transition took place, we saw two spots opposite each other (sometimes three) on the metal. They were 0.01 in., glowing red (approximately 1500°F) patches and they would move slowly. The movement is due to wear and the heating. They wear off from side to side; they acquire heat, and these two things come into balance. In fact, this is one of the few places where you can check the theory approximately. (Of course, Dow did an extremely good job). Because of the sensitivity of many of these interactions, it is difficult to predict exactly at what speed this thing is going to occur. But if you know the wear coefficient you can predict how fast the patches will creep around. Indeed, the speed at which they creep around is just about the same as we predicted.

Now I want to say something in favor of asperities. A patch is more like what one would think of as being a Hertzian contact between a ball and a flat. There are still asperities inside it.

J. R. Barber did some very provocative experiments. He made three little pins and put them in a bar and loaded them down against a drum. These pins would lift up alternately one after the other, symetrically. When one of the pins is a little longer than the rest, it would take all the load and heat up. Then it would wear off to the point where the next one would contact. It would then draw back and the next one would then take over the load. Thus the load is shifted around uniformly. In his model of what took place in the contact zone (he was interested in railroad wheel brakes at the time) these asperities would stick out and draw back like pins. In what we have been working with, we did not see that; but that does not mean it does not take place. There is another place for some needed research; namely, when do we get a sort of patch that moves along with asperities passing in and out of it and when do we actually get asperities drawing back and sticking out? I think Quinn's work at least matches that model.

T. F. J. QUINN, University of Aston: Do you see these at low speeds?

BURTON: Definitely. We get them at moderate speeds. I am talking about 2000 rpm, the kind of speed we have in an ordinary motor or a seal.

IX. WEAR IN PROCESSING

WEAR MECHANISMS IN METAL PROCESSING

M. C. Shaw

ABSTRACT

Many metal processing operations involve surfaces in sliding contact and therefore friction and wear are important considerations. This paper discusses the special conditions encountered in rpocessing and how they differ from those of ordinary sliders. Special emphasis is given to wear behavior and mechanisms that have been proposed to explain this behavior.

INTRODUCTION

The mechanical processing of materials involves the change of shape of a part which is usually accomplished in one or more of the following ways:

Joining (+)

Plastic deformation (o)

Material removal (-)

Molds, tools, or die surfaces are employed in most of these processes and consequently tribology plays an important role. Since the major objective of these processes is to accomplish a change in shape this usually involves a change in the apparent surface area of the part and this results in a more severe set of tribological problems than those associated with sliding contacts that are relatively lightly loaded.

In joining, material is added to a basic structure by welding or brazing macroscopic pieces, or even on an atomic scale as in electroforming. In plastic deformation processing material is neigher added nor removed but is moved from one point to another by plastic flow. In material removal operations unwanted material is removed by cutting or grinding on the microscopic scale, or by processes such as electrochemical machining on the atomic scale.

Tribology is most important in deformation and removal processing operations and the discussion here will be confined to these areas.

PLASTIC DEFCRMATION PROCESSING

Typical deformation processing operations are forgoing, extrusion, rolling, and drawing. Common to all of these processes is the movement of a fully plastic material over a harder elastically loaded tool surface. The fact that the entire subsurface is plastic is an important consideration that gives rise to the following departures from the behavior of ordinary friction sliders:

1. Amonton's law does not apply and the coefficient of sliding friction *decreases* with an increase in the mean contact pressure (load divided by apparent area of contact, A).

2. The new surface area generated is substantial.

3. Surface temperatures play an important role (particularly in hot forming operations that are carried out above the strain recrystallization temperature of the metal.)

4. The sliding speed and the rate of deformation play an important role.

5. The mean tool face pressure is higher than in the case of ordinary friction sliders.

The reason Amonton's Law does not hold has been discussed in reference (1). This is due to the fact that the real area of contact (A_R) is no longer proportional to the applied load when the areas of plastic flow beneath individual asperities uoin up. Figure 1 illustrates this point.

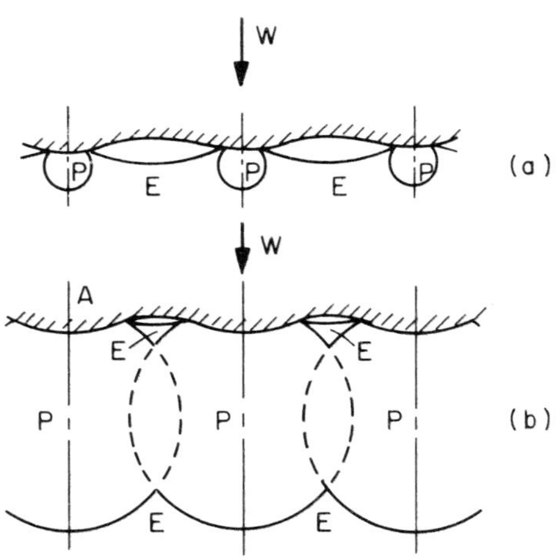

Fig. 1.—Plastic behavior of a) lightly loaded and b) haavily loaded surfaces. P = plastic and E = elastic regions.

Figure 1a is for a lightly loaded slider where only the tips of asperities are plastic while Figure 1b shows the onset of fully plastic subsurface flow. Figure 2 shows the variation of coefficient of friction for three regimes of A_R/A , which correspond to the following:

I Lightly loaded slider

II Heavily loaded slider

III Internal surface where $A_R = A$.

It is somewhat paradoxical that the coefficient of friction decreases with increase of pressure on the tool face. This is consistent however with

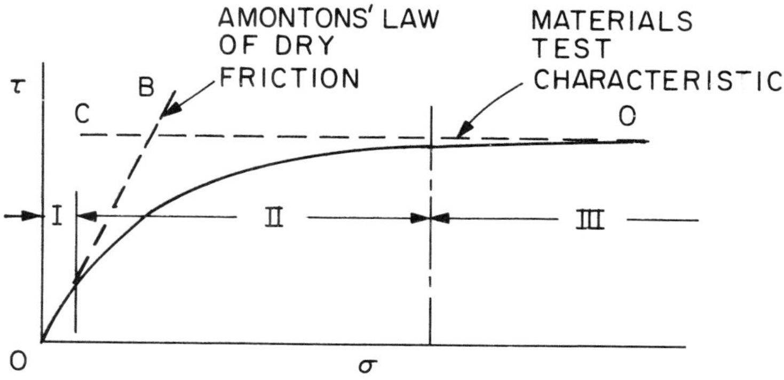

Fig. 2.—Variation in shear stress (T) with normal stress (s) Coefficient of friction f = T/s.

the fact that the coefficient of friction for a lightly loaded slider will be 0.1 to 0.4 while the same surfaces in a deformation process the coefficient of friction will be 0.01 to 0.04. This does not mean however that friction is less important in deformation processing than with lightly loaded sliders. The fact that the normal stress is very much higher on die surface than for ordinary sliding surfaces tends to offset the lower coefficients of friction associated with die surfaces.

A very important consideration in deformation processing is the constraint factor (c) since it has a major influence on the mean stress on the die surface and hence on the coefficient of sliding friction and further on the frictional energy associated with the process. The constraint factor c is defined as follows:

$$c = \frac{\text{mean stress on die face}}{\text{uniaxiaxal flow stress}} \quad (1)$$

The quantity in the denominator is the mean stress required to cause plastic flow of a completely unconstrained specimen such as the block of material shown in Figure 3. Figure 4 shows the manner in which c varies with the degree of constraint[2] as measured by the nondimensional ratio

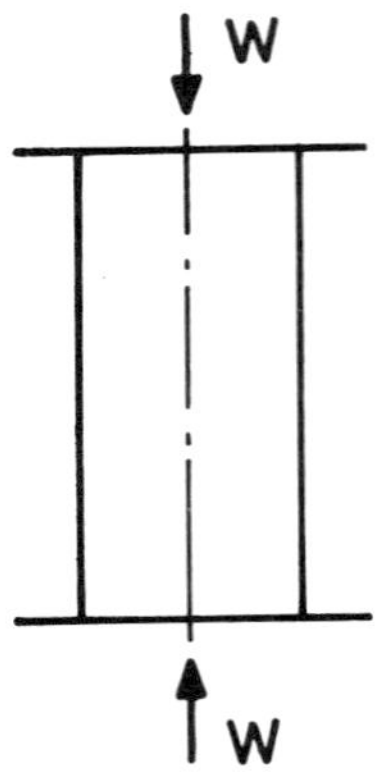

Fig. 3.—Uniaxial compression test.

h/b (where *h* = thickness of sheet and *b* = lateral extent of the punch). Figure 4 is for two dimensional (plane strain) punches that are about 10 times as wide (perpendicular to the paper in Figure 4) as dimension b.

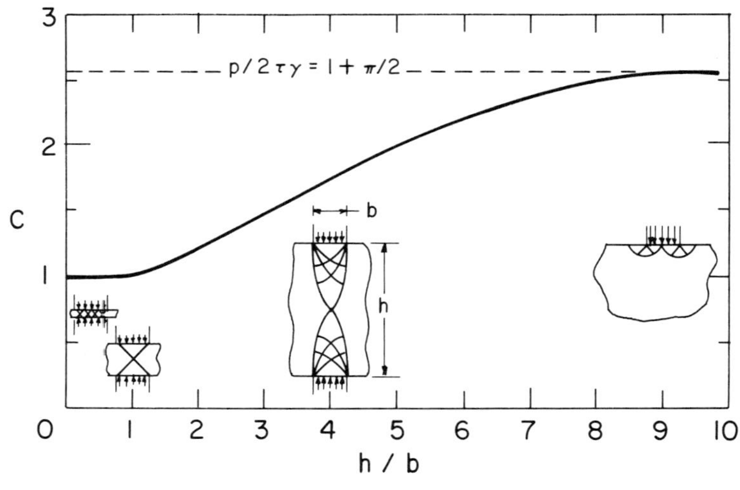

Fig. 4.—Variation of constraint factor (C) with ratio (h/b). (after Ref. 2)

When $h/b \to \infty$ this corresponds to a hardness test. The theoretical value for this case is 2.57 for a two-dimensional indentor [3] *and* 3.0 for an axysymetrical indentor[4]. The reason a constraint factor is involved is due to the fact that plastic flow is hampered when the plastic zone beneath the indentor is surrounded by elastic material (elastic constraint) or when it is caused to flow in a particular direction that depends on the geometry of the situation (flow constraint). Reference (5) discusses the differences associated with these two types of constraints.

A consequence of new area generated in materials processing operations is that the newly generated surfaces are highly active and tend to form strong bonds with the mating surface. This is not only due to the fact that a newly generated surface may be clean but also due to the fact that a new surface will tend to be out of equilibrium relative to electron density. An internal surface ($A_R = A$) has a certain desnity of electrons associated with it. These act in conjunction with positively charged ions to give rise to the cohesive forces that hold materials together. When a free surface is generated there will be an excess of electrons which leave the freshly generated surface. These are called exoelectrons and have been discussed in an interesting article by Rabinowicz[6].

After the new suraace has achieved its new equilibrium level of electron density the surface ions will have a slightly greater spacing than in an internal surface and a somewhat lower energy content. The difference in equilibrium energy level for an internal surface and a free surface is called surface energy. If a new generated surface contacts another surface before its new equilibrium electron density level has been reached by the emission of exoelectrons, the excess electrons present will tend to promote the formation of strong bonds with the mating surface. If some contaminating film (lubricant) is present in the vicinity of the newly generated surface it may spread over the newly formed surface and screen it from the mating surface to prevent the formation of strong bonds. A somewhat different function of *excess electrons* has been proposed in a recent paper[7] to explain the formation of spherical "wear" particles when extremely thin layers of metal are sheared from a surface at very high speeds.

The speed of sliding normally plays a negligible role for lightly loaded sliders and hence the Archard equation[8], which does not involve the speed of sliding, correlates well with experimental results. The Archard equation essentially states that the following nondimensional wear number[9] will be a constant

$$N_W = \frac{BH}{WL} \qquad (2)$$

Where B = Volume worn away

H = Hardness of softer member of sliding pair

W = Normal load

L = Sliding distance

This equation applies to abrasive wear as well as attritious wear[10], but only as long as thermal softening is not a factor. For lightly loaded sliders thermal softening is not encountered over a fairly wide range of speeds. However, in materials processing, speed normally plays an important r-le unless the duration of contact is so short that thermal softening does not occur as in some grinding operations.

In general, the consequences of new surface generation are far less severe in metal forming operations than in metal cutting. This is probably due to the fact that in most metal forming operations the new surface is formed by extending an existing surface rather than by the generation of a nascent surface, as in cutting. As long as the original surface is coated by a tenacious film, very little clean unscreened surface will be generated in metal forming operations. This is often achieved by coating the metal to be deformed with a porous phosphate or oxide that is impregnated with a surface active lubricant capable of extension during the metal working process. The result is that the rate of wear of forming tools is either much lower or less significant than the wear that occurs with metal cutting tools. This results in a tool life that is one or more orders of magnitude greater for forming tools than for cutting tools. Since tool wear is generally more important in cutting operations than in forming operations, the remainder of the discussion will be directed to wear mechanisms involved in metal cutting.

METAL CUTTING

Metal cutting tools remove unwanted metal by the process of concentrated shear shown schematically in Figure 5. As metal approaches the

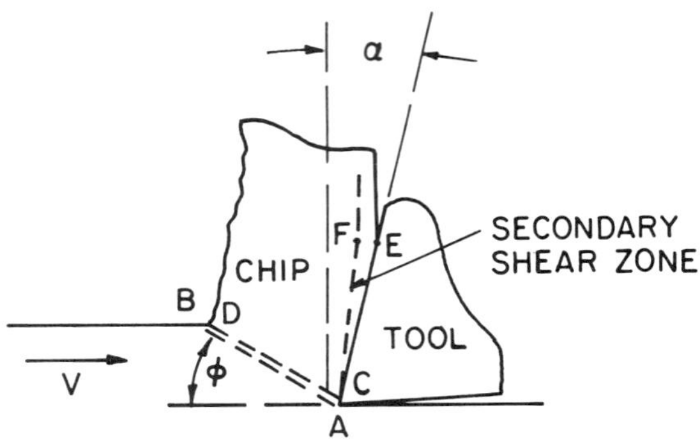

Fig. 5.—Schematic representation of metal cutting process.

cutting edge, essentially nothing happens until line AB is reached. Then, the material undergoes extensive shear strain as it moves from AB to CD. Beyond this point we have a chip which passes off essentially without further deformation. The chip leaves contact with the tool at point E and there is frequently some evidence of secondary shear (in a direction normal to that occurring in shear zone ABCD.) in the secondary shear zone AEF extending along the tool face.

There are, thus, two regions of importance that associated with the primary shear process (zone ABCD) and that associated with friction between chip and tool (zone AEF). Essentially all of the energy dissipated in cutting is associated with these two zones. About 3/4 of the total energy is involved in primary shear (zone ABCD) while 1/4 of the total energy is associated with secondary shear (in zone AEF).

It is important to realize that the two basic processes involved (primary shear and friction) cannot be considered independently. This may be seen by considering the chip as a free body (Figure 6). Resultant forces R and R^1 must be equal for a continuous cutting process. Hence, anything that occurs along the tool face to change the magnitude or direction of R^1 will cause a similar change in R and vice versa. This fact makes it difficult to identify cause and effect. For example, if an improved cutting fluid is used that lowers friction on the tool face this

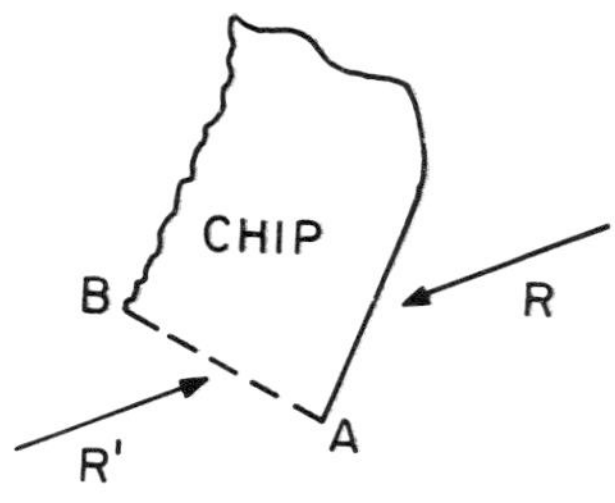

Fig. 6.—Free body diagram of chip.

will not only change the magnitude and direction of R^1 but will also cause a decrease in shear angle ϕ which in turn controls the mean shear strain associated with zone ABCD. The latter action will significantly reduce the specific primary shear energy, and since the primary shear process is energetically predominant, the main action could be assumed to originate with a change in the shear properties of the material instead of being associated with tool face friction. The reverse situation is also encountered in which a change in shear properties give rise to a change in the direction of R which appears to be due to a change in the coefficient of tool face friction. The primary action of carbon tetrachloride as a cutting fluid is very difficult to identify because of the above noted interaction between shear and frictional actions[11,12].

Another complication associated with cutting tool action is the so-called built-up edge (BUE). This is a wedge of metal that forms at the cutting edge and changes the effective rate angle α of the tool. The BUE is believed to form as a consequence of the temperature-shear strength behavior of steel in which strength is greater at a temperature of about 550 F than either at room temperature or temperature above 550 F. When the gradient of temperature from B to A (Figure 5) is appropriate, the metal within the chip may have a strength lower than that at the tool - chip interface due to thermal softening. When this occurs, shear takes place within the chip and a BUE forms. (Figure 7). An increase in cutting

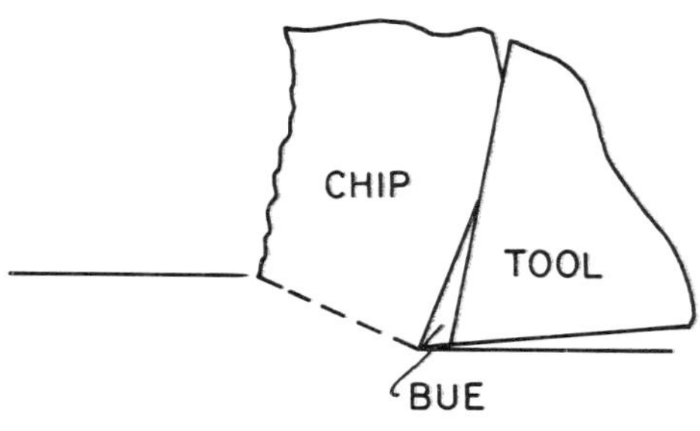

Fig. 7.—Schematic representation of metal cutting process with built-up edge (BUE).

speed will cause the plane of thermal softening to move closer to the tool face thus giving rise to a smaller (BUE). When the speed reaches a value where the temperature on the tool face is sufficiently high to give thermll softening there, the (BUE) disappears. A reduction in speed below that to give the masimum (BUE) will also tend to eliminate thermal softening of the chip, in which case the chip will shear closer to the tool face as a result of the stress gradient that is present in the chip. Thus, the size of the BUE is apparently due to a combination of thermal softening and the stress and temperature gradients across the chip.

The main complications associated with the BUE are due to its instability leading to a variation in its size and shape with time. In general, a BUE will increase the rate of tool wear due to the abrasive action of small pieces of highly work hardened material periodically shed by the BUE.

A very important aspect of BUE action is its influence on the roughness of the finished surface. This is caused by the fact that BUE grows slowly in size but decreases in size abruptly. The result is a characteristic surface roughness as shown in Figure 8.

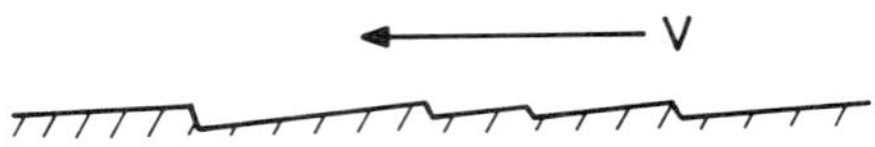

Fig. 8.—Characteristic surface roughness resulting from a built-up edge of variable size.

TYPES OF CUTTING TOOL WEAR

Wear in general has been classified as follows:

adhesive wear
abrasive wear
corrosive wear
fatigue

All of these are believed to be present in cutting situations in combination, the predominant wear mechanism depending upon cutting conditions. In addition to the above sources of tool failure, the following also pertain.

microchipping
gross fracture
plastic deformation

These however are usually readily identified and the solution is apparent. The fracture modes result from subjecting the tool to too high a cutting force, operating with too large a BUE or using a tool material that is too brittle. Plastic flow of the tool tip arises when the temperature is too high relative to the softening point of the tool material. When plastic flow occurs at the tool tip, tool clearance is lost, the temperature rises abruptly, and total tool failure occurs rather rapidly. The obvious solution to the latter difficulty is to use a lower cutting speed or a tool material that is more refractory. While cemented tungsten carbide is a very refractory material, even this material will suffer from plastic flow at the tool tip if the temperature is too high[13].

Cutting tools wear predominantly in different ways, depending on cutting conditions (principally cutting speed, V and undeformed chip thickness, t). Figure 9 due to Opitz[14] shows the principal types of tool wear, together with the approximate ranges of values of V and t where each type of wear is predominant. At low values of the product ($Vt^{0.6}$) tool wear consists predominantly of a rounding of the cutting point and a loss of tool sharpness. This is the main type of tool wear in the broaching operation where both V and t are small. As the product ($Vt^{0.6}$) increases, the predominant type of tool wear shifts as shown in Figure 9. The most important

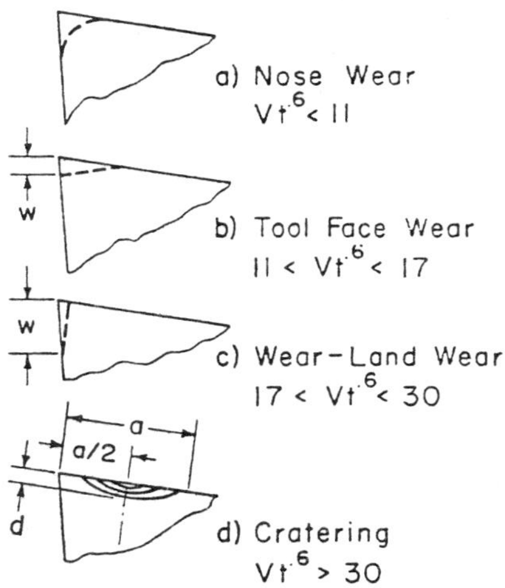

Fig. 9.—Types of tool wear for AISI 1045 steel as influenced by cutting speed (V) and feed (t) in a single point turning operation with a tungsten carbide tool. w = wear land, d = crater depth, a = chip-tool contact length. Depth of cut, 0.08 in. (Opitz,[14]).

types of tool wear for single point turning tools is wear land formation (9c) and crater formation (9d). Crater formation is apt to be more important than wear land formation in situations where cutting temperatures are high.

CUTTING TOOL TEMPERATURES

The results shown in Figure 9 are primarily due to the fact that the temperature on either the wear land or tool face vary as the product ($Vt^{0.6}$). This may be shown by performing a dimensional analysis with θ (tool face or tool flank temperature) as the main dependent variable. Table 1 lists the variables of importance. The product ($k\rho c$) for the work material is used in place of k (thermal conductivity) or ρc (volume specific heat) alone since it is the product that enters moving heat source problems[15]. After

TABLE I.—DIMENTIONAL ANALYSIS FOR CUTTING TEMPERATURE

Symbol	Quantity	Dimensions*
θ	Tool Temperature	θ
V	Cutting Speed	LT^{-1}
t	Undeformed Chip Thickness	L
u	Specific Cutting Energy	FL^{-2}
$k\rho c$	(Thermal Conductivity X Volume Sp. Ht)	$F^{2}L^{-2}T^{-1}\theta^{-2}$

*F = Force, L = length, T = Time, θ = Temperature

performing the dimensional analysis we have

$$\frac{\theta}{u}\left[\frac{k\rho c}{Vt}\right]^{0.5} = const \tag{3}$$

It is found experimentally that in single point turning operations[16]

$$u \simeq t^{-0.2} \tag{4}$$

combining equations (3) and (4)

$$\theta = \frac{const.}{t^{0.2}}\left[\frac{Vt}{k\rho c}\right]^{0.5} \sim V^{0.5}t^{0.3} \tag{5}$$

$$\text{or} \quad \theta^{2} \sim Vt^{0.6}$$

Thus, it is seen that the product ($Vt^{0.6}$) is proportional to θ^2. This suggests that the classification of wear types (Figure 9) due to Opitz[14] is of thermal origin.

TAYLOR TOOL LIFE EQUATION

Tool life (T) is generally found to vary with cutting speed (V) in accordance with the following equation initially suggested by F. W. Taylor [16]

$$VT^{n} = C \tag{6}$$

where n and C are empirical constants depending upon tool and work material and the undeformed chip thickness (t) which are held constant in equation (6). Representative values for n are given in Table II. Thus, for a high speed steel tool, tool life $T \sim V^{10}$ and from equation (5) $T \sim \theta^{20}$. This confirms the observation made previously that whereas speed is relatively unimportant for lightly loaded sliders, it plays a very important role for cutting tools. The above discussion also suggests that the major cause for the strong dependance of tool life (T) on cutting speed (V) is due to

TABLE II.—REPRESENTATIVE VALUES OF TAYLOR EXPONENT (n)

Tool Material	n
High Speed Steel	0.1
Tungsten Carbide	0.2
Ceramic (Al_2O_3)	0.4

tool temperature (θ).

The very high dependence of tool life on tool temperature ($T \sim \theta^{20}$) is consistent with other thermally induced chemical reactions. For example, in elementary chemistry a rule of thumb frequently cited is that the rate of a reaction doubles for each 10 C (18 F) rise in temperature. This is for aqueous chemistry where the temperature level is approximately 100 F (560 R). Thus, for this case

$$\frac{r_1}{r_2} = \frac{578}{560}^{x} = 2 \tag{6}$$

where r_1 and r_2 are two reaction rates and $x = 22$ is the appropriate exponent relating temperature and absolute temperature levels. This suggests that the reaction rate varies as the 22 power of the absolute temperature in the case of the simple rule of thumb from elementary chemistry. If we look on the attritious wear process for a cutting tool as a process involving chemical bonding, it is not unreasonable to expect a similar exponent. The closeness of the exponent 20 for a high speed steel tool ($T \sim \theta^{20}$) and 22 for the simple wet chemistry reaction rate situation suggests that the role of temperature in metal cutting wear is largely of chemical origin where the rate process will depend upon the amplitude of thermal vibration, which in turn is proportional to temperature.

By use of equation (5) the Taylor equation may be generalized as follows for cases where both V and t vary

$$VT^{n1}\ t^{n2} = C \tag{7}$$

where $n_2 > n_1$. This equation may in turn be generalized to include a variable depth of cut (b) as follows.

$$VT^{n_1}\ t^{n_2}\ b^{n_3} = C \tag{8}$$

where $n_1 < n_2 < n_3$

Typical values for n_1 , n_2 and n_3 for a high speed steel single point cutting tool are:

$$n_1 = 0.10 \quad n_2 = 0.17 \quad n_3 = 0.25$$

WEAR LAND TOOL WEAR

This is probably the most important type of wear for single point tools that are well matched to the material being machined. When the extent of the wear land is plotted versus cutting time at a constant rate of metal removal, curves such as those shown in Figure 10 are obtained. The wear rate is high at the beginning and at the end of the test.

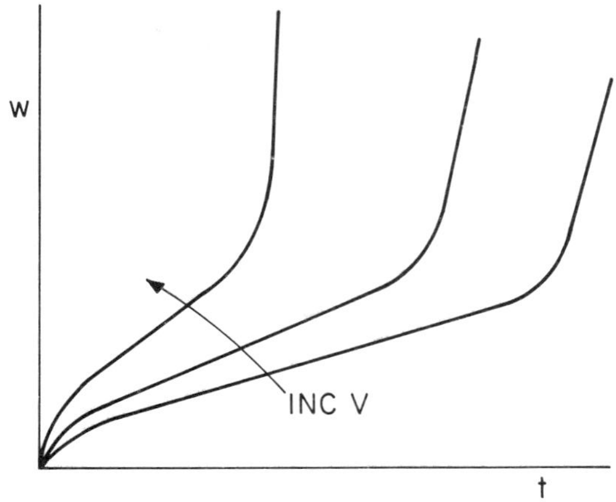

Fig. 10.—Typical wear land (w) versus cutting time (t) curves for different cutting speeds (V).

For a lightly loaded slider (which the wear land approximates), the real area of contact (A_R), and hence the wear rate, will be low as long as

$$\frac{W}{A} < H \qquad (9)$$

where W = load on wear land

A = apparent area of wear land

H = hardness of work material

This will not be the case at the beginning and end of the test. At the beginning A will be small and at the end H will be small due to thermal softening of the work material - *not the tool*. Hence, at the beginning and end of a wear land tool life test, we should expect a greater rate of wear than in the intermediate region, as shown in Figure 10.

Wear land wear is primarily attritious when the real area of contact (A_R) on the wear land is small relative to the apparent area (A), which is usually the case for the wear land.

Reference (18) discusses an interesting case in which HSS tool life, when cutting stainless steel, was significantly improved by addition of very finely divided Al_2O_3. While the presence of Al_2O_3 in a work material normally decreases tool life substantially, due to an increase in abrasive wear, (Al_2O_3 is one of the principal abrasive materials used in grinding wheels), the opposite is the case when the alumina particles are submicroscopic in size. The positive function of the alumina when finally divided is to make the work material more refractory so that the onset of thermal softening is postponed, thus making it possible for the tool to be used to a larger value of wear land before the onset of rapid wear (Figure 11).

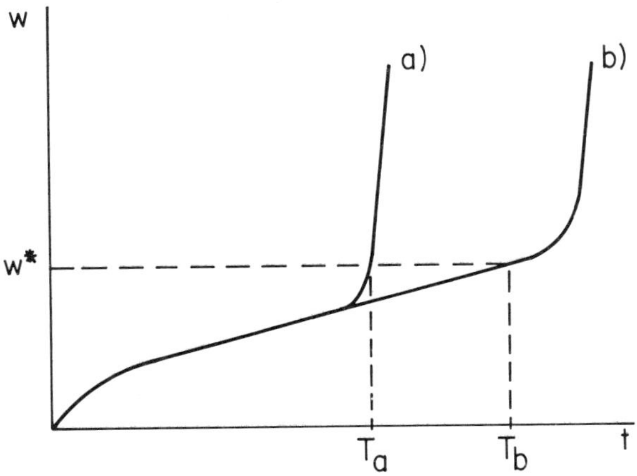

Fig. 11.—Different wear curves for a) work material that has a relatively low softening temperature and b) work material that is more refractory. In both cases T is the tool life corresponding to a certain value of wear land. (W).*

CRATER WEAR

The volume of wear that may be tolerated on the wear land is very much less than that for the allowable crater on the tool face. This is fortunate since the temperature on the tool face is very much higher and hence the volume rate of wear is much higher on the tool face. The maximum temperature on the tool face which occurs at about the midpoint of the contact length between chip and tool is usually sufficient to give thermal softening and a relatively large real area of contact for the tool face. The crater that forms has its maximum depth at the point of maximum temperature and, hence, the crater develops as shown in Figure 12 with its maximum point at approximately a constant distance from the cutting edge. The rate of crater formation depends on the stability of the constituents of the tool, the rate of diffusion of the products of decomposition and their influence on the strength of the work material.

When a tungsten carbide tool is used to cut a low carbon steel, the rate of crater formation is very high. This is because the iron on the surface of the chip has such a high affinity for carbon it causes the WC

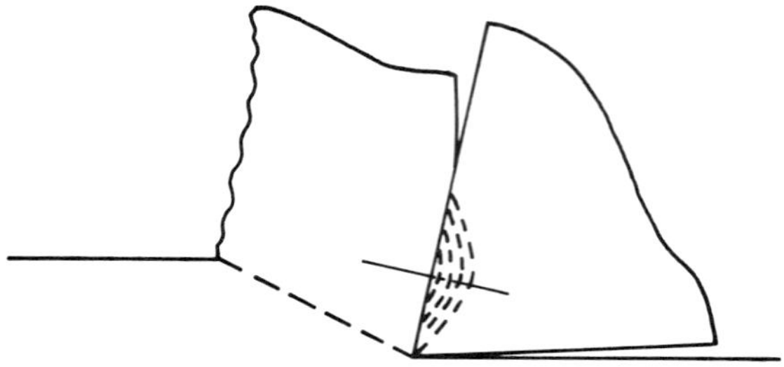

Fig. 12.—Manner in which crater develops on the tool face with its maximum depth at an approximately constant distance from the cutting edge.

particles to decompose, resulting in a high wear rate. This was not the case when mach·ning gray cast iron since the presence of excess carbon in the form of graphite prevents WC from being decomposed. Thus, it was not possible to machine low alloy steels with straight tungsten carbide tools for a whole decade after their introduction in hhe late 1920's. It was not until near the beginning of World War II that Mr. Philip M. McKenna found that a steel cutting grade of carbide could be produced by substituting TiC and TaC for some of the WC present in the original tools that could only be successfully used to machine gray cast iron and nonferrous metals[19]. The so-called steel cutting grades of cemented tungsten carbide revolutionized the machining operation and made it necessary for machine tool builders to design and manufacture higher speed, moee rigid machine tools of greater power.

SPECIAL DEOXIZED STEELS

An important development that was pioneered in Germany[20] and later extended in Japan[21] and the USA[22] is the use of special deoxidation methods to produce steel that has a lesser tendency to cause carbide tools to crater in high speed machining. This technique involves the use of calcium and ferro silicon as deoxidizing materials resulting in a relatively low melting ternary inclusion which spreads over the tool face in high speed machining and acts as a diffusion barrier.

HYDRODYNAMIC ACTION

It has been suggested[23] that hydrodynamic action is possible on the tool face. However, this is not a result of penetration of a liquid film of cutting fluid between chip and tool, but rather due to the formation of a wedge shaped liquid or semi-liquid layer between chip and tool due to thermal softening and spreading of an ingredient in the work material. The shape of such a liquid film will be as shown in Figure 13a. At first glance it may be concluced that the inclination present is such as to give a negative rather than a positive hydrodynamic pressure. However, as explained in reference (23) the resultant motion of the solid portion of the chip will be parallel to the tool face (*AE*) and not to the inclined surface of the

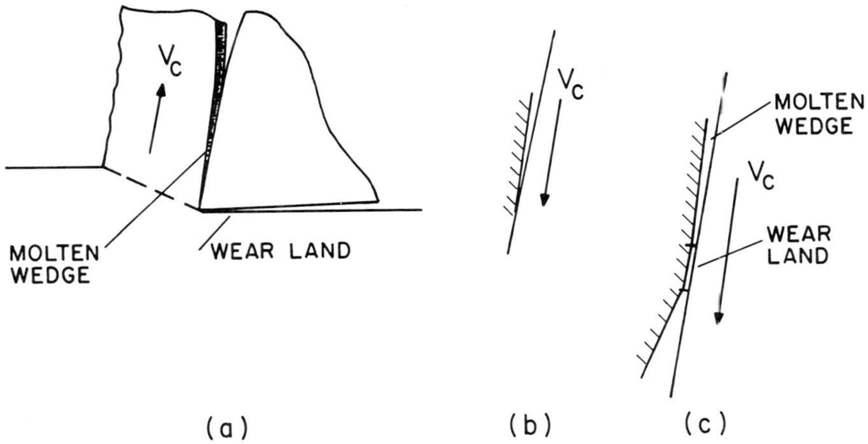

Fig. 13.—Hydrodynamic Pressure Generation in metal cutting a) cutting tool in operation with wedge shaped molten film of material between chip and tool, b) kinematical equivalent of a) and c) developed view of tool face - wear land combination.

molten layer (*AF*). This gives rise to a squeeze film action that more than offsets the sliding action associated with the negative inclination present to provide the generation of a positive hydrodynamic pressure between chip and tool. An alternative way of looking at this problem is to draw the kinematically equivalent diagram shown in Figure 13b where it is immediately clear that a positive pressure will be generated.

Figure 13c shows a developed view of the rake and clearance portions of the tool which suggests that the role of the wear land is to provide a composite bearing configuration and shroud action. Composite bearings are discussed in reference (24) as is the action of a parallel surface bearing corresponding to the situation on the wear land.

While the role of the inclusions present in specially deoxidized steels is generally believed to be that of a diffusion barrier for carbon, it could also tend to act as a thermal diffusion barrier just as the glass layer in the Sourjonet process used in hot extrustion of steel[(25)]. In this case a hot billet is coated with glass powder which metls to provide a continuous glass coating on hhe surface of the material to be extruded. An important function of the glass is to decrease heat transfer between the hot billet and the cold die. Similarly, the molten layer of oxide coming from the specially deoxidized steel would provide a poor thermal conducting layer that would tend to decrease heat transfer from chip to tool.

It has been reported that specially deoxidized steels tend not only to reduce crater formation on the tool face but also to reduce the rate of wear land formation. However, this is not believed due to a diffusion barrier action, since the temperature on the clearance face is very much lower than that on the tool face. A possible alternative explanation for the beneficial effect of oxide on the clearance surface is a hydrodynamic one - one associated with the liquid oxide that would leak past the cutting edge (Figure 13c).

It is possible that a hydrodynamic action is also responsible for the improved machinability of steels containing such free machining additives as lead and bismuth. These materials which are insoluble in steel will be squeezed from the chip to provide a thin wedge-shaped liquid layer on the tool face as in the case of the complex low melting oxides in specially deoxidized steels.

Analysis of the hydrodynamic aspects of metal cutting is extremely difficult since the inclination angle of the fluid wedge is not predetermined by the geometry of bearing surfaces but by a melting action that depends on the rate of heat generatinn, the rate of heat transfer and the rate of melting. From this one might expect the existence of an optimum cutting speed for each tool-work-inclusion system. If the cutting speed is too low, the energy level will be insufficient to cause melting; whereas if it is too high there will not be time for melting during chip-tool contact. This is obviously an optimization problem that is best approached experimentally rather than analytically.

It has been reported[20] that an optimum cutting speed does exist for specially deoxidized steels.

COATED TOOLS

In the late 1960's it was found that crater formation could also be retarded by the vapor phase deposition of a very thin (0.0002 in.) coating of TiC on a steel cutting grade carbide tool. It was later found that such coatings also tended to reduce the wear-land wear rate. Tool wear mechanisms are rarely of one type but generally consist of a combination of mechanisms. In the case of wear land wear there will be a component of wear associated with diffusion and it is this component that will be reduced by the presence of a thin TiC coating.

A TiC coating in contact with the iron chip is less likely to give up its carbon than WC. Use of such a coating avoids the tool weakening effect associated with relatively large additions of TiC or TaC to the tool in bulk.

Tools have also been coated with Al_2O_3 and this material is thought to function as a diffusion barrier which is one of the mechanisms believed responsible for the success of the specially deoxidized steels.

In addition to TiC, other coating materials in use include TiN, HfC and HfN. All of these appear to provide a more stable material in the presence of hot iron and at the same time act as diffusion barriers. A TiN coating gives lower tool face friction than TiC when cutting a low alloy steel. Although TiC is more stable than WC there will be some decomposition of TiC. The carbon that is released will be absorbed by the low carbon steel, thus strengthening it and causing an increase in tool-face friction. TiN does not appear to be as good a diffusion barrier as TiC, but when it decomposes the nitrogen released does not have as great a strengthening action on the steel as does carbon. This results in lower tool-face friction for a TiN-coating than for a TiC coating[26].

One carbide tool manufacturer markets a tool having three coating layers. Next to the carbide tool is a 0.0002 in. layer of TiC. This is followed by a 0.0008 in. layer of titanium carbonitride and then a 0.0002 in. layer of TiN. The TiC is apparently the ultimate diffusion barrier and the TiN layer provides coupling material for the inner and outer layers.

The subject of tool coating is a rapidly developing one and is akin to the solid lubrication of sliding surfaces. Important development in both of these fields are expected in the future.

CUTTING FLUIDS

Cutting fluids have a dual role - cooling and lubrication. In high speed cutting operations where the tool remains buried in the cut most of the time, such as in turning, the function of the fluid is primarily one of cooling. In low speed operations involving intermittent cutting, such as in broaching or tapping, lubrication is important. Both water base and oil base lubricants are used, the former are generally the better coolants while the latter are better lubricants. There are many secondary considerations associated with cutting fluids such as chip disposal, corrosion prevention, health and safety considerations, etc. A discussion of these aspects of cutting fluid technology is to be found in reference (27).

It is unlikely that anything but a vapor can penetrate the interface between chip and tool during a continuous cutting operation. This is due to the near perfect contact between chip and tool and the normally high speed motion of the chip counter to fluid penetration. From the latter point of view it is likely that any penetration will be from the side of the tool (i.e. parallel to the cutting edge).

GRINDING

Grinding is a material removal operation involving very small chips removed at very high speed. The tribological aspects of grinding operations are extremely difficult to interpret and anticipate since times involved are so short as to preclude equilibrium conditions. Since most of our technical experience involves equilibrium, we have little to go on in interpreting grinding mechanisms.

ACKNOWLEDGEMENTS

The author wishes to acknowledge a grant from the Office of Naval Research that has provided an opportunity of studying the Tribiological aspects of surfaces undergoing gross plastic flow. He should also like to thank Drs. Miller and Ellingsworth of ONR for their interest in his studies.

REFERENCES

1. Shaw, M.C., Ber, A. and Mamin, P.A., *American Society of Mechanical Engineers Transactions*, Vol. 82, 1960, p. 342.
2. Backofen, W.A., "Deformation Processing," Addison-Wesley, Reading, Massachusetts, 1972, p. 135.
3. Shaw, M.C. and DeSalvo, G.J., *American Society of Mechanical Engineers Transactions*, Vol. 92, 1970, p. 469.
4. Shaw, M.C. and DeSalvo, G.J., *American Society of Mechanical Engineers Transactions*, Vol. 92, 1970, p. 480.
5. Shaw, M.C. and DeSalvo, G.J., *Metals Engineering Quarterly*, Vol. 12, 1972, p. 1.
6. Rabinowicz, E., *Scientific American*, Vol. 237, 1977, p. 74.
7. Komanduri, R. and Shaw, M.C., *Philosophical Magazine*, Vol. 32, 1975, p. 711.

8. Archard, J.F., *Journal of Applied Physics,* Vol. 24, 1953, p. 981.
9. Shaw, M.C., *Wear,* Vol. 43, 1977, p. 263.
10. Shaw, M.C., *Chemical Technology,* July 1971, p. 432.
11. Shaw, M.C., *Wear,* Vol. 2, 1959, p. 217.
12. Shirakoshi, T., Komanduri, R. and Shaw, M.C., *American Society of Mechanical Engineers Transactions,* Vol. 100, 1978, p. 244.
13. Trent, E.M., in Proceedings of the 8th Machine Tool Design and Research Conference, Manchester, edited by S.A. Tobias and F.M. Roenigsberger, Pergamon Press, Part 2, 1967, p. 629.
14. Opitz, H., *Werkstattstechnik und Maschinenbau,* Vol. 46, 1956, p. 210.
15. Carslaw, H.S. and Jaeger, J.C., "Conduction of Heat in Solids," 2nd edition, Oxford University Press, London, 1959.
16. Shaw, M.C., "Metal Cutting Principles," Technology Press, MIT, Cambridge, Massachusetts, 1954.
17. Taylor, F.W., *American Society of Mechanical Engineers Transactions,* Vol. 28, 1907, p. 31.
18. Vilenski, D. and Shaw, M.C., *CIRP Annals,* Vol. 18, 1969, p. 623.
19. McKenna, P.M., U.S. Patent 2,113,353, 1938.
20. Opitz, H. and Koenig, W., *Industrie-Anzeiger,* Vol. 87, 1965, Part I, pp. 26, 46; Vol. 87, 1965, Part II, pp. 43, 845; and Vol. 87, 1965, Part III, pp. 51, 1033.
21. Working Group on Machinability in Japan (Chairman: T. Sata), *Bulletin of the Japan Society of Precision Engineers,* Vol. 3, No. 1, 1969, p. 1.
22. Tipnis, V.A. and Joseph, R.A., *American Society of Mechanical Engineers Transactions,* Vol. 93, 1971, p. 571.
23. DeSalvo, G.J. and Shaw, M.C., "Advances in Machine Tool Design and Research," Pergamon Press, Oxford, 1969, p. 961.
24. Shaw, M.C. and Macks, E.F., "Analysis and Lubrication of Bearings," McGraw-Hill, New York, 1949.
25. Sejournet, J., *Engineering,* Vol. 177, 1954, p. 463.
26. Rao, S.B., Kumar, K.V. and Shaw, M.C., *International Journal of Wear,* Vol. 49, 1978, p. 353.
27. Shaw, M.C., in Manufacturing Engineering Transactions, Society of Manufacturing Engineers, Dearborn, Michigan, Vol. 1, 1972, Paper No. 4.

DISCUSSION

N. P. SUH, MIT: I would like to make a comment for the benefit of the audience. Said Jahanmir in his Doctoral thesis here at MIT worked out exactly the state of stress as the tangential and normal loads slide by a reference point. In addition to the stresses set up during unloading, the material experiences stress below an asperity due to the loading by a nearby asperity also. As we follow the history of what happens in front of an asperity, we find that the crack tends to nucleate due to shear deformation which creates a tension around the inclusions. When we look at a large number of asperities moving tangentially, we observe the unloading tensile stress Prof. Shaw is referring to and the applied stress that together create a very complicated stress field. We find that the area behind the asperity is in complete elastic state of loading, even though in front of the asperity it is in complete plastic loading.

M. C. SHAW: I might mention that there is a very interesting film by Dr. Bush who did his Doctoral thesis at Hanover. He shows diamond and agate spheres sliding over a glass surface and he looks at it through the microscope and photographs it. There is clear evidence of plastic flow even with glass, because the deformed area is very small. Cracks form at the trailing edge. The explanation is that in unloading these spheres, the tensile stresses (normal to the surface) that develop form these horizontal cracks. To explain these subsurface cracks, it is not necessary to have a horizontal force. I therefore think you can explain the horizontal cracks based on the plastic flow that occurs during unloading.

QUESTIONER: I enjoyed your talk very much. In brittle materials, Lawn/ Wilshaw, and Evans recently observed radial cracks perpendicular to the surface during indentation. Upon unloading they will open up and lateral cracks almost parallel to the surface will form due to this stress condition. Numerous investigators have looked at the surface by sliding sharp objects or blunt indenters and they observed the same type of fracture. These radial cracks are elongated because of the sliding but you get the same type of crack distribution that you would from the indentation process.

SHAW: I think that instead of concentrating on the classical Hertz problem of a sphere with load vertical, we should have the load reversed and see what happens then.

SAME QUESTIONER: We should actually do it both ways. I didn't mean to imply that there is plastic deformation in the same sense that you are talking about, but there is a non-conservative process that takes place.

SHAW: I think that even with glass, if the load is light enough, we will observe plastic flow as in the case of Bush's experiment.

SAME QUESTIONER: I didn't make a distinction between blunt and sharp indentors. There are distinctions. You will get permanent impressions, but whether it is densification or plastic deformation is still being debated.

D. H. BUCKLEY, NASA: When you indent a sphere on a surface, have you tried repeated cycles of loading and unloading to see if the crack grows? Does the growth stop at some point after a number of cycles?

SHAW: We haven't done that yet. That is a very good suggestion. I might mention one instance related to shakedown stresses that may be of interest to the group. In some fatigue work where we had very heavily loaded material under the ball that was cyclically loaded the hardness apparently decreased with the number of cycles. In other words, the plastic indentation got bigger with each cycle up to a thousand cycles. It was approximately an exponential decay to a lower value of apparent hardness. However, when we measured the microhardness of the curved surface at the bottom of the indentation, the hardness was actually greater. The way I could explain this appears to be as follows. Actually the material itself is increasing in hardness but when you cycle a load on the sphere, relaxation of some of the residual stresses (that are normally there when you just load it once and unload it) takes place. In other words, it is a change in the residual stress pattern which is causing the apparent reduction in hardness, whereas the actual hardness is increasing because of strainhardening as the dent gets bigger. But I do not know whether that is related to what you are talking about. We did not study the influence of cycling on crack growth. That would be very interesting.

WEAR MECHANISMS IN MACHINING

S. Ramalingam

ABSTRACT

This discussion paper addresses the basic issues in tool wear during machining discussed in the review paper. The feasibility of adhesive, diffusive, and abrasive tool wear are examined in detail. Difficulties encountered in invoking adhesive wear and diffusive wear as formulated in literature are discussed. The data and arguments presented here stress the importance of abrasive wear in machining.

Physics of wear and wear mode change in tool materials prone to brittle-ductile transition are used to account for differences in abrasive wear in the crater and the flank wear regions of the cutting tool. Plausibility of tribo-chemical wear involving decomposition reactions is also examined.

INTRODUCTION

In metal removal by machining large specific surface areas are generated. The increase in specific surface area produced depends on the properties of the work and tool material, tool and process geometry, cutting conditions and so on. The significance of surface area production becomes clear when it is recognized that machining may be described as a process of helical or serial sectioning of the parent body. Characterized in this form, machining is seen to be a process of intensive, internal sampling of the work material by a mechanical sampling probe.

Structure sampling is carried out by a small part of the cutting tool which mechanically penetrates into and displaces the sampled material. Strains in the range of 1 to 5 and strain rates in excess of 10^4 sec^1 prevail during sampling. The sampling debris, the chip, remains in contact with the tool over a small distance. High normal stresses, shear stresses and elevated temperatures occur in this contact region due to the strains and strain rates imposed during sampling. The contact conditions are sufficiently severe to induce secondary shear in the chip in many instances.

The survival characteristics of the sampling probe under these sliding wear conditions depend on its structure and properties as well as its response to the thermal, mechanical and thermodynamic conditions encountered in course of sampling. Invariably material loss occurs from the sampling probe surfaces and when it reaches some limiting value, it is at the end of its useful life. This point determines the tool life. Process economy in machining depends on the rate of material loss from the sampling surface and the resulting tool life.

Metal removal by machining as recast here preserves all the phenomenology relevant to machining but allows one to focus on the structural questions connected with the work and tool material. It calls attention to the severe tool-work contact intensity peculiar to machining. It recognizes the fact that while the sampling probe is at a steady elevated temperature, the sampled material traversing across it only suffers transient heating, typically for a milli-second or less. That the "friction" encountered in machining is not adequately described by coulombic processes is also recognized. With this alternate modelling, the problem of tool wear and machinability can be examined from another perspective.

SOME GENERAL COMMENTS

Professor Shaw has enumerated and discussed the tool wear mechanisms in an earlier paper(1). Since under steady state machining conditions, tool failure is primarily by wear land formation and by crater formation (Figure 9, Reference 1), tool failure by fracture is not of interest. This discussion will address tool failure by adhesive wear, diffusive wear, abrasive wear, corrosive wear and by micro-chipping.

That tool wear process is thermally-activated is widely conceded. Some combinations of tool and work materials are more resistant to crater formations than others(1). Tool materials yielding higher crater resistance do not necessarily, and usually do not, exhibit higher flank wear resistance also. Intrinsically harder materials and tool materials which are thermodynamically more stable are generally known to permit higher metal removal rates(2). Little beyond this connected with tool wear can be supported with unambiguous physical evidence which can withstand rigorous scrutiny.

WORK MATERIAL STRUCTURE AND TOOL WEAR

In a machining process, there are two participants, a work material and a tool material. Dynamic interactions induced between these materials during machining generates tool wear. The importance of considering the material is seen when it is recognized that the temperatures generated in the thermally-activated tool wear originate as a result of the stresses and strains imposed on the work material to detach a portion of matter from it. Flow stress of the work material is hence an important parameter in tool wear processes.

The stresses and thermal fields imposed on the tool are generated by the shear dissipation in chip formation and by the frictional dissipation involved in overcoming flow restraint at the tool chip-interface. All of the power required to enforce metal removal are transmitted to the work material through a small tool-chip contact area.

Power dissipated in shear is larger than that to overcome flow restraint but proportionately more of it is transported away within the bulk of the chip. This dissipation is also farther removed from the tool surface than the frictional dissipation which occurs in direct contact with the cutting tool. Hence the thermal field generated in the thermally-activated wear process is dominated by the process at the tool-chip interface.

Chip formation is a problem in partially-constrained plastic flow(3) where the geometry is established dynamically. Boundary conditions involving stresses and contact length are so established that the resultant force R on the tool is equal to resultant R' acting across the shear region ABCD (Figures 5 and 6 of the paper by Shaw(1) in this volume). A decrease in flow restraint increases the shear angle to establish a new equilibrium cutting geometry(4) where smaller shear strains are obtained(5,6). Hence a decrease in apparent friction at the tool-chip interface not only decreases frictional dissipation but also lowers the thermal energy input into the tool from the shear zone. Large wear rate reductions are thus possible with a small decrease in tool-chip "friction".

According to the sampling model introduced here, tool-work contact intensity in machining is high. By incorporating small volumes of a second phase in the matrix of the work material, it should then be possible to modify the frictional processes at the interface quite effectively. Depending on the characteristics of the second phase incorporated it should be feasible to enhance or lower the tool wear in machining.

Free machining alloys attempt to, and succeed in lowering tool wear rate by the addition of such soft phases as lead, manganese sulfide, etc. In fine grain steels on the other hand, the additives such as aluminum, niobium and vanadium are usually present in their combined form as oxides, nitrides and other refractory phases. Tool wear rate is enhanced and lower machinability is then said to exist. Since the volume fraction is small but sufficient to induce changes in machinability, one should not expect changes in bulk mechanical properties in such circumstances.

It is seen that work materials can influence tool wear in machining both by their inherent resistance to chip formation (governed by flow stress) and by the dynamic interactions that can be generated by the second phase particles contained within their matrices. Ample evidence is available in the machining literature showing that (a) Tool wear rate is raised as the flow stress of the work material is increased (except by strain hardening), (b) Tool wear rate can be lowered by the judicious selection and incorporation of small quantities of a second phase within the material, (c) Tool wear rate is sensitive to the dispersion parameters associated with the second phase[7,8,9] and, (d) In the case of steels, tool wear rate is dependent on and can be altered by controlling the deoxidation practice[10,11].

EXPERIMENTAL STUDIES RELEVANT TO TOOL WEAR MECHANISMS

With the recognition that tool wear processes in machining depend on the properties of *both* the participants involved in this sliding contact wear problem, the relative importance and significance of the various wear mechanisms enumerated by Shaw[1] can be examined. As in any wear study, wear debris analysis can materially assist the identification of the dominant wear mechanism. In tool wear the total quantity of wear debris produced by the time the tool reaches the end of its useful life is less than 10^{-3} gm per mm of the width of cut. In typical laboratory machining studies, the width of cut ranges between 1.5 mm and 3.0 mm. The sub-milligram quantities of wear debris produced hence call for sophisticated wear debris analysis. Unless the wear debris is collected and concentrated, even EPMA (electron probe micro analysis) techniques are of marginal value.

Quantitative radio-active wear studies, auto-radiography, micro-chemical analytical studies and electron diffraction analysis are among the few experimental techniques that are adequate to obtain corroborating *physical evidence* to identify wear mechanisms when wear debris available is as small as in the present instance. A number of radioactive wear studies and auto-radiography studies connected with tool wear have been carried out[12-15]. Some micro-chemical analytical studies[16,17] have also been carried out. The available physical evidence can be summarized as follows: (a) The material detached from the tool surface during crater formation is transported away on the underside of the chip, (b) The material detached from the tool surface during wear land formation is transferred on to the work surface, and in some cases appears on the upper side on the chip, (c) Autoradiography and micro-chemical analytical studies indicate that substantial amount of wear debris is transported away as discrete wear debris,

(Professor Cook, based on his own auto-radiography finds a nearly uniform smear of tool material on the chip surface in the case of WC-Co cutting tools), and (d) The wear debris is weakly bonded to the chip surface to which it is transferred.

Preponderant *physical* evidence thus strongly favors the view that material loss occurs discretely with the largest wear debris dimension consistent with the dimensions of the individual carbide grains within these powder metal compacts.

Any wear mechanism advanced to account for tool wear during machining must (c) be thermally activable, (b) be consistent with work material interactions known to occur and discussed earlier, and (c) be able to account for discrete wear debris formation that is found to occur in almost all instances. There is one other requirement that must also be met. The wear mechanism must also be able to account for the distribution in tool life that is usually obtained under production machining conditions (when tool wear is primarily by crater formation life distribution is log-normal). With these requirements, the wear mechanisms enumerated[1] will now be examined.

ADHESIVE WEAR

Adhesive material transfer from one body to the other in two-body wear is well documented in literature. However, in almost every instance the material transfer is from the softer of the two bodies in contact. Tool material is invariably chosen to be harder of the two materials in machining. Thus, even if the softer of the two bodies, the work material, were allowed to transfer on to the tool surface, there is a fundamental difficulty in justifying how material loss occurs from the surface of the tool.

For material loss to occur from the surface of the harder body, cohesive failure must take place within its bulk and not failure at the interface between the adhered body and the tool material. This is possible, if the "weld" shear strength is high and the tensile fracture strength of the harder body is low. In the present case, the harder body, usually a cemented carbide, is weaker in tension. But the state of stress at and below the tool surface in the region of contact between the tool and the chip is bi-axial compression of considerable magnitude[18]. In the presence of this bi-axial compression, it is not realistic to expect cohesive failure within the tool material.

Zhilin and Tkachev[14] have moreover found that the wear debris from the tool is poorly bonded to the chip. Hence, adhesive wear, while it is consistent with discrete wear debris formation, is difficult to justify as a significant mechanism of tool wear.

The plausibility of adhesive tool wear can be more directly addressed, if the tool-chip interface can be observed during the machining process. Horne, Doyle and Tabor[19], using Sapphire cutting tools have observed the tool chip interface during a machining operation. By direct observation and by cinematography, they have been able to demonstrate that in the so-called "seizure zone" of the tool-chip contact region, there is a continuous movement of the chip material. This experimental observation that "seizure" does not take place also poses fundamental difficulties to the adhesive tool wear theory.

DIFFUSIVE WEAR

In tool wear literature diffusive matter transport is held to be a principal source of material loss from the tool surface during machining. There is hardly any direct evidence to support this wear mechanism. With the exception of the activation energy measurements of Cook[20] and

Nayak[16], there is no quantitative data. Even the identity of the diffusing atomic species is a matter of dispute.

Shaw[1] suggests that the principal diffusing species is Carbon. Cook[20,21] believes that diffusing Iron atoms control the wear rate. Trent[22] suggests that carbon, tungsten and cobalt diffusing simultaneously and with the same *mobility* into the chip control the wear rate.

Many relevant questions have not been considered in postulating the diffusive wear model. These need to be resolved before the applicability of the diffusive wear mechanism to tool wear can be admitted.

As mentioned earlier, while the tool is at a stationary elevated temperature, the work material only suffers transient heating. Consider a low carbon steel machined with a carbide cutting tool at 180 meters/minute with a chip thickness ratio of 0.66. The chip traverses across the tool at 2 meters/minute, and if the traverse time across the tool surface is taken to be 10^{-3} seconds (corresponds to a hot zone dimension of 2 mm) and the mean tool-chip interface temperature 800°C, the expected work material heating *rate* is between 1 and 8 x 10^5 °C/second.

The work material matrix initially consists of ferrite and cementite. As the material traverses across the "hot zone", if the ferrite is not transformed to austenite by the rapid heating, the ferritic phase is either supersaturated or nearly so. Carbon diffusion from the tool surface can then be not invoked. If the ferrite does transform, only massive transformation[23] can be expected leading to the formation of non-equilibrium austenite. The validity of applying equilibrium diffusion model to a couple where one of the participants is not in thermodynamic equilibrium is then open to question.

Furthermore the non-equilibrium austentite produced has a much larger specific interfacial area with cementite than with the carbide constituent in the tool material. The cementite is thermodynamically more unstable compared with the carbide, say WC, in the tool material (Free energy of formation of Fe_3C and WC at 298 K are +5.3 and -9.1 kcal/mol. respectively). It is therefore more appropriate to expect austenite to pickup carbon from the cementite than WC. This thermodynamic consideration casts considerable doubt on carbon diffusion model of tool wear.

Sectioned chips, even those produced at high cutting speeds, only exhibit deformed pearlitic structure when steels are machined. Thus, even the formation of non-equilibrium austenite considered in the last paragraph is very much open to question. Hence, the theoretical underpinning for the diffusion wear theory, the static work material - tool material diffusion couple data, is simply *not* relevant.

Published diffusive wear models assume that the non-zero concentration gradients across the interface provide the driving potential for diffusion. In polycomponent diffusion couples the rate of matter transport is governed by the gradients in chemical potential. Diffusion against a concentration gradient, the so-called "uphill diffusion" is feasible and has been shown to occur, especially for carbon diffusion in Fe-Si-C alloys. Diffusive wear models connected with tool failure have so far failed to recognize this.

When gray cast irons are machined, straight WC-Co tools do not fail by crater formation. Yet when spheroidal cast irons of the same total and combined carbon are machined, these same tools fail by crater formation. If carbon, iron, tungsten and cobalt, either singly or in some combination are responsible by a diffusive wear mechanism for tool wear, it is not clear what mechanism can be invoked to account for the extraordinary influence of *graphite type* on diffusional processes.

In the discussion of diffusive wear presented so far, the transverse velocity of the chip (velocity normal to the direction of diffusive matter transport) has been ignored. Since Horne, Doyle and Tabor by direct examination have confirmed continuous movement, this cannot be ignored.

In the diffusion couple involved in machining, the atomic mobility for any diffusing atom across the interface is many orders of magnitude lower than that mechanically impressed parallel to the interface. At such ratios of atomic mobilities, it is impossible to generate sufficient enrichment of atomic concentrations in the moving body to require the application of the transient diffusion equation to account for matter transport.

Tool wear under these conditions cannot be governed by the diffusion equation but by the atomic jump probability from the tool surface on to the moving chip surface. While this can be a thermally-activated process, it does not involve diffusive matter transport. The time-dependent recession of the tool surface giving rise to the formation of the crater is then governed by the atomic jump frequency from the stationary body to the moving one. Under these conditions, the crater formation is akin to a "dissolution" process rather than a diffusion process.

It is possible to attempt to rationalize the diffusive wear model by suggesting that the chip material is transferred to the tool surface with particle dimensions that are too small to be resolved by the optics used in the Horne, Doyle and Tabor[19] study. In intimate contact with the tool and at the prevailing tool-chip interface temperature, one or more atomic species may then be said to diffuse into the adhered body to account for the mass loss from tool surface. If such a two step diffusive wear model is invoked, it is difficult to account for the subsequent detachment of the adhered body. For reasons cited earlier, cohesive failure within the tool is unlikely. By diffusive transport, the adhered body must become solution hardened. Failure of the "weld interface" is then not permissible. Only layer by layer erosion of the adhered body can be visualized.

There are some additional consequences in the two-step diffusive wear model. Taking the tool as an infinite source and that tungsten diffuses from the tool surface into the chip, it is possible to estimate the time required for a specified depth of diffusion of tungsten, for example 14% tungsten, to a depth of one micron from the interface.

The diffusion parameters, diffusion constant and activation energy, for tungsten diffusion into a carbon steel (0.9% C; 0.9% Cr) have been determined by Gurevich, et al.[24] as 1.2×10^2 cm^2/sec., and 78 kcal/mol respectively. The activation energy for tungsten diffusion reported by Gurevich, et al. is consistent with the determined by Kieszniewski[25] in carbon steels containing 0.56% carbon and 0.23% carbon (80.3 and 77.5 kcal/mol respectively). The calculated times for diffusion of tungsten such that there is 14% tungsten one micron from the interface at a series of temperatures is as follows:

Temp °C	900	1000	1100
Time (sec)	7.2×10^3	5.2×10^2	55.2

Since Shirakashi and Usui[26], while machining a medium carbon steel with a carbide tool at 250 meters/minute report a peak tool-chip interface temperature of approximately 1150° to 1180°, computations up to 1100°C is sufficient for cutting at about 210 meters/minute used by Trent[22] in a recent study.

Trent[22] in the study referred to finds it possible to produce a 50 micron deep crater in 30 seconds while cutting at 213 meters/minute. The estimate presented here casts considerable doubt that even a two stage diffusive wear mechanism can account for the reported wear rate. The wear rate obtained is too high.

This discussion of the diffusive tool wear mechanism is closed by noting that there still is no unanimity of view regarding the diffusing species for a given tool-work pair. The thermodynamic foundations of the diffusive wear model are not satisfactory and are in need of closer scrutiny.

Despite years of experimental study, *hard evidence* is still lacking to support the postulated diffusive wear model. Available experimental data such as that of Trent[22] are insufficient to account for the observed wear rate. The model is inconsistent with preponderant auto-radiography and micro-chemical wear debris analytical data showing discrete wear debris formation. The diffusive wear model is unable to account either for the tool life scatter or the accelerated tool wear when parts per million quantities of aluminum, vanadium and niobium are added to steels (base chemistry of steel remains essentially unchanged and these additives do not form glassy phases to produce "protective" or anti-diffusion layers). Finally, the diffusive wear model does not attempt to even examine land formation (Flank wear) problems.

ABRASIVE WEAR

Of the wear mechanisms capable of discrete wear debris formation, only abrasive wear remains to be considered. The considerable indirect evidence tending to support the plausibility of abrasive wear has been summarized in recent papers[27-30]. Controlled machining studies carried out on powder metal compacts of iron and carbon containing varying volume fractions of abrasive phases (alumina and silica) strongly support the abrasive wear mechanism[7,8].

In abrasive wear theory, the hard (and usually) non-metallic inclusions contained in the work material are considered to indent (interpenetrate) the tool surface and to detach the interference volume during relative motion between the chip and the tool. The tool is considered to be at a steady elevated temperature during machining, while the non-metallic particles traversing across the tool surface are only considered to undergo transient heating during their passage across the shear zone. It is suggested that since traversing non-metallics only suffer transient heating, they will still be hard. But the tool, at a steady elevated temperature will have lost its hardness. Plastic indentation and abrasive wear are then not only possible but must occur. Discrete wear debris formation is attributed in the abrasive wear model to the dispersion of non-metallic species within the alloy matrix. Since machining involves intensive internal sampling, it is suggested that large volume fractions of non-metallics are not necessary to induce abrasive wear.

By examining the dislocation structure at and below the crater region of alumina-chromia tools, it has already been shown[31] that plastic flow can occur on hard tool material surfaces. Since alumina ceramics are more refractory than carbides, the possibility of and the occurence of plastic flow in carbides will have to be admitted. An analysis of the structure and properties of transition metal carbides also supports this view.

The transition metal carbides used to fabricate cutting tools rely on metallic binding for their transport properties, ionic binding for their crystal structure and covalent binding for their mechanical properties[32]. Recent band calculations[33] support the dominance of covalent binding invoked to account for the high intrinsic hardness of transition metal carbides. Despite their high room temperature hardness, these carbides lose their mechanical properties with increase in temperature more rapidly than even pure covalent solids such as silicon and germanium. Attention to this anamolous behavior initially reported by Williams[34] was called by Atkins and Tabor[35].

Elevated temperature hardness measurements by Platov[36-38], Dzhemelinski, et al.[39], Westbrook[40] and others confirm the precipitous drop in hardness with temperature. TiC, for example, with a room temperature hardness of 3200 kg/mm^2 possesses a hardness of only 400 to 600 kg/mm^2 at 800°C. It is such large decreases in hardness and flow properties of transition metal carbides at the tool-chip interface during machining that permit transiently heated abrasive inclusions in the work material to plow the tool surface to generate the wear debris. It is thus seen that crater wear can be satisfactorily accounted for by invoking abrasive wear.

Further analysis of the mechanical behavior of transition metal carbides at lower temperatures suggests that abrasive wear mechanism can also account for flank wear in machining. It should be noted that during any machining process, the flank temperatures are lower than those that prevail at the tool-chip interface by a few hundred degrees. "Indentation fracture" by traversing hard inclusions is then possible. The same process is also expected to occur at the crater surface at lower cutting speeds. In both instances, indentation fracture or micro-chipping will yield discrete wear debris. The physical basis for indentation fracture is as follows.

From the strain rate sensitivity observed in flow stress studies on TiC at elevated temperatures, Williams[41] has deduced that flow properties of transition metal carbides depend on Peierl's barrier climbing by dislocations (thermally activated kink motion). Activation volume measurements made at "low temperatures", 1000°C are of the order of 1000 $Å^3$, consistent with the Peierl's barrier[42]. The transition metal carbides, hence exhibit an exponential decrease in flow stress and hardness with increasing temperature. Such an exponential dependence of flow stress is also consistent with the creep studies of Dawhil and Frisch[43] on tungsten carbide compacts.

When any material exhibits a strong dependence of flow stress on temperature, a critical temperature must exist at which the stress required to enforce plastic flow equals that necessary to induce stress relaxation by fracture. This is the ductile-brittle transition temperature T_{DB}. At temperatures lower than T_{DB}, brittle fracture is the primary mode of failure. Many BCC materials exhibit this behavior at low temperatures. Due to the strong dependence of flow stress on temperature and the weak dependence of cleavage stress on temperature, transition metal carbides must also be expected to behave analogously. A definite ductile-brittle transition temperature can then be expected for each carbide. Toth[44] has estimated the brittle-ductile transition temperature of $TiC_{0.95}$ at 800°C, of $ZrC_{0.9}$ as 900°C and $NbC_{0.76}$ as 1000°C. It is, thus clear that flank wear occurs primarily in the brittle fracture temperature region of transition metal carbides.

When a brittle material is indented by a blunt body, hertzian cracks are formed if the contacts stresses are sufficiently large. If a tangential traction is applied to the indentor in addition to the normal force, the contact geometry is not greatly altered but the maximum radial tension at the trailing edge of the indentor is markedly enhanced[45]. If now the indentor is traversed, the circular symmetry of the conecrack is lost, the cone is developed farther, and the inelastic deformation encompasses larger volumes, extending the crack laterally[46]. Wear tracks are produced (due to micro-plasticity) with periodic transverse crack traces across the indentor track. Thus, as shown by Billinghurst, Brookes and Tabor[47], the sliding process can be a fracture-inducing process in brittle bodies leading to fragmentation and communition of the brittle solid whenever blunt indentors are made to traverse across their surfaces.

Lawn and co-workers[48] have carried out quantitative studies of indentation and fracture of brittle solids. They have verified that the critical load for ring crack initiation, when a brittle solid is indented depends on the indentor dimensions - the so-called "Auerbach's law". Powell and Tabor[49] have shown that Auerback's law also holds for titanium carbide compacts indented with hardened steel spheres. Experiments carried out under sliding contact[49] show that the fracture load is roughly proportional to the radius of the indentor. Thus, the feasibility of fragmentation and wear debris formation when brittle solids are subjected to indentor traverse (abrasion) is well established and the indentor dimensions have been shown to have a role in the process.

In the particular case of tungsten carbide, Buckley[50], by indenting and traversing a WC body with a 50 micron diamond stylus within a scanning electron microscope, was able to observe and show all the processes discussed in the previous paragraphs to occur at room temperature. By careful polishing and etching of worn carbide drill bits, Pons[51] has shown that indentation fracture by the abrasive wear mechanism also occurs in rock drilling with tungsten carbide tools.

It is therefore seen that abrasive wear is a viable tool wear mechanism at temperatures lower than T_{DB}. In this instance material loss occurs by fragmentation (micro-chipping) of the carbide due to indentation fracture. Since Auerbach's law has been found to hold, abrading body dimensions have a role in determining wear rate. In contrast with this, when abrasive wear occurs at temperatures greater than T_{DB}, due to geometric similarity in plastic plowing, indentor (abrading particle) dimensions have no effect on wear rate.

Machining tests on powder metal compacts containing silica and alumina[7,8] provide the principal experimental evidence in support of abrasive tool wear theory. Test results obtained when alumina containing compacts were machined[8] show that crater wear rate is independent of alumina particle size and depends only on the volume fraction of abrasive phase within the work material. The flank wear rate on the other hand is dependent both on the volume fraction and on the particle size. Machining test results would thus appear to support abrasive wear by plastic plowing for crater wear and by indentation fracture for flank wear. Additional and more detailed studies are necessary to establish the details of the abrasive wear process.

DECOMPOSITION REACTIONS

Although the abrasive wear mechanism is able to satisfactory account for both crater wear and flank wear, and is supported by machining tests on carefully prepared test samples[7,8], it cannot be invoked to account for the correlation between thermodynamic stability and tool life shown to exist by Suh, Cook and co-workers. The fine scale transfer of nearly uniform films of WC possessing "atomic dimensions" on to the chip surface when steels are machined with WC tools[20] is also difficult to explain. A problem of wear mechanism also arises when hypothetical, ultra pure materials containing no abrasive phases are machined. It is therefore necessary to postulate that other thermally activable mechanisms in addition to abrasive wear may also exist. One such may well be the decomposition reaction[59].

A decomposition reaction is formally defined by the chemical equation

$$A(solid) \rightarrow B(solid) + C(gas) \tag{1}$$

Due to the high temperatures at the tool-chip interface and the inability of the ambient atmosphere to penetrate into this contact region, an ideal, high temperature "reaction vessel" can be said to exist at the interstice between the tool surface and the sliding chip surface. This dynamic "reaction vessel" exists only in a transient form and due to the sliding motion, the reaction products (*B solid + C gas*) can be expected to be depleted continuously. Irreversible decomposition reactions may then occur to continuously decompose solid A. As a result, the surface of solid A recedes. If the decomposing solid is the tool material, surface recession rate of solid A is synonymous with crater formation rate.

When such a decomposition reaction occurs, the surface recession rate $\overset{\circ}{y}$ is given by:

$$\overset{\circ}{y} = J \frac{m_A}{\rho} \tag{2}$$

where J is the flux of A leaving the surface, m_A is its mass and ρ is its specific gravity. The flux of A depends on the decomposition rate, which in turn depends on the heat of reaction, reaction temperature and ambient pressure of gas C produced.

Flux J is proportional to $(P_V - P_C)$, where $P_V = 0$ is the initial partial pressure of C in the reaction vessel and P_C is the equilibrium partial pressure of C at the specified temperature. Flux J is then given by

$$J = - \frac{P_C}{(2\pi\ m_C\ k\ T)^{1/2}} \tag{3}$$

where m_C is the mass of C. Here k and T have their usual meaning. Surface recession rate than becomes:

$$\overset{\circ}{y} = - \frac{m_A}{\rho} \cdot \frac{P_C}{(2\pi\ m_C\ k\ T)^{1/2}} \tag{4}$$

Since the equilibrium partial pressure of C depends on the entropy and enthalphy change connected with the reaction, equation (4) may be written as:

$$\overset{\circ}{y} = - \frac{p}{T^{1/2}} \cdot exp\ (S/R) \cdot exp\ \frac{\Delta H}{RT} \tag{5}$$

where p accounts for all the constants in equation (4). Crater wear rate is then dependent on the heat of formation of the decomposing solid. To a first approximation, the temperature dependence is given by:

$$\frac{d[log(-\overset{\circ}{y})]}{d[1/T]} = - \frac{\Delta H}{R} \tag{6}$$

This preliminary analysis of the decomposition reaction suggests that thermodynamically more stable compounds will yield lower crater wear rates.

It is possible that due to the low thermodynamic stability of WC, WC-based tools do decompose rapidly to yield the thin films of WC on chip surfaces that were observed by Cook(20).

Although several studies of decomposition of TiC to yield Ti(gas) and C (graphite) have been carried out(52,53) at temperatures in excess of 2000°K and the vapor pressure of Ti(gas) determined, no such data are available at lower temperatures. Similar studies have also been carried out on ZrC(54,55), where decomposition is believed to occur congruently. It should be recognized that in all decomposition reactions, the reaction rate is very sensitive to the stoichiometry of the base compound. Thus, whether the tool wear is due to decomposition reactions or to the abrasive wear mechanism, carbides used in the tool materials need to be as close to stoichiometry as is feasible.

SUMMARY

The various mechanisms that have been advanced to account for tool wear in metal machining are examined in this discussion paper. A critical examination of the adhesive wear model and the diffusive wear model suggests that the physical foundations of both these models are weak. Clear-cut experimental evidence to support both these models is still lacking.

Tool wear studies carried out on Fe-C-SiO_2 and Fe-C-Al_2O_3 powder metal compacts strongly support the abrasive wear model. Radio-active tool wear studies and micro-chemical wear debris studies lend indirect support to the abrasive wear model. Detailed examination of the structure and properties of transition metal carbides show that abrasive wear model can be invoked to satisfactorily account for crater wear and flank wear of carbide cutting tools.

Tribo-chemical wear involving decomposition reactions is briefly examined. It is found that the need for high thermodynamic stability can be adequately accounted for by decomposition reaction model of tool wear. It would appear that the transfer of tungsten carbide as particles possessing "atomistic dimensions" can also be accounted for by the decomposition reaction. Attention is called to the lack of decomposition data for carbides in the temperature range of interest in metal machining.

We close by noting the following. The abrasive wear theory of tool wear supported by the powder metal compact machining studies rely on relative hardness reversal between the cutting tool carbide and the abrasive inclusions in the work material as a result of large energy dissipation during machining. Such a process has also been advanced to account for drill wear in rock drilling by Osborn[56]. Larsen-Basse[57] also invokes the same mechanisms to account for the high wear rate when rocks containing silica are drilled with carbides. Blombery and Perrott[58] also suggest that the rate controlling mechanism for drill wear in rock drilling may well be "localized crushing and transgranular fracture of the carbide grain beneath the path of indentation". The reader should recognize that in rock drilling, the loose drilling debris erodes the cobalt binder and thus releases the surface carbide grains from their residual compressive stresses to facilitate easier indentation fracture.

The abrasive wear model advanced here thus is seen to parallel that thought to operate in rock drilling. A unified theory of wear of carbides thus appears feasible.

ACKNOWLEDGEMENTS

The author wishes to thank Miss Melinda Wilson for the careful manuscript preparation. The author is also grateful to Drs. B. F. von Turkovich, P. Wright, and J. Watson for a number of interesting discussions. Some of the data obtained to substantiate the abrasive wear model was obtained in course of a research program funded by NSF Grant No. 75-09876. To Dr. Robert Reynik of the Materials Division of NSF, the author wishes to extend his thanks.

REFERENCES

1. Shaw, M.C., These Proceedings.
2. Suh, N.P., Cook, N.H., et al., in "NSF Hard Materials Research," National Technical Information Service Publication No. PB-221908, U.S. Dept. of Commerce, Washington, D.C., 1973, Vol. 2, p. 279.

3. von Turkovich, B.F., in "International Research in Production Engineering," American Society of Mechanical Engineers, New York, 1963, p. 26.
4. Merchant, M.E., *Journal of Applied Physics*, Vol. 16, 1945, p. 267.
5. Shaw, M.C., et al., *Journal of Engineering for Industry*, Vol. 83, 1961, p. 163.
6. Shaw, M.C., et al., *Journal of Engineering for Industry*, Vol. 85, 1963, p. 85.
7. Ramalingam, S. and Faulring, G., in Proceedings of VIth North American Metal Working Research Conference, University of Florida, Gainesville, Florida, 1978, p. 290.
8. Byrd, J.D. and Johnson, L., in Proceedings of VIth North American Metal Working Research Conference, University of Florida, Gainesville, Florida, 1978, p. 310.
9. Frohlke, M., in Proceedings of the International Symposium on Influence of Metallurgy on Machinability, Iron and Steel Institute of Japan, Tokyo, Japan, 1977, p. 41.
10. Weidtman, O., *Stahl und Eisen*, Vol. 56, 1936, p. 790.
11. Wicher, A., *Radex Rundschau*, 1965, No. 2, p. 432.
12. Merchant, M.E., et al., *American Society of Mechanical Engineers Transactions*, Vol. 75, 1953, p. 549.
13. Colding, B., *Acta Polytechnica Scandinavica*, Vol. 1, 1958, p. 1.
14. Zhilin, V.A. and Tkachev, V.N., *Fiziko-Khimicheskaia Mekhanika Materialov*, Vol. 10, No. 4, 1974, p. 80.
15. Nayak, P.N., Ph.D. Thesis, Massachusetts Institute of Technology, Cambridge, Massachusetts, 1966.
16. Uehara, K., et al., *Japan Society of Precision Engineers. Proceedings*, Vol. 42, 1976, p. 931.
17. Uehara, K., Kumagai, S. and Mitsui, H., 24th General Assembly of the International Institute of Production Engineering, Kyoto, Japan, 1974, Preprint.
18. Khishinami, T., Yokouchi, H. and Saito, K., *Japan Society of Precision Engineers. Proceedings*, Vol. 39, 1973, p. 1023.
19. Horne, J.G., Doyle, E.D. and Tabor, D., in Proceedings of Vth North American Metalworking Research Conference, University of Massachusetts, Amherst, Massachusetts, 1977, p. 237.
20. Cook, N.H., *Journal of Engineering for Industry*, Vol. 95, 1973, p. 931.
21. Cook, N.H., in "Influence of Metallurgy on Machinability," American Society for Metals, Metals Park, Ohio, 1975, p. 1.
22. Neerheim, Y. and Trent, E.M., *Metals Technology*, Vol. 7, 1977, p. 548.
23. Christian, J.W., "The Theory of Transformations in Metals and Alloys," Pergamon Press, Oxford, England, 1965, p. 602.
24. Gurevich, Ya. B., et al., *Metallovedenie i Termicheskaya Obrabotka Metallov*, No. 11, 1969, p. 52.
25. Kieszniewski, J., *Prace Instytutow Hutniczych*, Vol. 19, No. 4, 1967, p. 253.
26. Kitagawa, T., Shirakashi, T. and Usui, E., *Japan Society of Precision Engineers. Proceedings*, Vol. 42, 1976, p. 1178.
27. Ramalingam, S., *Materials Science and Engineering*, Vol. 29, 1977, p. 123.
28. Ramalingam, S. and Watson, J.D., in Proceedings of Vth North American Metalworking Research Conference, University of Massachusetts, Amherst, Massachusetts, 1977, p. 291.
29. Ramalingam, S. and Watson, J.D., in Proceedings of the International Symposium on Influence of Metallurgy on Machinability, Iron and Steel Institute of Japan, Tokyo, Japan, 1977, p. 67.
30. Ramalingam, S. and Watson, J.D., *Journal of Engineering for Industry*, Vol. 100, 1978, p. 201.

31. Ghate, B.B., et al., *American Ceramic Society. Bulletin,* Vol. 54, No. 2, 1975, p. 210.
32. Williams, W.S., *Progress in Solid State Chemistry,* Vol. 6, 1971, p. 955.
33. Williams, W.S., et al., *Physica Status Solidi.,* Vol. 75, 1976, p. 533.
34. Williams, W.S., *Journal of Applied Physics,* Vol. 33, 1962, p. 955.
35. Atkins, A.G. and Tabor, D., *Royal Society of London. Proceedings,* Vol. A292, 1966, p. 441.
36. Platov, A.B., *Soviet Powder Metallurgy and Metal Ceramics,* Vol. 1, 1962, p. 188.
37. Platov, A.B., *Soviet Powder Metallurgy and Metal Ceramics,* Vol. 2, 1963, p. 306.
38. Platov, A.B., *Soviet Powder Metallurgy and Metal Ceramics,* Vol. 4, 1965, p. 288.
39. Dzhemelinski, V.V., et al., in "Refractory Carbides," Consultants Bureau, New York, 1974, p. 403.
40. Westbrook, J.H., *Revue Internationale des Hautes Températures et des Réfractaires,* Vol. 3, 1966, p. 47.
41. Williams, W.S., *Journal of Applied Physics,* Vol. 35, 1964, p. 1329.
42. Williams, W.S., in 5th National Science Foundation Hard Materials Workshop, Brown University, Providence, Rhode Island, 1977, p. 116.
43. Dawhil, W. and Frisch, B., *Archiv für das Eisenhüttenwesen,* Vol. 36, 1965, p. 449.
44. Toth, L.E., "Transition Metal Carbides and Nitrides," Academic Press, New York, 1971, p. 173.
45. Hamilton, G.M. and Goodman, L.E., *Journal of Applied Mechanics,* Vol. 33, 1966, p. 371.
46. Culf, C.J., *Society of Glass Technologists. Journal,* Vol. 41, 1957, p. 157.
47. Billinghurst, P.R., Brookes, C.A. and Tabor, D., in "Physical Basis of Yield and Fracture," Institute of Physics and Physical Society, London, 1966, p. 253.
48. Lawn, B.R., et al., *Journal of Applied Physics,* Vol. 40, 1969, p. 4009.
49. Powell, B.D., and Tabor, D., *Journal of Physics D: Applied Physics,* Vol. 3, 1970, p. 783.
50. Buckley, D.H., *Wear,* Vol. 46, 1978, p. 19.
51. Pons, L., in "Anisotropy in Single Crystal Refractory Compounds," Vol. 2, edited by F.W. Vahldiek and S.A. Mersol, Plenum Press, New York, 1968, p. 393.
52. Fujishiro, S. and Gokcen, N.A., *Journal of Physical Chemistry,* Vol. 65, 1961, p. 161.
53. Coffman, J.A., et al., "Corbonization of Plastics and Refractory Materials Research," U.S. Air Force Report No. WADD-TR-60-646, 1963, Part II, p. 33.
54. McClaine, L.A., "Thermodynamic and Kinetic Studies for a Refractory Materials Program," U.S. Air Force Report No. ASD-TDR-62-204, 1964, Part III, p. 98.
55. Pollock, B.D., *Journal of Physical Chemistry,* Vol. 65, 1961, p. 731.
56. Osborn, H.A., *Powder Metallurgy,* Vol. 12, 1969, p. 471.
57. Larsen-Basse, J., *Powder Metallurgy,* Vol. 16, 1973, p. 1.
58. Blombery, R.I. and Perrott, C.M., *Australasian Institute of Metals. Journal,* Vol. 19, 1974, p. 266.
59. Suh, N.P., et al., 5th National Science Foundation Hard Materials Workshop, Brown University, Providence, Rhode Island, 1977 (verbal discussion).

DISCUSSION

K. C. TRIPATHI, IIT Research Institute: I think we are not asking the relevant question: what is the predominant and rate-determining process? In the work we have done, grinding wheel wear was lower when workhardened metals were ground than in the case of soft metals. Could you please explain the anomaly?

S. RAMALINGAM: Grinding is a complicated process. It has been shown previously that the specific energy of machining coldworked metals is lower than that of annealed material and the difference was found to be exactly equal to the energy put into the material during coldworking. What we are in effect doing by the coldworking operation is reducing the power dissipated during machining. Therefore, we are dropping the temperature which is a critical factor in machining. As a result, we will be able to obtain the high tool life. In the turning operation the tool wear is lower in the case of coldworked metals and that has been documented.

In the samples that we made by distributing different volume fractions of alumina we have identical bulk mechanical hardness. We varied the volume fraction by a small amount 0.2 to 0.4 percent. If we obtain such a large variation in tool wear, it stands to reason that abrasive wear is taking place. Furthermore, with leaded steel (say 0.1 percent lead) and the steel without lead which has the same bulk hardness, we get very large variation in tool wear. The phenomenon that is involved here is the modification of the process at the tool chip interface that changes the tool wear phenomena and the tool life.

QUESTIONER: Since you rule out diffusional processes, and by that I think also of microwelding processes between toll and chip, how do you explain the beneficial behavior of certain refractory carbide, nitride, and oxide catings? Can that be explained solely on the basis of thermal stability of these layers?

RAMALINGAM: That is what I would suggest. The principal observation we should pay attention to, when we rule out diffusion as a mechanism, is that with low chemical stability we always (with the exception of tungsten carbide) obtain wear patches -- discrete patches of material removed. And the spacing between the patches is as small in some instances as 10.0 μm. For a machining speed of 1 m/s the spacing between patches is 10.0 μm and the time involved is 10^{-6}s. It is impossible by any diffusional processes to remove in 10^{-6}s several thousand angstroms to produce a patch.

SAME QUESTIONER: Would you rule out any chemical interaction between the layers and the workpiece?

RAMALINGAM: I am very glad you raised this issue. Many of these layers have been detected using electron microprobe techniques which are inappropriate for this thin layer. I do not rule out the possibility that one might produce layers which preclude chemical decomposition reactions. I do not believe that they are diffusion barriers. They might be chemical barriers to prevent decomposition reactions.

M. C. SHAW, Arizona State University: I do not think anybody would deny that abrasive wear is important in machining. However, I think that the danger lies in claiming that it is the only mechanism or the key mechanism in a wide variety of situations. How do you explain that thin coatings of the order of only 0.5 μm, can have dramatic effects in terms of abrasive resisance. Secondly, why did we have to wait until 1940 to machine steel with carbide when the carbide that was used before that time was more abrasion resistant than the carbide we now use for machining steel. The point I am making is that the problem is very complex.

No single theory, such as an abrasive theory, is going to answer the question. It may be the predominant situation for one set of circumstances, but it certainly will not explain why we use triple carbide for steel and why the very thin layers of these diffusion barriers are so effective in promoting tool life in carbides.

RAMALINGAM: There is no data that is trustworthy in the literature documenting diffusion phenomena. In the case of graphitic cast irons, it is entirely feasible to machine with tungsten carbide but if you take the very same cast iron and change the free graphite into nodular form we are unable to machine it with the same tungsten carbide tool, despite identical chemistries. There again diffusion theories are in trouble. I would suggest that in the graphitic iron you do drop off graphite flakes on a hot surface. It burns there, produces a back pressure of carbon monoxide which prevents the decomposition of the tool. I do not imply that all wear is purely abrasive. I say in the upper end of the useful range, tool failure is either by abrasive mechanism or by a chemical stability mechanism. We cannot conceive of how one can account for an adhesive failure of the hard tool surface.

ABRASION MECHANISMS—DELAMINATION TO MACHINING

J. Larsen-Basse

ABSTRACT

The relationship between the abrasive wear rate of a metal and its hardness relative to the abrasive is examined in some detail, primarily in the light of experimental results obtained in the author's laboratory. It is shown that the abrasion process basically embraces mechanisms from machining to delamination wear, corresponding to abrasion by relatively hard and by relatively soft abrasives, respectively.

Hard abrasives act like cutting tools of rounded shape and remove metal in a complex plastic deformation process. The rate of metal removal depends closely on the plastic properties of the metal--flow stress, strain hardening, ductility, and strain rate sensitivity--and on the ability of the abrasives to indent and cut, as indicated by properties such as abrasive hardness, crushing strength and self-sharpening tendency.

Soft abrasives damage the metal surface by repeated loading and material is removed by surface and subsurface fatigue cracking, in a process which is similar to, or identical with, delamination wear. Hard and brittle particles in the metal surface may be removed by gradual fragmentation or by uprooting from the matrix.

The transition between the two regions is generally located at an abrasive hardness value of 20% above the metal surface hardness. However, soft abrasion mechanisms may be encountered for very hard abrasives under certain circumstances. These are basically conditions where the individual abrasive grains are insufficiently loaded for their cutting points to indent the metal to a sufficiently high plastic strain. They include abrasion by fine grit sizes, low humidity abrasion by certain abrasives, and fine grit or low load abrasion of cemented carbides.

INTRODUCTION

Shaw's review paper[1] has largely concentrated on the wear of cutting tools in metal machining operations, i.e., on situations involving dynamic contact between two materials of which one is much harder than the other. The purpose of the present paper is to give a view of the process of abrasive wear which shows that this process has many traits in common with machining. It also shows that as the hardness of the abrasive is decreased below 120% of the hardness of the metal surface the mechanism of material removal changes from a machining-type of mechanism to a contact fatigue-type of mechanism similar to delamination wear. The point of view presented is a rather personal one, being based primarily on work done in the author's laboratory and it should be pointed out that much work remains to be done, especially with relatively soft abrasives, before a complete picture of the abrasion process emerges.

THE EFFECT OF ABRASIVE HARDNESS

It is well known that the rate of metal removal in abrasive wear decreases drastically when the hardness of the metal, H_m, approaches the hardness of the abrasive, H_a [2-4]. The critical value has been estimated as $H_a/H_m = 1.3$ by Kruschov and Babichev[2] while Richardson[3] has used a value of 1.2. Since the bulk hardness value was used for H_m by the former investigators and the deformed surface hardness value was used by the latter, the results are probably in basic agreement. Tabor[5] has shown that the ratio between successive hardness values on the Moh's scale of scratch hardness for minerals is 1.2, i.e., that in order for one material to scratch another it must be 1.2 times harder. This obviously supports the findings of a radical change in wear rate at $H_a/H_m \approx 1.2$ and suggest a corresponding radical change in wear mechanism.

WEAR BY HARD ABRASIVES

Macroscopic Property Relations

Indentation hardness is the only commonly measured materials property which correlates well with the resistance to abrasion by hard abrasives. This general relationship has been known for many years and when Brinell developed his hardness test around the year 1900 he even substituted an abrasion test for the indentation test for materials which were too hard for measurements with a hardened steel indentor[6]. Rabinowicz[7] has presented a simple model of the abrasion process which shows that abrasion resistance should be proportional to metal hardness. The general validity of this approach was demonstrated by Kruschov[8]. He found that the abrasion resistance of pure metals and annealed steels is, indeed, proportional to their hardness, while linear relationships of lower slope were obtained for hardened and tempered steels, intersecting the line for the pure metals at the location of the data for the annealed state of the steel in question, as illustrated in Figure 1. Avient, et al.[9] pointed out that the hardness value to be used in this correlation really should be the hardness of the work hardened abraded surface, rather than the metal bulk hardness, and this has been confirmed by the present author[10-12]. However, it has been pointed out[11] that if the relationship between the two hardness values is reasonably similar for the different metals--and it generally is--then it really does not matter which of the hardness values is used.

Microscopic Observations

In two-body wear by hard abrasives the fixed abrasive plows a groove in the metal surface. The material from the groove is partly pushed into ridges at the sides of the groove and partly removed directly as a conventional-locking machining chip, see Figure 2. Due to the triaxial stress system it is possible to form chips for even quite brittle materials and it has been suggested that the mechanism more closely resembles extrusion than cutting[13]. In three-body abrasion, where loose abrasives are used, the material removal process is quite similar, except that the number of grooves formed is relatively small. Instead, the abrasives form craters in the metal surface and remove material in a backward extrusion

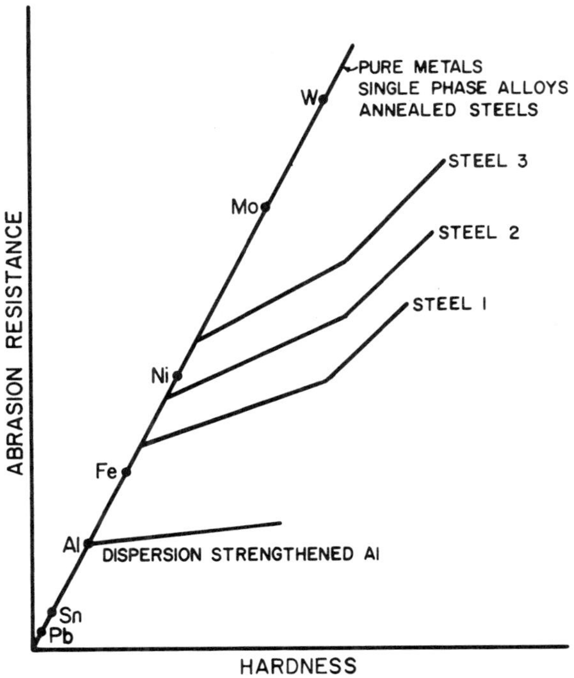

Fig. 1.—Sketch of the abrasion resistance-hardness relationship for various alloys for abrasion by hard abrasives. After Kruschov[2,8] *and Larsen-Basse*[10,11].

process. The same material removal mechanism is found even for quite brittle materials such as cemented carbides, see Figure 3. These alloys may lose some additional material by brittle fracture but this seems to be relatively rare.

The metal surface is work hardened by the metal removal process. The depth of deformation for normal abrasion of copper by silicon carbide abrasives has been determined by microhardness measurements to be about 80 μm with an extrapolated surface strain of around 0.25[14]. The average groove depth for the same conditions is approximately 15 μm[14,15]. This means that the subsurface material undergoes an average of 4-6 stress cycles with strain increasing from zero to 25% before it is removed as a chip at a much higher strain. And this means that material is removed in wear by hard abrasives in a process which essentially is severe plastic deformation.

Relationship with Plastic Flow Properties

If the cutting force in abrasion is divided by the cross sectional area of the grooves a specific cutting pressure, p, may be calculated. The calculation is more simply based on wear rate and cutting force:

$$p(Pa) = \text{cutting force } (N)/\text{wear rate } (m^3/m) \quad (1)$$

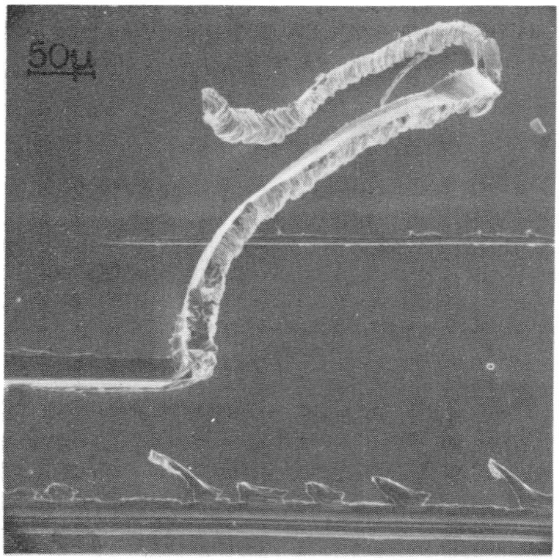

Fig. 2.—Formation of machining chip and pushed-up ridge in short-stroke abrasion of a polished copper surface by 120 μm silicon carbide abrasive papers.

and *p* may equally be considered as a specific work term (J/m^3). Since the cutting force does not vary greatly the specific cutting pressure is essentially proportional to the abrasion resistance. With this in mind, the following general model for the influence of plastic flow properties on resistance to abrasion by hard abrasives has been proposed(10), see Figure 4. The bulk hardness is approximately 3 times the flow stress at a strain of 8-10%(16) and the abraded surface hardness is presumably a similar

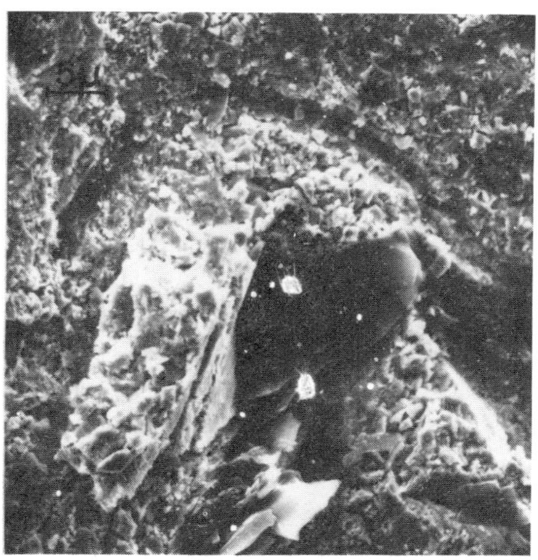

Fig. 3.—Crater formation and backward extrusion of heavily deformed material in three-body abrasion of WC-11 v/o Co alloy by 150 μm B_4C abrasives.

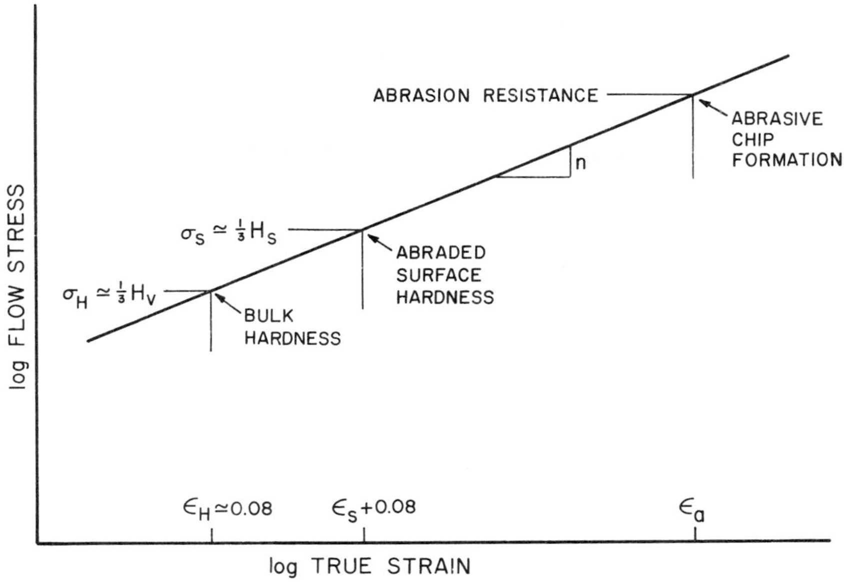

Fig. 4.—Sketch of the relationship between flow stress and strain and bulk hardness, abraded surface hardness, and abrasion resistance values for abrasion by hard abrasives. After Larsen-Basse[10].

measure of the flow stress at a strain corresponding to the surface deformation. The abrasion resistance is related to the flow stress at a strain of abrasion, ε_a, i.e., it is determined by the hardness, the coefficient of work hardening, n, and the ductility of abrasion, ε_a.

This model allows a qualitative explanation of the results shown in Figure 1[10]. Since n and ε_a do not differ greatly between different pure metals the abrasion resistance is proportional to bulk, or surface, hardness. For hardened and tempered steels n is lower and consequently the abrasion resistance increases less rapidly with hardness for these materials. By the use of bulk and surface hardness measurements and by assuming a deformed surface strain for steels equal to the 0.25 found for copper[14] a value of 4.8 was estimated for ε_a for tempered steels[12]. In a different approach[17] the model was used to explain the effect of sliding velocity as a strain rate effect and values of ε_a obtained in this case for very pure metals fell in the range 4.1-6.7. These ε_a-values appear reasonable in view of the severity of the chip formation process for tools with highly negative rake angles.

Correlations of abrasion resistance with microstructure parameters also indicate that a material's abrasion resistance is closely related to its flow stress at high levels of strain[10-12] and it seems reasonable to assume that the model has considerable validity.

WEAR BY RELATIVELY SOFT ABRASIVES

The most commonly occurring natural abrasive is quartz which is a hard abrasive relative to most metals, including hardened tool steels. This is probably one of the reasons that little interest has been shown in wear by soft abrasives. Another reason may be that it is much more difficult to obtain experimental data with soft abrasives. Actually, wear by soft abrasives probably occurs in many more operations than is generally realized. Examples include wear of cemented carbide bits in rock drilling or in machining of sand castings, wear of dentists' drills by calcium phosphate, and wear of hot work tool steels by iron oxide scale.

The wear rate for $H_a/H_m < 1.2$ is extremely low for ductile metals, while brittle materials such as cemented carbides show some material loss even in this range, as sketched in Figure 5. Richardson[3] found grooves on

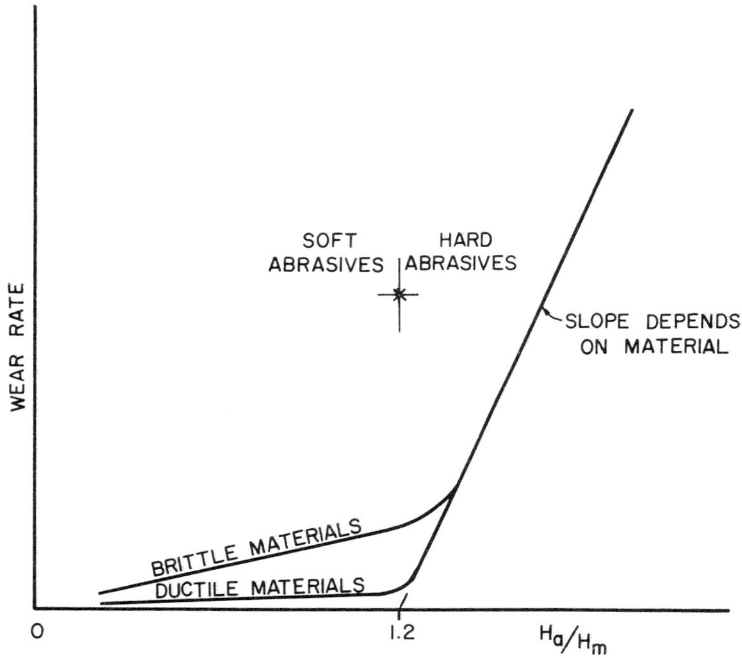

Fig. 5.—Sketch of the influence of relative abrasive hardness on wear rate of brittle and ductile materials. After Kruschov and Babichev[2] and Larsen-Basse and Tanouye[25].

abraded steel surfaces and suggested that soft abrasives also remove material by groove formation but that these abrasives are more susceptible to attrition and fracture and consequently give a low wear rate. Work in the author's laboratory[18] has confirmed the existence of grooves on abraded steel surfaces but has shown that these grooves are formed by carbides pulled from the matrix and rolled over the surface. A typical surface appearance is shown in Figure 6 which, in addition to the grooves, shows cracks formed perpendicular to the direction of sliding. The cracks extend below the surface, see Figure 7, and clearly resemble the contact fatigue-type of cracks described for delamination wear in metal-to-metal systems[19]. Similar cracks have been found in worn dental drills and in cemented carbides abraded by quartz. For the cemented carbides the process includes not only the contact fatigue mechanism[20] but also very small scale deformation--gradual extrusion of the binder and associated fatigue fracturing and

Fig. 6.—Surface of hardened tool steel abraded by glass abrasives.

removal of carbide grains as the supporting binder is extruded, see Figure 8.

Much work needs to be done in the area of wear by relatively soft abrasives in order to clarify effects of microstructure and environment, to determine crack nucleation and propagation mechanisms, etc.

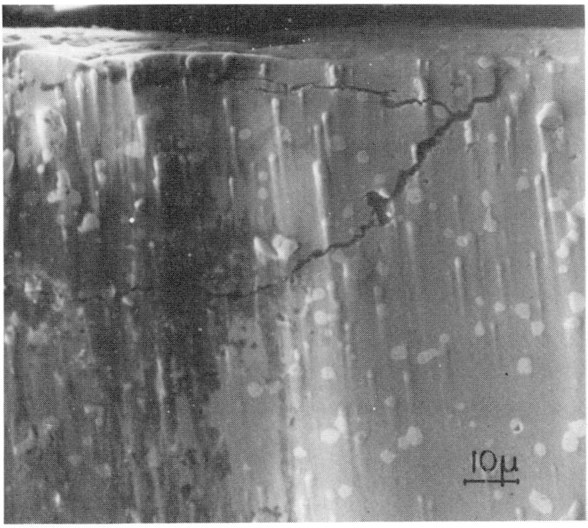

Fig. 7.—Section through the surface of hardened tool steel abraded by glass abrasives shown in Figure 6.

Fig. 8.—Section through the surface of WC-12.5 v/o Co alloy abraded by 100 μm quartz abrasives.

THE TRANSITION REGION

Closer study of the process of abrasion by hard abrasives shows that even in this case a large fraction of the abrasives are in contact fatigue-type of interaction with the surface rather than acting as cutting tools. There are generally sufficiently many cutting abrasives to remove the surface material relatively rapidly and the surface details which are typical of delamination and contact fatigue do not have sufficient time to become well developed. A few examples which illustrate this are given below.

It was found in abrasion of pure copper, aluminum and iron by silicon carbide abrasive papers[17] that the cutting work, w, is given by

$$w = \mu L + R \frac{\sigma_o \varepsilon_a^{n+1}}{n+1} (1 + K \ln v) \tag{2}$$

where μ is the coefficient of sliding friction between the abrasive and the metal surface, L is the applied load, R is the metal removal rate, K is a materials constant, and v is the sliding velocity. The first term on the right hand side represents the contribution from abrasive grits which are not removing material, i.e., the contribution from delamination-type of contacts. This term represents as much as 70-90% of the total even in this case where H_a/H_m falls in the range of 5-10 and the abrasive consequently is a very hard abrasive. The second term includes the plastic deformation in chip removal discussed above, with a velocity-independent and a strain rate-dependent component.

Another example is the effect of small grit size. As the size of hard abrasives is decreased past a certain value, which depends on metal hardness, specimen size, and applied load, the metal removal rate drops towards zero. The effect probably has a number of causes but the prime reason for the behavior is the increasing fraction of abrasive grains which do not indent the surface sufficiently to remove material, while they support an increasing fraction of the applied load[15,21], i.e., they are in delamination-type of contact with the surface.

A third example illustrates that abrasive hardness is not always a reliable indicator of the transition between machining and contact fatigue wear. It had been found that the metal removal rate for SiC abrading much softer metals increased strongly with increasing atmospheric humidity[22]. It was determined that the effect was due to moisture-assisted fracture of the abrasives which resulted in their continual self-sharpening[22,23]. For extremely dry conditions, on the other hand, the abraded surface showed almost no extrusion craters or grooves but instead a series of contact fatigue cracks[23].

A final example is drawn from the work on cemented carbides. These materials have bulk hardness values in the range 0.9-1.5 MPa while the hardness of the individual carbide grains is around 2.8 MPa. SiC abrasives of approximate hardness 2.4 MPa act as hard abrasives when the operating conditions allow them to indent the material like a bulk hardness indentor. For low loads or small grit sizes the individual abrasive grain contacts only one carbide grain at a time and it consequently acts as a soft abrasive under these conditions[24].

CONCLUSIONS

It has been demonstrated that the micro-chip formation process in abrasion of metals by relatively hard abrasives is closely related to the plastic flow properties of the metal. Abrasion by relatively soft abrasives, on the other hand, is very poorly understood. It is primarily a contact fatigue mechanism which is similar to, or perhaps identical with, the mechanism of delamination wear. It has also been demonstrated that the same mechanism is active in most situations involving abrasion by hard abrasives; it is usually not noticed because the chip forming process removes the surface material before noticeable surface fatigue damage appears. It is suggested that a closer study of the process of abrasion by soft abrasives and the effects of operating conditions, environment and microstructure could yield considerable engineering and scientific benefits.

ACKNOWLEDGEMENTS

This study was supported by the National Science Foundation under Grant No. DMR 76-17158. The SEM photographs were kindly provided by P. A. Tanouye and E. T. Koyanagi.

REFERENCES

1. Shaw, M.C., These Proceedings.
2. Kruschov, M.M., and Babichev, M.A., "Abrazivnoe Iznashivanie," Akademia Nauk SSSR, Moscow, 1970, p. 129.
3. Richardson, R.C.D., *Wear,* Vol. 11, 1968, p. 248.
4. Nathan, G.K. and Jones, W.J.D., *Institution of Mechanical Engineers. Proceedings,* Vol. 181, 1966-67, p. 215.
5. Tabor, D., *Journal of Physics D: Applied Physics,* Vol. 7, 1956, p. 159.
6. Wahlberg, A., *Iron and Steel Institute. Journal,* Vol. 59, 1901, p. 243.
7. Rabinowicz, E., "Friction and Wear of Materials," John Wiley and Sons, New York, 1965, p. 168.
8. Kruschov, M.M., in Proceedings of the Conference on Lubrication and Wear, Institution of Mechanical Engineers, London, 1957, p. 655.
9. Avient, B.W.E., Goddard, J. and Wilman, H., *Royal Society of London. Proceedings.* Series A, Vol. 258, 1960, p. 159.
10. Larsen-Basse, J., *Transactions American Society of Mining, Metallurgical and Petroleum Engineers,* Vol. 236, 1966, p. 1461.

11. Larsen-Basse, J., *Wear,* Vol. 12, 1968, p. 357.
12. Larsen-Basse, J. and Mathew, K.G., *Wear,* Vol. 14, 1969, p. 199.
13. Shaw, M.C., in Harold Armstrong Conference on Production Science in Industry, Melbourne, Australia, The Institution of Engineers, Australia, 1971, p. 1.
14. Du, F.O., M.S. Thesis, University of Hawaii, 1969.
15. Larsen-Basse, J., *Wear,* Vol. 11, 1968, p. 213.
16. Tabor, D., "The Hardness of Metals," Clarendon Press, Oxford, 1951, p. 104.
17. Larsen-Basse, J. and Tanouye, P.A., in "Wear of Materials 1977," edited by W.A. Glaeser, K.C. Ludema and S.K. Rhee, American Society of Mechanical Engineers, 1977, p. 194.
18. Premaratne, B., M.S. Thesis, University of Hawaii, 1977.
19. Suh, N.P., *Wear,* Vol. 44, 1977, p. 1.
20. Larsen-Basse, J. and Koyanagi, E.T., *Journal of Lubrication Technology,* American Society of Mechanical Engineers Transactions, Vol. 101, 1979, p. 208.
21. Larsen-Basse, J., *Wear,* Vol. 12, 1968, p. 35.
22. Larsen-Basse, J., *Wear,* Vol. 31, 1975, p. 373.
23. Dalmia, M.K., M.S. Thesis, University of Hawaii, 1974.
24. Larsen-Basse, J., Shishido, C.M. and Salem, L.K., in "Advances in Hard Material Tool Technology," edited by R. Komanduri, Carnegie-Mellon University, Pittsburgh, 1976, p. 231.
25. Larsen-Basse, J. and Tanouye, P.A., in "Advances in Hard Material Tool Technology," edited by R. Komanduri, Carnegie-Mellon University, Pittsburgh, 1976, p. 188.

DISCUSSION

K. C. TRIPATHI, IIT Research Institute: Would the absorption of water affect stress corrosion cracking of the silicon carbide grit?

J. LARSEN-BASSE: Yes, We measured the fracture strength values in a roller crushing device and found a tremendous difference in the fracture strength as a function of humidity. Also the grit fell into a large number of small fragments at high humidity. We also found that for aluminum, zinc, and glass the effect is much more pronounced and we take that to mean that there is also some effect of the surface properties of the material, an additional factor which I did not include in this paper.

M. C. SHAW, Arizona State University: From your paper one is inclined to infer that as long as we have a hardness ratio of 1.2 between the abrasive and the work piece, everything is fine. That is not so. A very important consideration is economics. If we want to grind a piece of steel with a wheel that was made of glass even though the hardness ratio was 1.2, the wear rate of the wheel would be so great that we would go out of business very quickly. In order to do things in a practical way in production, the hardness of the abrasive must be at least double, and preferably four times, the hardness of the work piece. Now, the hardest steel had hardness of about 1000 kg/mm^2. Aluminum oxide which is normally used to grind the steel has a hardness of the order of 2200 kg/mm^2. The abrasive really is not enough to do a good job even at a ratio of 2.0. We would do much better with cubic boron nitride (hardness $\sim$ 4500 kg/mm^2. In the case of tungsten carbide the hardness is of the order of 1400 kg/mm^2. To do a decent job on tungsten carbide we need a hardness of 6000 kg/mm^2 which means that we have to use diamond. I would say, therefore, that the ratio of the hardness practically should be at least double and preferably of the order of 4 to 5 for high metal removal rates in processing.

Another point is humidity. I think you said that the wear was reduced when the humidity was low. Are we talking about the wear of the abrasive or the workpiece? We might want the wear of the workpiece to go up because we want to machine it, whereas the wear of the abrasive should go down. We found the wheel wear and the forces of grinding correlated with humidity in the shop, and that the wheel wear was less when the atmosphere was dry as opposed to having a high humidity, say 50-70 percent.

LARSEN-BASSE: I think your point is well taken in that if we are looking for fast material removal we do need a hard abrasive. The wear rate of carbides versus hardness ratio showed a very rapid increase. That is, when we use cubic boron nitride we move up on the line and thus it checks very nicely.

SHAW: I think this work is very valuable for three-body wear (for rating hard facing materials and things like that) as opposed to production in the workshop. That is the point I am making.

LARSEN-BASSE: My point also was somewhat directed towards the testing that is being standardized for testing cemented carbides using alumina. I think we agree on the effect of humidity and we have the same sort of results.

WHEEL WEAR IN GRINDING AS A FUNCTION OF ENERGY CONSUMED

G. Werner

ABSTRACT

Prediction of the grinding wheel wear is a decisive prerequisite for the selection of favorable grinding conditions. Independent of the physical and chemical nature of wear, four mechanical criteria are used to describe the grinding wheel wear. These criteria are: average contact pressure per grain, sliding velocity, average contact time, and contact frequency. A comprehensive grinding wear function is derived relating these criteria to the energy consumed in the process. The model is in close agreement with results of wear tests and enables the determination of cost-optimal grinding conditions.

INTRODUCTION

In grinding, material is removed from the work piece as well as from the grinding wheel. The ratio of the work stock removal to the wheel wear, called G-ratio, is used as an efficiency measure of the grinding process with regard to grinding tool cost. To a certain extent, however, the wear on grinding wheels is necessary to achieve and maintain steady state operation, marked by constant grinding forces, sharp wheel surface, constant surface roughness of the work piece, and constant wheel wear.

For optimal steady-state grinding, knowledge of the relationship between wheel wear and the pertinent process parameters is required. By superimposing the wear criteria at individual cutting edges, a comprehensive wheel wear function can be derived as a function of the relevant process parameters.[1,2]

DEFINITIONS OF EDGE WEAR AND RADIUS WEAR

It is known from practice, especially in high-efficiency grinding, that both radius and edge wear increase progressively with metal removal rate, Z', which is proportional to the work speed, v_W, and the depth of cut, a, as shown in Figure 1.

Any comprehensive wear model should describe these relationships and

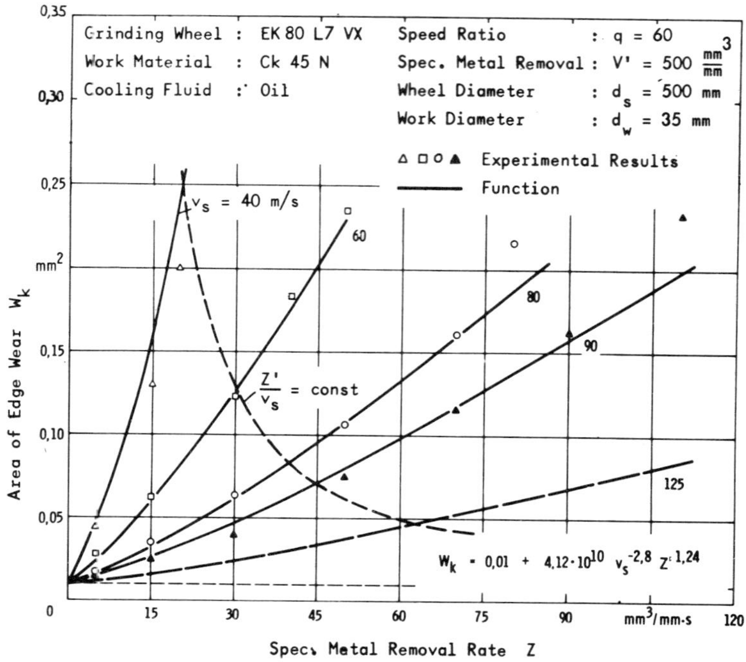

Fig. 1.–Application of Wear Model to Experimental Results in External Plunge Grinding.

should also consider the influence of the wheel diameter d_s, the work diameter, d_w, and the specific metal removal V'. Furthermore, it should also take into consideration wheel specification, work material, dressing method, cooling fluid, etc. which determine a specific grinding process.

In plunge grinding, the total wear area, W, is composed of the radius wear area, W_r, and the edge wear area, W_k, as shown in Figure 2. Both types of wear can be attributed to the same causes. As the mean life of the individual grits is shorter at the wheel edges, due to the diminished support, it leads to an edge rounding which is approximately circular.[1,2] The result is a greater radial wear intensity on the wheel edges.

In the literature up to the present time, the radius wear, Δr_s, has been specified by a dimension of length, while the edge wear, W_k, has been characterized by the wear cross section with the dimension of area. For exact reconditioning of the wheel profile, the grinding wheel must be dressed down to a depth of Δr_a, which is the sum of the radius wear, Δr_s, and the radius of the edge rounding, r_v (Figure 2):

$$\Delta r_a = \Delta r_s + r_v = \Delta r_s + \Delta r_k \qquad (1)$$

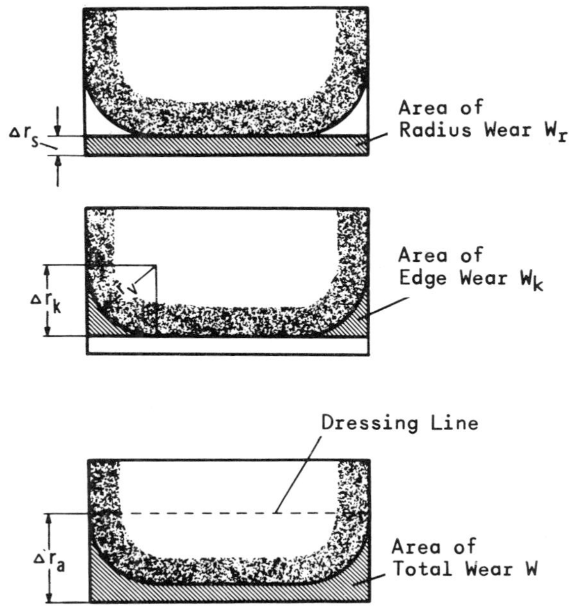

Fig. 2.—Definition of Grinding Wheel Wear.

where

$$\Delta r_s = W_r / b_s \tag{2}$$

$$\Delta r_k = W_k / (1-\pi/4)^{1/2} \tag{3}$$

b_s and W_r are grinding width and wear cross section, respectively.

EMPIRICAL DESCRIPTION OF GRINDING WHEEL WEAR

The basic wear profiles indicated in Figure 2 can only occur in plunge grinding where the wheel is continuously in contact with the work piece. In this case the radius wear, Δr_s, increases linearly whereas the edge wear, Δr_k, saturates with grinding time, t_g (Figure 3). The normal practice, however, is the intermittent plunge grinding operation as demonstrated in Figure 4, where the wheel plunges into a work piece until it reaches a depth of s_e. Then it is withdrawn and the same operation is performed with the next work piece. Even under these conditions, the radius wear, Δr_s, remains a linear function of the grinding time, t_g. The edge wear, Δr_k, however, approaches a constant value, Δr_k*. The reason for this behavior is that the lateral sections of the rounded wheel edges come into contact with the work piece much later than the straight cylindrical part of the wheel surface. This is also the reason why in practical plunge grinding the worn wheel edge shows an elliptical rather than a circular shape.

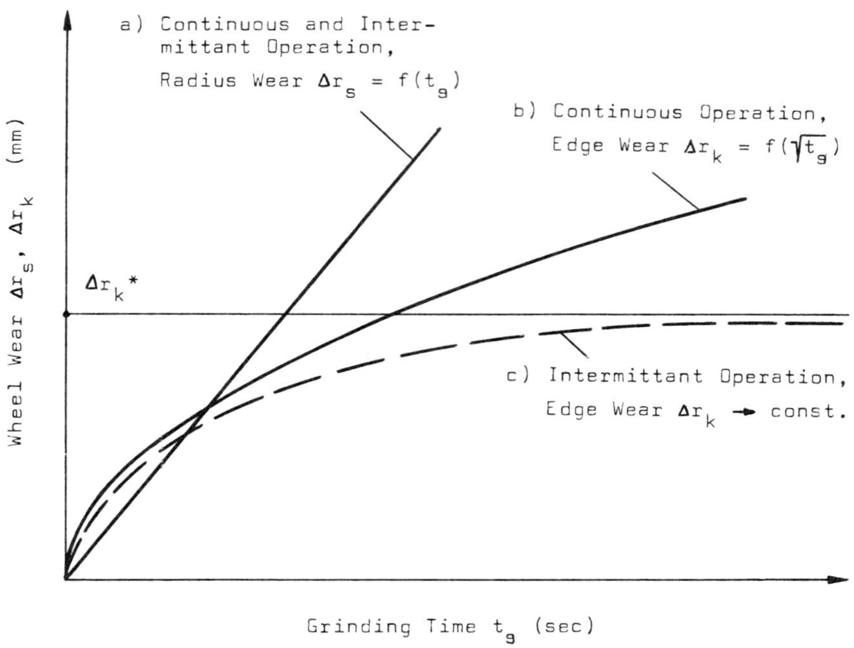

Fig. 3.—Wheel Wear Versus Grinding Time.

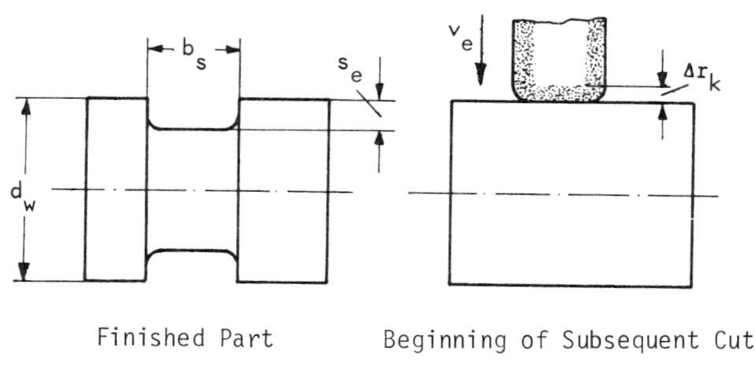

Fig. 4.—Intermittent External Plunge Grinding.

The relationship between grinding forces and wheel wear is an important subject. Figure 5 is a schematic of the principal relations. In the case of steady state wear (case a), when the number of cutting edges in the active wheel surface remains constant, the cutting force, F, is constant, and

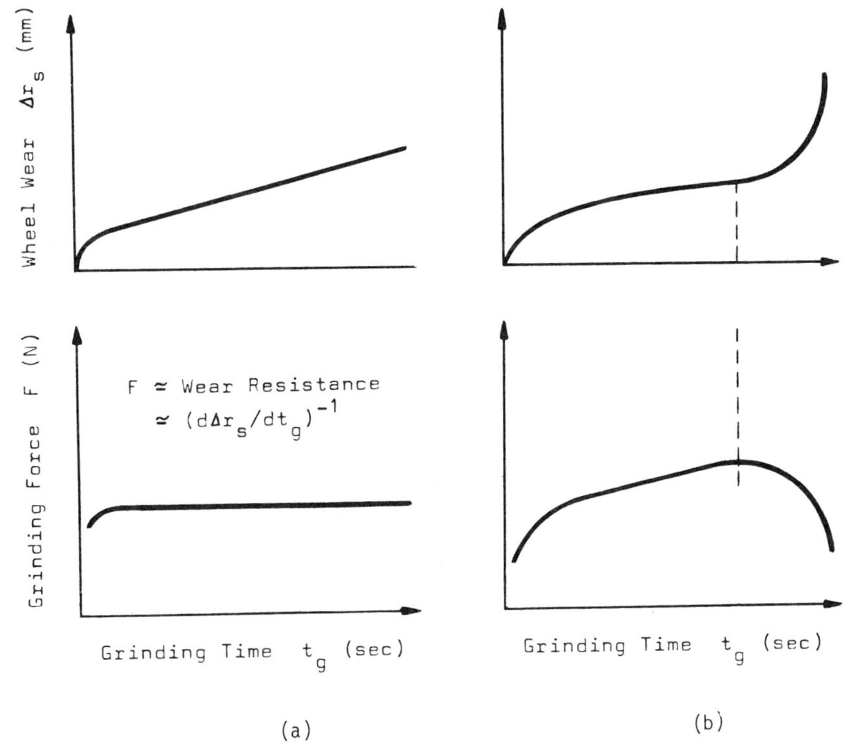

Fig. 5.—Relation Between Wear and Force in Grinding.

the radial wheel wear, Δr_s, increases linearly with grinding time with rapid increase at the beginning. During this phase the number of cutting edges increases with time resulting in greater wear resistance and higher grinding forces. The main reason for a non-steady state wear is that the wheel hardness is too high in relation to the average force per individual cutting edge. Consequently, the grits are held too long in the surface of the wheel and tend to develop excessive flattening which contributes to increasing grinding forces. When these forces reach a certain magnitude, the wheel surface structure collapses, resulting in progressively increasing wear and a simultaneous drop in grinding forces (case b). In practical grinding this is a signal for dressing the grinding wheel.

THE WEAR MODEL

When the grinding wheel cuts into the workpiece, several processes connected with friction and wear occur simultaneously (Figure 6). Local

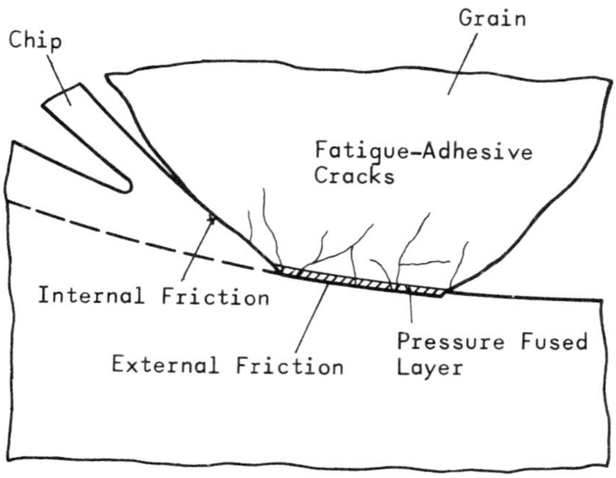

Fig. 6.—Friction and Wear on the Cutting Grain Surface.

contact areas are formed in the course of the motion between the two bodies. Thereby, particles are separated from one of the two surfaces depending on their mechanical properties, the environmental conditions, (such as lubricant and type of atmosphere) as well as such other parameters as pressure, friction coefficient, speed and temperature.[4,5]

During high-efficiency grinding with high metal removal rate, high quality grinding wheels are used. In this case, the largest amount of wear occurs at the cutting grains by the attrition of the outer layers.[6] The grains break off only when the individual cutting force increases as a result of excessive wear when the fixture of the grain in the bond is weakened, and when in addition, the grain contact area has been under alternating stress for a long time.

The friction process at the grain is characterized by the high pressure acting on the whole contact area and by the high contact temperature. These conditions lead to two different wear mechanisms that operate simultaneously.

The temperatures occuring with dry friction in the surface layers during the contact period lie between 1200° - 1800°C.[6] At these high temperatures oxidation and diffusion processes can cause great changes in the mechanical properties of the grain material within the affected volume. Then this layer is attrited by the mechanical stress which acts on the surface. The subsequent contact is also the pre-requisite for the formation of a new layer (Figure 5). In the literature[6] this process is known as pressure fusing. The layer thickness and therefore the wear intensity depend on the induced heat which increases with the contact time, the contact pressure and the speed.

With high efficiency grinding, a second wear mechanism - the fatigue of the crystalline bond caused by the mechanical and thermal alternating load - is of equal importance. In defect areas, which are present in every solid body in the form of lattice imperfections, grain boundaries, impurities, differences in hardness and internal stresses, fatigue cracks nucleate and grow with alternating load in the surface layer (Figure 6). The rate of formation of these cracks depends on the characteristics of the loaded materials, on the contact pressure and on the contact frequency.

The fine fatigue cracks which deepen and increase numerically with increasing contact frequency cause very small crystalline wear particles which are released from the surface if, under the effect of the external forces, the closure of a crack chain occurs. This so-called "fatigue-adhesive wear" is not identical with the grain break-off as mentioned above during which large portions of the grain are separated at the end of a wear cycle. However, such a macro-fracture process can be initiated and/or promoted in an advanced state of wear by fatigue effects. It should be remembered in the following considerations that the fatigue-adhesive wear increases with temperature and mechanical load, i.e., with contact time, the contact pressure and the relative sliding velocity.

Both types of wear, pressure fusing and fatigue wear, occur simultaneously. Thus, an interaction is possible as cracks can be closed by intense local heating. Therefore, a clear-cut separation of both wear processes is not possible and even not necessary because the intensity of both wear processes is equally affected by the three contact parameters: pressure, time and speed, determined by the grinding parameters and material characteristics.

The effect of coolants for reducing wear and friction should also be taken into consideration in this connection.[7] Lubrication is based on the fact that a third body is inserted between the two friction bodies and/or generated by chemical processes. In cases of high normal loads this third body shows minimal resistance to tangential displacements. The surface temperature is reduced as the friction work is diminished and therefore the application of a suitable lubricant reduces wear by pressure fusing. A total elimination of wear originating from temperature effects cannot be expected because the lubrication layers are themselves subject to wear[4] and lose their protective effect in the course of grain contact.

Similar reactions occur with adhesive-fatigue wear arising from alternating load. While the temperature is reduced, alternating normal forces which are less diminished despite lubrication, will still lead to the formation of fatigue cracks. The effect on these cracks may be even further increased due to the reduced temperature. Thus, it may be stated that the application of cooling lubricants in grinding can reduce wear but cannot fully eliminate it.

From the above description the following four criteria can be determined as the most important parameters for the average wear intensity at the individual cutting edge:[1,2]

1) contact force per grain, $\bar{F}_g \hat{=}$ contact pressure, p_s

2) relative sliding velocity, v_c

3) contact time, t_c

4) contact frequency, h_c.

In order to derive an explicit function for the grinding wheel wear, it is necessary to define these four criteria as a function of the grinding parameters and to assign them to the wear process, based on relevant physical relations. The basic kinematic and mechanical functions required for this task have been derived and described by the author earlier.[1,8] They result in the following expressions:

$$\bar{F}_g = \frac{F}{N} = K_1 \cdot \frac{v_w}{v_s}^{\varepsilon} \cdot \frac{a}{D}^{\varepsilon/2}, \quad 0 < \varepsilon < 1 \tag{4}$$

$$v_c = v_s \tag{5}$$

$$t_c = \frac{\ell s}{v_s} = \frac{1}{K_2} \cdot a.D^{1/2} \cdot v_s^{-1} \tag{6}$$

$$h_c = n \cdot t_g = \frac{v_s}{d_s \cdot \pi} \cdot t_g \tag{7}$$

Where F, total grinding force, N total number of active cutting edges, v_w is work speed, v_s is wheel speed, a depth of cut, $D = d_w \cdot d_s/(d_w \pm d_s)$ is the equivalent wheel diameter, d_w = work diameter, d_s = wheel diameter, $\ell = \sqrt{a \cdot D}$ = contact length between wheel and work piece, n = number of revolutions of grinding wheel, t_g = grinding time.

The average energy per cutting edge $\bar{u}_g$ can be expressed in terms of the previously mentioned four wear criteria.

$$\bar{u}_g = \bar{F}_g \cdot (v_c \cdot t_c \cdot h_c) \tag{8}$$

where the product $v_c \cdot t_c \cdot h_c$ represents the average frictional path of a cutting edge, and as a consequence the energy $\bar{u}_g$ is defined as the product of force times path of an individual edge. Using this fundamental physical model, and assuming that at steady state conditions the total grinding wheel wear is proportional to the average wear of the individual cutting edge, the wheel wear equation can be given as

$$\Delta r = k \cdot \bar{F}_g^{\ t} \cdot v_c^{\ i} \cdot t_c^{\ e} \cdot h_c^{\ h} \tag{9}$$

In the above, grinding wheel wear is interpreted as an exponentially-modified expression of the actual grinding energy. The modifying exponents (t,i,e,h) can approximately be assessed from friction and wear investigations [4,5,7]. They all must be positive; h has values near 1.0 because the edge wear increases proportionally with the contact frequency h_c; the exponent e represents the time-related influence of thermal energy on the edge wear and should take values clearly above 1.0; the exponent i reflects the influence of sliding speed and acquires values between 0.5 and 1.C depending on the properties of the contacting materials; and t describes the influence of the contact pressure and will attain values above 1.0. All other influences, like chemical reactions, lubrication and specific properties of the wheel structure are represented by the proportionality factor K, and to a minor degree in the specific numerical values of the four exponents.

The grinding time, t_g, can easily be expressed in terms of the specific material removal, V', and the specific material removal rate $Z' = a \cdot v_w$:

$$t_g = \frac{V'}{Z'} = \frac{V'}{a \cdot v_w} \qquad (10)$$

The equivalent wheel diameter D, is composed of the wheel and work diameter:

$$D = \frac{d_w \cdot d_s}{d_w \pm d_s} \qquad (11)$$

Here the plus sign refers to external grinding, whereas the minus sign is valid for internal operation. Inserting Equations (4) to (7) into Equation (9) and using equations (10) and (11) results in two basic equations for both the radial wear Δr_s and the edge wear Δr_k valid only for steady state wear conditions. For normal external grinding conditions the wheel diameter d_s is significantly larger than the work diameter d_w; therefore, with $d_s > 5d_w$ the resultant wear equations are:

$$\Delta r_s = k_s [v_w]^{m_s-h_s} [v_s]^{h_s+i_s-e_s-m_s} [a]^{e_s/2+m_s/2-h_s} [d_s]^{e_s/2-m_s/2-h_s} [V']^{h_s} \qquad (12)$$

$$\Delta r_k = K_k [v_w]^{m_k-h_k} [v_s]^{h_k+i_k-e_k-m_k} [a]^{e_k/2+m_k/2-h_k} [d_s]^{e_k/2-m_k/2-h_k} [V']^{h_k} \qquad (13)$$

The subscripts "s" and "k" indicate the respective model parameters for radial wear and edge wear. The exponents m_s and m_k are obtained from the product of the exponents ε and t of equations (4) and (9). Both equations have the same structure and reveal a rather complex interaction between the process variables (v_w, v_s, a, d_s, V') and the resulting wheel wear. If, for example, the wheel speed, v_s is changed, the wheel wear is influenced by all four wear criteria represented by the four exponential coefficients h,i,e and m. With increased wheel speed v_s, the wear intensity goes up due to increased contact frequency and higher speed. On the other hand, contact pressure and contact time are decreased resulting in a reduced wear intensity. These complex interrelations are described functionally for the first time by this model, which was proved to be in close accordance with practical wear measurements.[2]

APPLICATION OF THE WEAR MODEL

From Equations (3) and (10)

$$W_k = (1 - \pi/4) \,.\, \Delta r_k^{\ 2} \tag{14}$$

$$a = \frac{Z'q}{v_s} \tag{15}$$

where $q = v_s/v_w$ = speed ratio. Inserting these functions into equation (13) a special version of the wear model is derived, the variables of which represent those working parameters (q, v_s, Z' and V') which are applied most commonly in external plunge grinding. In addition an initial amount of wear A_k is taken into account, which is present in practice always due to the finite dimensions of the abrasive grains and due to the dressing process.

$$W_k - A_k = k_w \cdot [q]^{e_k - m_k} \cdot [v_s]^{2h_k + 2i_k - 3e_k - m_k} \cdot [Z']^{e_k + m_k - 2h_k} \cdot [V']^{2h_k} \tag{16}$$

This equation fits exactly the conditions of the practical wear investigations as presented in Figure 1. Assuming (from Reference 7) that $i_k = 0.5$, the three remaining exponents e_k, h_k, and m_k can by multiple regression be calculated from the measured values[2]:

$$\begin{aligned} i_k &= 0.5 \\ e_k &= 1.28 \\ m_k &= 0.96 \\ h_k &= 0.5 \end{aligned} \tag{17}$$

Inserting these values into the model equation (16) results in the following explicit wear model valid for the given practical results;

$$W_k - A_k = K_w \cdot q^{\,0.32}\, v_s^{\,-2.80}\, Z'^{\,1.29}\, V'^{\,1.0} \tag{18}$$

For a given combination of work material, grinding wheel, grinding machine and cooling fluid, the equation describes the edge wear W_k as a function of the main process parameters. W_k decreases significantly with increasing wheel speed, v_s, increases progressively with metal removal rate Z' and degressively with speed ratio, g, and is proportional to V', the amount of material removed per unit of grinding width.

The specific wear function (18) is compared with the practical measurements in Figure 1. It allows any intermediate value to be determined, and it also enables the extrapolation beyond the limits of the reproduced experimental results as demonstrated for the wheel speed v_s = 125 m/s.

Experiments[10] show that the effect of the depth of cut, a, is generally greater than that of the work speed, v_w, on wear. This becomes evident if the parameters a and v_w are varied such that the product $Z' = a \,.\, v_w$ remains constant as demonstrated in Figure 7 for surface plunge grinding of three different high-alloy steels with a resin-bonded grinding wheel. With $Z' = a \cdot v_w$ as a constant, the corresponding wear functions from equations (13), and (3) can be written as:

$$W_k - A_k = K_k \, [a]^{e_k - m_k} \qquad (19)$$

$$W_r - A_r = K_s \, [a]^{e_s - m_s} \qquad (20)$$

As the wear curves in Figure 7 increase with depth of cut, a, the exponential terms in (19) and (20) must be positive. This means that the time-and-temperature-related coefficients e_k and e_s must be greater than the force-related coefficients m_k and m_s. Furthermore, it is remarkable that the slope for the edge wear curve in Figure 7 is only half as large as that of the radius wear in all three cases.

These findings are of special interest for the interpretation of wheel wear at the extreme conditions of creep feed grinding operations. While the conventional method is characterized by small depths of cut, a, and large table speeds, v_w, creep feed grinding is a new method performed on

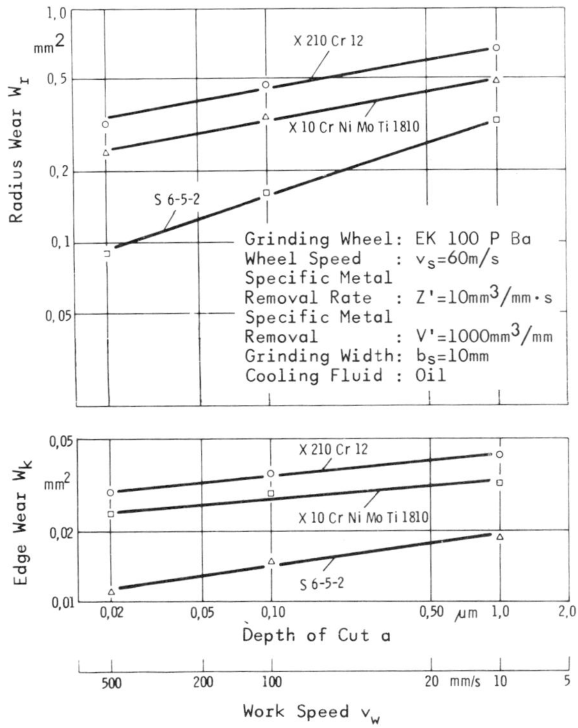

Fig. 7.—Radius and Edge Wear in Surface Grinding.

specially designed machines and characterized by very slow table speed, v_w, and extremely large depth of cut, a. Thus, deep profiles can be machined out of hardened steels in one single pass.[11]

The practical results in Figure 7 together with equations (19) and (20) indicate that at creep feed conditions wheel wear is higher for the materials tested. This effect is less pronounced for the edge wear W_k than it is for the radius wear W_r, resulting in a relatively good profile stability in creep feed grinding.

CONCLUSIONS

Superimposing the effects of the wear criteria - contact pressure, sliding velocity, contact time, and contact frequency - a comprehensive function for radius wear and edge wear of grinding tools is derived. This wear model can be interpreted as an exponentially modified expression of the grinding work, and it was proved to be in close accordance with the results of practical investigations.

The model, originally derived a few years ago,[1] is now being used in adaptive-control grinding processes to optimize external plunge grinding operations with regard to cost-optimal working condition.[12,13] Furthermore, it is a useful tool to analyze wear in grinding on the basis of the wheel and work-piece properties, and provides the base for functional descriptions of grinding cost and G-ratios.[2,12]

REFERENCES

1. Werner, G., Habilitation Thesis, Technical University, Aachen, 1973.
2. Werner, G., "Relation Between Grinding Work and Wheel Wear in Plunge Grinding," Technical Paper, MR75-610, Society of Manufacturing Engineers, Dearborn, 1975.
3. Guehring, K., Dissertation, Technical University, Aachen, 1967.
4. Kragelskii, I., "Friction and Wear" (German), Carl Hanser Verlag, Munich, 1971.
5. Rabinowicz, E., "Friction and Wear of Materials," John Wiley and Sons, New York, 1965.
6. Peklenik, J., Dissertation, Technical University, Aachen, 1957.
7. Shaw, M.C. and Young, C.T., *American Society of Mechanical Engineers Transactions*, July 1955, p. 645.
8. Malkin, S., Ph.D. Dissertation, Massachusetts Institute of Technology, 1968.
9. Werner, G., Dissertation, Technical University, Aachen, 1971.
10. Dederichs, M., Dissertation, Technical University, Aachen, 1972.
11. Werner, G., "Application and Technological Fundamentals of Deep and Creep Feed Grinding," Technical Paper, MR79-319, Society of Manufacturing Engineers, Dearborn, 1979.
12. Bierlich, R., Dissertation, Technical University, Aachen, 1976.
13. Werner, G., in Proceedings of the International Conference on Production Engineering, Japan Society of Precision Engineering, Tokyo, 1974, Part II, p. 64.

DISCUSSION

QUESTIONER: In our work with wheel manufacture we have shown that the addition of certain solid lubricants to grinding wheels increases the metal removal rate, reduces wheel wear and gives better finish to the part under given conditions of speed and pressure. I wonder how we could fit this into your model to optimize the solid lubricants we have been adding. Could it be done?

G. WERNER: It fits very well. By using a certain cutting fluid or by putting a solid lubricant into the wheel, or improved bonding, we influence the behavior of the four basic criteria. In the models I presented, cutting fluids, solid lubricants, and so on, are considered in terms of the specific numerical values of the model parameters.

THE ROLE OF TRIBOLOGY IN GRINDING

C. P. Bhateja

ABSTRACT

From a review of the grinding process the vital role of tribology in grinding is identified. It is shown that the two fundamental aspects of the grinding process, namely, the stock removal action of the grinding wheel and the surface texture of the ground workpiece are governed by tribological phenomena involving the interaction between the cutting surface of the wheel and the workpiece.

The wear of both the wheel and the workpiece in grinding, which involves almost every known tribological wear mechanism, is discussed. The influences of the various mechanisms of grinding wheel wear on the wheel's effectiveness as a metal removal tool as well as on the ground surface are identified. In view of the fact that ground surfaces in practice are perhaps the most commonly used in tribological applications, several processes inherent in grinding which control the topographical characteristics of these surfaces have been discussed. The fact that the surface texture - an important tribological characteristic of the grinding wheel - controls its apparent hardness in grinding which is a vital factor determining the grinding action of the wheel is considered to be especially significant.

INTRODUCTION

Of all metal removal and finishing processes in use today, grinding is perhaps the most interesting from the tribological point of view. The very grinding action of an abrasive wheel comes from the interaction of the cutting surface of the wheel with the workpiece, thus making grinding synonymous with Tribology - "the science and technology of interacting surfaces in relative motion and of related subjects and practices." Evidently, grinding is inherently a tribological situation, the objective of which is to produce abrasive wear of some work material by means of a composite, rotating, multipoint cutting tool having a complex topography of its cutting surface.

The above fundamental requirement of a grinding operation is apparently contrary to the usual purpose of the application of tribology, viz. to reduce wear of the two bodies in contact. A tribological analysis of the grinding process should be directed towards the identification of the conditions which promote abrasive wear and would therefore lead to some improvements in the metal removal potential of the process. In addition, the analysis should also be directed to help improve the process capability with regard to the economic production of certain desirable surface texture features on ground surfaces.

It is interesting to note here that in spite of the significant role of tribology in grinding as identified above, grinding has received surprisingly little attention from tribologists. Tribological investigations[1-5] which may be considered somewhat relevant to grinding have been mostly on the abrasive wear of metals in a 2-body or 3-body type abrasion system. The fundamental requirement common to abrasion and grinding is that both involve a hard, rough surface moving under pressure against a soft surface. In the majority of the studies of abrasion, material removal or "abrasive wear" has been assumed to occur through ploughing of the softer material by the hard asperities. Theoretical and empirical models of the abrasion process based on some known geometries of the asperities, have been proposed. A relationship of the form

$$Q = \frac{K\ F_n}{H} \qquad (1)$$

has been found to cover a wide range of abrasion situations,[6] where Q is the total volume of material displaced per unit distance travelled, F_n, the normal load, H, the hardness of the softer material and K, an abrasive wear constant.

From an examination of the grinding process, several major aspects of the process in which tribology plays vital roles in governing both the nature and the output of the grinding process may be identified. In fact the role of tribology in grinding is so significant that, as shown in Figure 1, tribology may be looked upon as the very medium which helps transform the process input elements and variables into the desired output parameters of a grinding operation. The following are some of the major aspects of the grinding process in which tribology is especially relevant:

a) The mechanism of metal removal in grinding.
b) The topography of the grinding wheel's cutting surface and its relationship with the workpiece surface.
c) The wear mechanisms of the grinding wheel.

These aspects of the grinding process will now be discussed in some detail together with some influences of tribology in other sections of the total grinding system.

TRIBOLOGICAL ASPECTS OF THE GRINDING PROCESS

The Mechanism of Material Removal

The linear tribological relationship of Equation 1 for abrasive wear has been confirmed in grinding.[7] However, contrary to the metal removal through ploughing assumed in abrasive wear, as shown qualitatively in Figure 2, the material removal may occur predominantly through either ploughing or chip removal, that is, cutting process depending upon the magnitude of the normal force intensity between the wheel and the workpiece in a grinding system. Naturally, for reasons of production economics and of part quality in grinding, the mechanism of cutting is usually preferred to ploughing. Therefore, the operating conditions are selected to induce and encourage the

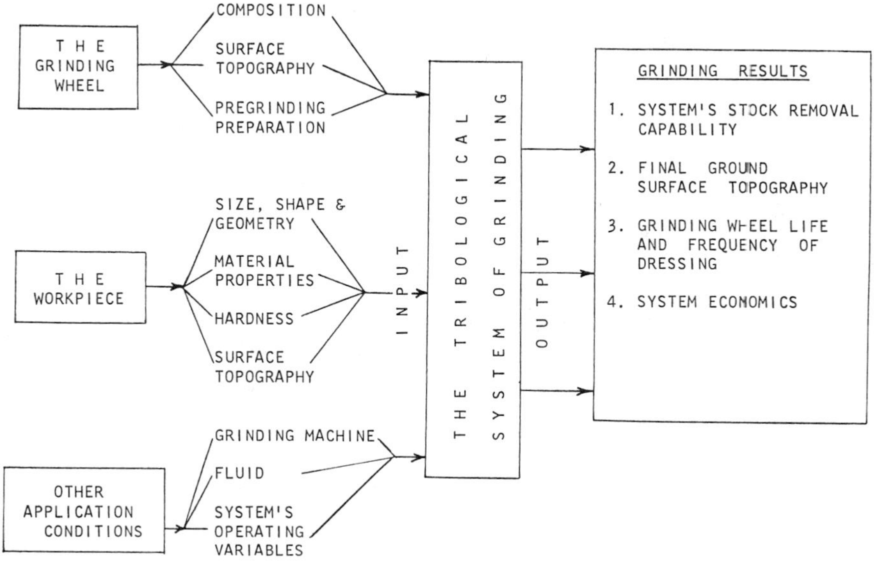

Fig. 1.—Tribology in grinding.

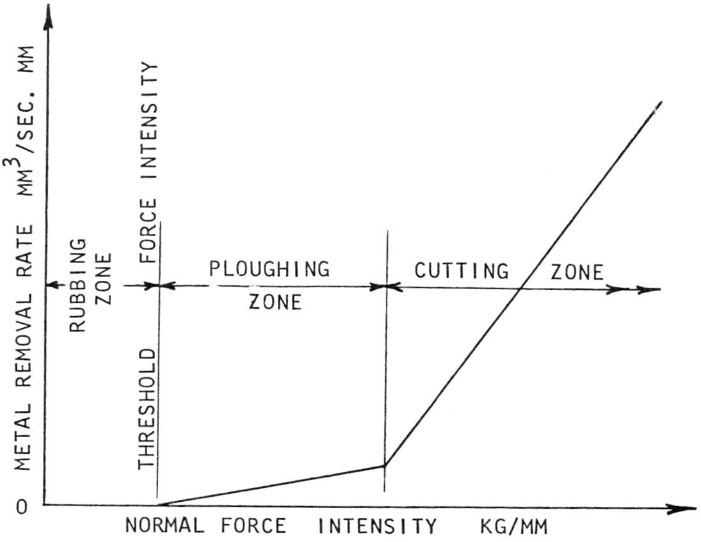

Fig. 2.—Metal removal - force intensity characteristic of a grinding system.

chip removal mechanism in grinding. It has been shown[8] that in spite of the presence of significant amounts of ploughing, the chip formation by the individual cutting edges on the grinding wheel surface predominates in grinding. Samples of grinding chips presented in the scanning electron photomicrographs of Figure 3 collected during the surface grinding of a bearing steel show strong resemblance to the chips in metal machining with a single point cutting tool.

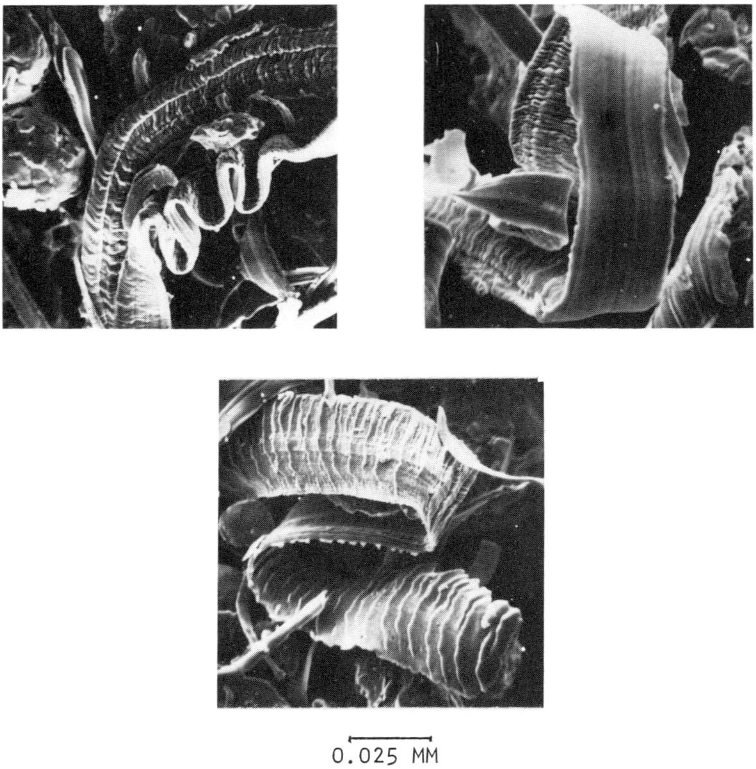

Fig. 3.—Grinding chips.

Even though grinding chips show marked primary shear which may consume as much as 75 percent of the total energy in a metal machining operation,[9] the typical "grooving" action of the grinding wheel cutting edges (Figure 4) seems to suggest that the ploughing and side displacement of work material may consume an appreciable portion of the total energy in grinding, particularly after some wear has developed on the grinding wheel surface.

(A) SHARP GRINDING WHEEL (B) WORN GRINDING WHEEL

125 μM

Fig. 4.—Ground surfaces showing the ploughing and side displacement of work-piece material in grinding.

In view of the predominance of the chip removal mechanism in grinding as shown by the grinding detritus of Figure 5, the majority of the comments regarding the balance of the forces on the chip due to the processes of primary shear and friction and the possible changes in these forces caused by the grinding fluids made by Shaw[9] for machining should also be valid for grinding. However, contrary to the common occurrence of the built-up edge in machining, owing to the large negative rake angles of the cutting edges in grinding and the high cutting speeds, a stable built-up edge is not common. In addition, the presence of the spherical particles in grinding detrius, which has been proposed[8] to be due to cooling under surface tension of the somewhat molten work material due to the intense heat in grinding, also indicates conditions not conducive to the stability of the built-up edge.

The high relative speeds between the cutting edges on the grinding wheel and the uncontaminated, freshly cut metal surface at the localized high temperature conditions in grinding naturally encourage wear of the work material through the adhesion mechanism. The deposition and growth of work material layer on the grinding wheel cutting edges, which would be a characteristic of adhesive wear,[10] has indeed been established in grinding.[11] Moreover, as would be common in such adhesive wear situations, the work material layers deposited on the cutting edges may be later transferred back onto the ground workpiece surface. Figure 6 shows that this process of adhesion and back transfer of work material increases as attritious type wear develops on the grinding wheel surface, it may occur to some extent even with a reasonably sharp grinding wheel.

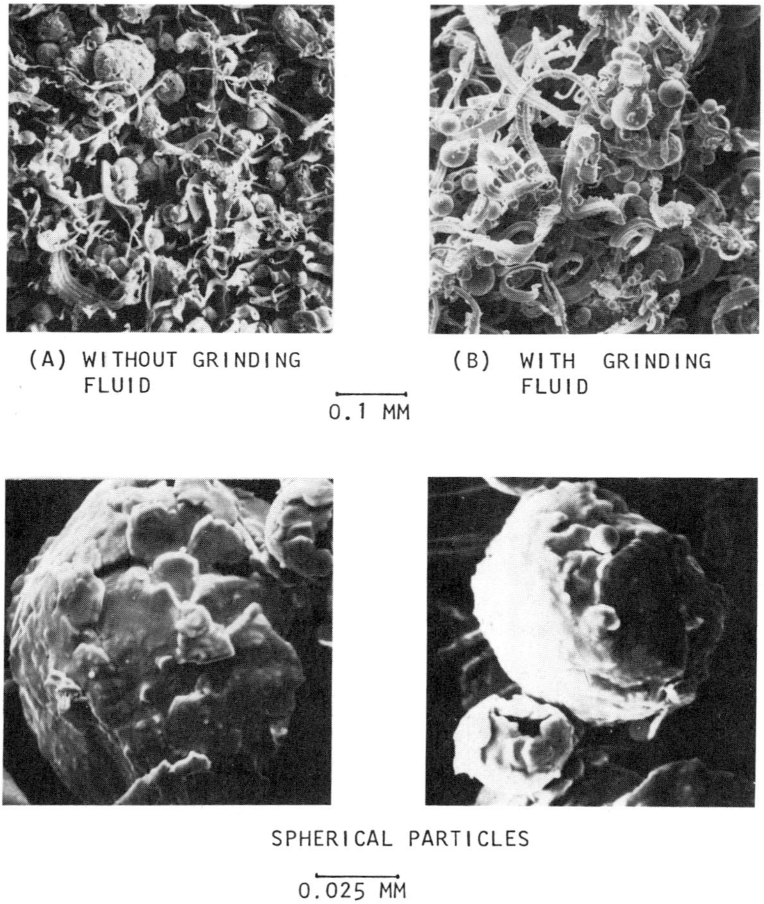

Fig. 5.—Grinding detritus showing chips and spherical particles produced in the surface grinding of steel.

The surface of a back transferred particle after the subsequent passage of grinding wheel cutting edges may be seen in Figure 6c. Apparently, the extrusion and smearing action of the cutting edges may cause some microcracks on these particles. These microcracks together with the repeated reworking of some particles trapped between wheel and workpiece and their possible work hardening (Figure 7) may in effect mean some superficial metallurgical damage of the ground surface. It is important to note here that in grinding these back transferred particles may often be work hardened and form loose oxidized abrasive particles in use. These particles may therefore be extremely undesirable on ground surfaces for critical applications in tribology.

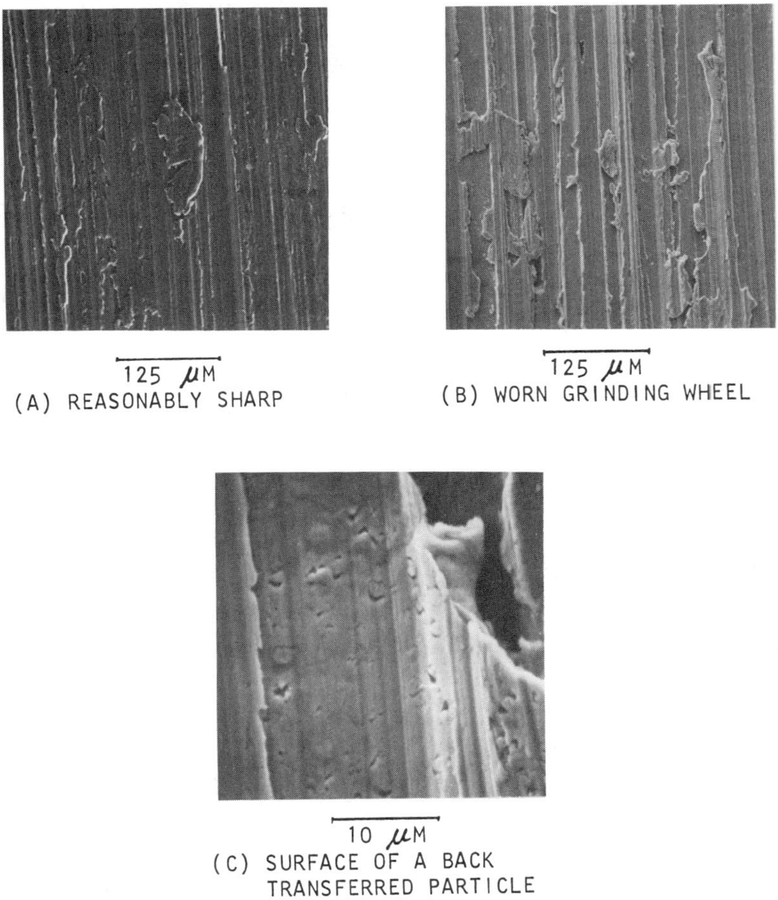

Fig. 6.—Back transferred particles on ground surfaces resulting from adhesive wear in grinding.

Surface Topographies of the Grinding Wheel and Workpiece and the Relationship Between Them

The characteristic of the grinding process which makes it synonymous with tribology is the interaction between the topographies of the grinding wheel and the ground workpiece surfaces. The ground surface, being the desired product of the grinding operation, is perhaps the most commonly used in tribological applications and therefore has far reaching implications in practice.[10] However, in the grinding process itself, the ground surface has been found to be closely related with the grinding action of the wheel itself under the operating conditions.[8,12,13,14,15] The ground surface can therefore be of tremendous value for both diagnostic and process evaluation purposes.

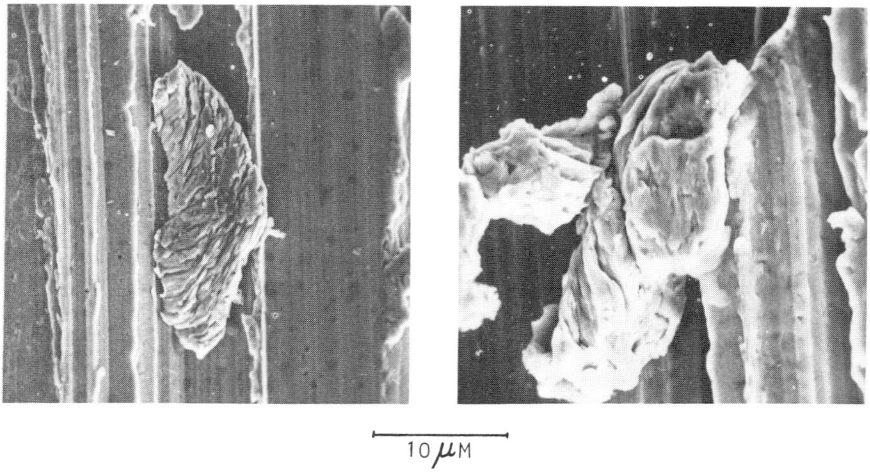

Fig. 7.—Examples of reworked back transferred particles on ground surfaces.

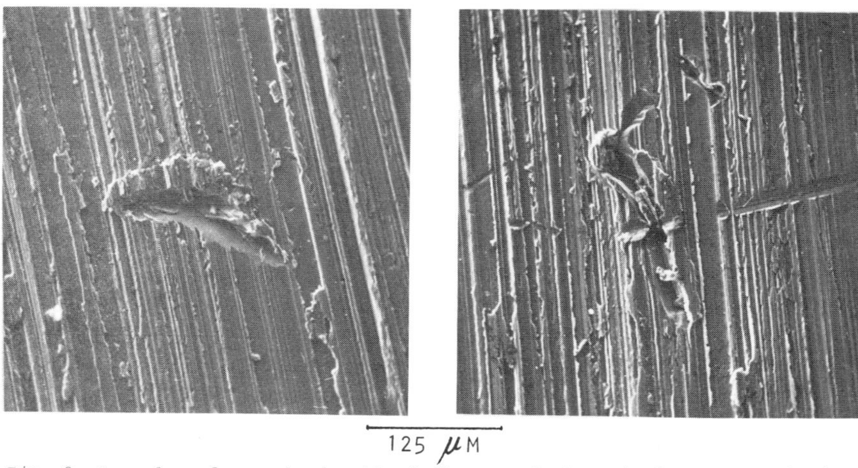

Fig. 8.—Ground surfaces showing the influence of the grit fracture mechanism of wheel wear.

The microscopic disposition of the ground surface has been shown to reflect some of the mechanisms of metal removal and wheel wear in grinding. Figure 8 presents the crater formation which may be caused by fragments of fractured abrasive grits during the initial rapid wear of the grinding wheel after dressing. The ability of the ground surface to reflect the influences of dressing and grinding wheel wear mechanisms is also demonstrated by the distributions of the asperity peaks and valleys and the bearing area characteristic of the workpiece surface profile presented in Figure 9. Apparently, when using single point diamonds for grinding wheel dressing, the use of coarser dressing leads results in the improvement of the bearing area of ground surfaces. Although, this fact may be used quite advantageously when grinding surfaces for critical load bearing applications in tribology, it

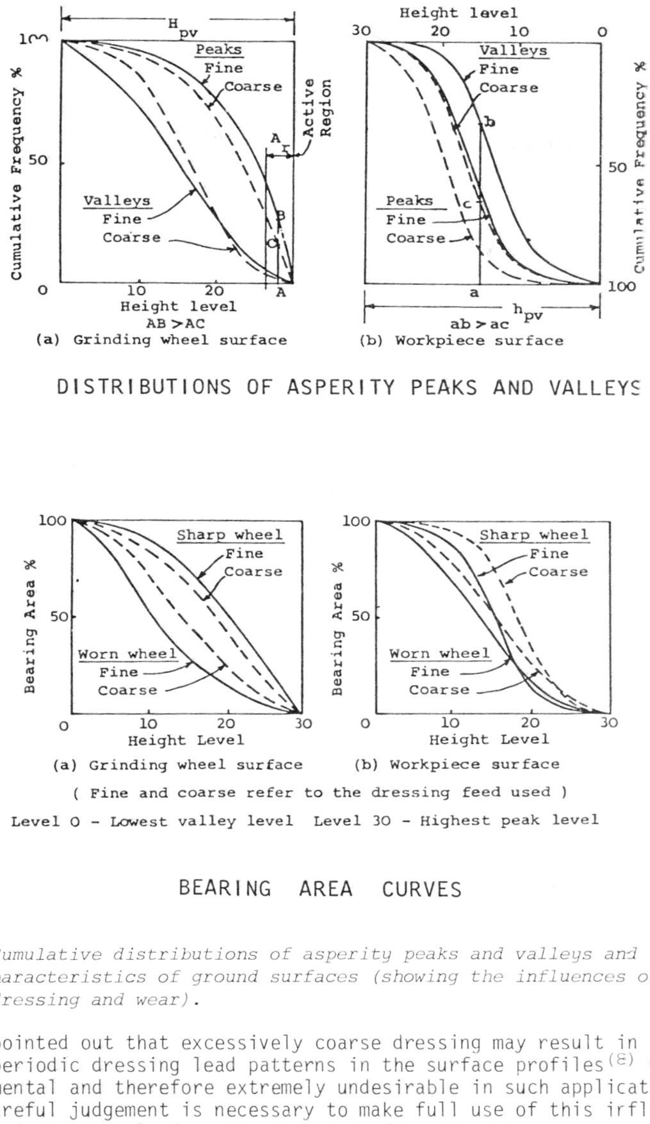

Fig. 9.—Cumulative distributions of asperity peaks and valleys and the bearing area characteristics of ground surfaces (showing the influences of grinding wheel dressing and wear).

must be pointed out that excessively coarse dressing may result in the appearance of periodic dressing lead patterns in the surface profiles(8) which can be detrimental and therefore extremely undesirable in such applications. Hence, careful judgement is necessary to make full use of this influence of the diamond dressing lead on the ground surface texture.

As to the grinding wheel surface topography, it is tribologically again an important characteristic of the wheel. It has been proposed(1-6) that cutting surface topography determines the inherent stock removal and surface finishing abilities of a grinding wheel. On the basis of the metal removal action of successive cutting profiles on the active surface of a grinding wheel, it has been suggested that two quantifiable functional parameters of the wheel, viz. the successive profile contribution characteristic and the metal removal function, may be derived from the wheel surface topography. One of the special merits of these parameters is their ability to reflect realistically some of the grinding behavior patterns of the wheel not accounted for otherwise. Such patterns include an increase in the apparent hardness of the grinding wheel and a reduction in the ground workpiece surface roughness with either an increase in the wheel's rotational speed or a decrease in the workpiece speed. Thus a wheel's surface topography - an important tribological feature of the wheel, affects its apparent hardness which is a vital grinding characteristic of the wheel. Moreover, it also accounts for the dynamic nature of the hardness of the grinding wheel.

The Wear Mechanisms of the Grinding Wheel

The wear of the grinding wheel is undoubtedly an important factor influencing its performance in any precision grinding operation. Although, generally the main emphasis in industrial grinding operations is on the amount of wheel wear, it is extremely important to remember that all wheel wear occurs through some tribological mechanisms which control directly the rate, the total amount and the uniformity or distribution of this wear on the wheel surface. Thus, the qualitative aspects of the grinding wheel wear are perhaps more important than its quantitative aspects which may well determine the performance of the wheel with regard to its ability to maintain size, consistency of the ground surface finish and, above all, the frequency of dressing which in turn determines the economics of the operation.

It has been shown[17] that the wear of the grinding wheel under conditions of constant material removal rate follows the pattern of Figure 10 and that the three stages of wear identified in the figure are associated with the predominance of different mechanisms of wheel wear. These mechanisms have been classified into two main types, namely fracture and attrition. Fracture includes both the partial or microfracture within the abrasive grit itself and the fracture of the bonding material resulting in a complete dislodgement of the whole grit. Attritious wear of the grinding wheel characterized by the slow, consistent and gradual dulling of the abrasive grit. Although attritious wear may be preferable to fracture type wear, it can result in a serious deterioration in the effectiveness of the grinding wheel as a metal removal tool.[7,18] It should also be noted here that these wear mechanisms also have some, what may be called as indirect influences.[19] These indirect influences are caused on the workpiece surface by the changes in the physical and topographical characteristics of the wheel surface resulting from the different wheel wear mechanisms (see Figure 9).

In grinding, although the attrition mechanism may account for only a small fraction of the measured wear, it usually affects the cutting action of the grits appreciably and therefore, is of considerable significance. Owing to the extremely high temperatures encountered in grinding[20] some of this attritious wear is only an apparent wear brought about by the deposition of a layer of the work material on the grinding wheel surface thus giving the appearance of dulling.

Certain chemical and diffusion type processes have also been identified[21,22] as two major mechanisms of wheel wear in grinding. For example, although both boron carbide and silicon carbide are harder than aluminum oxide, yet both dull more rapidly on steel than the latter. This difference is believed to be due to the greater solubility of the carbide abrasives in steel. On the other hand, when grinding glass aluminum oxide dulls at a rate too high to be explained by the relative hardness of the abrasive and glass.

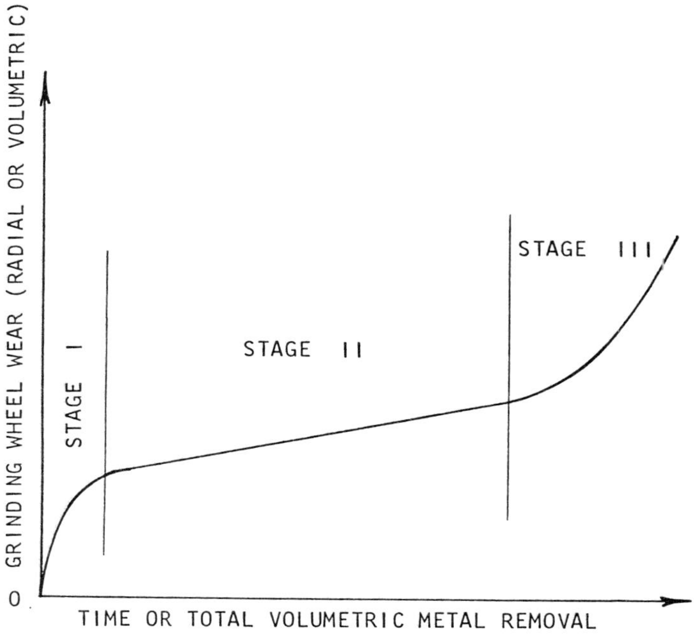

Fig. 10.—Typical pattern of grinding wheel wear.

The penetration hardness of glass is much lower than that of aluminum oxide. In fact, it is even lower than that of hard steel which, as just noted, is ground more efficiently with aluminum oxide abrasive. Again, the cause of the dulling seems to lie in the greater solubility of aluminum oxide in glass. Thus usually, because stock removal is the chief concern, aluminum oxide is used to grind steel and silicon carbide to grind glass. Recent studies of the apparent attritious wear of silicon carbide(23) have revealed some complex physical and chemical phenomena at work during the grinding of high cobalt and nickel alloys. It was found that the mechanisms involved may include some preferential removal of surface atoms on the abrasive by oxidation, dissociation of silicon carbide at high temperatures, diffusion of silicon into the work material and the formation of metal silicides and diffusion of carbon into the work material and formation of unstable metal carbides together with some cleavage type microfractures of silicon carbide crystals.

It is evident from the above therefore, that the chemical ard diffusion mechanisms of wheel wear influence heavily the selection of the right abrasive for a given work material. Generally, the abrasive wear of any work material does depend upon the hardness of the work material and it decreases with increasing hardness. This is consistent with the tribological relationship of Equation 1. Perhaps it is important to note here that in general the grinding of softer steels requires the use of harder grinding wheels and vice versa. This is attributed to the tendency of the softer materials to produce an early occurrence of bond fracture type wheel wear in grinding.(19)

The hardness of the work material as identified above is evidently an important tribological property of the work material which affects the grinding process significantly. Although the precise mechanisms may not have been identified, it has also been suggested[7] that work materials inherently may be classified into two different types for grinding purposes, namely easy-to-grind (ETG) and difficult-to-grind (DTG). The difficult-to-grind materials require larger threshold force intensities and higher ploughing-to-cutting transition force intensities[2] than the easy-to-grind materials. Examples of the ETG materials may include AISI 4150 and 52100 steels and those of the DTG materials include many steels in the M and T categories (e.g. M4 and T15), titanium alloys and high nickel steels.

TRIBOLOGY IN OTHER ASPECTS OF THE GRINDING SYSTEM

It is important to mention that the role of tribology in grinding is not limited to the localized interaction between the wheel and workpiece only. It is also vital to the success of the total grinding system in many other ways. An extremely important application of tribology occurs in grinding systems with friction type workpiece drives and work support mechanisms. Examples of these may include internal and external grinding systems with the workpiece supported on rolls or shoes and driven by magnetic chucks and carriers. It is widely known that the proper type and the positioning of the shoes to maintain a condition of part stability under the friction and grinding forces is essential in microcentric external grinding operations. Similarly in infeed or through feed centerless type production grinding operations, it is important to ensure a consistent part rotation under the control of the regulating wheel and the friction at the work rest blade to obtain good part geometry and roundness. As a matter of fact, it has been suggested[24] that improvements in workpiece roundness may be made simply by suitable control of the friction forces between the blade and the workpiece in relation to the force field of the grinding and regulating wheels. In addition, the abrasive wear of the workpiece support blade in a centerless grinding system determines the setup life and hence influences significantly the economics of the operation.

CONCLUSIONS

The various aspects of the grinding process which are significantly influenced by tribological phenomena have been identified as follows:

a) the mechanism of material removal
b) the interaction between the grinding wheel and ground workpiece surface topographies
c) the wheel wear mechanisms during grinding

These aspects have been discussed in some detail pointing out how tribology determines the output of a grinding operation.

It is also indicated that although grinding may be simply looked upon as the abrasive wear of the workpiece material, several apparent tribological anomalies exist in grinding. First, abrasive wear as generally referred to in tribology implies a lack of its desirability whereas in grinding, material removal is not just desirable, it is necessary. Second, contrary to the normal laws of tribology, softer steels, in general, tend to produce larger wear of grinding wheels than harder steels. Finally, unlike most tribological situations where the hardness of each of the two interacting surfaces is a static characteristic, the apparent hardness of a grinding wheel is a dynamic characteristic and it depends upon both the grinding conditions and the surface topography - an important tribological feature of the wheel.

REFERENCES

1. Khruschov, M.M. and Babichev, M.A., "Friction and Wear in Machinery, No. 12," translated from the Russian, American Society of Mechanical Engineers, New York, 1960, p. 5.
2. Avient, B.W.E., Goddard, J. and Wilman, H., *Royal Society of London. Proceedings*. Series A, Vol. 258, 1960, p. 159.
3. Rabinowicz, E., Dunn, L.A. and Russell, P.G., *Wear*, Vol. 4, 1961, p. 345.
4. Mulhearn, T.O. and Samuels, L.E., *Wear*, Vol. 5, 1962, p. 478.
5. Nathan, G.K. and Jones, W.J.D., *Wear*, Vol. 9, 1966, p. 300.
6. Halling, J., "Principles of Tribology," MacMillan, London, 1975, p. 101.
7. Hahn, R.S. and Lindsay, R.P., *Machinery*, July-November, 1971.
8. Bhateja, C.P., in Proceedings International Technical Conference, Chicago, Abrasive Engineering Society, 1975, p. 11.
9. Shaw, M.C., These Proceedings.
10. Neale, M.J., ed., "Tribology Handbook," John Wiley and Sons, New York, 1973, Section F6 by K.H.R. Wright.
11. Tsuwa, H., et al., *Japan Society of Precision Engineers. Journal*, Vol. 2, No. 1, 1966, p. 40.
12. Bhateja, C.P., et al., in Proceedings of the International Grinding Conference, Carnegie Press, Pittsburgh, Pa., April 1972, p. 685.
13. Bhateja, C.P., in Proceedings North American Metalworking Research Conference, Carnegie Press, Pittsburgh, Pa., May 1975, p. 498.
14. Bhateja, C.P., *CIRP Annals*, Vol. 26, Part 1, 1977, p. 333.
15. Lindsay, R.P., American Society of Mechanical Engineers Paper No. 72-WA/Prod-13.
16. Bhateja, C.P., *CIRP Annals*, Vol. 27, Part 1, 1978.
17. Krabacher, E.J., *Journal of Engineering for Industry*, Transactions ASME, Vol. 81, 1959, p. 187.
18. Hahn, R.S., *CIRP Annals*, Vol. 25, Part 1, 1976, p. 203.
19. Bhateja, C.P., et al., in Proceedings 12th International Machine Tool Design and Research Conference, Manchester, England, Macmillan Press Ltd., 1971, p. 535.
20. Outwater, J.O. and Shaw, M.C., *American Society of Mechanical Engineers Transactions*, Vol. 74, 1952, p. 73.
21. Letner, H.R., *Grinding and Finishing*, May 1955, p. 36.
22. Geopfert, G.J. and Williams, J.L., *American Society of Mechanical Engineers Transactions*, Vol. 81, 1959, p. 69.
23. Komanduri, R. and Shaw, M.C., *Journal of Engineering for Industry*, Vol. 98, No. 4, Nov. 1976, p. 1125.
24. Reeka, D., Doktor-Ingenieurs-Dissertation, Techniche Hochschule Aachen, February 1967.

DISCUSSION

G. WERNER, MIT: According to one hypothesis, the spherical particles are formed in grinding due to the high temperatures produced. Tiny metal particles solidify as small spheres with sizes ranging from several hundred micrometers to several micrometers. Incidently, this phenomenon is not confined to grinding only.

Your observation that surface roughness goes down with increasing wheel speed and goes up with work speed is similar to a model presented by Shaw in 1959 on titanium grinding. There is one difference; the surface roughness does not increase linearly.

C. P. BHATEJA: In the range of speeds that I have used, it is very much linear; but if we go to very high speeds I really do not know what happens. For speeds up to 200-300 feet per minute (say 6000 surface speed of the wheel itself) I think it is fairly linear. Now if you go to speeds like 500-600 or 1000 feet per minute which we use in industry, I am sure what the relationship would be. Perhaps non-linear.

WERNER: You said depth of cut has no influence on the surface roughness, at least if you go deep enough. Is that true?

BHATEJA: Yes. But I did not mention anything about the depth of cut as such in my model.

M. C. SHAW, Arizona State University: Regarding the spherical particle formation, we published one possibility about a year ago in Philosophical Magazine. And that has to do with surface energy approach to electron density.equilibrium for newly generated surfaces.

BHATEJA: I noticed this first in 1968. Since then I was hoping that someone would come up with an explanation. It is nice to hear that somebody has.

AN ANALYTICAL APPROACH TO TOOL WEAR PREDICTION

B. Kramer and N. P. Suh

ABSTRACT

An analytical technique that allows quantitative calculation of relative wear rates in high speed cutting from thermodynamic data is presented. The technique is based upon the identification of the rate controlling mechanism. It is shown to be an accurate predictor of wear rates for the important class of carbide cutting tools. Broad applications outside this class are indicated.

INTRODUCTION

The following analysis treats those wear processes in which dissolution of the tool material in the material being cut is required, the so-called "atomistic wear processes." Mechanical wear processes in which the tool material is removed as large particles consisting of many atoms are not treated. These wear processes are discussed elsewhere in this proceedings[1-3].

Trent[4,5] described wear by dissolution of the tool material in the material being cut qualitatively in his work. But he had difficulty in explaining the observed wear rates on the basis of the diffusion of tool constituents from the interface, and he settled on a mechanism in which melting occurs at the interface.

As can be seen herein, the flow properties and mechanics of the material in this region critically control the dissolution wear rate. However, the mechanics of the problem presents theoretical difficulties that make its solution inaccessible at the present time. In this paper, the critical quantities relating to the mechanics of the metal cutting process are inferred from the experimental data.

Autoradiographic studies by Cook and Nayak[6] clearly show that tungsten carbide-based tools wear by the dissolution of tungsten carbide in steel. Suh[7] has emphasized the importance of chemical stability in determining the solubility of the tool materials in the material being cut, and has suggested the free energy of formation as an index of chemical stability. The present work refines this treatment to allow for chemical reactions between the tool and the material being cut. For the first time, quantitative prediction of the relative wear rates of tool materials from their chemical properties is possible.

THEORETICAL ANALYSIS

With reference to Figure 1, the flux of tool material away from the tool may be expressed as

$$v_{wear} = K \left[-D \frac{\partial c}{\partial y} + C v_y \right]_{y = 0}$$

Where:

v_{wear} = the wear rate of the tool material (cm/sec)

D = the diffusivity of the slowest diffusing tool constituent atom in the material being cut (cm^2/sec)

c = the concentration of the tool material in the material being cut

C = solubility limit of the tool material in the material being cut

V_y = the bulk velocity of the chip material at the tool-chip interface

K = the ratio of the molar volumes of the tool and the workpiece materials

The first term is the diffusion flux. The driving force for diffusion is the concentration gradient $\partial c/\partial y$. The concentration gradient is strongly influenced by the velocity profile within the chip. In practice, it is known that the chip material in contact with the tool is slowed down as it passes along the tool face, forming the so-called "secondary shear zone." As a consequence, continuity considerations require that there be a compensating flow of chip material away from the tool. This flow is responsible

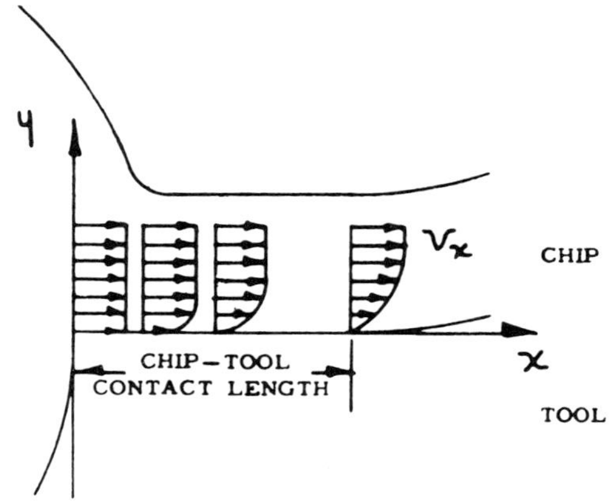

Fig. 1.—Schematic illustration of chip flow geometry.

for the second term in Equation (1), the flux due to the bulk flow of the chip material. It will be seen that this term predominates in most cases of practical interest.

THE MECHANICS OF CHIP FLOW

Available information suggests several features of chip flow. Near the tool tip, the chip slides along the tool face and there is a velocity discontinuity across the interface. Heat is generated at the interface. As can be seen in Figure 2, after Boothroyd[8], the locus of the points of maximum heat generation shifts from the chip-tool interface into the chip as the chip traverses the tool face, the transition occurring at the arrow on the figure. This would be expected if a transition from sliding to

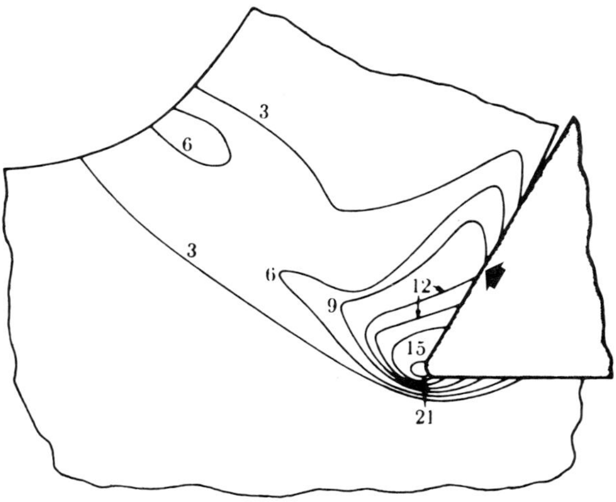

Fig. 2.—Heat generation within the chip (arbitrary units) after Boothroyd[8].

sticking occurred at this point. Similar behavior may be seen in Figure 3 for the locus of maximum temperature points.

It is reasonable to expect that the maximum deceleration of the chip material will occur in the region where the transition from sliding to sticking occurs. This will result in the greatest bulk velocity perpendicular to the interface and, consequently, in the greatest solution wear rate. Figure 4 shows a cross-section of the crater of a hafnium carbide-coated insert after penetration of the hafnium carbide surface layer. It can be seen that the steel chip begins to adhere to the tool near the maximum wear point.

There is no doubt that much can be learned from a detailed study of chip flow patterns at high magnification. Suh and Agustsson[9] have conducted such a study for the sliding wear of metals that helped to illuminate the near surface deformation patterns. A detailed electron microscopic study of the chip might yield similar informative results.

At the present time, it is impossible to predict the transition point from sliding to sticking from basic material properties. Even the related problem, prediction of the chip-tool contact length from basic material

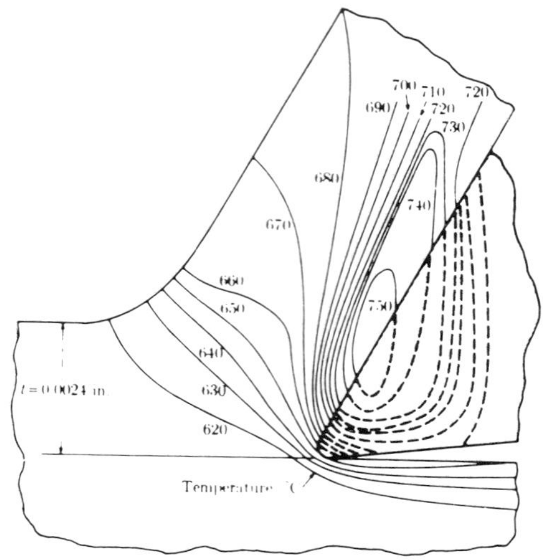

Fig. 3.—Temperature distribution within the cutting zone. The material being cut is mild steel, preheated to 611°C. After Boothroyd(8).

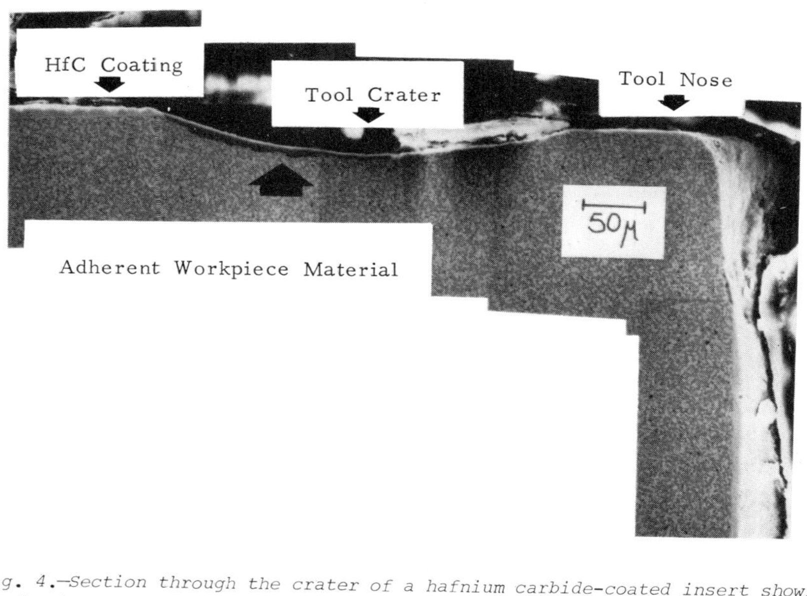

Fig. 4.—Section through the crater of a hafnium carbide-coated insert showing adhesion of workpiece material near the maximum wear point.

properties, has not been solved. These difficulties coupled with the general dearth of experimental determinations of those material properties that are known to be of importance, pose an awesome challenge to analysis. Numerical analysis is in order, and more than currently available computational power may be required.

ESTIMATION OF THE RELATIVE WEAR RATES

Because of our inability to determine v_y, only relative wear rates are calculated here. If v_y is similar for different tool materials under similar cutting conditions, the relative wear rate will depend *only* on the solubility of the tool material in the work and the molar volume ratio, K. Fortunately, relative wear rates provide all the information that is required to choose among the various potential tool materials for a given application.

In this study, solubilities have been estimated from thermodynamic properties. Approximate values of the solubilities of potential tool materials in body-centered cubic iron at 1,600 K, estimated from available data, are listed in Table 1. The techniques used to estimate these solubilities are discussed briefly in the following section. The reader is referred to Reference (12) for details concerning the techniques used in estimation.

It is believed that the characteristic short time of the cutting process does not allow sufficient time for the sluggish body-centered cubic to face-centered cubic transformation in iron to occur. While the highly strained material in the cutting zone defies proper structural description, it is believed that it is best approximated as metastable body-centered cubic iron.

If dissociation is required, at equilibrium:

$$\Delta G_{ij} = \Delta \bar{G}_i^M + \Delta \bar{G}_j^M$$

where:

ΔG_{ij} = the free energy of formation of the tool material,

$\Delta \bar{G}_i^M$ = the relative partial molar free energy of solution of component i, and

$\Delta \bar{G}_j^M$ = the relative partial molar free energy of solution of component j.

The relative partial molar free energy of solution of component i can be expressed in terms of the relative partial molar excess free energy of solution, the excess free energy:

$$\Delta \bar{G}_i^M = \Delta \bar{G}_i^{xs} + RT\ln c_i$$

TABLE I.—ESTIMATED AND REPORTED SOLUBILITIES OF POTENTIAL TOOL MATERIALS IN α-Fe AT 1,600°K

Potential Tool Material	Free Energy of Formation (cal/mole)	Estimated Equilibrium Concentration (solubility)	Exp. Results [11] Extra-polated to 1,600°K
ZrO_2	-190,300	3.60×10^{-8}	
Al_2O_3	-278,300	5.55×10^{-7}	
Ti_2O_3	-260,800	8.22	
TiO_2	-156,300	1.52×10^{-6}	
TiO	-91,020	1.40×10^{-5}	
HfN	-52,604	1.53×10^{-4}	
HfC	-49,122	1.97	
ZrN	-51,356	2.93	
$TiC_{.75}O_{.25}$	-52,395	5.42	
ZrC	-42,714	8.42	
TiN	-45,150	1.04×10^{-3}	
TaC	-34,604	1.41	2.1×10^{-3}
TiC(iron)	-39,520	1.86	6.1×10^{-3}
NbC	-32,236	2.01	6.8×10^{-3}
BN(graphitic)	-26,100	9.65	
WC	-8,144	--	2.6×10^{-2}
VC	-23,416	--	3.2×10^{-2}
TiC(nickel)	-39,520	2.24×10^{-2}	6.3×10^{-2}
Diamond	--	9.30	
Si_3N_4	-51,850	9.50	
β-SiC	-14,548	4.30×10^{-1}	

where:

c_i = the concentration of component i, and

$\Delta\bar{G}_i^{xs}$ = the relative partial molar excess free energy of solution of component i.

If the solution obeys Henry's Law, the excess free energy will remain constant with solute concentration at a given temperature.

Calculated constant values of the excess free energy for the tool constituents of interest are listed in Table 2. The assumptions and techniques employed in calculating these values are discussed in detail in Reference [10].

To calculate the solubility of TiC in α-iron, at equilibrium:

$$\Delta G_{TiC} = \Delta\bar{G}_{Ti}^{M} + \Delta\bar{G}_{C}^{M}$$

From Table 2:

$$\Delta\bar{G}_{Ti}^{M} = -6,900 + RT\ln c_{Ti} \qquad cal/mole$$

and

$$\Delta\bar{G}_{C}^{M} = +7,600 + RT\ln c_{C} \qquad cal/mole.$$

For TiC, $c_{Ti} = c_C$. Therefore, with reference to Table 1:

$$-39,520 = 700 + 2RT\ln C_{TiC}$$

TABLE II.–ESTIMATED EXCESS FREE ENERGIES OF SOLUTION OF TOOL CONSTITUENTS IN α-IRON.

Tool Constituent	$\Delta\bar{G}_i^{xs}$ (calories/mole)
Ti	-6,900
Ti (Ni)	-26,800
Zr	-5,000
Hf	-2,100
V	-9,100
Nb	-100
Ta	-200
Al	-10,700
Si	-16,700
B	-2,100
C	+7,600
C (Ni)	+11,600
O	-12,600
N	+5,700

Therefore:

$$C_{TiC} = 1.86 \times 10^{-3} \text{ at } 1{,}600°K.$$

Data concerning the solubility of refractory materials in the iron-group metals is scant. The data of Edwards and Raine[10] which is the only systematic study of the solid solubilities of the carbides in the iron-group metals, is reproduced in Table 3.

TABLE III.—THE SOLUBILITIES OF THE CARBIDES IN THE IRON-GROUP METALS AT 1,523° K

Carbide	Solubility in wt.% (mole %)		
	Cobalt	Nickel	Iron
WC	22(7.9)	12(3.9)	7(2.2)
TiC	1(1.0)	5(4.9)	<.5(<.5)
VC	6(5.6)	7(6.6)	1(.53)
TaC	3(.93)	5(1.6)	.5(.15)

Dawihl, et al.[11] cite conclusions made by Pfau and Rix in the German literature to the effect that, when tungsten carbide goes into solution in cobalt, tungsten atoms substitute for cobalt while carbon atoms fill interstitial sites. This will undoubtedly be the case for tool constituents dissolving in iron.

Table 4 shows a comparison of predictions of the relative wear rates of selected carbides with test results from inserts coated with these carbides by a patented process developed at MIT[13].

DISCUSSION

The discrepancy in the wear rate of TiC is explained by the pick-up of oxygen from the steel workpiece. As pointed out by Carson[14], the number of oxygen atoms per unit volume in normal steel exceeds the number per unit volume at normal atmospheric oxygen partial pressure. Suh, Carson, and Leung[15] have noted little difference in wear rate between sputtered coatings of $TiC_{.75}O_{.25}$ and TiC. The predicted wear rate of $TiC_{.75}O_{.25}$ is included in Table 4. Titanium and niobium are the only metals of interest that form stable monoxides.

The data clearly show that the wear of alumina is not controlled by the solubility of alumina in iron. This is encouraging for tool development since the mechanical properties of materials can be subjected to far greater manipulation than the chemical properties. The fact that the solution hardening of alumina increases tool life, suggests that plastic deformation operates in the wear process. Kane and Hasselman[16] show strong evidence of plastic deformation in alumina tools in steel cutting and suggest that dislocation pile-up may nucleate cracks. Carson[17] shows photographs of microcracking at sub-micron intervals in alumina-coated inserts. Subramanian[18] has also noted microcracks. The work of Suh and Fillion[19]

TABLE IV.—COMPARISON OF THEORETICAL PREDICTIONS WITH TEST RESULTS. WEAR RATE OF HfC IS TAKEN AS UNITY.

Carbide	Temperature (°K) 1600	1500	1400	1300	Test Results
HfC	1.0	1.0	1.0	1.0	1.0
TiC	7.65	8.82	10.6	12.8	2.59
$TiC_{.75}O_{.25}$	2.26	2.41	2.61	2.86	--
ZrC	4.44	4.87	5.47	6.20	5.74
TaC	6.33	7.19	8.39	9.98	10.4
	(9.43)	(10.7)	(12.5)	(14.9)	
NbC	9.13	10.6	12.7	15.6	16.7
	(31.0)	(36.0)	(43.1)	(52.8)	
WC	(107.0)	(153.0)	(215.0)	(332.0)	237

Terms without parentheses are calculated on the basis of the solubility estimated from thermodynamic properties. Terms in parentheses are calculated using the reported solubility (see Table 1).

on metal-bonded oxide tools is a new approach to improving the mechanical properties of oxides.

Accurate determinations of the diffusivities of the elements of interest in δ-iron would settle the issue of the role of diffusion in tool wear.

Lack of experimental data also makes it impossible to preclude the possibility that wear is limited by the kinetics of the separation of tool atoms from the tool lattice, a parameter related to the self-diffusivity of the tool material. This is the question of whether tool wear is an equilibrium process.

Explanation of the inversion in the relative wear rates of titanium carbide and tungsten carbide when cutting steel and nickel is extremely difficult if wear is controlled by the kinetics of separation. The phenomenon is explained by the greater solubility of titanium carbide in nickel. This result is expected from thermodynamic considerations and is due to the lower relative partial molar free energy of solution of titanium in nickel.

Thermodynamic analysis has broad application to cutting tool development outside of wear rate prediction, as has been demonstrated by truly revolutionary tool microstructures produced through spinodal decomposition reactions by Rudy(20). The major advances of the coming years in tool development will undoubtedly be through thermodynamic insight.

REFERENCES

1. Shaw, M.C., These Proceedings.
2. Ramalingam, S., These Proceedings.
3. Larsen-Basse, J., These Proceedings.
4. Trent, E.M., *Institution of Mechanical Engineers. Proceedings,* (A), Vol. 166, 1952, p. 64.
5. Trent, E.M., *Royal Society of London. Proceedings.* Series A, Vol. 212, 1952, p. 467.
6. Cook, N.H. and Nayak, P.N., "Development of Improved Cutting Tool Materials," Air Force Technical Report AFML-TR-69-185, 1969.
7. Naik, S. and Suh, N.P., *Journal of Engineering for Industry,* Vol. 97, 1975, p. 112.
8. Boothroyd, G., *Institution of Mechanical Engineers. Proceedings,* Vol. 177, 1963, p. 789.
9. Agustsson, G., S.M. Thesis, Massachusetts Institute of Technology, 1974.
10. Edwards, R. and Raine, T., in Plansee Proceedings, 1952, edited by F. Benesovsky, Metallwerk Plansee GmbH., Reutte/Tyrol, 1953, p. 232.
11. Dawihl, W. and Frisch, B., *Cobalt,* Vol. 22, 1964, p. 22.
12. Kramer, B., Ph.D. Thesis, Massachusetts Institute of Technology, June 1979.
13. Cook, N. and Kramer, B., "Tungsten Carbide Tools Treated With Group IVB and VB Metals," U.S. Patent 4,066,821, 1978.
14. Carson, W.W., Private Communication.
15. Suh, N.P., in Proceedings of the North American Metalworking Conference, McMaster University, Vol. 2, May 1973, p. 35.
16. Hasselman, D.P.H., et al., in "NSF Hard Materials Research," National Technical Information Service #PB-221908, Vol. 2, July 1973, p. 204.
17. Carson, W.W., et al., in "NSF Hard Materials Research," National Technical Information Service #PB-221908, Vol. 2, July 1973, p. 269.
18. Subramanian, K., Private Communication.
19. Suh, N.P. and Fillion, P., to be published.
20. Rudy, E., in "NSF Hard Materials Research," National Technical Information Service #PB-221908, Vol. 2, July 1973, p. 349.

DISCUSSION

J. T. BLACK, Ohio State University: Could you tell me a little bit more about solution wear?

B. KRAMER: The basic tenet is that tool materials are soluble in the work piece. The method for calculating the solubility is given in detail in the paper. Basically, the free energy of solution of the tool material is used as the activation energy for solution. Then the solubility of the tool material in the workpiece is calculated, and this will give a reasonable solubility if the free energy change is the controlling factor.

Now, since the chip is slowing down as it comes along the tool face, it must flow perpendicular to the tool surface in order to maintain continuity. Any tool material that dissolves at the interface will be carried away by this flow. The velocity of this flow could be estimated from the cutting data. For materials of interest at temperatures which are encountered, this velocity term dominates over diffusive terms and accounts for the wear.

M. M. BARASH, Purdue University: On a historical note, I think it was in 1951 that Trent published a classic paper explaining why tungsten-titanium carbide can cut steel and tungsten carbide cannot -- because of the difference in the solubility. It is rather surprising that for so many years nobody touched it.

D. H. BUCKLEY, NASA: I have a question about the calculation of free energy of formation. You have to take into consideration the relative concentration of the two species involved. How did you select the concentration here?

KRAMER: I calculated the relative concentrations from the free energy change.

BUCKLEY: You assumed the solution at the interface was an equilibrium situation. I would like to know how you arrived at that. I visualize the cutting operation as anything but an equilibrium from thermodynamic standpoint.

KRAMER: The interface between the cutting tool and the chip is a very small distance. I cannot see why equilibrium couldn't be established over these small distances.

BUCKLEY: Do you have any evidence to support this? I would infer just the opposite. If I could visualize any system at all as being non-equilibrium, it would be the cutting operation.

KRAMER: Well, equilibrium is a function of distances and times. Because of the small distances I feel equilibrium can be established. The evidence I have is that by doing this analysis and treating the problem as if equilibrium had been established, I get an excellent correlation between experimental results and predictions. I take that as confirmation of the fact that equilibrium was established.

D. TABOR, Cambridge University: The chip that is coming off is very highly distorted metal. Isn't it conceivable that many of the properties are rather different from the properties that you have been assuming based on textbook numbers about diffusivity, reactivity and so on? May this be a factor that changes the rate at which things occur at the interface?

KRAMER: In some cases it could be factor. In certain cases, e.g. high speed steels, the physical properties of the highly deformed material with a large number of voids and interstices may be important in the wear. However in these carbides the chemical properties are fairly similar, regardless of the mechanical state of the material.

R. VISWANATHAM, Martin Marietta Laboratories: Are most of your comments in the paper for coated tools?

KRAMER: Yes, the tools that were accessible to me for these tests were coated tools.

VISWANATHAM: Would you also take a similar approach for uncoated materials?

KRAMER: I would.

VISWANATHAM: Shall we then ignore completely the role of the binder? The reason I am saying this is because if the temperatures at the tool chip interface are indeed as high as you say, 1000° C and above, we are approaching temperatures where the stress rupture properties of the binder must have some role to play.

KRAMER: I was interested in an approach that would show why the various tools wear at different rates without complicating the problem. For materials with binders, you have to look at the preferential removal of the binder, the carbides, etc. and that is a complex problem. I was interested in looking at the chemical wear.

M. C. SHAW, Arizona State University: Our largest carbide manufacturer uses three layers of materials. I asked them about this and they were not clear about it. But they found that this worked better than just a titanium nitride. They used titanium nitride on the outer surface and on the inner surface, next to the carbide tool, they have titanium carbide and then an intermediate layer of carbonitride for coupling. They suggested after some inquiry that the titanium nitride gave lower friction between chip and tool but that titanium carbide was a better diffusion barrier.

So we set out to see if that was really the case. In diffusion couple experiments we anticipated, as suggested by Professor Tabor for example, that the material in the diffusion couple should be highly deformed. We built an apparatus with which we could study diffusion between a coated carbide and a work surface in a case where the work surface was plastically deformed in the experiment. We found that titanium carbide was a superior diffusion barrier to titanium nitride. We then did a series of experiments in which we measured the friction between the freshly machined work material and the coated tool surface and we found that titanium nitride gave a considerably lower friction coefficient than titanium carbide. Then we looked at the surface of the rubbed material using the Auger spectroscope and we found that in both cases there was evidence of carbon on the surfaces of the work piece when we had titanium carbide, and there was nitrogen in the case of titanium nitride.

The only conclusion we can come to is that the carbon that transfers from decomposition of the titanium carbide and transfers onto the asperities of the work material gave more strengthening than nitrogen does. So we came to the conclusion that the outer frictional characteristics of the titanium nitride as opposed to titanium carbide was due to the lower strengthening effect of the nitrogen that was released when the layer decomposed. It is an extremely thin layer that deposits on the material, but apparently it changes the interfacial characteristics of the material. The point I would like to make in this connection is that it is to be expected that maybe one coating is not enough. If we have more than one objective (as in the design of bearing materials in which we are interested in low shear strength in a superficial layer and high hardness in the bulk) we cannot get those in one material. Why not try to get the best with these coatings where you have low friction and high diffusion barrier. Titanium carbide for some reason stands as a better diffusion barrier for this particular application. The whole problem is very complicated, but I just mentioned this as I thought it might be of interest.

X. POLYMER WEAR

THE SLIDING WEAR OF POLYMERS: A BRIEF REVIEW

B. J. Briscoe and D. Tabor

ABSTRACT

This paper presents a short review of the sliding wear of polymers under the two headings of deformation wear and interfacial wear. The section on deformation wear mainly reports the work of other investigators and we have chosen a selection of classical and more recent papers to exemplify what we believe are the main features of this type of wear process. They involve primarily abrasive and fatigue wear mechanisms. The second main section deals with interfacial wear, the main emphasis being placed on adhesive or transfer wear. Here we cite mainly our own work and that of our colleagues. Naturally it is somewhat biased towards our own conception of the important features of this wear process, but we have tried to balance this with reference to other studies. This section also contains a brief summary of chemical wear but only in so far as it influences the nature of adhesive wear.

A proper understanding of polymer wear demands a detailed knowledge of the surface and bulk properties of polymers. Currently polymer science cannot provide this information and we must accept, for the present at least, rather tentative connections between polymer wear and material properties.

THE BASIC MECHANISMS OF WEAR

Many mechanisms are involved in the wear of polymers. Lancaster[1] has cited and discussed five mechanisms in the context of polymers; abrasive wear, erosive wear, fatigue wear, corrosive (or chemical) wear, and adhesive wear. Abrasive wear occurs when a hard sharp particle cuts or displaces material from the specimen. Erosive wear occurs in polymers when, for example, slurries are pumped through pipes or conduits. The process resembles abrasion and we shall not deal with it further. We may, however, note that in some cases high speed liquid drops can remove material from plastic surfaces (rain drops on radomes) and this is a type of erosion which is associated more with impact cracking than with abrasion. Fatigue wear, as its name implies, results from the repeated stressing of a given volume element of the specimen and, as we shall see, it often overlaps the abrasive process. Corrosive wear is usually associated with the influence of a chemically active medium in the presence of an applied stress. We shall deal with this only in the course of discussing other wear mechanisms. Finally adhesive wear involves strong intermolecular forces at the interface such that when sliding occurs, fragments of polymer are plucked out of the surface. These mechanisms have been studied in experiments deliberately devised to isolate them as individual wear processes. In most practical situations however, wear arises from a complex interaction of a number of these - and perhaps other - mechanisms.

In this review we shall consider the sliding wear of polymers under two broad headings based on our understanding of the origins of the friction force. Like wear, friction arises from many mechanisms, but to a first approximation we shall assume that two non-interacting basic processes are involved[2-4], Figure 1.

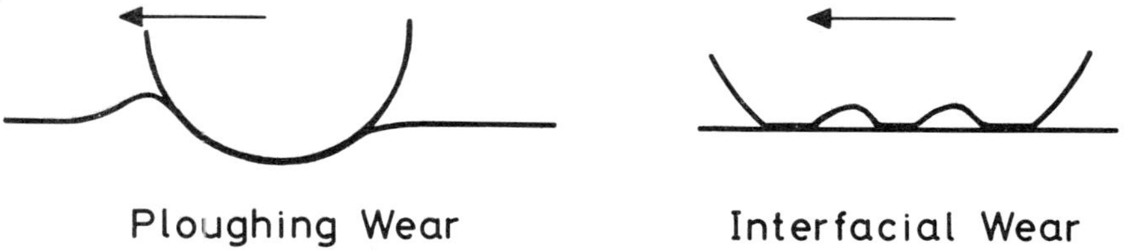

Fig. 1.—Two basic processes or classes of Wear.

The first energy dissipation process involves the ploughing or grooving of hard asperitites through the softer surface; the second process is the shearing of achesive junctions formed in the contact. In general the ploughing mechanism of friction involves relatively large volume deformations, small strains and rates of strain. The second component, the adhesive term involves much larger shear strains and rates of strain in a much thinner interfacial layer. In addition, as virtually all the mechanical work ultimately appears as heat[5], the surface layers are subjected to higher temperatures than the more extensive subsurface zone. For convenience, we may treat wear as arising from either the deformation or the interfacial shearing processes and neglect any mutual interactions. This is clearly an oversimplification and may in some cases lead to artificial distinctions and misleading conclusions.

It is also useful at this stage to classify polymers according to their response to the deformation in the contact regions.[6] This is done briefly in Table I. There are three basic types of deformation properties and failure modes; elastic or viscoelastic with tearing failure; plastic or viscous flow and necking failure; and brittle failure accompanied by fracture. These processes are generally influenced by the temperature and the rate of deformation; in addition several of them are sensitive to the chemical nature of the environment.

DEFORMATION WEAR

I. Elastomers

Schallamach[7-9] has studied the damage produced when a hard slider is dragged over the surface of a rubber specimen. Two types of behaviour are observed. If the slider is needle-like and penetrates deeply into the rubber the rubber surface is stretched as shown in Figure 2a and tensile

TABLE I.—DEFORMATION CHARACTERISTICS OF POLYMERS

Class	Example	T_g*/°C	Temp	State	Failure
Amorphous (un cross-linked)	PS	100	$>T_g$	viscous fluid	-
	PMMA	106	$<T_g$	glassy	Small strains to brittle fracture
Amorphous (cross-linked)	Rubbers		$>T_g$	elastic	High strains to fracture or tearing
	cispolybutadiene	-108			
	transpolybutadiene	-18			
	Resins	decompose below T_g	$<T_g$	glassy	Small strains to fracture
Semi-crystalline	PTFE	-	$>T_g$	plastic	Average (20%) strain to ductile failure (necking)
	PE	-85			
	PP	-			

* The glass transition temperature is a function of pressure and the rate of deformation. The values quoted are for low rates of strain and atmospheric pressure.

failure occurs at approximately right angles to the direction of sliding.[7] When the rubber frees itself from the slider it recovers elastically, and the tears are observed in the surface in the direction of sliding (figure 2b). A small lip is often formed above the level of the original specimen and some rubber may be detached during the sliding process. If on the other hand the slider is relatively blunt and does not penetrate deeply into the rubber, surface damage depends primarily on the frictional traction; this produces a much gentler distortion of the rubber.[10,11] Nevertheless at the rear of the contact region the frictional traction may produce a tensile stress large enough to rupture the rubber (figure 2c), an array of tears is generated at right angles to the direction of motion and a lip of rubber is "teased" out of the specimen ahead of each tear (figure 2d). Virtually no rubber is removed in this process though with subsequent traversals the lips may be torn away (see below). In practical situations these two processes merge into one another and abrasion results from a mixture of both. Evidently abrasion involves surface tearing and the availability of surface flaws. These investigations provide a very helpful approach to our understanding of the way in which rubber surfaces are damaged during abrasion; however they do not specify in detail how the wear fragments are actually detached.

In another set of experiments in which a rubber specimen was slid over fresh garnet abrasive paper, Schallamach showed that there is a close connection between the abrasive wear rate and the bulk tear strength of the rubber.[9] His data are shown schematically in Figure 3 where the velocity dependence of the reciprocal of the abrasive wear is compared with the rate dependence of the ultimate tensile strength. Ignoring the possible effect

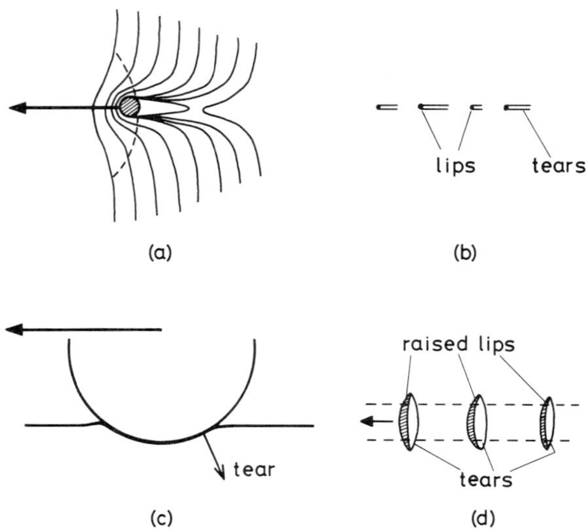

Fig. 2.—(a) and (b) Tearing of rubber by a needle (after Schallamach). (a) Deformation Contours. The rubber fails in tension behind the needle in a direction perpendicular to that of the maximum tensile stress (dotted line). (b) After the needle has passed-on the tear rotates through almost 90°; the surface is covered with intermittent tears parallel to the sliding direction with a small raised lip or some material may actually be removed. (c) and (d) Tearing of rubber by a sphere where surface traction due to friction is the main factor. (c) Profile showing that tear is generated at rear of contact region. (d) The tears are almost normal to the direction of sliding. A raised lip is "teased" out of the surface but no removal of material occurs.

of frictional heating the power law for the two processes is similar. These experiments were described by Schallamach as intrinsic abrasion. They were obtained by periodically changing the relative sliding direction during the wear measurement. If the wear direction is fixed, a rather different type of wear is observed, and "Abrasion Patterns" are observed on the rubber. The patterns appear as striations on the rubber surface normal to the sliding direction. In cross-section they have a saw-tooth geometry with the cutting direction of the saw in the opposite direction to the sliding direction. This unidirectional wear is greater than that observed in "intrinsic abrasion" and Schallamach suggested that it arose from "undercutting" effects. The tractive force simply pulls the teeth over, a crack propagates into the rubber and produces a lip which may be readily detached producing a large wear particle. This is a major source of wear in automotive tires. Recently Champ et al.[12] have studied this process in more detail using the two dimensional analogue (a razor blade) of the original Schallamach needle experiments. These authors modelled the stresses acting on the saw-tooth as a simple "trouser" test specimen under a cyclic load. In this way they were able to establish a relationship between the penetration of surface cracks in a bulk elastomer and the wear rate, the emphasis being on the generation of surface cracks rather than on the mechanism of wear-fragment formation. Lancaster has reviewed a similar approach based on a simplified empirical

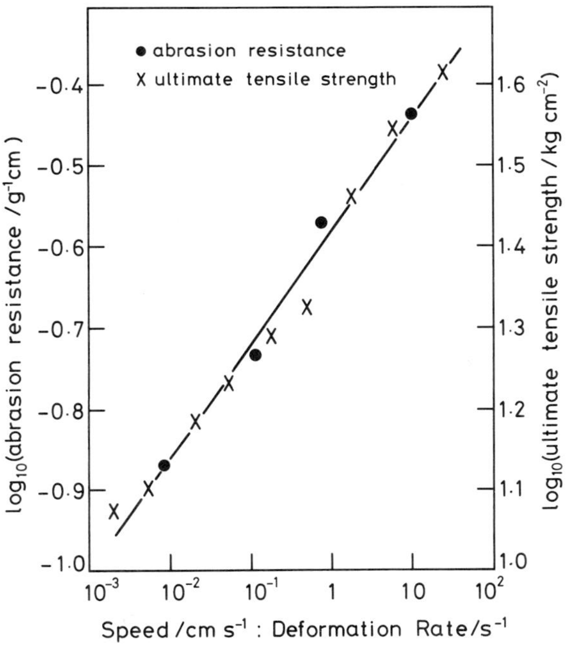

Fig. 3.—Log of Abrasion resistance (reciprocal of the wear rate) or log of U.T.S. plotted against the log of the Sliding velocity or the log of the Deformation rate for SBR rubber (Schallamach). The two sets of data have the same power law.

relationship for describing fatigue used by Kragelskii. As with Schallamach's and Champ's investigations this fatigue model does not specify either the periodicity of the pattern or the precise nature of the detachment process. It is also interesting that natural rubber, which strain crystallises in tearing tests does not appear to strain crystallise under the asperities. Champ et al. argue the rubber that is deformed by the asperities does not have time to crystallise during the transient deformation. It is possible that friction heating also plays a part. As we shall note later this is just one example where our understanding of bulk response is not completely mirrored in the behaviour of the polymer under contact conditions.

To summarize: although an extensive amount of work has been devoted to the abrasive wear and mechanical properties of rubbers the detailed nature of this wear process is still poorly understood. Nevertheless, it appears that this aspect of the wear of polymers is better understood than the other areas reviewed later in this paper.

II. Glassy Polymers

Polymers are relatively brittle at temperatures below their glass transition. This is also the case with highly cross-linked resins which do not exhibit thermally stable phase transitions. Under compressive point loading, however, appreciable plastic flow may occur prior to fracture[(16-18)] and radial and circumferential cracks may be formed. If a tangential stress is applied this greatly facilitates the formation of cracks.[(14-21)] With a typical inorganic glass[(20)], a critical mean Hertz contact pressure is required to create a cone crack. During sliding, however, there is generally an additional tensile component behind the contact and with a coefficient of friction of 0.3 this critical pressure is reduced to about one third of the static value. The onset of cracking in normal loading also depends on the coefficient of friction and the relative moduli of the contacting solids.[(21)] This damage due to tensile stresses in the surface may be regarded as similar to that observed by Schallamach when styli are slid over elastomers. These are fracture processes and are sensitive to stress crazing and cracking agents.[(23-25)] It is also worth noting that a ductile and brittle transition can be induced by the application of a hydrostatic stress.[(26-29)] This has been reported for polystyrene (*PS*)[(25)], polyimide (*PI*)[(27)], Polysulphone (*PSO*)[(28)] and Polymethylmethacrylate (*PMMA*)[(25)]. With *PMMA* and *PS* at room temperature the ductile - brittle transition occurs at 3×10^7 *Pa* in the absence of environmental effects. At low strain rates crazing occurs while at higher strain rates brittle fracture is observed.* In the presence of many fluids, for example silicon fluids, brittle behaviour is maintained up to much higher contact pressures. It is clear that the prediction of brittle failures at the surfaces of polymers, particularly in the presence of fluids, is a very uncertain exercise.[(30)]

Lancaster and Evans[(31)] have made a comprehensive study of polymer-fluid interactions in relation to wear in which they studied the effect of various solvents on the wear of *PMMA* and polyphenylene oxide (*PPO*). Under dry conditions (load *10N*, sliding velocity *ca 0.2 m/s*) the surface of these polymers showed extensive plastic flow indicating appreciable thermal softening. In addition the surface layers showed extensive delamination of the type described by Suh[(32)] and the wear debris consisted of clusters of lamellar plates. No transverse cracking was observed. The combination of high contact temperatures and pressures presumably inhibits brittle failure. However, Lancaster and Evans were able to show how certain liquids increased the wear of the *PMMA* and *PPO* by enhancing their susceptibility to crazing. Other solvents decreased the wear by perhaps plasticising the surface layers of the polymer, Figure 4. In certain fluids transverse striations were observed in the surface of the polymer. These features qualitatively resemble the abrasion patterns observed by Schallamach. These effects were not restricted to crazing and fracture enhancing solvents. The sliding wear of glassy polymers appears to involve extensive plastic deformations with perhaps thermal softening, although there is an indication that crazing and brittle failure may play a role particularly when thermal softening is not important. In contrast to semicrystalline polymers little transfer of polymer to the counterface is observed (see later).

High temperature polymers such as polyphenylene sulphide and epoxy resins might be expected to show more pronounced brittle failure particularly when they are mixed with hard fillers. These materials have been studied in composite form but the nature of the surface failure has not been reported in detail.

---

* The crazing process is still not fully understood but is associated with gentle tensile deformation across the interface. At high rates of strain gross brittle failure occurs and crazing is not so apparent. In rolling contact fatigue crazing is observed in subsurface regions (see later).

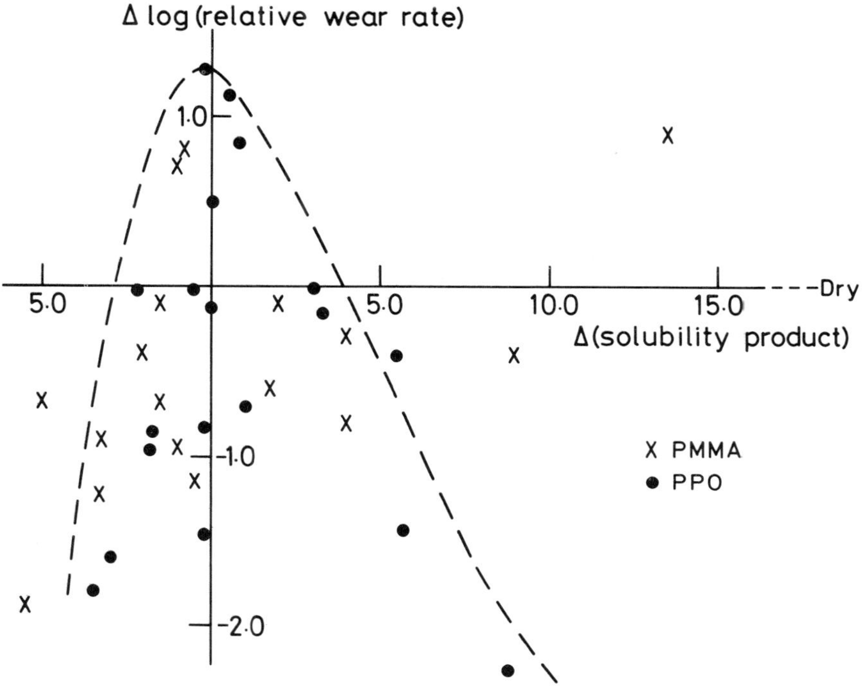

Fig. 4.—The effect of various fluids on the relative wear rate of PMMA. The relative wear rate is the difference between the dry and lubricated wear rates. The fluids are not named but are plotted according to their solubility parameters. The solubility parameter of PMMA is shown. Some fluids increase the wear, presumably by causing crazing, others reduce it, perhaps by plasticising the surface layers of the polymer (after Lancaster and Evans).

Surface tractions and thus excessive surface thermal and tensile stresses can be avoided in rolling contacts. Archard and Wannop[33] have investigated the rolling contact fatigue of PMMA and observed acute subsurfacing softening and subsequent failure at depths below the counterface consistent with calculations based on viscoelastic energy dissipated[34] in the bulk of the polymer beneath the contact. Similar mechanisms lead to the degradation of lignin in the fibrillation of wood pulp[35] and in the blistering of automobile tires.

III. Ductile Polymers

A hard sharp asperity may act as a miniature cutting tool when it is slid over the surface of a ductile polymer: wear produced in this manner may be described as abrasion. A simple theory[36] of abrasive wear may be developed if asperities are modelled by one parameter; their mean slopes, θ. For a conical indenter of slope θ, the frictional force F is

$$F = \frac{2}{\pi} \tan\theta$$

This energy is dissipated in deformation ahead of the asperity. If the area being deformed is, A; then

$$A \simeq 2w \tan\theta/\pi H$$

where H is some time-dependent hardness. If we assume that the volume of the deformed region is proportional to $A^{3/2}$ and that the probability of this volume creating a wear particle is K, then since the number of contact regions of area A per unit distance is proportional to $A^{-\frac{1}{2}}$, the wear per unit sliding distance is proportional to $A^{3/2} \cdot A^{-\frac{1}{2}}$ and we may write:-

$$Z = K\ w\ \tan\theta/H \qquad (1)$$

Equation (1) is a general case if θ is an average slope of the asperities. Whitehouse and Archard[37] have reviewed the significance of the mean slope θ and its relationship to other topological parameters. Tabor[38] has shown that

$$\tan\theta \simeq (\sigma/2R_{av})^{\frac{1}{2}} \quad \text{and} \quad \tan\theta = \sigma/\beta^* \qquad (2)$$

where σ, R_{av}, β^* are respectively the standard deviation of asperity heights, the average radius of the asperities and an autocorrelation distance. Whitehouse has reviewed these parameters and their properties elsewhere in this proceedings.

Before applying equation (1) to published data, a number of points are worth mentioning. First, this model applies to single traversals of asperities over the polymer surface. However, most abrasive experiments of the type we discuss here consist in sliding the polymer over a rough counterface. Thus, although in "single-traversal" or "single-pass" experiments the polymer traverses the counterface only once, many asperity interactions may occur at one and the same region of the polymer so that it may be subjected to fatigue as well as to micro-cutting. Multiple traversals over the same wear track enhance the fatigue process. We shall deal with this later. In the present context the more important feature is that wear debris accumulates on the rough counterface significantly reducing its effectiveness as an abrading surface. Invariably this leads to lower wear. Figure 5 shows this effect for high density polyethylene (*HDPE*) sliding on abrasive papers.[39] Secondly as θ, β^* and R_{av} are generally obtained from surface profilometry the absolute values of θ are often independent of the scale of the roughnesses.[40,41] Finally we note that this approach, like that described for the fatigue wear of elastomers, does not specify the details of the wear mechanism.

Figure 6 shows data from Hollander and Lancaster[42] for the single traversal abrasive wear of nylon, polyethylene and polypropylene. Clearly the wear is proportional to (σ/R_{av}). Recently Warren and Eiss have studied the influence of mean slope and its standard deviation on the high-load single-pass abrasive wear of polyvinylchloride (*PVC*), nylon 6:6 and polychlorotrifluoroethylene. Using scanning electron microscopy they observed that when polymer was transferred to asperities on the counterface, it did so by depositing polymer behind the asperity. The upper surface of this deposit made a reasonably constant angle with the horizontal datum for a given polymer. The greater the angle the less the wear as monitored by neutron activation analysis. This work also confirms the suggestion, originally proposed by Lancaster[44], that there is a minimum asperity slope below which no abrasive wear is observed. In addition these authors also found the now quite well-established relationship between the abrasive wear

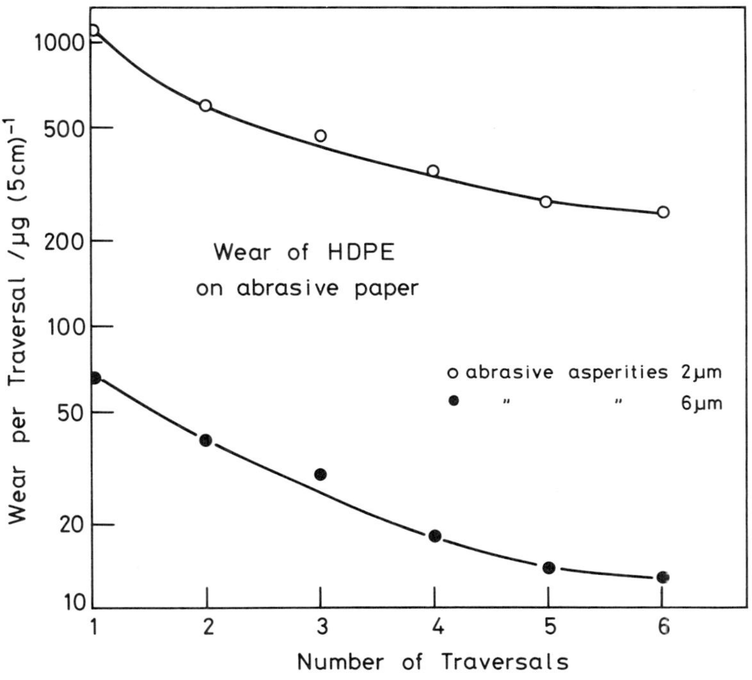

Fig. 5.—Wear against the number of traversals (see text) over the same wear track on abrasive paper (Briscoe, Pogosian and Tabor).

rate and the reciprocal of the work to rupture of the polymer. This correlation was probably first proposed by Ratner and his collaborators.[45] Figure 7 shows data taken from Lancaster[1], where the single pass wear is plotted against the reciprocal of work to rupture for various polymers. The agreement is very good and somewhat surprising when one realizes that no attempt has been made to compare the deformations under comparable conditions of hydrostatic stress, strain rate and with polymers of similar morphology. The K factor in equation (1) may be considered as containing the reciprocal of the work to rupture. It is interesting in view of the general relationship between wear rate and the rupture energy, to note that the failure process is probably similar for many polymers which exhibit ductile abrasive wear. Unfortunately equation (1) also follows from a simple model of brittle fatigue failure as might occur in glasses and rubbers. The different gradients of PMMA and the ductile polymers[42] when wear rate is plotted against σ/R_{av} as shown in Figure 6 might reflect this difference in failure processes. However, in general, it has proved extremely difficult to differentiate between abrasive and fatigue wear. In some cases environmental sensitivity (see earlier) provides an indication of the nature of the failure process. The experiments described above refer to single traversals. Most practical applications involve multiple traversals and in addition many polymers used in practice contain comparatively hard and abrasive fillers. The effective surface topography often changes markedly during the initial stages of wear. Often the transfer of polymer debris and

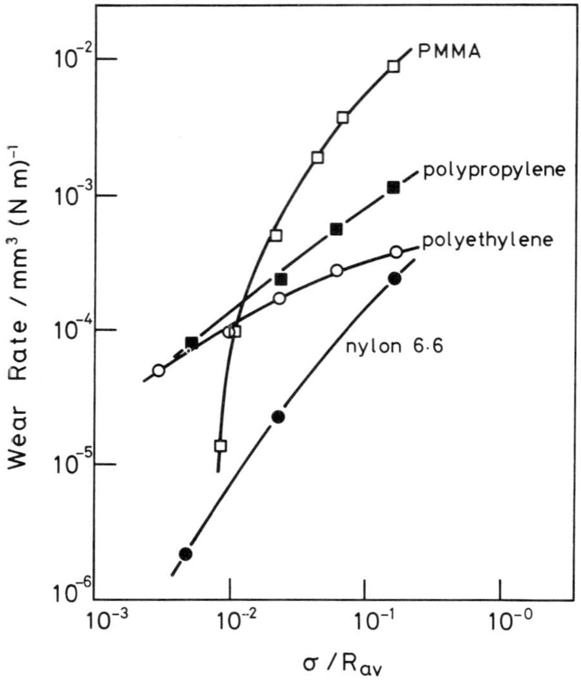

Fig. 6.—The unlubricated wear rate as a function of the mean asperity slope (σ/R_{av}) for nylon, polyethylene, polypropylene and PMMA.

polishing by the fillers reduces the roughness.[46] It has been reported that the wear in a prolonged experiment may be minimised by the choice of a suitable initial counterface roughness. For PTFE this roughness has been quoted as being about 0.5 μm c.l.a. (see later).

A rather different type of 'wear' which involves large volume deformations is caused by gross thermal softening and the subsequent loss of dimensional stability. Tanaka and Uchiyama[47] have made a comprehensive study of surface melting and its influence on the friction and wear of a number of semi-crystalline polymers. For example in the case of high density polyethylene sliding at 1 m/s under a load of 1 Kg against steel the molten layer is about 15 μm in thickness. Similar experiments by Kar and Bahadur[48] show that, at comparable sliding speeds, surface melting occurs with HDPE and polyoxymethylene but not with PTFE. This is not entirely in agreement with an earlier paper by Tanaka et al.[49] on the high-speed transfer of PTFE.

In Tanaka and Uchiyama's work where surface melting of HDPE is well established they were able to show in a semiquantitative way that the frictional force can be accounted for by the viscous flow of the polymer in the molten layer. Two mechanisms of wear are suggested when repeated traversals are made over the same wear track. In the case of low density polyethylene a relatively thick coherent layer of molten polymer is transferred

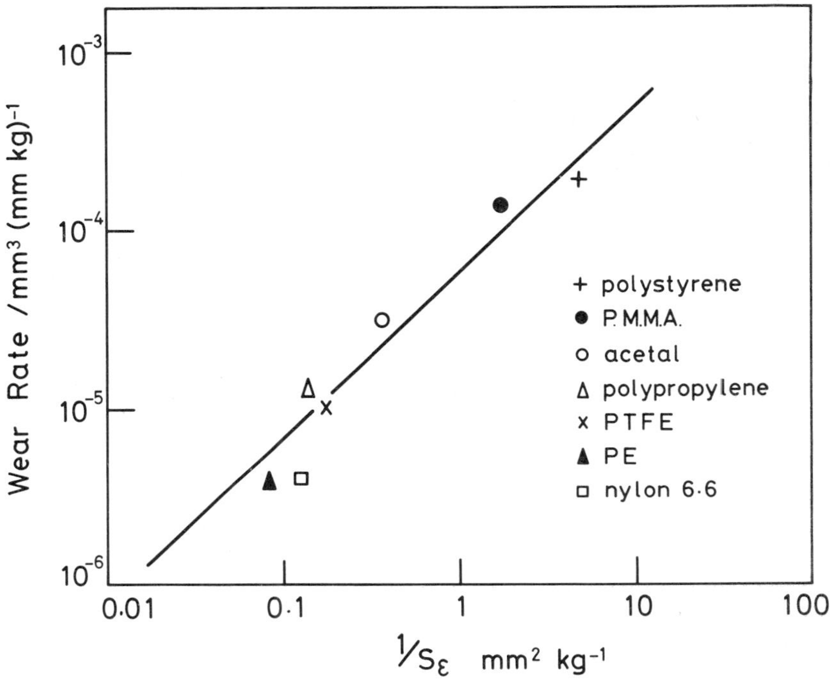

Fig. 7.—Wear rate as a function of the reciprocal of the work to rupture for various ductile polymers.

to the counterface. Other polymers such as high density polyethylene, polyacetal, nylon 6:6, and polypropylene transfer a much thinner molten layer. The first mechanism of wear involves the removal of this layer by subsequent passes of the slider. In this respect it is similar to the mechanism of adhesive wear to be described in the next section although the transferred layers are probably thicker when surface melting takes place. A second novel wear process, tentatively suggested by these workers, involves the extrusion of molten polymer out of the contact region. We have examined this process with a glassy polymer, PMMA, and observed the formation of extruded polymer at the rear of the contact. This extruded material has an interesting layered structure and it is attached to the rear of the polymer specimens (Figure 8).

INTERFACIAL WEAR

In the previous section we have considered energy dissipation processes which occur primarily within the bulk of the polymer beneath the interface. The nature of the countersurface and the surface of the polymer has been considered in the context of surface roughness and thermal conductivity but no direct mention has been made to the adhesion between surfaces except in the case of the tearing of elastomers and the transverse cracking of glass where surface tractions may be involved. Such tractions arise from

Fig. 8.—The wear of PMMA at high sliding speeds. The sliding direction is clockwise. The molten polymer layers accumulate at the rear of the contact and form layered or fibrillar debris.

the adhesion between the surfaces. In these wear processes surface flaws and the surface morphology are also certainly of importance. In this section we will consider those wear processes that may be regarded as mainly interfacial in origin. There seem to be two types: adhesive or transfer wear and chemical wear. These processes involve relatively thin layers of the polymer surface and as a consequence higher strains, strain rates and temperatures are involved than in the regime of deformation wear.

A. "Adhesive" or "Transfer" Wear

A good deal has been written on adhesive wear. In summary, the process involves the formation of adhesive junctions between the polymer and the counterface followed by bulk failure in the surface layers of the polymer in response to the applied shear stress. Polymer is consequently transferred to the counterface and repeated traversals of the polymer specimen over this layer detach the film which is then ultimately displaced from the contact. A further layer is transferred, the process is repeated and the polymer gradually wears away. The process may be rather more complex. For example, there is evidence for multiple film transfer upon previously transferred layers during repeated traversals of the polymer specimen. There are a number of important features in this mechanism that may be discussed separately. They include (I) the nature of the initial adhesion between the polymer and the counterface; (II) the mode of failure at the junction and criteria for film transfer; (III) the structure and thickness of the transferred layer; (IV) the bonding between the transferred film and the counterface; (V) the mechanism of film removal and the displacement of the debris from the contact region. We will deal with these aspects in turn.

I. The initial adhesion between the polymer and the countersurface

With organic polymers the adhesion arises from two sources. One is Coulombic electrostatic forces. The importance of this type of bonding has been the subject of dispute.[50-53] The other process involves van der Waals forces which are present irrespective of the nature of the interacting atoms.[54] There may also be in addition hydrogen bonding between the polymer and oxygenated or hydrated layers on the countersurface. If the two surfaces adhere strongly the work of adhesion may be very much greater than the thermodynamic work of adhesion because of the large amount of work involved in ductile and visco-elastic losses. This applies to polymers and elastomers above their glass transition temperature. With these materials the work to separate the polymer from the interface may be greater than the energy required to tear out a lump of polymer. As a result appreciable transfer may occur. On the other hand, with polymers or rubbers below their glass transition temperature, the adhesive failure is largely brittle and the work of adhesion and the tendency to produce polymer transfer will be very much smaller.

II. Junction failure

Under the application of a tangential stress, the adhesive junctions, as indicated above will fail either at the original interface or within the polymer. With clean elastomers on smooth surfaces the elastomeric junctions appear to fail at the interface itself and no coherent transfer occurs. Some transfer at a submicroscopic level may be formed but most experimental methods are incapable of detecting the polymer itself as distinct from some low molecular weight impurity. This point is reminiscent of the arguments which surround the distinction between cohesive and interfacial failure of adhesives. Elastomers wear by "large" volume fatigue or tearing and sometimes by chemical degradation. Glassy polymers also do not appear to transfer polymer to the counterface except in isolated patches. At high sliding speeds surface heating may increase the interfacial temperatures above the glass transition temperature and then quite gross, but again patchy, transfer is observed.[54] These observations are consistent with the low adhesion of glassy materials. Highly cross-linked resins often appear to form thin coherent transferred layers but, as mentioned above, this material may be low molecular weight impurities which diffuse to the surface or they may be the residue of thermal degradation.

III. Transferring polymers: the nature and structure of the transferred film

Ductile polymers (semi-crystalline polymers above their glass transition temperatures) often transfer bulk polymer to the counterface although during the transfer process there are marked changes in morphology sometimes accompanied by a certain amount of chain scission.[55] When transfer occurs there appear to be two types of behaviour which depend upon the polymer and the sliding conditions. The most general ("normal") behaviour is the formation of a "lumpy", relatively thick, *ca. 0.1* to *1.0 μm*, film which contains quite highly oriented material.[56-65] This behaviour is found in low density polyethylene, polypropylene and nylon 6:6. Good transfer is aided by clean substrates and there is evidence that mild surface roughness *ca 0.5 μm* or less, may improve the transfer.[66,67] However structural studies on these layers are difficult. The most widely used method which is electron diffraction is restricted because of the ease of radiation damage in polymers.

The other class of behaviour is only shown, as far as we know, by high density polyethylene (*HDPE*), *PTFE*, and as the more recent experiments by Kar and Bahadur show by polyoxymethylene.[68] In this case the transferred film may be much thinner *ca. 10 nm* to *50 nm*[60] and more highly orientated with the major axis of the polymer chains aligned in the direction of sliding.[60,69,70] This rather special behaviour of *HDPE*, *PTFE* and *POM* appears to be restricted to low speeds, intermediate temperatures and smooth counterfaces.[57] Outside these, as yet, ill defined limits "normal" lumpy transfer is thought to occur. At higher sliding speeds, as pointed out earlier, there is evidence for frictional melting of *HDPE* and of *POM* but not of *PTFE*.[48]

The mechanism of thin film transfer is not clear. It certainly involves large plastic shear strains and we may say that transfer occurs when the interfacial shear strength is greater than that of the bulk shear strength of the polymer close to the interface. Pooley and Tabor have shown that prior to thin film transfer reorientation of a surface layer of the polymer occurs for *HDPE* and *PTFE*. It is likely that this material will have a lower shear strength than the isotropic polymer.[71,72] Reorientation of surface layers, prior to transfer is not as important with "normal" ductile polymers which show lumpy transfer and higher friction. These observations are consistent with the observed thin film shear strengths of polymer films. For a given set of contact conditions the shear strengths of *HDPE* and *PTFE* are less than those of nylon, low density polyethylene and polypropylene which are again less than those of glassy polymers such as *PMMA* and polystyrene.[72] In the case of the glassy polymers little reorientation is likely and true interfacial sliding may take place.[73] These observations also broadly agree with a number of interesting experiments which have been reported, where various polymer combinations have been slid over one another. Transfer takes place from the polymer with the lowest cohesive energy density[64] or critical surface tension of wetting.[63] In general reorientation and segmental mobility favour transfer and this is reflected in the relatively poor transfer of radiation cross-linked polymers and ultrahigh molecular weight polyethylenes.

In the transfer mechanism energy is dissipated within relatively narrow zones; it is possible that adiabatic shear may be involved. Adiabatic shear bands are seen in certain metals[74] when they are deformed at very high rates of strain and the process is also thought to occur in some polymer fibre drawing processes.[75] The mechanism of transfer seems to have something in common with fibre drawing. Both seem to involve a reorientation of lamellae and the subsequent alignment of amorphous material as well as the pulling out of chains from the crystalline regions. What is difficult to account for in the transfer process is the observed range of film thickness. It may be controlled by the position of adiabatic shear planes and thus related to the thermal properties of the polymers. Tanaka et al.[62], however, have suggested that the thickness may be governed by the lamella size of the crystalline portion of the polymer. In the work of Pooley and Tabor, where they observed film thicknesses much smaller than those of the lamellae, to some extent conflicts with this view. In addition they found that morphology had little effect on the friction and the transfer.

It is evident from this discussion that the factors controlling film thickness are not yet established. The way in which the film is laid down may be determined by some characteristic morphological dimension or by the position of an adiabatic shear plane. What is certainly required for transfer is a fair degree of drawability with its associated requirements of segmental mobility. In the section below we will be considering the use of fillers in improving the wear of polymers. We will note here that fillers, particularly those with high aspect ratios, may significantly hinder the reorientation process as well as changing the morphology of the pclymer. Steward[76] has shown for example, that mica-filled high density polyethylene requires a significantly longer distance to establish a low friction regime than the pure polymer. This suggests that the fillers, with their relatively low mobility, interfere with the alignment of the polymer chain into directions of easy shear. We might guess that these effects might be more important with platelets and fibres. If reorientation is a prerecuisite for transfer, fillers may reduce the rate of transfer as well as modifying the polymer-counterface interactions as outlined below.

IV. The bonding between the transferred layers and the counterface

In the absence of chemical degradation the bonding will be Coulombic and van der Waals. The electrostatic forces in this case seem to be larger than those in (I) as can be seen if a transferred layer is stripped from a counterface. Whether this frictional electrification significantly enhances the adhesion of the film is unknown. However, the adhesion of these films is generally quite poor. The films can often be removed with adhesive tape or during cellulose acetate replication prior to transmission electron microscopy. The films are also highly strained and wetting them with aqueous detergents may detach the film as well as causing it to curl at its edges.[60] It is not surprising that multiple traversals of the polymer can readily displace the film. During these multiple traversals it is not established how much additional film transfer can occur on top of previously deposited films. The work of Pooley and Tabor[60] suggests that this is not an important effect at low speeds with *PTFE* and *HDPE*. At higher speeds where frictional heating and adiabatic shear may be more important, thicker films form with these polymers but it is not clear whether these layers are characteristic of "normal" transfer or multiple thin film transfer. Work in our laboratory tentatively indicates the latter can occur.

Later work[39], suggests that the adhesion of the first layer to the counterface is of critical importance in governing the long-term wear of *HDPE* and *PTFE*. The evidence indicates that certain inorganic fillers help to reduce the wear of polymer by increasing this adhesion. This enhanced adhesion mechanism seems to be restricted to *HDPE* and *PTFE* and the fillers do not seem to significantly reduce the other mechanisms of wear. A number of experiments demonstrate this enhanced film adhesion. First the transferred polymer layer formed against a ferrous counterface by a glass or carbon-filled *PTFE* composite is extremely difficult and often impossible to detach from the counterface by the simple means described above. Second, this transferred layer (or one like it formed from *HDPE* in the presence of lead and copper oxides as fillers) will act as a low wear counterface for *pure HDPE* until the film is eroded away after which the wear rate of pure *HDPE* is observed. Typically these fillers will reduce the wear of *PTFE* and *HDPE* by a factor of 10^3.[39,77] Fillers do not appreciably change the total energy dissipated (the friction force) or markedly alter the thermal properties of the specimen and it has been suggested that they act as stress or temperature intensifiers in the contact. This is consistent with the observation that those fillers which cause abrasion of the counterface are also effective in reducing the wear of the polymer composite.[78] How this local intensification of stress and temperature improves the adhesion of the transferred film is not properly understood. In the case of *PTFE* composites there is no strong evidence for significant chemical effects. However, early work by Pratt[77,79] noted the formation of coloured debris when *PTFE*/Copper/Lead composites were slid against steel. A similar type of debris is seen when mica-filled *PTFE* is slid against lead bronze. Again Pocock and Cadman[80] have studied the thermal stability of *PTFE* composites and detected some chemical reactions that might be responsible for chemical bonding. More recent work by Steward[76] has failed to identify any pronounced catalytic effects using similar composites. On the other hand, in the case of *HDPE* filled with lead and copper oxides the evidence for chemical effects is stronger. This is based on the observed synergistic action of lead and copper oxides in reducing the wear of the polymer. These fillers are thought to promote a mild gas phase oxidation of the polymer. This aspect is dealt with further in the next section. In addition to possible contributions from valence bonding, the fillers acting as temperature intensifiers in the contact zone may locally melt the polymer, annealling out stresses and creating more intimate contact between the film and the counterface.

It should be stated that another school of thought exists on the role which fillers play in reducing wear. Tanaka[81] and Arkles et al.[82] suggest that a filler-rich surface layer is formed at the interface and the particles, which have low wear rates, effectively support the load and separate the polymer from the counterface. The polymer acts as a boundary lubricant. Excess filler concentrations are generally observed at the surface of the composite, and with some types of fillers, the wear of the filler particles can be readily observed, Figure 9. The two mechanisms are not mutually exclusive and both may play a part in the wear process. Other factors may also be important. In an earlier section we noted that there is evidence that a certain scale of surface roughness gives a minimum wear. This roughness may act as the critical stress intensifier which strengthens film adhesion without producing excessive abrasion. In addition, many effective fillers abrade the counterface and the fillers may thus act to provide an optimum surface roughness. Recent work[73] on glass filled compo-

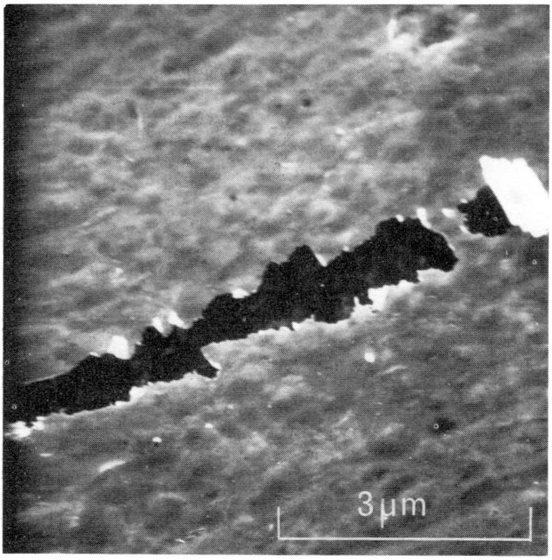

Fig. 9.—The worn surface of a PTFE composite filled with glass spheres. The glass spheres show pronounced wear.

sites indicates that the filler particles can, given sufficient time, produce a counterface roughness close to the optimum value almost irrespective of the initial counterface roughness.

The use of fillers is a valuable method for improving the wear resistance of polymers when they are slid against reasonably smooth surfaces and when their transferred layers are formed on the counterface. These fillers seem to be less useful in reducing deformation wear such as abrasion and fatigue. The mechanism of filler action is not well understood but the requirement of some gentle abrasion of the counterface is a good empirical rule. There do, however, appear to be some exceptions. For example, Arkles et al.[83] have shown that two high temperature polymers, polyphenylene sulphide and polyimide, are very effective fillers for *PTFE*. Although these workers did not report data on counterface abrasion, we presume that it did not occur. Nevertheless, factors such as the concentration, shape, size, aspect ratio and the relative hardness of the filler and the counterface can affect the wear rate. Direct correlation between these variables and the wear rate is often difficult to establish because of the breakdown and accumulation of the filler particles[78,81] in the contact zone and because of the work hardening of the counterface.[84] Many of these composites show pronounced environmental sensitivity which can sometimes be attributed to specific interactions with the filler particles such as the oxidative wear of carbon. In other cases, for example at high humidities, it seems that the mechanism of enhanced adhesion is severely inhibited. The development of polymer bearings which can operate effectively in aqueous media is an area of renewed research interest. Unfortunately the film transferring polymers which are used as dry bearings and sometimes as marine bearings are intrinsically unsuitable for this purpose.

V. The mechanism of film removal

Little systematic work has been carried out on this complex problem. In some cases patches of the film are pulled out of the surface of the transferred layer by the tractive stress[39], Figures 10 and 11. This film debris can sometimes be observed adhering to the surface of the polymer. This process of "back transfer" is important and it is often responsible for the influence of contact geometry on the wear rate. This is particularly so when the debris cannot be readily displaced from the contact region as in, for example, a journal bearing. Ahroni[85] has suggested that the detached layers may form into rolls prior to displacement. This behaviour may be seen in the micrograph of a *PTFE* transferred film, Figure 11. Roll formation was originally proposed as a mechanism for the wear of rubbers by Reznikouski and Brodskii[86], but this may be the result rather than the cause of wear. Fatigue followed by delamination[32] may be a useful model for the detachment process but clearly the removal process is extremely complex.

B. Chemical Wear

High temperatures are generated at sliding contacts. Ductile polymers such as *HDPE*, *PTFE*, nylon and polypropylene and brittle polymers such as *PMMA* and polystyrene melt or soften without significant chemical decomposition. Resins, certain high polymers and elastomers suffer appreciable chemical degradation without becoming viscous. For these materials chemical wear is important. For example we have all seen the blue smoke evolved from automobile tires. Here some softening occurs, as is evidenced by the transfer of rubber to the road.

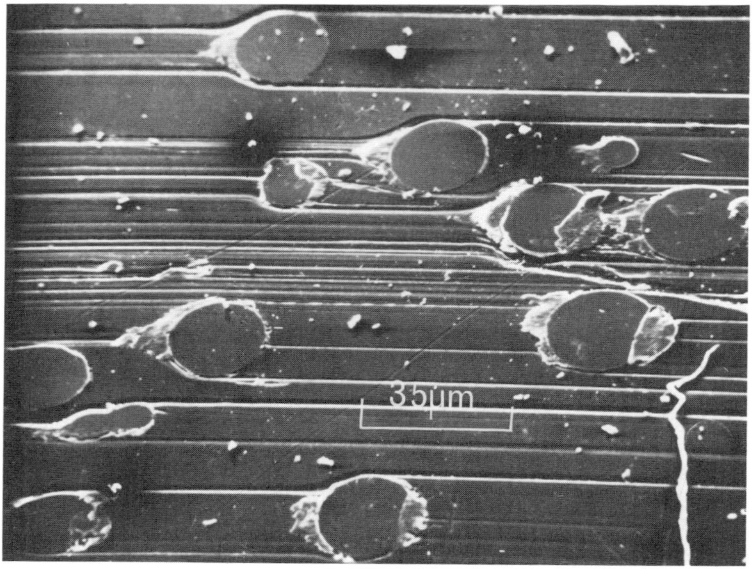

Fig. 10.—The buckling of a PTFE transferred layer. The layer was formed by a carbon-filled PTFE composite sliding against steel. Sliding direction left to right.

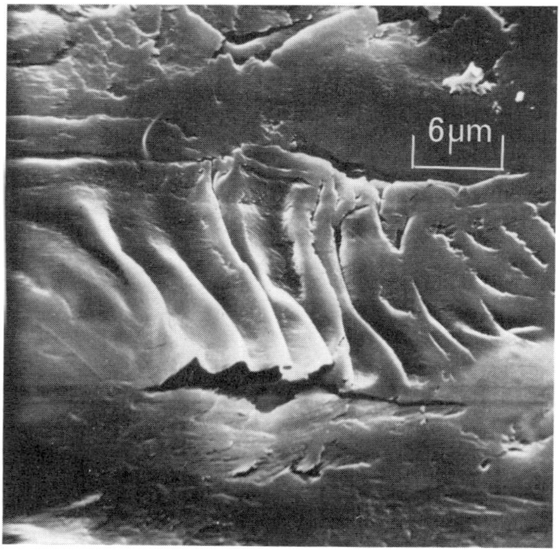

Fig. 11.—A lesion in a transferred film presumably caused by abrasion. A roll of film is formed at the right of the picture.

Systematic studies of chemical wear are few and they take different forms. Richardson[87] has briefly reviewed this field. Rhee and his collaborators[88] have applied the approach established by Quinn for the oxidative wear of ferrous metals to the study of the chemical wear of a class of brake materials based on high temperature resins. The contact temperature may be estimated as a function of the sliding variables and the chemical wear rate is then described by an Arrhenius equation. The activation energies calculated correspond to those typical of chemical degradation. A rather different type of experiment involves sliding polymers and composites in vacuum and monitoring the concentration and type of gaseous species evolved using mass spectroscopy. Buckley and Johnson[89] have studied the chemical decomposition of filled and pure *PTFE* in sliding contacts using Auger and mass spectoscopy. With the pure polymer the major products were CF^+ and CF_2^+ ions. With the glass filled polymer the major product was F^+ indicating that higher temperatures and stresses existed in the contact. This is in agreement with the intensification-of-stress arguments described earlier. Copper fillers produce much lower concentrations of CF^+ and CF_2^+ indicating that there is less stress concentration with this filler. Copper-filled *PTFE* has a higher wear rate than that of glass-filled *PTFE* at similar filler concentrations. Professor Belyi's laboratory in Gomel has carried out similar experiments but on less stable polymers such as substituted polycarbonates and polyalkylmethacrylates. They suggest that the gas evolved in these cases may provide effective "gas" lubrication of the polymer. More recent work[90] on *PTFE* shows that the wear by *chemical decomposition* may be very large in vacuum and the loss of material may be comparable with the *adhesive wear* observed under open laboratory conditions. Apparently the catalytic activity of clean metal counterfaces can significantly increase chemical decomposition. This type of catalytic decomposition may be responsible for the high rates of wear that are observed with filled *PTFE* when sliding occurs in very dry pure inert gases.[91] Here presumably the protective oxide films naturally present on the counterface are worn away.

An alternative approach which provides some insight into chemical wear is to examine the thermal decomposition of polymers and composites in various environments using differential scanning calorimetry (DSC) and gravimetric analysis. For example, Hauser[92] has shown that the particular combination of lead oxide and copper oxide fillers which gives a minimum wear of *HDPE* also promotes a mild low temperature decomposition of the polymer. Whereas the role of the lead oxide is uncertain, it seems that copper, or one of its oxides, acts as a catalyst for the gas phase oxidation of polyethylene. For example, copper is known to catalyse the oxidation of polyethylene-coated copper conductors.[93] Similarly Pocock and Cadman[80] used DSC to investigate the oxidation of *PTFE* and showed that copper had a catalytic effect. In contrast Steward[76] using a thermal balance has found little enhanced reactivity with glass and carbon fillers though they are very effective in reducing the wear of *PTFE*. Clearly these DSC and thermogravimetric studies must be interpreted with some caution. The role of clean surfaces and their catalytic effects may render invalid any comparison between static tests and sliding wear experiments. In addition it is possible that large stresses, or rather large strains, may themselves reduce the activation energy for the decomposition[94] at the interface.

Various workers[95] have indicated that mild oxidation can improve the adhesion of polymers to metals. This may well account for the enhanced adhesion of transferred layers in the presence of fillers. The evidence for chemical effects in improving the adhesion of transferred films is still somewhat inconclusive.

CONCLUSIONS

The wear of polymers involves a variety of complex and interacting processes. It is at least as complex as the wear of metals and perhaps more so because of the wide range of mechanical properties found in polymers, their marked dependence on temperature and strain rate and the pronounced environmental sensitivity of their failure processes. In another sense they are less complex because their surfaces are less susceptible to chemical and mechanical damage. In comparison with metals and high strength solids their wear is thus more influenced by mechanical than surface properties. This is one of their virtues in dry bearing applications.

At a fundamental level we can often account for the wear of polymers by considering their mechanical failure characteristics in bulk specimens. We can, for example, correlate the wear of rubbers with their fatigue life. The energy input during sliding creates a certain amount of fatigue damage. However the connection between damage and weight debris cannot be given. We cannot for example, predict the size of the wear particles. The same situation holds for abrasive and chemical wear of brittle and ductile polymers. The mechanism of transfer wear is even less well developed. The subject of polymer wear is thus more of a speculative art than a definitive science.

REFERENCES

1. Lancaster, J.K., in "Polymer Science," edited by A.D. Jenkins, North Holland Publishing Company, London, 1972, Chapter 14.
2. Bowden, F.P. and Tabor, D., "Friction and Lubrication of Solids," Clarendon Press, Oxford, 1950, Part I; 1964, Part II.
3. Briscoe, B.J. and Tabor, D., in "Polymer Surfaces," edited by D. Clark and J. Feast, John Wiley and Sons, New York, 1978, p. 1.
4. Briscoe, B.J. and Tabor, D., *British Polymer Journal,* Vol. 10, 1978, p. 74.
5. Rumford, Count Benjamin, *Journal de Chemie,* Vol. 1, 1798, p. 9.
6. See, for example, Ward, I.M., "The Mechanical Properties of Polymers," John Wiley and Sons, London, 1972.
7. Schallamach, A., *Journal of Polymer Science,* Vol. 9, 1952, p. 385.
8. Schallamach, A., in "The Chemistry and Physics of Rubber-like Substances," edited by L. Bateman, MacLaren and Sons, London, 1963, Chapter 13.
9. Schallamach, A., *Wear,* Vol. 1, 1958, p. 384.
10. Ratner, S.B. and Ferberova, I.I., *Soviet Plastics,* Vol. 9, 1960, p. 51.
11. Reznikovskii, M.M. and Brodskii, G.I., in "Abrasion of Rubbers," edited by D.I. James, MacLaren and Sons, London, 1967, p. 14.
12. Champ, D.H., Southern, E. and Thomas, A.G., in "Advances in Polymer Friction and Wear," edited by H. Lee, Vol. 5A of *Polymer Science and Technology,* Plenum Press, New York, 1974, p. 133.
13. Kraghelskii, I.V. and Napomnyashchii, E.F., *Wear,* Vol. 8, 1965, p. 303.
14. Schallamach, A., *Journal of Applied Polymer Science,* Vol. 12, 1968, p. 281.
15. James, D.I., ed., "Abrasion of Rubbers," MacLaren and Sons, London, 1967.
16. Bartenev, S.M., Razumouskaya, I.V. and Sanditov, D.S., *Journal of Non-Crystalline Solids,* Vol. 1, 1969, p. 388.
17. Puttick, K.E., Smith, L.S.A. and Miller, L.E., *Journal of Physics D: Applied Physics,* Vol. 10, 1977, p. 617.
18. Hirst, W. and Howes, M.G.J.W., *Royal Society of London. Proceedings.* Series A, Vol. 311, 1969, p. 429.

19. Gilroy, D.R. and Hirst, W., *Journal of Physics D: Applied Physics*, Vol. 3, 1969, p. 1784.
20. Billinghirst, P.R., Brooks, C.A. and Tabor, D., in Proceedings of the Conference on the Physical Basis of Yield and Fracture, Oxford, edited by A.C. Stickland, Institute of Physics and Physical Society, London, 1966, p. 253.
21. Powell, B.D. and Tabor, D., *Journal of Physics D: Applied Physics*, Vol. 3, 1970, p. 783.
22. Johnson, K.L., O'Connor, J.J. and Woodward, A.C., *Royal Society of London. Proceedings*. Series A, Vol. 334, 1973, p. 95.
23. Rehbinder, P. and Shchukin, R., *Progress in Surface Science*, Vol. 3, Part 2, 1974.
24. Westwood, A.R.C., *Journal of Materials Science*, Vol. 9, 1974, p. 1871.
25. Matsushige, K., Radcliffe, S.V. and Baer, E., *Journal of Materials Science*, Vol. 10, 1975, p. 833.
26. Pugh, H.L.D., Chandler, E.F., Holliday, L. and Mann, J., *Polymer Engineering and Science*, Vol. 11, 1971, p. 463.
27. Bhateja, S.K. and Pae, K.D., *Journal of Polymer Science. Part B: Polymer Letters*, Vol. 10, 1972, p. 531.
28. Sauer, J.A., Bhateja, S.K. and Pae, K.D., Proceedings-3rd Inter-American Conference on Materials Technology, Rio de Janeiro, Brazil, 1972.
29. Jones Parry, E. and Tabor, D., *Journal of Materials Science*, Vol. 8, 1973, p. 1510.
30. Kambour, R.P., *Journal of Polymer Science, Part D: Macromolecular Reviews*, Vol. 7, 1973, p. 1.
31. Evans, D.C. and Lancaster, J.K., "Polymer-Fluid Interactions in Relation to Wear," Royal Aircraft Establishment Technical Report 76099, Farnborough, U.K., 1976.
32. Suh, N.P., *Wear*, Vol. 25, 1973, p. 111.
33. Wannop, G.L. and Archard, J., *Institution of Mechanical Engineers. Proceedings*, Vol. 187, 1973, p. 615.
34. Greenwood, J.A., Minshall, H. and Tabor, D., *Royal Society of London. Proceedings*. Series A, Vol. 259, 1961, p. 408.
35. Atack, D. and May, W.D., *Pulp & Paper Magazine of Canada*, Vol. 59, No. C, 1958, p. 265.
36. Kraghelskii, I.V., "Friction and Wear," Butterworths, London, 1965.
37. Whitehouse, D. and Archard, J., *Royal Society of London. Proceedings*. Series A, Vol. 316, 1970, p. 97.
38. Tabor, D., *Wear*, Vol. 32, 1975, p. 269.
39. Briscoe, B.J., Pogosian, A. and Tabor, D., *Wear*, Vol. 27, 1974, p. 19.
40. Whitehouse, D., Private Communication.
41. Briggs, G.A.D. and Briscoe, B.J., *Journal of Physics D: Applied Physics*, Vol. 10, 1977, p. 2453.
42. Hollander, D.E. and Lancaster, J., *Wear*, Vol. 25, 1973, p. 155.
43. Warren, J.H. and Eiss, N.S., Jr., in "Wear of Materials, 1977," edited by W.A. Glaeser, K.C. Ludema and S.K. Rhee, American Society of Mechanical Engineers, New York, 1977, p. 494.
44. Lancaster, J.K., *Plastics & Polymers*, Vol. 41, 1973, p. 297.
45. Ratner, S.B., Farberova, I.I., Radyukervick, O.V. and Lure, E.G., *Soviet Plastics*, Vol. 7, 1964, p. 37.
46. Lancaster, J.K. and Ciltrow, J.P., *Wear*, Vol. 16, 1970, p. 357.
47. Tanaka, K. and Uchiyama, Y., in "Advances in Polymer Friction and Wear," edited by H. Lee, Vol. 5B of *Polymer Science and Technology*, Plenum Press, New York, 1974, p. 499.
48. Kar, M.K. and Bahadur, S., in "Wear of Materials, 1977," edited by W.A. Glaeser, K.C. Ludema and S.K. Rhee, American Society of Mechanical Engineers, New York, 1977, p. 501.
49. Tanaka, K., Uchiyama, Y. and Toyooka, S., *Wear*, Vol. 23, 1973, p. 153.

50. Derjaguin, B.V., *Research (London)*, Vol. 8, 1955, p. 70.
51. Skinner, S.M., Savage, R.L. and Rutzler, J.E., *Journal of Applied Physics*, Vol. 24, 1953, p. 438.
52. Von Harrach, H.G. and Chapman, B.N., in Proceedings of the International Conference on Thin Films, Venice, Elsevier, Amsterdam, 1972, Vol. II, p. 157.
53. See, for example, Tabor, D., *Journal of Colloid and Interface Science*, Vol. 58, 1977, p. 2.
54. Steward, M.D. and Briscoe, B.J., Unpublished Data.
55. Arkles, B.C. and Schireson, M.J., *Wear*, Vol. 39, 1976, p. 177.
56. Belyi, V.A., Sviridyonok, A.I., Petrokovetz, A.I. and Savkin, V.G., *Nauka i Tekhnika*, 1976, p. 3.
57. Makinson, K.R. and Tabor, D., *Royal Society of London. Proceedings*. Series A, Vol. 281, 1964, p. 49.
58. Bowers, R.C., Clinton, W.C. and Zisman, W.A., *Lubrication Engineering*, Vol. 9, 1953, p. 204.
59. Steijn, R.P., *Wear*, Vol. 12, 1968, p. 193.
60. Pooley, C.M. and Tabor, D., *Royal Society of London. Proceedings*. Series A, Vol. 329, 1972, p. 251.
61. Bowers, R.C., Clinton, W.C. and Zisman, W.A., *Modern Plastics*, Vol. 31, No. 6, 1954, p. 131.
62. Tanaka, K., Uchiyama, Y. and Toyooka, S., *Wear*, Vol. 23, 1973, p. 153.
63. Sviridyonok, A.I., Belyi, V.A., Smurugov, V.A. and Savkin, V.G., *Wear*, Vol. 25, 1973, p. 301.
64. Jain, V.K. and Bahadur, S., in "Wear of Materials, 1977," edited by W.A. Glaeser, K.C. Ludema and S.K. Rhee, American Society of Mechanical Engineers, New York, 1977, p. 487.
65. Tanaka, K. and Uchiyama, Y., in "Advances in Polymer Friction and Wear," edited by H. Lee, Vol. 5B of *Polymer Science and Technology*, Plenum Press, New York, 1974, p. 499.
65a. Rhee, S.K. and Ludema, K.C., in "Wear of Materials, 1977," edited by W.A. Glaeser, K.C. Ludema and S.K. Rhee, American Society of Mechanical Engineers, New York, 1977, p. 482.
65b. Belyi, V.A., Kragelskii, I.V., Savkin, V.G. and Sviridyonok, A.I., in "Wear of Materials, 1977," edited by W.A. Glaeser, K.C. Ludema and S.K. Rhee, American Society of Mechanical Engineers, New York, 1977, p. 532.
66. Swikert, M.A. and Johnson, R.L., "Simulated Studies of Wear and Friction in Total Hip Prostheses Components with Various Ball Sizes and Surface Finishes," NASA, TN D-8174, March 1976.
67. Dowson, D., Challan, J.H., Holmes, K. and Atkinson, J.R., in Leeds-Lyon Symposium on Tribology, Leeds, September 1976, Paper 4 (iv), p. 99.
68. Kar, M.K. and Bahadur, S., *Wear*, Vol. 30, 1974, p. 337.
69. Bonfield, W., Edwards, B.C. and Markham, A.J., *Wear*, Vol. 37, 1976, p. 383.
70. Briscoe, B.J., Pooley, C.M. and Tabor, D., in "Advances in Polymer Friction and Wear," edited by H. Lee, Vol. 5A of *Polymer Science and Technology*, Plenum Press, New York, 1974, p. 191.
71. Briscoe, B.J. and Tabor, D., *American Chemical Society Preprints*, Vol. 21, No. 1, September 1976, p. 10.
72. Briscoe, B.J. and Tabor, D., *Journal of Adhesion*, Vol. 9, 1978, p. 145.
73. Briscoe, B.J. and Tabor, D., *American Society of Lubrication Engineers. Transactions*, Vol. 17, 1974, p. 158.
74. Winter, R. and Field, J.E., *Royal Society of London. Proceedings*. Series A, Vol. 343, 1975, p. 399.
75. See, for example, Morton, W.E. and Hearle, J.W.S., "Physical Properties of Textile Fibres," Heinemann, London, 1975.
76. Steward, M.D. and Briscoe, B.J., Unpublished Data.

77. Pratt, G.C., "Recent Developments in PTFE-Based Dry Bearing Materials and Treatments," Proceedings-5th Lubrication and Wear Convention, Institution of Mechanical Engineers, London, 1967.
78. Briscoe, B.J., Steward, M.D. and Groszek, A., *Wear,* Vol. 42, 1977, p. 99.
79. Pratt, G.C., in "Transactions and Journal," Plastics Institute, London, Vol. 32, 1964, p. 255.
80. Pocock, G. and Cadman, P., *Wear,* Vol. 37, 1976, p. 129.
81. Tanaka, K., in "Wear of Materials, 1977," edited by W.A. Glaeser, K.C. Ludema and S.K. Rhee, American Society of Mechanical Engineers, New York, 1977, p. 532.
82. Speerschneider, C.J. and Li, C.H., *Wear,* Vol. 5, 1962, p. 392.
83. Arkles, B., Theberge, J. and Schireson, M., *American Society of Lubrication Engineers. Transactions,* Vol. 33, 1977, p. 33.
84. Lancaster, J.K., *Wear,* Vol. 20, 1972, p. 315.
85. Aharono, S.M., *Wear,* Vol. 25, 1973, p. 309.
86. Reznikouskii, M.M. and Brodskii, G.I., in "Abrasion of Rubbers," edited by D.I. James, MacLaren and Sons, London, 1967.
87. Richardson, M.O.W., in "Advances in Polymer Friction and Wear," edited by H. Lee, Vol. 5B of *Polymer Science and Technology,* Plenum Press, New York, 1974, p. 787; see also *Wear,* Vol. 17, 1971, p. 89.
88. Liu, T. and Rhee, S.K., in "Wear of Materials, 1977," edited by W.A. Glaeser, K.C. Ludema and S.K. Rhee, American Society of Mechanical Engineers, New York, 1977, p. 552; and references therein.
89. Buckley, D.H. and Johnson, R.L., NASA Report 69060, 1969.
90. Wilkins, W. and Kranz, O., *Wear,* Vol. 15, 1970, p. 215.
91. Summers-Smith, D. and Hillman, B., Private Communication.
92. Hauser, H., Private Communication.
93. Chan, M.G. and Allara, D.L., *Journal of Colloid and Interface Science,* Vol. 47, 1974, p. 697.
94. See Davis, L.A., Baughman, R.H. and Pampillo, C.A., *Journal of Polymer Science. Part A-2: Polymer Physics,* Vol. 11, 1973, p. 2441.
95. Packham, D.E., "The Adhesion of Polyethylene to High-Energy Surfaces," Adhesion Conference, City University, London, March 19, 1970.

DISCUSSION

G. HARDY, Celanese Research Company: I believe it is implicit in the excellent papers I just heard that we should pay more attention to the fact that the mechanical and other properties of high polymers are extraordinarily sensitive to temperature in a range that very often does not concern the metallurgists. These effects can be of great importance because the properties of the material at the interface where the sliding is actually occurring will be quite different from the bulk properties. In many cases, very thin layers of adsorbed materials on the surface of a polymer (even on a metal) can affect its frictional behavior. The small temperature rise that occurs during rubbing is going to desorb these layers and perhaps wetting and adhesion can occur where it would not have otherwise. More importantly these polymers will change their rheological properties drastically with a very small temperature rise. They will melt if they are semi-crystalline and when they resolidify again they maynot be in the crystalline state they were in before melting. We will have a material of rubbery properties, quite different from semi-crystalline polymer we started with. Finally there are orientation effects in both crystalline and non-crystalline polymers. If the plane of orientation of the high polymer is parallel to the gross interface, it leads to a substantial drop in the energy for fracture. It is common to observe a preferred fracture plane parallel to the interface due to the tractions that are exerted during sliding. I think that all of these items have to be considered very carefully before we can interpret the results of polymers sliding against each other or sliding against metals.

B. J. BRISCOE: I agree.

S. GANESH, Bendix Research Laboratory: You showed a plot for high density polyethylene where drastic reduction in the wear rate with the addition of 30 percent lead oxide and 5 percent copper oxide was observed. Is this improvement independent of the counterface material? Is it independent of the load, sliding speed and so forth? What is the mechanism by which you achieve this improvement?

BRISCOE: We spent a great deal of time trying to answer problems of this type. The choice of the filler depends on the counterface against which the polymer is slid, the sliding conditions, and particularly the environment and the surface topography. This material was modeled on a commercial material. It is very effective for rubbing against smooth, ferrous surfaces at low humidities. That particular composite batch is not produced any more because it would not work on rough surfaces at intermediate humidities.

As regards the mechanism of wear, the answer is: we do not know. In my talk I tried to make the point that the fillers in some way increase the energy dissipation locally within the transferred film, and thus enhance the adhesion of the transferred film to the counterface thereby reducing wear. What is the mechanism of that process? Is it simply melting? Reduction in the area of contact? Reduction in the stored elastic strain? Or is it perhaps a modification of the surface topography? We do not have the answers. We know from adhesive technology that if we can create specific valance bonds between polymer and substrates by mild oxidation, for example, we can enhance the strength of the adhesive junction. With that th[illegible]ng, one wonders whether the fillers can modify locally the chemistry of the [illegible]ce of the transferred film and the substrate to generate strong valance [illegible]mical bonds. This work was put together because we considered that copper acts as catalyst for the oxidation of the high density polyethylene, and lead is an agent for the transport of oxygen. In that way we hoped to establish, by changing the environment and the relative ratio of these materials, strong chemical effects. They must not be too large or we will get gross chemical degradation. If the temperatures are too high one will catalytically corrode the material. In a companion paper by Dr. Cadman, et al. hypotheses of this sort of mechanism are tested. They used very sophisticated analytical tools to prove or disprove the point that there is local degradation of the polymer which might in some way facilitate the enhanced adhesion. The fillers intensify the energy dissipation, increase the temperature, and in that way, accelerate mild chemical reaction which can improve the adhesion of the film.

D. A. RIGNEY, Ohio State University: Could you tell us something about what is going on below the surface? Could you compare what you see in polymers with what is seen in the case of metals?

D. TABOR: The answer really depends on the system. There are systems in which you get subsurface damage due to the generation of high shear stresses just as you do beneath the Vickers and Brinell indentations. Underneath those regions we get high shear stresses and that will produce heating which will produce failure. For example, Archard has shown that in the rolling of polymer discs, we can actually get subsurface damage due to energy dissipation at the region of maximum shear stress below the surface. The other type of experiment, of course, is by Tanaka; he showed that at high sliding speeds, we can get surface melting and the smearing of a relatively thick film on the counterface. Therefore, the answer depends on the system. We can get quite appreciable damage extending beyond the surface layers.

S. A. KARPE, DTNSRDC, Annapolis: I would like to make a comment based on some of the work I have done on the friction and wear of polyethlenes. In some of the figures presented by the authors we see just P.E. We heard reference to pure High Density Polyethylene. In many of the papers in the literature there is very little description of what material is used. Polyethylene, or polyethylene-like can be as different as stainless steel and mild steel. I suggest that we should identify the polymer carefully.

Now a question. I would like to know the influence of mechanical interlocking at the sliding interface instead of pure London or van der Waal forces, on film transfer. What surface finish is best suited for minimizing wear?

BRISCOE: I agree that it would be desirable to specify more carefully not only the molecular weight, but also the molecular weight distribution of the material and perhaps its thermal and strain history, orientation, and so forth. These are certainly important. I think, in addition, it is very important to specify the nominal purity of the material, as many materials contain oxidation inhibitors. Clearly, the presence of an oxidation inhibitor may well modify the chemistry at the interface. To answer your question, our opinion is that the molecular weight influences dynamic response of the material. Polyethylene is a good example. At low molecular weight, it is essentially a ductile material. If one radiation cross-links it, or uses ultra-high molecular weight materials, because of the entanglements in the long-range order, one would turn it into a material of rather different mechanical properties and it would show some features of brittle response.

KARPE: I would like to say one thing about the brittleness of ultra-high molecular weight polyethylene which has a molecular weight of about a million: it is a ductile material, not brittle.

BRISCOE: I said that we changed its features from ductile towards brittle. In this context the polymers do not behave with wholly brittle response. What I am saying is that it will in some way modify the transfer process. It is more difficult for the material to show the necessary segmental mobility to allow transfer and adhesion to occur. Certainly one would not expect to go into brittle regimes of wear.

TABOR: The only bit of work I know which has been concerned with molecular structures in relation to the frictional properties is this. We did some work, not on high density polyethylene but on PTEE. The interesting thing we found was that PTFE has a capacity for adhering to clean surfaces not too strongly, but just strongly enough to draw out a very thin film which is the order of perhaps 20 Å thick in which the molecules are all oriented in the direction of sliding parallel to the surface. Now this film is the one on which subsequent sliding occurs and gives very low friction. We then tried some changes both in molecular weight and in the chemistry of the PTFE. We put in bulky side groups -- such as hexafluoropropylene and so on. As soon as we put bulky side groups into the chain, we found that we lost the ability to produce the thin transferred film and low friction. Apparently what was important in the low frictional properties of PTFE is the ease of drawing of the molecular chains. It looked as if a smooth molecular profile was an important feature in providing a material that gave thin film transfer and low friction. We have not done that systematically with polyethylene, but certainly the low density polyethylene has straggling side groups. It shows high friction and transfer in lumps. As soon as you get rid of those straggling side groups you get a material which gives you smooth transfer and relatively low friction.

N. HOLLAND, CTI Cryogenics: I am concerned with a very practical problem, namely, wear and friction characteristics in a reciprocating seal in an inert atmosphere such as helium gas. Was any research done in this field? If so, would you please comment on it?

BRISCOE: This is a very interesting area of research. The reason is that these materials -- polymers and their composites -- are often used in conditions in whcih there are no lubricants. One does not want to use a metal, for example, when he is pumping pure rare gasses. These environments have very low oxygen concentrations and very low humidities. A number of cases are recorded in the commercial literature that are not in the scientific literature. When sliding systems are inert, that is, any oxygen or water vapor has been scavenged from the system, there is a pronounced increase in the wear of the material. There is much speculation as to whether this can actually be generated in a laboratory experiment. We have not been able to do it. My opinion is that in equipment such as pumps and gas compressors working for protracted periods in the presence of dry argon, active ingredients such as oxygen and water vapor are scavenged from the system. One approaches the situation which is very similar to ultra high vacuum with clean surfaces. Transition from transfer mode to a gross form of chemical corrosive wear of the material is possible. The amount of debris is far more severe in the case of filled systems than in the unfilled systems. This supports the argument that temperature and stress in the contact are intensified by the fillers. When the surfaces are too clean and too active, degradation of the polymer takes place.

MECHANISM OF FILM FORMATION IN POLYMER-METAL SLIDING

M. K. Kar and S. Bahadur

ABSTRACT

Sliding experiments between a polymer disc with radial abrasion marks and a steel indentor were performed at a low speed to study the mechanism of film formation at the rubbing surface. The extraction replicas from the sliding tracks were examined in a transmission electron microscope. It was found that thin films were stretched across abrasion grooves in case of the semicrystalline polymer polypropylene. In the case of amorphous polycarbonate, however, fragmented films were found on the polymer sliding surface. Electron diffraction revealed that the films belonged to the respective polymeric materials. On the basis of these studies along with others reported earlier by the authors, the possible wear mechanisms are suggested for low speed sliding. These involve shearing across amorphous regions between lamellae and drawing of the latter into thin films for semicrystalline polymers and deformation of nodular structure for amorphous polymer.

The possibility of softening of the polymer surface during high speed sliding is also discussed.

INTRODUCTION

In the sliding of polymers against metal and glass surfaces, the wear of polymers seems to be dominated by the transfer of softer polymeric material to the harder counterface material. The material transferred may be in the form of lumps or thin films, as reported by several workers.[1-6] The transfer of Polytetrafluoroethylene (*PTFE*) to clean glass surfaces has been reported by Makinson and Tabor[1] and Pooley and Tabor.[2] The latter workers also observed the lumps of high density polyethylene on sliding steel surfaces. Among others who detected that thin *PTFE* films are produced in sliding were Bowers et al.[3] and Steijn.[4] They used transmission electron microscopy and electron diffraction to identify the films. Brainard and Buckley[5] have also observed using Auger emission spectroscopy the transfer of a uniform and continuous film of *PTFE* to a metal surface. According to Tanaka et al.[6], the formation of thin *PTFE* films is due to the slippage between crystalline slices of the banded structure of *PTFE*. Tanaka and Miyata[7] have also reported the presence of thin films of semicrystalline polymers other than *PTFE* when sliding was performed between abraded surfaces of these polymers and a clean glass plate. However, there was no attempt made to identify these films by electron diffraction. Based on their recent studies using transmission electron microscopy, Kar and Bahadur[8] reported the formation of thin films of the polymeric material in the case of sliding between high density polyethylene, polyoxymethylene, *PTFE* and a steel indentor. In all the cases, the films were identified by electron diffraction and were found to be of the polymeric material involved in sliding.

The present work is a follow up of the above studies[8] with respect to two more polymers--semicrystalline polypropylene and amorphous polycarbonate. The mechanisms for the formation of film in low speed sliding have been proposed. In addition, some observations have also been reported for medium speed and high speed sliding conditions.

EXPERIMENTAL

Sliding under low speed conditions was performed between polymer discs inscribed with radial abrasion marks (figure 1) and a steel indentor in a tribometer described elsewhere in detail.[9] With the disc rotating about its center, the sliding direction was at 90^0 to the abrasion marks. The sliding conditions chosen consisted of a load of *1600 g*, a speed of *0.002 m/sec* and a time of *20 min*. Replicas suitable for transmission electron microscopy were made from both the sliding track and out-of-the track portions. The repliaction technique involved shadowing the polymer surface in a vacuum with *Ge*, sputtering with *C*, depositing a thin layer of a 3% aqueous solution of polyacrylic acid and removal of the dried layer from the polymer surface which was later floated on distilled water to dissolve the polyacrylic acid. The final replica thus consisted of shadowed *Ge* with a backing of carbon. The replicas were examined in a transmission electron microscope to investigate the mechanisms of film formation under low speed sliding conditions.

Sliding under medium speed (*0.5 m/sec*) and high speed (*2.5 m/sec* and

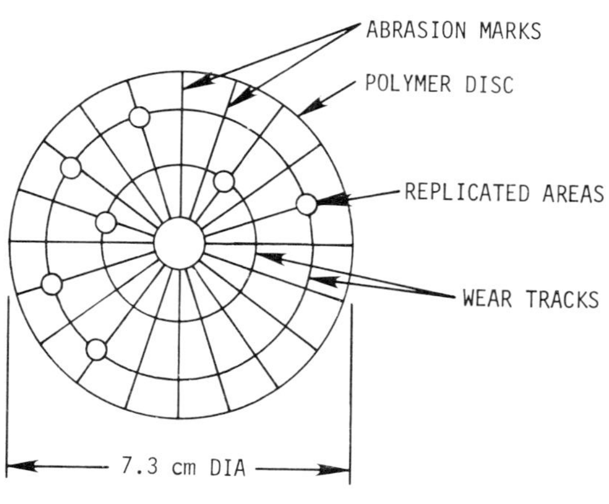

Fig. 1.—Sketch of polymer disc showing the abrasion marks, wear tracks and replicated areas.

higher) conditions was performed in a pin-and-disc type of wear apparatus. Here the circular end of a stationary, cylindrical polymer pin was loaded against the cylindrical periphery of a rotating steel disc.

RESULTS AND DISCUSSION

Sliding at a low speed of *0.002 m/sec* was performed on polypropylene and polycarbonate discs. Except for the formation of a shallow track, there were no other effects visible on the surface by optical microscopy. As such, extraction replicas from the polymer sliding surface were prepared and examined in a transmission electron microscope. Referring to the micrograph for polypropylene in Figure 2, it is seen that thin films are bridging across the abrasion groove. A selected area electron diffraction pattern obtained from the circled location in the film is inserted in the top left corner of the micrograph. Since the monoclinic form is more common than the hexagonal form for isotactic polypropylene[(10)], the indexing of the diffraction pattern was performed presuming a monoclinic crystal structure. With increasing diameters of diffraction spots, the corresponding planes are (110), (130), (041), (150), (060) and (200). The diffraction from these planes has earlier been reported for polypropylene by Binsberger and DeLange[(11)] and Bahadur.[(12)]

Contrary to the results obtained by the authors for *PTFE*, polyethylene and polyoxymethylene in the previous studies[(8)] and polypropylene in the present work, films stretching across the abrasion grooves could not be observed when the replicas of polycarbonate (Bisphenol-A) sliding surfaces were examined in a transmission electron microscope. In this case, merely

Fig. 2.—Transmission electron micrograph of polypropylene sliding surface and electron diffraction pattern from the encircled portion in the top-left corner. Sliding conditions: time 20 min. Load 1600 g; speed 0.002 m/sec.

the fragmented films appear to be scattered around on the surface (figure 3). The electron diffraction pattern, given in a corner of the micrograph, was obtained from one of these films. It shows sharp diffraction spots. The appearance of such diffraction spots from an amorphous polymer may raise some questions. However, research in recent years on the morphology of amorphous polymers has shown that these polymers consist of small domains (about *30-100* Å) with local order in them. These domains are described in terms of the nodular structure in an amorphous polymer.[13] The random coil structure concept used in the past to explain the morphological structure of amorphous polymers is losing ground rapidly. To date, nodular structure has been observed in several polymers, namely, polycarbonate, poly(ethylene terephthalate), natural rubber, isotactic and atactic polystyrene and poly(methyl methacrylate).[13,14] It is possible that severe deformation at the polymer sliding surface orients these locally ordered regions of nocules thereby creating an induced crystal structure. If it is so, a diffraction pattern of the type obtained in this work would be expected. Yeh and Geil[15] observed that when a thin film of poly(ethylene terephthalate) was stretched (drawn), the nodules aligned themselves in short rows in the direction of drawing. The electron diffraction pattern from such a film indicated that a three-dimensional crystal structure had developed.

The examination of transmission electron micrographs of the sliding surfaces of semicrystalline polymers, viz., PTFE, high density polyethylene, polyoxymethylene in the previous work[8] and polypropylene in the present study, led us to conclude that low speed sliding between a smooth metallic indentor and an abraded polymer surface produces thin polymer films. The latter are stretched across abrasion grooves. These films are produced by

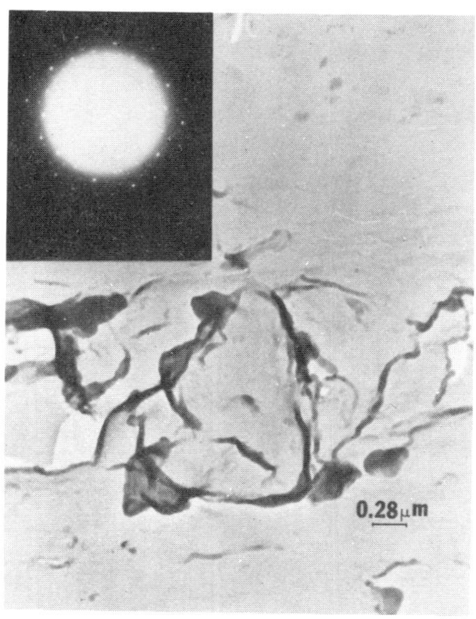

Fig. 3.—Transmission electron micrograph of polycarbonate sliding surface showing fragmented films and electron diffraction pattern from one such film in the top-left corner. Sliding conditions: same as in Figure 2.

drawing, caused by the stress applied in the direction of sliding due to the frictional resistance. Since the films have an estimated thickness of a few Å and are highly oriented, it is anticipated that these are produced as a result of shearing in the amorphous regions between lamellae. This is illustrated in Figure 4(a) which shows two adjacent lamellae consisting of folded molecules separated by the amorphous region which serves as a weak connection between the two. A strong adhesive bond between the metallic indentor and polymer surface produces shearing action in the polymer substrate so that a sheet of the polymeric material roughly equal in thickness to the interlamellar spacing is produced. This is subjected to the stretching action in the direction of sliding.

Thin films were found even on the sliding surface of polycarbonate which is an amorphous polymer. They were highly fragmented and exhibited a high degree of orientation in the direction of sliding. The films are formed due to severe deformation produced by adhesive bonding at the sliding interface which possibly results in the shearing of nodules followed by the alignment of molecules in the direction of sliding. The mechanism of film formation in the sliding of an amorphous polymer is explained in Figure 4(b). Here the molecules A, B and C in the nodule before deformation get

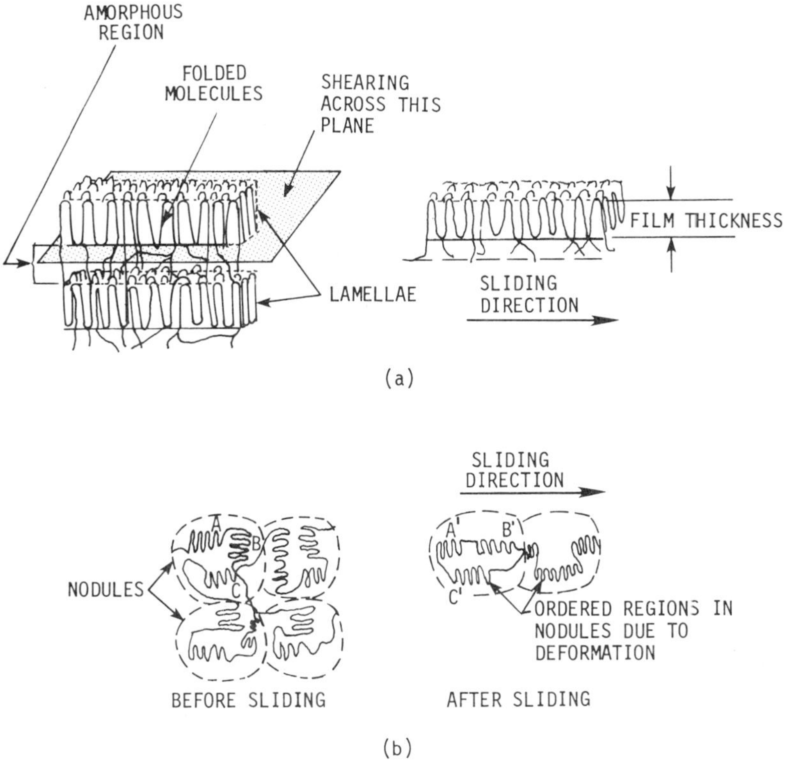

Fig. 4.—Schematic of the proposed wear model; (a) crystalline polymers, (b) amorphous polymers.

aligned in the direction of sliding to A', B', and C' after deformation, thereby producing a better order in their arrangement.

Contrary to the case of low speed sliding described above, irregularly shaped polymer wear particles could be seen escaping from the interface when sliding was performed at a medium speed of *0.5 m/sec* in a pin-and-ring type of apparatus. The shape and size of these wear particles was studied by optical microscopy. It was followed by a theoretical analysis considering two situations--a wear particle may escape from the sliding interface or may get entrapped between high points of the sliding surfaces. In the latter case, the particles formed would get bonded again to the previously formed layer resulting in the agglomeration of worn material. Depending upon the criterion used, the thickness of the ellipsoidal polymer wear particle was found to vary by a factor of about 6-12. The thickness of the *PTFE* wear particle, estimated from this analysis, was found to lie roughly between *5000 Å* and *55,000 Å*, which agrees with the values reported by Briscoe et al.[16] The details of this analysis are yet to be published.

The experiments involving high speed sliding (*2.5 m/sec* and higher) between high density polyethylene and polyoxymethylene pins and a steel disc at the authors' laboratory showed drawing of softened fibers through the rear edge of polymer pins.[8] Such a phenomenon has also been reported by Tanaka and Uchiyama.[17] The differential thermal analysis revealed lowering in the melting point of the fibrous material[8] which indicates that the worn material had undergone some changes in its morphological structure.

In a recent study to be published later, involving the measurement of the temperature rise at the sliding interface, it was found that the temperatures just prior to the failure of high density polyethylene and polyoxymethylene pins sliding at high speeds against the metal disc were *114°C* and *168°C*, respectively. Since these temperatures are close to the melting points of the respective polymers, it appears that the failure of the polymer specimens occurred due to thermal softening. On the other hand, the measured interface temperature for *PTFE* sliding against a metal disc was well below the melting point of the polymer. These observations further support our earlier conclusions[8] that softening results in catastrophic wear failure of high density polyethylene and polyoxymethylene pins in high speed sliding. However, such is not the case for *PTFE*.

CONCLUSIONS

Transmission electron microscopy of the abraded sliding surfaces of several polymers revealed that thin, oriented films, several hundred Å in thickness, are produced in low speed sliding of polymeric materials against a smooth metal surface. The films of semicrystalline polymers are large and seem to bridge across the abrasion groove more or less continuously, whereas those of the amorphous polycarbonate are highly fragmented. The probable mechanism of film formation for semicrystalline polymers is shearing across the amorphous region between lamellae and drawing of the lamella into an oriented film. In the case of an amorphous polymer, the film is produced due to the deformation of nodular structure which produces an ordered arrangement of the molecules.

In the case of high speed sliding, there is enough evidence to support the conclusion that the likely mode of catastrophic wear failure is thermal softening for high density polyethylene and polyoxymethylene but not so for

PTFE.

Depending upon whether the particle is able to escape from the interface or whether it gets adhesively bonded to the sliding surface, the analysis shows that the thickness of wear particles formed under medium speed sliding conditions may vary by a factor of 6-12.

ACKNOWLEDGEMENT

This research was supported by the Engineering Research Institute of Iowa State University.

REFERENCES

1. Makinson, K.R. and Tabor, D., *Royal Society of London. Proceedings.* Series A, Vol. 281, 1964, p. 49.
2. Pooley, C.M. and Tabor, D., *Royal Society of London. Proceedings.* Series A, Vol. 329, 1972, p. 251.
3. Bowers, R.C., Clinton, W.C. and Zisman, W.A., *Modern Plastics,* Vol. 321, 1954, p. 131.
4. Steijn, R.P., *Wear,* Vol. 12, 1968, p. 193.
5. Brainard, A.W. and Buckley, D.H., *Wear,* Vol. 26, 1973, p. 75.
6. Tanaka, K., Uchiyama, Y. and Toyooka, S., *Wear,* Vol. 23, 1973, p. 153.
7. Tanaka, K. and Miyata, T., *Wear,* Vol. 41, 1972, p. 383.
8. Kar, M.K. and Bahadur, S., *Wear,* Vol. 46, 1978, p. 189.
9. Bahadur, S., "Mechanism of Dry Friction in Deformation Processing of Polymers," ERI Project 896-S, Final Report, NSF Grant GK-27345, Engineering Research Institute, Iowa State University, Ames, Iowa, May 1973.
10. Natta, G. and Corrandini, P., *Nuovo Cimento,* supplement to Vol. 15, 1960, p. 40.
11. Binsberger, F.L. and DeLange, B.G.M., *Polymer,* Vol. 9, 1968, p. 23.
12. Bahadur, S., *Journal of Materials Science,* Vol. 10, 1975, p. 1425.
13. Neki, K. and Geil, P.H., *Journal of Macromolecular Science. Part B. Physics,* Vol. 8, 1973, p. 295.
14. Geil, P.H., "Polymeric Materials Relationship Between Structure and Mechanical Behavior," American Society for Metals, Metals Park, Ohio, 1974, p. 119.
15. Yeh, G.S.Y. and Geil, P.H., *Journal of Macromolecular Science. Part B. Physics,* Vol. 1, 1967, p. 251.
16. Briscoe, B.J., Pogosian, A.K. and Tabor, D., *Wear,* Vol. 27, 1974, p. 19.
17. Tanaka, K. and Uchiyama, Y., in "Advances in Polymer Friction and Wear," edited by L.H. Lee, Vol. 5B, Plenum Press, New York, 1974, p. 499.

DISCUSSION

S. GANESH, Bendix Research Laboratories: You gave an estimate of the film thickness and said that it is about the order of lamellae thickness. I believe it is about 300Å to 400Å. How did you estimate that?

M. K. KAR: The films are curled at the edges of the grooves. We measured the curled edge and found that to be about 300 to 400 angstroms. It is a technique which had been used earlier by Steijn.

GANESH: As it was so close to the lamellar distance I was wondering how much confidence you have in the estimate?

KAR: Well, I would say that it is a very crude technique. It has some validity, but it is not a perfect way to measure the thickness.

J. S. WOLLMAN, C. S. Draper Laboratory: I am interested in the material that you picked up on the slider. You said that the melting point was lower than the bulk material. It seems to me there are two possibilities: one is leaching of the low melting fraction during the sliding process and the other is that the material is highly stressed. Did you repeat melting point measurement? I think that would differentiate between those two.

KAR: We repeated the measurements quite a few times and it was reproducible.

WOLLMAN: That would indicate that you were extracting a lower melting fraction in the wear process. Do you concur with that?

KAR: Because of the heating and cooling action at the interface the morphological structure was changed altogether. That is why we got the reduction in the melting point.

B. J. BRISCOE, Imperial College: I would like to comment on this problem of the film thickness. Historically, it was suggested by Makinson and Tabor that there was some relationship between the film thickness as observed under these conditions and the morphology of the polymer. In later papers, other things were found to be important. Having thought about this problem and looked for analogies in bulk response, as so often one tries to do in this work, we can see the closest analogy is fiber drawing. As you all probably know there is a good deal of literature on the changes in the morphology of semi-crystalline polymers when they are subjected to very high strains. The models of drawing essentially involve the stretching of the amorphous region first and then the drawing out of the cystalline lamellae, and then there are all sorts of long-range order problems which lead to scission. So what we would like to think of is that the film thickness is not governed by the morphology but by some process which is involved in fiber drawing. If one looks at fiber drawing phenomena, particularly at high rates of strain, he finds that adiabatic shear (because of the defects that occur in this sort of system) is very likely. If one is prepared to take the analysis a little further he arrives at the possibility that this layer is governed by the way in which energy is dissipated during this process. It could be considered as a type of adiabatic shear of thermal local weakening. One need not invoke very large temperatures. Now if one does this analysis (of course, there are so many parameters that he can get essentially what he likes), the figures are consistent at least with the spectrum of film thickness values that one sees. I think the data on thermal conductivity can be put into this sort of argument. It is a very complicated problem.

S. BAHADUR: The thickness of the drawn film is going to depend upon a number of factors. We all know that if the rate of drawing is smaller, the thickness of the drawn film will also be smaller. So we fully realize that the thickness is going to depend upon the sliding speed. We have tried to give a little thought to it and all we can guess is that probably the smallest thickness that we can come up with is equal to the interlamellar thickness - not the lamellar thickness. There is another possibility: some unfolding of the molecules, or at least partial unfolding may take place. But all the evidence that we have based on the crude measurement technique suggests that we are possibly in that region. I mean that we are getting a thickness of the film which might be 400 to 800 angstroms which is, in effect, the interlamellar thickness.

D. A. RIGNEY, Ohio State University: From the wear debris how do you distinguish which particles have come loose right away and which have been reprocessed by being stuck in there and transferred and spread out? Is it the thickness of the particles, aspect ratio, appearance of the surface of the particle, or some combination of this?

BAHADUR: We could not distinguish between the two. We have carried out a large number of measurements on worn particles by changing velocity, load, and so forth. We have found that there was a large distribution, and it is very hard to say whether a particle that just formed has been separated from the interface of the particle or it came from the interface as a result of agglomeration. We estimated the thickness and tried to measure it and there seemed to be some correspondence. That is the best I can say about it. There is no way we can tell what sort of history the particle has gone through just by looking at the wear particles.

BEN SU, MIT: Do you have an estimate of the total strain produced in the transfer film? With that amount of the total strain, do you think the orientation in the transfer film is really due to the orientation of the molecules or is it only the ordering of the crystal-like lamellae in the sliding orientation? Can this strain actually be compared to the very high strain in the fiber drawing process?

BAHADUR: We realize that in a sliding motion there are very high strains induced and the smaller the thickness of the film coming out, the higher the strain rate to velocity ratio. But we have also to realize that there is a special situation in this fiber drawing or the extrusion and that is the hydrostatic stress state. So I do not think that you can try to connect the two things together directly. All we can say is we are aware that there are high strain rates involved, and we are able to see some of the extruded films. The thickness of the film turns out to be of the order of interlamellar dimension. Can we say that there is some unfolding or is it intermolecular arrangement? Of course, both processes occur in the extrusion. There is some arrangement between the molecules taking place which leads to the ordered arrangement and at the same time there is some unfolding of the molecules occurring; amorphous regions are being stretched. As a matter of fact a multiplicity of processes is going on and the end result is oriented film. That is the best I can tell you.

QUESTIONER: You mentioned the possibility that the structure of the amorphous polymer that you worked with was nodular and attempted to explain the events that you observed on that basis. I would like to remind you that in the last five years in several laboratories around the world it has been shown that results of neutron diffraction studies on several amorphous polymers, in particular polycarbonate, are not consistent with the idea of a nodular supramolecular structure in amorphous polymers.

BAHADUR: Well, all I can say is that we have some faith in the approach that has been taken by Prof. Geil at Case Western Reserve University. You may be familiar with the fact that Geil and his associates have been doing a lot of work and they have been pleading for nodular structure.

KAR: Professor Geil has reported that if a polycarbonate or PMMA sample is drawn we get an orientation effect. He explained that also in terms of this nodular structure. Of course that drawing process was not sliding; it was drawn mechanically just to see what sort of effect is produced because of the drawing. So it is merely a conjecture.

SLIDING WEAR MECHANISMS OF POLYMERS

M. Clerico

ABSTRACT

The wear mechanisms of two polymeric composites and three commercially pure polymers sliding against a metal were experimentally investigated by varying the sliding distance, normal force, and sliding speed.

Microscopic observations showed that the mode of wear of the two composite materials was almost the same. Subsurface deformation, crack nucleation at the matrix-hard particle or matrix glass fiber interface, crack propagation parallel to the surface at a depth corresponding to the friction coefficient, and cracks shearing to the surface were found in both composites. Furthermore, the wear particles were very often thin sheets. These observations suggest that the wear mechanism is delamination similar to that observed in metals. Since film transfer greatly influences friction and wear, prediction of wear of polymeric composites must take into account such a film in addition to the crack nucleation and propagation mechanisms.

Microscopic observations of surfaces and subsurfaces of pure polymers showed that wear was caused by continuous deformation together with thermal softening or melting of the surface.

INTRODUCTION

The paper by Briscoe and Tabor is a very comprehensive review of the literature on the sliding wear of polymers. From this review it follows that, although many wear mechanisms have been proposed for ductile polymers, no distinction has been made to date between the wear process of ductile polymers and that of the same polymers filled with fibers or particles of some other material. The purpose of this paper is to describe the wear mechanisms in polymeric composites[1,2] in terms of sliding distance, normal load, sliding speed and the wear effects on the subsurfaces. Also, new results on the wear behavior of commercially pure polymers are reported.

It was previously found[1,2] that the wear of polymer composites is essentially a fracture process due to cyclic loading, not confined just to the contacting asperities. The wear mode of these composites is similar to that of the delamination theory proposed by Suh[3]. The wear rate depends on the friction coefficient, flow stress, microstructure of the materials, etc. If the temperature due to the high normal load and sliding speed is large, the film transfer may decrease the friction coefficient and thus the wear rate.

Experiments on ductile polymers without fibers or particles also disclose that the wear process is not confined to the contact asperities. They deform continuously due to the surface traction. In addition, it is likely that an extrusion process of molten polymer out of the contact region takes place[4]. Adhesion, as well as abrasion, theory may be valid for polymer wear only at the start of the sliding.

Film transfer is the most striking phenomenon which differentiates the polymer from the metal in friction and wear[5-7]. The role played by this film is not clearly understood to date. It is claimed that the film can smooth out the metal counterface and reduce both friction and wear rate, as the sliding condition changes from that of polymer sliding on metal to one of polymer sliding on polymer. The friction and wear behavior is also affected by the film thickness which is related to the temperature and thus to the sliding speed and normal load. Knowledge of the chemistry of the film, and the nature of degradation products (due to the very high temperature in local spots of the polymeric sliding surfaces and the presence of metallic counterface), might be of help in understanding its role in friction and wear.

Thus, an understanding of polymer wear must take into account the spreading and the subsequent removal of any transferred film and its chemical changes, in addition to mechanisms like fracture due to cyclic loading and continuous deformation, together with surface thermal softening or melting.

EXPERIMENTAL

Ultramid A 4, Ultramid A 3R (U.N.), Ultramid A 3WG5 (F.N.) (BASF, West Germany) are all nylon 6/6. The first one was commercially pure; the second one, specially prepared for reducing friction and wear, contains particles of different size harder than the matrix; Ultramid A 3WG5 (F.N.), used for bearing cages, is filled with randomly oriented glass fibers (25% - wt). Hostaform C 9021, and Hostaform C 9021 K (Hoechst-West Germany), specially prepared for reducing friction and wear, are unfilled acetal copolymers. Other details about the materials are given in Table 1.

TABLE I.—PROPERTIES OF MATERIALS

Material	Specific Gravity	Hardness (Rockwell)	Yield Strength (daN/mm^2)	Tensile Modulus (daN/mm^2)
Ultramid A4	1.14	82(R_L)	5.5	170
Ultramid A3R (U.N.)	1.11	51(R_H)	5	165
Ultramid A3WG5 (F.N.)	1.28	75(R_H)	10	700
Hostaform C9021	1.41	74(R_H)	7.3	350
Hostaform C9021K	1.44	73(R_H)	7.2	350
Steel 38NiCrMo4UNI (S)	7.8	66.5(R_A)	85	21,000
Bronze B 14UNI (B)	8.7	41(R_A)	13	11,000

Specimens were prepared by machining from extruded bars. The surface roughness was varied by grinding as well as by turning in order to get preliminary information about the influence of surface texture.

The tests were conducted using a modified Amsler apparatus (Figure 1) in a standard laboratory atmosphere. Cylindrical polymer specimens were slid against metal cylinders. Specific sliding factors were 1.662 for the metal and 2.511 for the polymer. The apparent contact area, A, and the maximum normal stress, σ_{max} (both depending on the normal load) were approximately determined from the Hertz's formulae. The test conditions are given in Table 2.

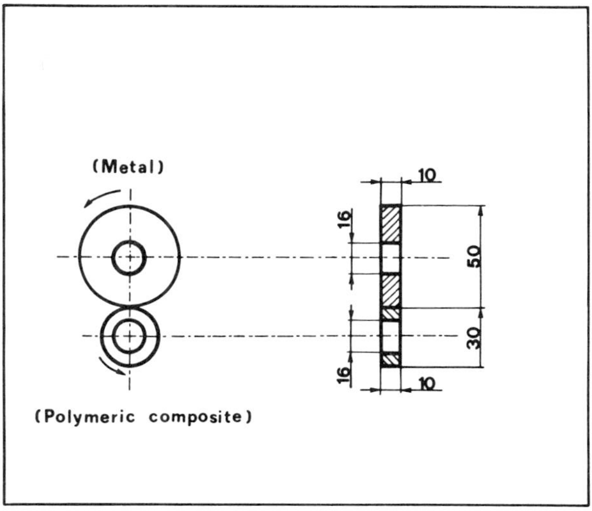

Fig. 1.—Operating conditions and specimen dimensions.

RESULTS

Pure Nylon

Figure 2 shows the effect of load on the friction coefficient and wear of pure nylon[8]. At higher load, the wear increases very rapidly with the normal load whereas the frictional force increases up to a maximum, beyond which it decreases.

Worn surface and subsurface characteristics are shown in Figure 3. Though its mechanism is not clear yet, wear seems to be caused by continuous deformation together with surface thermal softening or melting[3,4].

Composites

General observations

The initial surface texture of the metal cylinders does not significantly affect the friction and wear results. Profilometer tracings after the test show that all the sliding surfaces became smoother, especially if they are free from wear particles. Values of the normal force (5,15,25 daN)

TABLE II.-TEST CONDITIONS

Materials of the Pairs	Sliding Distance (10^4m)	Sliding Speed (ms^{-1})	Normal Load (daN/mm^2)	Apparent Contact Area (mm^2)	Maximum Normal Stress (daN/mm^2)
Ultramid A4 Steel	5.5	0.75	2 to 25	3.1 to 11	8 to 2.8
Ultramid A3R Steel or Ultramid A3R Bronze	11	0.25 0.50 0.75	5,15,25	5.0,8.6,11.1	1.2,2.1,2.8
Ultramid A3WG5 Steel or Ultramid A3WG5 Bronze	11	0.25 0.50 0.75	5,15,25	2.4,4.2,5.4	2.5,4.4,5.7
Hostaform C9021 Steel or Hostaform C9021K Steel	5.5	0.75	2 to 40	2.2 to 10	1.1 to 5.1

for tests on the polymeric composites were chosen from the results of reference 8 (Figure 2), where different types of wear (low, medium, high) according to the normal load and the contact temperature were postulated.

The steady state friction coefficient and the wear volume of both the nylon composites sliding against steel versus the sliding distance for a load of 5 daN and for the three tested speed levels are shown, as an example, in Figure 4. The wear volume of nylon as a function of sliding distance does not show the three characteristic stages, i.e., the initial stage of large wear rate, the intermediate stage of imperceptible change in volume and the last stage of catastrophic wear. It seems that there are just two stages, as if the normal load is large enough to produce catastrophic wear after the short initial period of "running-in", when the metal asperities deform and cut off the polymer asperities. A normal load of 5 daN is indeed large enough to cause sizable wear in pure nylon (Figure 2). Finally, although there is no substantial difference in the wear rates of U.N. against bronze and against steel, the wear of F.N. against bronze is much more catastrophic than that of F.N. against steel.

Although bronze always wears when it is against F.N. (it wears like the counterpart), a small amount of nylon adheres to its sliding surfaces. A polymeric film also transfers to the steel surface which also slightly wears.

The wear rate of U.N. sliding against steel (Figure 5) and bronze increases with decreasing friction coefficient, whereas, that of F.N. decreases. Large loads and/or high sliding speed produce very high contact temperatures, sizeable wear rate, and low friction coefficients (Figure 6). The actual temperature in the contact regions must be so high that the nylon (probably also the glass) transferred to the metal surfaces softens and develops chemical compounds with metal which decrease friction. The glass fiber composite at high temperature seems to act in an analogous way to some E.P. additives that give low friction but not low wear.

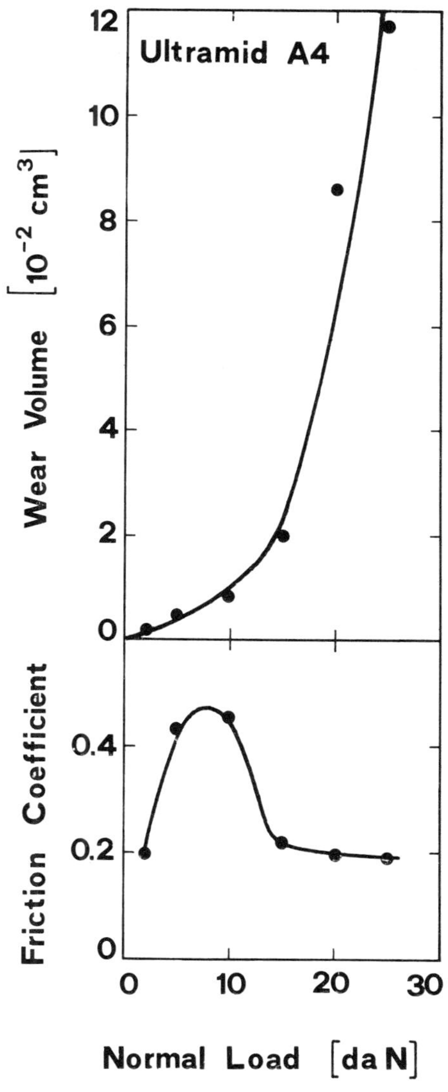

Fig. 2.—Wear volume and friction coefficient of Ultramid A4 versus normal force[8].

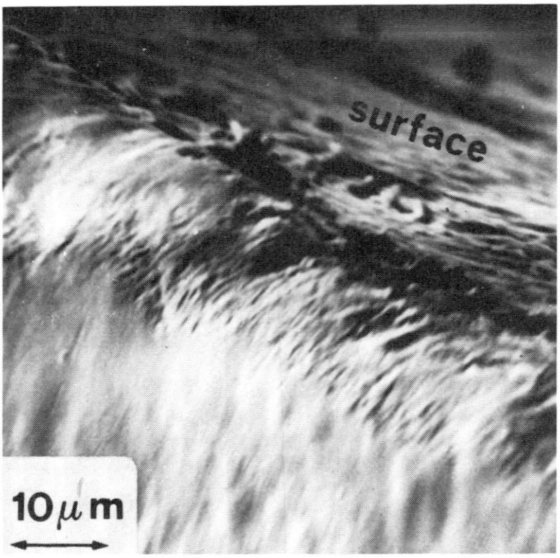

Fig. 3.—Worn surface and subsurface of Ultramid A4.

Characteristics of worn surfaces

SEM micrographs of worn surfaces are shown in Figures 7 and 8. Polymer surfaces (Figures 7a, 8a) become very smooth; the initial asperities are completely smeared over the surface and wear tracks are plowed by the hard asperities on the surface. The worn surfaces also exhibit wear craters (Figure 7b), wear sheet formations (Figure 7c), and cracks orthogonal to the sliding direction (Figures 8a and c). Polymeric wear particles containing glass fibers adhere to F.N. surfaces sliding against steel (Figure 8c).

SEM micrographs of bronze and steel counterfaces show that large amounts of polymer adheres to steel (Figures 7d, 8d); if the polymers is F.N. (Figure 8d), the film holds broken glass fibers. Their edges are made round like stones in the ocean. The small amount of polymer adhering to the bronze counterface (Figure 7b) forms a very thin and uniform film. The wear as well as the friction is more than in the composite-steel pairs. The relatively smooth surfaces of the metal and the wear tracks on the bronze surface (after eliminating polymeric debris) demonstrate that both of the metal specimens wear, though different extents.

Characteristics of subsurfaces

Subsurfaces of both of the polymers show cracks (Figure 9) which nucleate (Figure 10) at matrix-hard particle or matrix-glass fiber interface. These cracks propagate parallel to the surface (Figure 9) at a depth which depends on the normal load and sliding speed. When a critical length is reached, they shear to the surface producing wear sheets (Figure 11). Both polymeric composites behave the same way.

The proportion of fibers beneath the F.N. sliding surfaces does not increase during the wear process as was found by Tanaka[9] on carbon-filled PTFE or polyacetal. Bronze and steel counterfaces (Figures 12 and 13) wear by the delamination mechanism[3]. Crack propagation depths normalized to

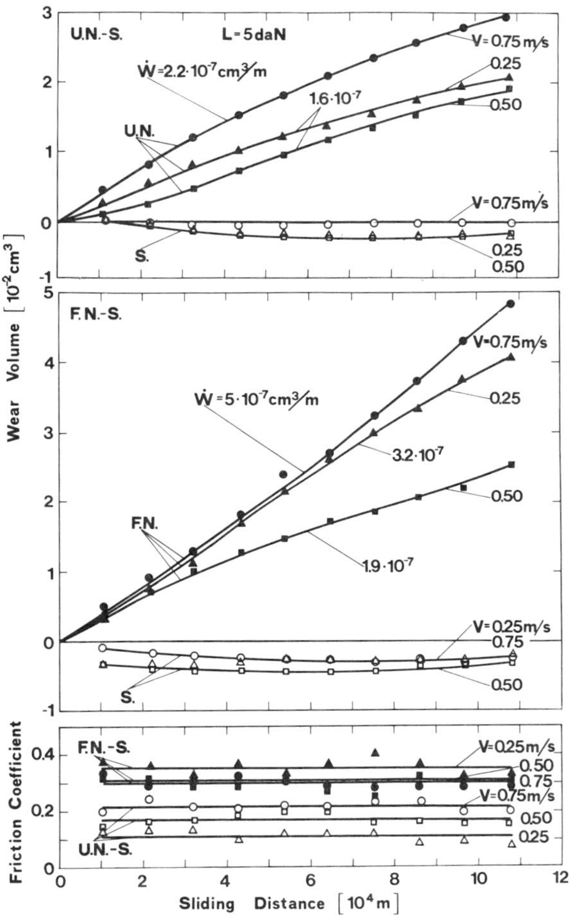

Fig. 4.—Wear volume and friction coefficient of steel-nylon composite pairs versus sliding distance[1].

the contact half length versus friction coefficient are plotted in Figure 14.

Wear particles

Wear particles of different material and shape were collected: parts of the harder particles found in the U.N. matrix, squashed glass fibers or nylon pieces (Figure 7d), filamentary nylon particles, sheets from U.N., F.N. (Figures 15b, c and 16a), bronze (Figure 16b) and steel. The metal particles are seldom free from polymeric film as well as the nylon particles from metal residue. In the ferrographs shown in Figure 16, the brown-black color indicates polymer and the red-brown particles are metal covered by polymer film.

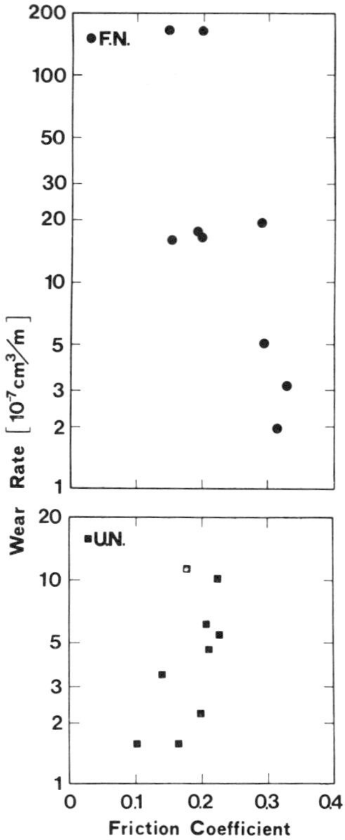

Fig. 5.—Wear rate of steel-composite pairs versus friction coefficient(1).

UNFILLED ACETAL COPOLYMERS

The wear and friction coefficient versus normal force of the tested unfilled acetal copolymers sliding against steel (Figure 17) are quite similar to that of pure nylon (Figure 2). The polyacetal Hostaform C 9021 K wears less than the Hostaform C 9021, whereas its friction coefficient is negligibly larger. They cannot compete with nylon for the low wear.

Worn surfaces exhibit a layer with evident marks of shearing (Figure 18), similar to the sea waves generated by the wind, as if the frictional heat generated during sliding accounts for the surface softening or melting of polymer. The layer thickness seems to depend upon the normal force, it varies from 3 to 5 μm for a normal force of 5 daN, increases till about 10 μm at 20 daN (Figures 19-22). The material of the surface layer is more brittle and more deformed by shearing than the bulk material. There are not cracks underneath the layer as in polymeric composites, just deformation (Figures 21, 22). It is likely that softened or molten polymer is extruded out of the contact region (Figure 21).

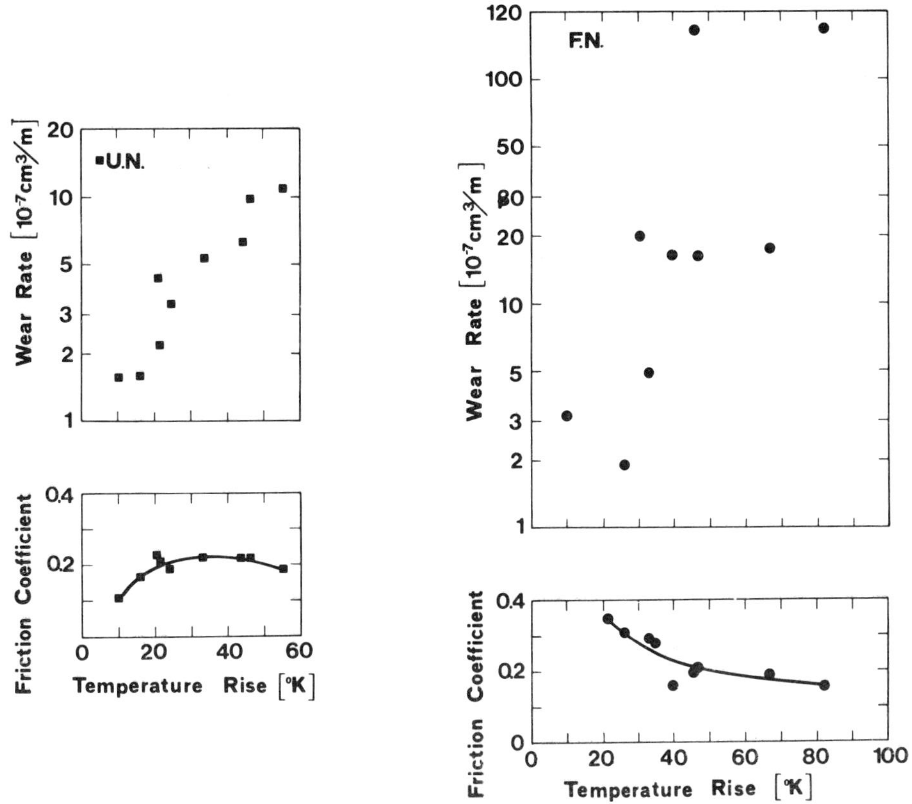

Fig. 6.—Wear rate and friction coefficient of steel-composite pairs versus temperature rise[1].

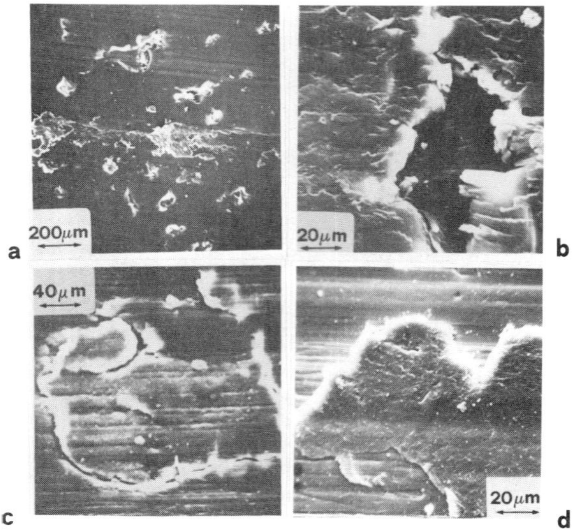

Fig. 7.—Worn surfaces of U.N. - metal pairs (the normal force was 15 daN): (a) Worn surface topography of U.N. sliding against bronze at 0.75 m s^{-1}(2); (b) crater in U.N. sliding against steel at 0.75 m s^{-1}; (c) sheet formation in U.N. sliding against steel at 0.25 m s^{-1}; (d) worn surface of steel (the sliding speed was 0.50 m s^{-1}).

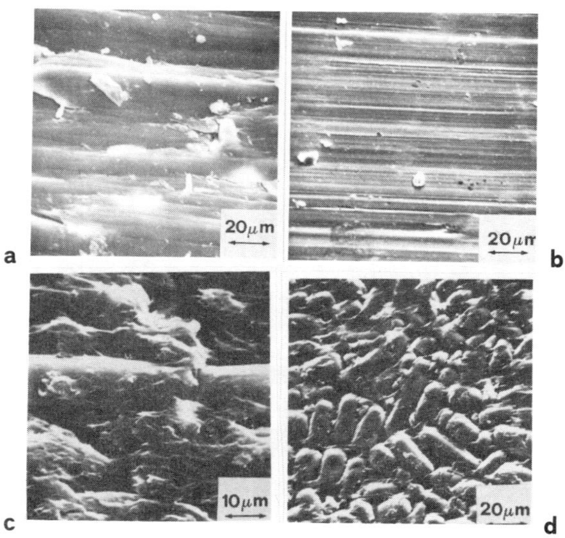

Fig. 8.—Worn surface of F.N. - metal pairs (the normal force was 15 daN and sliding speed was 0.25 m/s^{-1}): (a) F.N. surface and (b) bronze counterface; (c) F.N. surface and (d) steel counterface.

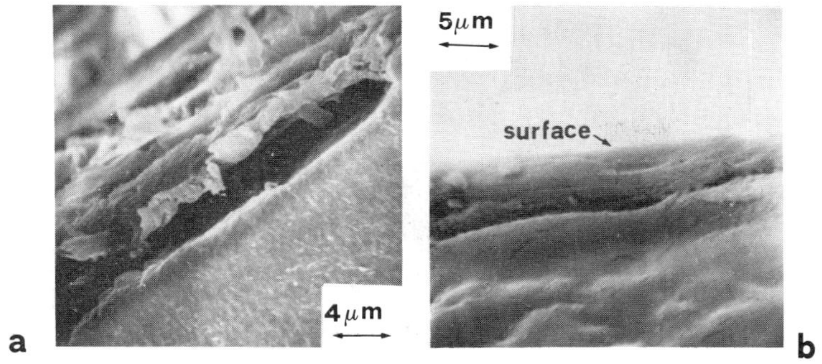

Fig. 9.—Crack propagation in F.N.: (a) sliding against steel[1] (the normal force was 5 daN, sliding speed was 0.50 m s^{-1}); (b) sliding against bronze[2] the normal force was 25 daN, sliding speed was 0.50 m s^{-1}.

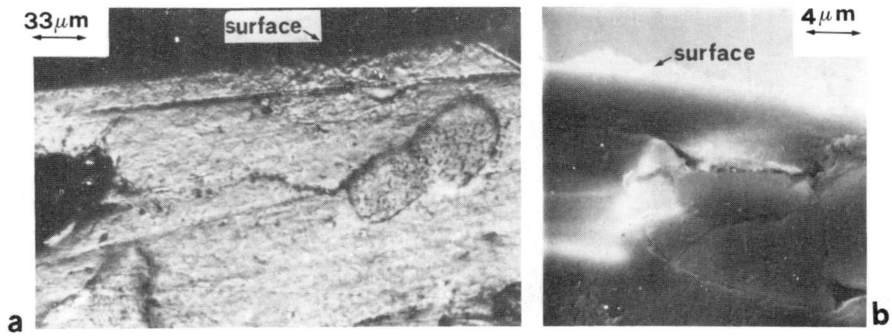

Fig. 10.—Crack nucleation in polymeric composites sliding against bronze: (a) U.N., (b) F.N. The normal force was 25 daN, sliding speed was 0.75 m s^{-1}[2].

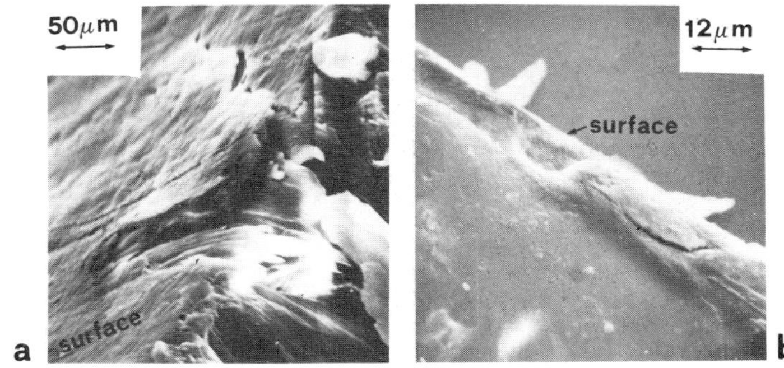

Fig. 11.—Sheet formation in the same specimens of the previous figure[2].

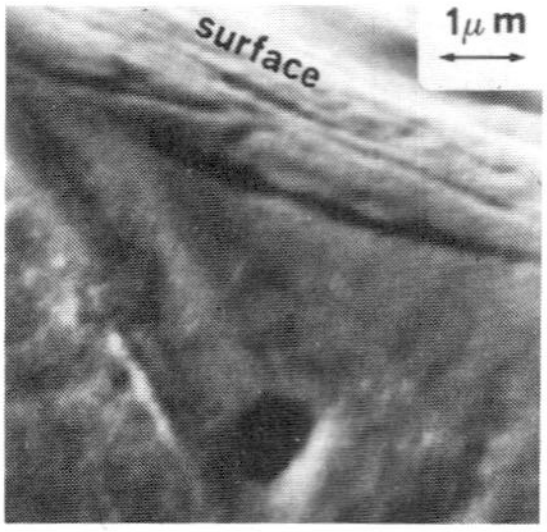

Fig. 12.—Crack propagation in steel sliding against U.N. The normal load was 15 daN, sliding speed was 0.50 m s^{-1}(1).

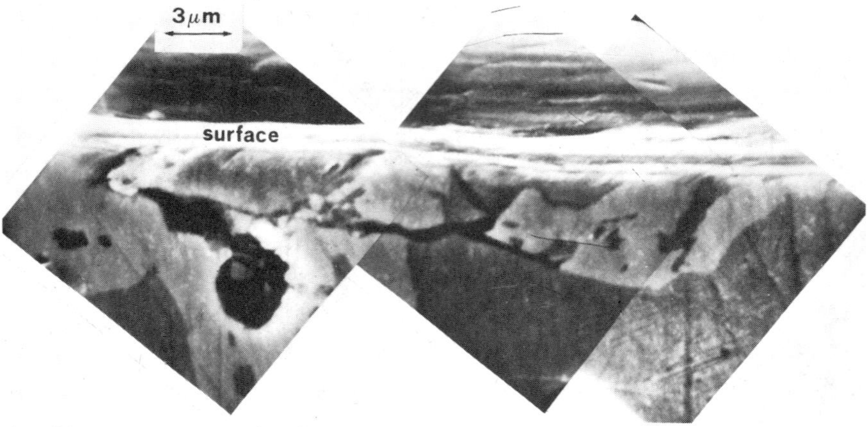

Fig. 13.—Crack propagation in bronze sliding against U.N. The normal load was 25 daN, sliding speed was 0.75 m s^{-1}(2).

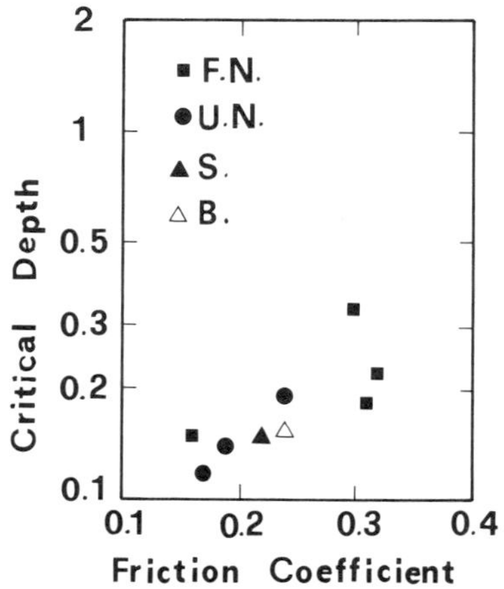

Fig. 14.—The crack propagation depth as a function of friction coefficient(2).

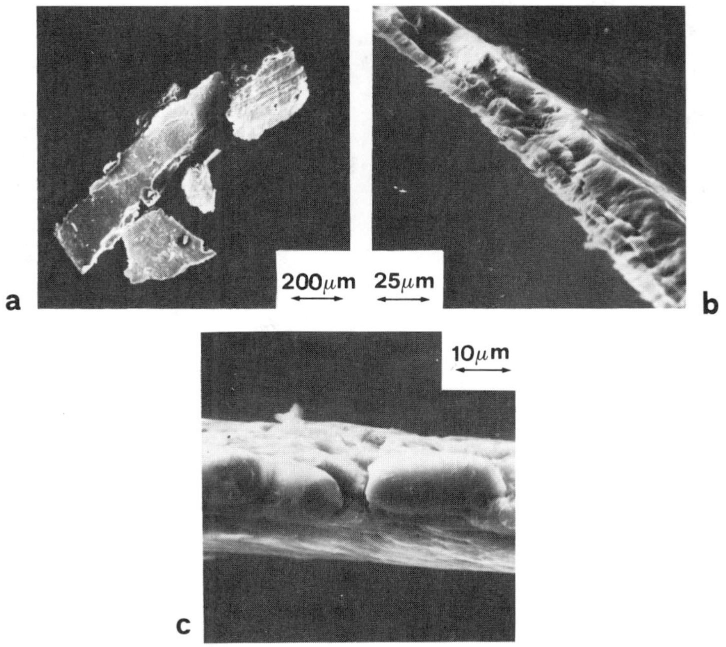

Fig. 15.—Wear particles generated from F.N. specimens sliding against steel. The normal force was: (a) and (c) 15 daN, (b) 25 daN. The sliding speed was: (a) 0.5 m s^{-1}, (b) 0.75 m s^{-1}, (c) 0.25 m s^{-1}.

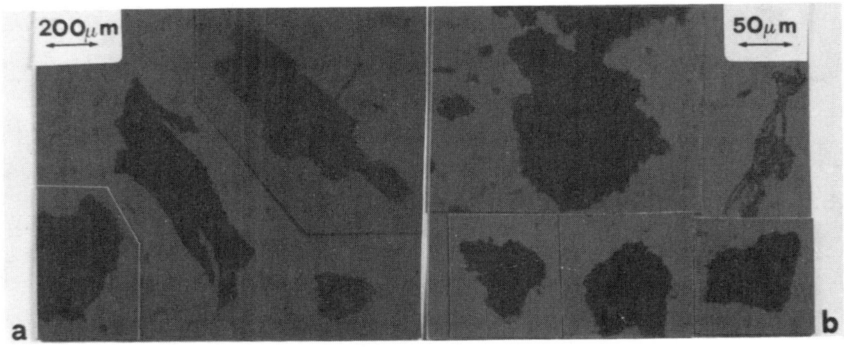

Fig. 16.—Ferrographs of wear particles. (a) F.N. sliding against steel. The normal force was 15 daN, sliding speed was 0.25 m s^{-1}, (b) Bronze covered by polymeric film. The normal force was 15 daN, sliding speed was 0.25 m s^{-1}.

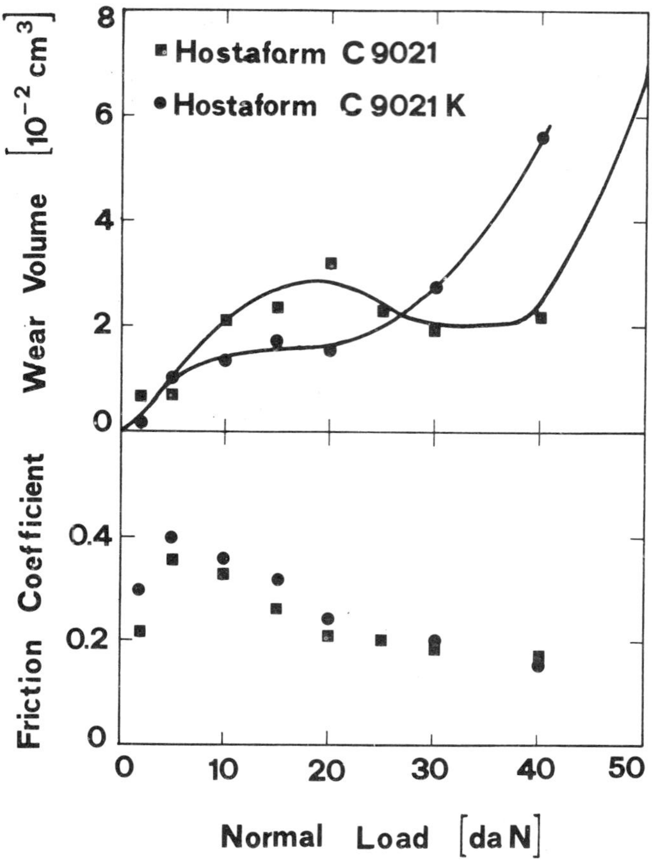

Fig. 17.—Wear volume and friction coefficient of acetal copolymer versus normal load.

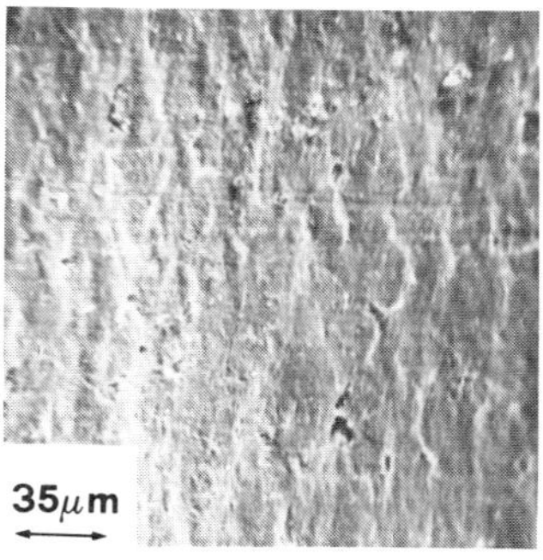

Fig. 18.—Worn surface of Hostaform C 9021 K. The normal force was 20 daN.

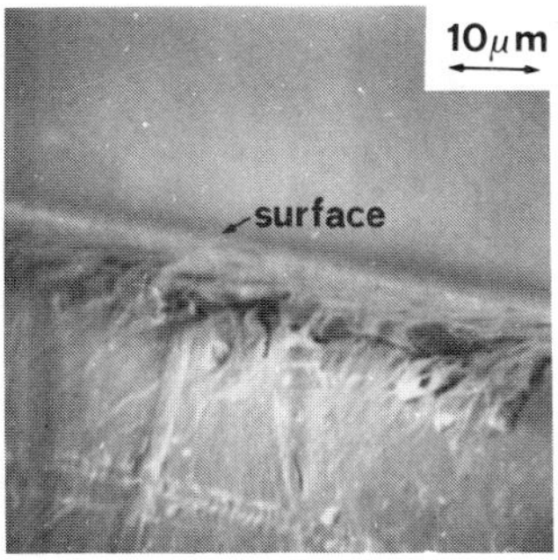

Fig. 19.—Subsurface of Hostaform C 9021. The normal force was 5 daN.

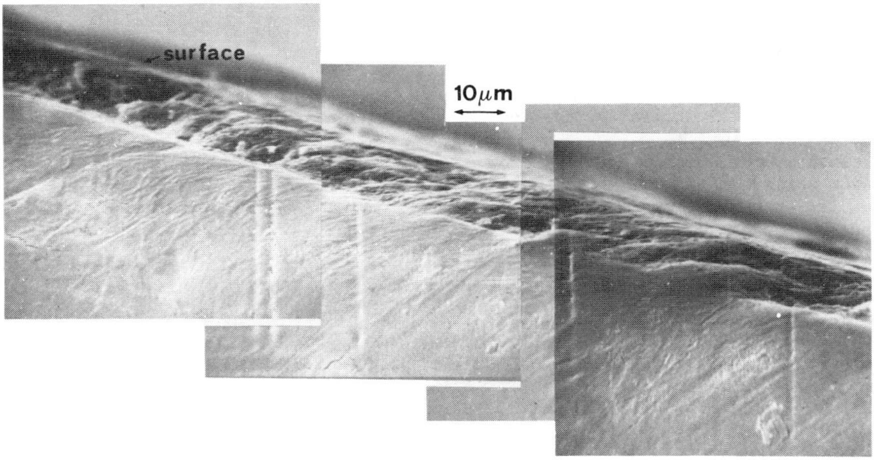

Fig. 20.—Subsurface of Hostaform C 9021. The normal force was 20 daN.

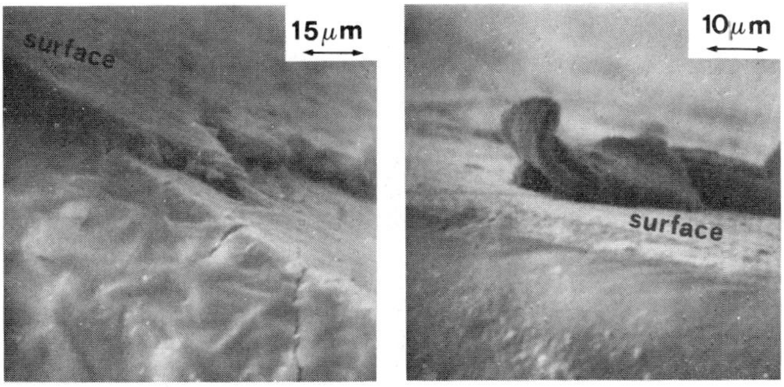

Fig. 21.—Subsurfaces of Hostaform C 9021 K. The normal force was 5 daN.

Fig. 22.— Subsurface of Hostaform C 9021 K. The normal force was 20 daN.

CONCLUSIONS

The two polymeric composites essentially exhibit the same wear mechanisms: subsurface deformation, crack nucleation at the matrix-hard particle interface or matrix-glass fiber interface, crack propagation parallel to the surface at a depth from it correlated to the friction coefficient, and crack shearing to the surface. The final result is a thin wear sheet.

Bronze counterfaces also wear (about the same as composites filled with glass fiber) whereas steel wears negligibly compared to F.N. Metal subsurfaces show substantially the same kind of damage found in composite subsurfaces.

These results lead to the conclusion that wear mechanisms of semicrystalline composites are similar to the delamination wear of metals.

Differences in the behavior seem to derive from the amount and the nature of the forming film which can decrease friction more than wear.

Continuous deformation dissipates energy at the sliding surface contact. As a result, the surface material softens or melts depending on the normal force. Extrusion of molten polymer out of the contact region seems to be involved in wear process of unfilled nylon 6/6 and polyacetals.

ACKNOWLEDGEMENTS

The work reported in this paper was partially supported by C.N.R. (Italian National Research Council) under contracts No. 75/0131807 and No. 76/0163207. The author is grateful to Dr. V. Patierno - C.R. F. FIAT for his help in making the micrographs.

REFERENCES

1. Clerico, M., in 4th National Conference of the Society for Theoretical and Applied Mechanics, Florence, Italy, Vol. 3, 1978, p. 53.
2. Clerico, M. and Patierno, V., *Wear,* Vol. 53, 1979, p. 279.
3. Suh, N.P., *Wear,* Vol. 44, 1977, p. 1.
4. Tanaka, K. and Uchiyama, Y., in "Advances in Polymer Friction and Wear," edited by H. Lee, Vol. 5B, Plenum Press, New York, 1974, p. 499.
5. Richardson, M.O.W., *Wear,* Vol. 17, 1971, p. 89.
6. Sviridynok, A.I., Bely, V.A., Smurugov, V.A. and Savking, V.G., *Wear,* Vol. 25, 1973, p. 301.
7. Tanaka, K. and Miyata, T., *Wear,* Vol. 41, 1977, p. 383.
8. Clerico, M., *Ingegneria Meccanica,* Vol. 19, 1970, p. 41.
9. Tanaka, K., *Journal of Lubrication Technology,* Vol. 99, 1977, p. 408.

DISCUSSION

G. F. HARDY, Celanese Research Company: Am I right in assuming that your composite plastic specimens were cut from extruded rods?

M. CLERICO: Yes, they were extruded.

HARDY: Did you examine the specimens with a microscope to see if the fibers were randomly oriented? Is it possible that you have a preferred orientation of fibers relative to sliding direction?

CLERICO: They were not examined very carefully. It is possible that fibers in polymer extruded rods are not quite random. Also, the fiber orientation should have some effect on friction and wear. I would like to determine this effect.

HARDY: I must say it is very difficult to get truly random fibers in that kind of material.

BEN SU, MIT: In wear tests on polymers which generate very small particles, we usually found that these particles were redeposited on the surface. Especially with fibers, all these particles may accumulate between fibers and press together. I was wondering how you could differentiate between these back transferred particles and the wear sheet that was supposed to have generated from the original polymer surface.

CLERICO: I have made micrographs showing also this kind of particles; some of them are still adhering to the surface. Since they melt in addition to press together, their aspect is quite different.

APPLICATION OF ADVANCED LIGHT SCATTERING TECHNIQUES TO THE STUDY OF POLYMER PROPERTIES

J. M. Schnur, P. Schoen and S. L. Wunder

ABSTRACT

It has been well established that a polymer's propensity to wear is related to its toughness or viscoelastic properties. Unfortunately, the specific variables contributing to a material's 'toughness' are neither well understood nor easily measured. This presentation will discuss the types of relaxation phenomena and relaxation times that are important to the viscoelasticity of a material, their molecular basis, and the various optical techniques available for the characterization of these properties. Specifically the application of Raman, Brillouin, and correlation light scattering techniques to the characterization of solid polymeric materials will be discussed with particular emphasis placed on the relationship between types of information one can obtain from these light scattering experiments and the strength of polymeric materials.

INTRODUCTION

Single component polymers can vary in molecular weight, dispersity, crystallinity and morphology, all of which affect material strength and wear. Indeed, a polymer can manifest a range of macroscopic properties (density, tensile strength, etc.) while its chemical composition, molecular weight and molecular weight distribution are kept constant. This is possible because the microscopic properties of molecular conformation and orientation profoundly affect macroscopic properties. Thus if we are to understand fully the mechanisms of strength and wear in polymers we must be able to probe them on the molecular level.

The response of a polymeric material is dependent on the frequency of the applied load, making it important in polymer characterization to measure material properties over a wide frequency range. Information on time scales not measurable by available techniques is typically obtained by making use of the time/temperature superposition principle (which implies equivalence of frequency and temperature for some systems); dynamical mechanical measurements made over a limited frequency range at a series of temperatures are converted to plots spanning a wide range of frequencies at a single temperature. It is clearly preferable to be able to measure a material's response in real time, thereby avoiding the assumptions inherent in time-temperature superposition. Recently, progress has been made towards this goal with the application of light scattering techniques to the problem of polymer characterization.

Information about the internal dynamics of a polymer is obtained from the frequency changes light undergoes when it is scattered. A single frequency f_o of light from a laser strikes a sample and is scattered with a new frequency or frequencies. The change in frequency Δf is characteristic of some process in the sample. Thus Δf may be a doppler shift caused by reflection of the light from a moving particle, or else it may be the frequency of a sound wave in the sample or of a molecular vibration.

Light scattering can be divided under the broad subheadings of correlation, Brillouin, and Raman scattering according to the ranges of the frequency shifts involved. Figure 1 indicates the frequency ranges spanned and the molecular processes expected to occur within those frequency domains.

Correlation spectroscopy deals with the lowest frequencies detectable by the light scattering technique. The name, correlation, refers to the electronic technique of processing the detected optical signal. The resulgint correlation function can be fourier transformed to give its frequency spectrum (the range is roughly *0.1 Hz* to *1 MHz*). As indicated in Figure 1 this range corresponds to the slow, diffusive motions of large polymer chains and chain segments. The technique is used to observe the Brownian motion and rotation of polymer molecules in solution (light scattered from the moving particles is doppler shifted) and slow relaxations in solids near the glass transition, T_G. It also can be used to determine thermal conductivity, viscosity, and other transport properties through the local density fluctuations these processes cause.

At higher frequencies interferometric means can be used to measure frequency shifts. The lower frequency limit is about *1-10 MHz* (depending on laser frequency stability). The upper limit is usually about *100 GHz* although there is no reason in principle why one could not work at frequencies several orders higher than this. Figure 1 shows that this range deals mainly with Brillouin scattering, i.e. scattering due to sound waves. The technique determines the elastic moduli (or in fluids, the compressibilities) governing sound propagation. In turn these moduli depend on high frequency collision-induced relaxations. In fluids the tumbling of smaller molecules, depending upon high frequency viscosity, falls into this range, as does the vibrational motion of small molecules and molecular segments in crystals and plastic crystals.

In Raman scattering, dealing with the highest frequency shifts, 10^{10} to 10^{14} *Hz*, spectral information is measured by diffraction grating. In this regime, light scattering is a true molecular probe, as shown in Figure 1, revealing the motions of atoms relative to each other, and sensitive to crystallinity and molecular conformation. In the lower end of this region are found the so-called longitudinal "acoustic" modes (LAM's) whose frequencies are inversely proportional to the length of straight molecular chain segments in polymers. There is currently much interest in LAM's because of the information they yield about molecular conformations.

Optical measurements have the additional useful properties: (1) we can see the part of the sample that is being probed by the light beam, (2) the beam can be focused to a spot as small as a few microns if desired for looking at very small samples or regions of samples, and (3) light scattering can be done remotely on samples located inside pressure cells, tempera-

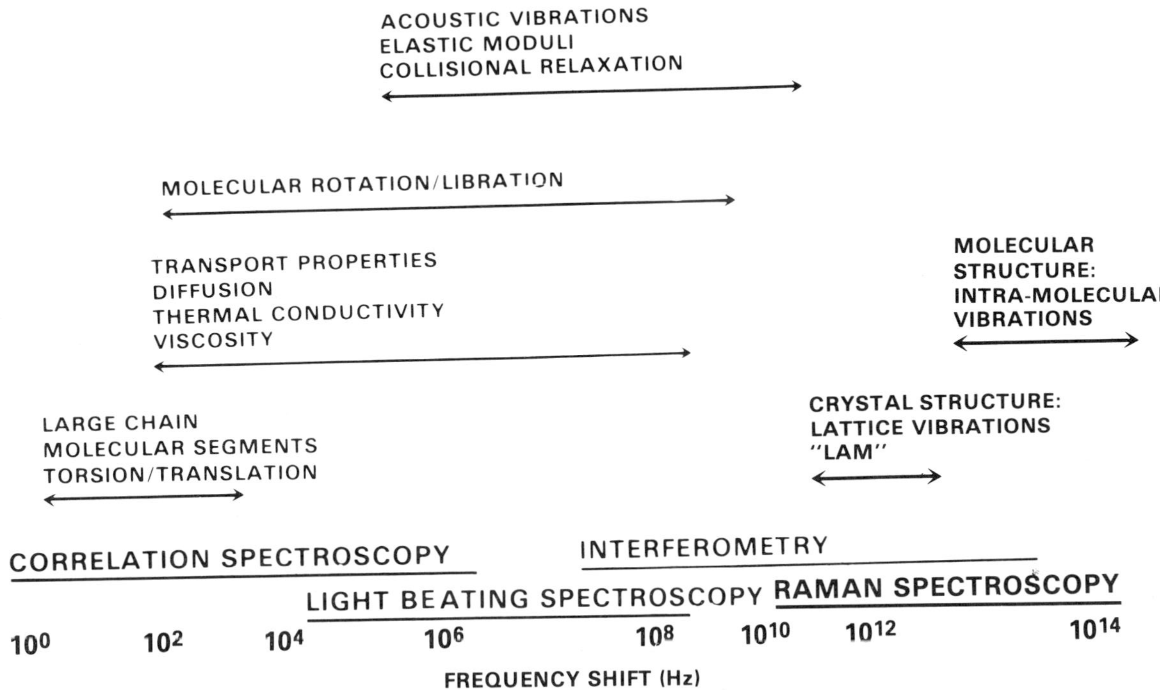

Fig. 1.—Range of frequency shifts encountered in correlation, Brillouin and Raman spectroscopy and the excitation frequencies in various samples which these techniques may be used to measure.

ture baths, or even at great distances such as the tops of factory smokestacks...

The typical sample has many processes occurring within it which cause light scattering so that there is usually a great deal of information available in a spectrum. Interpretation of this information is the chief task of the spectroscopist. Light scattering is used in combination with X-ray diffraction, neutron scattering, deuteration and other isotopic substitution, chemical labeling and substitution, computer modeling and other techniques to classify and identify spectral features seen in different circumstances in different compounds and classes of compounds. Since the discipline of light scattering is mature, a considerable library of information is now available that can be brought to bear upon new problems. The art of sorting out the details of a spectrum, drawing upon this pool of techniques and past work, has advanced considerably. Today there is the potential of utilizing this experience in the pursuit of a fuller understanding of the observed macroscopic properties of polymeric materials.

CORRELATION SPECTROSCOPY

The correlation technique[1,2] has been used in light scattering for about fifteen years. The technique has had broad application to problems both biological and physical. Among the first uses were investigations of transport properties of fluids near their critical points[2-4], and of the diffusive motions (translational and rotational) of large macromolecules and viruses for the purpose of determining their sizes and shapes.[1,3,5] Another type of experiment has been measuring chemical reaction rate constants by observing fluctuations in the fluorescence of reacting species.[6] Liquid crystal dynamics and viscoelastic constants have also been observed.[1]

Correlation spectroscopy has been used to measure doppler shifts caused by blood flow in the arteries of living animals[7], to determine particle velocities in wind tunnels and jets[8], and to obtain the viscosities of lubricants by measuring the velocity with which a slug falls through the viscous liquid.[9] This last technique is quite new and extends the range of viscosities which can be measured by the falling slug method by one or two orders of magnitude (to about 10^8 poise), while reducing the time required for the measurement from hours to minutes. To determine still higher viscosities (up to 10^{11} poise) correlation measurement of the distrubution of fluid structural relaxation times (the diffusive "Mountain mode") has been shown to give good results in lubricants subjected to very high pressures and characterizes the dynamic viscosity over a range of frequencies.[10] Figure 2 shows the distribution of relaxation times for a pair of lubricants whose short time lubricating abilities differ sharply because of the difference in their short time relaxation behavior.

A substantial amount of the work done by correlation spectroscopy has been in the area of determining relaxation times for polymer chains in dilute solution.[11] The relaxation times for depolarized light scattered in the forward direction from polymer molecules were measured and extrapolated to infinite dilution. The relaxation times obtained were compared initially to the predictions of the Rouse-Zimm model and found in general to disagree, except for the longest wavelength internal motions.[12] It was found, again contrary to prediction, that the local diffusive relaxation of fluctuations in position of small polymer segments in a large random coil

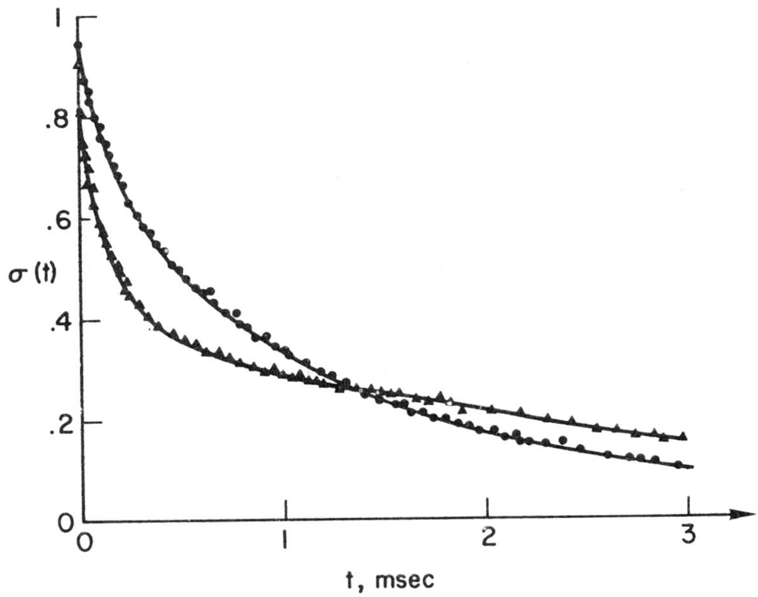

Fig. 2.—Correlation functions σ(t) for a polyphenyl ether (O) at 1.2 kbar pressure and for a short chain methyl phenyl siloxane (Δ) at 2.1 kbar. The two curves illustrate relaxation functions for liquids which have the same low frequency, long time scale viscosity. The latter polymer will quickly relax a suddenly applied stress whereas the former will not.

(Permission of C. Montrose (ref. 10) to reproduce this figure is gratefully acknowledged.)

resembled that of independent small molecules. Recent measurements by Caroline and Jones[13] show good agreement with an improved theoretical model.

All of the above cases have involved fluids or suspensions of particles in fluids. Use of a correlator with solid materials is rare because of the strong stray light scattering from such samples. However, recently the method has been used to measure low frequency motions in *PMMA*[14,15] as a function of temperature near the glass transition. Two relaxation times were observed of order *.1 sec* and *.01 sec*. The shorter time changed as a function of temperature below T_g. Its activation energy was found to be lower than that assigned to the side-chain molecular reorientations from dielectric and NMR measurements. The internal motion observed from the correlation spectroscopy experiments was thus thought to arise from a coupling of this side-chain motion with the low activation energy process of torsional oscillations of the chain segments about their equilibrium positions. The longer relaxation time observed was temperature insensitive below T_g but merged with the backbone main chain relaxation at T_g. The measurements of this internal relaxation mode above T_g in the bulk phase by correlation spectroscopy agreed reasonably well with those obtained by mechanical and dielectric measurements. The relaxation below T_g, which has not been observed by other techniques, was thought to represent a rearrangement of free volume or of "configurational entropy."

BRILLOUIN SPECTROSCOPY

Brillouin spectroscopy[16] is light scattering caused by high frequency (*10 MHz* to *100 GHz*) sound - i.e. random thermal oscillations which are always present in a sample with a non-zero temperature. The mechanism of scattering resembles that of Bragg X-ray diffraction, with a sound wave causing diffraction instead of crystal layers. As with Bragg scattering, the scattering angle determines the sound wavelength observed, but since the sound wave is moving, the scattered light is doppler shifted in frequency relative to the incident frequency. From the doppler frequency shift one can determine the velocity of sound. The absorption of sound is determined from frequency broadening of the scattered light. In turn one can then find the viscoelastic moduli of the sample and related transport properties, such as viscosity, thermal conductivity, and diffusion coefficients.[1,17] We note here that processes at such fast rates - on the nanosecond time scale - are difficult or impossible to obtain by ultrasonic methods and that X-ray and neutron techniques which do go to such frequencies are considerably less accurate and more difficult, expensive, and damage the sample.

A very large amount of work has been done on a wide variety of fluids and solids to determine their moduli, transport properties, and relaxation times. We will restrict our attention to a few examples chosen from the field of polymers. Energy in such systems is exchanged between molecular internal vibrations and external translational motions.[18] Processes which occur faster than such energy exchange encounter considerably different compressibilities and viscosities, for instance, than slower processes do.[1,19]

Paraffin oils have been examined in depolarized Brillouin scattering.[20] The tumbling of such chain molecules produces a spectral line whose width is related to viscosity. Linewidths were measured as a function of chain length and temperature and the values of the rotational relaxation time were compared with those obtained from flow birefringence. The values were similar but not identical, indicating a possible change in paraffin chain flexibility with longer chains. The relaxation times exhibited an Arrhenius temperature dependence and yielded rate-activation energies which had a chain length dependence.

Brillouin spectra have been obtained for poly (dimethyl siloxane) (*PDMS*) of molecular weight 7.7×10^4 and reveal four relaxation peaks in the hypersonic absorption as a function of temperature.[21] This data with microwave and dielectric relaxation times show Arrhenius type behavior for higher temperatures and WLF behavior below, with a sharp break at ~ 180^oK. The authors suggest this to be the Debye temperature of

The amorphous polymer poly (methyl acrylate) (*PMA*) appears to be one of the few solids in which a frequency dependent hypersonic relaxation has been observed.[22] Frequency dispersion in solid polymers often occurs at considerably lower frequencies than are encountered in Brillouin scattering so that the parameters obtained, such as the velocity, are generally the infinite frequency values. For *PMA* the glass temperature is ~ 6^oC. Relaxation is observed at temperatures above 85^oC in both the hypersonic velocity and attenuation, and the size of the relaxation phenomenon is frequency dependent. Using a single relaxation time theory the infinite frequency velocity and the relaxation time were determined. The value of the activation energy (*7 Kcal/mole*) suggests the relaxation is of the β type (side chain).

The Landau Placzek (LP) ratio - the ratio of the elastically scattered (zero frequency shift) light intensity to the Brillouin intensity - is often used as a probe of solid polymer behavior under stress. The Brillouin intensity should be relatively constant, as it is an intrinsic property of the material. The elastic scattering on the other hand is sensitive to impurities, domain boundaries, strains, inclusions, etc. Mitchell and Guillet[23] and Coakley et al.[24] have studied the effects of annealing upon the LP ratio and they have suggested that as the temperature of a sample is dropped the naturally occurring fluctuations in the material are "frozen" in below the glass temperature. The LP ratio then is related to the amount of strain energy stored in the glass.

One of the difficulties of Brillouin scattering in amorphous solids is "contrast," the ability of the spectrometer to distinguish the Brillouin lines, which are weak, from the nearby elastic scattering peak, which is very strong. In recent years considerable progress has been made in improving contrast by the technique of multipassing[25] which increases contrast by many orders of magnitude. Dil et al.[26] and, very recently, Sandercock[27] have reported use of 5 and 7 pass systems to observe Brillouin scattering in metal surfaces. Since the penetration depth of light into a metal surface is small, the technique yields elastic moduli for the region within a few hundred angstroms of the surface. Other phenomena such as the propagation of Lamm acoustic waves in thick films also can be observed.[27]

RAMAN SCATTERING

In the Raman scattering process[28,29] the motion of an atom relative to its neighbors in a molecule or lattice, causes fluctuations in the atomic or molecular polarizability. Since the first observation of the effect almost 50 years ago Raman spectroscopists have used a variety of techniques to determine the nature of the molecular motions which give rise to the Raman spectral lines. This has involved the use of infrared spectroscopy, crystallography, symmetry, substitution of isotopes of different masses onto molecules, and calculations using computer models of molecules and molecular force fields.

One body of work of particular interest in the polymer field has been the application of Raman spectroscopy to the study of crystalline polymer morphology. Many polymers crystallize by folding their long chains into short segments which lie side by side like stacks of soda straws or sheaves of wheat, forming lamellae. The frequency shift of one of the banks of the Raman spectrum, the LAM, is a function of the lamellar thickness. Initial work was done on polyethylene and the n-paraffins.[30,31] The latter were used to determine the proportionality constant between the Raman frequency shift and the inverse of the straight chain length as a function of the number of carbon atoms. Frequency shifts obtained from polyethylene samples having lamellar morphology could then be correlated with lamellar thickness assuming that the Raman band corresponded to chain segments whose fold length was that of a similar n-paraffin.[31] For polyethylene single crystals where the tilt angle between the chains and the lamellar surface could be determined independently, lamellar thicknesses determined by the Raman method agreed fairly well with those obtained using small angle X-ray scattering (SAXS).[32]

Comparison has been more difficult for samples crystallized from the melt, since the tilt angle is generally unknown. Recent work has focused on the effects of the gauche (twisted) segments in the amorphous loop region[33], and of conformational[34] and mass defects[35] on the frequency of the LAM. This data has yielded information on the nature of the chain fold in polyethylene.

Despite initial difficulty, LAM's have now been observed in the helical polymers polyethylene oxide[36], isotactic polypropylene[37,38], polyoxymethylene[38] and in the copolymers (random) tetrafluorethylene - hexafluoropropylene[39] and (block) polyethylene oxide - polypropyleneoxide[40], making it possible to use this Raman band in the study of the crystallization and structure of polymer lamellae. It should be noted that the advantage of the Raman scattering technique is that it does not depend[33] on the regularity of the lamellar stacking as is necessary for SAXS. This has permitted the observation of double lamellar populations.[32]

If the polymer chain is regarded as a uniform rod the proportionality constant between the LAM frequency and the inverse of the chain length is the sound velocity along the rod, $v = (E/\rho)^{\frac{1}{2}}$, where E is Young's modulus and ρ is the density. Thus, if an estimate of the tilt angle in the lamellar crystal is made, SAXS long period measurements can be combined with Raman frequency shifts to determine the ultimate Young's modulus of a polymer chain. Reports thus far[38,39] indicate that estimates made in this way agree with those determined from inelastic neutron scattering, but are substantially larger than those determined from wide angle X-ray scattering.

One particular class of problem which has received a great deal of theoretical attention for the past 30 years or so has been the mode structure of long chain molecules of the polyethylene type.[41,42] Considerable progress has been made in this effort and, of particular interest to us, not only have the normal modes of the ideal straight chain been identified but also the effects upon the spectra of chain kinking have begun to be understood. In addition to the LAM's which give information on the length of straight segments between kinks, spectral features have been identified (see Figure 3) which reveal the relative number of gauche and trans bonds (the "optical skeletal" bands)[43], and the ordering of chains and chain segments relative to each other (the "methylene stretching" bands[44] and crystal field splitting[45]). These bands have been used to investigate the effects of annealing on polymer lamellae thicknesses[46], the relative elastic moduli of crystalline and amorphous polymers[47], the effects upon chain order in model biological membrane chain molecules caused by phase changes and other perturbations[48], the effect of high pressure upon polymer structure and crystallinity[49], and the effect of high pressure upon chain kinking in the liquid alkanes.[50] In the latter case it has been seen that high pressure (up to *20 kbar*) causes short chain alkanes (heptane, octane) to kink up and longer chains (polyethylene) to straighten out to the "super extended" chain phase. Work of this nature is continuing now to try to determine the relation between lubricant chain conformation and "lubricity," polymer structure and strength, and the influence upon these properties of perturbations such as temperature, pressure, chain length and dispersity, sample history, etc.

CONCLUSION

We have discussed briefly correlation, Brillouin, and Raman spectroscopies and have attempted to demonstrate in a general way their use and

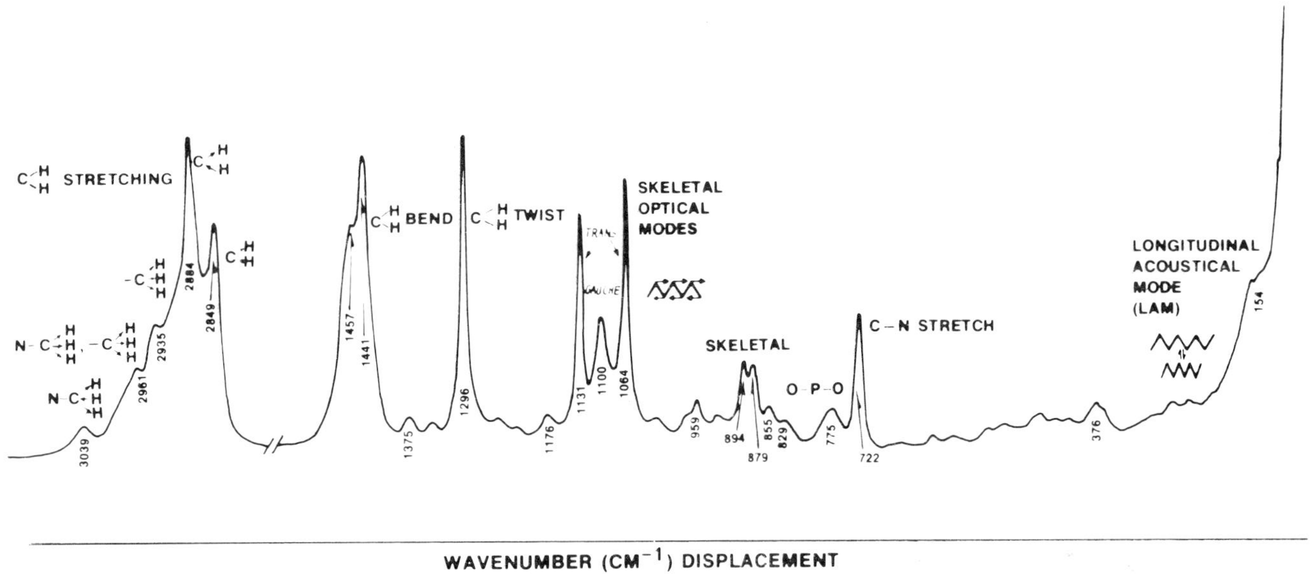

RAMAN SPECTRUM OF DIPALMITOYL PHOSPHATIDYL CHOLINE (EXCITATION SOURCE; Kr LASER AT 647 nm)

Fig. 3. Raman spectrum of the model biological membrane molecule, dipalmitoyl phosphatidyl choline, which shows many of the features typical of paraffin and polyethylene molecules. Each spectral band is labeled with the type of atomic motion causing it and the numbers give the frequency shifts in wavenumbers. (1 wavenumber = 3 x 10^{10} Hz). The LAM at 154 wavenumbers is fairly weak in this sample whereas in polyethylene and the n-paraffins it is quite strong.

range of applicability. We have tried to show some of their unique abilities while at the same time emphasizing the need to employ them in concert with other spectroscopies and mathematical modeling techniques. We believe that light scattering is unusual and exciting because it is both new and old; new in the sense of new and dramatically improved instrumentation and lengthening list of materials and fields in which it is finding use, and old in the sense of being a mature science with a great store of information and technology available to help with experimentation and interpretation.

ACKNOWLEDGEMENT

We gratefully acknowledge the support of the Office of Naval Research for some of the work reported herein.

REFERENCES

1. Cummins, H.Z. and Pike, E.R., eds., "Photon Correlation and Light Beating Spectroscopy," NATO Advanced Study Institute, Capri, Italy, 1978; also Plenum Press, New York, 1974.
2. Benedek, G.B., in "Polarisation, Matière et Rayonnement; Livre de Jubilè en l'honneur du Professeur A. Kastler," edited by the French Physical Society, Presses Universitaires de France, Paris, 1969.
3. Cummins, H.Z. and Swinney, H.L., *Progress in Optics*, Vol. 8, 1970, p. 133.
4. Chu, B., "Laser Light Scattering," Academic Press, New York, 1974.
5. Cummins, H.Z., Carlson, P.D., Herbert, J.J. and Woods, G., *Biophysical Journal*, Vol. 9, 1969, p. 518.
6. Elson, E.L. and Magde, D., *Biopolymers*, Vol. 13, 1974, p. 1; and Magde, D. and Elson, E.L., *Biopolymers*, Vol. 13, 1974, p. 29.
7. Riva, C., et al., *Investigative Ophthalmology*, Vol. 11, 1972, p. 936.
8. Birch, A.D., Brown, D.R., Thomas, J.R. and Pike, E.R., *Journal of Physics D: Applied Physics*, Vol. 6, L71, 1973.
9. Jackson, D.A. and Bedborough, D.S., *Journal of Physics D: Applied Physics*, Vol. 11, Section L1, 1978.
10. Drake, P.W., et al., *Journal of Chemical Physics*, Vol. 67, 1977, p. 1969.
11. Berne, B.J. and Pecora, R., "Dynamic Light Scattering," John Wiley and Sons, New York, 1976, p. 182.
12. Lee, W.I. and Schurr, J.M., *Chemical Physics Letters*, Vol. 23, 1973, p. 603.
13. Caroline, D. and Jones, G., *American Physical Society. Bulletin*, Vol. 23, 1978, p. 272.
14. Jackson, D.A., et al., *Journal of Physics C: Solid State Physics*, Vol. 6, Section L55, 1973.
15. Cohen, C., Sankur, V. and Pings, C.J., *Journal of Chemical Physics*, Vol. 67, 1977, p. 1436.
16. Fleury, P.A. and Boon, J.P., *Advances in Chemical Physics*, Vol. 24, 1973, p. 1.
17. Mountain, R.D., *Reviews of Modern Physics*, Vol. 38, 1966, p. 205.
18. Herzfeld, K.F. and Litovitz, T.A., "Absorption and Dispersion of Ultrasonic Waves," Academic Press, New York, 1959.
19. Lao, Q.H., Schoen, P.E. and Chu, B., *Journal of Chemical Physics*, Vol. 64, 1976, p. 3547.
20. Champion, J.V. and Jackson, D.A., *Molecular Physics*, Vol. 31, 1976, p. 1159.

21. Lindsay, S.M., Adshead, A. and Shepherd, I.W., *Polymer,* Vol. 18, 1977, p. 862.
22. Huang, Y.Y., et al., in "Light Scattering in Solids," Proceedings of the International Conference in Paris, 1971, p. 488; also *Flammarion,* 1971.
23. Mitchell, R.S. and Guillet, J.E., *Journal of Polymer Science. Part A-2; Polymer Physics,* Vol. 12, 1974, p. 713.
24. Coakley, R.W., et al., *Journal of Applied Physics,* Vol. 47, 1976, p. 4271.
25. Sandercock, J.R., in "Light Scattering in Solids," Proceedings of the International Conference in Paris, 1971, p. 9; also *Flammarion,* 1971.
26. Dil, J.G. and Brody, E.M., *Physical Review B (Solid State),* Vol. B14, 1976, p. 5218.
27. Sandercock, J.R., *American Physical Society. Bulletin,* Vol. 23, 1978, p. 387.
28. Szymanski, H.A., ed., "Raman Spectroscopy," Plenum Press, New York, 1967.
29. Craver, C.D., ed., "Polymer Characterization-Interdisciplinary Approaches," Plenum Press, New York, 1971.
30. Schaufele, R.F., *Journal of Chemical Physics,* Vol. 49, 1968, p. 4168.
31. Peticolas, W.L., et al., *Applied Physics Letters,* Vol. 8, 1971, p. 87.
32. Dlugosz, J., et al., *Polymer,* Vol. 17, 1976, p. 471.
33. Hsu, S.L. and Krimm, S., *Journal of Applied Physics,* Vol. 48, 1977, p. 4018.
34. Reneker, D.H. and Fanconi, B., *Journal of Applied Physics,* Vol. 46, 1975, p. 4144.
35. Fanconi, B. and Crissman, J., *Journal of Polymer Science. Part B: Polymer Letters,* Vol. 13, 1975, p. 421.
36. Hartley, A., et al., *Polymer,* Vol. 17, 1976, p. 355.
37. Hsu, S.L., Krimm, S., Krause, S. and Yeh, G.S.Y., *Journal of Polymer Science. Part B: Polymer Letters,* Vol. 14, 1976, p. 195.
38. Rabolt, J.F. and Fanconi, B., *Journal of Polymer Science. Part B: Polymer Letters,* Vol. 15, 1977, p. 121.
39. Rabolt, J.F. and Fanconi, B., *Polymer,* Vol. 18, 1977, p. 1258.
40. Hartley, A.J., et al., in Proceedings of the 5th International Conference on Raman Spectroscopy, Schulz, Freiberg, 1976, edited by E.D. Schmid, Plenum Press, p. 496.
41. Tasumi, M. and Shimanouchi, T., *Journal of Molecular Spectroscopy,* Vol. 9, 1962, p. 261.
42. Schachtschneider, J.H. and Snyder, R.G., *Spectrochimica Acta,* Vol. 19, 1963, p. 85, and also Vol. 19, 1963, p. 117.
43. Lippert, J.L. and Peticolas, W.L., *National Academy of Sciences. Proceedings,* Vol. 68, 1971, p. 1572.
44. Larsson, K., *Chemistry and Physics of Lipids,* Vol. 10, 1973, p. 165; and Larsson, K. and Rand, P., *Biochimica et Biophysica Acta,* Vol. 326, 1973, p. 245.
45. Boerio, F.J. and Koenig, J.L., *Journal of Chemical Physics,* Vol. 52, 1970, p. 3425.
46. Koenig, J.L. and Tabb, D.L., *Journal of Macromolecular Science. Part B. Physics,* Vol. B9, 1974, p. 141.
47. Peterlin, A., et al., *Journal of Polymer Science. Part B: Polymer Letters,* Vol. 9, 1971, p. 583.
48. Gaber, B.P. and Peticolas, W.L., *Biochimica et Biophysica Acta,* Vol. 465, 1977, p. 260; Priest, R.G. and Sheridan, J.P., in "Liquid Crystals and Ordered Fluids," edited by J.F. Johnson and R.S. Porter, Plenum Press, New York, 1978, p. 209.
49. Wu, C.K. and Nicol, M., *Journal of Chemical Physics,* Vol. 58, 1973, p. 5150; and *Chemical Physics Letters,* Vol. 24, 1974, p. 395.
50. Schoen, P.E., Priest, R.G., Sheridan, J.P. and Schnur, J.M., *Nature,* Vol. 270, 1977, p. 412.

DISCUSSION

R. A. DASKIVICH, General Motors Research Laboratories: I was happy to see that at the end of the talk you brought up the subject of a material that is of some interest in engineering -- polyethylene. In working with this material, what approximately would be your signal to noise ratio, or the effect of signal to noise ratio that you have to play with?

J. M. SCHNUR: We are working with molecular weights ranging from 16,000 to about one million. We are really concerned about purity in specifying material properly. In the materials that we annealed for a very long time, some decomposition occurred although the DSC gave us purities of 99.5% which is bringing our signal to noise ratio down to about 3 to 4 for these low frequency modes. However, in the French samples which were not annealed, we found some very interesting characteristics. We were able to get a signal to noise ratio of 20 to 100 with the interferometer coupled to the double monochromator. The signal counts were quite low and we had to count for a long time, but the signal to nosie ratio was quite good. If we do not use the interferometer, in both cases we do not see anything at all. We are enhancing contrast from 10^{12} to 10^{13} with the double monochromater by itself, to about 10^{17} with the interferometer coupled in there; that is, 10^5 times increase in contrast.

X-RAY PHOTOELECTRON SPECTROSCOPIC STUDIES OF P.T.F.E. AND GRAPHITE FLUORIDE SURFACE INTERACTIONS

P. Cadman and G. M. Gossedge

ABSTRACT

Polytetrafluoroethylene (P.T.F.E.) and graphite fluoride polymers have possible applications as dry bearings due to their low coefficient of friction and "unreactive" nature. Reactions of these polymers with tin, cadmium and indium, by either heating the mixture of metal/polymer together at temperatures up to 500°C, or evaporation of the metals onto the clean polymer surface, have been found to produce fluoride ions and elemental carbon. Rubbing P.T.F.E. onto both nickel and stainless steel also produces fluoride ions.

Diffusion of P.T.F.E. molecules through the surface layers of the polymers occurs at a significant rate and may be important in the mode of action of bearings of this material. P.T.F.E. in two differing forms has been found transferred to the counterface after rubbing with polymer in both vacuum and air.

INTRODUCTION

The paper by Briscoe and Tabor[1] has comprehensively reviewed the current state of knowledge of the tribological interactions of polymers and has highlighted the areas where many questions remain unanswered. In this paper we should like to discuss further, certain aspects mentioned in that review: namely, the mechanism of polymer transfer, the occurrence and importance of possible chemical reactions between fluorinated polymers and metallic counterfaces, especially in regard to P.T.F.E. and graphite fluoride.

Relatively little attention has been shown in the elucidation of the importance of possible chemical reactions that may occur in tribological processes even though the increasing utilization of polymers[2,3] and polymer based composites[4] as solid lubricants has generated a need for such basic information. Pratt[5] believes the action of lead/bronze/P.T.F.E. composites is connected with the reaction that occurs between this polymer and lead or lead oxides above 300°C. X-ray photoelectron spectroscopic (X.P.S.) examination of lead - P.T.F.E. mixtures heated at 500°C in nitrogen atmosphere showed that partial decomposition of the P.T.F.E. had occurred with the concomitant production of fluoride (F^-) ions.

Polymer-metal interactions such as suggested by Pratt[5] have been found at elevated temperatures[6] (less than 500°C) such as might exist on the atomic scale at surfaces during rubbing. These reactions were found by heating mixtures of the polymer with metals and metal oxide powders in a Differential Scanning Calorimeter (D.S.C.) and are shown in Table 1. X-ray photoelectron spectroscopic examination of the residues from D.S.C. showed the presence of fluoride ions and the visual appearance of the residue was that of a black carbonaceous mass.

Recently[7] we have found other reactions of metals with P.T.F.E. by heating the mixtures of the powders in aluminum pans (1:1 ratio by volume) in the D.S.C. to 497°C in nitrogen atmosphere. These results are shown in Table 2. Thermal events which cound not be assigned to phase transitions were exotherms at 390°C for cadmium/P.T.F.E. Indeed one such reaction of cadmium with P.T.F.E. when heated in gram quantities under a vacuum was so exothermic that it caused violent explosions[8].

Another fluorinated polymer which has a low coefficient of friction is graphite fluoride. This polymer is prepared by the action of fluorine on graphite at temperatures above 400°C. The unsaturation and planar structure of the graph-te is changed to a puckered ring structure with fluorines substituted alternately above and below the rings[9]. Mixtures of metals with graphite fluoride were heated in the D.S.C. under similar conditions to those above. Reactions were found for cadmium/graphite fluoride, tin/graphite fluoride and for indium/graphite fluoride (Table 3).

The black residues left from the reactions of P.T.F.E. and graphite fluoride with the metals mentioned above were examined by X.P.S. The presence of fluoride (F^-) ions and elemental carbon was found together with undecomposed polymer and the metal. Such spectra are shown in Figure 1.

TABLE I.—BINARY SYSTEMS CONTAINING METAL AND P.T.F.E.

Metal*	Thermal Behavior	Comments
Cu (75%)	P.T.F.E. transition peak only only	No reaction occurring between Cu and P.T.F.E. up to 500°C or after soak at 450°C
Sn (65%)	Sn melting peak and P.T.F.E. transition peak	No reaction between Sn and P.T.F.E. scanning up to 500°C
Pb (65%)	Pb melting peak and P.T.F.E. transition peak superimposed	No reaction between Pb and P.T.F.E. scanning up to 500°C after soak at 450°C
Ag (70%)	P.T.F.E. transition peak only	No reaction between Ag and P.T.F.E. scanning to 500°C or after 450°C soak
Fe (60%)	P.T.F.E. transition peak only	No reaction between Fe and P.T.F.E. scanning up to 500°C
Zn (55%)	Zn melting peak and P.T.F.E. transition peak only	No reaction between Zn and P.T.F.E. scanning to 500°C or after soak at 450°C
Sn (72%)	Large exotherm after prolonged soak at 450°C	A reaction is evident at this temperature between Sn and P.T.F.E.

*Percentage in brackets is metal composition.

TABLE II.—BINARY SYSTEMS CONTAINING METAL AND P.T.F.E. HEATED IN D.S.C.

Metal	Thermal Events	Comments
Al	P.T.F.E. transition	No reaction up to 430°C
Cd	Cd mpt. + P.T.F.E. transition exotherm at 390°C	Reaction at 390°C black residue
Cr	P.T.F.E. transition	No reaction up to 430°C
Cu	"	"
Fe	"	"
Ni	"	"
Mg	"	"
Sn	Sn mpt. P.T.F.E. transition exotherm at 400°C	Reaction at 400°C black residue
Zn	mpt. Zn P.T.F.E. transition	No reaction up to 430°C

TABLE III.—BINARY SYSTEMS CONTAINING METAL AND GRAPHITE FLUORIDE HEATED IN THE D.S.C.

Metal	Thermal Events	Comments
Ag	None	No reaction up to 500°C
Al	"	"
Cd	Exotherm 410°C mpt. Cd	Reaction at 410°C black residue
Cr	None	No reaction up to 500°C
Cu	"	"
Fe	"	"
Ni	"	"
Pb	Pb mpt.	"
Sn	Exotherm 460°C mpt. Sn	Reaction at 460°C black residue
Zn	Zn mpt.	No reaction up to 500°C
In	In mpt. Exotherm at 300°C	Reaction at 300°C black residue

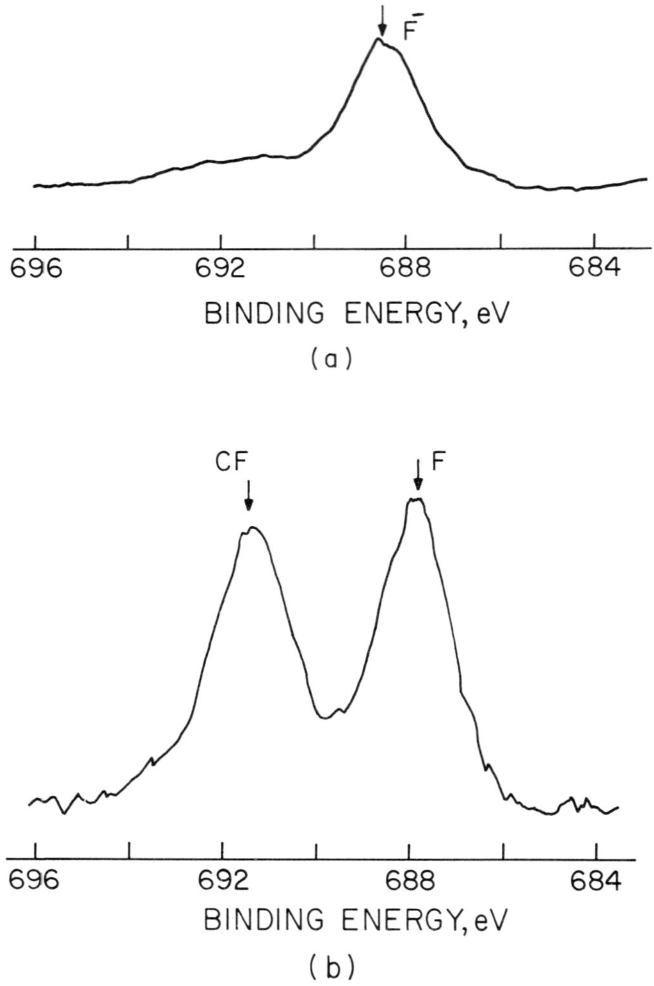

Fig. 1.—(a) Cd-PTFE DSC residue, F(1s) spectrum.
(b) Sn-CF DSC residue, F(1s) spectrum. Heating Cd and Sn with PTFE and graphite fluoride, respectively, in the DSC produces chemical reactions and concomitant fluoride formation.

To further illustrate the chemical reactivity of these binary systems and to eliminate any possible interference/catalytic effects of surface contaminants; tin, cadmium, and indium were separately evaporated onto a "clean" polymer surface. "Clean" polymer surfaces were prepared by scraping with a razor blade. X.P.S. characterization of these surfaces prepared in such a manner have shown that less than about 5% of monolayer impurities are present after such treatment[(7)]. An atomically clean layer of the metals was deposited onto the polymer surfaces. This was done for tin and indium under ultra high vacuum (10^{-8} Torr) in a preparation chamber directly attached to the X.P. spectrometer. (For practical reasons the cadmium was evaporated in Edwards evaporator at 10^{-6} Torr.) Table 4 shows the results of these evaporations. Fluoride ions were again detected and were most probably present at the metal-polymer interface.

Although experimental conditions in the above results were such as might possibly occur in tribological situations, such reactions have also been found for iron-graphite films run to failure[12] where fluoride ions have been detected. We have detected the presence of fluoride ions at the metal polymer interface[10] when P.T.F.E. was rubbed against stainless steel and nickel counterfaces under atmospheric conditions. Transfer of P.T.F.E. to the metal counterface occurs (detected by X.P.S. examination) and obscures the interfacial region. Argon ion etching of the surface film removed the polymer layer but unfortunately depth profiling by the etching itself produced fluoride ions and elemental carbon. Instead the polymer layer was removed by careful scraping with a razor blade. X.P.S. examination showed the presence of fluoride at the interface as well as P.T.F.E. on the metal surface (Figure 2).

Similar treatment for graphite fluoride rubbed onto stainless steel and nickel under atmospheric conditions also revealed metallic fluoride at the metal-polymer interface.

The relative importance of such processes in adhesion, friction and wear of bearings involving such polymers is still uncertain, but the occurrence of such chemical reactions should at least be considered.

Clean P.T.F.E. (prepared and characterized as above) has been rubbed separately against lead and tin counterfaces in U.H.V. These metals were evaporated on to cleaved basal plane graphite substrates at less than 10^{-8} Torr and the surfaces of the atomically clean and unoxidized metal.

TABLE IV.—EVAPORATED METALS ON A POLYMER SUBSTRATE

Metal	Polymer	Metal Fluoride F(1s)* Peak Position (eV)
Tin	P.T.F.E.	4
"	$(CF_{1.0})n$	3.5
Cadmium	P.T.F.E.	6.5
"	$(CF_{1.0})n$	6
Indium	P.T.F.E.	4
"	CF	4
Silver	P.T.F.E.	Fluoride not found
Lead	"	4.5
Gold	"	Fluoride not found

*Position at higher K.E. relative to P.T.F.E. F(1s).

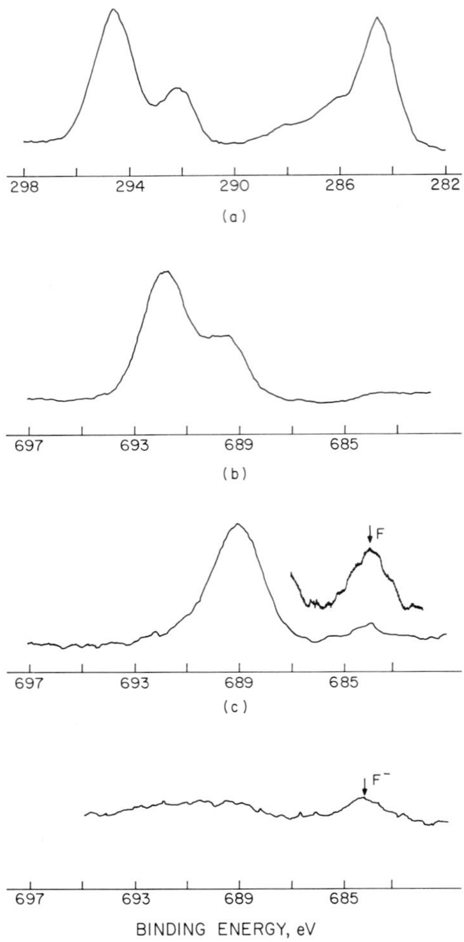

Fig. 2.—Stainless steel rubbed with PTFE.

(a) Initially rubbed with PTFE, C(1s) spectrum, (L.H.S. of peaks x 3.3). Left peak is PTFE peak. Other peaks are C(1s) contaminator, shifted due to oxide.

(b) Initially rubbed with PTFE, F(1s) spectrum. Left peak does not move with applied bias. Right peak moves with applied bias. (L.H.S. of peaks x 1).

(c) Rubbed with PTFE and scraped with razor blade, (x 3.3 on L.H.S. of peak). Blown up peak arrowed F (x 10 on L.H.S. of peak and also peak below it).

(d) After rubbed with PTFE and scraped with razor blade, sample is argon etched. Peak arrowed F remains, (x 3.3 on L.H.S. of peak).

P.T.F.E. transfers very readily onto such surfaces. X.P.S. examination showed the presence of two differing forms of P.T.F.E. on the metal counterface[11] when an electrical bias of about 40 volts was applied across the sample (Figure 3). One pair of carbon (1s) and fluorine (1s) peaks moves with the bias and one pair does not. We shall call these two types of P.T.F.E. as "charging" and "non-charging" although the terms are relative.

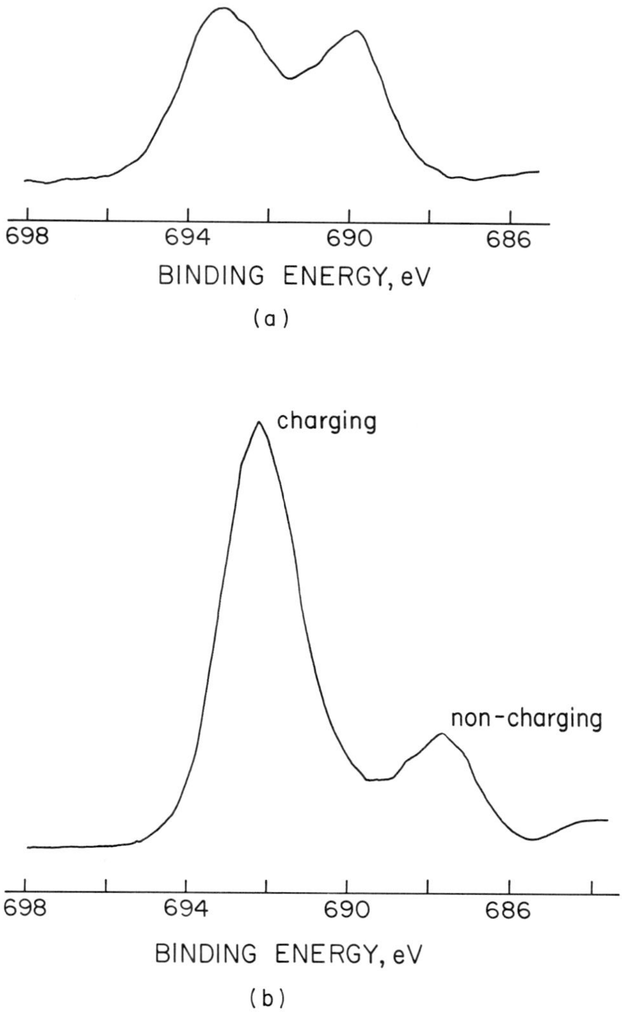

Fig. 3.—Lead film rubbed with PTFE under UHV conditions.

(a) Initial rubbing with PTFE, F(1s) spectrum. Rubbing metals with PTFE produces two polymer species (charging and non-charging) on the counterface.

(b) Additional rubbing with PTFE, F(1s) spectrum. Left peak arrowed as "charging", right peak arrowed as "non-charging".

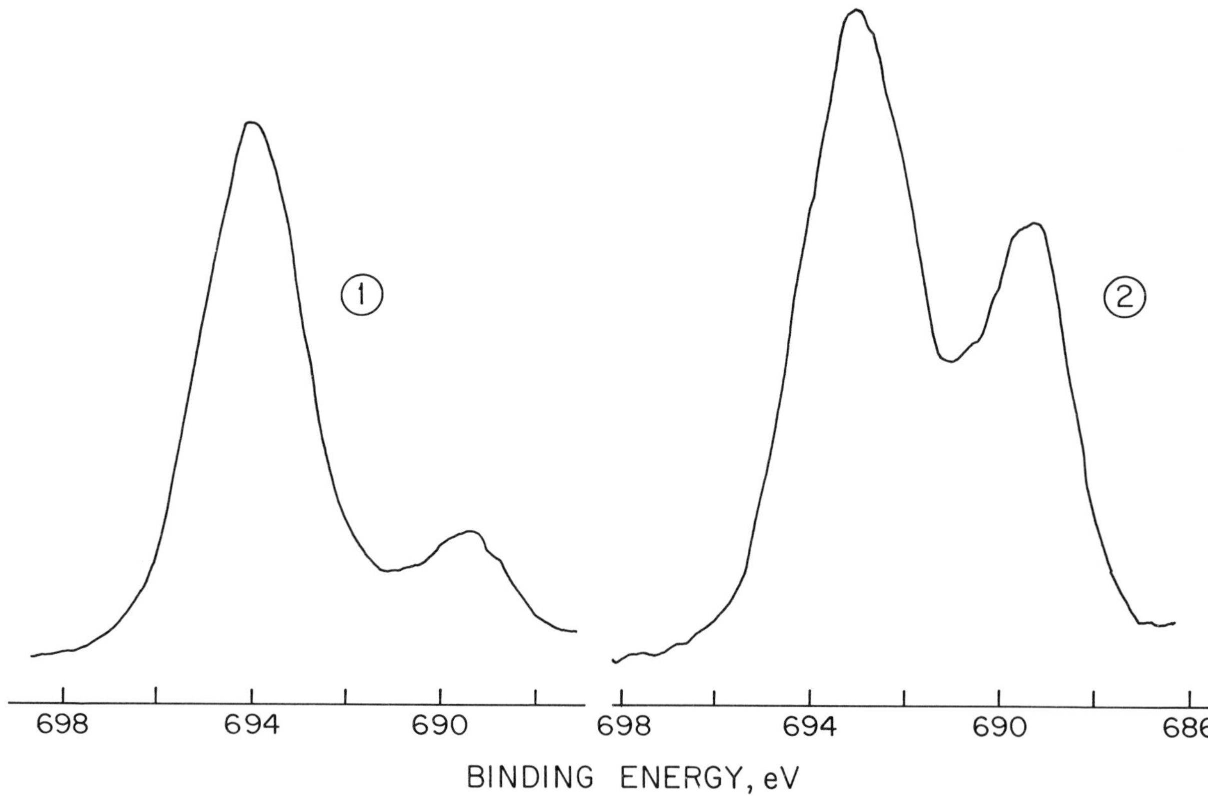

Fig. 4.—Angular studies of PTFE species on lead film, F(ls) spectrum. First set of peaks headed (45°), second set of peaks headed (75°). First set of peaks intensity x 3.3 on L.H.S.

It is to be expected that the "non-charging" P.T.F.E. is in more intimate electrical contact with the metal such as in the form of thin layers while the "charging" P.T.F.E. is in less close electrical contact with the metal and perhaps is in the form of lumps. These types of transfer have been envisaged by Tabor[13] and by Tanaka et al.[14] occurring in the polymer transfer mechanism. Variation of the angle of the sample w.r.t. the electron energy analyzer enhanced the signal from the non-charged layer relative to the charge one in support of such an assignment (Figure 4). The relative intensities of the metal substrate together with a knowledge of the "average" depth from which the electrons escape in P.T.F.E.[15] are compatible with a thickness of the layer of the order of 60 Å, well within the range postulated earlier[13,14]. There was no firm evidence for any chemical reactions in these cases but the thickness of the polymer layer made the detection of fluoride ions difficult.

Removal of the samples from the spectrometer and application of the "Cellotape test" showed that this film was quite strongly adherent to the metal counterface.

The mode of action of P.T.F.E. composite bearing materials is thought to be due to the fact that P.T.F.E. transfers to the counterface on rubbing and so interaction is P.T.F.E. on P.T.F.E. rather than P.T.F.E./metal. In our experiments involving argon ion etching of P.T.F.E. it was noticed that although this process preferentially removed the fluorine from $-CF_2-$ groups leaving residual carbon and -CF, on standing the X.P.S. signals associated with P.T.F.E. gradually reappeared over several hours, suggesting that the polymer molecules were labile enough to diffuse to the surface at a significant rate at room temperature. Such a process may have important significance in the regeneration of P.T.F.E./P.T.F.E. surfaces in bearings. Whether these polymer molecules are low molecular weight species present in the surface layers as suggested by Schonhorn[16] in adhesion studies, low molecular weight species produced by chain scission from the X-rays used for analysis, or even high molecular weight species which are more labile than may have been thought, is not clear at present.

REFERENCES

1. Briscoe, B.J. and Tabor, D., These Proceedings.
2. Richardson, M.O.W., *Wear,* Vol. 17, 1971, p. 89.
3. Bowden, F.P. and Tabor, D., "Friction and Lubrication of Solids," Clarendon Press, Oxford, Part I, 1950, p. 164; Part II, 1964, p. 216.
4. Pascoe, M.W., *Tribology,* October 1973, p. 184.
5. Pratt, G.C., *Institution of Mechanical Engineers. Proceedings,* Vol. 181, 1966-67, p. 58.
6. Cadman, P. and Pocock, G., *Wear,* Vol. 37, 1976, p. 129.
7. Cadman, P. and Gossedge, G.M., Unpublished Work.
8. Cadman, P. and Meadows, B., Unpublished Work.
9. Rudorff, W., *Advances in Inorganic Chemistry and Radiochemistry,* Vol. 1, 1954, p. 230.
10. Cadman, P. and Gossedge, G.M., *Journal of Materials Science,* Vol. 14, 1979, in press.
11. Cadman, P. and Gossedge, G.M., *Wear,* Vol. 54, 1979, p. 211.
12. Atkinson, I.B., Waghorne, R.M. and Swift, P., *Wear,* Vol. 37, 1976, p. 123.
13. Pooley, C.M. and Tabor, D., *Nature Physical Science,* Vol. 237, 1972, p. 88.
14. Tanaka, K., Uchiyama, Y. and Toyooka, S., *Wear,* Vol. 23, 1973, p. 153.
15. Cadman, P. and Gossedge, G.M., *Journal of Electron Spectroscopy,* Vol. 13, 1978, p. 1.
16. Schonhorn, H. and Ryan, F.W., *Journal of Adhesion,* Vol. 1, 1969, p. 43.

DISCUSSION

D. TABOR, Cambridge University: I would like to say that I liked the way you have tackled this problem. But I have a question. If you were to take the wear properties of a piece of PTFE against two metals, one which gives no chemical change and the other which gives this change in chemistry, do you find any substantial difference in the wear properties or not? One needs to know all the things you have been describing, but one would also like to know what effect it has on the wear properties of that combination.

P. CADMAN: Well, it is something we could perhaps do together. Your set-up in your laboratory for these wear properties we always set up to examine the sort of interfacial regions. If we come together perhaps we could answer that question.

D. H. BUCKLEY, NASA: You mentioned that your thin PTFE film is approximately 60Å. Maybe I missed it, but what technique did you use to establish that?

CADMAN: Very approximately. There were some other measurements on these fluorinated polymer films. We actually measured what we call escape depth. This is an average, if you want to put it very crudely, of how far beneath the surface the electrons come from. With Auger we know it is between five and ten angstroms perhaps with XPS it is probably more. I could discuss this afterwards because it is not something I could answer in a minute. We measured these escape depths for polymers like polyethylene, polysterene, PTFE, and graphite fluoride. We found it to be of the order of 30 to 60 angstroms, although I might say there is some controversy over these values at the moment.

XI. WEAR MONITORING

MONITORING OF WEAR

V. C. Westcott

ABSTRACT

Using Ferrography it is possible to monitor wear by analyzing debris generated by the wear process. Techniques used to identify various alloys and the use of morphology of the wear particles in indicating the mode of wear are discussed. Ferrography is more accurate than other monitoring methods since it is only concerned with the product of wear rather than the entire wearing system.

INTRODUCTION

It is fair to say that one of the most difficult problems in the design, operation and maintenance of machinery is the assessment of wear. If the factors which influence wear rate are to be evaluated accurately, it is necessary to test mechanisms and even complete machines under realistic conditions. Although the problem is different for actual machines, as distinguished from wear test rigs, in both types of tests accurately evaluating the nature and amount of wear has remained a problem.

With the exception of measurement of the radio-activity of lubricants as the result of the wearing away of parts that have been made radioactive artificially, such evaluations have been based on the measurement of the dimensions, weight loss, or the size of the wear scar of the worn part. Such measurements have two inherent disadvantages: they are based on the measurement of a large object in order to evaluate a small quantity and they generally require the disassembly of the machine. As a result, the measurements are usually made only after a prolonged test that provides enough wear. Therefore, only a few measurements can be made during a test. This is analogous to evaluating the performance of a power house in terms of weekly readings of a watt-hour meter. Usually there are insufficient data to detect and study the "break-in" of the parts and the onset of failure modes. Equally important is the fact that each disassembly and reassembly of a machine results in a new "break-in" cycle, so that a test often cannot be followed progressively. It is this fact that results in such wide variations, particularly in the case of rolling fatigue.

The development of Ferrography has made it possible to base measurements of wear on the debris rather than on the part. It is analogous to measuring the size of a saw cut from the sawdust rather than from the tree.

The quantitative assessment of wear from the debris in the lubricant requires:

1. The existence of a functional relationship between the amount of debris in the oil and the total wear (or the wear rate depending on the mean lifetime of the particles).

2. The ability to obtain a representative sample of the oil.

3. The ability to recover the wear particles from the oil and distinguish them from associated non-wear particles either by selectively precipitating only the wear particles or by distinguishing them from non-wear debris after precipitation.

4. The ability to measure the quantity of wear debris.

5. The ability to distinguish the types of wear particles to identify the part wearing and the mode of wear.

The following paragraphs take up each of these issues in the order they are listed.

THE RELATIONSHIP BETWEEN THE CONCENTRATION OF WEAR DEBRIS IN THE OIL AND TOTAL WEAR OR WEAR RATE

It is natural to assume that as a machine wears, debris enters the oil and the concentration of debris in the oil increases indefinitely as the wear progresses. In other words, as wear continues the oil would have a progressively higher concentration of wear particles in it. However, this is usually not the case.

For example, an oil filter would set an upper limit to the concentration. As the concentration increases, the rate of loss of particles to the filter increases until a concentration is reached where the loss rate equals the generation rate. The more effective the filter, the sooner such an equilibrium is reached.

There are other ways in which particles are lost and, in fact, all machines exhibit such an equilibrium even when there is no filter present. Particles are lost by a variety of mechanisms such as settling to the bottom of the oil sump, sticking in gelatinous coatings on the parts, being ground down into smaller and smaller particles by the moving parts of the machine and by being swept out with oil that is lost to the system.

If it is assumed that the probability of a particle being lost is independent of the presence of other particles, the loss rate for a particular type and size of particle is proportional to the concentration of that type and size. Once a general equilibrium is reached, an equilibrium for all of the sizes and types being considered, it is possible to write an expression for the loss rate.

Assuming that the volume of oil circulating in the machine is constant, the loss rate under a given set of circumstances may be expressed as:

$$Loss\ rate = KCV$$

where c is concentration in grams/m^3

v is volume in m^3

K is proportionality constant of the machine expressed in grams/gram/sec or sec^{-1}

If the machine is operating under steady state conditions and the wear rate, i.e. the particle generation rate, is constant, the concentration of particles in the oil will rise until equilibrium is reached. At ecuilibrium,

$$Wr = KVC$$

where Wr is wear rate expressed in grams/sec.

Therefore, the wear rate is proportional to the concentration of wear particles in the oil.

When first started up, machines that have been serviced with relatively particle-free oil show an exponential increase in the concentration of wear particles, the slope of which is dependent on the average lifetime of a particle in the oil. Time constants for typical machines range from a few minutes to several hours. For example, the time constant for a jet engine might be twenty minutes.

Figure 1(a) illustrates the relationship of particle concentrations of 1 to 2 μm particles vs time for machine that has been broken-in and operates in a normal benigh wear mode. The oil was clean and relatively particle free at the start. Such a simple concentration vs time relationship is not always exhibited. Frequently the machine will go through a break-in period in which the waaring surfaces conform to their mating members and the particle concentration shows a rapid rise followed by a fall oo equilibrium, as illustrated in Figure 1(b).

A consequence of the relationship between loss rate and concentration is that machines that have long equilibrium time constants are those machines that operate with a high concentration of particulate matter in the oil. Oridinarily machines that operate with few particles in the oil have a short probable particle life and must, therefore, reach an equilibrium quickly. The actual wear rate of hhe machines does not play a part in determining the time to reach equilibrium.

OIL SAMPLING

In order to obtain a sample of the lubricant that is representative of the wearing condition it is necessary to consider two general questions - the time the oil is taken and the place from which it is taken.

If the oil sample is taken while the machine is running, gooc mixing can be obtained. Even the largest particles will not have time tc settle. In general, it is good practice to obtain "live" samples if practicable. However, there is one complication: the particle density in the oil raiidly follows any changes in the operating parameters of the machine. For example, gas turbines have been observed to have more particles in the oil during acceleration to the maximum throttle setting than during deceleration toward idle. Significant changes can be observed to occur in a matter of a minute or two. One explanation of this behavior is that such engines "hide" oil

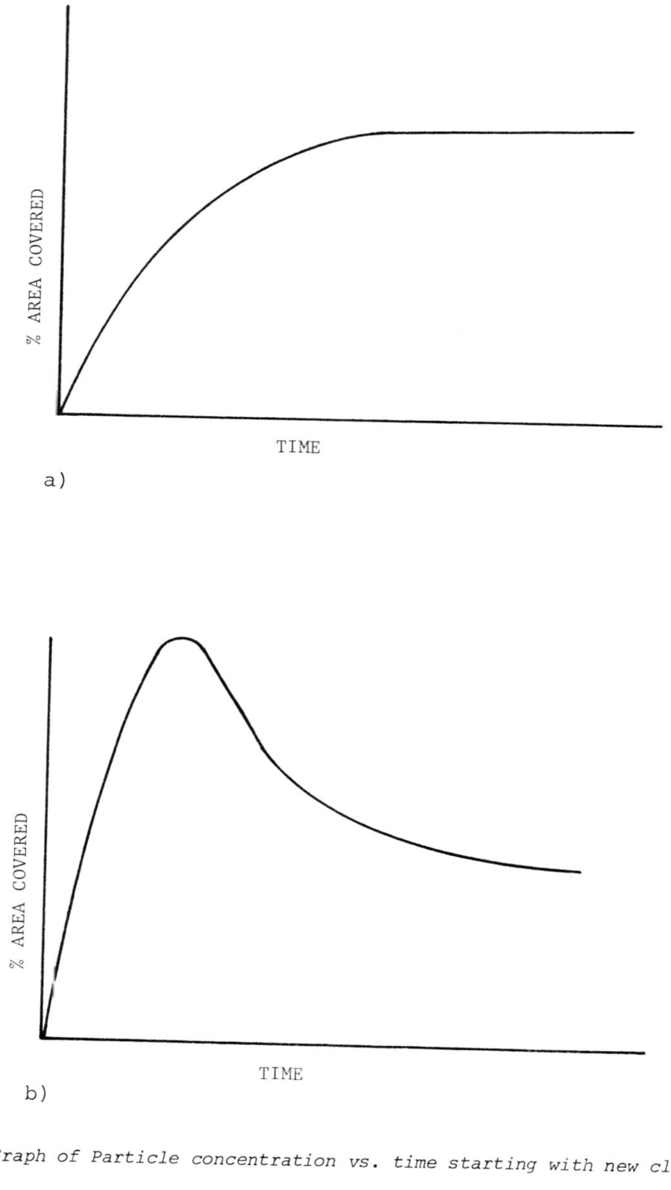

Fig. 1.—Graph of Particle concentration vs. time starting with new clean oil.

in cavities ir the engine. The amount of oil hidden depends on the engine speed and may be as high as 30 to 50% of the total volume of oil in the tank. In such a case the rapid return of the oil on slow down influences the particle concentration. Figure 2 is a recording of the mass of oil in a gas turbine tank as a function of time. It is to be expected that similar

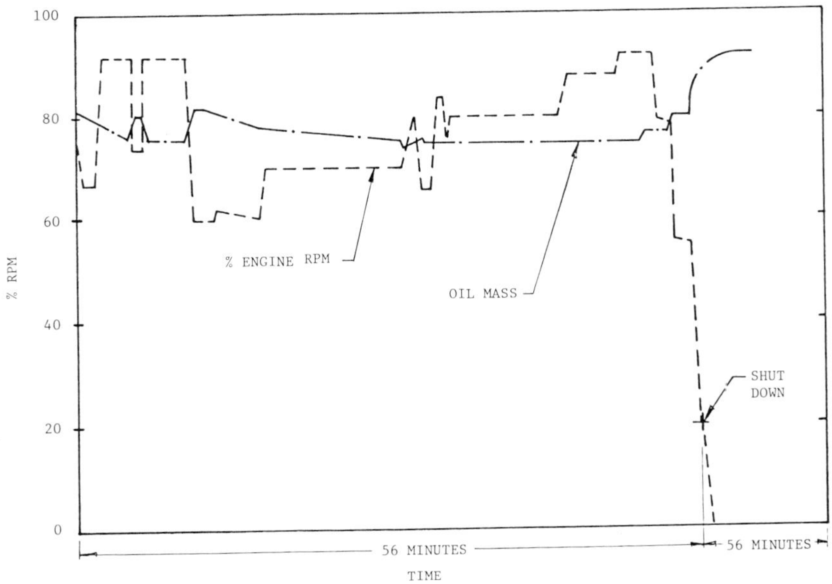

Fig. 2.—Plot of oil mass and engine R.P.M. vs. time for a gas turbine.

complexities may be experienced in other machines but in general those that have been observed are of a lesser magnitude than with gas murbines.

If the oil is to be sampled during running, a simple T connection to an oil return line from a sump may be sufficient. However, it is important to allow a small quantity of oil to flow before taking the sample so that trapped particles are flushed away.

The other alternative is to sample after the machine is shut down. Usually the oil is taken from the oil tank. Figure 3 illustrates several methods of doing this. The use of a permanently installed stand pipe is to be preferred because the possibility of human error is minimized. Caution should always be exercised to prevent scraping up particles from the bottom of the tank since they may be very old and unrepresentative of current operating conditions.

When test rigs are sampled it is often feasible to remove all of the oil at each sample time. For example, tests have been run in which the oil, about 200 cc, has been changed every 15 minutes. In such a case the particle concentration is not necessarily in equilibrium and it may be feasible to measure the total wear in the interval rather than the wear rate.

The time between machine shut down and the taking of the sample is usually not critical providing the sample is taken at a point sufficiently below the surface[1].

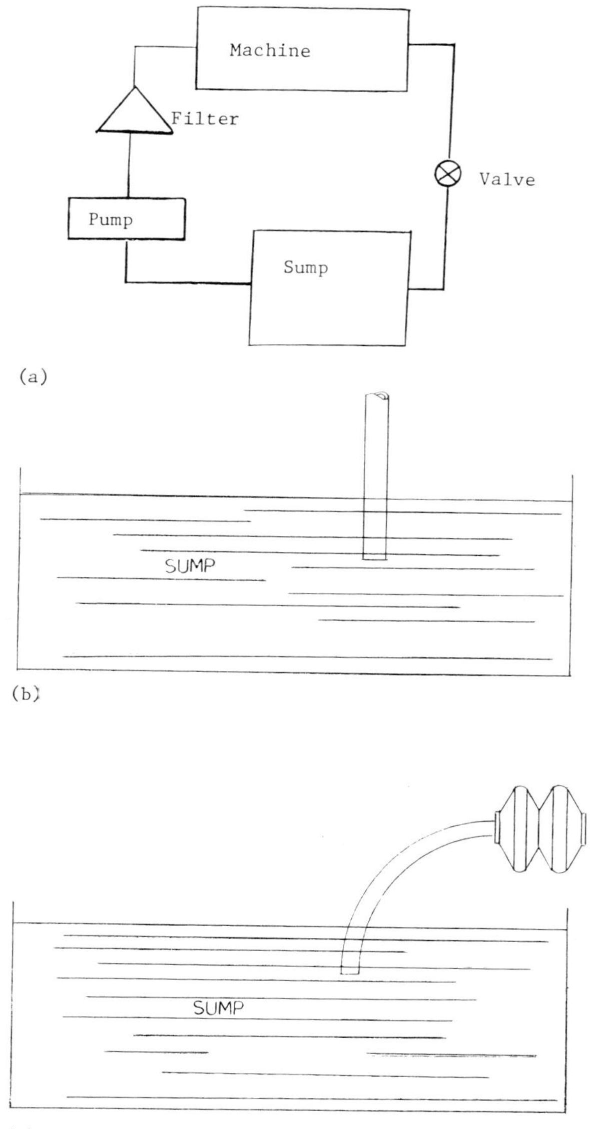

Fig. 3.—Oil Sampling Methods. (a) Sampling from a valve. Be sure to allow oil to flush out valve before taking sample. (b) Use clean glass tube or clean metal tube to sample from middle of sump. Use thumb on top of tube to create partial vacuum to support fluid column in tube. (c) Plastic tube and plastic bottle with bellows.

THE SEPARATION OF WEAR PARTICLES FROM NON-WEAR PARTICLES

In many cases in the laboratory, and in most cases in the field, wear particles are mixed with a miscellany of particles from other sources, such as dust from the air. In the case of some types of farm machinery, for example, the quantity of siliceous particles sucked into the engine may exceed the number of wear particles by several times. Figure 4(a) is a photomicrograph of the surface of a membrane filter through which 2 cc of oil from an automatic transmission has been filtered. Even this relatively clean oil leaves a deposit of metallic wear particles. Figure 4(b) shows the entry deposit of a Ferrogram after processing 2 cc of oil from the same source(1). The siliceous contaminants have passed over the Ferrogram leaving only ferrous particles or particles that have rubbed against ferrous surfaces. In addition, plastic particles and particles of friction polymer deposit as a result of ferrous metal or oxides embedded in the matrix. Figure 4(c) shows friction polymer that was so deposited.

MEASURING WEAR DEBRIS

The quantity of wear particles on a Ferrogram is determined by measuring the fractional area covered by particles at different positions along the Ferrogram. If the fractional area covered by the particles is measured in a given field of view in the microscope and the average height of the particles determined with the focusing mechanism, the volume of the particles in view may be determined.

Figure 5 shows a cross calibration with a Quantimet television type particle analyzer. Since the television system is absolute in the sense of counting the area of particles that are indicated (by the equipment) it is possible to study the performance of the densitometer when viewing various types of particles.

The quantity of particles that are present in a given field of view on a Ferrogram are measured in terms of the percent area covered by the particles.

The measurement is made in terms of the amount of light that is blocked by the particles. If the particles are opaque, as they are in the case of metals, the reduction in light intensity is proportional to the fraction of the area covered by the particles. An optical densitometer mounted on the microscope is calibrated to read directly in percent area covered by the particles.

As mentioned above, the product of the average height of particles in any one field of view times the area covered by particles can be used to determine the volume of particles present. If this is done over the whole area of a Ferrogram, the total volume present may be estimated. This is usually very tedious; but fortunately, in most cases, more than 90% of the volume of wear particles present is located at the entry deposit; so that if the volume of the entry deposit is estimated, the total volume present can be judged. The measuring procedure is discussed in Reference 2.

In actual use it is seldom of value to estimate the volume of particles directly, the percent area covered by selected particles being a more versatile index of the concentration of wear particles.

When machines are running in a stable mode, most of the wear particles are of a type called rubbing wear particles. They are thin flakes, usually

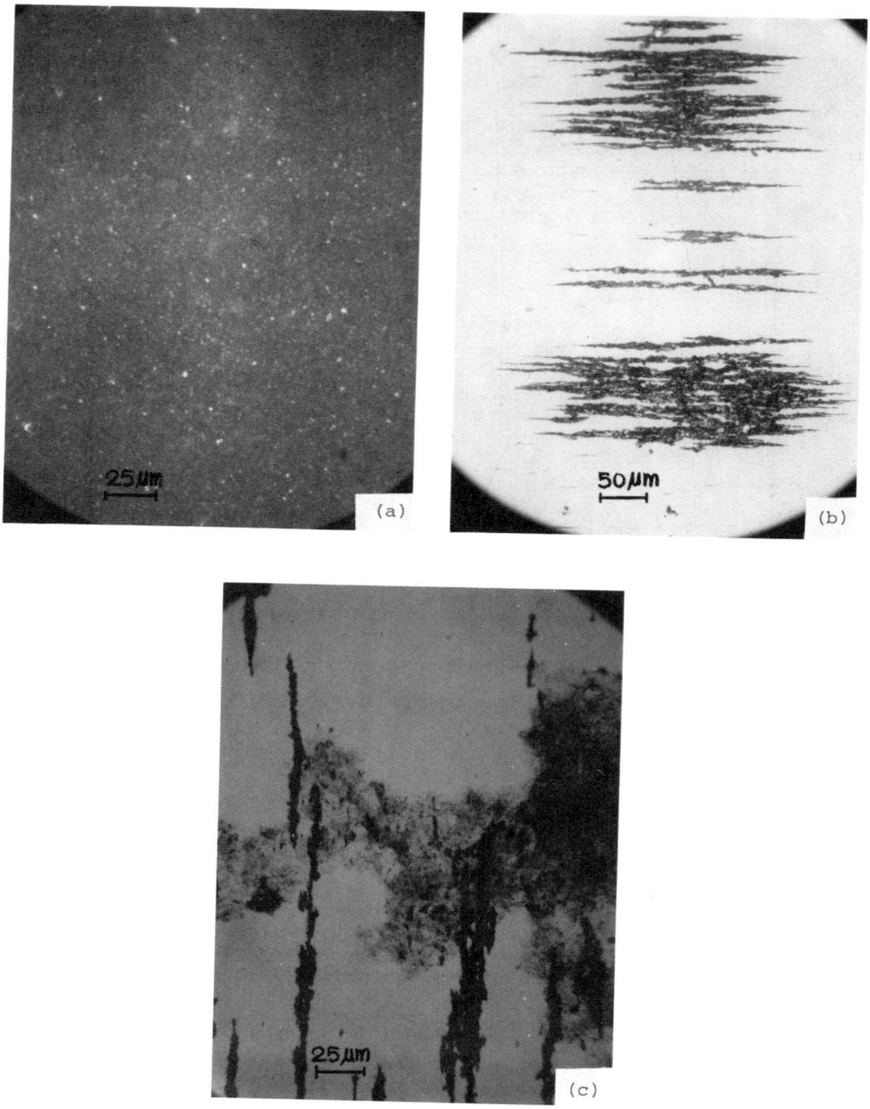

Fig. 4.—(a) Phctomicrograph of membrane filter through which 2 cc of automatic transmission oil has been filtered. (b) Entry deposit of a Ferrogram. (c) Ferrogram of Friction polymer.

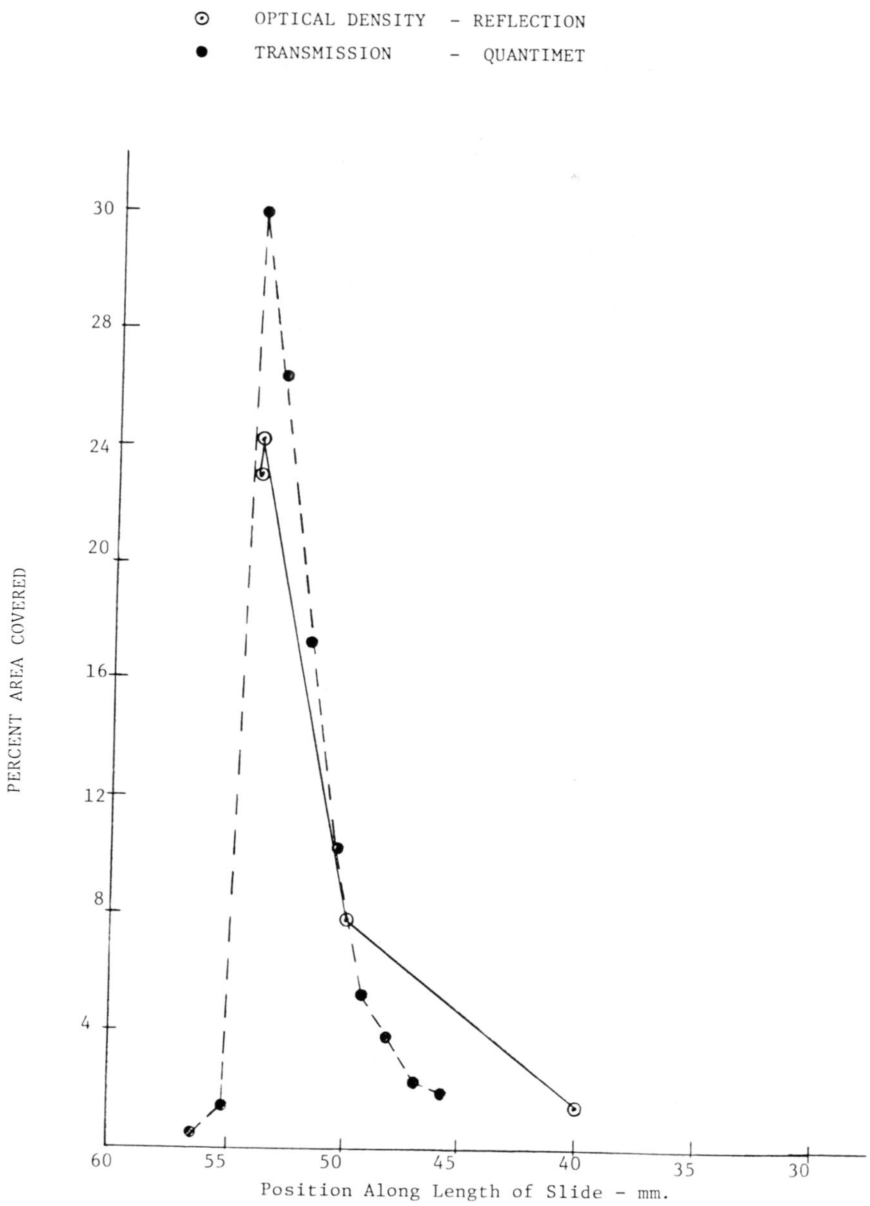

Fig. 5.—Comparison between optical density measurement and quantimet entry region.

having a maximum dimension between 5 and 15 micrometers for steel. Typical rubbing wear particles are shown in Figure 6. Their morphology has been compared with corn flakes and indeed they are often curved like corn flakes.

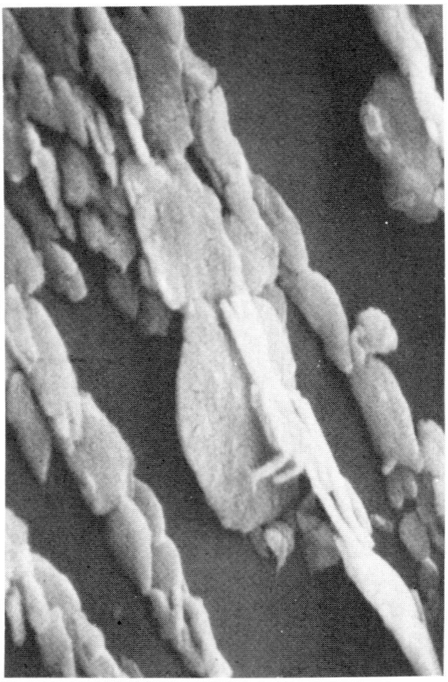

Fig. 6.—SEM photomicrograph of rubbing wear particles.

The rubbing wear particles are the result of the sliding of metal surfaces under hydrodynamic or boundary lubrication[3]. Sliding wear in the two regimes of wear, Regime 1 and 2, results in these thin wear particles. They are generated in the shear mixed layer. The flakes are composed of crystallites having a range of order of about 20 nm (200 Å) or less. It is hypotheiszed that the original grain structure of the metal has been broken down and the surface layer mixed by the shearing action of the opposing surfaces.

The shear mixed layer can migrate tangentially thousands of times the thickness of the layer. In this respect the metal smears like butter and is observed to cover over scratches and other details of the original surface topography, Figure 7.*

If the load on the opposing surfaces is increased or if the oil film becomes too thin, the sliding surfaces begin to adhere and much larger particles are generated, Figure 8. Such particles, from steel surfaces, often have maximum lengths of 150 to 200 μm and are sometimes 50 to 100 μm wide

---

*The author wishes to acknowledge the courtest of Prof. Nam P. Suh for supplying the original photograph.

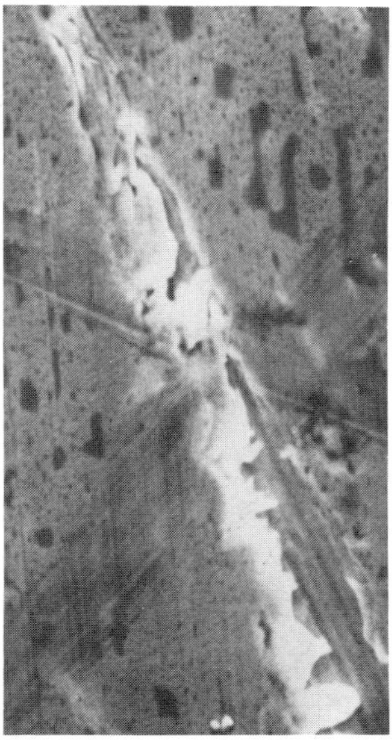

Fig. 7.—Photomicrograph of shear mixed layer covering over a scratch.

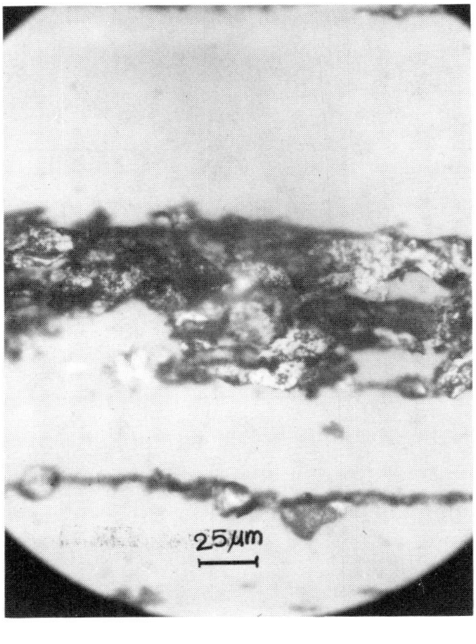

Fig. 8.—Regime 3 wear particles.

and 5 to 10 μm thick. They are generally longer than they are wide and the corresponding cavities in the wearing surface from which they are torn are seen to be longer in the direction of motion. The particles are torn out by the motion.

Two regimes of oxidative wear and a very severe wear mode in which the metal fails in volume can also be exhibited with sliding motion.

It will be noted that all of the severe wear regimes, 3, 4, 5 and 6, along with the serious fatigue wear modes, generate particles that are much larger in dimension, particularly in their volume, than the rubbing wear particles of regimes 1 and 2. Therefore, the sudden appearance of much larger particles in the oil is symptomatic of the inception of a severe wear mode.

An empirical quantity referred to as the severity of wear index I_S is obtained from area covered readings on a Ferrogram. The index has the property that it increases rapidly at the inception of a severe wear mode. Thus:

$$I_S = (A_L + A_S)(A_L - A_S) = A_L^2 - A_S^2$$

where A_L = percent area covered at the entry deposit of the Ferrogram (See Figure 9).

A_S = percent area covered at 50 mm (Fig. 9).

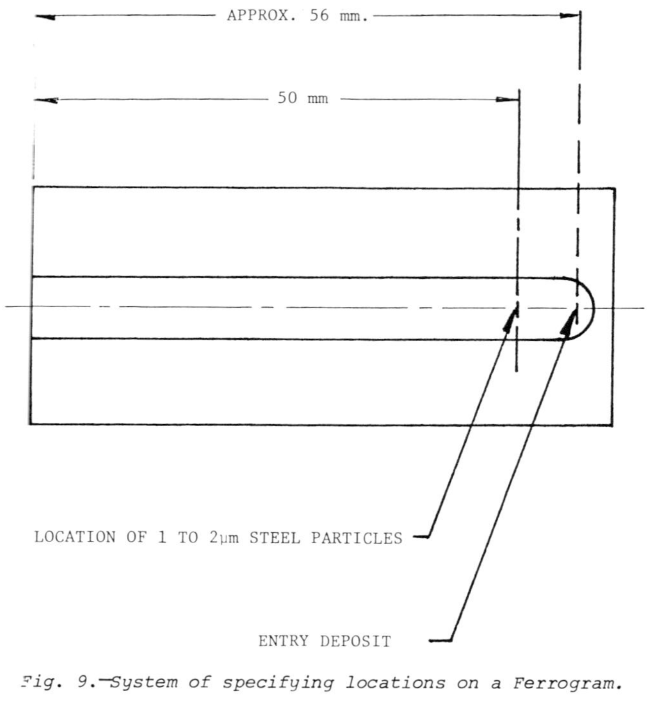

Fig. 9.—System of specifying locations on a Ferrogram.

For steel the major dimension of particles at the location of A_L is usually 5 μm or larger. At 50 mm, the location of the A_S reading, the sizes of the particles are usually between 1 and 3 μm.

The term $(A_L + A_S)$ is indicative of the concentration of wear particles while the term $(A_L - A_S)$ reflects the bias toward large particles. The quantity $(A_L + A_S)$ represents the intensity or amount of the wear rate while $(A_L - A_S)$ is a measure of its severity.

The wear mechanisms may be such that $(A_L + A_S)$ is large but $(A_L - A_S)$ approaches zero or is even negative. Such is the case if the machine has a high wear rate but does not generate large particles. This means that no surface is exfoliating large pieces of metal. The wear is, therefore, distributed and is of a general nature. Examples of such wear occur when sand enters the oil and when the oil becomes corrosive.

On the other hand, if the wear rate is high and if one or only a few surfaces are contributing the wear, large particles must be generated in order to result in the high volume of wear. In such a case A_L will be larger than A_S , and the product of the two terms of I_S will be large.

The severity of wear index is unfortunately not independent of the dilution of the oil and it is, therefore, necessary to take into account the amount by which the oil is diluted before making a Ferrogram. It is common practice to dilute the oil sample sufficiently before making a Ferrogram to prevent overloading the Ferrogram with particles. As a machine is failing it may be necessary to dilute the oil by 10, 100 or even 1000 times. In such cases A_L and A_S are reported as the readings observed on the optical densitometer multiplied by the dilution factor. The result is that A_L and A_S may be larger than 100%. On occasion the severity of wear index may climb into the thousands.

BREAK-IN WEAR

After assembly, when a machine first starts to run, very large irregular particles enter the oil. Undoubtedly some of these are the direct result of machining operations. Others are generated as high points are stripped or broken off, Figure 10(a). Even the rubbing wear particles are larger than normal, e.g. about 15 μm max.

In addition, a variety of long particles from the shear mixed layer are seen, Figure 10(b). These particles are generated on the top of grinding marks and other long parallel ridges. They are not generated after break-in.

As the machine wears, the generation rate of these larger particles drops precipitously. The quantities A_L , A_S and I_S , drop. I_S decreases rapidly and usually assumes a low value characteristic of the design of machine.

The concentration of wear particles as expressed by A_L and A_S and the severity of wear index I_S indicates quantitatively the wear state of the machine, but it does not contain all the information to determine how a machine is wearing or what its condition is. The problem is that the usual machine is much too complex to be adequately quantified by two or

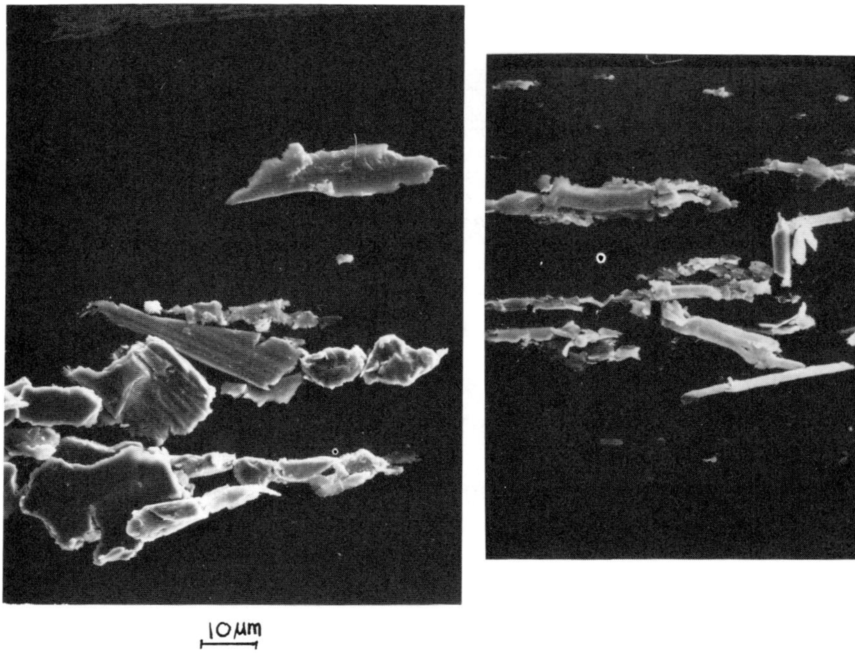

(a) (b)

Fig. 10.—(a) Severe wear particles, break-in
(b) Long break-in wear particles.

three numbers. Ordinarily the quantities A_L , A_S and I_S are used to determine whether further analysis is desirable and worth the effort. In those instances a Ferrogram is made and examined to determine the types of particles present. This is the area where the recognition of the types of particles is valuable.

Too much has been published on these matters to make it desirable to go into the details here, but it can be pointed out that certain classes of particles make it possible to distinguish rapid catastrophic wear from severe but less dangerous wear and even benign wear. This ability is invaluable in preventing unnecessary tear fowns of expensive machines and, of course, the prevention of teardowns adde to reliability by eliminating the infant mortality rate associated with the reassembly of machines.

IDENTIFYING STEEL, HIGH ALLOY STEEL, CAST IRON

Ferrograms may be heated to cause certain classes of ferrous wear particles to be distinguished. For example, the Ferrogram may be heated for

90 seconds on a hot plate at 330° C thus causing all low alloy steel particles to turn bright blue. Cast iron, stainless steel, certain other categories of ferrous particles and nickel remain close to their original color. Further heating of the Ferrogram to a higher temperature for 90 seconds causes cast iron particles to become speckled blue somewhat like an Easter egg while the low alloy steel particles return to their original color[12]. By estimating the percentage of each particle type present in combination with measuring the percent area covered by all metal particles in a given field of view, it is possible to track the amount of wear of each of low alloy steel, cast iron, nickel, stainless steel and certain other ferrous metals. Aluminum and bronze behave in a characteristic manner when present on Ferrograms that have been heated.

The technique of heating Ferrograms has been adapted to the monitoring of diesel engines. Here the wear of low alloy steel parts such as the shafts, cams, etc., has been separated from wear of the cylinder liners[5].

Frequently, specific types of wear such as fatigue spalling of gears and ball bearings, generate particles that are characteristic of the mechanism. Gears generate fatigue particles that often have scuff marks on their surface. Ball bearings, on the other hand, may generate tiny steel spheres ranging from 1 to 5 μm in diameter in micro cracks that precede fatigue spalling. In some cases such spheres are generated by the millions.

Cutting wear particles represent one of the easiest types to classify and, if present in quantity, are the harbingers of failure because one surface must penetrate another to generate them. They are not confined to wear by abrasives but may be generated when any hard surface interpenetrates another, as might occur when gears are out of line. When abrasive particles lodge in the soft metal of a journal bearing they may cut the shaft and generate cutting wear particles.

Curiously the presence of abrasives such as sand does not always result in cutting wear. If the abrasive rolls, or shatters, rubbing wear may be increased. This is a much less dangerous condition than the generation of cutting wear.

The interpretation of Ferrograms is too extensive a subject to be discussed in detail here, but a class of organic particles deserve special attention. In addition to metallic wear particles, the oil often contains copious quantities of various organic wear products. They are deposited because steel wear particles and particles of the oxides of iron become embedded in them. Their type and quantity are often of great significance in diagnosing the operation of a machine. Friction polymers, debris from gaskets including the abrasive fillers in gaskets, worn plastic parts and gel-like deposits from the plasticizers in plastic hoses are seen. In one case the presence of particles of a birefringent plastic signaled the wear of a plastic gear in the machine. This was a significance because prior to the detection of the particles of plastic it was not recognized that such parts were in the machine. Extensive literature about the particles of wear exists and photographs of the most significant types has been embodied in an atlas of wear particles[6]. Work continues on this project and additional sections of the Atlas will be available in the future.

QUANTITATIVE FERROGRAPHY

It is tempting to try to compare readings of the density of particles on a Ferrogram (usually expressed in percent area covered) with other methods of measuring the quantity of wear debris in oil especially spectrometric readings. Usually such attempts fail, primarily for two reasons. The first is that Ferrogram readings represent the concentration of specific sizes of particles. For the same total weight of particles per unit volume of oil, the readings change if the size distribution changes. Secondly, in the case of direct comparison with spectrometers it is found that spectrometers are less sensitive to metals in the form of large particles than to the smaller particles. Generally speaking the spectrometers undervalue metal in the form of particles larger than 3 to 5 μm in major dimension. On occasion, however, most of the wear particles of a metal may be large. This is the case when, for example, a gear train contacts its aluminum housing. In such cases the strong bias of the size distribution of the aluminum particles toward large sizes (> 10 μm) will cause the spectrometer readings to be low by a substantial ratio, sometimes by as much as an order of magnitude. On the other hand when the wear particles are small and particularly if they are of the rubbing wear type (flakes) gravimetric and spectrometric estimates of the particle concentration of steel, agree closely(7).

The question arises as to how repeatable Ferrographic readings may be. Figure 11 shows a series of D. R. readings taken on oil samples, each of which was taken from the same bottle of oil. However, such repeatability

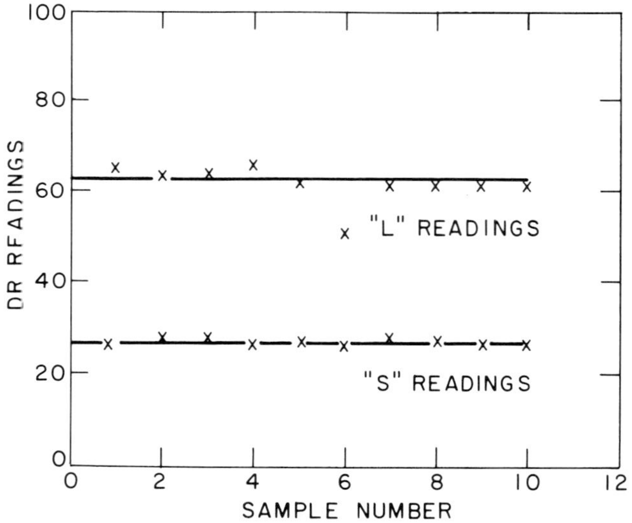

Fig. 11.—Successive readings of samples taken from the same bottle. (Full scale = 100 DR units).

is not obtained easily. It is vitally important to understand how to ensure that the wear particles are in fact dispersed evenly in the container from which the samples are taken. Frequently particles adhere to the walls of the bottle, trapped in a gel or wax. Heating the oil to 65° C, combined with vigorous shaking will disperse the particles adequately in most cases. Without heating virtually no amount of shaking will dislodge some of the particles. Once an oil sample is heated it may usually be used for one or two days at room temperature without further heating.

Figure 12 shows the results of a gravimetric calibration of a D.R. Ferrograph. The lines shown are the result of a linear regression analysis

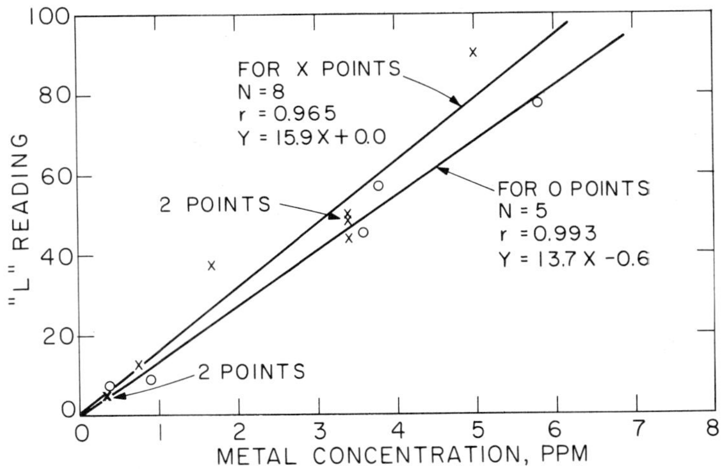

Fig. 12.—Gravimetric calibration DR Ferrograph.

of two sets of readings corresponding to two independent dilution sequences. Wear particles from a transmission were first separated using the D.R. Ferrograph as a separator. They were washed to remove other particles so that they could be weighed and redispersed in oil. Specific concentrations were obtained by diluting samples of this oil.

The "L" readings are the result of the deposit of the largest particles, mostly larger than 5 μm. The "S" readings, Figure 13, are the result of measuring the deposit in the location where particles of 1 to 2 μm are located. The scatter of readings is greater for the large particles than for the small ones, partially as the result of the smaller number of large particles resulting in poorer statistics for the population. Fewer of the larger particles are needed to generate the readings and, therefore, random variations in this smaller number produces a larger scatter of the "L" readings. The divergence of the two lines implies that the number of large particles differed in the two samples used for the dilutions.

The "S" readings vs. concentration on the other hand shows much more uniformity which provides evidence that the scatter in the "L" readings is a genuine reflection of the populations since the same electronic apparatus was used to measure both sizes.

Figure 14 shows a similar calibration of an emission spectrometer vs. gravimetrically determined concentration. The data were obtained on samples submitted to a standard oil analysis (S.O.A.P.) station. More details about the method of sample preparation, results, etc., are given in Ref. 7.

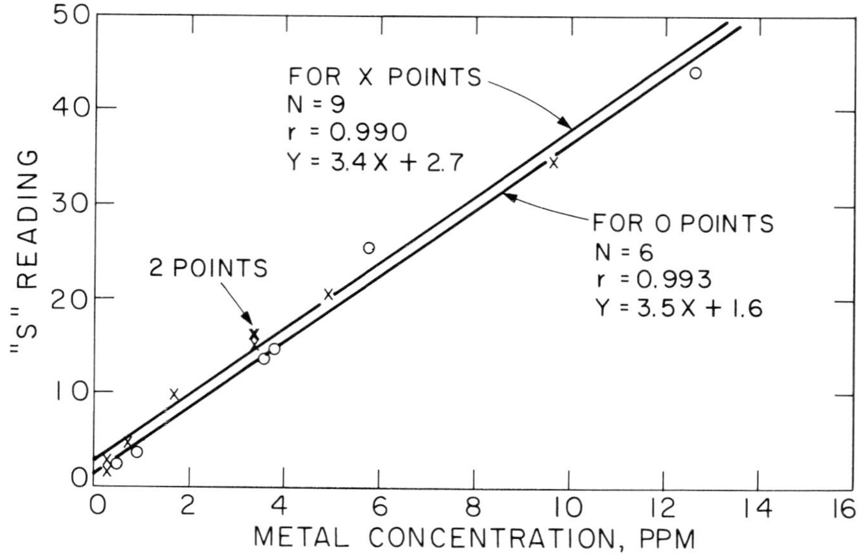

Fig. 13.—Gravimetric Calibration.

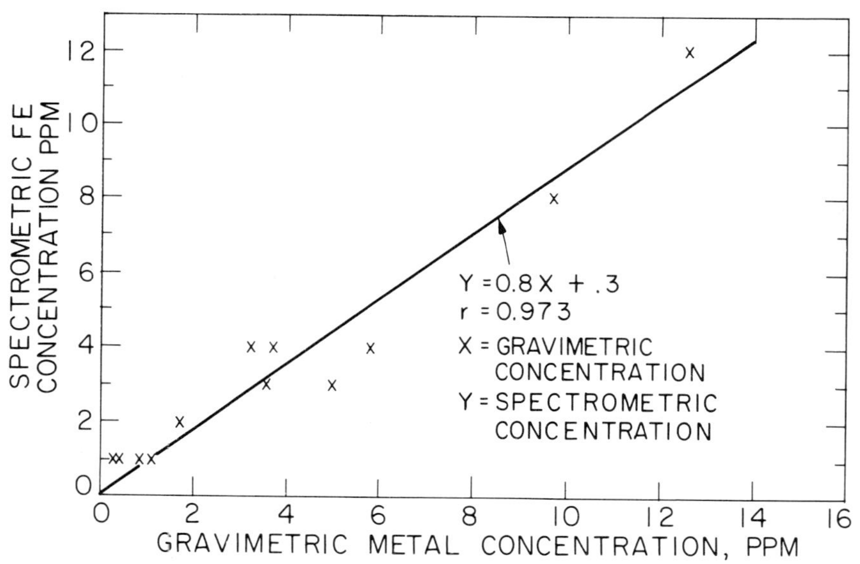

Fig. 14.—Spectrometric vs. Gravimetric Concentration.

SUMMARY

In summary, the quantitative monitoring of wear is feasible. The quantative size distribution and concentration of the wear particles is significant in determining the state of the machine and any likely failures.

The types of particles present is a function of the type of part wearing and its wear mode. In a sense, the quantitative size distribution reflects how fast the machine is wearing and how severe the wear is. The direct observation of the particles, on the other hand, gives information on the type of wear and often on the kind of part that is wearing. Frequently the direct observation of the particles is crucial in determining what action is required to correct the machine or whether the machine should remain in service.

REFERENCES

1. Scott, D., Seifert, W.W. and Westcott, V.C., *Scientific American,* Vol. 230, No. 5, May 1974, p. 88.
2. Pocock, G. and Gadd, P., *Wear,* Vol. 35, 1976, p. 161.
3. Reda, A.A., Bowen, E.R. and Westcott, V.C., *Wear,* Vol. 34, p. 261.
4. Barwell, F.T., et al., *Wear,* Vol. 44, 1977, p. 163.
5. Hoffman, M.V. and Johnson, J.H., "The Development and Application of Ferrography to the Study of Diesel Engine Wear," to appear in *Wear.*
6. "Wear Particle Atlas," Final Report under Contract No. N00156-74-C-1682 for Naval Air Engineering Center, Lakehurst, New Jersey, July 1976.
7. Anderson, D.P. and Silva, R.S., "The D.R. Ferrograph, Design, Calibration and a Field Application," to be published in *Lubrication Engineering.*

DISCUSSION

A. W. RUFF, NBS: In an engine or a mechanical system we are not only interested in wear volume but also in the particle size distribution. Did you find any problem in recovering large wear particles from complicated systems? There may be sinks where the particles may be trapped from the lubricant.

V. C. WESTCOTT: That is absolutely the case. Indeed, one has to consider the effects of the filter. Filters that are capable of removing particles larger than 3 μm are quite efficient. When we use these filters the density of the larger particles is extremely low, as one simply gets the particles that are on their way to the filter. Nevertheless, by increasing the size of the oil sample we get a representative sample for the analysis. This is done very often and it works. The other thing that one may do is to use a sampling filter. The practice should be to use a small high gradient magnetic filter (such as the one developed by Sala Magnetics) so that one can quickly remove the element and recover the wear particle from the element. We are presently working on that and we hope to get a proper filter. There are, of course, questions such as how one should connect the filter to the oil system.

K. TRIPATHI, IIT Research Institute: When we work metals by hotrolling, grinding, and so on, we produce an oxide film of the order 15 to 50 Å thick. I am wondering if these oxides you are seeing are, in fact, produced as you take the debris out of the oil and expose it to air.

WESTCOTT: No, I do not think so. These particles are much thicker than anything you have mentioned. I do not think that we are looking at that thin surface oxide at all. I believe it is too thin to produce interference colors and we would ordinarily not see it.

TRIPATHI: Interference colors can be produced with 5 to 10Å thick films.

WESTCOTT: If you were to produce interference colors it is evident that the wave length of the light is in the visible spectrum. Since the refractive index of red iron oxide is about 2.5, I would be very surprised if you could get interference colors from films much less than a quarter wave length of visible light.

MONITORING OF MACHINE WEAR BY USED OIL ANALYSIS

J. H. Johnson

ABSTRACT

The overall factors that need to be considered in the oil analysis of debris in operating machines are reviewed. Particularly significant is the fact that the filter of a machine contains the most significant debris that characterizes the machine's mode of wear. A comparison of the parameters that spectroscopy, Ferrography and radioactive tracer wear measurement methods detect is outlined. Ferrography is shown to be an advantageous measurement method since it is able to separate particles by size. The issues related to the use of Ferrography are discussed.

BACKGROUND

The advancement of technology has resulted in lubrication systems that contain complex wear processes. With the development of systems such as gas turbines, reciprocating internal combustion engines, hydraulic systems, gear boxes, etc., monitoring the wear trends of the internal components in these systems has become necessary. Much of the earlier wear research, due to the complex nature of the subject and the lack of sensitive oil analysis procedures, has been limited to a simple bench system or component analysis basis. Often this research has been conducted with the parts or lubricants highly stressed beyond their normal operating conditions. Ferrography now provides engineers with a new tool to analyze the debris in complex lubrication systems under realistic conditions.

As background information, it is useful to summarize the types of engineering situations where it is desired to study the effect of machine design and operating variables on wear. These types of situations are as follows:

1. Laboratory performance tests
2. Laboratory durability tests
3. Field engineering tests - performance and durability
4. Field data from typical user operations.

It is also useful to break down the various variables of a machine that affect the wear rates and the resulting debris analyzed in the oil. These variables are as follows:

1. Design variables, i.e., bearing-loads, diameters, materials, lengths, etc.

2. Operating variables, i.e., loads, speeds, oil temperatures, etc.

3. Filtration system design, i.e., type of filter, filter media, full-flow or by-pass, primary-secondary filters, etc.

4. Lubricants, i.e., viscosity characteristics, additives, engine or machine performance characteristics, etc.

From an engineering viewpoint, there is concern about two basic characteristics in a complex lubricated machine system. First, it is desirable to achieve low friction and hence high efficiencies. Second, there is a requirement of long life, with a high degree of reliability, i.e., no machine failure due to abnormal wear at any of the wearing surfaces. Oil monitoring is a useful tool to help achieve these characteristics.

In considering the problem of oil monitoring, four types of particle contaminants are present in various systems. These are as follows:

1. The metal wear particles
2. Organic wear particles, i.e., friction polymers, gasket debris, plastic parts, etc.
3. Silica particles from dust ingestion
4. Exhaust soot particles in diesel engines.

Any oil analysis method must be sensitive to the particles that affect the performance and durability of machines. Oil analysis is carried out for three basic reasons. First, to monitor the condition of the lubricant/additives. Second, to monitor the condition of the metal wearing surfaces in the machine. Third, to monitor any outside contaminants that might affect either the lubricant or the machine's wear surfaces. Table 1 shows various oil analysis methods in relation to the types of debris or additives that are being monitored.(1)

In a general sense, oil monitoring will find application in three broad cases. First, for developing improved bearing, lubricant, and filtration system performance by using either component bench or full scale short term laboratory tests. Second, to indicate maintenance of the machine, i.e., oil replacement, additive additions, oil or air filter replacement, air leaks, coolant leaks, abnormal bearing wear, etc. Third, the onset of wear rates that will lead to a catastrophic failure which will in turn lead to costly machine downtime or to extensive damage to a greater portion of the machine than the part that will fail. Oil analysis to date has largely been applied to the last two cases although Ferrography opens up the possibility of oil monitoring for short term laboratory tests. It is this application of Ferrography that will be discussed in greater detail with particular emphasis on the monitoring of the metallic wear debris in diesel engines. This paper will largely expand on and be complementary to the basic review paper by Westcott.(2) The effect of a filter in the machine's lubrication system needs more discussion; and in fact, the filter stores (with loss of time memory) the debris that has the most important information for monitoring the mode of wear of the various machine surfaces, i.e., the large (>5 μm) particles.

TABLE I.—OIL ANALYSIS METHODS IN RELATIONSHIP TO WHETHER MACHINE (M) OR LUBRICANT (L) IS MONITORED

Oil Analysis Method	Contaminates: Metal Wear Particles (Wear)	Organic Wear Particles	Silica Particles	Exhaust Soot Particles or Fuel	Additive Depletion
Insolubles Content (ASTM Method D 893)	M	M		M	L
Acidity (TAN) (ASTM Method D 664)					L
Alkalinity (TBN) (ASTM Method D 664 and D 2896)					L
Elemental Analysis* (Emission Spectrograph Atomic Absorption and X-Ray Absorption)	M		M		
Chemical Identification (Infrared Spectroscopy)					L
Viscosity Determination				M	L
Dispersancy Characteristics					L
Ferrography	M	M	M		
Particle Counting	M				

*Other than C, H, and O.

In summary, the need to develop standardized oil sample collection procedures including Ferrogram preparation, analysis and parameter calculation procedures for various applications is outlined. The information presented here is drawn from our publications on the application of Ferrography to the study of diesel engine wear in the laboratory under short term performance tests.(3-5)

THE MACHINE AS A COMPLEX PARTICLE GENERATOR

The diesel engine (or any modern oil lubricated machine) is an intricate system with many wear surfaces making wear analysis of the system even more complex. Figure 1 shows the engine in schematic form as a series of wear particle generators with various particle production rates (given as mass flow rates $-\dot{m}_i$). Therefore, the total production rate $\dot{m}$ is given as

$$\dot{m} = \sum_{i=1}^{N} \dot{m}_i \qquad (1)$$

where i = the individual surfaces and
N = total number of surfaces generating metallic particles.

Examples of these particle generating surfaces include cylinder liner, piston rings, main bearings, crankshaft, camshaft, valve guides, etc. In

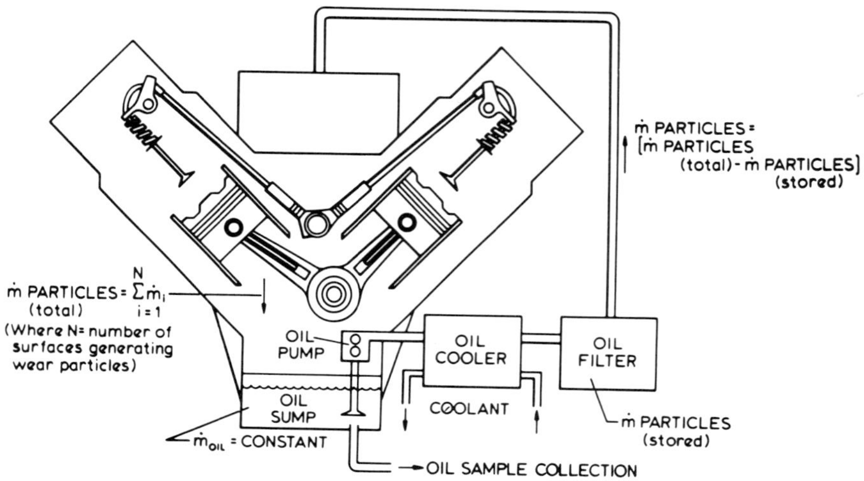

Fig. 1.—Schematic of the diesel engine as a particle generator including the concept of the filter as a particle storage reservoir.

addition, non-metallic particles can also be added at several locations in the system. Particulate matter contained in the exhaust products (soot) can be absorbed into the oil film on the cylinder wall. These particles are largely carbonaceous in composition, but have also been known to contain various acids formed during the combustion process. Dust (primarily silica) can also be added to the oil system via the intake air in much the same manner as the soot particles. If exhaust gas recirculation (EGR) is used as a means of NO_x control, additional soot particles also have the potential of being absorbed into the oil during the intake stroke. With EGR, exhaust gases are re-introduced into the engine which increases the potential of particle loading of the oil. Without EGR, primarily "clean" air is in contact with the oil film and engine parts. The soot particles can adsorb various additives in the oil such as zincdithiophosphates (ZDP's) which affects the life of the lubricant and causes higher valve train wear in diesel engines as in gasoline engines.[6] This paper, although studying particles found in the oil system, will concentrate only on the metallic particles generated at the various wear interfaces.

The mass flow rates, of course, are made up of a vast assortment of sizes and shapes of metallic particles and, therefore, have a characteristic size distribution of particles for each wear surface. Equation (2) accounts for the range of particle sizes:

$$m_i = \sum_{j=1}^{K} m_j \tag{2}$$

where j = the various particle size categories and $j=1$ represents the soluble metal.
K = the total number of particle size categories.

Ideally, the particles when mixed with the oil (particle carrier) by the constant circulation of the oil, yield a homogeneous mixture of particles from all the wear surfaces. Assuming constant particle production rates, the resulting equation for system particle production is simply a function of time as indicated below:

$$\dot{m} = \sum_{i=1}^{N} \dot{m}_i = \sum_{i=1}^{N} \sum_{j=1}^{K} \dot{m}_{ij} =$$

$$f(t)\Big|\ \text{operating variables} = \text{const.} \tag{3}$$

Retaining the assumption of the homogeneous mixture of wear particles in the oil, the concentration of particles may be calculated for any given instant of time.

For an idealized case, it can be assumed that the $\dot{m}$ is constant with time and the various particle production rates are constant. In an actual application, however, the oil consumption (particle and oil removal) and oil additions (particle-free oil additions) have a great deal of influence on the concentrations that are seen to exist. Break-in of new surfaces or parts, start-up and shut-down and the various operating variables have also been found to have an effect on the actual particle concentration indicated in the oil. The particles also have a tendency to settle or plate out either as deposits or sludge during actual engine operation or during engine shut-down. Another system, probably the most influential on particle concentration, is the filtration system. Due to the close machining tolerances in a majority of the engine components and the general size of the wear particles produced, the interaction of the particles in the wear interfaces has a tendency to cause greatly accelerated and often catastrophic wear. Oil filtration has, therefore, become an accepted concept in diesel and other engines.

With the use of filters comes another complication for our idealized system analysis. The general oil filter acts as a particle removal mechanism and, as expected, is a function of many factors. A wide variety of filters are commercially available, ranging from the very crude filter, catching only the largest of particles, to the very efficient, removing everything in the oil including particles down into the sub-micron size range and often the oil additive package. Proper selection of filter is often very complicated with a number of aspects to be considered. One aspect of importance to our ideal system analysis is the particle removal efficiency of the filtration system. The particle removal efficiency is a function of the filter matrix material, shapes and sizes of particles, types of particle materials and the flow rate of the oil through the matrix. Relating this to our idealized system:

Particle Removal Efficiency =

f(particle material, oil flow rates, particle size category -j, and filter matrix material) (4)

and would be expected in general to be a decreasing function for smaller and smaller particle sizes as shown in Figure 2.

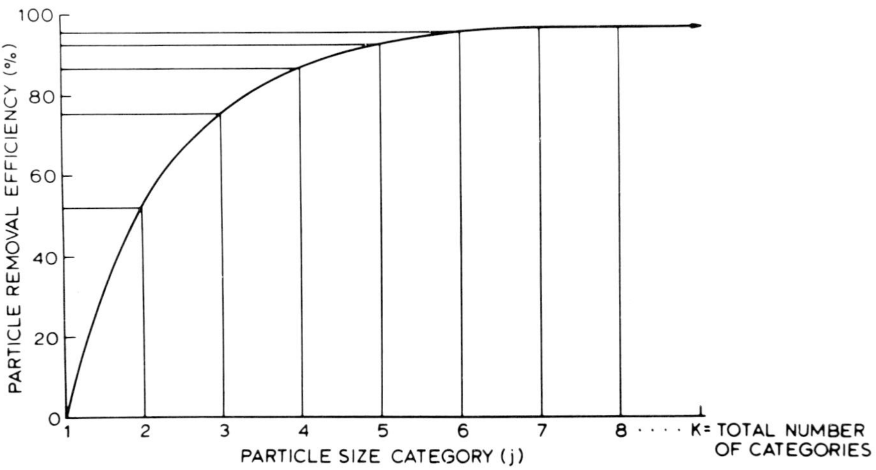

Fig. 2.—Particle filtration efficiency versus particle size.

From the complete system shown in Figure 1, the filter removes some fraction of the particles for each pass of the oil through the filter. For some large particle size categories, it would remove practically all particles. The fact that the large particles are removed from the oil by the filter is also important from the viewpoint of oil analysis since the important particles for understanding the wear situation and predicting impending failure are the large particles (> 5 μm). The filter in effect removes and stores important information about the particles being generated at the individual wear surfaces. This causes a loss in sensitivity for the oil analysis methods that measure the particle concentration suspended in the oil. It is also possible that those particles that are not removed are then circulated through the engine, again subject to various grinding-up and breaking-up processes as the particles pass through the wear interfaces. When the particles are again pumped through the filter, they are then subject to a certain probability that they will be removed. Each time they are successful in passing through the filter, they have a higher probability of staying suspended in the oil, since in general, filtering efficiency decreases with decreased particle size. Improved filtration of small particles would also tend to cause further losses in oil analysis sensitivity. These time varying processes are quite complex and consequently an equilibrium particle concentration (by size and chemical composition) might never occur in a machine/oil system as complex as the diesel engine.

From the discussion of the idealized wear particle generator, one can quickly realize how complicated wear analysis of an actual operating diesel engine or other machine can become. Any change in the operating variables has the potential to change the nature of the wear process. Considerable caution must be taken therefore in interpreting even the most carefully controlled laboratory experiment dealing with the various wear phenomena. Because of this, the methods of wear measurement must be very accurate and extremely sensitive to changes in the wear conditions. One must interpret the data from the various wear measurement methods in order to obtain meaningful results.

The next logical step in this discussion is the wear measurement techniques and how they relate to the study of diesel engine wear. As mentioned earlier, the four most common wear measurement methods used in wear research, both in the laboratory and in field wear studies, are the radioactive tracer method, particle counters, spectroscopy and Ferrography. Each of these methods will be discussed in the following section. A comparison of the advantages and disadvantages of each technique when used in measuring diesel engine wear is also included.

FERROGRAPHY AS COMPARED TO OTHER OIL MONITORING METHODS

The different quantities of wear debris measured by spectroscopy, Ferrography and the radioactive tracer method are illustrated in Figure 3 which shows the basic generated concentration versus time of a given element from a number of surfaces of a wearing system. Spectroscopy measures in principle this total concentration and it can be seen that this one number contains the summation of all the concentrations from the individual surfaces. In addition, all of the size distribution information is contained in this one concentration number. This concentration would in practice be reduced by the lack of sensitivity of spectroscopy to the particles larger than 1 μm in size. In addition, the filter can significantly reduce the total measured concentration because of the removal of the larger size particles.

Figure 3 also shows in schematic form the advantages of the Ferrographic oil analysis technique in that it is possible to analyze the larger (> 1 μm) particles from the individual surfaces and to give a size ordered distribution of the particles. A disadvantage of Ferrography is that methods are only available to quantify these particles by taking percent area covered readings on the glass slide rather than the more basic parameter of particle concentration. Another disadvantage of Ferrography is that it does not measure the soluble metal. These two disadvantages are not serious since the large metallic particles are felt to be the most important information relative to the interface wear mode prediction. The filter will also affect the sensitivity of Ferrography due to the reduction of the concentration of the larger particles.

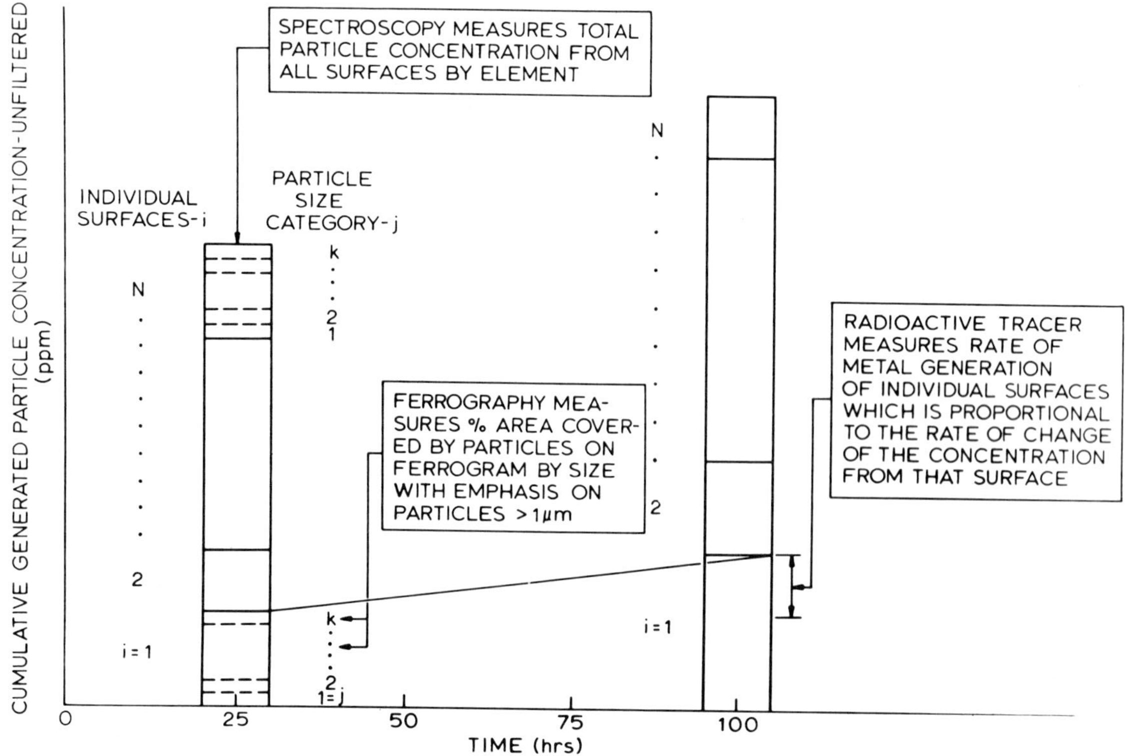

Fig. 3.—Schematic illustrating what spectroscopy, Ferrography and the radioactive tracer wear measurement methods detect.

Figure 3 shows how the radiotracer method is able to measure the individual surface wear rate. Its main disadvantage is that no size distribution data are generated. Its other disadvantage is related to the fact that the filter must be removed from the system and this could result in increased wear rates over a filtered system.

The particle counters would basically measure the number of particles per volume of oil in each size category with particles from all surfaces summed together in this number. The particle counts would only be available for size categories j = 7 - 10 since present particle counters usually measure particle sizes greater than 1 μm because they are based on the scattering and attenuation of a very sharply defined beam of light focused through a sample cell. Counters for diesel engines or other machines that have soot particles in large numbers would appear to be difficult to use because of the large counts that would be contributed by the carbonaneous material. The other disadvantage of counters is that they do not discriminate the particle counts by surface.

FURTHER AREAS OF RESEARCH AND DEVELOPMENT FOR FERROGRAPHY

One of the areas that needs additional research and development is accurate and repeatable quantitative data from the analytical Ferrography. The basic appraoch is as follows:(4,5)

1. Measure percent area covered versus distance down Ferrogram, i.e., entry - 55, 50, 40, 30, 20 and 10 mm locations.

2. Calculate a severity index (I_S) based on the area covered at the entry (A_L) and the 50 mm location (A_S).

3. Calculate the Area Under the Curve (AUC) as a measure of the total debris in the oil sample.(5)

4. Heat the slide (HFA - Heated Ferrogram Analysis) for 90 seconds at 625°F and measure the approximate fraction of the area covered by the cast iron (brown), low carbon and/or alloy steels (blue) and the unoxidized metals such as lead, aluminum, copper and tin.(4,5,7)

5. Assess the quantity of various types of particles using four descriptive words, i.e., none, few, moderate, heavy.

The direct reading and real time Ferrographs only use fiber optics to read the percent area covered at the entry (large particle) and the small particle locations of the precipitator tube.

All these approaches to quantitative Ferrography need further definition. Data for items 4 and 5 above are fairly well defined in the literature and can be developed into fairly standardized engineering practices. The assessment of the quantity of various types of particles is dependent on the volume of oil passed across the slide or the amount of dilution of clean oil used to get an acceptable density of particles on the slide. This raises a basic need in Ferrography: to define a "standard oil sample volume" (SOV) when using it on practical machines since various situations result in various particle concentrations in the oil.

It is important to make a basic assessment of what fundamentals are in effect when the concentration of particles per volume of oil and the total volume of oil passed across the slide is varied at a constant oil volume/fixer volume ratio. We can hypothesize a general family of curves of particle concentration versus the total volume of oil passed across the analytical Ferrogram as shown in Figure 4. From our research we feel that

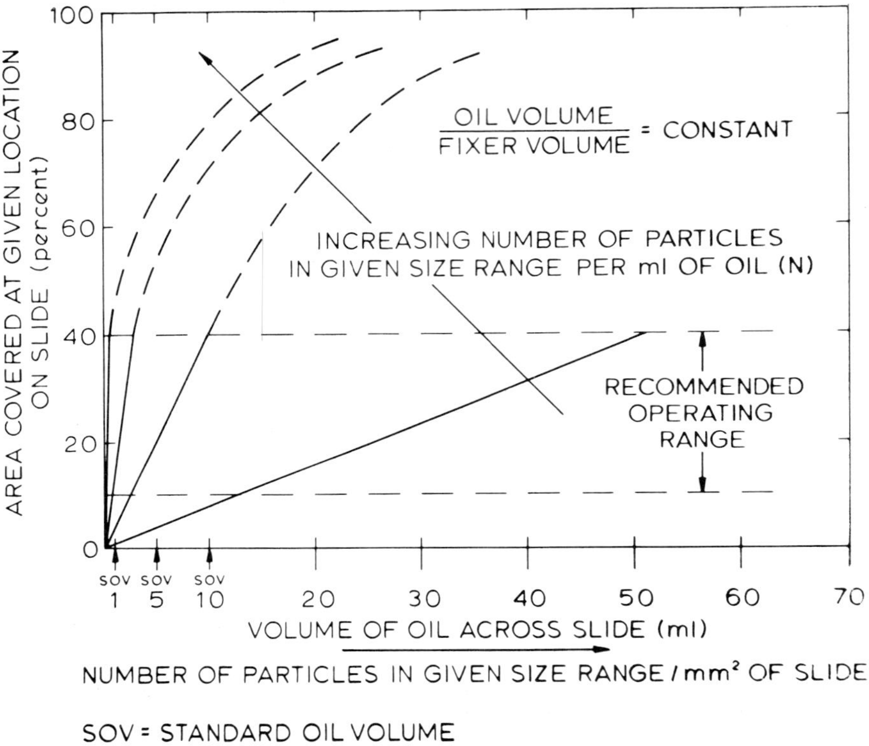

Fig. 4.—Schematic of percent area covered at given location on slide for Ferrography as a function of the volume of oil passed across the slide for various concentrations of particles in the oil.

the region 10-40% is a good recommended range for operating since it appears that it is linear up to 40% and the lower limit of 10% provides good accuracy and precision relative to background zeroing problems and instrument zero drift. As the volume of oil passed across the slide increases, the number of particles per unit area of the slide in a given size range increases. This size range is based on the field of view of the microscope and the magnetic drag forces on the particles. As the particle concentration or the total oil volume across the slide increases so that the percent area is greater than 40%, a non-linear response occurs. This is due to the piling up of particles on top of each other and the possible deposition of smaller particles at each slide location than would occur with lower than 40%. It is for this reason that a "standard oil sample volume" (SOV) must be used for various machine applications for determining the percent area covered along the slide and in calculating a severity index (I_S) from these readings. The SOV for the high concentration systems should be such that the percent area covered readings will not be greater than approximately 60% (slightly greater than linear region) and for low concentration systems such that percent area covered readings will not be less than approximately 4% (greater than background and zero drift effects).

To support the need for SOV's, two cases of the use of Ferrography will be outlined along with the basic equations that describe the percent area covered versus the volume of oil across the slide. Experiments with a large diesel oil sample run for 12 hours in the VT-903 engine test set-up were performed.(5) The oil/fixer volume ratio was set at 3/1 and the volume of the oil used to make Ferrograms was 6, 9, 12, 18, 24, and 36 ml. Figures 5 and 6 shows these data at the entry location ($A_L \approx 55$ mm) and at the 50 mm locations (A_S) including the least squares fit through the data. It should be noted that these data indicate an intersection point at the zero oil volume. Figures 7 and 8 show similar type of data with the real time Ferrograph for a bearing and disc tests with a 25 μm filter.(8)

We can then write equations describing these experimental results and develop equations describing the dilution of highly concentratec debris samples with clean uncontaminated oil. These equations can then be used in conjunction with SOV's to calculate percent areas covered at the standard oil sample volume and in turn a severity index at the SOV. This procedure will provide the means to develop percent area covered readings and severity indexes that are comparable, repeatable and accurate for various oil sample particle concentrations and various oil volumes across the slide.

Therefore,

$$A = M\,V_O + b \qquad (5)$$

$$A_L = M_L\,V_O + b_L \qquad (6)$$

$$A_S = M_S\,V_O + b_S \qquad (7)$$

where A = percent Area Covered at location on slide

A_L = percent Area Covered at entry ($\approx$ 55 mm)

A_S = percent Area Covered at 50 mm

M = Slope of oil volume curve, %/ml

M_L = Slope of oil volume curve at entry %/ml

M_S = Slope of oil volume curve at the 50 mm location, %/ml

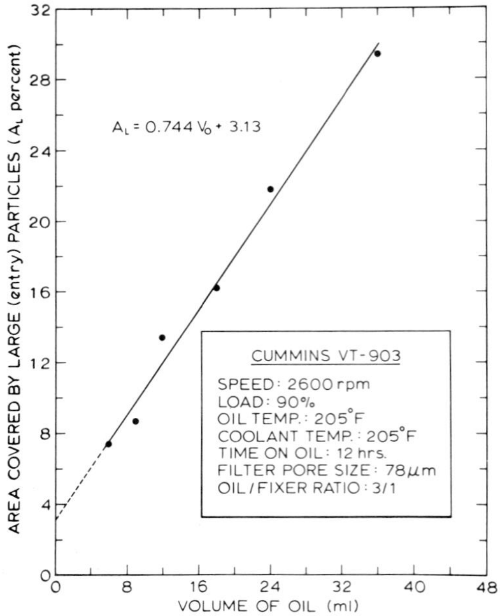

Fig. 5.—Percent area covered at entry (≃ 55 mm) as a function of oil volume across slide for Cummins VT-903 diesel engine.

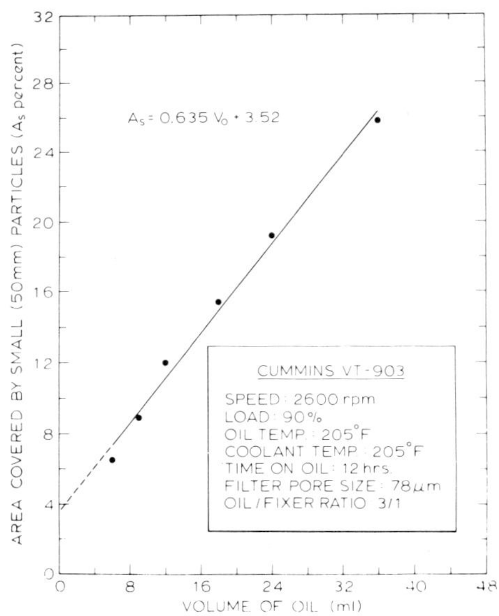

Fig. 6.—Percent area covered at 50 mm location as a function of oil volume across slide for Cummins VT-903 diesel engine.

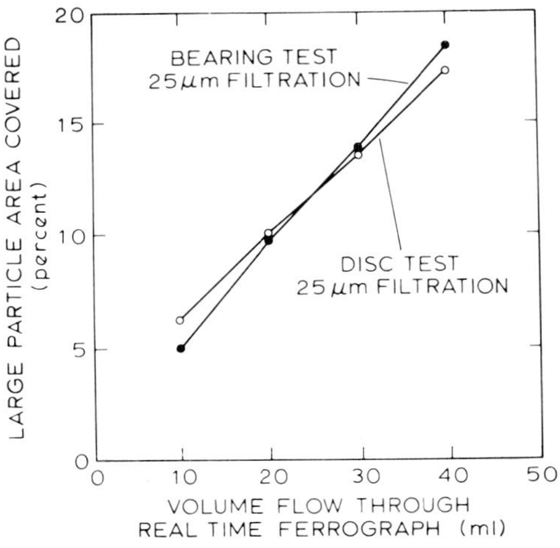

REFERENCE: Effectiveness of the Real Time Ferrograph and other Oil Monitors as related to Oil Filtration, Report No. NAPC-PE-2, D. Poposhev and R. Valori

Fig. 7.—Percent area covered for large particles versus volume of oil flow through real time Ferrograph.(8)

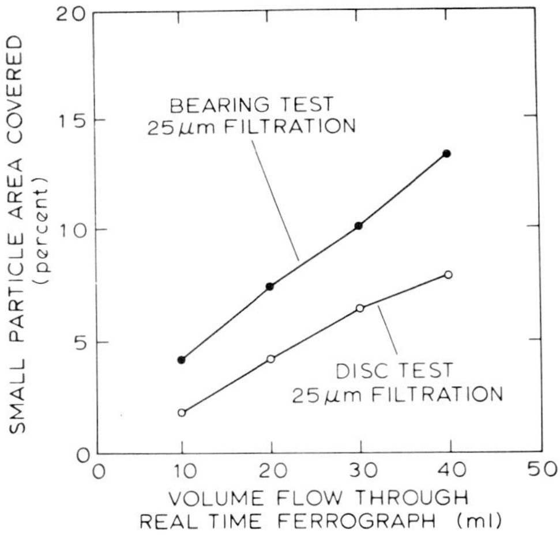

REFERENCE: Effectiveness of the Real Time Ferrograph and other Oil Monitors as related to Oil Filtration, Report No. NAPC-PE-2, D. Poposhev and R. Valori

Fig. 8.—Percent area covered for small particles versus volume of oil flow through real time Ferrograph.(8)

b = Constant (A intercept at $V_o = 0$), percent

b_L = Constant (A_L intercept at $V_o = 0$), percent

b_s = Constant (A_s intercept at $V_o = 0$), percent

V_o = Volume of oil passed across slide or through precipitator tube in direct reading or real time Ferrograph, ml

The slope of the curve, M, will be a function of the oil viscosity, the oil/fixer volume ratio, the particle size distribution, the types of particles, and probably several other variables. The remaining discussion will concentrate on applying these equations.

There are basically two cases that need to be considered. The first is the case of highly concentrated particles that at the minimum feasible oil volume (3 ml) would result in percent area covered readings above 40%. This is referred to as Case I. This oil sample needs to be diluted with clean oil so that the diluted concentration brings the percent area covered reading below 40%. If it is assumed that the undiluted particle concentration of the total range of particle sizes or an individual size category can be expressed as

$$N_U = n/V_{OU} \tag{8}$$

where n = Number of particles (total or individual size category -j) in oil sample

V_{OU} = Volume of undiluted oil in sample

If this oil sample is now diluted with clean uncontaminated oil, the following equation holds

$$N_D = n/(V_{OU} + V_{OD}) \tag{9}$$

where V_{OD} = Volume of clean oil in diluted sample.

Dividing Equation 8 by Equation 9 gives

$$\frac{N_U}{N_D} = 1 + \frac{V_{OD}}{V_{OU}} \tag{10}$$

Figure 9 shows in schematic form this case and is an extension of Figure 4. The general procedure involves finding by trial and error the volume of clean oil (V_{OD}) required to bring the concentration of particles down so that the minimum feasible quantity of oil can be run and the percent area covered readings are within the 10-40% range. An example is shown in Figure 9 where an undiluted oil sample concentration (N_U) would give a response above 40%. A dilute mixture of 2 ml of the contaminated (used) oil was mixed with 10 ml of clean oil to give N_U/N_D ratio of 6. From this diluted oil sample a Ferrogram was prepared using 3 ml of the diluted oil and 1 ml of fixer. This mixture gave a 33% area covered reading at the entry location. Next, it was assumed for simplicity that the constant b_L was equal to zero. This calculation could have been carried out with a value of b_L from Figure 5. The slope M_{LD} was then calculated as $M_{LD} = 11\%/ml$. The percent area covered at a SOV of 1 ml was then calculated to

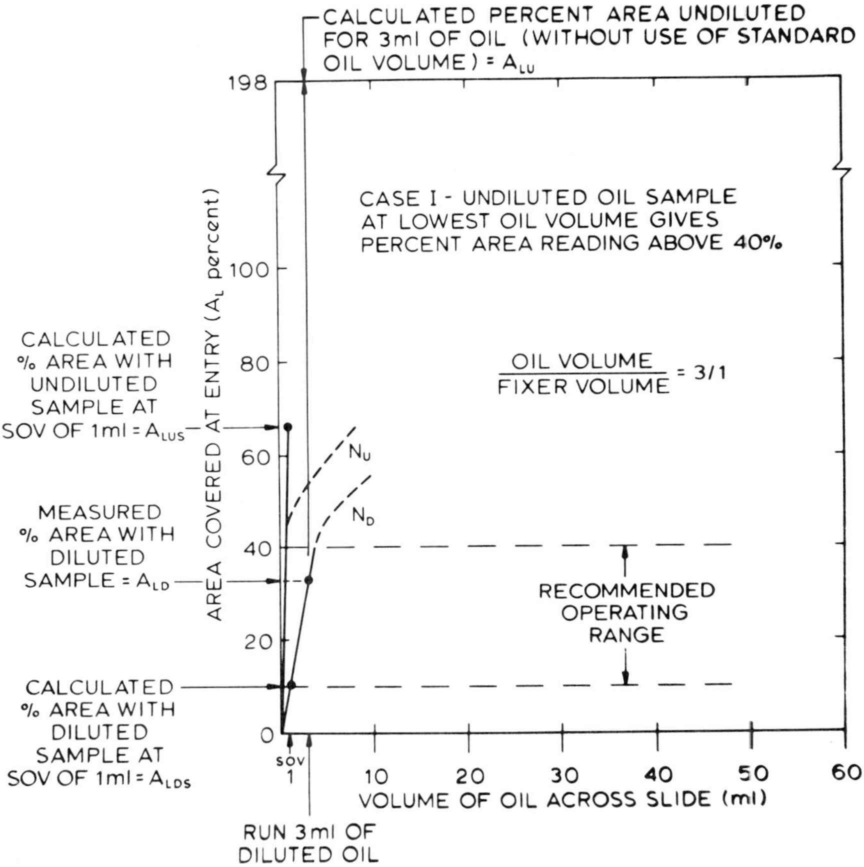

Fig. 9.—Percent area covered at entry versus volume of oil across slide for Case I, i.e., undiluted oil sample at lowest oil volume gives percent area covered reading above 40%.

be 11%. This is an equivalent or diluted standard oil volume value of percent area covered at entry for 1 ml of oil across the slide. The undiluted standard area covered at entry is then calculated assuming a linear relation between concentration and percent area covered, or which is computed to be 66%.

This value is above the 40% linear region and slightly beyond the 60% maximum recommended earlier and points out why a SOV of .3 or .5 ml might be required for some high particle concentration systems.

It is useful to consider the case where the concentrations of the undiluted samples are below 10% area at the minimum oil volumes. For this case, the volume of the oil passed across the slide is increased until a value of A_L is reached that is greater than 10 and less than 40%. An example of this case is shown in Figure 10. For this case 30 ml of oil was

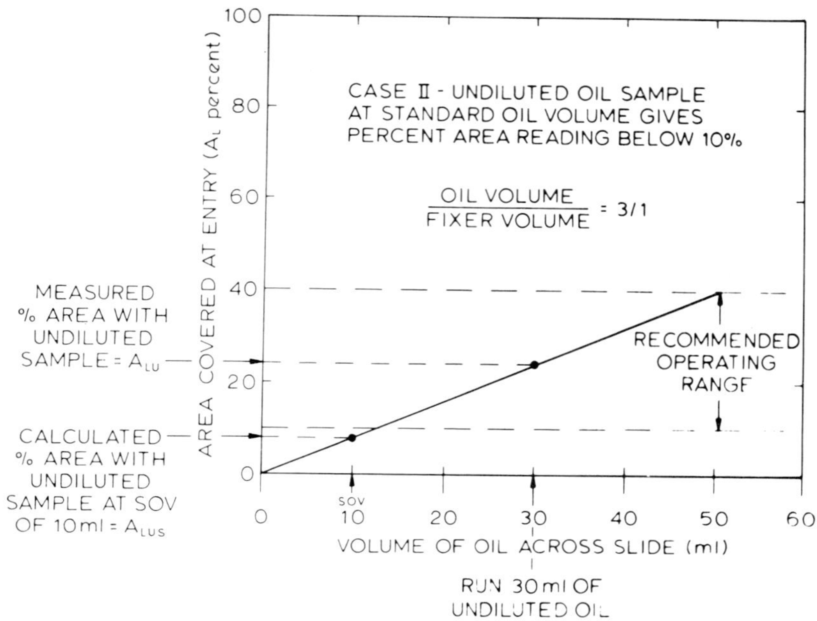

Fig. 10.—Percent area covered at entry versus volume of oil across slide for Case II, i.e., undiluted oil sample at minimum oil volume gives percent area covered reading less than 10%.

run across the slide and a percent $A_{LU} = 24$ was determined. A SOV of 10 ml was used for this calculation since for this type of case, a low SOV such as 1 ml would put the A_{LU} within the region of background and zero drift uncertainty. Therefore, the slope assuming $b_L = 0$ is determined to be

$$M_{LU} = \frac{24}{30} = .8\%/ml \text{ and } A_{LUS} = M_{LU} V_{OS} = .8(10) = 8\%$$

This is the equivalent or standard oil volume percent area covered at entry (A_{LUS}) where V_{OS} = Standard Oil Volume, ml. The slope of the curve could also have been used to represent the debris concentration and for this case it would have been 0.8%/ml.

This same procedure can then be used down the complete slide to get percent A readings at a SOV. The A_L and A_S readings can then be used to calculate a severity index (I_S) at the SOV. This brings up another concern in the literature and that is the large number of different severity indexes. Hopefully, a standard I_S can be agreed to in a standardized Ferrography procedure. Several of the I_S's in the literature are as follows:

$$I_s = \frac{A_L}{A_s} \tag{11}$$

$$I_S = A_L(A_L - A_S) \tag{12}$$

$$I_S = (A_L + A_S)\ (A_L - A_S) = A_L^2 - A_S^2 \tag{13}$$

We have standardized on the form of I_S shown in Equation (13) and found it satisfactory. Reference 8 also expresses a preference for Equation 13.

Once a percent A curve at the SOV versus the slide length has been determined the Area Under the Curve (AUC) can then be calculated as a measure of the total debris on the slide at the SOV. The AUC data have been found to generally correlate quite linearly to spectroscopic concentration data for iron as was indicated in Reference 4. The linear correlation is believed to be a result of the generally constant thickness of the wear particles at a given Ferrography distance for different engine operating conditions (thickness ranged from 1.0 μm at entry to 0.2 μm at the 10 mm location) and for the range of particles studied by Ferrographic techniques (particle sizes ranging from 0.1 μm to 7.5 μm). Therefore, it was concluded that a factor for our diesel engine work (approximately a constant) in the AUC to account for the thickness of the particles was not necessary for developing comparative AUC data. The AUC parameter from Ferrography is felt to be a number proportional to the total iron particle concentration and in conjunction with the I_S parameter provides useful quantitative data from Ferrography. Further work needs to be carried out to see if a particle thickness factor should be developed at each slide location to multiply by the percent area covered reading in order to determine a parameter that is proportional to the volume of the particles. The area under the curve (AUC) would then be the area under the curve of the multiple of percent A x particle thickness summed from entry to 10 mm.

Other areas of further research in Ferrography are: (a) the comparison of the percent area covered readings of the analytical Ferrograph to those of the direct-reading (DR) Ferrography[10] (b) the development of procedures to correct percent area covered readings for oil additions and oil consumptions.[11,12]

As mentioned earlier, the other important variable that must be considered in Ferrography is the effect of the oil filter on the percent area covered readings. The AUC data are really the data of interest in determining what the effects of filtration are on particle concentrations in the oil. From our recent laboratory diesel engine studies the AUC data with and without a filter for operating the engine for 6 hours at a given engine operating condition is shown in Figure 11. It can be seen that there is approximately thirteen times more material deposited on the Ferrogram from an oil sample taken from the system operated without filtration as from the filtered lubrication system. Reference 8 has shown that the Real Time Ferrograph is ineffective in indicating failure for both the disc scoring and bearing fatigue tests when the filtration system ratings are lower than 40 μm. Additional research work should be carried out to overcome this limitation of Ferrography.

We have recently completed a diesel engine laboratory test set-up to overcome this problem. This test set-up includes a High Gradient Magnetic Separator (HGMS) which will be used in the by-pass mode to collect metal wear particles as shown in Figure 12.[13] The experimental approach will be to operate the engine for a fixed time period at a given operating

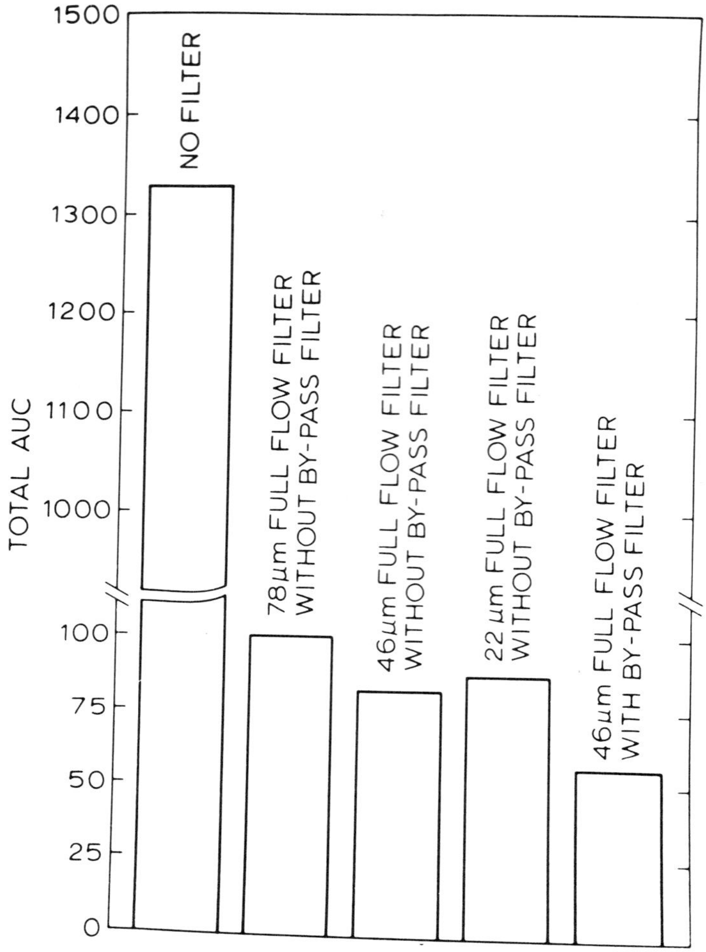

Fig. 11.—Comparison of the total AUC data with and without filters for the same moderate wear condition (2600 rpm/70% rated load/180°F inlet oil temperature/180°F outlet coolant temperature - Cummins VT-903 diesel engine).

condition. The HGMS magnetic field is then shut-off and fresh oil is back flushed through the HGMS for a fixed time period. The particles that were stored during engine operation are then suspended in the fresh oil. A sample of the oil is then used for the preparation of a Ferrogram. Note that this approach provides the means to study the large particles (> 3 μm) generated by the engine while still protecting the engine with highly filtered oil.

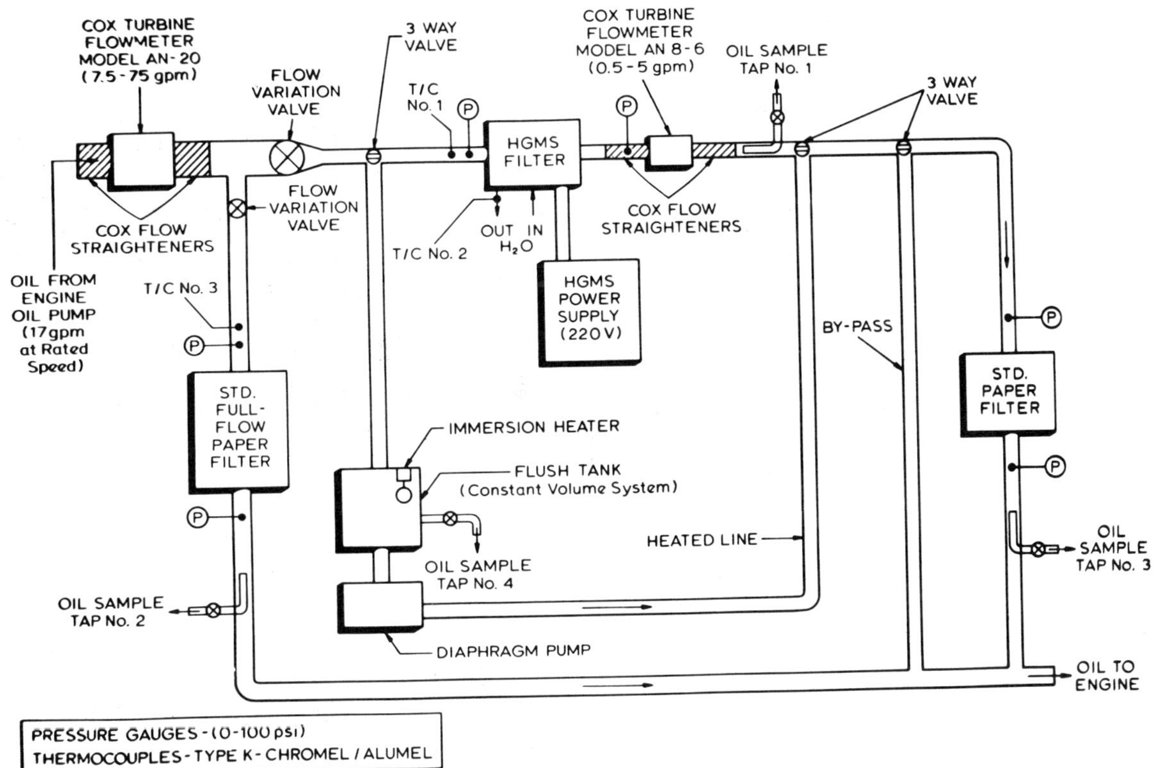

Fig. 12.—Michigan Technological University High Gradient Magnetic Separator (HGMS) oil filter test set-up.

STANDARDIZED FERROGRAPHY TEST PROCEDURE

It is clear that Ferrography is a powerful research and engineering oil analysis technique. It complements other machine and lubricant oil analysis tests as shown in Table 1 and extends the ability to study the metallic debris by size distribution. Its measurement sensitivity is affected when increased levels of filtration are used. Other methods are also affected in a similar way. Ferrography has the ability to be semi-quantitative in an indirect way by the use of the percent area covered readings at various locations on the slide. These data can then be translated to additional parameters such as the severity index (I_S), Area Under the Curve (AUC) or volume of particles at the entry location. It can also be used to observe the type and quantity of particles for the purpose of identifying the mode of wear taking place in the machine. This approach to analysis is outlined in the Wear Atlas[14] and is an excellent guide for this aspect of Ferrography. It is an extremely sensitive method which can be used to study oil samples with high debris concentrations or low debris concentrations from normally wearing machines. The ability to study these low concentration samples quantitatively is only possible by increasing the oil flow across the slide while the high concentration samples can only be studied by diluting the used oil samples with clean oil as proposed in this paper. It is also proposed that Standard Oil Volumes (SOV) be developed so that the data can be made comparable and physically meaningful.

Researchers can obviously use Ferrography to develop meaningful and useful data to further extend the theoretical and experimental understanding of the oil analysis area of Tribology. Ferrography is a powerful analytical technique to study machine generated wear debris without complete disassembly of the machine. To extend Ferrography to a routine user and engineering analysis method clearly needs additional research and development and will require the development of a standardized test procedure which will develop the percent area covered data to be accurate, repeatable and comparable between laboratories.

This test procedure needs to contain detailed guidelines for: 1) Taking the oil sample, i.e., size and type bottle, how full, etc., 2) Preparing the slide, i.e., proper oil to fixer ratio for various viscosity oils, dilution of high concentration oil samples to be within the linear and accurate 10-40% area covered region, increased oil volume across the slide for low concentration oil samples, heating and shaking of oil sample bottle, etc., 3) Standard means, locations, and magnification to read percent area covered including accurate and repeatable zeroing procedures, 4) Standardized oil samples and standard slides for checking calibration and accuracy of complete system, 5) Calculation procedures to correct readings to the SOV's including guidelines to standardize on SOV's for given applications so as to minimize need to correct data to the SOV, 6) Procedures for calculating a severity index (hopefully only just one), 7) Procedures for calculating a AUC, 8) Procedures for carrying out the HFA, 9) Procedures and guidelines for assessing the quantity of the various types of particles using the standardized data sheet and the four descriptive words, i.e., none, few, moderate, heavy, 10) Specific procedures for taking (exposing) good Polaroid pictures, 11) Procedures for plotting and doing trend analysis with data including guidelines on interpretating the data, 12) Procedures to correct data for oil addition and oil consumption, 13) Guidelines for the proper use of the DR Ferrograph in relation to the analytical Ferrograph including calculation procedures to get percent area covered and a severity index from the DR Ferrograph.

Once a draft procedure has been prepared, it then must be used in an extensive interlaboratory test program using known standardized oil samples and slides and unknown oil samples from a variety of different type machines and of different particle concentrations. The oil samples should also be taken from systems with different levels of filtration which in turn will result in various particle size distributions. This cooperative program should use a standard reporting format.

With draft procedure also comes the need to develop an accurate and extensive data base which can be used as input for developing better guidelines for interpreting the oil analysis data from Ferrography.

ACKNOWLEDGEMENTS

I would like to express my appreciation to Kysor of Cadillac and the Office of Naval Research for their support of our diesel wear research program. Michael Hofman and David Anderson have contributed data and stimulated several of the ideas presented in the paper. Their help is greatly appreciated.

REFERENCES

1. Asseff, P.A., "Used Engine Oil Analysis," SAE Paper No. 770642, Society of Automotive Engineers Fuels and Lubricants Meeting, Bartlesville, Oklahoma, June 7-9, 1977.
2. Westcott, Vernon C., These Proceedings.
3. Bolis, David A., Johnson, John H. and Daavetilla, Donald A., "The Effect of Oil and Coolant Temperatures on Diesel Engine Wear," SAE Paper No. 77086, Society of Automotive Engineers International Automotive Engineering Congress and Exposition, Detroit, Michigan, February 28-March 4, 1977.
4. Hofman, M.V. and Johnson, J.H., *Wear,* Vol. 44, No. 1, 1977, p. 183.
5. Hofman, M.V. and Johnson, J.H., "The Development and Application of Ferrography to the Study of Diesel Engine Wear," SAE Paper No. 780181, Society of Automotive Engineers Congress and Exposition, February 27-March 3, 1978.
6. Rounds, Fred G., "Carbon: Cause of Diesel Engine Wear?," SAE Paper No. 770829, Society of Automotive Engineers Passenger Car Meeting, Detroit, Michigan, September 26-30, 1977.
7. Barwell, F.T., et al., *Wear,* Vol. 44, 1977, p. 163.
8. Popgoshev, D. and Valori, R., "Effectiveness of the Real Time Ferrograph and Other Oil Monitors as Related to Oil Filtration," Report No. NAPC-PE-3, Naval Air Propulsion Center, November 1977.
9. Anderson, Daniel P. and Silva, Robert S., "The D.R. Ferrograph Design, Calibration and a Field Application," Paper No. 78-AM-IA-1, American Society of Lubrication Engineers Meeting, Dearborn, Michigan, April 17-20, 1978.
10. Hofman, M.V., MS Thesis, Michigan Technological University, 1977.
11. Scheller, Karl and Eisentrant, Kent J., "Statistical Analysis of Wear Metal Concentration Measurements in Oil: Calculation of Significant Wear Metal Production Rates," MFPG 26th Meeting, Chicago, Illinois, May 17-19, 1977.
12. Land, Malcolm L., Winer, Ward O. and Schwarz, Charles F., "A New Look at Wear Metal Analysis," SAE Paper No. 770085, Society of Automotive Engineers International Automotive Engineering Congress and Exposition, Detroit, Michigan, February 28-March 4, 1977.

13. Anderson, D., M.S. Thesis, Michigan Technological University, 1979.
14. Bowen, E.R. and Westcott, V.C., "Wear Particle Atlas," Prepared by Foxboro/Trans-Sonics, Inc. for the Naval Air Engineering Center under Contract No. 156-74-C-1682, July 1976.

DISCUSSION

QUESTIONER: You said that the correlation between SOAP and Ferrography is very good and for that reason you think it can be extrapolated to total metal content in the lube system. I think experience has shown that SOAP generally disregards large particles. The reason is not so much the sampling effect which can be taken into account by taking a sample at the right time with the right technique. It is that the number of large metal particles would be very small and the probability of trapping them in a given sample is very low. Therefore, I think this large particle end of the spectrum influences the total metal content data to a very large degree. SOAP has this potential weakness and I do not think that you can extrapolate to the metal content in the lube system by looking only at the small particle end of the spectrum.

J. H. JOHNSON: I do not really see that there is a correlation with SOAP. If you have a wear situation that is producing a lot of small particles that are in the grinding-up processes, I think the area under the curve is a parameter that you can get out of Ferrography which can guide us. I do not want to say that it exactly correlates with SOAP. I would guess that there are too many complexities in the spectrometric methods in terms of the response to large particles and sample collection.

E. A. SAIBEL, Army Research Office: I would like to make a short comment which I think will not be out of place in this audience. That is, there is another reason, and a very good one, for knowing the distribution of sizes and densities of the wear particles. One of the effects of the particles, like the effect of additives, is to make the fluid non-Newtonian. Now as far as I know most calculations on bearings -- pressure distributions friction, oil flow and so on -- are based on linear viscosity formulas and these are obviously inadequate. Now if we have a knowledge of the particles -- the distribution of size and so on -- I believe that it is possible to work out these important factors in an engine. After all we do have to know the load-carrying capacity and we should be able to calculate the friction. Just giving a number for viscosity is not enough for those calculations.

F. T. BARWELL, University College of Swansea: I am addressing the question of sump filtration. Professor Johnson was not particularly disturbed by this question because he takes oil sample immediately upstream of the filter. Basically if you consider steady-state conditions you are collecting the particle generated in that interval of time represented by clean oil leaving the filter going through the engine. As point of fact, the filter will allow some smaller particles to go on circulating. So you have possibly got a background or a noise. I think those particles are deposited well down the slide in the Ferrogram and if we are considering the first two spots they are not particularly important. Would you not agree that the right place to take a sample is immediately before the filter.

Now as far as Dr. Saibel's point was concerned, I am not particularly worried about the effect of the particles on the viscosity. But I do consider it very important that their size should be small in relationship to the various clearances in the engine. Therefore I think it is important to relate the calculations to some information about particles.

JOHNSON: You will notice that we take the sample from the sump, that is before the filter. We observed 15 to 20 μm particles in an unfiltered system. With a filter the particle size was 3 to 4 μm. You can still work with Ferrography, I think, and get some information. Presently we are working with a magnetic filter where we hope to back wash and collect these particles and suspend them in new oil to get a better feeling of what we are really losing.

I agree with Dr. Saibel. I think that the wear particles might act like additives. As there is a high concentration in a diesel engine, they may affect viscosity, and load-carrying capacity. Unfortunately, we have not done enough experiments to prove it conclusively.

J. W. WILSON, Mobil Research and Development: I think it is important to realize that the severity index is an empirical value and not statistically sound. If you have a wide range of distribution particles, you seem to be producing useful information by accident. However, it is important to realize that automotive transportation lubricants comprise only about 50 percent of the lubricants that are used. When we get into something like steam turbine oils we observe uniform particles. In that case the severity index is not a useful parameter. Therefore, statistical handling of the data is much needed in the immediate future.

D. GODFREY, Chevron Research: You briefly mentioned the corrosive wear effect. I think it should be emphasized more because in Ferrography only magnetic materials are identified. In many diesel engines and other engines after the break-in, the mode of wear is considered to be essentially corrosive which produces sulphates or sulphides of iron. Ferrography may not be able to identify these particles. My suggestion is that in any standardization we should adopt both Ferrography and other techniques for non-magnetic particles.

JOHNSON: All these methods are really complementary and I think we have to be careful in the choice of the technique, application, the kind of information we obtain, and so on.

V. C. WESTCOTT, Foxboro/Trans-sonics: May I just make a very short comment on the corrosive wear products. It is quite common on Ferrograms to see substantial fractions of these particles. They are converted to oxides and chlorides. They are sufficiently magnetic so that they do precipitate. I think that what we lose are the solubles. Seifert and I examined a number of oil samples by filtering them through 0.45 μm filters. We found that both the spectrographic and Ferrographic methods indicated virtually zero iron content. We have examined a jet engine in Sweden in which a sizeable amount of soluble iron was found, but I think that is the only case we have found.

MONITORING WEAR IN HYDRAULIC SYSTEMS

R. K. Tessmann

ABSTRACT

Life and reliability have always been important considerations in the application of hydraulic systems. As component loading, system investments, and maintenance costs have risen, the need for highly reliable and long-lasting hydraulic systems has become very important. In order to accurately assess the probable life of such systems, it is necessary to measure the wear rate exhibited by the system under various conditions.

The intent of this paper is to discuss the concepts and viewpoints of a review paper, entitled "Monitoring of Wear," by Vernon C. Westcott, as those ideas apply to hydraulic systems. The important topics covered are: (1) the relationship between wear rate and debris concentration, (2) system sampling, (3) wear debris recovery, (4) wear debris measurement, and (5) interpretation of wear debris analysis. While there are many other aspects of wear monitoring which could be discussed, it is felt that these are very critical subjects.

INTRODUCTION

The need for high-pressure, high-performance hydraulic systems has focused considerable attention on the life and reliability of such systems. Because of their extremely high power-to-weight ratio and versatility, hydraulic systems are gaining widespread application. In addition, modern technology has improved both the power-to-weight ratio and the operational efficiency. Unfortunately, the initial cost of modern hydraulic components is high, and this cost added to that of maintenance, downtime, and repair has created a great demand for hydraulic systems which can survive the rigors of field operation for long periods of time. The field service life of hydraulic components is assessed through the degradation in their performance. For example, a hydraulic pump is considered "worn out" when it no longer provides sufficient flow at desired operating pressure.

Wear is one of the most costly phenomena which occurs during the use of devices in the world today. The U.S. Navy has reported (1) that the cost of wear control in aircraft and surface ships is approximately 2/3 the cost of the fuel. In many wear processes, especially those occurring in oil-wetted components, abrasive particles serve to accelerate the deleteriour attack upon critical internal surfaces. In some cases, engineers have learned to harness this phenomenon for the benefit of industry. For example, various manufacturing processes (such as grinding, lapping, sanding, etc.) make good use of abrasive particles in controlling the rate of material removal and the texture of the finished surface. However, when such processes are undesirable, as is the case in a hydraulic system, abrasive particles become a contaminant.

Three basic modes of failure have been defined for hydraulic system components [2,3] - transient, catastrophic, and performance degradation. Transient failures are characterized by a temporary malfunction (such as an excessive pressure overshoot or a momentary hesitation in response). Normally, such unsatisfactory performance is caused by the presence of contaminant particles lodged in critical clearances of a component, causing increased drag forces or a temporary configuration change between mating parts. Catastrophic failure, on the other hand, occurs suddently and is of a permanent nature. This type of failure may be typified by a complete locking of moving parts. When the condition occurs because of the presence of particulate contamination, it is referred to as *contaminant lock*[4]. However, a catastrophic failure can also result from a loss of a hydrostatic or hydrodynamic balance accompanied by a breakdown in the fluid film, causing severe surface contact.

While both transient and catastrophic failures certainly reduce the life and reliability of hydraulic systems, it is almost impossible to treat them as wear processes. Performance degradation is the only basic failure mode which predominantly occurs as a result of a wear process. Such a failure results when the performance of a hydraulic component is impaired to such an extent that it must either be repaired or replaced. The characteristic feature of this failure mode is the gradual but persistent deterioration of critical surfaces within the component, as evidenced by a similar loss in performance capabilities.

The destruction of critical surfaces within the hydraulic component will produce two related phenomena. First of all, the leakage paths within the component will be enlarged; and, second, wear debris will be generated. The debris generated during the destruction of these surfaces will be added to the particulate matter already entrained in the system fluid. Therefore, it should be obvious that wear in fluid power components can be measured by a change in the performance or the amount of debris generated. In order to assess performance degradation, the component must be operated in a controlled manner while critical performance parameters are evaluated.

Wear analysis of debris generation is not a new idea; however, the development of the Ferrograph [5] has made it possible to study the characteristics of the debris as well as to evaluate its concentration. Useful interpretation of a wearing situation by use of the debris entrained in the fluid relies upon the following five aspects [6]:

1. The functional relationship between debris concentration and wear rate
2. Fluid sampling
3. Wear debris recovery
4. Wear debris measurements
5. Interpretation of wear debris measurements

This paper discusses each of these aspects as they apply to hydraulic systems.

DEBRIS CONCENTRATION IN TERMS OF WEAR RATE

In modern hydraulic systems, particulate contamination is extremely harmful. Therefore, a diligent effort is normally made to remove such injurious material. There are three sources of particulate matter in most fluid systems. Contaminant particles can be ingested by the system from the external environment, through seals, breathers, etc. In addition, even with extreme care, manufacturing residue can be left in the system during fabrication. Finally, the wear of components will introduce particles into the system - normally called *generated contaminant*.

Since a constant contamination addition from one or all of these sources would cause a progressively increasing contamination level, a filter is used in a hydraulic system to control the particle concentration. Therefore, the system filter is a controller, just as a relief valve is a pressure controller. In other words, the amount of contaminant present in a hydraulic system is a function of the material being introduced and the material being removed.

By assuming that the only particles being added to a fluid system are produced by wear, a mathematical relationship can be derived for the concentration of wear debris. Such an analytical expression is based upon a fundamental material balance[7,8]. On an idealistic basis, such a balance could be written as follows:

$$G_s(t) = G_o V_s + \int g_w \, dt - \int g_R \, dt \qquad (1)$$

where: $G_s(t)$ = debris concentration at any time, t, upstream of the filter (mg/litre)

V_s = system volume

G_o = initial debris concentration

g_w = wear rate (mg/min)

g_R = debris removal rate (mg/min)

This equation states that the concentration of wear debris at any time, $G_s(t)$, measured upstream of the system filter times the system volume equals the initial concentration, G_o , times the system volume plus the amount of debris added due to wear minus the amount of debris removed by the filter. Equation 1 can be simplified by realizing that the debris removal rate is a function of the efficiency of the filter, which can be developed analytically as follows:

$$g_R = (G_s(t) - G_d(t) Q \qquad (2)$$

where: G_d = debris concentration at any time, t, downstream of filter (mg/litre)

Q = flow through filter (litres/min)

and:

$$G_d(t) = (1 - \varepsilon)\ G_s(t) \tag{3}$$

where: ε = filter efficiency

Substituting Eq. 3 into Eq. 2 and simplifying produces:

$$g_R = \varepsilon Q\ G_s(t) \tag{4}$$

Then, substituting Eq. 4 into Eq. 1 yields the following expression:

$$G_s(t)\ V_s = G_o\ V_s + \int g_w\ dt - \int \varepsilon Q\ G_s(t) dt \tag{5}$$

Dividing through by the system volume, V_s, integrating and rearranging produces the differential equation which describes the wear debris concentration in a hydraulic system as follows:

$$\frac{d\ G_s}{dt} + \varepsilon \frac{Q}{V_s} G_s = \frac{g_w}{V_s} \tag{6}$$

If it is further assumed that the parameters of Eq. 6 are time invariant, then it is possible to obtain a closed-form solution, as shown below:

$$G_s = \frac{g_w}{\varepsilon Q} \left[1 - e^{-\frac{\varepsilon Q}{V_s} t}\right] \tag{7}$$

From this equation, it can be seen thtt the final debris concentration achieved after the system has reached equilibrium is directly proportional to the wear rate and inversely proportional to the product of the filter effienency times the flow rate. The speed at which the debris concentratinn attains a steady value is a function of the filter efficiency and the ratio of flow divided by the system volume or the turnover rate. Thus, it should be apparent that two hydraulic systems which have exactly the same wear rate would exhibit different debris concentrations if the filter efficiency were different between them.

Equation 7 can be graphically represented, as shown in Figure 1. As can be seen in this figure, it was assumed that the system was initially filled with *"clean"* fluid. Therefore, the initial concentration, G_o, was taken as zero. In cases where the system fluid contains significant amounts of debris so that the initial concentration cannot be ignored, Eq. 6 will still apply; but, the solution given by Eq. 7 will not be accurate. The curve shown in Fig. 1 has assumed a constant wear rate for the system of interest. In some cases, such an assumption is not correct. For

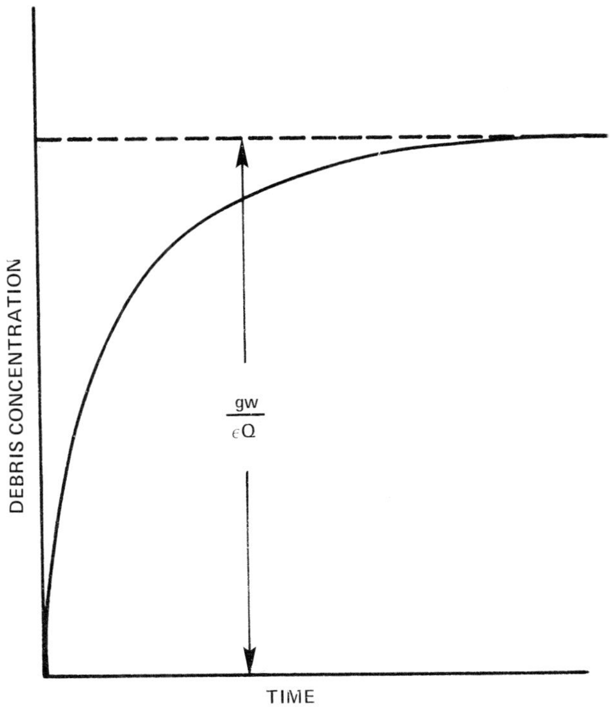

Fig. 1.—Debris Concentration as a function of time for a filtered system.

example, new systems will exhibit a break-in period during which the internal surfaces of the components are aligning themselves. The wear rate may be quite high at the beginning of this period and decrease exponentially throughout the break-in phenomenon. Here again, Eq. 6 applies, but a solution under this constraint will be different than that given by Eq. 7.

FLUID SAMPLING

Without a doubt, fluid sampling is the most important and often overlooked aspect of fluid analysis. In order to obtain a meaningful interpretation of a wear situation through Ferrographic analysis or any other fluid analysis technique, it is necessary to have a representative sample. In this case, representative means that the contamination level in the sample is the same as the system fluid during operation. Basically, there are three critical considerations in acquiring a representative fluid - the condition of the container into which the sample is placed, the time the sample is taken, and the place from which it is removed.

The degree of cleanliness required of the sample containers[9] is associated with the contamination level of the system fluid. That is, if a high concentration of contaminant is in the fluid, then a small amount in the container will never be noticed. In addition, the degree of cleanliness of the sample containers depends upon the type of analysis which will be conducted. For example, since the Ferrograph is capable of separating component wear debris from environmental dust, some of this dust in the sample container is of little concern. However, if a particle count is to be made, such a container would produce erroneous results.

Basically, the International Organization for Standardization (ISO) has defined two simple types of sampling methods - *dynamic* and *static*. Dynamic fluid sampling is defined as the extraction of a sample of fluid from a turbulent section of a flow stream. Conversely, static fluid sampling is defined as the extraction of a sample of fluid from a fluid at rest. Although it is possible to obtain a representative sample through static sampling of the reservoir fluid, great care must be taken to insure against particle settling. A device for acquiring a reservoir sample is shown in Figure 2. The ISO recommends that the dynamic sampling method[10] be

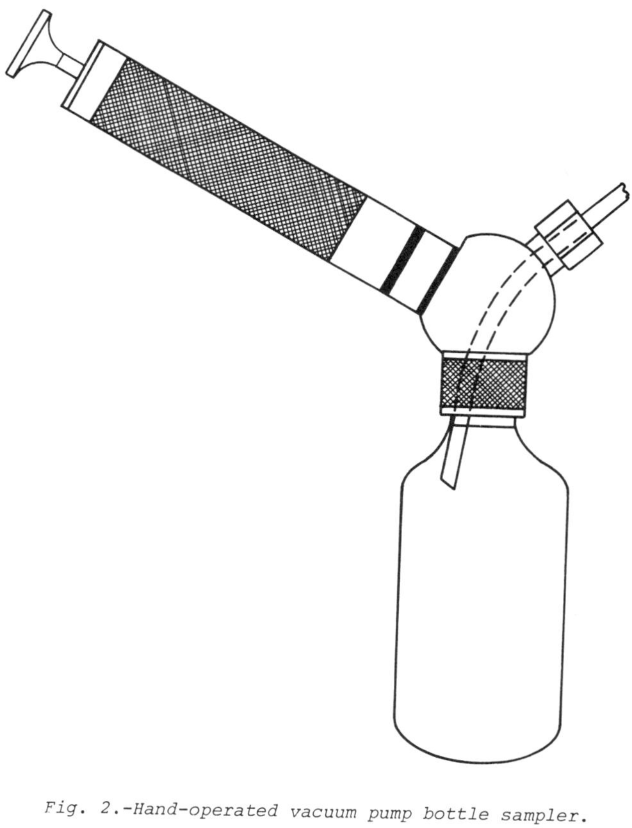

Fig. 2.-Hand-operated vacuum pump bottle sampler.

employed to obtain a representative sample. A typical field type sample device is shown in Figure 3 from the ISO standard.

WEAR DEBRIS RECOVERY

In most hydraulic systems, the amount of extraneous material entrained in the oil far exceeds the concentration of the wear debris. Therefore, if a conventional laboratory membrane is utilized to filter the contamination from a sample of fluid extracted from a hydraulic system, the extraneous particles (environmental dust, filter fibers, friction polymers, bits of rubber, etc.) would cover the wear debris, making any analysis difficult if not impossible. As another example, the contaminant sensitivity of hydraulic components is a critical consideration in their selection and use[11]. In order to determine the contaminant wear resistance of a hydraulic pump, a test is conducted using a controlled contaminant level of 300 milligrams per litre of various particle size ranges of test dust obtained through the classification of AC Fine Test Dust. In order to analyze the contaminant wear process induced during this test, the wear debris must be adequately separated from the test dust.

Ferrographic oil analysis has been successfully applied to examine the contaminant wear of hydraulic pumps. Typical results from these studies are shown in Figures 4 and 5[12]. Figure 4 shows the results of using the same gear pump tested at various contamination levels, while Figure 5 shows the data acquired using various pump outlet pressures. The most important aspect to note from these figures is that the Ferrographic analysis method was sufficiently sensitive and discriminatory to show small changes in wear rates caused by parameter variations.

WEAR DEBRIS MEASUREMENTS

The Ferrographic oil analysis method has been successfully applied to the measurement of wear debris from hydraulic components and systems. To the author's knowledge, no other method is currently available which has enjoyed anywhere near the degree of success as the Ferrograph. Therefore, it is only natural that a section concerned with the measurement of wear debris from hydraulic systems would be restricted to a discussion of the Ferrographic technique.

Of primary importance to the use of any instrument is data scatter or repeatability. In any case of the Ferrograph, two types of data are obtained. Quantitative data are obtained through optical desnity readings from the Ferrogram. The preparation of a Ferrogram is adequately discussed in Refs. (5) and (6) and will not be addressed here. The second type of information is subjective in nature and acquired by visual observations of the debris particles through the use of the features available on the bichromatic microscope. Therefore, repeatability is only a consideration in the quantitative data derived through the use of the optical densitometer.

In order to appraise the repeatability and saturation characteristics of the Ferrographic method, a sample of fluid was extracted from a hydraulic system in which contaminant wear had been induced. Three Ferrograms were made from each of the following sample volumes - 0.25, 0.50, 0.75, 1.0, 2.0, and 3.0 millilitres. The density readings obtained from these 18 Ferrograms (at the 54 millimetre position) are shown in Figure 6. These data

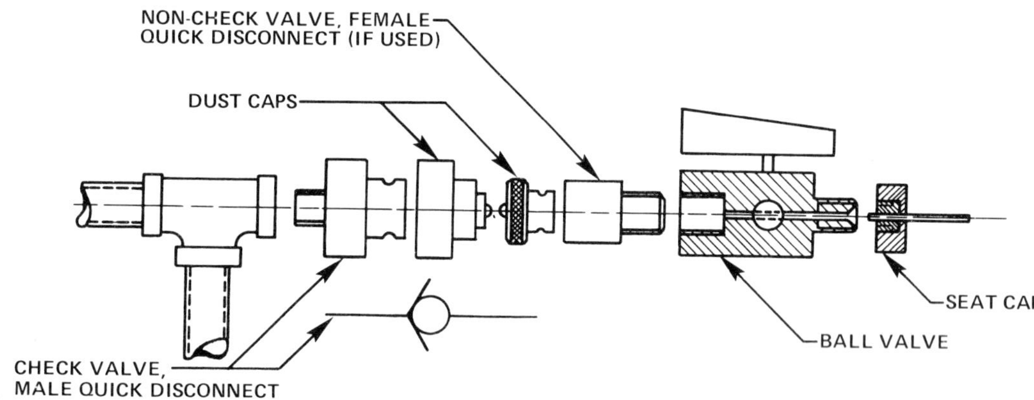

Fig. 3.-Typical field type sampling device.

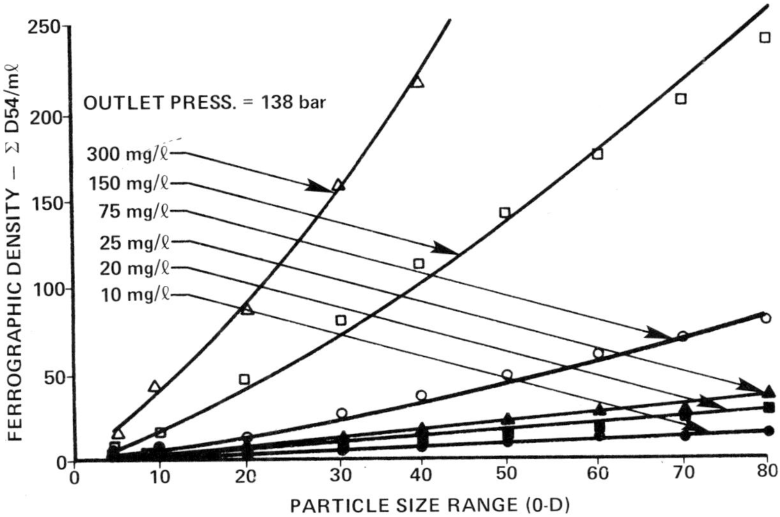

Fig. 4.-Summary of pump test results at various contaminant concentrations with a pump outlet pressure of 138 bars.

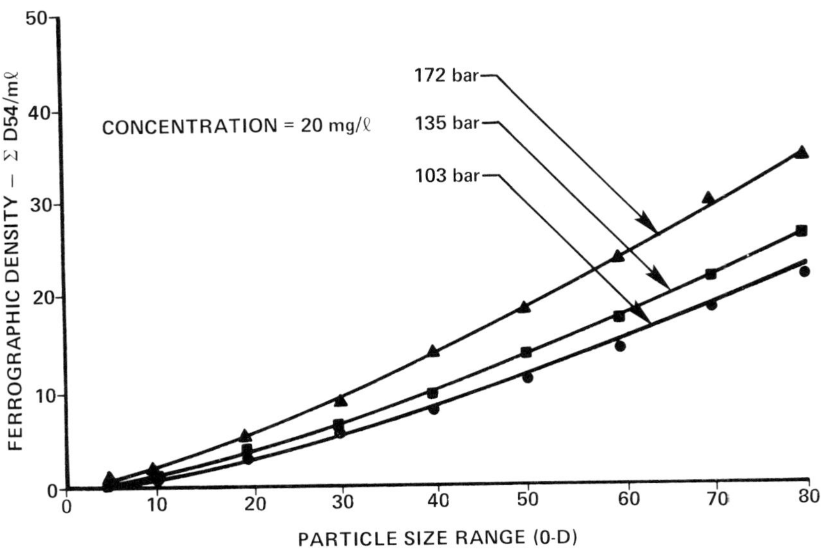

Fig. 5.-Summary of pump test results at various outlet pressures with a contaminant concentration of 20 milligrams/litre.

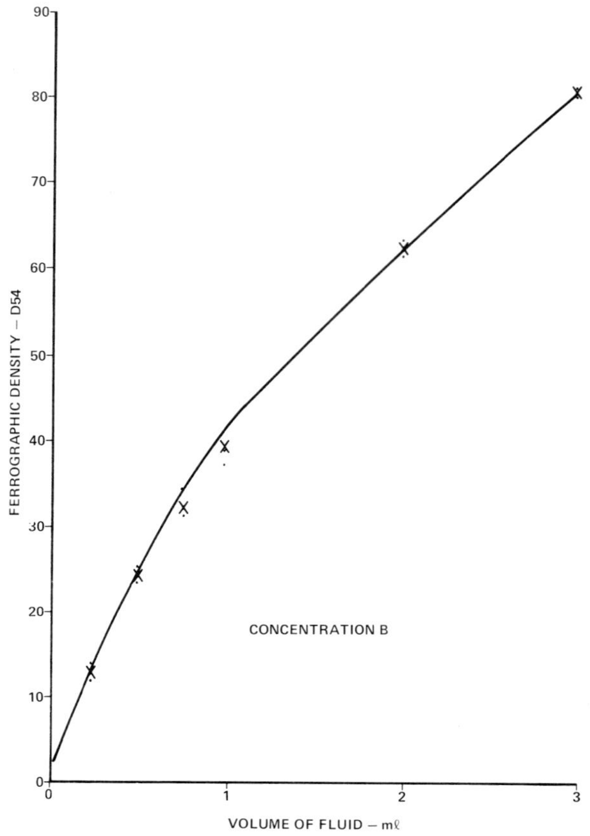

Fig. 6.—Repeatability and saturation characteristics of ferrographic technique.

clearly show that some type of saturation occurs at a density reading of about 40 at the 54 millimetre position. In addition, the maximum coefficient of variation exhibited by these data is about 5% (using the range to estimate the standard deviation).

INTERPRETATION OF WEAR DEBRIS

It is the firm belief of this author that all wear measurements should be interpreted in terms of performance degradation. This is not a simple task, since a device is seldom considered *"worn out"* by a user until its performance is no longer satisfactory. In some hydraulic components, a relationship has been derived and verified which expresses wear debris concentration in terms of the degradation in a critical performance parameter. This has not been attempted on all hydraulic components, but it is believed that the task is very similar.

Basically, it is necessary to rationalize that the result of any wearing process in a hydraulic component is the destruction of critical surfaces. The secondary result of this destruction is both debris generation and performance degradation. Therefore, if the generation of debris can be mathematically related to the amount of surface destruction, then the amount of surface wear can be used to calculate performance changes. While the transformation can be expressed in words fairly simply, it is most difficult to achieve. Howev er, until such relationships are derived and verified, scientists and engineers must continue with parameter influenced measurements and trend analysis.

CONCLUSION

The development of the Ferrograph has been a tremendous breakthrough in the area of wear in hydraulic systems and components. Up to the introduction of this powerful technique, about the only reliable method employed to study wear in fluid power components was direct performance parameter changes. It is felt that a succinct relationship can be derived to describe the interaction between wear rates, degris concentration, and hydraulic system operating parameters. Such an expression renders debris analysis of hydraulic systems a feasible method of diagnosis.

Tribologist must begin to equate himself with the medical doctor. The fluid system is the patient, and they must learn what parameters to monitor in order to make a successful diagnosis. The medical doctor has heartbeat, blood pressure, skin color, blood counts, etc. to use in the diagnosis of the human system. All of the critical parameters of a fluid system must be discovered. In addition, norms must be developed so tribologists know when a parameter is too high or too low. Finally, the professional must learn what to do when an unsatisfactory diagnosis is made. That is, if the wear rate is too high, tribologists must develop *"cures"*.

ACKNOWLEDGEMENT

The author would like to acknowledge the support and guidance which has been received from the Office of Naval Research and the Naval Air Engineering Center. In addition, the support, guidance and encouragement received from the various members of the fluid power industry through the Hydraulic System Diagnostics Project are gratefully acknowledged.

REFERENCES

1. "Wear Control in Naval Aircraft," Proceedings of the 4th Annual Meeting on Materials and Processes, Aeronautical Analytical Rework Program, Naval Air Development Center, Warminster, Pennsylvania, December 2-4, 1975.
2. Farris, J.A., "The Importance of Silt Removal Filtration in Mobile Hydraulic Systems," Paper No. 690606, Society of Automotive Engineers, New York, New York, 1969.
3. Kirnbauer, E.A., "Contamination Control for Hydraulic Systems - Recent Developments in the United States," 2nd Fluid Power Symposium, Guildford, England, January 1971.

4. Surjaatmadja, J.B. and Fitch, E.C., "The Characteristics of Contaminant Lock in Fluid Components," Annual Report No. 10, Basic Fluid Power Research Program, Fluid Power Research Center, Oklahoma State University, Stillwater, Oklahoma, October 1976.
5. Scott, D., Seifert, W.W. and Westcott, V.C., *Scientific American*, Vol. 230, No. 5, May 1974, p. 88.
6. Westcott, V.C., These Proceedings.
7. Fitch, E.C. and Tessmann, R.K., "Practical and Fundamental Descriptions for Fluid Power Filters," Paper No. 730796, SAE Transactions, Society of Automotive Engineers, New York, 1974.
8. Tessmann, R.K., "The Effect of System Configuration on Filtration Performance," Paper No. P75-23, 9th Annual Fluid Power Research Conference, Fluid Power Research Center, Oklahoma State University, Stillwater, Oklahoma, 1975.
9. American National Standard, "Procedure for Qualifying and Controlling Cleaning Methods for Hydraulic Fluid Power Fluid Sample Containers," ANSI B93.20-1972, American National Standards Institute, 1972.
10. International Standard, "Hydraulic Fluid Power - Particulate Contamination Analysis - Extraction of Fluid Samples from Lines of an Operating System," ISO 4021-1977 (E).
11. "Method for Establishing the Flow Degradation of Hydraulic Fluid Power Pumps When Exposed to Particulate Contaminant," NFPA Recommended Standard T3.9.18-1976, National Fluid Power Association, Milwaukee, Wisconsin, 1976.
12. Tessmann, R.K. and Fitch, E.C., "Contaminant-Induced Wear Debris for Fluid Power Components," Tribology 1978 - Materials Performance and Conservation, University College of Swansea, April 3-4, 1978, Institution of Mechanical Engineers, Wales.

DISCUSSION

QUESTIONER: In the tests that you are doing you do not pick out particles in the range 70 to 80 μm. Rather you are using 0 to 80 μm particles. As a result you do not know what the concentration of 70 to 80 μm particles is and what influence that particular size has on wear rate.

R. K. TESSMANN: First of all, we do know the concentration of particles as a function of the particle size for the whole range. There is a more fundamental consideration that you are missing entirely. How many practical systems are subjected to 70 to 80 μm particles? In hydraulic systems, about 75 percent of the contaminants are environmental contaminants. They come in through breathers and seals. The particles have full distribution. A little bit of analytical interpretation tells us what was caused by 70 to 80 μm particles if we know what happens with 0 - 70 μm and 0 - 80 μm particles. When we ran a full distribution over 1000 to 1500 hrs., we found that the estimate was 50 percent lower than the experimental values. I think that is great.

SAME QUESTIONER: But unless you are sure that the size distribution and level of contamination and so on are the same as in practice, you cannot predict the efficiency of the pump from your laboratory test results.

TESSMANN: Well, we use nine different contamination levels and I will guarantee you that at least one of them will be close to the one in the field. Of course, every practical system will give a different particle size distribution, depending on the operating conditions, filters, and so on.

SAME QUESTIONER: If we break the size distribution down to narrow size ranges, we can then superimpose the results to get any size distribution we want.

TESSMANN: We worked for twelve years trying to do that without much success. The problem becomes extremely difficult when a narrow size range is used. It is also prohibitively expensive.

J. L. TEVAARWERK, University of Waterloo: I would think that most industries would have a fairly standard distribution of particles. We could possibly classify them with a mean value and a standard deviation for which the test would be valid. All we have to do is to take a sample of that particular concentration and distribution.

TESSMANN: Yes, you are absolutely right. But the standard distribution is only the forcing function of what we have in our system. We also have to consider what the seal does, what the breather does and what the filter does. For example, if we test ten filters we will not get the same performance out of any one of them. We get stochastic distribution of performances.

W. NEEDELMAN, Pall Corporation: How are you filtering out materials like copper alloys that seem to be coming out of hydraulic pumps? You mentioned that filters do have a big effect on the type and distribution of wear particles. Are you considering doing any work when there is a filter in the system?

TESSMANN: The point I am trying to make is not how to measure contamination wear. Rather the object is to show that under a heavily contaminated environment the Ferrograph produced very important information.

NEEDELMAN: In many systems the predominantly wearing particles are steel. Hydraulic systems seem to be different.

TESSMANN: I have seen several articles published which say that very little steel particles come out of the hydraulic system. I have been running tests for four years and in my experience steel particles came out of everything I have ever tested. These include piston pumps, gear pumps and so on.

CONDITION MONITORING OF GAS TURBINE BY FERROGRAPHIC TREND ANALYSIS

D. Scott, P. McCullagh and G. Mills

ABSTRACT

The application of Ferrographic trend analysis to machinery condition monitoring has been investigated. The most informative method was found to be cumulative plots of the total wear and the severity of wear on the same graph. Convergence of the curves gives an early warning of uncharacteristic trends which should be investigated by full Ferrographic analysis; a cross-over of the curves indicates that urgent action is required. Monitoring of engines on the test-bed gives an indication of subsequent service performance. Evidence of deterioration of components in the gas stream can be indicated from Ferrographic oil analysis procedure. Monitoring of both lubricant washed components and components in the gas stream may be most informative.

INTRODUCTION

Westcott[1] has emphasized that wear assessment is a difficult problem in the design and operation of modern complex machinery and in the development of lubricants essential for the successful operation of such machinery. He has described the use of Ferrography[2] to monitor wear by examining and measuring the debris generated by the wear process.

To aid the application of Ferrographic analysis to the condition monitoring of gas turbines, a program of trend analysis has been carried out[3] to assess various methods of using data provided by the DR and the Real Time Automated Ferrographs, and to investigate by analytical Ferrography and complementary techniques the significance of trends in values of specific parameters. As trend analysis depends on recording values of significant parameters to indicate uncharacteristic changes, it is essential to monitor an engine throughout its complete service life. Samples of lubricant similar to those taken for SOAP were obtained periodically from 100 gas turbine engines of a single type used in the fleet of a single operator. Engines on the test bed prior to service installation were also sampled, followed by sampling in service to assess the value of test bed monitoring as a guide to expected service performance.

As monitoring wear and deterioration of non-lubricant washed vital components such as turbine discs, seals and blades in gas turbines is considered an even more difficult problem than the monitoring of lubricant washed components, a simple technique of sampling wear particles deposited from the gas stream on to surfaces in the gas path has been developed for use with Ferrographic analysis procedure.[4] Several of the engines in the Ferrographic trend analysis program were monitored from both the lubricant and the gas path.

Monitoring of Lubricant Washed Components

From the investigations the most informative method of condition monitoring was found to be cumulative plots of the total wear and the severity of wear from DR Ferrograph readings on the same graph.[3] For a satisfactory engine there appears to be a steady increase in both cumulative total wear $(D_L + D_S)$ and cumulative severity of wear $(D_L - D_S)$ as shown in Figure 1. For a suspected deteriorating engine, Figure 2 shows that initially there is a steady increase in both total wear and severity of wear, then the severity of wear curve tends to rise toward the total wear curve before eventually intersecting the latter just before both increase rapidly in gradient. It was found that engines which performed satisfactorily in service gave low DR Ferrograph readings on the test bed. A few engines which gave high DR Ferrograph readings on the test bed continued to give progressively high readings in service prior to premature in-flight shut down.

From the exploratory trend analysis it appears that a satisfactory engine provides steady, well-separated curves for total wear and severity of wear. When the severity of wear curve tends to approach the total wear curve, it indicates an uncharacteristic trend, and full Ferrographic analysis should be carried out to investigate possible causes.

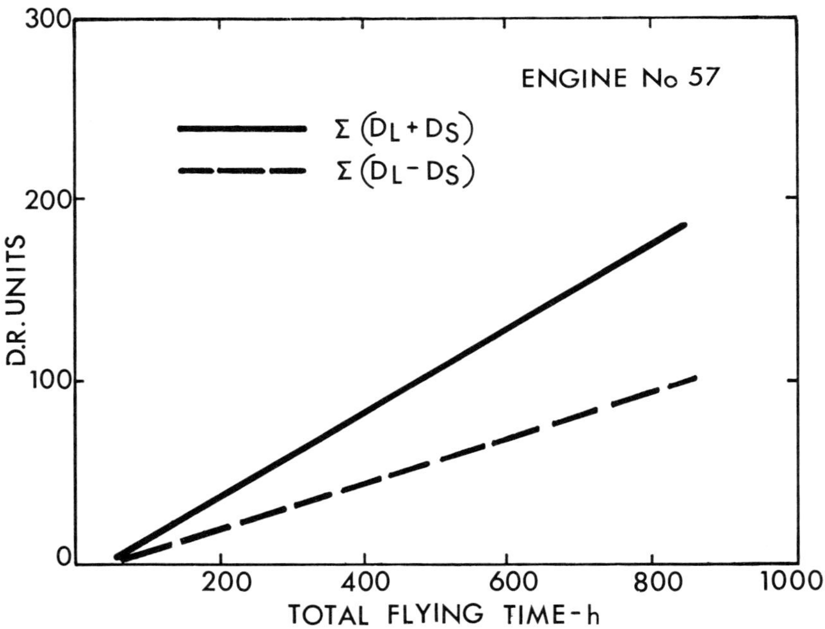

Fig. 1.—Trend plot of cumulative total wear and severity of wear for a satisfactory engine.

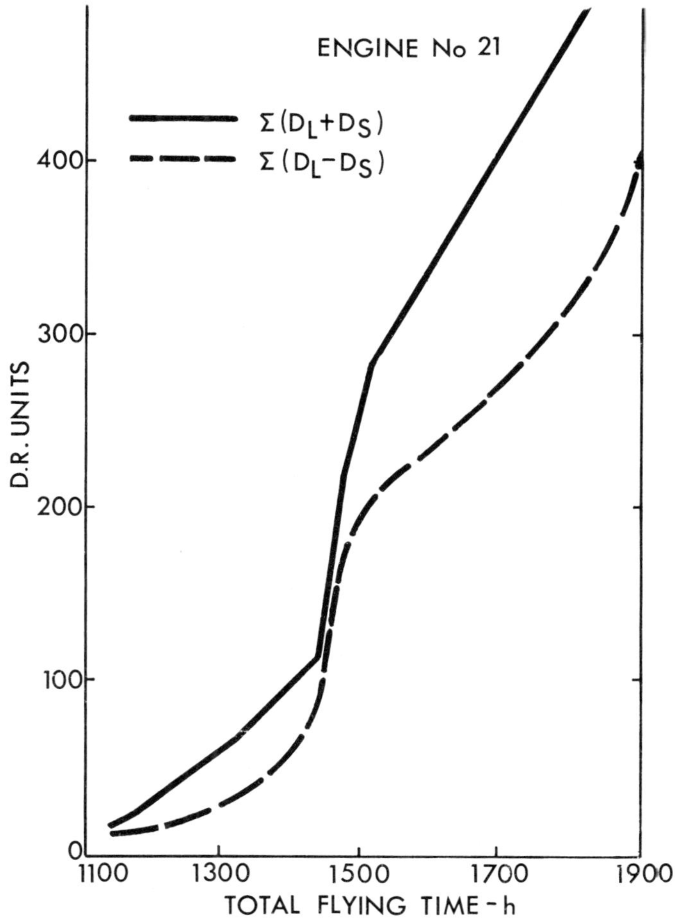

Fig. 2.—Trend plot of cumulative total wear and severity of wear for a deteriorating engine.

Ferrograms from satisfactory engines generally contained small rubbing wear particles, Figure 3, and Ferrograms from lubricant samples frcm the same engines on the test bed were similar.

With deteriorating engines, Ferrograms from the initial satisfactory running period showed principally small rubbing wear particles. However, a Ferrogram prepared from an oil sample at the point of intersection of the total wear and severity of wear curves showed massive cutting wear (Figure 4), indicative of a change in wear mode and impending failure. Ferrograms from lubricant samples from an engine on the test bed which gave a poor short service life showed fine cutting wear particles. The size of these increased to severe cutting wear particles and large severe wear particles just prior to engine withdrawal, Figure 5.

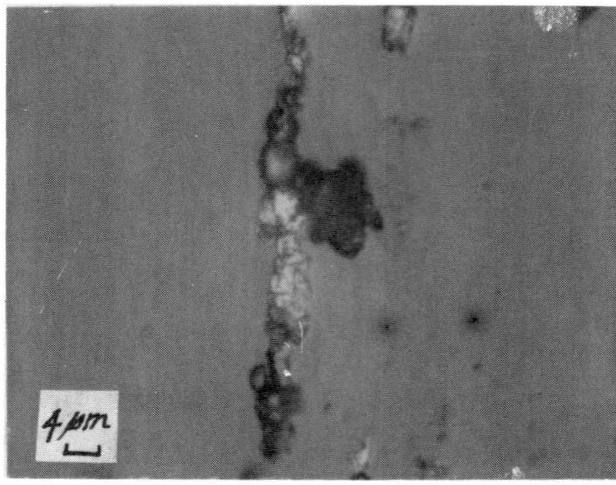

Fig. 3.—Small rubbing wear particles from a satisfactory engine.

Fig. 4.—Massive cutting wear particles from a deteriorating engine.

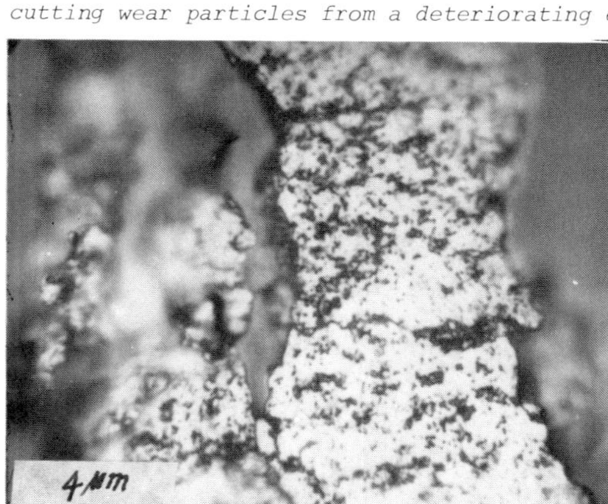

Fig. 5.—Large score wear particles from an engine just prior to failure.

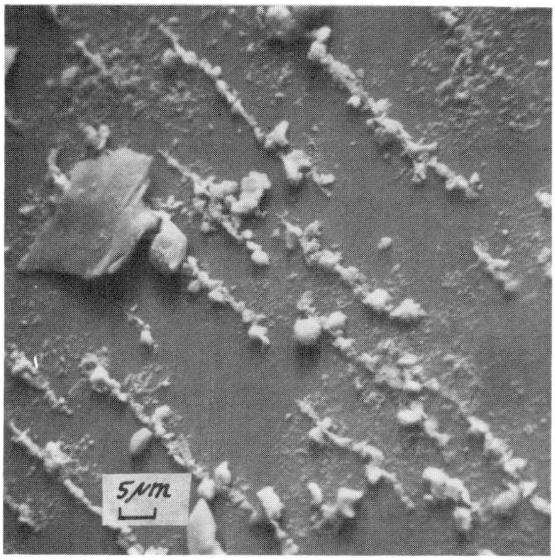

Fig. 6.—Typical debris collected from the gas stream of a deteriorating engine.

Fig. 7.—Typical debris collected from the gas path of a deteriorating engine.

Summarized data indicates that DR Ferrograph readings should provide satisfactory operating bands of wear. In the type of engine monitored, it appears that in satisfactory condition the normally quoted severity of wear index (I_o) should be below a limit of 150. Generally the readings for D_L and D_S as well as the total wear $(D_L + D_S)$ and the severity of wear $(D_L - D_S)$ should be in the ratio of 2:1. If it falls below 2 and approaches unity, the curves converge and a more severe wear mode is indicated and a full Ferrographic analysis is advised.

Wear Particles From the Gas Stream

Particles carried by the gas stream of a gas turbine engine, which impinge on surfaces contacted such as the tail pipe, can be removed with adhesive tape.[(4)] Solution of the adhesive on the tape, or even the tape itself, by a suitable chemical provides a liquid sample containing debris which can be precipitated by Ferrography. Samples were obtained from several of the engines being monitored, by pressing adhesive tape onto the inner surface exhaust outlets.

Figure 6 shows typical debris collected from the tail pipe of a deteriorating engine by adhesive tape. This engine was eventually removed for suspected turbine blade damage. The debris consisted of miscellaneous types of particles which by X-ray energy analysis were of varying composition indicative of turbine blade and seal material.

Figure 7 shows debris collected from the tail pipe of a different deteriorating engine which was withdrawn from service and found to have gearbox trouble. It is interesting to note that much of the debris was of analysis indicative of gear-box material including large flake-like particles of Al-based material similar to that found in the oil samples. It thus appears that debris from deteriorating oil-washed components has leaked into the gas path.

CONCLUSIONS

DR Ferrographic trend analysis appears potentially attractive as a simple, economic means of machinery condition monitoring. Convergence of the curves for cumulative total wear and severity of wear give an early warning of some uncharacteristic trend which should be investigated; cross-over of the curves indicates that urgent action is required. Monitoring of engines on the test bed can give an indication of service performance. Monitoring of both lubricant-washed components and components in the gas stream appears to be the most informative, as deterioration of one can greatly influence the rate of wear of the other.

ACKNOWLEDGEMENTS

This paper is published by permission of the Director of The National Engineering Laboratory of the Department of Industry. It is Crown copyright and reproduced by permission of the Controller, Her Britannic Majesty's Stationery Office.

REFERENCES

1. Westcott, V.C., These Proceedings.
2. Bowen, F.R.,et al., *Tribology International,* Vol. 9, No. 3, 1976, p. 109.
3. Scott, D., McCullagh, P.J. and Campbell, G.W., *Wear,* Vol. 49, 1978, p. 373.
4. Scott, D. and Mills, G.H., *Wear,* Vol. 48, No. 1, 1978, p. 201.

DISCUSSION

QUESTIONER: In figure 2 the point where convergence took place was at about 1400 hours. Then the curves were together for about 200 hours and then diverged again. The duration for which these two stayed together was nearly 200 hours and then afterwards the engine ran another 400 hours. Was the engine reconditioned at this point? Or, did the engine operate another 400 hours until it was taken off the wing after the indication that it would cause problems?

D. SCOTT: Several thousand samples collected over a period of time from the various engines were delivered together for analysis. When the trend of the engine in question was established the operators were notified but the engine was eventually shut down in flight having sustained considerable damage which necessitated very expensive overhaul.

A. BEERBOWER, UCSD: I am interested in the convergence of the curves. Is this a reflection of the shape of the particles. Do you still put some credence in the characteristic shapes of particles as a diagnostic tool in the determination of the mechanism of wear.

SCOTT: The trend plots are prepared from simple DR readings and when the trend of results justifies further investigation analytical Ferrography is carried out. Convergence of the curves is usually due to an increase in the number of large particles generated.

BEERBOWER: Does that mean that you still use morphology of the particle as a diagnostic tool for wear mechanism?

SCOTT: Yes. Particle size usually indicates the severity of wear and particle morphology indicates the operative wear mode.

BEERBOWER: So it is your early warning system.

SCOTT: Yes. The simple DR Ferrograph and trend analysis indicates any uncharacteristic trend which is investigated by analytical Ferrography to provide information on the initiation of a more severe wear mode and the component which is in distress.

R. K. TESSMANN, Oklahoma State University: Let me add a comment in support of DR Ferrograph. If we are not able to bring the Ferrographic technique out of the laboratory into the field as a real time monitoring device, it will remain an academic tool. For example, what I am doing with the Ferrogram now is essentially academic. The industrial sponsors are now particularly uneasy about the fact that it takes 30 minutes to get a single reading from the Ferrograph.

SCOTT: One objective of the simple DR Ferrograph is to allow condition monitoring on site by relatively unskilled personnel. The automated on-line Ferrograph is also being developed for continuous condition monitoring.

D. H. BUCKLEY, NASA: What filtration was in the system and from where do you take the samples? When did it occur to you that the engine was in a relatively unstable condition?

SCOTT: The normal filtration system was in use and samples were taken before the filter. The slope of the trend graphs indicates the rate of debris generation and any acceleration in wear rate which should be investigated.

BUCKLEY: How about the sampling technique?

SCOTT: The sampling technique is important and requires careful consideration. This will be particularly so for the on-line Ferrograph. However, for the results reported samples identical with those used for SOAP were used.

QUESTIONER: How did you collect these samples from the gas stream?

SCOTT: For the exploratory investigation adhesive tape was pressed on to the interior of the tail pipe after flight and thermocouple sheaths were rinsed clear of impinging debris after removal from the engine. Other methods have been considered and a small removable filter in the thermocouple sheath may be employed.

As we had several thousand samples sent from the engines in service the Ferrographic analysis was well behind the sampling.

DEBRIS ANALYSIS OF EROSIVE AND ABRASIVE WEAR

A. W. Ruff

ABSTRACT

Debris particles have been recovered from dry abrasion and erosion studies of AISI 1015 steel specimens. The size and morphology of the debris particles have been related to the worn surface topography. In both cases it appears that the principal wear mechanism involves removal of plastically deformed, exposed material from the lips of impact craters (erosion) or surface grooves (abrasion). Theoretical models that consider cutting processes on flat surfaces do not appear to properly consider the steady state morphology actually involved. The advantage of obtaining information from studies of both worn surfaces and debris particulates has been demonstrated.

INTRODUCTION

Analysis of wear debris is frequently carried out in connection with lubricated wearing systems in order to determine the system condition and the wear rate. However, the techniques used there can also be applied to a wide range of different wear problems as reported by Westcott[1] (for example, artificial implants and prosthetic devices) and by Scott et al.[2] on aircraft turbine engines using particulates recovered from the exhaust system. This paper will present some results obtained at NBS concerning the dry wear processes of solid particle impact erosion and two-body scratching abrasion on a mild (AISI 1015) steel. The experiments were carried out to examine the mechanisms of erosion and abrasion under similar conditions, and further, to compare the microscopic processes involved in debris formation.

Wear debris analysis develops an additional source of information concerning the wear process beyond that provided by wear surface analysis and subsurface examination. It is important to study the process by which material is actually removed from the wearing surface as contrasted to the processes that lead to only surface damage and deformation. Relatively little characterization has been done of debris resulting from steady state, multiple particle erosion, according to a recent survey.[3] However, information developed from debris analysis can furnish guidance for theoretical developments and an imporved understanding of erosion. Studies of abrasively worn surfaces produced under various wear conditions have been reported previously, including discussions of possible wear mechanisms and theories of wear rates.[4-6] Under two-body wear conditions with the abrasive particles

Fig. 1.—Abrasive particles of 50µm Al_2O_3 used in erosion and abrasion studies.

rigidly fixed during contact with the opposing surface, the contact interactions involve local processes of cutting (material chip removal), plowing (material displacement and deformation), and a load bearing function. It is reported that cutting contacts are a small proportion of the total number of contacts, from 12%[5] to about 50%.[7] Further, even though surface groove formation is the prominent process used in theoretical modeling, it has been estimated that less than 40% of the groove volume is removed as wear debris.[8] It seems likely that careful study of debris can provide further insight into the details of the abrasive wear mechanisms involved.

EXPERIMENTAL

The abrasive used in these studies was a commerical grade of alumina, Al_2O_3, of nominal size 50µm. A photograph showing the shape and angularity of a collection of these abrasive particles is shown in Figure 1. Fragmentation of some of the particles during erosion and wear tests is known to have occurred from the debris analysis results (to be described). The hardness of the particles (about 2100 kg/mm^2 [20 GPa]) was considerably greater than that of the steel specimens (about 100 kg/mm^2 [980 MPa]). The erosion equipment used has been described previously.[9] It consisted of a pressurized gas supply, a particle mixing chamber, and a delivery nozzle. Specimens were positioned at a working distance of 1 cm from the nozzle and exposed for a period of 2 min to a particle concentration of 1 g/l of nitrogen gas. The particle velocity was 40 m/s at an attack angle of 30 deg. The two-body abrasion experiments were carried out in reciprocal dry sliding on a surface of abrasive particles embedded in epoxy adhesive. A load of 2.2N and a velocity of about 5 cm/sec were used. After sufficient abrasive wear to achieve steady state conditions, the debris particles were removed from the Al_2O_3-epoxy surface using a clean lubricating oil wash. The oil-debris mixture was then magnetically separated to obtain the debris particles. In the case of the erosion experiments, a portion of the abrasive particles was collected after impacting the specimen surface. That dry mixture of abrasive particles

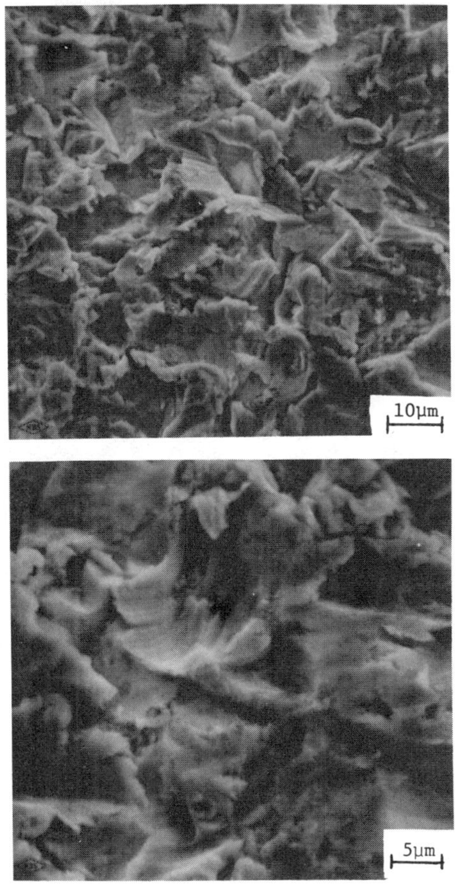

Fig. 2.—Scanning electron micrographs of the eroded surface on 1015 steel specimens. Individual particle impact craters can be identified.

and metal erosion debris was then mixed with clean oil in order to recover the metallic debris using the ferrographic technique. The wear debris collections (on glass slides) were subsequently coated with a thin Au-Pd layer in a sputtering system and examined in a scanning electron microscope. Stereo-pairs of photographs were taken (stereo angle of 7 deg) in many cases in order to determine the size and shape of both debris particles and surface features after wear and erosion tests.

RESULTS AND DISCUSSION

The appearance of the eroded surfaces of the 1015 steel specimens was generally that shown in Figure 2. Two significant aspects of the observed

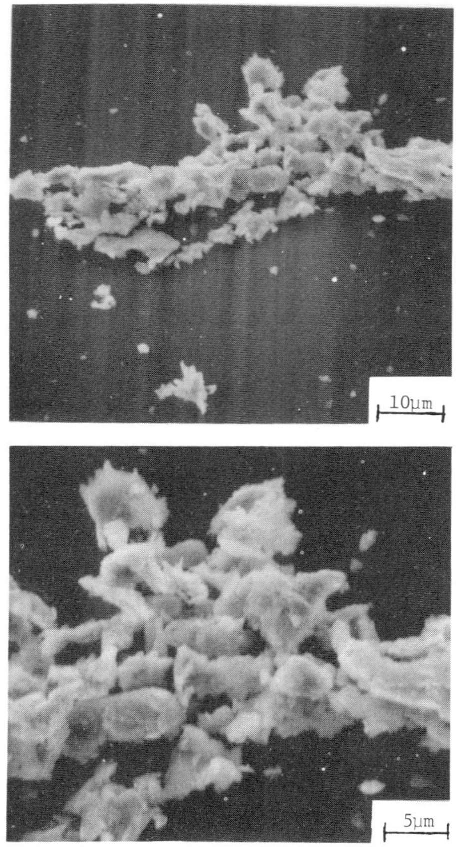

Fig. 3.—Metallic erosion debris particles recovered after exposure using the ferrographic method.

surface morphology were: (1) impact crater sites with distinct walls or lips of plastically displaced metal, and (2) an overall surface layer of deformed, distorted material containing embedded abrasive particle fragments along with many small craters and cavities. Metal debris particles recovered after erosion had the appearance shown in Figure 3, where the clustering of individual particles results from the magnetic field associated with the recovery method. The debris particles were about 1 μm to 5 μm in size. Stereo viewing of the debris revealed that the particles were generally smaller in the third dimension, that is, they were somewhat disc-shaped. Comparison of the metallic debris with the eroded surface morphology suggests that many debris particles arise from the deformed material comprising the edges of the impact craters. One characteristic feature of the debris particles was small, ragged edges that suggested a ductile tearing process associated with their formation. As far as this particular steel was concerned, there were no particle

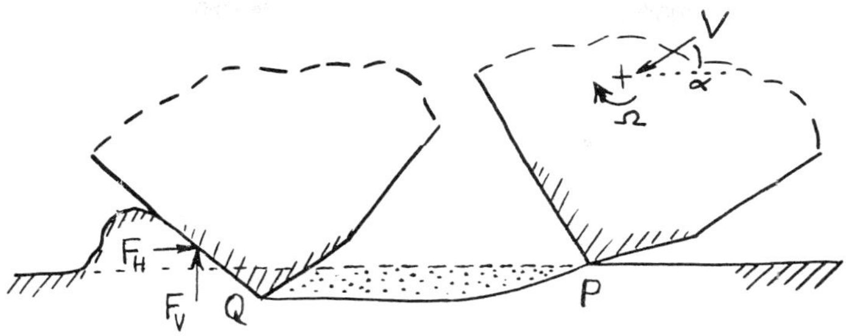

Fig. 4.—Schematic illustration of particle impact during erosion. Particle with velocity, V, and attack angle, α, impacts surface at point P. Material is displaced (shaded region) until particle leaves surface at point Q.

features that suggested that brittle fracture processes were involved.

Theories of ductile material erosion(10) generally depict the processes of plowing and cutting of surface material by the incident particle as shown schematically in Figure 4. Undoubtedly many of the incident particles interact in this way. However, from the size and shape of the wear particles, it seems likely that the affected volume of material (shaded area, Figure 4) is deformed and displaced rather than removed as debris from the target surface. Subsequent particle impacts with the more exposed crater lip material probably produce most of the debris recovered in these experiments, based on observations of the debris size and shape.

The abraded specimen surfaces had a characteristic appearance that can be seen in Figure 5. Long grooves representing displaced metal were found aligned in the sliding direction. Plastically deformed metal comprised the lips of material that were produced at the edges of these grooves. Stereo photographs of the surface topography suggested that much of this lip material was being removed subsequently in the form of wear debris. Studies of the recovered debris particles (Figure 6) indicated that their size and shape was consistent with the nature of the exposed surface material. Only relatively few examples of long chips of metal were found (e.g., Figure 6b). The predominant size and shape of the wear particles suggested that, much as in the case of the erosion damage, the exposed (and deformed) surface features were a principal source of the wear debris.

In the case of two-body abrasive wear, most theories(11) involve a rigid particle cutting process as depicted in Figure 7. The groove volume of material (shaded) is equated with removed material, as wear debris. However, observation of the surfaces and the debris particles in these experiments involving 1015 steel suggest that the groove material is, for the most part, deformed and displaced into more exposed positions above the surface. Subsequent abrasive particle motion across the surface then removes that material as debris particles. Preliminary studies of particle size distributions in these two modes of wear have been carried out. Figure 8 indicates that a difference exists under present conditions. The erosion impacts tends to produce a somewhat larger mean particle size and a lower frequency of particles 1 μm and smaller. Further studies of debris size and shape characteristics are in progress.

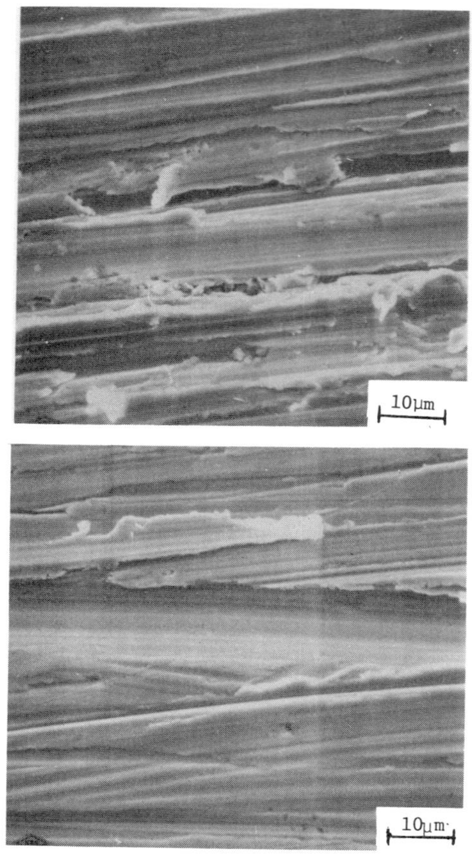

Fig. 5.—Scanning electron micrographs of the two-body abraded surface on 1015 steel specimens illustrating the surface deformation and topography produced.

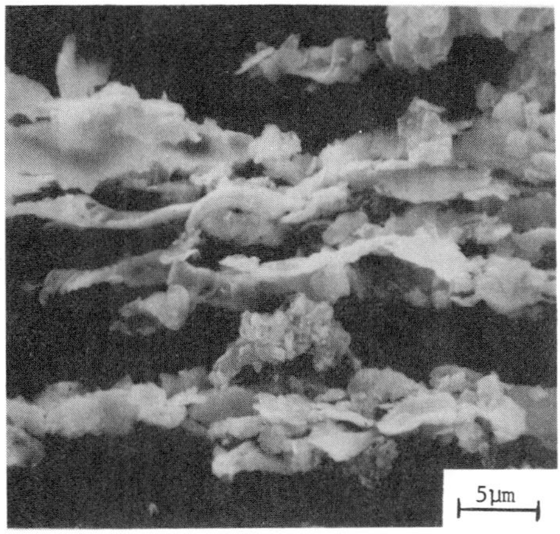

(a)

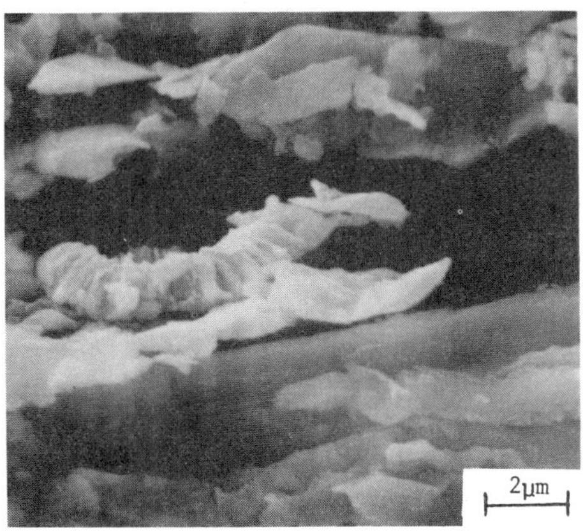

(b)

Fig. 6.—Metallic wear debris particles recovered from the dry two-body abrasion studies using the magnetic recovery method.

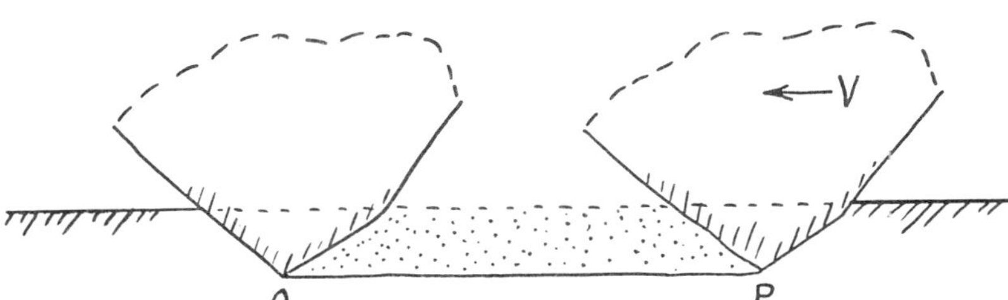

Fig. 7.—Schematic illustration of the abrasive particle wearing geometry during abrasion.[11] The relative velocity, V, in the presence of a normal load causes a volume of displaced material (shaded) to be formed over the length PQ.

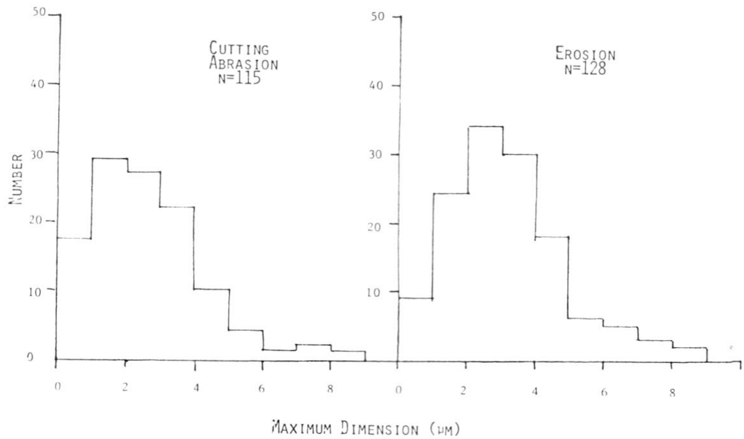

Fig. 8.—Debris particle size distributions obtained from cutting abrasion and erosion studies on 1015 steel.

SUMMARY

Debris particles have been recovered from dry abrasion and erosion studies of AISI 1015 steel specimens. The size and morphology of the debris particles have been related to the worn surface topography. In both cases it appears that the principal wear mechanism involves removal of plastically deformed, exposed material from the lips of impact craters (erosion) or surface grooves (abrasion). Theoretical models that consider cutting processes on flat surfaces do not appear to properly consider the steady state morphology actually involved. The advantage of obtaining information from studies of both worn surfaces and debris particulates has been demonstrated.

REFERENCES

1. Westcott, V.C., These Proceedings.
2. Scott, D., et al., These Proceedings.
3. Ruff, A.W. and Wiederhorn, S., in "Treatise on Materials Science and Technology: Erosion," edited by C.M. Preece, Academic Press, New York, 1979, in press.
4. Moore, M.A., *Wear,* Vol. 27, 1974, p. 1.
5. Mulhearn, T.O. and Samuels, I.E., *Wear,* Vol. 5, 1962, p. 478.
6. Larsen-Basse, J., *Wear,* Vol. 12, 1968, p. 35.
7. Larsen-Basse, J., *Wear,* Vol. 11, 1968, p. 213.
8. Stroud, M.F. and Wilman, H., *Journal of Physics D: Applied Physics,* Vol. 13, 1962, p. 173.
9. Young, J.P. and Ruff, A.W., *Journal of Engineering Materials and Technology,* American Society of Mechanical Engineers Transactions, Vol. 99, 1977, p. 121.
10. Finnie, I., *Wear,* Vol. 19, 1972, p. 81.
11. Rabinowicz, E., "Friction and Wear of Materials," John Wiley and Sons, New York, 1965, p. 168.

DISCUSSION

QUESTIONER: Could you elaborate on the method by which you recovered the particles and the Ferrogram preparation technique?

A. W. RUFF: I do not want to go into all the details here. Basically the particles were suspended in oil after the experiments. We had suspension of both the metallic and abrasive particles and then it was run in a routine way. We had a lot of abrasive particles on the Ferrogram with transferred metal on them. This is also an advantageous technique for looking at abrasive fragmentation and metal transfer because it deposits at least some abrasives that have received a transfer film of iron.

The sampling problem is very complicated. Normal filtering techniques or sieving techniques would give a different population. If one is interested in mechanisms in relation to surface topography, it is necessary to capture as much of the metallic debris as possible.

QUESTIONER: Did you find any difference in the amount of abrasive on the Ferrogram in the different modes?

RUFF: Well, in these experiments the stress per particle was not controlled. What we want to do is to compare abrasion and errosion with the same stress per particle. The higher stress levels would involve more metal transfer and possibly more fragmentation of the abrasive. I could not conclude anything useful right now, but that is of interest to us.

D. SCOTT: You are comparing abrasion with erosion. Abrasion seems to be a very inefficient means of metal removal. Can you comment on this?

RUFF: I think that there is a lot of information in the literature on this. There is probably a need to do some more experiments under simple conditions. We hope to make a comparison in the same system under the same loading conditions. It appears that they are similar mechanisms; they both involve ploughing as well as cutting. The theories that I am familiar with are not complete in that they describe, to a large extent, the surface topography that develops but not the actual production of wear debris. In many instances abrasion is one percent effective and erosion is even less than that particularly for low particle velocities. The missing aspect is that the wear mechanism really has not been studied. I think one needs to develop models that really describe the process of wear debris formation.

XII. WEAR PREVENTION

STATUS OF WEAR PREVENTION

M. B. Peterson, M. J. Devine and A. J. Koury

ABSTRACT

A review of the literature in the field of wear has been carried out to determine areas where additional information is necessary to improve our ability to predict and prevent wear in service. Areas covered are: understanding the effects of certain variables, materials, design, wear prediction techniques, and environmental effects.

INTRODUCTION

Wear is an applied science. Developments should eventually be measured against their contribution to prevention of wear of service equipment. Accordingly, it is often useful to review our ability to prevent or correct excessive wear of a given piece of equipment. Such an exercise not only illustrates the practical barriers to wear prevention but also defines areas where additional research and development is necessary. This might be called service-oriented research, where the studies are directed not to solve specific problems but to fulfill a general need for information.

The authors, for a period of three years, were involved in a wear control program for naval aircraft. The purpose of this program was to assess the status of wear control technology and to make recommendations for improvements. Specifically, the aim was to determine what new developments in the field of wear could be usefully applied to reduce the maintenance costs and extend the service life of naval aircraft.

The program consisted of several phases. In the first phase, information was gathered relating to the cost and consequences of wear in order to determine the magnitude of the problem. Secondly, a technology review was undertaken to accumulate recent technical and materials developments which might be applied to wearing aircraft components. Visits were then made to naval facilities to isolate areas of potential improvement to which these developments might be applied. Once component problem areas were identified, solutions were proposed, evaluated, and introduced into service. Although the objective of this program was to prevent wear of specific components, the results of the program serve a second objective, the evaluation of the status of wear prevention technology. This is the subject of the present paper. Our ability to solve service problems is evaluated here.

In making this evaluation, due attention must be paid to practical considerations which very often override technical solutions. Economics, timing, and uncertainty are strong motivating forces for maintaining the "status quo".

WEAR COSTS AND CONSEQUENCES

It is recognized that improvements can be made in almost any manufactured product--especially in products which see service for 20 years, as many aircraft do. During this service life, many new technical developments are conceived and commercialized which were not available during the design of the aircraft. The question to be answered is whether such improvements are significant enough to justify the retrofit; or, put another way, whether the problems are severe enough to justify a detailed analysis. In order to answer such questions, certain cost information was assembled for illustrative purposes. In addition, other qualitative information was assembled which illustrated the consequences of wear. These data are presented in the following sections.

Cost of Wear - Attack Aircraft (A Case Study)

In order to obtain quantitative wear cost data, all maintenance actions on one aircraft were acquired through one tour of duty. This included both organization level maintenance and eventual rework.

The costs associated with malfunctions for this aircraft for a two-year period 1974 and 1975, are shown in Table 1. It can be seen that the wear costs total $197 per flight hour during the selected period. Fuel costs for the same period were approximately $376 per flight hour. Seventy percent of these costs for wear were incurred at the squadron level, 66 percent of all costs being associated with labor and 34 percent with materials. Almost all of the costs were either in the airframe or avionics systems. Only about 5 percent were engine related.

*TABLE I.—OPERATIONAL WEAR COSTS**

	$/Year	$/FH
Component Maintenance (unscheduled)	$ 40,400	$140
Component Maintenance	16,450	57
Total (Wear)	$ 56,850	$197
Total (Fuel)	$216,000	$376

*575 Flight Hours.

TABLE II.—COMPONENT REWORK

Malfunction	Rework Hours	Remove/Install. Hours	Total
Wear	201.03	248	$11,226
Cracks	122.52	292	10,363
Corrosion	121.00	213	8,350
FOD/Torn	46.28	106	3,807
Test	26.09	27	1,327
Electrical	22.54	---	564
Clean	5.16	---	129

The rework costs for the same aircraft are given in Table II. In Table II the total rework hours for components have been categorized into seven major malfunctions along with the hours required to remove those components from the aircraft. For wear, a total of 449.03 hours was required. Assuming a total of $25 per hour, the total cost of component wear was $11,226. The cost due to cracks was $10,363.

Similarily, data were obtained on the costs of malfunctions on the aircraft itself (as distinguished from components which are removed); from engineering changes made to overcome some deficiency in the design; and the materials purchased to accomplish the rework. These data showed essentially the same trend as that shown in Table II; that wear is the predominant rework cost, followed closely by cracking (fatigue and corrosion). Costs of other malfunctions were significantly lower. A summary of all costs is given in Table III. Thus, the total wear costs associated with this aircraft during its two-year tour of duty (exclusive of the tire wear) is $235 per flight hour which is almost two thirds the fuel costs. The main conclusion drawn from the data was that a good deal of money is spent each year on wear and some steps should be taken to control it. Some further analyses of these data are useful. The total cost of ownership of such an aircraft is $5842 per flight hour of which $4329 is the cost of operation and $1513 for amortizing the aircraft. Of the $4329, $883 is for personnel, $505 for spare parts, $62 for engine repair, $308 for depot maintenance, $820 for component rework, $768 for fuel and other expenable items and $971 for indirect costs associated with this aircraft. Thus, the total cost of maintenance is approximately $2000, of which wear is roughly 11%. The total cost of wear is small in comparison with the cost of ownership; however, this is the major malfunction cost and should be reduced as far as possible. This cost of wear can be increased substantially if down time is included. The aircraft mission is programmed so that the aircraft is operational at least 70% of the time. If it is not, then a high price has been paid for a service not rendered. For this aircraft, down time amounts to approximately $2000/day. Thus, for each day the aircraft is "down" greater than 30% of the time, the total cost of wear must be increased by $2000. For this aircraft 1076 maintenance actions were performed for wear at an average of 3.3 hours per action, equalling 3540 hours.

TABLE III.—TOTAL COST OF WEAR*

	$/Year	$/FH
Operational		
Unscheduled	$ 40,400	$140.00
Scheduled	16,450	57.00
Overhaul		
Component	5,633	20.00
Discrepancies	488	1.70
Tech Directives	3,700	12.87
Materials		3.30
Bearings	612	---
Seals	86	---
Gears	42	---
Hinges	10	---
Total (Wear)	$ 67,600	$235.00
Total (Fuel)	$108,000	$376.00

* 2-Year Period

Thus, in a 24-month period, it was maintained for wear 20% of the time the 3.3 hours representing only one man. Since a number of maintenance actions can be performed at the same time, it could not be determined if the 30% was exceeded.

Another cost factor should be mentioned. If the aircraft were lost due to a wear failure, the remaining down time left in the aircraft must be added.

As seen, maintenance for wear is expensive. However, it is not just the cost of the part and its replacement, but also the fact that a maintenance staff must be available at all times to respond. This increases the direct cost of wear. Even though the economics of the situation itself suggests that better wear control be exercised, there are several other reasons which in themselves justify action.

First of all, for various reasons parts and materials are in short supply. It has been found that equipment is out of service for longer periods of time awaiting parts. This increases the amount of down time for a given malfunction. Because of this, wear control is more significant for complex *purchased* parts than for simple, easily repaired parts.

Secondly, worn parts cause many secondary problems such as increased vibration which leads to fatigue damage, shock loading, misalignment, and accelerated wear. Worn parts also cause a deterioration in performance which, although acceptable, is certainly not desirable.

High wear costs, even though economically justify a change, are often not a sufficient reason for change. The change itself costs a large amount of money since the supply inventory must be changed, the maintenance manuals revised, procedures modified, etc. There is also a large resistance to change because it introduces major new uncertainty and risk. It is often much safer and less expensive to live with a poor, well-understood system than to introduce a better ill-defined one.

In addition, labor cost is much higher in comparison to materials. Accordingly, it is often deemed cheaper to consistently change parts than to try to resolve a problem.

In any case, wear prevention is a policy decision and two approaches are possible: (1) accept a certain level of wear and provide sufficient numbers of parts and manpower to maintain an acceptable level of machinery availability, or (2) reject wear and apply new technology for improved reliability. The choice of a course of action is an economic decision, but it depends to a large extent on the confidence in the technology.

TYPES OF WEAR

Wear, as previously stated, is not a single process but a variety of processes which can take place independently. To understand wear and its control, each of these processes must be isolated and studied independently. In the past, wear has been categorized in a variety of different ways. Most lists, however, include the following: adhesion, abrasion, corrosion, fatigue, fretting, erosion, cavitation, attrition, gouging, and lubricated wear. Actually, these are not independent types of wear, but rather processes which occur under a defined set of circumstances. For example, erosion occurs when hard particles, fluids, or gases strike the surface of a solid. Abrasion occurs when particles or a rough surface scratches another surface. For the purposes of research and understanding phenomena, it is felt more advisable to classify wear according to the mechanism of particle removal.[1] Such a classification is as follows:

Adhesion and transfer	Surface fracture
Corrosion film wear	Surface reactions
Cutting	Tearing
Plastic deformation	Melting
Fatigue	Electromechanical

Here the major types of wear are listed in the left hand column while those on the right refer to special cases. A brief description of each wear process is given in Table IV.

TABLE IV.—TYPES OF WEAR

Adhesion and Transfer	Material welds at sliding asperity tips, is transferred to the harder member, grows in subsequent encounters and is eventually removed by fracture or fatigue.
Corrosion Film Wear	A film formed by reaction with the environment or the lubricant is removed by sliding.
Cutting	A sharp particle or asperity cuts a chip.
Plastic Deformation	The surface is worked plastically. Cracks form, grow, and coalesce forming wear particles.
Surface Fracture	If nominal stress exceeds the fracture stress of a brittle material, particles can be formed by fracture.
Surface Reactions	One material dissolves or diffuses into another.
Tearing	Elastic material can be torn by a sharp indenter.
Melting	High velocities can cause wear by melting.
Electrochemical	The difference in potential on the surface due to a moving fluid can cause a material to go into solution.
Fatigue	The surface is worked elastically. Microcracks form, grow, and coalesce forming wear particles.

In the following list, some of the conventional wear descriptions are given in terms of the suggested types of wear:

Abrasion	Cutting and deformation
Fluid erosion	Deformation and fatigue
Solid erosion	Cutting and deformation
Cavitation	Deformation and corrosion
Lubricated wear	Adhesion and corrosion
Dirt abrasion	Deformation and fatigue
Filing	Cutting
Sanding	Cutting and deformation

Unlubricated metal wear	Adhesion and deformation and corrosion
Unlubricated plastics wear	Adhesion and deformation
Unlubricated rubber wear	Tearing and fatigue
Carbon wear	Fatigue
Wear of Ceramics	Fracture

The above list of wear types is not considered final and would be modified as new insights are obtained. However, they represent a logical classification of the basic wear processes which can be isolated, studied, and related to the basic material properties and processes.

Reduction of Adhesive Wear

Adhesive wear usually results from an inadequate supply of lubricant to the contact area or from a poor choice of materials for the operating conditions. If the wear debris shows large metallic particles and the surfaces are damaged with material transferred from one surface to another, then the amount of lubricant reaching the contact area should be investigated. Often with large contact areas, sufficient lubricant passages are not provided and unlubricated regions result. This initiates the failure process. Thus, the first step is either more lubricant or better distribution. If this does not appear to be the cause of the high wear, improved materials should be considered.

Adhesive wear is the result of high attractive forces of the atoms composing each of the sliding surfaces. In sliding, two surface asperities come into contact, adhere, and are deformed under the action of the normal and tangential stresses. As this process continues (if the bond is strong enough), a break will occur in the weaker member and a particle will be transferred to the harder surface. The stronger the bond, the greater the probability of forming a particle; the greater the ductility of the softer metal, the larger that particle will be. This transfer does not produce a wear particle; however, in subsequent passes, portions of the film are removed from the surface. The greater the rate of transfer, the greater the rate of wear. Thus, adhesive wear rates are proportional to the bond strength and the ductility. It is not possible to calculate actual bond strengths between metals; however, research has indicated that certain techniques may be used to indicate weak bonds. The two most widely used techniques are solubility and atomic size. If phase diagrams indicate low solubility, adhesion will be low. Rabinowicz[(2)] has shown that under such circumstances wear will also be low as indicated by his data shown in Table V. Generalizing on these results and applying it to common aircraft materials of aluminum and steel would indicate that material of limited solubility in those metals would give minimum adhesive wear. Such metals are listed in Table VI from test results of Reference 3.

Of the *soluble* materials, a series of high-speed, high-load sliding tests (100 ft/sec, 60 lbs., 120 psi) were run using a variety of pure metals against steel. It was found that the least surface damage resulted with CU, W, and Mo.[(4)]

TABLE V.—WEAR COEFFICIENTS

Metal Pair	Solubility	Wear coefficient (X 10^{-4})
Cu-Pb	No	.10
Ni-Pb	No	.21
Fe-Ag	No	.68
Fe-Pb	No	.69
Al-Pb	No	1.4
Al-Zn	Yes	3.9
Al-Fe	Yes	1.0
Fe-Cu	Yes (Min)	19.0
Al-Al	Yes	30
Mg-Mg	Yes	36
Fe-Mg	Yes (Min)	38
Fe-Fe	Yes	77
Cu-Ni	Yes	81
	Yes	130
	Yes	290

Low sliding speed tests at light loads on iron showed that the materials which gave the least damage were the following: Mg, Ag, Sn, An, Pb, and Bi.

It should be noted, however, that these materials may be poor for other forms of wear. For other techniques of selecting low adhesion materials, the reader is referred to Reference 1.

In many instances wear prevention techniques are directed toward increasing the hardness (lower ductility) and too little attention is given to improved lubrication and improved materials.

Prevention of Corrosive Wear

Only a very thin, solid film is needed for effective lubrication. If more is formed, it is quickly worn away. This is illustrated in Figure 1 where the wear rate is plotted against film thickness. With very thin films, appreciable metal contact and adhesive wear result.

TABLE VI.—INSOLUBLE PAIRS WITH GOOD OR FAIR SCORE RESISTANCE

Aluminum	Steel
Cd	C
In	Ag
Pb	Cd
Bi	In
Zn	Sn
	Pb
	Bi

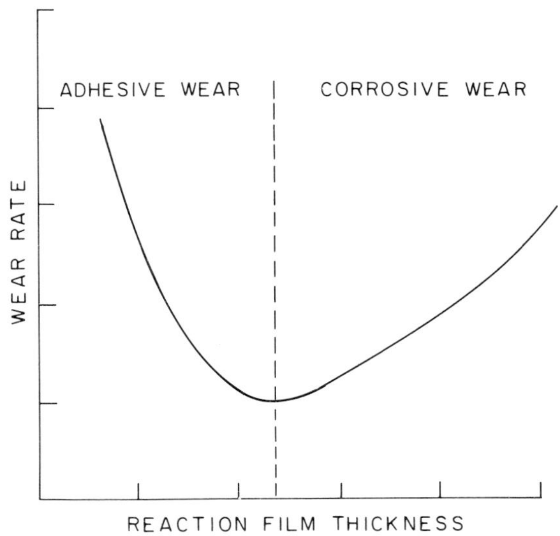

Fig. 1.—Effect of reaction film thickness on wear.

With increasing film thickness, the adhesive wear component is reduced until the optimum film thickness is obtained. Greater film thickness means greater wear. As previously discussed, it is believed that all of the excess film is removed (above the optimum) during each sliding cycle. Thus, the wear rate would be proportional to the rate of film formation. To reduce this wear rate, one must reduce the rate of reaction. The most generally used approach is to use more corrosion resistant materials or surface treatments; however, less reactive lubricants will have the same effect. Here it is not necessary for the treatment to increase the surface hardness; thus, many simple corrosion resistant treatments may be useful in reducing wear.

Prevention of Cutting Wear

Cutting wear is usually the result of adhesive dirt particles in the fluid. If such are identified in the oil, the first step is to either prevent their entry or remove them by more rigorous filtration. If this is not possible, then improved materials are suggested.

In cutting wear a hard particle or asperity cuts a chip from the wearing surface. As sliding takes place between the surfaces, two independent deformation processes can occur; either a chip is cut, or material is plastically or elastically deformed about the particle.

If a chip is cut, exceedingly high rates of wear will result as previously discussed. Fortunately, a chip will be cut only under very special conditions shown by Equation (7) Reference 1. High-friction materials and lubricants mean less cutting wear since fewer cutting edges are involved.

Also, to cut a chip the particle or asperity must be firmly attached to the surface. Rolling particles have from 0.01 to 0.10 times the wear of rigid particles. Furthermore, to cut a chip, the material must not be blunted by the sliding process; materials and lubricants which dull the cutting edge will suffer less wear.

As previously shown, the most important factor is the hardness of the cutting agent. To cut, it must have a hardness greater than 1.2 times the material being cut. Since SiO_2 has a hardness of approximately $1000\ H_v$, the wear resistant surface should have a hardness of *1200*. There are very few metals and surface treatments which can reach this level of hardness. Chrome plate (if carefully controlled) is one. Others are boriding, plasma sprayed cermets like tungsten carbide-cobalt and ceramics, and certain forms of electroless nickel. More use should be made of such materials in abrasive situations.

Various correlations have been attempted between abrasive wear and material properties (hardness, modulus, hardness-modulus ratio, strain energy, plastic work, etc.) without much success when a large variety of metals and alloys are considered. However, for pure (homogeneous) materials (ceramics, metals, plastics), *cutting wear resistance is directly proportional to hardness*. Thus, to reduce cutting wear, higher hardness materials should be selected. Although it cannot be used in the selection process, it should be remembered that it is not the static hardness but rather the dynamic hardness after sliding (cold work) which is important.

Hardness becomes less significant with duplex structures since hardness is only resistant to indentation while cutting requires independent penetration of both the hard and soft phases. It has been shown that if the cutting particles are large, the overall or average hardness applies; while if the particles are small, structure effects become significant. Important material factors under these conditions are the relative hardness of the matrix and the included phase, the spacing, and the bond between them. Application of these principles generally means fine grain, small particle size, fibrous structures are more wear resistant to cutting.

Deformation Wear Reduction

Deformation wear consists of continual plastic working of the surface until cracks form, grow, and coalesce; and a wear particle is produced. If the stresses are high, wear tracks are produced, the plastic working is severe, and large wear particles can be produced in only a few cycles. In this case the wear particles are ridges or lips from the edge of the wear tracks. At lower stresses (still in the plastic range) plastic working produces small surface spalls within the contact area. This suggests that lower loads or larger contact areas be used if this type of wear is identified.

Very few fundamental studies have been conducted to obtain definite correlation with material properties; however, it seems clear that there should be a correlation between deformation wear and some fracture mechanics property sucn as the rate of crack growth, fracture toughness, or elongation at fracture.

Such a correlation has been established for plastics[5]. The wear rate is plotted against the product of tensile strength and elongation (a crude estimate of the toughness). It is shown that the wear resistance does increase as the toughness increases.

The best materials depend very much on the operating temperature;

however, at room temperature the following order of increasing wear rate was found:

Polyethylene
Nylon 6/6
Polytetraflouroethylene
Polypropylene
Nylon 11
Polyactal

A correlation has not been obtained for metals; it is generally agreed that high toughness is necessary to resist deformation wear. Values for this product (strength x elongation) are given for some common metals in Table VII. Although this list corresponds qualitatively to the deformation wear resistance as shown by their use in certain applications, more precise information would be desirable based upon some stress/time impact cycle.

There are several sets of conditions where deformation wear predominates in machinery:

(1) In dirt abrasion where wear or metal transfer has dulled the abrasive particles so it will not cut.

(2) During conditions of impact loading between the two surfaces.

(3) During conditions of erosion, either fluid or solid, with high angles of impact.

(4) In well-lubricated sliding where a surface with smooth, hard asperities slides against a softer material (minimum transfer and cutting wear).

Under these conditions where deformation wear predominates, it is a common fallacy of wear control to try to improve the wear resistance by increases of hardness. Usually this has just the opposite effect.

Prevention of Fatigue Wear

Fatigue wear is the result of repeated vibratory stressing of the surface of a body. Microcracks form, usually at inclusions, and are transmitted throughout the surface. Eventually these cracks coalesce and wear particles can be formed[6]. Although the overall stress is within the elastic range, the stresses at the crack tip are intensified and fracture can take place. This type of wear is more prevalent in rolling contacts such as bearings and gears; however, it is becoming increasingly apparent that fatigue wear is significant in sliding contacts where other more severe types (cutting, adhesion, plastic deformation) have been eliminated. Furthermore, fatigue wear requires a given number of stress cycles and often predominates after a part has been in service a long period of time. Since the fatigue wear rate is greater than the adhesive wear rate, the prediction of its onset would be a very valuable contribution to wear prediction and wear prevention.

The important fact is that there has been found a definite correlation between the fatigue resistance of the material in fatigue tests and in wear tests on different rubbers.

TABLE VII.—METAL PROPERTIES

Metal	(Strength x Elongation ($X 10^6$ psi)
Stainless Steel 304 (Annealed)	5.1
12 Mn Steel	4.5
Martensitic SS 431 (Heat Treated)	4.1
A-286	3.6
AISI 4340 (Hardened and Tempered)	3.0
Nitralloy 135	2.8
13 Carbon Steel	2.3
AISI 1020 Steel	2.3
AISI 440C (Annealed)	1.5
AISI 440C (Hardened and Tempered)	.57
Cobalt Alloy L650	8.7
200 Ni (annealed)	3.2
Inconel x 750	3.4
Inconel 718	4.7
1100 Al (Annealed)	.52
7075 Al (Annealed)	.56
Cu (annealed)	1.6
Al Bronze (Hardened)	2.9
Mg	6.1
Ti6Ai6Va	1.9

This not only confirms the presence of fatigue wear but shows that fatigue resistant materials can be used in fatigue wear situations.

Effect of Hardness on Wear Prevention

It is very common to make contacting surfaces of similar metals (i.e., Steel on steel, aluminum on aluminum) and the same hardness (e.g., Rc 45 on Rc 45). It has been found that this is undesirable and to be avoided for a number of different reasons.

The abrasive wear of metal combinations with different kinds of abrasive was shown by Hirano and Ura(7) to depend on the metals' relative hardnesses. By making a hardness difference between the surfaces of at least 50 points Vickers or Brinell (or about 3 to 10 points Rockwell C) either up or down, the wear of both metal surfaces can be significantly decreased. This is an important finding for it is relatively easy to change the hardness of steel parts by this amount. It is doubtful, however, that the benefit would be obtained with aluminum alloys as they are precipitation hardened and do not obey the same rules as steels.

Another reason for a hardness difference of at least 50 points Vickers is to prevent surface distress. Welsh(8) studies the surface distress of steel rollers and found that a positive hardness difference with the surface of 50 to 60 points Vickers develops with many cycles of operation. This can lead to surface distress and rippling unless one of the surfaces exceeds the other in initial hardness by this amount. This phenomenon has also been observed in gears and is now explainable on this basis. It would be good practice, therefore, to require all heavily-loaded surfaces (cams, latches, hinges, rollers, etc.) to have a hardness difference between mating surfaces of at least 50 points Vickers.

The effect of the relative hardness of bushing and bolt materials on the wear of nonlubricated friction pairs was the subject of a study by Matveevskii and Khrushchov[9]. They established that if the bushing and bolt materials are of the same hardness, the wear of both elements sharply increases--reaching the maximum possible wear values. This study again shows that metals of the same hardness should not operate together as a friction pair.

The dry rubbing of steel surfaces generally exhibits one of two kinds of adhesive wear phenomena: severe wear in which large metallic fragments come from the wearing surfaces, giving rise to a high degree of surface damage and a high wear rate; and mild wear in which the surfaces wear by the formation of a tenacious oxide layer and its subsequent removal in the form of very small oxide particles, thereby producing negligible surface damage and a very low wear rate.

Exhaustive and thorough research on the adhesive wear of carbon steels[10] has shown that the prevention of the transition from mild wear to severe wear in rubbing components and thereby the prevention of a catastrophic wear situation depends on maintaining a minimum surface hardness. This hardness represents the ability of the surface to successfully resist the failure of its oxide film. Welsh found that the critical Vickers hardness needed to maintain the mild wear form of wear was related to carbon content as follows:

Percentage Carbon	Critical Vickers Hardness
0.12	>457
0.34	>433
0.52	>436

As a Vickers hardness of 460 corresponds to a Rockwell C Hardness of 47, by requiring only one of the two rubbing steel surfaces to have a minimum of Rc 50, it should be possible to prevent the occurrence of catastrophic severe wear.

This means that *both* surfaces do not have to be heat treated to high hardness. As long as the oxide on only one of the surfaces resits failure, severe wear is prevented.

A study to prove the validity of this proposition was carried out by Calabrese, Peterson, and Chiang[11]. They obtained data for combinations of the alloy steels SAE 4340, SAE 4140, and the tool steel FS02. They found that mild wear could be obtained if one of the surfaces was hardened to Rc 50 or better, while the other surface had a minimum hardness of Rc 30.

By *not* heat treating both surfaces to high hardness, significant savings can be realized.

In many friction pairs, one component moves while the other experiences no movement. Some examples are hinges and hinge pins and bushings and bolts. If such a system is to be made of materials with different hardnesses, the question of which component to make harder.

Tests (for aircraft hinge pin purposes) on steel over a range of hardnesses from 24 Rc to 48 Rc, comparing pairs consisting of soft bushings and hard pins and for pairs consisting of hard bushings and soft pins, show that the total wear of the former is approximately three times greater than that of the latter[(12)]; i.e., it is more advantageous to combine *a hard bushing with a soft pin* than a soft bushing with a hard pin.

Filtration

Most hydraulic systems, pneumatic systems, reciprocating engines and compressors, bearing systems, and other machinery have their flowing lubricating oil and air filtered to remove abrasive contaminants. The concentration and size of the abrasive particles allowed to pass these filters is important to the wear life of these systems. However, the degree of filtration selected was usually done quite arbitrarily without real attention being paid to the consequences of this selection.

Only recently [(13-16)] has the meaning of filtration quantitatively emerged. Tests showed that the abrasion of chrome plated surfaces, typical of those found in hydraulic systems, *rapidly diminishes with decreasing abrasive particle size when the abrasive is smaller than 5 to 7 microns.* The abrasion caused by 5 micron particles was five times greater than that caused by one micron particle. Filtering to obtain the smallest, practically-achievable contaminent particle size is, therefore, highly advantageous.

These studies also show the importance of abrasive contaminant concentration (by weight%). Above an extremely small critical concentration, a significant increase in abrasive concentration only leads to a modest increase in wear rate. For example, tests on a series of particle sizes from 1 to 10 microns found that an increase of eight fold in abrasive concentration (from 0.05 to 0.4 wt. percent) caused only a doubling of wear rate.

This explains the common industrial observation that in many cases the clogging or removal of filters from machinery does not seem to have any effect. In these situations, clearly the minimum concentration of abrasive present in the fluid when the filter is operative already exceeds the minimum required for abrasion. The logical action in such a situation is to either greatly improve the filtration system or dispense with it entirely.

NEW SURFACE TREATMENTS FOR WEAR CONTROL

A review of the surface treatments in use at the rework facilities has indicated that primarily chrome plate, electroless nickel, and anodizing are used for most wear resistant applications. Plasma spraying is also used but it is limited to a few engine components and has not found the general use of which it is capable. This capability should be expanded to include many other possible applications where improved wear resistance is desirable.

Over the years a large number of materials and treatments have been developed for wear control. These are quite well known to most materials engineers and are described in detail in reference sources such as the ASM Metals Handbook. However, in recent years a number of new treatments have been proposed which have certain advantages over previously used techniques either with regard to their ease of application or their superior performance. An extensive review of the technical and commercial literature was made to select such treatments. These are described in the following sections. The intent here is merely to describe the treatment.

Teflon Filled Anodizing

Anodizing has been used to impart wear resistance to aluminum and other light metals. Recently a treatment has been developed and commercialized whereby the pores in the hard anodize are filled with teflon. In this way a solid lubricant is combined with a wear resistant surface. It can be used with or without a fluid lubricant and provides some degree of corrosion prevention. It is generally used on lightly-loaded aluminum surfaces such as hinges, sliding tracks, and guides. This treatment is commercially available from a number of sources.

Boron Diffusion Coatings

Industry is finding increasing uses for boron diffusion treatments as both a wear control treatment and as an aid for lubrication. It is almost exclusively used on carbon steels where case depths from 0.5 to 10 mils can be obtained. Since it is a high-temperature diffusion process, the parts must be heat treated (where necessary) after the wear treatment is applied.

Microhardness from 70 to 80 Rc can be obtained. Recent research has indicated that in lubricated sliding, boride surfaces become polished which increases the probability of fluid film lubrication. It is best used for lubricated sliding on applications where nitriding might be considered. Several different treatments are used.

Sprayed Molybdenum

One of the present authors (M.J.D.) first pointed out the advantages of the use of molybdenum when lubricated sliding surfaces are being considered. Since that time others have noted its effectiveness in unlubricated sliding at high temperatures as protection against fretting damage, for improved brake material, and for gun barrel erosion. The advantages of molybdenum and its alloys are their good resistance to galling (due to lack of ductility) combined with the fact that under conditions of sliding they form lubricating oxide films. Facilities for spraying molybdenum are readily available. It should be used primarily for the reduction of adhesive wear since high hardness levels are not obtained. The following references give further evidence [17-20].

Chromium Plate

Chromium plating is widely used for wear resistance purposes and has been so used for many years. Several recent developments indicate that under certain conditions, higher wear resistance can be achieved. Studies show that the hardness and the wear resistance of chrome plate vary considerably depending upon the plating condition. A maximum hardness of 1186 Kg/mm^2 (80 Rc) was obtained at a temperature of 113 F and a current density of 4 $amps/in^2$. The significant point here is that this hardness level is about 1.2 times that of SiO_2 which as previously stated should eliminate cutting wear. Secondly, channel type chrome overcomes the non-wetting characteristics of chromium plating and provides good surface oil retention which is important to low-wear-rate, long-life components (hydraulic systems, for example). Soviet work[21] has shown that the channels have an optimum number of islands per area for minimum wear rate. This should be an important quality control specification for deviation from the optimum leads to significant increases in wear. The optimum is 12 to 25 $areas/mm^2$.

Current military specifications call for plating hardness of 800 to 900 Kg/mm^2. Higher hardness will, of course, make the plating less ductile and could affect other properties. However, this may not be significant in certain applications.

Since chrome plate is widely used and understood, this approach merits further consideration. Evaluations are necessary to determine the impact resistance and adhesion of such chrome and if it could be consistently plated for wear applications. It could be then evaluated in those applications where abrasion is a problem (bushings, slat tracks, actuators, and valves).

Electroless Plating

Electroless nickel plating is widely used in industry and at the Naval Rework Facilities for wear and corrosion protection. This treatment depends upon phosphorous as a hardening agent. There are several new developments in the field of electroless plating which may find application in the rework of aircraft.

Electroless nickel-boron has been developed and introduced commercially. Its advantage is that the deposits are harder than Ni-P and can reach levels of greater than 1200 Kg/mm^2[(22)]. Thus, it may find application where abrasion resistance is needed. It can be applied to a wide variety of substrates.

Electroless cobalt baths have been developed[(23)]. For sliding applications, cobalt is known to have better sliding characteristics than nickel. This is attributed to its hexagonal structure. Thus, it may be more wear resistant than the corresponding nickel bath.

Spark Hardening

Using a small, inexpensive, bench type device, a thin film of carbide can be applied to a metal surface. Essentially the film is formed by vibrating a rod of the carbide against the surface while an arc is struck between the carbide and the metal surface. A coating up to two mils thick is applied with a hardness of 1500 Vickers. It is used mostly on tool surfaces, but should be equally applicable to all metals. A study by Welsh[(24)] indicates that it does have superior wear resistance under sliding (adhesive) wear conditions using a coated pin sliding against a steel disk. Spark hardened steel, titanium, and nickel gave lower wear under these conditions than case-carburized steel, high speed steel, stellite 100, and sprayed Ni-Cr-B. This technique should prove useful in many special situations where it is desirable to apply an in-shop wear resistant coating, for example, on small surfaces such as are found in the fuel or hydraulic systems on aircraft.

It was also felt that it might be possible to use this technique to apply soft metal coatings in the field. For example, it might be used to apply corrosion resistance cadmium coatings to the fasteners of aircraft and in the filling of corrosion pits on aluminum.

The overall approach of spark coatings is very attractive for field work. At the present time it is being primarily used for carbide coating of tools. There are no technical reasons why it cannot find wider application. Much more work would be necessary to establish the correct conditions of application for each metal combination and their value in preventing and controlling wear.

Teflon 954-101

These products referred to as Teflon S are a single coat finish based upon fluorocarbon resins and modifiers dispersed in an organic solvent. The coating can be applied to clean aluminum and steel. Best results, however, are obtained when these metals are grit blasted or with chromate treated aluminum or phosphated steel. The main advantage of this coating is the ease of application and its low temperature cure (450°F). It can be applied by air spray, electrostatic spray, or dipping. The resulting product is a film approximately one mil thick which gives adequate resistance to sliding wear (10^6 cycles at 50 psi) and good corrosion resistance (1000 hours in salt spray) on grit blasted aluminum. These results suggest its use on simple sliding applications such as *hinges* and *latches* exposed to the atmosphere.

Powder Coating (Nylon 11)

An investigation of plastic powder coatings has indicated that plasma or electrostatic sprayed nylon 11 had considerable potential as a wear and corrosion resistant coating for aluminum[25].

Nylon 11 has a low melting temperature, 367°F, high toughness, and low friction coefficient. The low melting temperature allows it to be applied to aluminum and magnesium without overaging. The further advantage of this method of coating is that it eliminates the use of solvent carriers.

The application of Nylon 11 has been propsed to solve many critical aircraft problems such as:

. Spline wear
. Surface fatigue
. Dirt and sand abrasion
. Wheel corrosion
. Antenna corrosion/erosion
. Tie down chains

Maintenance Tapes

Recently a wide variety of adhesive backed tapes became commercially available. They consited of a variety of filled and unfilled plastic materials. Although they are not necessarily designed to be wear resistant, they might be usable for many simple, lightly-loaded applications in aircraft rework. In most instances the cost of application is almost negligible compared to applying a solid film lubricant or plating. Accordingly, an extensive industrial survey was made to select those designed for sliding application. Unfortunately, wear data for such products are almost nonexistent although other properties are given. Such information is necessary before they can be considered for a wide variety of potential applications.

Maintenance Putties

The simplest technique for rebuilding a worn surface is to use a putty-type material which can be applied by hand, is air dried, and then finished to the desired size. These materials are generally epoxies filled with metals, carbides, and solid lubricants. As was the case for the tapes, wear data was almost non-existent. With these data many new applications could be considered.

Formaflex

Formaflex is a technique for applying tungsten carbide coatings on irregular surfaces. A flexible sheet of teflon containing the carbide and a self-brazing nickel alloy powder is attached to the surface and heated to high temperatures. The resultant coating is a carbide-nickel composite from 5 to 6 mils in thickness with porosity levels of less than five percent. The advantage of this coating technique is that a metallurgical bond is formed between the coating and the substrate. Accordingly, it can be used on flexible substrates when bond failures are prevalent. It is applicable to all metals which withstand the high forming temperatures (not aluminum, titanium, or magnesium).

Co-Deposition

Probably one of the most significant advances in the field of wear resistant coatings is being made in the area of co-deposited platings. Basically, the process consists of suspending small particles (.01 to 10 micron) in conventional electroplating baths, introducing agitation to keep the particles suspended, and plating in the normal manner. Using this tehcnique, platings such as nickel, copper, nickel chromium, and cobalt have been deposited with a variety of additives such as MoS_2[26], teflon[27], graphite[28] barium sulfate[29], silicon carbide[30], diamond[31], refractory metals[32], zinc sulfide[33], and others. Limited evaluation has shown that these composites yield lower friction, increased load capacity, and reduced wear. Future developments should see platings (adhesion resistant metals) made abrasion resistant with hard particles (carbides) containing lubricants (MoS_2, teflon, and graphite). Such coatings have the further advantage of being applied with conventional plating equipment at low temperatures so the mechanical properties of the substrates are not destroyed. Another advantage is that the coating can be tailored to the actual application. Further development of such coatings would be very desirable.

Solid Intermetallic (Tribaloy)

A series of metals have been developed primarily for use in fuels, hydraulic fluids, water and synthetics. They can also be used effectively, in certain applications, unlubricated to temperatures. These materials consist of an intermetallic compound (Co-Mo-Si) in a metallic base (bronze, nickel, stainless steel, molybdenum, cobalt, etc.) producing a wide variety of compositions. The material can be fabricated into solid shapes by casting, or by powder metallurgy techniques, or sprayed on the surface as a coating. The intermetallic compound can also be used as an additive to plastics or metal plating solution for co-deposition. A composite consisting of the intermetallic in molybdenum has been found to be effective as a high temperature brake material. In a nickel base, effective sliding was found against aluminum and anodized aluminum in water and oils at high temperatures.

Teflon Filled Polyurethane

Polyurethanes are finding increasing use as wear resistant films for applications like sand erosion where large amounts of sliding are not involved. Recently, to provide a low friction surface for sliding applications, teflon has been added. The resulting film has found application as an antichaff coating on commercial and military aircraft. It generally has a white color, although other colors can be formulated. It could find many areas of use in military aircraft.

Lubricant Adjunct Coatings

It is well documented in the lubrication literature that a variety of coatings will greatly improve lubrication. Such coatings allow better running in of the surfaces to obtain conformal geometries, hold lubricant in the contact area, reduce wear and metal transfer, and increase the load capacity. They are particularly beneficial for use in fluids which are poor lubricants. In spite of the many advantages, the practice is not widely used by industry and should be expanded. Some specific examples are as follows:

Application	Reference
Use of wax impregnated phosphate films as a fretting inhibition.	34
Use of sulfiding to increase load capacity and extend lubrication.	35
Use of soft metal flims to reduce wear and increase load capacity in hydrocarbons, silicones, and diesters.	36
Phosphating of aluminum to extend lubrication of aluminum.	37
Use of tin plating to prevent piston scuffing.	38
Use of molybdenum as an aid to lubrication.	39
Soft metals diffused into aluminum and titanium surfaces to improve lubrication.	40

Improved Nitriding

In the past ten to fifteen years considerable work has been carried out which has resulted in improved nitriding. One of the most significant has been in the ability to nitride stainless steel and nickel base alloys. Although the case depth is not as thick as that which can be obtained with the nitralloys, it is sufficiently deep (several mils) to protect the surface from sliding surface damage. The process consists of diffusing nitrogen into the surface at a temperature of approximately 1000°F. Case hardness of 58 Rockwell C can be obtained for treatment times of 6 to 48 hours.

AIRCRAFT COMPONENTS FOR IMPROVED WEAR CONTROL

The selected aircraft components that exhibit high wear rates and thus require rework are shown in Table VIII.

These components then became candidates for the application of the improved wear control techniques and treatments previously discussed. Not all components were studied or studied to the same degree. Fasteners and erosion surfaces were not considered since they were the subjects of other investigations.

In the following sections each of these components is considered separately. The main purpose of the review is to consider the adequacy of the

TABLE VIII.—COMPONENTS REQUIRING IMPROVED WEAR CONTROL

Hinges
Hydraulic Actuators
Bushings
Servo Valves
Fairing Surfaces
Splines
Door Flanges and Latches
Bearings
Seals
Fits

technology to reduce wear in this given situation. In the original work[41], of course, the emphasis was on selecting and evaluating specific solutions.

To reduce wear in a given application the following steps are necessary:

(1) Identify the type of wear and its cause.
(2) Determine the operating parameters of the system.
(3) Propose technical solutions which have some degree of creditability.
(4) Determine practicality of the solutions.
(5) Select a method for efficiently affecting a change in an established rigid system.

TABLE IX.—WEAR PREVENTION LIMITATIONS

	Causes	Wear Type	Materials	Technology	Practical
Hinge	*				*
Actuator					*
Fairing				*	
Hydraulic Valves				*	*
Splines	*	*	*		
Flanges				*	
Bushings	*			*	
Seals					*
Valves	*			*	
Fits					*

Hinges

Piano type rolled hinges are used in a variety of aircraft locations. Essentially they consist of an aluminum hinge and a stainless steel pin. The pin can be cadmium plated, chrome plated, or sprayed with a solid film lubricant. The aluminum hinge can be anodized for additional protection. In service the hinges are lubricated with VVL-800 at frequent intervals. Often this lubrication is neglected or is ineffective. As a result, many hinges are either seized, corroded, broken, or worn when the aircraft arrives for rework. They must be cut off the aircraft and replaced, a process which requires a large amount of time. Examination of the surface indicates that adhesive wear was taking place; however, there was considerable evidence of salt, dirt, and wear particle contamination in the hinge clearance. In this instance the most logical approach was to provide for more effective permanent lubrication in the form of a coating. Many such coatings are available; however, the dilemma is that most of these would be rather useless if the initial cause of wear was the dirt which destroyed the meager lubrication that existed. Because of this uncertainty, one desired both a hard coating and a lubricating one. Since the pin is small (90 mil diameter) and flexible, a hard coating is not practical and attention must be directed to the aluminum hinge. However, a uniform complex coating inside a long small hole poses many application problems. After some evaluations it was found that teflon-filled anodizing performed satisfactorily; however, the best solution turned out to be using a spray can of teflon on the pin. The rather thick coating, inaccurately applied, acted as a seal and kept the lubricant in and the dirt out.

For this application the dilemma was not a lack of technology or material availability but rather a lack of understanding of the wear sequence and the practical limitations of coating hinges.

Hydraulic Actuators

Hyrdraulic actuators represent a major rework expense. The actuators in either the control system or in the brake system are subjected to abrasive wear due to small particles of dirt which become trapped in the clearance between the piston and the cylinder or between the piston rod and the end cap. Actually, the trapped dirt particle makes a scratch in the cylinder which allows leakage. This situation is most prevalent with aluminum cylinders rather than steel. Present practice requires that the cylinder be honed to remove the scratches. If the cylinder is then out of tolerance, it is usually discarded; however, some rework facilities have rebuilt the cylinder with electroless nickel or installed a liner. This problem is common to most aircraft and is accentuated by the fact that these actuators are complex configurations. This means that they are expensive and have long waiting times for procurement. Thus, the cause of wear and the type of wear could be determined. The materials and the technology were also adequate.

The criteria for coating selection was that previously discussed. To prevent scratching a coating of hardness 1100 to 1200 Vickers would be needed. A survey was made and the following were selected:

- Tungsten carbide (fine grain - low porosity)
- Chromium plate (applied to insure maximum hardness)
- Electroless nickel boride.

A second approach would be to apply a soft coating on the cylinder to embed the dirt particle as is usually done with journal bearings.

Electroplated tin is a possibility.

The real dilemma of this application was the mechanics of applying, grinding, and lapping coatings on the inside diameter of a small (2" to 4") long (1' to 4') tube. Although this can be done at outside facilities there is considerable reluctance (because of purchase procedures and short lead times) to start an actuator flow. Although they cost several thousand dollars each, it is cheaper to purchase new ones.

Fairing Wear

There are many locations on an aircraft where one surface slides against another; for example, on the A7, where the horizontal stabilizer rubs against the fuselage. This is not just a cosmetic problem since the removal of the paint can cause severe corrosion in that area. A survey was made of commercially available products which could be simply applied in the field. The previously described maintenance tapes were particularly directed to this application. Thus, the polycohr tape which had the longest wear life in the tests was suggested for these kinds of wear situations. A second approach which seemed feasible but was not evaluated was the use of a teflon filled polyurethane paint. This is often used by the airlines to prevent skin damage and is now being used on the F14 aircraft. It is, however, somewhat more difficult to apply.

This is an example of a straightforward problem. Adhesive wear and galling results when aluminum rubs on aluminum. Since the conditions are not severe it appeared that the solution would be adequate; however, because of a lack of wear data, the life of the solution was in doubt.

Hydraulic Spool Valves

Hydraulic valves require close clearances for proper operation. Even small amounts of wear will cause leakage due to the high pressure and improper operation. This affects the safety of the aircraft. A literature search was conducted and discussions held with various valve manufacturers. In addition, new and used valves were inspected. The wear problem concerns two parts: the spool, which slides in a cylinder; and the flapper pin, which moves the spool. As the spool moves, grooves on its circumference open and close ports which direct the fluid. The flapper pin has two hemispherical tips which fit into a groove to move the spool. The movement of the flapper is controlled electrically. Wear takes place on the lands and on the tips of the flapper pins. The literature search revealed considerable information concerning the operation of the valve, but almost nothing relating to the materials or their relative wear behavior. A metallurgical analysis revealed that the material was within specifications as to composition (440C stainless) and hardness (60Rc). Studies of the wear tracks showed highly polished surfaces, and wear rates were estimated to be typical of the lubricated condition. Thus, it appears that the wear is not an unusual or failure type situation but typical of these materials under these conditions. Accordingly, new materials or improved wear treatments are necessary if better wear resistance is desired.

Discussions with the manufacturers and processors of valve parts resulted in the following information. The spool and sleeve must be of the same material to avoid differential thermal expansion problems. They are lapped to fit; the edges are then ground for ports. Thus, thin coatings would be ground off and brittle coatings would leave an uneven edge. A surface treatment applied after grinding and lapping would be most desirable, however, this must not distort the spool or change the surface dimensions. Although nitriding would improve the wear resistance of the 440C, intricate experiments would be needed to determine if it could be successfully applied. A more feasible approach for the present valve would be to change the material to either T15 tool steel or carbides. Adequate attention, however, must be given to determine if these materials have sufficient corrosion resistance. The increased costs of the carbides could be justified based upon the present rework expense. Changing the basic materials of the valve would require requalification of the system paid for by the one suggesting the change. This is extremely expensive but would be justified if there were reasonable assurances of success. Since these (wear rates of these materials under these conditions) were not available there would be little interest in the evaluation.

Splines

Wear of splines is one of the major wear problems in naval aircraft. Efforts are being devoted to develop new spline geometries; but in the interim, some technique is necessary to improve the wear behavior of existing configurations. Research and evaluation of spline lubrication and wear technique is being investigated by Southwest Research Institute[(42)]. Accordingly, it was not studied in this investigation. However, based upon the review of wear techniques, certain recommendations were made.

The problem arises due to misalignment of the spline. As the spline shaft turns, a fretting action takes place on the spline teeth which results in wear if suitable lubrication is not provided. The amount of wear is directly proportional to the amount of misalignment so any reduction in misalignment is desirable. However, splines are supposed to compensate for misalignment so other solutions are required. Adequate lubrication is necessary, so low-viscosity lubricants are desirable. However, the most satisfactory solution has been to use wear-resistant coatings and treatments.

With splines, the fact is usually ignored that wear takes place by different modes and that different treatments are required for each mode. From visual examinations of spline wear and from reviews of the literature, it is concluded that the following treatments are most appropriate for the conditions described:

(1) *Lubricant Depletion*. If the lubricant is unable to reach the fretting contact area, adhesive wear and galling will result. The surfaces will appear severely damaged with transferred metal visible. In such circumstances, it is desirable to use treatments to hold the lubricant on the surface (sand blast, phosphate, etc.) or adhesion-resistant materials (silver plate, chrome plate, carbides, etc.) which can withstand periodic lubricant interruption.

(2) *Minimum Slip*. If the motions are small, it may be desirable to add an elastic coating (nylon 11) where all the slip motion may be taken elastically.

(3) *Abrasive Wear*. If the slip motion generates abrasive oxide films, these can accelerate wear. Materials harder than their oxide films are beneficial here (nitriding, boriding, carbides, etc.) or corrosion resistant materials or treatments (electroless nickel).

(4) If torsional or other vibrations impose an impact loading between the spline teeth, a coating with sufficient impact resistance must be selected.

It is clear that for splines the cause and type of wear and its relation to the motions in the system are unknown. As long as this situation exists solutions will be good guesses.

There are applications in splines and other components where impacts result. Hard coatings are of no use since they readily fracture. Here a treatment is needed to add toughness to the surface. Although hard facing can be used, it is too thick for many applications. Some diffusion treatments to toughen the existing surface would be most welcome.

Door Flanges

Aircraft access doors are made of aluminum and close against a thin flange of aluminum, mounted on the aircraft skin. During operation of the aircraft, fretting can occur between these surfaces and severe wear results. This is especially true when the door becomes loose due to hinge wear. Inspection of a number of aircraft showed that in a number of instances the flange was almost completely worn through. When this happens, the only repair technique is to remove and replace either the skin or the structure.

In this application the motions are severe due to the flexing of the structure so damping of the motions is not possible. Either an elastic gasket material or a lubricated coating is desirable. A wide variety of materials are available but there is insufficient information to determine which will be satisfactory. Once a satisfactory coating is selected there appears to be no serious practical problem in its application. This, of course, is always easy to say before the coating is selected.

Bushings

Bushings are used in a variety of aircraft applications such as wing folds, landing gears, flaps, etc. They are usually made of 4340 steel or aluminum bronze sliding against chrome-plated steel. Many of these are replaced at rework due to excessive wear. The wear usually results from lack of lubrication or dirt abrasion. The problem is essentially the same as that for hinges except that the bushings are much easier to replace. Wear data has been developed for the existing materials combinations and for selected other materials and some improved materials have been suggested. For an improvement the material must be tough enough to resist the high impact loads of the application, be hard enough to resist abrasion from dirt, and be adhesion resistant. Although the desirable properties to resist a single kind of wear can be estimated, not enough is known to optimize for a combination of types of wear. Furthermore, not enough information is known of the causes of wear to determine which type predominates. Practical problems exist in that maintenance people are accustomed to replacing bearings and see little advantage in improved wear behavior.

Seals

Seals, or rather leaking seals, are the main reason for removal of aircraft from service. Leakage is caused by seal wear but it is also caused by a variety of other causes such as misalignment, distortion, improper dynamic operation, deterioration of the material, coking and clogging by the lubricant, corrosion and probably several other reasons. Accordingly, at rework all seals are automatically changed so most wear problems are avoided. Where wear problems exist it is usually the result of abrasion by dirt or corrosion of the seal face. Since the seals are purchased as a unit, it is not easy to make simple material changes and one usually finds a different type of seal more suitable to the existing situation.

Pneumatic Valves

A number of installations have reported wear of high-temperature pneumatic valves. Very little attention has been given to this problem in design. The wear problem concerns dirt abrasion of the aluminum housing by the stainless steel flapper. Ideally, the aluminum should be coated to prevent its damage using one of the same coatings recommended for aluminum actuators. However, a simpler solution, more practical for rework, would be to coat the flapper with a soft coating such as silver plate so that it wears rather than the aluminum. In rework, it would then be necessary only to replate the flapper. However in this case, as in most others, insufficient information is available on the nature of the problem or on the wear rates of materials to predict success. It becomes more or less "propose and evaluate".

Fits

One of the most untalked about wear problems concerns the fits or component locating surfaces. They are not supposed to move but do, as a result of vibration. Since they can't be lubricated (that would make them slip more) and they can't be locked (they must be disassembled), other solutions must be sought. The cause of wear is usually adhesion because one of the surfaces is a soft structural material. Many damage-resistant coatings exist but they are difficult to apply to a small area of a large structural part. As a result, liners are used which can be increased in size to compensate for wear. Thus, wear prevention is limited by practical considerations.

SUMMARY

Considerable knowledge exists on the nature of wear and how various materials behave under specific wearing conditions. The types of wear are known and materials properties which resist these types are generally known. A lot of qualitative information exists on the effect of certain variables on wear but quantitative information is lacking. Furthermore, a great variety of materials and coatings are available for use in wear prevention. However, one encounters a great deal of difficulty in making practical use of these data when confronted by a specific wear problem. From examining a variety of wear problems the reason for this dilemma is somewhat clearer. It appears that too little is known of the causes of wear and the conditions under which wear takes place in a variety of applications. In many cases the nature of the motions are not sufficiently understood or the specific sequence of events which lead to excessive wear are not clear. What is needed are more component studies or service wear examinations under adverse operating conditions so that a specific chain of events can be defined. For a given component for which there are wear problems it is desirable to know what forms of wear exist and what are the conditions which cause them to exist. Does wear debris generated in the bearing represent a significant form of wear or does it not. If it does, then it would be easy to develop a material with soft wear debris. It is this type of vast uncertainty which limits adequate wear control.

However, even where this can be done, other technology barriers exist. For any given proposed solution sufficient knowledge of wear rates is not available so that success can be assured. It becomes an endless matter of evaluating possible solutions in test rigs (whose extrapolation is questionable) or in the application itself (which can be disastrous). More specific information is needed on how changes in a given material property affect its wear rate so that appropriate compromises can be made to accommodate different forms of wear and the different requirements of materials (corrosion resistance, strength, etc.)

Even in the best of situations, where known treatments will cure known diseases, there are many practical limitations which must be overcome. One of the main ones is that alternate solutions for wear prevention and compensation have been developed. These include readily changeable parts, inserts, liners, etc. Their economics and effect is known and planned. Supply systems are filled with required parts and people know that certain parts must be changed at times. It is all arranged and it has a great deal of inertia. To make a change takes some major crisis.

Other practical limitations usually include such things as initial costs, machining difficulties, thermal expansions, and lack of adequate internal facilities. Purchase costs are usually much more than the cost of the item.

One of the main material limitations is that most of the wear problems are caused by dirt. The hard surfaces necessary to resist this type of wear are brittle and are damaged by impact loading. Since most wear surfaces have impact loads, solutions are very limited. Some new approaches to abrasive wear resistance are needed.

REFERENCES

1. Anonymous, "Aircraft Maintenance and Support Costs," Special Recap Report, Naval Air Development Center, Warminster, Pennsylvania.
2. Anonymous, "Proposal for a Wear Control Handbook," American Society of Mechanical Engineers, New York.
3. Peterson, M.B. and Murray, S.F., in "Boundary Lubrication - An Appraisal of World Literature," edited by F.F. Ling, E.E. Klause and R.S. Fein, American Society of Mechanical Engineers, New York, 1969, Chapter 3, p. 19; Chapter 9, p. 145.
4. Rabinowicz, E., "The Friction and Wear of Materials," John Wiley and Sons, New York, 1966.
5. Davies, R., in "Handbook of Mechanical Wear," edited by C. Lipson and L.U. Colwell, University of Michigan Press, 1961.
6. Ho, T.L. and Peterson, M.B., "Development of Aircraft Brake Materials," NASA CR 134663, March 1974.
7. Lancaster, J.K., *Institution of Mechanical Engineers. Proceedings,* Vol. 183, Part 3, 1968, p. 98.
8. Suh, N.P., *Wear,* Vol. 25, No. 1, 1973, p. 111.
9. Hirano, F. and Ura, A., in "Tribology in Iron and Steel Works," Iron and Steel Institute, London, 1970, p. 163.
10. Welsh, N.C., in Proceedings of the Conference on Lubrication and Wear, Institute of Mechanical Engineers, London, 1957, p. 701.
11. Matveevskii, R.M. and Kruschchov, M.M., *Vestnik Mashinostroeniia,* No. 7, 1959; references by Krylov, K.A., in "Friction and Wear in Machinery," Vol. 15, American Society of Mechanical Engineers, New York, 1964, p. 80.

12. Welsh, N.C., *Royal Society of London. Philosophical Transactions.* Series A, Vol. 257, 1965, p. 31.
13. Calabrese, S.J., Peterson, M.B. and Chiang, T., "Effect of Hardness on the Wear and Surface Damage of Steel," MEI Report 67TR49, July 1967, prepared for Watervliet Arsenal under Contract No. DA30-144-AMC-1473(W).
14. Garkunos, D.N., in "Friction and Wear in Machinery," Vol. 14, American Society of Mechanical Engineers, New York, 1962, p. 81.
15. Maev, E., in "Friction and Wear in Machinery," Vol. 19, American Society of Mechanical Engineers, New York, 1964, p. 51.
16. Finkin, E.F., in "Evaluation of Wear Testing, Special Technical Publication 446," American Society for Testing and Materials, Philadelphia, 1969, p. 55.
17. Babichev, M.A., in "Friction and Wear in Machinery," Vol. 14, American Society of Mechanical Engineers, New York, p. 1.
18. Lomakin, V.S., in "Friction and Wear in Machinery," Vol. 11, American Society of Mechanical Engineers, New York, p. 39.
19. Devine, M.D., Lamson, E.R. and Bowen, J.H., in Proceedings USAF Aerospace Fluids and Lubricants Conference, April 1963, p. 317.
20. Taylor, D.E. and Waterhouse, R.B., *Wear,* Vol. 20, No. 3, 1972, p. 401.
21. Peterson, M.B., in "Handbook of Mechanical Wear," edited by C. Lipson and L. Colwell, University of Michigan, 1961, p. 16.
22. IIT Research Reported in April 1972, "Materials Engineering," Molybdenum and Tungsten for Gun Erosion, Vol. 75, p. 13.
23. Garkunov, D.C. and Poliakov, A.A., in "Friction and Wear in Machinery," Vol. 11, American Society of Mechanical Engineers, New York, 1956, p. 98.
24. Weightman, R.F. and Pearlstein, F., "Hardness and Wear Resistance of Electroless NiP and NiB Deposits," DDC Report AS A002904, November 1974.
25. Pearlstein, F. and Weightman, R.F., *Electrochemical Society. Journal,* Vol. 121, 1974, p. 1023.
26. Welsh, N.C. and Watts, P.E., *Wear,* Vol. 5, No. 3, 1962, p. 289.
27. Koury, A.J., Conte, A.A., Jr. and Devine, M.J., in Proceedings of the 2nd North American Conference on Powder Coatings, Toronto, Canada, March 1972, p. 164.
28. Vest, C.E. and Bazzarre, D.F., *Metal Finishing,* Vol. 65, No. 11, November 1967, p. 52.
29. Plastiques, S.A., "Sulfide Composite Coatings," French Patent 1,316,-700, May 1963.
30. Martin, W.M., *Metal Finishing,* Vol. 11, 1965, p. 447.
31. Tomaszewski, T.W., Tomaszewski, L.C. and Brown, T.H., *Plating,* Vol. 56, 1969, p. 1234.
32. Kedward, E.C., *Cobalt,* Vol. 53, 1973, p. 27.
33. Bellis, H.E., "Nickel or Cobalt Wear-Resistant Compositions and Coatings," U.S. Patent 3,674,447, July 1972.
34. Hirakis, E.C., "Protection of Niobium," U.S. Patent 3,057,048, October 1962.
35. Ertel, M., "Sulfide Codeposition Coatings," U.S. Patent 2,999,789, September 1961.
36. Waterhouse, R.B. and Allery, M., *Wear,* Vol. 8, 1965, p. 112.
37. Gregory, J.C., *Wear,* Vol. 9, 1966, p. 249.
38. Wolfe, G.F., *Lubrication Engineering,* Vol. 19, 1963, p. 38.
39. Murphy, J.A., "Surface Preparation and Finishes for Metals," McGraw-Hill, 1971, p. 406.
40. "Properties and Selections of Metals," American Society for Metals, 1961, Vol. 1.
41. Devine, M.J., Lamson, E.R. and Bowen, J.H., in Proceedings of the USAF Conference on Aerospace Fluids and Lubricants, 1963, p. 317.

42. Shapiro, A. and Gisser, H., *American Society of Lubrication Engineers Transactions,* Vol. 6, 1963, p. 40.
43. Peterson, M.B., Gabel, M.K., Devine, J.J. and Minuti, D.V., "Wear Control for Naval Aircraft Components," ARP/SLP Report, Naval Air Development Center, Warminster, Pennsylvania.
44. Weatherford, W.D., Valtierra, M.L. and Ku, P.M., *American Society of Lubrication Engineers Transactions,* Vol. 9, No. 2, 1966, p. 171.

DISCUSSION

J. G. HANOOSH, Foster-Miller Associates: I would like to make a comment. Beyond just disseminating information, other fields can also contribute materials for different wear applications. For example, materials developed for injection molding of glass-filled plastics have provided highly abrasive wear-resistant materials for coal gasification systems, essentially off-the-shelf. So if you have a problem you should also look into other technological areas which may initially seem to have no bearing on your problem but may provide useful information.

M. B. PETERSON: That is correct. Indeed, careful classification of the types of wear allows you to do this. For instance, instead of your calling a certain kind of wear "fretting" which many may not understand, you call "fretting" fatigue wear (and it turns out that is what it is), then the whole field of fatigue opens up to us. It may not be applicable as a wear prediction technique, but we certainly know what materials are fatigue-resistent. By classifying wear in a certain way rather than by calling it by some obscure name, we can readily make use of this kind of information.

S. A. KARPE, DTNSRDC, Annapolis: If I call it fretting several other fields, adhesion, abrasion, surface fatigue, etc., open up for us. What do we do then?

PETERSON: Well, what I am trying to say is that under certain kinds of motion, we can identify a certain kind of wear based on the way the particle is removed. In fretting the material is wearing by fatigue and thus you identify it as fretting fatigue.

KARPE: It does not wear by fatigue; it wears by many other modes. Fretting defines just a type of motion -- reciprocal sliding. Maybe we should define wear by the state of stress, mode of motion and the environment instead of calling it adhesive wear, abrasive wear, or fatigue.

PETERSON: We can also work out a system based on what you proposed. My main point however is not what we call it. Somehow we should agree on what we are going to call things and stop changing back and forth.

O. KNOTEK, University of Aachen: As you indicated there are problems in introducing a new solution. We should think of coatings during the design of the product itself not just as a tool for repair and maintenance.

PETERSON: Yes, of course, the time to solve the problem is when you build the airplane. Then you could use appropriate materials and work out a whole methodology of predicting wear before the fact.

E. S. SPROLES, Rensselaer Polytechnic Institute: I would like to comment that in the area of fretting and fretting fatigue the important parameter is often not the amount of material removed. In fretting wear, the worst problem is that the fretting initiates a fatigue crack which causes the part to fail regardless of how much material has been removed.

PETERSON: That is right. What I have tried to do in wear prevention is to define wear as the removal of unwanted material by chemical or mechanical action. Failure processes are quite different. Under this sort of classification you can get fretting-fatigue failures and you can get wear; you have to decide which one is important in a given situation.

H. CZICHOS, BAM, Berlin: In the aircraft wear prevention program, what monitoring techniques are mainly used and for what components in the aircraft?

PETERSON: As far as I know, spectrographic oil analysis, magnetic plugs and measurement of temperatures of the fluid are employed.

DISCUSSION OF STATUS OF WEAR PREVENTION

E. F. Finkin

ABSTRACT

The economic factors which control the present status of wear prevention are identified and discussed.

INTRODUCTION

The economic penalty that people are willing to pay by ignoring a branch of technology depends on their perception of the magnitude of the potential economic benefit which can be achieved by using that technology. For example, structures were made very thick with little detailed stress analysis, as this was the conventional approach. Now the optima design evolves thorough stress analysis involving highly complex computer programs. This is an example in which the perception of the economic benefit of utilizing a branch of technology -- stress analysis -- has come to be widespread.

Could the same thing happen to wear technology? Perhaps. Wear technology stands roughly in the same position that stress analysis did seventy years ago. The fundamental principles are reasonably known and much detailed knowledge exists. This knowledge is, however, by and large ignored. The heart of the status of wear prevention, and all technology for that matter, is therefore economic.

For a third of a century the world has sustained a substantial rate of economic growth. To support and participate in this economic growth industry has focused its attention on expansion. This meant emphasis on product development, emphasis on addition of manufacturing capacity, and emphasis on adoption and development of improved manufacturing practices. By contrast, those activities which are not growth oriented, but rather aim to preserve and extend the life of existing assets and reduce their operating costs have received less attention.

As economic growth decreases, profit margins shrink. As a direct consequence, the emphasis on cost reduction increases. Presumably, the consequence of this is that the relative importance of activities focused on the future and on the present will shift more towards the latter.

PRODUCT DEVELOPMENT

Consumer products, machinery and transportation equipment are designed and developed to perform a function at a reasonable cost. Reasonable cost is determined by the marketplace. Secondary issues include life, reliability, and operating cost, and the norms for these also are generally set by the marketplace. It is when performance measured against one of these norms is inadequate that development attention is focused on it. This inherently means secondary attention will be applied to such fields as fatigue, corrosion, wear and vibration.

The present state of affairs reduces product development cost and reduces the time needed to commercialize a product. Unfortunately, it also increases operating costs compared to what they could be if development attention had been paid to these fields, by reducing product life, decreasing reliability, and increasing maintenance, downtime costs and parts replacement costs.

Is this trade-off satisfactory? It must be at present, or the market wouldn't cause it. The reason the situation is economically satisfactory is that it reduces business risk. Firms developing consumer products, machinery and transportation equipment do not really know how many of the specific items being developed will sell. If none is sold the development costs are a total loss. To minimize the business risk due to sales uncertainty, a product must be brought to market for the lowest reasonable expenditure. This inherently implies minimizing attention to all technological areas secondary to the purpose of the product. Wear, therefore, will always be a step-child discipline.

AFTER THE FACT

What happens when a product is already commercialized? If the value of the product is high enough (e.g., fleets of ships, airplanes or bulldozers), the owner may have sufficient economic incentive to reduce operating costs to warrant his devoting money to this purpose. Such was the case with Mr. Peterson's studying ship engines and naval aircraft.

Even there the resources available for re-engineering the product, to properly take wear into account, were so insignificant compared to the resources devoted to developing the product, that little could be accomplished -- or could hope to be accomplished.

Mr. Peterson has been seeking an answer to the question of why wear specialists could not accomplish more in solving the wear problems they identified in naval aircraft. The answer to this question is that knowledge of principles is not enough. Resources (men, money, equipment and time) are needed and these are not forthcoming in sufficient magnitude.

ON THE STATUS OF WEAR PREVENTION

P. M. Ku

*ABSTRACT**

Many wear processes take place simultaneously rather than independently, complicating the analysis of wear problems. The wear of many selected parts are discussed to illustrate the point.

INTRODUCTION

The authors have stated that "wear is not a single process but a variety of processes which can take place independently." Viewed independently, the various wear processes or modes, as classified by the authors, appear logical and easy to comprehend. However, even in controlled laboratory experiments, it is not unusual for several wear processes to occur *simultaneously,* and to isolate and study them *independently* is not necessarily a straightforward matter. Broadly speaking, different wear modes are affected differently by the contact geometry; the lubrication process; the physical and chemical properties of materials, surfaces, lubricant, and environment; as well as the operating variables. Many of these variables are mutually interacting. What is even more exasperating is that some of the wear modes may more than simply compete, but also interact with one another. Thus post-mortem examination of the failed specimens, without the benefit of careful and often sophisticated monitoring of the wear history, may not yield complete insights. The complexity of the problem is such that the "systems approach" would appear to be in order. But systems approach is difficult to apply at least at this juncture, since so many of the causes and effects have not been satisfactorily isolated and quantified. For these reasons, the authors' broad, critical appraisal of available wear data answers a real need in wear technology, and their plea for development of additional systematic wear data is equally timely.

When it comes to the practice of wear control in a design office, the problem is more challenging than wear study in the research laboratory. The designer seldom has complete freedom in design, nor has he control over how the equipment will eventually be used and serviced. Even if he does, the basic information available is, as said before, incomplete and the stakes

*By editors.

are high, so that the tendency against innovation is understandable.

The maintenance engineer must somehow cope with the legacy of the designer, and be further handicapped by other constraints so eloquently enumerated in the paper, to such an extent that his option has been characterized by the authors as a "policy decision." This is essentially what it is, although some policy makers might prefer to regard it as being based on a "life-cycle cost analysis."

All this is fine if we know enough about the inputs and the input-output functions for wear and life-cycle cost analyses; but we certainly do not at this time. Meanwhile, machines do break down and need to be repaired one way or another. It is against this background that the authors have performed a worthy service in sorting out some general guidelines and in relating their experience in dealing with a variety of specific aircraft service problems. It is hoped that the remains which follow will help to emphasize some of the points the authors have made or implied.

LUBRICATION PROCESS

The lubrication process to be provided a machine element is basically the prerogative of the designer, but the maintenance engineer does have a role in ensuring that adequate lubrication is maintained in service.

Where other considerations are not overriding, the designer probably prefers the use of recirculating liquid lubrication for the critical, heavy-duty machine elements. Whether it be conformal bearings such as plane or journal bearings, or counterformal bearings such as rolling-element bearings, cams, or gears, the maintenance of a minimum fluid film thickness sufficient to prevent rubbing of surface asperities and with solid contaminants is always helpful. This tends to eliminate most mechanical types of wear (i.e., except corrosive and electrochemical, if present), and greatly minimize though not necessarily eliminate surface fatigue altogether. Thus, when liquid-lubricated bearings have failed, it is useful to make sure that an adequate supply of lubricant, which is required to maintain an adequate fluid film and to provide cooling, has not been somehow impeded in service.

Some of the actions over which the designer often has no direct control are the "upgrading" of equipment to perform more severe tasks, and the "streamlining" of maintenance procedure to lengthen the overhaul period or dispense with some maintenance operations. Have these steps been such as to encourage the accumulation of deposits so as to cause lubricant starvation or inadequate cooling, or to help retain at the rubbing sites wear debris generated there and solid contaminants entering from outside? Can a change to a different oil help? Are filtration and sealing adequate? Can some means of facilitating oil circulation help? These are some ways to prevent wear short of a major design change, rather than accepting wear as inevitable.

MATERIALS AND SURFACE TREATMENTS

Of course, not all bearings are fluid-lubricated, for a variety of reasons. Indeed, many bearings, by virtue of their functions or locations, need not or cannot conveniently be so--and that is especially true of the aircraft for which weight is at a premium. Such bearings operate in the boundary or dry wear regimes, thus material compatibility, surface treatment, and hardness control are, as emphasized by the authors, of paramount importance. In fact, even fluid-lubricated bearings must encounter transient operating conditions when fully-developed hydrodynamic or elastohydrodynamic lubrication is not achieved. Thus, even here, material compatability, surface treatment, and hardness control are also important.

The authors have referred to the effect of hardness differential on rubbing wear rate. In the practice of this technique, it is sometimes advantageous to allow the more expensive or less accessible part of a rubbing part to wear minimally, and let the less expensive or more easily replaced part to serve as the sacrificial member. In this manner, the time and cost of replacements may be less in the long run.

MISALIGNMENT

Misalignment is virtually impossible to avoid even in original equipment, and inevitably gets worse in service. Yet its effect on wear is enormous. In machine elements of "open configuration" such as most gears and cams, misalignment alters the contact stress distribution but not the imposed load, unless some unusual dynamic conditions are involved. In machine elements of "closed configuration" such as roller bearings and spline couplings, the danger of increased imposed load due to misalignment is ever present, and this aggravates the contact stress distribution even more. The increased maximum contact stress generally tends to decrease rubbing wear life more or less linearly if a constant wear coefficient is assumed, but it decreases surface fatigue life by a powerful exponential function. There is no question that every repair shop should keep a sharp eye on misalignment and have the proper means to measure and minimize it.

A not-infrequent misconception that this discussor has observed is the indiscriminate use of the term "average contact stress." When any misalignment is present, average contact stress is basically meaningless; it is the local, maximum contact stress that controls the lubrication and wear behavior.

The discussor has encountered cases, when excessive wear due to misalignment is experienced, that proposals are made to increase the face width of spur gear, increase the length of bearing journal, or increase the length of spline coupling--all with the thought of increasing the "contact" area and reducing the average contact stress. It is clear that, if the misalignment is not changed, such a scheme will not significantly alter the contact stress distribution in bearings of open configuration, and will only aggravate the contact stress distribution in bearings of closed configuration. Indeed, if a bearing of closed configuration is otherwise adequate in strength (and this is not infrequent) and if the misalignment cannot be easily corrected, shortening the bearing should be considered to reduce the extra loading caused by the misalignment and thereby reduce the maximum contact stress. For bearings of both open and closed configurations, the more expensive means of crowning may also alleviate the problem. In some other cases, a more flexible or elastic member may be introduced into the system to accommodate the imposed misalignment.

SPLINE COUPLINGS

Spline couplings are widely used to transmit power or drive accessories, or to operate components within accessories. Their principal advantages are mechanical simplicity, compactness and lightweight, and low cost. Although spline couplings have a reputation of being able to accommodate some misalignment, their closed configuration is such that tooth fracture and fatigue may occur if misalignment is large, and fretting wear is inevitable even if the misalignment is seemingly insignificant. The authors have referred to an early work[1] from this discussor's laboratory. References 2-6 report some other interesting findings. It is neither possible nor necessary to discuss our work in detail here, but a few salient points will be made to round out the authors' perceptive treatment.

In principle, the best way to lubricate a spline coupling is by a recirculating oil system. This not only provides good boundary lubrication, but also enables the wear debris and hard oxides to be carried away from the contact zone and be subsequently removed by filtration. Under these conditions, the wear problem is generally minimal. This is indeed usually done for splines located in the engine compartment or adjacent to it.

We have found that with splines operating in liquids such as an oil or a fuel, wear reduction can be achieved quite effectively by improving the liquid's boundary-lubrication capability; by using suitable antioxidants; by suitable choice of metal combination and relative hardness; and by certain surface treatments.[2,4,5] Of course, changing the liquid composition has limited practical application because the liquid, be it an oil or a fuel, has other requirements to satisfy. Still, any engine oil is a much better performer than a grease. In any case, metal and hardness changes and surface treatments can readily be accomplished in the maintenance process.

The advantages of recirculating liquid lubrication notwithstanding, the lubrication of "interface splines" i.e., those used to drive accessories, is a problem of different practical character. The problem is serious because interface splines are so numerous (typically about 100) on a modern aircraft.[3] The use of oil recirculation entails considerable penalties in terms of weight and mechanical complexity. On new aircraft, a reasonable compromise is to use the "wet pad," which is, in effect, a recirculating oil system (with circulation facilitated by centrifugal action) of limited reservoir capacity. However, on many existing aircraft, retrofitting of interface splines with wet pads is not always possible, mainly due to space or other limitations. Such splines are usually grease-lubricated, and considerable studies on greases, metal combinations and relative hardnesses, and metallic coatings, have shown that their effects are much more modest than in liquid lubrication.[6] This is because the vigorous fretting action raises the temperature of grease in the very limited clearance space between the spline teeth, so that the grease actually solidifies. Grease solidification prevents the metal surfaces from intimate contact with fresh grease and thus impairs effective boundary lubrication. It also retains the wear debris and hard oxides at the rubbing sites to cause abrasive wear.

Solid-film lubricants have proved no help at all, mainly due to the vigorous fretting action involved.[5] Plastic coatings tend to separate and tear from the metal surfaces due to the fretting action,[3] and we have not seen conclusive evidence of their benefit for spline application.

It should be emphasized that the real culprit of spline wear is misalignment. Due to the closed configuration, the detrimental effect of misalignment on spline wear life is exponential.[5] Accordingly, any scheme that reduces the misalignment or reduces the effective slip amplitude will help greatly. One very effective way is to interpose a high-strength, self-lubricating plastic "muff" (such as one made of a polyimide) between the metal members, and otherwise dispense with lubrication altogether.(6) With such a scheme, and even with uncrowned metal and plastic teeth, the muff divides the total imposed misalignment between two rubbing interfaces, and thus effectively minimizes wear at both interfaces. Moreover, wear of the metal teeth is virtually nil, and the plastic teeth experience mostly plastic deformation rather than rubbing wear. This plastic deformation serves to accommodate the imposed misalignment, i.e., to reduce the effective misalignments at both rubbing interfaces more than a metal muff can provide. Finally, if the muff ever needs replacement, it is a simple matter to do so.

The use of a plastic muff can be a very effective maintenance scheme for relatively large spline units (about 3-cm outer diameter and over). Where still larger units can be accommodated, a coupling with a specially formed plastic muff and a crowned inner spline, which gives even better performance, is now accepted practice.(7) For smaller units, the use of a plastic outer spline to engage a metal inner spline appears to be a practical compromise.(6)

Of course, where high torque-transmitting capacity is required, and at high operating temperatures, the use of a plastic spline component is not viable. For such applications, the use of a metal muff can provide partial relief, particularly if it is designed to take up the major share of the wear and thus serve as an easily replaceable item.(6)

CONCLUSION

It would be remissness not to emphasize once again that the authors have performed a noble service in presenting their views in general and their experiences in aircraft maintenance in particular. Despite the numerous constraints they have encountered, improvements have been made. One cannot measure their efforts on an absolute scale, for there is none. Any tangible improvement, however arrived at, is a success.

Although the authors' goal is practical, their approach is as fundamental as current knowledge appears to indicate. The implications of their work is far-reaching.

REFERENCES

1. Weatherford, W.D., Valtierra, M.L. and Ku, P.M., *American Society of Lubrication Engineers Transactions,* Vol. 9, 1966, p. 171.
2. Weatherford, W.D., Valtierra, M.L. and Ku, P.M., *Journal of Lubrication Technology,* Vol. 90, 1968, p. 42.
3. Valtierra, M.L., Brown, R.D. and Ku, P.M., "A Critical Survey and Analysis of Aircraft Spline Failures," Final Report, Contract N00156-70C-2156, Naval Air Engineering Center, Philadelphia, Pennsylvania, 1971.
4. Valtierra, M.L., Pakvis, A. and Ku, P.M., *Lubrication Engineering,* Vol. 31, 1975, p. 136.
5. Ku, P.M. and Valtierra, M.L., *Journal of Engineering for Industry,* Vol. 97B, 1975, p. 1257.
6. Valtierra, M.L. and Ku, P.M., *Lubrication Engineering,* Vol. 35, No. 1, January 1979, p. 9.
7. "Circular Spline and Adapter Details, Engine Driven Accessories," Military Standard MS14169(AS), 1976.

WEAR PREVENTION

O. Knotek

ABSTRACT

The dependence of wear resistance on various properties of materials used for wear prevention is discussed. The volume fraction of the hard phase and the corrosion resistance are important in many cases. Methods of application of wear resistant surface coatings are discussed. Because of the numerous possibilities (diffusion, PVD, CVD welding, etc.) cost is a major consideration in the use of coatings against wear and corrosion. Therefore, welding processes present the best solution with respect to metals and alloys. Examples of wear resistance on the welding parameters of several alloys are given. Most hard alloys used for welding (spraying) are based on Fe-Cr-C, Co-Cr-W-C and Ni-Cr-B-Si. The structure of recently developed boron free hard nickel alloys and electroless nickel boron alloys are discussed. The increase of wear resistance by boronizing is shown in the case of boronized cobalt alloys and cemented carbides.

INTRODUCTION

If a wear resistant material is not used by itself for a tool or machine part, wear prevention is accomplished by a combination of two or more different materials. The wear resistance of materials is based mainly on the resistance against (a) chemical reactions (wear occurs by continuous removal of reaction products), (b) the diffusion of the wearing agent into the base material (cratering of machining tools), and (c) plastic deformation, chipping and fatigue caused by the wearing operation. At least one of these three basic wear phenomena can be observed in many cases. Wear resistance is a property of the whole system, and it cannot be described by a single property like the hardness.

It is clear, therefore, that wear protection has to be based on those three basic possibilities. For instance, corrosion is observed in many wear problems and many wear resistant materials are indeed good corrosion resistant materials.

SURFACE COATINGS

Coatings can be used as diffusion barrier (e.g. Ti(C,N) coating on carbide tools). The protection itself can be done by diffusion or by the application of coatings. For a long time various welding operations were the preferred techniques for coating against wear. The vacuum metallurgy and electronics industry have most recently made new coating procedures possible. These physical vapor deposition (PVD) and chemical vapor deposition (CVD) processes can be operated very well. The coatings are thinner than welded layers but the possibilities are vast. The properties of the substrate may, however, be influenced by the coating operation (heat generated by welding). This has to be considered in the coating process.

In CVD, a gas reaction on the surface of a heated substrate forms the coating desired. Hard compounds like carbides, nitrides, borides and oxides are produced. The most common CVD coatings are TiC/TiN and Al_2O_3 on carbides and steels. The problem of most CVD-processes is the high substrate temperature, which limits the substrate material, the reaction of the reactive gas mixture (i.e. $TiCl_4 + CH_4 \rightarrow TiC + 4\ HCl$) with the substrate and the gasflow. The new developments are in the direction of lower reaction temperatures.

The most simple PVD process is the evaporation. Here, as in the case of all other CVD and PVD processes the coating achieved is thin and should not exceed a few tens of micrometers. For wear resistant coating the adhesion of a simple evaporated coating is not sufficient for many cases. A glow discharge has to be added and, in many cases, the substrate is heated. Electronbeam guns permit evaporation of high melting compounds.

An example for the use of evaporation or plasmatron sputtering is the turbine blade coating of MCrAlY alloys ($M = Ni, Cr, Fe$) against corrosion and wear. The plasma spraying of same alloy powders is in use also.

Recent techniques employ higher substrate temperatures, reactions in the gaseous phase, acceleration of the gas particles in the direction of the substrate. The results are the ARE process[1] and the ion plating. Dense TiC-coatings of more than 100 μm can be produced by ARE on steel. Evaporated titanium reacts with acetylene according to the reaction $2Ti + C_2H_2 \rightarrow 2TiC + H_2$. The substrate temperature is far below that of the common CVD-process.

The evaporated titanium particles can be pushed by ionized argon at in an electric field. This "ionplating" results in better adhesion, primarily by biasing the substrate. But the coating rate is low and the control of the process is not easy. If a reaction is introduced to form hard compounds ($2Ti + N_2 \rightarrow 2TiN$) it works well with one reaction product (Figure 1).[2] If a second product appears, difficulties grow quickly ($2Ti + C_2H_2 \rightarrow 2TiC + H_2$) because of disturbance of the electric field applied. But various tools are coated now by ionplated TiN. A scheme of structures achieved[3] is given in Figure 2.

To date, for many cases welding processes permit the best solution for metals and alloys. This is the most important area of practical use. The amount of wear tolerated at many machine parts is larger than the coating thickness achieved by several other coating methods. The chemical deposition of chromium and boron nickel produces economic results in cases where it can be applied.

Diffusion processes using carbon, nitrogen, aluminum, and chromium, are

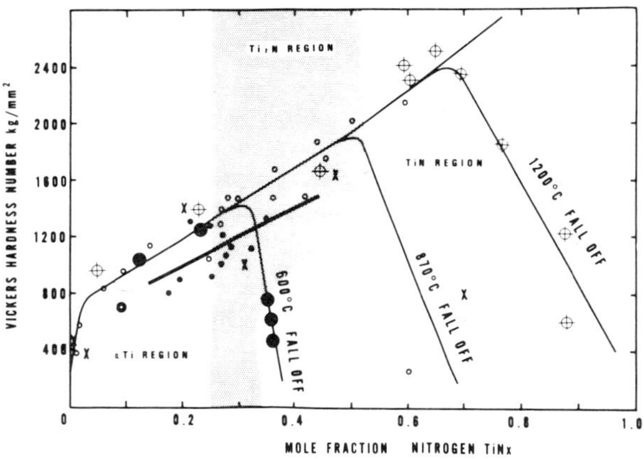

Fig. 1.—Hardness of Ion plated TiN.

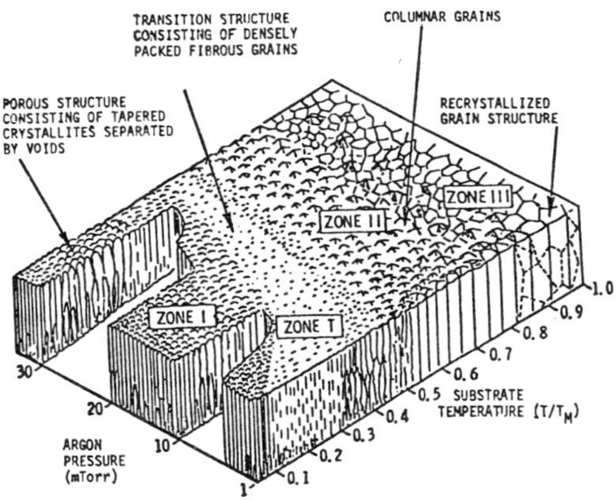

Fig. 2.—Schematic Microstructure of PVD Coatings.

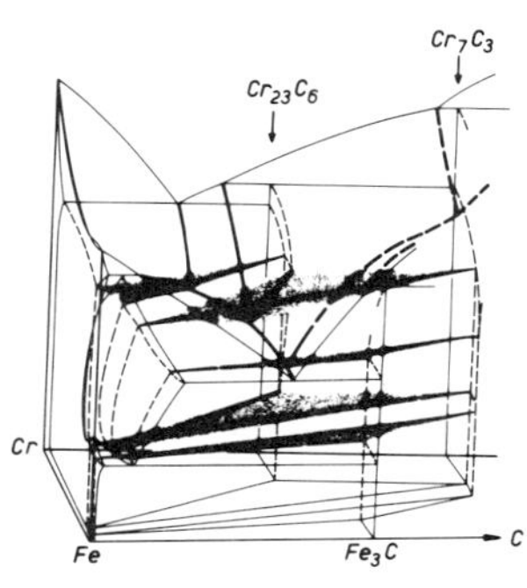

Fig. 3.—System Fe-Cr-C.

also common. The diffusion of boron was introduced recently. So the following discussion of wear prevention will deal with welding and briefly with chemical (electroless) deposition of nickel boron and boron diffusion on special substrates.

In the early days hardfacing by welding was used for repair and maintenance. Within the past decade, surface treatment of components and in particular the hard facing of engineered parts has become commonplace with the manufacture of original equipment. Parallel to these developments, the economics of the welding operation has changed markedly due to increased labor costs and to the difficulty of having a sufficient number of skilled welders to achieve reproducible results of hardfacing. Generally, the hardfacing process needs more welding experience than the joining by welding. This has caused the development of automatic and semi-automatic welding equipment which permit high deposition rates and an increase in the performance-cost ratio.

For hardfacing alloys a proven scheme is based on the matrix metals iron, cobalt and nickel. Tables 1-3 give the composition of common iron, cobalt and nickel hard facing alloys.[4] Hard phases are formed by addition of carbon (in Fe, Co) or boron (in Ni). Examples of the ternary and quarternary systems [5-7] are shown in Figures 3-5. Intermetallic Laves phases are used as hard phases in special cobalt and nickel base allows for powder spraying also.

The welding process plays an important role in the formation of the structure of the welded deposit via cooling rate. It is impossible to transform every hardfacing alloy in every kind of consumable due to their inferior mechanical properties. Many of the alloys shown above can be produced only as cast rods or powders.

The welding processes employed for hardfacing are:

(a) gas torch welding
(b) manual arc welding
(c) submerged arc welding
(d) gas shield welding
(e) open arc welding
(f) thermal spraying
(g) fusion treatment
(h) plasma spraying
(i) transferred arc plasma
(j) flame plating
(k) deposition processes (electroslag)

Table 4 shows an example of the relation between welding process, dilution, coating thickness and kind of consumable used. Figures 6 and 7 show structures of iron base alloys at different welding conditions. Wear resistance of welded overlays are shown in Figure 8.

The volume fraction of hard phase is very important for the wear resistance in a welded overlay. Figures 9 and 10 indicate the dependence of wear resistance on the volume fraction in cobalt-based alloys.[8] Often there is no proportional dependence and the best wear resistance is not achieved for the highest hard phase concentration.

In many applications the behavior of hard alloys at elevated temperatures is more important than the room temperature behavior. (Figure 11).

TABLE I.—IRON BASE HARDFACING ALLOYS

	Composition wt % (rest iron)					
	C	Cr	Mn	Si	Others	Welding Method
1	0.1	0.8	1		Ti	E UP MIG
2	0.15	1	2		Ti	E UP MIG
3	0.1	1.5	1	1	0.9Mo	UP OA MIG
4	1	2	2	1.6		E
5	0.1	3	1			E UP MIG
6	0.1	5	1.5		0.6Mo 0.5v	E UP MIG
7	0.3	6	1.5	1.1	0.8Mo	UP OA
8	0.5	9	0.5	3		E UP
9	0.5	9	0.5		0.5MoIV	E UP
10	0.5	9	2	1.4	1.8Mo	UP
11	2	13				A E WIG
12	0.75	15.5	6	1.75		UP OA
13	1	17			0.5Mo	E
14	1	17			16Mo 6.25Co 1.9v	AE
15	0.1	18	6		8Ni	WIG MIG E
16	5	20			10V	A WIG E
17	2	23.5	2	2		UP OA
18	3.3	28	1	0.75	3Co 0.75B	UP OA
19	0.15	29		0.5	9 Ni	H UP CA MIG
20	3	30				AE
21	3.5	32		2		UP OA
22	4	34	0.5	1		E
23	3.75				10Mo	A
24	0.6		16	1.75	4.5Ni	E UP CA

TABLE I.—IRON BASE HARDFACING ALLOYS (CONT.)

	Composition wt % (rest iron)					
	C	Cr	Mn	Si	Others	Welding Method
25	0.3	2.3			2Co 0.3V 8.8W	A E WIG
26	0.65	1.6			0.7Mo 1,3V 0.5W + Ta/Nb	A E WIG
27	0.8	3.7			8.5Mo 1V 1.8W	A E WIG MIG UP
28	0.8	4.3			4.8Co 0.9Mo 1.6V 18W	A E WIG
29					60-70W_2C WC	A E OA
30					80WC	E
31					70WC 10Cr_3C_2	E

A: Gas Welding

E: Electric Arc Welding

UP: Electroslag Welding

OA: Open Arc Welding

WIG:
MIG: Inert Gas Welding

Figures 12 and 13 show results of corrosion and wear resistance tests and the influence of the alloying elements.

Due to the availability of different matrix metals, formation of hard phases within, and the variety of welding processes, hardfacing by welding permits the solution of a variety of wear problems. On the one hand, there are the powder welding techniques with low heat influence to the base metal (less prone to martensitic tempering or few machining problems after surfacing) and on the other hand, high deposition rates almost in the range of casting. There are even techniques where solid carbide powder is put into the molten pool and so on. The joining presurfaced parts to a component is also used.

TABLE II.—COBALT BASE HARDFACING ALLOYS

	Composition wt % (Rest Cobalt)							
	C	CR	W	Mo	Ni	Nb	Others	Welding Method *
1	1	26	5					A E WIG (MIG)
2	1.8	27	9					A E WIG (MIG)
3	2.5	29	13					A E WIG
4	1.7	26	12		20			WIG
5	2.2	25			15			WIG
6	1.8	30		3.5	8	5	1.5Cu	WIG
7	Matrix Nr. 6						30WC	WIG
8	Matrix Nr. 6						40WC	WIG
9	1.7	27	6				30Fe	

* See Table I.

TABLE III.—NICKEL BASE HARDFACING ALLOYS

				Composition wt % (Rest Nickel)				
	C	Cr	Mo	W	B	Si	Others	Welding Method*
1	0.1	15.5	16	3.7			5.3Fe	E UP MIG
2	0.6	15			3.5	4	4Fe	A
3	0.4	12			2.5	3.5	3.5Fe	A
4	0.3	10.5			2.8	4	3.5Fe	A
5	Matrix Nr. 4						70-80W_2C/ WC	A
6	Matrix Nr. 4						80WC 10Fe	WIG

* See Table I.

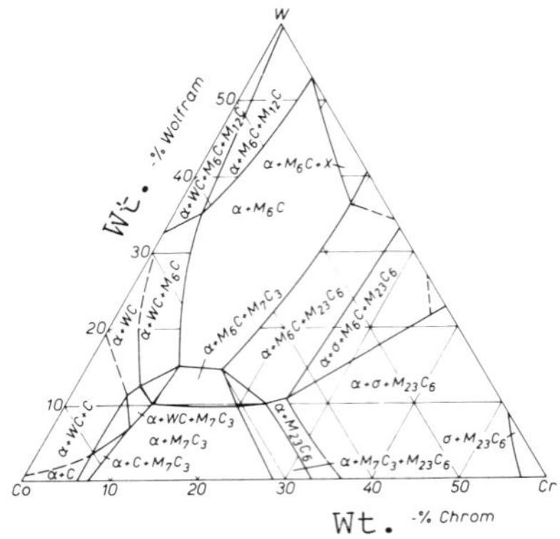

Fig. 4.—Co-Cr-W-1C-system.

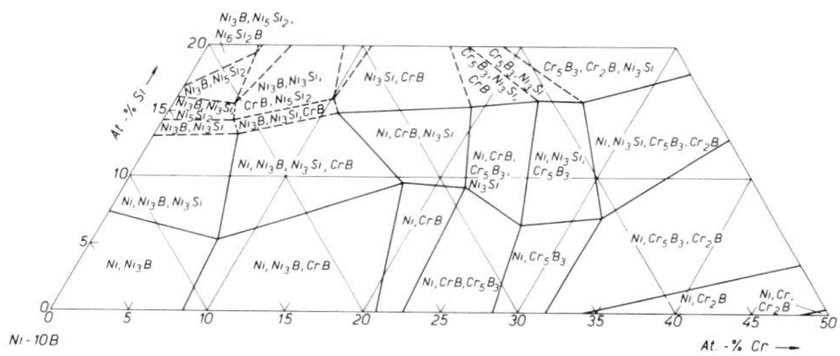

Fig. 5.—Ni-Cr-2B-Si-system.

Fig. 6.—Microstructure of an Fe-Cr-C hard facing alloy.

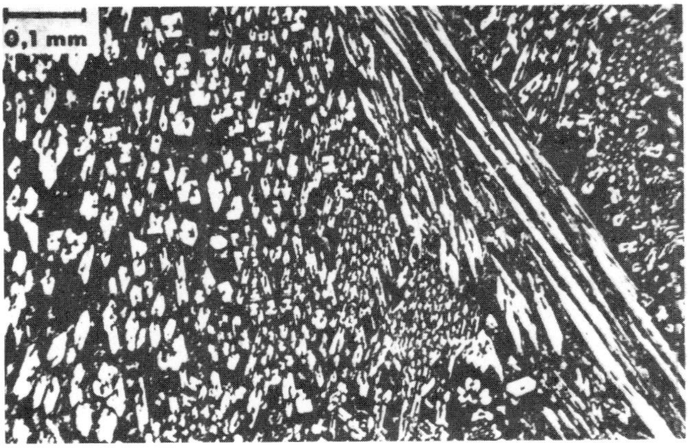

Fig. 7.—Microstructure of an Fe-Cr-C hard facing alloy.

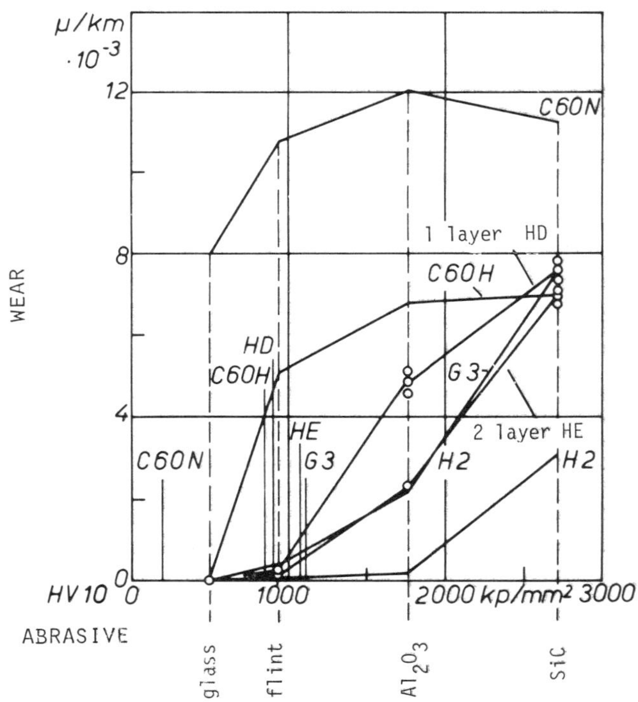

Fig. 8.—Wear resistance of welded alloys.

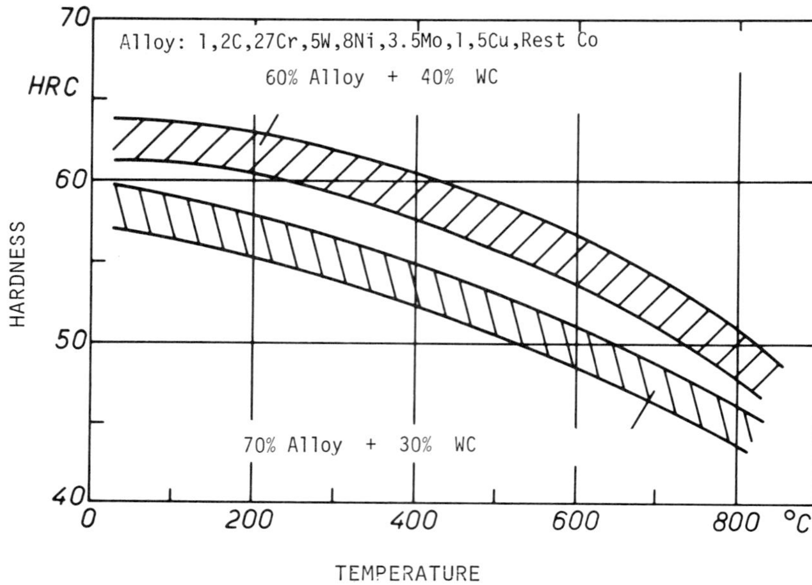

Fig. 9.—Hot hardness of two cobalt base alloys.

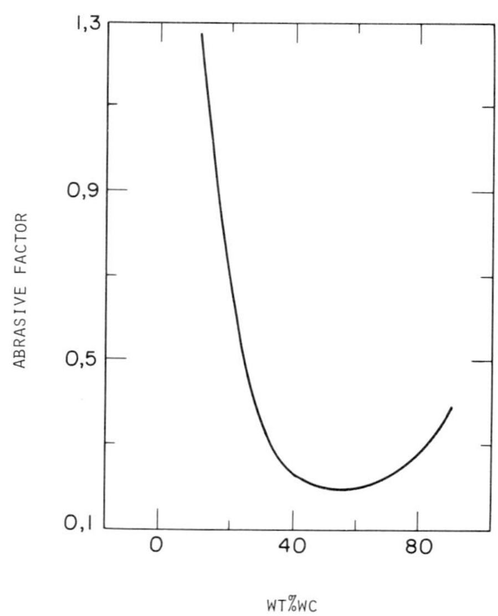

Fig. 10.—Wear resistance of tungsten carbide - cobalt base alloys.

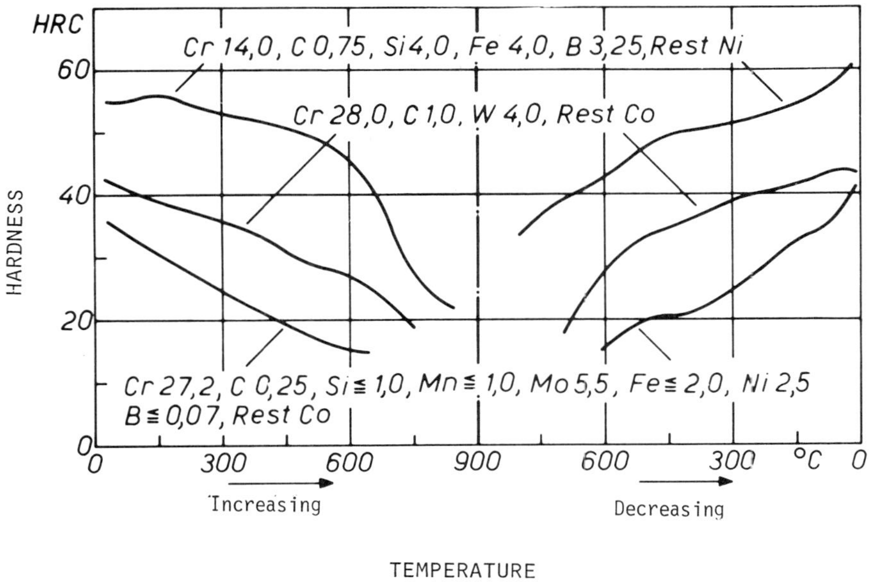

Fig. 11.—Hot hardness of cobalt and nickel base alloys.

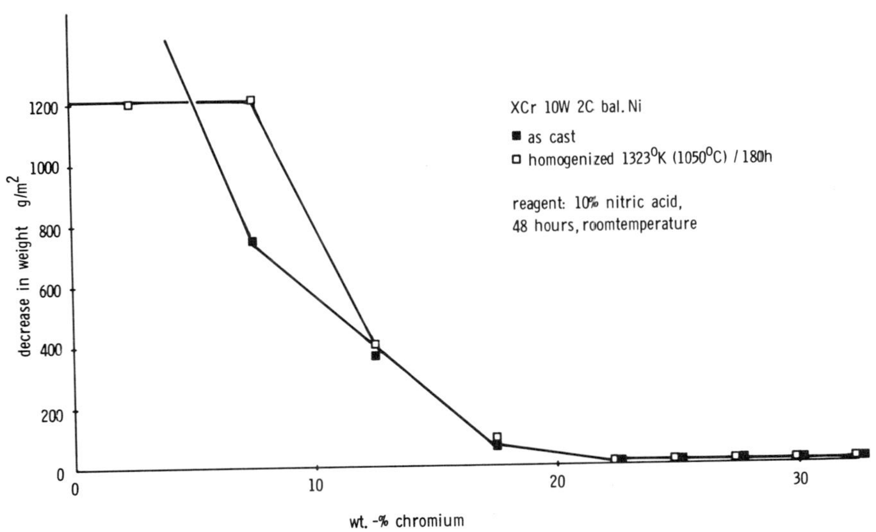

Fig. 12.—Corrosion resistance of Ni-Cr-W-C alloys.

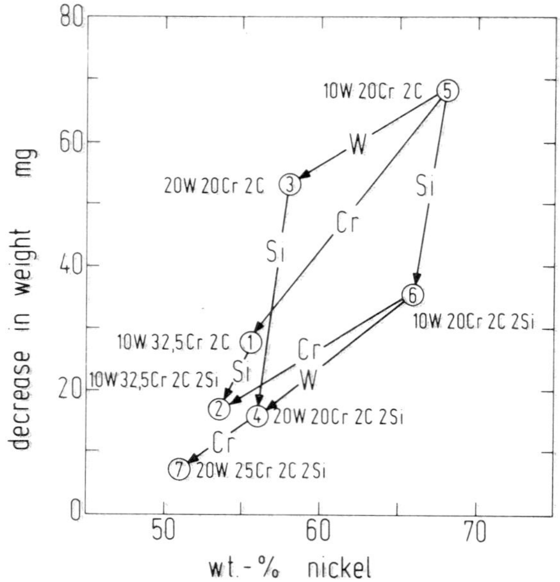

Fig. 13.—Wear resistance of Ni-Cr-W-C alloys.

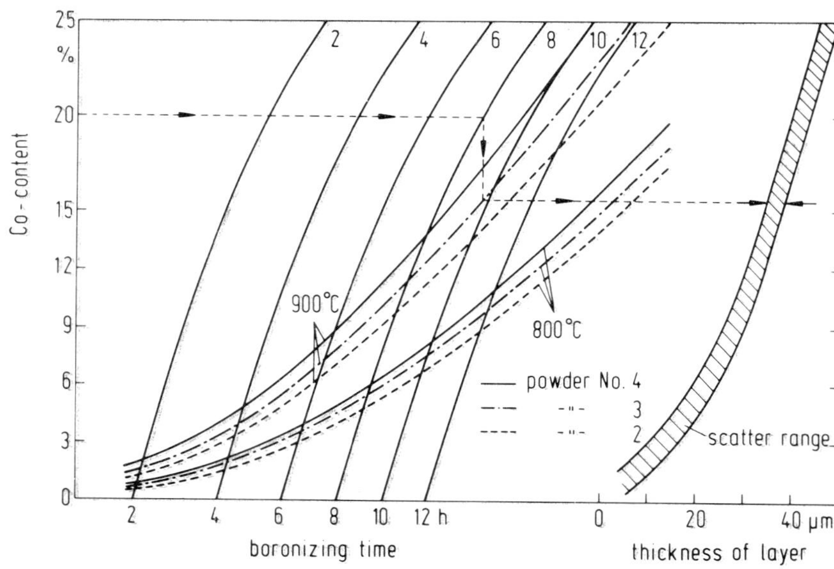

Fig. 14.—Nomogram for estimation of boronized layer thickness on WC-CO cemented carbides.

TABLE IV.—HARDFACING WITH CO-ALLOYS

Process	Deposition Rate kg/h	Dilution %	Minimum Thickness of Depcsition mm
Gas	1.8	1	0.75
WIG	2.3	10	2.0
Plasma	3.2	2	0.25
MIG	5.5	30	3.0
MIG + Cold Wire	25.0	20	4.5

ELECTROLESS DEPOSITION

The electroless chemical deposition of nickel and nickel phorphor are well known. Although the deposition of nickel boron is ten years old, the break through came only recently. After deposition the nickel matrix contains about 4%B by weight. This boron forms Ni_3B during the following heat treatment. The treatment can be done at 50-320°C from 2 hours up to 40 weeks. A hardness of 1200 - 1300 Hv can be achieved by a 2 h treatment at 310°C Maximum hardness of 1600 to 2000 HV was observed. The formation of Ni_3B in the deposit during the heat treatment has been proved. We think that during deposition the NiB_{12} is formed and stabilized by impurities which are present. The pure NiB_{12} is unstable. A little increase of temperatures of only 50°C initiates the formation of the stable Ni_3B. One of the most interesting findings is the mobility of boron at this low temperature.

BORONIZING

Boronizing can be carried out in a gaseous atmosphere, in a liquid (molten salts) and by powder packing or brushing a paste. We chose the last two possibilities because of the easy handling.[10] Mixtures of boron carbide, silicon carbide, amorphous boron and potassium boron fluorate (activator) were used at temperatures of 800, 900, 1000°C. Boronizing time was 4,8,12h. Pure cobalt, hard cobalt alloys (i.e. Co-Cr,Mo) and cemented carbides were boronized. The layer on pure cobalt (900°C, 8h) consists of Co_2B.

For boronizing cemented carbides the limit of temperature is about 1000°C due to the cobalt-boron eutectic. Up to 1000°C the carbide phases are not boronized. Cemented WC shows a formation of WB (HV 3700) at 1200°C. At 1400° and 3 h exposure, WB and W_2B_5 could be observed. At boronizing temperatures of 1000°C the cobalt binder phase is changed only. Three zones were found: CoB, C_2B and CO_3B. At temperatures from 900°C to 1000°C an additional zone containing WCoB is formed. A nomogram for estimation of boronized layers on cemented carbides is given in Figure 14. Figure 15 gives the hot hardness of a boronized steel, cemented carbides and Figure 16 the microhardness of boronized layer on WC-CO-carbides.

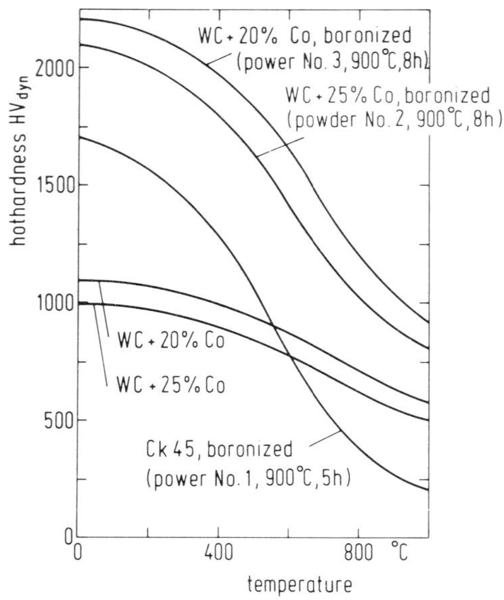

Fig. 15.—Hot hardness of boronized steel cemented carbide and boronized cemented carbide.

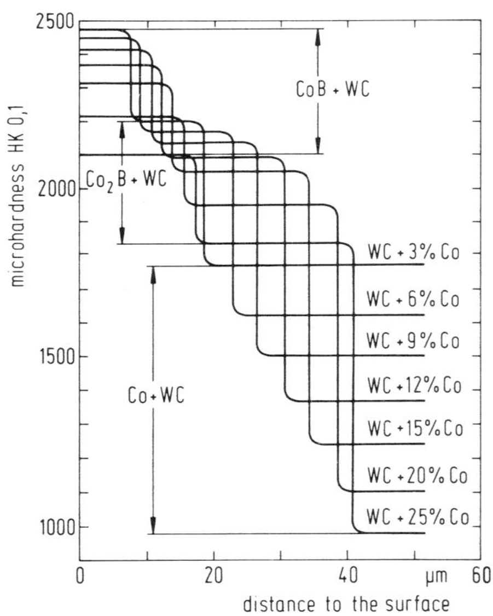

Fig. 16.—Microhardness of boronized layer on WC-Co-carbides as a function of the distance to the surface and the cobalt content.

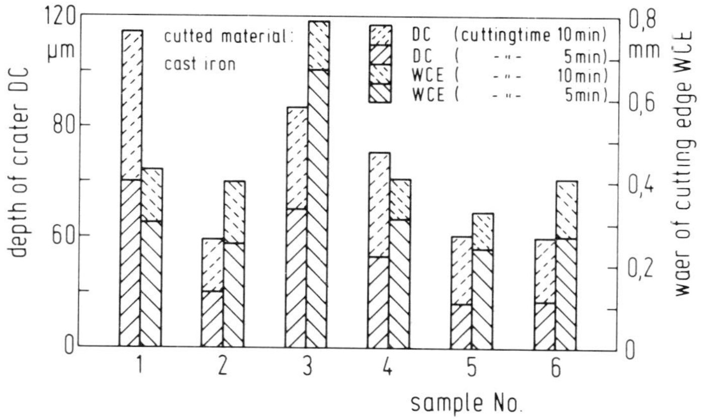

Fig. 17.—Crater and flank wear of boronized TiC coated and untreated cemented carbide.

A variety of commercial cemented carbides were boronized, CVD coated or used untreated for machining cast iron. The results are given in Figure 17. The crater and flank wear is similar for boronized and CVD coated tools.

REFERENCES

1. Bunshah, R.F. and Raghuram, A.C., *Journal of Vacuum Science and Technology,* Vol. 9, 1972, p. 1385.
2. Hill, R.J., Scheuermann, G. and Lucariello, R., *Thin Solid Films,* Vol. 40, 1977, p. 217.
3. Moschan, B.A. and Demichishin, A.V., *Fizika Metallov i Metallovedeniye,* Vol. 28, 1969, p. 653.
4. Knotek, O., Lugscheider, E. and Eschnauer, H., "Hartlegierungen zum Verschleibschutz," Verlag Stahleisen, 1975, p. 38.
5. Bungardt, K., Kunze, E. and Horn, E., *Archiv für das Eisenhüttenwesen,* Vol. 29, 1958, p. 193.
6. Knotek, O., Seifahrt, H. and Kieffer, R., *Archiv für das Eisenhüttenwesen,* Vol. 11, 1968, p. 869.
7. Knotek, O., Lugscheider, E. and Reimann, H., Proceedings of the 2nd Conference on Structure of Property Relationship in Thick Films and Bulk Coatings, San Francisco, 1975.
8. Knotek, O. and Lugscheider, E., *Verein Deutscher Ingenieure,* VDI-Berichte, Nr. 194, 1973, p. 161.
9. Knotek, O., Lugscheider, E. and Wichert, W., Proceedings of the International Conference on Metallurgical Coatings, San Francisco, 1978.
10. Knotek, O., Lugscheider, E. and Leuschen, K., *Thin Solid Films,* Vol. 45, 1977, p. 331.

DISCUSSION

I. AITCHISON, Atomic Energy, Canada: Your presentation is very interesting. Everybody seems to know how to deposit Stellite 6. But the nickel-based alloys seem hard to deposit. We are interested in these alloys for nuclear applications where we cannot use cobalt. I wonder if you have any remarks on the technology and the practical difficulties of applying nickel-based alloys.

O. KNOTEK: The reason for the development of nickel-based alloys is to avoid boron and in some cases cobalt in the nuclear power generation equipment. Generally speaking, nickel and cobalt are chemically very similar. But with cobalt we always form very complex carbides. The solubility of silicon in nickel is much higher than in cobalt. Therefore, we can play a little bit more with the carbon-silicon ratio in the nickel alloy than with a cobalt alloy. We can go up to 5 percent silicon in the nickel alloys.

This alloy was prepared by plasma melting, sand casting and machining. Then it was cleaned in a pickling bath which contained hydrofluoric acid. No other alloy would withstand that. But high silicon nickel alloy can withstand that. On the other hand, silicon stabilizes different carbides in the as cast condition. M_2C and sometimes M_3C_2 carbides are stabilized which may give different properties to the plasma sprayed coating. We made cast alloys and used them for turbines and heavy motors in Europe. It compares very well with Stellite 6. So far we were unable to make powders of this alloy at our university.

XIII. BOUNDARY LUBRICATION

REVIEW OF USEFULNESS OF NEW SURFACE ANALYSIS INSTRUMENTS IN UNDERSTANDING BOUNDARY LUBRICATION

D. Godfrey

ABSTRACT

The literature from 1967 to 1977 was reviewed to determine the usefulness of new surface analytical tools in boundary lubrication. The scanning electron microscope permitted significant advances in understanding fatigue and delamination wear, diffusion of alloying elements and the nature of corrosive wear. Electron probe microanalysis, electron spectroscopy for chemical analysis, and Auger electron spectroscopy revealed the concentration and distribution of elements and fragments of compounds from lubricating oil additives in boundary lubrication films on metal surfaces. Other instruments such as low energy electron diffraction showed the transfer of elements during adhesive contacts and the exchange of elements during sliding.

INTRODUCTION

The purpose of this paper is to assess the present understanding of boundary lubrication (BL). The author has chosen to determine what useful knowledge tribologists have gained in the last ten years by the use of modern analytical instruments, such as the microprobe. The literature has been surveyed back through 1967, and the results have been classified and evaluated. The most useful instrument was the scanning electron microscope which has provided new concepts on the shape of surfaces and wear fragments.

In the last ten years, there have been BL reviews by Tabor(1), Godfrey(2), Campbell(3) (with 210 references), Clayfield and Galvin(4), Fein(5), Beerbower(6,7), and Rowe(8). The writing of another similar review seemed unnecessary if not repetitious. All reviewers have concluded that we do not know the details of the mechanisms in BL. In particular, more research is needed to reveal the conditions of the formation and the nature of films on the areas of real contact.

A concise summary of the state of knowledge of BL as revealed by the reviewers may be useful.

1. BL is a complex phenomena involving metallurgy, surface roughness, adsorption, chemisorption, corrosion, catalysis, temperature effects, and time of reaction. In practical machines, BL rarely occurs in a pure state

and is mixed with or intermittently replaced by hydrodynamic lubrication.

2. The most important factor in BL is the formation of a surface film on metals to minimize damage during solid-to-solid contacts.

3. Film formation is determined by the chemistry of the film former and the chemical properties of the surface. Films can be composed of adsorbed long chain molecules, chemisorbed soaps such as iron stearate, deposited solids such as zinc sulfide, resins or "friction polymer," laminar solids, and plastics.

4. Lubrication effectiveness is determined by the physical properties of the film including thickness, shear strength or hardness, cohesion, adhesion, melting point, and the solubility of the film in the base oil.

5. Other environmental species such as oxygen, water, and competing surfactants affect film formation.

6. The mode of operation of machines with surfaces in relative motion determines the success of BL. Speed, load, and rate of loading with or without break-in, constant load, cyclic loading, temperature, rate of heating or cooling, reciprocation or unidirectional sliding, all affect performance.

The complete characterization of a thin film on a boundary-lubricated surface is a difficult and expensive task. The principal difficulties are the recognition of the load carrying part of the film, the identification of the composite structure of a complex-layered film, and the recognition and characterization of amorphous material.

It is clear from the literature that the use of any one surface analytical technique is insufficient to characterize a surface[9,10]. For example, the use of diffraction only would miss all amorphous material in a film. The answer is to use many techniques of analysis.

METHODS OF SURFACE ANALYSES USEFUL IN BL

There was no attempt, and the author was not qualified, to review the methods of surface analyses in detail. The highlights of the application of some newer selected methods to tribology will be given.

Others have reviewed surface analysis techniques. Quinn wrote a paper[11] and a book[12] on physical techniques in wear with emphasis on electron microscopy as well as on electron and X-ray diffraction. Haltner[13] reviewed techniques for studying and characterizing solid surfaces. In 1976, Furey and Eiss put on an advanced ASLE course on the "Physical, Chemical, and Topographical Characterization of Surfaces." The lectures included chemical characterization by energy emission. The book, "Characterization of Solid Surfaces," by Kane and Larrabee[14] was used for the course and is a good reference for details of many techniques to be discussed. Buckley, a pioneer in the use of surface analysis instruments in tribology, recently reviewed the methods for the study of wear[15].

A few examples from the author's files of analyses of surfaces related to BL are presented.

TECHNIQUES WHICH REVEAL STRUCTURE

X-Ray Diffraction (XRD)

This older useful method reveals the structure of crystalline material from which compounds can be identified. When a good diffraction pattern is obtained, the identification of the compound can be positive. No other technique provides such an analysis. A modified technique using glancing angle of X-rays is useful for surface films in situ. XRD can be used to identify wear fragments, corrosion products, deposits, films stripped or scraped off or pulled off with tape. X-ray crystallographers can now analyze micrograms of material. XRD will also identify the presence of amorphous material.

Electron Diffraction (ED)

A diffracted electron beam gives structure and composition of crystalline material similar to XRD. The shorter wave length of electrons compared to X-rays allows its use for smaller crystallites and thinner films. ED can be used on a film in place on a surface by directing the beam at a glancing angle. Films removed from the surface can be examined by transmission selected area electron diffraction. ED is also useful in the study of wear fragments and corrosion products. One danger in ED is that the intense beam may alter, if not destroy, some organic-type specimens. Another problem is that spot or incomplete patterns may be obtained, some of which cannot be identified with certainty.

Low Energy Electron Diffraction (LEED)

LEED utilizes a low energy electron beam which is diffracted by molecules regularly arranged on a surface. Usually, the surface is a single crystal in the shape of a pin point and examined at high temperature in a hard vacuum. LEED reveals the location of atoms and the structure of surfaces even less than a half a monolayer. LEED has been used to study the structural changes and kinetics of interaction of iron and water and the fundamentals of tribological surfaces. In one case, wear experiments on cermets were observed in action in an SEM[16]. LEED requires very good vacuum with idealized uncontaminated specimens. There may be difficulties in interpretating the spot patterns.

TECHNIQUES WHICH REVEAL TOPOGRAPHY AND SHAPE

Scanning Electron MIcroscopy (SEM)

SEM provides photographs of dramatic three-dimensional quality because of its great depth of field and wide range of magnification. A focused electron beam is caused to scan a specimen which emits secondary electrons. The emitted electrons strike a collector from which an image can be obtained. The depth of field is 300 times greater than the light microscope. SEM was compared back to back with electron and optical microscopy[17]. Eyre and Dutta[18] described the method and its varied use in wear studies. A large working distance permits the analysis of large specimens such as gear teeth. If a specimen is too large for an SEM apparatus, replicas of the surface can be used. The replica technique has been used to follow the wear of gear teeth during a run[19].

SEM also provides the opportunity to view a whole wear area, for example, wear scars on balls from a four-ball machine. Often one can learn more from this low magnification macrograph than from a high magnification micrograph of one small area, such as obtained from the electron microscope. SEM has been used with closed-circuit television to observe wear in action in a vacuum[20].

Figure 1 demonstrates the great depth of field of SEM at high magnification. The pits are due to the absence of graphite on the surface of nodular cast iron in an engine. The work suggested that oil cavitation removed the graphite. Figure 2 shows the Hureaulite crystals of a commercial phosphate coating on a new cast iron camshaft.

METHODS WHICH GIVE CHEMICAL ANALYSIS OF SURFACES

Electron Probe Microanalysis (EPMA)

X-rays emitted from an element which is bombarded with an electron beam provide information on the location and concentration of the element on and in surface layers. EPMA does not give direct information on the compounds in a film; however, associations of elements are supporting evidence. For example, the finding of iron associated with oxygen would certainly suggest iron oxide. The element identification and distribution data can be in the form of total counts on a computer or as a spectrum on graph paper or on film as white dots on a black background. In BL these X-ray images are very useful in associating elements with particular microfeatures shown by SEM. For example, the element sulfur may be associated with a stress corrosion crack. Rough surfaces present a problem, and the method is essentially qualitative for BL but can be semiquantitative for thicker films or deposits.

Figure 3 is an example of the use of EPMA. In this case, it identified inclusions in metal. The specimen is the cast iron foot of a new commercial hydraulic valve lifter. It has been shown that cracks originated at these inclusions and led to contact fatigue. The correspondence of the black particles in the sample current image and the element X-ray images shows that the inclusions are a manganese silicide and not graphite as might have been expected in cast iron. Figures 4 and 5 further exemplify the combined use of SEM and EPMA, showing salt as the cause of the corrosive wear (rusting) of the cast iron sides of a piston ring from a locomotive run on a seacoast line. Figure 4 shows the X-ray images of sodium and chlorine in the pits. Figure 5 shows the element tracings across the ring and the intimate association of sodium and chlorine. Figure 6 shows how the microprobe can help determine if scuffing has occurred. The X-ray image of iron shows that iron is present on the chromium-plated surface of the piston ring, proving transfer from the cast iron cylinder wall.

Electron Spectroscopy for Chemical Analysis (ESCA)

ESCA determines the elements on a surface by measuring their binding energies. A surface is irradiated with X-rays. Electrons are emitted from the atoms in the surface and collected and separated according to their energy. From the energies, the binding energy is deduced to identify the elements emitting the electrons. The binding energies vary slightly with chemical state of the element. These chemical shifts allow identification of valence state of the element and thus its possible association with other elements. For example, ESCA will show the oxidation state of iron. Another example in BL is determining the form of sulfur on a lubricated surface. The technique will determine if the sulfur is in the form of elemental sulfur, sulfate, sulfide, or as a sulf-oxide. A problem is determining the association of elements, for example, if zinc, iron and sulfur as

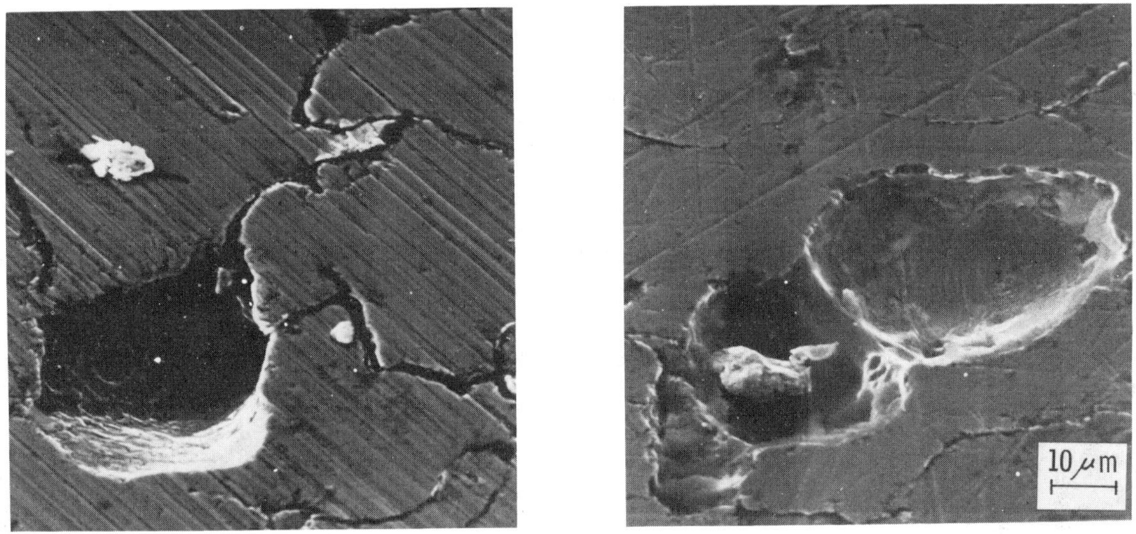

Segment of Cylinder Wall

Thin Film Cavitation Specimen

Fig. 1.—Scanning electron micrographs of cast iron specimens.

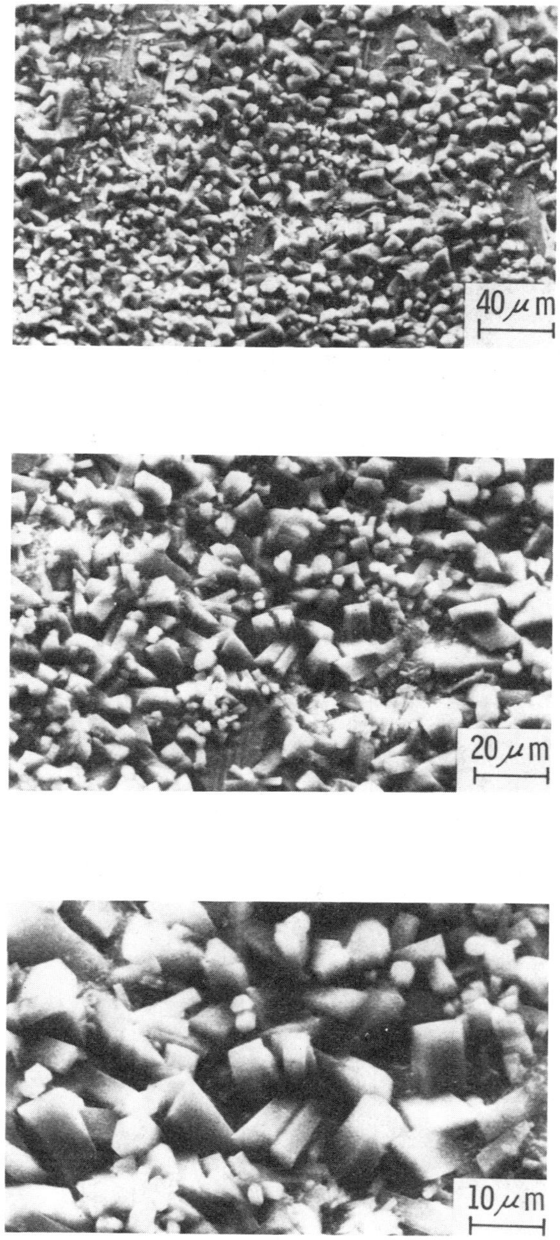

Fig. 2.—Scanning electron micrographs of new phosphated cam.

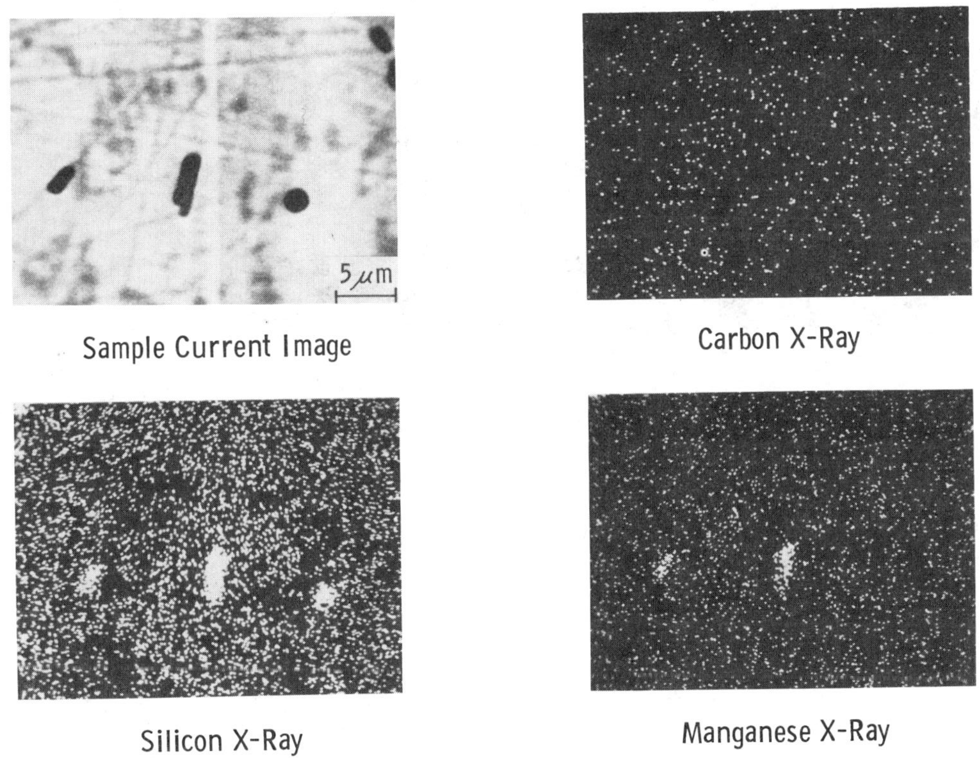

Fig. 3.—Electron microprobe analysis of new hydraulic valve lifter, polished.

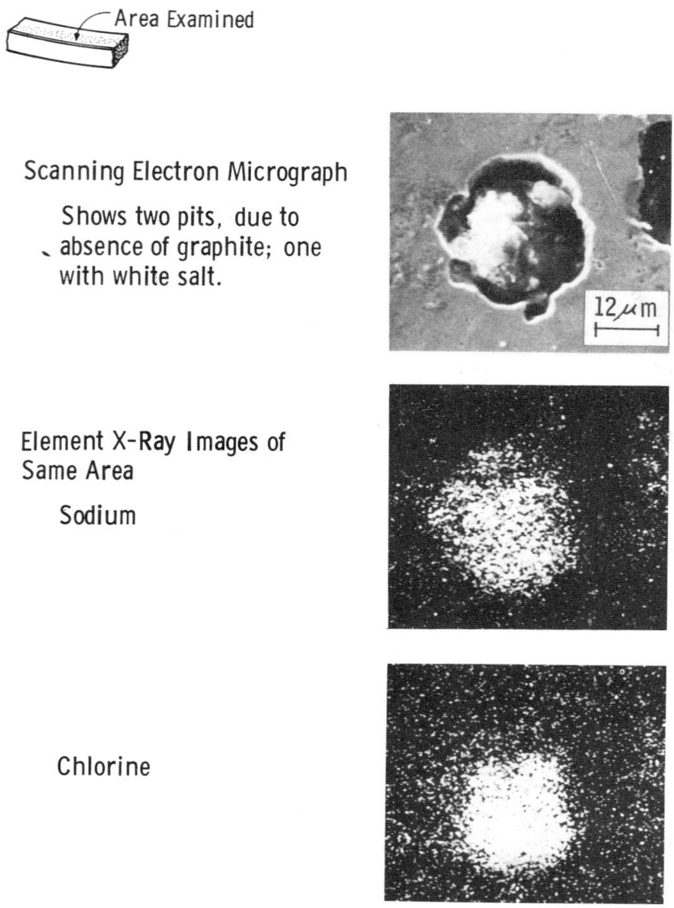

Fig. 4.—Scanning electron microprobe analyses of top side of locomotive piston ring.

sulfide, are present, one cannot say if the compound is zinc sulfide or iron sulfide. ESCA analyzes only the extreme outermost layers of surface material, 2 nm. Contamination is a problem. The beam covers large areas because the specimen is excited by flooding the surface with X-rays and, therefore, the spatial resolution is poor. The method is nondestructive and gives useful chemical information but offers poor elemental sensitivity. Details of analytical problems with ESCA are given in Reference 21. Depth profiles and thickness can be determined with ESCA by removing layers by ion bombardment. An ESCA method has recently been reported(23) which has a resolution of 15-20 μm. It uses a small electron beam focused on aluminum foil with the specimen on the backside. This technique would be suitable for a stripped film.

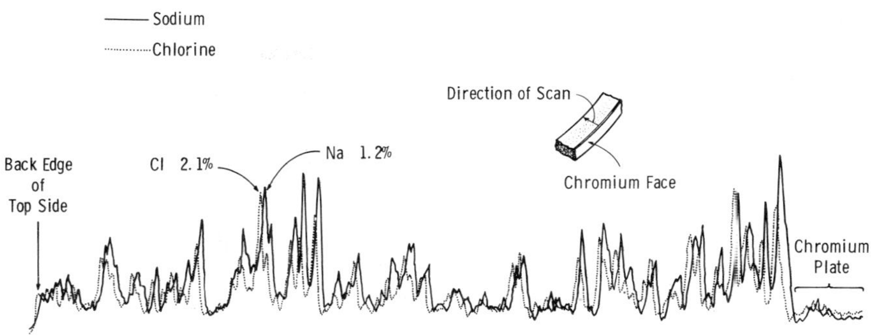

Fig. 5.—Electron microprobe element tracings across top side of worn piston ring.

Figure 7 shows an example of the sue of ESCA in film analysis related to BL. The specimen is the surface of the lobe of a used cast iron cam run in an automobile for approximately 50,000 miles with an oil containing sulfur in the additives. The spectra presented show that sulfur in the surface layers is present as a sulfide, possibly a metal sulfide. Also present were two species of oxidized sulfur, possibly as sulfite, and as sulfate. The results are similar to those in Reference 22. The specific chemical compounds could not be identified.

Auger Electron Spectroscopy (AES)

AES is a method which permits analysis of the surface layers for very small areas of a surface. The method analyzes all elements heavier than helium. The surface is irradiated with a small finely focused electron beam which provides good spatial resolution down to 1 μm. AES can show the change in the ratio of elements on a surface. Also, as in ESCA, ion bombardment is available for depth profiling and elemental imaging. AES can be combined with LEED and SEM.

In BL, it is particularly useful for detecting the important elements -- carbon, sulfur, nitrogen, phosphorus, and oxygen. A detailed description of this and other methods is given in References 23 and 24. One disadvantage of AES is that the intense electron beam may damage the surface being analyzed.

Scanning Auger Microscopy (SAM)

SAM is a technique of scanning with AES and ideal for the detailed characterization of a nonhomogeneous surface, such as a worn surface. In SAM, the scanning electron beam is a small spot and thus small areas such as a film on a microplateau can be analyzed. Thus, SAM provides high resolution in the form of SEM imaging combined with elemental mapping.

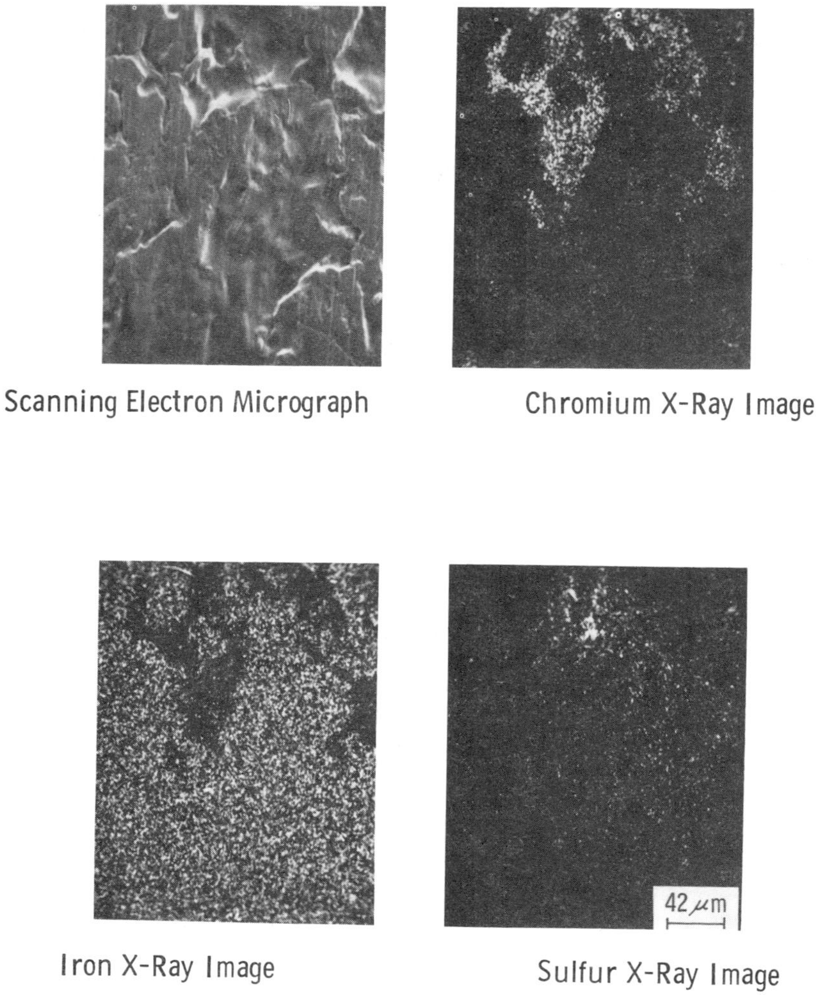

Fig. 6.—Electron microprobe analysis of scuffed area on face of chromium plated piston ring.

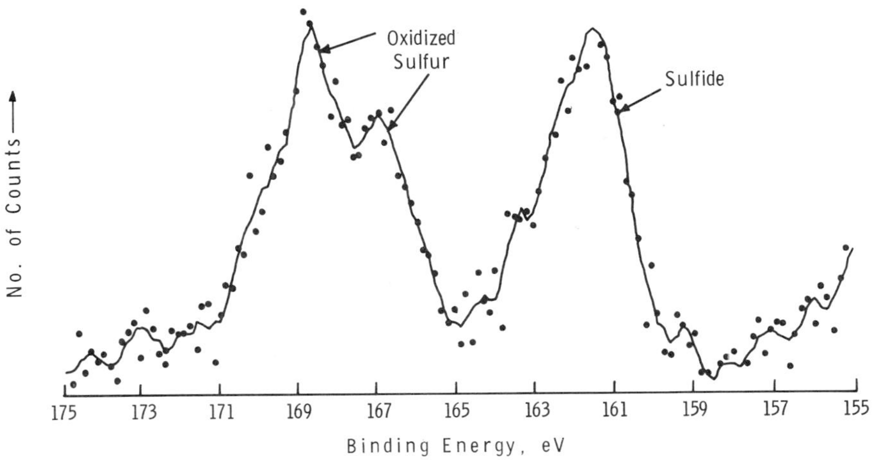

Fig. 7.—Electron spectroscopy for chemical analysis. Film on cast iron cam from automobile run 50,000 miles.

Secondary Ion Mass Spectrometry (SIMS)

SIMS is an older technique more widely used in other fields. Its features are high elemental sensitivity and the ability to detect hydrogen and molecular fragments. The surface is irradiated by a beam of incident ions which destroys the film on the surface. However, less than 1/100 of a monolayer provides a usable signal. Depth profiles are possible. The chief advantage is analysis for fragments of compounds such as sulfate SO^{-2} or phosphate PO^{-3} from a small area. A disadvantage is that the analysis is not quantitative.

Energy Dispersive X-Ray Fluorescence (EDXRF)

This method detects elements heavier than sodium by exciting the specimen with X-rays and detecting the element's characteristic X-ray. It is rapid, nondestructive, and capable of simultaneous multielement analysis in the ppm range. The method can be quantitative if standards are used. X-rays are emitted from the top 10 μm of the film so surface sensitivity is usually not great enough for depth profiling. EDXRF is very useful in BL for detecting elements in metals, in films on metals, in corrosion products, in oils, and in wear fragments. It is particularly useful in identifying oxygen, phosphorus, sulfur, and chlorine. Some instruments allow analysis of various sizes and shapes of specimens in a few minutes' time.

Figure 8 is a demonstration of the use of EDXRF for film analysis. A steel strip was immersed in hot oil to promote film formation by additives in the fresh experimental crankcase oils. The results show that the ratio of phosphorus, sulfur, and zinc in the films formed by the two oils were different.

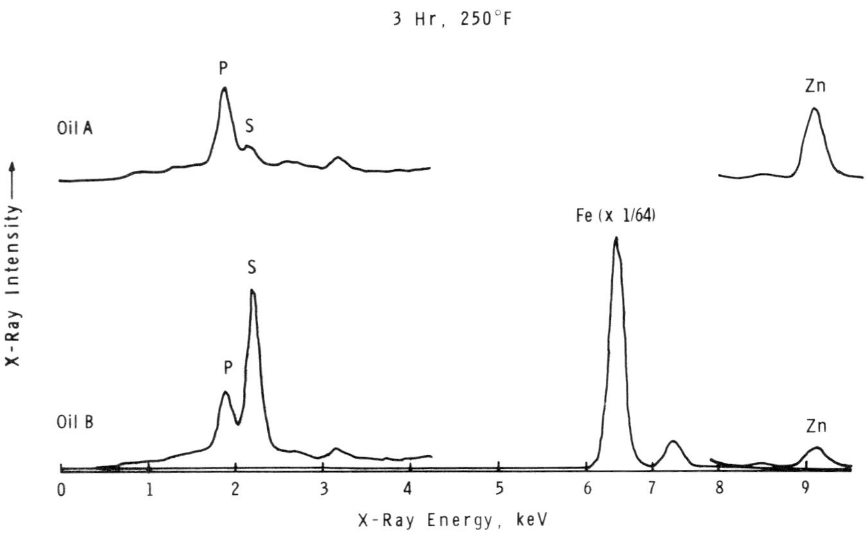

Fig. 8.—Energy dispersive X-ray fluorescence of films formed on iron during immersion in crankcase oils.

Reflection Infrared Spectroscopy (RIS)

In this method, the infrared beam is reflected several times between two parallel polished metal specimens on which the unknown film is present. The method is also known as attenuated total reflectance infrared techniques. It was first applied to BL by Francis and Ellison[25].

Inelastic Electron Tunneling Spectroscopy (IETS)

The detection of whole organic molecules adsorbed on metal surfaces is not provided by any of the methods already described. IETS, however, can do this by measuring the vibrational spectrum of monolayers or organic molecules on a metal oxide surface. For example, it can detect fatty acid monolayers important in BL[26]. The specimen must be sandwiched between two metal films and measured at liquid helium temperatures. There are no known applications to BL, but it appears promising.

Field Ion Microscopy (FIM)

In this method, the polished tip of a needle-like single crystal is subjected to very high potential. The image is projected at high magnification and gives the locations of molecules on the crystalline planes with a resolution of 2-3 nm. In tribology, it is most suitable for the study of adhesion due to contact.

Other Methods

Other methods of surface analysis have been used in other fields of science but no papers were found where they were applied to tribology. These include: ion microprobe, secondary ion spectroscopy, scanning electron stimulated desorption, pyrolysis mass spectrometry, ultraviolet photoemission spectroscopy, Mössbauer spectroscopy, electron energy loss spectroscopy, ion scattering spectroscopy, and neutron activation.

RESULTS

Adhesion

A combination of LEED and AES has been used to study the transfer of clean metals to clean iron or nickel during normal contact in vacuum(15). The amount of transfer and adhesion is related to the physical properties of the metals and the presence of oxide films. AES has also been used in combination with a sliding experiment to monitor rider metal transfer to a disk in vacuum(27). In the first passes, silver, aluminum, and copper were transferred to the oxidized steel disk. EPMA has been used to measure transferred metal during sliding(28).

Wear Mechanisms

SEM is widely used in tribology to show new, worn, or damaged surfaces, pits, cracks, corrosion, and delamination. SEM has been an important tool in the development of the delamination theory of wear because it shows cracks and lifting of fragments not otherwise visible(29). In one case, involving the study of lubricated machined surfaces(30), the author concludes that SEM is more indicative of the true condition of the surface than are surface roughness measurements. SEM shows detailed features, such as pits, grooves, and dimples, not revealed by a surface roughness apparatus. The SKF Rolling Bearing Damage Atlas(31) has used many SEM's to show surface features. Also, SEM has revealed the size and shape of crystals of phosphate coatings(32) which affects antiscuff properties of metals. (See Figure 2.)

SEM has been used for the topography of wear scars on balls from four ball machine lubricant tests(33). SEM clearly reveals the details of the failed surfaces in relation to the whole scar.

SEM has permitted the use of taper microsection which exposes the surface, subsurface material, and the base metal structure all in one photograph. This clever technique has been demonstrated by Eyre and his co-workers(18,34,35) and in one case, shows the running-in and the depth and extent of the white carbide phase formed during scuffing of cast iron piston rings. Eighty-three SEM and microprobe photographs are used to show the effect of cylinder bore finish on piston ring scuffing(36). One set shows chromium particles from the ring embedded in and associated with abrasion furrows in the cast iron bore. The surface detail exemplified by this study was not obtainable by any other technique. One or more element EMPA line scans are superimposed on SEM's of taper sectioned specimens to show the unlubricated wear of aluminum alloys(37). The change in alloy concentration on the surface, the subsurface and the bulk metal was seen in one photograph. This is a useful combination of techniques.

The adhesion of polytetrafluoroethylene (PTFE) to metals during normal contact and subsequent sliding wear by embedded metal particles in the PTFE, has been revealed using FIM, AES, and SEM(16). The authors demonstrate how

PTFE can machine an aluminum disk. SEM has shown microscopic pockets in rolled aluminum that had been formed by a very viscous lubricant. The aluminum oxide film was ruptured[38].

SEM has revealed that wear of wet and dry human cartilage in a pendulum machine is due to detachment of fibrous strands on the surface[39].

Wear Fragments

SEM is superior to optical microscopy for revealing the nature of wear fragments. Side-by-side photographs are shown in Reference 17. SEM has revealed spherical metal particles in wear debris from rolling contact fatigue[40], cavitation[40], fretting[41], and pitted gears[42]. Spherical particles are now established as a symptom of fatigue. Wear fragments collected from a pin on ring apparatus run in an ester oil were identified by XRD as a mixture of αFe_2O_3, Fe_3O_4, and FeO. The condition represents poor lubrication and a severe form of oxidative wear[10]. EPMA and SEM have been used to characterize wear fragments collected from oil out of rear axles run for 50,000 miles[43]. The author again showed the superiority of SEM over optical microscopy and the elements associated with the fragments. Titanium dioxide from gear-marking compounds and some silicon dioxide contaminants were found. Wear fragments from unlubricated cast iron, sliding under mild conditions, were found to be flat plates of Fe_3O_4[44] by SEM and XRD. SEM of wear fragments has provided an understanding of the wear of steel[45].

CHEMISTRY OF FILMS IN BL

Oxygen

Oxygen is an important element in BL. Microanalytical techniques have shown that dissolved oxygen reacts with the super refined mineral oil and the metal surface to form organometallics which reduce friction and wear[46]. XRD has also been useful[47] in deducing the temperatures during unlubricated sliding of iron surfaces by analysis of the oxides found in the wear fragments. EPMA was used with SEM in Reference 48 to show the presence of islands of less oxidized iron on a worn oxidized surface. The X-ray element tracing was matched with the SEM to clearly show the change in the amount of oxide with location. In a wear study[49] of MgO sliding on steel, EPMA has shown that a solid solution of FeO and MgO had formed on the sliding surface.

Sulfur

Forbes first reported the use of SEM and EPMA[50,51] in BL with the distribution of sulfur from various sulfur compounds (and oxygen from air) on four-ball wear scars. This work reveals the risks of analyzing rough scuffed surfaces where elements are in nonrubbing, depressed areas. The material in depressions may be nonfunctional varnish-like deposits from the lubricant rather than load carrying reaction products in films on the plateaus. In similar work[52], sulfur concentration is increased only with "seizure" and reaches concentrations up to 30% of the surface. Films from sulfur and chlorine containing additives on four-ball wear scars are examined by AES[53]. The higher load carrying sulfur additives form thicker films than the chlorine additive, and the sulfur content increases with load. Sulfur on scuffed four-ball scars has been found to occur on the high plateaus and as iron sulfide (with iron oxide) by glancing XRD[52]. XRD of surface films formed by sulfur compounds in the presence of air are composed of a mixture of iron sulfide and iron oxide[54,55]. Films of mixed composition protect the metal from excessive sulfidation and provide better lubrication. Buckley was the first and only user of AES[56,57] for sliding surfaces. He has demonstrated that sulfide films on iron surfaces are displaced by oxygen and that sliding inhibits the adsorption of H_2S and methyl mercaptan to an iron surface.

ESCA (labelled XPS in the paper) has been used to analyze unfailed films on Falex machine steel pins run in 16 different organo-sulfur compounds in oil[58]. All the additives produced the same metal sulfide. Sulfide concentration showed a correlation with wear. Other elements from the additives were present.

Phosphorus

EPMA aligned with SEM's of surfaces run in 1% tricresyl phosphate in mineral oil has shown iron and phosphorus in the film. XRD of the film formed on a hot wire in the same oil identified the film as a mixture of iron phosphates[59]. In another study with a series of amine phosphates[60], EPMA has indicated low concentrations of phosphorus on four-ball scars. EPMA has been used effectively to resolve the composition of antiwear films formed by dialkyl phosphates[61]. The films contain iron-organo-phosphite formed by hydrolysis. Under more severe sliding conditions, further hydrolysis occurs to form an iron-inorgano-phosphite film. XRD also indicates that iron phosphates are the crystalline constituent in films formed in various organo phosphate compounds in the hot wire method[62] and in sliding experiments[63]. The microprobe is used to associate film composition with film failure during sliding in a four-ball lubricant tester[64,65]. The authors show that iron oxide and iron phosphate from tricresyl phosphate promote high load carrying capacity by smoothing a surface and enhancing elastohydrodynamic lubrication.

The nature of the crystals in iron manganese phosphate coatings on steel have been revealed by SEM. The size and shape prove to be related to antiscuff performance[66].

Wear scars on Timken blocks run in sulfur-phosphorus type EP oils have been examined by EPMA[67]. In the early high wear stages of the run, the upper part of the film was rich in sulfur; and the lower part was rich in phosphorus. In the later low wear stages of the run, the element concentrations are reversed. The reason for the reversal is not clarified.

ED is used to show the chemical composition of wear fragments formed during sliding of tensile stressed and nonstressed specimens[68]. The unstressed fragments from the base mineral oil were composed of a mixture of FeO and Fe_3O_4, whereas with tensile stress present, the fragments were only Fe_2O_3. The tensile stress promotes oxidation corrosion. With TCP present, Fe_3O_4 is found for both conditions. With dibenzyl disulfide present, a mixture of Fe_3O_4 and FeS_2 is found for both conditions. For all oils, wear is greater with the tensile stress imposed than without.

Zinc Organo Dithiophosphate (ZDTP)

The films formed by the antiwear additives zinc organo dithiophosphates is of great interest in practical lubrication. ZDTP is probably the most abundantly used lubricating oil additive in the world even though its mechanism is not known with certainty. The authors in Reference 10 suggest that the additive is decomposed because of the distribution of zinc, phosphorus, and sulfur in the film is different than in the additive.

Neutron activation has been used with chromatography to study the antiwear properties of zinc dihexyldithiophosphate[69]. The authors conclude that a phosphite type of impurity originally in the commercial material or formed during its decomposition, is the effective antiwear agent. EPMA combined with SEM has been used to determine the additive elements in films on the scars of the four-ball machine[70]. The data show that the zinc and phosphorus (and a small amount of sulfur) are present in the film under mild as well as severe conditions of sliding.

Rounds[71-74] used X-ray fluorescence to monitor element concentration in films on various steel ball specimens run and heated in ZDTP and other sulfur-, chlorine-, and phosphorus-containing lubricating oil additives. His results suggested that ZDTP reaction products react with the metal surface to form the antiwear film. ESCA has also been used to identify the compounds formed on steel surfaces by EP additives in oil[22]. Zinc dialkyldithiophosphate forms a polymeric film which occurs in patches on the surface. Also, the concentration of zinc, sulfur, and phosphorus decreases as the film is etched away. When a specimen worn in ZDTP is subjected to tensile stress, wear is decreased. XRD has been used to identify the film and corrosion products[75].

Chlorine

The exact mechanism of lubrication with chlorine additives is also unknown. The role of chlorine in BL has been studied by sliding clean iron in vacuum with traces of organic chlorides. AES analysis has been used to monitor the chlorine, oxygen, and carbon in the films. Results show the effects of chlorine concentration in the film, chemisorption of chlorine compounds, and polymerization on sliding[57].

Solid Lubricants

The use of solid lubricants such as molybdenum disulfide, graphite, and polytetrafluoroethylene, when finely divided and dispersed in oil is in the regime of BL. EPMA, SEM, and EDXRF have been used to identify MoS_2 on engine parts[76]. These tests as well as infrared reflection spectroscopy and ED were used on films on Falex and LFW-1 specimens. By deduction, the author concludes that MoS_2 was present on engine valve parts. In the same paper, it is shown that zinc pyrophosphate forms a phosphate film the same as does zinc dialkyl dithiophosphate, and $Ca(OH)_2$ in trifluoropropylmethyl siloxane forms a film of CaF_2.

Alloy Element Diffusion

EDXRF has been used to show the diffusion of alloying elements in bearing materials due to sliding[77]. The data show that in an automobile journal bearing consisting of a lead-tin alloy overlay on a second layer of a copper alloy, running causes tin to diffuse into the second layer and form the intermetallic compound Σ-Cu_3Sn (by XRD). The Σ-Cu_3Sn is associated with high wear and seizure. EPMA has shown that tin segregated from a copper-tin alloy and concentrated on the surface[78].

Polysiloxanes

The BL ability of polysiloxanes is of interest because of the attractive stability of the fluid. Reflectance infrared was used to demonstrate the growth of films on steel at high temperatures and with a static charge[79,80].

Corrosion

In a study of the corrosion of copper-lead bearings, the authors[81] have shown that SEM can resolve which metal phase is being corroded. Because of the great depth of field, the SEM provides photographs that are more readily interpreted than light micrographs; for example, one can clearly see that the lead phase is missing.

Friction Polymer

The role of "friction polymer," a material associated with sliding surfaces during BL, is of current interest. Friction polymer from glycol derivatives which was shown to reduce wear has been pyrolyzed in a mass spectrometer and has revealed a mass of 294 and high thermal stability[82]. In measuring the durability of monolayers of stearic acid, an "agglomerate" has been found on the sliding surfaces which affected friction and wear[83]. The material reduces friction, is tenacious and insoluble in acetone, and has a melting or decomposition point greater than 280°C. The authors conclude that it was a polymer from the fatty acid. SEM and ESCA have been useful in identifying undesirable stains, possibly "friction polymer," on cold rolled aluminum[84,85]. The stains consist of patchy films of a "gluey" polymer in which wear fragments and aluminum are entrapped.

SUMMARY OF ADVANCES IN BL ATTRIBUTABLE TO NEW ANALYTICAL TECHNIQUES

1. The use of SEM, ESCA, and EDXRF have contributed the following advances in BL.

   a. Improved understanding of the nature and defects in new surfaces including the differences in phosphate coatings.

   b. The diffusion of elements in and near sliding surfaces.

   c. The existence of spherical wear fragments from fatigue, platelets being lifted during delamination wear, and loosened fibrous fragments from cartilage wear.

   d. The nature of corrosion of copper lead bearings.

   e. Films formed by static immersion are not the same composition as films formed during sliding, and in the latter they occur in microscopic patches.

2. ESCA, AES, and LEED contributed to understanding BL.

   a. Sulfur in films is sometimes in the form of sulfites and sulfates as well as sulfides and varies in concentration with the depth in the film, and the severity of sliding.

   b. Oxygen can convert sulfide films to oxide.

DISCUSSION

Despite some significant advances, particularly in the fields of SEM, ESCA, and EDXRF, the contributions of the new analytical instruments to BL has been less than expected. The results have demonstrated the distribution of elements in a film, but the full composition has not been resolved. The problem may be that most tribologists including the author have not yet applied them to definitive experiments or have not asked the right questions.

The author recommends the analysis of films on unfailed sliding surfaces from practical machines such as a gear set, thus relating composition to proven performance.

ACKNOWLEDGEMENT

Mr. K. L. Hall provided the example of ESCA analysis.

REFERENCES

1. Tabor, D., in "Surface and Colloid Science," edited by E. Matjevic, Vol. 5, Wiley-Interscience, New York, 1972, p. 245.
2. Godfrey, D., in "Interdisciplinary Approach to Friction and Wear," edited by P.M. Ku, NASA SP-181, 1968, p. 335.
3. Campbell, W.E., in "Boundary Lubrication, an Appraisal of World Literature," edited by F.F. Ling, et al., American Society of Mechanical Engineers, New York, 1969, p. 87.
4. Clayfield, E.J. and Galvin, G.D., in Tribology Convention 1969, Proceedings 1968-1969, Institution of Mechanical Engineers, London, Vol. 183, Part 3P, 1969, p. 152.
5. Fein, R.S., in "Interdisciplinary Approach to the Lubrication of Concentrated Contacts," edited by P.M. Ku, NASA SP-237, 1970, p. 489.
6. Beerbower, A., *American Society of Lubrication Engineers Transactions,* Vol. 14, No. 2, 1971, p. 90.
7. Beerbower, A., "Boundary Lubrication," Dept. of Army, AD 747 336, 1972.
8. Rowe, C.N., in "Interdisciplinary Approach to Liquid Lubricant Technology," edited by P.M. Ku, NASA SP-318, 1973, p. 527.
9. Godfrey, D., *American Society of Lubrication Engineers Transactions,* Vol. 5, No. 1, 1962, p. 57.
10. Coy, R.C. and Quinn, T.F.J., *American Society of Lubrication Engineers Transactions,* Vol. 18, No. 3, 1975, p. 163.
11. Quinn, T.F.J., *Tribology International,* Vol. 3, No. 4, 1970, p. 198.
12. Quinn, T.F.J., "The Application of Modern Physical Techniques to Tribology," Van Nostrand Reinhold Co., New York, 1971.
13. Haltner, A.J., in "Boundary Lubrication, An Appraisal of World Literature," edited by F.F. Ling, et al., American Society of Mechanical Engineers, New York, 1969, p. 39.
14. Kane, P.F. and Larrabee, G.B., eds., "Characterization of Solid Surfaces," Plenum Press, New York, 1971.
15. Buckley, D.H., in "Wear of Materials - 1977," edited by W.A. Glaeser, K.C. Ludema and S.K. Rhee, American Society of Mechanical Engineers, New York, 1977, p. 12.
16. Brainard, W.A. and Buckley, D.H., *Wear,* Vol. 26, No. 1, 1973, p. 75.
17. Bowen, R., et al., *Tribology International,* Vol. 9, No. 3, 1976, p. 109.
18. Eyre, T.S. and Dutta, K., *Lubrication Engineering,* Vol. 31, No. 10, 1975, p. 521.
19. Anderson, S., *Tribology International,* Vol. 10, No. 4, 1977, p. 206.
20. Brainard, W.A. and Buckley, D.H., *American Society of Lubrication Engineers Transactions,* Vol. 19, No. 4, 1976, p. 309.
21. Bird, R.J. and Galvin, G.D., *Wear,* Vol. 37, No. 1, 1976, p. 143.
22. Riggs, W.M., et al., *Industrial Research,* Vol. 19, No. 9, 1977, p. 84.

23. Trietz, N., *Journal of Physics E: Scientific Instruments,* Vol. 10, 1977, p. 573.
24. Thompson, M., *Talanta,* Vol. 24, 1977, p. 399.
25. Francis, S.A. and Ellison, A.H., *Journal of Chemical and Engineering Data,* Vol. 6, No. 11, 1961, p. 83.
26. Hansma, K., *Physics Reports,* Vol. 30C, 2, 1977, p. 147.
27. Buckley, D.H. and Pepper, S.V., *American Society of Lubrication Engineers Transactions,* Vol. 15, No. 4, 1972, p. 252.
28. Aronstein, J., *American Society of Lubrication Engineers Transactions,* Vol. 14, No. 4, 1971, p. 322.
29. Suh, N.P. and Coworkers, "The Delamination Theory of Wear," Elsevier Sequoia S.A., Lausanne, 1977, p. 70.
30. Becker, H.C., "Lubrication," Vol. 61, 1975.
31. Tallian, T.E., et al., "Rolling Bearing Damage Atlas," SKF, King of Prussia, Pa., 1974.
32. Perry, J. and Eyre, T.S., *Wear,* Vol. 43, No. 2, 1977, p. 185.
33. Czichos, H. and Kirschke, K., *Wear,* Vol. 22, No. 3, 1972, p. 321.
34. Eyre, T.S. and Maynard, D., in Tribology Convention 1972; London, September 27-28, 1972, Institution of Mechanical Engineers, London, 1973, p. 62.
35. Eyre, T.S. and Dutta, K.K., "Piston Ring Scuffing," Institution of Mechanical Engineers, London, 1975, p. 125.
36. Carver, D.P. and Johnson, K.A., "Piston Ring Scuffing," Institution of Mechanical Engineers, London, 1975, p. 165.
37. Beesley, C. and Eyre, T.S., *Tribology International,* Vol. 9, No. 2, 1976, p. 63.
38. Ratnager, D.D., Cheng, H.S. and Schey, J.A., *Journal of Lubrication Technology,* Vol. 96, No. 4, 1974, p. 591.
39. Clarke, I.C., Constini, R. and Kendi, R.M., *Journal of Lubrication Technology,* Vol. 97, No. 3, 1975, p. 358.
40. Scott, D. and Mills, G.H., *Wear,* Vol. 16, No. 3, 1970, p. 234.
41. Hurricks, P.L., *Wear,* Vol. 27, No. 3, 1974, p. 319.
42. Cummins, R.A., Doyle, E.D. and Rebeccki, B., *Wear,* Vol. 27, No. 1, 1974, p. 115.
43. Cantley, R.E., *Lubrication Engineering,* Vol. 30, No. 10, 1974, p. 498.
44. Eyre, T.S. and Wilson, F., *Lubrication Engineering,* Vol. 29, No. 2, 1973, p. 65.
45. Quinn, T.F.J. and Woolley, J.L., *Lubrication Engineering,* Vol. 26, No. 4, 1970, p. 312.
46. Klaus, E.E., Tewksbury, E.J. and Bose, A.C., in Proceedings of the Japan Society of Lubrication Engineers-American Society of Lubrication Engineers International Lubrication Conference, Tokyo, June 9-11, 1975, edited by T. Sakurai, Elsevier Scientific Publishing Company, New York, 1976, p. 39.
47. Quinn, T.F.J., in Tribology Convention 1969, Proceedings 1968-1969, Institution of Mechanical Engineers, London, 1969, p. 129.
48. Endo, K., Komai, K. and Shiomi, H., *Wear,* Vol. 30, No. 3, 1974, p. 285.
49. Sugeta, T. and Suzuki, K., *Wear,* Vol. 45, No. 1, 1977, p. 57.
50. Forbes, E.S., in "Lubrication and Wear: Fundamentals and Applications to Design," Institution of Mechanical Engineers, London, Vol. 182, Part 3A, 1967-1968, p. 383.
51. Allum, K.G. and Forbes, E.S., *American Society of Lubrication Engineers Transactions,* Vol. 11, No. 2, 1968, p. 162.
52. Coy, R.C. and Quinn, T.F.J., in Tribology Convention 1972, London, September 27-28, 1972, Institution of Mechanical Engineers, London, 1973, p. 62.
53. McCarroll, J.J., et al., *Nature,* Vol. 266, Issue 5602, 1977, p. 518.

54. Godfrey, D., *American Society of Lubrication Engineers Transactions*, Vol. 5, No. 1, 1962, p. 57.
55. Tomaru, M., Hironaka, S. and Sakurai, T., *Wear*, Vol. 41, No. 1, 1977, p. 117.
56. Buckley, D.H., *American Society of Lubrication Engineers Transactions*, Vol. 17, No. 3, 1974, p. 206.
57. Buckley, D.H., *American Society of Lubrication Engineers Transactions*, Vol. 17, No. 1, 1974, p. 36.
58. Baldwin, B.A., *American Society of Lubrication Engineers Transactions*, Vol. 19, No. 4, 1976, p. 335.
59. Weigand, H. and Broszeit, E., *Wear*, Vol. 21, No. 2, 1972, p. 289.
60. Forbes, E.S., et al., *Wear*, Vol. 18, No. 4, 1971, p. 269.
61. Forbes, E.S. and Battersby, J., *American Society of Lubrication Engineers Transactions*, Vol. 17, No. 4, 1974, p. 263.
62. Sakurai, T. and Sato, K., *American Society of Lubrication Engineers Transactions*, Vol. 13, No. 4, 1970, p. 252.
63. Godfrey, D., *American Society of Lubrication Engineers Transactions*, Vol. 8, No. 1, 1965, p. 1.
64. Tomura, M., Hironaka, S. and Sakurai, T., *Wear*, Vol. 41, No. 1, 1977, p. 141.
65. Sethuramiah, A., Okabe, H. and Sakurai, T., *Wear*, Vol. 26, No. 2, 1973, p. 187.
66. Wilson, R.W., *Tribology International*, Vol. 2, No. 3, 1969, p. 166.
67. Masuko, A., Hirata, M. and Watanabe, H., *American Society of Lubrication Engineers Transactions*, Vol. 20, No. 4, 1977, p. 304.
68. Hamaguchi, H., Sato, K. and Sakurai, T., in Proceedings of the Japan Society of Lubrication Engineers-American Society of Lubrication Engineers, International Lubrication Conference, Tokyo, June 9-11, 1975, edited by T. Sakurai, Elsevier Scientific Publishing Company, New York, 1976, p. 386.
69. Barton, D.B., et al., *American Society of Lubrication Engineers Transactions*, Vol. 16, No. 3, 1973, p. 161.
70. Forbes, E.S., Allum, K.G. and Silver, H.B., in Tribology Convention 1969, Proceedings 1968-1969, Institution of Mechanical Engineers, London, 1969, p. 35.
71. Rounds, F., *American Society of Lubrication Engineers Transactions*, Vol. 11, No. 1, 1968, p. 19.
72. Rounds, F., *American Society of Lubrication Engineers Transactions*, Vol. 15, No. 1, 1972, p. 54.
73. Rounds, F., *American Society of Lubrication Engineers Transactions*, Vol. 16, No. 2, 1973, p. 141.
74. Rounds, F., *American Society of Lubrication Engineers Transactions*, Vol. 18, No. 2, 1975, p. 79.
75. Sakurai, T., et al., *American Society of Lubrication Engineers Transactions*, Vol. 17, No. 3, 1974, p. 213.
76. Holinski, R., *American Society of Lubrication Engineers Transactions*, Vol. 18, No. 4, 1975, p. 263.
77. Oda, Y., Kimura, M. and Nakajima, K., *Wear*, Vol. 20, No. 2, 1972, p. 159.
78. Roberts, A.G. and Cameron, A., *American Society of Lubrication Engineers Transactions*, Vol. 18, No. 4, 1975, p. 270.
79. Jemmett, A.E., *Tribology International*, Vol. 1, No. 4, 1968, p. 237.
80. Jemmett, A.E., *American Society of Lubrication Engineers Transactions*, Vol. 16, No. 3, 1973, p. 233.
81. Weetman, D.G., et al., SAE Preprint 760559.
82. Zaslavsky, Y.S., et al., *Wear*, Vol. 30, No. 2, 1974, p. 267.
83. Georges, J.M., et al., *Wear*, Vol. 42, No. 2, 1977, p. 217.
84. Tripathi, K.C., *Tribology International*, Vol. 8, No. 4, 1975, p. 146.
85. Tripathi, K.C., *Lubrication Engineering*, Vol. 33, No. 12, 1977, p. 630.

DISCUSSION

J. S. WOLLAM, C. S. Draper Laboratory: I would like to know how you analyze for boron. Also you talked about the measurement of the hardness of a thin film. That is a very tricky area and I would like to hear more about that.

D. GODFREY: We used scanning Auger to measure boron. It is much more sensitive than the electron microscope. The hardness of the films was measured by scratch hardness test.

S. A. KARPE, DTNSRDC, Annapolis: I feel it is quite difficult to measure physical properties of thin films because you are forming the films on a substrate which may control these properties. There is no point in measuring the shear strength without doing some friction tests. In other words, it is more appropriate just to evaluate the properties in service instead of trying to measure properties of isolated films.

GODFREY: I agree with your comments, but in an isolated state I think they could be measured. For example, if the film is on a surface of interest, take the part and slide riders over it, and measure its coefficient of friction. At light loads it should represent the shear strength of the film.

R. DASKIVICH, G. M. Research Laboratories: I agree with you that we should now be looking for the physical properties of films. At present the SEM is an extremely useful tool. The thing we sometimes forget is that the SEM is at best a semi-quantitative device. Are there any attempts being made to calibrate, if you will, so as to quantify film thicknesses we are measuring?

GODFREY: I welcome your impression on that. There is no place in the tribology literature where this attempt has been made.

DASKIVICH: There was a small attempt in that direction a few years ago at NBS.

GODFREY: Was it published?

DASKIVICH: I am not sure. The project was on plating teeth.

D. H. BUCKLEY, NASA: All these instruments have limitations. They are all man-made and man is not perfect. We should recognize the limitations of those instruments and use them. If properly used they can supply a considerable amount of information that is useful in interpreting what is on the surface which is very important for us.

You suggest that we should measure the physical properties -- physical properties of what? We do not have the tools that tell us what is on the surface -- What its composition is, what its chemistry is, and what the physical properties are.

In one of your figures you showed cavitation damage with cracks. What is the mechanism of fracture?

GODFREY: Those were not cracks; those were the exposure of the flake type graphite. This was a mixture of graphite in the nodular form and the flake type; it was exposure of the edge of the flakes.

Now in regard to your first point, my philosophy is that we should look at the films on the surfaces of practical machines where a failure has not occurred. Because they demonstrate the desirable property, they are the good films. I do not believe in looking at scuffed surfaces. We should look at surfaces which show low wear, low friction, or whatever good property you want. So I have gotten away from static specimens, immersing coupons in oil and so on. I would like to find a low wear specimen out of a practical machine-tooth, piston ring, a segment of a cylinder wall, raceway of a ball-bearing and so forth.

A. BEERBOWER, UCSD: We seem to have two kinds of tribologists. Those who look at the surfaces and throw away the particles, and those who look at the particles and throw away the bearings. We seem to be throwing away half the evidence all the time.

GODFREY: I think I mentioned very briefly that wear fragments represent attrited films in the case of boundary lubrication. They represent the reaction products from the surfaces. Sometimes it is easier to see their composition (if you can collect them) than the film on the surface.

E. RABINOWICZ, MIT: I have always felt that research is not a rational activity anyway. Godfrey makes clear that in research expensive instruments drive out cheap instruments. Gresham's Law that bad money drive out good money is demonstrated here.

QUESTIONER: You talked about the observation of films in the SEM with energy dispersive analysis and electron microprobe. But we are interested in the actual surface and not in the bulk film.

GODFREY: I do not agree that that top 20 angstroms, or the monolayer is necessarily the important factor in boundary lubrication. I think we should look at the whole film. As the asperities ride on the film, elasticity and the plasticity of the film to give, relax and resist penetration are important.

P. ASHLEY, ALCAN, Canada: I find that Mr. Godfrey is directed to information which has been obtained over fifty or sixty years of good wet chemistry. No doubt about it. But for those of us who are working with materials like aluminum, aluminum alloys and titanium, these particular instruments that you said really do not show you very much, are of great value right now. The reason is that we are able to get information faster than we ever did with your wet methods. And for that reason alone, I find that that kind of instrumentation is worth the investment even though expensive.

GODFREY: I do not mean to degrade those instruments for basic wear phenomena, but I just do not see their usefulness in boundary lubrication.

N. P. SUH, MIT: Is the organometallic compound a chelated compound? If it is, is there a possibility that it may undergo a transition as it gets squeezed between the surfaces?

GODFREY: I do not think it is a chelating compound. Its lubrication mechanism is very clearly related to its decomposition. It just starts popping out various things that are useful to the surface -- a little hydrogen sulphide, phosphoric acid-type derivatives that will phosphate the surface to a degree. In the end, when it is all decomposed, it is precipitating zinc sulphide. Basically the decomposition is responsible for its lubricity.

I. GOLDBLATT, Exxon Research: Much of the chemical research has been done to try to identify what kind of film is formed. Clearly, that is not very interesting because the important thing is to have the film there, rather than to know what the particular film is. The intent of the chemist should be not merely to identify a compound, but to know and to understand how it forms. By understanding how a compound forms, he should be able to relate this to other potential reactions, and then modify those reactions. Unfortunately, much of the chemical research that has been done in tribology has not been of this kind. I contend that there is a lot of work that should and could be done in that area to further the area of tribology.

GODFREY: That is the justification for my work, i.e., to determine what is on the surface so that the chemists can make better additives that will form the good films more readily. But in the tribology literature there is nothing like that revealed.

I. L. SINGER, NRL: When I started to investigate these tools for studying antiwear additives, I asked a chemist when it was first realized that an iron phosphate film was formed. He estimated it to be fifteen years. According to my literature search, the first time iron sulphide film was identified positively on the surface in the antiwear regime was about two years ago. Speculations were made as to what kind of film was formed based on inferences from chemical reactions. Furthermore, a number of studies about eight years ago were performed which identified sulphur and sulphide films by X-ray diffraction, but those were very thick films. The antiwear films which are 10 or 50 angstroms were never fully identified until ESCA and XPS came into wide use.

GODFREY: I am sorry I disagree with that. Larson, Williams and others clearly identified iron sulphide by X-ray diffraction a long time ago.

D. SCOTT, Paisley College of Technology: I do not think you should be too pessimistic about the use of these instruments. Professor Rabinowicz says that expensive instruments replace the cheap ones. They do not really, you know. One thing people forget about these expensive instruments is that they help improve the cheaper and useful instruments. For instance, once people can see something in the electron microscope they realize that they will be able to see something of that size in the optical microscope. And today we have refined optical microscopes which people never thought about years ago. An example is a polarizing microscope to identify particles.

ON THE FUNDAMENTALS OF BOUNDARY LUBRICATION

T. Sakurai

ABSTRACT

Mechano-chemical activity of worn surface is an important, as well as interesting, problem in the field of boundary and extreme pressure lubrication. This paper discusses various techniques employed in the research on the subject. The techniques include monitoring the amount of exo-electron emission, ESCA, and examination of the reactions of EP additives and vinyl monomers with the worn surfaces.

Dr. Godfrey has reviewed the usefulness of new surface analytical tools in boundary lubrication (BL). He pointed out that the results of surface analysis are very important to know the chemical reaction between mating surfaces in BL.

The purpose of this paper is mainly to discuss the chemistry of films formed in BL. Godfrey demonstrated that the results of XRD and ED of surface films formed by sulfur compounds in the presence of air were composed of a mixture of iron sulfide and iron oxide.[1] He pointed out that oxygen is an important element in BL.

It has been revealed by many researchers[2-5] that oxygen plays an important role as one of reactants in BL and EP lubrication (EPL). In sulfurization reaction, the presence of oxygen considerably affects the reaction between metal surfaces and sulfur compounds.

In the hot wire method, the sulfurization reaction has been retarded remarkably in the presence of iron oxide films (~ 1000A in thickness) on the iron surface in argon atmosphere. On the other hand, when the iron wire was covered with thinner oxide films, it was severely sulfurized in Ar.[6] These sulfurization experiments were conducted at the same temperature. Such interesting results have been obtained in the dynamic condition by Buckley.[7,8] He has demonstrated by using AES that sulfide films on iron surfaces are displaced by oxygen.

The tribochemical reaction seems to be not the same as the chemical reaction under the static condition because the reaction temperature of compounds with metals may be lowered during sliding because of the higher activation for the reaction system. From the DTA results, Meyer has demonstrated that the decomposition temperature of organic sulfide is decreased when the activated iron powder is used.(9)

It has been found that the adsorption of surfactants on metal surfaces is more important in BL. Rowe has revealed that the adsorption energy is theoretically related to wear in BL.(10) The author has demonstrated by using the autoradiography that C14-stearic acid adsorbed predominantly on iron sulfide surface and not on iron oxide surface.(11) From the results of the measurements of cumulative heat of adsorption, it was found that some surfactants adsorbed strongly on FeS and Fe_3O_4 surfaces compared to α-Fe_2O_3 surface as shown in Figure 1. The effect of adsorption of stearic acid onto the various surfaces on wear(12) is shown in Table 1. The predominant effect

*TABLE I.—EFFECTS OF ATMOSPHERE AND SURFACTANT ON WEAR**

	Wear Volume, X $10^{-6}mm^3$							
Atmosphere	In Air				In Argon			
Additive	None†	Stearic Acid	Stearyl Alcohol	Methyl Stearate	None†	Stearic Acid	Stearyl Alcohol	Methyl Stearate
Sulfidized ball	19.1	0.2	8.2	11.5	12.5	22.6	22.5	22.5
Steel ball	19.8	5.5	17.3	13.9	15.7	6.6	6.2	10.3

* Ball-on-disk friction machine, sliding distance 8.5m, 20°C, 1 kg., 0.1 weight per cent additive.

† Base oil: hydrofinished oil, viscosity 32.2 x $10^{-6}m^2/s$ (32.2 cSt) at 37.8°C, 5.3 x $10^{-6}m^2/s$ (5.3 cSt) at 98.9°C. All experiments were repeated three times. The average values are shown.

of stearic acid on the wear of sulfidized balls could be explained by its strong adsorption and possible reactivity. These results also suggest that oxygen has a pronounced role of reducing the wear of sulfidized balls with surfactants compared with that of the unsulfidized balls.

It seems that surface defects may be created by mechanical working. The tensile stress promotes oxidation corrosion and the amount of wear is greater than without tensile stress for all EP additives applied when a tensile stress is imposed on the specimen.(13)

From the results of film analysis by EPMA and observation of surface roughness, it was found that some EP additives promote the high load carrying capacity by smoothing a surface and enhancing partial elastohydrodynamic lubrication.(14)

There is an additional result in which the presence of oxygen is very important in BL and/or EPL. Many friction machines are employed to estimate the performance of EP oils. A four-ball machine is generally used for the

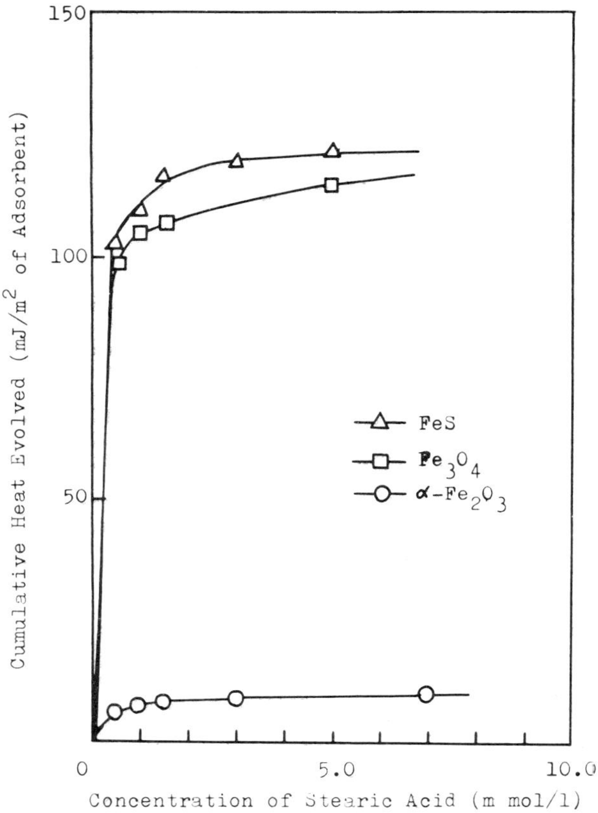

Fig. 1.—Cumulative Heats of Adsorption of Stearic Acid from n-heptane onto FeS, α-Fe_2O_3, and Fe_3O_4 powders.

estimation of EP oils. As shown in Table 2, the different results are obtained even though the same friction machine and the same EP oils are used when two different testing methods are applied for Shell four-ball machine.[15]

These results may indicate the different chemical reactions occurred depending on the contents of oxygen on interface during sliding. The contents of oxygen between lubricated surfaces seems to vary in BL conditions, depending on pressure, temperature, film thickness and capability of oils as an oxygen carrier.

It is clear that in severe conditions, lubrication should be kept in the lubrication regime I or II which is indicated by Salomon.[3] EP oils may be required to have following properties: anti-wear, oxygen carrying capacity, capability of smoothing a surface, suitable decomposition temperature and mild corrosivity for metals.

Finally, precise analysis of surface films will be required to under-

TABLE II.—THE ORDER OF THE LOAD CARRYING CAPACITY OF EP OILS

Testing Method	Order of Load Carrying Capacity	Environments
Step-load test*	DPDS>DBDS>ES	Air
Standard test	ES >DBDS >DPDS	Air
Standard test	ES ≃DBDS >DPDS	Ar

* The conformed surface was prepared in a lubricant without additives before testing.

DPDS; diphenyl sulfide, DBDS: dibenzyl disulfide
ES: elementary sulfur

stand the chemical reactions in BL, especially in dynamic conditions. In the BL system, the chemical reaction between solid surfaces and lubricants occurs to form metal salt films or metal complex films. These films such as sulfide, chloride or phosphide are generally more stable than any physically or chemically adsorbed film. However, the chemisorption of surfactants on lubricated surfaces may be important for the effective BL. The surface films are believed to function by their properties of low shear strength but of high melting point. The mechanism of EP lubrication is that the films on metal surface reduce metal-to-metal contact by their smoothness and that shear occurs within the film. However, the mechanism of BL or EPL has not been fully understood. To approach the mechanism of BL or EPL, it may be important to know the chemical components of surface films and to analyze the kinetics of chemical reaction during sliding.

REFERENCES

1. Godfrey, D., *American Society of Lubrication Engineers Transactions*, Vol. 5, 1962, p. 57.
2. Fein, R.S., "Interdisciplinary Approach to the Lubrication of Concentrated Contacts," NASA SP-237, edited by P.M. Ku, 1970, p. 489.
3. Salomon, G., *Lubrication Engineering*, Vol. 32, 1976, p. 570.
4. Czichos, H., *Wear*, Vol. 22, 1972, p. 321.
5. Begelinger, A. and de Gee, A.W.J., *Wear*, Vol. 22, 1972, p. 337.
6. Tomaru, M., Hironaka, S. and Sakurai, T., *Wear*, Vol. 41, 1977, p. 117.
7. Buckley, D.H., *American Society of Lubrication Engineers Transactions*, Vol. 17, 1974, p. 206.
8. Buckley, D.H., *American Society of Lubrication Engineers Transactions*, Vol. 17, 1974, p. 36.
9. Meyer, K., Keil, G. and Berndt, H., in Eurotrib '77, October 3-5, 1977, Dusseldorf, Band I 58, Preprint, p. 1.
10. Rowe, C.N., *American Society of Lubrication Engineers Transactions*, Vol. 9, 1966, p. 101.
11. Sakurai, T., et al., *American Society of Lubrication Engineers Transactions*, Vol. 5, 1962, p. 67.
12. Yahagi, Y., Hironaka, S. and Sakurai, T., *American Society of Lubrication Engineers Transactions*, Vol. 21, 1978, p. 231.

13. Hamaguchi, H., Sato, K. and Sakurai, T., in Proceedings JSLE-ASLE International Conference, Tokyo, 1975, edited by T. Sakurai, Elsevier Scientific Publishing Company, New York, 1976, p. 386.
14. Sethuramiah, A., Okabe, H. and Sakurai, T., *Wear*, Vol. 26, 1973, p. 187.
15. Tomaru, M., Hironaka, S. and Sakurai, T., *Wear*, Vol. 41, 1977, p. 141.

DISCUSSION

T. F. J. QUINN, University of Aston: Oxygen in the air and lubricant are important in the wear behavior. Do you have any idea about the temperatures in your experiments?

T. SAKURAI: I have no idea. It is very difficult to estimate the temperatures.

D. GODFREY, Chevron Research Company: I would like to add a remark regarding the effect of oxygen. It is very important to use deaerated oil because just using argon over an oil system does not get rid of the dissolved oxygen in the oil. The amount of dissolved air in the oil amazes me. I think we would get larger differences in the additive rating by deaerating the oil.

ROLE OF MECHANOCHEMICAL ACTIVITY IN BOUNDARY LUBRICATION

Y. Tamai

ABSTRACT

Mechanochemical activity of worn or sliding surfaces is one of the most important problems of boundary and extreme-pressure lubrication. Exo-electron emission phenomenon seems to give the required information on this activity. Application of the exo-electron emission was made for milling of silica, aluminum and iron powders. The mechanochemical reaction of milled powder with vinyl monomer was examined in comparison with exo-electron emission. By means of ESCA evidence of mechanochemical activity in the reaction between steel and organic sulfur compound after mild cutting is also presented.

INTRODUCTION

In the previous paper Mr. Godfrey reviewed and pointed out the usefulness of modern tools to analyze solid surface in order to increase the understanding of the mechanism of boundary lubrication. In general his opinions are correct. Therefore, the aim of this discussion paper is to add another method, i.e., exo-electron measurement to determine mechanochemical activity of solid surface during or after milling and an example of ESCA application to show the role of mechanochemical phenomena in boundary or extreme pressure lubrication.

EXO-ELECTRON MEASUREMENT

Exo-electron, discovered by Kramer,(1) is considered to play a significant role in mechanochemical activity. Exo-electron emission is measured in many ways, for example, by a Geiger-Müller counter of open-window type.(2) Using this device we have conducted research in which milled powders of silica, aluminum and iron were brought into contact with vinyl monomers of acrylonitrile (AN), methyl methacrylate (MMA), styrene (ST) and vinyl acetate (VAc), and the activity was monitored by exo-electron emission from these solids. The details are as follows.

The powders (about 100 mesh) of 15 g were vibromilled in a cylindrical glass jar (120 mm dia. and 120 mm height) containing 25 glass balls (12 mm dia.) at 1200 rpm with an eccentricity of 5 mm for 1 hour (for exo-electron measurement) and 2 hours (for chemical activity examination). Both samples were divided into two parts, and kept in air after working for 10 minutes and 120 minutes, respectively.

Exo-electron emission was measured with an open-window Geiger counter illustrated in Figure 1 at 30°C with argon (150 torr) and methanol (10 torr) quenching. The cathode voltage was 1260 V and the grid voltage was 100 V.

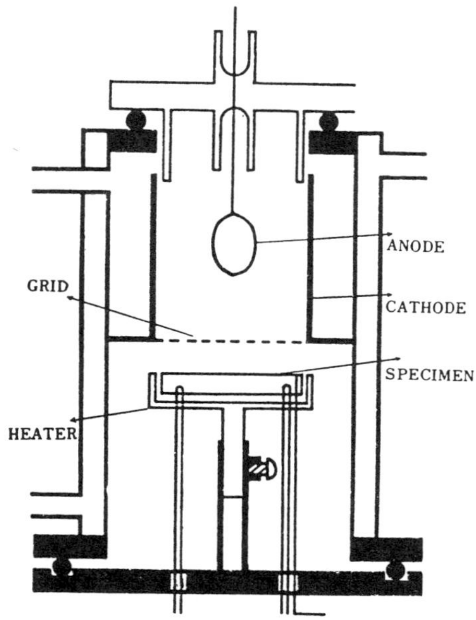

Fig. 1.—Schematic of Exo-electron Counter.

The powders were put in the specimen dish. The results are listed in Table 1. The emission intensity decreased when the powder was exposed to air.

TABLE I.—EXO-ELECTRON EMISSION INTENSITY FROM MILLED POWDERS

Powder	Emission Intensity (c/s, 30°C) (after 10 min)	(after 120 min)
Aluminum	50	15
Silica	5	2.5

The mechanochemical activity was determined by polymerization of vinyl monomer on the powder surface. The powder was immersed after working for 2 hours, into a liquid pool of monomer at 20°C. The wetted powders were taken out and dried under high vacuum and the weight increase of the powder was measured. The results are shown in Figure 2 for silica, aluminum and iron powders and methyl methacrylate, and in Figure 3 for silica powder and four vinyl monomers together with benzene.

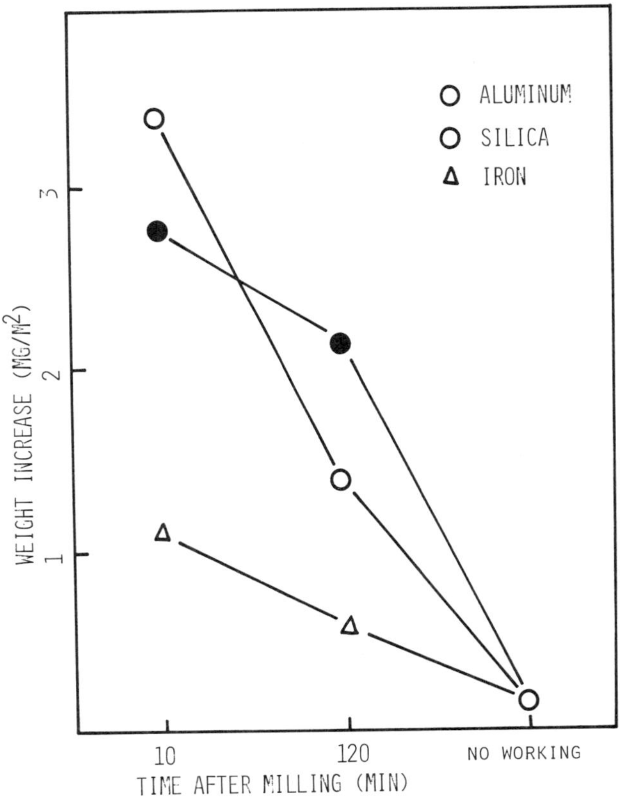

Fig. 2.—Mechanochemical Polymerization of Methyl Methacrylate on Milled Powders.

It is clear that the measured activity is not due to high temperature created by frictional heating and high pressure at the colliding contact, because the activity was tested after milling. From Figure 2 the activity decreased with time after milling, suggesting some relation to the exo-electron emission which also decreased after milling. From Figure 3 the order of polymerization is in agreement with the order of anion sensitivity in polymerization, which also suggests the role of exo-electron in this reaction.

In an earlier work by Smith,[3] the role of exo-electron was already con-

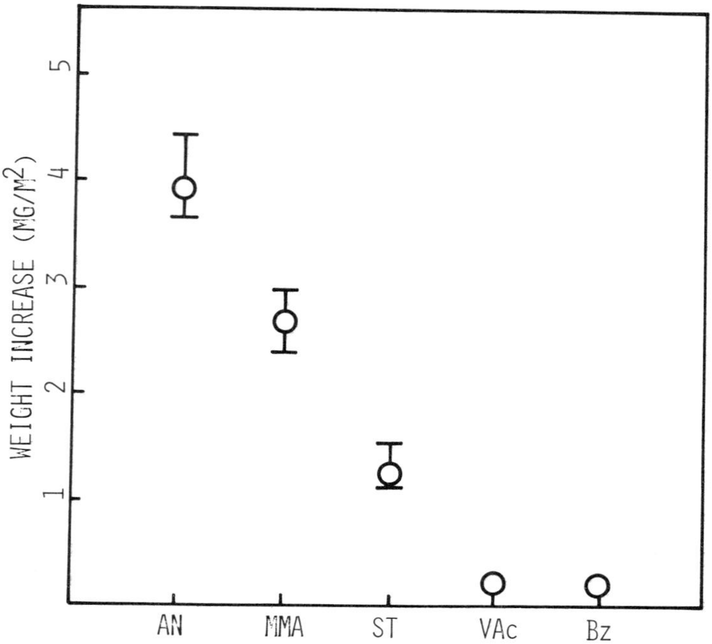

Fig. 3.—Mechanochemical Polymerization of Vinyl Monomers on Milled Silica.

sidered as accelerating soap formation of fatty acid with metal. Goldblatt[4] also imagined exo-electron to produce anion radical of condensed aromatics in explaining the effect of aromatics for wear reduction. The author[5] reported the relation between exo-electron emission and adsorption of hydrocarbon onto steel surface. Recently so-called friction polymer covering of sliding surface is attracting more attention introducing potential polymer-forming additives.[6] In this sense, the exo-electron measurement should be one of the powerful means of surface analysis.

As another example of mechanochemical activity, the reaction between mineral oil solution of di-t-dodecyl polysulfide and mild steel was investigated under light cutting at the speed of 45 m/min and of 15 mm width and 0.1 mm depth. Two kinds of specimens were prepared. The one was cut in the flow of mineral oil containing 1% of sulfur compound. The other was cut and lubricated with plain mineral oil. In both cases, the cut surface of the steel rod showed no color change and the chemical analysis of the surface was impossible with usual methods.

After cutting, the first specimen was immersed in the sulfur containing oil for 10 minutes at 20°C and then washed with cyclohexane. The second specimen was first dipped in the plain oil for 5 minutes and then followed by a 10 minute sulfur-containing oil immersion and successive cyclohexane washing in the same way as for the first.

These specimens were brought under ESCA examination. The spectrum for sulfur is shown in Figure 4. According to Bird[7] simple thermal reaction

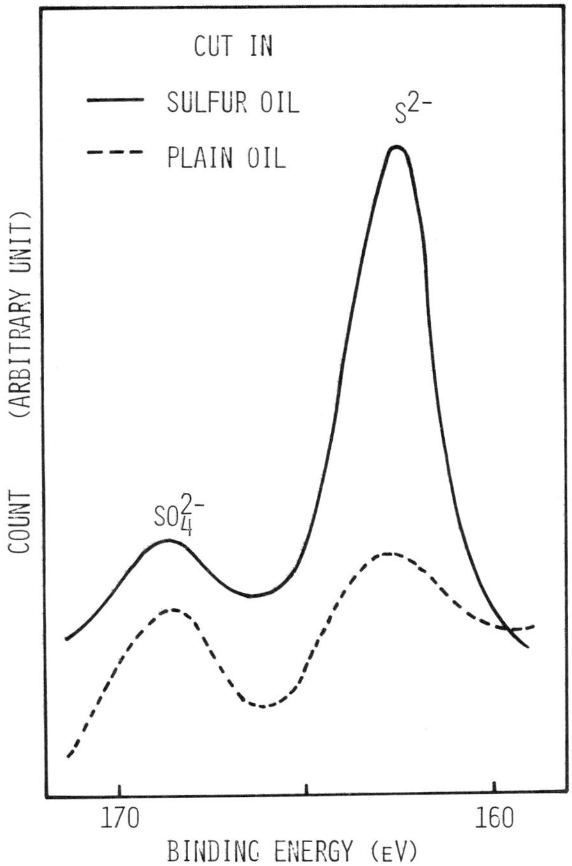

Fig. 4.—ESCA Spectrum of Worked Steel Immersed in Sulfur Compound Solution.

cannot give the existence of S^{2-}. The present results indicate that the mechanochemical reaction does exist in cutting and this activity remains even after machining, which is quite similar to the results on vinyl polymerization. This shows again the importance of mechanochemical phenomena in boundary lubrication.

REFERENCES

1. Kramer, J., *Z. Phys.*, Vol. 145, 1949, p. 739.
2. Grunberg, L., Scott, D. and Wright, K.H.R., *Journal of Physics D: Applied Physics*, Vol. 12, 1961, p. 134.
3. Smith, H.A. and McGill, R.M., *Journal of Physical Chemistry*, Vol. 61, 1957, p. 1025.
4. Goldblatt, I.L., *Industrial & Engineering Chemistry Product Research and Development*, Vol. 10, 1971, p. 270.
5. Tamai, Y., Suzuki, M. and Momose, Y., Preprint - Symp. Div. Petrol. Chem. ACS 13, 1968, B159.
6. Furey, M.J., *Wear*, Vol. 26, 1973, p. 369.
7. Bird, J. and Calvin, G.D., *Wear*, Vol. 37, 1976, p. 143.

DISCUSSION

E. S. SPROLES, Rensselaer Polytechnic Institute: It is proposed that exoelectrons are the cause of the formation of friction polymer over the precious metal surfaces. Would you comment on that.

Y. TAMAI: Indeed the purpose of my experiment was to show how important exoelectron emission is in friction polymer formation. Yes, I agree with that hypothesis.

D. H. BUCKLEY, NASA: Do you think it is the mechanical activity that produces the exoelectrons or the generation of a clean surface?

We did some experiments in a vacuum chamber, using single crystals of aluminum with (110) orientation and magnesium with (0001) orientation. Exoelectron emission was monitored from the surface of annealed and strained crystals while looking at the surface by Auger also. There was no exoelectron emission from both surfaces -- the strained and the unstrained surface. When we bled a small amount of oxygen into the system, we observed the exoelectron emission go up and both surfaces have indentical emission rates. Simultaneously we observed an increase in the concentration of oxygen by Auger spectroscopy. After the large peak exoelectron comes off you get a smaller peak which is due to a secondary exoelectron effect. I believe that the first peak is associated with adsorption of the oxygen on the surface giving rise to exoelectron emission whether the surface is strained or unstrained. The secondary peak is the chemical shift in Auger spectroscopy for the aluminum and the formation of aluminum oxide. The first peak is due to adsorption and the second one is a chemical compound formation of the surface. None is due to mechanical working at all.

TAMAI: My experiment was done after cold working and the chemical reaction occurred after that. I guess your experiment with exoelectron is almost the same as ours. You performed the experiment under high vacuum but we prepared the sample by cutting or grinding to remove the surface oxide which was originally there.

SELF-GENERATED VOLTAGES UNDER BOUNDARY LUBRICATED CONDITIONS

I. L. Goldblatt

ABSTRACT

A program has been carried out to determine the relationship between Self-Generated Voltages and wear under Boundary Lubricated Conditions. The voltages generated between wearing members are shown to be unrelated to such factors as fluid conductivity or fluid viscosity. These voltages are shown to be associated with the wear process itself. A relationship has been obtained between wear and Self-Generated Voltages and has been found for both a steel-on-steel system and an aluminum-on-steel system under dry air blanketing. Under these conditions, at low wear, the measured voltages are several millivolts in magnitude, and they decrease as wear increases.

A preliminary mechanism to explain how the voltages are generated under Boundary Lubricated Conditions is discussed. It is suggested that two primary electrochemical processes occur under Boundary Lubrication Conditions, an electron transfer process and an oxidation process.

INTRODUCTION

Tribology is a technology largely based upon empiricism and, as such, currently possesses only a limited capability to predict what will occur when operating conditions are varied. Recently several models have been proposed which have been used to explain a portion of the body of tribology.[1-6] These models fall into two broad categories. Some are purely chemical, involving primarily the lubricant, and others are metallurgical or physical, often involving a mathematical description of the phenomenological process being modeled.

In order to adequately describe the real nature of tribology, a model incorporating both approaches will be necessary. However, in order to do this, relationships between the metal surfaces and lubricants are required, and these relationships could well be defined by electro-chemical processes which take place at the surface involving both the lubricant and the metal. Such understanding may then be used to relate the purely chemical models to the metallurgical and physical transformations which have been described.

It will be shown that a relationship between self-generated voltages and wear may exist which could serve to fill this role. Furthermore, it will be shown that a detailed investigation of some of the chemical parameters involved in the lubrication process tend to support the idea that self-generated voltages might be related to fundamental electrochemical processes which can in turn be related to the wear mechanism.

EXPERIMENTAL

The experimental program has been carried out using the Ball-on-Cylinder device, which has been described previously.[7,8] The device, Figure 1, consists of a stationary ball which is loaded onto a rotating cylinder.

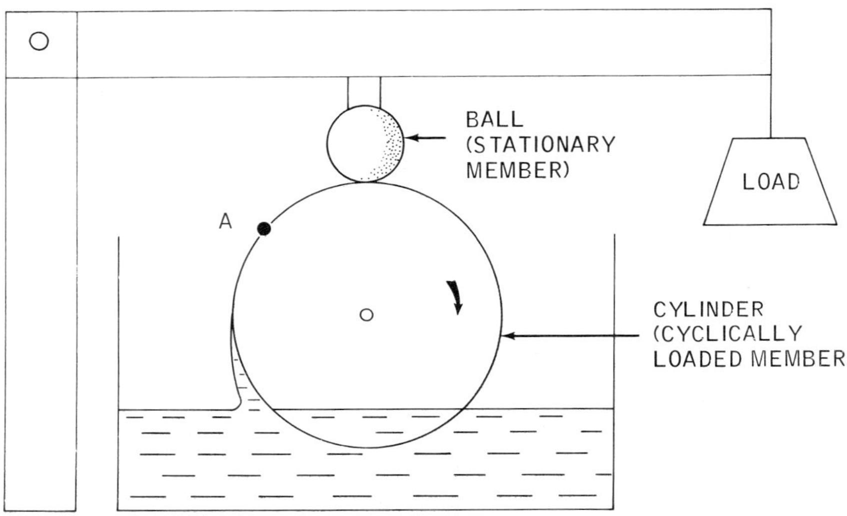

Fig. 1.—Schematic Diagram of the Ball-on-Cylinder Device.

Both are enclosed in a small environmental control chamber, which under lubricated conditions, serves as the lubricant reservoir. This reservoir may be blanketed with various atmospheres through the use of an attached gas inlet system. Heaters, embedded in the base of the reservoir, permit operation up to bulk fluid temperatures of about 200°C.

The fluids which have been employed include several highly refined saturated petroleum white oils, a lightly refined bydrocarbon base stock, several ester base stocks, as well as several pure hydrocarbons, Table 1. An antistatic agent, ASA-3, a proprietary antistatic agent sold commercially, was also used for some tests.

Two combinations of materials have been employed in this study, a 52100 steel ball (R_c 62-64) sliding on a 52100 steel cylinder (R_c 20) and an AMS 2017 aluminum ball (R_b 58) sliding on a 52100 steel cylinder.

TABLE I.—LUBRICANTS USED IN THE TEST PROGRAM

Lubricant	Per Cent Concentration Saturates	Unsaturates + Aromatics	Viscosity (mPa.S)
Decane	100		0.859
Dodecane	100		1.37
Tetradecane	100		2.06
Hexadecane	100		3.09
LVWO (1)	100		2.43
IVWO (2)	100		17.7
HVWO (3)	99.8	0.2	35.8
HVLRO (4)	65	35	37
Toluene		100	0.536
Mesitylene		100	0.855
1-Methylnaphthalene		100	2.69
1-Chloronaphthalene		100	3.14
Di-2-Ethylhexyl Sebacate	100% Ester		17.1
Hercolube J	100% Ester		39.4

(1) LVWO = Low Viscosity White Oil.
(2) IVWO = Intermediate Viscosity White Oil.
(3) HVWO = High Viscosity White Oil.
(4) HVLRO = High Viscosity Lightly Refined Oil.

Electric Potential Measurements

In order to measure the magnitude and polarity of the Self-Generated Voltages (SGV's), the ball and cylinder were thoroughly insulated from each other, from ground, and from the rest of the Ball-on-Cylinder device. The test balls were insulated by mounting in special adaptors fabricated either from "Lexan" (polycarbonate) or "lucite" (polymethacrylate). Electrical contact with the ball was then effected using 0.07 mm stainless steel wire. A special collar to lock the balls was machined from nylon. Steel spindles, shielded with a nylon sheath 0.083 mm thick, served to insulate the cylinders. Electrical contact with the cylinders was obtained through a conducting set screw via a mercury contact cell. The ball mounting adaptor and shielded steel spindle are shown in Figure 2.

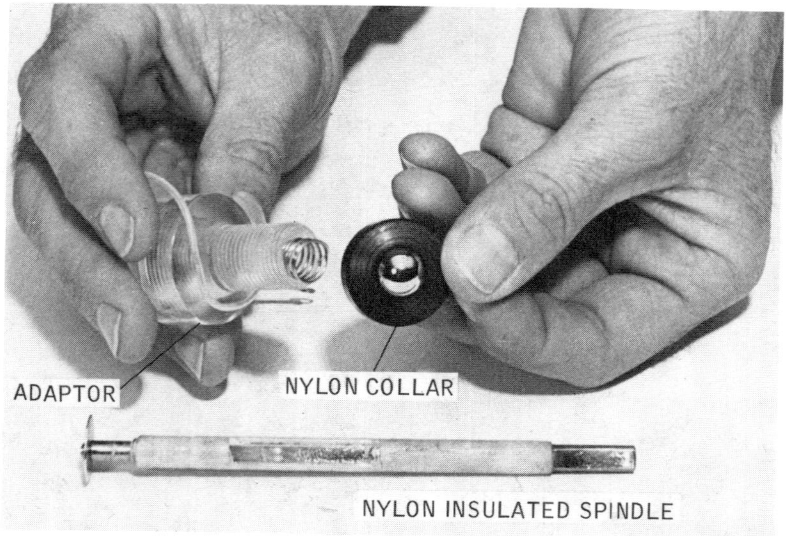

Fig. 2.—The Ball Adaptor and Nylon Covered Steel Spindle for Insulating the Ball and the Cylinder.

Measurements of potential were made using a Keithly Model 602 Solid State Electrometer. The output from the Keithly was fed to a Hewlett-Packard Model 7100B Moseley Strip Chart Recorder.

It should be noted that it is not necessary to establish the absolute polarity of either of the test members relative to ground. Only their relative polarity is required. The convention that will be adopted for reporting the polarities of the SGV's will be the polarity of the cylinder relative to the ball.

Experimental Method

The procedure followed throughout this program consists of starting the cylinder rotating and at the same time purging the lubricant and blanketing the test chamber. After purging the system for about 15 minutes, the ball is loaded by dead weight upon the cylinder. For tests in inert atmospheres, the test chamber is blanketed for 30 minutes and the surrounding dry box filled with a dry nitrogen atmosphere. Test duration, in general, is 32 minutes, during which time the friction force and the SGV's are recorded continuously. At the conclusion of the experiment, the cylinder is stopped, the contact potential between the ball and the cylinder measured, and the ball unloaded. The ball is then removed and is replaced by a new ball. It is retained for visual examination and wear measurement (wear scar diameter).

The sample reservoir is then emptied of its contents, in the case of lubricated studies; the chamber is thoroughly rinsed with hydrocarbon solvent such as hexane or heptane, and is air dried. Next, the cylinder is moved along its axis by about 1 mm to another track position, and a sample of fresh lubricant is introduced. After completion of about 8 to 12 tests, the cylinder is removed, and the surface profile is recorded, from which the track wear is assessed.

RESULTS - GENERAL SYSTEM VARIABLES

The experimental program was designed to investigate the relationship which may exist under boundary lubricated conditions between wear and Self-Generated Voltage (SGV). In order to relate the SGV's to wear models currently in use, it was decided to determine the influence of certain operating and chemical parameters upon both SGV's and wear. Included among these parameters are load, moisture (water), temperature, metallurgy, atmosphere (oxygen) and chemical nature of the lubricant.

Prior to investigating these parameters, however, it was necessary to investigate the influence of several other germane nonreadily controlled or measured general system variables, namely, lubricant conductivity, fluid motion and lubricant film thickness. Each of these variables could influence the voltages generated and thus confound the detection of relationships between wear and SGV.

Before describing the experimental results in detail, a brief description of the general characteristics of the time dependent variation of the SGV's will be presented.

General Characteristics of the Time Dependence of the Self-Generated Voltages (SGV's)

In this study, two general types of time dependences for the SGV's have been recognized. Both are characterized, usually, by an initial break-in period, during which the magnitude of the SGV increases (Figure 3). Differences in behavior can be encountered after this initial period. The first type of response, illustrated by the results obtained in dry air, is typified by a brief decrease in the magnitude of the SGV after the break-in period, in this example to 1.2 mv (Figure 3). There is then a gradual approach towards a higher asymptotic value of SGV, in this case about +4. 5 mv after 32 minutes. The second type of behavior is exhibited under wet blanketing in which an initial rise in the SGV to about 0.9 mv is observed. This falls to about 0 mv, increases to about 0.3 mv, then decreases to the point where the polarity of the cylinder relative to the ball changes. The SGV at the end of a 32 minute test period is about -1.5 mv.

The behavior of other lubricant and/or metallurgical systems are, in general, similar to one of the above examples. The actual magnitude and polarity will depend, however, upon the particular system under investigation.

General System Variables

Influence of Fluid Conductivity

The first parameter to be investigated was lubricant conductivity. Lubricant conductivity is recognized as being important in the static charging of flowing liquids since fluid conductivity influences the charge relaxation process and thereby the magnitude of charge buildup. In order, therefore,

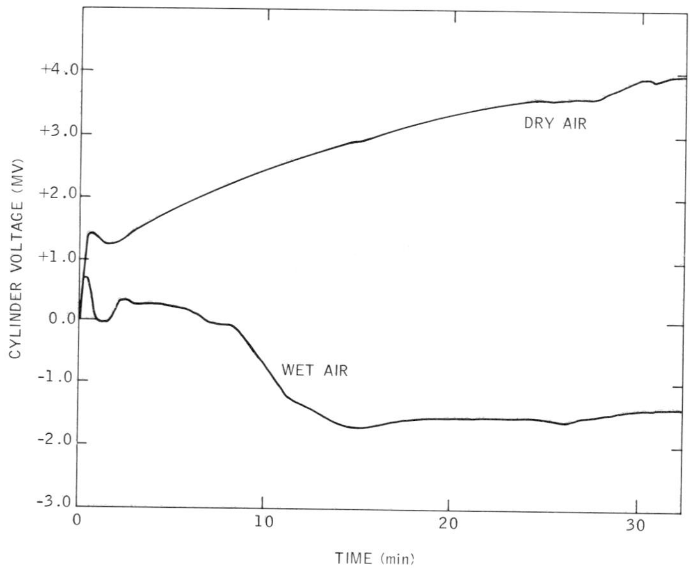

Fig. 3.—Time Dependent Variation of the Self-Generated Voltage (Test Conditions: Ball-on-Cylinder Device, HVWO, 1,000 g Load, 240 rpm, 25°C, 32 minutes, 52100 Steel-on-Steel, Atmospheres as Noted).

to establish the relationship between SGV and wear under boundary lubricated conditions, it is of importance to determine if changes in charge relaxation or fluid conductivity contribute to the measured voltage.

Evaluation of the influence which fluid conductivity has upon the voltage measured is best performed under experimental conditions in which the wear parameters remain essentially constant while only conductivity is varied. Such conditions may best be achieved by using very potent commercial antistatic agents at very low concentrations (under 20 ppm).

By using several different concentrations of the proprietary antistatic agent ASA-3 in a low viscosity white oil, the fluid's conductivity may be altered by about five orders of magnitude, while its wear characteristics are not changed, as may be seen from the measured cylinder and ball wear (Table 2). Correspondingly, the magnitudes and polarities of the measured voltages at the 500 g load used in the tests are similar for all of the fluids at the conclusion of the 32 minute test. Traces showing the time variation of the average measured voltages for the runs listed in Table 2 are presented in Figure 4. From the data presented, it is clear that there is no substantial difference between the measured voltages for the oils under similar test and wear conditions.

Fluid motion effects - streaming potential

The next aspect to be considered is what voltages might be generated due to the flow of oil between a separated ball and cylinder, i.e., streaming

TABLE II.—THE INFLUENCE OF FLUID CONDUCTIVITY MODIFIED USING ASA-3 ANTI-STATIC ADDITIVE UPON THE MEASURED VOLTAGE

Additive to Bayol 35	Load (g)	Conductivity (Picosiemens/m)	Ball Wear Volume (cm^3)(2)	Cylinder Wear Volume (cm^2)	Measured Voltage (mv)
None	500	0.017	$3.77x10^{-7}$	0.08×10^{-5}	+ 0.5
0.05 ppm ASA-3	500	4.59	$2.64x10^{-7}$	0.08×10^{-5}	+ 0.6
0.4 ppm ASA-3	500	50.51	$2.64x10^{-7}$	0.12×10^{-5}	+ 0.6
1.5 ppm ASA-3	500	215.0	$3.77x10^{-7}$	0.09×10^{-5}	+ 0.2
15 ppm ASA-3	500	2066.0	$2.41x10^{-7}$	$0.008x 10^{-5}$	+ 0.7

(1) Test Conditions: Ball-on-Cylinder device, Loads as noted, IVWO as base fluid, 240 rpm, Dry Air Atmosphere, 25°C., 32 Minute, 52100 Steel-on-52100 Steel.
(2) Ball Wear Volume calculated as $7.73 \times WSD^4 x10^{-2}$ (cc).

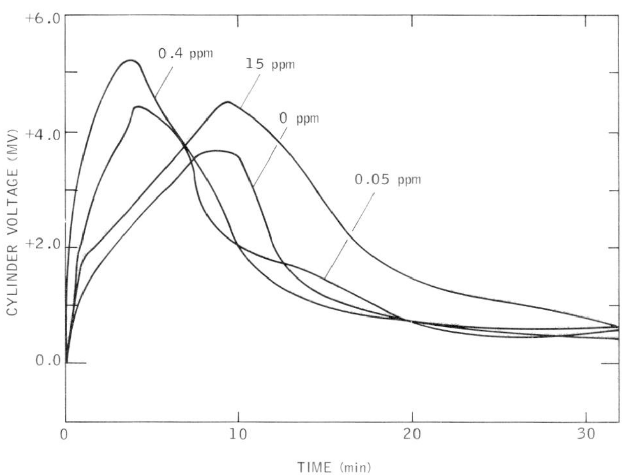

Fig. 4.—Influence of the Antistatic Agent ASA-3 Upon the Time Dependent Variation of the Measured Voltages of the LVWO (Test Condit ons: Ball-on-Cylinder Device, 500 g Load, Dry Air Atmosphere, 240 rpm, 25°C, 32 minutes, 52100-on-52100 Steel).

potential effects. The magnitude of the streaming potential effect was estimated by measuring the potentials developed during tests in which oil flowed between relatively large ball and cylinder separations of about 0.55

mm and 0.25 mm (Figure 5).

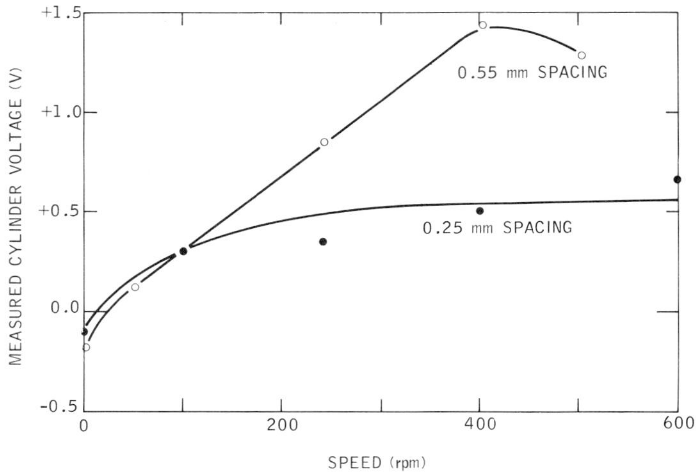

Fig. 5.—Measured Voltage Generated Due to Flow of Fluid Between a Separated Ball and Cylinder (Test Condition: Ball-on-Cylinder Device, Hercolube J, Dry Air Atmosphere, 25°C, 52100 Steel-on-52100 Steel).

These measured voltages at the higher speeds were found to compare favorably with the values calculated using the Equation below.[9]

$$V_{axis} - V_{wall} = SR^2/4\ \varepsilon\varepsilon_o \qquad (1)$$

V = Voltage (V)

S = charge density (C/m^3)

R = distance between core of fluid and wall (taken in this case as the distance between the ball and the cylinder)

ε_o = the absolute dielectric constant of a vacuum (8.85×10^{-12} A·S/V·m)

ε = the relative dielectric constant (∿2.0)

Since the equation above appears to predict the magnitude of the streaming potential due to the flow of relatively thick films, the magnitude of the streaming potentials to be expected using the various fluids employed in this program may also be computed using this equation. In general, it is found that the streaming potentials computed are less than 10^{-2} mv. This value is small as compared with the measured voltages. Hence, the recorded voltages are probably not related to streaming potentials.

Lubricant film thickness

Another parameter which may affect the measured voltages is lubricant film thickness. Experimentally, it was decided to vary lubricant viscosity and thereby film thickness, and thus probe the influence which lubricant film thickness has upon measured voltages. This has been done using two white oils, the LVWO and the IVWO, having viscosities of 2.43 mPa·S and 17.7 mPa·S, respectively, and mixtures thereof. The metallurgical combination selected in order to ensure nearly constant wear rate over the range of film thicknesses (or viscosities) investigated is an aluminum ball on a steel cylinder. It is clear (Table 3) that the measured voltages are essentially independent of film thickness and/or viscosity over a reasonably broad range of the film thicknesses. As may be seen from Table 3, both ball and cylinder wear are virtually identical for all of these fluids.

From the above series of tests, it is evident that the measured voltages, under conditions in which wear rates (and presumably wear mechanism) are similar, are not affected by changes in film thickness.

RESULTS - TEST VARIABLES

It has been demonstrated that the voltages generated under boundary lubricated conditions are not due to fluid conductivity, streaming potential or changes in fluid film thickness. The voltages generated are therefore, presumably, associated with surface damage and wear.

Gathered in Table 4 are the calculated wear and measured SGV's for tests with 52100 steel under dry air blanketing at room temperature using several base stocks, including paraffinic hydrocarbons, aromatic type hydrocarbons, mixtures of aromatics and paraffins, the HVWO, the HVLRO and esters.

Included in Table 4 is a parameter labeled the Surface Fatigue Demerit Rating (SFDR) which reflects the nature and character of the track surface profile. This rating system is illustrated in Figure 6. A relatively smooth track, showing no apparent cracking is assigned a demerit rating of 1. One which exhibits some damage is assigned a value of 2. A surface profile which exhibits extensive "cracking," is jagged and non-symmetric, is assigned a value of 3. A rating of 1^+ indicates slight damage within the wear track.

Presented graphically in Figures 7 and 8 are the relationships between SGV's and both cylinder and ball wear. In general, the magnitudes of the SGV's decline rapidly with an increase in wear under dry air blanketing.

A relatively large body of data has also been gathered under conditions of wet air blanketing, Table 5. The data show no general trend. It should also be noted that the polarities (and magnitudes) of the SGV's for the HVWO are different from those for the other base stocks investigated under this atmospheric environment.

The relationships between wear and SGV seem to exist at elevated temperatures also. Increased temperature, under dry air blanketing, leads, in general to increased wear and either relatively more negative or unchanged, negative, SGV's, Table 6. This data is also included in Figures 7 and 8.

Data are also available for aluminum-on-steel using the HVWO, HVLRO and mixed aromatic-LVWO base stocks under dry air blanketing (Table 7). The general relationships which emerge using this combination of materials, Figures 9 and 10 for cylinder and ball wear, respectively as a function of SGV's, are generally similar to those presented in Figures 7 and 8, namely that higher SGV's give rise to lower wear.

TABLE III.—MEASURED WEAR AND MEASURED VOLTAGES AS A FUNCTION OF RELATIVE FILM THICKNESS[1]

Lubricant	Load (g)	Viscosity (mPa·S)	Relative Film Thickness ho/ho'[2]	Ball Wear Scar Diameter (mm)	Ball Wear Volume (cc)	Track Cross Sectional Area (cm^2)	Cylinder Wear Volume (cc)	Measured Voltage (mv)
LVWO	500	2.43	1.0	0.62	11.42×10^{-7}	0.88	1.16×10^{-5}	+ 0.3
50:50 LVWO:IVWO	500	5.6	1.81	0.55	7.07×10^{-7}	0.90	1.19×10^{-5}	+ 0.45
30:70 LVWO:IVWO	500	8.2	2.37	0.50	4.83×10^{-7}	0.63	0.83×10^{-5}	+ 0.6
15:85 LVWO:IVWO	500	11.3	2.98	0.50	4.83×10^{-7}	0.53	0.70×10^{-5}	+ 0.51
IVWO	500	17.7	4.10	0.48	4.10×10^{-7}	0.52	0.69×10^{-5}	+ 0.33

(1) Test Conditions: Ball-on-Cylinder Device, Mixtures of LVWO and IVWO, 240 rpm, 25°C, 32 Minutes, Wet Air Atmosphere, Aluminum-on-Steel.

(2) $ho/ho' = (\eta o/\eta o')^{0.725}$ as per Cheng, H. S., "A Numerical Solution of the EHD Film Thickness in an Elliptical Contact," Trans. ASME, J. Lub. Tech., 92, pgs. 155-162, (assumed nearly circular contact).

TABLE IV.—MEASURED WEAR AND SELF-GENERATED VOLTAGES UNDER DRY AIR BLANKETING FOR AISI 52100 STEEL BALLS SLIDING ON AISI 52100 STEEL CYLINDERS(1)

Lubricant	Load (g)	Wear Volume Ball (cc)	Wear Volume Cylinder (cc)	Surface Fatigue Demerit Rating	Self-Generated Voltage (mv)
HVWO	200	0.08×10^{-7}	0.11×10^{-5}	1	+6.3
Decane	500	36.69×10^{-7}	0.05×10^{-5}	2	+0.21
Dodecane	500	20.77×10^{-7}	0.18×10^{-5}	1+	+0.20
Tetradecane	500	12.18×10^{-7}	0.12×10^{-5}	1	+0.24
LVWO	500	3.17×10^{-7}	0.08×10^{-5}	1	+0.50
Hexadecane	500	2.64×10^{-7}	0.26×10^{-5}	3	+1.65
1-Methylnaphthalene	500	1.16×10^{-7}	0.50×10^{-5}	1	+3.0
HVWO	500	0.41×10^{-7}	0.24×10^{-5}	1	+4.9
HVLRO	500	0.18×10^{-7}	0.24×10^{-5}	1	+2.6
20/80 Toluene/LVWO	1000	12.18×10^{-7}	0.33×10^{-5}	3	+0.15
20/80 Mesitylene/LVWO	1000	10.02×10^{-7}	0.22×10^{-5}	1	+0.40
LVWO	1000	6.10×10^{-7}	0.30×10^{-5}	1	+0.18
HVWO	1000	0.71×10^{-7}	0.50×10^{-5}	1	+4.3
HVLRO	1000	0.30×10^{-7}	0.40×10^{-5}	1	+2.3
20/80 1-Methylnaphthalene/ LVWO	1000	1.16×10^{-7}	0.17×10^{-5}	2	+3.0
20/80 1-Chloronaphthalene	1000	8.16×10^{-7}	0.42×10^{-5}	1	+4.0
LVWO	2000	27.17×10^{-7}	0.45×10^{-5}	3	-0.05
20/80 1-Methylnaphthalene/ LVWO	2000	4.83×10^{-7}	1.11×10^{-5}	3	-0.15
20/80 1-Chloronaphthalene/ LVWO	2000	16.53×10^{-7}	1.83×10^{-5}	3	+0.8
HVWO	2000	0.92×10^{-7}	0.54×10^{-5}	1+	+0.55
Di-2-Ethylhexyl Adipate	2000	0.71×10^{-7}	1.25×10^{-5}	1	+0.20
Hercolube J	2000	0.63×10^{-7}	0.50×10^{-5}	3	+1.50
HVLRO	2000	1.61×10^{-7}	1.15×10^{-5}	3	-0.06
20/80 1-Chloronaphthalene/ LVWO	4000	31.66×10^{-7}	2.31×10^{-5}	3	+0.45
HVWO	4000	3.77×10^{-7}	2.35×10^{-5}	3	-0.30
HVLRO	4000	4.10×10^{-7}	2.82×10^{-5}	3	-0.24

Test Conditions: Ball-on-Cylinder Device, Loads as noted, 240 rpm, 25°C, 32 min.

The influence of oxygen availability upon the development of SGV's was investigated by studying systems under Argon or partially inerted atmospheres, Table 8. Oxygen availability is clearly necessary for the development of SGV's since, in its absence, the SGV's are all nearly zero, even though wear may be substantial.

DISCUSSION

Relationships Between SGV's and Wear

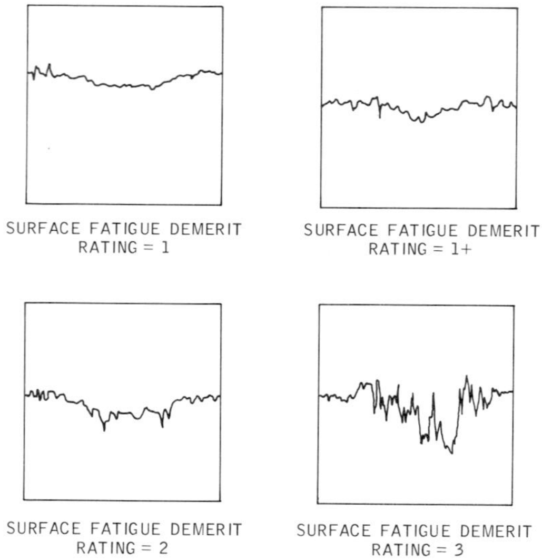

Fig. 6.—Surface Profiles Which Illustrate the Surface Fatigue Demerit Rating System.

With increased wear on either the ball or the cylinder, there is a general decline in the magnitudes of the SGV's for the steel-on-steel system under dry air blanketing, Figures 7 and 8 and for the aluminum-on-steel, Figures 9 and 10.

The relationship between ball wear and SGV's for the steel-on-steel system in Figure 8 is particularly striking. The relationship between cylinder wear and SGV's for the steel-on-steel system in Figure 6 is not as clear, particularly because low values of cylinder wear may be associated with either high or low values for the SGV's. However, it should be noted that the systems which comprise the data base clustering in the low cylinder wear - low SGV's region, all exhibit large values for ball wear. This characteristic, high ball wear, low cylinder wear is typical of systems undergoing severe corrosive wear. There is, therefore, reason to believe that the systems which comprise this set of data are different from the rest of the data presented in Figure 7. The smooth curve drawn in Figure 7, therefore, disregards these values. In so doing, the relationship between SGV's and cylinder wear for the steel-on-steel system is reasonably good.

For the aluminum-on-steel system, there is also a general trend that increased wear leads to lower SGV's. There are again several systems which lie substantially away from the smooth curves drawn in Figures 9 and 10, namely, those for the LVWO, and the 20/80 blends of toluene and mesitylene in the LVWO. The reason for this failure is apparent, however, upon elemental analysis, performed using EDAX, of the aluminum balls employed in these particular tests (Figure 11). The major elemental component at the surface of the wear scar in these cases is found to be iron rather than the aluminum metallurgy of the ball (Figures 11a-11c). Presumably, considerable metal transfer from the cylinder to the ball has taken place, essentially convert-

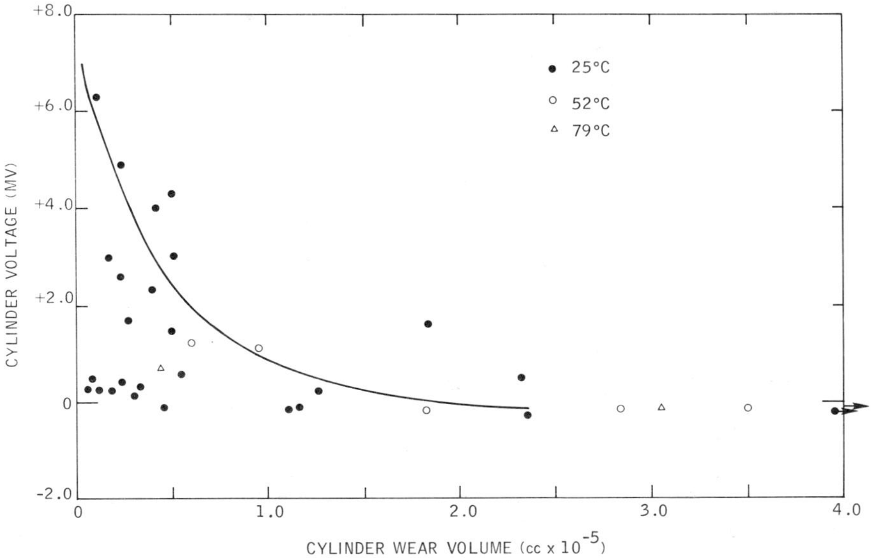

Fig. 7.—The Relationship Between Cylinder Wear and Self-Generated Voltages for the Steel System Under Dry Air Blanketing (Test Conditions: Ball-on-Cylinder Device, 240 rpm, 32 minutes, Loads as Noted in Tables 4 and 6.

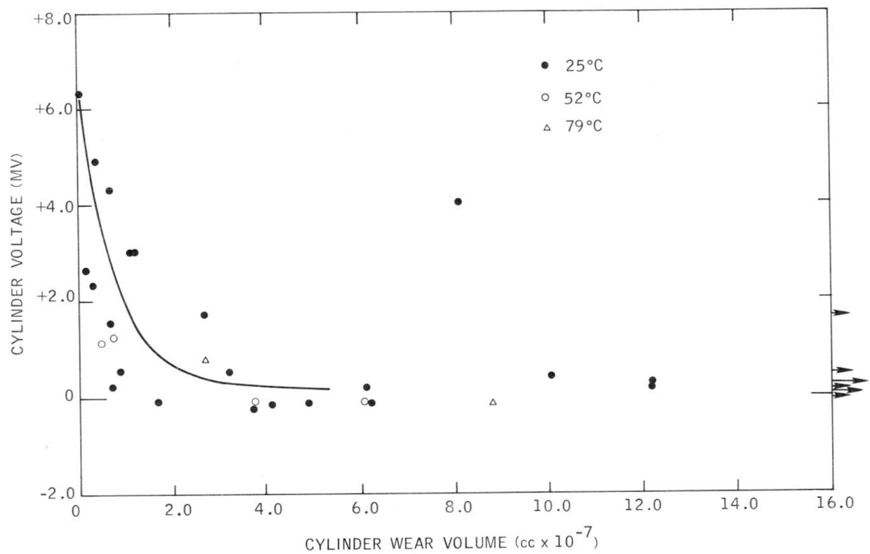

Fig. 8.—The Relationship Between Ball Wear and Self-Generated Voltages for the Steel System Under Dry Air Blanketing (Test Conditions: Ball-on-Cylinder Device, 240 rpm, 32 minutes, Loads as noted in Tables 4 and 6).

TABLE V.—MEASURED WEAR AND SELF-GENERATED VOLTAGES UNDER WET AIR BLANKETING FOR AISI 52100 STEEL BALLS SLIDING ON AISI 52100 STEEL CYLINDERS[1]

Lubricant	Load (g)	Wear Volume Ball (cc)	Wear Volume Cylinder (cc)	Surface Fatigue Demerit Rating	Self Generated Voltage (mv)
HVWO	200	0.48×10^{-7}	0.08×10^{-5}	1	-0.84
Hexadecane	250	24.46×10^{-7}	0.04×10^{-5}	1+	+0.09
HVWO	350	0.92×10^{-7}	0.12×10^{-5}	1	-1.50
LVWO	500	10.70×10^{-7}	0.33×10^{-5}	1	+0.20
Hexadecane	500	44.28×10^{-7}	1.06×10^{-5}	3	+0.09
HVWO	500	1.16×10^{-7}	0.24×10^{-5}	1	-1.80
1-Methylnaphthalene	500	6.10×10^{-7}	0.37×10^{-5}	1	+2.9
HVLRO	500	0.63×10^{-7}	0.13×10^{-5}	1	+2.8
Di-2-Ethylhexyl Sebacate	500	0.26×10^{-7}	0.15×10^{-5}	+	+0.69
HVWO	700	1.45×10^{-7}	0.22×10^{-5}	1	-2.19
1-Chloronaphthalene	1000	83.67×10^{-7}	0.59×10^{-5}	1	+0.6
LVWO	1000	24.46×10^{-7}	0.95×10^{-5}	3	+0.15
20/80 Toluene/LVWO	1000	44.28×10^{-7}	0.32×10^{-5}	3	+0.0
20/80 Mesitylene/LVWO	1000	44.28×10^{-7}	0.79×10^{-5}	2	+0.01
20/80 1-Methylnaphthalene/ LVWO	1000	15.58×10^{-7}	0.59×10^{-5}	1	+0.18
20/80 1-Chloronaphthalene/ LVWO	1000	241.87×10^{-7}	0.59×10^{-5}	3	+0.02
HVWO	1000	1.98×10^{-7}	0.33×10^{-5}	1+	-1.38
HVLRO	1000	0.63×10^{-7}	0.69×10^{-5}	1	+0.66
Di-2-Ethylhexyl Sebacate	1000	0.30×10^{-7}	0.24×10^{-5}	+	+0.57
HVWO	1500	1.61×10^{-7}	0.53×10^{-5}	1+	-0.90
LVWO	2000	323.32×10^{-7}	163.68×10^{-5}	3	+0.06
HVWO	2000	2.41×10^{-7}	1.02×10^{-5}	1+	+0.15
HVLRO	2000	1.80×10^{-7}	0.90×10^{-5}	1	+0.46
Di-2-Ethylhexyl Sebacate	2000	0.48×10^{-7}	0.25×10^{-5}	1	+0.24
Di-2-Ethylhexyl Adipate	2000	0.92×10^{-7}	0.71×10^{-5}	3	+0.06
Hercolube J	2000	0.81×10^{-7}	0.49×10^{-5}	3	+1.41
1-Chloronaphthalene	4000	160.29×10^{-7}	6.07×10^{-5}	3	0.0
HVWO	4000	6.10×10^{-7}	4.54×10^{-5}	1+	+0.30
HVLRO	4000	3.46×10^{-7}	1.97×10^{-5}	1+	+0.09
Di-2-Ethylhexyl Azelate	4000	1.45×10^{-7}	2.55×10^{-5}	1+	+0.24

Test Conditions: Ball-on-Cylinder Device, Loads as noted, 240 rpm, 25°C, 32 minutes.

ing the test metallurgy from aluminum-on-steel to steel-on-steel. Indeed, the SGV's measured for the above named systems are close to those noted for the steel-on-steel case, Table 4. Typical surface analysis of the worn balls from some of the other test systems are shown in Figures 10d-10f where it is noted that in contrast with Figures 11a-11c the dominant element observed is aluminum. Since the three special cases discussed above represent circumstances closer to steel-on-steel wear rather than aluminum-on-steel, it is readily understood why they do not correlate with the relationships presented

TABLE VI.—INFLUENCE OF TEMPERATURE UPON THE MEASURED WEAR AND SELF-GENERATED VOLTAGES UNDER DRY AIR ATMOSPHERES FOR AISI 52100 STEEL BALLS SLIDING ON AISI 52100 STEEL CYLINDERS(1)

Lubricant	Load (g)	Temperature (°C)	Wear Volume Ball (cc)	Wear Volume Cylinder (cc)	Surface Fatigue Demerit Rating	Self-Generated Voltage (mv) (2)
HVLRO	500	25	0.18×10^{-7}	0.24×10^{-5}	1	+2.6
HVLRO	500	52	0.63×10^{-7}	0.60×10^{-5}	2	+1.2
HVLRO	500	79	1.30×10^{-7}	0.13×10^{-5}	1+	erratic
HVWO	1000	25	0.71×10^{-7}	0.50×10^{-5}	1	+4.3
HVWO	1000	52	0.48×10^{-7}	0.95×10^{-5}	1	+1.11
HVWO	1000	79	2.64×10^{-7}	0.44×10^{-5}	2	+0.78
HVLRO	4000	25	4.10×10^{-7}	2.82×10^{-5}	3	-0.24
HVWO	4000	25	3.77×10^{-7}	2.35×10^{-5}	3	-0.30
HVLRO	4000	52	6.10×10^{-7}	3.50×10^{-5}	3	-0.18
HVWO	4000	52	3.77×10^{-7}	1.81×10^{-5}	3	-0.21
HVLRO	4000	79	8.75×10^{-7}	3.05×10^{-5}	3	-0.15
HVWO	4000	79	19.64×10^{-7}	10.23×10^{-5}	3	-0.19

(1) Test Conditions: Ball-on-Cylinder Device, 240 rpm, 32 minutes.

(2) The zero for the SGV's in the steel-on-steel system is temperature dependent and follows the relationship SGV (MV) = $0.209 - 6.96 \times 10^{-3}$ T (°C) over the temperature range 25°C - 160°C. SGV's have been temperature corrected.

TABLE VII.—MEASURED WEAR AND SELF-GENERATED VOLTAGES UNDER DRY AIR BLANKETING FOR ALUMINUM BALLS SLIDING ON STEEL CYLINDERS[1]

Lubricant	Load (g)	Wear Volume Ball (cc)	Wear Volume Cylinder (cc)	Surface Fatigue Demerit Rating	Self-Generated Voltage (mv)
HVWO	200	$0.63x10^{-7}$[2]	$<0.09x10^{-5}$[2]	1	+10.8
20/80 Toluene /LVWO	500	$1.16x10^{-7}$	$0.11x10^{-5}$	1	+5.8
HVLRO	500	$0.63x10^{-7}$	$0.11x10^{-5}$	1	+3.6
LVWO	1000	$2.41x10^{-7}$	$0.26x10^{-5}$	3	+0.45
20/80 Toluene/LVWO	1000	$2.64x10^{-7}$	$0.13x10^{-5}$	1+	+0.48
20/80 Mesitylene/LVWO	1000	$2.41x10^{-7}$	$0.29x10^{-5}$	3	+0.45
20/80 1-Methylnaphthalene/ LVWO	1000	$2.41x10^{-7}$	$0.17x10^{-5}$	1	+2.6
HVWO	1000	$1.45x10^{-7}$[2]	$<0.12x10^{-5}$[2]	1	+4.9
HVLRO	1000	$1.03x10^{-7}$	$0.13x10^{-5}$	1	+3.1
HVWO	2000	$2.90x10^{-7}$[2]	$0.18x10^{-5}$[2]	1	+3.8
HVLRO	2000	$3.77x10^{-7}$	$0.30x10^{-5}$	1	+1.9
20/80 1-Chloronaphthalene/ LVWO	4000	$15.58x10^{-7}$	$1.74x10^{-5}$	3	+0.24
HVLRO	4000	$4.83x10^{-7}$	$0.59x10^{-5}$	1	+0.7
HVWO	4000	$4.83x10^{-7}$	$0.46x10^{-5}$	1	+0.7

([1]) Test Conditions: Ball-on-Cylinder Device, 240 rpm, 25°C, 16 minute tests (see Note 2).
([2]) Measured Wear after 32 minutes of Testing, Self-Generated Voltages as read at 16 minutes.

in Figures 9 and 10. Disregarding these three values, the smooth curve relationships presented in Figures 9 and 10 are fairly good.

For the steel-on-steel system under wet air blanketing, Table 5, it is apparent that little correlation exists between wear and the SGV's. There are two difficulties associated with this wet air system. The first is that the SGV's at low wear for the HVWO are much lower in magnitude than (and opposite in polarity from) the other systems investigated. The second difficulty stems from the fact that several of the systems investigated undergo

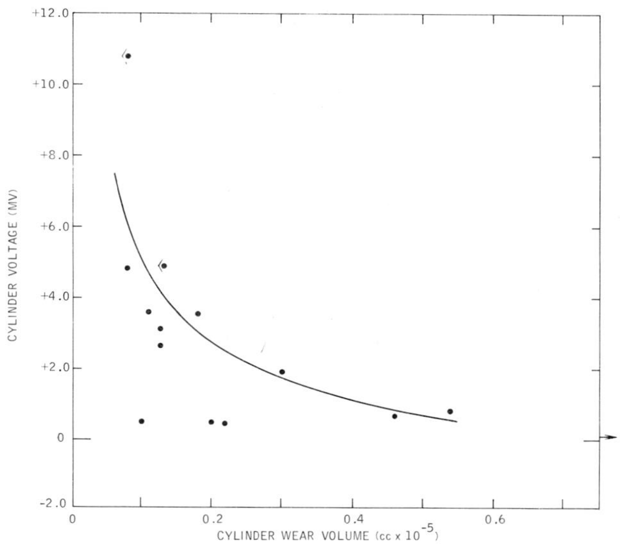

Fig. 9.—The Relationship Between Cylinder Wear and Self-Generated Voltages for an Aluminum-on-Steel System Under Dry Air Blanketing (Test Conditions: Ball-on-Cylinder Device, 240 rpm, 25°C, 16 minute tests, See Table 7 for Applied Loads and Details Concerning Wear Measurements).

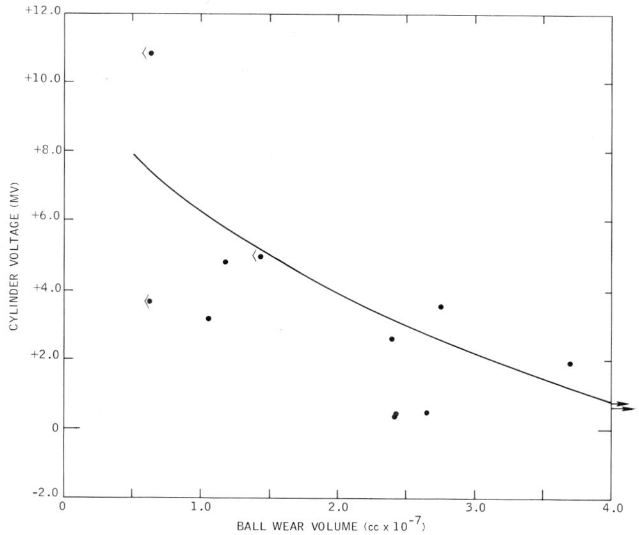

Fig. 10.—The Relationship Between Ball Wear and Self-Generated Voltages for an Aluminum-on-Steel System Under Dry Air Blanketing (Test Conditions: Ball-on-Cylinder Device, 240 rpm, 25°C, 16 minute Tests, See Table 7 for Applied Loads and Details Concerning Wear Measurements).

TABLE VIII.—MEASURED WEAR AND SELF-GENERATED VOLTAGES UNDER BOTH INERTED AND OXYGENATED ATMOSPHERES FOR THE HVWO USING AISI 52100 STEEL BALLS SLIDING ON AISI 52100 STEEL CYLINDERS[1]

			Argon			Air		
Load	Metallurgy (Ball/Cylinder)	Time (min)	Ball Wear Volume (cc)	Cylinder Wear Volume (cc)	Self-Generated Voltage (mv)	Ball Wear Volume (cc)	Cylinder Wear Volume (cc)	Self-Generated Voltage (mv)
				DRY CONDITION				
200	Aluminum/Steel	32	--	--	--	$0.63x10^{-7}$	$0.13x10^{-5}$	+14.1
250	Aluminum/Steel	32	$1.98x10^{-7}$	$0.23x10^{-5}$	-0.19	--	--	--
500	Aluminum/Steel	32	$7.07x10^{-7}$	--	-0.08	$1.98x10^{-7}$	$0.22x10^{-5}$	+ 2.5
500	Steel/Steel	36	$0.22x10^{-7}$	$0.30x10^{-5}$	+0.06	$0.40x10^{-7}$	$0.24x10^{-5}$	+ 4.9
500	Steel/Steel	360	$0.22x10^{-7}$	$0.21x10^{-5}$	+0.18	--	--	--
2000	Steel/Steel	36	$0.81x10^{-7}$	$0.66x10^{-5}$	0.00	$0.92x10^{-7}$	$0.54x10^{-5}$	+ 0.55
				WET CONDITION				
200	Steel/Steel	32	--	--	--	$0.48x10^{-7}$	$0.08x10^{-5}$	- 0.84
350	Steel/Steel	32	--	--	--	$0.92x10^{-7}$	$0.12x10^{-7}$	- 1.5
500	Steel/Steel	32	$0.30x10^{-7}$	$0.20x10^{-5}$	+0.05	$1.16x10^{-7}$	$0.24x10^{-5}$	- 1.8
2000	Steel/Steel	32	$0.63x10^{-7}$	$0.91x10^{-5}$	+0.03	$2.41x10^{-7}$	$1.02x10^{-7}$	+ 0.15

Test Conditions: Ball-on-Cylinder Device, 240 rpm, 25°C.

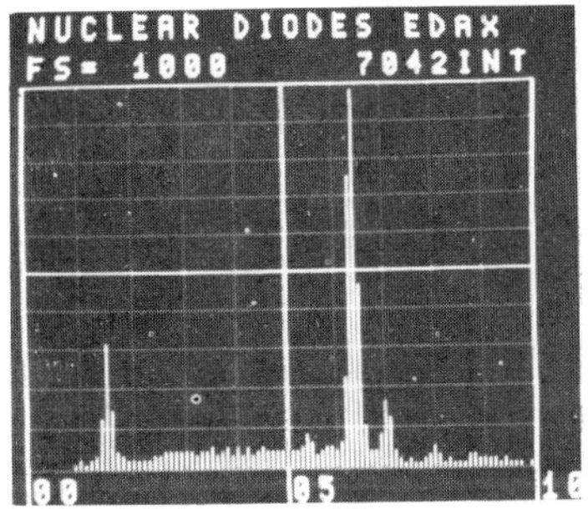

11a. TOLUENE

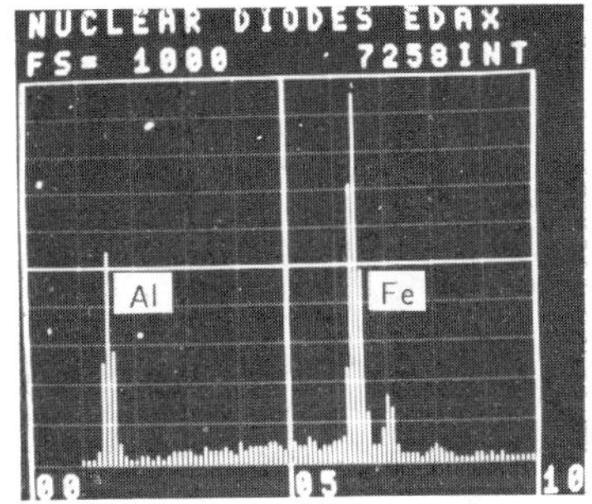

11b. MESITYLENE

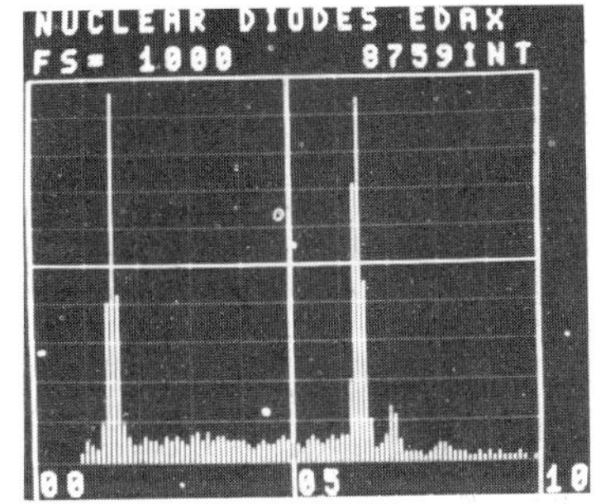

11c. LVWO

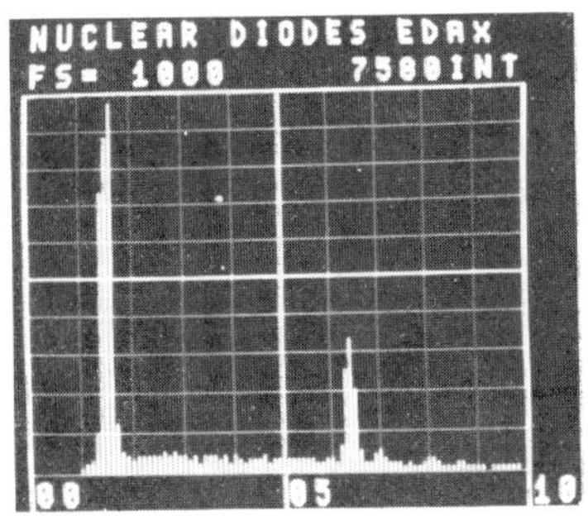

11d. 1-METHYLNAPHTHALENE

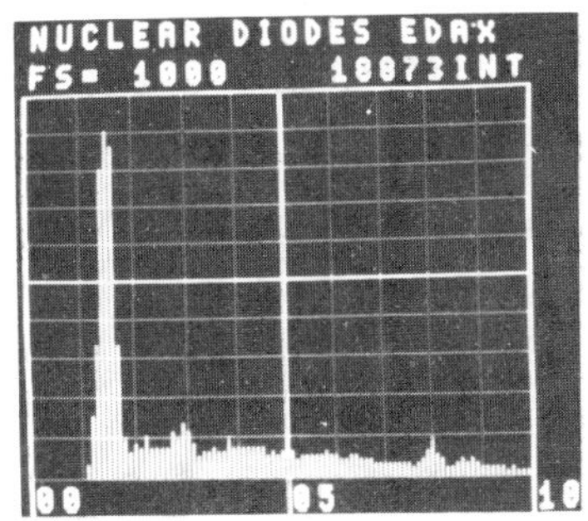

11e. HVWO

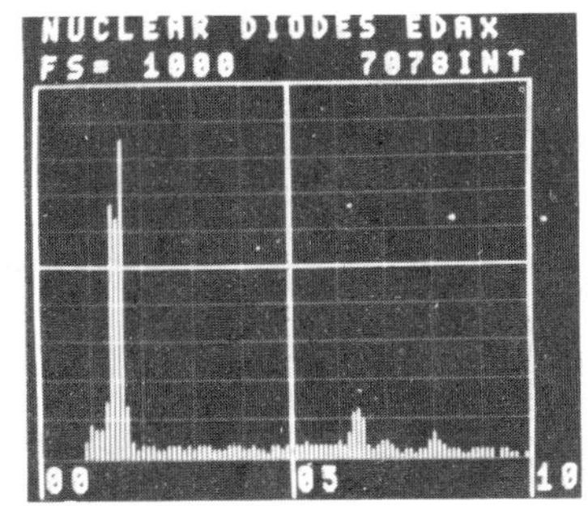

11f. HVLRO

Fig. 11.—Energy Dispersive Analysis by X-Rays (EDAX) Analysis of Wear Scars of Aluminum Balls for Tests with Various Base Stocks for Aluminum-on-Steel Tests Under Dry Air Blanketing (Test Conditions: Ball-on-Cylinder Device, 240 rpm, 25°C, 16 minute Tests, See Table 7 for Applied Loads and Details Concerning Wear Measurements).

severe corrosive wear, and hence, for example, low cylinder wear may be associated with low SGV's. It is, therefore, obvious that because of the commingling of competitive wear mechanisms under the wet air, steel-on-steel conditions, it would be difficult to obtain a general, unambiguous relationship between wear and SGV's.

Based upon the evidence above, it appears that a relationship between wear and Self-Generated Voltages exists under dry air blanketing conditions. Under dry air conditions for the steel-on-steel system, it was pointed out that there are instances in which low cylinder wear is associated with low SGV's. If these data are excluded from the smooth curves presented in Figures 7 and 8, the relationship which is suggested is seen even more clearly, Figures 12 and 13. It should be noted, however, that only a limited portion of the data obtained under wet air blanketing may be fit to this relationship.

Relationships Between SGV and Wear Mechanism

Mechanism of voltage generation under boundary lubricated conditions

It is also of interest to determine if relationships between wear, wear mechanism and SGV's exist. It is of particular interest to determine whether surface fatigue wear, a process in which microcracks form either at, or near, the surface and thereafter propagate due to the repeated stress cycling, leads to the development of potential differences between the stationary and cyclically loaded members of a lubricated system. This would be consistent with, and therefore tend to confirm that the electron transfer process, depicted schematically below is an important controlling primary electrochemical process for surface fatigue wear.

$$M \rightarrow M^{+\chi} + \chi\, e^{-} \qquad (2)$$

$$Lubricant + e^{-} \rightarrow (Lubricant\cdot\cdot e)^{-} \qquad (3)$$

In the above scheme, M is the metal surface, and the reaction is written as liberating "χ" electrons. It is suggested that these "χ" electrons are removed from the free electron-cloud in the metal. They need not be removed totally from the surface. They need only be placed into a relatively bound state, thus essentially demetallizing the surface. In this scheme, therefore, metal ions need not be formed and, therefore, need not enter into solution. As a result of such transfer of electrons from an unbound metallic state to a surface bound state or at times even to a free state, within an adsorbed layer, a surface may appear to act as an electron source and, as a consequence, would assume a negative polarity relative to a non-reacting surface.

Based upon the above scheme, the cylinder which is under cyclic stress, and is undergoing surface fatigue wear is expected to assume a potential which is negative relative to the ball. Indeed, it is found that the cylinder potential becomes negative relative to the ball with increased wear and that at the higher loads studied under dry blanketing, the cylinder potentials are often negative. Furthermore, it is found that the SFDR's generally increase as the cylinder potential becomes relatively more negative. This is consistent with the assumed relationship between surface fatigue wear and SGV's generated as a consequence of the electron transfer process.

It is noted, however, that increased wear does not always lead to rela-

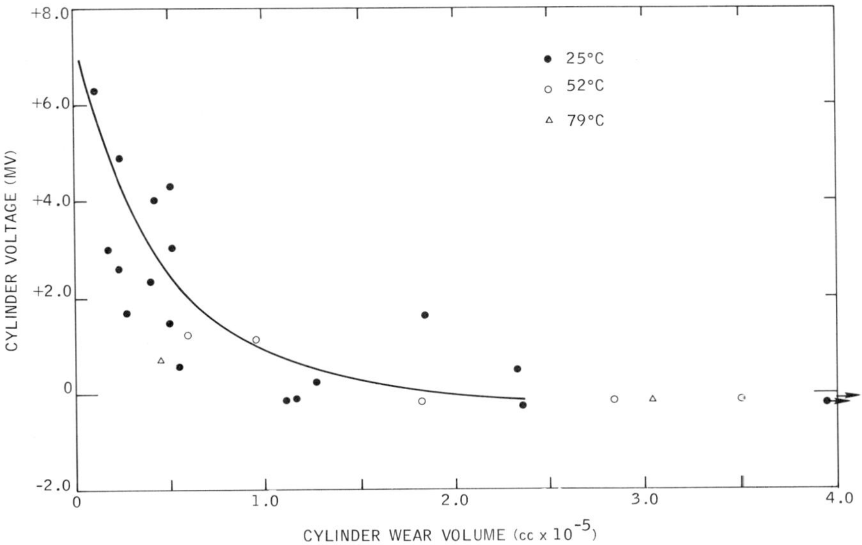

Fig. 12.—The Relationship Between Cylinder Wear and Self-Generated Voltages for the Steel System Under Dry Air Blanketing, Excluding Fluids Exhibiting High Ball Wear.

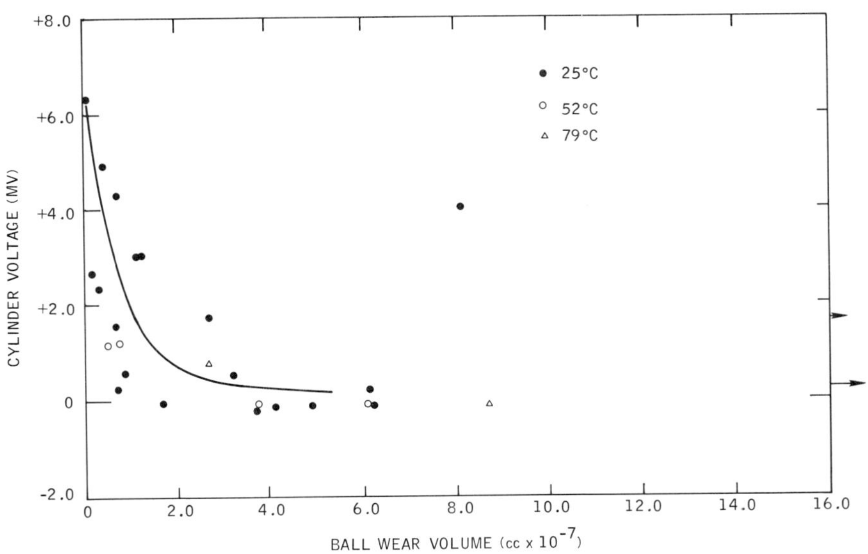

Fig. 13.—The Relationship Between Ball Wear and Self-Generated Voltages of the Steel System Under Dry Air Blanketing, Excluding Fluids Exhibiting High Ball Wear.

tively more negative potentials. This suggests that the scheme by which SGV's are developed under boundary lubricated conditions may be more complex than described above.

The idea that the single reaction scheme above is not sufficient to explain how SGV's may be developed under boundary lubricated conditions is furthermore supported by the fact that under similar operating conditions the SGV's under wet air blanketing are in general of different polarity from those observed under dry air blanketing. The fact that the polarities of the cylinder are negative at light loads under wet air blanketing, a condition assumed to be associated with enhanced oxidative or corrosive wear, and then become relatively positive at higher loads indicates that an oxidative electrochemical process should also be included in the reaction scheme. Such an oxidative electrochemical process leading to negative potentials of the reacting surface would probably be controlled by the electrochemical processes outlined below, reactions 4 and 5.

$$M \rightarrow M^{a+} + ae^- \qquad (4)$$

$$2ae^- + \frac{a}{2}\, O_2 + aH_2O \rightarrow 2aOH^- \qquad (5)$$

Electrons are liberated via reaction 4, the metal oxidation step. The surface undergoing this process assumes a negative polarity since it behaves as an electron sink. The electrons may be consumed via a reduction step such as reaction 5. Under conditions in which the metal ions are not free to migrate away from the surface, an oxide/hydroxide film would grow on the metal surface. The net consequence of this oxidation-reduction process would be to cause the reacting test member to assume a negative polarity relative to the nonreacting surface.

There are, therefore, at least two primary electrochemical processes, the electron transfer and the oxidation processes, whereby SGV's may be developed under boundary lubricated conditions. These as indicated above, are associated with two competitive fundamental wear processes, the surface fatigue wear process and the corrosive wear process, respectively. Therefore, in a wearing system both primary electrochemical processes may be operative simultaneously. Furthermore, it might be expected that the primary electrochemical process responsible for the cylinder wear and, therefore, the cylinder potential would be different from that encountered on the ball. The potential difference which is measured between the ball and the cylinder is, therefore, the resultant of the single cell potentials for each member. Since systems, in general, do not exhibit a single unique wear mode, it is readily understood that the correlation between wear and SGV's, the latter a function of the combined single cell processes, would exhibit some scatter. Furthermore, a clear relationship between SGV's and any single wear mechanism would be difficult to establish.

Oxygen effects upon SGV's

The fact that Argon tests lead to near zero values for the SGV's may also be explained in the following way. It may be assumed that a thin insulating oxide film is necessary in order to measure the SGV's. With such a film as might be formed under oxygenated atmospheres, a pseudostationary level for the SGV may be measured. However, under inerted atmospheres, the oxide films are sufficiently thin so that voltage discharge between the test members occurs frequently, and a measurable pseudoequilibrium value for the SGV is not established. Since SGV's cannot be measured under these con-

TABLE IX.—SELF-GENERATED VOLTAGES FOR SYSTEMS TESTED UNDER SIMILAR OPERATING PARAMETERS EXHIBITING SIMILAR WEAR

Lubricant	Load (g)	Wear Volume Ball (cc)	Wear Volume Cylinder (cc)	Surface Fatigue Demerit Rating	Self-Generated Voltage (mv)	Table No.
Metallurgy: 52100 Steel-on-52100 Steel						
1-Methylnaphthalene	500	1.16×10^{-7}	0.50×10^{-5}	1	+3.0	4
HVWO	500	0.41×10^{-7}	0.24×10^{-5}	1	+4.9	4
HVLRO	500	0.18×10^{-7}	0.24×10^{-5}	1	+2.6	4
20/80 1-Methylnaphthalene/LVWO	1000	1.16×10^{-7}	0.17×10^{-5}	2	+3.0	4
HVWO	1000	0.71×10^{-7}	0.50×10^{-5}	1	+4.3	4
HVLRO	1000	0.30×10^{-7}	0.40×10^{-5}	1	+2.3	4
HVWO	2000	0.92×10^{-7}	0.54×10^{-5}	1+	+0.55	4
Di-2-Ethlyhexyl Adipate	2000	0.71×10^{-7}	1.25×10^{-5}	1	+0.20	4
Hercolube J	2000	0.63×10^{-7}	0.50×10^{-5}	3	+1.50	4
HVLRO	2000	1.61×10^{-7}	1.15×10^{-5}	3	-0.06	4
HVWO	4000	3.77×10^{-7}	2.35×10^{-5}	3	-0.30	4
HVLRO	4000	4.10×10^{-7}	2.82×10^{-5}	3	-0.24	4
Metallurgy: 52100 Steel-on-Aluminum						
20/80 1-Methylnaphthalene/LVWO	1000	2.41×10^{-7}	0.17×10^{-5}	1	+2.6	7
HVWO	1000	1.45×10^{-7}	0.12×10^{-5}	1	+4.9	7
HVLRO	1000	1.03×10^{-7}	0.13×10^{-5}	1	+3.1	7
HVWO	2000	2.90×10^{-7}	0.18×10^{-5}	1	+3.8	7
HVLRO	2000	3.77×10^{-7}	0.30×10^{-5}	1	+1.9	7
HVWO	4000	4.83×10^{-7}	0.59×10^{-5}	1	+0.7	7
HVLRO	4000	4.83×10^{-7}	0.46×10^{-5}	1	+0.7	7

ditions, no relationship to wear can be established, even though electron transfer processes may still be important.

Several reports have appeared describing the relationship between exoelectron emission and wear in general and fatigue in particular.[(10-12)] One striking similarity between the two phenomena of exoelectron emission and of SGV's is that, in general, both require the presence of air (or oxygen) for their observation. Electron activation during the oxidation step is used to rationalize the need for oxygen in order to observe exoelectron emission. Perhaps, therefore, an alternate explanation for the absence of SGV's in inert atmospheres is that oxygen is necessary in order to activate and free the electrons via chemical reaction.

Effects of chemical parameters upon SGV's

Another set of data consistent with the relationship between wear mechanism and SGV's is obtained by determining the effects of variations in the chemical nature of the lubricant while maintaining the test system under otherwise similar conditions. A selection of data, taken from those presented in Tables 4 and 6, for several different base stocks which exhibit similar (although not identical) wear under identical operating conditions, Table 9, indicates that molecules which are better electron acceptors exhibit relatively more negative cylinder potentials.

The HVWO is the poorest electron acceptor and it, in general, exhibits the highest relative potentials. The polycyclic aromatics (present for example in the HVLRO) which behave as strong electron acceptors appear to lead to the lowest relative potentials. Of the esters tested, Di-2-Ethylhexyl Adipate, leads to potentials intermediate between the HVWO and the aromatics while the ester of pentaerythritol alcohol, Hercolube J, leads to more positive potentials. Of the two tested, the Hercolube J is expected to be the poorer electron acceptor at a freshly abraded surface because it is more sterically hindered and hence cannot adsorb upon the metal surface as strongly as the other ester. Possibly unexpected is that the Hercolube J leads to more positive potentials than the HVWO, perhaps reflecting some additional chemical influences at the surface. Thus, it appears that the relative magnitude (and polarity) of the SGV's are associated with an electrochemical process.

ACKNOWLEDGEMENTS

This work was sponsored by a grant from the Office of Naval Research under Contract Number N00014-75-C-1080. The author wishes to thank Dr. R. S. Miller and Mr. G. L. Harting for many valuable and interesting discussions and suggestions.

REFERENCES

1. Suh, N.P., *Wear,* Vol. 25, 1973, p. 111.
2. Suh, N.P., Jahanmir, S., Abrahamson, E.P., II and Turner, A.P.L., *Journal of Lubrication Technology,* Vol. 96, 1974, p. 631.
3. Jahanmir, S., Suh, N.P. and Abrahamson, E.P., II, *Wear,* Vol. 23, 1974, p. 235.
4. Fleming, J.R. and Suh, N.P., *Wear,* Vol. 44, 1977, p. 39.
5. Goldblatt, I.L., *Industrial & Engineering Chemistry Product Research and Development,* Vol. 10, 1971, p. 270.
6. Goldblatt, I.L., *American Society of Lubrication Engineers Transactions,* Vol. 16, 1973, p. 150.
7. Tao, F.F. and Appeldoorn, J.K., *American Society of Lubrication Engineers Transactions,* Vol. 11, 1968, p. 345.
8. Goldblatt, I.L., in "The Relationship Between Engine Oil Viscosity and Engine Performance," SAE Special Publication, SP 416, 1977, p. 47.
9. Klinkenberg, A. and Van der Minne, J.L., "Electrostatics in the Petroleum Industry," Elsevier, Amsterdam, 1958, p. 57.
10. Baxter, W., *Journal of Applied Physics,* Vol. 44, 1973, p. 608.
11. March, P.A. and Rabinowicz, E., *American Society of Lubrication Engineers Transactions,* Vol. 20, 1977, p. 315.
12. Ferrante, J., *American Society of Lubrication Engineers Transactions,* Vol. 20, 1977, p. 328.

DISCUSSION

P. ASHLEY, ALCAN Research Center, Canada: When you were running the aluminum on steel or vice-versa, did you look at the steel surface? It is very seldom you can run aluminum on steel and not transfer the aluminum to the steel.

GOLDBLATT: I was running an aluminum ball on a steel cylinder. I looked at the steel cylinder and I did not get the transfer of aluminum to the steel under those conditions.

N. SAKA, MIT: Did you measure the friction coefficient simultaneously?

GOLDBLATT: Yes, we do that routinely.

SAKA: Was there a large difference in friction coefficient?

GOLDBLATT: No. It was basically very similar. It was about 0.13 ± 0.02, except under conditions of extreme wear when the friction coefficients are substantially higher.

QUESTIONER: Can you apply the potential in the opposite direction and change the wear rate?

GOLDBLATT: Indeed you can. This sort of thing has been demonstrated under conditions in which there was a conducting fluid such as water. We have been able to do it with a lubricant also.

R. A. DASKIVICH, G. M. Research Laboratories: Every series of experiments has base line experiments. What is your reference system?

GOLDBLATT: As I pointed out, the high viscosity white oil under wet air conditions performs somewhat differently from all the other fluids. It is very sensitive to any impurities and gross changes in metallurgy and so on. I used that system as a reference system to make sure that the test conditions were reasonably well produced.

XIV. ELASTOHYDRODYNAMIC LUBRICATION

FUNDAMENTALS OF ELASTOHYDRODYNAMIC CONTACT PHENOMENA

H. S. Cheng

ABSTRACT

Elastohydrodynamic phenomena in lubricated concentrated contacts are reviewed in three parts. The first part covers major analytical and experimental developments in full-film EHD were the surfaces are completely separated by a lubricant film. Emphasis is given to the mechanism of film generation, film prediction, and other secondary effects on film formability. Correlations are presented between calculated film thickness and that measured by capacitance and optical methods. Principal features in film and pressure distributions in line and point contacts are described, and recent experimental confirmations on film and pressure profiles are discussed. Attention is drawn to effects of surface and film temperature rise on film thickness, pressure, and friction in the conjunction. Detailed discussions are given to the important developments leading to successful predictions of traction behavior in sliding EHD contacts. The second part reviews developments on partial-EHD where the load is shared between fluid pressure and asperity contacts. Attention is given to surface characterization and methods of determination of the average hydrodynamic pressure and asperity contact pressure in the conjunction. Discussions are also given to proposed mechanisms of lubrication breakdown in partial-EHD. The last part gives a brief review on micro-elastohydrodynamic lubrication which deals with the local film thickness and pressure fluctuations around an asperity or furrow.

1. INTRODUCTION

In a relatively short period of a quarter of a century, elastohydrodynamic lubrication (EHD) has grown from a relatively unknown and obscure subject to a well-recognized regime of lubrication in the field of tribology. In this past period, some have regarded EHD as the most important development in tribology since Reynolds work on hydrodynamic lubrication. Yet, others have considered that the developments in EHD have grown out of proportion, and they appear to serve little purpose other than to enable some academecians to perpetuate their own kind. Regardless where EHD would eventually settle in the history of tribology, present trends indicate that EHD theories already have been used not only as a research tool for studying lubrication performance but also as a tool to establish new design criteria in lubricated contacts. As the field expands, there is a continuing need to review the understanding of major variables in EHD lubrication, which include detailed film, pressure, temperature, and traction distribution in lubricated Hertzian contacts. There are at least two major comprehensive reviews contributed by Dowson[1] and by McGrew, et al.[2]

The objectives of this presentation are to update the recent findings on EHD fundamental phenomena, and to discuss the possible relations between EHD and the failure processes in lubricated concentrated contacts.

In hydrodynamic lubrication, the deformation of bearing surfaces is normally negligible in comparison to the film thickness, and one only needs to solve the Reynolds equation for the pressure distribution to determine the lubrication performance. In lubrication problems involving either low modulus bearing materials or counter formal contacts, the surface deformation can approach and often exceed the lubricant film thickness. Under these conditions, the surface deformation must be considered as an unknown quantity which must be solved together simultaneously with the Reynolds equation. Examples of lubrication problems involving low modulus materials include compliant journal bearings, elastomeric seal lubrication, tire hydroplaning problem, foil bearings, lubrication in synovial joints, etc. While the examples of compliant bearings bring out many interesting aspects of the EHD problem, they are not the primary reason for the rapid development in this field. The main factor behind the surge in EHD development is its application to lubrication in concentrated contacts which, in many cases, seem to limit the performance and life of mechanical equipment. For this reason, papers included in this review will be confined mainly to EHD in concentrated contacts.

Work on lubrication in EHD contacts can be further classified into two main categories, namely, full-film EHD, and partial-film EHD. In full-film EHD, the average film thickness everywhere in the concentrated contact is assumed to be well above three times the composite r.m.s. surface roughness of both contacting surfaces; as depicted in Figure 1a. Theories of full-film EHD are based on the assumption of smooth surfaces, and are substantiated by experiments with highly polished surfaces separated by a thick lubricant film. Developments on full-film EHD are reviewed in detail in Section 3. Unlike hydrodynamic sliding bearings for which full film lubrication is customarily considered as a necessary condition for smooth performance, EHD contacts can operate quite satisfactorily even if the film thickness falls below three times the composite surface roughness. In fact, it is believed that the majority of EHD contacts do operate in the regime of partial-film EHD in which a thick lubricant film is interrupted occasionally by the presence of asperities. The condition of partial-film EHD is depicted in Figure 1b. Because of the difficulty in incorporating the surface roughness effects, progress on partial-film EHD has been slow in spite of early recognition of its importance by Christensen[3] and Tallian et al.[4] Recent advancements in surface roughness characterization by Williamson,[5] Nayak,[6] and Whitehouse and Archard,[7] have triggered further progress in this area. A detailed coverage of partial-film EHD is reviewed in Section 4.

Full-film EHD studies enable one to determine conditions for ideal lubrication performance, and partial-film EHD studies provide information on surface roughness effects on average film thickness and pressure. However, they do not fully provide the basis for understanding the mechanisms of lubrication breakdown where there are severe asperity interactions. For this understanding, it might be more fruitful to study the local fluctuations of pressure and film thickness around a single asperity as it passes through the conjunction. Such problems are known as micro-elastohydrodyna-

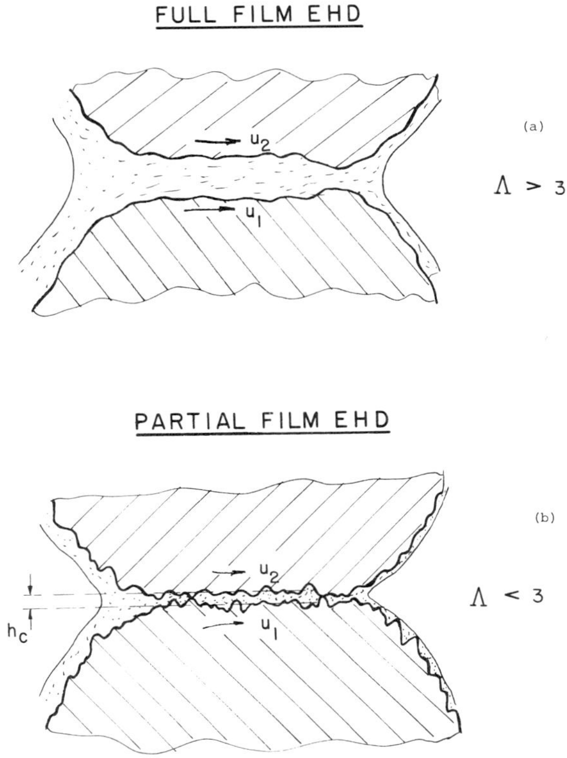

$$\Lambda = \frac{h_c}{(\sigma_1^2 + \sigma_2^2)^{1/2}} = \frac{h_c}{\sigma}$$

Fig. 1.—Full-Film and Partial-Film EHD Contacts.

mic lubrication. Developments in this area are recent, and they are discussed in Section 5.

2. EHD GEOMETRY

Considering a general case of a lubricated Hertzian contact between two bodies as shown in Figure 2, the geometry at the point of contact is characterized by the principal radii, R_{x1}, R_{y1} for body 1 and R_{x2}, R_{y2} for body 2. In general, the principal planes containing R_{x1} and R_{x2} may not coincide; however, for most EHD contacts, such as roller or ball bearings and gears, the principal radii R_{x1} and R_{x2} do lie in the same plane. These surfaces can be classified as convex, concave, or saddled, depending upon whether R_x and R_y are both positive, both negative, or mixed. Thus, contacts between two convex surfaces such as roller-inner race contacts or gear teeth contacts for external spur gears are counter-formal. Contacts between one

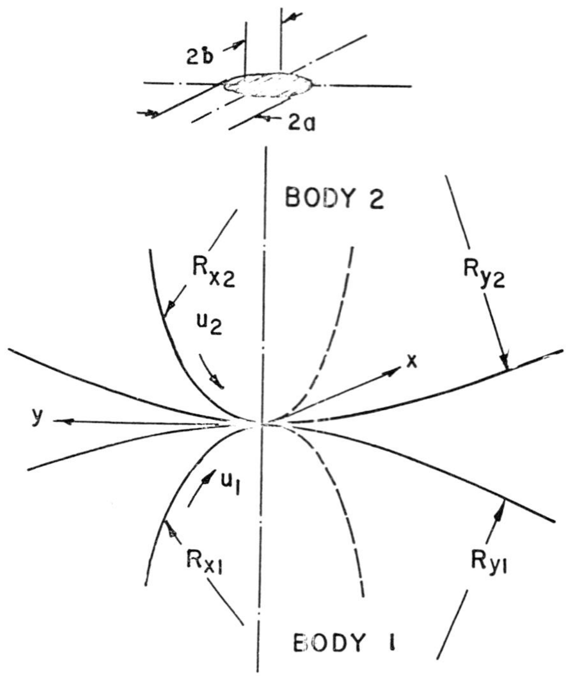

Fig. 2.—Geometry of an EHD Contact.

convex and one concave surface such as roller-outer race contacts or gear teeth contacts for internal spur gears are conformal. For contacts between one convex and one saddled surface such as deep-groove ball-inner race contacts, the result is conformal in one direction and counterformal in the perpendicular direction.

In these heavily loaded contacts, the normal load is distributed over an elliptical area approximately conforming to the Hertzian ellipse of a dry contact as shown at the top of Figure 2. In most cases, the direction of rolling is along the minor axis of the ellipse with surface velocities u_1 and u_2 for bodies 1 and 2, respectively. The degree of sliding is measured by a quantity known as the slide to roll ratio denoted by $S = 2(u_1 - u_2)/(u_1 + u_2)$ with $S = 0$ corresponding to pure rolling and $S = 2$ to simple sliding.

3. FULL-FILM EHD

3.1 Problem Description

The behavior of an EHD contact, such as that shown in Figure 1a, operating in the thick-film regime can be characterized by the distributions of film thickness, lubricant pressure, surface stress, surface temperature, and film temperature, within the elliptical conjunction. The major task in full-film EHD is the accurate prediction of these quantities with a given set of input data which include:

u_1, u_2	- surface speeds
W	- contact load per unit width
E_1, E_2	- Young's moduli
ν_1, ν_2	- Poisson's ratios
ρ_1, ρ_2	- density of the solids
c_1, c_2	- specific heat of the solids
k_1, k_2	- thermal conductivity of the solids
μ_o	- inlet viscosity of the lubricant
α, β, γ, etc.	- lubricant parameters relating viscosity as a function of pressure and temperature
ρ_f, c_f, k_f	- density, specific heat, and thermal conductivity of the lubricant
$R_{x1}, R_{x2}, R_{y1}, R_{y2}$	- principal radii at the contact.

After two decades intensive efforts, the EHD phenomenon in the full-film regime is now reasonably well understood. Even though in reality most concentrated contacts are elliptical or point contacts, their basic phenomena do not differ greatly from those developed for cylindrical or line contacts. In fact, most of the significant results were obtained with two-dimensional analysis for line contacts. Reviews in this area will include earlier work describing the mechanism of film generation and other major results on the behavior of film shape, pressure, temperature, and traction.

3.2 Film Generation

3.2.1 Rigid roller lubrication

Consider the case of two rigid rollers shown in Figure 3. The lubricant transported at any point along the film is

$$\left(\frac{u_1+u_2}{2}\right) h - \frac{h^3}{12\mu}\frac{dp}{dx} \tag{1}$$

Let the lubricant flow rate be $1/2(u_1+u_2)h^*$, where h^* is the gap where dp/dx is zero, then the continuity of flow gives

$$\frac{dp}{dx} = 12\mu\left(\frac{u_1+u_2}{2}\right)\frac{(h - h^*)}{h^3} \tag{2}$$

It is seen that on the left of h^*, dp/dx is positive resulting in a hydrodynamic pumping action. On the right of h^*, dp/dx is negative resulting in a sealing action. Thus, given h^*, one can readily solve Equation (2) for the pressure distribution hence the load and film thickness relationship. This was accomplished by Martin[8] for an isovicous lubricant. The result

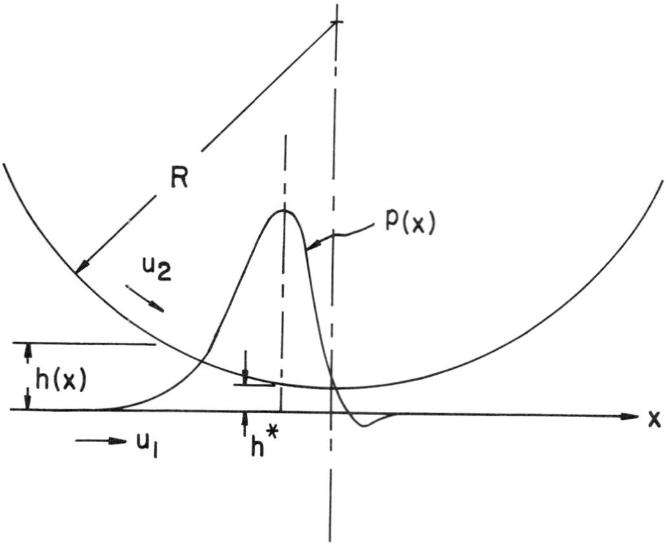

Fig. 3.—Lubrication Between Rigid Rollers.

was that the film thickness was far too small for roller dimensions comparable to gears and rolling element bearings.

After Martin's work, many believed that the increase of viscosity with pressure can give higher load capacity hence greater film thickness. Including exponential pressure-viscosity relation

$$\mu = \mu_o e^{\alpha p} \tag{3}$$

in Equation (2), one obtains

$$\frac{1}{e^{\alpha p}} \frac{dp}{dx} = 6\mu_o \left(u_1+u_2\right) \frac{h - h^*}{h^3} \tag{4}$$

By introducing a reduced pressure

$$q = \frac{1 - e^{-\alpha p}}{\alpha}$$

Equation (4) can be simplified to

$$\frac{dq}{dx} = 6\mu_o \left(u_1+u_2\right) \frac{h - h^*}{h^3} \tag{5}$$

Equation (5) can be readily solved for q which leads to p and the load capacity. This was accomplished by Gatcombe,[9] Block,[10] and McEwen.[11] The results showed some improvement in film thickness, but it is far from being enough to suggest a full-film hypothesis in roller lubrication.

3.2.2 Line contact film thickness analyses

It was Grubin(12) who assumed the inlet film shape h(x) conforming to a deformed Hertzian shape, and provided the first convincing evidence of full-film EHD lubrication.

Essentially, Grubin succeeded in solving Equation (5) for a Hertzian shape shown in Figure 4. He determined h_o for the reduced lubricant pres-

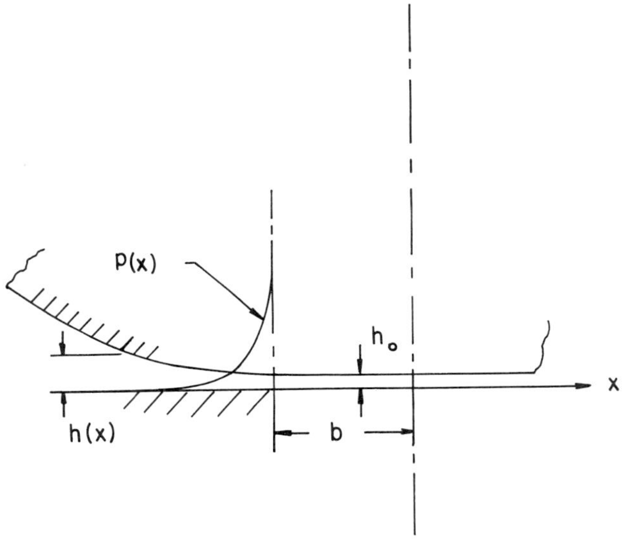

Fig. 4.—Inlet Geometry of Grubin-Type EHD Line Contact.

sure q to reach $1/\alpha$ (or $p \to \infty$) at $x \to -b$, that is at the entrance of the Hertzian conjunction. Once $e^{-\alpha p}$ approaches zero, μ becomes very large in the conjunction. A very minute change in film thickness in the conjunction would cause enormous change in pressure. Therefore, for heavily loaded contacts, the gap in the conjunction is practically uniform. Equation (5) was solved by Grubin using the Hertzian inlet film profile.

$$h(x) = h_o + \frac{4}{\pi}\frac{W}{E_D}\left(|\bar{x}|\sqrt{\bar{x}^2-1} - \ln\left(|\bar{x}| + \sqrt{\bar{x}^2-1}\right)\right) \quad (6)$$

where $1/E_D = 1/2\ (1-\nu_1^2)/E_1 + (1-\nu_2^2)/E_2$ and $\bar{x} = x/b$. Empirical fitting yields the following equation for the film thickness.

$$\frac{h_o}{R} = 1.95\frac{(GU)^{8/11}}{(\bar{W})^{1/11}} \quad (7)$$

where $G = \alpha E_D$, $U = \frac{\mu_o u}{E_D R}$, $u = \frac{1}{2}(u_1+u_2)$,

$\overline{W} = \frac{W}{E_D R}$, and $R = R_{x1}R_{x2}/(R_{x1}+R_{x2})$

The accuracy of Grubin's formula for predicting h_0 holds well for a reasonably wide range of conditions. It becomes less accurate when one of the following conditions occurs: 1) G < (lubricant with a small pressure-viscosity dependence or material of low modulus), 2) U is large enough to cause a significant loss of viscosity due to inlet shear heating, 3) Insufficient lubricant supply at the inlet to cause a starvation effect.

Solution of h_0 for the full range of G from 0 to 5000 was obtained by Cheng[13] and more recently by Ford.[14] The effect of lubricant compressibility was also included in Reference 13.

When U is large, the reduction of viscosity due to shear heating must be taken into account by solving the energy equation simultaneously with Equation (5) in the inlet region. This was achieved by Greenwood and Kauzlarich,[15] Cheng,[16] and more recently by Murch and Wilson.[17] Results led to a thermal reduction factor, ϕ_T, which can be multiplied into the Grubin film thickness for the actual h_0. It was shown in Reference 16 that ϕ_T varies slightly with W, and the lubricant parameter. As a first approximation, one may use the curve shown in Figure 5 to determine the value of ϕ_T.

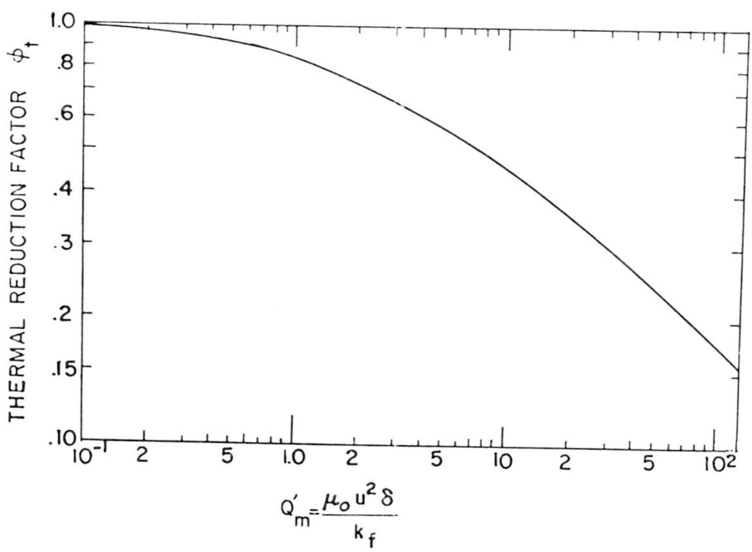

Fig. 5.—Thermal Reduction Factor.

The starvation effect introduces still another reduction factor of the

Grubin film thickness. The starvation reduction factor is shown by Wolveridge and Archard[18] as a function of the distance between the inlet film meniscus and the entrance edge of the Hertzian conjunction. Figure 6 shows the variation of the reduction factor with the starvation parameter.

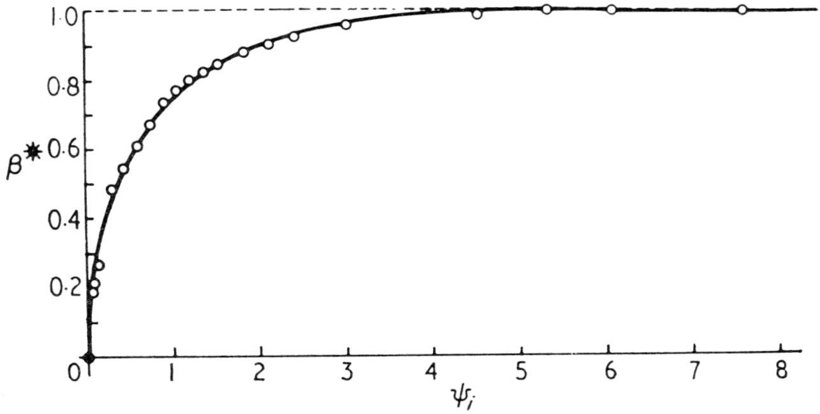

Fig. 6.—Elastohydrodynamic Theory Expressed in Practical Form (Ref. 18).

3.2.3 Line contact film thickness measurement

Measurements of film thickness in line contacts have been reviewed thoroughly by Archard.[19] Only a brief coverage of the significant results will be given here.

The first successful experimental confirmation of Grubin's film prediction can be credited to Crook[20] who deduced the film thickness from the measured capacitance between two discs. The second significant film measurement was by Sibley and Orcutt.[21] With the exception that Sibley and Orcutt's results show a slight load dependence, their data are in good agreement with Crook's capacitance data.

After Crook, further capacitance film measurements were made by Archard and Kirk,[22] Christensen,[23] and Dyson, Naylor, and Wilson.[24] The refined technique used in Reference 24 produced an excellent correlation with a line contact film thickness formula given by Dowson and Higginson.[25] More recently, optical film thickness for line contacts were also made available by Wymer and Cameron.[26] The agreement between the predicted film with that measured by either capacitance or optical techniques can be seen in Figure 7.

3.2.4 Point contact film thickness analysis

The Grubin type of analysis can also be carried out for spherical contacts with a circular conjunction. This solution was provided by Archard and Cowking.[27] They solved the two-dimensional Reynolds equation outside the circular conjunction region for a film thickness distribution compatible to the Hertzian solution for an unlubricated contact. They used a Grubin type boundary condition, $q = 1/\alpha$, around the circumference of the circular

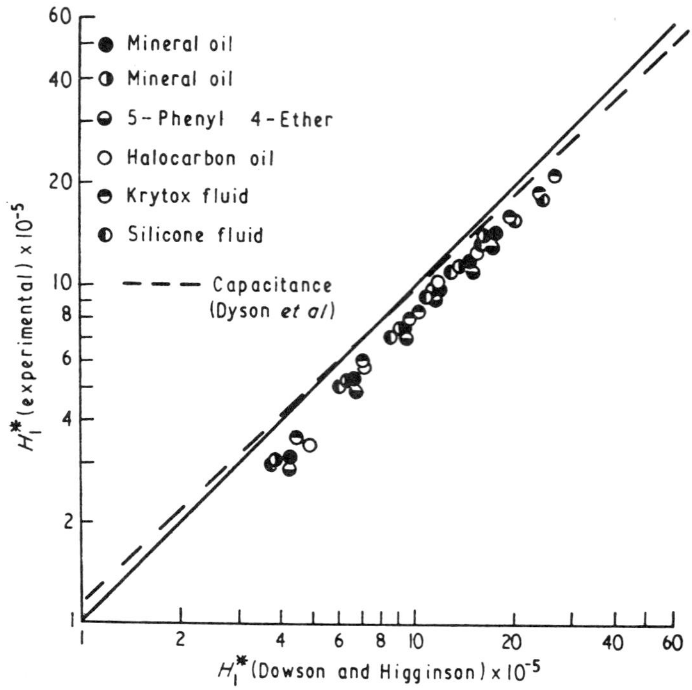

Fig. 7.— Minimum Thickness at Exit and Capacitance Measurements Compared with Dowson and Higginson Theory (Ref. 26).

conjunction. Their results gave the following film thickness formula for circular point contacts in terms of maximum Hertzian pressure p_o.

$$\frac{h_o}{R} = 1.37\ (GU)^{0.74}\left(\frac{p_o}{E}\right)^{-0.22}$$

For point contacts with an elliptical conjunction, analytical solution is not feasible, and numerical solutions have been contributed by Cameron and Gohar[28] and Cheng.[29] The results in Reference (29) gave

$$\frac{h_o}{R} = C\ (GU)^{n_1}\left(\frac{p_o}{E}\right)^{n_2} \qquad (9)$$

where C, n_1, n_2 are tabulated for four values of ellipticity ratios of the conjunction, a/b, (b - semi-axis in the direction of motion, a - semi-axis normal to the motion).

$\frac{a}{b}$	C	n_1	n_2
5	1.625	0.74	-.22
2	1.560	0.736	-.209
1	1.415	0.725	-.174
0.5	1.145	0.688	-.066

Predictions of h_0 using Equations (8) and (9) would gradually become inaccurate when the rolling speed becomes excessive. In the high speed region, both the inlet heating effect and the starvation effect can cause a reduction in h_0. The inlet heating effect for elliptical point contact has not been investigated; however, it is believed that the thermal reduction factor obtained for line contacts may also be applicable for point contacts.

More recently, full computer solutions for point contacts were made available by Hamrock and Dowson[30-33] for flooded as well as for starved contacts. Based on their calculations, the film thickness formulas for the flooded contacts appear as

$$H_{c,F} = 2.69\, U^{0.67} G^{0.53} W^{-0.067} \left(1 - 0.61 e^{-0.73k}\right) \quad (10)$$

$$H_{min,F} = 3.63\, U^{0.68} G^{0.49} W^{-0.073} \left(1 - e^{-0.68k}\right) \quad (11)$$

where

$H_{c,F} = h_{c,F}/R_x$

$h_{c,F}$ = central film thickness for flooded contacts

$H_{min,F} = h_{min,F}/R_x$

$h_{min,F}$ = minimum film thickness for flooded contacts

k = elliptical parameter, $k = 1.03\,(R_y/R_x)^{0.64}$

$R_x = R_{x1}R_{x2}/(R_{x1}+R_{x2})$; $R_y = R_{y1}R_{y2}/(R_{y1}+R_{y2})$

$W = w/(ER_x^2)$; w = total load

For the starved contacts, the formulas are

$$H_{c,s} = H_{c,F}\left(\frac{m-1}{m^*-1}\right)^{0.29} \quad (12)$$

$$H_{min,s} = H_{min,F}\left(\frac{m-1}{m^*-1}\right)^{0.25} \quad (13)$$

where the subscript s refers to the starved contacts, m is the distance of the inlet meniscus from the center of the contact, and m* is the inlet distance required for achieving the flooded conditions. m* can be expressed as

$$m^* = 1 + 3.06\left[\left(\frac{R_x}{b}\right)^2 H_{c,F}\right]^{0.58} \quad (14)$$

where b is the semiminor axis of the elliptical conjunction in the rolling direction.

So far, in most of the starvation analysis, the location of the inlet miniscus is considered as known. However, in reality this quantity is not known beforehand, and is dependent on the lubricant supply rate and the system configuration. The reduction in film thickness due to starvation, treating rolling element bearings as a system, has been studied in considerable detail by Chiu et al.[34] They concluded that the starvation effects in most rolling element bearings are considerably greater than the inlet heating effects.

3.2.5 Point contact film thickness measurement

Film thickness in point contacts using capacitance method was reported by Archard and Kirk[22] and Archard[19] with a cross cylinder apparatus. They found the side leakage effects are not large in point contact. Thus, line contact theories can be used to estimate the central film thickness for point contacts.

The introduction of optical technique in measuring film thickness[28] enables many to gather extensive data for point contact film thickness and to make correlations with the point contact film analysis. Good agreement were found by Wedevan,[35] Westlake and Cameron[36] and others. Typical results given in Reference 36 are shown in Figure 8.

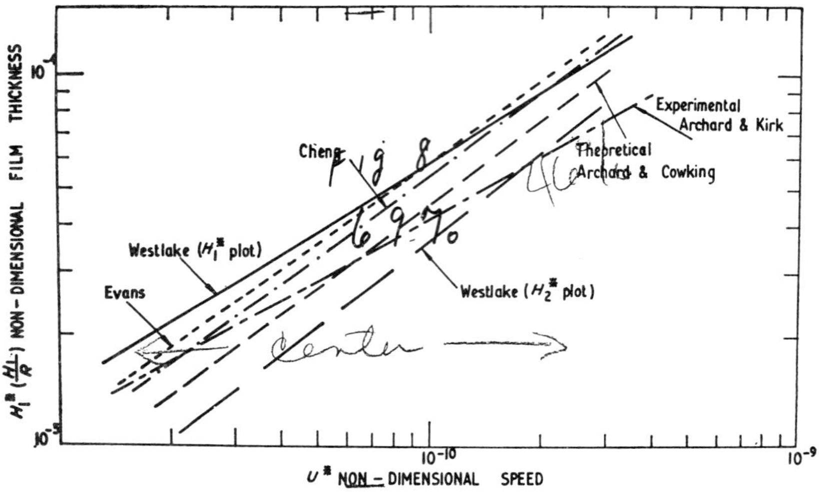

Fig. 8.—Theoretical and Experimental Central Film Thicknesses Versus Speed Parameter U (Ref. 36).*

Extensive film thickness data by X-ray transmission were also obtained with crowned rollers by Parker and Kannel.[37] The X-ray film data showed a much stronger load dependence than that predicted by EHD analysis at maximum Hertzian pressures beyond 1 GP_a(150,000 psi). This disagreement was explained by Gentle et al.[38] as possibly due to a combination of thermal and surface roughness effects. The point contact EHD film theory at extreme pressures (up to maximum Hertzian pressure equal to 2 GP_a(300,000 psi)) was also validated in Reference 33 by using a sapphire disk and a tungsten carbide ball.

3.3 Film Shape and Pressure Distributions

3.3.1 Line contact analyses

Grubin's method can predict a film thickness at the inlet region, but it is not capable of yielding a detailed film thickness and pressure distribution. For these details, it is necessary to solve the coupled Reynolds and deformation equations.

Because of the inherent difficulty associated with the high compliance and the non-linear pressure-viscosity relation, these two coupled equations have been proven to be not too feasible for either analytical solutions or conventional numerical methods. However, limited success have been achieved by Dorr,(39) Weber and Saalfeld,(40) and Stephenson and Orsterle.(41) These solutions are limited to maximum Hertzian pressure well below 0.14 GP_a (~ 20,000 psi), which is one order of magnitude lower than the pressures in gears and rolling element bearings.

The first numerical solution for a limited number of heavily loaded cases was reported by Petrusevich.(42) The full numerical results did not appear until quite a few years later when Dowson and Higginson(43) succeeded in obtaining the full EHD numerical solution.

Their results revealed, for the first time, the main features of the film and pressure profiles as functions of dimensionless speed, load, and lubricant parameters. The most distinctive feature of their solution is the existence of a sharp pressure spike accompanied with a film constriction at the conjunction exit, as shown in Figure 9. At first, these striking re-

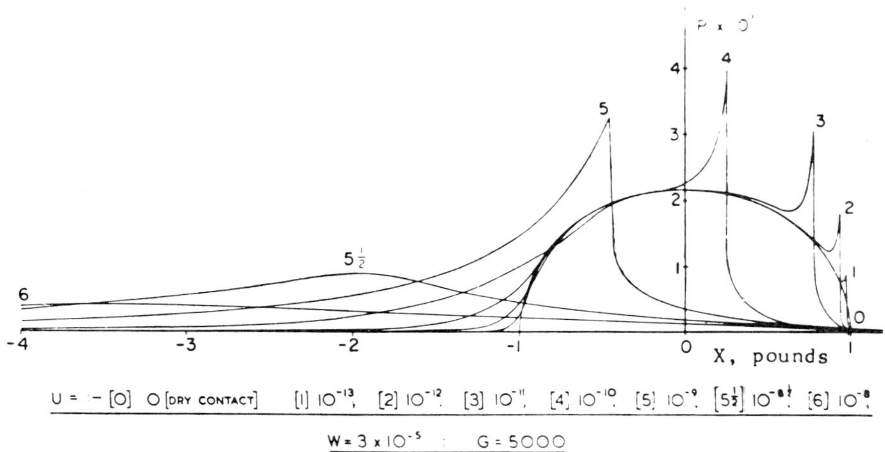

Fig. 9.—Pressure Distributions for a Compressible Lubricant (Ref. 43).

sults were received with considerable skepticism particularly among machine designers. Not very much later, results of Dowson and Higginson were confirmed by Archard, Gair and Hirst.(44) Similar features were also found later in other isothermal EHD numerical solutions.(45-48)

The minimum film thickness at the constriction was shown to be approximately 70-75% of the inlet uniform thickness. Dowson and Higginson's data provided the following well-known empirical formula for predicting the minimum film thickness.

$$\frac{h_{min}}{R} = 2.65 \frac{G^{0.54} U^{0.7}}{W^{0.13}} \qquad (15)$$

3.3.2 Line contact film shape measurement

The first positive experimental evidence of the exit constriction was seen in the circumferential film profile measured by the X-ray technique.[49] This was later followed by a much more satisfactory film profile measurement by the optical interferometry due to Gohar and Cameron.[28] Figures 10 and 11 show a series of optical film profiles for line contacts measured by Wymer and Cameron.[26] While their measured nominal film shows good corre-

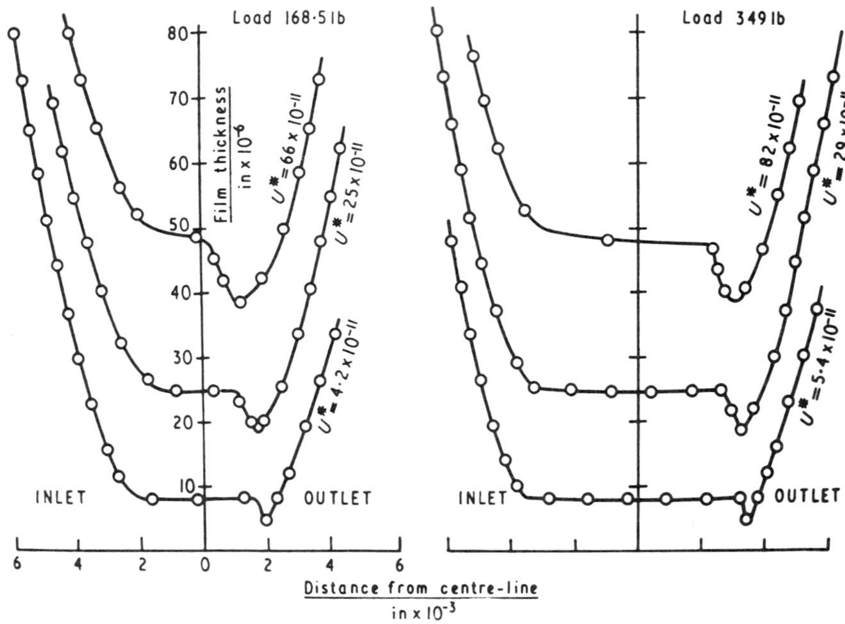

Fig. 10.—Film Profiles at Mid-Point in Direction of Rolling (Ref. 26).

lation with EHD theories, the measured ratio of minimum film to nominal film appear to be considerably smaller than .7 to .75 as predicted by EHD analysis. The reason for this discrepancy is still unknown. Their results also show some interesting features of film variation at the edges of roller contact as shown in Figure 11. These features are currently not predictable analytically.

3.3.3 Line contact pressure measurement

Early pressure measurements were conducted by Dowson and Longfield[50]

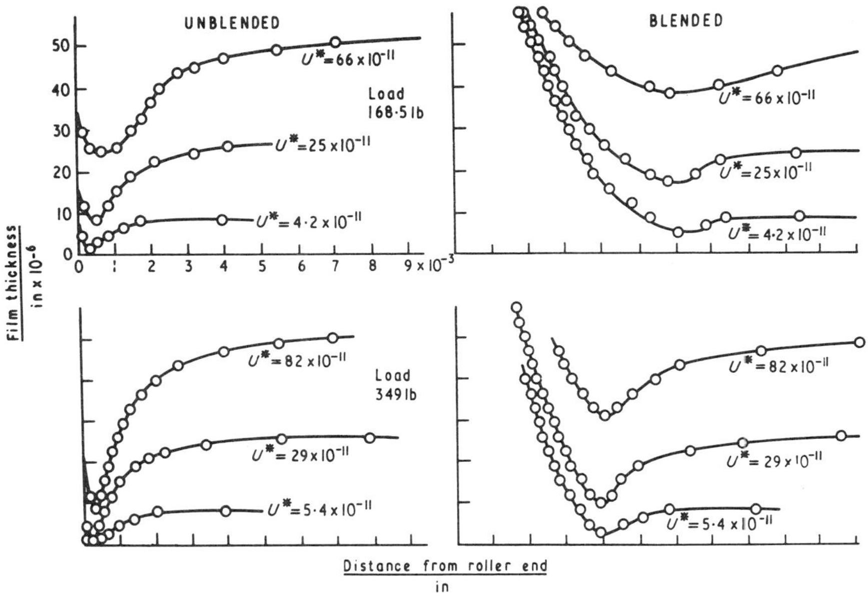

Fig. 11.—Film Profiles Through End Closure in Axial Direction (Ref. 26).

by using a large rotating disc in a simple sliding contact. Because of the large area of contact in a conformal contact, no secondary pressure peaks were detected. Subsequent attempts were made by Orcutt(51) and Kannel(52) in measuring pressure profile in heavily loaded contacts by means of a tiny vapor deposited manganin strip. The confirmation of secondary pressure pikes in EHD line contacts can be seen in Figure 12 taken from Reference 52. The trend of the effect of load on the location of the pressure spike agrees with that predicted analytically.

More recently, Hamilton and Moore(46) obtained even more convincing evidence of the EHD secondary pressure spike by using a similar technique used by Orcutt and Kannel. As shown in Figure 13, the observed pressure peak appears to be more pronounced than those shown in Figure 12.

3.3.4 Point contact numerical solution

Even though the film shape of EHD point contact was measured optically (28) as early as 1966, successful numerical solutions for point contacts did

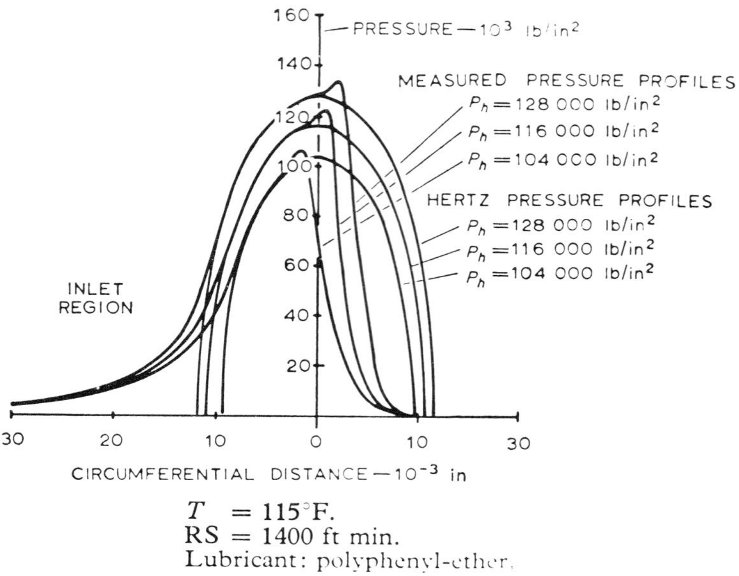

Fig. 12.—Measured Variation in Pressure Profile with Disc Loading (Ref. 52).

not appear until 1970 when Jacobson[(53)] succeeded in obtaining the first numerical solution for spherical point contacts treating the lubricant as a solid-like substance inside the conjunction. Unfortunately, his results did not lend themselves to a direct comparison with existing point contact film thickness formulas. However, limited comparisons were made in Reference 53 between the author's own optical film measurement and the calculated film shape. Close agreement was found.

The second numerical work on point contact film and pressure shape was contributed by Ranger[(54)] who obtained results up to 0.53 GP_a(~ 80,000 psi). Convergence difficulties were found beyond this load. His results substantiated all aspects of optical film shape measurements with only one exception. Ranger's results of central film thickness show a load dependence contrary to that observed in practice. The reason is yet to be found.

A more complete set of point contact numerical solutions appeared recently in a series of papers by Hamrock and Dowson.[(30-33)] They employed a direct iterative procedure, and solved the two-dimensional Reynolds and elasticity equations. Figures 14(a) and (b) show typical contours of film thickness and pressure in a circular contact. The pressure and film profile along the central strip reveal the same features as those found in the line contact theories. The minimum film thickness, in most cases, occurs at the sides of the contact.

Hamrock and Dowson's results confirm quite well the film thickness formulas based on the existing Grubin-type point theories. They also provide a much needed complement to the optical EHD film measurements. The impact

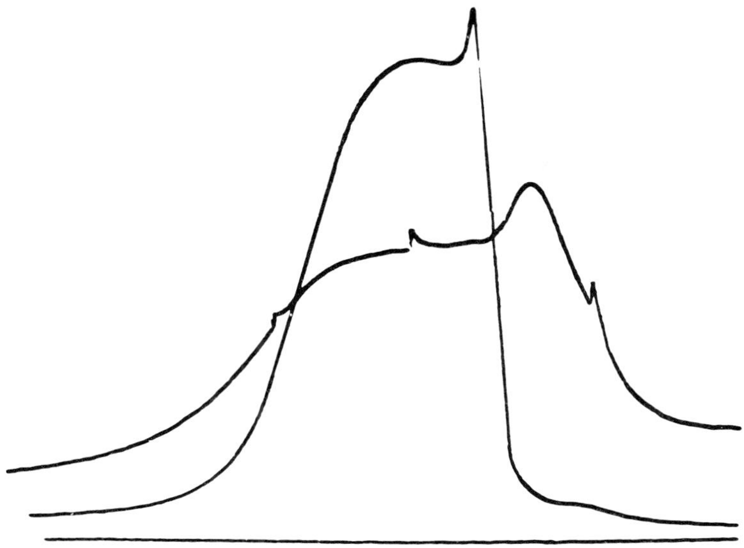

Fig. 13.—Synchronized Pressure and Shape Photograph Obtained from a Gauge with Very High Resolving Power (width of pressure sensing element 7.6μm). Load 100.3 kN/m, speed 1.30 m/s. (Ref. 46)

of these point contact numerical solutions may even be greater if future interest will be directed to asperity lubrication. Since asperities are most conveniently modeled as elliptical contact, point contact numerical solutions will be immediately applicable. This point will be expounded further in Section 5 on micro-elastohydrodynamics.

3.3.5 Point contact film shape measurement

In studying the film shape in point EHD contacts experimental efforts have preceeded the theory. This was largely due to the optical interferometric technique which was explored initially by Archard and Kirk[55] and later fully developed by Gohar and Cameron.[28] For studying the film shape in point EHD contact, the method involves observation of the interferometric map at the contact between a highly polished steel ball and a transparent plate. By identifying the successive fringes, a constant thickness contour can be mapped.

The development of lubricant film in a point contact as the rolling speed increases can be shown in Figure 15 taken from Reference 28. It is seen that the minimum film for heavily loaded cases occurs at the two sides instead of at the exit. The ratio of the minimum film thickness to the central film thickness obtained by Foord et al.[56] and Westlake and Cameron[36] is lower than 70-75% predicted by the line contact theory, particularly for the thin film cases.

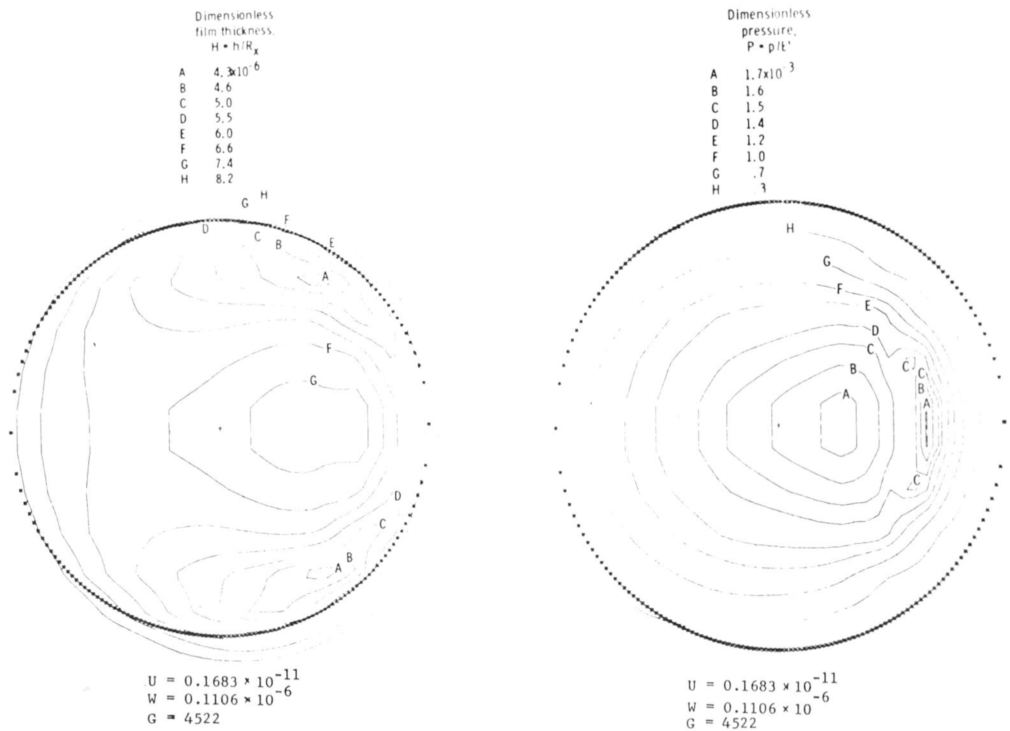

(a) (b)

Fig. 14.a.—Contour plot of dimensionless film thickness. (Ref. 32)
b.—Contour plot of dimensionless pressure. (Ref. 32)

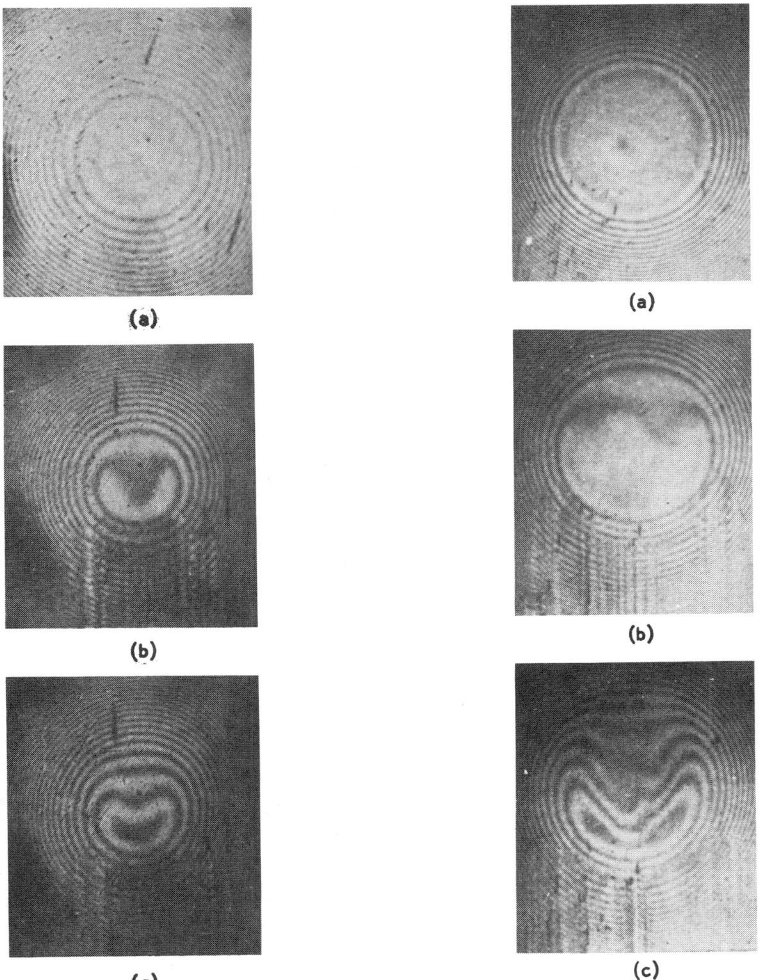

(a) P = 2 lb U = 0 cm/sec
(b) P = 2 lb U = 2 cm/sec
(c) P = 2 lb U = 7 cm/sec

(a) P = 16 lb U = 0 cm/sec
(b) P = 16 lb U = 3.2 cm/sec
(c) P = 16 lb U = 28.6 cm/sec

Fig. 15.—Rolling Point Contacts (Ref. 28).

3.4 Film Thickness Chart

The minimum film thickness of EHD line contacts may be determined from a survey diagram contributed by Moes.[57] He demonstrated that the film parameter h_{min}/R, the speed parameter $\mu_0 u/E'R$, load parameter, and the lubricant parameter α^*E' can be regrouped to form an implicit relation among only three independent parameters. Thus only one family of curves is needed to relate the film thickness parameter with other parameters. As shown in Figure 16, this family of curves covers a wide range of loads, speeds, and lubricant parameters. It includes the Martin's results as an asymptote for the rigid/isoviscous case, and Herrebruch[58] results as an asymptote for the elastic/isoviscous case.

3.5 Temperature

3.5.1 Thermal analyses

Since viscosity is strongly influenced by temperature, thermal effects are expected to play a role in EHD performance particularly for sliding contacts. The energy equation in Hertzian contacts was considered first by Crook[59] and by Bell et al.[60] Crook developed a simplified theory in determining the mid-film temperature by neglecting the surface temperature rise, the viscous heat dissipation due to the Poiseulle flow, and the heat convected by the lubricant. His method yields a reasonably accurate mid-film temperature within a moderate speed range. Crook's theory has been used widely in estimating the sliding friction in EHD contacts. Bell et al.[60] also developed an analytical solution of the energy equation which revealed the relative significance of heat due to sliding, rolling, and compression. Both theories indicated that lubricant film temperature rises significantly even for cases of moderate slip. These results suggested a need to include the energy equation in the full-EHD solution.

The numerical solutions of the full thermal-EHD problem were contributed by Cheng and Sternlicht[45] and Dowson and Whittaker.[61] It was hoped that the incorporation of temperature in the full-EHD analysis might have led to a significant reduction of the pressure spike. Results in both analyses indicated that at least for EHD contacts lubricated with mineral oils the basic features of the pressure spike and the exit protrusion are not significantly altered by the thermal effects.

3.5.2 Temperature measurement

Measurement of surface temperature in EHD contact has proven to be a rather difficult task. Orcutt[51] used a platinum wire as the temperature transducer, and obtained temperature profiles for a moderately loaded line contact.

A continuous effort has been exerted by Kannel[62,63] in improving the surface temperature measurement in line contacts using a titanium wire deposited over a silica layer on the disc surface. Figures 17 and 18 show the effects of load and speed on the surface temperature. In general, the temperature level is considerably higher than that measured by Orcutt.

Recently, Bartz and Ehlert[64] has also used vapor deposited transducers to obtain extensive data on pressure, temperature, and film thickness profiles. The qualitative features of their measurements confirm the re-

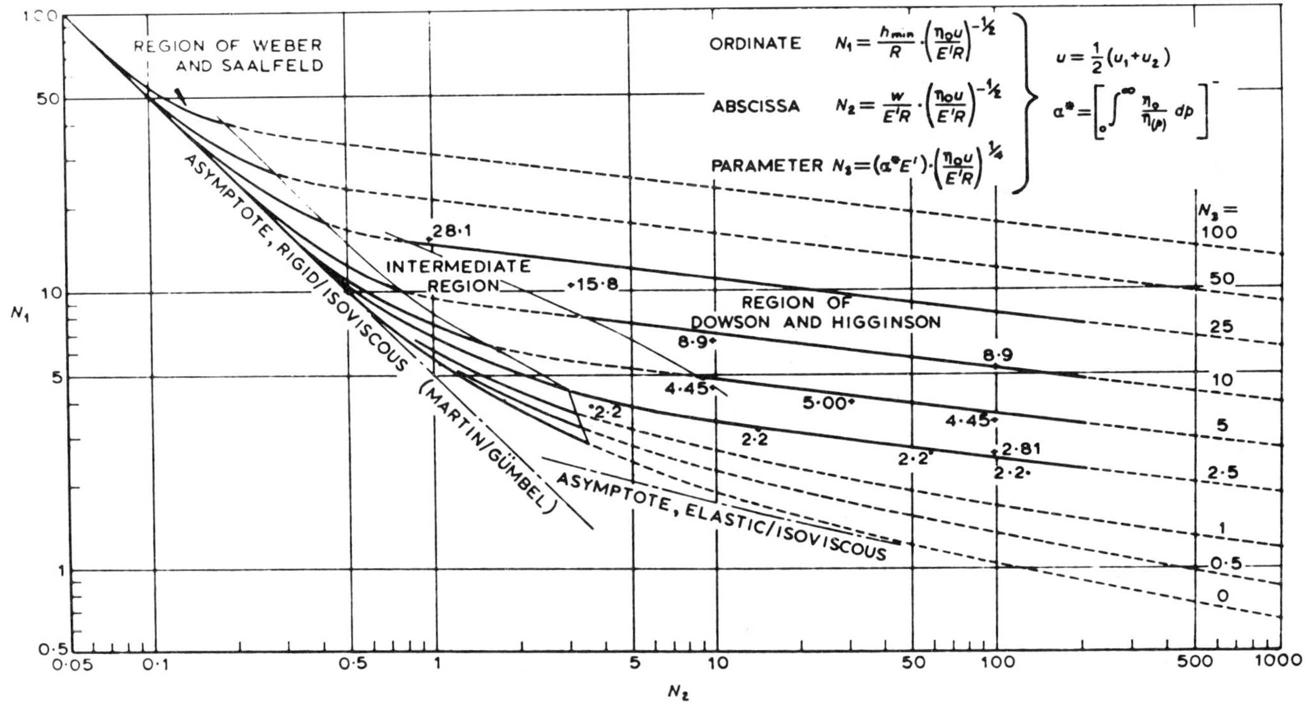

Fig. 16.—Survey Diagram for Incompressible Isothermal Elastohydrodynamic Lubrication (Ref. 57).

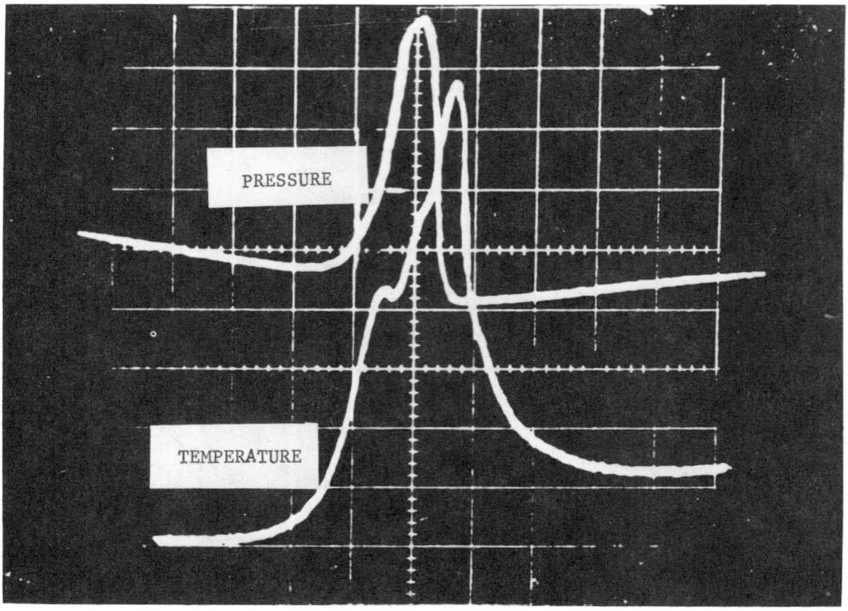

V = 2,500 rpm, Slip = 1 m/sec, P_h = 1,000 MN/m^2 (150,000 psi), T = 2C/ div.

Fig. 17.—Pressure-Temperature Traces for Synthetic Paraffinic Lubricant (XRM-109) (Ref. 63).

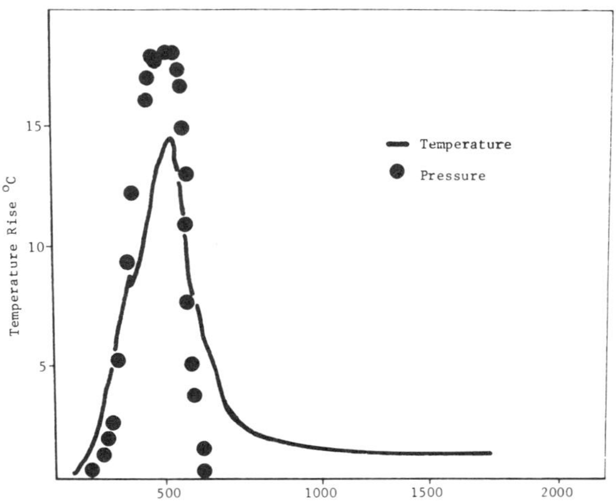

P_h = 1000 MN/m^2 (-150,000 psi), V = 2500 rpm (.077 m disks), slip = 1 m/sec, lubricant TL-7165

Fig. 18.—Corrected Temperature Profile for Steel Disks (Ref. 63).

sults obtained by previous investigators. However, the surface temperature measurements for pure rolling cases are considerably higher than those obtained by others.

While the technique of using vapor deposited probes for surface temperature measurement is yet to be perfected for a satisfactory agreement with calculated temperature, an entirely new technique has been developed by Nagaraj and Winer[65] in measuring the surface as well as film temperature in circular contacts using an infrared probe. At least for the surface temperature measurement, the data obtained so far has been remarkable. Figures 19 and 20 show traces of surface and film temperature along the center strip of the circular contact. Figure 21 shows the comparison between the measured temperature and that predicted from the Jaeger-Archard[66] Agreement is far better than the limited correlations between the titanium or platinum wire temperature data and the circulated temperature data for line contacts.

3.6 Friction

Sliding friction in concentrated contacts is perhaps one of the most intriguing aspects of EHD. The basic features of sliding friction are revealed most clearly in the friction curves taken from a two-disc machine.[67, 68] A typical family of friction curves, taken from Reference 68 is shown in Figure 22. In the low-slip or linear region, the friction increases linearly with slip similar to the behavior of a Newtonian fluid. As slip increases, the friction would gradually taper off and would eventually flatten out. This is labeled as the non-linear region because the stress is believed to be no longer governed by the linear constituative relations. The region beyond is characterized by a decreasing shear stress with an increasing slip, and is labeled as thermal region because of the strong thermal influence on the fluid properties at high sliding speeds.

Behavior of sliding friction in these regions are quite different, and are reviewed separately in the following sections.

3.6.1 Low-slip friction

A simplified view of the lubricant behavior as it passes through the high pressure conjunction with a minute slip can be depicted in Figure 23. An elemental fluid, represented by a rectangle ABCD at the inlet, deforms into a parallelogram A'B'C'D' at the exit due to a slightly faster velocity of the bottom surface. Part of the strain ε, shown in Figure 23, is elastic and recoverable. The remainder due to viscous flow is permanent. The ratio of the recoverable to the permanent strain can be estimated by the quantity $\mu U/Gb$, known as Deborah Number. If this number is very large, the traction will be governed by elastic deformation. At the other extreme, the traction will follow the law of viscous flow.

Crook[67] and Dyson[69] have employed the linear viscoelastic model to correlate with the measured effective viscosity, or the initial slope of the tractive-slip curve. The agreement was not satisfactory.

One major question in the Maxwell viscoelastic model is whether the viscosity and shear modulus would reach an equilibrium value when the fluid is subjected to a high pressure during a short time interval. The transient viscosity model was explored by Harrison and Trachman[70] with remarkably

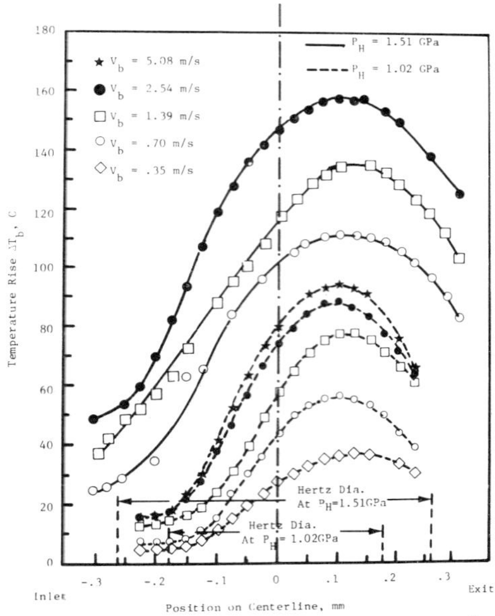

(Smooth ball: .011 µm R_a, Fluid N1, T_{bath} = 40 C, V_{sa} = 0).

Fig. 19.—Ball Surface Temperature Rise Along Centerline Versus Speed (Ref. 65).

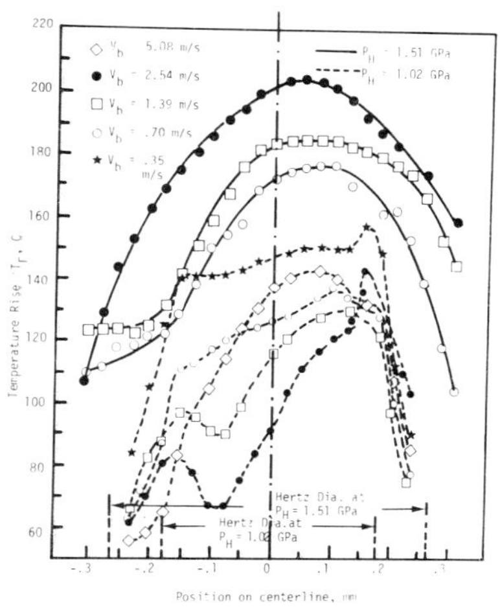

(Smooth ball: .011 µm R_a, Fluid N1, T_{bath} = 40 C, V_{sa} = 0)

Fig. 20.—Fluid Temperature Rise Along Centerline Versus Speed (Ref. 65).

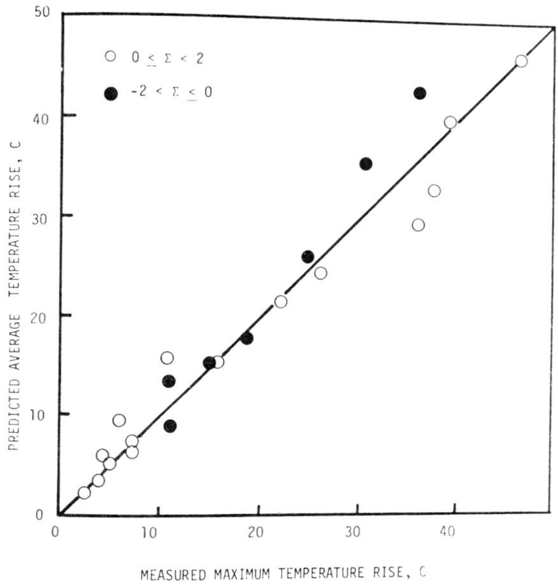

Fig. 21.—Comparison of Predicted Average and Actual Maximum Ball Surface Temperature Rises for L > 5 (Ref. 65).

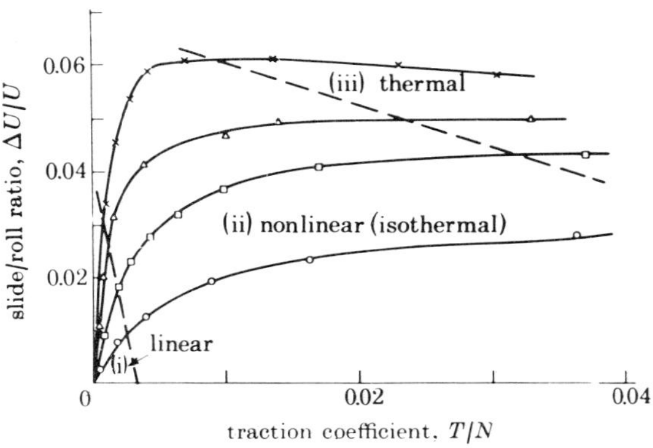

Fig. 22.—Typical traction curves measured on a two-disk machine in line contact (from Johnson and Cameron[68]), at varying mean contact pressures, $\bar{p}$: x, 1.03; Δ, 0.68; , 0.51; o, 0.40 GPa.

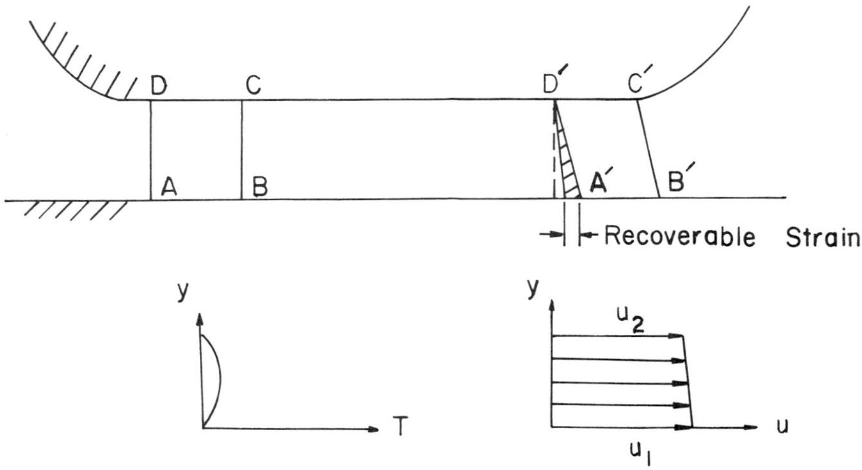

Fig. 23.—Characteristics of Low-slip Traction.

good correlation with the measured effective viscosity by Johnson and Cameron.[68] This was accomplished, however, without considering the shear elastic strain.

More recently, Johnson and Tevaarwerk[71] devised a spin-roll point contact rig to discern whether the fluid under high pressure would deform like an elastic solid or viscous liquid. They have shown some convincing evidence that the oil under a high pressure and a short transient time tends to deform more as a solid. However, the values of shear modulus deduced from the low-spin tests are about one order of magnitude below the known equilibrium shear modulus under high pressure.[72] They attributed that this discrepancy in shear modulus is due to transient effects under a suddenly applied pressure.

Very recently, Montrose et al.[73] developed perhaps the most complete viscoelastic model to describe the low-slip traction behavior. In this model, the shear elastic strain is considered in the same manner as the conventional viscoelastic theory. However, instead of using the actual local pressure for the viscosity a fictive pressure is used in order to account for the transient viscosity effects similar to that considered in [71]. The fictive pressure is determined from an exponentially decaying structural relaxation function. Preliminary results appear to be quite promising. It is quite possible that a more complete correlation with the observed traction using Montrose's model might provide an answer to why there is such a large discrepancy in the shear modulus in Johnson and Tevaarwerk's work.

To summarize, the low-slip friction can be characterized by the conventional Maxwell viscoelastic model. The deformation of lubricant in the conjunction under high pressure and at high speed is predominantly elastic. However, nonrecoverable viscous flow is not totally negligible because of the inability of the fluid viscosity to respond fully to the local pressure. The model described by Montrose et al.[73] appears to hold the best promise for accurate prediction of low-slip friction in EHD contacts.

3.6.2 High-slip friction

As the sliding velocity increases, the slope of the friction curve gradually decreases to zero, and, in many cases, becomes slightly negative at very high sliding speed. Under these conditions, the fluid undergoes a large shear strain usually in the mid-plane of the lubricant film, as depicted in Figure 24. An elemental fluid ABCD is severely sheared to the shape A'B'C'D' as it reaches the exit section.

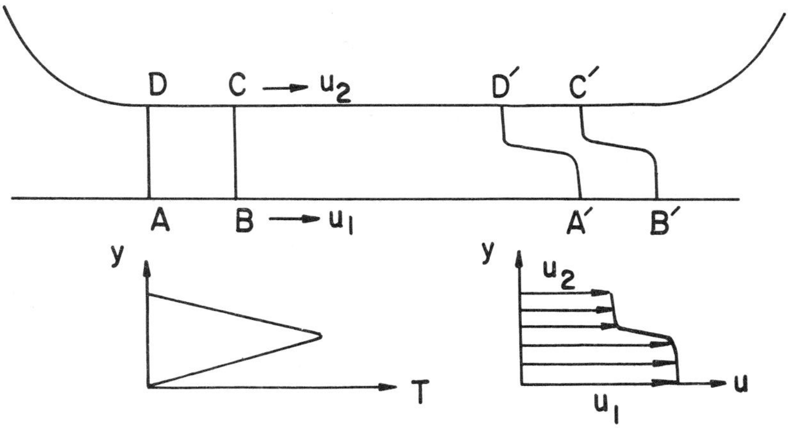

Fig. 24.—Characteristics of High-slip Traction.

Crook first predicted the trend of the friction curve in the high-slip region for moderately loaded contacts ($P_{Hz} < 0.5$ GP_a) on the basis of a Newtonian liquid. Results agree with the measured friction only in trend, but not in magnitude particularly at the friction peak. The same conclusion was also found in (74) based on a thermal-EHD numerical solution. The inadequacy of the Newtonian model was finally firmly established by Johnson and Cameron(68) in Figure 25, which shows that in heavily loaded contacts, the Newtonian model would yield a friction force almost one order of magnitude higher than the measured value.

The use of a non-Newtonian model for friction in high-sliding region was first analyzed by Bell et al(75) on the basis of a shear dependent Ree Eyring liquid. Later, Dyson(69) postulated that the liquid possesses a limiting shear stress under a large strain and strain rate. He provided a semi-empirical function of pressure and temperature. As demonstrated in (76), prediction of high-slip friction using Dyson's formula appears to work well over a wide range of conditions.

The limiting shear behavior at high sliding speed was explained by Smith(77) on the basis of a plastic solid. This concept is gradually gaining more acceptance because most high sliding EHD contacts operate at a high

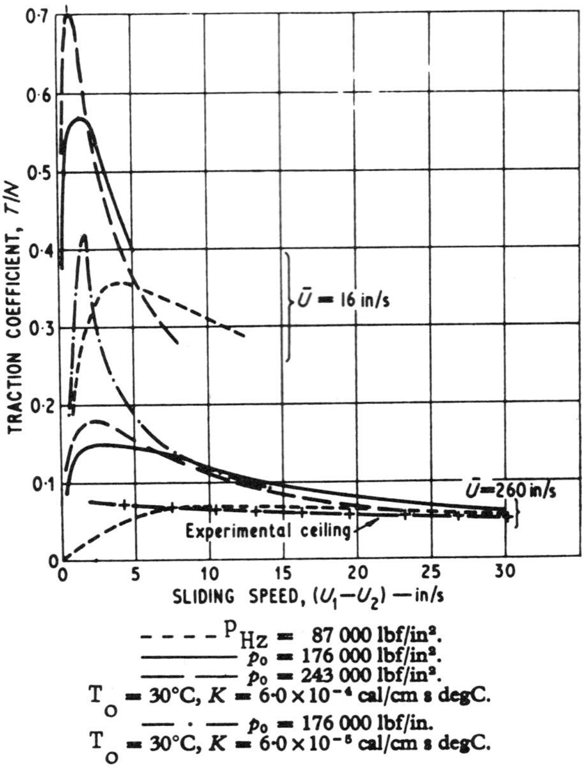

Fig. 25.—Variation of Traction Coefficient with Sliding Speed Calculated by Crook's Theory, at Various Contact Pressures (Ref. 68).

pressure level where the lubricant most likely would shear like a solid rather than a non-linear liquid.

Johnson and Tevaarwerk[(71)] proposed a most interesting friction model which is capable of describing linear and nonlinear viscous, linear and nonlinear viscoelastic, as well as elastic/plastic behavior in a single equation. This model is depicted in Figure 26 and is expressed as

$$\dot{\gamma} = \frac{1}{G}\frac{d\tau}{dt} + F(\tau) = \dot{\gamma}_e + \dot{\gamma}_v$$

$$F(\tau_e) = \left(\frac{\tau_o}{\eta}\right) \sinh\left(\frac{\tau_e}{\tau_o}\right)$$

where $\dot{\gamma}$ = *total strain rate*

$\dot{\gamma}_e$ = *elastic strain rate*

$\dot{\gamma}_v$ = *viscous strain rate*

τ_e = *equivalent stress*

τ_o = *representative stress, a fluid property*

η = *viscosity*

G = *shear modulus*

τ_o, η, and G are fluid properties to be deduced from traction tests. Limited

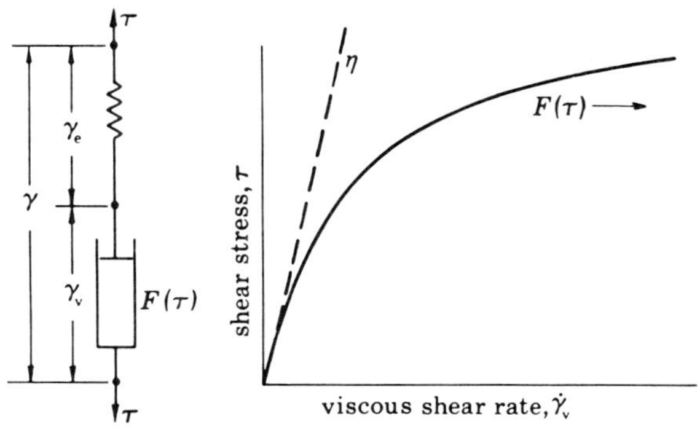

Fig. 26.—Non-linear Maxwell fluid with zero-shear-rate viscosity and infinite-rate shear modulus G.F(τ) denotes the nonlinear viscous function (Ref. 71).

values of these properties are given in Reference 71 for five oils. Good correlations were found with data taken from spin-tests as well as side-slip tests.

Very recently, Winer et al.[78] developed a new method to measure the yield stress and shear modulus of glassy lubricants under high hydrostatic pressure without reliance on any EHD traction data. Preliminary results are very encouraging. It appears before long there will be considerable primary data of the fluid properties available to predict high-slip traction based on the elastic/plastic solid concept.

3.7 Summary (Full-Film EHD)

For heavily loaded EHD contacts in ordinary rolling element bearing, gears, and cams, the film thickness is essentially uniform over most of the Hertzian conjunction. The uniform gap is reduced significantly along the exit edges to form a minimum film either at the central exit edge or at the two sides. Optical film measurements have verified that present EHD theories are remarkably accurate in predicting this uniform gap for a wide range of speed and for load as high as 2.5 GP_a(350 ksi). However, some discrepancies still remain between the calculated and measured minimum film thickness at the trailing edge or at the two sides. Formulas are also available to determine the reduction in film thickness due to inlet shear heating effects, but its accuracy is yet to be fully substantiated. Film reduction due to a starved inlet can also be predicted accurately if the inlet meniscus position is given.

- The pressure distribution in moderately loaded EHD contacts conforms largely to a Hertzian distribution except that it contains a well-publicized spike in the exit region. This pressure peak can now be considered as an experimentally confirmed phenomenon. However, the measured peak sharpness is much reduced possibly because of the inability of the lubricant to reach the extreme shear stress at the peak. Temperature effect was found to have little effect on the sharpness of the peak.

- Friction in EHD contacts increases initially linearly with slip in the low-slip region. The initial slope of the friction curve is predictable on the basis of a Maxwell substance. For speeds and loads encountered in most Hertzian contacts in practice, the lubricant deformation is found to be largely elastic. Both the viscosity and shear modulus show signs of not being able to reach its equilibrium value under a rapidly applied pressure. Method is available to predict the transient effective viscosity based on the structure relaxation concept. No method is yet available to predict the transient shear modulus.

- Friction in the high-slip region can be modelled as a shear dependent fluid or, in more heavily loaded cases, as elastic/plastic solids. Prediction of friction can be made for elliptical EHD contacts for a reasonable range of sliding, rolling, and spinning speeds based on fluid constants either deduced from EHD friction experiments or from another independent primary experiment.

- Both surface and film temperature predictions based on Newtonian fluid is not sufficiently accurate in a sliding EHD contact. Models, including a shear thinning viscosity and a pressure-temperature dependent yield stress for the lubricant in the glassy state, are needed to determine the temperatures.

4. PARTIAL-EHD

Partial-EHD is recognized as the regime where the average film thickness becomes less than three times the composite surface roughness. Under this condition, the local film would become extremely thin or interrupted at the tip of tall asperities. Understandings of the lubrication process in partial-EHD are essential because a majority of Hertzian contacts in gears, cams, and rolling element bearings operate in this regime. A complete understanding of this problem would go a long way to predict the performance and failure of these contacts.

As suggested by Tallian,[79] who contributed the first theory on partial-EHD, the quantities which describe the phenomena in partial-EHD may include:

a. the average film thickness
b. the ratio of the EHD load to the asperity load
c. the average friction.

The main tasks in partial-EHD are to devise methods to predict the above quantities with given input data on contact geometry, material, lubricant, operating conditions, and surface roughness parameters.

4.1 Surface Characterization

For years, the roughness of bearing surfaces has been characterized by a single parameter, namely, the standard deviation of the roughness amplitude from a mean plane. This is commonly known as σ, the r.m.s. value of roughness. The value σ gives an indication of the roughness level, but not the roughness texture. For instance, two surfaces may have the same σ, but entirely different asperity shape. One surface may have long sharp asperities, and the other deep valleys.

The surface roughness texture can be described by two statistical parameters, namely, the height distribution and the auto-correlation function, a.c.f. The height distribution has been studied extensively by Williamson.(5) It was found that for surfaces finished by an abrasive process the height distribution is approximately Gaussian. For running-in surfaces, the height distribution is slightly skewed from Gaussian.

Auto-correlation function is essentially a measure of the wave length structure of a surface profile in a given direction. It is defined as

$$R_x(\lambda) = \frac{1}{\ell}\int_0^{\ell} \delta(x)\cdot\delta(x+\lambda)\,dx \tag{16}$$

where λ is the correlation length, δ is the height function along the x direction, and $R_x(\lambda)$ is the a.c.f. in the x direction. Whitehouse and Archard(7) concluded that the a.c.f. for most engineering surfaces is an exponentially decaying function. Others(80-82) have used a linear function as an approximation for the a.c.f.

Many engineering surfaces have roughnesses which are directionally oriented, i.e., it contains long wave lengths in one direction and short wave lengths in the normal direction. The directional properties of roughness can be conveniently described by a surface pattern parameter γ, first introduced by Kubo and Peklenik.(83) It is defined as the ratio of x and y correlation length

$$\gamma = \frac{\lambda_{0.5\ x}}{\lambda_{0.5\ y}} \tag{17}$$

where $\lambda_{0.5\ x}$ is defined as the correlation length at which the a.c.f. of the profile is 50% of the value at the origin. γ may be interpreted as the length-to-width ratio of a representative asperity. Purely transverse, isotropic, and purely longitudinal roughness patterns have $\gamma = 0, 1, \infty$ respectively. Surfaces with $\gamma >$ longitudinally oriented.

For determining partial-EHD performance, the following surface roughness parameters are required for each surface.

1. σ - r.m.s. surface roughness

2. height distribution function

3. $\lambda_{0.5\ x}$, $\lambda_{0.5\ y}$ - 50% correlation length in x and y direction

4. a.c.f. - auto correlation function

4.2 Average Film Thickness

The average film thickness as affected by pure longitudinal or transverse roughness was first explored by Johnson et al.[83] for pure rolling contact based on Christensen Stochastic Theory.[84] They developed a Grubin type solution, valid for $\sigma \ll h$ only, and concluded that for EHD contacts, where the load is primarily carried by the fluid pressure, the effects of roughness on the average film thickness is minimal.

The cases for $h/\sigma > 3$ and for rolling and sliding contacts were studied by Berthé[85] and by Chow and Cheng.[86] The results show that:

1. For pure rolling contact with pure transverse roughness, the average film thickness is higher than that predicted by the smooth surface EHD theory. This effect is greatly enhanced as h/σ approaches three. For sliding contacts with one surface smoother than the other, the roughness effect is enhanced if the smoother surface is faster, and retarded if the smoother surface is slower.

2. For pure longitudinal roughness, the average film thickness is lower than that predicted by the smooth surface theory. Superimposing of sliding on rolling has little influence on the roughness effect for pure longitudinal surfaces.

Very recently, Patir and Cheng[82] developed a method known as the average flow model to handle surface roughnesses of any arbitrary surface pattern parameter γ. This method enables one not only to investigate the effect of more practical surface roughness texture, but also to extend the results to h/σ below 3 where part of the load is shared by asperity contacts. Figure 27 depicts the flow pattern for longitudinally oriented ($\gamma > 1$) and isotropic roughness ($\gamma=1$). Typical results are shown in Figure 28 where the actual film thickness ratio $H_o=h_o/\sigma$ is plotted against the film parameters $\Lambda = h_{smooth}/\sigma$. It is interesting to note that the roughness effect for a realistic longitudinally oriented roughness, $\gamma=6$, is considerably smaller than the pure longitudinally roughness $\gamma=\infty$. The case of isotropic roughness only shows a slight increase in the average film compared to the smooth surface theory. Even for the case of $\gamma = 1/6$, the effect is not as overwhelming as that suggested by the pure transverse roughness theory.

4.4 Asperity Load to EHD Load Ratio

The average asperity contact pressure in a partial-EHD contact is a function of the ratio of the compliance to the composite surface roughness h/σ. Here the compliance is the distance between the two mean planes based on the undeformed surfaces. Such functions are generally known as the load compliance relationships. For Gaussian surfaces, Tallian[79] has derived the asperity load as a function of h/σ for both plastically or elastically deformed asperities. Typical load sharing ratios were also presented in [79] for a constant maximum EHD pressure.

The load sharing ratios in circumferential ground EHD contacts (longitudinal roughness) can be obtained by a full numerical solution given by Chent and Dyson[87] for discs with known surface roughness characteristics.

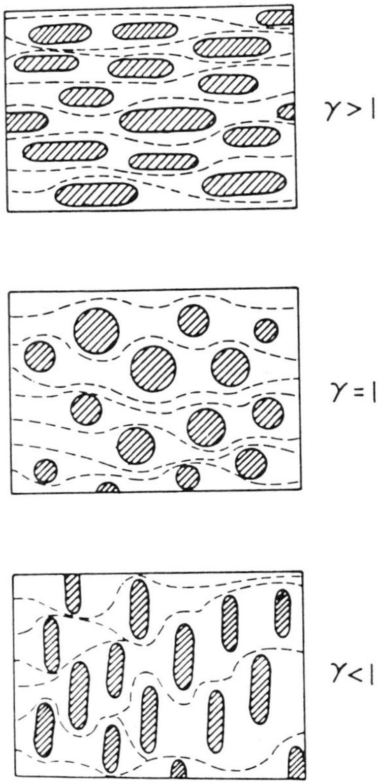

Fig. 27.—Typical contact areas for longitudinally oriented, isotropic and transversely oriented rough surfaces (Ref. 82).

4.5 Average Friction

Once the ratio of the asperity load to the fluid pressure load is determined, the total frictional force in partial-EHD, as suggested in [79] can be readily evaluated by

$$F = \mu_a Q_a + \mu_{EHD} Q_{EHD}$$

where F = *total frictional force*

μ_a, μ_{EHD} = *the coefficients of friction for the asperity load and hydrodynamic load respectively*

Q_a, Q_{EHD} = *asperity, hydrodynamic loads*

For most partial-EHD contact, the value of μ_a is believed to be between 0.1 0.2. The value of μ_{EHD} can be taken from frictional coefficients for full-film EHD contacts.

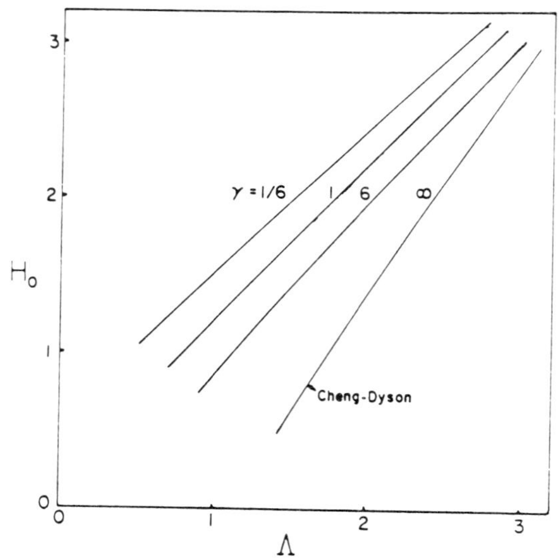

$P_0/E' = .003$, $K' = .003$, $U_{sr} = 0$, $C_3 = 0.$

Fig. 28.—Comparison of center film thickness obtained for rough surfaces with the EHL film thickness obtained for smooth surfaces.

4.6 Partial-Film EHD Breakdown

When the load is increased to a certain level in a sliding partial-EHD contact, a transition would take place, and is associated with a sudden increase in friction and wear rate. This transition is sometimes known as scuffing, or the change from mild to severe wear.

Mechanisms of scuffing in partial-EHD sliding contacts have been reviewed by Dyson,(88) and are still considered as an unresolved problem. A widely accepted explanation for this transition is the critical temperature concept which states that scuffing occurs when the maximum surface temperature of a contact exceeds the critical temperature of a given lubricant. This concept appears to be in harmony in most gear failure tests, but in discord with a number of major disc experiments.

Recently, an alternative explanation(87,89) has been given for partial-EHD scuffing. It is based on the postulate that scuffing would occur if the average lubricant pressure P_{EHD} reach a state where

$$1 - e^{\alpha P_{EHD}} \ll 1.$$

This hypothesis appears to agree well with the scuffing data for at least one set of disc experiments,(90) but does not seem to be in accord with simple sliding scuffing tests.

4.7 Summary (Partial-EHD)

- Surface roughness parameters affecting the performance of partial-EHD include: σ_1, σ_2, the r.m.s. roughness of the two surfaces; γ, the asperity length to width ratio; the height distribution, and the autocorrelation functions in two mutually perpendicular directions.

- Methods are now available to predict the average film thickness in partial-EHD contacts based on the average flow model approach for three-dimensional surface roughness. Both the transverse oriented ($\gamma=1/6$) and isotropic roughness ($\gamma=1$) tend to raise the average film from that determined by the smooth surface theory. Longitudinal oriented roughness seems to have little effect on the average film thickness.

- Ratio of asperity load to EHD load in a Hertzian conjunction is predictable by numerical solution for partial-EHD contacts. However, the accuracy of this load-sharing ratio depends critically on the validity of the load and compliance relation which is somewhat uncertain particularly for deeply penetrated asperities.

- Partial-EHD breakdown can result from an excessive temperature on one of the contacting surfaces or from the loss of pressurized viscosity in the valleys. Successful correlation have been found with both mechanisms.

5. MICRO-ELASTOHYDRODYNAMIC LUBRICATION

Micro-EHD(91) deals with the local pressure and film fluctuations around asperities and furrows within a macro-EHD conjunction. Problems in this area can be classified into three aspects: 1) the normal approaching of a single asperity at the inlet section; 2) the sliding of a single asperity; 3) asperity-asperity collision.

For pure rolling EHD contacts, local pressure and film thickness distribution at asperities are governed mainly by the normal approach action. As a deep ellipsoidal asperity approaching the opposing surface, the lubricant is first pressurized at the asperity center to a piezo-viscous condition, i.e. $e^{-\alpha p} \to 0$. Subsequently, the lubricant is entrapped to form a central pocket as it travels through the Hertzian conjunction, as shown qualitatively in Figure 29. The asperity film thickness for the lubricant to become piezo-viscous is predictable for both longitudinal and spherical asperities,(91-93) but the entrapped film thickness is currently not predictable and a more sophisticated normal approach analysis is required.

For pure transverse ridges, the normal approach action in a pure rolling process causes the lubricant in the Hertzian region ahead of the asperity to reach the piezo-viscous state. Subsequently, the asperity becomes nearly frozen together with the lubricant and is transported through the Hertzian conjunction as an integral unit.(94) The micro-EHD film h^* corresponding to the lubricant reaching the piezo-viscous condition is predictable by a Grubin-type analysis for surfaces with a transverse ridge at the inlet.(91)

For sliding EHD contacts, micro-EHD film thickness h^* is controlled largely by the sliding entrainment of the lubricant at the inlet of an asperity. For transverse asperities, a lower limit of h^* can be estimated by using the classical EHD film thickness formulas to a sliding asperity in a

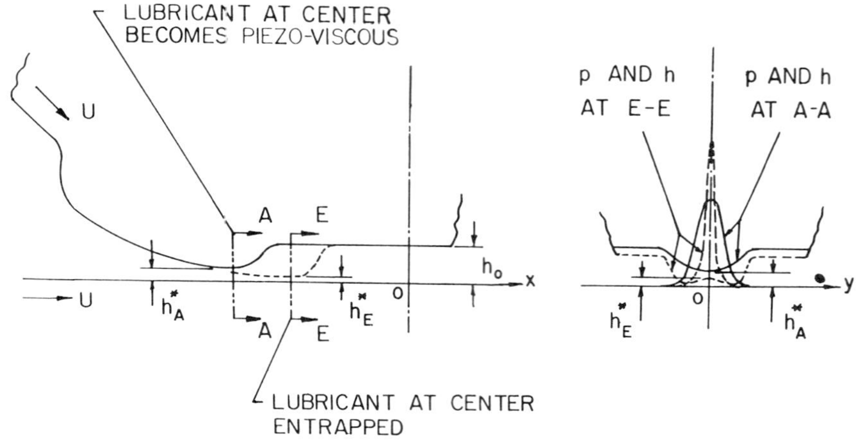

Fig. 29.—Normal Approach of a Longitudinal Asperity at the inlet of an EHD Contact (Ref. 91).

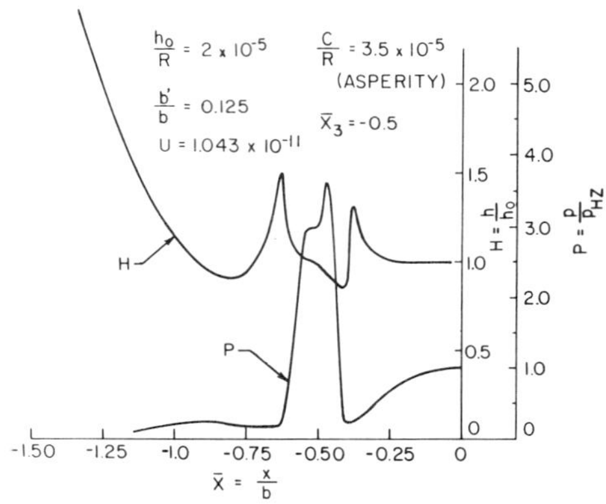

Fig. 30.—Pressure and Film Thickness Around a Sliding Asperity (Ref. 91).

low pressure ambient. For longitudinal asperities, very little is available to estimate the minimum film thickness in the micro-EHD contact. Further work is needed.

Micro-EHD of a sliding deep stationary transverse ridge or furrow can be studied by using an existing EHD computer code.(13) Figure 30 shows that typical micro-EHD pressure and film characteristics at an asperity contain similar features found in the classical EHD theories.

For a pair of transverse asperities colliding in a lubricant of low ambient pressure the micro-EHD film thickness can be estimated reasonably well with existing theories.(80-81) If the collision takes place in a high pressure ambient, h* would increase but cannot be predicted quantitatively at the present time.

REFERENCES

1. Dowson, D., *Institution of Mechanical Engineers. Proceedings*, Vol. 180, Part 3B, 1965-66, p. 7.
2. McGrew, J., et al., "Elastohydrodynamic Lubrication, Phase I," Technical Documentary Report AFAPL-TR-70-27, Air Force Aero Propulsion Laboratory, November 1970, p. 1.
3. Christensen, H., *Institution of Mechanical Engineers. Proceedings*, Vol. 180, Part 3B, 1965-1966, p. 147.
4. Tallian, T.E., McCool, J.I. and Sibley, L.B., *Institution of Mechanical Engineers. Proceedings*, Vol. 180, Part 3B, 1965-66, p. 169.
5. Williamson, J.B.P., in "Interdisciplinary Approach to Friction and Wear," NASA SP-181, edited by P.M. Ku, 1968, p. 143.
6. Nayak, P.R., *Journal of Lubrication Technology*, Vol. 93, No. 3, July 1971, p. 398.
7. Whitehouse, D.J. and Archard, J.F., *Royal Society of London. Proceedings.* Series A, Vol. 316, 1970, p. 97.
8. Martin, H.M., *Engineering*, Vol. 102, 1916, p. 199.
9. Gatcombe, E.K., *American Society of Mechanical Engineers Transactions*, Vol. 67, 1945, p. 177.
10. Blok, H., *Institute of Petroleum. Journal*, Vol. 38, 1952, p. 673.
11. McEwen, E., *Institute of Petroleum. Journal*, Vol. 38, 1952, p. 646.
12. Grubin, A.N. and Vinogradova, I.E., "Central Scientific Research Institute for Technology and Mechanical Engineering," Book No. 30, D.S.I.R. Translations, Moscow, No. 337, 1949.
13. Cheng, H.S., *Journal of Lubrication Technology*, Vol. 94, No. 1, January 1972, p. 35.
14. Ford, R.A.J., Ph.D. Thesis, University of London, March 1975.
15. Greenwood, J. and Kauzlarich, J., *Journal of Lubrication Technology*, Vol. 95, No. 4, October 1973, p. 417.
16. Cheng, H.S., "Calculation of Elastohydrodynamic Film Thickness in High-Speed Rolling and Sliding Contacts," Mechanical Technology Inc. Technical Report MTI-67TR24, May 1967, p. 1.
17. Murch, L.E. and Wilson, W.R.D., *Journal of Lubrication Technology*, Vol. 97, No. 2, April 1975, p. 212.
18. Wolveridge, P.E., Baglin, K.P. and Archard, J.F., *Institution of Mechanical Engineers. Proceedings*, Vol. 185, 1970-71, p. 1159.
19. Archard, J.F., *Institution of Mechanical Engineers. Proceedings*, Vol. 180, Part 3B, 1965-1966, p. 17.
20. Crook, A.W., *Royal Society of London, Philosophical Transactions*, Series A254, 1961, p. 223.
21. Sibley, L.B. and Orcutt, F.K., *American Society of Lubrication Engineers Transactions*, Vol. 4(2), 1961, p. 234.
22. Archard, J.F. and Kirk, M.T., *Journal of Mechanical Engineering Science*, Vol. 6, 1964, p. 101.
23. Christensen, H., *Acta Polytechnica Scandinavica. Mechanical Engineering*, Series 15, 1963.
24. Dyson, A., Naylor, H. and Wilson, A.R., *Institution of Mechanical Engineers. Proceedings*, Vol. 180, Part 3B, 1965-1966, Paper 10, p. 119.
25. Dowson, D. and Higginson, G.R., *Journal of Mechanical Engineering Science*, Vol. 2, No. 3, 1960, p. 188.
26. Wymer, D.G. and Cameron, A., *Institution of Mechanical Engineers. Proceedings*, Vol. 188, 1974, p. 221.
27. Archard, J.F. and Cowking, E.W., *Institution of Mechanical Engineers. Proceedings*, Vol. 180, Part 3B, 1965-1966, p. 47.
28. Gohar, R. and Cameron, A., *American Society of Lubrication Engineers Transactions*, Vol. 10, 1967, p. 215.

29. Cheng, H.S., *Journal of Lubrication Technology*, Vol. 92, No. 1, January 1970, p. 155.
30. Hamrock, B.J. and Dowson, D., *Journal of Lubrication Technology*, Vol. 98, No. 2, April 1976, p. 223.
31. Hamrock, B.J. and Dowson, D., *Journal of Lubrication Technology*, Vol. 98, No. 3, July 1976, p. 375.
32. Hamrock, B.J. and Dowson, D., *Journal of Lubrication Technology*, Vol. 99, No. 2, April 1977, p. 264.
33. Hamrock, B.J. and Dowson, D., *Journal of Lubrication Technology*, Vol. 99, No. 1, January 1977, p. 15.
34. Chiu, Y.P., et al., "Exploratory Analysis of EHD Properties of Lubricants," SKF Technical Report No. AL 72P010, 1972.
35. Wedevan, L.D., Ph.D. Thesis, University of London, March 1970.
36. Westlake, F.J. and Cameron, A., *Institution of Mechanical Engineers. Proceedings.*
37. Parker, R.J. and Kannel, J.W., "EHD Film Thickness Between Rolling Discs With a Synthetic Paraffinic Oil to 589 K," NASA Technical Note D-6411, 1970.
38. Gentle, C.R., Duckworth, R.R. and Cameron, A., *Journal of Lubrication Technology*, Vol. 97, July 1975, p. 383.
39. Dörr, J., *Ingenieur-Archiv*, Vol. 22, No. 3, 1954, p. 171.
40. Weber, C. and Saalfeld, K., *Zeitschrift fuer Angewandte Mathematik und Mechanik*, Vol. 34, Nos. 1-2, 1954, p. 54.
41. Orsterle, J.F. and Stephenson, R.R., *American Society of Mechanical Engineers Transactions*, Vol. 5, No. 2, 1962, p. 365.
42. Petrusevich, A.I., *Izvestiya Akademia Nauk SSSR* (OTN), Vol. 2, 1952, p. 209.
43. Dowson, D. and Higginson, G.R., *Journal of Mechanical Engineering Science*, Vol. 1, No. 1, 1959.
44. Archard, G.D., Gair, F.C. and Hirst, W., *Royal Society of London. Proceedings.* Series A, Vol. 262, 1961, p. 51.
45. Cheng, H.S. and Sternlicht, B., *Journal of Basic Engineering*, Vol. 87, No. 3, 1965, p. 695.
46. Hamilton, G.M. and Moore, S.L., *Royal Society of London. Proceedings.* Series A, Vol. 322, 1971, p. 313.
47. Rodkiewicz, C.M. and Srinivansan, V., *Journal of Lubrication Technology*, Vol. 94, No. 4, October 1972, p. 324.
48. Rohde, S.M., *Royal Society of London. Proceedings.* Series A, Vol. 343, 1975, p. 315.
49. Kannel, J.W., et al., "A Study of the Influence of Lubricants on High-Speed Rolling-Contact Bearing Performance," Technical Report No. ASD-TR-61-643, Air Force Aero Propulsion Laboratory, 1964, Part IV.
50. Dowson, D. and Longfield, M.O., in Proceedings of Lubrication and Wear Convention, Institution of Mechanical Engineers, 1963, Paper 3, p. 27.
51. Orcutt, F.K., *American Society of Lubrication Engineers Transactions*, Vol. 8, 1965, p. 381.
52. Kannel, J.W., *Institution of Mechanical Engineers. Proceedings*, Vol. 180, Part 3B, 1965-1966, p. 135.
53. Jacobson, B., *Acta Polytechnica Scandinavica. Mechanical Engineering.* Series 54, 1970.
54. Ranger, A.P., Ph.D. Thesis, University of London, March 1974.
55. Archard, J.F. and Kirk, M.T., in Proceedings of Lubrication and Wear Convention, Institution of Mechanical Engineers, 1963, Paper 15, p. 181.
56. Foord, C.A., et al., *Institution of Mechanical Engineers. Proceedings*, Vol. 184, Part 1, 1969-70, p. 487.
57. Moes, I.H., *Institution of Mechanical Engineers. Proceedings*, Vol. 180, Part 3B, 1965, p. 244.

58. Herrebrugh, K., *Journal of Lubrication Technology,* Vol. 90, No. 1, January 1968, p. 262.
59. Crook, A.W., *Royal Society of London, Philosophical Transactions,* Series A, Vol. 254, p. 237.
60. Bell, J.C., Kannel, J.W. and Allen, C.M., *Journal of Basic Engineering,* Vol. 86, No. 3, 1964, p. 423.
61. Dowson, D. and Whittaker, B.A., *Institution of Mechanical Engineers. Proceedings,* Vol. 180, Part 3B, 1965-66, p. 57.
62. Kannel, J.W. and Bell, J.C., in Elastohydrodynamic Lubrication 1972 Symposium, The Institution of Mechanical Engineers, Paper C 24/72, p. 118.
63. Kannel, J.W. and Zugars, F.F., "The Role of Temperature in EHD," Battelle Columbus Laboratory Technical Report G-6306, Columbus, Ohio, October 1976.
64. Bartz, W.J. and Ehlert, J., *Journal of Lubrication Technology,* Vol. 98, October 1976, p. 500.
65. Nagaraj, H.S., Sanborn, D.M. and Winer, W.O., *Wear,* Vol. 49, No. 1, July 1978, p. 43.
66. Jaeger, J.C., *Royal Society of New South Wales. Journal and Proceedings,* Vol. 56, 1942, p. 203; also Archard, J.F., *Wear,* Vol. 2, No. 6 (October 1959), p. 438.
67. Crook, A.W., *Royal Society of London, Philosophical Transactions,* Series A, Vol. 255, 1963, p. 281.
68. Johnson, K.L. and Cameron, R., *Institution of Mechanical Engineers. Proceedings,* Vol. 182, 1967-68, p. 307.
69. Dyson, A., *Royal Society of London, Philosophical Transactions,* Vol. 266, No. 1170, 1970, p. 1.
70. Harrison, G. and Trachman, E.G., *Journal of Lubrication Technology,* Vol. 95, No. 4, October 1972, p. 306.
71. Johnson, K.L. and Tevaarwerk, J.L., *Royal Society of London. Proceedings.* Series A, Vol. 356, 1977, p. 215.
72. Barlow, A.J., et al., *Royal Society of London. Proceedings.* Series A, Vol. 327, 1972, p. 403.
73. Montrose, C.J., Moynihan, C.T. and Sasake, H., "Dynamical Shear and Structural Viscoelasticity in EHD Lubrication," Vitreous State Laboratory Tech. Report, July 1977.
74. Cheng, H.S., *American Society of Lubrication Engineers Transactions,* Vol. 8, No. 4, October 1965, p. 397.
75. Bell, J.C., *American Society of Lubrication Engineers Transactions,* Vol. 5, 1962, p. 160.
76. Trachman, E. and Cheng, H.S., in Elastohydrodynamic Lubrication 1972 Symposium, Leeds, England, April 1972, The Institution of Mechanical Engineers, 1972, Paper 37, p. 142.
77. Smith, F.W., *Journal of Basic Engineering,* Vol. 87, 1965, p. 170.
78. Winer, W.O., Private Communication, 1977.
79. Tallian, T.E., *Wear,* Vol. 21, 1972, p. 49.
80. Fowles, P.E., *Journal of Lubrication Technology,* Vol. 91, 1969, p. 464.
81. Fowles, P.E., *Journal of Lubrication Technology,* Vol. 93, July 1971, p. 383.
82. Patir, N. and Cheng, H.S., *Journal of Lubrication Technology,* American Society of Mechanical Engineers Transactions, Vol. 100, No. 1, January 1978, p. 12.
83. Johnson, K.I., Greenwood, J.A. and Poon, S.Y., *Wear,* Vol. 19, 1972, p. 91.
84. Christensen, H., *Institution of Mechanical Engineers. Proceedings,* Vol. 184, Part 1, No. 55, 1969-1970, p. 1013.
85. Berthe, D., D.Sc. Thesis, University of Lyon, 1974.

86. Chow, L.S.H. and Cheng, H.S., *Journal of Lubrication Technology*, Vol. 98, No. 1, January 1976, p. 117.
87. Cheng, H.S. and Dyson, A., *American Society of Lubrication Engineers Transactions*, Vol. 21, No. 1, January 1978, p. 25.
88. Dyson, A., *Tribology International*, Vol. 8, No. 2, April 1975, p. 77.
89. Dyson, A., *Institution of Mechanical Engineers. Proceedings*, Vol. 190, 1976, p. 52.
90. Bell, J.C., Dyson, A. and Hedley, J.A., *American Society of Lubrication Engineers Transactions*, Vol. 18, No. 1, 1975, p. 62.
91. Cheng, H.S., in Proceedings of 4th Leeds-Lyon Symposium on Lubrication, April 1977.
92. Christensen, H., *Royal Society of London. Proceedings*. Series A, Vol. 266, 1962, p. 312.
93. Christensen, H., *Journal of Lubrication Technology*, Vol. 92, January 1970, p. 145.
94. Lee, K.M. and Cheng, H.S., "The Effect of Surface Asperity on the Elastohydrodynamic Lubrication," NASA CR-2195, February 1973.

ADDITIONAL ASPECTS OF ELASTOHYDRODYNAMIC LUBRICATION

B. J. Hamrock

ABSTRACT

Professor Cheng† is to be congratulated for doing an outstanding job in presenting an up-to-date review of the varying aspects of elastohydrodynamic lubrication. The purpose of the present paper is to add some material which was not covered in Professor Cheng's paper, in particular, some recent work on elastohydrodynamic lubrication of materials of low elastic modulus and on hydrodynamic lubrication. Both topics will be applicable for contacts with any ellipticity parameter (ranging from a circular contact to a line contact). These two studies will be combined with some previous work to supply film thickness contours for the different regimes of lubrication as a function of the elasticity and viscosity parameters. From these studies the lubrication regime and the minimum film thickness within the contact can be easily determined.

NOMENCLATURE

a semimajor axis of contact ellipse

b semiminor axis of contact ellipse

E modulus of elasticity

E' $\dfrac{2}{\dfrac{1-\nu_A^2}{E_A}+\dfrac{1-\nu_B^2}{E_B}}$

ε elliptical integral of second kind

F normal applied load

G dimensionless material parameter, $\alpha E'$

H dimensionless film thickness, h/R_x

†These proceedings.

$\hat{H}$ dimensionless film thickness, $H(W/U)^2$

h film thickness

k ellipticity parameter, a/b

m dimensionless inlet distance

m^* dimensionless inlet distance at boundary between fully flooded and starved conditions

R effective radius

r radius of curvature

U dimensionless speed parameter, $u\eta_0/E'R_x^2$

u surface velocity in rolling direction, $(u_A + u_B)/2$

W dimensionless load parameter, $F/E'R_x^2$

α pressure-viscosity constant

β R_y/R_x

η_0 atmospheric viscosity

ν Poisson's ratio

Φ $\left[1 + \frac{2}{3B}\right]^{-1}$

Subscripts

A solid A

B solid B

E elastic

F fully flooded conjunction

I isoviscous

min minimum

R rigid

S starved conjunction

V viscous

x,y coordinate system

INTRODUCTION

Figure 1 points out the difference between hydrodynamic and elastohydrodynamic lubrication (EHD). Hydrodynamic lubrication normally deals with conformal contacts while elastohydrodynamic lubrication deals with non-conformal contacts. Elastohydrodynamic lubrication can be divided into materials of low and high elastic modulus. Note the maximum pressure is three orders of magnitude less for materials of low elastic modulus as compared to that of high elastic modulus. Also, the minimum film thickness is the smallest for EHD of materials of high elastic modulus and next smallest for EHL of materials of low elastic modulus and the largest for hydrodynamic lubrication contacts.

Lubricant in elastohydrodynamic lubrication of *high* elastic modulus materials enters a typical contact with atmospheric pressure, is subjected to a maximum pressure of approximately 300,000 psi, traverses a passage one thousand times longer than its height, and is then ejected into the atmosphere within a time span of about 10^{-4} or 10^{-5} second. The viscous character of the fluid changes drastically while passing through the contact, going from an easy flowing liquid to a pseudosolid and back to an easy flowing liquid in less than a fraction of a millisecond.

Lubricant in elastohydrodynamic lubrication of *low* elastic modulus materials enters a typical contact with atmospheric pressure and is subjected only to a maximum pressure of 3×10^2 psi due to the elastomeric nature of the material. For such materials the distortions are large even with light loads. Another feature of EHD of *low* elastic materials is the negligible effect of pressure on the viscosity of the lubricating fluid. Therefore the viscous character of the fluid does not change while passing through the contact.

EHD FOR MATERIALS OF LOW ELASTIC MODULUS

The work presented by Professor Cheng related only to materials of *high* elastic modulus (e.g., steel). The work that I would like to present in this section is for materials of *low* elastic modulus (e.g., nitrile rubber). Engineering applications in which elastohydrodynamic lubrication is important for *low*-elastic-modulus materials include seals, human joints, tires, and elastomeric-material machine elements.

The problem of fully flooded line contacts has been solved theoretically for *low*-elastic-materials by Herrebrugh(1), Dowson and Swales(2), and Boglin and Archard(3). The solutions of References 1 and 2 were obtained numerically and are based on simultaneous solutions of the hydrodynamic and elasticity equations; the analytical solution of Reference 3 relied on the assumption of a simplified form for the film shape in the contact region. Biswas and Snidle(4) used the approach of Reference 3 to solve the point-contact situation. Reference 5 presents, to the best of the author's knowledge, the first complete numerical solution of the problem of fully flooded isothermal elastohydrodynamic lubrication of elliptical contacts for low-elastic-modulus materials. Reference 6 extends the work of Reference 5 by studying the effect of lubricant starvation on pressure and film thickness within the conjunction. This section of the present paper will utilize the work of References 5 and 6.

Reference 5 produces the following dimensionless minimum film thickness for fully flooded elliptical contacts for materials of low elastic modulus

$$\left\{H_{min,F}\right\}_{I,E} = 7.43\left(1 - 0.85\, e^{-0.31\, k}\right) U^{0.65} W^{-0.21} \tag{1}$$

The subscript I,E refers to the isoviscous elastic lubrication condition which is the condition that exists in dealing with materials of *low* elastic modulus.

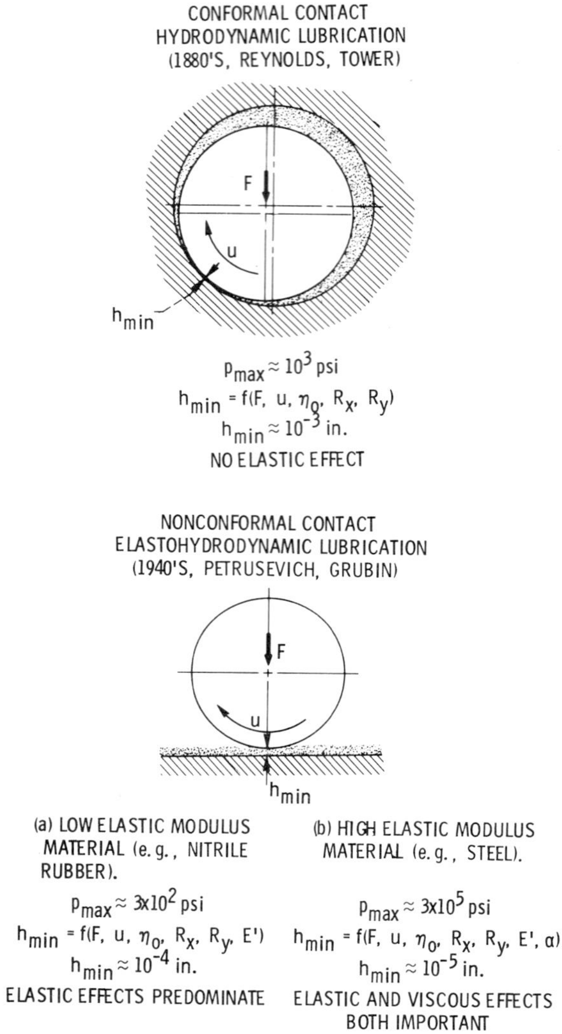

Fig. 1.—Hydrodynamic and elastohydrodynamic lubrication.

It is interesting to compare the equation for materials of *low* elastic modulus (eq. 1) with the corresponding equation found in Reference 7 for materials of high elastic modulus, namely:

$$\{H_{min,F}\}_{V,E} = 3.63(1 - e^{-0.68k})U^{0.68}W^{-0.073}G^{0.49} \tag{2}$$

The powers of U in equations (1) and (2) are quite similar, but the power of W is much more significant for *low*-elastic-modulus materials. The expressions showing the effect of the ellipticity parameter is of exponential form in both equations, but with quite different constants.

A major difference between equations (1) and (2) is the absence of a material parameter in the expression for the minimum film thickness for low-elastic-modulus materials. The reason for this is the negligible effect of pressure on the viscosity of the lubricating fluid for low-elastic-modulus materials.

When the elliptical contact becomes starved, the dimensionless minimum film thickness for materials of low elastic modulus (6) can be written as

$$\{H_{min,S}\}_{I,E} = \{H_{min,F}\}_{I,E}\left[\frac{m - 1}{m^* - 1}\right]^{0.22} \tag{3}$$

where m is the distance of the inlet miniscus from the center of the contact, and m^* is the inlet distance required for achieving the fully flooded condition. From Reference 6 m^* can be expressed as

$$m^* = 1 + 1.07\left[\left(\frac{R_x}{b}\right)^2 \{H_{min,F}\}_{I,E}\right]^{0.16} \tag{4}$$

where b is the semiminor axis of the elliptical contact.

To make it easier to calculate the semiminor axis of the contact ellipse (b), the elliptical integral of the second kind ($\bar{\varepsilon}$), and the ellipticity parameter, the approximate expressions developed in Reference 8 will be used and are given below as

$$b = \left[\frac{6\varepsilon F}{\pi k E'\left(\frac{1}{R_x} + \frac{1}{R_y}\right)}\right]^{1/3} \tag{5}$$

$$\varepsilon = 1 + \frac{3R_x}{5R_y} \tag{6}$$

$$k = 1.03\left(\frac{R_y}{R_x}\right)^{0.64} \tag{7}$$

where

$$\frac{1}{R_x} = \frac{1}{r_{Ax}} + \frac{1}{r_{Bx}} \tag{8}$$

$$\frac{1}{R_y} = \frac{1}{r_{Ay}} + \frac{1}{r_{By}} \tag{9}$$

The approximate expressions (eqs. 5-7) enable one to easily calculate these terms within 3 percent accuracy without resorting to charts or numerical methods.

To explain more fully what happens to the film thickness in going from a fully flooded to a lubricant starvation condition for materials of low elastic modulus, Figures 2 and 3 are presented. Figure 2 represents a fully flooded condition and Figure 3 a severely starved condition. In these figures the symbol + indicates the center of the Hertzian contact. Because of the way the coordinates are made dimensionless, the actual contact ellipse becomes a Hertzian circle regardless of the ellipticity parameter. The Hertzian contact circle is shown in these figures by asterisks. At the top of these figures the contour lable and its corresponding value is given. The inlet region is to the left and the exit region is to the right.

It is observed in these figures that the central portion of the film-thickness contours has become parallel for the severely starved condition (Figure 3). The minimum film thickness area is closer to the exit region for the severely starved condition. Note also that the values of the film thickness contours for the severely starved condition (Figure 3) are much lower than those of the fully flooded condition (Figure 2).

DIMENSIONLESS GROUPING

The dimensionless group $\{H, U, W, G, \text{and } k\}$ used in equations (1) and (2) have served a very useful purpose in aiding the understanding of the results found in References 5-7. Several authors (e.g., References 9 and 10) have noted that the examination of the dimensionless representation of governing equations shows the above set of dimensionless group can be reduced by one without any loss of generality. Although Johnson's paper(10) does not consider elliptical contacts, it does state what the nondimensional parameters would be, namely,

Dimensionless film thickness

$$H_{min} = H_{min} \left(\frac{W}{U}\right)^2 \tag{10}$$

Dimensionless viscosity parameter

$$g_1 = \frac{GW^3}{U^2} \tag{11}$$

Dimensionless elasticity parameter

$$g_3 = \frac{W^{8/3}}{U^2} \tag{12}$$

The ellipticity parameter (k) remains a parameter as discussed in equation (7). Therefore, the reduced dimensionless group is $\{\hat{H}_{min}, g_1, g_3, \text{and } k\}$.

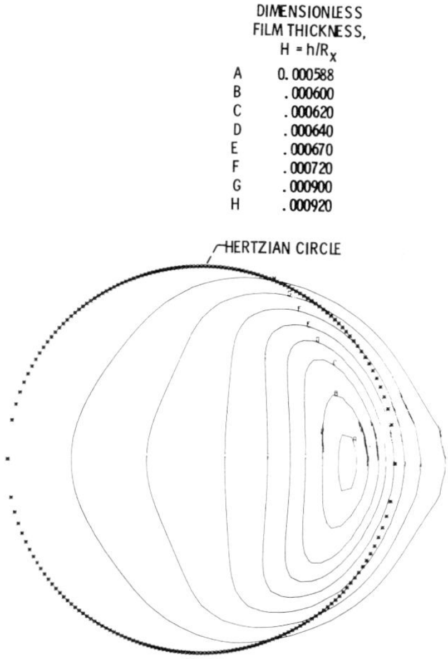

Fig. 2.—Contour plot of dimensionless film thickness in a fully flooded conjunction for materials of low elastic modulus.

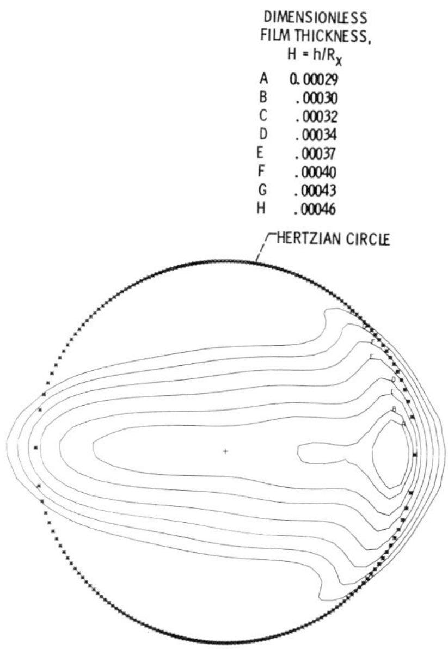

Fig. 3.—Contour plot of dimensionless film thickness is a severely starved conjunction for materials of low elastic modulus.

FILM THICKNESS IN ELLIPTICAL CONTACTS FOR FOUR LUBRICATION REGIMES

Reference 11 makes use of the reduced dimensionless grouping to give the film thickness equations for four fluid film lubrication regimes found in elliptical contacts. These regimes are distinguishable by the influence or lack of influence of elastic and viscous effects. The film thickness equations for the respective regimes come from earlier theoretical studies on elastohydrodynamic (5 and 7) and hydrodynamic (12) lubrication of conjunctions of elliptical form.

Isoviscous Rigid Regime

The influence of geometry on the hydrodynamic film separating two rigid solids was investigated in Reference 12. It was found that the minimum film thickness had the speed, viscosity, and load dependence as the classical Kapitza[13] solution. However, the incorporation of Reynolds condition resulted in an additional geometry effect. Therefore, from Reference 12 the dimensionless minimum film thickness for the isoviscous rigid lubrication regime can be written as

$$\{\hat{H}_{min}\}_{I,R} = 128\beta\Phi^2 \left[1.68 \tan^{-1} \frac{\beta}{2} + 0.13\right]^2 \tag{13}$$

where

$$\beta = \frac{R_y}{R_x} \approx \left(\frac{k}{1.03}\right)^{1/0.64} \tag{14}$$

$$\Phi = \left[1 + \frac{2}{3\beta}\right]^{-1} \tag{15}$$

In equation (13) note that the dimensionless film thickness is strictly a function of the geometry of the contact (R_y/R_x).

Viscous Rigid Regime

Making use of Blok[14] while adding a geometry effect for an elliptical contact as discussed in Reference 11, the minimum film thickness for the viscous rigid regime can be written as

$$\{\hat{H}_{min}\}_{V,R} = 1.66\, g_1^{2/3} (1 - e^{-0.68k}) \tag{16}$$

Isoviscous Elastic Regime

Expressing equation (1) in terms of the reduced dimensionless group ($\hat{H}_{min}$, g_1, and k) from Reference 11 the dimensionless minimum film thickness for the isoviscous elastic regime can be written as

$$\{\hat{H}_{min}\}_{I,E} = 8.70\, g_3^{0.67} (1 - 0.85\, e^{-0.31\, k}) \tag{17}$$

Viscous Elastic

Expressing equation (2) in terms of the reduced dimensionless group (H_{min}, g_1, g_3, and k) the minimum film thickness for the viscous elastic regime can be written as

$$\{\hat{H}_{min}\}_{V,E} = 3.45\, g_1^{0.49} g_3^{0.17} (1 - e^{-0.68\, k}) \tag{18}$$

Having defined the dimensionless minimum film thickness for the four lubrication regimes in equations 13, 16-18 these equations were used to develop the dimensionless film thickness contours for the different regimes of lubrication. Reference 11 shows these maps on a log-log grid of dimensionless viscosity parameter and dimensionless elasticity parameter for ellipticity parameters of 1, 2, 3, 4, and 6. Figure 4 shows a sample of these results for an ellipticity parameter of 1. The four lubrication regimes are clearly indicated in this figure. In this figure we see that

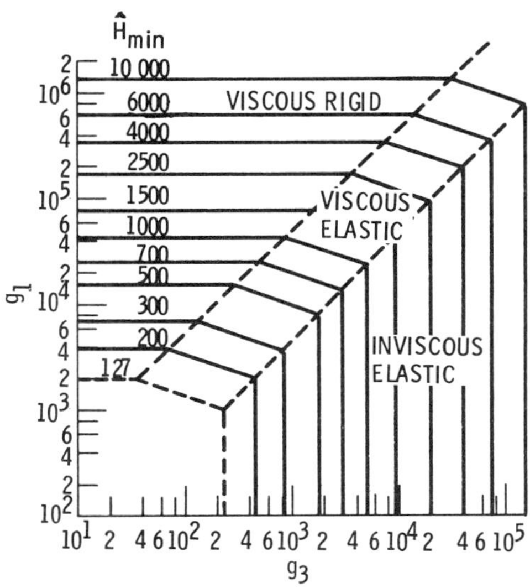

Fig. 4.—A map of the regimes of lubrication with dimensionless film thickness ($\hat{H}_{min}$) contours on a log-log grid of the dimensionless viscosity parameter (g_1) and the dimensionless elasticity parameter (g_3) for an ellipticity parameter of one ($k = 1$).

the transition between the viscous rigid and viscous elastic is nearly smooth whereas the transition from the isoviscous elastic and viscous elastic is not as good. One might speculate that the transition from the isoviscous elastic and viscous elastic might produce slightly lower dimensionless film thicknesses than those in Figure 4. Similar results are shown for other ellipticity parameters in Reference 11.

Therefore, given an ellipticity parameter (k), the value of the viscosity parameter (g_1), and the elasticity parameter (g_3), one can determine the lubricating regime as well as an approximate value of the dimensionless minimum film thickness. Knowing the lubrication regime a more accurate value of $\hat{H}_{min}$ can be obtained using the appropriate dimensionless minimum film thickness equation ((13), (16), (17), or (18)).

REFERENCES

1. Herrebrugh, K., *Journal of Lubrication Technology,* Vol. 90, 1968, p. 262.
2. Dowson, D. and Swales, P.D., in Proceedings of the 4th International Conference on Fluid Sealing, British Hydromechanics Research Association, 1969, Vol. 2, Paper 1, p. 1.
3. Baglin, K.P. and Archard, J.F., in Symposium on Elastohydrodynamic Lubrication, Institution of Mechanical Engineers, 1972, p. 13.
4. Biswas, S. and Snidle, R.W., *Journal of Lubrication Technology,* Vol. 98, 1976, p. 524.
5. Hamrock, B.J. and Dowson, D., *Journal of Lubrication Technology,* Vol. 100, 1978, p. 236.
6. Hamrock, B.J. and Dowson, D., *Journal of Lubrication Technology,* Vol. 101, 1979, p. 92.
7. Hamrock, B.J. and Dowson, D., *Journal of Lubrication Technology,* Vol. 99, 1977, p. 264.
8. Brewe, D.E. and Hamrock, B.J., *Journal of Lubrication Technology,* Vol. 99, 1977, p. 485.
9. Moes, H., *Institution of Mechanical Engineers. Proceedings,* Vol. 180, Part 3B, 1965-66, p. 244.
10. Johnson, K.L., *Journal of Mechanical Engineering Science,* Vol. 12, 1970, p. 9.
11. Hamrock, B.J. and Dowson, D., Fifth Leeds-Lyon Symposium on Tribology, Elasto-Hydrodynamics and Related Topics, September 19-22, 1978.
12. Brewe, D.E., Hamrock, B.J. and Taylor, C.M., *Journal of Lubrication Technology,* Vol. 101, 1979, p. 231.
13. Kapitza, P.L., *Zhurnal Tekhnicheskoi Fiziki,* Vol. 25, 1955, p. 747.
14. Blok, H., *Institute of Petroleum. Journal,* Vol. 38, 1952, p. 673.

DISCUSSION

H. P. EVANS, University College of Cardiff: I like to ask what you have done about the variation of the film thickness with the ellipticity parameter, k. I would like to point out k is independent of the load; it depends on the geometry of the situation. I find it difficult to believe that the variation of film thickness with the shape of the contact is going to be independent of the load and the speed.

B. J. HAMROCK: In changing the ellipticity you consider what effect it has by keeping the other parameters constant.

EVANS: Yes, I agree with you. What I am getting at is that you are assuming that the effect of it is going to be the same for all U, u, etc.

B. J. HAMROCK: The ellipticity parameter is the last one you consider. You first study what the effect of dimensionless speed is. You can do that directly without affecting the other parameters such as load and materials. Then you consider the effect of geometry. You incorporate that effect into established exponents for these. You then try to find out an expression that would best fit this information. We went through a number of different forms and we found that indeed the elliptical form seemed to fit the data much better.

TOPICS RELATED TO THE FUNDAMENTALS OF EHD PHENOMENA

J. Pirvics

ABSTRACT

This material comments and expands on Professor Cheng's survey of recent developments in the analysis of elastohydrodynamic contacts. Application within general mechanical system simulation software is noted. Some analytic advances in "soft" EHD contact performance assessment are described and comments are offered on recent efforts in the description of the "elastic" interaction. A brief note is also made of a correction to the published literature on asperity load sharing.

INTRODUCTION

Professor Cheng's paper† is a timely and comprehensive update of the state of the art in EHD contact phenomena. Additional references, detailing more investigations, within the categories noted under the heading of concentrated contacts, are available, but do not redefine the substance of the matter he presents.

The following contains brief discussions of some topics which are related to "The Fundamentals of EHD Phenomena". The material:

- takes note of the implementation of existing EHD analyses within mechanical system simulation software
- describes some analytic advances in "soft" EHD contact assessment
- comments on recent efforts in the analysis of the "elastic" interaction, and
- advises of a recent correction, as yet not in the public literature concerning the computation of the asperity-borne portion of a concentrated contact load.

APPLICATION WITHIN SYSTEM SIMULATION SOFTWARE

The past years, during which effort has been devoted to the detailing of individual EHD contact behavior, have also witnessed the emerging

†These proceedings.

confidence to place such models into mechanical system simulation computer programs. The complete range of contact behavior analysis from full to partial films, under varying thermal, load, and lubricant supply conditions have been used. Reference [1] contains papers by Gupta, Sibley and Pirvics which note programs containing EHD analysis for rolling contacts.

An example of such software is SHABERTH. The mechanical system addressed starts with a flexible shaft which is supported by a number of bearings. The *SH*aft *BEAR*ing analysis logic is *TH*ermally coupled to lubricant flows, contact heat generation and housing heat dissipation. System performance is computed with consideration of detailed individual contact behavior.

A particular example of such system analysis by software is that by Crecelius et al.[2] for the input pinion of a helicopter transmission. The type of contact behavior encountered is seen by examining the nodal maps used by SHABERTH, Figures 1 and 2. Successful performance simulation is found in corroboration with experimental evidence. Table 1 displays the steady state nodal temperature map. Figure 3 illustrates the time transient temperature behavior of a particular outer raceway within a ball bearing after loss of lubrication.

Further confidence in the validity of current EHD contact behavior assessment is found in the experimental/analytic work performed by Coe and Zaretsky[3] on high speed ball bearing performance.

The correlations obtained have established confidence in the analytic tools to guide hardware commitment. This has been done for thrust carrying, flange loaded, cylindrical rolling element bearings. The prototype generated and its performance are reported by Morrison et al.[4]. Ceramic rolrolling element contact performance has been explored similarly within a mechanical system simulation to reveal solutions for high speed, high temperature use.

ANALYSIS OF "SOFT" EHD CONTACTS

Cheng's review of EHD fundamentals restricts itself to the efforts which relate to the study of concentrated contacts. Other developments, within the broader definition of the discipline, merit attention. Specifically the behavior of compliant surface fluid film bearings is mentioned here.

Benjamin[5] has addressed the general three dimensional, finite boundary, elasticity problem. A Green's function approach is used in the analysis and it results in a procedure which eliminates the problem occasioned by mathematical singularities present when Poisson's ratio approaches 0.5. The elasticity formulation has been coupled with a variable film thickness Reynolds' equation and the resulting set of nonlinear equations solved. The specific applications considered by Benjamin address thrust as well as self-acting journal bearings, Figure 4. Both have a compliant layer on the bearing surface. The computed deflection profiles reproduced in Figure 5 reveal the conservation of material volume which is characteristic of deformed elastomeric layers. The self sealing and thus self imposed lubricant flow restriction as well as elevated load carrying capacity are clearly displayed for this elastohydrodynamic bearing. Figures 6 and 7 are offered as examples of available simulation of EHD performance within such practical applications as main shaft bearings of ships.

Pirvics and Castelli[6,7] formulated and explored inertial and viscoelastic effects of finite thickness compliant surface layer EHD behavior in slider and journal configurations, Figure 8. Elasticity and fluid mechanics representations are coupled and solved by a multidimensional Newton-Raphson iteration on Fourier coefficients. Numerical computation indicates that layer inertia effects become significant as the shear wave speed defined

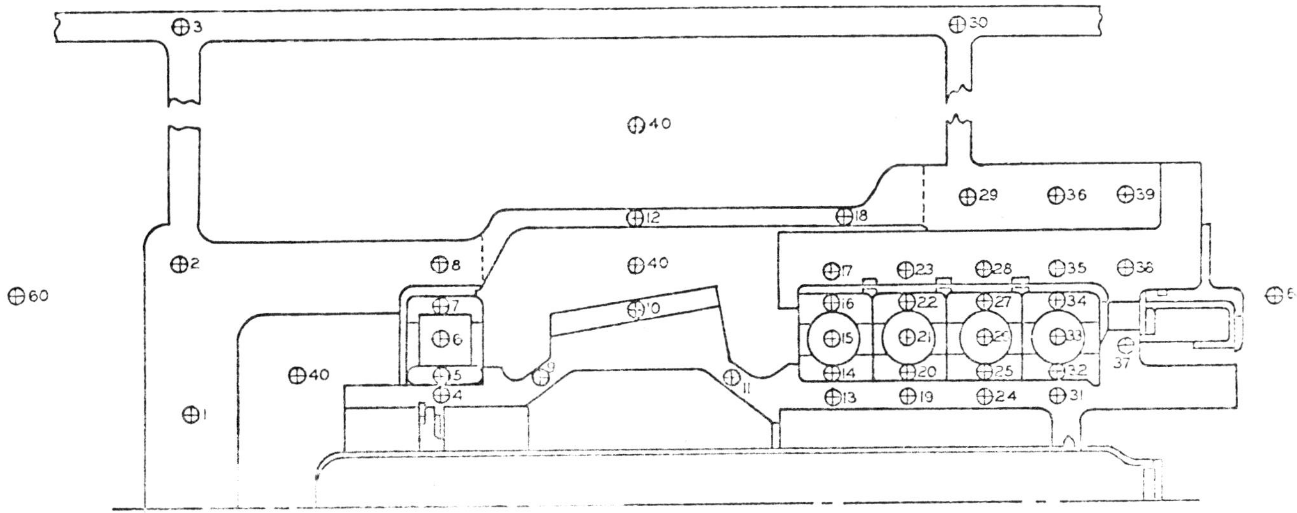

Fig. 1.—Calculated and experimental temperatures at time T = 0 after termination of lubricant supply. CH53 helicopter nose gear box.

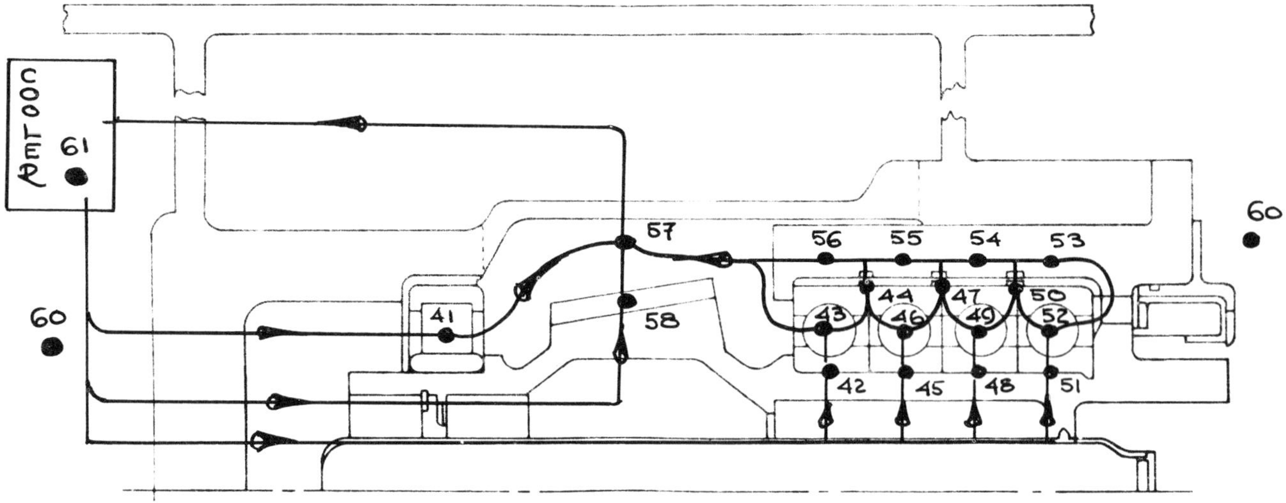

Fig. 2.—Lubrication node map. CH53 helicopter nose gear box.

TABLE I.—STEADY STATE NODAL TEMPERATURE

Node No.	Calc. Temp. °C	Measured Temp. °C	Node No.	Calc. Temp. °C	Measured Temp. °C	Node No.	Calc. Temp. °C	Measured Temp. °C	Node No.	Calc. Temp. °C	Measured Temp. °C
1	25		12	87		23	103		34	102	92/91
2	58		13	87		24	86		35	99	
3	76		14	96		25	94		36	96	
4	119		15	109		26	108		37	111	
5	121	116	16	104	105	27	102	102/92	38	96	
6	117		17	103	105	28	100		39	95	
7	105	106	18	94		29	97		40	89	
8	96		19	86		30	77				
9	112	97	20	95	102	31	88		60	24	
10	116		21	108		32	96				
11	107		22	104	102/105	33	109				

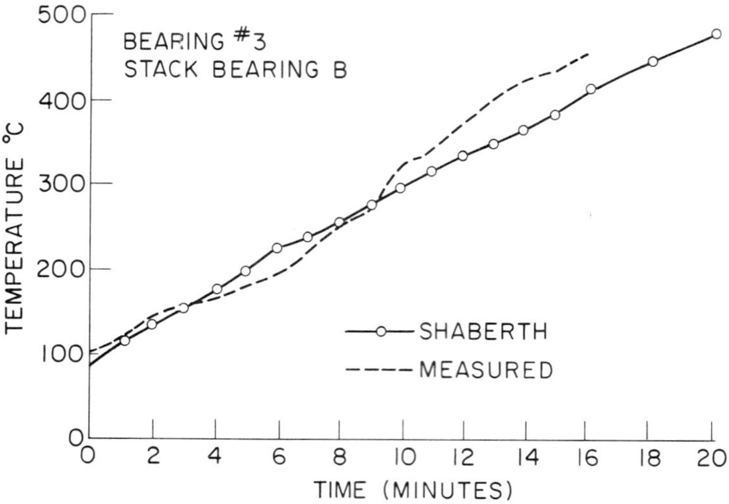

Fig. 3.—Outer raceway temperature.

Mach number exceeds 0.5. Particular attention is directed to the extent of the displacement profile outside of the defined load zone. Implications are raised for successive slider coupling through unrelaxed elastomer deformations in these "soft" elastohydrodynamic bearings, Figure 9.

THE "E" IN EHD

Most current analyses of EHD concentrated contacts employ a Hertz definition of the interacting body problem and assume purely elastic behavior to exist within an unstrained homogeneous material. Many practical materials, geometries and loading conditions do not respect this desire for computation simplicity. Some additional recent efforts which address the "elastic" part of elastohydrodynamic lubrication are noted in the following.

Asperity-film and asperity-asperity interactions produce tangential forces. These tangential loads interact with those in the normal direction to redefine contact shape as well as subsurface stress strain states. Tallian, et al.[12] comprehensively investigated the effects of asperity slope and surface traction. The results are noteworthy for their ability to relate traction coefficients to experimentally observed metallographic changes in the subsurface. Specifically, a rotation of the Martin angle of deformation bands is observed to vary with film thickness/roughness ratio. This rotation is related to an effective traction coefficient. The authors are presently examining the altered non-"Hertzian" stress-strain state to refine the accepted formulation for bearing life to include traction/surface effects in the contact.

A continuing effort, seeking characterization for the formation of residual stress/strain states and their interaction with surfaces has been underway by Chiu[8-11]. Figure 10 illustrates a half pane subjected to changes in the applied moving load. Here, only a 10% increase in the maximum "Hertz" pressure creates a trailing subsurface plastic deformation band where none had existed previously. The zone, Figure 11, has linear dimensions in the order of the contact area and begins within a quarter of the contact width beneath the surface.

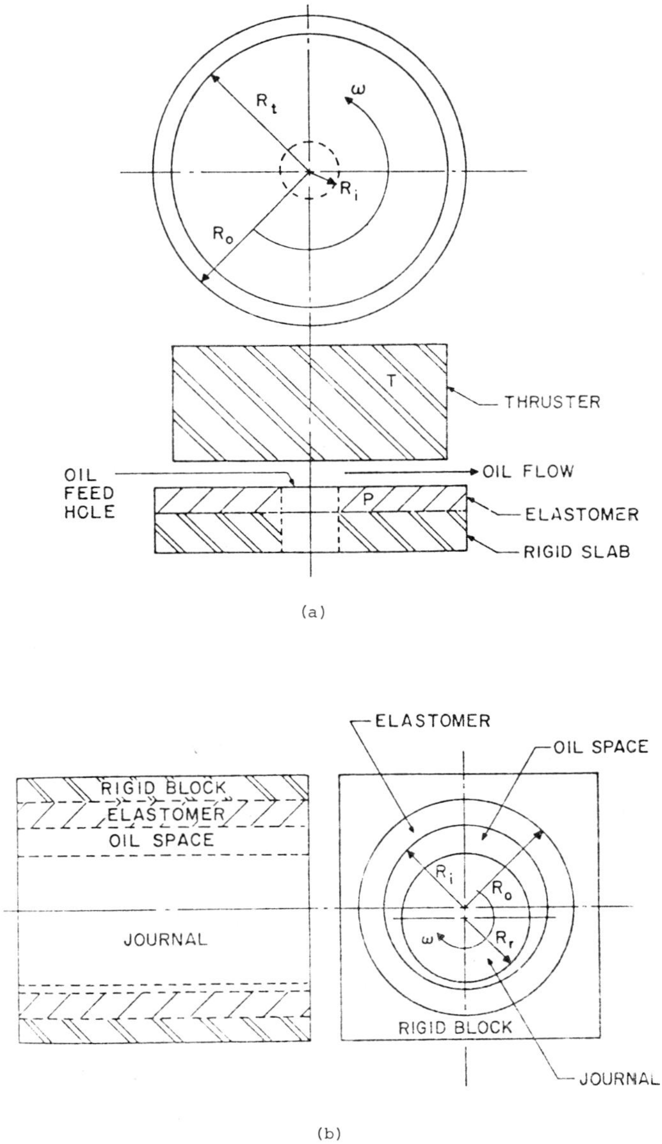

Fig. 4.—(a) Thrust bearing geometry. (b) Journal bearing geometry.

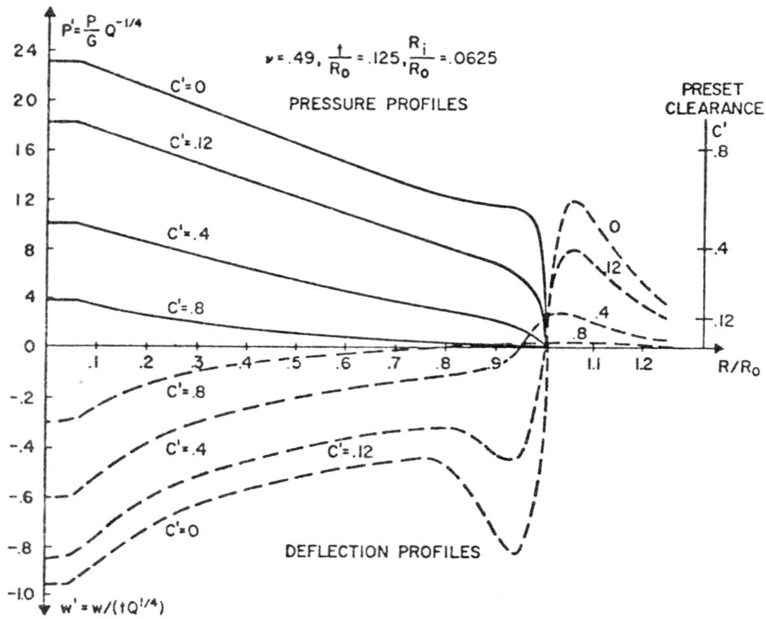

Pressure Profiles ————
Deflection Profiles — — — —

Fig. 5.—Sample pressure and deflection profiles for various values of dimensionless preset clearance C'.

When a moving concentrated contact has passed, leaving such a strained state, the free surface for subsequent load passage is permanently altered. Figure 12 illustrates the permanent displacement from a given reference plane within a half space containing a strained element. Dilatational as well as unidirectional strain effects are presented. These results indicate a procedure for computation of material "shakedown" due to the repeated passage of a concentrated load.

ASPERITY/EHD FILM LOAD RATIO COMPUTATION - A CORRECTION

Tallian(13), within the literature cited by Cheng, prescribed computation procedures for calculating asperity to EHD contact load ratios. A correction(14), as yet not in the public literature, has been formulated.

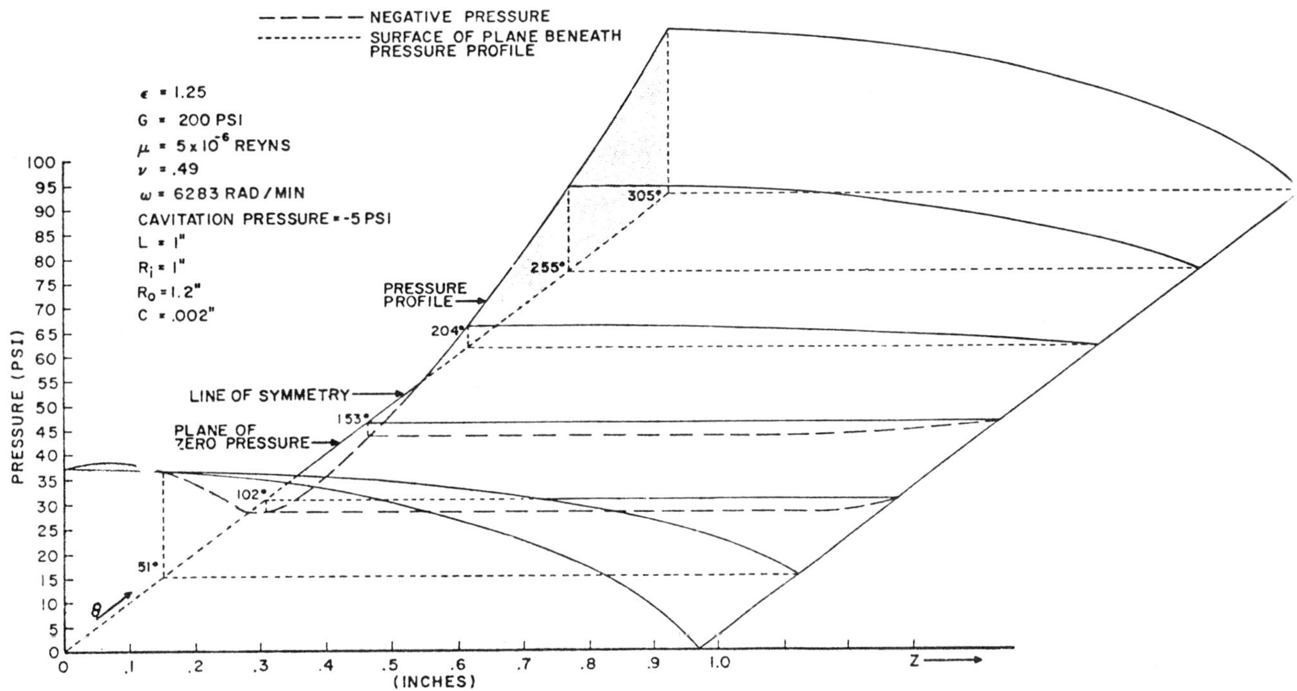

Fig. 6.—Pressure profile for finite elastomer compliant surface journal bearing.

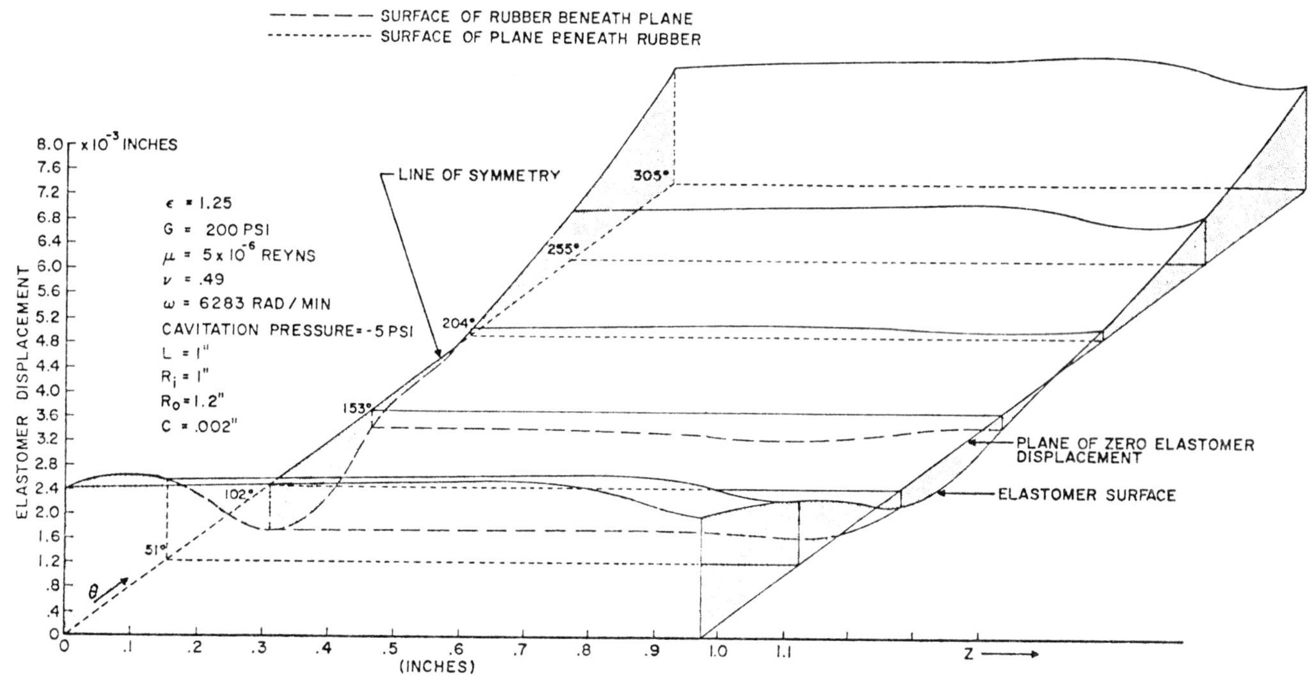

Fig. 7.—Elastomer displacement profile for finite elastomer compliant surface journal bearing.

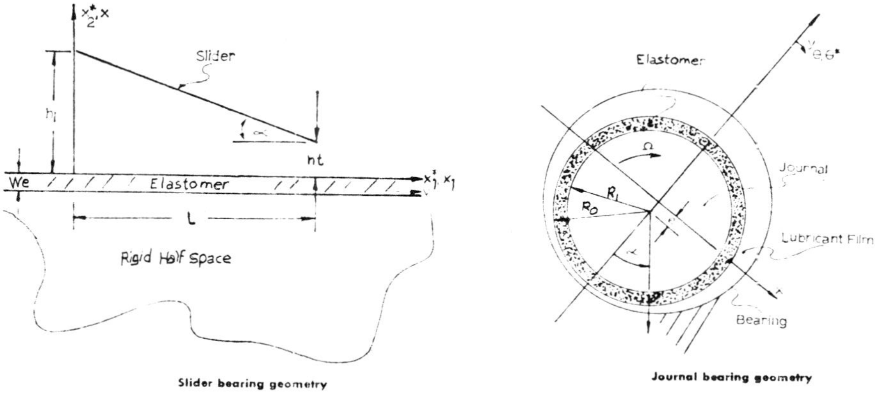

Fig. 8.—Slider and journal bearing configurations.

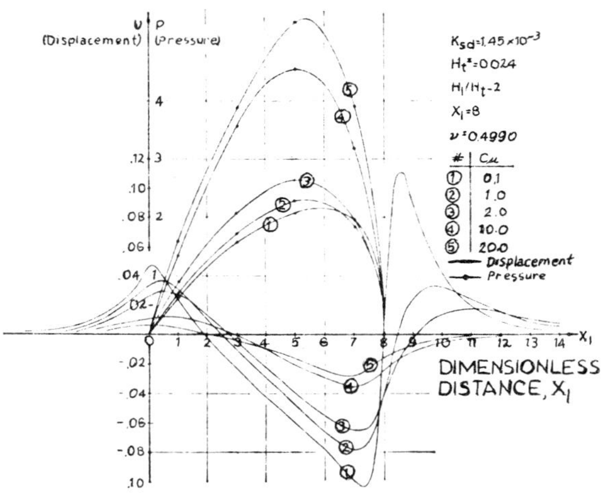

Fig. 9.—Viscoelastic effect of damping.

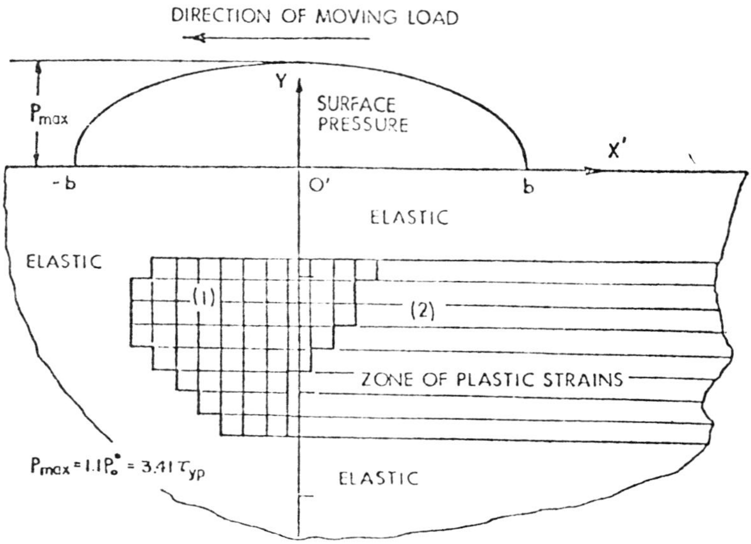

Fig. 10.—A coordinate system (x',y) moving with load and discretizing (rectangular and semi-infinite) elements for the plastic zone in a half plane.

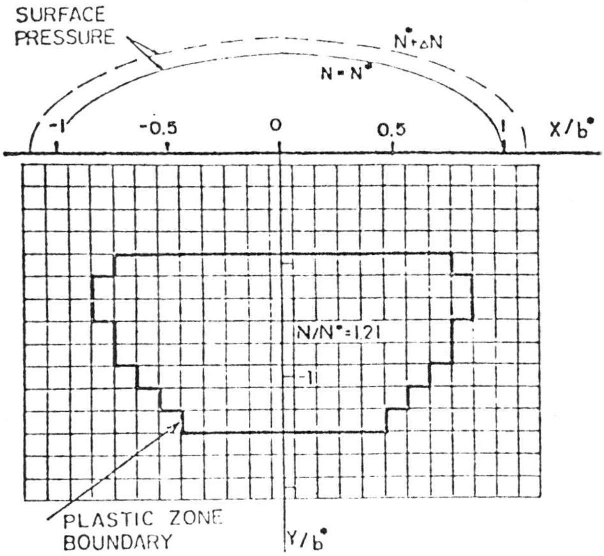

(a)

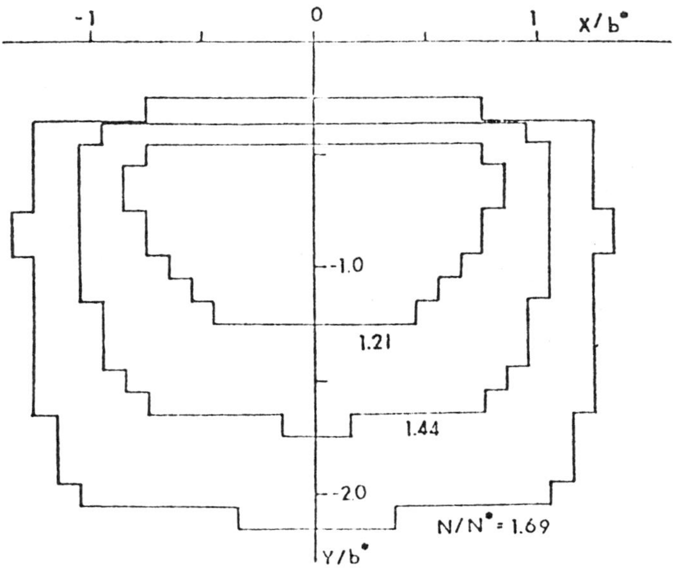

(b)

Fig. 11.—(a) Arrangement of square elements and computed plastic zone under an applied normal load N/N = 1.21. (b) Boundaries of plastic zone at three applied loads.*

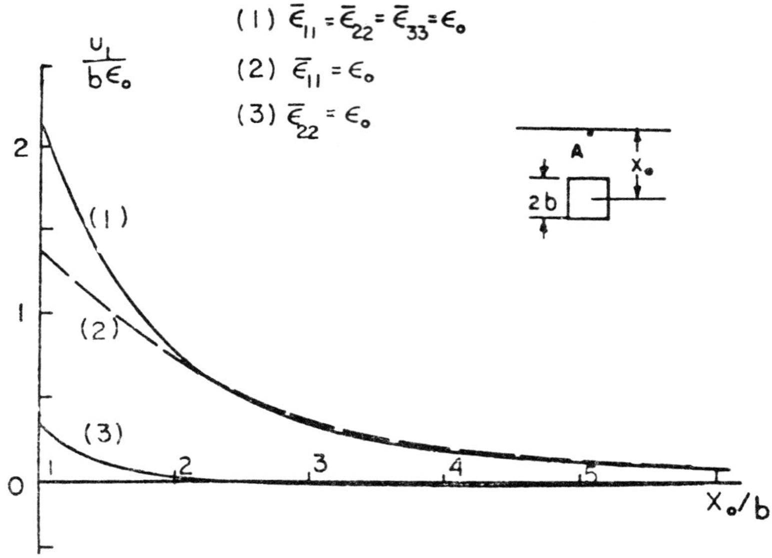

Fig. 12.—Surface deformation due to residual strain.

The author of that work has revised his original manuscript to conclude that the total load carried by the Hertz contact should be expressed as the *sum* of

a. the area integral of the contact pressure causing asperity deflection *plus* the EHD pressure acting over *all* asperity contact areas, and

b. the area integral of EHD pressures acting on areas within the Hertz contact that are not occupied by asperity contacts.

The correction also stipulates that the *sum* of the ambient film and asperity deflection pressures is to be used in computing the normal load in calculating boundary lubrication traction. In consideration of micro-EHD effects *without* boundary lubrication, the pressures of macro and micro-EHD are to be superimposed to generate viscous traction. Here *macro*-EHD traction must be calculated *only* over areas *not* covered by asperity contacts.

This change in computation assumes obvious significance for thin films and high slip regimes. Practical experience offering lower than theoretically predicted indications of sliding in some rolling bearing applications substantiates the validity of the corrections proposed.

REFERENCES

1. Kennedy, F.E. and Cheng, H.S., "Computer-Aided Design of Bearings and Seals," American Society of Mechanical Engineers, New York, 1976, pp. 19, 95.
2. Crecelius, W.J., Liu, J.Y. and Pirvics, J., "Predictions of the Time-To Failure of the CH-53 Helicopter Power Input Module After Loss of Lubrication," USA Ballistic Research Laboratories, 1978, in preparation.
3. Coe, H.H. and Zaretsky, E.V., "Predicted and Experimental Performance of Jet-Lubricated 120-Millimeter-Bore Ball Bearings Operating to 2.5 Million DN," NASA TP1196, April 1978.
4. Morrison, F.R., Pirvics, J. and Crecelius, W.J., *Journal of Lubrication Technology,* Vol. 101, 1979, p. 164.
5. Benjamin, M.K., Ph.D. Thesis, Columbia University, New York, 1969.
6. Pirvics, J. and Castelli, V., *Journal of Lubrication Technology,* Vol. 95, 1973, p. 372.
7. Pirvics, J. and Castelli, V., *Journal of Lubrication Technology,* Vol. 95, 1973, p. 363.
8. Chiu, Y.P., *Journal of Applied Mechanics,* Vol. 49, 1977, p. 587.
9. Chiu, Y.P., *Journal of Applied Mechanics,* Vol. 45, 1978, p. 302.
10. Chiu, Y.P., "On the Solution of Stresses and Displacements of a Half Plane and Layer Containing Localized Initial Strains," to be published.
11. Chiu, Y.P., in Proceedings of Ninth Southeastern Conference of Theoretical and Applied Mechanics, May 1978, p. 79.
12. Tallian, T.E., Chiu, Y.P. and Van Amerongen, E., *Journal of Lubrication Technology,* Vol. 100, April 1978, p. 155.
13. Tallian, T.E., *Wear,* Vol. 21, 1972, p. 49.
14. Tallian, T.E. and McCool, J.I., "Corrections to the Paper 'The Theory of Partial EHD Contacts,'" SKF Report AL78Q004.

DISCUSSION

J. L. TAVAARWERK, University of Waterloo: I congratulate you for your excellent presentation considering the complexity that one can get into in these elastic-plastic topics. You showed the time-temperature trace. Where does most of the heat come from? Is it from the churning of the lubricants in these systems or is it actually in the slip and spin losses in the contact?

J. PIRVICS: The case shown indicates behavior in the absence of lubrication, however, slip, spin, churning and drag effects are considered in the software. It has been shown, particularly by some of the NASA work which has just come out, that the assumed volume of lubricant within the bearing cavity plays an extremely significant role in the correlation between experiment and theory in predicting temperatures. Correlation is very sensitive to churning.

TEVAARWERK: This is what I expected.

PIRVICS: The effect of the lubricant present within a bearing cavity upon heat generation is in a gray area which can be bounded because experiments are available which give an idea about the temperatures generated. You can bound the input for analysis, but by no means is it an accurate calculation.

TEVAARWERK: But for the first order approximation one could just simply look at the churning losses.

PIRVICS: It varies from running condition to running condition. But you can have as much as half of it going into churning.

TEVAARWERK: So the statement sometimes is made that by reducing the slip losses in EHD one can make contributions to the energy conservation. It is a bit of a fifth order effect.

PIRVICS: I would not say a fifth order.

TEVAARWERK: But it can be a secondary order.

PIRVICS: Yes.

H. S. CHENG, Northwestern University: I am really very delighted to see that some of the things that we have been talking about are now really incorporated in these series of programs that you have at SKF. I am interested to know whether you have other programs besides SHABERTH in your series that deal with gears and other elements.

PIRVICS: We have programs that deal with general systems such as SHABERTH. We have specific programs which go into the details of starvation effects. We also have programs which look at particular bearing types. I estimate there are twenty or thirty different programs which go into various degrees of detail concerning the EHD contact itself.

CHENG: One particular aspect which I am interested in is the starvation effects. According to Y. P. Chiu, it comes in much earlier in the rolling element bearing systems than inlet heating effects.

PIRVICS: Yes, there is a program which specifically addresses that and the starvation effects are included within SHABERTH.

THICK FILM AND TRANSIENT ELASTOHYDRODYNAMIC LUBRICATION PROBLEMS

S. M. Rohde

ABSTRACT

The mathematical formulation of the EHD problem is discussed together with methods for obtaining approximate solutions. Research related to the normal approach, transient, and "soft" EHD problems is reviewed. Particular emphasis is placed on compliant hydrodynamic bearings.

INTRODUCTION

In the previous review paper, Professor Cheng has given an illuminating account of some of the research efforts that have been underway in the area of concentrated contact elastohydrodynamic lubrication (EHD). Indeed, that area has developed rapidly during the past twenty years.

In parallel with that development, several other related EHD areas have also progressed rapidly. These areas include the normal approach EHD problem, the more general transient EHD problem, and the class of "soft" EHD problems (steady state or transient). These problems arise in the study of the lubrication of gear teeth, the lubrication of natural and artificial biological joints as well as mechanism joints, compliant slider and journal bearings, elastomeric seals, and the viscous hydroplaning of tires. It is precisely the significance of these areas from an engineering point of view which has motivated the development of this area of EHD.

This paper will review research related to the normal approach, transient, and "soft" EHD problems. Before doing so, and because this class of lubrication problems is so intriguing from analytical point of view, the general mathematical formulation of the EHD problem will first be briefly discussed together with methods for obtaining approximate solutions. The reader who is not interested in such "details" may proceed directly to Section 3 where the steady state compliant hydrodynamic bearing problem is discussed. In Section 4 the normal approach problem is reviewed together with the more general transient EHD problem.

Although a considerable amount of research has been directed toward the solution of these problems, more work remains to be done. Some of the areas for future research will be outlined in the appropriate sections of this paper. Finally, the author wishes to emphasize the fact that no attempt was made to make this review complete. The author's intent was, rather, to give the reader a "flavor" for this branch of EHD and some of its applications.

MATHEMATICAL FOUNDATIONS OF ELASTOHYDRODYNAMICS LUBRICATION THEORY

Elastohydrodynamic lubrication theory is concerned with the solution of a complex fluid-solid interface problem for the creeping flow of a viscous fluid between deformable boundaries. The combined continuity and momentum equations for this configuration is the famous Reynolds equation of hydrodynamic lubrication theory. For an incompressible lubricant having constant properties across the thickness of the fluid film, Reynolds equation can be written as:

$$\nabla \cdot \frac{h^3}{12\mu} \nabla p = \nabla \cdot \frac{(U_1 + U_2)}{2} h + \frac{\partial h}{\partial t}. \tag{1}$$

In equation (1) p is the pressure generated in the fluid film, $U1$ and $U2$ are the tangential velocities of the top and bottom bearing surfaces, μ is the lubricant viscosity, and h is the film gap. The region over which equation (1) is to hold is the bearing surface area and the boundary condition on p is (usually) that p has a specified value on the boundary of the region.

In equation (1), $\partial h/\partial t$ represents the rate at which the bearing surfaces are approaching each other. For steady state operation this term vanishes identically whereas for dynamically loaded bearings, it is precisely this term which governs the pressure levels and allows the resultant of the pressure on the bearing surface to remain in equilibrium with the applied load at each instant of time.

If h is a given function of position and time, we see that, depending upon whether μ is a function of pressure or not, a linear or nonlinear boundary value problem results, respectively. If, however, the film thickness, h, also depends upon the film pressure, a highly nonlinear (typically an integrodifferential equation) results. The solution of this equation is the central mathematical problem in EHD studies.

Before proceeding to outline methods to solve the EHD problem, we first note that although the deformations of the bounding bodies in EHD theory may be large when compared to the fluid film thickness under consideration, they are small when compared to the dimensions of those bodies. Hence linear elastic theory is applicable and we may write the deformation field, d, due to the pressure field, p, as:

$$d = L\, p \tag{2}$$

where L is a linear operator. The simplest form of L is a constant, i.e., we assume a Winkler solid. In any event, the film thickness h can be written as

$$h(x,y,t) = h_g(x,y,t) + d = h_g(x,y,t) + L\, p \tag{3}$$

The function h_g does not depend upon pressure, and for a large number of cases can be written as

$$h_g(x,y,t) = h_o(t) + h_p(x,y) \tag{4}$$

where h_p is the undeformed geometry of the gap and h_o can be viewed as determining the nominal separation (and can be negative). By substituting (3) into (1) we observe that the character of the equation may change, i.e., an initial value problem in pressure results for the nonsteady state case (completely analogous to the corresponding situation for an unsteady gas lubricated bearing).

Turning now to the actual solution of the EHD problem, we first consider the steady state problem. The most natural iteration scheme for that problem is the so called "direct" method in which an assumed value of film thickness, h_o, is substituted into Reynolds equation and a pressure distribution p_o is obtained. From p_o a film thickness h_1 is obtained using (3) and the process is repeated. Schematically this process may be written as

$$h_{n+1} = f_1(h_n) \tag{5}$$

or

$$p_{n+1} = \overline{f}_1(p_n)$$

Unfortunately, this process typically diverges as EHD effects become important. To overcome this convergence difficulty, numerous investigators introduce a damping factor, β, and write for example:

$$h_{n+1} = \beta f_1(h_n) + (1-\beta)h_n \tag{6}$$

Introduction of such a damping factor can, indeed, lead to an increased range of convergence as shown, for example, by Stephenson and Osterle[1]. The manner in which a damping factor works can be illustrated by a very simple example. Suppose we wish to solve the equation $x = \alpha x,\ \alpha \neq 1$. We set $f(x) = \alpha x$ and write

$$x_{n+1} = f_n(x_n) = \alpha^{n+1}\ x_o \tag{7}$$

if $x_o \neq 0$, we see that $x_n \to 0$, the solution, if $|\alpha|<1$. On the other hand, if $|\alpha|>1$ or $\alpha = -1$ the scheme diverges or oscillates, respectively. Although this is a trivial example it does reflect the type of behavior we obtain using a scheme such as (5) to solve EHD problems. In the latter cases, however, the relationship between the operating conditions and geometry and the convergence is not as transparent as in our example. Now let us introduce a damping factor into simple problem. We obtain

$$x_{n+1} = \beta f(x_n) + (1-\beta)x_n = \beta\alpha x_n + (1-\beta)x_n = [1 + \beta(\alpha-1)]x_n$$

$$= [1+\beta(\alpha-1)]^{n+1}x_o = \gamma^{n+1}x_o \tag{8}$$

Observe that for $\alpha>1$ there is no positive value of β which allows $|\gamma|<1$. Hence the scheme would diverge. In this case, however, we can set $\beta < 0$ and recover convergence. This amounts to weighing previous estimates more heavily ($1-\beta>1$) and probably accounts for the reason that schemes such as presented by Huebner and Oh[2] converged. On the other hand if $\alpha<-1$ we see that $\beta>0$ can always be chosen so that $|\gamma|<1$. In some EHD problems the corresponding "α" is probably changing from iteration to iteration or (and) $\alpha<<-1$ so β must be chosen to be extremely small as Dowson and Hamrock[3] have done.

A critical aspect of this iteration scheme is the question of when to terminate the iteration. Typically, the iteration is terminated when

$$e_n = ||\ p_{n+1} - p_n\ || \leq \varepsilon\ ||\ p_n\ || \tag{9}$$

where $||\ ||$ is a norm, e.g., the integral of the square of the function in brackets and $\varepsilon>0$ is a small number. If our iterative process is converging slowly as it typically is in EHD direct iterative approaches in which we are forced to set $\beta\approx 0$, we should replace (9) by something of the form

$$\frac{e_n}{1-\dfrac{e_n}{e_{n-1}}} < \varepsilon\ ||\ p_{n+1}\ || \tag{10}$$

Expression (10) accounts for the slow convergence as the reader can verify by studying a geometric series with a ratio near one. By using (10), we will not terminate the process too soon.

Due to the slow convergence, or lack thereof, of the direct method for EHD problems other methods were devised. For steady state dimensional problems the inverse iteration scheme devised by Dowson and Higginson[4] was found to give satisfactory results. This scheme, however, makes essential use of the fact that the one dimensional Reynolds equation can be inverted. This, as well as certain other assumptions, preclude the use of the method for two dimensional problems.

One of the most powerful schemes for solving nonlinear equations of virtually any type is Newton's method and its variations. Several authors have used the Newton-Raphson method (Newton's method applied to systems of algebraic equations) to the algebraic problem resulting from the discretization of (or part of) the EHD problem. In particular we cite the work of Cheng and Sternlicht[5] and Pirvics and Castelli[6]. More recently Rohde and Oh[7] have applied Newton's method directly to the steady state EHD equations. Linear integrodifferential equations were obtained for the pressure correction term. A sequence of these linear problems are then solved (usually) by numerical methods. In particular, finite element methods have been found to lead to efficient computational schemes[7].

Another method which has been found to be effective for a class of EHD problems is the use of regular perturbation methods. These methods, when applicable, allow simple expressions (e.g., polynomials) for the lubricated contact performance characteristics to be derived. Unfortunately, however, the range of applicability of these techniques is rather limited as discussed by Dowson and Taylor[8] and Rohde[9].

Turning now to the numerical treatment of the transient problem, we remark that far less work has been done in this area. As mentioned previously, we note that the character of the EHD problem changes in this case. Recent work done by Rohde et al.[10] outlines an effective scheme for treating this class of problems. Essentially all time derivatives are first discretized at time t. Implicit representations are used. The resulting nonlinear integrodifferential equation for the pressure distribution at time t is then solved by Newton's scheme previously discussed. The nominal separation can either be a-priori given or $h_o(t)$ can be determined by specifying the load on the bearing at each instant of time. The latter requires the "adjoining" of a "constraint" equation to the EHD system and solving simultaneously for $p(x,y,t)$ and $h_o(t)$ (or $h_g(x,y,t)$). Inclusion of this constraint causes no difficulties. For additional discussion and details pertaining to the solution of lubrication problems, and in particular EHD problems, the reader is referred to Rohde[11].

One area in the mathematical and numerical analysis of EHD problems which requires more research effort, is the treatment of free boundaries. Such problems typically arise in cases in which a nominal diverging film gap is present. The effective incorporation of Reynolds boundary condition in such problems remains an open question, It is the author's contention, however, that this problem should be attacked from the point of view of variational inequalities.[46]

COMPLIANT HYDRODYNAMIC BEARINGS--STEADY STATE OPERATION

Compliant hydrodynamic bearings are of practical interest for a number of reasons. First, in a large number of fluid film bearing installations, hydrodynamic pressures are sufficiently high to distort the bearing components--although the latter are not intended to deform. Such situations exist, for example, in internal combustion engine bearings, and heavily loaded slider and thrust bearings. In other cases a compliant bearing geometry is intentionally incorporated into the design. This may be done to obtain favorable bearing dynamic response characteristics, to attain a high level of insensitivity to dirt in the lubricant or to non-ideal bearing geometries, to drastically reduce the lubricant flow requirements, or to reduce the cost to manufacture. Regardless of whether intentionally designed into the bearing or not, the resulting hydrodynamic bearing--in which the bearing components deform--leads to an EHD problem.

In 1964 Carl[12] published the results of an experimental investigation which showed that elastic distortions of journal bearings can be important even at relatively low pressures (14 MPa.). Carl found that the oil film extent increased, the peak film pressure was reduced, and the point of minimum film thickness moved toward the point of maximum film pressure because of elastic distortions. At about the same time Higginson[13] examined the EHD bearing problem for a journal bearing with a thin elastic liner. An infinitely long bearing was assumed. Furthermore, the bearing shell was assumed to behave as a Winkler solid, simplifying the problem considerably. Among the results he obtained, was the result that the coefficient of friction was reduced due to the deformation of the bearing shell.

A significant contribution to the study of the EHD journal bearing problem was made by O'Donoghue et al.[14]. They obtained solutions for the EHD journal bearing problem assuming an infinitely long bearing and a bearing housing which is infinite in extent (in the radial direction). The latter assumption, they claim, is valid for cases in which the outside diameter of the bearing shell is greater than about 3.5 times the shell diameter. This assumption, however, does seem a bit restrictive, particularly from the viewpoint of automotive engine bearing design. Nevertheless, their results are extremely interesting and reflect experimentally observed trends. Figures 1 through 5 show some of their results. Note how the presence of elastic deformations allows the eccentricity ratio to go beyond one. For a given load on the bearing, we also note that the minimum film thickness is *increased* and the peak film pressure is reduced. Hence from these results one may conclude that elasticity effects can be beneficial for this class of problems as they are in concentrated contact situations. Future research in this area should, perhaps, be directed toward purposely incorporating compliance into journal bearing housing design.

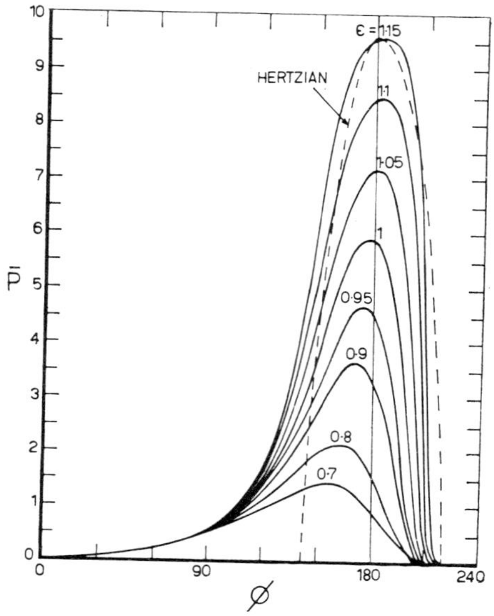

Fig. 1.—Pressure distributions for elastic bearing (Ref. 14).

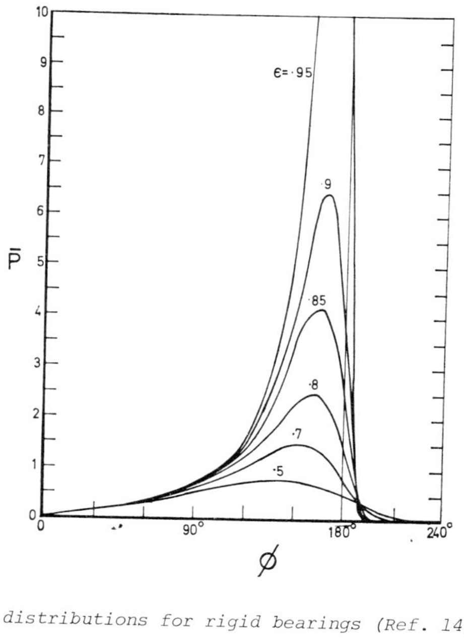

Fig. 2.—Pressure distributions for rigid bearings (Ref. 14).

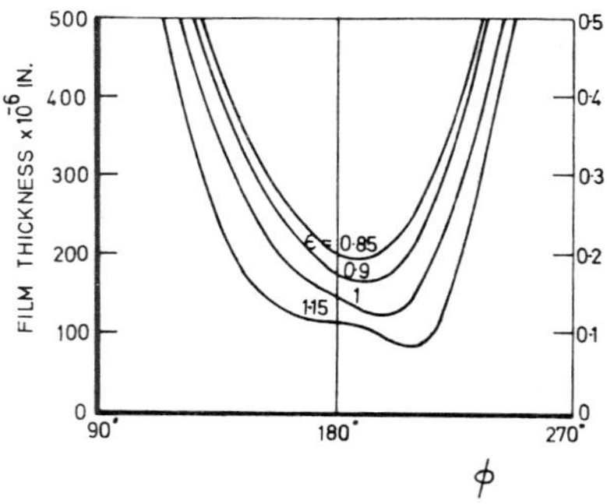

Fig. 3.—Film thickness for elastic bearing (Ref. 14).

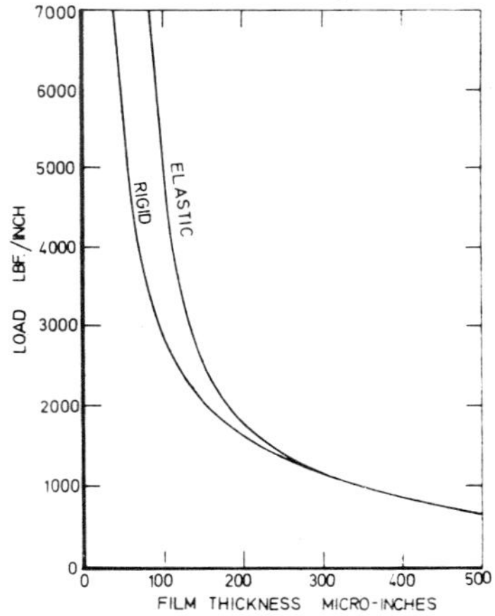

Fig. 4.—Load versus minimum film thickness (Ref. 14).

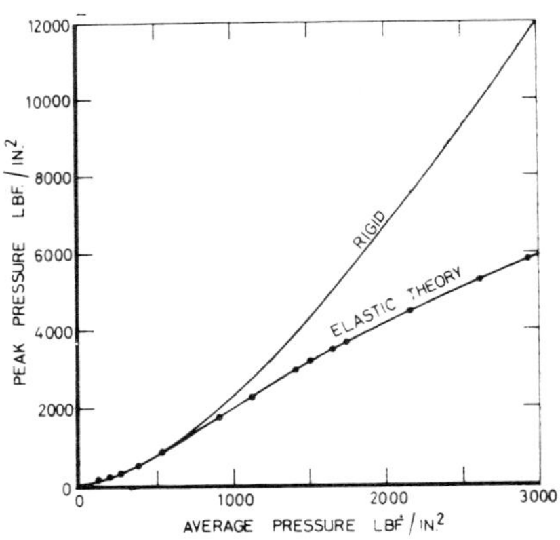

Fig. 5.—Peak pressure versus average pressure (Ref. 14).

One of the first analytical treatments of a *finite* compliant journal bearing was performed by Benjamin and Castelli[15]. Their analytical investigations, directed toward bearings having a compliant elastomeric layer as shown in Figure 6, led to interesting results such as a rapid decrease in attitude angle with increasing eccentricity ratio as shown in Figure 7. The latter result, they report, is due to a "pooling" of the fluid near the minimum film thickness. A large portion of[15] is devoted to a discussion of the numerical procedures used. Essentially the EHD problem is solved using a Newton-Raphson scheme.

At about that time finite element methods (FEM) were blossoming. These methods were thus naturally applied to lubrication problems beginning with Reddi[16,17]. The paper by Huebner and Oh[2] contains an application of FEM methods to the finite EHD journal bearing problem. A three dimensional finite element model of the geometry shown in Figure 8 was constructed. Although somewhat idealized, this geometry is canonical to the geometry of a connecting rod in a reciprocating engine. The coupling of the elasticity and fluid equations described in their paper followed a direct iterative scheme and thus required many iterations and a considerable amount of computer time for convergence. Nevertheless, several interesting results were obtained. A typical result is shown in Figure 9. This result is in qualitative agreement with previous work. Unfortunately no minimum film thickness data was presented.

More recently Fantino et al.[18] have constructed a finite element model for a connecting rod and have coupled that model to the fluid film equation (again directly). They obtain the interesting result that, for a given load, although the eccentricity ratio may increase beyond one, the minimum film thickness does *not* vary significantly from the value corresponding to a rigid bearing. The minimum film thickness, however, *decreases* due to elasticity effects as shown in Figures 10 and 11. The fact that the film thickness corresponding to the rigid and elastic cases are close, fortunately, adds credibility to standard bearing design practices.

An analysis of a compliant journal bearing operating with a compressible lubricant was given by Oh and Rohde[19]. They considered a thin foil shell as shown in Figure 12. Higher order finite elements were used together with their Newton scheme[7] to obtain approximate solutions. Numerical results were presented for a number of different operating conditions and geometrical parameters. A decrease in attitude angle over the corresponding values for the rigid bearing case was again found and offered as a possible explanation for the improved dynamic stability characteristics associated with these bearings. The configuration considered can be regarded as a more complex form of the classical foil bearing problem. The latter class of problems, although EHD, will not be discussed here. The reader is referred to[20-23] for additional information.

Like journal bearings, slider and thrust bearing components can deform due to pressure and (or) thermal loads. Such situations may again exist either intentionally as with elastomeric bearings or unintentionally as in the design of heavily loaded or (and) high speed thrust bearings.

Experimental and analytical investigations of a compliant hydrostatic thrust bearing were almost simultaneously published by Castelli et al.[24] and Dowson and Taylor[8]. Both sets of investigators studied the configuration shown in Figure 13. The motivation for studying this configuration is provided by the fact that an entrapment (and hence reduced flow) "phenomenon" is present in this configuration. Furthermore, this configuration is simpler to analyze and has practical significance in the design of hydrostatic bearings. In[24] a complete elastic layer model was studied as well as a column or Wirkler model whereas in[8] only the latter model was considered. In both cases, however, reasonable agreement between theory and experiment was obtained as shown in Figures 14 and 15. More recently the

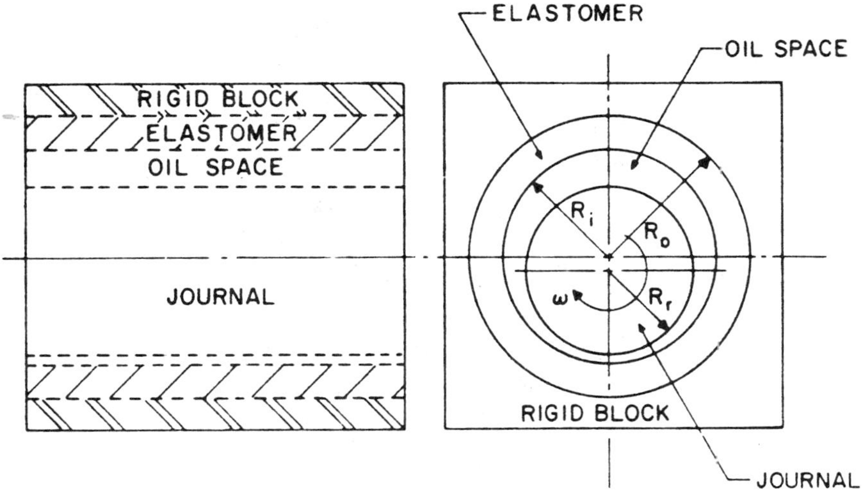

Fig. 6.—Journal bearing geometry (Ref. 15).

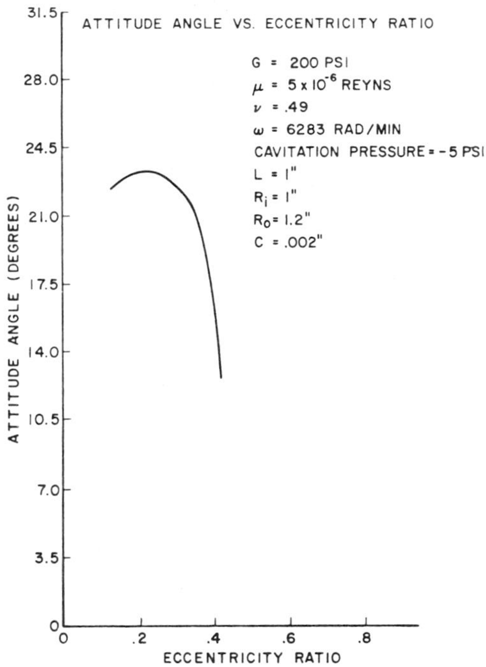

Fig. 7.—Infinite elastomer compliant surface journal bearing attitude angle versus eccentricity ratio (Ref. 15).

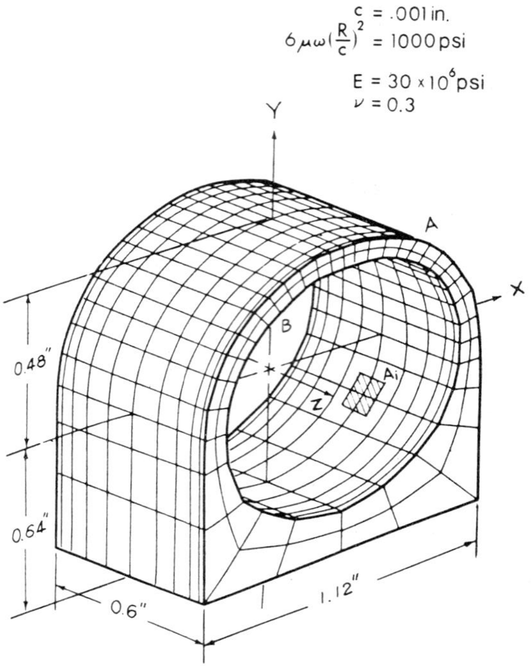

Fig. 8.—Bearing structure and coordinate system (Ref. 2).

EHD problem corresponding to a heavily loaded hybrid bearing was studied by Rohde et al.[25]. A typical result from that investigation is shown in Figure 16.

Although the fact that slider and thrust bearings can distort in operation has been known for a long time, it is only within the past fifteen or twenty years that invesitgators have been able to quantitatively characterize these distortions. An early contribution in this direction was provided by Sternlicht et al.[26] who analyzed a centrally pivoted thrust bearing pad. Thermal and elastic effects were included in their analysis. A plate model was used for the bearing components. From the cases considered by the authors, they concluded that elastic deflections of the pads can be several times larger than the minimum film thickness and must be included in a realistic analysis.

A discussion of a more complete numerical scheme to treat the tilting pad EHD problem (including thermal effects) is given by Castelli and Malanoski[27]. Included in their paper are some charts for designing tilting pad thrust bearings including thermal and elastic effects.

In a number of practical situations, deformations of the bearing components due to thermal loadings are significant. An extreme case which demonstrates the significance of thermal deformations is provided by the parallel surface (or centrally pivoted) thrust bearing. Fogg[28] first observed that parallel sliders can support a load. This phenomenon, termed the "Fogg effect" stimulated much research into fluid film bearing behavior. Many years later, Hahn and Kettleborough[29] showed that Fogg's results can be explained as resulting from thermal distortions of the bearing

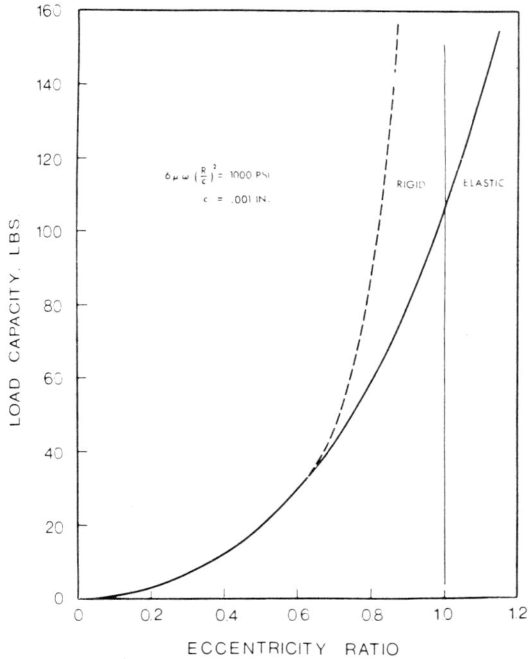

Fig. 9.—Load capacity for a bearing with a rigid or elastic housing (Ref. 2).

components. Hahn and Kettleborough considered the full thermohydrodynamic (THD) solution for an infinitely long slider bearing. (By THD we indicate that in the analysis: a) temperature variations *across* the fluid film thickness are considered as well as along the film extent, and b) heat transfer to the bearing components was accounted for.) They also conclude that the effect of thermal distortions is greatly reduced in the case in which the slider profile has a nominally converging region.

The first complete thermoelastohydrodynamic (TEHD) analysis of a *finite* slider bearing was provided by Rohde and Oh[(30)]. The configuration they analyzed is shown in Figure 17. Both mechanical and thermal distortions were included in their analysis. Plate models and semi-infinite half-space models were used to compute the deformations of the bearing components. For the cases they considered, the thermal effects dominated the elasticity effects as shown in Figure 18 and Table 1. These results are in complete agreement with the results of Reference 29.

Recent experimental verification of the load generating mechanism in parallel surface thrust washers can be found in the paper by Taniguchi and Ettles[(31)] who studied parallel surface sector bearings. Included in their paper are results from a thermo-elastic analysis similar to that contained in [(26)]. Reasonable agreement between theoretical and measured minimum film thicknesses were obtained as shown in Figure 19. The differences, however, between the film pressure distribution resulting entirely from analysis, and that obtained by substituting the measured film thicknesses into the film equations as shown in Figure 20, cause one to speculate as to the relative significance of the chamfer and the thermal distcrtions.

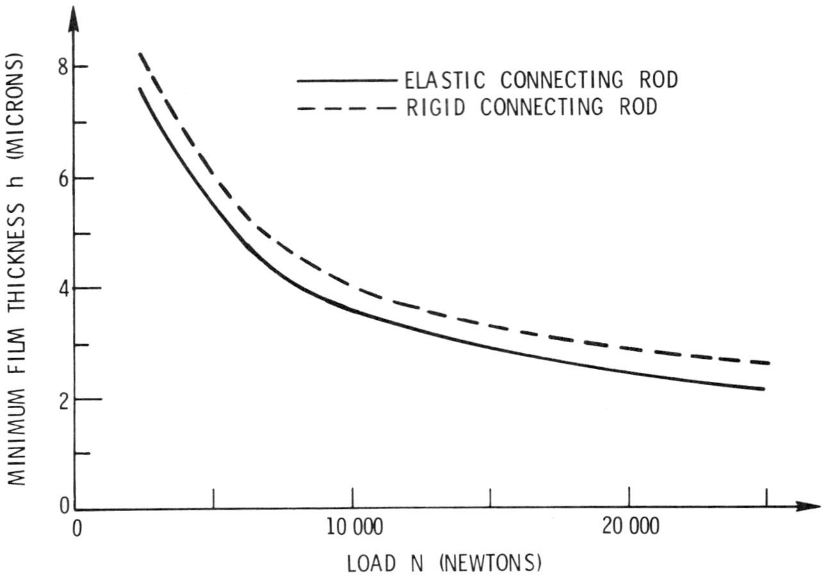

Fig. 10.—Minimum film thickness versus load (Ref. 18).

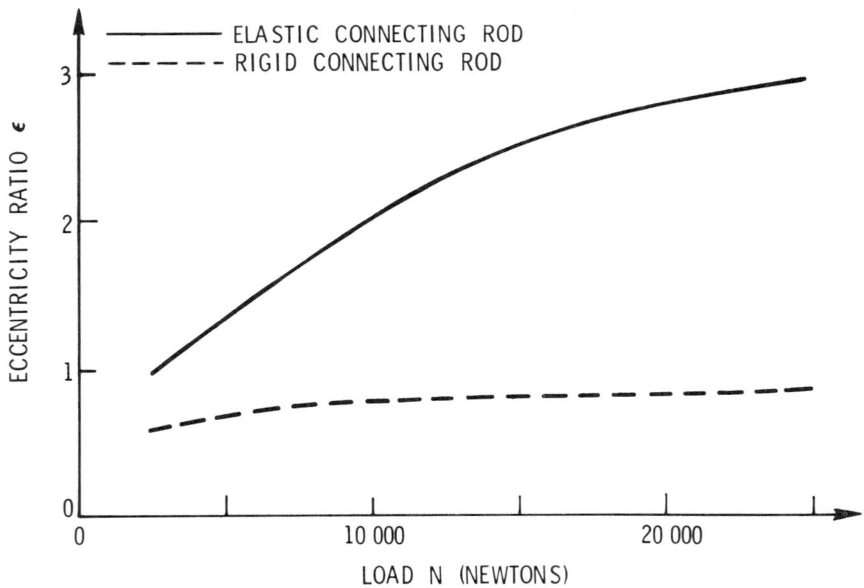

Fig. 11.—Eccentricity ratio versus load (Ref. 18).

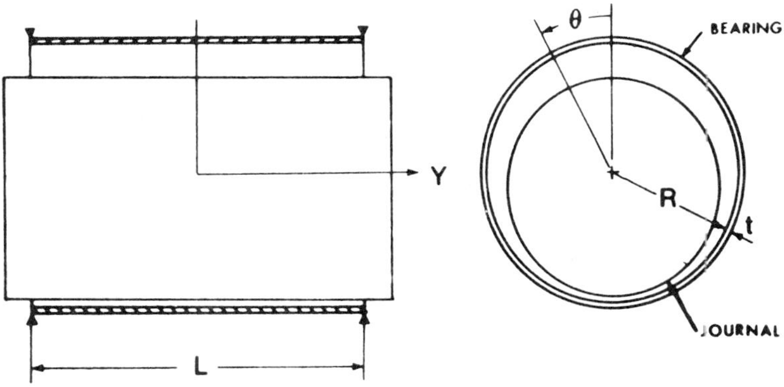

Fig. 12.—Schematic diagram of the shell bearing analyzed (Ref. 19).

In a subsequent paper, Ettles[32] outlines an analysis technique (or rather a computer program) for predicting the performance of sector tilting pad thrust bearings including the effects of thermal and elastic distortions. The analytical results are compared to experimental results obtained by Robinson[33]. Figures 21 and 22 show typical results obtained in this study. The improved agreement between theoretically and experimentally determined minimum film thicknesses as the unit pressure is increased is somewhat surprising, particularly in view of the increasing disparities in temperature rise.

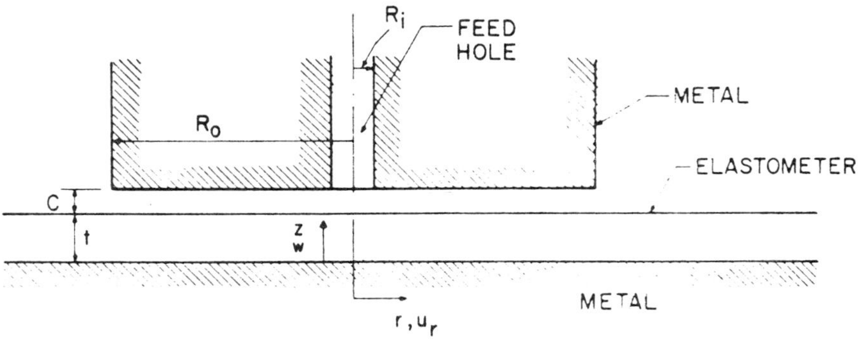

Fig. 13.—Bearing geometry (Ref. 24).

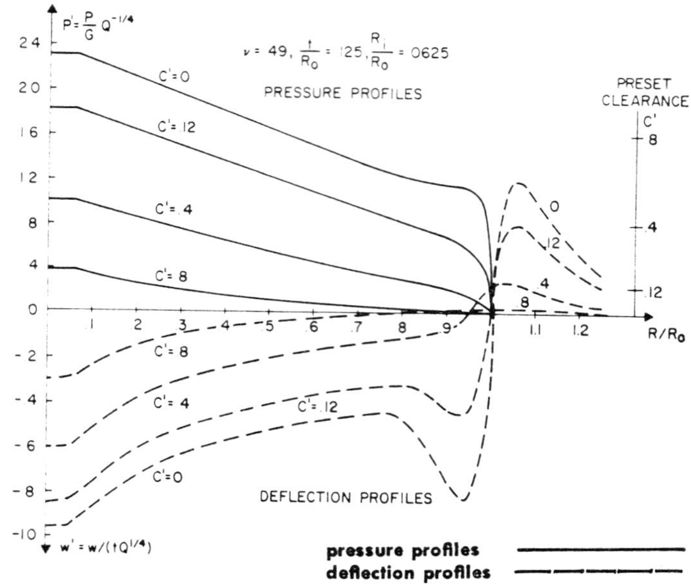

Fig. 14.—Sample pressure and deflection profiles for various values of dimensionless preset clearance C' (Ref. 24).

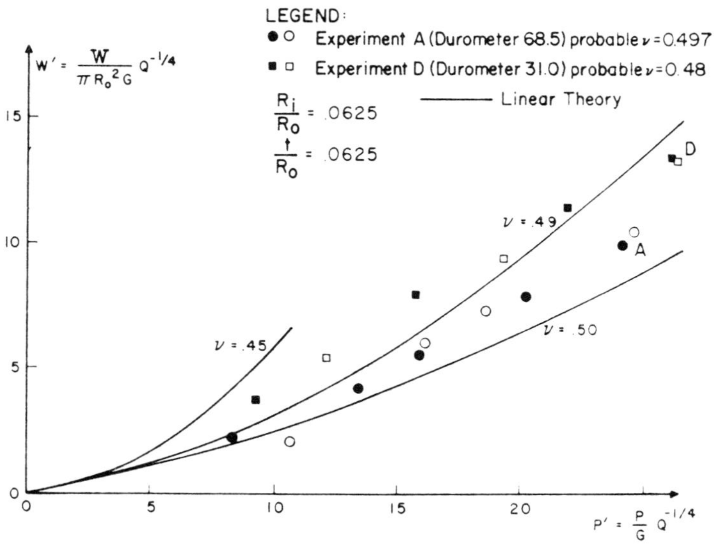

Fig. 15.—Load versus recess pressure (Ref. 24).

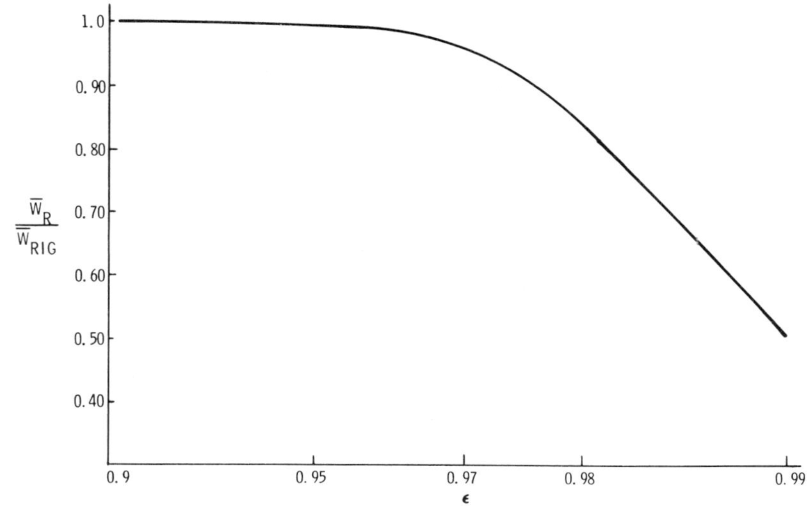

Fig. 16.—Effect of surface deformation on load capacity as a function of eccentricity ratio, $\overline{U}$ = 2.78, P_2 = 33.33 (Ref. 25).

An extensive experimental investigation of the performance of tilting pad compliant surface thrust bearings was reported by Rightmire et al.[34] A compliant elastomeric layer was affixed to a rigid backing in their experiments as shown in Figure 23. Pressure and film thickness traces were both taken for a wide variety of speeds, loads, and pivot locations. From their investigation, the authors conclude that compliant surface pivoted-pad thrust bearings can support loads at much lower speeds than can be realized with rigid shoes, can start and stop under load, can self-align, and can provide vibration attenuation.

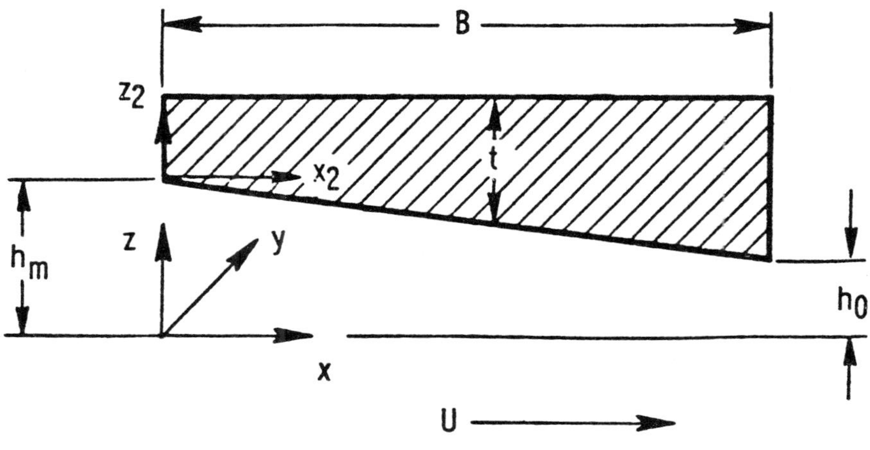

Fig. 17.—Schematic diagram of a slider bearing (Ref. 30).

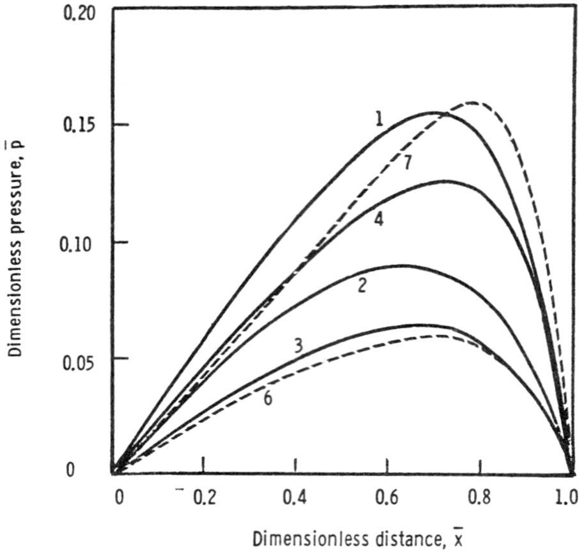

Fig. 18.—Centerline pressure distribution (Ref. 30).

TABLE I.—DESCRIPTION OF CASES CONSIDERED

Case	Fluid Solid Bndry Cond	Film Temp	Mechanical Dist Incl	Thermal Dist Incl	Solid Model	Load Var	Center of Press
1	iso	iso	no	no	rigid	0.068	0.58
2	iso	var	no	no	rigid	0.041	0.56
3	thd	var	no	no	rigid	0.029	0.57
4	iso	iso	yes	no	inf	0.056	0.59
5	iso	var	yes	no	inf	0.037	0.56
6	thd	var	yes	no	inf	0.027	0.58
7	iso	iso	yes	n0	plate	0.065	0.62
8	iso	var	yes	no	plate	0.040	0.58
9	thd	var	yes	no	plate	0.030	0.60
10	thd	var	no	yes	plate	0.029	0.57
11	thd	var	yes	yes	plate	0.030	0.60

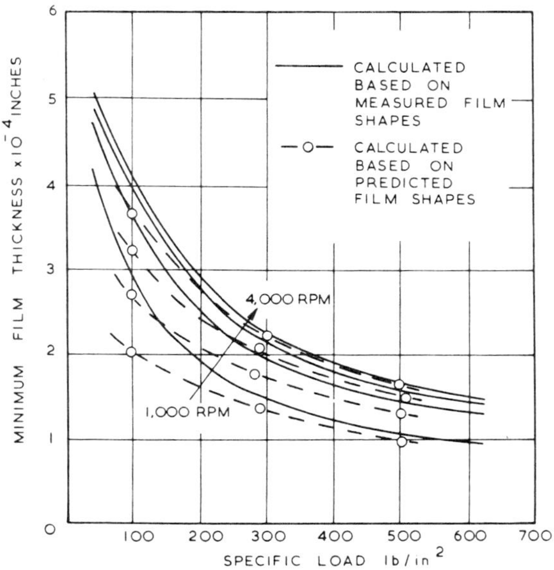

Fig. 19.—Variation of theoretical and experimental minimum film thickness with load for bearing W3 (Ref. 31).

In recent thesis by Eidelberg[35], a finite element approach to the poro-EHD problem is presented. A slider like geometry is considered. Both the geometry and the physical parameters considered are representative of those found in natural biological joints (see Figure 24). Numerous results for different parameter values are given by the author. These results extend those presented by Eidelberg and Booker[36] for the rigid porous problem.

In summary, although a good deal of work has been done in the compliant hydrodynamic bearing area, more remains to be done. In particular, more effort is needed to understand and to quantify the effects of deformations on bearing performance. More realistic geometries should be considered in the analyses. Ultimately, the results of such research efforts should be directed toward the design process.

NORMAL APPROACH AND TRANSIENT EHD PROBLEMS

Research related to the normal approach EHD problem began with the early work of Christensen[37]. He considered the problem of the normal approach of two infinitely long cylinders separated by a viscous fluid. An analysis and numerical scheme were presented for this problem. The rate of deformation term $(\partial d/\partial t)$, however, was omitted from the analysis, simplifying the problem. Results obtained showed the development of 'bumps" in the film thickness distribution which tend to move toward the ends of the contact region as the cylinders are moved toward each other. Simultaneously, the resulting pressure distribution approaches a Hertzian distribution. These results are shown in Figure 25. Qualitative verification of the analytical predictions was provided by experimental results obtained by the author. Since those experiments were performed using a ball rather than cylinder, qualitative agreement is all that one can expect.

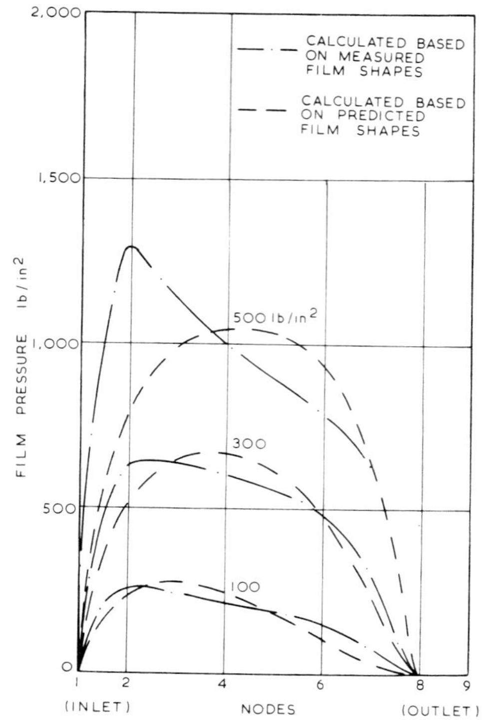

Fig. 20.—Film pressure vs. position (Ref. 31).

Herrebrugh[38] gave an elegant formulation of the normal approach problem as an integral equation in the film thickness. Assumptions similar to (37) were made and results were presented only for the isoviscous case. An interesting discovery by the author is the existence of a bifurcation point for solutions to this problem.

The same year Christensen[39] analyzed the normal approach of spherical bodies. His previous analysis and conclusions for cylinders were thus extended to spheres. In addition he reports a "transition" film thickness where the behavior changes from "rigid" to elastic. For values of film thickness above the transition film thickness, he concludes that elastic distortions can be neglected. Furthermore, in this region, the effects of pressure on viscosity were found to be more important than elastic distortion effects.

The first investigation which included the deformation rate term in the normal approach problem was conducted by Lee and Cheng[40]. They considered the normal approach of infinitely long cylinders. Arbitrary variations of viscosity and density with pressure were accounted for in their analysis. In addition, a considerably more effective numerical scheme was incorporated than had been used previously. The latter allowed results for a larger range of parameters to be studied. In particular, Lee and Cheng were able

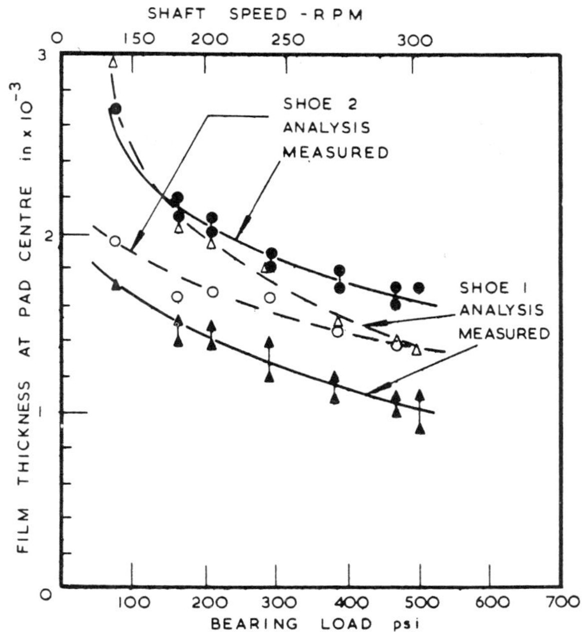

Fig. 21.—Comparison of film thickness measured at pad center with analytical result (Ref. 32).

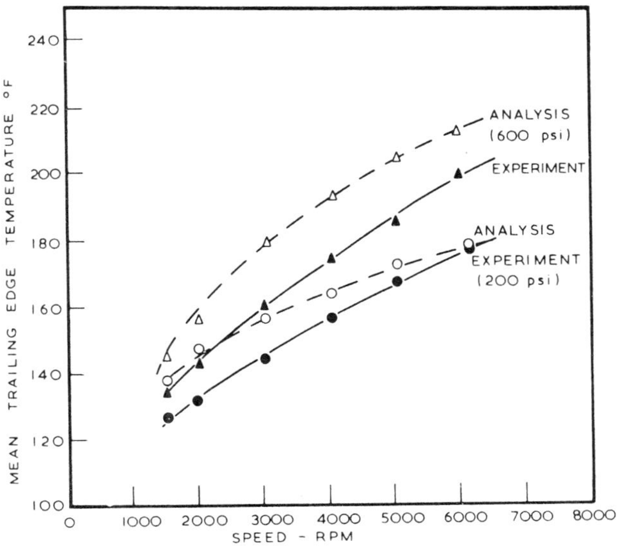

Fig. 22.—Comparison of mean trailing edge temperature with analysis (Ref. 32).

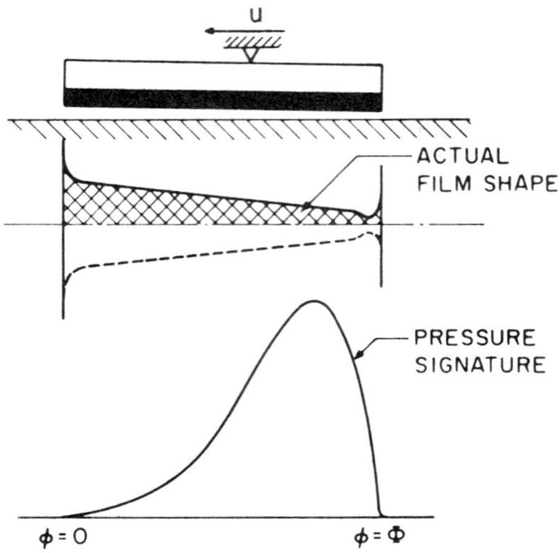

Fig. 23.—Bearing schematic (Ref. 34).

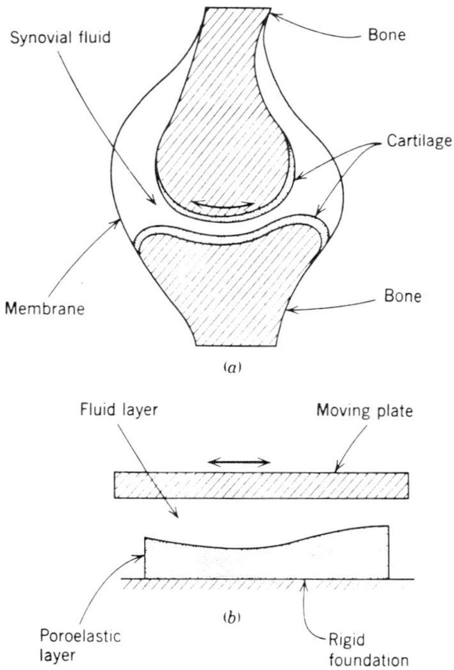

Fig. 24.—The geometry of a natural joint (19). (a) actual joint, (b) idealized model (Ref. 35).

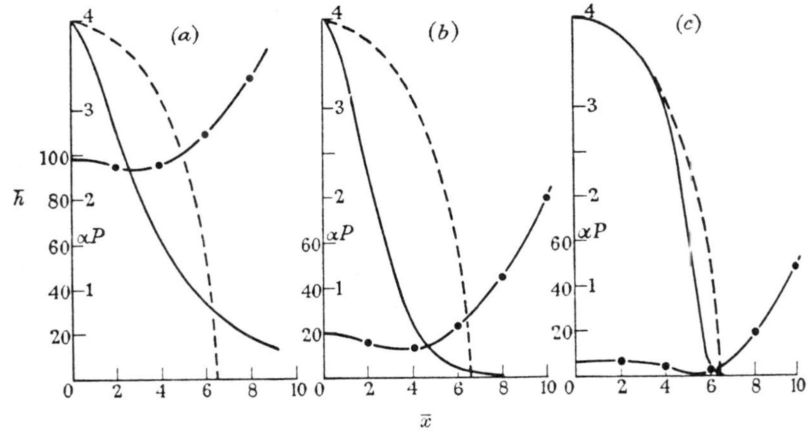

Fig. 25.—Pressure distributions and film shapes: ———, pressure, - - - -, Hertzian pressure; -●-●-, film shape. (a) h_o = 99; (b) h_o = 20; (c) h_o = 7 (Ref. 37).

to study the final stages of normal approach. Significant results obtained by the authors indicate that, due to the presence of the deformation rate term, a pocket of fluid is entrapped in the center region of the contact as shown in Figure 26. The pressure in this center region does not decrease in the final stages of descent as reported previously, but rather begins to

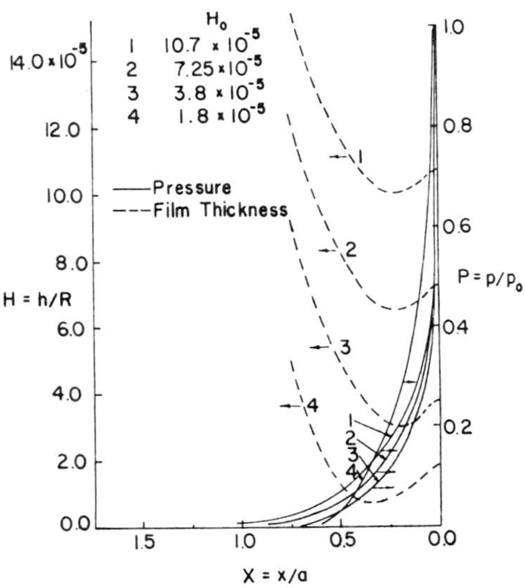

Fig. 26.—Pressure and deformation profiles, straight exponential lubricant, G = 3180, $p_o = 1.5 \times 10^3$ psi (Ref. 40).

rise again as the film thickness is further reduced. Hence the authors conclude that the pressure profile does *not* approach the Herzian distribution.

The fact that the deformation rate term can become significant was also shown by Conway[41]. He studied the problem of a rigid roller on an elastic layer (assumed to behave as a Winkler foundation) and concluded that the deformation rate term can become important in the case of high approach velocities, thin films, or oils whose viscosity increases substantially with pressure.

The normal approach problem is a special case of the more general transient EHD problem. The latter class of problems arise in the study of seals, biological joints, and the viscous hydroplaning of tires. Motivated by these applications, some recent research has been concentrated in this latter area.

The work of Rohde et al.[10] is specifically directed toward the more general transient EHD problem. Figure 27 shows the geometry they considered: a rigid indenter moving into a deformable (soft) half-space separated by a thin layer of lubricant. Because deformation rate terms are included, the associated mathematical problem is explicitly shown to lead to an *initial value* problem. The authors considered the case in which the indenter velocity was *a-priori* given as well as the cases in which the indenter velocity (as a function of time) was computed so that either the center point film pressure or the load supported by the film has a specified time history. Among the results found by the authors, was an entrapment of fluid as shown in Figure 28 and a dramatic increase of the rate at which the indenter approaches the half-space due to the elasticity effects. Figure 29 shows a typical result of this kind. These results are in qualitative agreement with the experimental data of Gaman et al.[42].

In a subsequent paper by Browne et al.[43], the specific application of the above methods to the viscous hydroplaning of tires was made. In particular, the slip term was included in their formulation. The deformable

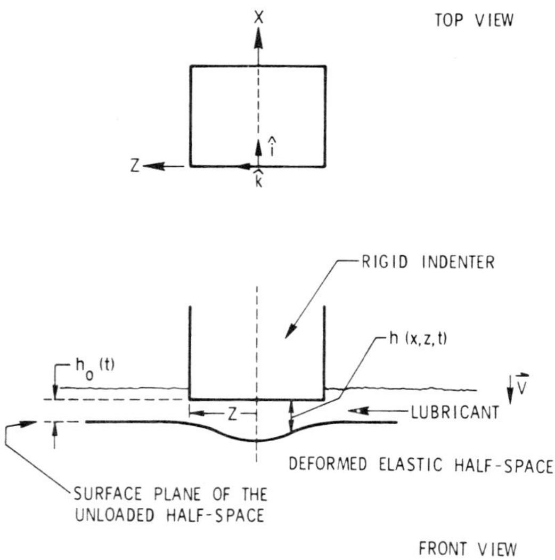

Fig. 27.—Elastohydrodynamic squeeze film geometry (Ref. 10).

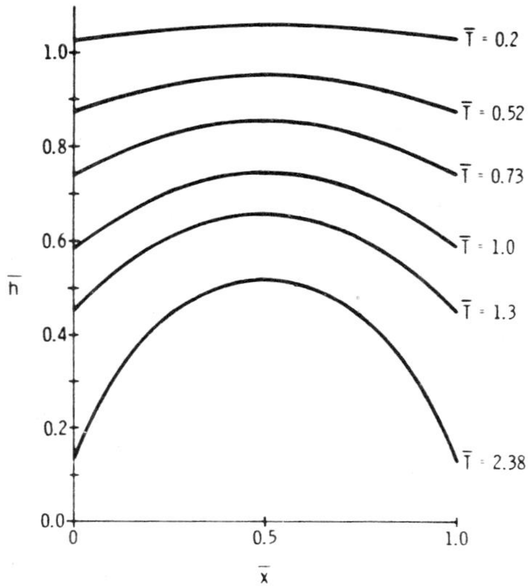

Fig. 28.—Centerline film profile as a function of time, ramp center point pressure; scale = 1.2, $P_{00}/T = 0.1$ (Ref. 10).

solid was assumed to be a three dimensional rectangular parallelopiped which modelled a tire tread "element" as shown in Figure 30. The elastic behavior of the tread element was shown to be capable of dramatically altering analytical traction predictions. The effect of slip, on the other hand, was shown to be quite small unless one is considering speeds of over 100 mph as shown in Figure 31.

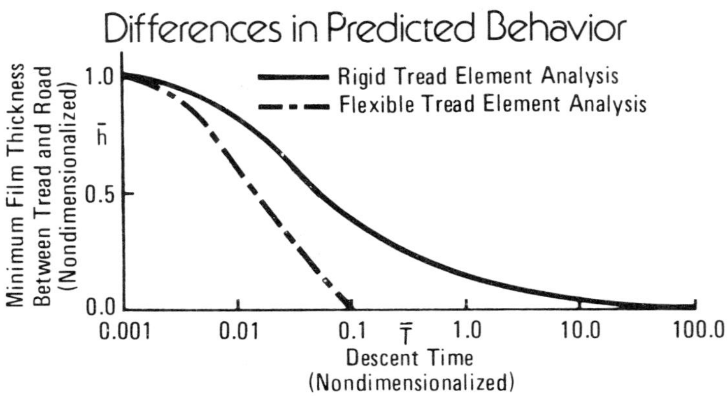

Fig. 29.—(Ref. 43) A comparison of time histories of minimum fluid film thickness, obtained with the rigid and flexible element modesl. Shown is the vastly different descent behavior predicted by the two models.

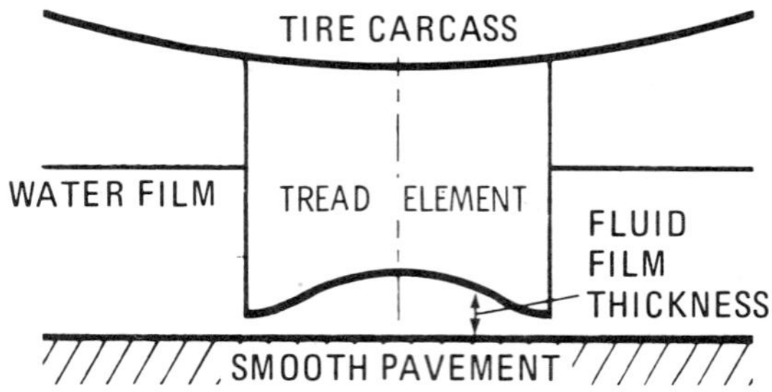

Fig. 30.—(Ref. 30) As the tread element descends, fluid is trapped between its deformed surface and the pavement surface.

A further extension of the above analysis by Whicker et al.[44], considered the motion of a squeezing, rotating, sinking element descending through a viscous fluid. The configuration studied is shown in Figure 32. The analytical results obtained were compared with the experimental results of Moore[45]. Figure 33 shows this comparison together with the results of an ad-hoc theory given by Moore. As can be seen reasonable agreement was found.

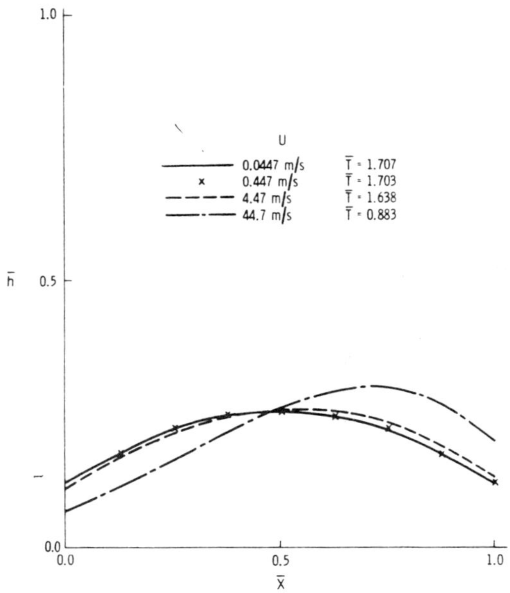

Fig. 31.—Centerline film thickness distribution as a function of slip velocity U; h_{min} = 5.5 x 10^3, W = 0.4, D = 117 (Ref. 43).

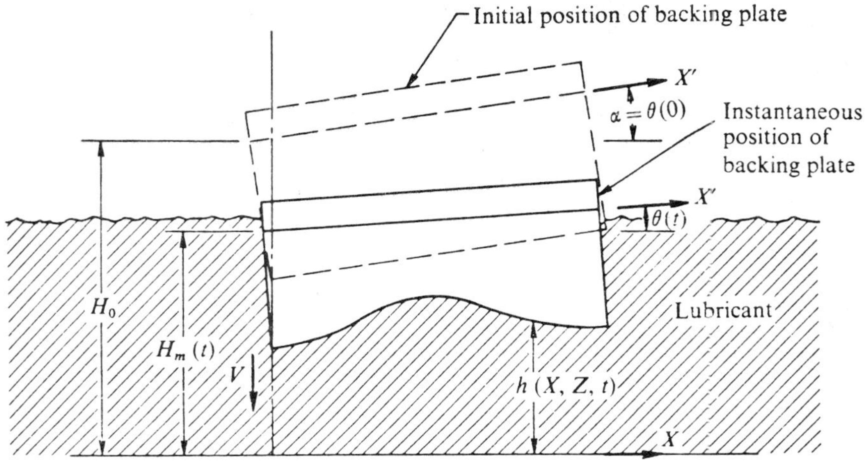

Fig. 32.—Motion of sinkage element (Ref. 44).

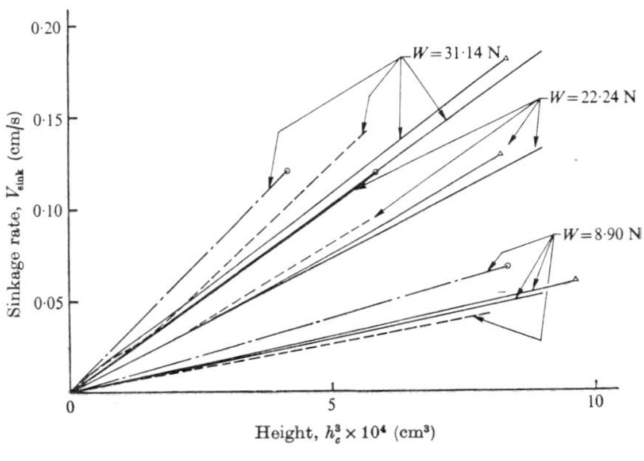

Fig. 33.—Sinkage velocity: comparison of theory with experiment. ——, theory (Browne et al.); ▲——▲, inertial theory (Browne et al.); - - -, experiment (Moore); ○—— - ——○, theory (Moore). Values of all parameters except W (Ref. 44).

Future work in the area of transient EHD problems should concentrate on the application and extension of previously derived methods and techniques to various areas. In particular, applications to the lubrication of natural and artificial biological joints should be pursued. Likewise, the squeezing out of an entrapped pocket of fluid in situations such as the viscous hydroplaning of tires should be studied.

REFERENCES

1. Stephenson, R.R. and Osterle, J.F., *American Society of Lubrication Engineers Transactions,* Vol. 5, 1962, p. 2.
2. Oh, K.P. and Huebner, K.H., *Journal of Lubrication Technology,* Vol. 95, 1973, p. 342.
3. Dowson, D. and Hamrock, B., *Journal of Lubrication Technology,* Vol. 98, 1976, p. 223.
4. Dowson, D. and Higginson, G.R., *Journal of Mechanical Engineering Science,* Vol. 1, 1959, p. 1.
5. Cheng, H.S. and Sternlicht, B., *Journal of Lubrication Technology,* Vol. 87, 1965, p. 695.
6. Pirvics, J. and Castelli, V., *Journal of Lubrication Technology,* Vol. 95, 1973, p. 363.
7. Rohde, S.M. and Oh, K.P., *Royal Society of London. Proceedings.* Series A, Vol. 343, 1975, p. 315.
8. Dowson, D. and Taylor, C., *Journal of Lubrication Technology,* Vol. 89, 1967, p. 237.
9. Rohde, S.M., *Wear,* Vol. 27, 1974, p. 273.
10. Rohde, S.M., Whicker, D. and Brown, A.L., *Journal of Lubrication Technology,* Vol. 98, No. 3, 1976, p. 401.
11. Rohde, S.M., General Motors Research Publication, GMR-2279, 1976, to appear in "Lectures in Tribology," McGraw-Hill, 1980.
12. Carl, T.E., in Proceedings 2nd Conv. Lub. and Wear, Institution of Mechanical Engineers, 1964, p. 100.
13. Higginson, G.R., *Institution of Mechanical Engineers. Proceedings,* Vol. 180, Part 3B, 1965.
14. O'Donoghue, J., Brighton, D.K. and Hook, C.J.K., *Journal of Lubrication Technology,* Vol. 89, No. 4, 1967, p. 409.
15. Benjamin, M.K. and Castelli, V., *Journal of Lubrication Technology,* Vol. 93, No. 1, 1971, p. 191.
16. Reddi, M.M., *Journal of Lubrication Technology,* Vol. 91, No. 3, 1969, p. 524.
17. Reddi, M.M. and Chu, T.Y., *Journal of Lubrication Technology,* Vol. 92, No. 3, 1970, p. 502.
18. Fantino, B., Frene, J. and Godet, M., "Etude des deformations des coussinets de bielles," to be published.
19. Oh, K.P. and Rohde, S.M., *Journal of Lubrication Technology,* Vol. 99, No. 1, 1977, p. 75.
20. Heuer, D.F. and Collins, R.A., Tech. Rep. AFAPL-TR-73-56, Air Force Aero Propulsion Lab, 1973.
21. Licht, L., Branger, M. and Anderson, W.J., *Journal of Lubrication Technology,* Vol. 96, No. 2, 1974, p. 215.
22. Barlow, E.J. and Wildmann, M., *Journal of Lubrication Technology,* Vol. 90, 1968, p. 640.
23. Baumann, G.W., *Journal of Lubrication Technology,* Vol. 93, 1971, p. 457.
24. Castelli, V., Rightmire, G.K. and Fuller, D.D., *Journal of Lubrication Technology,* Vol. 89, 1967, p. 510.
25. Rohde, S.M., Whicker, D. and Browne, A.L., *American Society of Lubrication Engineers Transactions,* Vol. 21, No. 3, 1978, p. 264.

26. Sternlicht, B., Carter, G.K. and Arwas, E.B., *Journal of Applied Mechanics,* Vol. 28, 1961, p. 179.
27. Castelli, V. and Malanowski, S.B., *Journal of Lubrication Technology,* Vol. 91, 1969, p. 634.
28. Fogg, A., *Institution of Mechanical Engineers. Proceedings,* Vol. 155, 1946, p. 49.
29. Hahn, E.J. and Kettleborough, C.F., *Journal of Lubrication Technology,* Vol. 90, 1968, p. 233.
30. Rohde, S.M. and Oh, K.P., *Journal of Lubrication Technology,* Vol. 97, No. 3, 1975, p. 450.
31. Taniguchi, S. and Ettles, C., *American Society of Lubrication Engineers Transactions,* Vol. 18, No. 4, 1975, p. 299.
32. Ettles, C., *American Society of Lubrication Engineers Transactions,* Vol. 19, No. 2, 1976, p. 153.
33. Robinson, C.L., Ph.D. Thesis, University of London, 1971.
34. Rightmire, G.K., Castelli, V. and Fuller, D.D., *Journal of Lubrication Technology,* Vol. 98, 1976, p. 95.
35. Eidelberg, B.E., Ph.D. Thesis, Cornell University, Ithaca, New York, 1976.
36. Eidelberg, B.E. and Booker, J.F., *Journal of Lubrication Technology,* Vol. 98, No. 1, 1976, p. 175.
37. Christensen, H., *Royal Society of London. Proceedings.* Series A, Vol. 266, 1961, p. 312.
38. Herrebrugh, K., *Journal of Lubrication Technology,* Vol. 92, 1970, p. 292.
39. Christensen, H., *Journal of Lubrication Technology,* Vol. 92, 1970, p. 145.
40. Lee, K.M. and Cheng, H.S., *Journal of Lubrication Technology,* Vol. 95, 1973, p. 308.
41. Conway, H.D., *Journal of Lubrication Technology,* Vol. 95, 1973, p. 391.
42. Gaman, I.D.C., Higgonson, G.R. and Norman, R., *Wear,* Vol. 28, 1974, p. 345.
43. Browne, A.L., Whicker, D. and Rohde, S.M., *Tire. Sci. Tech.,* Vol. 3, 1975, p. 215.
44. Whicker, D., Browne, A.L. and Rohde, S.M., *Journal of Fluid Mechanics,* Vol. 78, Part 2, 1976, p. 247.
45. Moore, D.F., *Journal of Fluid Mechanics,* Vol. 20, 1964, p. 321.
46. Rohde, S.M. and McAllister, G.T., *International Journal of Engineering Science,* Vol. 13, 1975, p. 841.

DISCUSSION

E. A. SAIBEL, Army Research Office: This is really a very short comment to put things in historical context. In 1957 Osterle and I did a paper in which we had the interaction of hydrodynamic and elastic effects. We had to work out the problem of the elastic half space with the slider bearing on it. It was fun for me because it was the first usage of the digital computer in our locality. Osterle was elected to go down to the Mellon Bank at two in the morning and work until six in the morning to get the things done. I enjoyed that.

And the other thing is that in Herb Cheng's review this morning he left out what I think was the first paper on the treatment of the film as a Maxwell fluid. T. S. Chow and I did that about ten years ago. Maybe when the Big Book is written someday it will be included in that.

S. M. ROHDE: It was not a deliberate intention to delete your name from the brief review. Because of space limitation, no attempt was made to write a complete review, but rather to give a flavor for some of the work that has been done. We apologize.

TOWARDS A REFINED SOLUTION OF THE ISOTHERMAL POINT CONTACT EHD PROBLEM

H. P. Evans and R. W. Snidle

ABSTRACT

The existing full numerical solutions of the point contact elastohydrodynamic lubrication problem are reviewed. It is concluded that, while these solutions are important in establishing the viability of a straightforward iterative technique of solving the problem, further refinement of the numerical techniques is required before the point contact problem can be said to have as firm a theoretical basis as the corresponding line contact problem. As a contribution towards this effort some of the recent techniques are described.

A technique for the evaluation of elastic deformation is described with which it is possible to achieve high accuracy when using the relatively coarse finite difference grids which must be used in this problem. Higher order finite difference methods for the solution of the hydrodynamic equations have been developed for the same reason and are also described in detail.

Some of the problems which occur in the straightforward iterative method of solution are described and the importance of convergence criteria is demonstrated. Results are presented showing typical pressure distributions and film shapes for high elastic modulus conditions when using an exponential viscosity/pressure relationship. The use of more realistic viscosity/pressure laws such as that due to Roelands is discussed in relation to work which is now in progress.

NOMENCLATURE

c	Parameter in pressure viscosity relationship
d	Elastic deformation
E	Elastic modulus
E'	Effective elastic modulus
G	Non-dimensional material parameter
h	Film thickness
h_o	Film thickness at coordinate origin
K	Parameter in pressure-viscosity relationship
p	Pressure
p_o	Maximum Hertzian pressure for dry contact
q	Transformed (reduced) pressure
r	Radial coordinate
R	Radius of sphere in sphere-plane contact
u	Hydrodynamic velocity of surfaces
U	Non-dimensional speed parameter
W	Non-dimensional load parameter
x, y	Cartesian coordinate
α	Pressure coefficient of viscosity in Barus equation
$\overline{\alpha}$	Equivalent α value for non exponential pressure-viscosity relationship
η	Viscosity
η_o	Viscosity at zero pressure
ν	Poisson's ratio
ρ	Density

1. INTRODUCTION

In his review paper on the fundamentals of elastohydrodynamic contact phenomena Professor Cheng has discussed both past and present developments in the field against the background of line contact EHD. It is, of course, appropriate that line contact should be the reference condition since most of the experimental work has been done using this configuration and, in addition, the corresponding fundamental theory of line contact EHD is now well understood and accepted. In many engineering situations, however, it must be admitted that nominal contact between lubricated machine elements occurs not along a line but at a single point. The nominal contacts in spherical roller bearings and hypoid gear sets are obvious examples of single point contacts. Even in line contacts the nominal contact area becomes divided into multiple point contacts under conditions of partial EHD, which, as Professor Cheng points out, is probably the regime under which the majority of EHD contacts operate in engineering practice.

A fundamental understanding of point contact EHD is therefore important in both extending EHD theory to full film point contacts and in contributing to our understanding of partial EHD in the general case.

In this discussion paper our aims are, first, to add to Professor Cheng's detailed review of some of the recent developments in the area of fundamental theory of point contact EHD and, second, to indicate the ways in which some of the theoretical problems involved in a full solution to the point contact problem might be tackled in the future.

2. REVIEW OF EXISTING NUMERICAL SOLUTIONS

In this section we examine some of the assumptions made and techniques adopted in the two full solutions to the point contact EHD problem obtained by Ranger et al.[(1)] and Hamrock and Dowson[(2-5)].

These solutions are similar in that both use a straightforward re-iterative scheme in which an initially assumed pressure distribution is modified in the light of the hydrodynamically generated pressure which corresponds to the deformed film shape resulting from the application of the assumed pressure to the elastic, lubricated surfaces. For point contact solutions this type of approach is adopted because of the obvious difficulties in finding an inverse solution of the hydrodynamic equations considering side leakage. Although both solutions referred to above use the same basic approach, there is one quite important difference between them as can be seen from Figure 1 which shows simplified flow diagrams for the two solutions. In Ranger's solution the elastic deformation is calculated only once per main (outer) iteration whereas in the solution of Hamrock and Dowson the calculation of elastic deformation is included in an inner loop and must therefore be calculated more than once per main iteration. Ranger's approach would appear to be preferable on the grounds that since calculation of elastic deformation is the most time-consuming aspect of the problem, it is desirable to minimize the number of times the calculation is performed.

The two solutions also differ in the way elastic deformation is calculated. Ranger replaces the pressure distribution by overlapping "pyramidal" elements and calculates the contribution to the deformation at a point due to a pressure pyramid by analytical means. Hamrock and Dowson use a similar, if much more straightforward, technique and calculate the deformation arising from blocks of uniform pressure.

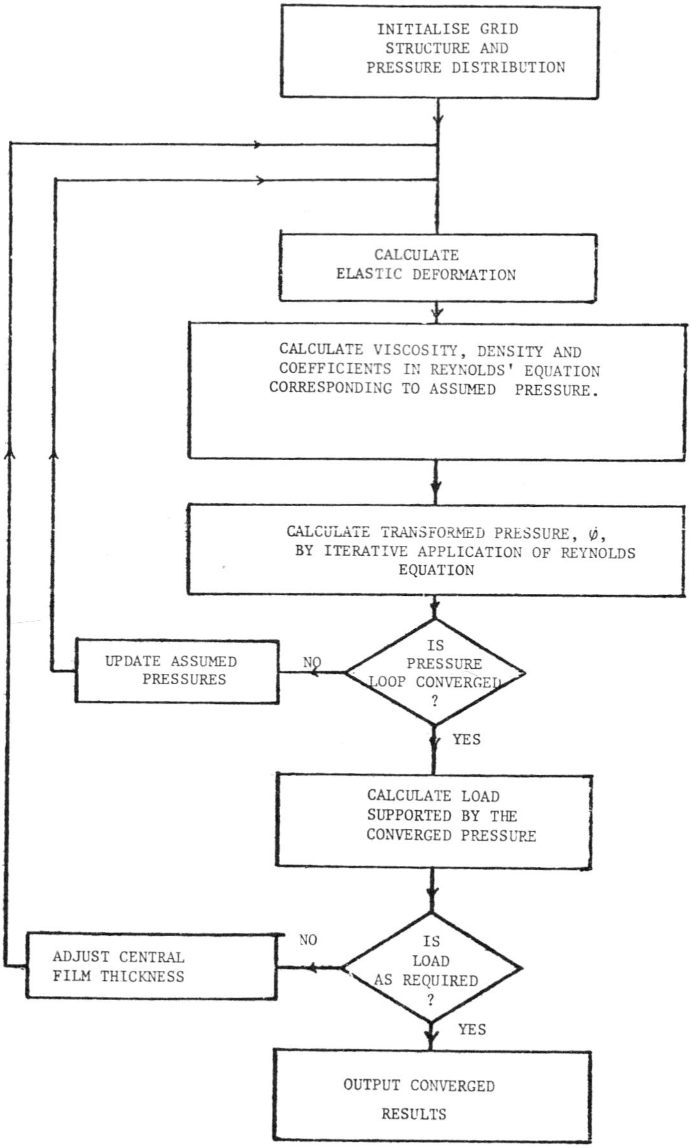

(a)

Fig. 1a.—Simplified flow diagrams for Hamrock and Dowson.

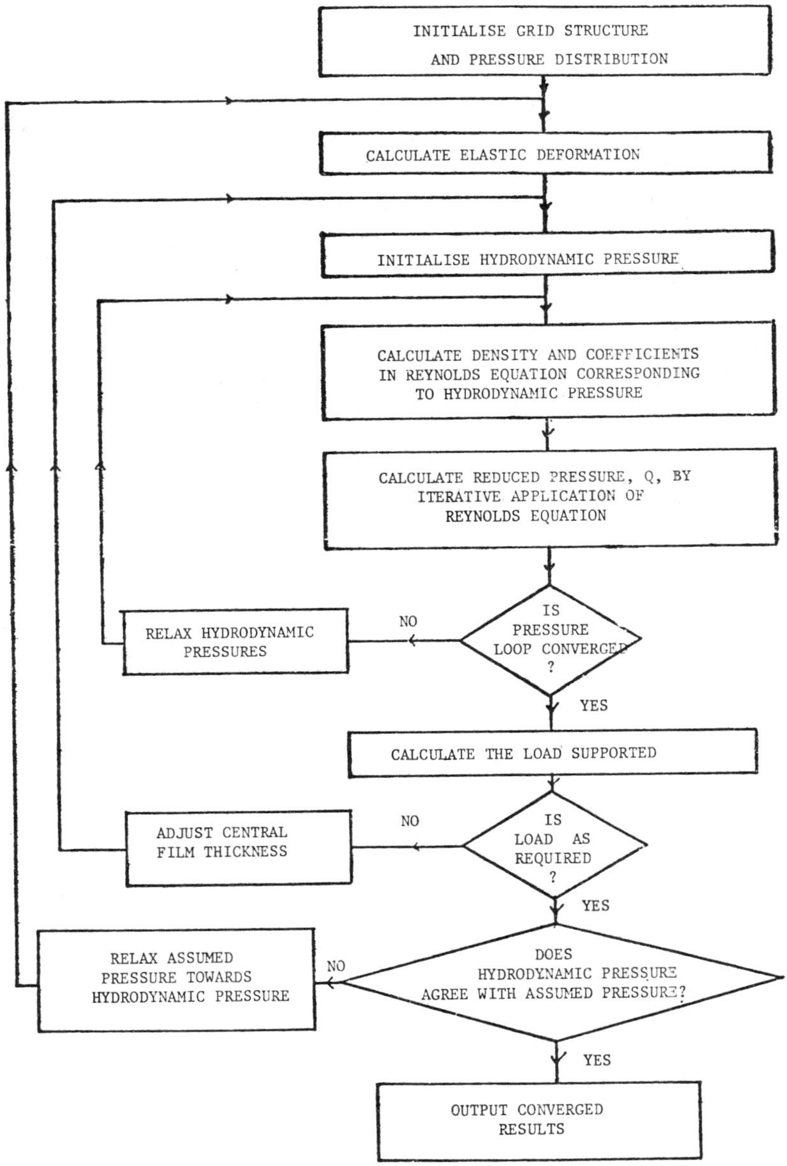

(b)

Fig. 1b.—Ranger's method of solution.

In both solutions the Reynolds equation with side leakage is solved by performing Gauss-Seidel relaxation on transformed values of pressure. Ranger uses the reduced (i.e. isoviscous) pressures and Hamrock and Dowson adopt the Voghelpol transformation. The smoothing of the pressures achieved by these transformations is desirable because of the pressure dependence of viscosity and because of the steep pressure gradients which occur near the exit from the contact. The reduced pressure transformation gives a greater degree of smoothing, but care is needed because of the prediction of infinite pressures when the reduced pressures reach a value of $1/\bar{\alpha}$ (where $\bar{\alpha}$ is defined in Section 4).

Ranger assumed an exponential viscosity/pressure relationship whereas Hamrock and Dowson adopted the Roelands[6] equation. In both these solutions a second order (5 point) finite difference representation of transformed pressures was adopted in the Gauss - Seidel relaxation procedure.

As will be discussed later, because of the time required to perform evaluation of elastic deformation it is necessary to adopt relatively coarse finite difference meshes (much coarser, for example, than the meshes which have been used in the solution of line contact EHD problems).

An irregular grid spacing in the x direction is used by Ranger with a grid spacing that is nowhere finer than a/12 where a is the corresponding Hertzian radius for dry contact. An even grid is used in the y direction with a spacing of a/7. The determination of elastic deformation is straightforward in Hamrock and Dowson's method and they are able to use an even grid with spacing of a/13 in the x direction and a/5 in the y direction.

In Ranger's solution convergence is considered acceptable, when

$$\frac{\Sigma\ Pi^{old} - Pi^{new}}{\Sigma\ Pi^{new}} < 0.01. \qquad (1)$$

The convergence criterion utilized by Hamrock and Dowson is not clear, but comparison of the convergence criteria of the two methods is likely to be misleading since, as can be seen from the flow diagrams of Figure 1, the iterative procedures adopted by them are not equivalent.

Both solutions give converged results for conditions corresponding to situations in which significant deformation occurs. Ranger's results are limited to less severe conditions than those obtained by Hamrock and Dowson. This is probably due, first, to the fact that Ranger uses the Barus, exponential pressure-viscosity relationship, which, as will be discussed later, represents a very demanding test of the ability of the solution to converge and second, possibly due to a difference in the convergence criteria adopted.

Where the two solutions do not agree, however, is in the film thickness relationship which has been derived from a number of solutions covering a range of conditions.

The Ranger result if

$$\frac{ho}{R} = 1.393\ U^{0.571}\ W^{0.048}\ G^{0.477} \qquad (2)$$

while the Hamrock and Dowson formula for the comparable case of a circular contact is

$$\frac{ho}{R} = 1.899\ U^{0.67}\ W^{-0.067}\ G^{0.53} \tag{3}$$

Both formulas give film thicknesses which are comparable with the earlier Grubin-type formula obtained by Archard and Cowking

$$\frac{ho}{R} = 1.4\ (UG)^{0.74}\ W^{-0.074} \tag{4}$$

In these expressions the non-dimensional speed, load and material parameters U, W and G are defined as follows

$$U = \frac{u\ \eta o}{E'R} \qquad W = \frac{L}{E'R^2} \qquad G = \alpha E'$$

where α is the pressure-viscosity coefficient (or its equivalent, $\overline{\alpha}$ in Hamrock and Dowson's case), L is the normal applied load, and E' is the effective elastic modulus given by

$$\frac{1}{E'} = \frac{1}{2}\left\{\frac{1-\nu_1^2}{E_1} + \frac{1-\nu_2^2}{E_2}\right\} \tag{5}$$

The positive load exponent in Ranger's formula is unexpected particularly as the method of elastic displacement calculation utilized can be expected to be more accurate than that used by Hamrock and Dowson. The authors feel that this anomalous result is probably due to 'side-starvation' occuring in Ranger's solutions at the lighter loads considered. This point will be given further attention in Section 8.

The conclusions to be drawn from the pioneering solutions of Ranger et al. and Hamrock and Dowson are first, that the forward iterative method of solution does appear to work and second, that further work needs to be done to refine the solutions available and, in particular, to remove the anomaly between the two film thickness formulas which have so far been obtained.

In the following sections we describe some of the techniques which the authors have developed with the aim of producing a more accurate solution to the point contact elastohydrodynamic lubrication problem.

3. ACCURATE EVALUATION OF ELASTIC DEFORMATION

The elastic deformation of the lubricated surfaces (Figure 2) at a point (x_1, y_1) is given by

$$d(x_1, y_1) = \left\{\frac{1 - \nu_1^2}{\pi E_1} + \frac{1 - \nu_2^2}{\pi E_1}\right\} \iint_A \frac{p(x,y)\ dx\ dy}{\left[(x-x_1)^2 + (y-y_1)^2\right]^{\frac{1}{2}}} \tag{6}$$

where A is the region over which the pressure, p, acts.

There are two main problems in dealing with this integral in a numerical solution. The first problem is due to the singularity at $x = x_1$, $y = y_1$. The second problem is a purely computational one. If we assume, for example, that we require the elastic deformation at all points on a square $n \times n$ finite difference grid, then the time required to compute equation (6) is proportional to n^4.

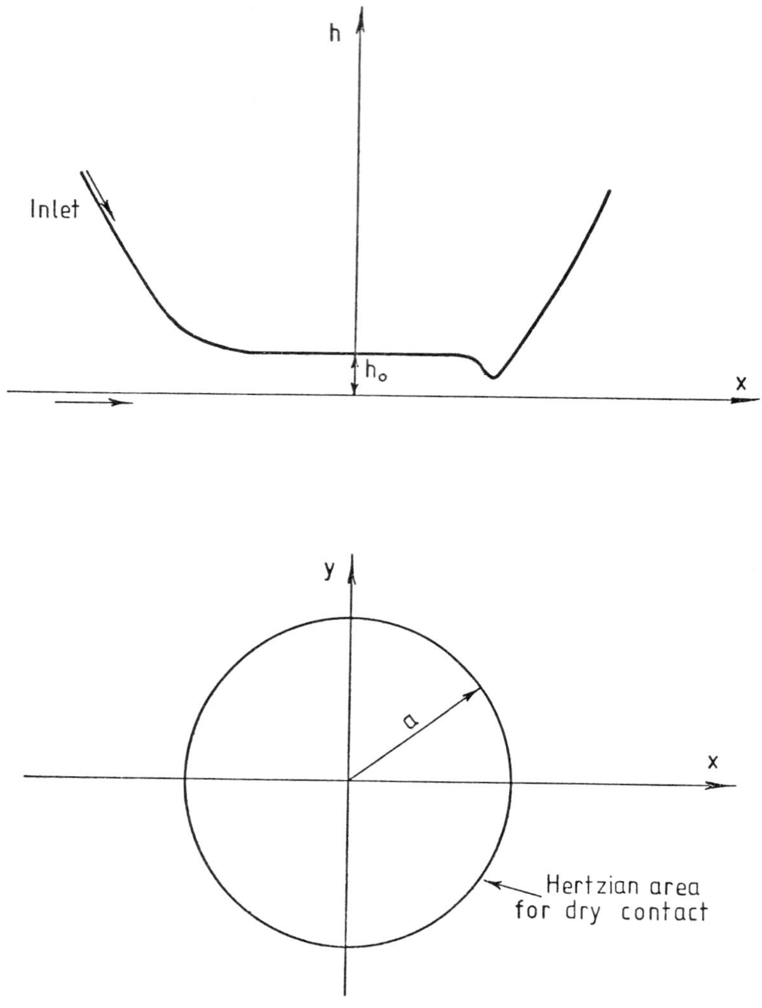

Fig. 2.—Schematic arrangement of circular elastohydrodynamic contact showing coordinate system.

In practice this means that relatively coarse finite difference grids must be used, otherwise unacceptably long computational times are required. The accepted method of dealing with the singularity problem is to replace the pressure function $p(x,y)$ with some analytical expression in the region of the singularity such that the integral can be obtained in closed form. The form of pressure function used by Hamrock and Dowson (blocks of constant pressure) and Ranger et al. ("pyramids") have already been described. Recently, Biswas and Snidle[12] have developed more accurate, second order, techniques for the evaluation of the elastic deformation which are more appropriate when using the unavoidable coarse grid structure.

At each point where the elastic deformation is required the evaluation of the integral contained in equation (6) is performed by partitioning region A into two regions B and C so that

$$d(x_1, y_1) = \left(\frac{1 - \nu_1^2}{\pi E_1} + \frac{1 - \nu_2^2}{\pi E_2}\right) \left\{ \iint_B \frac{p(x,y)\,dxdy}{\left[(x-x_1)^2 + (y - y_1)^2\right]^{\frac{1}{2}}} + \iint_C \frac{p(x,y)\;dxdy}{\left[(x-x_1)^2 + (y-y_1)^2\right]^{\frac{1}{2}}} \right\} \tag{7}$$

C is a rectangular area containing the singular point (x_1,y_1) and is bounded by lines drawn through its neighbouring grid points, and B is the remainder of region A. In region B the integral is evaluated by normal numerical cubature using the bivariate form of Simpson's rule. The second integral is obtained by expressing the pressure function as an analytical function of x and y and integrating by formal methods.

We consider the case where the singular point is at the origin so that the integral over region C becomes

$$\int_{-a_1}^{a_2} \int_{-b_1}^{b_2} \frac{p(x,y)}{(x^2+y^2)^{\frac{1}{2}}}\, dy\; dx \tag{8}$$

The pressure function is approximated by a bi-quadratic polynomial of the form

$$p(x,y) = \sum_{i=0}^{2} \sum_{j=0}^{2} C_{ij} x^i y^j \tag{9}$$

The nine coefficients C_{ij} are determined uniquely by the value of $p(x,y)$ at the singular point (the origin) and the eight nodal points on the boundary of C. Thus the integral becomes

$$\sum_{i=0}^{2} \sum_{j=0}^{2} C_{ij}\; I_{ij}$$

where

$$I_{ij} = \int_{-a_1}^{a_2} \int_{-b_1}^{b_2} \frac{x^i\, y^j}{(x^2+y^2)^{\frac{1}{2}}}\, dy\; dx \tag{10}$$

The nine integrals I_{ij} can be determined analytically and the problem is simplified if either $a_1 = a_2$ or $b_1 = b_2$, when integrals involving $i = 1$ in the former case and $j = 1$ in the latter become zero.

The choice of a grid structure for the solution of the point contact problem requires careful consideration. It is advantageous to adopt a uniform mesh due to the increased accuracy of finite difference approximations to second derivatives (encountered in the hydrodynamic part of the problem) for such a mesh. However, the conflicting requirements of fine resolution in the exit half of the contact in order to resolve the steep pressure gradients which occur there, and a long inlet region which is necessary to establish the fully flooded boundary condition, make it impractical to use an even grid spacing in the direction of motion (x direction). Similar arguments apply in the y direction also since it has been found that the position of the side boundaries can be critical, particularly under less severely loaded conditions. The adoption of non-uniform grid spacing in both x and y directions has therefore been found inevitable if a high accuracy solution is to be obtained with the given computer resources available to the authors.*

The grid structure which is being used at the present stage of development of the solution is shown in Figure 3. Elastic deformation is calculated and the hydrodynamic equations are solved at the intersections of the lines. The grid is uniform in the y direction and non-uniform in the x direction.

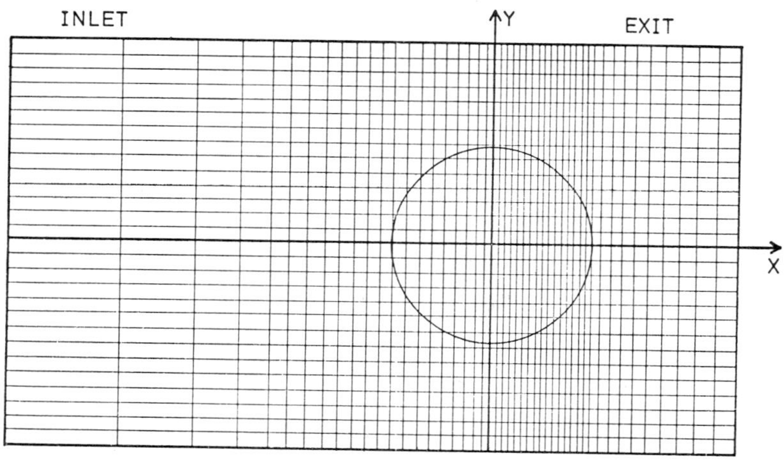

Fig. 3.—Finite difference mesh. Circle is Hertzian contact region for dry contact at given load.

---

* The computer available to the authors is a CDC 7600 with a normal maximum single run time of 1200 seconds.

A more refined grid, which is being implemented at the present time, has non-uniform divisions in both x and y directions. This refinement has been introduced so that wider side boundaries can be used thus avoiding "side-starvation" at the lighter loads and higher speeds.

4. REYNOLDS EQUATION

The generation of hydrodynamic pressure is governed by the following form of Reynold's equation

$$\frac{\partial}{\partial x}\left(\frac{\rho h^3}{\eta}\frac{\partial \rho}{\partial x}\right)+\frac{\partial}{\partial y}\left(\frac{\rho h^3}{\eta}\frac{\partial \rho}{\partial y}\right) = 12\ u \frac{\partial}{\partial x}(\rho h) \tag{11}$$

Because the viscosity, η, usually varies very rapidly with pressure under high elastic modulus conditions, it is difficult to obtain a stable solution to the equation in this form. The usual way of dealing with Reynold's equation where viscosity/pressure effects are present is to apply a transformation such that transformed pressures vary less rapidly. The Voghelpol transformation, $\phi = ph^{3/2}$, is an attractive way of achieving this aim and has been used by Hamrock and Dowson. An alternative transformation, which was used by Ranger et al. and has been adopted by the present authors, is the reduced pressure transformation

$$q = \eta o \int_o^p \frac{dw}{\eta(w)} \tag{12}$$

The effective pressure viscosity coefficient, $\bar{\alpha}$ is given by $\underset{p\to\infty}{Lt}\left(\frac{1}{q(p)}\right)$. With this transformation we obtain the isoviscous form of Reynold's equation

$$\frac{\partial}{\partial x}\left(\rho h^3 \frac{\partial q}{\partial x}\right)+\frac{\partial}{\partial y}\left(\rho h^3\frac{\partial q}{\partial y}\right)= 12\ \eta o\ u \frac{\partial}{\partial x}(\rho h) \tag{13}$$

The transformed, or reduced, pressures are therefore the isoviscous pressures and since the calculation of the reduced pressures does not involve a calculation of the viscosity, the transformation leads to a more stable method of calculating pressures. The dependence of density upon pressure remains, of course, but this effect is much less of a problem and can be included in the analysis quite simply by adding a density pressure iteration in the evaluation of q values. It is useful if a functional relationship can be found between q and p. In the case of the exponential viscosity/pressure law

$$\eta = \eta o\ exp\ (\alpha p) \tag{14}$$

we obtain, from equation (12) the well known relationship

$$q = \frac{1}{\alpha}\ (1 - e^{-\alpha p}) \tag{15}$$

A more physically realistic, and now widely accepted formula is the single parameter Roelands viscosity/pressure law[6]

$$\eta = \frac{A}{\gamma}\left(\frac{\gamma\eta o}{A}\right)^{(1+ P/B)^Z} \tag{16}$$

where $A = 0.001\ Ns/m^2$, $B = 1.961 \times 10^8 Nm^{-2}$ and the constant $\gamma - 10^{1.2}$ is the same for nearly all fluids. If $\eta(p)$ is given by equation (16) then equation (12) cannot be integrated to yield a simple functional relationship. However, if the two parameter viscosity/pressure law of the form

$$\eta = \eta o\ (1 + Cp)^k \tag{17}$$

is used (which can be fitted very closely to the Roelands law as is shown in Figure 4) then a simple functional relationship between q and p does exist

$$q = \frac{1}{C(k-1)} \left(1 - \frac{1}{(1+Cp)^{k-1}} \right). \tag{18}$$

Values of the reduced pressure can therefore be calculated from equation (13) and the true pressures are then

$$p = \frac{1}{C} \left([1 - qC(k-1)]^{1/(1-k)} - 1 \right) \tag{19}$$

Equation (13) may be solved by the Gauss-Seidel method with over-relaxation.

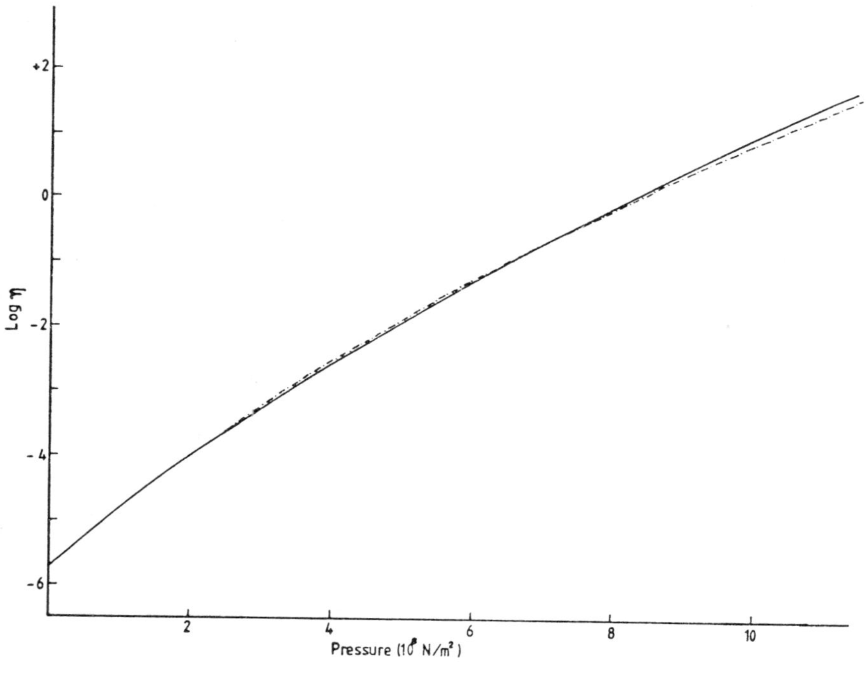

Fig. 4.—Roelands pressure viscosity relation as used by Hamrock and Dowson (solid line) with example of relation of the form $\eta = \eta o(1+Cp)^k$ (broken line). $\eta o = o.o411\ Ns/m^2$ $Z = 0.67$ (Roelands) $C = 6.145 \times 10^{-6}\ m^2/N$, $K = 24$.

In order to achieve high accuracy with the relatively coarse finite difference grids which must be used in the problem, a fourth order finite difference representation is used. Equation (13) may be written as

$$\nabla^2 (\rho h^3 q) = q \nabla^2 (\rho h^3) - \rho h^3 \nabla^2 q + 24\eta_o u \frac{\partial(\rho h)}{\partial x} \tag{20}$$

where $$\nabla^2 = \frac{\partial^2}{\partial x^2} + \frac{\partial^2}{\partial y^2} \tag{21}$$

Values of q, ρh and $\rho h^3 (=f$, say) are specified at the nine nodes of the finite difference grid shown in Figure 5. We can then write Taylor series expansions for the function f at each of the points 1,2,3, and 4 about the point 0. For example

$$f_1 = f_o - a \frac{\partial fo}{\partial x} + \frac{a^2}{2} \frac{\partial^2 fo}{\partial x^2} - \frac{a^3}{6} \frac{\partial^3 fo}{\partial x^3} + \frac{a^4}{24} \frac{\partial^4 fo}{\partial x^4}$$

$$- \frac{a^5}{120} \frac{\partial^5 fo}{\partial x^5} + \frac{a^6}{720} \frac{\partial^6 fo}{\partial x^6} + \ldots\ldots\ldots \tag{22}$$

This gives four equations from which we can obtain expressions for $\frac{\partial fo}{\partial x}$ and $\frac{\partial^2 fo}{\partial x^2}$, in which terms higher than the fourth order are neglected. Corresponding expressions are obtained for the y direction and we finally obtain a computing equation for q of the form

$$q_o = \frac{\sum_{n=1}^{8} H_n(\rho_n h_n^3 - \rho_o h_o^3) q_n - 12 \eta_o u \sum_{n=0}^{4} (F_n \rho_n h_n)}{\sum_{n=0}^{8} (H_n \rho_n h_n^3) - 2 H_o \rho_o h_o^3} \tag{23}$$

where $F_i = F_i(a,b,c,d)$ and $H_i = H_i(a,b,c,d,\Delta)$ (a,b,c,d, and Δ are the dimensions of the grid as shown in Figure 5).

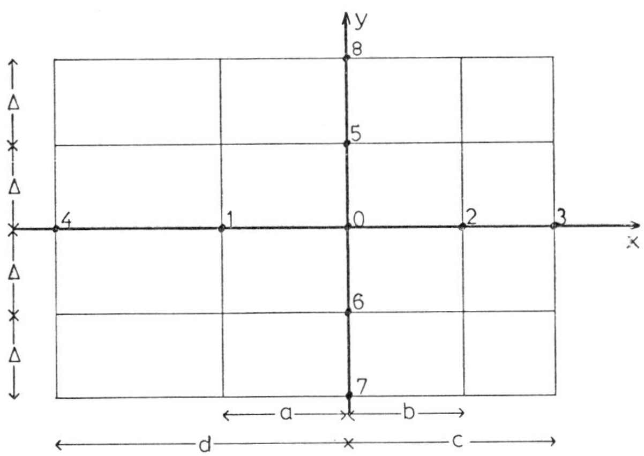

Fig. 5.—Finite difference notation used for solution of Reynolds' equation.

5. ITERATIVE SOLUTIONS

The iterative technique which the authors have adopted for the simultaneous solution of the elasticity, hydrodynamic, viscosity/pressure and the density/pressure equations is based upon those of Stephenson and Orterle[10] Sternlicht, Lewis and Flynn[11] (line contact) and Ranger et al.[1] (point contact).

The first step in the numerical solution, after deciding upon the operating conditions of load, speed, atmospheric viscosity, viscosity/pressure relationship, elastic constants and geometry, is to set up the finite difference mesh as described earlier. An initial or first guess for the pressure distribution (p_S) is then established in one of two ways. The first method is to use values from a previous solution at the same load. If these values are not available, an approximation to the expected pressure distribution is used. This approximation corresponds to a modified Hertzian distribution for dry contact. A starting pressure distribution which has been used in some of our earlier work is shown in Figure 6. Also shown is the corresponding centerline film shape calculated from the elasticity equation. It can be seen that this starting pressure gives a film shape which has approximately the right characteristics of an elastohydrodynamic film.

Having established an initial pressure and film shape, the film thickness must then be initialized and, when values from a previous solution are not available, the value predicted by the approximate heavy-load formula of Archard and Cowking[8] is used.

The steps involved in obtaining a converged solution are then as shown in the flow diagram in Figure 7.

It will be noted that the application of the elasticity equation to find the film shape is required only once per main (outer) iteration. This is desirable from the purely computational point of view since evaluation of the deformation integral is the most costly (in terms of computing time) aspect of the solution.

In general the tolerances on convergences are reduced as the main iterative loop converges. Overall convergence is measured by the sum of absolute pressure residuals divided by the sum of pressures and is taken to be acceptable if less than 1%.

i.e. $$\frac{\Sigma \mid Ps - Phydro \mid}{\Sigma P_s} < 0.01 \qquad (24)$$

This criterion must be satisfied over the computing region as a whole and, separately, over the area of the Hertzian circle for dry contact.

Other convergence requirements on inner loops at final convergence are as follows

Load 5×10^{-3}

Transformed pressures in Gauss-Seidel iteration 2×10^{-6}

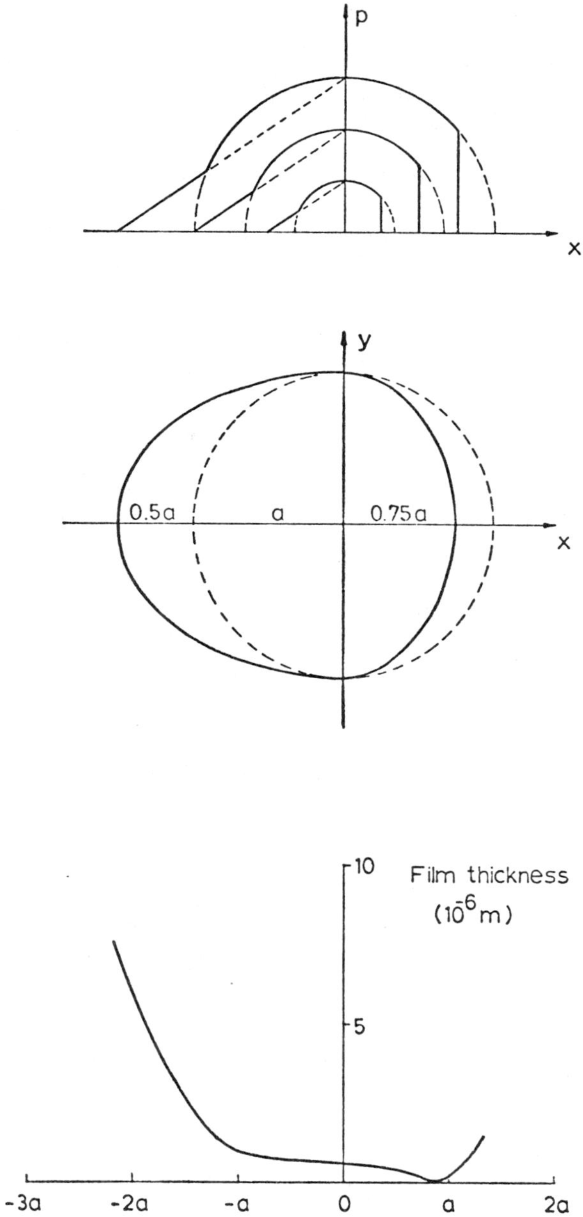

Fig. 6.—Modified Hertzian pressure distribution used as initial pressure and example of corresponding film shape.
$\nu = 0.3$, $E = 10.8\text{x}10^{10}$ N/m^2, $R = .0254$m. $L = 120$N.

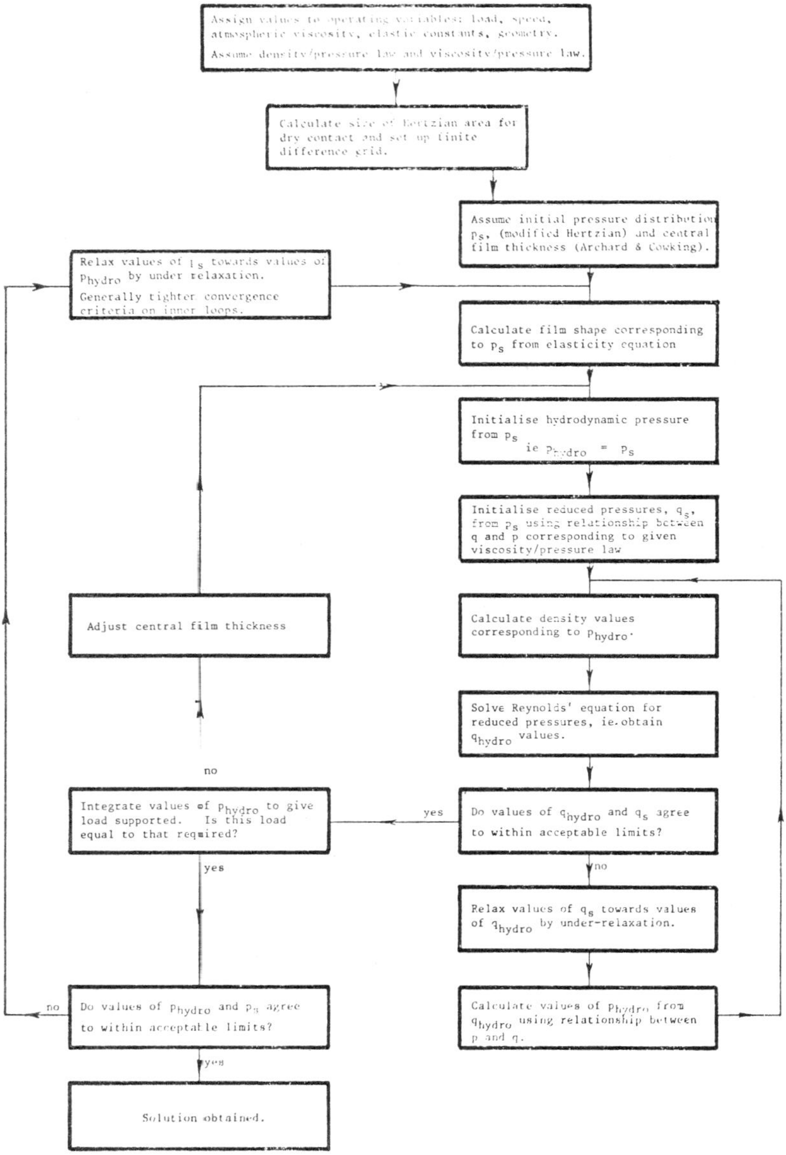

Fig. 7.—Flow diagram for numerical solution of isothermal point contact EHD problem.

6. EXAMPLE OF NUMERICAL RESULTS

Examples of some of the preliminary results obtained from the numerical analysis are shown in Figures 8 and 9. The main geometrical and materials variables have been chosen to correspond to the experimental work of Snidle and Archard[12] and the following values have been assumed.

$$R = 0.0254 \text{ m}$$

$$E_1 = E_2 = 10.80 \times 10^{10} \text{ N/m}^2$$

$$\nu_1 = \nu_2 = 0.3$$

$$\eta_o = 0.5159 \quad \text{Ns/m}^2$$

In the preliminary results presented in this paper an exponential viscosity/pressure relationship has been used, but work now in progress uses the power-law fit to the Roelands relation described earlier. Numerical results are presented corresponding to two different conditions of load and speed representing relatively lightly loaded and relatively heavily loaded conditions, respectively.

Figure 8 shows the center line pressure distribution and film shape for lightly loaded conditions. Corresponding isobars and film thickness contours are shown in Figure 9. Figure 10 shows center line pressure and film thickness variations for a more heavily loaded case and Figure 11 shows the corresponding isobars and film thickness contours.

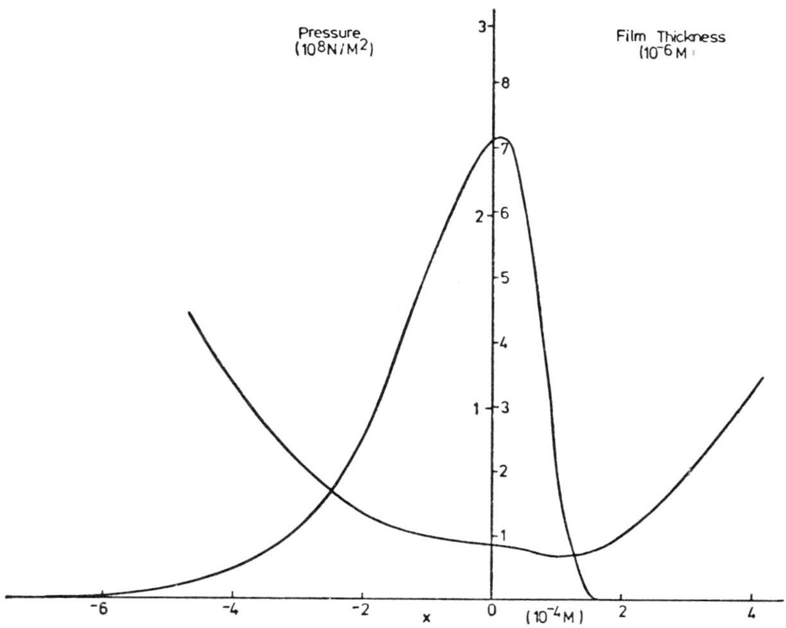

Fig. 8.—Centre line pressure distribution and film profile.
$u = 0.5$ m/s, $L = 15$N, $\alpha = 10^{-8}$ m²/N.

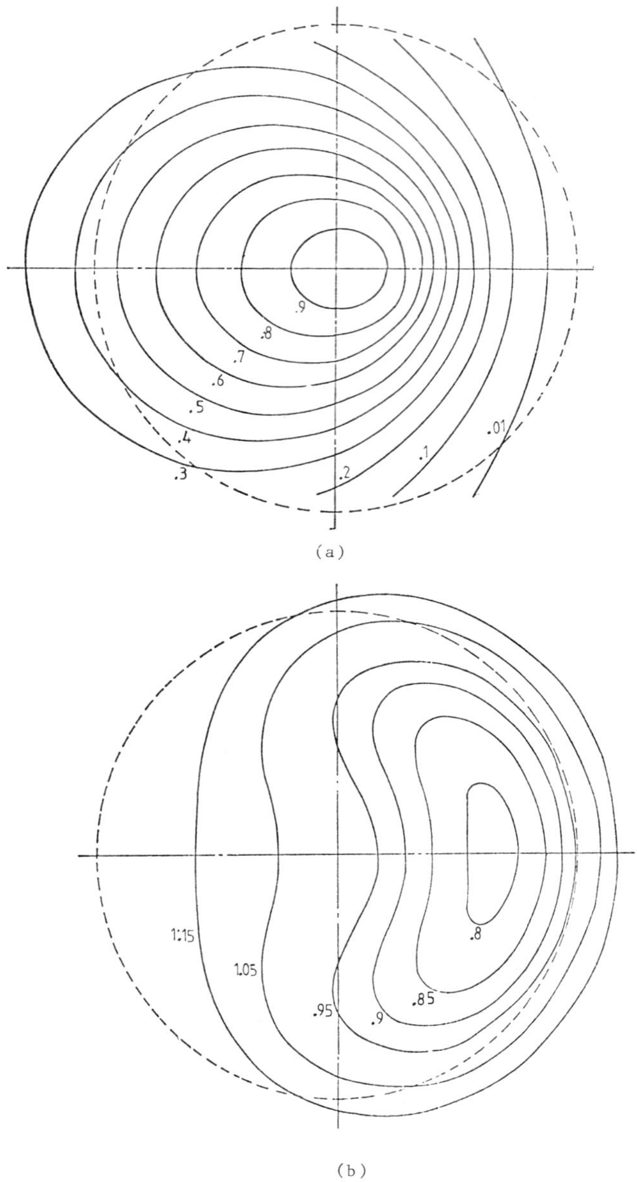

Fig. 9.—Isobars and film thickness contours.
$u = 0.5$ m/s, L - 15N, $\alpha = 10^{-8}$ m^2/N.
(a) Pressure, figures indicate values of P/Po, (b) Film thickness, figures indicate values of h/h_o. Broken circle indicates corresponding Hertzian area for dry contact.

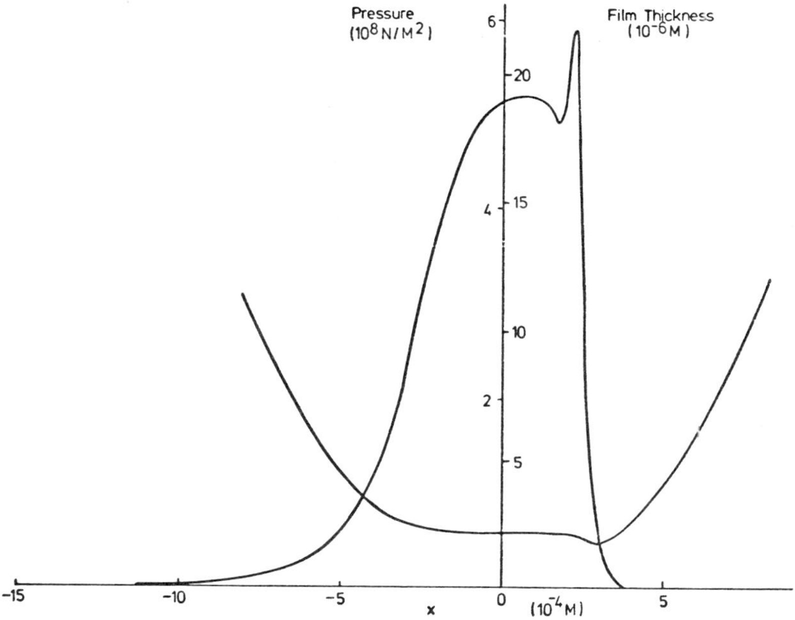

Fig. 10.—Centre line pressure distribution and film profile.
$u = 2$ m/s, $L = 120$ N, $\alpha = 10^{-8}$ m^2/N.

7. CONVERGENCE PROBLEMS

At higher values of load and effective α values the iterative scheme eventually fails to converge. It remains possible to obtain rough convergence (to within 10%, approximately) of pressure values, but the pressure distribution in the heavily loaded part of the contact tends to oscillate giving rise to the type of center line pressure distribution shown in Figure 12. The reason for the oscillations or "wobbles" in the pressure curve would seem to arise from the combination of an almost parallel film and high values of the viscosity. This can be seen from the integrated form of Reynolds' equation for flow in one direction

$$\frac{dp}{dx} = 12\, u\eta \frac{(h - h^*)}{h^3} \tag{25}$$

Since η is very high, it follows that any errors in the film shape represented by $h - h^*$ will become greatly magnified when values of pressure are calculated. This also emphasizes the need for high accuracy elastic deformation methods such as we have aimed to develop.

Those problems are particularly evident when attempting to use realistic values of the viscosity/pressure coefficient in the Barus relationship. When using the Roelands relationship (or the power-law relationship which is a good fit to the Roelands-law) the problem is less serious because of the way the viscosity/pressure relationship levels off at the higher pressures.

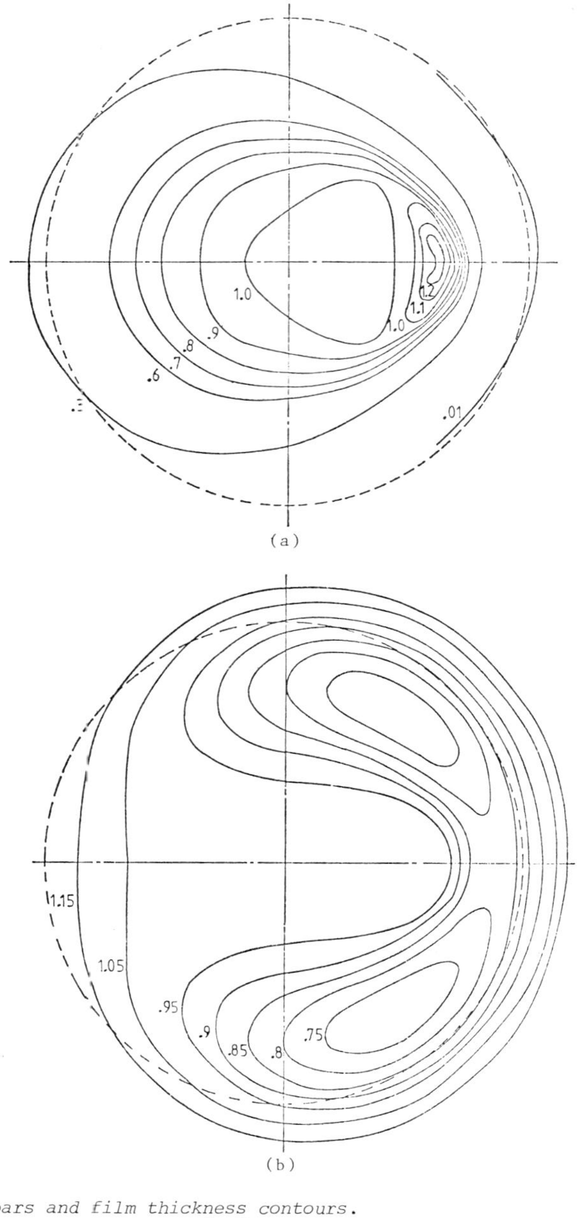

Fig. 11.—Isobars and film thickness contours.
$u = 2$ m/s, $L = 120$ N, $\alpha = 10^{-8}$ m^2/N.
(a) Pressure, figures indicate values of p/p_o, (b) Film thickness, figures indicate values of h/h_o. Broken circle indicates corresponding Hertzian area for dry contact.

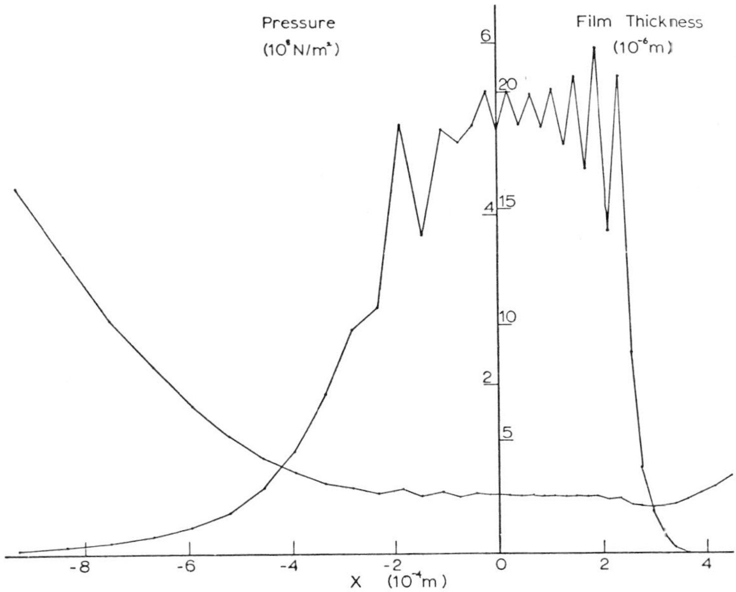

Fig. 12.—Centre line pressure distribution and film profile for roughly converged solution (10%). $u = 2$ m/s, $L = 120N$, $\alpha = 1.5 \times 10^{-8}$ m^2/N.

The use of the Barus law in this type of solution is a particularly demanding test of the method and explains, for example, why the results obtained by Ranger et al. (using the Barus law) correspond to less severely loaded conditions than the results given by Hamrock and Dowson (using the Roelands law).

Apart from using viscosity/pressure laws which correctly give lower values of viscosity at the higher pressures, other methods of improving the stability of the solution are being examined and must be pursued if the range of conditions for which solutions are possible is to be extended.

In Figure 13 the center line pressure distribution and film shape are shown at four stages in the development of a converged (to 1%) solution from the initial Hertzian-type pressure distribution. It is seen that the film shape becomes accurately established quite rapidly relative to the pressure distribution, which is still being changed at the 1% level. This stability of the film shape can be misleading, as can be seen from Figure 12. In this example the film thickness is stable at the shape shown but the corresponding pressure field is far from realistic and is neither converged nor proceeding toward convergence.

This is, of course, an extreme example, but it illustrates the danger of using coarse grids and slack convergence criteria, both of which may seem to explain a lack of smoothness of the pressure fields in cases where the film shape seems well established, whereas in reality the numerical method is incapable of producing a converged solution.

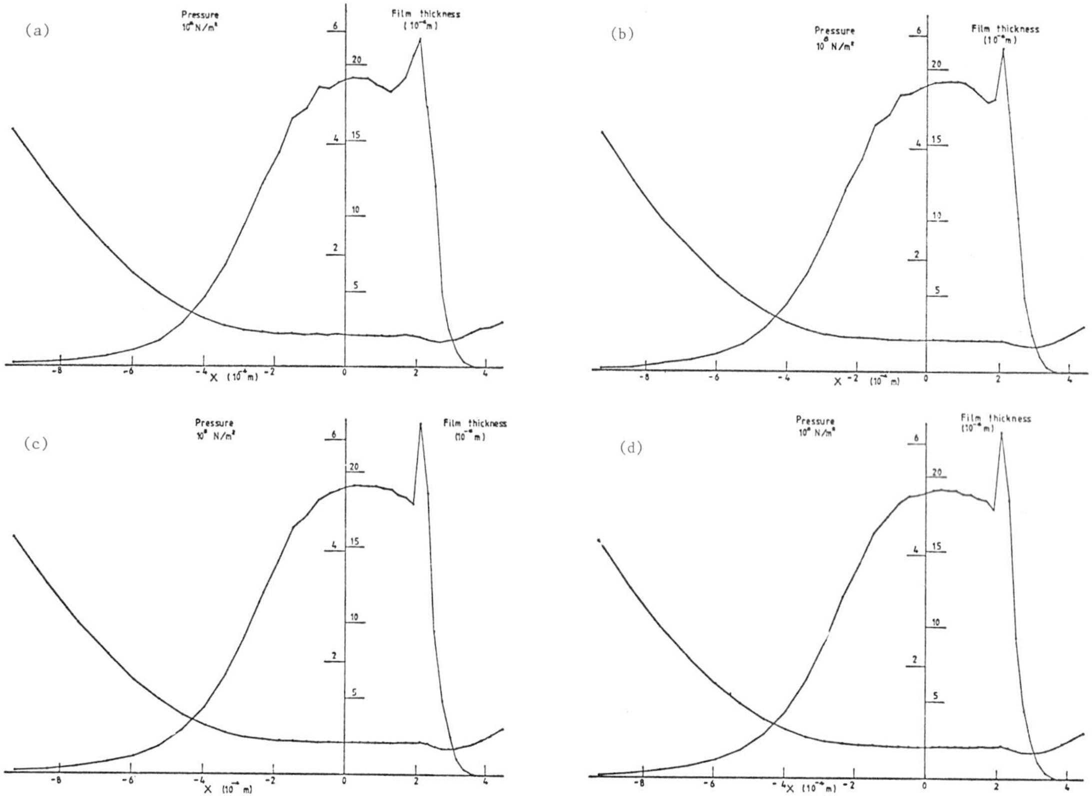

Fig. 13.—Center line pressure distribution and film profile at four stages of convergence of solution. $u = 2$ m/s, $L = 120N$, $\alpha = 10^{-8}$ m^2/N. (a) 10%; (b) 5%; (c) 2%; (d) 1%.

8. SIDE STARVATION

The importance of the inlet boundary condition and the consequences of starvation in the case of line contact EHD are well understood[13]. In point contact EHD the inlet boundary extends around the sides of the contact region and, in numerical solutions in which a rectangular zero pressure boundary is assumed, it is clear that the position of the side boundaries of the computing region is of importance, and while a specific numerical solution with given side boundary positions may have some physical relevance, it is clear that the influence of the position of the side boundaries needs to be understood before comparisons are made between, say, values of film thickness under different operating conditions. For while starvation due to the side boundaries being too close may be apparent at a light load or high speed the effect may disappear at heavier loads or lower speeds.

The effect of varying the position of the side boundary is shown in Figure 14. The film thicknesses plotted were produced by a model using a coarser grid in the y direction (5 points per Hertzian radius), and would seem to indicate that under the conditions considered the side boundary needs to be up to four times the Hertzian radius from the center line before the assymptotic film thickness is produced. As the side boundaries are moved closer to the center line the central film thickness is seen to decrease. This effect is least pronounced at high loads and low rolling speeds.

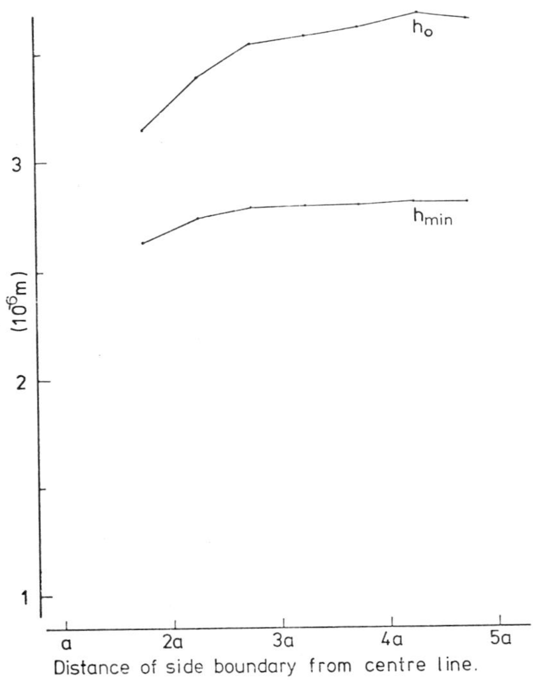

Fig. 14.—Effect on central and minimum film thickness of varying position of zero pressure side boundary. $u = 5$ m/s, $L = 120N$, $\alpha = 10^{-8}$ m^2/N.

The side boundaries used in Ranger's model are at a distance 1.5 times the Hertzian radius from the center line, a distance which is indicated by Figure 14 to be small enough to be capable of causing a significant side-starvation effect on the film thickness. The result of this is that film thickness values at lighter loads were probably under-estimated in Ranger's work and this is, in the authors' view, the most likely explanation of Ranger's unexpected film thickness formula which predicts an increase of film thickness with load.

It is hoped this matter will be resolved by using a grid structure which has variable spacing in both x and y directions. A model incorporating such a grid is currently being developed.

Hamrock and Dowson, although using a side boundary which is comparable to Ranger's (i.e. 1.6 times the Hertzian radius from the center line), appear to have avoided side-starvation effects by basing their film thickness formula on results obtained at heavier loads.

9. DISCUSSION AND CONCLUSIONS

In this discussion paper we have tried to review, in greater detail than has been possible in Professor Cheng's overall review of the subject, some of the recent work in the area of numerical solutions to the point contact EHD problem. The solutions obtained by Ranger et al. and Hamrock and Dowson have been of great importance in establishing the viability of the forward iterative method of solving this difficult problem. Now that the basic approach has been proven, further work needs to be carried out to refine the techniques developed and to try to extend the range of conditions for which closely converged solutions are possible.

It would seem to be an ambitious hope to produce solutions to the point contact problem which are as accurate and refined as the numerical solutions obtained for the line contact case, particularly at the heavier loads which are encountered in engineering practice. This must be our aim, however. We should regard the solutions of Ranger et al. and Hamrock and Dowson as being the first step towards the "refined" type of solution.

We have mentioned some of the shortcomings of the existing solutions. In particular we have pointed to the need to improve the accuracy of calculation of elastic deformation which is the major shortcoming of the Hamrock and Dowson work. We have also suggested that because relatively coarse finite difference grids are unavoidable (given the maximum computer power presently available) higher order finite difference methods are generally indicated. The adoption of second order methods for elastic displacement calculations and fourth order finite difference methods for the solution of Reynolds' equation, as used by the authors, are therefore desirable and are felt to be an important contribution towards obtaining a more accurate solution.

It is also hoped that the development of higher accuracy finite difference techniques will help to extend the range of solution by overcoming the problem of oscillations or "wobbles" in the pressure distribution near the center of the EHD contact at high loads. The results shown in Figure 13 indicate that much more closely converged solutions will help to overcome this problem.

The choice of a viscosity/pressure relationship is also important in determining the range of loads for which stable solutions can be found.

The Barus relationship is a particularly demanding test of the type of solutions described because of the extremely high viscosities which occur near the center of the contact when this relationship is used. The incorporation of viscosity/pressure laws such as the Roelands into the solutions are therefore desirable not only because of their greater physical realism but also because it should be easier to obtain a numerical solution when these laws are used. This is a rare but happy example of greater physical realism going hand in hand with an easier theoretical solution.

The forward iterative technique of solutions should obviously be pursued and the authors are further developing the techniques described in this paper. However, more radical methods of solution may be needed if solutions to the point contact problem are to be obtained under the most severe conditions which are encountered in practice. In this respect it may be worth attempting to exploit some of the techniques which were developed for the solution of the line contact problem. In particular it would seem possible to treat the heavily loaded part of the contact, where side leakage of the lubricant is minimal, as a series of elemental line contacts under heavily loaded conditions. The inverse techniques adopted for the solution of the line contact problem could therefore be used in this region and the inlet region (where side leakage is important) treated separately.

REFERENCES

1. Ranger, A.P., Ettles, C.M.M. and Cameron, A., *Royal Society of London. Proceedings.* Series A, Vol. 346, 1975, p. 227.
2. Hamrock, B.J. and Dowson, D., *Journal of Lubrication Technology,* Vol. 98, No. 2, 1976, p. 223.
3. Hamrock, B.J. and Dowson, D., *Journal of Lubrication Technology,* Vol. 98, No. 3, 1976, p. 375.
4. Hamrock, B.J. and Dowson, D., *Journal of Lubrication Technology,* Vol. 99, No. 2, 1977, p. 264.
5. Hamrock, B.J. and Dowson, D., *Journal of Lubrication Technology,* Vol. 99, No. 1, January 1977, p. 15.
6. Roelands, C.J.A., Druk V.R.B., Groningen, Netherlands, 1966.
7. Cheng, H.S., *Journal of Lubrication Technology,* Vol. 98, 1976, p. 228.
8. Archard, J.F. and Cowking, E.W., *Institution of Mechanical Engineers. Proceedings,* Vol. 180, 1965-66, p. 47.
9. Biswas, S. and Snidle, R.W., *Journal of Lubrication Technolcgy,* Vol. 99, 1977, p. 313.
10. Stephenson, R.R. and Osterle, J.F., *American Society of Lubrication Engineers Transactions,* Vol. 5, 1962, p. 365.
11. Sternlicht, B., Lewis, P. and Flynn, P., *Journal of Basic Engineering,* Vol. 83, 1961, p. 213.
12. Snidle, R.W. and Archard, J.F., in Elastohydrodynamic Lubrication Symposium, Leeds, 1972, Institution of Mechanical Engineers, London, p. 5.
13. Wolveridge, P.E., Baglin, K.P. and Archard, J.F., *Institution of Mechanical Engineers. Proceedings,* Vol. 185, 1970-71, p. 81.

DISCUSSION

QUESTIONER: I would like to make a few points here. First of all, it is not clear from a mathematical point of view that a higher order method is necessarily more accurate. One can get into trouble for a variety of reasons. For example, you are assuming pressure distributionis parabolic, and you put three points through it and you have a pressure distribution somewhere that can be extremely negative. Secondly, it is not clear that the nine-point formula is better than a five-point formula. You have to remember that the truncation error depends on a higher derivative of the function. In a lot of these lubrication problems, these higher derivatives do not exist. Even with a simple thing like a step slider bearing you can get into that problem; just because there is a nominal h to the 19th power out in front does not mean that you are going to get a better approximation. Another point is this: a more refined solution really should be lloking at the technique for obtaining the solution rather than the details. And I think we need to develop techniques like generalized Newton methods. You had a very good point about the convergence criteria and in a lot of these EHD problems things may converge very slowly, especially if you are using direct iterative kinds of approaches. One might liken this to a geometric series with the geometric ratio near one. What is happening is that you take the nth term and it is very small, but the nth term does not represent the area.

H. P. EVANS: I would like to take up two points. One is about this oscillator when it breaks down. It seems to me, by looking at the things before it breaks down and after it breaks down, that the hydrodynamic part of the solution is no longer able to correct; any oscillation along the x-axis of the pressure distribution is being repeated by the next prediction. By contrast in the results that converge, if you have a wobble on it, then what comes out of the Reynold's Equation is a wobble but out of phase. I think the breakdown has something to do with the hydrodynamic part of the solution.

On the nine point representation, the question of accuracy has certainly been discussed in the mathematical literature. There is no general agreement about that. But certainly in a lot of cases five-point is better than nine-point, especially if you can take a lot of points. We have tried it with the five-point representation but did not get a smooth pressure distribution out of that.

J. PIRVICS, SKF Industries: This is a suggestion. I think this work would benefit from reading the available material. A review was performed back in 1968 by Professor Castelli and myself. I think that the methods presented there indicate convergence criteria for non-linear convergences, techniques for solving Reynold's Equations, step variations, discrete supply sources, and so on. I think that an examination of available literature would have extracted you from a lot of the difficulty in your calculations.

SPIN TRACTION PREDICTION

J. L. Tevaarwerk

ABSTRACT

The spin traction is predicted using the recently proposed traction models. Material parameters for the equations are extracted from the zero spin traction curve and the predicted results are compared with experimental results from a point contact traction machine. It may be observed that elastic effects in the fluid play a major role in correctly predicting the spin traction. Also examined are the effect of forward pressure center shift and compressional viscoelasticity on the spin traction prediction.

NOMENCLATURE

Symbol	Explanation	Units
a	radius of the Hertzian contact area	m
B	non-dimensional strain rate in the elastic/plastic model	-
G	local elastic shear modulus of the fluid	N/m^2
$\bar{G}$	average elastic shear modulus of the fluid over the contact area	N/m^2
h	mean separation of the Hertzian contact planes	m
N	normal load on the contact	N
m	initial linear slope on the traction curves	-
P	local Hertzian pressure in the fluid	N/m^2
s	disposable parameter in the Ree-Eyring viscosity model	m^2/N
T	traction force (area integral of the shear stress)	N
U	velocity in the principal rolling direction	m/s
V	velocity perpendicular to the principal rolling direction	m/s

x,y,z	coordinates	m
α	pressure viscosity coefficient for the Barus equation	m^2/N
γ	local shear strain	-
$\mathring{\gamma}$	local shear strain rate	1/s
$\mathring{\gamma}_{ij}$	local shear strain rate tensor	1/s
$\mathring{\gamma}_e$	equivalent local shear strain rate = $\sqrt{\tfrac{1}{2}(\mathring{\gamma}_{ij}\cdot\mathring{\gamma}_{ij})}$	1/s
μ	maximum vlaue of the traction coefficient	-
τ	local shear stress	N/m^2
τ_e	equivalent local shear stress = $\sqrt{\tfrac{1}{2}(\tau_{ij}\cdot\tau_{ij})}$	N/m^2
τ_c	local limiting shear stress	N/m^2
τ_o	representative stress in the Ree-Engineering viscosity model as used by Johnson and Tevaarwerk ($= \frac{1}{s}$)	N/m^2
τ_{ij}	local shear stress tensor	N/m^2
$\bar{\tau}_c$	average limiting shear stress	N/m^2
η	local fluid viscosity	Ns/m^2
η_o	inlet fluid viscosity	Ns/m^2
ω	spin velocity	rad/s

INTRODUCTION

When counter conformal surfaces are loaded together, the stresses occurring at the contact zone are high enough to cause the metal surfaces to deform elastically. This, together with the effect of a hydrostatic pressure on the viscosity, lead to the formation as what is known as an elastohydrodynamic film of oil between the moving surfaces. This mode of lubrication is common in gears, rolling element bearings, cams and tappets, and traction drives. In traction drives, the fluid film transmits the tractive force from one element to the next. Development of the elastohydrodynamic lubrication (EHD) theory, is based upon the assumption that at high pressures and shear rates, the behaviour of lubricants remains of the Newtonian form. This has led to a theory which accurately predicts the mean separation of the moving surfaces by the oil film. Successful predictions of the EHD oil film thickness were first made by Grubin,(1) and in a more complete computer analysis by Dowson and Higginson,(2) and by Archard, Gair and Hirst.(3) Experimental verifications of these predicted film thicknesses came from Crook,(4) Sibley and Orcutt,(5) Dyson et al.,(6) Archard and Kirk(7) and others.

An adequate theory of EHD must also be capable of correctly predicting the frictional tractions at the opposing surfaces. It is in this area that EHD theory has proved to be inadequate. When the oil film in an EHD contact is being sheared, a number of conditions exist which could influence the properties to give an entirely different response compared to the influence of the individual conditions. In a typical EHD situation the time a fluid element is subjected to the pressure varies between 10 μs to 1 ms and is rarely more than 1 ms. Studies of fluid properties in a conventional viscometer can never be conducted in so short a time. Also some fluid properties tend to rise exponentially with pressure so there is a limit to the pressure at which these steady state properties may be determined in a reasonable time. Besides the above problems of transit time and experimental time limits, the rates of shear strain as found in the EHD contact are impossible to match because of serious heating effects in conventional viscometers. Shear heating does not become prohibitive in the EHD contact.

In order to gain insight into the EHD shear properties we must conduct our experiments under EHD conditions and the simplest way to do this is to study the traction behaviour in disc machines. The information which must be derived from the disc machine traction results, therefore, is the dependence of the properties on pressure; temperature and transit time and their interdependence; but foremost, a suitable constitutive equation which requires the minimum number of fluid properties and which describes the EHD traction behaviour under any conceivable strain distribution. This task is formidable and progress has been slow, not through lack of attempts, but because of the limited number of ways in which the EHD shear can be studied.

DISC MACHINE TRACTION CURVES

As opposed to the EHD film thickness, which is mainly governed by the conditions in the inlet, the frictional tractions are determined by the fluid properties within the EHD contact area. As Crook[8] showed, we may regard the EHD region as forming a parallel-plate viscometer, see Figure 1, and estimate the fluid properties from the known film thickness and the observed traction curves. If the film thickness and load are maintained constant the traction v. slip curves have the familiar form as shown in Figure 2. Crook[8] showed that this traction consists of two contributions; the rolling tractions caused by velocity gradients in the inlet region, and the sliding tractions caused by the shearing of the oil in the contact zone. Owing to the exponential response of the shear properties with pressure, the rolling traction constitutes a very small portion of the total traction.

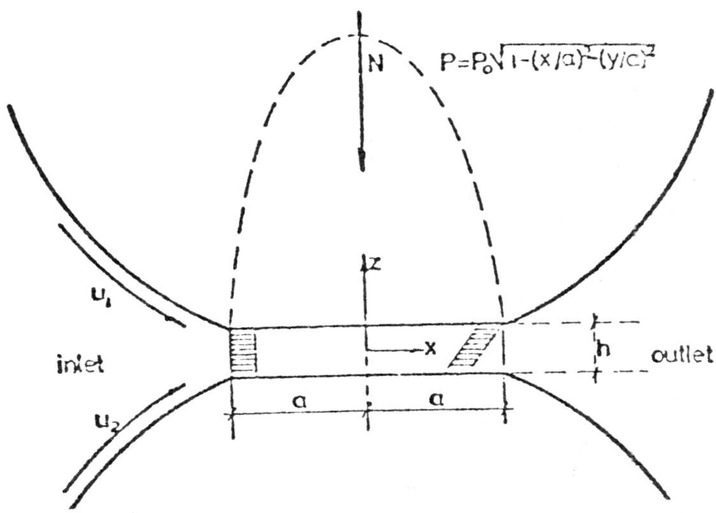

Fig. 1.—Idealized EHD contact showing a film of oil under shear.

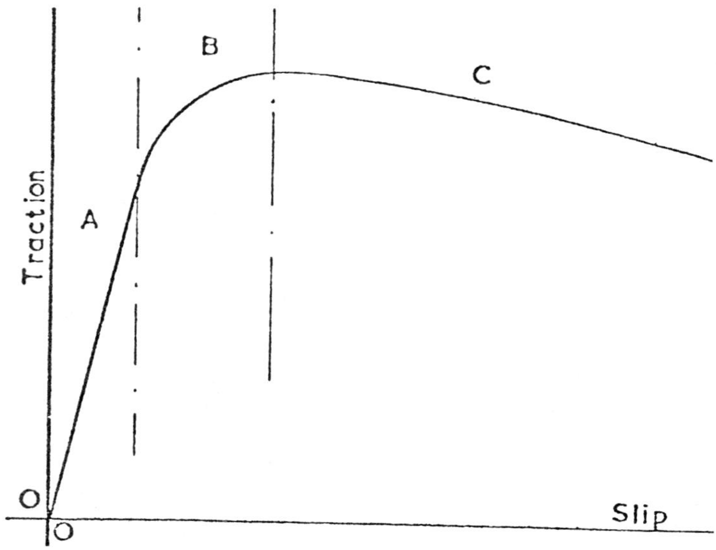

Fig. 2.—Typical traction curve with the three commonly designated regions. Region A, small strain linear region; Region B, isothermal non-linear region; Region C, thermally influenced region.

The traction curves as observed can be divided into three separate regions and studies of each have been made in the past. Region A is known as the linear small strain region where the traction increases linearly with slip. Thermal effects do not play any part in the observed traction behaviour. Region B is the non-linear region where the traction increases less rapidly as the slip increases. Towards the end of this region thermal effects become significant. Region C is known as the thermal region and any observed traction behaviour is greatly influenced by the effect of shear heating. The traction curves obtained from the disc machine experiments are strongly dependent upon the operating conditions of speed, pressure and temperature. They are also dependent upon the degree of spin that is present on the contact as can be seen from Figure 3, taken from Gaggermeier.(9) Spin occurs in many of the machine elements dependent upon EHD and it is very important that fluid traction models correctly predict the influence of the spin on the traction.

EXISTING TRACTION MODELS

When the slip is increased beyond the linear region the traction increases less and less rapidly, region B in Figure 2. The first thought is that this may be caused by shear heating effects and that the non-linear behaviour is simply showing the temperature dependence of the linear slope lubricant properties. Crook(8) calculated the shear temperature and from the small strain results deduced the shape of the traction curves with the thermal effects. Although the calculated variations of traction with slip were qualitatively similar to those observed experimentally, it was found that the rate of heat generation was insufficient to cause the traction to drop to the observed levels. Similar analyses done by Cheng,(10) Bell,

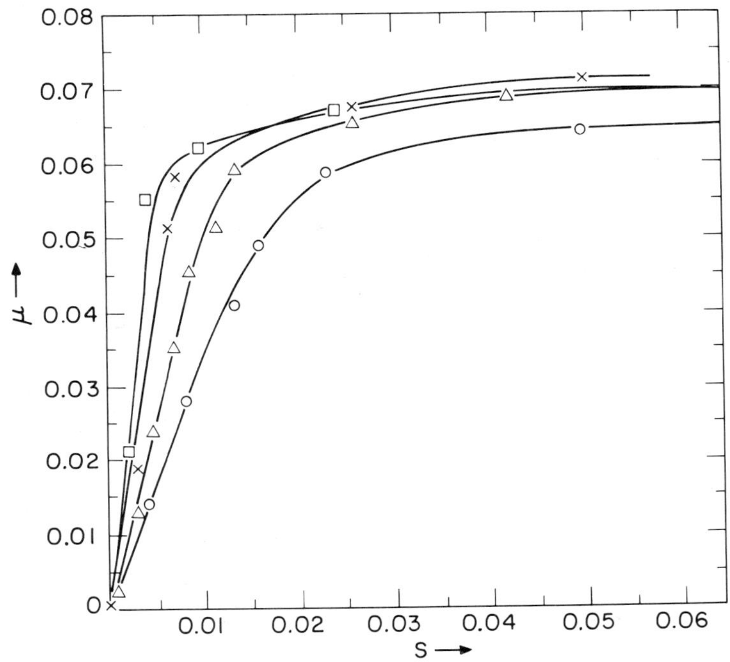

Fig. 3.—The influence of spin on the traction curves. These curves represent the traction resulting from longitudinal slip. Po = 1.24 GPa, U = 4.19 m/sec and temperature 50°C. Spin $\frac{\omega\sqrt{ab}}{U}$; □, o.o; χ, 6.2 x 10^{-3}; Δ, 1.28 x 10^{-2}; O, 4.13 x 10^{-2}. (Figure 56-2 from Gaggermeier.)[9]

Kannel and Allen[11] and Johnson and Cameron,[12] led to the same conclusion.

Because of the above evidence a number of investigators looked at possible non-linear models to predict the shape of the linear region A and the nonlinear region B. For reasons of comparison we will still look at the linear and nonlinear regions separately and in particular the deformation mechanisms.

The first of the traction models suggested was that due to Bell, Kannel and Allen.[11] They used the Ree-Eyring model of viscosity which has the form:

$$s\eta \mathring{\gamma} = \sinh(\tau s) \qquad (1)$$

For small strain rates this equation becomes linear and it predicts a simple Newtonian viscosity for the fluid. At larger stresses the stress strain relationship is towards shear thinning, see Figure 4a. Bell, Kannel and Allen[11] found that while qualitative agreement was found, the equation could not exactly predict the entire isothermal nonlinear behaviour.

Dyson[13] instead suggested that the nonlinear behaviour was caused by viscoelastic effects in the fluid. He pointed out that the shear deformation

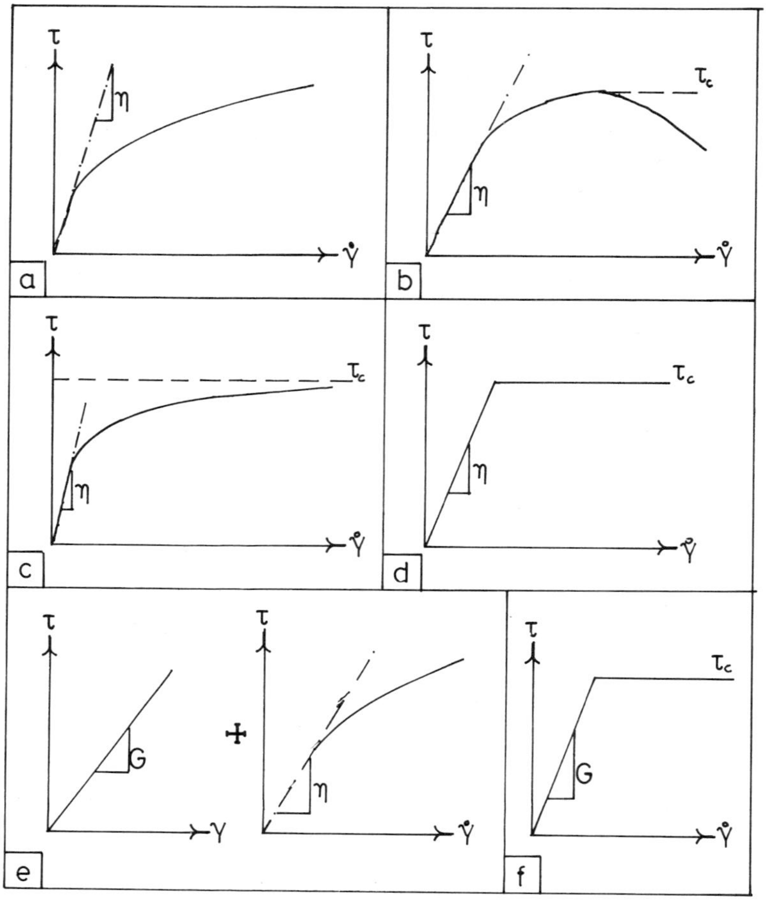

Fig. 4.—Various proposed stress/shear strain rate relationships for fluids under EHD conditions. 4(a), Bell, Allen and Kannel; 4(b), Dyson visco-elastic model; 4(c), Trachman and Cheng; 4(d), Lingard viscous/plastic model; 4(f), Limiting case of the Johnson and Tevaarwerk model used at high pressures.

of an element is accompanied by translation and rotation of the element. It is necessary therefore to use convected derivatives in the constitutive equations. Dyson[13] used the steady state large strain rate solution for a Maxwell viscoelastic model as derived by Fromm.[14] This solution gives a relation in shear stress and strain rate as follows:

$$\tau = \eta \mathring{\gamma} \, [1+(\frac{\eta}{G} \mathring{\gamma})^2]^{-1} \tag{2}$$

The resulting stress strain rate relationship is given in Figure 4(b). The initial slope of this model is of a viscous nature as the equation indicates. It should be recognized here that the non-linear behaviour in the traction curve is not an intrinsic property of the material but rather as a consequence of the entire stress system. Dyson found that the model had several drawbacks and in particular the fact that the shear stress reaches a maximum and then drops. Based on isothermal arguments this is unacceptable and he therefore modified the Fromm[14] solution as indicated in Figure 4(b). After the maximum point in the stress strain curve the stress continues like a limiting shear stress model would. Another drawback of this model is that at the maximum shear stress point the recoverable elastic shear strains are 50%. These high levels have not been observed anywhere, even in amorphous solids (except for rubber).

An outgrowth of the Dyson model was used by Trachman and Cheng.[15] They suggested a different form to the traction equation which did not have the undesirable isothermal drop in traction at high shear rates. Their equation is of the form

$$\tau = \frac{\eta \mathring{\gamma}}{1 + \eta \mathring{\gamma}/\tau_c} \qquad (3)$$

and is shown in Figure 4(c). For small strains the euqation predicts a viscous linear relationship and at large values of strain, the stress reaches the limiting value of τ_c. The equation predicts a smooth transaction from linearly viscous to the limiting plastic shear stress. Lingard[16] suggested a similar fluid model except that in his model the stress strain rates would stay Newtonian until the maximum shear stress was reached. After that point the fluid would behave like a solid and exhibit a limiting shear stress. This model is shown in Figure 4(d). With such a traction model, a smooth transition between the linear region on the traction curve and the final limiting shear may be obtained provided that not all the fluid yields at the same instant. This can be accomplished by letting the viscosity vary with pressure as suggested by the Barus equation; $\eta = \eta_0 e^{\alpha P}$ and by letting the limiting shear stress be a constant fraction of the applied pressure. Both Lingard[16] and later Gaggermeier[9] used this model with a great deal of success on simple slip/spin traction curves. The most recently suggested traction model is that due to Johnson and Tevaarwerk.[17] The equation they proposed is capable of describing linear elastic or viscous, nonlinear viscous and in the limit elastic/plastic-like behaviour. The model is shown in simple mechanical terms in Figure 4(e) and is expressed as

$$\frac{1}{G}\frac{d\tau_{ij}}{dt} + \frac{\tau_{ij}}{\tau e} F(\tau e) = \mathring{\gamma}_{ij} \qquad (4)$$

The dissipative function may take on any form consistent with experiment, but Johnson and Tevaarwerk found that an adequate function was that given by the Ree-Eyring viscosity model

$$F(\tau_e) = \left(\frac{\tau_o}{\eta}\right) \sinh\left(\frac{\tau_e}{\tau_o}\right) \qquad (5)$$

where τ_o = representative stress.

For normal operating conditions this equation has three fluid parameters; G, η and τ_o, each of which may be determined from the traction curve. The initial small strain rate slope is either viscous or elastic depending upon the so-called Deborah number. The Deborah number is the ratio of the relaxation time of the fluid, given by η/G, and the transit time a/u. At Hertzian pressures in excess of 1 GPa it was found that many of the commonly used fluids in EHD were giving an elastic small strain response. In the limit for very high pressures the constitutive equation reduces to that for an elastic/plastic like solid as was shown by Tevaarwerk.[18] The large strain rate behaviour is like that of a plastic solid, see Figure 4(f).

DEGREE OF FIT TO TRACTION CURVES

It is interesting to note that, with the exception of the Ree-Eyring model as proposed by Bell, Kannel and Allen,[11] all the models show the same behaviour at the higher shear stress end. Differences only occur in the transition between linear small strain and the limiting behaviour and in the nature of the small strain rate linear region. This is viscous for the first four models, while for the Johnson and Tevaarwerk model, it may be viscous or elastic. If the limiting shear stress model is used for the dissipative element then the linear region of this model is elastic in nature. Any differences that may occur in using the various models to predict the shape of the traction curve will therefore only be caused by the initial linear and transition regions of the models. It is for this reason that only two of the viscous/plastic models will be used together with the elastic/plastic model to observe to what degree the models predict the actual traction curves without and with spin.

Results for the Simple Slip Traction Curves

As an example traction curve, the results reported by Tevaarwerk[18] will be used in the analysis. This traction curve, see Figure 5, was obtained on a point contact machine under side slip. The fluid used here was a specially formulated traction fluid known as Hydra-Torque Medium. The three traction models that are used to fit this traction curve are the Trachman and Cheng model, the Lingard model and the Johnson and Tevaarwerk model. Regardless of the interpretation given to the various fluid parameters they are in fact disposable constants in the equations and may be used to obtain a best fit for the given traction curve.

The resulting traction curves obtained from the curve fit are shown in Figure 5 for comparison. For the Trachman and Cheng model and for the Lingard model the viscosity was allowed to vary as given by the Barus equation

$$\eta = \eta_0 e^{\alpha P} \tag{6}$$

where the local pressure was taken to be that given by the Hertzian pressure distribution and the value of α was selected to obtain the correct slope in the initial linear region. The critical shear stress τ_c was taken to be directly proportional to the local pressure:

$$\tau_c = \mu P \tag{7}$$

The constant μ was then selected to give the best overall fit to the data. To obtain the actual traction curves, numerical integration of the equations over the contact area was employed.

For the Johnson and Tevaarwerk model, averaged properties were used and the two mutual parameters $\bar{G}$ and $\bar{\tau}_c$ were extracted from the actual traction curve. For the averaged properties over the contact area the elastic/plastic model may be integrated analytically (see Tevaarwerk)[18] and the resulting

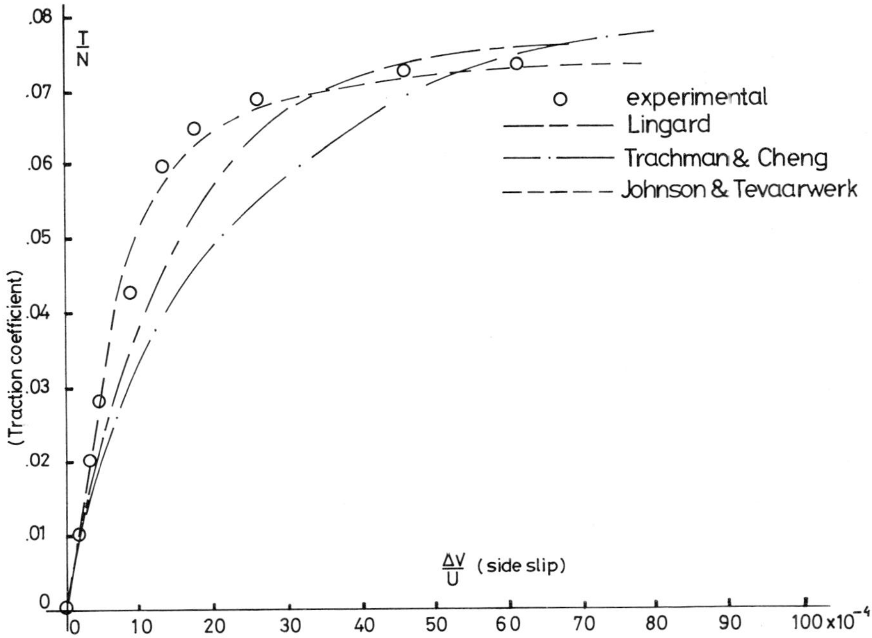

Fig. 5.—Side slip traction curve showing the experimental data points and the degree of fit obtained with the three traction models. Experimental information: Po = 1.3 GPa; U = 1.87 m/sec; a = 4.05 x 10^{-4}m; inlet temperature 17°C; N = 445 Newtons; ηo = 8 x 10^{-2} Pas; h = 9.33 x 10^{-7}m. Fluid Hydra Torque Medium. Disposable constants used for the models have the following values:

Lingard model: α = 1.15 x 10^{-8} m^2/N, μ = .101
Trachman and Cheng: α = 1.16 x 10^{-8} m^2/N, μ = .150
Johnson and Tevaarwerk: m = 66.7, μ = .077

equation is as follows:

$$\frac{T}{N} = \frac{\mu}{\pi}\left[2\cos^{-1}\left(\frac{1}{2B}\right) + \frac{8B}{3}\left(1 - \sqrt{1-\left(\frac{1}{2B}\right)^2}\right) - \frac{1}{3B}\sqrt{1-\left(\frac{1}{2B}\right)^2}\right]$$

$$\text{for} \quad B = \frac{\bar{G}}{\bar{\tau}_c}\frac{a\,\Delta v}{h\,U} \geq \frac{1}{2} \tag{8}$$

$$\text{and} \quad \frac{T}{N} = \mu\,\frac{8}{3\pi}B \quad \text{for} \quad B \leqslant \frac{1}{2}$$

The value of $\bar{G}$ may be obtained from the initial slope m of the traction curve

$$G = \frac{3}{8}\, m\, \frac{hN}{a^3} \qquad (9)$$

where N is the normal load on the circular contact of radius a.

The closeness of the fit for the elastic/plastic model is remarkable when one considers the simplicity of it. This, however, does not directly indicate that the model correctly describes the fluid shear behaviour. Any polynomial expression with some plausable explanation for the coefficients may be acceptable as a fluid model. Also, by invoking shear heating the higher strain rate regions can be made to fit better than shown here. This technique was employed by both Trachman and Cheng, and Lingard. A real test of the correctness of the fluid model comes, however, when one alters the strain distribution over the contact area without altering the pressures, velocity, temperature and film thickness. This may be done by introducing spin on the contact area.

Traction Prediction with Spin

Spin occurs in many mechanical elements that are dependent upon elastohydrodynamic lubrication. It arises when the two axes of rotation of the contacting bodies are not parallel to the plane of contact. The resulting motion is essentially one of twist about the line of centers of the contacting bodies. Under the assumption that the film thickness in the contact area is constant and of value h, then the resulting strain rate distribution for a circular contact is as shown in Figure 6(a). When additional side slip is introduced of increasing intensity, the resulting total slip distribution for the centerline of a circular contact is as shown in Figure 6(b). The ability to change the strain rate distribution over the contact area is very useful in traction model analysis since all the other major variables are not altered by the introduction of spin on the contact. In Figure 6(b) the difference at low side slip values is the sign of the spin, being either positive or negative. At larger values of side slip the effect of spin on the total slip is not significant, especially if the lubricant in the contact zone behaves in a plastic manner. At these high values of side slip one would expect the same amount of traction regardless of the value of the spin. In the second experiment reported here, the spin introduced an amount of slip at the contact edge equal to:

$$\frac{\omega a}{U} = \pm 3.28 \times 10^{-3}$$

(spin is positive according to the right hand rule on the ijk triad). This value was kept constant for the two spin experiments reported here while the side slip was increased from zero, giving pure spin, to the maximum value obtainable on the machine. The resulting experimental traction data is shown in Figure 7 for both positive and negative spin. The conditions of temperature, pressure and speed were the same as for the results in Figure 5.

The three traction models that were used in fitting the simple side slip traction curves were also used in predicting the traction curves resulting from the different strain rate distributions. Due to the nonlinear nature of the traction model equations and the fact that the material at high stress behaves like a plastic solid, slight alterations in the equations of Trachman and Cheng and Lingard are necessary to take this into account.

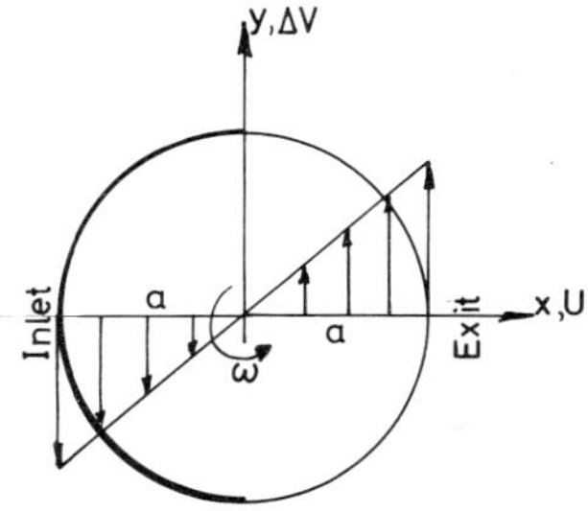

(a)

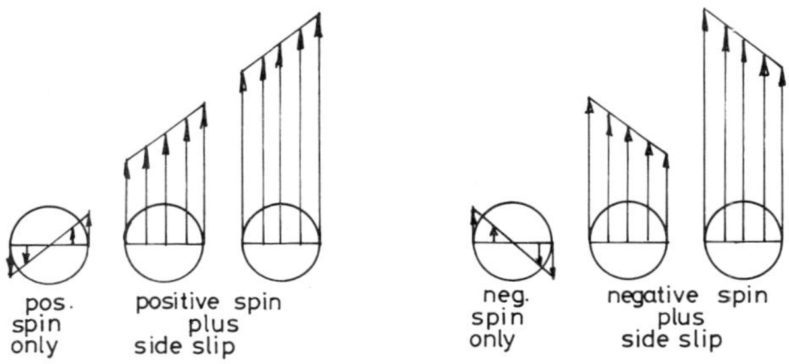

(b)

Fig. 6.—(a) Resulting slip distribution the x axis due to positive spin on the contact. (b) Resulting slip distribution on the x axis due to spin and side slip.

The equation may be modified as follows -

For the Trachman and Cheng model:

$$\tau_{ij} = \frac{\eta\mathring{\gamma}_e}{1+\eta\mathring{\gamma}_e/\tau_c} \left(\mathring{\gamma}_{ij}/\mathring{\gamma}_e \right) \tag{10}$$

and for the Lingard model:

$$\tau_{ij} = \eta\mathring{\gamma}_{ij} \text{ when } \eta\mathring{\gamma}_e < \mu P$$

$$\text{and}$$

$$\tau_{ij} = \mu P\, \mathring{\gamma}_{ij}/\mathring{\gamma}_e \text{ when } \eta\mathring{\gamma}_e = \mu P \tag{11}$$

where $\quad \mathring{\gamma}_e = \sqrt{(\mathring{\gamma}_{ij}\cdot\mathring{\gamma}ij)/2}$

These equations were then solved for the shear stress in the side slip direction and simple integration produced the resulting traction curves as a function of side slip. Exactly the same fluid parameters were used as in the simple side slip traction curve predictions and the resulting traction

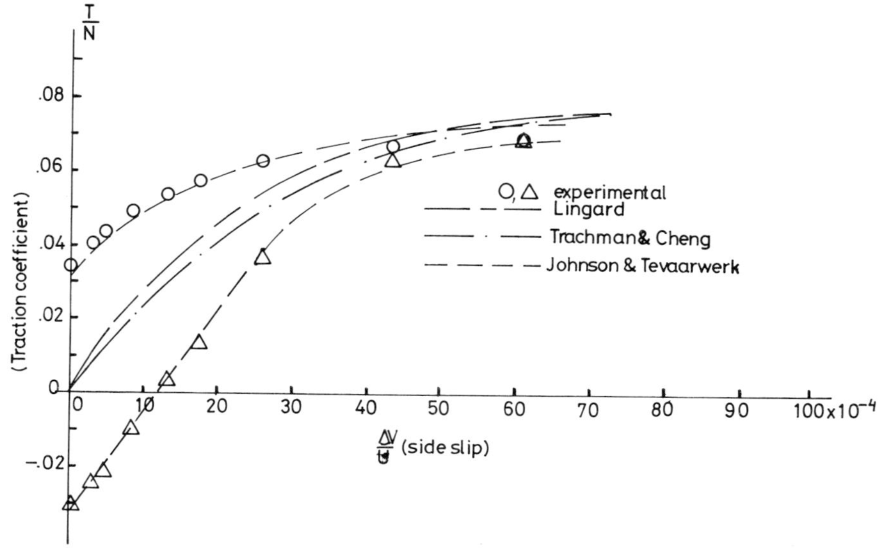

Fig. 7.—Combined side-slip and spin traction curves and the degree of fit obtained by using the various traction models. Experimental data: Spin $\frac{\omega a}{U}$ *:* $O = -3.28 \times 10^{-4}$, $\Delta = 3.28 \times 10^{-4}$. *The rest of the parameters are the same as in Figure 5.*

curves are shown in Figure 7.

DISCUSSION

From the results in Figure 7 it may be seen that for the Lingard model and the Trachman and Cheng model the predictions in the small side slip regions are very poor indeed. Not only do the results appear independent of the value of the imposed spin, but also the values are incorrect. At the higher side slip values the predictions do come closer to the experimentally observed traction values. This leads one to suggest that the problem with the Lingard and Trachman and Cheng models lies with the initial linear region. In the Johnson and Tevaarwerk model for this pressure it is taken as elastic-like, while for the other two models, it is viscous. When considering the point of zero side slip in this experiment one may examine why the elastic/plastic model does give rise to some traction while any simple viscous/plastic model cannot. In the elastic/plastic model the stress is dependent upon the shear strain while for the viscous/plastic it is dependent upon shear strain rate. For the pure spin case this gives rise to shear strain distribution as shown in Figure 8(a) for the center line stress in a circular contact. The traction, being the integral of the stress over the contact area, will be non-zero for the elastic/plastic model while it will be zero for the viscous/plastic models.

There are two effects known in elastohydrodynamics which may alter this picture somewhat. They are the time dependence of fluid properties and their response to sudden increases and decreases in pressure. This is im-

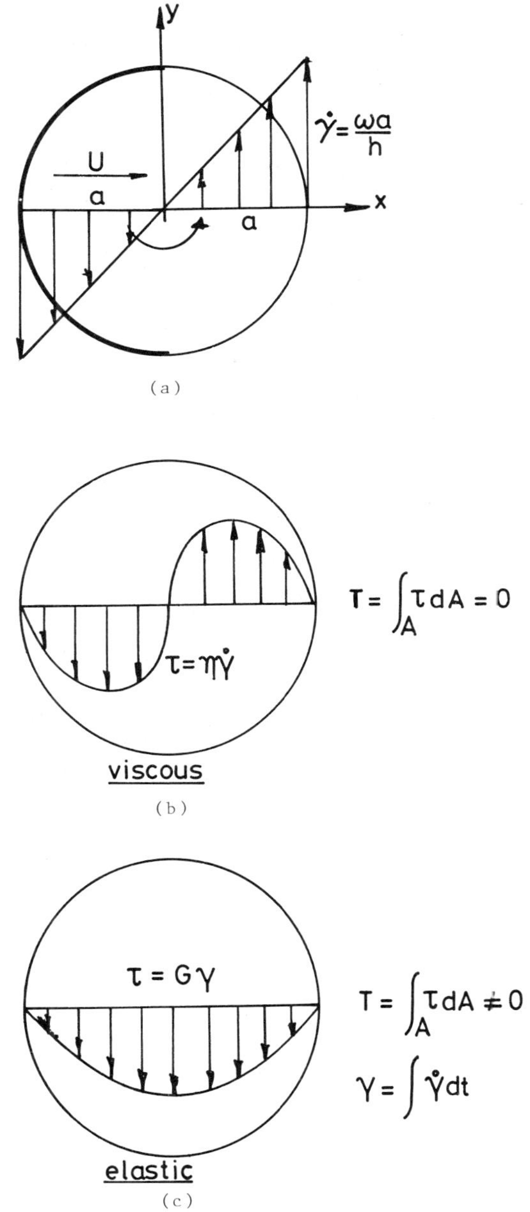

Fig. 8.—(a) Shear strain rate distribution due to spin on the x axis. (b) Shear stress distribution on the x axis for a viscous material. The net resulting traction will be zero. (c) Shear stress distribution on the x axis for an elastic material. The net resulting traction will not be zero and will be negative for positive spin.

portant in EHD since the fluid element is subjected to very high pressures in short times. The second effect is that due to the alterations in the Hertzian pressure distribution when a fluid is drawn into the constriction zone. Their effect on the spin traction curves in the initial linear region will now be examined. Trachman and Cheng[15] suggested that a fluid viscosity could not respond instantaneously to the applied pressure and that there would be a certain time delay involved. They suggested a variation of viscosity with time as shown in Figure 9(a). This variation, together with the assumed Hertzian pressure distribution, will produce an asymmetric viscosity distribution over the contact area, shown schematically in Figure 9(b). The average viscosity over the back part of the contact will be larger than that in the front. It will produce a shear stress distribution under conditions of spin as shown in Figure 9(c), and will result in a positive spin traction for positive spin at zero side slip; opposite to the observed spin traction for elastic-like material response. This form of spin traction response is in disagreement with the results reported here and those of Johnson and Roberts[19] which clearly show a negative spin traction with positive spin. Clearly then, the effect of compressional delays on the viscosity in the contact cannot produce the correct response to the observed spin traction data. The second effect that may influence the initial spin traction is that due to the forward movement of the actual pressure distribution as compared to the Hertzian pressure distribution. This may be seen in Figure 10(a), reproduced from Dowson and Higginson.[20] This means that the fluid viscosity will be higher in the frontal region of the Hertzian contact than in the rear. This can give rise to negative spin traction with positive spin on a viscous material. In order to calculate the effect of this forward pressure the contact was modelled as shown in Figure 10(b). The pressure distribution is still assumed to the Hertzian but the maximum value occurs a distance δ before the line of centers. Spin takes place about the line of centers and this then will result in a negative net spin traction due to positive spin. To evaluate the exact amount of spin-traction resulting from this phenomena, the maximum pressure for the Hertzian distribution was taken to coincide with the pressure center of the actual pressure distribution. The amount that the pressure center is forward of the line of centers was calculated by Hamrock[21] and is given by:

$$\delta = 4.25\, a (g_V)^{.022} (g_e)^{-.35} \tag{12}$$

where g_V and g_e are the Johnson[22] viscosity and elasticity parameters. For the spin/side slip experiment as reported here, δ/a amounts to 3.2%. This offset was introduced in the spin distribution for the circular contact and the Lingard viscous/plastic model was used to show the net resulting spin traction due to this effect. The results of the calculations are shown in Figure 10(c). While there is some influence of the pressure center forward movement, it is not sufficient to account for the difference between the observed and the predicted traction values.

CONCLUSION

From the evidence presented here it appears that traction models used for predicting the effect of spin on the traction curves require an initial elastic-like response. This is especially true for traction prediction in traction drives where moderately high pressures are encountered. The Johnson and Tevaarwerk model correctly predicts the shape of the traction curves with and without spin, and it is based on simple elastic/plastic like behaviour of the fluid in the constriction zone. The use of viscous/plastic

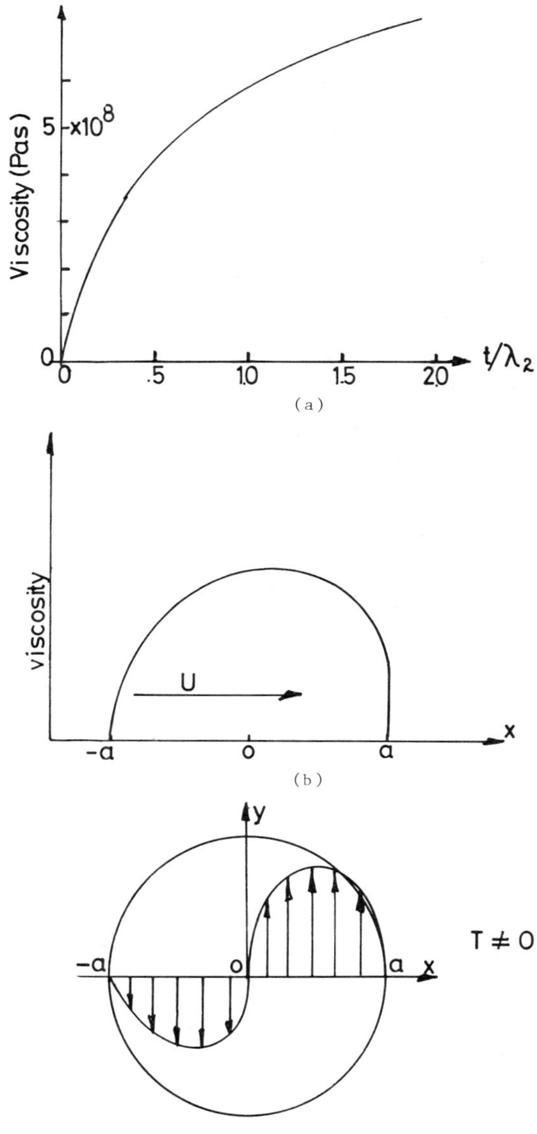

Fig. 9.—(a) Variation of viscosity with time following an applied pressure step as suggested by Trachman and Cheng.[15] *(b) Resulting non-symmetric viscosity distribution due to the viscoelastic time delay model. (c) Shear stress distribution on the x axis due to spin only for the viscosity distribution as shown in Figure 9(b). The net resulting traction will be positive for positive spin.*

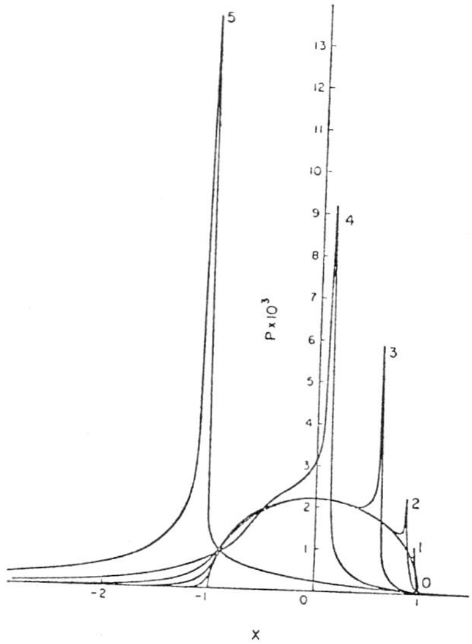

(a)

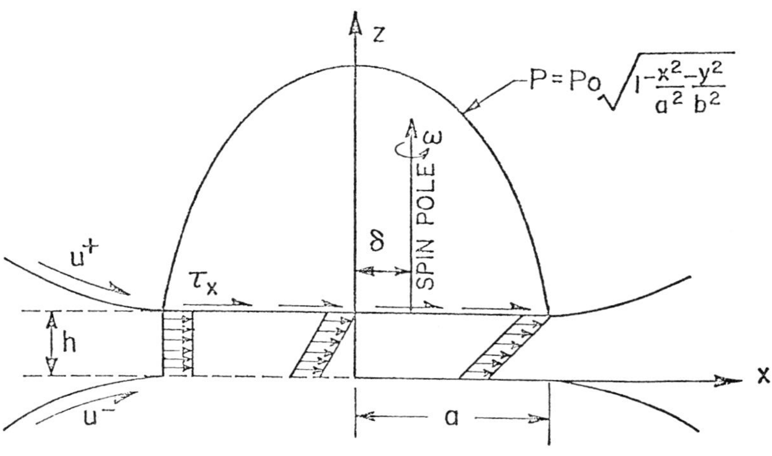

(b)

Fig. 10.—(a) Pressure distribution for an incompressible lubricant. $W = 3 \times 10^{-5}$, $G = 5000$, $U = (0)0$, (1) 10^{-13}, (2) 10^{-2}, (3) 10^{-11}, (4) 10^{-10}, (5) 10^{-9}. (b) Idealized contact zone with allowance for pressure center offset.

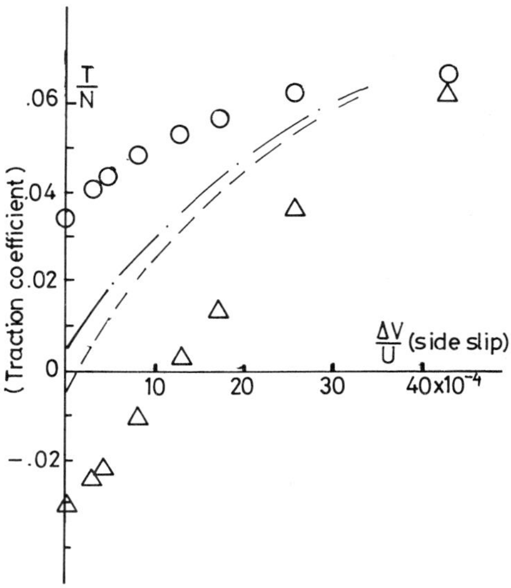

Fig. 10(c).—Comparison between the experimental[(c)] *results, as shown in Figure 7, and the predicted spin/side slip traction of the viscous/plastic Lingard model. In these calculations allowance was made for the effect of forward pressure center.*

models together with viscoelastic time delay effects or the influence of the forward movement of the pressure center cannot account for the observed spin traction results.

ACKNOWLEDGEMENTS

The author would like to thank the Canadian National Research Council for their support under grant number A4214.

REFERENCES

1. Grubin, A.N. and Vinogradova, I.E., Book #30, D.S.I.R. Translation, Moscow, 1949, No. 337.
2. Dowson, D. and Higginson, G.R., *Journal of Mechanical Engineering Science*, Vol. 1, No. 1, 1959, p. 6.
3. Archard, G.D., Gair, F.C. and Hirst, W., *Royal Society of London. Proceedings*. Series A, Vol. 262, 1961, p. 51.
4. Crook, A.W., *Royal Society of London, Philosophical Transactions*, Series A, Vol. 250, 1958, p. 387.
5. Sibley, L.B. and Orcutt, F.K., *American Society of Lubrication Engineers Transactions*, Vol. 4, No. 2, 1961, p. 234.
6. Dyson, A., Naylor, H. and Wilson, A.R., *Institution of Mechanical Engineers. Proceedings*, Vol. 180, Part 3B, 1966, p. 119.
7. Archard, J.F. and Kirk, M.T., *Royal Society of London. Proceedings*. Series A, Vol. 253, 1961, p. 52.

8. Crook, A.W., *Royal Society of London, Philosophical Transactions*, Series A, Vol. 255, 1963, p. 281.
9. Gaggermeier, H., Ph.D. Thesis, Technische Universitaet Muenchen, 1977.
10. Cheng, H.S., *American Society of Lubrication Engineers Transactions*, Vol. 8, No. 4, 1965, p. 397.
11. Bell, J.C., Kannel, J.W. and Allen, C.M., *Journal of Basic Engineering*, Vol. 86, 1964, p. 423.
12. Johnson, K.L. and Cameron, R., *Institution of Mechanical Engineers. Proceedings*, Vol. 182, Part 1, No. 14, 1967.
13. Dyson, A., *Royal Society of London, Philosophical Transactions*, Series A, Vol. 258, 1965, p. 529.
14. Fromm, H., *Zeitschrift fuer Angewandte Mathematik und Mechanik*, Vol. 28, No. 2, 1948, p. 43.
15. Trachman, E.G. and Cheng, H.S., Conference Publication, Institution of Mechanical Engineers, Vol. 9, 1972, p. 142.
16. Lingard, S., *Tribology International*, Vol. 7, 1974, p. 228.
17. Johnson, K.L. and Tevaarwerk, J.L., *Royal Society of London. Proceedings*. Series A, Vol. 356, 1977, p. 215.
18. Tevaarwerk, J.L., Ph.D. Thesis, University of Cambridge, 1976.
19. Johnson, K.L. and Roberts, A.D., *Royal Society of London. Proceedings*. Series A, Vol. 337, 1974, p. 217.
20. Dowson, D. and Higginson, G.R., "Elastohydrodynamic Lubrication," Pergamon Press, Oxford, 1966, p. 90.
21. Hamrock, B.J., Private Communication, 1977.
22. Johnson, K.L., *Journal of Mechanical Engineering Science*, Vol. 12, No. 1, 1970, p. 9.

DISCUSSION

J. H. HUTTON, Shell Research, England: This paper is an excellent contribution to this Conference. I would like to ask Professor Tevaarwerk just how general his conclusions are. Does he think that there will be elastohydrodynamic conditions in which viscous models just would not work, for example?

J. L. TEVAARWERK: All I would say, from the sort of traction rise that I have seen and the pressures and speeds that are employed there, is that one is likely to make a mistake. He will make it with a viscous plastic model, not with an elastic-plastic model. One can be in transition over a large range (not having just purely elastic response, but also viscous response) before this model goes up significantly. I think the real strength of the model lies in its simplicity and the fact that it can be integrated analytically on a non-thermal basis.

H. S. CHENG, Northwestern University: I seem to be convinced that the traction and temperature in EHD contact in most cases go hand in hand. You seem to show that you can go a long way by isothermal analysis. At what stage (particularly wiht gear contacts), must the temperature come in? Can you venture to guess what?

TAVAARWERK: I am working on a very simple model on that. The temperatures can be significant but what saves the day a bit here is the elastic-plastic behavior. First of all the elastic effects are going to store energy; they are not going to dissipate it. The results by Johnson and Cameron showed that the effect of temperature on plastic properties is not nearly as significant as it is in viscosity. We tend to think of exponential drops in viscosity. In the plastic situation that does not happen. I have difficulties relating the simple isothermal theory to these large spin tractions. I can get it if I assume properties which are measured at that large spin and then I can show that, in fact, the temperature variation between almost zero side slip and a lot of side slip is not very great. It is just that the initial spin seems to raise the temperature there which drops the properties down. But I think the plastic properties are not nearly as sensitive to the temperatures as the viscosity is.

A VISCOELASTIC FREE VOLUME THEORY OF TRACTION IN ELASTOHYDRODYNAMIC LUBRICATION

D. M. Heyes and C. J. Montrose

ABSTRACT

The behavior of a lubricant passing between heavily loaded rolling contacts is investigated using an elastohydrodynamical model. The liquid's free volume is used to illustrate and calculate the effects of the short transit time of the fluid in the contact zone. The results agree well with those of a previously formulated theory based on fictive pressure. The conditions which maximize traction between the rollers have been determined. The calculated traction efficiency for many lubricants is, typically, twice as large as that obtained experimentally.

INTRODUCTION

At the present there is a great interest in the development of lubricants that can efficiently transmit a shearing force from one rolling element to another, thereby driving the latter.[1] This property, known as traction, is improved by those lubricants that transiently (in $\approx 10^{-4}$ s) become glass-like when subjected to high pressures (>1GPa) in the contact region. The traction coefficient, C_T, is a convenient measure of the efficiency of this process and is defined as the ratio of the tangential to normal force per unit length along each roller.

Any theoretical treatment of traction at high pressures must consider the elastic deformation of the rollers as well as the behavior of the fluid between them. Such an elastohydrodynamical (EHD) model has been developed recently.[2] The model is confined to those conditions for which the tractive force increases linearly with the difference in the roller speeds, i.e., in the 'low slip' region of the so-called "traction curves." It assumes that the fluid passes between rollers that are flattened in the contact region.

The main purpose of this article is to propose an alternative (free volume) description of the fluid in this region, whose basis is less empirical than the (fictive pressure) approach of the previous work,[2] hereafter referred to as I. The dependence of the average viscosity on the lubricant's physical properties and the operating conditions are also examined in order to maximize the traction coefficient.

THE MODEL

The details of the model have been explained elsewhere(2) and will only briefly be outlined here.

The tractive force depends on the average state of stress of the lubricant in the contact region and it is convenient to calculate this by following the progress of a thin slice of the lubricant as it passes between the driving and driven rollers, whose circumferential velocities are U_1 and U_2, respectively. This is illustrated in Figure 1.

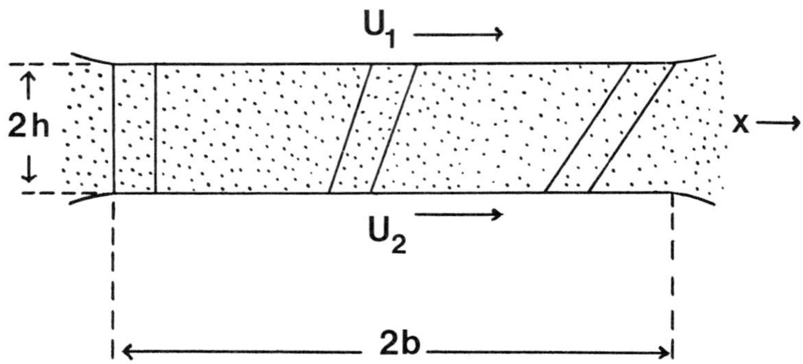

Fig. 1.—The deformation of a fluid element as it passes through the EHD contact zone between two rollers with circumferential velocities U_1 and U_2 ($U_1>U_2$).

Part of the response of a lubricant to a pressure increment occurs instaneously while the remainder takes time to manifest itself. These correspond to glass-like behavior, involving changes in the average nearest neighbor distance, and to liquid-like behavior which involves changes in molecular configuration.

We assume that the important behavior of a thin slice of lubricant in the high pressure contact can be characterized in terms of some representative property of the fluid, $\xi(t)$, the time dependence of which is given by

$$\xi(t) = \xi(0) + \int_0^t dt'\ \gamma(t')\ \left(\frac{\partial\xi(t')}{\partial P}\right)\ dP/dt'$$

$$+ \int_0^t dt'(1-\gamma(t'))\ \left(\frac{\partial\xi(t')}{\partial P}\right)\ dP/dt'\ \{1-\phi[\int_{t'}^t\ dt''/\tau(t'')]\} \qquad (1)$$

In this equation $t = 0$ is the instant at which the fluid element enters the contact region; $0 < \gamma\ (t') < 1$. The instantaneous value of $\frac{\partial\xi(t)}{\partial P}$ is taken to be that for an equilibrium system having a value of ξ equal to $\xi(t')$.

The second term on the right of Equation (1) denotes those changes in ξ which occur instantaneously on the application of a pressure change. The third term represents those changes in ξ which take time to be realized in a manner governed by a normalized, i.e., $\phi(0) = 1$, structural relaxation function, ϕ, which is formulated in terms of a structural relaxation time τ

and whose argument $\int_{t'}^{t} dt''/\tau(t'')$ has been generalized (from simply $(t-t')/\tau$ for an equilibrium system). That is, we assume that the non-equilibrium nature of the fluid element as it passes through a series of structural states can be described simply by allowing the characteristic relaxation time to be itself time dependent.

Eq. (1) can be evaluated numerically by assuming the changes in ξ occur in steps of height $\Delta\xi_i$ produced by pressure increments ΔP_i, at the times t_i

$$\xi_k = \xi_0 + \sum_{i=1}^{k} \gamma_i \Delta\xi_i + \sum_{i=1}^{k} (1-\gamma_i)\Delta\xi_i (1-\phi(\sum_{j=i}^{k} \Delta t_j/\tau_j)) \qquad (2)$$

where ξ_k is the value of ξ at t_{k+1}, after k pressure increments; γ_i is the fraction of $\Delta\xi_i$ which takes place instantaneously; $\Delta t_j = t_{j+1} - t_j$; τ_j is the structural relaxation time at t_j.

The structural relaxation function can be written as a linear combination of L exponentials having a spread of relaxation times. Thus,

$$\xi_k = \xi_0 + \sum_{i=1}^{k} \gamma_i \Delta\xi_i + \sum_{i=1}^{k} (1-\gamma_i)\Delta\xi_i (1- \sum_{\ell}^{L} g_\ell \exp(- \sum_{j=i}^{k} \Delta t_j/\tau_j^\ell)) \qquad (3)$$

where g_ℓ is the weight of the ℓth relaxation function

$$\exp(- \sum_{j=i}^{k} \Delta t_j/\tau_j^\ell) \text{ and}$$

$$\sum_{\ell}^{L} g_\ell = 1 \qquad (4)$$

Eq. (3) can be rewritten in the form below, for computational convenience:

$$\xi_k = \sum_{\ell}^{L} g_\ell \; \xi_k^\ell \qquad (5a)$$

where $\xi_{k+1}^\ell = \xi_k^\ell + (\xi_0 + \sum_{i=1}^{k+1} \Delta\xi_i - \xi_k^\ell)(1-\exp(-\Delta t_{k+1}/\tau_{k+1}^\ell))$

$$+ \gamma_{k+1} \Delta\xi_{k+1} \exp(-\Delta t_{k+1}/\tau_{k+1}^\ell) \qquad (5b)$$

In I the fictive pressure, $P_f(t)$, was chosen as the representative property, that is, the structure of the fluid element being that of the lubricant in equilibrium at a pressure $P = P_f(t)$. In this case, Eq. (5b) becomes

$$P_{f,k+1}^\ell = P_{f,k}^\ell + (\sum_{i=1}^{k+1} \Delta P_i - P_{f,k}^\ell)(1 - \exp(-\Delta t_{k+1}/\tau_{k+1}^\ell)) \qquad (6)$$

The fictive pressure is a measure solely of the configurational or time dependent changes of the fluid. Any instantaneous changes are assumed to be obtainable from the applied pressure P(t).

In contrast, by using the free volume of the element $V_f(t)$, as the representative property, both instantaneous and time dependent changes can be explicitly incorporated in the model.[2] In this case, Eq. (5b) becomes

$$V^{\ell}_{f,k+1} = V^{\ell}_{f,k} + (-\sum_{i=1}^{k+1} \kappa^{0}_{i} V_i \Delta P_i - V^{\ell}_{f,k})(1 - exp(-\Delta t_{k+1}/\tau^{\ell}_{k+1}))$$

$$- \kappa^{\infty}_{k+1} V_{k+1} \Delta P_{k+1} exp(-\Delta t_{k+1}/\tau^{\ell}_{k+1}) \quad (7)$$

where κ^0_i and κ^∞_i are the equilibrium and instantaneous compressibilities immediately before the pressure increment ΔP_i. The total volume and free volume at t_i, V_i and V_{fi}, respectively are related as follows

$$V_{fi} = V_i - V_c \quad (8)$$

where V_c is the close-packed volume of the element, which is assumed density independent.

A fundamental problem is: how are the structural and shear relaxation times and hence the form of $\xi(t)$ to be calculated? The structural and shear relaxation times, $\tau(t)$ and $\tau_s(t)$, respectively, depend on the state of the fluid element. The fictive pressure approach only yields information on the relaxational (time dependent) structural changes. Out of necessity, the influence of the instantaneous and time dependent changes of the fluid on the relaxation times were incorporated in the fictive pressure calculations in an empirical way[2]

$$\tau(t)/\tau(0) = \tau_s(t)/\tau_s(0) = exp\ (\alpha\ (x\ P(t) + (1-x)P_f(t))) \quad (9)$$

where $P(t)$ is the time dependent applied pressure and α is the pressure-viscosity coefficient; $0 < x < 1$. When the fluid is subjected to an infinitely slowly applied pressure, then $P_f(t) = P(t)$ and Eq. (9) reduces to an experimentally supported relationship giving the ratio of the shear viscosity at $P(t)$ to that at zero applied pressure.[2] The applied pressure component in Eq. (9) is designed to incorporate instantaneous changes. This form was adopted in the spirit of a similar fictive temperature relationship[3] which has been found empirically to have success in glass annealing studies. The disadvantage of Eq. (9) is that x is experimentally difficult to determine. The relaxation times can be directly related to free volume using another commonly used relationship[4]

$$\tau_s(t)/\tau_s(0) = exp\ (B\ V_c\ (1/V_f(t) - 1/V_{fo})) \quad (10)$$

where $B \approx 1$ and V_{fo} is the free volume of the inlet fluid element. In this case there is less arbitrariness involved in determining $\tau_s(t)$ and $\tau(t)$. The relaxation times are calculated from the free volume after its instantaneous change and assumed constant during the following time step.

In order to compare the fictive pressure and free volume approaches it is reasonable to adjust the parameters so that they give the same $\tau_s(t)$ for an infinitely slowly applied pressure, that is,

$$\tau_s(t)/\tau_s(0) = exp\ (\alpha\ P(t)) = exp\ (BV_c(1/V_f(t) - 1/V_{fo}) \quad (11)$$

which implies that $\kappa_0(V) = \frac{\alpha}{B}\ (1 - \frac{V_c}{V})\ (\frac{V}{V_c} - 1)$ (12)

This is a reasonable form for the equilibrium compressibility, $\kappa_0(V)$, because it decreases as the free volume is expelled on the application of increasing pressure. Also, Eq. (11) can be rearranged to give an expression for the pressure dependence of the density, ρ, which agrees with that found experimentally;[5] that is,

$$\frac{\rho}{\rho_0} = 1 + \frac{AP}{1+CP} \tag{13}$$

where

$$A = (1-\rho_0 V_c + (V_c^2 - V_c V_0)\rho_0^2)\ \alpha / V_c \rho_0 \tag{14a}$$

and

$$C = (V_c V_0 - V_c^2)\ \rho_0\ \alpha / V_c \tag{14b}$$

$(B=1.0)$

where V_0 and ρ_0 are the total volume and density of the fluid element at atmospheric pressure.

An increase in V_{f0} increases κ_0, A and C; that is, the fluid is more compressible when it contains a larger atmospheric pressure free volume. This is illustrated in Figure 2, which shows the volume dependence of the equilibrium compressibilities for model liquids with V_{f0}/V_0 of 0.15 and 0.3, and whose zero applied pressure values are 0.53 and 2.57 GPa^{-1}, respectively (α = 20 GPa^{-1}). The former is a typical value for many natural and synthetic lubricants; for example 0.33 GPa^{-1} for 5P4E[6] and 0.95 GPa^{-1} for MLO-71-6[7] at atmospheric pressure and 22°C. The A and C values given by Eq. (14) for $V_{f0}/V_0 = 0.15$ are 0.529 and 0.300 GPa^{-1}, respectively while for $V_{f0}/V_0 = 0.3$ they are 2.571 and 0.600 GPa^{-1}. In practice, A and C values of 0.581 and 1.680 GPa^{-1} produce a good fit for many lubricants.[5] It is reassuring that the experimental V_{f0}/V_0 of the lubricants 5P4E (0.16 ± 0.02) and MLO-71-6 (0.10 ± 0.02), obtained from extrapolation of log vs ρ^{-1} curves, are closer to the smaller test value above.

At atmospheric pressure $0.25 < \gamma < 0.75$;[8] for example, the values for 5P4E[6] and for MLO-71-6[7] are 0.75 and 0.73 at atmospheric pressure, respectively. This ratio (κ_∞/κ_0) decreases with increasing pressure, e.g., by 13% at 0.1 GPa and 21% at 0.2GPa for 5P4E; the MLO-71-6 values are 36% and 50% respectively. Nevertheless, γ was assumed to be pressure independent in these calculations to maintain the simplicity of the model.

The effective viscosity, $\mu(t)$, of the fluid element is defined to be the instantaneous shear stress divided by the shearing rate $\dot{\varepsilon}$, and may be calculated by following the shear stress in the fluid element, as follows

$$\mu(t) = G_\infty \int_0^t dt'\ \phi_s \left[\int_{t'}^{t} dt''/\tau_s(t'')\right] \tag{15}$$

where G_∞ is the shear modulus (assumed pressure independent), ϕ_s is the shear stress relaxation function and τ_s is the shear stress relaxation time. The latter can be calculated using Eq. (10).

The average effective viscosity is defined below.

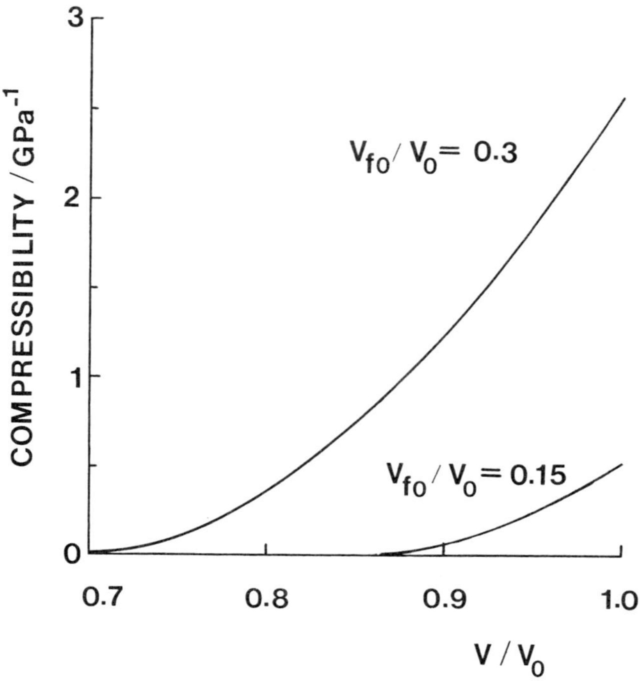

Fig. 2.—The dependence of the equilibrium compressibility on the total volume of the liquid, V, using Eq. (12). The values of the zero applied pressure free volume, V_{f0}, are shown on the figure.

$$\overline{\mu} = \frac{1}{t_0} \int_0^{t_0} dt\ \mu(t) \tag{16}$$

where t_0 is the transit time of the element between the rollers.

The traction coefficient, C_T, can be calculated using the expression:(2)

$$C_T = \frac{8R}{\pi h E} [\overline{\mu}/\frac{1}{2}\ G_\infty\ t_0]\ G_\infty\ \frac{\Delta U}{U} \tag{17}$$

where $R = R_1R_2/(R_1 + R_2)$ is an effective radius for the roller pair of radii R_1 and R_2;

$E = E_0/(1-\nu^2)$ where E_0 is Young's modulus of the rollers and ν is their Poisson's ratio;

ΔU is the difference in circumferential velocity of the rollers (the slip speed) and $U(= (U_1 + U_2)/2)$ is called the rolling speed;

h is half the film thickness; $\dot{\varepsilon} = \Delta U/2h$.

RESULTS AND DISCUSSION

The choice of typical input parameters has been discussed elsewhere(2) and, unless stated, the following values were used in the calculations:

$G_\infty = 2GPa;$

$\eta_0 = 1\ Pa\ s;$

$\tau_s(0) = \eta_0/G_\infty;$

$\alpha = 20\ GPa^{-1};$

$\tau(t)/\tau_s(t) = 3;$

$$\phi(t/\tau) = exp(-t/\tau)^\beta; \qquad (18)$$

$$\phi_s(t/\tau_s) = exp(-t/\tau_s)^\beta \text{ where} \qquad (19)$$

$\beta = 0.5;$

$t_0 = 10^{-4}s;$

The applied pressure profile was of an elliptical Hertzian form,

$$P(x) = P_{HZ}[1-(x/b)^2]^{1/2} \qquad (20)$$

where 2b is the width of the contact zone in the x direction ($0<x<2b$); see Figure 1.

The first objective was to assess the effects of changing γ between 0.25 and 0.75 and V_{f0}/V_0 between 0.15 and 0.3 using a selection of P_{HZ} ($\leq$3GPa). Table 1 gives values of $\bar{\mu}$, the mean free volume, $\bar{V}_f/V_0$, mean density, $\bar{\rho}/\rho_0$, and the minimum free volume V_f^m/V_0 of the fluid element.

The average effective viscosities of the free volume and fictive pressure calculations agree well; for example, the $\gamma = 0.5$ results are within 14% of those of the $x = 0.5$ fictive pressure computations. The agreement between the two approaches is not significantly altered on changing γ and V_{f0}/V_0 (the values of $\bar{\mu}$ always being within 12% of the $\gamma = 0.5$ results). Figures 3 and 4 show the $\mu(t)$ of the $V_{f0}/V_0 = 0.15$ computations with $P_{HZ} = 0.75$ and $1.5\ GPa$. The effective viscosity more rapidly ascends when γ is large because the film can more readily respond to the applied pressure. However, for $t > 90\ \mu s$ the structural (and hence shear) relaxation times decrease more rapidly for the same reason. The result is that the area under each $\mu(t)$ profile (and hence $\bar{\mu}$) is relatively insensitive to γ. The $\bar{\mu}$, V_f^m/V_0 and $\bar{V}_f/V_0$ for $V_{f0}/V_0 = 0.3$ are very similar to those of $V_{f0}/V_0 = 0.15$ using the same P_{HZ}. This arises because of the $\exp(V_c/V_f(t))$ dependence of the structural and shear relaxation times. V_c is ≈ 1 in both cases and consequently the relaxation times become much larger than the transit time for similar free volumes (the easily excludable free volume having been expelled within $\simeq 10\ \mu s$).

Only a few per cent more free volume can be excluded in the $V_{f0}/V_0 = 0.3$ case. The free volume and $\mu(t)$ profiles of Figure 5 further illustrate this phenomenon. The time dependent reduced densities, $\rho(t)/\rho(0)$, of the $V_{f0}/V_0 = 0.15$, $\gamma = 0.5$ calculations are illustrated in Figure 6. The maximum density increase is less than 15% for $P_{HZ} < 3GPa$. The average densities

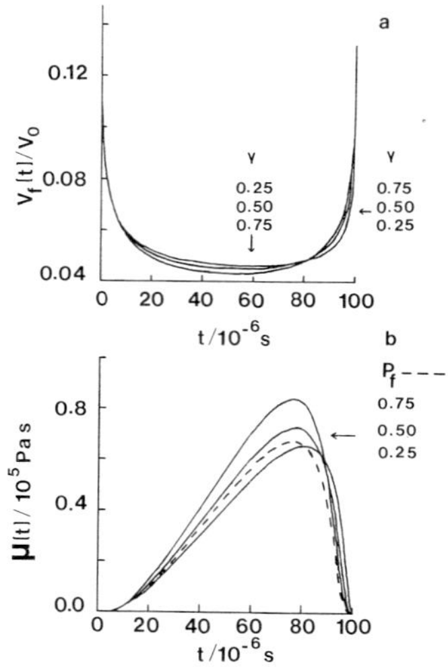

Fig. 3.—The response of the fluid element under the conditions: $P_{HZ} = 0.75$ GPa, $G_\infty = 2$ GPa, $t_0 = 100$ μs, $\eta_0 = 1$ Pa s, $\alpha = 20$ GPa^{-1}, $\tau/\tau_s = 3$ and ϕ and ϕs are given by Eqs. (18) and (19) with $\beta = 0.5$, $V_{f0}/V_0 = 0.15$. (a) Reduced free volume $V_f(t)/V_0$ and (b) effective viscosity, $\mu(t)$. The effects of changing γ are shown. The fictive pressure effective viscosity ($x = 0.5$) is given for comparison (broken curve).

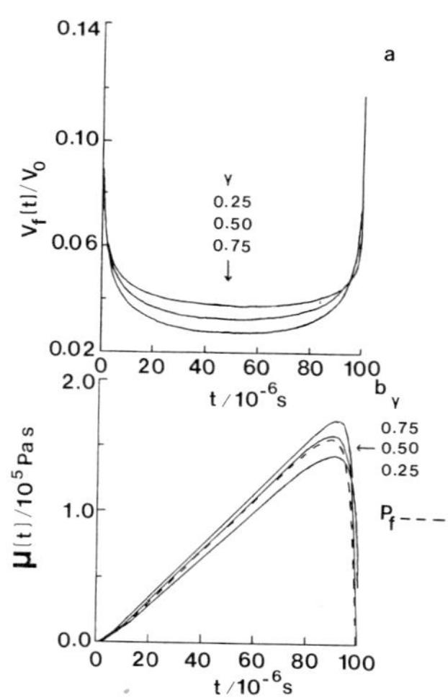

Fig. 4.—The time dependence of the (a) free volume and (b) effective viscosity, of a fluid element in the contact zone for the conditions: $P_{HZ} = 1.5$ GPa, other parameters are as for Figure 3. The dependence of the results on γ are shown. The effective viscosity calculated using the fictive pressure model ($x = 0.5$) is given for comparison (broken curve).

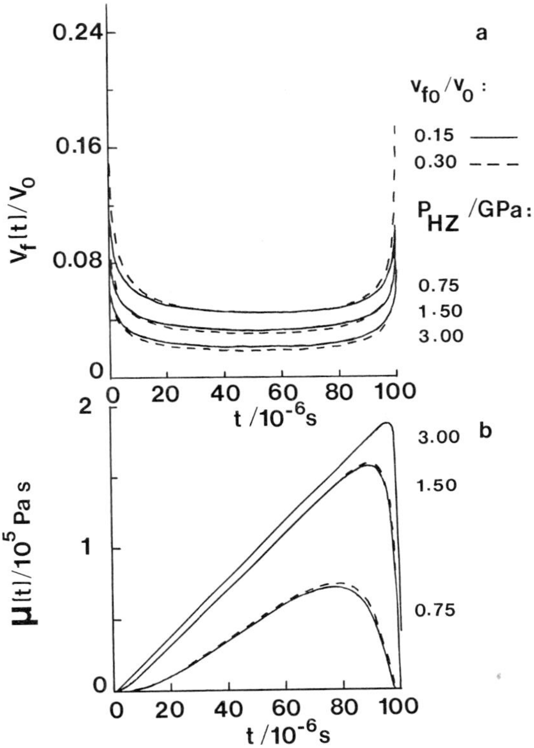

Fig. 5.—The effect on the free volume and shear response of varying the peak Hertzian pressure P_{HZ} (values of which are given on the right of the figures). $V_{f0}/V_0 = 0.15$, ______; $V_{f0}/V_0 = 0.30$, -----. Other parameters are those used for Figure 3.

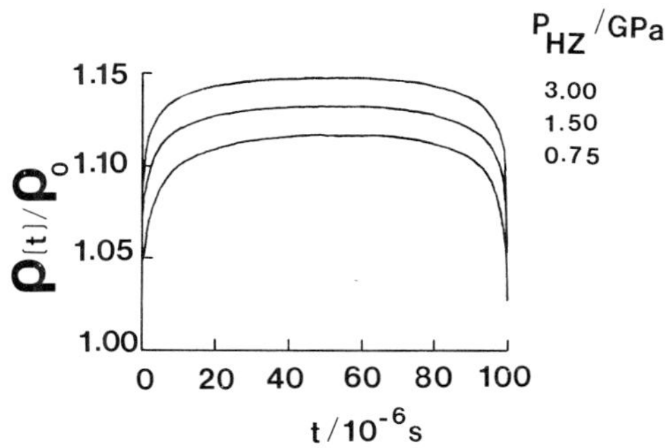

Fig. 6.—The change in density of the fluid element as it passes through the contact region as a function of P_{HZ}, the values of which are shown on the figure; other conditions are those listed for Figure 3.

of the contact zones only deviate significantly from those expected from instantaneous responses to the applied pressure when $P_{HZ} > 0.75$ GPa (see Figure 7).

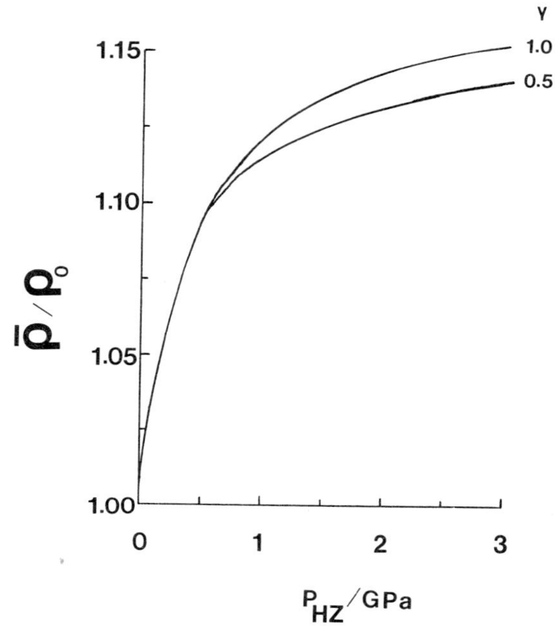

Fig. 7.—The dependence of the average lubricant density in the contact region on the maximum Hertzian pressure. The other parameters are the same as those used to calculate Figure 6.

The combination $\gamma = 0.5$ and $v_{f0}/v_0 = 0.15$ was chosen for succeeding calculations because $v_{f0}/v_0 = 0.15$ gives more experimentally reasonable compressibilities than $v_{f0}/v_0 = 0.3$. The same choice of γ and x yield similar $\bar{\eta}$ (see Table 1). The value $\gamma = 0.5$ was chosen for the succeeding calculations because of the close agreement of their $\mu(t)$ with those of the comparable (x=0.5) fictive pressure computations (see Figures 3-4) and is a reasonable extrapolation of the low pressure values to high pressures.

The traction coefficient C_T, defined by Eq. (17), is proportional to the reduced traction slope, $\bar{\mu}/\frac{1}{2}\, G_\infty\, t_0$, or χ. The latter can be calculated directly from the computations. However, χh^{-1} is perhaps more useful in assessing the true effects of varying the EHD conditions. It has been found empirically[2,9] that,

$$h \propto \alpha^{0.6} \, (\eta_0/t_0)^{0.7} \, P_{HZ}{}^{0.45} \tag{21}$$

where

$$t_0 = 2b/u \tag{22}$$

and

$$b = 4R\, P_{HZ}/E \tag{23}$$

TABLE I.—THE EFFECTS OF γ, x AND P_{HZ} ON THE EHD CALCULATED STATES OF THE FILMS. $G_\infty = 2$ GPa. $\eta_0 = 1$ Pa s, $\tau/\tau_s = 3$, $t_0 = 10^{-4}$s, $\alpha = 20$ GPa^{-1}. RELAXATION FUNCTIONS ARE DEFINED USING EQ.S (18) AND (19) WITH $\beta = 0.5$.

		Free Volume				Fictive Pressure	
P_{HZ}/GPa	γ	V_f^m/V_0	$\overline{V}_f/V_0$	$\bar{\rho}/\rho_0$	$\bar{\eta}/10^5$ Pa s	x	$\bar{\eta}/10^5$ Pa s
$V_{f0}/V_0 = 0.15$							
0.75	0.25	0.0464	0.0531	1.107	0.353	0.25	0.311
0.75	0.50	0.0451	0.0529	1.108	0.383	0.50	0.349
0.75	0.75	0.0432	0.0521	1.109	0.433	0.75	0.406
0.75	1.00	0.0408	0.0507	1.111	0.495	1.0	0.476
0.75	0.50	0.0451	0.0529	1.108	0.383	0.50	0.349
1.50	0.50	0.0329	0.0391	1.125	0.866	0.50	0.845
3.00	0.50	0.0211	0.0266	1.141	0.992	0.50	0.984
$V_{f0}/V_0 = 0.30$							
0.75	0.50	0.0448	0.0554	1.325	0.398	0.50	0.349
1.50	0.50	0.0309	0.0382	1.355	0.876	0.50	0.845
3.00	0.50	0.0188	0.0245	1.381	0.995	0.50	0.984

($E = 0.23 \times 10^3$ GPa for hardened steel[9])

Arbitrarily, a reference film half-thickness, h_0, has been defined using the liquid properties and conditions, $\alpha = 20\ GPa^{-1}$, $\eta_0 = 1\ Pas$, $P_{HZ} = 1.5\ GPa$ and $t_0 = 10^{-4} s$. The quantity, χ', where

$$\chi' = (\frac{h_0}{h})\ \chi \tag{24}$$

is useful in assessing the effects on C_T of changing the input parameters. These investigations are discussed below.

Maximum Hertzian Pressure and Pressure-Viscosity Coefficient

An increase in the maximum Hertzian pressure results in larger average structural and shear relaxation times and the film behaves in a more solid-like manner because the time scale of the structural changes becomes large when compared with the transit time. This is evident from the $\mu(t)$ of Figure 5 which approach the solid limit[2]

$$\mu(t) = G_\infty\ t \tag{25}$$

with increasing P_{HZ}.

A high P_{HZ} decreases h, all other parameters being kept constant. Both χ and h change on increasing P_{HZ} and thus increase the traction coefficient. However, wearing of the contact surfaces becomes a problem as h is decreased. If a pressure independent t_0, is ensured by compensating changes in b with those in U, using Eq. (22), then h *increases* with pressure. Hence, the traction coefficient will maximize at a P_{HZ} which will depend on the bulk properties of the lubricant (see Table 1). The effects on χ of changes in P_{HZ} are also exhibited by altering α in the same proportions. This arises because P(t) and α always appear as products in the formulae which determine the relaxation times in the free volume and fictive pressure calculations. Again, assuming a pressure independent t_0, then χ' also varies with α in a similar way to the same fractional changes in P_{HZ} because of the similar dependence of h on α and P_{HZ} in Eq. (21).

Transit Time

It is possible to increase χ by reducing t_0 (i.e., increasing the rate of revolution of the driving roller). This arises because the pressure increments take place in a shorter time and consequently the film is less able to respond structurally to the applied pressure history. As a result the lubricant acts in a solid-like fashion for a larger fraction of the transit time. However, Table 2 illustrates that, even for an order of magnitude change in t_0 ($P_{HZ} = 1.5$ GPa) χ varies by less than 10%. At these high P_{HZ} where $\chi \approx 1$ the main effect of an increased rolling speed is to increase h and thus reduce C_T.

Inlet Viscosity

The inlet viscosity is proportional to the initial shear and structural relaxation times. Thus the higher the inlet viscosity the less well is the

TABLE II.—THE EFFECTS OF CHANGING THE MAXIMUM HERTZIAN PRESSURE, TRANSIT TIME, PRESSURE VISCOSITY COEFFICIENT AND INLET SHEAR VISCOSITY ON THE REDUCED TRACTION SLOPES. $V_{f0}/V_0 = 0.15$, $\gamma = 0.5$. THE OTHER PARAMETERS ARE AS FOR TABLE I.

			Free Volume				Fictive Pressure	
P_{HZ}/GPa	$t_0/10^{-4}$	α/GPa^{-1}	η_0/Pa s	V_f^m/V_0	χ	χ'	χ (x = 0.5)	χ'
0.75	1.0	20.0	1.0	0.0451	0.383	0.523	0.349	0.477
1.50	1.0	20.0	1.0	0.0329	0.866	0.866	0.845	0.845
3.00	1.0	20.0	1.0	0.0211	0.992	0.726	0.984	0.720
1.50	1.0	10.0	1.0	0.0451	0.383	0.580	0.349	0.529
1.50	1.0	40.0	1.0	0.0211	0.992	0.655	0.984	0.649
1.50	0.4	20.0	1.0	0.0335	0.889	0.468	0.869	0.457
1.50	4.0	20.0	1.0	0.0437	0.828	2.185	0.805	2.124
1.50	1.0	20.0	10^{-2}	0.0300	0.721	18.102	0.694	17.4
1.50	1.0	20.0	10^{2}	0.0364	0.956	0.038	0.942	0.038

lubricant able to follow the applied pressure changes. Consequently, although the minimum free volume occurs for the low η_0, the average structural and shear relaxation times are larger in the high η_0 cases. The lubricant behaves in a glassy fashion for a larger fraction of the transit time. Unfortunately, the film thickness increases with η_0 (see Eq. (21)) and so a high η_0 ($\approx 10^2$ Pa s) is not a means to improve traction (see Table 2) when $\chi \approx 1$ at high P_{HZ} (> 1 GPa).

Relaxation Time Ratio τ/τ_S

The structural development of the lubricant in the contact zone is arrested by increasing τ/τ_S. This is evident from a reduction in the average expelled free volume of the film as τ/τ_S increases. The shear relaxation time is calculated using Eq. (10) and is hence independent of the choice of τ/τ_S. Thus $\bar{\mu}$ decreases with increasing τ/τ_S, although a $1 < \tau/\tau_S < 10$ range only changes $\bar{\mu}$ by a few per cent when $P_{HZ} = 1.5$ GPa.

Effect of Changing the Form of the Relaxation Functions

Each time step the average relaxation times, τ_S and τ, are calculated from the instantaneous state of the element. In this section, three ways in which the structure and stress can relax are considered; that is, three forms of the relaxation functions,

$$\phi^{(1)}(t/\tau) = \exp(-t/\tau) \tag{26}$$

$$\phi^{(2)}(t/\tau) = \exp(-(t/\tau')^{\beta}) \tag{27}$$

$$\phi^{(3)}(t/\tau) = a \exp(-(t/\tau')^{\beta}) + (1-a)\exp(-(t/\tau'')) \tag{28}$$

The τ' and τ'' are defined so that each relaxation function has the same average relaxation time τ, where

$$\tau = \int_0^{\infty} \phi^{(i)}(t/\tau)\,dt \qquad \text{where } i = 1,\ 2 \text{ or } 3$$

$\phi^{(2)}(t/\tau)$ and $\phi^{(3)}(t/\tau)$ may be thought of as a linear combination of exponentials with a spread of relaxation times, τ_i, about τ. By setting $a = 0.8$ and $\tau'' = 200\,\tau'$, $\phi^{(3)}(t/\tau)$ has a long time tail in the (1-a) term which is compensated for by more highly weighted short range components than those of $\phi^{(2)}(t/\tau)$. The combinations

I $\quad \phi(t/\tau) = \phi^{(1)}(t/\tau)$

$\quad\quad \phi_S(t/\tau_S) = \phi^{(1)}(t/\tau_S)$

II $\quad \phi(t/\tau) = \phi^{(2)}(t/\tau)$

$\quad\quad \phi_S(t/\tau_S) = \phi^{(2)}(t/\tau_S)$

III $\quad \phi(t/\tau) = \phi^{(2)}(t/\tau)$

$\quad\quad \phi_S(t/\tau_S) = \phi^{(3)}(t/\tau_S)$

IV $\phi(t/\tau) = \phi^{(3)}(t/\tau)$

$\phi_S(t/\tau_S) = \phi^{(3)}(t/\tau_S)$

were adopted and their $V_f(t)$ and $\mu(t)$ are illustrated in Figure 8. Reduced

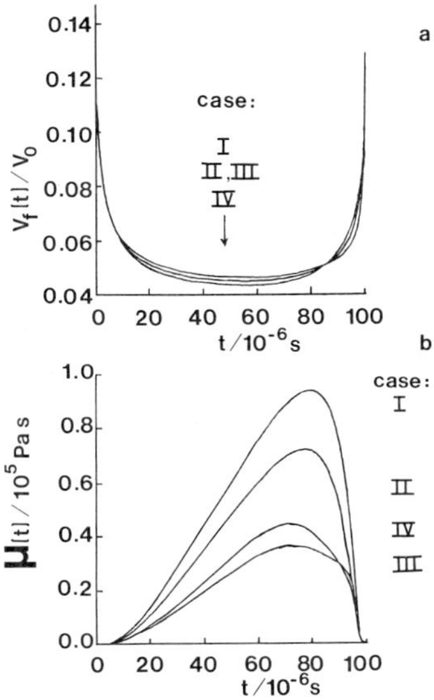

Fig. 8.—The dependence of the response of the system on different structural and shear relaxation functions (the four cases shown are described at the end of Section III), (a) free volume versus time; (b) effective viscosity versus time. The other conditions are as for Figure 3.

traction coefficients are given in Table 3.

It is evident from Figure 8 and Table 3 that those structural and shear relaxation functions with rapidly relaxing components ($\tau_i < \tau$) are more capable of responding to the applied pressure profile and consequently have a smaller V_f^m. However, their χ's are the smallest because of more rapid shear stress relaxation in these cases. A comparison between the $\mu(t)$ of II and III further illustrates this point. The only difference between these two calculations is that the shear relaxation function components of III are more heavily biased towards the low τ_i and large τ_i extremes. The result of this is that $\mu(t)$ of III rises less rapidly than that of II, on the application of pressure, but it also decays less rapidly for $t > 90\mu s$ because of the long time tail in $\phi_S(t/\tau)$. The χ of II is greater than that of III despite this latter effect.

To summarize, the traction properties of the lubricant can be enhanced by having a structural relaxation function with short time response components combined with a shear relaxation function without this bias.

TABLE III.—THE EFFECTS OF THE FORM OF THE RELAXATION FUNCTIONS ON THE REDUCED TRACTION SLOPES. $P_{HZ} = 0.75$ *GPa,* $t_0 = 10^{-4}$, $\eta_0 = 1.0$ *Pa s,* $G_\infty = 2$ *GPa,* $\alpha = 20$ GPa^{-1}, $\tau/\tau_s = 3$. $\phi^{(1)}$, $\phi^{(2)}$ *AND* $\phi^{(3)}$ *ARE DEFINED BY EQS. (26)-(28), RESPECTIVELY, TAKING* $\beta = 0.5$, $a = 0.8$ *AND* $\tau'' = 200\ \tau'$, $V_{f0}/V_0 = 0.15$.

Case	Free Volume			Fictive Pressure	
	V_f^m/V_0 ($\gamma = 0.5$)	χ	χ'	χ ($x = 0.5$)	χ'
I $\phi = \phi^{(1)}(t/\tau)$					
$\phi_s = \phi^{(1)}(t/\tau_s)$	0.0464	0.495	0.677	0.444	0.607
II $\phi = \phi^{(2)}(t/\tau)$					
$\phi_s = \phi^{(2)}(t/\tau_s)$	0.0451	0.383	0.523	0.349	0.477
III $\phi = \phi^{(2)}(t/\tau)$					
$\phi_s = \phi^{(3)}(t/\tau_s)$	0.0451	0.212	0.289	0.194	0.265
IV $\phi = \phi^{(3)}(t/\tau)$					
$\phi_s = \phi^{(3)}(t/\tau_s)$	0.0435	0.244	0.333	0.225	0.307

COMPARISON BETWEEN EXPERIMENTAL AND THEORETICAL TRACTION COEFFICIENTS

A number of reduced traction coefficients, $C_TU/\Delta U$, of lubricants have been computed using the free volume EHD model and are compared with the experimental values in Table 4. Most of the calculated $C_TU/\Delta U$ are at least

TABLE IV.—COMPARISON BETWEEN EXPERIMENTAL AND THEORETICAL TRACTION COEFFICIENTS

Lubricant	T /°C	p_{HZ} /GPa	U /ms^{-1}	t_0 /10^{-4}s	h /μm	b /μm (Eq. (23))	$C_TU/\Delta U$	
							expt.	theory
5P4E[a]	27	0.816	0.24	24.0	1.1[h]	273.6	47.1	82.2
(a poly-	27	0.385	2.20	1.21	2.4[h]	128.9	6.7	25.1
phenyl	27	0.385	0.24	11.1	1.1[h]	128.9	27.0	39.2
ether)	27	0.816	2.13	2.73	2.4[h]	273.6	27.0	39.6
5P4E[b]	126.7	0.827	22.9	0.24	0.64[h]	277.2	13.53	0.8
MIL-L-7808[c]	48.9	0.827	22.9	0.24	0.39[g]	277.2	0.31	1.41
L63/1266[d] (paraffinic)	20	0.494	0.89	3.72	0.8[h]	165.6	2.20	14.02
L63/1271[e] (naphthenic)	20	0.494	0.65	5.12	0.9[h]	165.6	17.2	28.91
L63/1271[f]	27	0.56	0.60	6.26	0.6[h]	187.7	24.8	43.6

a. α = 45 GPa^{-1}, η_0 = 1.9 Pa s (Ref. 5), G_∞ = 0.47 GPa, β = 0.65

G_∞ / GPa = 0.926 - 0.0168 T/°C (Ref. 6), τ/τ_s = 17.5 (Ref. 12)

b. α = 10.9 GPa^{-1}, η_0 = 9 x 10^{-3} Pa s (Ref. 11), G_∞ = 0.24 GPa from G_∞^{-1} = 1.1438 + 0.023523 T/°C (Ref. 6)

c. α = 10.9 GPa^{-1}, η_0 = 9 x 10^{-3} Pa s (Ref. 11), G_∞ = 0.3 GPa (Ref. 9)

d. α = 26.3 GPa^{-1}, η_0 = 0.22 Pa s (Ref. 13), G_∞ = 0.3 GPa (Ref. 9)

e. α = 33.1 GPa^{-1}, η_0 = 0.31 Pa s (Ref. 14), G_∞ = 0.3 GPa (Ref. 9)

f. α = 3.13 GPa^{-1}, η_0 = 0.18 Pa s (Ref. 13), G_∞ = 0.3 GPa (Ref. 9)

g. The film thickness was calculated using Eq. (21) (Ref. 9)

h. The film thickness was obtained from the paper.

Other parameters used in the calculations are those given at the beginning of Section III.

two times larger than experiment. The reason for this is not obvious. These experiments were performed at pressures and rolling speeds where the effective viscosity should change little as γ approaches 1 (see Table 1). Consequently, it is not realistic to interpret this phenomenon in terms of the lubricant's inability to respond to the applied pressure while it is in the contact region. Also, at these experimental slip speeds ($\simeq 10^{-3}$m s^{-1}) it is not expected[15] that the shearing rate in a large part of the film should be less than its nominal value of $\Delta U/2h$; another possible cause for the low experimental C_T. There is recent evidence[16] that the shearing and compression of the lubricant in the inlet zone can cause a significant increase in its temperature. This would reduce the average effective viscosity of the lubricant and hence could be a cause of the discrepancy between experiment and this isothermal theory.

CONCLUSIONS

Tables I-III show that there is good agreement between the effective viscosities of the free volume and fictive pressure traction computations that incorporate the same proportion of instantaneous and time dependent components in each model's relaxation behavior. Therefore, the conclusions of this work are similar to those reported previously.[2] That is, traction can be enhanced as follows. Assuming a minimum film thickness then C_T can be increased by applying a larger pressure to the rollers and choosing a lubricant with a large pressure-viscosity coefficient. Any increase of the film thickness can be avoided by decreasing U.

The main effect of increasing the inlet viscosity at high P_{HZ} (> 1 GPa) is to increase h and consequently decrease traction. Order of magnitude changes in the ratio of the average structural to shear relaxation times have a small effect on C_T. The traction coefficient is increased for those lubricants that have a structural relaxation function with a short time response bias and a shear relaxation function without such short time behavior.

REFERENCES

1. Johnson, K.L. and Tevaarwerk, J.L., *Royal Society of London. Proceedings*. Series A, Vol. 356, 1977, p. 215.
2. Montrose, C.J., Moynihan, C.T. and Sasabe, H., "Dynamical Shear and Structural Viscoelasticity in Elastohydrodynamic Lubrication," Technical Report No. 6, Naval Research Contract No. N00014-75-C-0856, July 1977, submitted to ONR Arlington, Va. (submitted to *Journal of Lubrication Technology* for publication).
3. Debolt, M.A., Easteal, A.J., Macedo, P.B. and Moynihan, C.T., *American Ceramic Society. Journal*, Vol. 59, 1975, p. 16.
4. Macedo, P.B. and Litovitz, T.A., *Journal of Chemical Physics*, Vol. 42, 1965, p. 245.
5. Hirst, W. and Moore, A.J., *Royal Society of London. Proceedings*. Series A, Vol. 344, 1975, p. 403.
6. Dill, J.F., Drake, P.W. and Litovitz, T.A., "The Applicability of Light Scattering Spectroscopy to the Measurement of the Viscoelastic Parameters of Lubricants," Technical Report No. 2, Naval Research Contract No. N00014-67-A-0377-0018, March 1974, p. 28, submitted to ONR Arlington, Va.
7. Dill, J.F., "Viscoelastic Properties of a Perfluoroalkylpolyether Lubricant (MLO-71-6)," Technical Report No. 3, N00014-67-A-0377-0018, January 1975, submitted to ONR Arlington, Va.

8. Fein, R.S., *Journal of Lubrication Technology,* Vol. 89, 1967, p. 127.
9. Harrison, G. and Trachman, E.G., *Journal of Lubrication Technology,* Vol. 95, 1972, p. 306.
10. Smith, C.O., "The Science of Engineering Materials," Prentice-Hall, New Jersey, 1969, p. 396.
11. Walowit, J.A. and Smith, R.L., *Journal of Lubrication Technclogy,* Vol. 98, 1976, p. 607.
12. Dill, J.F., Drake, P.W. and Litovitz, T.A., *American Society of Lubrication Engineers Transactions,* Vol. 18, 1975, p. 202.
13. Adams, D.R. and Hirst, W., *Royal Society of London. Proceedings.* Series A, Vol. 332, 1973, p. 505.
14. Hirst, W. and Moore, A.J., *Royal Society of London. Proceedings.* Series A, Vol. 337, 1974, p. 101.
15. Trachman, E.G. and Cheng, H.S., "Rheological Effects on Friction in Elastohydrodynamic Lubrication," NASA Contract Rep. CR-2206, Washington, D.C., March 1973, p. 64.
16. Miller, R.S., Private Communication.

DISCUSSION

QUESTIONER: The density response to pressure that you showed was of an instantaneous nature, at least judging by the shape of the curves that are perfectly symmetrical about the center of the contact. I think that there is a certain time dependence in the density response. It is not a fully symmetrical reversible-type response at all. I would expect asymmetrical density curve about the contact.

The other thing that I would like to add is that you calculated the slopes from this sort of viscoelastic model and came up with predictions that were twice as high as measured values. This has also been observed by Professor Johnson at Cambridge University several times. In a disc machine under these high pressures the shear stress-strain rate behavior cannot be limited to the film only. The discs themselves respond significantly and influence the traction slopes. In fact the pressures you quoted show that simple correction factors dominate the initial slope of these traction curves. What you are seeing there, a factor of two, is nothing surprising to those who are familiar with the compliance of the discs.

The last thing I would like to comment on is the following. Johnson and Roberts took the Maxwell model which has one relaxation time. That was done for mathematical simplicity and also because they wanted to look at the first order effects only. Now if you want to include secondary effects I think you should include all of them. Film thickness over the contact cannot be taken as flat. You must also consider the sort of initial inlet slope where the pressure starts to rise, the effect of the pressure in the initial inlet, and the constriction at the back. All these are very important secondary effects.

D. M. HEYES: Well, in response to the first question, the density is not a very good quantity to illustrate the asymmetry of the response. It does show a slight lag in decay in the outlet region; this is more evident in the free volume plots. In fact, if you compare the spontaneous density response to that calculated using a value of γ equal to 0.5 then one can see a significant difference on a graphical plot at pressures above 0.75 GPa. Only then do the time dependent effects become significant.

A. BEERBOWER, UCSD: How did you get the fully compressed volume? Is it obtained from your experiments?

HEYES: We can get the fully compressed volume from the experimental values of the viscosity versus pressure plots. I took the values which agreed with a certain number of good traction fluids.

XV. SYSTEMS APPROACH

FUNDAMENTALS OF THE SYSTEMS APPROACH TO TRIBOLOGY AND TRIBO-TESTING

H. Czichos

ABSTRACT

The introductory review paper by Mr. Douglas Scott, clearly indicates the economic impact and the extremely broad scope of tribology. While the review paper deals with the present state of the art of the various topics of the interdisciplinary field of tribology, ranging from surface studies and the mechanisms of lubrication and wear to computer aided design and machinery condition monitoring, questions arise as to how to link the various sub-topics in order to achieve unified fundamentals of the entire field of tribology.

This contribution attempts to answer these questions by applying the system concept to tribology. It is shown that the different levels of complexity of tribological problems can be identified by proper choice of a "systems envelope." The system concept is then used to study the influence of friction and wear processes on the "structure" and "function" of mechanical systems in connection with tribo-induced changes of the properties of mechanical components and questions of mechanical equipment reliability. Finally, a methodology is outlined which may be used for tribo-testing and for treating practical friction and wear problems systematically.

INTRODUCTION

In the previous paper, Mr. Scott has dealt with various important topics of tribology including surface studies, mechanisms of wear, lubrication, tribo-engineering materials, computer aided design, machinery condition monitoring, etc. His review once again shows the economic impact and the interdisciplinary nature as well as the extremely broad scope of this subject matter.

While the various sub-topics of tribology will be discussed separately in detail in the following papers, some questions may arise:

1. Is it possible to combine the different aspects of tribology to develop a unified picture?

2. Is it possible to develop a convenient methodology to deal systematically

with practical friction and wear problems?

The following attempts to answer these questions by explaining briefly the system concept and its application to tribology and tribo-testing.

The System Concept and Its Application to Tribology

The idea of systems approach to tribology is not new. Experienced research investigators, system designers and analysts, and test and service engineers are generally aware that a mechanical system composed of many relatively moving tribological components is exceedingly complex and a complete understanding of its behaviour requires a multidisciplinary or systems effort.[1] However, without the guidance of a formalized concept and technique, the so-called systems approach in the hands of the inexperienced may frequently prove to be incomplete, and the conclusions drawn may thus be misleading and risky. On the other hand, for the experienced practitioner, the systems approach appears to be nothing else but "organized technological common sense."

The first step in the application of the system concept is to identify the subject under discussion with the proper location of a so-called "systems envelope" or "control surface." By proper choice of the systems envelope it is possible to identify and to resolve the different levels of complexity of a given problem.[2] This may be explained by an example.

In Figure 1, various levels of complexity of tribological subjects and of corresponding tribo-testing approaches are shown schematically. In the

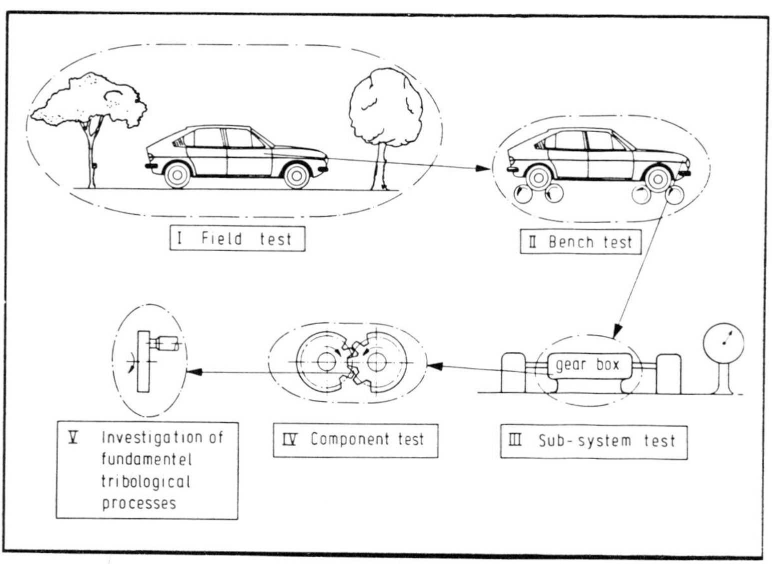

Fig. 1.—Identification of tribological subjects by choice of a systems envelope.

first case, the systems envelope is located very widely around the subject under discussion, namely driving a car. It is obvious that in this case the behaviour of the system results through dynamic interactions of the vehicle, the driver, the road, and the atmospheric conditions. If the systems envelope is located narrower around the vehicle, the behaviour of the system may be tested in a bench test. In locating the systems envelope still narrower, a sub-system may be identified, as for example a gear box. Next, the behaviour of the basic components of the gear box, namely the two gears, may be studied. Finally, the elementary contact, friction, lubrication, and wear processes between two relatively moving surfaces may be investigated by a pin-on-disc system in a laboratory test.

This simple example clearly shows how the different levels of complexity can be identified and resolved by proper choice of a systems envelope. Clearly, at the different levels of complexity (or hierarchy) different operating variables and parameters must be taken into consideration. Further, from Figure 1 it is obvious that different types of testing of tribological subjects may be distinguished, ranging from field tests to fundamental laboratory friction and wear test and surface investigations.

The Description of a Tribo-mechanical System

Having identified the system under discussion by proper choice of a systems envelope, it is then possible to describe and to compile its various aspects in a systematic manner.[3] To illustrate this, consider a typical tribo-technical system, namely a gear box, as shown in Figure 2.

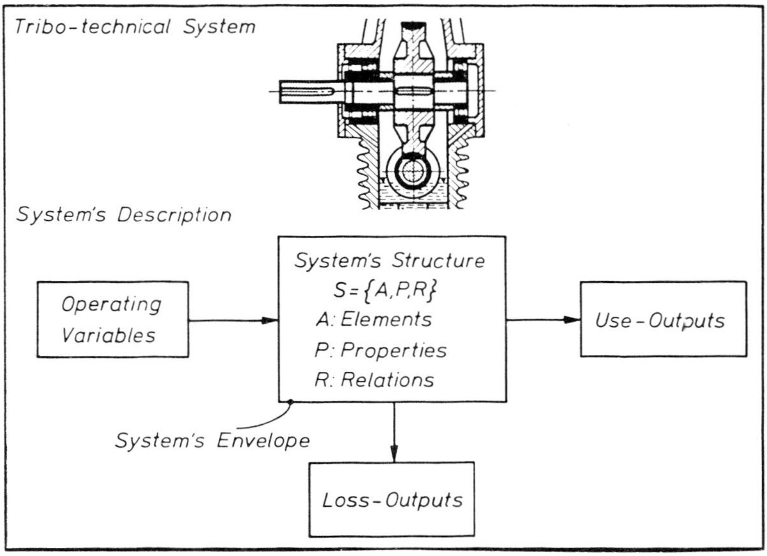

Fig. 2.—Analysis and description of a tribo-technical system.

As described above, the two partners which form the "tribologically interacting surfaces," i.e., gear 1 and gear 2, can be hypothetically separated from their environment by the proper choice of a systems envelope. All material components of the system are within this envelope and are part of the so-called "structure" of the system. The system structure consists of the elements of the system (A), their relevant properties (P) and their interrelations (R) described formally by the set

$$S = \{A, P, R\}.$$

The inputs of the operating variables are transformed through the structure of the system into outputs which are used technically: the use-outputs. Simultaneously, as a consequence of the tribological interactions between the elements of the system, loss-outputs occur, denoted in summary by the terms friction loss and wear loss. The way in which the inputs are transformed into outputs determines the technical function of the system.

This example shows that a detailed description of a tribo-mechanical system can be achieved with the following steps:

(1) Systems function
- a) separate the system from its environment by the proper choice of a systems envelope,
- b) compile all inputs and outputs,
- c) describe the functional input-output relations.

(2) Systems structure
- a) identify the "elements" (or material components) of the system,
- b) characterize the interrelations and interactions between the elements (i.e., the contact, friction, and wear processes)
- c) specify the relevant properties of the elements.

Based on this systems procedure it is possible to study and to compile the basic influencing factors and mechanisms relevant to the function and the structure of mechanical systems in which friction and wear processes occur[3].

Function of Tribo-mechanical Systems

The various engineering systems in which friction and wear processes occur can be easily classified according to their function in considering the pertinent inputs and outputs. A broad classification is given in Table 1. Invariably, motion is a characteristic of any tribo-mechanical system. This motion may constitute a transfer of work, materials or information. In some instances the purpose of a system may be to change a rate of motion or to eliminate it altogether. It is also often desired to restrict motion, i.e., to reduce the number of degrees of freedom a machine element may possess. In other instances materials are not merely moved but also changed in state or form. Mechanical devices which produce or transfer information are still common, but are being steadily replaced by devices in which there is little or no mechanical motion, for example the replacement of mechanical clocks by digital electronic clocks.

Having classified in Table 1 the various tribo-mechanical systems according to their external function, the question of the internal "structure" of these systems is to be discussed.

TABLE I.—CLASSIFICATION OF TECHNICAL FUNCTIONS OF TRIBO-MECHANICAL SYSTEMS

Inputs and outputs needed for technical function		Primary technical function of the system	Examples
Main inputs {X}	Main outputs {Y}		
Motion + Work	Motion	Guidance of motion Coupling of motion Annihilation of motion	Bearings Clutches Brakes
	Work	Power transmission (mech., hydr., pneum.)	Gears
	Information	Generation of information	Clocks; Cams and followers
		Reproduction of information	Data transducer (audio, video; tape or record)
Motion + Materials	Materials	Transportation	Wheel/rail Pipeline
		Forming of materials	Wiredrawing

Structure of Tribo-mechanical Systems

As described above, the structure of a tribo-mechanical system is given by the systems elements (i.e., the material components of the system), their relevant properties and their interrelations, described formally by the set S = {A, P, R}. If the systems envelope is drawn as closely as possible around the origin of a friction and wear process, it appears that in most tribological systems four different basic elements are involved in the friction and wear processes, as shown schematically in Figure 3. For a simple sliding system, the components that form the pair of the "interacting surfaces in relative motion," are the moving element (1) and the stationary element (2). The other two basic elements are the lubricant (3) (if any) and the environment (4). These main elements are linked to others or may be composed of sub-constituents. For example, the element (3), the lubricant, may consist of a base oil and additives.

The tribological interactions between the elements of a mechanical system, i.e., the contact, friction, lubrication, and wear processes are of paramount interest in the description of any tribo-mechanical system. In Figure 4, the basic tribological processes which are known today are expressed in the form of schematic diagrams for systems of increasing complexity, i.e., increasing number of interacting elements.

In an ultrahigh vacuum, the simplest tribological system consists only of the two interacting partners (1) and (2). The main possible tribological relations between moving and stationary elements are given by contact deformation, surface fatigue, abrasion and adhesion. In air, i.e., under dry frictiön condition these mechanisms are supplemented by interactions between the moving and the stationary partners (1) and (2) and the atmosphere (4). Through these interactions tribo-chemical reactions result. Finally,

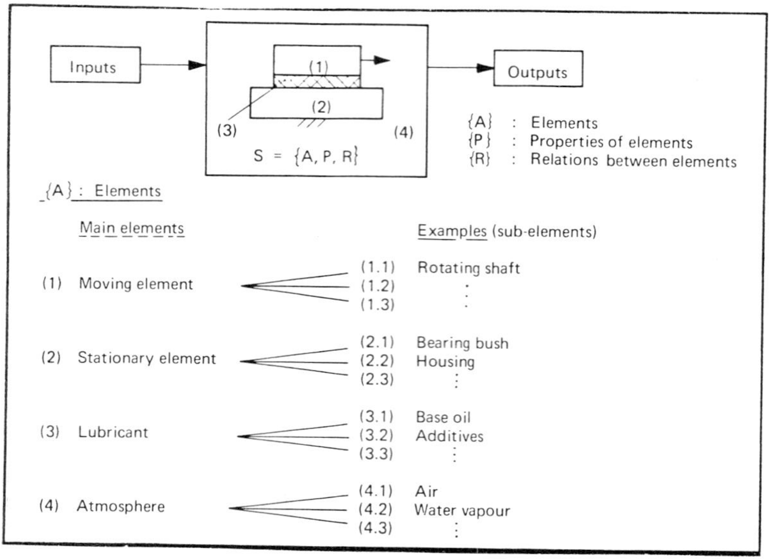

Fig. 3.—Schematic representation of basic elements of tribo-technical systems.

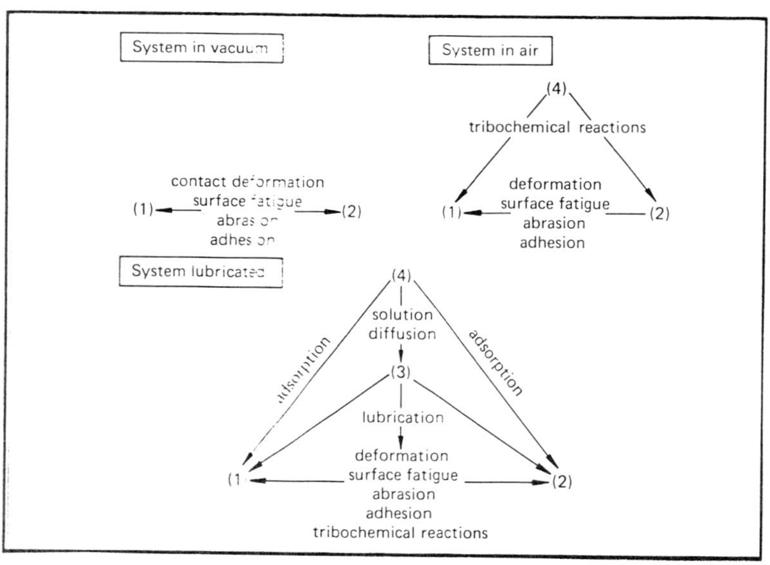

Fig. 4.—Schematic representation of tribological interactions between system components.

in a lubricated system tribological processes between the elements are given by interactions between all four basic elements. In this case, the direct (contact) interactions between moving and stationary elements are prevented through the different mechanisms of lubrication. Depending on the thickness of the lubricating film, different lubrication regimes result.

Also here, interactions between (4) and (3) with (1) and (2) should be taken into account. For instance the diffusion of atmospheric oxygen into the lubricant (4)→(3), followed by oxidation processes between the lubricant and the moving and stationary partners (3)→(1), (2), can distinctly influence the mechanisms of mixed and boundary lubrication.

Influence of Tribological Processes on the Structure and the Function of Mechanical Systems

Having compiled the various aspects of tribological systems in the terms of their internal structure and external function, the question arises, how the tribological processes, i.e., the friction and wear processes, influence and disturb the behaviour of a given mechanical system. This is illustrated schematically in Figure 5.

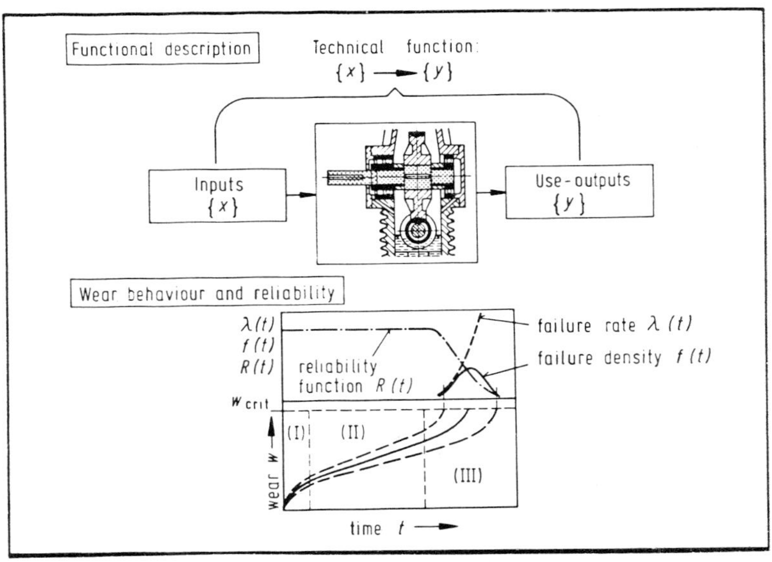

Fig. 5.—Influence of wear-induced changes of systems structure on function and reliability.

In the upper part of Figure 5, a typical mechanical system, namely a gear box, is shown schematically. The technical function of the system is to transform certain inputs, namely angular velocity and torque, into outputs which are used technically. The technical function can then be described formally as a transformation of the inputs into the outputs via a

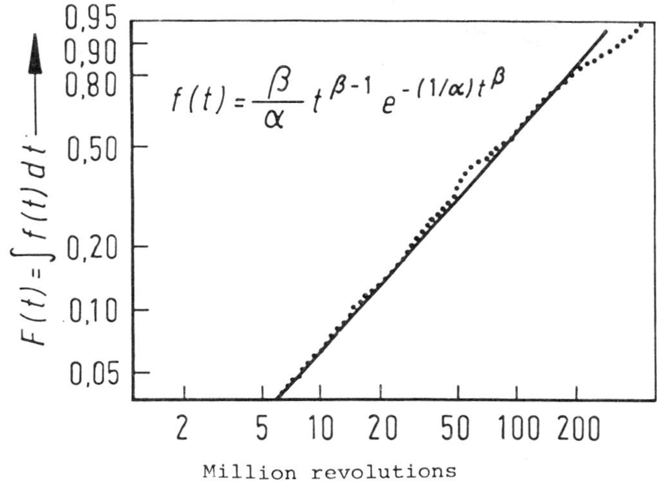

Fig. 6.—Failure probability of ball bearings; Weibull distribution (after (5)).

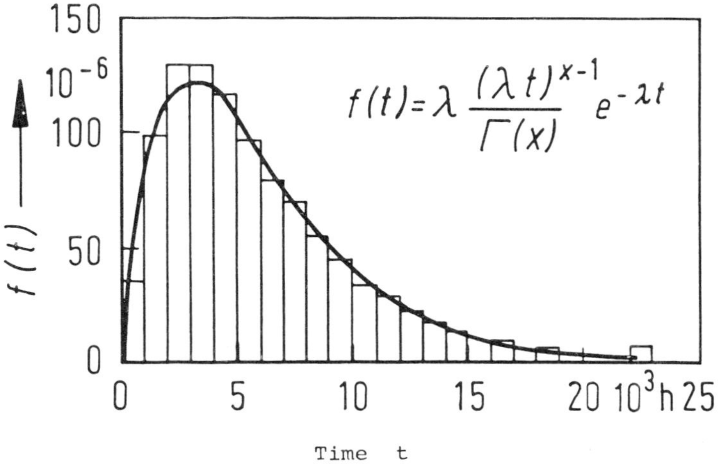

Fig. 7.—Density function of the failure of Diesel engine pistons; Gamma distribution (after (6)).

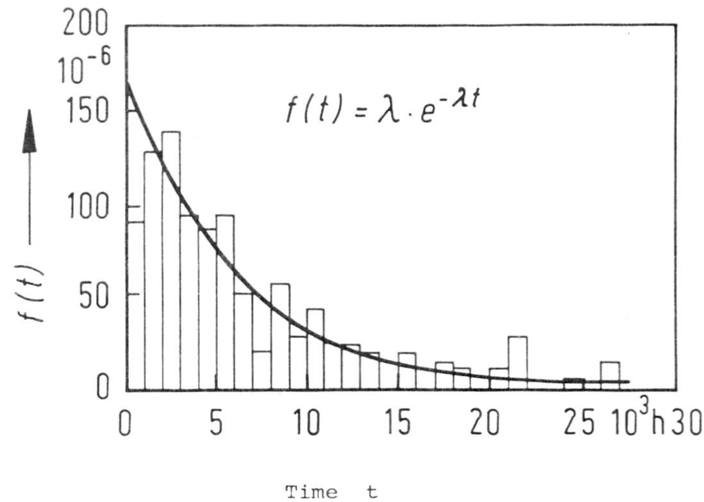

Fig. 8.—Density function of the failure of Diesel engine control units; Exponential distribution (after [6]).

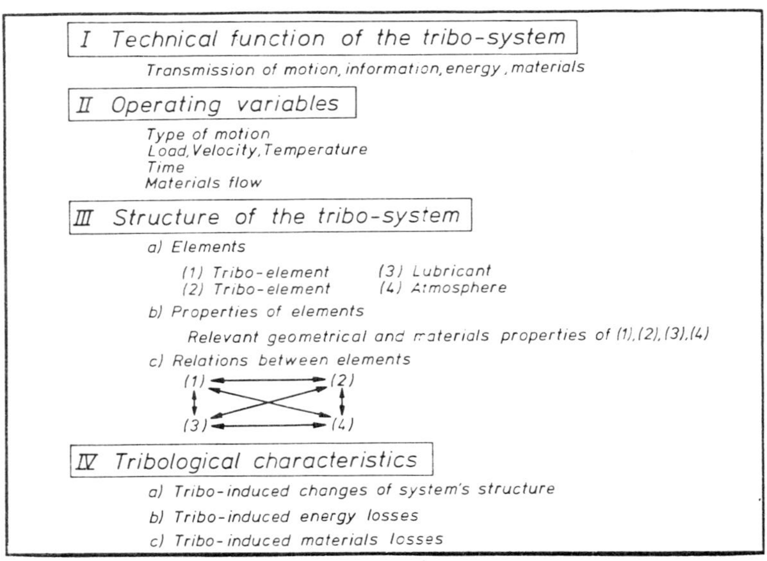

Fig. 9.—Main groups of system parameters.

In addition, the system concept has been used in developing a methodology for tackling systematically a given practical friction and wear problem. A couple of case studies of tribological problems which have been solved successfully with the help of the systems methodology were published recently.[9] The case studies range from material and lubricant selection procedures to wear-reduction attempts and failure analyses. It follows, that for a unified description of tribological subjects as well as in attempts to solve practical friction and wear problems, both the functional and the structural aspects must be considered systematically.

REFERENCES

1. Ku, P.M., editor, "Interdisciplinary Approach to Friction and Wear," NASA SP-181, 1968, Preface.
2. Salomon, G., *American Society of Lubrication Engineers Transactions*, Vol. 17, 1974, p. 295.
3. Czichos, H. and Mølgaard, J., *Wear*, Vol. 44, 1977, p. 247.
4. Czichos, H., *Materialprüfung*, Vol. 20, 1978, p. 33.
5. Bergling, G., *Kugellager-Zeitschrift*, Vol. 51, 1976, p. 1.
6. Fleischer, G., *Wissenschaftliche Zeitschrift TH Magdeburg*, Vol. 16, 1972, p. 289.
7. Czichos, H., *Wear*, Vol. 41, 1977, p. 45.
8. de Gee, A.W.J., *Wear*, Vol. 36, 1976, p. 33.
9. Czichos, H., "Tribology - A Systems Approach to the Science and Technology of Friction, Lubrication and Wear," Elsevier, Amsterdam-New York, 1978, p. 300 ff.

DISCUSSION

R. DASKIVICH, G. M. Research Laboratories: In the figure giving examples of Weibull and gamma distribution, you mentioned that the figure applies for a diesel engine piston. Could you explain what would be included in the system envelope that you would term diesel engine piston?

H. CZICHOS: These examples are taken from the literature. They have not been worked out in our laboratory. I would just like to show that there are correlations between the observed failure distributions and the friction and wear processes. For instance, if you find the Weibull distribution, in most cases you can relate that to a fatigue mechanism. Instantaneous failure, which doesn't depend on the history, is related to an exponential distribution. The examples are chosen to illustrate these relationships.

DASKIVICH: I understand the reason behind it. I want to underline the fact that the piston is a complex mechanism which may wear by all the known mechanisms.

CZICHOS: Right, I agree. In most cases it is not possible to correlate it with a single process. It may be an overlap of different wear processes.

DASKIVICH: All right.

N. P. SUH, MIT: I agree with the comment that several mechanisms operate simultaneously; however, I disagree with the view that all the mechanisms are equally important. If we look at the data, there is always a rate-controlling process. This rate-controlling process often causes the failure. I think in any engineering system it is very important that we recognize what the rate-controlling process is. To say that all processes are important is often not correct.

H. CZICHOS: This was intended to be a general survey. Of course, under special situations the specific mechanism is important.

R. B. WATERHOUSE, University of Nottingham: I felt like I was looking through a microscope where Prof. Czichos increasingly changed the power and focused on those things that are important and I think that is part of the strategy we have to use. This kind of approach would be beneficial to bridge that canyon between the researcher and the designer.

AN INTRODUCTION TO TRIBO-ENGINEERING

N. Ohmae

ABSTRACT

The concept of tribo-engineering system is introduced and the dynamics of the system is studied by an autoregression model. The results of the analysis and simulation suggest the possibility of tribological process control.

INTRODUCTION

Systems analysis is increasingly used in many fields of science and technology. In the field of tribology, a multidisciplinary approach through systems analysis has received considerable attention recently. The research at IRG-OECD[1-6] emphasized the need for systems analysis in tribology. Salomon[1] described the tribological system from the level of complexity, while Czichos applied a black-box model to a tribological system.[2,7] Although these systems analyses provide a new outlook on tribology, their predictive power to practical problems is presently limited.

The term "tribo-engineering" is introduced in this paper. It is defined as the engineering-based discipline for practicing tribology. Systems analysis, in the strict sense, is an applied science based primarily on scientific and mathematical disciplines rather than on empirical or trial and error methods.[8-10] However, at the present stage of progress, tribology is still an empirical science. Therefore, the problem of primary importance is to make effective use of resources of the knowledge that the tribologists have. The object of tribo-engineering should be to treat tribological problems as a whole by systems analysis rather than in parts.

In this paper, a comparison between practical tribological systems and test tribological systems is made in order to discuss tribo-engineering. Then, an autogregression (AR) model is applied to a dynamic tribological system. The model simulation, and the stability of the AR model are discussed. The development of the tribological process is examined by removing compulsory inputs from the simulator. Tribo-control is also discussed.

TRIBOSYSTEM

Figure 1 shows a schematic of cost-reliability relationship when a manufacturer produces machinery parts. Manufacturing cost increases with increasing reliability, while replacement cost due to tribological problems decreases with reliability. The total cost becomes minimum at a certain level of the reliability. Recent advances in theoretical studies of the reliability and its applications[11,12] should be extended to the tribological system in the near future.

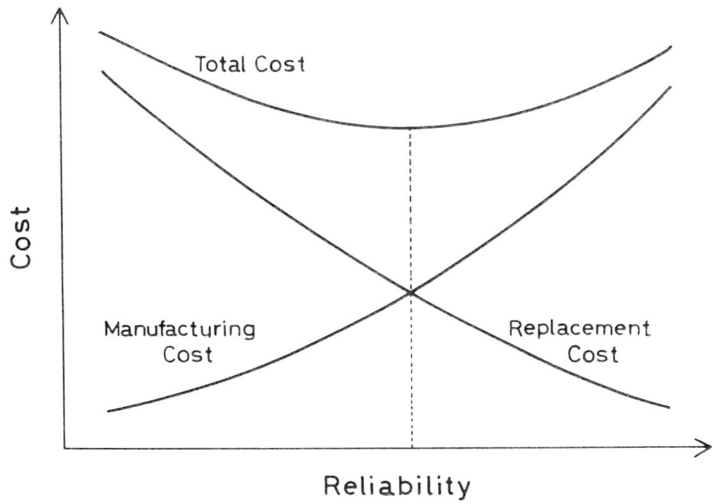

Fig. 1.—Cost-reliability relationship when developing a new machinery component.

User-decision also is an important factor. When the reliability of the system is high, the user will rapidly adopt the system. In contrast, the utility will decrease with reliability. A typical example is the commercial airplane. When the airplane is highly reliable, those who want to use this airplane will increase. However, the high reliability of the airplane may result in a reduction of capacity such as seats, etc. Therefore, the maximum profit is expected at an intermediate level of reliability. (Figure 2)

The tribological system is in fact a multi-object system.[13] This concept is shown in Figure 3. The value of the system can be evaluated in accordance with the levels of performance, cost, reliability, time and flexibility. Technology assessment can only be done from the intangible viewpoint, since some of these factors cannot clearly be expressed by figures. The technical value of the system, in general, may simply be a function of performance and cost. Cost of the system rapidly increases with increasing complexity. Performance, however, saturates in the range of high complexity (see Figure 4). In such a case, the system value is determined by the ratio

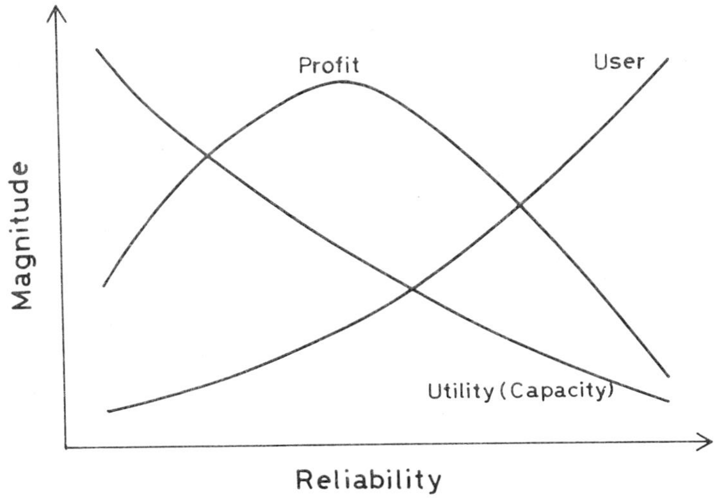

Fig. 2.—Profit of manufacturer under the influence of user-decision.

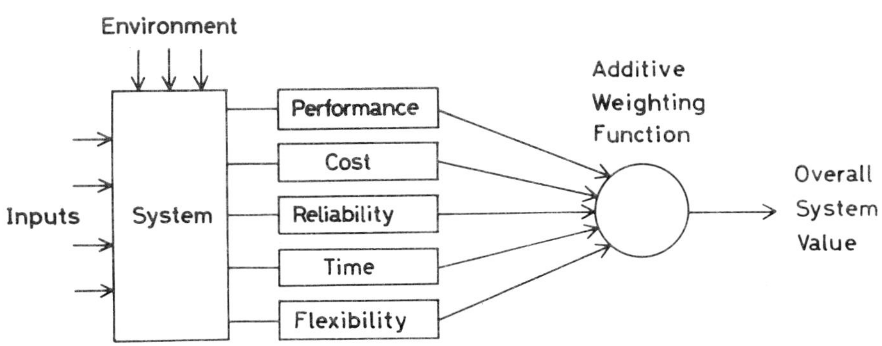

Fig. 3.—Structure of multi-object system and its overall system value (ref. 13).

of performance to cost.

The solution of system value can also be discussed by the relationship between the new and old, as shown in Figure 5.[14] If the cost/time to build

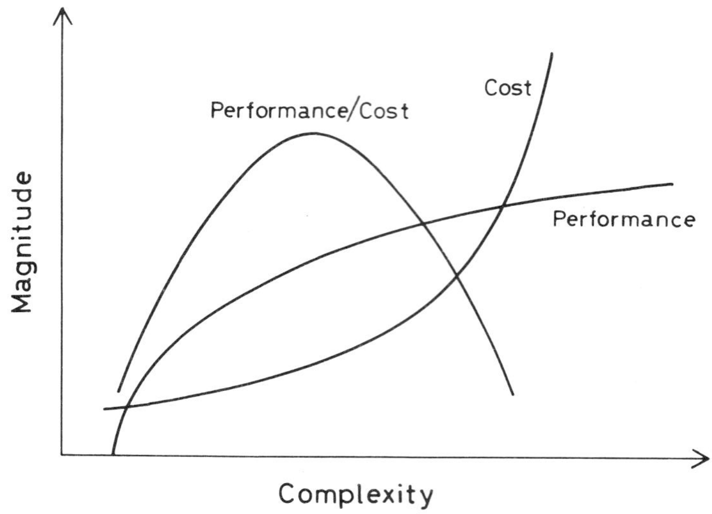

Fig. 4.—Performance/cost as a function of complexity (ref. 13).

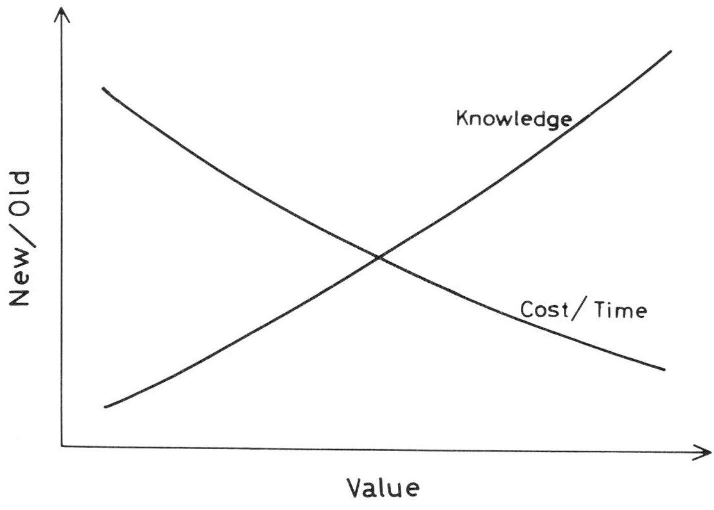

Fig. 5.—Value of the system as determined by knowledge and cost/time (ref. 14).

a new system is high, then the value may be low. On the other hand, if the cost/time to use the new system is low, or if the new knowledge is highly significant compared to the old knowledge, then the value of the system is high.

PRACTICAL AND TEST SYSTEMS

Two different tribological systems can be defined; one is a practical system (PS), and the other a test system (TS). The former may be an actual tribological system in industry while the latter is a tribological testing system in the laboratory. Czichos[15] has proposed the use of a checklist to characterize the similarities of the two systems. He also suggested the possibility of reproducing PS by properly selecting the test conditions of TS.

Tribo-engineering is the overall system management which considers three elements: input, output and process (see Figure 6). Since fairly

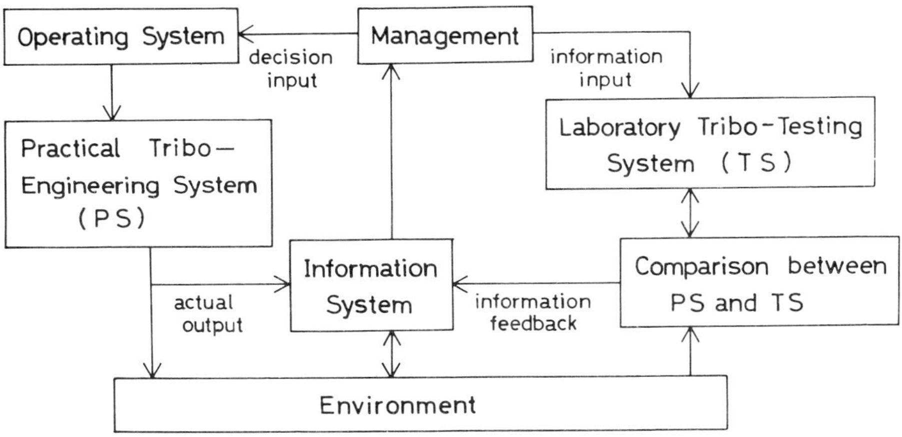

Fig. 6.—Systems management and a concept of tribo-engineering system.

extensive data are available in PS, the first step is to design an information system. The actual outputs from PS are collected in the information system. The information system evaluates this output-performance, and then transmits the resultant information to the systems management. This systems management, of course, means a user-decision. In case the limit of control is exceeded, a correction should be operated through systems management. The second step is an exchange of information between PS and TS. The information from TS is compared with that from PS through the information system. Then on the basis of this result, the user can correct with the use of a control device such as the mini-computer when the system needs correction. This correction means a transmission of what is determined, and becomes an input to PS (decision-making). Furthermore, a feedback of information towards TS is made for modification by iterating this procedure. One of the prototypes will be a tribo-meter in which such a process control system as mini-computer is utilized.

TRIBOLOGICAL CONTROL

Figure 7 shows an interaction, at a boundary, between the system and its environment. In a tribological process, in general, wear and friction are identified as non-controlled outputs (loss-outputs).[2] In a practical

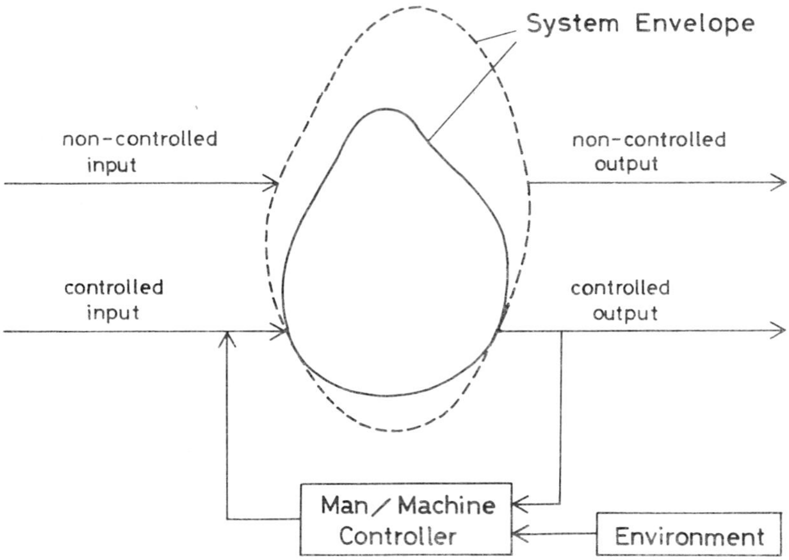

Fig. 7.—System and its environment. Note that the system envelope behaves actively.

sense, by adding a restrictive condition to the system, non-controlled outputs can be converted to controlled outputs. In other words, when a system is inactive, it is possible to prevent tribological damages by pre-determining a suitable protective technique of a surface at man/machine interface. However, when a system behaves actively (usually it does), it is quite difficult to control the loss-outputs. Thus, to control tribological processes, both the analysis of systems inter-structure and the simulation of non-controlled inputs become primary problems to be solved.

AUTOREGRESSION MODEL OF A TRIBO-PROCESS

Autoregression (AR) model* has been employed in a variety of process controls. In this section, we shall attempt to analyze a dynamic tribological process using the AR model. Let us consider an n-dimensional vector $\underline{x}(k)$. Assuming that a stationary time series of $\underline{x}(k)$; $k=1,\ldots,N$ is generated from an Mth order multi-dimensional AR model, we have

$$\underline{x}(k) = \sum_{m=1}^{M} A_M(m)\ \underline{x}(k-m) + \underline{\varepsilon}(k) \qquad (1)$$

where $A_M(m)$ is an (n x n) matrix of unknown parameters, and $\underline{\varepsilon}(k)$ n-dimensional white noise. In an AR model, it is of critical importance to determine M, the order of the model. One of the best methods, at the present progress

* For the details of AR model, refer to appendix.

in systems methodology, is the determination from final prediction error (FPE) proposed by Akaike.[16,17] FPE is of the form

$$FPE(M) = (1 + \frac{Mn + 1}{N})^{n} (1 - \frac{Mn + 1}{N})^{-n} det(D_M) \quad (2)$$

An example of the tribological system discussed here is the sliding between mild steel and fiber reinforced plastics. The static systems analysis of this combination of materials was reported previously.[18] Friction coefficient x_1, temperature x_2 (measured at 1 mm below sliding surface) and specific wear rate x_3 are set as typical state variables. Based on the measured data, the system identification was done using multi-dimensional AR model. The results of computer calculations are shown below:

1 order;

$$\begin{bmatrix} x_1 (k) \\ x_2 (k) \\ x_3 (k) \end{bmatrix} = A_1(0) + A_1(1) \begin{bmatrix} x_1 (k-1) \\ x_2 (k-1) \\ x_3 (k-1) \end{bmatrix} \quad (3)$$

2 order;

$$\begin{bmatrix} x_1 (k) \\ x_2 (k) \\ x_3 (k) \end{bmatrix} = A_2(0) + A_2(1) \begin{bmatrix} x_1 (k-1) \\ x_2 (k-1) \\ x_3 (k-1) \end{bmatrix} + A_2(2) \begin{bmatrix} x_1 (k-2) \\ x_2 (k-2) \\ x_3 (k-2) \end{bmatrix} \quad (4)$$

3 order;

$$\begin{bmatrix} x_1 (k) \\ x_2 (k) \\ x_3 (k) \end{bmatrix} = A_3(0) + A_3(1) \begin{bmatrix} x_1 (k-1) \\ x_2 (k-1) \\ x_3 (k-1) \end{bmatrix} + A_3(2) \begin{bmatrix} x_1 (k-2) \\ x_2 (k-2) \\ x_3 (k-2) \end{bmatrix}$$

$$+ A_3(3) \begin{bmatrix} x_1 (k-3) \\ x_2 (k-3) \\ x_3 (k-3) \end{bmatrix} \quad (5)$$

When unknown parameter $A_M(m)$ is used for predicting the model, the error due to this methodology should be carefully examined. As indicated earlier, FPE was used for this purpose. Figure 8 shows the variation of FPE(M) with respect to M. FPE(M) is minimum at $M=1$. This result indicates that a high predicting accuracy of the system model is expected at $M=1$.

Now that the state variables in the tribological process are able to be reproduced within an AR simulator, the behavior and development of the system can be studied. The turbulence of the system is observed by admitting a compulsory input x_1 (friction coefficient) to the AR simulator. In this case, a white noise of normalized random numbers (mean 0, variance 0.01) is added to the compulsory input. Thick line in Figure 9 shows the compulsory input x_1. Thin lines show the development of the system after the compulsory input is restrained. With respect to x_1, x_2 and x_3 show a stable pur-

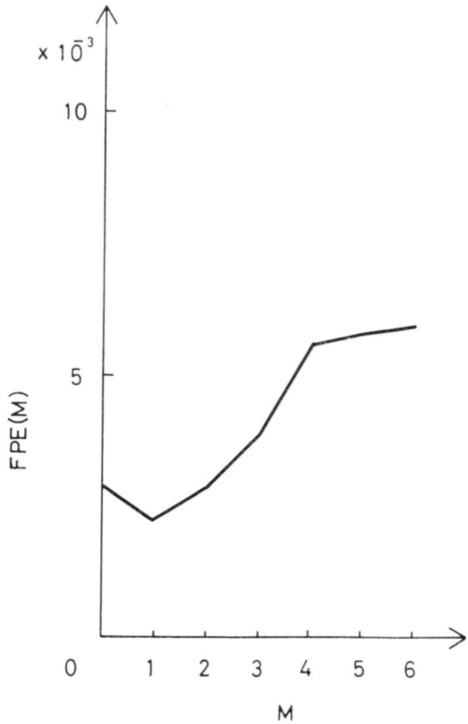

Fig. 8.—Variation of FPE(M) with respect to the order of the AR model, M.

suit at *M=1*, while unstable pursuit at *M=2* and *3*. This result implies that a considerable amount of error is involved in $A_M(m)$ at *M=2* and *3*, and is consistent with the minimal FPE(M) obtained at *M=1* shown in Figure 8. At *M=3* the predicting accuracy of $A_M(m)$ seems rather low, but a tendency of curve fitting is recognized. This may be caused by the fact that the order of the model is high.

Figure 9 also shows the development of the system after a restriction of the compulsory input is removed at locations a, b, c and d. The purpose of this simulation was to examine the effect of abrupt stimuli on the system, since tribological system is an open system which interacts with its environment, as pointed out earlier. It is therefore possible to discuss about the system behavior at "off-control." At *M=1* each variable shows a stable convergence, and a turbulence in the system monotonically decreases with time. In contrast at *M=2*, each variable diverges and the turbulence increases with time. At *M=3* the system develops slowly towards a convergence, which may be similar in manner to a damped oscillation. Thus at *M=1* and possibly at *M=3* the system develops towards a favorable state even if a turbulence generates. This may be explained by a feed-back function which the system itself has. That is, the convergent property is clearly maintained. However at *M=2*, the system develops towards an unfavorable state. In such a case, it is necessary for the operator to control the system. If

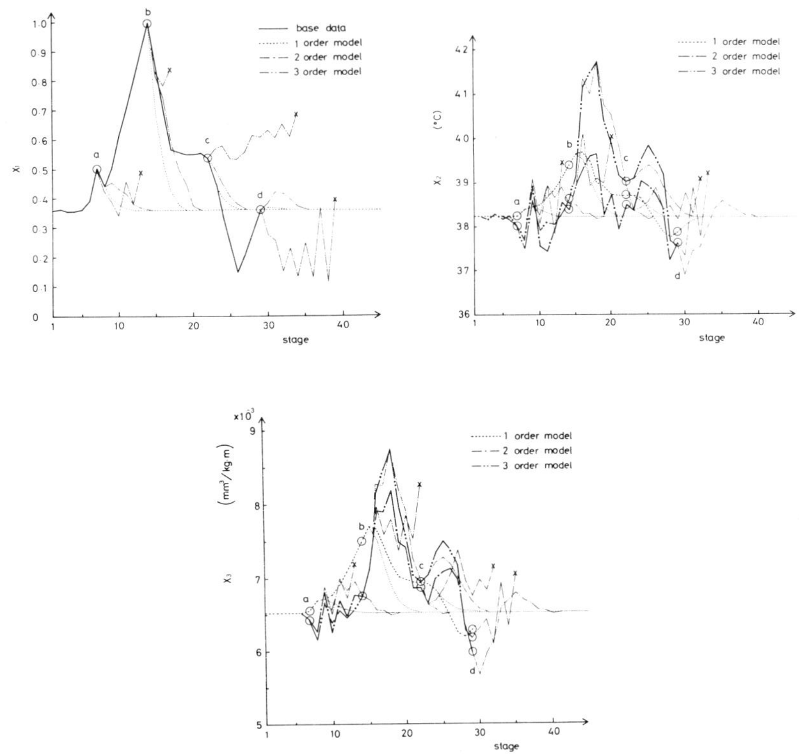

Fig. 9.—Tribological processes of state variables x_1 (friction coefficient), x_2 (temperature) and x_3 (specific wear rate). Thick line of x_1 shows compulsory input, and thin lines the behavior of the process after removing a restriction of compulsory input at a, b, c, and d.

this kind of turbulent condition continues, the system will proceed to a catastrophe. When the turbulence drives the system towards an unfavorable state surface protective treatments will serve as an effective technique for the prevention of tribological damages. On the contrary, when the stability of the AR model is well evaluated on the basis of FPE, the surface treatment selected for the above purpose may cause an adverse result.

ACKNOWLEDGEMENTS

I wish to thank Masao Yukumoto for carrying out much of the AR modelling and simulation. Sincere thanks are due Dr. H. Tamura for stimulating discussions on AR analysis, and to Professor T. Tsukizoe and Mr. S. Ishikawa for helpful comments. The computations were carried out with NEAC 2200 (model 700) at the computation center in Osaka University. This work was partially supported by the Grant-in-Aid for Scientific Research (A) from the ministry of Education, Japan.

APPENDIX

1) One-dimensional AR model

The stationary time series is defined as a series of values from such a stochastic process that the probability density function does not depend on time (stationary process) and that the set mean equals to the time mean with a probability of unity (ergodicity). Let us assume that a stationary time series of $\{x(k);\ k=1,\dots,N\}$ is generated from M-order AR model. Then we have

$$x(k) = \sum_{m=1}^{M} a(m)\cdot x(k-m) + \varepsilon(k) \qquad (1)$$

where a(m) is unknown parameter and $\varepsilon(k)$ a white noise with a mean of 0 and a variance of σ^2. That is,

$$E\varepsilon(k)\varepsilon(\ell) = \delta_{k,\ell}\sigma^2$$
$$E\varepsilon(k) = 0 \qquad (2)$$

where E denotes an operator representing a set mean, and $\delta_{k,\ell}$ the Kronecker's delta. Such an a(m) that minimizes the unknown noise variance σ^2 satisfies a following equation of

$$Ex(k)x(k-\ell) = \sum_{m=1}^{M} a(m)Ex(k-\ell)x(k-m);\ \ell=1,\dots,M \qquad (3)$$

When an autocovariance function of $x(\cdot)$ is known, i.e.

$$\Gamma_{xx}(k+\ell,\ k) \triangleq Ex(k+\ell)x(k) \qquad (4)$$

a(m) is obtained from equation (3). We will have from an assumption of stationary process

$$\Gamma_{xx}(k,\ k-\ell) = \Gamma_{xx}(\ell)$$

and from that of ergodicity

$$C_{xx}(\ell) \triangleq \frac{1}{N} \sum_{k=1}^{N-\ell} x(k+\ell)x(\ell) \tag{5}$$

where $C_{xx}(\ell)$ is an unbiased consistent estimator.
Using $C_{xx}(\ell)$ and $c_{xx}(m-\ell)$ instead of $Ex(k+\ell)x(k)$ and $Ex(k-\ell)x(k-m)$ respectively, we obtain

$$C_{xx}(\ell) = \sum_{m=1}^{M} \hat{a}(m)C_{xx}(\ell-m);\ \ell=1,\cdots,M \tag{6}$$

Normal equation (6) is solved by simple recurrence formulae shown below. Let $\hat{a}_p(m);\ m=1,\cdots,p$ be an estimator of parameter of p-order AR model and $\hat{\sigma}_p^2$ an estimator of σ^2. Then, $\hat{a}_{p+1}$ and $\hat{\sigma}_{p+1}^2$ of p+1 order AR model is written as

$$\begin{aligned} \hat{a}_{p+1}(p+1) &= (\hat{\sigma}_p^2)^{-1}\{C_{xx}(p+1) - \sum_{m=1}^{P} \hat{a}_p(m)\, C_{xx}(p+1-m)\} \\ \hat{a}_{p+1}(m) &= \hat{a}_p(m) - \hat{a}_{p+1}(p+1)\hat{a}_p(p+1-m);\ m=1,\cdots,p \\ \hat{\sigma}_{p+1}^2 &= \hat{\sigma}_p^2\{1-\hat{a}_{p+1}(p+1)^2\} \end{aligned} \tag{7}$$

Finally we obtain

$$\hat{\sigma}^2 = C_{xx}(o) - \sum_{m=1}^{M} \hat{a}(m)C_{xx}(m) \tag{8}$$

2) Multi-dimensional AR model

Consider n-dimensional vector $\underline{x}(k)$, every component of which satisfies a stationary time series. Let us assume that a stationary time series of $\{\underline{x}(k);\ k=1,\ldots,N\}$ is generated from an M-order multi-dimensional AR model. Then,

$$\underline{X}(k) = \sum_{m=1}^{M} A(m)\underline{x}(k-m) + \underline{\varepsilon}(k) \tag{9}$$

where A(m) is an (nxn) matrix of unknown parameters, and $\underline{\varepsilon}(k)$ n-dimensional white noise, i.e.

$$\begin{aligned} E\underline{\varepsilon}(k)\underline{\varepsilon}(\ell)^T &= \delta_{k,\ell}D \\ E\underline{\varepsilon}(k) &= \underline{0} \end{aligned} \tag{10}$$

To least square estimate the unknown parameter of multi-dimensional AR model is to determine such an A(m) that independently minimizes each component of unknown noise variance-covariance matrix D. In other words, the least square estimation can be done by minimizing (i,j) element of D_{ij} in the equation of

$$D = \Gamma(0) + \sum_{m=1}^{M} \sum_{\ell=1}^{M} A(m)\Gamma(m-\ell)A(\ell)^T$$

$$- \sum_{m=1}^{M} \{A(m)\Gamma(m)^T - \Gamma(m)A(m)^T\} \tag{11}$$

where $\Gamma(m)$ is a variance-covariance matrix of $\underline{x}(\cdot)$, i.e.

$$\Gamma(m) \triangleq E\underline{x}(k)\underline{x}(k-m)^T \tag{12}$$

$$= \Gamma(-m)^T \tag{13}$$

By partially differentiating D_{ij} with $A_{pq}(m)$, which is the (p,q) element of A(m), and by letting this equation be zero, we will have

$$\Gamma_{rq}(m) = \sum_{\ell=1}^{M} \sum_{s=1}^{n} \Gamma_{qs}(m-\ell)A_{rs}(\ell),$$
$$r=1,\cdots,n;\ q=1,\cdots,n;\ m=1,\cdots,M;\ s=1\cdots,n. \tag{14}$$

where $\Gamma_{rq}(m)$ is the (r,q) element of $\Gamma(m)$.

This simultaneous n^2xM equation is written as a following matrix equation of

$$\Gamma(\ell) = \sum_{m=1}^{M} A(m)\Gamma(m-\ell);\ \ell=1,\cdots,M \tag{15}$$

When estimating A(m) in actual case, we should solve the normal matrix equation of

$$C(\ell) = \sum_{m=1}^{M} \hat{A}(m)C(m-\ell);\ \ell=1,\cdots,M \tag{16}$$

where $C(\ell)$ is the unbiased consistent estimator of $\Gamma(\ell)$ and written as

$$C(\ell) = \frac{1}{N} \sum_{k=1}^{N-\ell} \underline{x}(k+\ell)\underline{x}(k)^T \tag{17}$$

Then, the estimator of D is

$$\hat{D} = C(0) - \sum_{m=1}^{M} \hat{A}(m)C(m)^T \tag{18}$$

Normal matrix equation (16) is calculated from recurrence formulae given by equation (19). When $\hat{A}_p(m);\ m=1,\cdots,\ p$ and $\hat{D}_p$ of p-order AR model are known, the estimators of (p+1)-order AR model are written as

$$E_p = C(p+1) - \sum_{m=1}^{M} \hat{A}_p(m)C(p+1-m)$$

$$F_p = C(0) - \sum_{m=1}^{M} B_p(m)C(m)$$

$$
\begin{aligned}
\hat{D}_p &= C(0) - \sum_{m=1}^{M} \hat{A}_p(m)C(m)^T \\
\hat{A}_{p+1}(p+1) &= E_p F_p^{-1} \\
\hat{A}_{p+1}(m) &= \hat{A}_p(m) - \hat{A}_{p+1}(m)B_p(p+1-m);\ m=1,\cdots,p \\
B_{p+1}(p+1) &= E_p^{T}.\hat{D}_p^{-1} \\
B_{p+1}(m) &= B_p(m) - B_{p+1}(p+1)\hat{A}_p(p+1-m);\ m=1,\cdots,p
\end{aligned}
\tag{19}
$$

3) Final prediction error (FPE)

Akaike has proposed the use of FPE when a finite order AR model is applied to the n-dimensional stationary time series of $\{x(k);\ k=1,\cdots,N\}$. To simplify the problem, let the mean of $\{\underline{x}(k)\}$ be $\underline{0}$. Then the series of residuals,

$$
\underline{\varepsilon}(k) = \underline{x}(k) - \sum_{m=1}^{M} A_M(m)\underline{x}(k-m) \tag{20}
$$

is regarded practically as a white noise with a mean of $\underline{0}$. The order of the model, M, is an integer to minimize the equation (21).

$$
FPE(M) = \left(1+\frac{Mn+1}{N}\right)^n\left(1-\frac{Mn+1}{N}\right)^{-n}\det(D_M) \tag{21}
$$

D_M is an estimated covariance of $\underline{\varepsilon}$ and can be solved using an estimator of parameter matrix, $\hat{A}_m(m)$, in equation (18).

REFERENCES

1. Salomon, G., *American Society of Lubrication Engineers Transactions*, Vol. 17, 1974, p. 295.
2. Czichos, H., *American Society of Lubrication Engineers Transactions*, Vol. 17, 1974, p. 300.
3. Czichos, H. and Salomon, G., in Proceedings of the Japan Society of Lubrication Engineers-American Society of Lubrication Engineers International Lubrication Conference, Tokyo, edited by T. Sakurai, Elsevier, Amsterdam, 1976, p. 428.
4. Mølgaard, J. and Czichos, H., in Proceedings of Wear of Materials, 1977, St. Louis, Missouri, edited by W.A. Glaeser, et al., American Society of Mechanical Engineers, New York, 1977, p. 30.
5. Czichos, H., in Proceedings of Eurotrib '77, Düsseldorf, Gesellschaft für Tribologie, Homberg, F.R. Germany, 1977, Nr. 17.
6. Salomon, G., *Lubrication Engineering*, Vol. 34, 1978, p. 353.
7. Czichos, H., "Tribology," Elsevier, Amsterdam, 1978, p. 29.
8. White, H.J. and Tauber, S., "Systems Analysis," W.B. Saunders, Philadelphia, 1969, p. 1.
9. Porter, W.A., "Modern Foundations of Systems Engineering," MacMillan, New York, 1966, p. 1.
10. Rudwick, B.H., "Solving Management Problems, A Systems Approach to Planning and Control," Wiley-Interscience Publication, John Wiley and Sons, New York, 1979, p. 13.

11. Crow, L.H. and Shimi, I.N., in "The Theory and Applications of Reliability," edited by C.P. Tsokos and I.N. Shimi, Vol. 2, Academic Press, New York, 1977, p. 299.
12. Abdel-Hameed, M., in "The Theory and Applications of Reliability," edited by C.P. Tsokos and I.N. Shimi, Vol. 1, Academic Press, New York, 1977, p. 397.
13. Chestnut, H., "Systems Engineering Methods," John Wiley and Sons, New York, 1967, p. 10.
14. Hansteen, H.B., Miller, R.E. and Wahlstrom, S.O., in "Large Engineering Systems," edited by A. Wexler, Pergamon, Oxford, 1977, p. 19.
15. Czichos, H., *Wear,* Vol. 41, 1977, p. 45.
16. Akaike, H., *Institute of Statistical Mathematics. Annals,* Vol. 22, 1970, p. 203.
17. Akaike, H., *Institute of Statistical Mathematics. Annals,* Vol. 26, 1974, p. 363.
18. Ohmae, N., Yukumoto, M. and Tsukizoe, T., Analysis of Systems Structure in the Wear of FRP, Proceedings of Eurotrib '77, Düsseldorf, Gesellschaft für Tribologie, Homberg, F.R. Germany, 1977, Nr. 57.

DISCUSSION

H. CZICHOS, BAM: Your analysis looks a bit complicated at first glance. To what actual system does this analysis apply?

N. OHMAE: This is only an example of the tribological system. The results were obtained from the static system of the sliding between mild steel and fiber-reinforced plastics.

XVI. DOCUMENTATION OF TRIBOLOGY LITERATURE

SCOPE AND SUBDIVISION OF THE FIELD OF TRIBOLOGY WITH REGARD TO DOCUMENTATION AND INFORMATION

K. Kirschke

ABSTRACT

A classification system for tribology subjects is given for the purpose of documentation of the published literature.

INTRODUCTION

The comprehensive term tribology embraces friction and wear, in all their aspects, including lubricants, lubrication, bearing design and selection of materials. Its importance is growing with technological progress which is connected with ever heavier demands on the materials used. Tribology has to attend to higher velocities, higher pressures, extreme temperatures, environmental conditions, etc.

Basic and applied research in tribology have been intensified, and it has become obvious that an effective treatment of tribological problems necessitates the cooperation of physicists, chemists, metallurgists and mechanical engineers. A systematic, effective and profitable collection of the results of basic research work and of investigations into matters of industrial application is the underlying concept of the scope and classification of the whole field of tribology.

SCOPE AND SUBDIVISION OF THE FIELD OF TRIBOLOGY

Many practical problems cover a wide area ranging from mathematical theories to processing techniques. Basic research and engineering experiments often have the same objective; one, being primarily of a physical approach comprising the investigation of the friction and wear characteristics under conditions that permit the observation of individual mechanisms. The results of such experiments are to be regarded merely as information on the respective mechanisms and as contributions toward a basic understanding of friction and wear phenomena. The technological approach consists of the investigation of the friction and wear characteristics of real machines and machine parts under actual operation conditions.

The underlying considerations for the subdivision of the tribological literature are explained in the following. Basically, various possibilities lend themselves to a systematic subdivision. Such subdivision could proceed from basic research aspects, i.e., from friction and wear mechanisms. But phenomena, or materials, or design and operational characteristics could also be adopted. Considering the various possibilities, the conclusion will be arrived at that it is not fruitful to classify the complex field of wear, friction, and lubrication according to a single system. It is impossible to devise a generally acceptable method of classifying the host of published information covering the field according to a single principle. However, the search for a suitable subdivision has been guided by the need to devise a classification system of sufficient flexibility which will permit all publications relating to the entire field to be included. Limited access to only a part of the published information would be counter to the need for an exchange of information between experts in all the fields involved. A plant engineer, for instance, will in his daily work be confronted with the problem of friction, wear, and lubrication of machine parts under complex operating conditions. He will approach it with the remedies known in engineering practice. He should also have at his disposal the most essential findings in the research field relating to his sphere of work. In contrast, the researcher in the laboratory works under idealized conditions over which he has a fair amount of control, with optimum experimental facilities at his disposal. If his work is to be successful, however, he should also be acquainted with the conditions which may give rise to friction and wear in practical operation. It is therefore essential for the system to cover all aspects of the problem without any priorities. Bearing the above in mind, it has been considered expedient to create a system in which the various fields are side by side.

Upon this consideration, the following classification system (Table 1) has been devised by the Documentation Center of BAM for the field of tribology, which has proved useful for documentation purposes for over ten years. It is safe to assume that it is suitable for other applications as well.

TABLE I.—DIVISION AND SUBDIVISION OF SUBJECTS

1. General, monographs, reviews

2. Experimental methods and equipment

3. Tribological fundamentals, friction and wear: laws, mechanisms, effects
   - 3.1 Mechanical and acoustical phenomena (contact formation, contact stresses, deformation, hardening and softening, fatigue, crack formation, material removal, lattice waves, sound emission)
   - 3.2 Chemical, physico-chemical and thermal phenomena (tribochemical reactions, activation, diffusion, adhesion, material transfer, heat development, structural change, residual stresses)
   - 3.3 Electrical and optical phenomena (contact potential, contact resistance, frictional electrification, exoelectron emission, triboluminescence)

4. Appearance of damage

5. Wear and friction under specific stressing conditions
   - 5.0 Resting and preliminary displacement in static friction
   - 5.1 Sliding, including stick-slip
   - 5.2 Rolling, rolling with sliding
   - 5.3 Spin
   - 5.4 Impact
   - 5.5 Sliding against grains
   - 5.6 Fretting, fretting corrosion

5.7 Wear caused by flowing media
 5.71 Fluid erosion, abrasive erosion
 5.72 Impact erosion
 5.73 Cavitation erosion, impingement of drops
 5.74 Electroerosion
5.8 Thermal wear, ablation

6. Materials and combinations of materials (with and without intermediate materials)
6.1 Metals
 6.11 Metal vs. metal
 6.12 Metal vs. mineral
 6.13 Metal vs. polymer
 6.14 Metal vs. composite
6.2 Minerals, ceramics
 6.21 Mineral vs. metal see 6.12
 6.22 Mineral vs. mineral
 6.23 Mineral vs. polymer
 6.24 Mineral vs. composite
6.3 Polymers (plastics and rubber)
 6.31 Polymer vs. metal see 6.13
 6.32 Polymer vs. mineral see 6.23
 6.33 Polymer vs. polymer
 6.34 Polymer vs. composite
6.4 Composites (glass-fibre reinforced materials, cermets, laminated materials etc.)
 6.41 Composite vs. metal see 6.14
 6.42 Composite vs. mineral see 6.24
 6.43 Composite vs. polymer see 6.34
 6.44 Composite vs. composite
... .

6.8 Textile materials, leather
6.9 Other materials and combinations of materials (incl. wood, paper, shoe soles, floors, granular materials, biological materials, dental materials)

7. Influential factors and control parameters
7.1 Material dependent factors
 7.11 Strength, residual stresses
 7.12 Composition, texture, structure
 ....
 7.19 Other material characteristics (incl. compatibility)
7.2 Surface dependent factors
 7.21 Topography of surfaces
 7.22 Surface layers (adsorbed films, oxide layers)
 7.23 Surface treatment (mechanical and heat treatment, electropolishing)
 7.24 Surface coatings
  7.241 Inorganic coatings
   7.2411 Electroplating, chemical plating, hot-dip coating, vapor depositing
   7.2412 Surfacing, sintering, casting, cladding, spray coating
   7.2413 Glazing, enamelling
   7.2414 Diffusion coating (carburizing, nitriding, boriding, chromising, alumenising, etc.)

- 7.2415 Chemical conversion coating (phosphating, anodizing, etc.)
- 7.242 Organic coatings (incl. paint coating, printing inks)
- 7.3 Operating conditions
  - 7.31 Motion
    - 7.311 Running distance, running time, orientation of motion, start and stop periods
    - 7.312 Velocity
  - 7.32 Shear force, load
    - 7.321 Static loading
    - 7.322 Dynamic loading (espec. oscillations)
  - 7.33 Temperature
  - 7.34 Environment
    - 7.341 Atmosphere, vacuum
    - 7.342 Liquid media
      - 7.3421 Liquified gases
      - 7.3422 Other liquids (acid, alkaline, and salt solutions, molten salts, water, etc.)
- 7.4 Other influential factors
  - 7.41 Electric potential, current passage, magnetic field
  - 7.42 Solid intermediate particles, dust, contaminants
  - 7.43 Irradiation
  - 7.44 Geometric factors, effects of design

8. Lubrication
  - 8.1 Thick-film lubrication, espec. hydrodynamic lubrication
  - 8.2 Thin-film lubrication (boundary lubrication, elastohydrodynamic lubrication)
  - 8.3 Lubrication in special environments
    - 8.31 Lubrication in vacuum and/or under space conditions
    - 8.32 Lubrication as affected by aggressive liquids incl. water
  - 8.4 Lubrication at extreme temperatures
  - 8.5 Lubrication under radiation and/or in reactor technique
  - 8.6 Extreme pressure lubrication
  - 8.7 Lubricants, lubricant testing
    - 8.70 Lubricant testing
    - 8.71 Liquid lubricants, greases
    - 8.72 Additives
    - 8.73 Solid lubricants
  - 8.8 Lubrication systems and methods
  - 8.9 Hygienic and microbiological problems

9. Fields of application, technical designs and processes
  - 9.0 Life, reliability and maintenance of technical systems
  - 9.1 Machinery and components: selection and performance
    - 9.11 Bearings
      - 9.111 Sliding bearings
        - 9.1111 Liquid film and grease-lubricated sliding bearings
        - 9.1112 Gas-lubricated sliding bearings
        - 9.1113 Dry sliding bearings
        - 9.1114 Magnetic and MHD bearings
      - 9.112 Ball and roller bearings
      - 9.113 Sliding/rolling hybrid bearings
      - 9.114 Foil bearings
      - 9.115 Elastomeric bearings
      - ....

9.119 Special bearings
- 9.1191 Precision bearings (instrument bearings, gyroscopic devices, clock and counter bearings)
- 9.1192 Large bearings for antennas and telescopes
- 9.1193 Engineering joints
- 9.1194 Human and animal joints
- 9.1195 Construction and bridge bearings

9.12 Transmissions
- 9.121 Toothed gears
- 9.122 Chain drives, belt drives, rope drives
- 9.123 Friction drives
- 9.124 Crank drives, cam drives
- 9.125 Hydraulic transmission, automatic transmissions
- 9.126 Screw drives

9.13 Couplings, clutches and brakes

9.14 Seals

9.15 Machines, sets. Vehicles and vessels
- 9.150 Vehicles and vessels
  - 9.1501 Vehicles
  - 9.1502 Watercrafts
  - 9.1503 Aircrafts, spacecrafts
- 9.151 Piston-type engines, pumps and compressors
  - 9.1511 Cylinders, pistons, piston-rings
  - 9.1512 Connecting rods, crank-shafts (incl. bearings)
  - 9.1513 Carburetors, valves, governors
- 9.152 Turbines, turbomachines, centrifugal pumps, blowers, propellers
- 9.153 Hydraulic and pneumatic systems
- 9.154 Machine tools (guides, slideways, spindles, etc.)
- 9.155 Textile machinery
- 9.156 Printing machinery, paper machinery
- 9.157 Chemical engineering, food technology
- 9.158 Measuring instruments, optical instruments, cinematography, phonography, business machines, data processing
- 9.159 Electric plants
  - 9.1591 Electric motors
  - 9.1592 Electrical contacts and switches

9.16 Wheel and rail

9.17 Tire and road

9.18 Specific industries
- 9.181 Comminution, dressing
- 9.182 Mining (incl. well drilling) and metallurgy
- 9.183 Construction, quarrying, earth movement, agriculture, forestry
  - 9.1831 Equipment, wearing parts
  - 9.1832 Foundations
- 9.184 Mechanical conveying, handling, and storage (pipings, fittings, bins, means of transportation)
- 9.185 Engineering for power

9.19 Other machine elements and products

9.2 Manufacturing processes. Tools

9.21 Forming
- 9.211 Metals
  - 9.2111 Casting
  - 9.2112 Forging, upsetting

9.2113 Rolling
9.2114 Drawing
9.2115 Extrusion
9.2116 Deep drawing
9.2117 Other metal forming processes
9.212 Polymers (plastics, rubber)
9.213 Minerals and ceramics, powder metallurgy
9.214 Tablets, foodstuffs
....
9.23 Cutting
9.231 Shear cutting, machining
9.232 Grinding, lapping, polishing, abrasive cutting
9.233 Jet cutting
9.24 Surface finishing (blasting, brushing, roller burnishing)
9.25 Joining
9.251 Screwing, riveting clamping, forcing-in
9.252 Friction welding, ultrasonic welding
9.253 Other joining techniques
9.26 Coating
9.3 Materials testing
9.4 Rock mechanics, geomechanics

10. Basic information on solid surfaces and contacts
10.1 Solid surfaces
10.11 Morphology
10.12 Strength, surface tension and energy, chemical reactions, environmental effects
10.13 Surface analysis techniques
10.2 Contacts
10.21 Area of contact, deformation of contact, heat transfer
10.22 Joint character, adhesion

THE IMPORTANCE OF A DOCUMENTATION AND INFORMATION SERVICE

Scientific and technological work is, as a rule, published to make essential findings available to experts. The steady intensification of basic and applied research has led to rapid progress in science and technology and this in turn has triggered off a considerable increase in publications each year.

Because of this growth, the capacity of the human memory no longer suffices to have all knowledge concerning even a small sector of a special subject. Even search and retrieval of required information in literature involves considerable difficulties. Scientists are no longer in a position to scan regularly all publications in their field of work. Also, card indexing for all topics must be regarded as too time-consuming. Besides, such card catalogues may not suffice when unexpected problems arise. It follows that it is more effective to assign this task, for a specific field, to a single and competent institution working for the benefit of all scientists concerned. Otherwise recent findings will not receive the due attention and the application they deserve. Considerable funds are wasted again and again because findings already in existence are not known at the start of a project. Useless duplication of work results in considerable loss of time and money. According to conservative estimates, between 30 and 85 percent of man-hours are thus wasted every year in research and development.

Considerable disadvantage and cost will be caused, if publications are not available when they are needed, if one does not know that they exist and if one cannot find out whether there are relevant publications. For some years it has been realized that this is an essential problem of our time. The promotion of science must include the furtherance of information and documentation services. For the individual scientist, work on a research project is usually complete with the publication of the findings. He turns to new problems. In many fields this has led to a situation where more research work is done and published, yet the results receive less and less attention by the experts. It is, however, not sufficient that experiments and findings merely exist, printed in some journal or presented at some symposium. They should also be easily accessible and retrievable. The more troublesome the search for published knowledge is, the less it is used. The only remedy is to set up a continuous and reliable documentation for every special branch. Without a fairly complete documentation for every special branch. Without a fairly complete documentation service the gap between the growing number of yearly publications and the diminishing possibility of getting to know them will be widening. Scientists working for the documentation service thus play an important part as mediators.

Greater knowledge and its early application constitutes an advantage for every enterprise in industrial competition. The importance of technical literature, the methods of its application and the possibilities of using it should be pointed out to scientists during their training.

In addition, the existence of a documentation service, available to all scientists concerned, leads to an improved coordination of work resulting from better mutual information. Progress requires research, and research without the application of findings by others is inconceivable today. Many still fail to realize the importance that technical literature has gained in recent decades.

As a consequence, in allocating funds for research and development projects, an appropriate sum should be set aside for the purpose of tapping all available sources of information. In the USA five percent of the research funds are allocated for information purposes, and in the Soviet Union scientists have already called for ten percent to be used for this purpose. Experts consider an amount of up to 20 percent of the funds granted appropriate for information purposes[(1)].

In fact, it is strange that, on the one hand, a scientist is expected to publish the findings of his research work. (In many cases he is rated according to the number of his publications.) Yet, on the other hand, nobody cares much about whether he himself has taken into account the relevant publications of other experts, and to what extent his work is based on the current state of knowledge attained through an extensive search of literature. In research and development, rationalization and increased efficiency are possible only if the organizations promoting research projects have considered more than hitherto the various aspects of documentation and have made a precondition for the allocation of research funds the stipulation that existing literature should be taken sufficiently into account. Projects could then frequently be started at a far more advanced level.

According to the Weinberg report, a scientist should be both a research scientist and an information scientist, unless, as in a big working group, one member is exclusively concerned with information work.

After giving this general report about the prevailing situation, it can be said that it is indeed time for all those persons holding top positions in big state-run scientific institutions or important companies to recognize the decisive task of documentation and the compiling of information for the future and to realize the necessity of contributing, at least on a moderate level, their share to make it work.

THE DOCUMENTATION SERVICE TRIBOLOGY

The "Documentation Tribology," edited by Bundesanstalt für Materialprüfung (BAM) [Federal Institute for Materials Testing], Unter den Eichen 87, D-1000 Berlin 45 (West-Germany), is an organized reference guide to the world's literature on tribology in all its aspects, published in single volumes at regular intervals. The documentation aims at providing a comprehensive bibliography of relevant titles, and therefore, all materials available as sources are scanned for references. Hence this is a very comprehensive work of reference which quite obviously contains the bulk of the world's known published material on tribology(2). Volume 13 (1976) for example covers 6000 references. The publication references are grouped into sections and subsections by their subject matter. Within each section items are arranged in alphabetical order by authors. The specified subdivision of subjects and a subject index facilitate the literature search. The subscriber to this documentation is in a position to get quick information on the state of knowledge concerning a special subject when the need arises.

The way references are quoted makes the documentation suited for international use; wherever the title is non-English, an English translation is given. If the title is not given in the original language, the respective language is put in brackets. Beside the original source, abstracts are frequently referred to facilitating quick information. In addition, reference is made to translations of papers available elsewhere.

FURTHER DEVELOPMENT OF THE DOCUMENTATION SERVICE

The utilization of a computer will make it feasible for the Documentation Centre to improve and facilitate the possibilities of employment of the Documentation Tribology. Besides the Subject Index which is already at hand, a series of other Indexes will be submitted. Shortly, an Author Index, a Standards Index, A Report and Paper Number index and a Conference Index will be available.

In view of the numerous findings and reports on relevant experiments in technical literature, it should not be left at listing and classifying all published information. In the wake of rationalization, also in the fields of research and development, the listing and classifying should be followed by an evaluation of literature with the aim to present the state of knowledge in an applicable manner(3).

REFERENCES

1. Lutterbeck, E., *Verein Deutscher Ingenieure*, No. 33, August 13, 1969.
2. *Industrial Lubrication and Tribology*, Vol. 28, No. 3, 1976, p. 113.
3. Kirschke, K., *Lubrication Engineering*, Vol. 31, No. 7, July 1975, p. 362.